Biotechnology

Biotechnology

Applying the Genetic Revolution

Authors

David P. Clark
Department of Microbiology
Southern Illinois University
Carbondale, Illinois

Nanette J. Pazdernik
Math and Science Division
Southwestern Illinois College
Belleville, Illinois

AMSTERDAM • BOSTON • HEIDELBERG • LONDON
NEW YORK • OXFORD • PARIS • SAN DIEGO
SAN FRANCISCO • SINGAPORE • SYDNEY • TOKYO
Academic Press is an imprint of Elsevier

Elsevier Academic Press
30 Corporate Drive, Suite 400, Burlington, MA 01803, USA
525 B Street, Suite 1900, San Diego, California 92101-4495, USA
84 Theobald's Road, London WC1X 8RR, UK

This book is printed on acid-free paper. ♾

Library of Congress Cataloging-in-Publication Data
Biotechnology : applying the genetic revolution / editors, David P. Clark, Nanette Pazdernik.
p. ; cm.
Includes bibliographical references and index.
ISBN 978-0-12-175552-2 (hardcover : alk. paper) 1. Biotechnology–Textbooks.
2. Genetic engineering—Textbooks. 3. Molecular biology—Textbooks. I. Clark, David P. II. Pazdernik, Nanette.
[DNLM: 1. Biotechnology–methods. 2. Genetic Engineering. 3. Molecular Biology.
4. Recombinant Proteins. QU 450 B6158 2009]
TP248.2.B55145 2009
660.6—dc22
2008017422

British Library Cataloguing in Publication Data
A catalogue record for this book is available from the British Library

ISBN 13: 978-0-12-175552-2

For all information on all Elsevier Academic Press publications
visit our Web site at www.elsevierdirect.com

Printed in China
08 09 10 9 8 7 6 5 4 3 2 1

This book is dedicated to Donna. —DPC

This book is dedicated to my children and husband. Their patience and understanding have given me the time and inspiration to research and write this text. —NJP

CONTENTS

PREFACE

Biotechnology has made the world a different place. Biotechnology has made it possible to identify the genetic causes behind many different inherited diseases. Biotechnology has made it possible for people to survive to a much higher population density by providing more food per acre. The advent of modern molecular biology and genetics has advanced our understanding of the genomes of a wide range of organisms from viruses and bacteria to trees and humans. The application of this knowledge has revolutionized the sciences, changing them from a descriptive nature to a variety of disciplines that provide new products such as drugs, vaccines, and foods.

Biotechnology has opened doors to making proteins with new functions, and even new biochemical pathways with altered products. With new proteins and new biochemical pathways, it seems only logical to find ways to incorporate the new functions into crops, into animals, and, it is hoped, into people with genetically based illnesses. Only a short time ago, agriculturists largely relied on green fingers to get good yields; today they use green fluorescent protein to assess gene expression in transgenic crops. The ability to make such direct changes will result in major changes for the future. Will biotechnology find the proverbial fountain of youth by identifying the molecular changes that cause us to age or develop cancer? Will it change the way we treat diseases? Will the way we wage war change with the development of new biological agents?

Biotechnology: Applying the Genetic Revolution explains how the information from the genetic revolution is being used to answer some of these questions. It informs the reader about the many avenues where biotechnology has changed the original field of study. The first few chapters provide a clear and concise review of the basics of molecular biology. These topics are explained in more detail in the first book of this series, entitled *Molecular Biology: Understanding the Genetic Revolution*. This review will take the student through the basics, including DNA structure, gene expression, and protein synthesis, as well as survey the variety of organisms used in biotechnology research. The student is then presented with the basic methodologies used in biotechnology research. Chapter 3 explains how nucleic acids are isolated, cloned into humanmade genetic vehicles, and then reinserted into one of the model organisms for in-depth analysis. The next two chapters discuss in more detail various techniques that have been developed to investigate the function of genes. Chapter 4 focuses on DNA, dealing with both *in vivo* and *in vitro* synthesis of DNA and the polymerase chain reaction. Chapter 5 focuses on RNA, explaining antisense technology, RNA interference, and ribozymes. Familiarity with these chapters is critical to understanding the rest of the textbook.

The remaining chapters focus on different fields of research, presenting some of the ways the genetic revolution has irreversibly changed these areas. Chapter 6 begins this approach by presenting newer techniques to generate antibodies for genetic research and for creating new vaccines. Chapter 7 delves into a different realm, one based on the nanoscale. This chapter evaluates how molecular biology will be changed by the ability of scientists to work in the nanoscale world. It discusses how scientists are using novel nanoscale structures to deliver drugs, identify biological molecules *in situ*, and manufacture antibacterial materials. The chapter illustrates how nanobiotechnology exploits the self-assembly property of DNA to create nanodevices. It shows how DNA can physically control the shape of proteins. This new field of research is intimately intertwined with molecular biology and will only become a stronger component of molecular biology courses in the future.

The next section returns to the more familiar world of genomics and proteomics. These chapters emphasize the applied aspect of these topics and discuss the medical applications of advances in genomics and proteomics. The proteomics chapter includes a variety of techniques used to isolate and characterize proteins, including the more recent developments in mass spectrometry. Proteomics provides a nice segue to the next chapter, which surveys how proteins are studied by expressing them in various organisms and cultured cells. The creation of proteins with novel properties by protein engineering follows.

Because single genetically modified proteins have their limitations, Chapter 12 moves from the lab to the environment and presents the emerging field of metagenomics. This approach bypasses the traditional method of identifying new genes one at a time from model organisms in the laboratory. Instead, metagenomics skips directly to isolating genomic sequences from the environment without identifying the organism from which they originate. The investigation of novel gene functions continues in Chapter 13. Biochemical pathways may be altered using recombinant DNA technology, and this chapter presents a few of these novel pathways. Construction of novel proteins and biochemical pathways is pointless unless they can be inserted into plants and animals. So the next two chapters present the student with the latest advances in creating transgenic plants and animals.

The next block of chapters focuses on the medical arena. First, in Chapter 16, the molecular basis for inherited defects is examined. This leads into the following chapter on gene therapy. Several chapters then present the molecular basis of cancer, a selection of noninfectious diseases, such as erectile dysfunction, diabetes, and obesity, and then aging. Last, molecular biology has made huge strides in our understanding of bacterial and viral diseases. In Chapter 21 and Chapter 22, the student will learn how bacteria and viruses exploit our cellular machinery to cause disease. The latest research on the unusual prion diseases, such as mad cow disease and Creutzfeldt-Jakob disease, is also covered. Chapter 23 builds on the knowledge of bacterial and viral pathogenesis to present a survey of biowarfare and bioterrorism.

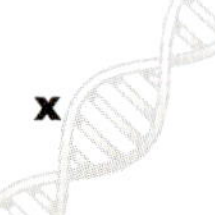

Chapter 24 surveys how the field of forensics has been altered by the genetic revolution. The way criminals are identified via molecular biology has changed the penal system irreversibly. New cases, old cases, and unsolved cases are all now examined with DNA testing which is more accurate and reliable than previous identification methods. The use of DNA in criminal investigation has even spawned popular television series that showcase these advances and their effect on society. As the book comes to a close in Chapter 25, the subject of bioethics is presented. Rather than discussing scientific methodology, this chapter asks questions about the role of these methodologies in society. Should we use the genetic revolution to clone a human, create transgenic crops, do research on human stem cells? Should our genetic identity be open to the public domain?

Biotechnology: Applying the Genetic Revolution demonstrates many different ways in which advances in technology and the revolution in molecular biology have merged. The combined ability to process large volumes of information along with analyzing our bodies and other organisms with infinitesimal precision has and will continue to change our society, our ethics, and our personal surroundings. This book gives the student a basic knowledge of some of those changes that have already occurred, with the hope that they will be able to apply this knowledge toward future advances.

ACKNOWLEDGMENTS

We would like to thank the following individuals for their help in providing information, suggestions for improvement and encouragement: Laurie Achenbach, Rubina Ahsan, Phil Cunningham, Donna Mueller, Dan Nickrent, Holly Simmonds, and Dave Pazdernik. Especial thanks go to Alex Berezow and Michelle McGehee for writing the questions and to Karen Fiorino for creating most of the artwork.

MODERN BIOTECHNOLOGY RELIES ON ADVANCES IN MOLECULAR BIOLOGY AND COMPUTER TECHNOLOGY

Traditional biotechnology goes back thousands of years. It includes the selective breeding of livestock and crop plants as well as the invention of alcoholic beverages, dairy products, paper, silk, and other natural products. Only in the past couple of centuries has genetics emerged as a field of scientific study. Recent rapid advances in this area have in turn allowed the breeding of crops and livestock by deliberate genetic manipulation rather than trial and error. The so-called green revolution of the period from 1960 to 1980 applied genetic knowledge to natural breeding and had a massive impact on crop productivity in particular. Today, plants and animals are being directly altered by genetic engineering.

New varieties of several plants and animals have already been made, and some are in agricultural use. Animals and plants used as human food sources are being engineered to adapt them to conditions that were previously unfavorable. Farm animals that are resistant to disease and crop plants that are resistant to pests are being developed in order to increase yields and reduce costs. The impact of these genetically modified organisms on other species and on the environment is presently a controversial issue.

Modern biotechnology applies not only modern genetics but also advances in other sciences. For example, dealing with vast amounts of genetic information depends on advances in computing power. Indeed, the sequencing of the human genome would have been impossible without the development of ever more sophisticated computers and software. It is sometimes claimed that we are in the middle of two scientific revolutions, one in information technology and the other in molecular biology. Both involve handling large amounts of encoded information. In one case the information is humanmade, or at any rate man-encoded, and the mechanisms are artificial; the other case deals with the genetic information that underlies life.

However, there is a third revolution that is just emerging—nanotechnology. The development of techniques to visualize and manipulate atoms individually or in small clusters is opening the way to an ever-finer analysis of living systems. Nanoscale techniques are now beginning to play significant roles in many areas of biotechnology.

This raises the question of what exactly defines biotechnology. To this there is no real answer. A generation ago, brewing and baking would have been viewed as biotechnology. Today, the application of modern genetics or other equivalent modern technology is usually seen as necessary for a process to count as "biotechnology." Thus, the definition of *biotechnology* has become partly a matter of fashion. In this book, we regard (modern) biotechnology as resulting in a broad manner from the merger of classical biotechnology with modern genetics, molecular biology, computer technology, and nanotechnology.

The resulting field is of necessity large and poorly defined. It includes more than just agriculture: it also affects many aspects of human health and medicine, such as vaccine development and gene therapy. We have attempted to provide a unified approach that is based on genetic information while at the same time indicate how biotechnology has begun to sprawl, often rather erratically, into many related fields of human endeavor.

CHAPTER 1

Basics of Biotechnology

ADVENT OF THE BIOTECHNOLOGY REVOLUTION

Biotechnology involves the use of living organisms in industrial processes—particularly in agriculture, food processing, and medicine. Biotechnology has been around since the dawn of time, ever since humans began manipulating the natural environment to improve their food supply, housing, and health. Biotechnology is not limited to humankind. Beavers cut up trees to build homes. Elephants deliberately drink fermented fruit to get an alcohol buzz. People have been making wine, beer, cheese, and bread for centuries. All these processes rely on microorganisms to modify the original ingredients (Fig. 1.1). Over the ages, farmers have chosen higher yielding crops by trial and error, so that many modern crop plants have much larger fruit or seeds than their ancestors (Fig. 1.2).

FIGURE 1.1 Traditional Biotechnology Products Bread, cheese, wine, and beer have been made worldwide for many centuries using microorganisms, such as yeast.

FIGURE 1.2 Teosinte versus Modern Corn Throughout history, people have improved many plants for higher yields. Teosinte (smaller cob and green seeds) is considered the ancestor of commercial corn (larger cob; a blue-seeded variety is shown). Courtesy of Wayne Campbell, Hila Science Camp.

The reason we think of biotechnology as modern is because of recent advances in molecular biology and genetic engineering. Huge strides have been made in our understanding of microorganisms, plants, livestock, as well as the human body and the natural environment. This has caused an explosion in the number and variety of biotechnology products. Face creams contain antioxidants—supposedly to fight the aging process. Genetically modified plants have genes inserted to protect them from insects, thus increasing the crop yield while decreasing the amount of insecticides used. Medicines are becoming more specific and compatible with our physiology. For example, insulin for diabetics is now genuine human insulin, although produced by genetically modified bacteria. Almost everyone has been affected by the recent advances in genetics and biochemistry.

Mendel's early work that described how genetic characteristics are inherited from one generation to the next was the beginning of modern genetics (see Box 1.1). Next came the discovery of the chemical material of which genes are made—**DNA (deoxyribonucleic acid)**. This in turn led to the central dogma of genetics; the concept that genes made of DNA are expressed as an **RNA (ribonucleic acid)** intermediary that is then decoded to make **proteins**. These three steps are universal, applying to every type of organism on earth. Yet these three steps are so malleable that life is found in almost every available niche on our planet.

Biotechnology affects all of our lives and has altered everything we encounter in life.

Box 1.1 Gregor Johann Mendel (1822–1884): Founder of Modern Genetics

As a young man, Mendel spent his time doing genetics research and teaching math, physics, and Greek to high school children in Brno (now in the Czech Republic). Mendel studied the inheritance of various traits of the common garden pea, *Pisum sativum*, because he was able to raise two generations a year. He studied many different physical traits of the pea, such as flower color, flower position, seed color and shape, and pod color and shape. Mendel grew different plants next to each other, looking for traits that mixed together. Luckily, the traits he studied were each due to a single gene that was either dominant or recessive, although he did not know this at the time. Consequently, he never saw them "mix." For example, when he grew yellow peas next to green peas, the offspring looked exactly like their parents. This showed that traits do not blend in the offspring, which was a common theory at the time.

Next Mendel moved pollen from one plant to another with different traits. He counted the number of offspring that inherited each trait and found that they were inherited in specific ratios. For example, when he cross-pollinated the yellow and green pea plants, their offspring, the F_1 generation, was all yellow. Thus the yellow trait must dominate or mask the green trait. He then let the F_1 plants produce offspring, and grew all of the seeds. These, the F_2 generation, segregated into 3/4 yellow and 1/4 green. When green seeds reappeared after skipping a generation, Mendel concluded that a "factor" for the trait—what we call a gene nowadays—must have been present in the parent, even though the trait was not actually displayed.

Mendel demonstrated many principles that form the basis of modern genetics. First, units or factors (now called genes) for each trait are passed on to successive generations. Each parent has two copies of each gene but contributes only one copy of the gene to each offspring. This is called the **principle of segregation**. Second, the **principle of independent assortment** states that different offspring from the same parents can get separate sets of genes. The same phenotype (the observable physical traits) can be represented by different genotypes (combinations of genes). In other words, although a gene is present, the corresponding trait may not be seen in each generation. When Mendel began these experiments, he used purebred pea plants—that is, each trait always appeared the same in each generation. So when he first crossed a yellow pea with a green pea, each parent had two identical copies or **alleles** of each gene. The green pea had two green alleles, and the yellow pea had two yellow alleles. Consequently, each F_1 offspring received one yellow allele and one green allele. Despite this, the F_1 plants all had yellow peas. Thus yellow is dominant to green. Finally when the F_1 generation was self-pollinated, the F_2 plants included some that inherited two recessive green alleles and had a green phenotype (Fig. A).

Mendel published these results, but no one recognized the significance of his research until after his death. Later in life he became the abbot of a monastery and did not pursue his genetics research.

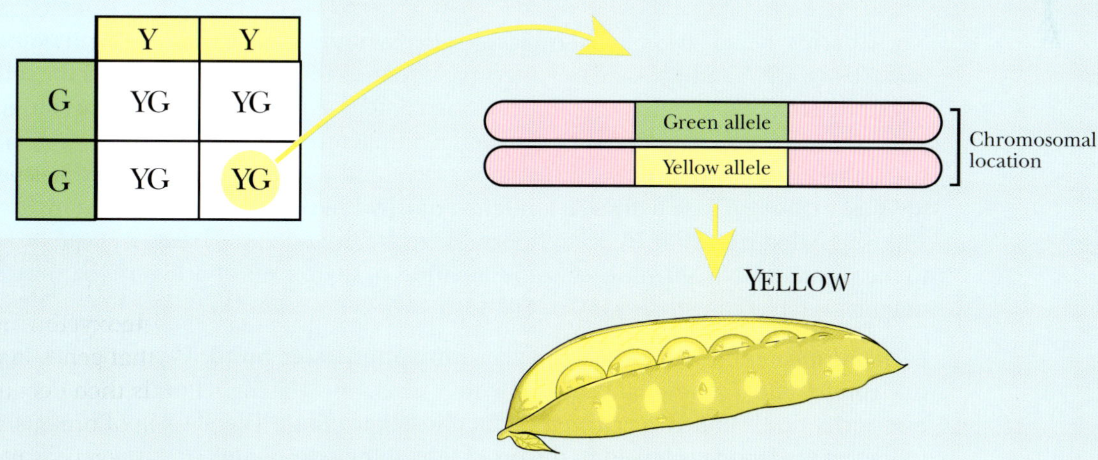

FIGURE A Relationship of Genotype and Phenotype
(A) Each parent has two alleles, either two yellow or two green. Any offspring will be heterozygous, each having a yellow and a green allele. Since the yellow allele is dominant, the peas look yellow.

(*Continued*)

Box 1.1 Gregor Johann Mendel (1822–1884): Founder of Modern Genetics—cont'd

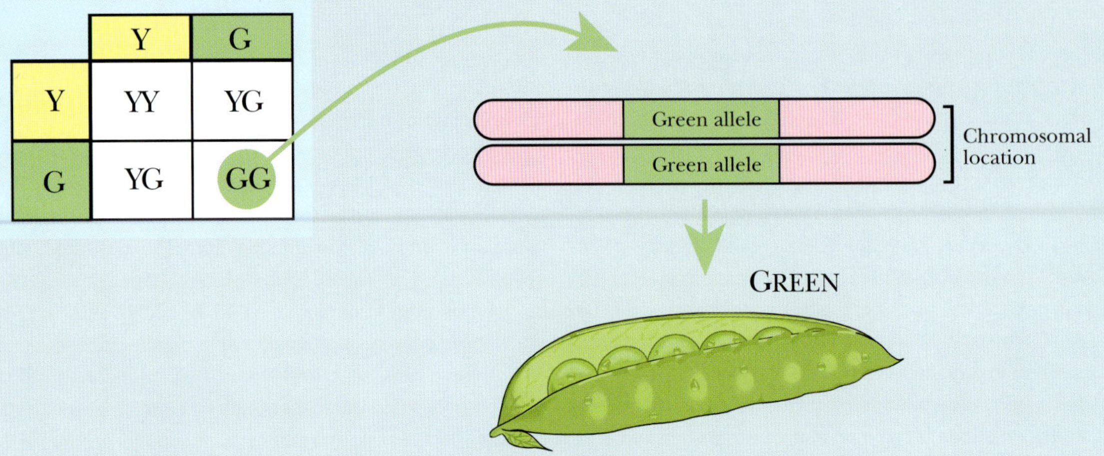

FIGURE A Relationship of Genotype and Phenotype, cont'd
(B) When the heterozygous F_1 offspring self-fertilize, the green phenotype re-emerges in one-fourth of the F_2 generation.

CHEMICAL STRUCTURE OF NUCLEIC ACIDS

The upcoming discussions introduce the organisms used extensively in molecular biology and genetics research. Each of these has genes made of DNA that can be manipulated and studied. Thus a discussion of the basic structure of DNA is essential. The genetic information carried by DNA, together with the mechanisms by which it is expressed, unifies every creature on earth and is what determines our identity.

Nucleic acids include two related molecules, deoxyribonucleic acid (DNA) and ribonucleic acid (RNA). DNA and RNA are polymers of subunits called **nucleotides**, and the order of these nucleotides determines the information content. Nucleotides have three components: a **phosphate group**, a five-carbon sugar, and a nitrogen-containing **base** (Fig. 1.3). The five-carbon sugar or **pentose** is different for DNA and RNA. DNA has **deoxyribose**, whereas RNA uses **ribose**. These two sugars differ by one hydroxyl group. Ribose has a hydroxyl at the 2′ position that is missing in deoxyribose. There are five potential bases that can be attached to the sugar. In DNA, guanine, cytosine, adenine, or thymine is attached to the sugar. In RNA, thymine is replaced with uracil (see Fig. 1.3).

Each phosphate connects two sugars via a **phosphodiester bond**. This connects the nucleotides into a chain that runs in a 5′ to 3′ direction. The 5′-OH of the sugar of one nucleotide is linked via oxygen to the phosphate group. The 3′-OH of the sugar of the following nucleotide is linked to the other side of the phosphate.

The nucleic acid bases jut out from the sugar phosphate backbone and are free to form connections with other molecules. The most stable structure occurs when another single strand of nucleotides aligns with the first to form a double-stranded molecule, as seen in the DNA **double helix**. Each base forms hydrogen bonds to a base in the other strand. The two strands are **antiparallel**, that is, they run in opposite directions with the 5′end of the first strand opposite the 3′end of its partner and vice versa.

The bases are of two types, **purines** (guanine and adenine) and **pyrimidines** (cytosine and thymine). Each base pair consists of one purine connected to a pyrimidine via hydrogen bonds. Guanine pairs only with cytosine (G-C) via three hydrogen bonds. Adenine pairs only with thymine (A-T) in DNA or uracil (A-U) in RNA. Because an adenine-thymine (A-T) or adenine-uracil (A-U) base pair is held together with only two hydrogen bonds, it requires less energy to break the connection between the bases than in a G-C pair.

The double-stranded DNA takes the three-dimensional shape that has the lowest energy constraints. The most stable shape is a double-stranded helix. The helix turns around a central axis in a clockwise manner and is considered a **right-handed helix**. One complete turn is 34 Å in length and has about 10 base pairs. DNA is not static, but can alter its conformation in response to various environmental changes. The typical conformation just described is the

FIGURE 1.3 Nucleic Acid Structure
(A) DNA has two strands antiparallel to each other. The structure of the subcomponents is shown to the sides.

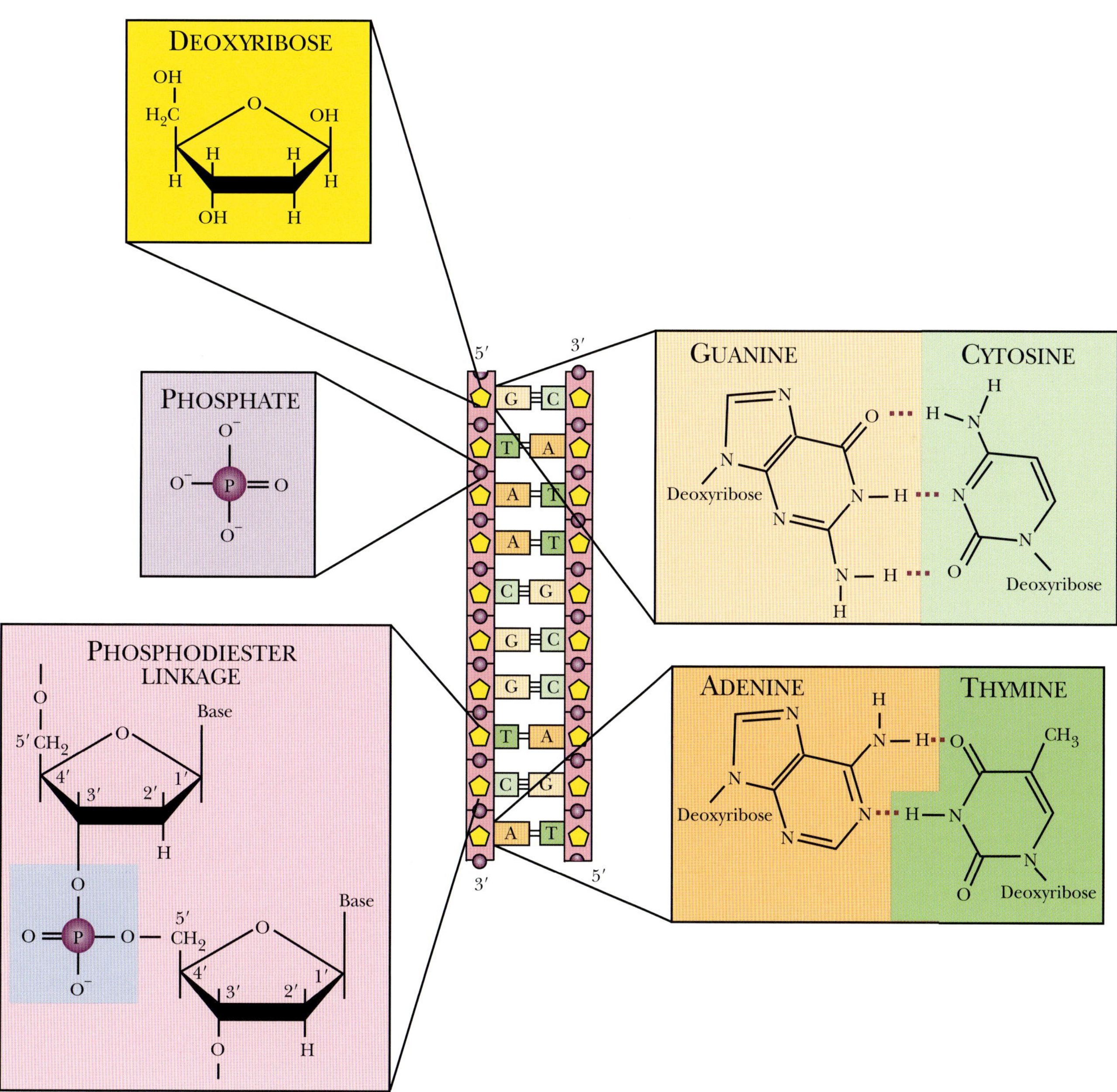

(Continued)

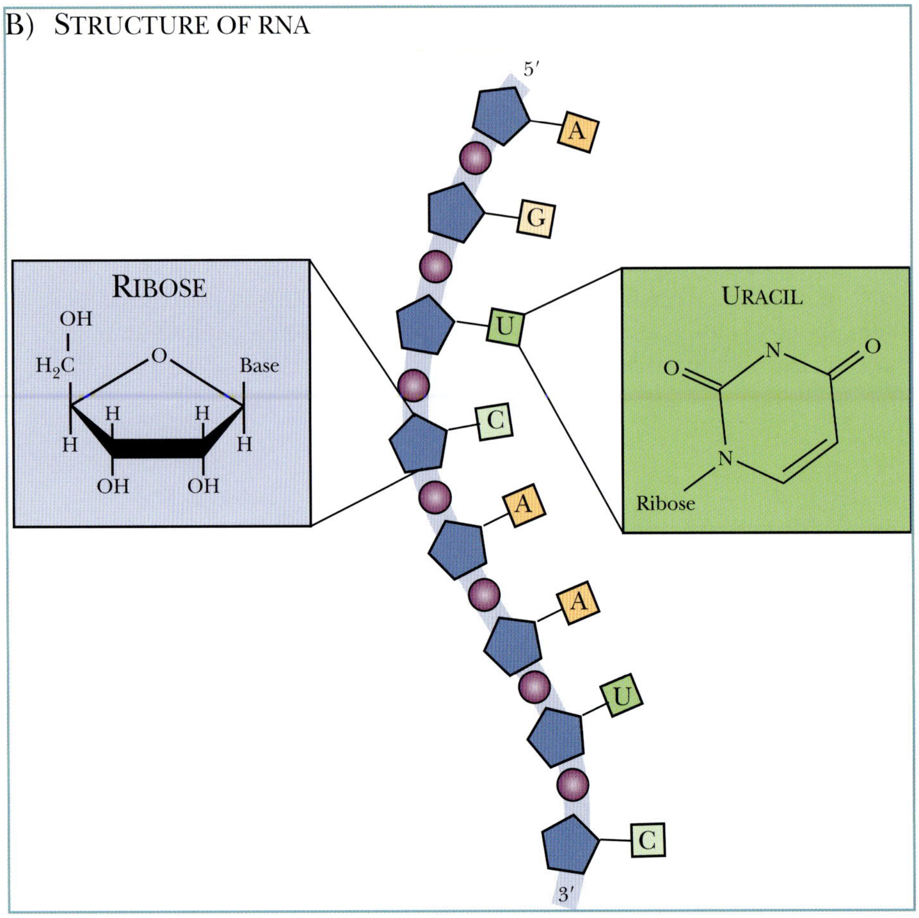

FIGURE 1.3, cont'd
(B) RNA is usually single-stranded and has two chemical differences from DNA. First, an extra hydroxyl group (-OH) is found at the 2′ position of ribose, and second, thymine is replaced by uracil.

B-form of DNA and is most prevalent in aqueous environments with low salt concentrations. When DNA is in a high-salt environment, the helix alters, making an **A-form** that has closer to 11 base pairs per turn. Another conformation of DNA is the **Z-form**, which has a left-handed helix with 12 base pairs per turn. In this form, the phosphate backbone has a zigzag conformation. These two forms may be biologically relevant under certain conditions.

> DNA and RNA are both structures with alternating phosphate and sugar residues linked to form a backbone. Base residues attach to the sugar and project out from the backbone. These bases can base-pair with another strand to form double-stranded helices.

PACKAGING OF NUCLEIC ACIDS

Bacteria have just a few thousand genes, each approximately 1000 nucleotides long. These are carried on a chromosome that is a single giant circular molecule of DNA. A single DNA double helix with this many genes is about 1000 times too long to fit inside a bacterial cell without being condensed somehow in order to take up less space.

In bacteria, the double helix undergoes **supercoiling** to condense it. Supercoiling is induced by the enzyme **DNA gyrase**, which twists the DNA in a left-handed direction so that about 200 nucleotides are found in one supercoil. The twisting causes the DNA to condense. Extra supercoils are removed by **topoisomerase I**. The supercoiled DNA forms loops that connect to a protein scaffold (see Fig. 1.4).

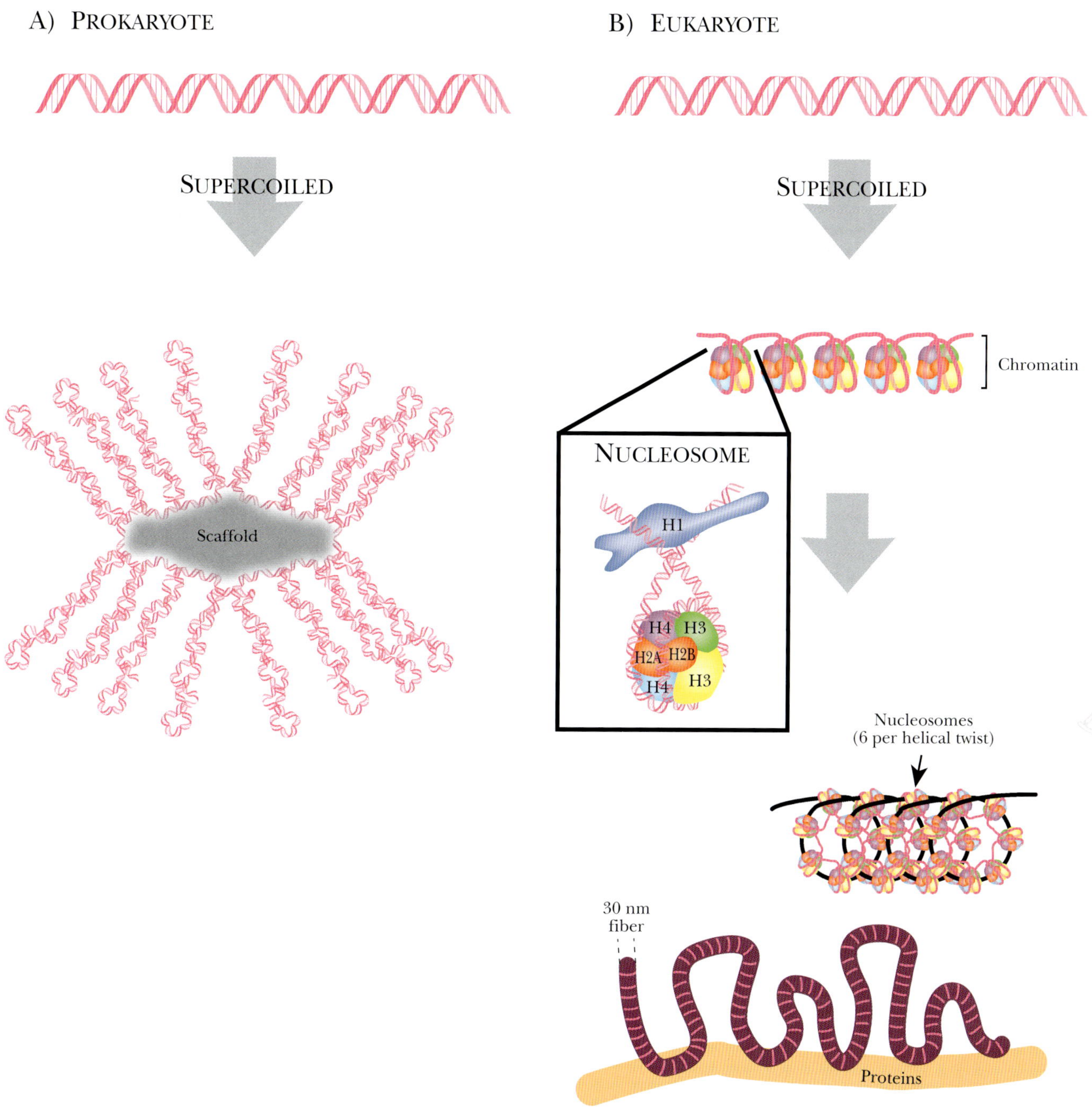

FIGURE 1.4 Packaging of DNA in Bacteria and Eukaryotes
(A) Bacterial DNA is supercoiled and attached to a scaffold to condense its size to fit inside the cell. (B) Eukaryotic DNA is wrapped around histones to form a nucleosome. Nucleosomes are further condensed into a 30-nm fiber attached to proteins at MAR sites.

In humans and plants, vastly more DNA must be packaged, so just adding supercoils is not sufficient. Eukaryotic DNA is wound around proteins called **histones** first. Histones have a positive charge to them, and this neutralizes the negatively charged phosphate backbone. DNA plus histones looks like beads on a string and is called **chromatin**. Each bead or **nucleosome** has about 200 base pairs of DNA and nine histones, two H2A, two H2B, two H3, two H4, and one H1. All the histones form the "bead" except for H1, which connects the beads by holding the DNA in the linker region. The histones are highly conserved proteins that are found in all eukaryotes and, in simplified form, in archaebacteria. Histone tails stick out from the nucleosome and are important in regulation. In regions of DNA that are expressed, the histones are loose, allowing regulatory proteins and enzymes access to the DNA. In regions that are not expressed, the histones are condensed, preventing other proteins from accessing the DNA (this structure is called **heterochromatin**).

Chromatin is not condensed enough to fit the entire eukaryotic DNA genome into the nucleus. It is coiled into a helical structure, the **30-nanometer fiber**, which has about six nucleosomes per turn. These fibers loop back and forth, and the ends of the loops are attached to a protein scaffold or chromosome axis. These attachments occur at **matrix attachment regions (MAR)** and are mediated by **MAR proteins**. These sites are 200–1000 base pairs in length and have 70% A/T. The structure of A/T-rich DNA is slightly bent, and these bends promote the connection between proteins in the matrix and the DNA. Often, enhancer and regulatory elements are also found at these regions, suggesting that the structure here may favor the binding of protein activators or repressors. This structure refers to chromosomes during normal cellular growth. When a eukaryotic chromosome readies for mitosis and cell division, it condenses even more. The nature of this condensation is still uncertain.

DNA must be condensed by supercoiling and wrapping around nucleosomes to form chromatin, and finally attached to protein scaffolds in order to fit into the nucleus.

BACTERIA AS THE WORKHORSES OF BIOTECHNOLOGY

DNA is the common thread of life. DNA is found in every living organism on earth (and even in some entities that are not considered living—see later discussion). Only a tiny selection of these organisms has been studied in the molecular biology laboratory. These few generally have special traits or features that make them easy to grow, study, and manipulate genetically. By now, each model organism has had its entire genome sequenced. The model organisms are used both as a guide to understand other related organisms not investigated in detail and for various more practical biotechnological purposes.

Bacteria live everywhere on the planet and are an amazing part of the ecosystem. There are an estimated 5×10^{30} bacteria on the earth, with about 90% of these living in the soil and the ocean subsurface. If this estimate is accurate, then about 50% of all living matter is microbial. Bacteria have been found in every environmental niche. Some bacteria live in icy lakes of Antarctica that only thaw a few months each year. Others live in extremely hot environments such as hot sulfur springs or the thermal vents at the bottom of the ocean (Fig. 1.5). There has been great interest in these extreme microbes because of their physiological differences.

For example, *Thermus aquaticus*, a bacterium from hot springs, can survive at temperatures near boiling point and at a pH near 1. Like others, this bacterium replicates its DNA using the enzyme **DNA polymerase**. The difference is that *T. aquaticus* DNA polymerase has to function at high temperatures and is therefore considered **thermostable**. Molecular biologists have exploited this enzyme for procedures like **polymerase chain reaction** or **PCR** (see Chapter 4), which is carried out at high temperatures. Other bacteria from extreme environments provide interesting proteins and enzymes that may be used for new procedures. Hydrothermal vents found on the ocean floor have revealed a fascinating array of novel organisms (see Fig. 1.5). Water temperatures in different vents range from 25 °C to 450 °C.

FIGURE 1.5
Hydrothermal Vent Tubeworms
These hydrothermal vent tubeworms from the Pacific Ocean get energy from symbiotic bacteria that live inside them. Courtesy of National Oceanic & Atmospheric Administration/National Undersea Research Program (NURP).

Bacteria are highly evolved into every niche of the planet and provide researchers with many unique properties.

ESCHERICHIA COLI IS THE MODEL BACTERIUM

Although extreme bacteria are interesting and useful, more typical bacteria are the routine workhorses for research in molecular biology and biotechnology. The most widely used is *Escherichia coli*, a rod-shaped bacterium about 1 by 2.5 microns in size. *E. coli* normally inhabits the colon of mammals including humans (Fig. 1.6). *E. coli* is a gram-negative bacterium that has an outer membrane, a thin cell wall, and a cytoplasmic membrane surrounding the cellular components. Like all prokaryotes, *E. coli* does not have a nucleus or nuclear membrane, and its chromosome is free in the cytoplasm. The outer surface of *E. coli* carries about 10 flagella that

A)

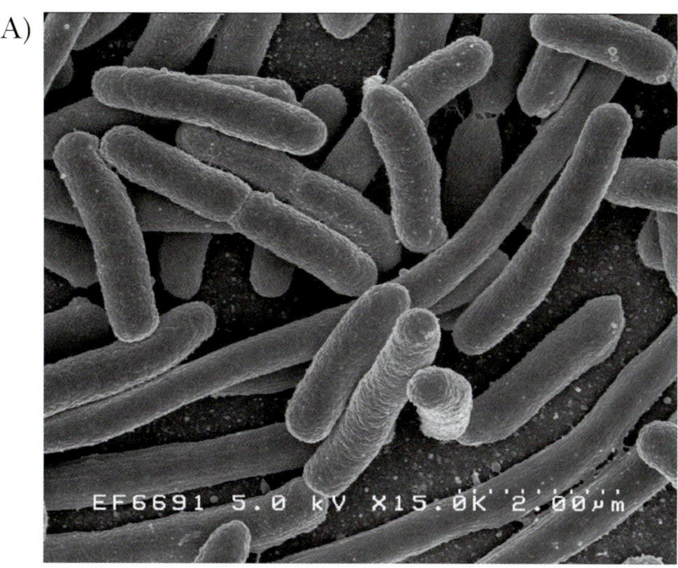

B) GRAM-NEGATIVE (e.g., *E. coli*)

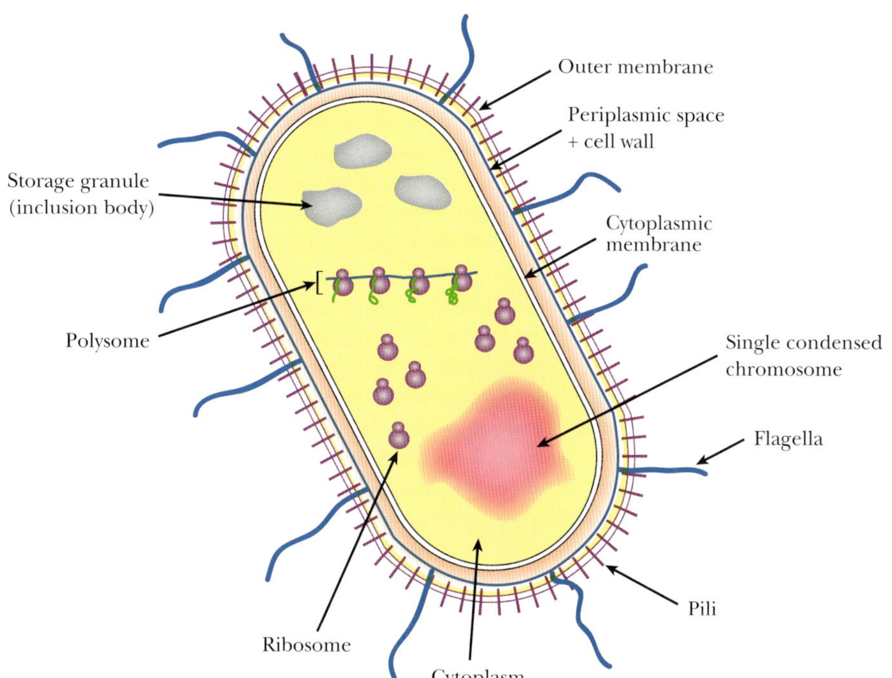

FIGURE 1.6
Subcellular Structure of *Escherichia coli*
(A) Scanning electron micrograph of *E. coli*. The rod-shaped bacteria are approximately 0.6 microns by 1–2 microns. Courtesy of Rocky Mountain Laboratories, NIAID, NIH. (B) Gram-negative bacteria have three structural layers surrounding the cytoplasm. The outer membrane and cytoplasmic membrane are lipid bilayers, and the cell wall is made of peptidoglycan. Unlike eukaryotes, no membrane surrounds the chromosome, leaving the DNA readily accessible to the cytoplasm.

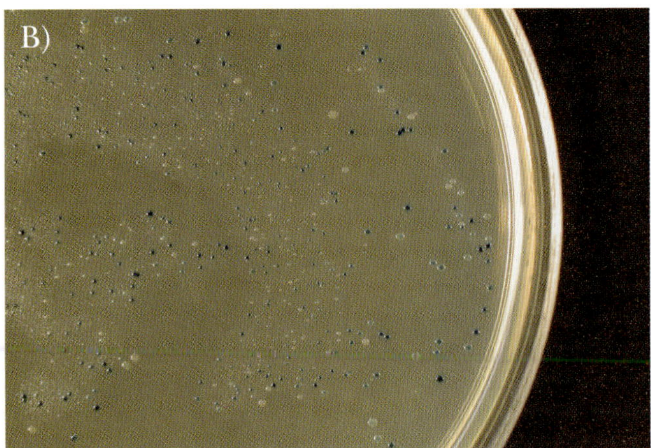

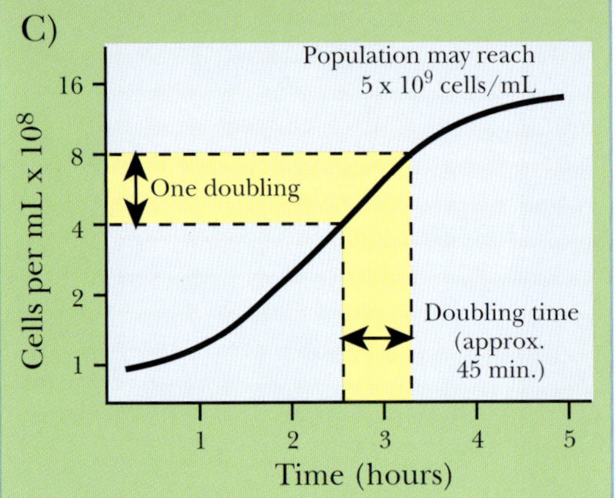

FIGURE 1.7 Bacteria are Easy to Grow
(A) Bacteria growing in liquid culture. (B) Bacteria growing on agar. This photo shows a mixture of bacterial colonies from the blue/white method for screening plasmid insertions—see Chapter 3 and Fig. 3.15 for a full explanation.
(C) Fast-growing bacteria can double in numbers in short periods. Here, the number of bacteria double after approximately 45 minutes and reach a density of 5×10^9 cells/mL in about 5 hours.

propel the bacteria to different locations, and thousands of pili that allow the cells to attach to surfaces.

Although the media often report about *E. coli*–contaminated food, *E. coli* is usually harmless. However, occasional strains of *E. coli* are pathogenic and secrete toxins that cause diarrhea by damaging the intestinal wall. This results in fluid being released into the colon rather than being extracted. *E. coli* O157:H7 is a particularly potent pathogenic strain of *E. coli* with two toxin genes that can cause bloody diarrhea. It is especially dangerous to young children, the elderly, and those with compromised immune systems.

Bacteria provide many advantages for research. Bacteria have growth characteristics that are very useful when large numbers of identical cells are needed. A culture of bacteria can be grown in a few hours and can contain up to 10^9 bacterial cells per milliliter. Growth can be strictly controlled, that is, the amount and types of nutrients, temperature, and time may all be adjusted based on the desired result. *E. coli* are so easy to grow that they can grow in mineral salts, water, and a sugar source. The cells can be grown in liquid cultures or as solid cultures on agar plates (Fig. 1.7). Liquid cultures can be stored in a refrigerator for weeks, and the bacteria will not be harmed. Additionally, bacteria can be frozen at −70 °C for 20 years or more, so different strains can be maintained without having to constantly culture them. *E. coli* are normally grown in air, but can grow anaerobically if an experiment requires that oxygen be eliminated.

Bacteria are single-celled organisms. The cells in a bacterial culture are identical in contrast to mammalian cells where even a single tissue contains many different types of cells. Each *E. coli* has one circular chromosome with one copy each of about 4000 genes. This is significantly fewer than in humans, who have two copies each of about 25,000 genes on 46 chromosomes. This makes genetic analysis much easier in bacteria (Fig. 1.8).

Escherichia coli is the model bacterial organism used in basic molecular biology and biotechnology research. The organism is simple in structure, grows easily in the laboratory, and contains very few genes.

MANY BACTERIA CONTAIN PLASMIDS

Because many different types of bacteria are found in every environment, competition for nutrients and habitat occurs regularly. Many bacteria compete using a form of biological warfare and secrete toxins, called **bacteriocins**, which kill neighboring bacteria. For example, nisin, a bacteriocin from *Lactococcus lactis*, kills other food-borne pathogens such as *Listeria monocytogenes* and *Staphylococcus aureus*. *E. coli* also produce bacteriocins, called **colicins**. *Bacteriocin* is a general term, whereas *colicin* specifically refers to toxins produced by *E. coli*. (Sometimes *colicin* is used as a general term, but this is not strictly correct.) *E. coli* makes different types of colicins, such as colicin E1 or colicin M, to kill neighboring cells. Colicins act by two main mechanisms. Some puncture the cell membrane, allowing vital cellular ions to leak out, and destroying the proton motive force that drives ATP production. Others encode nucleases that degrade DNA and RNA. These toxins do not affect their producer cells because the cell that makes the toxin also makes an **immunity protein** that recognizes the toxin and neutralizes it.

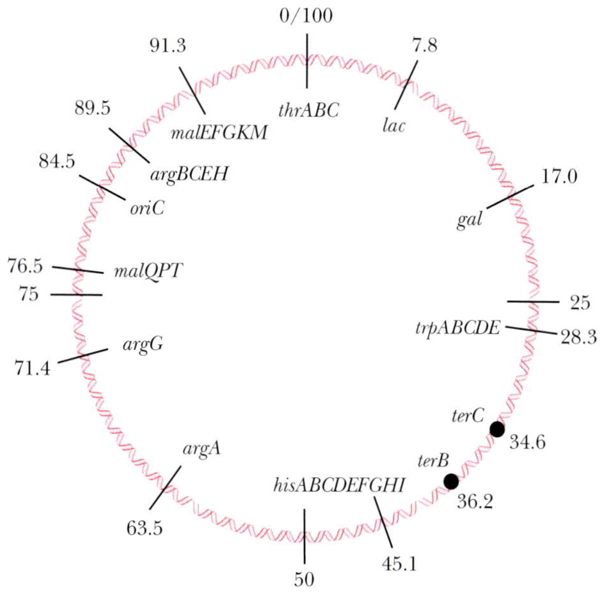

The ability to make colicin is due to the presence of an extrachromosomal genetic element called a **plasmid**. These are small rings of DNA that exist within the cytoplasm of bacteria and some eukaryotes such as yeast. A colicin-producing plasmid has several genes: the gene for the colicin, the gene for the immunity protein, and genes that control plasmid replication and copy number. In addition, all plasmids contain an **origin** for DNA replication. When the host cell divides, the plasmid divides in step (Fig. 1.9). These colicin plasmids are used extensively for molecular biology. The colicin genes have been removed, and the remaining segments have been greatly modified so that other genes can be expressed efficiently in bacteria. The resulting **recombinant plasmids** are the crux of all molecular biology. All the modern advances in biotechnology started with the ability to express **heterologous** proteins in bacteria (see Chapter 3 for cloning vectors).

FIGURE 1.8 The *E. coli* chromosome
The *E. coli* chromosome is divided into 100 map units, arbitrarily starting at the *thrABC* operon. Various genes and their locations are shown. The replication origin (*oriC*) and termination zone (*terB* and *terC*) are indicated.

> Another useful trait of *E. coli* is the presence of extrachromosomal elements called plasmids. These small rings of DNA are easily removed from the bacteria, modified by adding or modifying genes, and reinserted into a new bacterial cell where new genes can be evaluated.

OTHER BACTERIA IN BIOTECHNOLOGY

Other bacteria besides *E. coli* are used to produce biotechnology products. *Bacillus subtilis* is a gram-positive bacterium that is used as a research organism to study the biology and genetics of gram-positive organisms. *Bacillus* can form hard spores that can survive almost indefinitely. It is also used in biotechnology. For industrial production, secreting proteins through the single membrane of gram-positive bacteria is much easier than secreting them through the double membrane of gram-negative bacteria; therefore, *Bacillus* strains are used to make extracellular enzymes such as proteases and amylases on a large scale.

Pseudomonas putida is a bacterium that normally lives in water. It is a gram-negative bacterium like *E. coli*, but is commonly used in environmental studies because it is able to degrade many aromatic compounds (see Chapter 13). *Streptomyces coelicolor* is a soil bacterium that is gram positive. This organism degrades cellulose and chitin, and also produces a large number of different antibiotics. Another example of a common industrial microorganism is *Corynebacterium glutamicum*, which is used to produce L-glutamic acid and L-lysine for the biotechnology industry.

> Many different bacteria are used for biotechnology research because of their unique qualities.

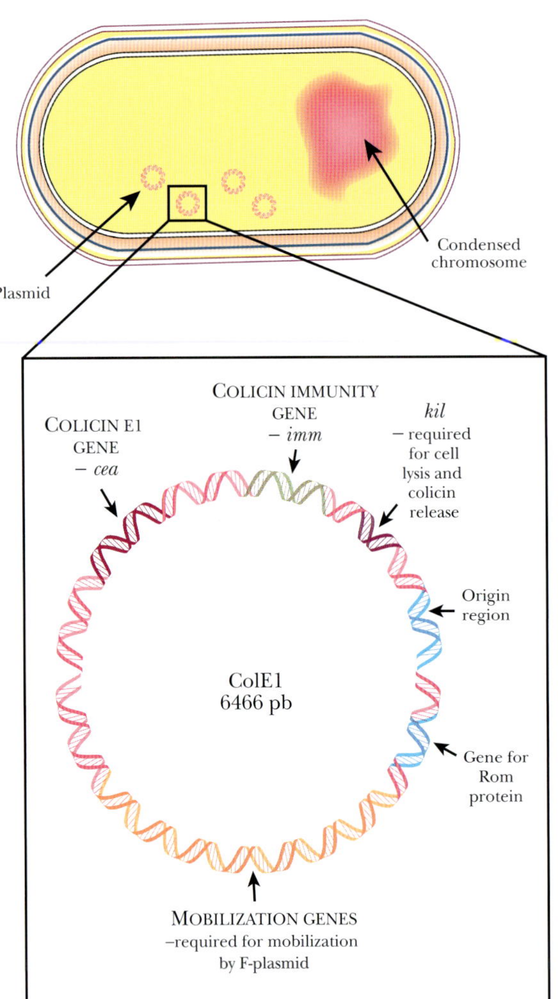

FIGURE 1.9
Plasmids Encode the Genes for Colicin
ColE1 plasmids are extrachromosomal DNA elements that are maintained by bacteria for producing a toxin (*cea* gene). They also carry genes for toxin release and immunity. These plasmids have been modified to carry genes useful in genetic engineering.

BASIC GENETICS OF EUKARYOTIC CELLS

Most eukaryotes are **diploid**, that is, they have two homologous copies of each chromosome. This is the case for humans, mice, zebrafish, *Drosophila*, *Arabidopsis*, *C. elegans*, and most other eukaryotes. Having more than two copies of the genome is extremely rare in animals, and only one rat from Argentina has been discovered with four copies of its genome. On the other hand, many plants, especially crop plants, are **polyploid** and contain multiple copies of their genomes. For example, ancestral wheat has seven pairs of chromosomes (i.e., its diploid state = $2n = 14$), whereas the wheat grown for food today has 42 chromosomes. Thus modern wheat is hexaploid. Domestic oats, peanuts, sugar cane, white potato, tobacco, and cotton also have four to six copies of their genome. This makes genetic analysis very difficult!

In animals, there is a division between **germline** and **somatic** cells. Germline cells are the only ones that divide to give haploid descendents. Diploid germline cells give rise to haploid gametes—the eggs and sperm that propagate the species. After mating, the two haploid cells fuse to become diploid (forming the zygote). Somatic cells, on the other hand, are normally diploid and make up the individual. Any mutations in a somatic cell disappear when the organism dies, whereas a mutation in a germline cell is passed on to the next generation (Fig. 1.10).

If a somatic cell is mutated early in development, all the somatic cells derived from this ancestral cell will receive the defect. Suppose this ancestral cell is the precursor of the left eye and that this defect prevents the manufacture of the brown pigment responsible for brown eyes. The right eye will be brown, but the mutant left eye will be blue (Fig. 1.11). Blue eyes are not due to blue pigment; they simply lack the brown pigment. People or animals with eyes that don't match are unusual but not incredibly rare. Such events are known as **somatic mutations**. They are not passed on to the offspring. Nonetheless, mutations in somatic cells can cause severe problems, as they are the cause of most cancers (see Chapter 18).

In plants, the division between germline and somatic cells is less distinct, because many plant cells are **totipotent**. A single plant cell has the ability to form any part of the plant, reproductive or not. This is not true for the majority of animal cells. Nevertheless, many animal cells do have the potential to form several different types of cells. A cell able to differentiate into multiple cell types is called a **stem cell**. Research on embryonic stem cells has become a hot political topic because of the potential ability to form an embryo. However, researching adult stem cells holds much promise. For example, researchers are hoping to identify stem cells that can form new neurons, so that patients with spinal cord injuries can be cured.

Eukaryotic cells are more complex than bacteria and contain multiple copies of the genome within a cell. Eukaryotic cells are also specialized, that is, some cells are for reproduction, some cells are stem cells that can differentiate into somatic cells, and some cells are specialized in function and shape.

YEAST AND FILAMENTOUS FUNGI IN BIOTECHNOLOGY

Fungi are incredibly useful microorganisms in the world of biotechnology. Anyone who has grown mold on a loaf of bread understands the ease with which these are cultured. Fungi are traditionally used in food applications. Yeasts are used in baking and brewing and other fungi in cheese-making, mushroom cultivation, and making foods such as soy sauce. Cheese production uses a variety of fungi. For example, a mold called *Penicillium roqueforti* makes the blue veins in cheeses such as Roquefort, and *Penicillium candidum, Penicillium caseicolum,* and *Penicillium camemberti* make the hard surfaces of Camembert and Brie cheeses. Soy sauce is made from soybeans that are fermented with *Aspergillus oryzae.*

Fungi are responsible for the production of many industrial chemicals and pharmaceuticals. The most famous is penicillin, which is manufactured by *Penicillium notatum,* in large tanks called **bioreactors**. Citric acid is a chemical additive to food that occurs naturally in lemons. It gives the fruit their sour taste. Rather than extracting citric acid from lemons, it has been manufactured since about 1923 by culturing *Aspergillus niger.*

Much like bacteria, yeast has a twofold purpose in biotechnology. It offers many of the same advantages as bacteria with the additional advantage of being a eukaryote. Yeasts are also important for production of biotechnological products. The most common research strain of yeast is brewer's or baker's yeast, *Saccharomyces cerevisiae.* This is the same little creature that makes the alcohol in beer and makes bread soft and fluffy by releasing carbon dioxide bubbles that get trapped in the dough.

Yeast is a single-celled eukaryote that has its cellular components compartmentalized (Fig. 1.12). Like all eukaryotes, yeasts have their genomes encased in a **nuclear envelope**. The nucleus and cytoplasm are separated, but they communicate with each other through gated channels called **nuclear pores**. *Saccharomyces cerevisiae* has 16 linear chromosomes that have **telomeres** and **centromeres**, two features not found in bacteria. The yeast genome was the first eukaryotic genome sequenced in its entirety. It has 12 Mb of DNA with about 6000 different genes. Unlike higher eukaryotes, yeast genes have very few intervening sequences or **introns** (see Chapter 2). Outside the nucleus, yeast has organelles including the endoplasmic reticulum, Golgi apparatus, and mitochondria to carry out vital cellular functions.

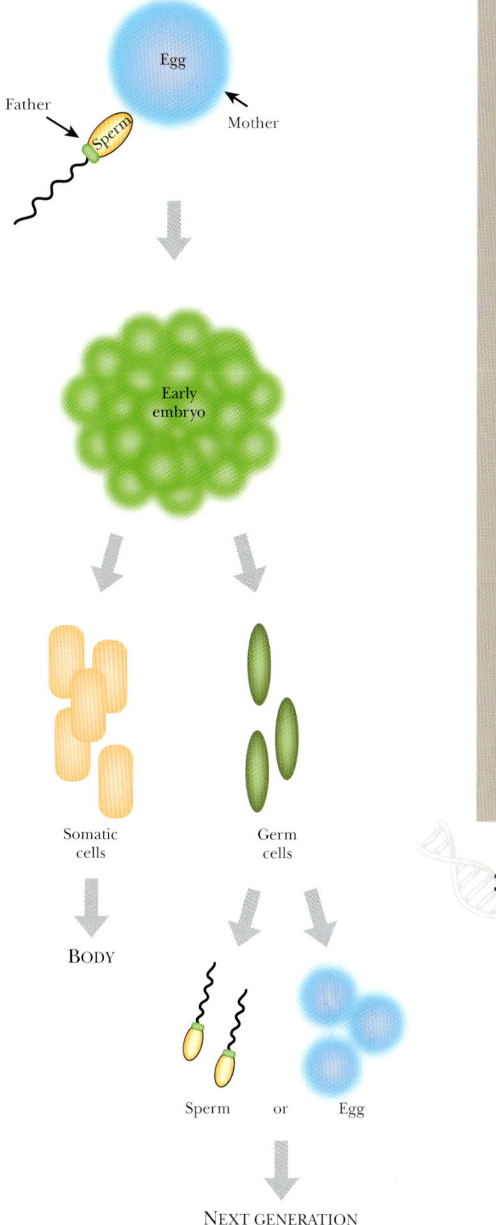

FIGURE 1.10
Somatic versus Germline Cells
During development, cells either become somatic cells, which form the body, or germline cells, which form either eggs or sperm. The germline cells are the only cells whose genes are passed on to future generations.

Like bacteria, yeast grow as single cells. A culture of yeast has identical cells, making genetic and biochemical analysis easier. The culture medium can either be liquid or solid, and the amount and composition of nutrients can be controlled. The temperature and time of growth may also be controlled. Under ideal circumstances, yeast doubles in number in about 90 minutes, as opposed to *E. coli*, which doubles in 20 minutes. Although slower than bacteria, the growth of yeast is fast in comparison to other eukaryotes. Like bacteria, yeast cells can be stored for weeks in the refrigerator and may be frozen for years at −70 °C.

Much like bacteria, some yeast cells also have extrachromosomal elements within their nuclei. The most common element is a plasmid called the **2-micron circle**. Like the chromosomes of all eukaryotes, the DNA of this plasmid is also wound around histones. This element has been exploited as a cloning vector (see Chapter 3) to express heterologous genes in yeast. The plasmid has two perfect DNA repeats (*FRT* sites) on opposite sides of the circle. The plasmid also has a gene for **Flp protein**, also called **Flp recombinase** or **flippase**. This enzyme recognizes the *FRT* sites and flips one half of the plasmid relative to the other via DNA recombination (Fig. 1.13). Flippase

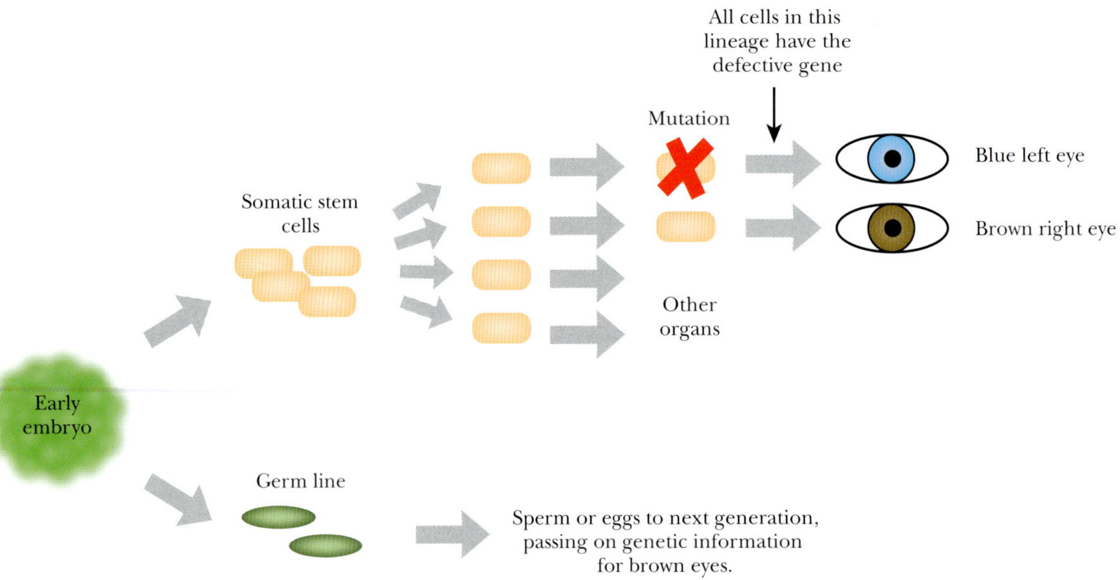

FIGURE 1.11 Somatic Mutations
The early embryo has the same genetic information in every cell. During division of a somatic cell, a mutation may occur that affects the organ or tissue it gives rise to. Because the mutation was isolated in a single precursor cell, other parts of the body and the germline cells will not contain the mutation. Consequently, the mutation will not be passed on to any offspring.

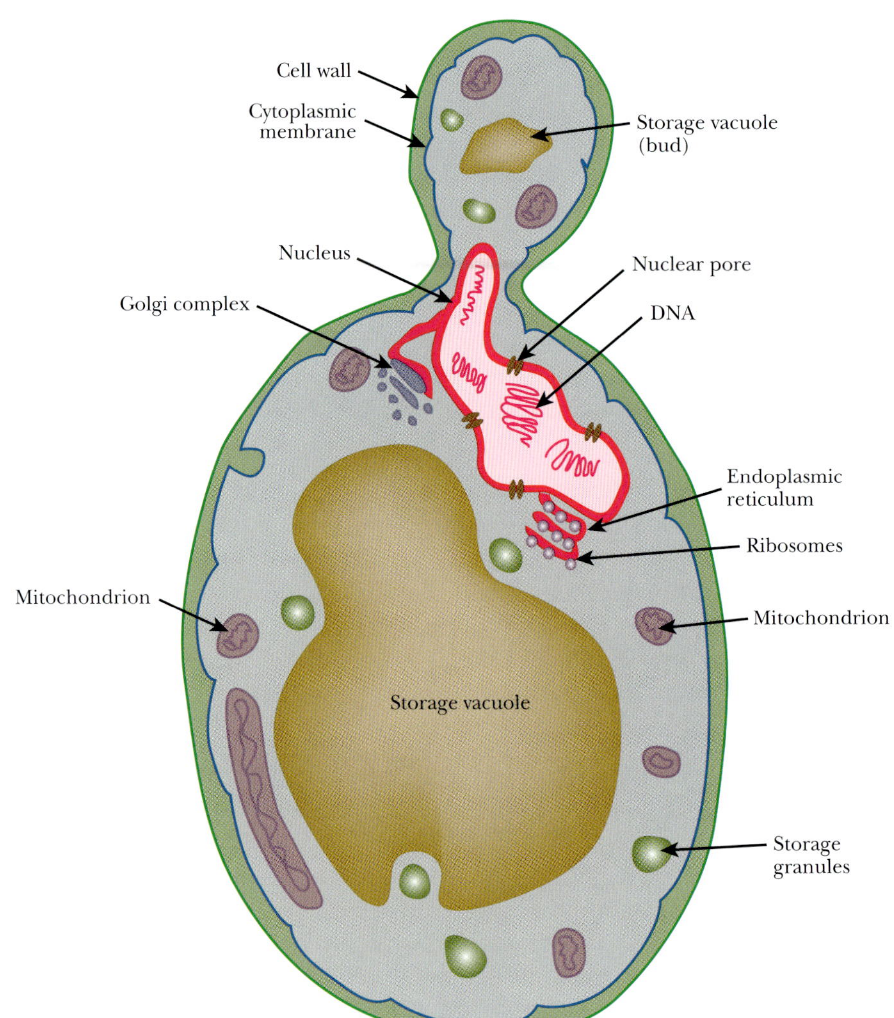

FIGURE 1.12 Structure of Yeast Cell
This yeast cell, undergoing division, is starting to partition components into the bud. Eventually, the bud will grow in size and be released from the mother (lower oval), leaving a scar on the surface of the cell wall.

recombines any DNA segments carrying *FRT* sites, no matter what organism they are in. Consequently, flippase is used in transgenic engineering in higher organisms (see Chapter 15). In plants, a related system, Cre (recombinase) plus *LoxP* sites, is used in a similar way (see Chapter 14).

Yeast offer a variety of advantages to biotechnology. They are single-celled organisms that grow fast. Yeast are eukaryotes with chromosomes that have telomeres and centromeres, like the human genome.
Yeast cells have extrachromosomal elements similar to plasmids that are used to study new genes.

YEAST MATING TYPES AND CELL CYCLE

Yeast cells grow and divide by budding. Cellular organelles such as mitochondria and some cellular proteins are partitioned into the growing **bud**. Finally, mitosis creates another nucleus and when the bud has reached a sufficient size, the new daughter cell is released, leaving a scar on the surface of the mother cell. Budding creates genetically identical cells because the genome divides by mitosis.

Yeast has diploid and haploid phases in its life cycle, greatly simplifying genetic analysis. Most yeast found in the environment is diploid, having two copies of its genome. Under poor environmental conditions, yeast can undergo meiosis, creating four **haploid spores**, called **ascospores**, contained within an **ascus**. These are released to find a new environment with more nutrients. If the spores find a better environment, they germinate. In the laboratory, the haploid cells can be isolated and grown separately, but in the wild, haploid cells quickly fuse with another, forming diploid cells again (Fig. 1.14). This life cycle allows individual genes to be followed during segregation and inheritance patterns to be analyzed much as with Mendel's peas. However, the shorter life cycle of yeast allows greater numbers to be analyzed.

Just as meiosis creates haploid male and female gametes in humans, meiosis in yeast creates haploid cells of two different mating types. Because they are structurally the same, rather than calling them male and female, the yeast mating types are **a** and **α**. Fusion may only occur between different mating types, that is, only an **a** plus an **α** cell can merge forming a diploid. Each mating type expresses a distinct mating pheromone that binds to receptors on the opposite mating type. The pheromones are secreted into the environment. For example, when an **a** cell encounters the **α** pheromone, a cell surface receptor, the **α** receptor, binds the **α** pheromone, readying the yeast for fusion. Conversely, when **α** cells encounter an **a** pheromone, the cell surface **a** receptor binds the **a** pheromone and readies the cell for mating. The two cells fuse, combining two different genomes into one. The exchange of genes during sex is important for evolution, as it forms new genetic combinations that may have an advantage in different environments.

Yeast cells control mating type through a genetic locus, called the ***MAT* locus**, for mating type. This segment of DNA has two genes that are transcribed in opposite directions. In haploid **a** cells the locus genes are *MAT***a1** and *MAT***a2**. In **α** cells the locus genes are *MAT***α1** and *MAT***α2**. During meiosis, the genes at this locus are exchanged by recombination so that half the spores have *MAT***a1** and *MAT***a2** and the other half, *MAT***α1** and *MAT***α2**. The gene products control expression of the mating pheromone and receptors in the haploid cells.

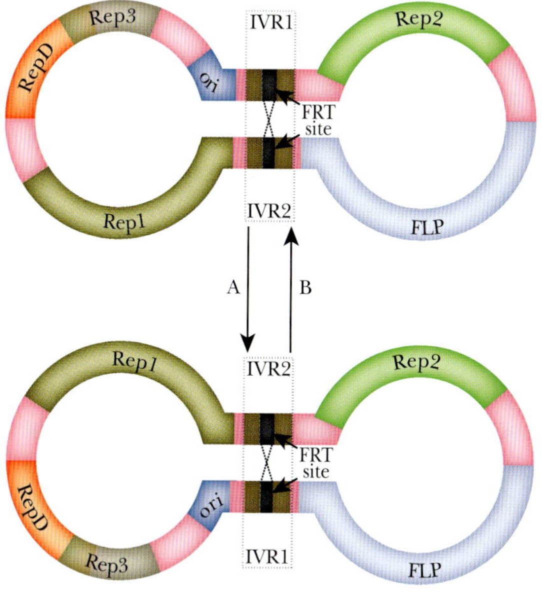

FIGURE 1.13 The 2-Micron Plasmid of Yeast
Two different forms of the 2-micron plasmid are shown. The enzyme Flp recombinase recognizes the *FRT* sites and recombines them, thus flipping one half of the plasmid relative to the other half.

Diploid yeast will also form genetic clones by budding when plenty of nutrients are available for growth.

Yeast, like other eukaryotic organisms, can create new genetic combinations with sexual reproduction. The two forms of haploid yeast are **a** and α, which are functionally equivalent to male and female. These mate to form a new genetically unique diploid cell.

MULTICELLULAR ORGANISMS AS RESEARCH MODELS

Single-celled creatures offer many advantages, but understanding human physiology requires information about cellular interactions. Although single-celled organisms interact with each other, this is not the same as multicellular organisms where one cell is surrounded by other cells on all sides. The location of cells affects both their role and development. The cells in our hair follicles are different from our skin cells. Bone cells differ drastically from the long nerve cells of our spinal cord. Much basic work on cellular interactions, development of multicellular organisms, and understanding cellular physiology in different tissues has been done on the roundworm *Caenorhabditis elegans*. Although this is a multicellular organism, it is still relatively simple compared to mammals or other vertebrates.

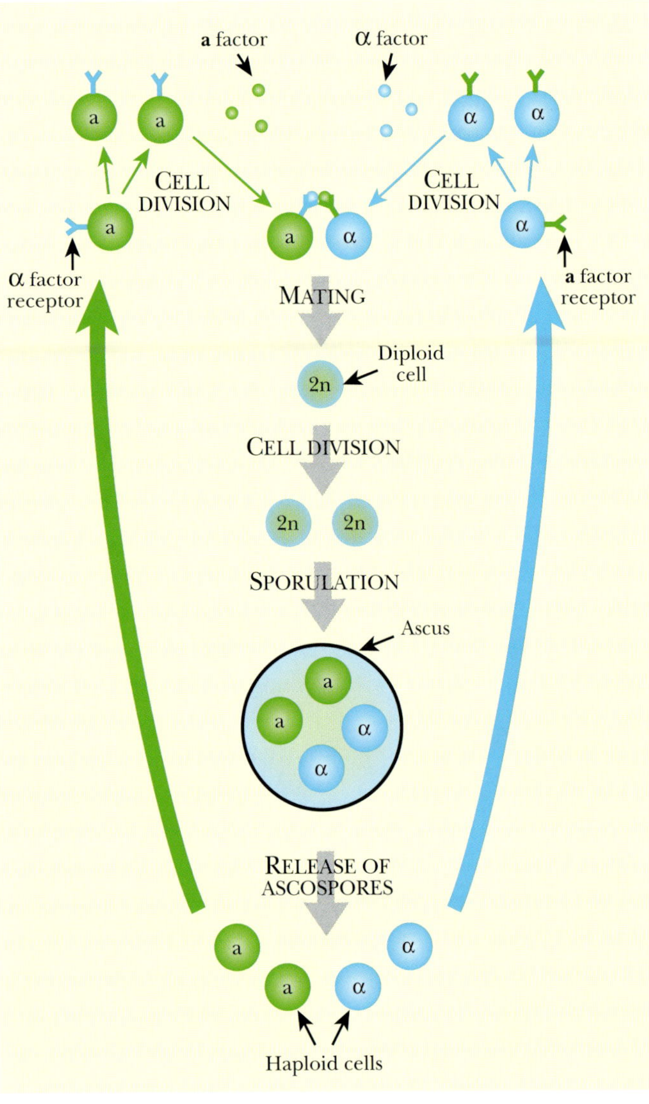

FIGURE 1.14
Alternating Haploid and Diploid Phases of Yeast
Haploid cells come in two different forms, **a** and **α**. These express mating pheromones, **a factor** and **alpha (α) factor**, which attract the two forms to each other. When the pheromones bind to receptors on the opposite cell type, the two haploid cells become competent to fuse into a diploid cell. Diploid cells sporulate under growth limiting conditions. Otherwise, the diploid cells form genetic clones by budding.

Caenorhabditis elegans, a Small Roundworm

C. elegans is a small roundworm that is found in soil, particularly rotting vegetation, where it feeds on bacteria (Fig. 1.15). There are two sexes, a self-fertilizing hermaphrodite and a male, allowing genetic studies on both self- and cross-fertilization. The body is shaped as a simple nonsegmented tube that is encased in a cuticle layer to prevent dehydration. Inside *C. elegans*, there are 959 somatic cells, which include more than 300 neurons. The head has many sense organs that respond to taste, smell, temperature, and touch, but no eyes. There is a nerve ring that serves as the brain and a nerve cord that runs down the back of the body. The digestive system consists of a pharynx followed by intestine and anus. There are 81 muscle cells that control the sinusoidal movement of the worm around its environment. The reproductive system occupies the largest volume within the worm. In the hermaphrodite, the tail is long and tapered, whereas the male has a blunt end. The hermaphrodite has a vulval opening where it lays eggs. The sperm cells come either from itself or from a male *C. elegans* in a sexual encounter.

C. elegans has many advantages for molecular biology and genetics. These creatures are transparent and can be studied in real time using various fluorescent techniques. They have many physiological characteristics similar to higher animals. For example, they undergo programmed cell death, and the genes involved are similar to genes found in humans (see Chapter 20). *C. elegans* are used to study development, aging, sexual dimorphism, alcohol metabolism, and many other phenomena that apply to humans.

The life cycle of *C. elegans* is conducive to research. One generation lasts about 3 days. First, a sperm and egg cell fuse, and a single-celled egg develops within the hermaphrodite's body. After the egg hatches, the larval stages begin. There are four stages, with L4 considered the adult. The sexual organs develop at the transition between L2 and L3, and sperm and egg cells develop in late L3, before the transition from L3 to L4 (Fig. 1.16). Genetic analysis is also helped because *C. elegans* is a hermaphrodite. Self-fertilization allows homozygous organisms to be easily generated. This is useful when mutants are isolated.

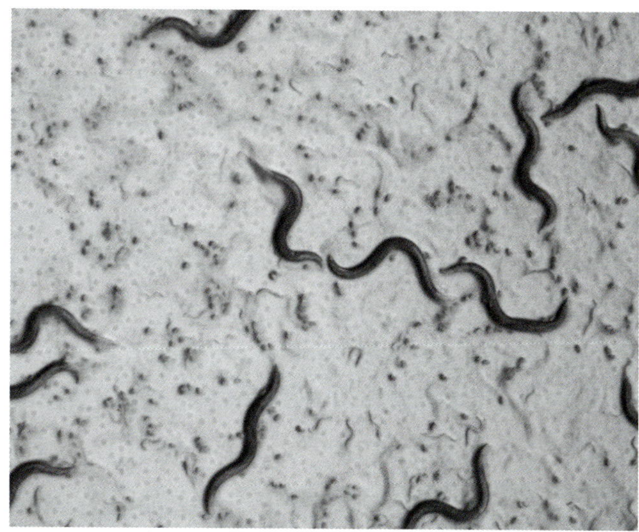

FIGURE 1.15 ***Caenorhabditis elegans***
Courtesy of Jill Bettinger, Virginia Commonwealth University, Richmond, VA.

C. elegans is a model multicellular eukaryotic organism. Biotechnology research uses this organism because it is easy to grow, it is transparent, and it is a hermaphrodite, so it can create either genetic clones or novel genetic organisms.

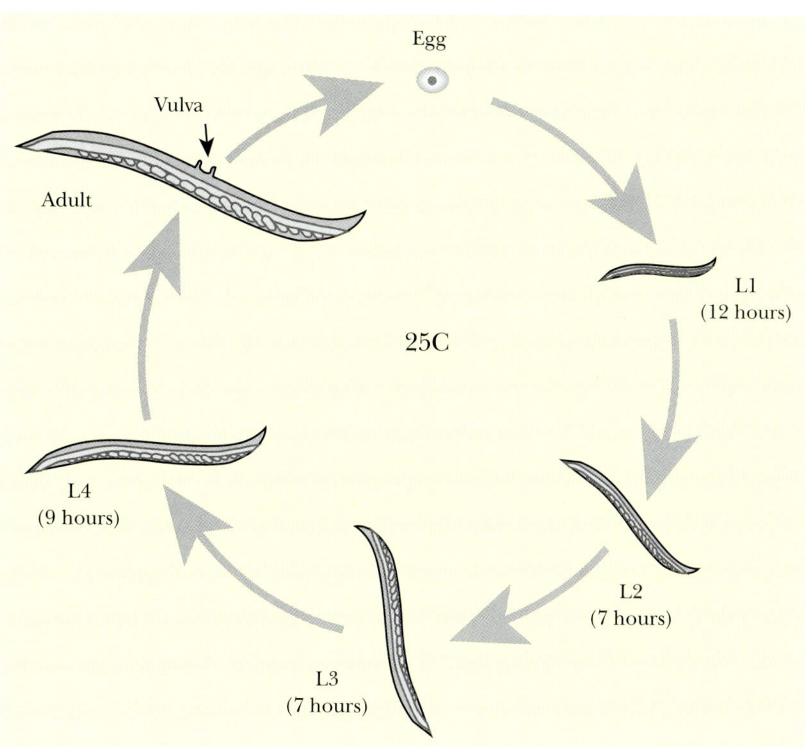

FIGURE 1.16 Life Cycle of *Caenorhabditis elegans*
When the *C. elegans* sperm fuses with an egg, a small worm develops (L1). The larva goes through multiple stages until it reaches the sexually mature adult phase. *C. elegans* has six different chromosomes: five autosomes and one X chromosome. The worms are diploid, with two sets of chromosomes. When the embryo has two X chromosomes, it becomes a hermaphrodite. If the embryo has only one X, it becomes a male, but males make up only 0.05% of a normal population. The genome is 97 Mb and was completely sequenced in 1998. Approximately 27% of the genome is coding sequence with about 19,000 genes, more than 900 of which are RNA coding genes. The average gene contains five introns and is about 3000 base pairs long. Intronic DNA accounts for 26% of the total genome. The remaining 47% of the genome is intergenic and noncoding.

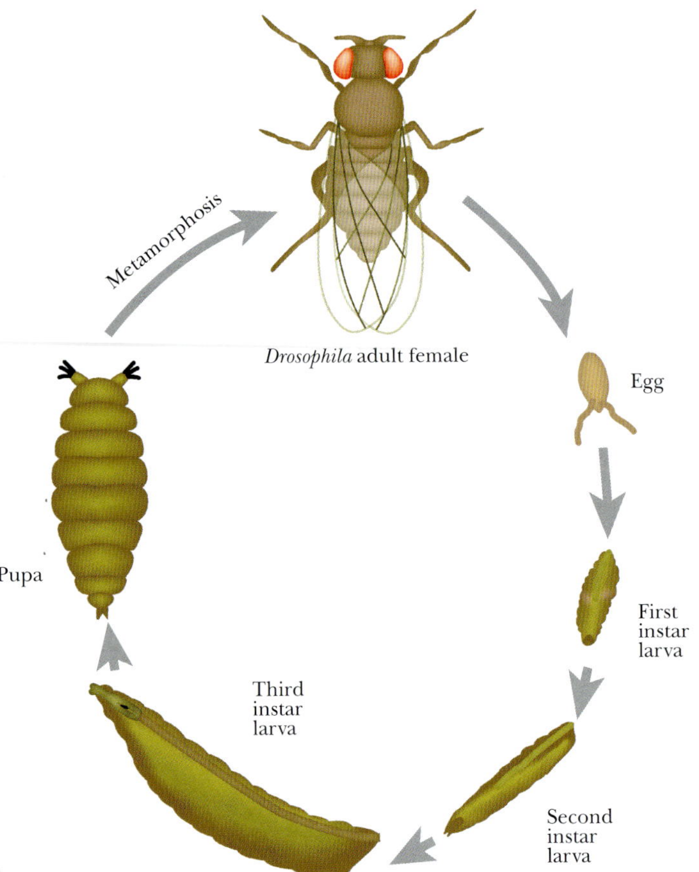

FIGURE 1.17 Life Cycle of *Drosophila melanogaster*
Drosophila fruit flies start as a tiny egg that develops into a worm (*maggot*). After a series of larval stages, the worm forms a pupa where the adult form develops.

Drosophila melanogaster, the Common Fruit Fly

Another multicellular model organism widely used because of its genetics is *Drosophila melanogaster*, usually referred to simply as *Drosophila*, the common fruit fly. This insect is about 3 mm in length and can often be found around rotting fruit. These flies are easy to grow and maintain in a lab. They need a food source and are kept in a bottle capped with cotton so they cannot escape. Their entire life span is 2 weeks and starts with an egg about 0.5 mm in length (Fig. 1.17). The embryo develops in one day in the egg and hatches into a worm-like larva. There are three larval instars that develop 1 day, 2 days, and 4 days after hatching. Each instar grows and eats continuously and molts to form the next instar. The third larval instar forms a pupa that is immobile. The pupa usually clings to the side of the glass flask, where it stays for about 6 days. During this time the larva transforms into the winged adult fly. Wings, legs, antenna, segmented bodies, eyes, and hair are formed.

The main focus of *Drosophila* research is genetics. Many different mutations are available, from simple changes such as longer or shorter body hairs, to dramatic mutations where body segments are duplicated. That is, some mutants of *Drosophila* have four wings or extra legs. Studying these mutants has identified many genes that determine basic body patterns in *Drosophila*, and based on homology, humans too. The genome of *Drosophila* has been sequenced and has 165 Mb of DNA, divided among three autosomes and the X/Y sex chromosomes. There are a predicted 12,000 genes in the genome.

During the rapid growth of the larval stages, the number of cells actually remains fairly constant. The size of the cells does increase dramatically, though. In order for these large cells to work, a large amount of extra protein and mRNA needs to be made, and the chromosomes duplicate hundreds of times to provide multiple gene copies. Although they duplicate, they do not divide but stay attached to each other, creating thick **polytene chromosomes** (Fig. 1.18). These are so large they can be visualized under a light microscope. The polytene chromosomes have characteristic banding patterns, with each section of each chromosome being unique. The banding pattern allows some mutations to be localized. For example, a deletion that causes white eyes (in the adult) would alter the banding pattern on the corresponding polytene chromosome. Thus the mutation can easily be mapped to its chromosome location.

The true sexual reproduction of *Drosophila* allows genetic manipulations, and the complex alterations that occur in the pupal to adult fly metamorphosis are two key characteristics that are studied by researchers.

Zebrafish Are Models for Developmental Genetics

The small zebrafish (*Danio rerio*) is a simple vertebrate used in molecular biology research. It is a common fish found in pet stores for keeping in freshwater aquariums. The qualities that have made it so prevalent as a pet also make it attractive for research. It is easy to maintain and breed in an aquarium. A wide variety of mutations exist, which makes the fish handy for genetics research. The adult is about an inch long with black horizontal stripes down its body (Fig. 1.19). The mother lays about 200 eggs at one time, so many offspring can be studied after one mating. Embryonic development occurs outside the mother. The embryos are completely transparent, so that the effects of mutations that affect embryo development can be seen with ease. Moreover, different cells can either be destroyed or moved to new locations, and the effect on development can be traced. Such experiments are wonderful for deciphering the effect of position on cellular development. The embryos develop from one single cell to a tiny fish in about 24 hours, so that studies of development can be done relatively quickly.

The zebrafish genome has been sequenced. There are 25 pairs of chromosomes with a haploid genome size of 1700 Mb of DNA. About 75% of zebrafish DNA is homologous to humans. Thus, when a new gene function is identified in the fish, it suggests possible roles for corresponding human genes.

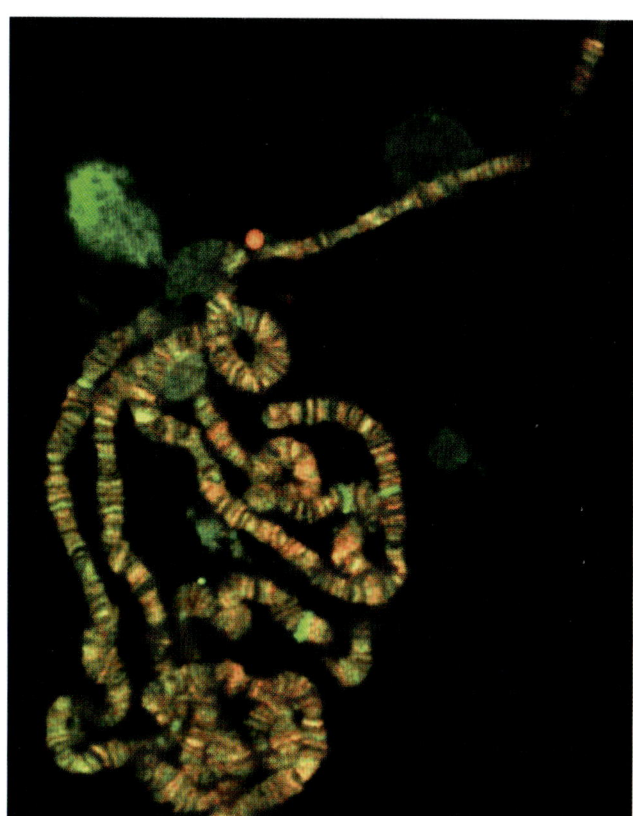

FIGURE 1.18 Polytene Chromosome Fluorescent staining of polytene chromosome from *Drosophila.*

Zebrafish are key organisms to study development of embryos because they have live babies that develop outside the mother.

Mus musculus, the Mouse, Is Genetically Similar to Humans

The model organism most closely related to humans is the mouse. The mouse genome has about 2600 Mb of DNA on 20 different chromosomes. Less than 1% of the genes have no human gene counterpart, so mouse genetics relates to humans very readily. Mice are easy to manipulate genetically, and animals with one or more genes inactivated (knockout mice) are fairly easy to generate (see Chapter 15). In addition to genetic deletions, extra genes can be inserted and expressed in the mouse, giving **transgenic** animals (see Chapter 15). The effect of such genetic manipulations on growth, development, or physiology can be determined.

FIGURE 1.19 The Zebrafish, *Danio rerio* This fish is used as a model vertebrate to study genetics, cell biology, and developmental biology.

Researchers consider mice very similar to humans because they have so many genes in common.

ANIMAL CELL CULTURE *IN VITRO*

Another way to approximate human physiology is by studying mammalian cells cultured *in vitro* (Fig. 1.20). Many different cell lines have been generated from humans and monkeys,

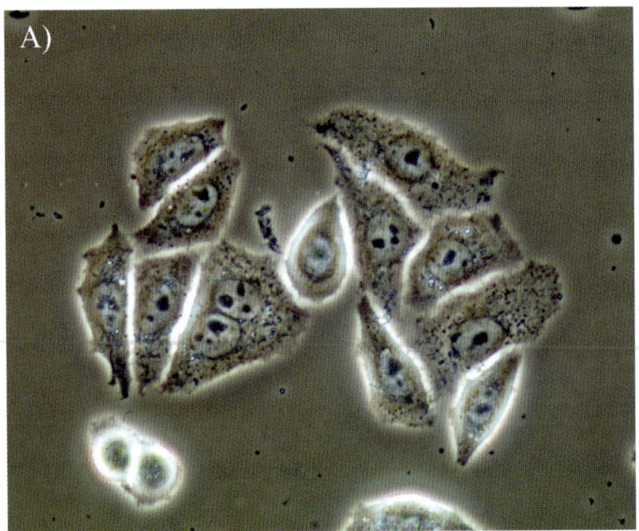

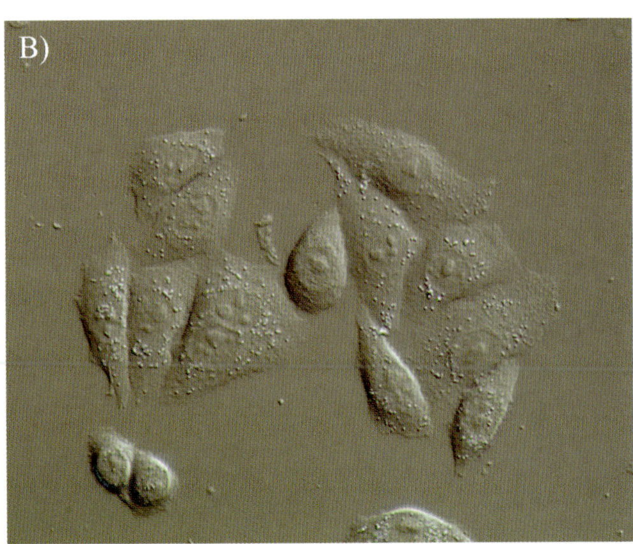

FIGURE 1.20 Human HeLa Cells Grown *in Vitro*
HeLa cells were taken from the tumor of Henrietta Lacks, a woman suffering from cervical cancer, in the 1950s and have been cultured continuously ever since. (A) Viewed under phase contrast. (B) Viewed under differential interference contrast. Courtesy of Michael W. Davidson, Optical Microscopy Group, National High Magnetic Field Laboratory, Florida State University, Tallahassee, Florida.

and these can be grown in plastic dishes or flasks using culture media containing growth factors and nutrients. Cell lines must be maintained at 37°C and require an atmosphere rich in carbon dioxide. **Adherent cell lines** stick to and divide on the plastic dishes, whereas **suspension cells** grow and divide in liquid culture. Most cell lines are one particular type of cell from a particular tissue, and many different cell lines have been grown from kidney, liver, heart, and so forth. The original cell lines cannot divide in culture forever. Primary cells, as they are called, can be maintained for only a short time. Using cancer cells overcomes this limitation since cancer cells do not stop dividing (see Chapter 18 for discussion). These cell lines are immortal and can, in principle, be grown under the correct circumstances forever.

The best aspect of using cultured human cells is the ability to do genetic studies. Different genes can be expressed in cultured cells, and their effect on cellular physiology can be determined. In addition, gene deletions or mutations can be examined. Although the cells are not inside a human, their genetic complement is identical to those in the whole organism. Cultured mammalian cells are also important for production of recombinant proteins, which are then isolated and purified for medicine, research, and other biotechnology applications.

Cell lines have also been developed from insects (Fig. 1.21). They are primarily used to express heterologous proteins for the biotechnology industry. Insect cells are preferred to mammalian cells because they require fewer nutrients for growth, surviving in media free of serum. Mammalian cells require serum from fetal cows, which is very expensive and in limited supply. Insect cells also grow at lower temperatures without carbon dioxide, and therefore do not require special incubation chambers.

Insect cells are used in research to study viruses that are transmitted between insects and plants, as well as cell signaling pathways. Insect cell lines are primarily derived from *Spodoptera frugiperda* (fall armyworm), *Trichoplusia ni* (cabbage looper), *Drosophila melanogaster* (fruit fly), *Heliothis virescens* (tobacco budworm), the mosquito, and others. The most common cell lines are those from ovarian tissue of *S. frugiperda*, which include Sf9 and Sf21 cells, those from embryonic cells from *T. ni*, which include the "High Five" cell lines, and those from late-stage *Drosophila* embryos, which include Schneider S2 cells.

Studying cells in a dish rather than an organism provides the researcher with another way to study genes. The cell lines are useful for genetic manipulations such as expressing new genes or deleting existing genes.

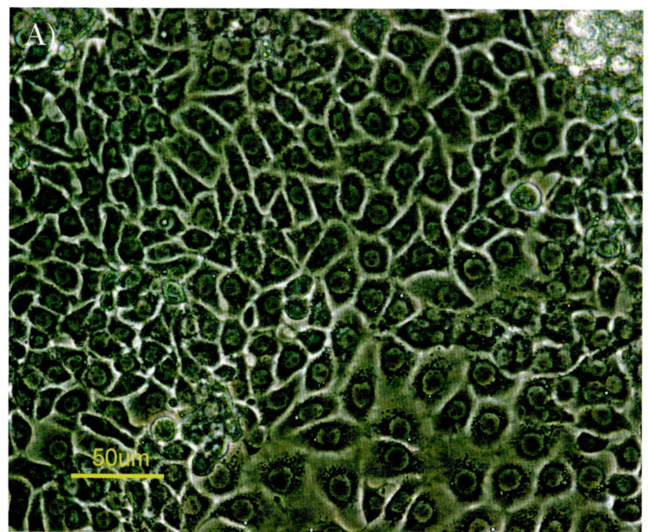

FIGURE 1.21 Insect Cells in Culture
(A) HvT1 cells from tobacco budworm testes are strongly attached to the surface of the dish. (B) TN368 cells from cabbage looper ovary are only loosely attached. Courtesy of Dwight E. Lynn, Insect Biocontrol Lab, USDA, Beltsville, MD.

ARABIDOPSIS THALIANA, A MODEL FLOWERING PLANT

The model organism most widely used in plant genetics and molecular biology is the weed *Arabidopsis thaliana*, wild mustard weed or mouse ear cress (Fig. 1.22). Plant research has typically lagged behind research on humans, but there is extensive interest. Growing different crops to feed the world population is incredibly important, and much money is invested in research on the crops most used for food, such as rice, soybean, wheat, and corn. These plants have huge genomes, and most are polyploid—even hexaploid (such as wheat). Therefore, a model organism is essential to learn the basic biology of plants. *Arabidopsis* has much the same responses to stress and disease as crop plants. Moreover, many genes involved in reproduction and development are homologous to those in plants with more complex genomes.

Arabidopsis has many convenient features. First, it is easily grown and maintained in a laboratory setting. The plant is small and grows to match its environment. If there is plenty of space and nutrients, the plant can grow to over a foot in height and width. If the environment is a small culture dish in a lab, the plant will grow about 1 cm in height and width. At either size, the plant forms flowers and seeds. An entire generation from seed to adult to seeds is finished in 6–10 weeks, which is relatively quick for a plant. (Note that for corn or soybeans, only one generation can occur in the span of a summer.) In *Arabidopsis*, many seeds are produced on each plant, so aiding genetic analysis. Much like yeast, *Arabidopsis* can be maintained in a haploid state.

Arabidopsis has a small genome for a plant, containing only five chromosomes with a total of 125 Mb of sequence. The genome was completely sequenced in 2000, allowing researchers to identify about 25,000 genes and important sequence features. Rice has also been sequenced and has an estimated 40,000 to 50,000 genes. This tops the number of predicted human genes, and so rice (and doubtless many other plants) may be more "advanced" than us lowly humans.

Plant research also relies on a model organism to study. *Arabidopsis thaliana* is used because of its size, ease of growth, and small genome.

FIGURE 1.22
Arabidopsis thaliana
The plant most used as a model for molecular biology research is *A. thaliana*, a member of the mustard family (Brassicaceae). Courtesy of Dr. Jeremy Burgess, Science Photo Library.

VIRUSES USED IN GENETICS RESEARCH

Viruses are entities that border on living. But unlike genuine living organisms, viruses cannot survive outside a host organism. Viruses are pathogens that invade host cells and subvert them to manufacture more viruses. Viruses are simple in principle and yet very powerful. They consist of a protein shell called a **capsid** surrounding a genome made of RNA or DNA. The particle is called a **virion** and, unlike a living cell, has no way to make its own energy or duplicate its own genome. The virus relies on the host to do this work.

Viruses come in many different types and can inhabit every living thing from bacteria to humans to plants. Viral diseases in humans are extremely common, and most cause only minor symptoms. For example, when rhinovirus invades, the victim ends up with a runny nose and other cold symptoms and usually feels miserable for a few days. However, viruses do cause a significant number of serious diseases, such as AIDS, smallpox, hepatitis, or the recently identified Ebola.

When viruses invade bacteria, the infected bacteria usually die. Bacterial viruses are called **bacteriophage** or **phage**, and they normally destroy the bacterial cell in the process of making new viral particles. When bacteria grow on an agar plate, they form a hazy or cloudy layer (a bacterial **lawn**) over the top of the agar. If the culture of bacteria is infected with bacteriophage, the viruses eat holes or **plaques** into the bacterial lawn, leaving clear zones where all the bacteria were killed.

Bacteriophages, like other types of viruses, have the following stages of their life cycle (Fig. 1.23):

- **(a)** Attachment of the virion to the correct host cell
- **(b)** Entry of the virus genome
- **(c)** Replication of the virus genome
- **(d)** Manufacture of new virus proteins
- **(e)** Assembly of new virus particles
- **(f)** Release of new virions from the host

Not every virus kills the host cell, and in fact, many viruses have a latent phase where they lie dormant within the cell, not producing any proteins or new viruses. **Latency**, as it is called in animal cells, is also called **lysogeny** when referring to bacteria. (In contrast, the phase of viral growth where the host cell is destroyed is called the **lytic phase**.) Sometimes a virus becomes latent by inserting its genome into the genome of the cell. The viral genome integrates into a host chromosome and remains inactive until some stimulus triggers it to reactivate. The integrated virus is called a **provirus** (or a **prophage** if the virus invades bacteria).

The great variety of viruses can be divided into groups based on capsid shape or the type of genome. The three major shapes are spherical, filamentous, and complex. Spherical viruses actually have 20 flat triangular sides and are thus icosahedrons. Complex viruses come in various shapes, but some have legs that attach to the host cell, a linear segment that injects the DNA or RNA genome into the host, and a structure that stores the viral genome. This type of complex virus is common among bacteriophages, several of which are widely used in molecular biology research. Bacteriophage T4, lambda, P1, and Mu all look like the Apollo lunar landers (Fig. 1.24).

Viral genomes are varied in size, but all contain sufficient genetic information to get the host cell to make more copies of the virus genome and make more capsid proteins to package it. At the very least, a virus needs a gene to replicate its genome, a gene for capsid protein, and a gene to release new viruses from the host cell. Bacteriophage Qβ infects bacteria; its entire genome is only 3500 base pairs, and the entire genome consists of only four genes. On the other hand, large complex viruses may have more than 200 genes that are used at different

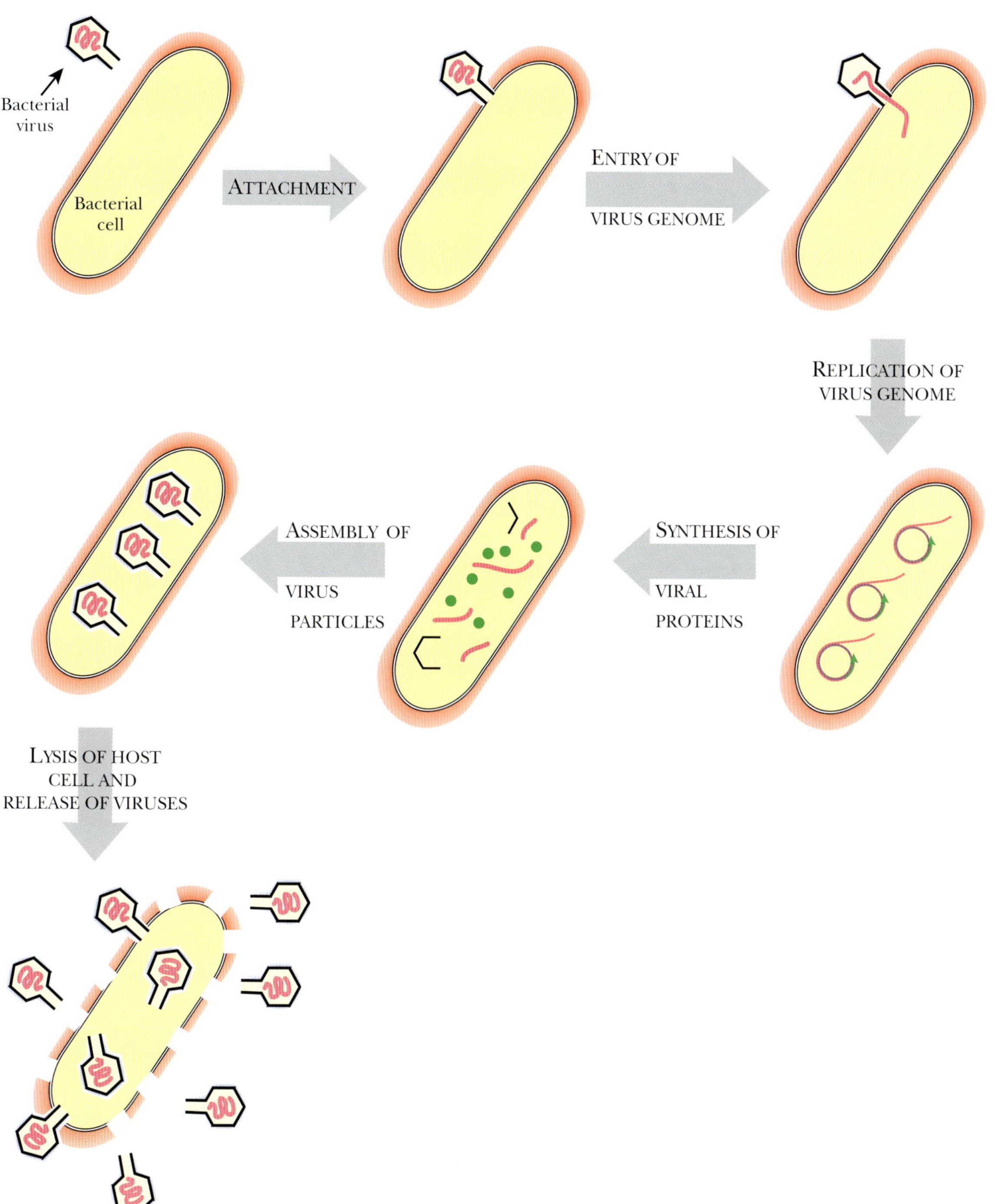

FIGURE 1.23 Virus Life Cycle
The life cycle of a virus starts when the viral DNA or RNA enters the host cell. Once inside, the virus uses the host cell to manufacture more copies of the virus genome and to make the protein coats for assembly of virus particles. Once multiple copies of the virus have been assembled, the host cell bursts open, allowing the progeny to escape and find other hosts to invade.

times after infecting the host cell. The genes are then divided into categories based on when they are active. Some genes are considered **early genes** and are active immediately after infecting the host, whereas **late genes** are active only after the virus has been inside the host cell for some time.

Viral genomes are either made from DNA or RNA, can be double-stranded or single-stranded, and can be circular or linear. When viruses use a single strand of RNA as genome,

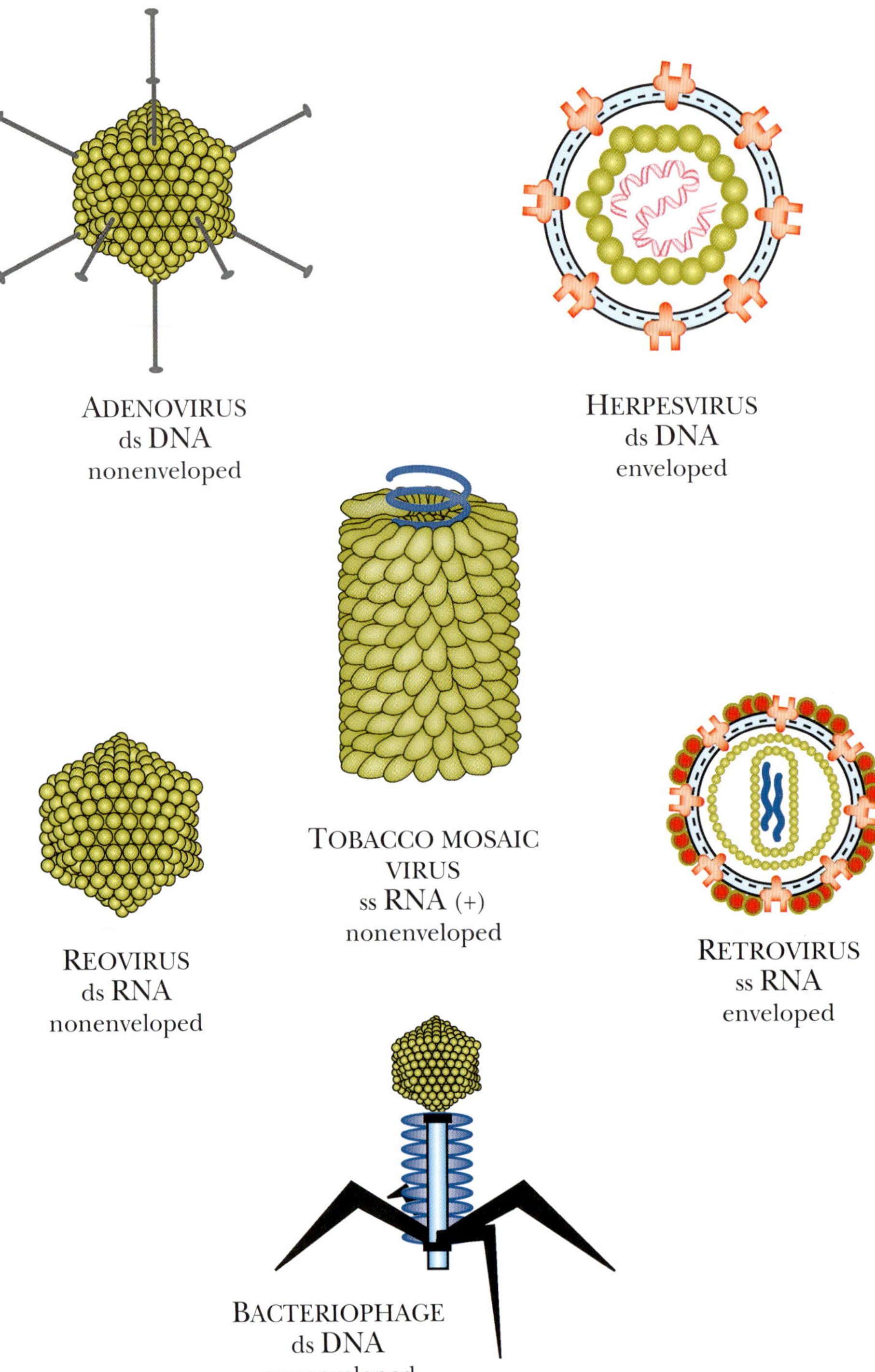

FIGURE 1.24
Examples of Different Viruses
Viruses come in a variety of shapes and sizes that determine whether the entire virus or only its genome enters the host cells.

this can either be the **positive** or **plus (+) strand** or the **negative** or **minus (–) strand**. The positive strand corresponds to the coding strand and the negative strand to the template strand (see Chapter 2). When a positive-strand RNA virus injects its genome into the host, the RNA can be used directly as a messenger RNA to make protein. If the RNA virus has a negative-strand genome, the RNA must first be converted into double-stranded form (the **replicative form** or **RF**). Then each strand is used: the negative strand is used as a template to make more positive-stranded genomes, and the positive strand is used to make proteins.

Some viruses actually use both RNA and DNA versions of their genome (Fig. 1.25). **Retroviruses** infect animals and include such members as HIV **(human immunodeficiency**

virus). The genome inside a retrovirus particle is a single-stranded RNA that is converted to DNA once it enters the host. **Reverse transcriptase** is the enzyme that manufactures the DNA copy of the RNA genome and is used extensively in molecular biology and genetic engineering (see Chapter 3). Once the DNA copy is made, it is inserted into the host DNA using two repeated DNA sequences at the ends called **long terminal repeats (LTRs)**. Once integrated, the retrovirus becomes part of the host's genome. This is why there is no complete cure for acquired immunodeficiency syndrome (AIDS); the host can never rid itself of the retroviral DNA once it becomes integrated. The viral genes then direct the synthesis of new viral particles that infect neighboring cells. Reverse transcriptase is an example of a viral gene product that is synthesized by the host and packaged inside the virions for use in the next infection cycle.

Retroviral genomes have three major genes, *gag, pol*, and *env*, as well as several minor genes. The *tat* and *rev* genes regulate the expression of the other retroviral genes. *Nef, vif, vpr*, and *vpu* encode four accessory proteins that block the host cell's immune defense and increase the efficiency of virus production. *Gag, pol*, and *env* each give single mRNA transcripts that encode multiple proteins. *Gag* encodes three proteins involved with making the capsid. *Pol* gives three proteins, a **protease** that digests other proteins during particle assembly, reverse transcriptase that makes the DNA copy of the genome, and an **integrase** that integrates the viral DNA into the host chromosome. *Env* codes for two structural proteins; one forms the outer spikes and the other helps the virus enter the host cell.

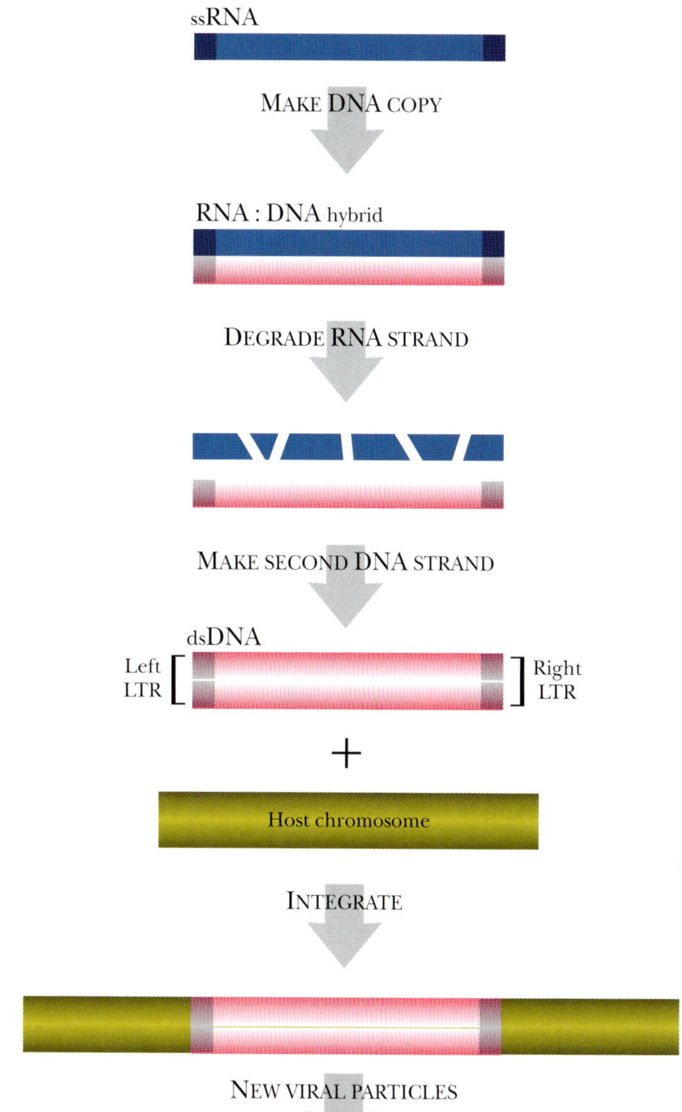

FIGURE 1.25
Retroviral Life Cycle
Retroviral genomes are made of positive RNA. Once the RNA enters the host, a DNA copy of the genome is made using reverse transcriptase. The original RNA strand is then degraded and replaced with DNA. Then the entire double-stranded DNA version of the retrovirus genome can integrate into the host genome.

Viruses are used extensively in biotechnology research because they specialize in inserting their genome into the host genome. They subvert the host into expressing their genes and making more copies of themselves. Researchers exploit these characteristics to study new genes, to alter the genomes of other model organisms, and to do gene therapy on humans.

SUBVIRAL INFECTIOUS AGENTS AND OTHER GENE CREATURES

We have used the term **gene creatures** to refer to various genetic entities that are sometimes called subviral infectious agents. These creatures exist, but are not considered living because none of them can produce their own energy, duplicate their own genomes, or live independent of a host. The main advantage a virus has over a gene creature is the ability to survive as an inactive particle outside the host cell. Gene creatures are not normally found outside the host cell.

Satellite viruses are defective viruses. They can either replicate their genome or package their genome into a capsid, but they are unable to do both by themselves. Satellite viruses rely on a

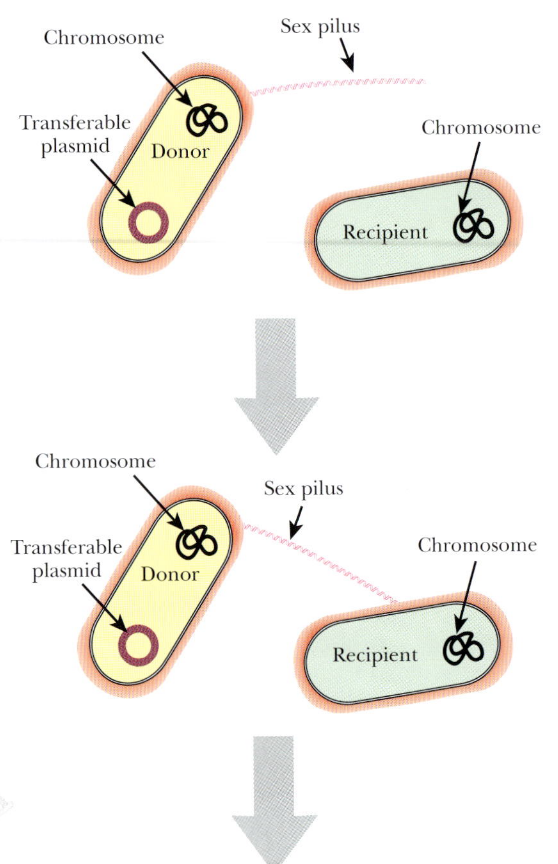

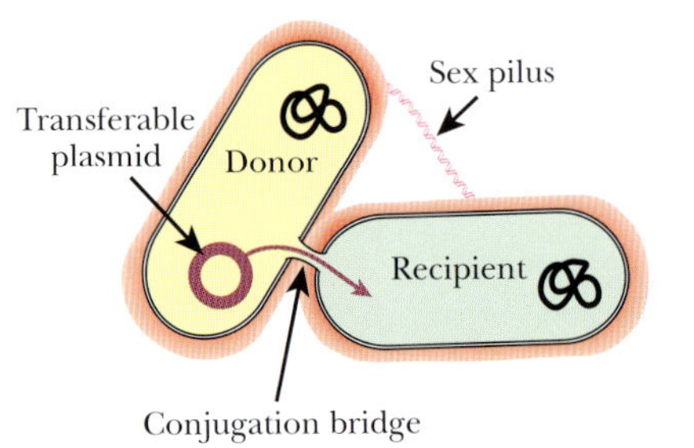

FIGURE 1.26
Conjugation in *E. coli*
Bacteria with a transferable plasmid can make a sex pilus that attaches to a recipient cell. When the two cells touch, a conjugation bridge forms, and a copy of the plasmid transfers from one cell to another. If the plasmid is integrated into the donor cell genome, segments of genomic DNA may also pass across.

helper virus to supply the missing components or genes. For example, hepatitis delta virus (HDV) is a small single-stranded RNA satellite virus that infects the liver. Its helper is hepatitis B virus. Bacteriophage P4 is a satellite virus that infects *E. coli*. It is a double-stranded DNA virus that can replicate as a plasmid or integrate into the host chromosome, but it cannot form virus particles by itself. It relies on P2 bacteriophage to supply the structural proteins. P4 sends transcription factors to the P2 genome to control expression of the genes it pirates.

Gene creatures also include genetic elements that may be helpful to the host. For example, the plasmids of *E. coli* and yeast are genetic elements that cannot produce their own energy and rely on the host cell to replicate their genome. They cannot survive outside a host cell. These traits qualify plasmids as gene creatures. Like viruses and satellite viruses, plasmids are **replicons**, that is, they have sufficient information in their genome to direct their own replication. Plasmids may confer positive traits to the host. For example, plasmids can provide antibacterial enzymes, such as bacteriocins, that help their host compete with other bacteria for nutrients (see earlier discussion). Plasmids may carry genes for antibiotic resistance, thus allowing the host bacteria to survive after encountering an antibiotic. Plasmids may confer virulence, making the host bacteria more aggressive and deadly. Finally, some plasmids contain genes that help the host degrade a new carbon source to provide food.

Plasmids are usually found as circles of DNA, although some linear plasmids have been found. Plasmids come in all sizes, but are usually much smaller than the bacterial chromosome. The genes on plasmids are often beneficial to the host. Because the plasmid coexists within the cytoplasm of the host cell, it does not generally harm its host.

The F plasmid is found in some *E. coli*, and it is about 1% of the size of the chromosome. It was named "F" for fertility because it confers the ability to mate. F plasmids can transfer themselves from one cell to the next in a process called **conjugation** (Fig. 1.26). The plasmid has genes for the formation of a specialized pilus, the F-pilus, which physically attaches an F^+*E. coli* to an F^- cell. After contact, a junction—the conjugation bridge—forms between the two cells. During replication of the F plasmid, one strand is cut at the origin and the free end enters the cytoplasm of the F^- cell via the conjugation bridge. Inside the recipient a complementary strand of DNA is made and the plasmid is recircularized. The other strand of the parent plasmid remains in the original F^+ cell and is also duplicated. Thus, after conjugation, both cells become F^+. Occasionally, the F plasmid integrates into the host chromosome. If an integrated F plasmid is transferred to another cell via conjugation, parts of the host chromosome may also get transferred. Therefore, bacteria can exchange chromosomal genetic information through conjugation.

Another gene creature that is very useful in biotechnology is the **transposable element** or **transposon**. This genetic element is merely a length of DNA that cannot exist or replicate as an independent molecule. To survive, it integrates into another DNA molecule. **Mobile DNA** or **jumping genes** are two terms used to describe transposons. When the transposon moves from one location to another, the process is called **transposition**. Unlike plasmids, transposons lack an origin of replication and are not considered replicons. They can only be replicated by integrating themselves into a host DNA molecule, such as a chromosome, plasmid, or viral genome. Transposons can move from site to site within the same host DNA or move from one host molecule of DNA to another. If a transposon loses its ability to move, its DNA remains in place on the chromosome or other DNA molecule. This is what comprises much human "junk DNA" and is discussed in Chapter 8.

Transposons come in several varieties and are classified based on the mechanism of movement. Transposons have two inverted DNA repeats at each end and a gene for **transposase**, the enzyme needed for movement. Transposase recognizes the inverted repeats at the ends of the transposon and excises the entire element from the chromosome. Next, transposase recognizes a **target sequence** of 3 to 9 base pairs in length on the host DNA. The transposon is then inserted into the target sequence, which is duplicated in the process. One copy is found on each side of the transposon. When a transposon is completely removed from one site and moved to another, the mechanism is **conservative transposition** or **cut-and-paste transposition** (Fig. 1.27). This leaves behind a double-stranded break that must be repaired by the host cell. Several cellular mechanisms exist to make this type of repair.

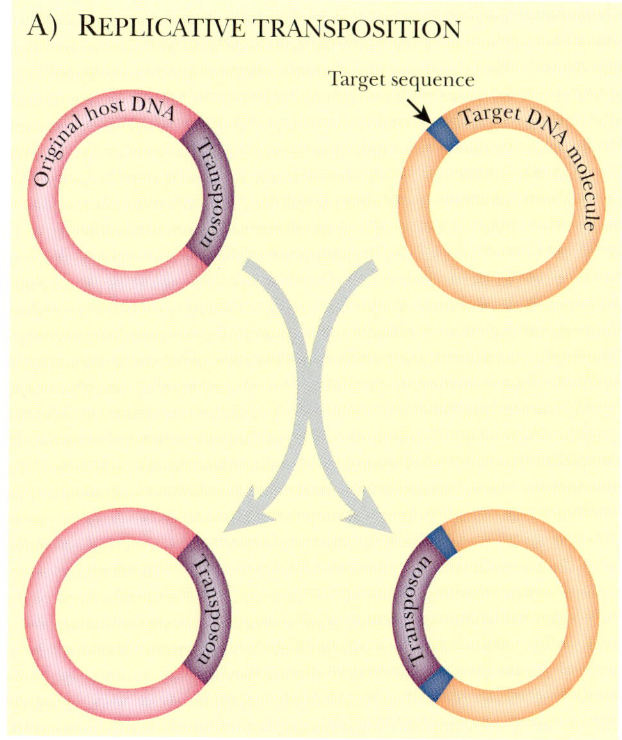

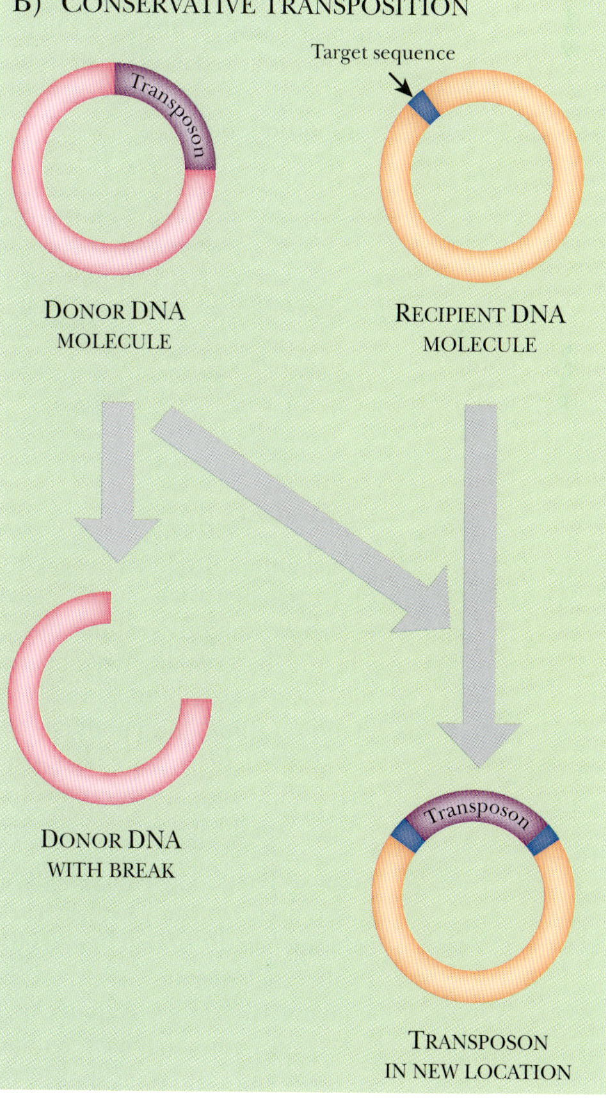

FIGURE 1.27 Transposons Move by Replicative or Conservative Transposition
(A) Replicative transposition leaves the original transposon in its original place, and a copy is inserted at another site within the host genome. (B) During conservative transposition, the original transposon excises from its original site and integrates at a different location.

An alternative mechanism is **replicative transposition**, where a second copy of the transposon is made. **Complex transposons** use this method. Much as before, transposase recognizes the inverted repeats of the transposon. However, in this case it only makes single-stranded nicks at the ends. Transposase then makes two single-stranded nicks, one at each end of the target site. Each single DNA strand of the transposon is joined to one host strand at the target site. This creates two single-stranded copies of the transposon. The host responds to such single-stranded DNA regions by making the second, complementary strand of the transposon. This gives two copies of the transposon. Notice how the transposon itself does not replicate. It tricks the host into making the replica.

Transposon movement can cause problems for the host. When the transposon moves, there is a potential for insertions, deletions, and inversions in the host DNA. If two copies of a transposon are found on a plasmid and the target sequence is on the host chromosome, a segment of the plasmid (flanked by the transposons) may be inserted into the host DNA. More generally, when multiple transposons are near each other, the ends of two neighboring but separate transposons may be used for transposition. When the two ends move to a new location, the DNA between them will be carried along. Whole genes or segments of genes may be deleted from the original location in this process. Conversely, regions of chromosome may become duplicated. If transposons are active and move often, the genome will become very damaged and the host cells often commit suicide (see Chapter 20). Because the transposon will be destroyed along with its host, many transposons move only rarely. Controlling their movement preserves their existence within the genome and keeps the host cell from committing suicide.

Gene creatures is a term to describe genetic elements that exist within the confines of a host cell, yet are separate from the original host genome. Some gene creatures include satellite viruses, plasmids, and transposons.

The plasmid is a unique gene creature because it confers positive traits such as resistance to antibiotics, bacteriocins, and the ability to transfer genetic material between two cells.

Transposons do not contain origins for their independent replication as do plasmids. These elements subvert the cell to make their copies by inducing breaks in the genome.

Summary

This chapter introduces the variety of different organisms used to study genes useful for biotechnology. Each organism, even the lowly gene creatures, is based on DNA. DNA and RNA have unique structures that ensure their survival and existence in all facets of life. Each structure has a backbone of alternating phosphate molecules with sugar residues. In DNA, the sugar, deoxyribose, is missing a hydroxyl group on the 2′ carbon. The bases, which attach at the 1′ carbon, form pairs so that adenine joins with thymine and guanine joins with cytosine. These pairs are held together with hydrogen bonds that induce the two backbones to twist into a double-stranded helix. In RNA, the sugar, ribose, has one extra hydroxyl group, and the base thymine is replaced with uracil.

Many different organisms are used in biotechnology research, and they have a particular trait that is useful to study new genes. Bacteria are genetic clones that are easily grown and stored for long periods of time. Two key traits are their simple genomes and availability of plasmids to alter their genetic makeup. Although useful, bacteria are prokaryotes and differ greatly from humans. Therefore, eukaryotic model organisms are also used for research. Yeasts are single-celled eukaryotes that have similar traits to human cells, such as multiple chromosomes, a nucleus, and various organelles. In addition, yeasts also have plasmids in which extra genes can be added to study in a model organism. Finally, the chapter outlines the key traits of multicellular organisms from barely visible roundworms such as *C. elegans* to mice, cultured human, animal, and insect cells, and the model plant organism, *Arabidopsis*.

Besides real organisms, research in biotechnology relies on gene creatures such as viruses, transposons, and plasmids. These genetic vehicles are critical to manipulating the genome of the model organisms. In fact, viruses may be the key to accomplishing gene therapy in humans also.

Viruses are used as vehicles to inject foreign DNA into a host cell. Transposons are also used to deliver new genes into the host DNA. Plasmids are used for the same purpose, but do not work in higher organisms and, therefore, are restricted to cultured cells, yeast, and bacteria. The use of gene creatures and model organisms is key to biotechnology research.

End-of-Chapter Questions

1. Which statement best describes the central dogma of genetics?
   - **a.** Genes are made of DNA, expressed as an RNA intermediary that is decoded to make proteins.
   - **b.** The central dogma only applies to yellow and green peas from Mendel's experiments.
   - **c.** Genes are made of RNA, expressed as a DNA intermediary, which is decoded to make proteins.
   - **d.** Genes made of DNA are directly decoded to make proteins.
   - **e.** The central dogma only applies to animals.
2. What is the difference between DNA and RNA?
   - **a.** DNA contains a phosphate group, but RNA does not.
   - **b.** Both DNA and RNA contain a sugar, but only DNA has a pentose.
   - **c.** The sugar ring in RNA has an extra hydroxyl group that is missing in the pentose of DNA.
   - **d.** DNA consists of five different nitrogenous bases, but RNA only contains four different bases.
   - **e.** RNA only contains pyrimidines and DNA only contains purines.
3. Which of the following statements about eukaryotic DNA packaging is true?
   - **a.** The process involves DNA gyrase and topoisomerase I.
   - **b.** All of the DNA in eukaryotes can fit inside of the nucleosome without being packaged.
   - **c.** Chromatin is only used by prokaryotes and is not necessary for eukaryotic DNA packaging.
   - **d.** Eukaryotic DNA packaging is a complex of DNA wrapped around proteins called histones, and further coiled into a 30-nanometer fiber.
   - **e.** Once eukaryotic DNA is packaged, the genes on the DNA can never again be expressed.
4. Which statement about *Thermus aquaticus* is false?
   - **a.** *T. aquaticus* was isolated from a hot spring.
   - **b.** The DNA polymerase from *T. aquaticus* is used in molecular biology for a procedure called polymerase chain reaction (PCR).
   - **c.** The DNA polymerase from *T. aquaticus* is able to withstand very high temperatures.
   - **d.** *T. aquaticus* can survive high temperatures and low pH.
   - **e.** *T. aquaticus* is found in the frozen lakes of Antarctica.

(Continued)

5. Which statement about *Escherichia coli* is not correct?
   a. *E. coli* is called "the workhorse of molecular biology."
   b. *E. coli* can grow in a simple solution of water, a carbon source, and mineral salts.
   c. All *E. coli* strains are pathogenic, and therefore must be handled accordingly.
   d. The chromosome of *E. coli* consists of one circular DNA molecular containing approximately 4000 genes.
   e. All of the above answers are correct.

6. Plasmids from bacteria can be described by which of the following statements?
   a. Plasmids provide an advantage to the host bacterium to compete against non-plasmid-containing bacteria for nutrients.
   b. Plasmids are used as a molecular biology tool to express other genes efficiently in the host bacterium.
   c. Plasmids are extrachromosomal segments of DNA that carry several genes beneficial to the host organism.
   d. Plasmids have their own origin of replication.
   e. All of the above statements describe plasmids.

7. Which of the following statements is not correct about the usefulness of fungi in biotechnology research?
   a. Fungi produce the blue veins in some types of cheeses.
   b. Yeast is responsible for the alcohol in beer and for bread rising.
   c. Fungi are called "the workhorses of molecular biology."
   d. The 2-micron circle is a useful extrachromosomal element from yeast that can be utilized in molecular biology research.
   e. Fungi produce many industrial chemicals and pharmaceuticals.

8. What mechanism does yeast utilize to control mating type in the cells?
   a. Yeast is only able to reproduce through mitosis.
   b. The MAT locus in the yeast genome contains two divergent genes that encode for the pheromones **a** and α, along with the pheromone receptors.
   c. The mating type of yeast is determined by pheromones called **b** and β.
   d. There are no mechanisms to control mating type in yeast because all of the cells are structurally the same.
   e. Yeast mating types are generally referred to as either male or female.

9. Which of the following yeast cellular component is typically not found in bacteria?
   a. centromeres
   b. telomeres
   c. nuclear pores
   d. nuclear envelope
   e. All of the above are found in yeast and not bacteria.

10. Identify the statement about multicellular model organisms that is correct.

a. *C. elegans* has been used extensively to study multicellular interactions partly because the creature can reproduce by self-fertilization (genetic clones) or sexually (novel genetic organisms).

b. Based on homology, research on *Drosophila* mutants has identified genes in the human genome responsible for body patterns.

c. The zebrafish, or *Danio rerio*, are used to study developmental genetics because the embryonic cells are easily destroyed or manipulated and the effects can be observed within 24 hours.

d. The mouse is a model organism for studying human genetics, physiology, and development because less than 1% of the genes in the mouse genome have no genetic homology in humans.

e. All of the statements are correct.

11. What is the main advantage for studying cells in culture rather than in a whole organism?

a. Cell lines in culture are easily manipulated genetically to introduce new genes or delete other genes.

b. Cell lines are not very stable and therefore, it is more advantageous to study cells within the organism itself.

c. There is no advantage to studying cells in a cell line rather than in a live organism.

d. The information obtained from studying cell lines as opposed to live organisms is not relevant to what happens *in vivo*.

e. None of the above is the main advantage.

12. Why is *Arabidopsis thaliana* used as a model organism for plant genetics and biology?

a. *Arabidopsis* responds to stress and disease similarly to important crop plants such as rice, wheat, and corn.

b. *Arabidopsis* is easy to grow and maintain in the laboratory.

c. The genome of *Arabidopsis* is relatively small compared to other plants.

d. The generation cycle of *Arabidopsis* is shorter than most other crop plants and produces many seeds for further study.

e. All of the above statements are reasons for using *Arabidopsis* as a model organism.

13. Why are viruses significant to biotechnology?

a. They are able to insert their genome into the host genome, thus integrating genes in the process.

b. Viruses can be used to alter the genomes of other organisms.

c. Reverse transcriptase, an enzyme used in molecular biology, is encoded in a retroviral genome.

d. Viruses play an important role in delivering gene therapy to humans.

e. All of the above statements are reasons why viruses are significant to biotechnology research.

(*Continued*)

14. Which statement best describes the F plasmid?
- **a.** F plasmids contain genes for formation of a specialized pilus that forms a conjugation bridge between two cells for the purpose of transferring genetic material.
- **b.** The F plasmid does not have an origin of replication and can therefore not replicate itself.
- **c.** The primary host for the F plasmid is *Saccharomyces cerevisiae*.
- **d.** The F plasmid is not important for biotechnology research.
- **e.** All of the above statements describe the F plasmid.

15. Which of the following elements is important in biotechnology research?
- **a.** transposons
- **b.** F plasmid
- **c.** satellite viruses
- **d.** plasmids
- **e.** all of the above

Further Reading

Altun ZF Hall DH (Eds) (2002–2006). WormAtlas. http://www.wormatlas.org

Clark DP (2005). *Molecular Biology: Understanding the Genetic Revolution.* Elsevier Academic Press, San Diego, CA.

Garcia-Hernandez M, Berardini TZ, Chen G, Crist D, Doyle A, Huala E, Knee E, Lambrecht M, Miller N, Mueller LA, Mundodi S, Reiser L, Rhee SY, Scholl R, Tacklind J, Weems DC, Wu Y, Xu I, Yoo D, Yoon J, Zhang P (2002). TAIR: A resource for integrated *Arabidopsis* data. *Funct. Integrative Genomics* **2**(6), 239.

Glick BR, Pasternak JJ (2003). *Molecular Biotechnology: Principles and Applications of Recombinant DNA.* 3rd ed. ASM Press, Washington, DC.

Lynn DE (2002). Methods for maintaining insect cell cultures. *J. Insect Sci. 2.9.* Available online: http://insectscience.org/2.9

McCarroll L, King LA (1997). Stable insect cell cultures for recombinant protein production. *Curr Opin Biotechnol* **8**, 590–594.

Sprague J, Bayraktaroglu L, Clements D, Conlin T, Fashena D, Frazer K, Haendel M, Howe D, Mani P, Ramachandran S, Schaper K, Segerdell E, Song P, Sprunger B, Taylor S, Van Slyke C, Westerfield M (2006). The Zebrafish Information Network: The zebrafish model organism database. *Nucleic Acids Res* **34**, D581–D585.

CHAPTER 2

DNA, RNA, and Protein

THE CENTRAL DOGMA OF MOLECULAR BIOLOGY

Two essential features of living creatures are the ability to reproduce their own genome and manufacture their own energy. In order to accomplish these feats, an organism must be able to make proteins using information encoded in its DNA. Proteins are essential for cellular architecture, giving the cell a particular shape and structure. Proteins include enzymes that catalyze reactions used to make energy. Proteins control cellular processes like replication. Proteins provide channels in the membrane for cells to communicate with each other or share metabolites. Making proteins is a key operation for all living organisms.

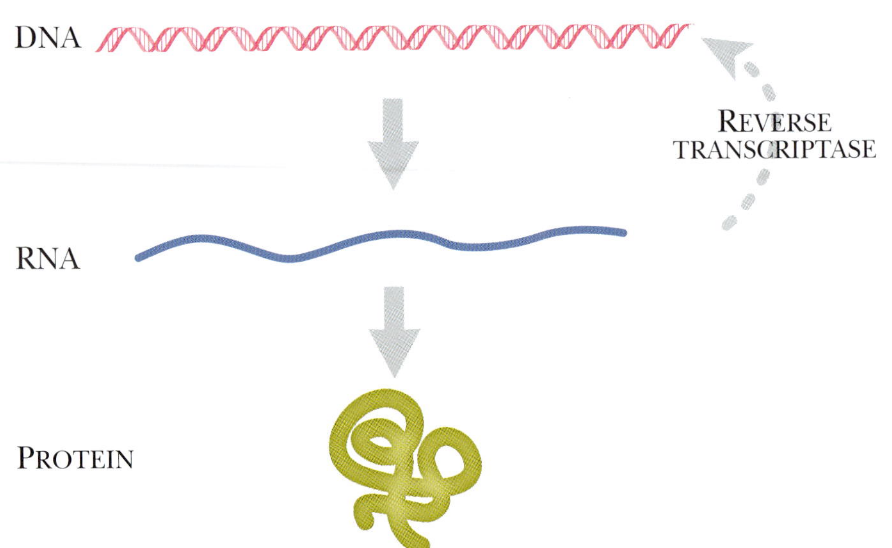

FIGURE 2.1 The Central Dogma
Cells store the genetic information to function and replicate in their DNA. When a protein is needed, DNA is transcribed into RNA, which in turn, is translated into a protein.

The **central dogma** of molecular biology states that information flows from DNA to RNA to protein (Fig. 2.1). First, this chapter focuses on how RNA is made from DNA in a process called **transcription.** Next, the mechanisms used to control transcription are discussed. We then discuss how particular RNA molecules called **mRNA** or **messenger RNA** are used to make protein in a process called **translation.** Hopefully, by examining these processes, the reader will gain an understanding of the complexity involved in engineering cells for the purposes of biotechnology.

The central dogma of molecular biology is that DNA is transcribed into RNA, which in turn is translated into proteins.

TRANSCRIPTION EXPRESSES GENES

Gene expression involves making an RNA copy of information present on the DNA, that is, transcribing the DNA. Making RNA involves uncoiling the DNA, melting the strands at the start of the gene, making an RNA molecule that is complementary in sequence to the **template** strand of the DNA with an enzyme called **RNA polymerase**, and stopping at the end of the gene. The newly made RNA is released from the DNA, which then returns to its supercoiled form.

An important issue in transcription is identifying the right gene. Which gene needs to be decoded to make protein? There are different types of genes. Some are **housekeeping genes** that encode proteins that are used all the time. Other genes are activated only under certain circumstances. For instance, in *E. coli*, genes that encode proteins involved with the utilization of lactose are expressed only when lactose is present (see later discussion). The same principle applies to the genes for using other nutrients. Various inducers and accessory proteins control whether or not these genes are **expressed** or made into RNA and will be discussed in more detail in upcoming sections.

The final product encoded by a gene is often a protein but may be RNA. Genes that encode proteins are transcribed to give messenger RNA, which is then translated to give the protein. Other RNA molecules, such as tRNA, rRNA, and snRNA, are used directly (i.e., they are not translated to make proteins). Some RNA molecules, such as large-subunit rRNA, are called **ribozymes** and can catalyze enzymatic reactions (see Chapter 5). Most of the time though, genes ultimately code for a protein via an mRNA intermediate. The coding region of a gene is sometimes called a **cistron** or a **structural gene** and may encode a protein or a nontranslated RNA. (The term *cistron* was originally defined by genetic complementation using the *cis/trans* test.) In contrast, an **open reading frame (ORF)** is a stretch of DNA (or the corresponding RNA) that encodes a protein and therefore is not interrupted by any stop codons for protein translation (see later discussion).

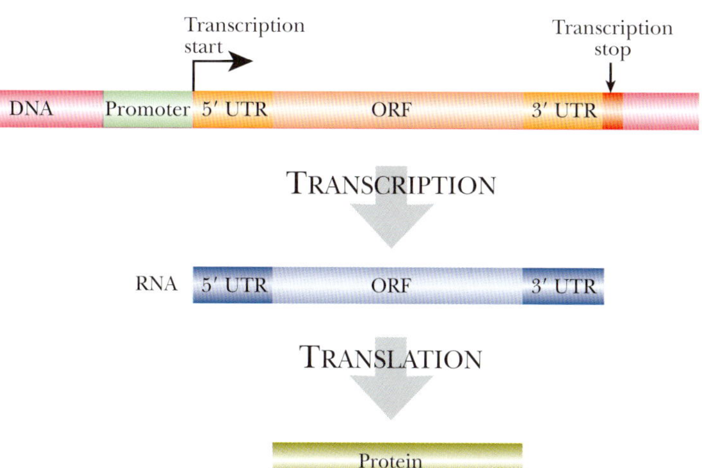

FIGURE 2.2 The Structure of a Typical Gene
Genes are regions of DNA that are transcribed to give RNA. In most cases, the RNA is translated into protein, but some RNA is not. The gene has a promoter region plus transcriptional start and stop points that flank the actual message. After transcription, the RNA has a 5′ untranslated region (5′ UTR) and 3′ untranslated region (3′ UTR), which are not translated; only the ORF is translated into protein.

The next issue is finding the start site of the gene. Every gene has a region upstream of the coding sequence called a **promoter** (Fig. 2.2). RNA polymerase recognizes this region and starts transcription here. Bacterial promoters have two major recognition sites: the −10 and −35 regions. The numbers refer to their approximate location upstream of the transcriptional start site. (By convention, positive numbers refer to nucleotides downstream of the transcription start site and negative numbers refer to those upstream.) The exact sequences at −10 and −35 vary, but the consensus sequences are TATAA and TTGACA, respectively. When a gene is transcribed all the time or **constitutively**, then the promoter sequence closely matches the consensus sequence. If the gene is expressed only under special conditions, **activator proteins** or **transcription factors** are needed to bind to the promoter region before RNA polymerase will recognize it. Such promoters rarely look like the consensus.

Just after the promoter region is the **transcription start site**. This is where RNA polymerase starts adding nucleotides. Between the transcription start site and the ORF is a region that is not made into protein called the **5′ untranslated region (5′ UTR).** This region contains translation regulatory elements like the ribosome binding site. Next is the ORF, where no translational stop codons are found. Then there is another untranslated region after the ORF, known as the **3′ untranslated region (3′ UTR).** Finally comes the termination sequence where transcription stops.

Bacterial RNA polymerase is made of different protein subunits. The **sigma subunit** recognizes the −10 and −35 regions and the **core enzyme** catalyzes the RNA synthesis. RNA polymerase only synthesizes nucleotide additions in a 5′ to 3′ direction. The core enzyme has four protein subunits, a dimer of two α proteins, a β protein, and a related β' subunit. The β and β' subunits form the catalytic site, and the α subunit helps recognize the promoter. The 3D structure of RNA polymerase shows a deep groove that can hold the template DNA, and a minor groove to hold the growing RNA.

Genes have a transcriptional promoter, where RNA polymerase attaches to the DNA and begins making an RNA copy of the template strand. The RNA has three regions: The 5′ UTR contains information important for making the protein, the ORF has the actual coding region translated into amino acids during translation, and the 3′ UTR contains other important regulatory elements.

MAKING THE RNA

In bacteria, once the sigma subunit of RNA polymerase recognizes the −10 and −35 regions, the core enzyme forms a **transcription bubble** where the two DNA strands are separated from each other (Fig. 2.3). The strand used by RNA polymerase is called the template strand (aka **noncoding** or **antisense**) and is complementary to the resulting mRNA. The core enzyme adds RNA nucleotides in the 5′ to 3′ direction, based on the sequence of the template strand of DNA. The newly made RNA anneals to the template strand of the DNA

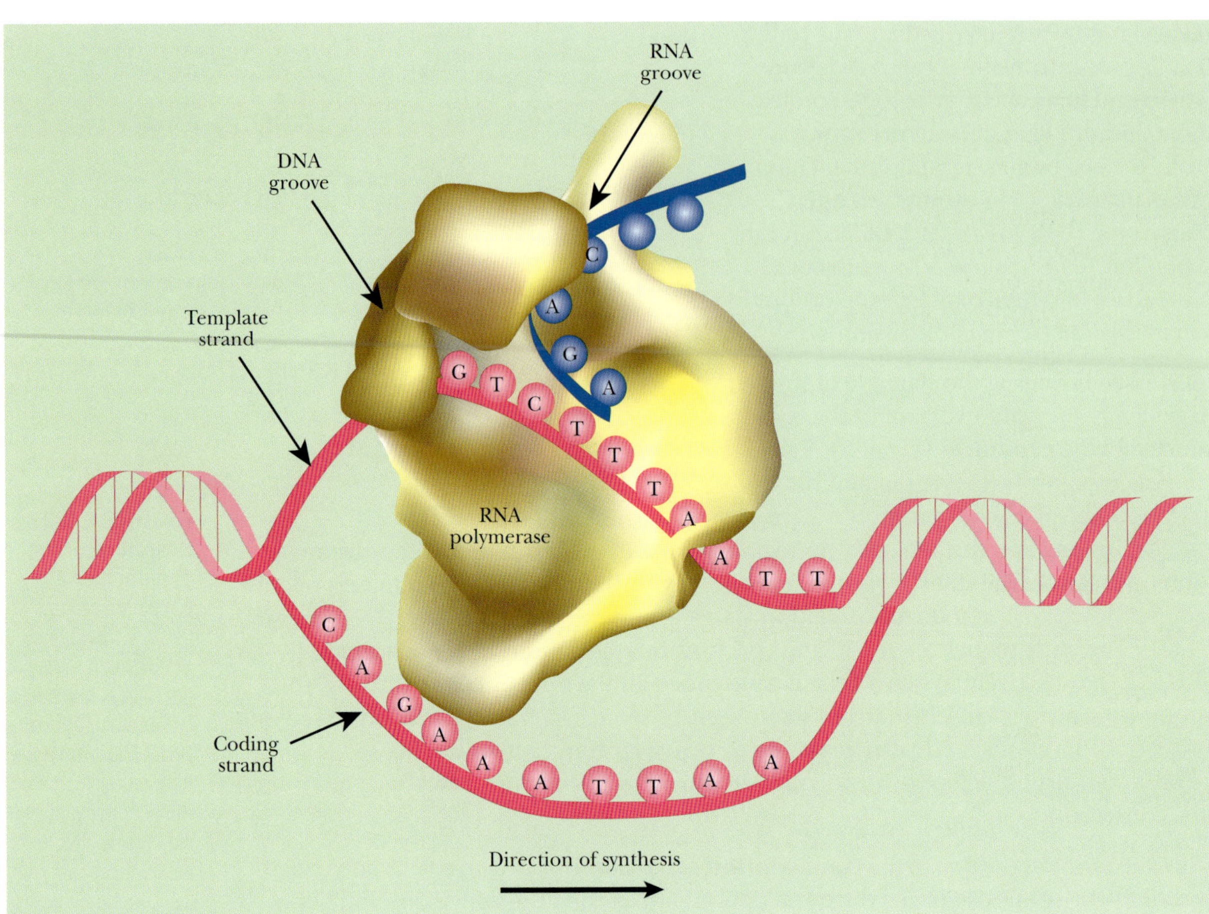

FIGURE 2.3 RNA Polymerase Synthesizes RNA at the Transcription Bubble
RNA polymerase is a complex enzyme with two grooves. The first groove holds a single strand of DNA, and the second groove holds the growing RNA. RNA polymerase travels down the DNA, adding ribonucleotides that complement each of the bases on the DNA template strand.

via hydrogen bonds between base pairs. The opposite strand of DNA is called the **coding strand** (aka **nontemplate** or **sense strand**). Because this is complementary to the template strand, its sequence is identical to the RNA (except for the replacement of thymine with uracil in RNA).

RNA synthesis normally starts at a purine (normally an A) in the DNA that is flanked by two pyrimidines. The most typical start sequence is CAT, but sometimes the A is replaced with a G. The rate of elongation is about 40 nucleotides per second, which is much slower than replication (~1000 bp/sec). RNA polymerase unwinds the DNA and creates positive supercoils as it travels down the DNA strand. Behind RNA polymerase, the DNA is partially unwound and has surplus negative supercoils. Just as during replication, DNA gyrase and topoisomerase I either insert or remove negative supercoils, respectively, returning the DNA back to its normal level of supercoiling (see Chapter 4).

RNA polymerase makes a copy of the gene using the noncoding or template strand of DNA. The RNA has uracils instead of thymines.

TRANSCRIPTION STOP SIGNALS

RNA polymerase continues transcribing DNA until it reaches a termination signal. In bacteria, the **Rho-independent terminator** is a region of DNA with two inverted repeats separated by about six bases, followed by a stretch of A's. As RNA polymerase makes these sequences, the two inverted repeats form a hairpin structure. The secondary structure causes RNA polymerase to pause. As the stretch of A's is transcribed into U's, the DNA/RNA hybrid molecule becomes unstable (A/U base pairs only have two hydrogen bonds). RNA polymerase "stutters" and then falls off the template strand of DNA in the middle of the A's.

Bacteria also have **Rho-dependent terminators** that have two inverted repeats but lack the string of A's. **Rho (ρ) protein** is a special helicase that unwinds DNA/RNA hybrid double helices. Rho binds upstream of the termination site in a region containing many cytosines but very few guanines. After RNA polymerase passes the Rho binding site, Rho attaches to the RNA and moves along the RNA transcript until it catches RNA polymerase at the hairpin structure. Rho then unwinds the DNA/RNA helix and separates the two strands. The RNA is then released.

Transcription terminates either in a Rho-independent manner or in a Rho-dependent manner.

THE NUMBER OF GENES ON AN mRNA VARIES

Bacterial and eukaryotic chromosomes are organized very differently. In prokaryotes, the distance between genes is much smaller, and genes associated with one metabolic pathway are often found next to each other. For example, the lactose **operon** contains several clustered genes for lactose metabolism. Operons are clusters of genes that share the same promoter and are transcribed as a single large mRNA that contains multiple structural genes or cistrons. Thus these transcripts are called **polycistronic mRNA** (Fig. 2.4). The multiple cistrons are translated individually to give separate proteins. In eukaryotes, genes are often separated by large stretches of DNA that do not encode any protein. In eukaryotes, each mRNA only has one cistron and is therefore called **monocistronic mRNA**. If a polycistronic transcript is expressed in eukaryotes, the ribosome only translates the first cistron, and the other encoded proteins are not made.

Bacterial mRNA transcripts have multiple open reading frames for proteins in the same metabolic pathway. Eukaryotes tend to have only one open reading frame in a single mRNA transcript.

EUKARYOTIC TRANSCRIPTION IS MORE COMPLEX

Several differences between eukaryotic and prokaryotic transcription occur because eukaryotic cells are more complex. The simple fact that eukaryotic mRNA is made in a nucleus makes the process more involved than bacterial transcription, but this is only one of the differences.

In contrast to the single RNA polymerase in prokaryotes, eukaryotes have three different RNA polymerases that each transcribe different types of genes. **RNA polymerase I** transcribes the eukaryotic genes for large ribosomal RNA. These two rRNAs are transcribed as one long mRNA that is cleaved into two different transcripts, the 18S rRNA and 28S rRNA. These are used directly and not translated into protein. **RNA polymerase III** transcribes the genes for tRNA, 5S rRNA, and other small RNA molecules. **RNA polymerase II** transcribes the genes that encode proteins and has been studied the most.

Starting transcription of eukaryotic genes is more complex than in bacteria. The layout of the eukaryotic promoter is much different. RNA polymerase II needs three different regions, the **initiator box**, the **TATA box**, and various upstream elements that bind proteins known as transcription factors. The initiator box is the site where transcription starts and is separated by about 25 base pairs from the TATA box. The upstream elements vary from gene to gene and aid in controlling what proteins are expressed at what time.

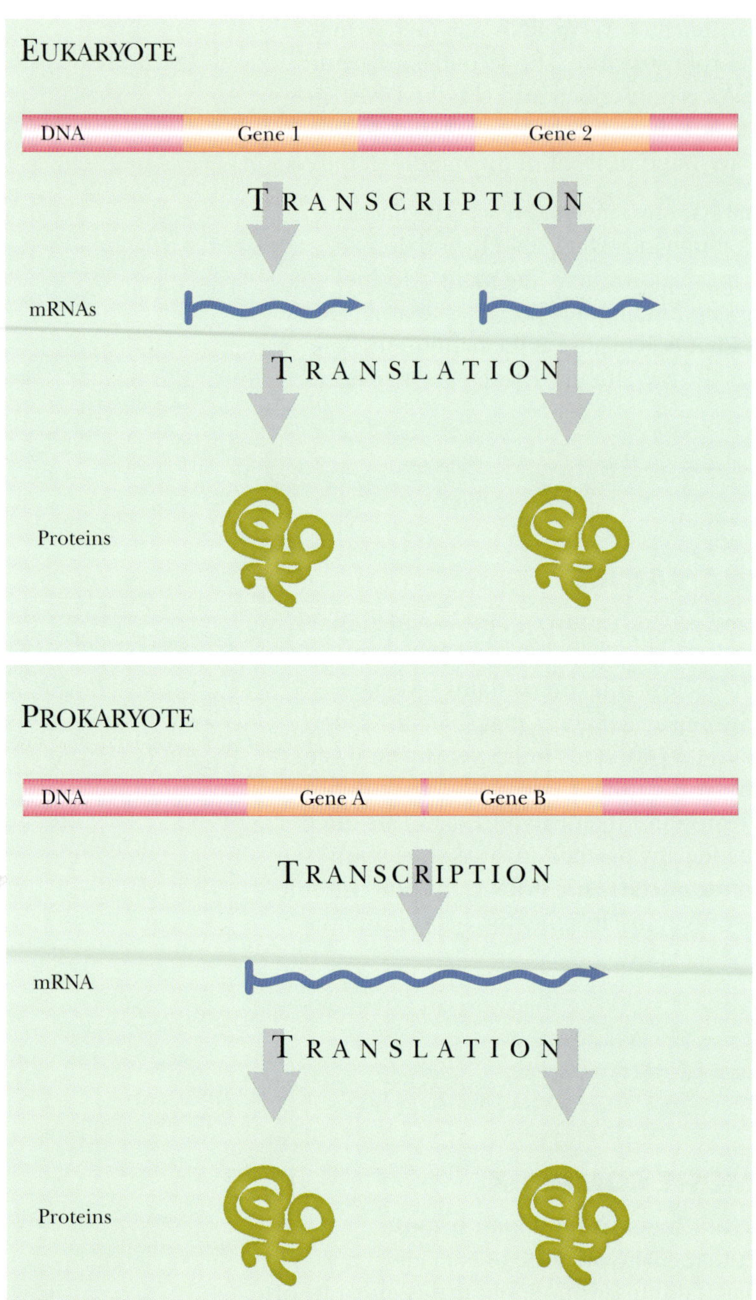

FIGURE 2.4 Monocistronic versus Polycistronic Eukaryotes transcribe genes in single units, where each mRNA encodes for only one protein. Prokaryotes transcribe genes in operons as one single mRNA, and then translate the proteins as separate units.

Many proteins are involved in positioning eukaryotic RNA polymerase II at the transcriptional start site (Fig. 2.5 and Table 2.1). RNA polymerase II requires several **general transcription factors** to initiate transcription at all promoters. In addition, **specific transcription factors** are needed that vary depending on the particular gene (see later discussion). The **TATA binding factor** or **TATA box factor** (TBF) recognizes the TATA box. This factor is used by all three RNA polymerases in eukaryotes. For RNA polymerase II, TBF is found with other proteins in a complex called TFIID. (For the other RNA polymerases, TBF associates with different proteins.) After this complex binds, TFIIA and TFIIB bind to the promoter, which then triggers the binding of RNA polymerase II. RNA polymerase is associated with TFIIF, which probably helps it bind to the promoter. Once RNA polymerase II has bound to the promoter, it still requires TFIIE, TFIIH, and TFIIJ to initiate transcription. In particular, TFIIH phosphorylates the tail of RNA polymerase II, which allows it to move along the DNA. As RNA polymerase II leaves the promoter, it leaves behind all of the general complexes except TFIIH.

Bacterial RNA polymerase can function with a promoter containing no upstream elements. But in eukaryotes the upstream elements are essential to RNA polymerase II function, and a promoter with no upstream elements is extremely inefficient at initiating transcription. These elements are from 50 to 200 base pairs in length and vary based on the gene being expressed. They bind regulatory proteins known as specific transcription factors, as opposed to the general transcription factors shared by all promoters that use RNA polymerase II. For example, the specific transcription factors Oct-1 and Oct-2 proteins bind only to the Octamer elements. Oct-1 is found in all tissues, whereas Oct-2 is only found in immune cells. A plethora of specific factors exist that are beyond the scope of this discussion.

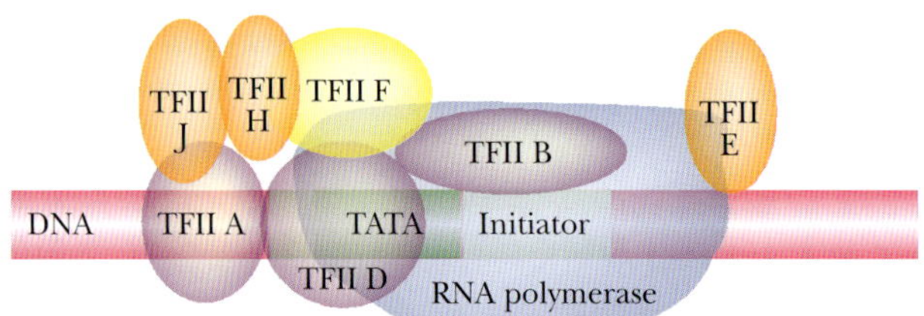

FIGURE 2.5 Eukaryotic Transcription
Many different general transcription factors help RNA polymerase II find the TATA and initiator box region of a eukaryotic promoter.

Table 2.1	General Transcription Factors for RNA Polymerase II
TBF	binds to TATA box, part of TFIID
TFIID	includes TBP, recognizes Pol II specific promoter
TFIIA	binds upstream of TATA box; required for binding of RNA Pol II to promoter
TFIIB	binds downstream of TATA box; required for binding of RNA Pol II to promoter
TFIIF	accompanies RNA Pol II as it binds to promoter
TFIIE	required for promoter clearance and elongation
TFIIH	phosphorylates the tail of RNA Pol II, retained by polymerase during elongation
TFIIJ	required for promoter clearance and elongation

Eukaryotes have three different RNA polymerases that transcribe different genes. RNA polymerase II requires many different proteins to transcribe a gene.

Eukaryotes require specific transcription factors to initiate gene transcription. These are highly abundant, and each gene has a different set of factors that regulate its transcription.

REGULATION OF TRANSCRIPTION IN PROKARYOTES

In prokaryotes, a variety of activator and **repressor** proteins control which genes get transcribed. The activators and repressors all work by binding to DNA in the promoter region and either stimulating or blocking the action of bacterial RNA polymerase. In *E. coli*, about 1000 of the 4000 total genes are expressed at one time. Activator proteins work by **positive regulation**; in other words, genes are expressed only when the activator gives a positive signal. In contrast, repressors work by **negative regulation**. Here the gene is expressed only when the repressor is removed. Some repressors block RNA polymerase from binding to the DNA; others prevent initiation of transcription even though RNA polymerase has bound.

Regulation of transcription is complex, even in simple prokaryotes. Many genes are controlled by a variety of factors. Some operons in bacteria have multiple repressors and activators. Less often, regulatory proteins may block elongation either by slowing the actual rate of elongation or by signaling premature termination. Conversely, a few **antiterminator proteins** are known that override termination and allow genes downstream of the termination site to be expressed.

Prokaryotes use both positive regulation, where activator proteins signal RNA polymerase to transcribe the gene, or negative regulation, where the transcription factor inhibits RNA polymerase.

Prokaryotic Sigma Factors Regulate Gene Expression

Prokaryotic RNA polymerase has the sigma subunit, which recognizes the promoter first and binds the catalytic portion of the enzyme (the core enzyme). There are many different sigma subunits, and each one recognizes a different set of genes. The σ 70 subunit, or RpoD, is the most commonly used form. It recognizes most of the housekeeping genes in *E. coli*. During the stationary phase, when *E. coli* is not growing rapidly, σ 38, or RpoS, activates the necessary genes. (Sigma subunits are named either by σ plus their molecular weight or by Rpo [for RNA polymerase] plus their function: D = default, S = stationary, etc.)

Another sigma factor, RpoH, or σ 32, activates genes needed during heat shock. Normally *E. coli* grows at body temperature, 37 °C, and stops growing at temperatures much above 43 °C. At such higher temperatures proteins begin to unfold and are degraded. RpoH activates expression of **chaperonins** that help proteins fold correctly and prevent aggregation. RpoH also activates **proteases** that degrade proteins too damaged by the heat to be saved.

The transcription and translation of RpoH depend on temperature. When *E. coli* grows at a normal temperature, very few misfolded proteins are present. DnaK (a chaperonin) and HflB (a protease) are found in the cytoplasm but have very few bad proteins to "fix"; therefore, they bind to RpoH and degrade it. They even degrade partially translated RpoH protein. When high temperatures promote unfolding and aggregation of proteins, DnaK and HflB bind to the aberrant proteins and no longer destroy RpoH. Now the sigma factor initiates transcription of other genes associated with heat shock.

Sigma (σ) subunits are transcription factors that associate with prokaryotic RNA polymerase and control which genes are transcribed.

Lactose Operon Demonstrates Specific and Global Activation

Many genes require specific regulator proteins to activate transcription via RNA polymerase. Some of these proteins exist in two forms: active (binds to DNA in promoter region) and inactive (nonbinding). The forms are interconverted by small **signal molecules** or **inducers** that alter the shape of the protein. For example, the inducer *allo*-lactose controls the lactose operon.

The lactose or *lac* operon is well characterized because it was the first to be studied. It has a promoter upstream of three structural genes, *lacZ*, *lacY*, and *lacA*. The *lacZYA* genes are transcribed as a polycistronic message. Upstream of the promoter is the gene for LacI protein, the *lac* operon repressor (Fig. 2.6). The *lacZ* gene encodes **β-galactosidase**, which cleaves the disaccharide lactose into galactose and glucose. The *lacY* gene encodes **lactose permease**, which transports lactose across the cytoplasmic membrane into the bacteria. Finally, the *lacA* gene encodes the protein **lactose acetylase**, with an unknown role. The promoter has a binding site, *lacO*, for the repressor protein, which overlaps the binding site for RNA polymerase. This region is also known as the **operator**, and when the repressor binds, RNA polymerase cannot transcribe the operon.

There is also a binding site for **CRP protein (cyclic AMP receptor protein)**, also known as **CAP (catabolite activator protein)**. This is a global regulator that activates transcription of many different operons for using alternate sugar sources. It is active when *E. coli* does not have glucose to utilize as an energy source.

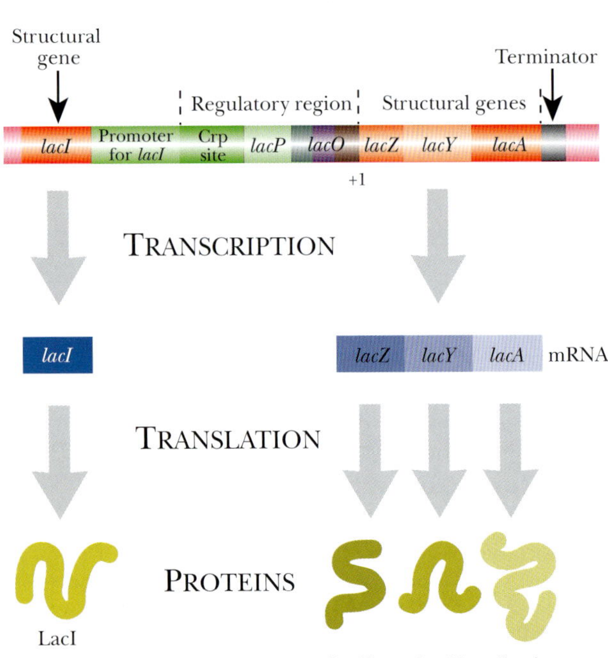

FIGURE 2.6 Components of the *lac* Operon
The *lac* operon consists of three structural genes, *lacZYA*, which are all transcribed from a single promoter, designated *lacP*. The promoter is regulated by binding of the repressor at the operator, *lacO*, and of Crp protein at the Crp site. Note that in reality, the operator partly overlaps both the promoter and the *lacZ* structural gene. The single *lac* mRNA is translated to produce the LacZ, LacY, and LacA proteins. The *lacI* gene that encodes the LacI repressor has its own promoter and is transcribed in the opposite direction from the *lacZYA* operon.

The environment controls whether or not the lactose operon is expressed (Fig. 2.7). When *E. coli* has plenty of glucose, then the lactose operon is turned off (as well as other operons for other sugars such as maltose or fructose). When glucose is present, levels of a small inducer, **cyclic AMP (cAMP)**, are low. If *E. coli* runs out of glucose, the levels of cAMP increase, and bind to Crp, the global regulator. Crp, in turn, dimerizes so that it can bind to the Crp sites in various promoters, such as the lactose operon. Crp binding alone will not activate the lactose operon; lactose must also be present to activate transcription. If lactose is available, β-galactosidase catalyzes a side reaction, converting some lactose into *allo*-lactose. This acts as an inducer and binds to the tetrameric LacI repressor protein. This releases the repressor from the promoter. The lactose operon is expressed only when both glucose levels are low and lactose is present. The control relies on two inducer molecules: cAMP binds to

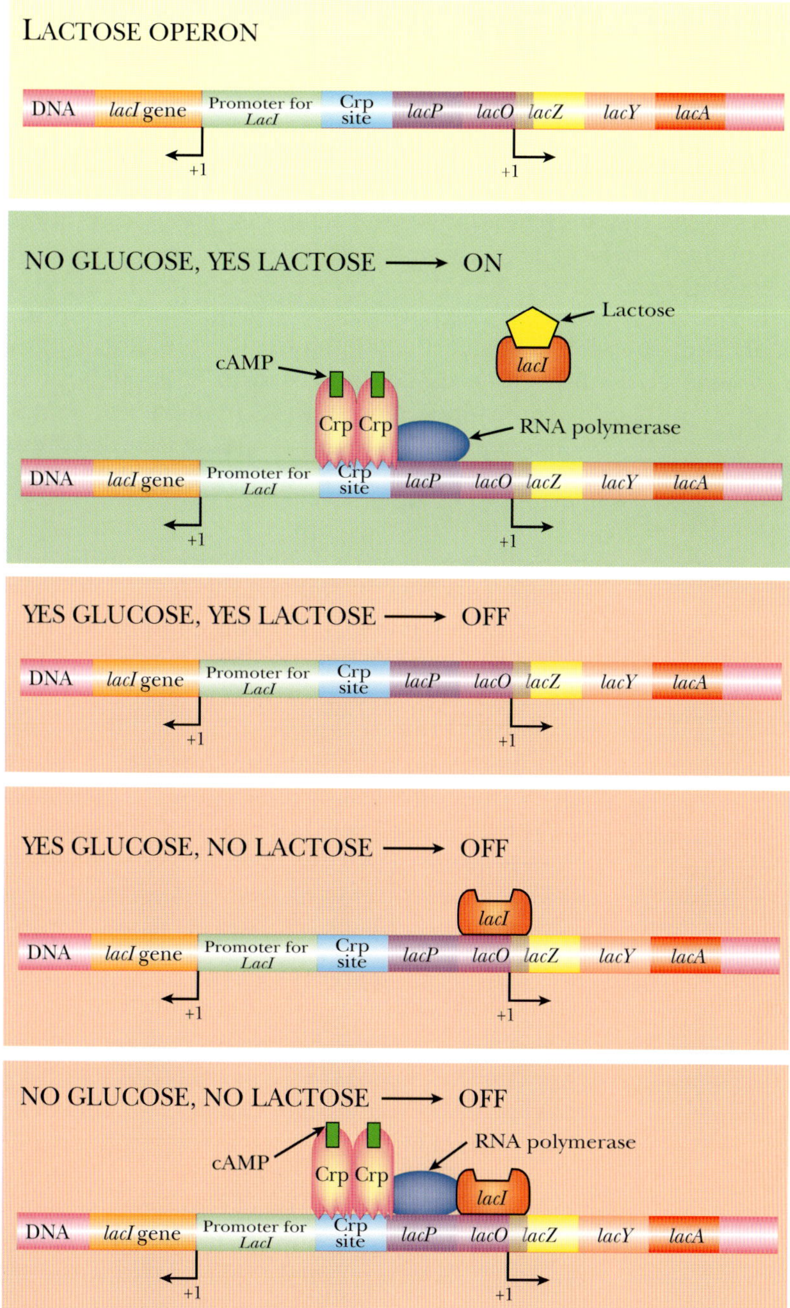

FIGURE 2.7 Control of Lactose Operon

The lactose operon is turned on only when glucose is absent but lactose is present. When glucose is available, the global activator protein, Crp, does not activate binding of RNA polymerase. When there is no glucose, Crp binds to the promoter and stimulates RNA polymerase to bind. The lack of lactose keeps LacI protein bound to the operator site and prevents RNA polymerase from transcribing the operon. Only when lactose is present is LacI released from the DNA.

the global activator, Crp, and *allo*-lactose binds to the specific repressor, LacI. One control is global (Crp) because it controls many different operons, and one control is specific (LacI) because it only regulates the lactose operon.

Many researchers use the lactose promoter to control expression of other genes. In the lab, *allo*-lactose is not used to induce the promoter. A **gratuitous inducer**, **IPTG (isopropyl-thiogalactoside)**, is used (Fig. 2.8). IPTG is not cleaved by β-galactosidase because its two halves are linked through a sulfur rather than oxygen. Since it is not metabolized, IPTG does not have to be added continually throughout the experiment (as would be the case for a natural inducer).

The *lac* operon is important to understand because its inducers and regulators are used to control new genes that are engineered into model organisms.

Control of Activators and Repressors

A variety of mechanisms controls gene activators and repressors. In some cases, the repressor or activator binds to the promoter of its own gene and controls its own transcription; this is called **autogenous regulation**.

Many activators and repressors rely on activation by small molecules, as for Crp and lacI. In some cases, a repressor needs a **co-repressor** in order to be active. For example, ArgR, represses the arginine biosynthetic operon when arginine is present. Arginine is a co-repressor and ensures that the bacteria do not make the amino acid when it is not needed.

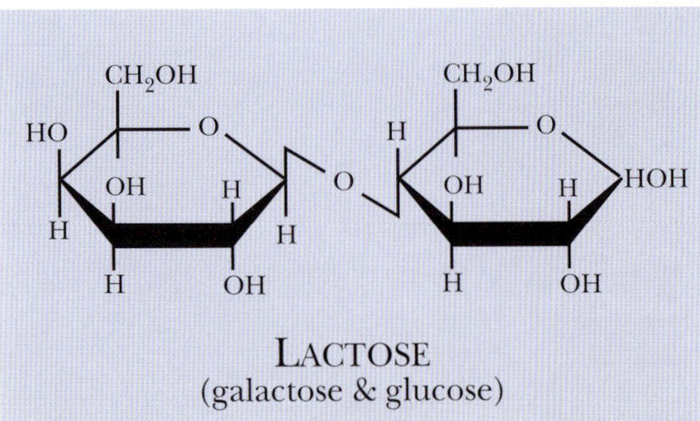

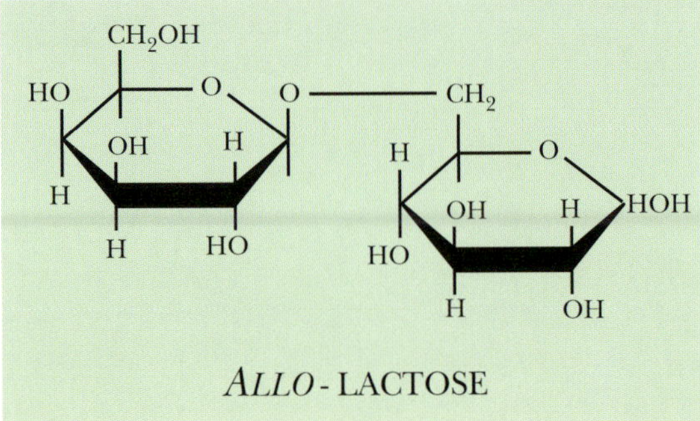

ISOPROPYL-β-D-THIOGALACTOSIDE
(IPTG)

FIGURE 2.8
Structures of Lactose, *allo*-Lactose, and IPTG
IPTG is a nonmetabolizable analog of the lactose operon inducer, *allo*-lactose. β-galactosidase cannot break the sulfur linkage, and therefore, does not cleave IPTG in two.

In many cases, adding different groups, such as phosphate, methyl, acetyl, AMP-, and ADP-ribose, covalently modifies activators or repressors. The **two-component regulatory systems** of bacteria transfer phosphate groups from a sensor protein to a regulator protein (Fig. 2.9). The first protein, the **sensor kinase**, senses a change in the environment and changes shape. This causes the kinase to phosphorylate itself using ATP. The phosphate group is then transferred to the regulator protein (an activator or repressor), which changes shape to its DNA binding form. The phosphorylated regulator then binds to its recognition site in the target promoter. This either stimulates or represses transcription of the operon.

There are many different two-component systems in bacteria that respond to a variety of environmental conditions. For example, when there is low oxygen, the ArcAB system modifies gene expression to compensate. The ArcB protein is the sensor kinase, and it has three phosphorylation sites. The ArcA regulator has only one site for phosphorylation. The phosphate group is passed from one site to the next in a **phosphorelay** system, ultimately regulating transcription of the genes. These types of phosphorelays are very common, particularly in eukaryotes where there are often more than two components.

Prokaryotic regulators are controlled by different mechanisms so that the gene of interest is expressed only when necessary.

REGULATION OF TRANSCRIPTION IN EUKARYOTES

Just as the initiation of transcription is more complex in eukaryotes, so is its control. The mechanisms to regulate which gene is expressed at what time are very complicated. The fact that eukaryotic DNA is wound around histones hinders many proteins from binding to the DNA, delaying access by activators and repressors. In addition, the nuclear membrane prevents the access of most proteins to the nucleus. Consider the complexity of the human body, with its multiple tissues. Each gene in each cell needs to be expressed only when needed, and only in the amount needed. In addition to normal organ functions and to changes during development, the environment has a huge impact on our bodies, and changes in gene expression help us adapt. Overall, the numbers and types of controls for gene expression are staggering. Other eukaryotes such as mice, rats, *Arabidopsis*, *C. elegans*, and even the relatively simple yeast have similarly complex control systems.

There are many different transcription factors for eukaryotic genes, yet they all have at least two domains: one binds to DNA, and the other binds to some part of the transcription apparatus. The two domains are connected, yet may function when separated from each other (Fig. 2.10). If the DNA binding domain of one transcriptional regulator is connected to the activation domain of another, the hybrid protein will work, each part retaining its original characteristics. That is, the DNA binding domain will bind the same sequence as it did before, and the activation domain will activate transcription as before. This property can be exploited when trying to identify protein-to-protein interactions with newly characterized proteins in the yeast two-hybrid screen (see Chapter 9).

Transcription factors work via an assembly of many proteins called the **mediator complex** (Fig. 2.11). These proteins receive all the signals from each of the activator proteins, compile the message, and transmit this to RNA polymerase II. The mediator contains about 20 different proteins, most of which make up the core. The presence of other proteins may vary depending on the cell or organism. These accessory proteins were originally thought to be co-activators or co-repressors since their presence varies based on tissue.

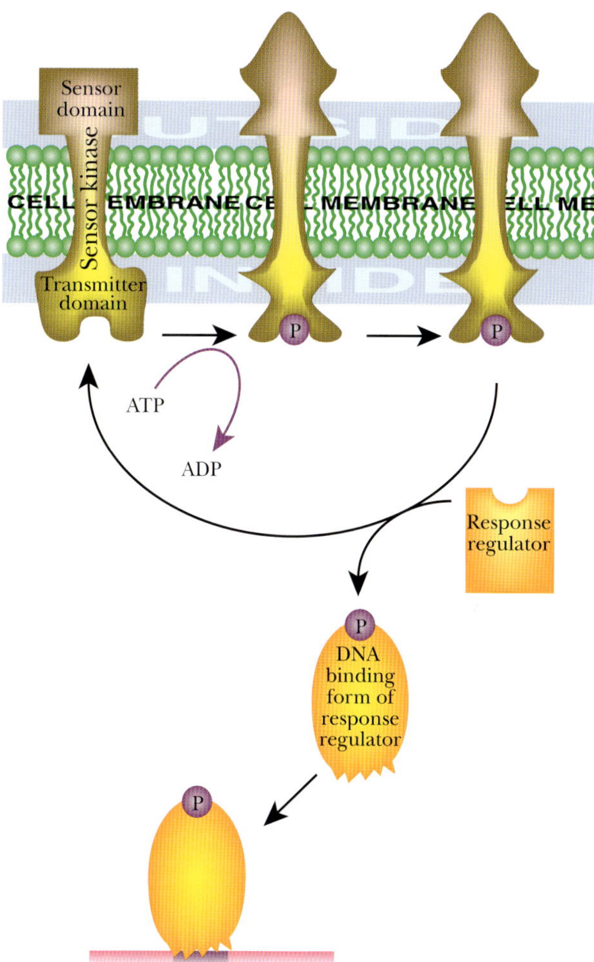

FIGURE 2.9 Model of Two-Component Regulatory System
The two-component regulatory system includes a membrane component (sensor kinase) and a cytoplasmic component (regulator). Outside the cell, the sensor domain of the kinase detects an environmental change, which leads to phosphorylation of the transmitter domain. The response regulator protein receives the phosphate group, and as a consequence, changes configuration so as to bind the DNA.

The mediator complex sits directly on RNA polymerase II waiting for information from activators or repressors. These may bind to regions just upstream of RNA polymerase II and the mediator complex. However, eukaryotic transcription factors may also bind to DNA sequences known as **enhancers** that may be thousands of base pairs away from the promoter. Even so, the regulatory proteins bind directly to the mediator complex. The enhancer elements work in either orientation, but affect only genes that are in the general vicinity. The prevailing theory is that the DNA loops around so that the enhancer is brought near the promoter.

Insulators are DNA sequences that prevent enhancers from activating the wrong genes. Insulators are placed between enhancers and those genes they must not regulate. The **insulator binding protein (IBP)** recognizes the insulator sequences and blocks the action of enhancers that are on the far side of the insulator (see Fig. 2.11). Insulator sequences may be controlled by methylation. When the DNA sequence is methylated, IBP cannot bind and the enhancer is allowed to access promoters beyond the insulator.

The eukaryotic transcription factor has two domains. The DNA binding domain binds to the DNA at the promoter and the activator domain has sites for initiating RNA polymerase action.

Eukaryotic transcription factors control gene expression by binding to the mediator complex. Insulator sequences prevent transcription factors from binding to the wrong promoter. Enhancer sequences are far from the promoter but may loop around to directly bind to the mediator complex.

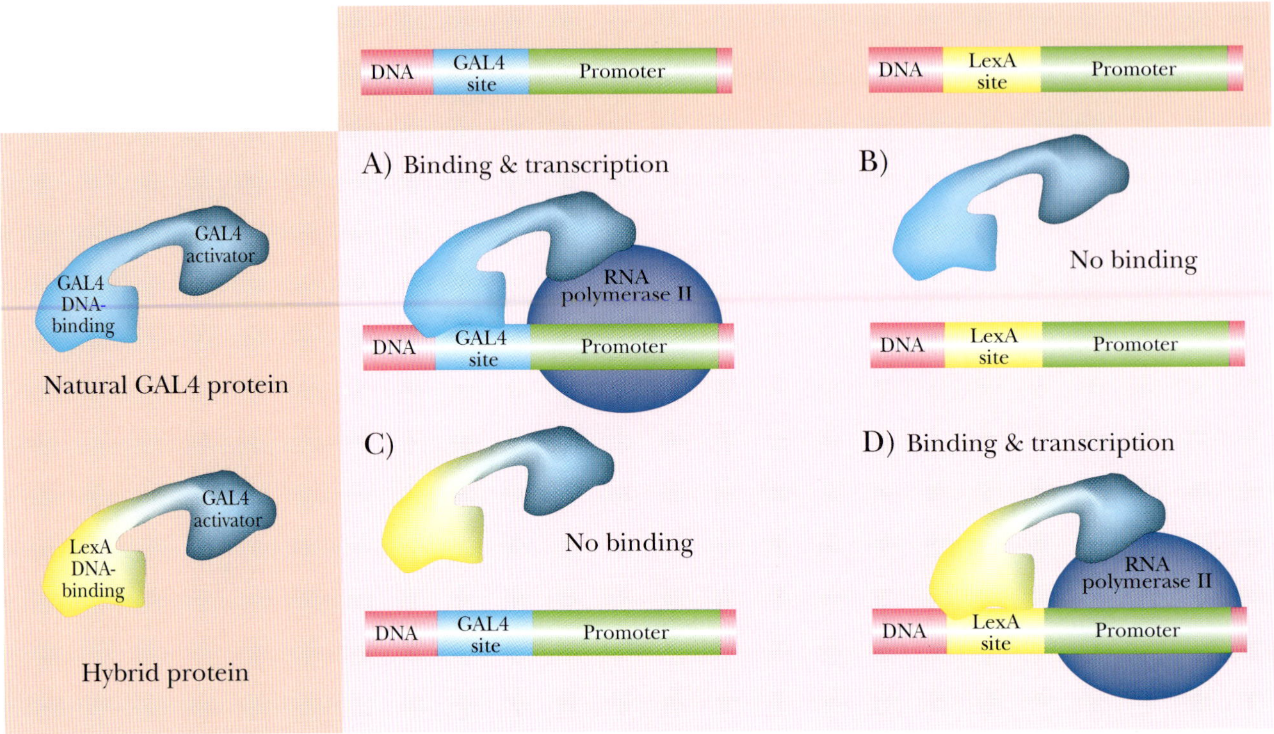

FIGURE 2.10 Transcription Factors Have Two Independent Domains
(A) One domain of the GAL4 transcription factor normally binds to the GAL4 DNA recognition sequence and the other binds the transcription apparatus. (B) If the LexA sequence is substituted for the GAL4 site, the transcription factor does not recognize or bind the DNA. (C) An artificial protein made by combining a LexA binding domain with a GAL4 activator domain will not recognize the GAL4 site, but (D) will bind to the LexA recognition sequence and activate transcription. Thus, the GAL4 activator domain acts independently of any particular recognition sequence. It works as long as it is held in close contact with the DNA.

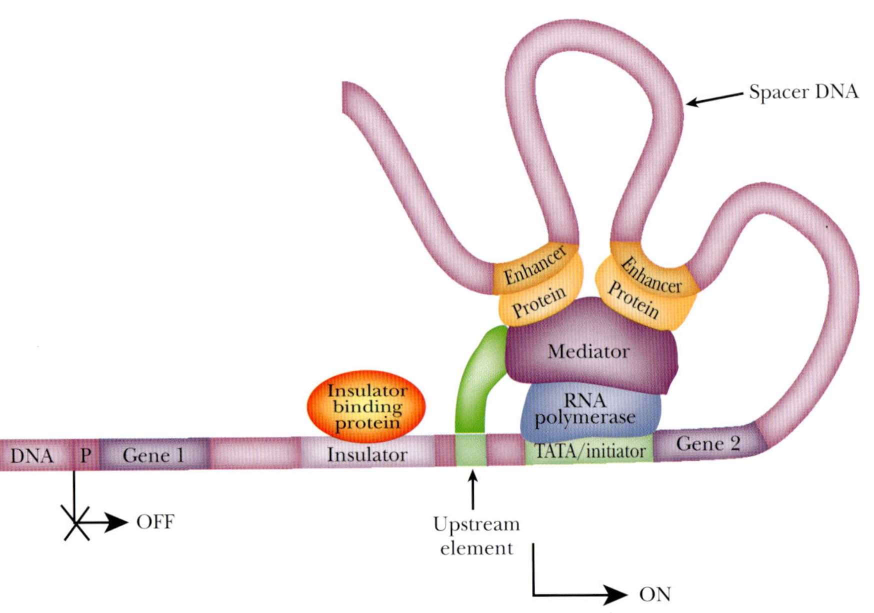

FIGURE 2.11 Enhancer and Insulator Sequences
Enhancer elements are found many hundreds of base pairs from the gene they control. They bind specific proteins that interact with the mediator complex by looping the DNA around. In this example, the enhancer proteins interact only with gene 2 because the insulator binding protein prevents their access to gene 1.

Eukaryotic Transcription Enhancer Proteins

An example of a eukaryotic transcription factor that works as an activator in some cell types is **AP-1 (activator protein-1)**. This affects a variety of genes and responds to a wide range of different stimuli. The most potent stimulators of AP-1 include growth factors and UV irradiation. Growth factors stimulate cell growth, and UV irradiation kills cells, so the genes activated by AP-1 are quite varied. The complex effects of this single transcription factor are still being investigated.

AP-1 is a dimer of two proteins, Fos and Jun, and recognizes the palindromic sequence 5′-TGACTCA-3′ (Fig. 2.12). AP-1 belongs to a family of DNA binding proteins called **bZIP proteins**. The proteins each have a leucine zipper, which brings the partners

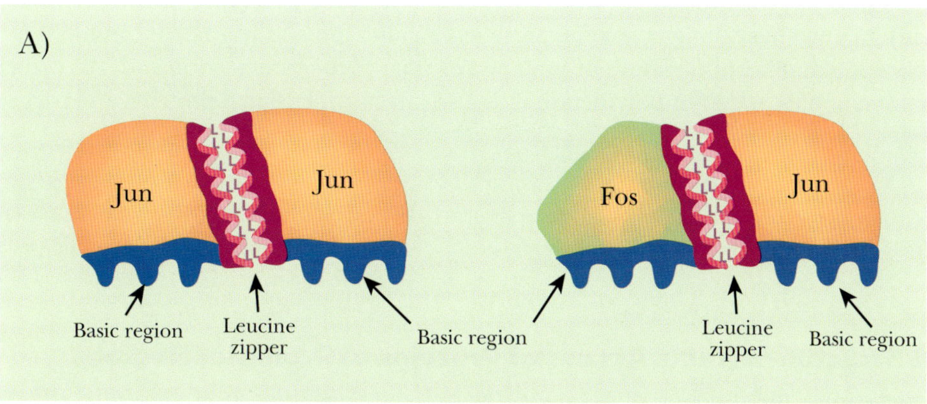

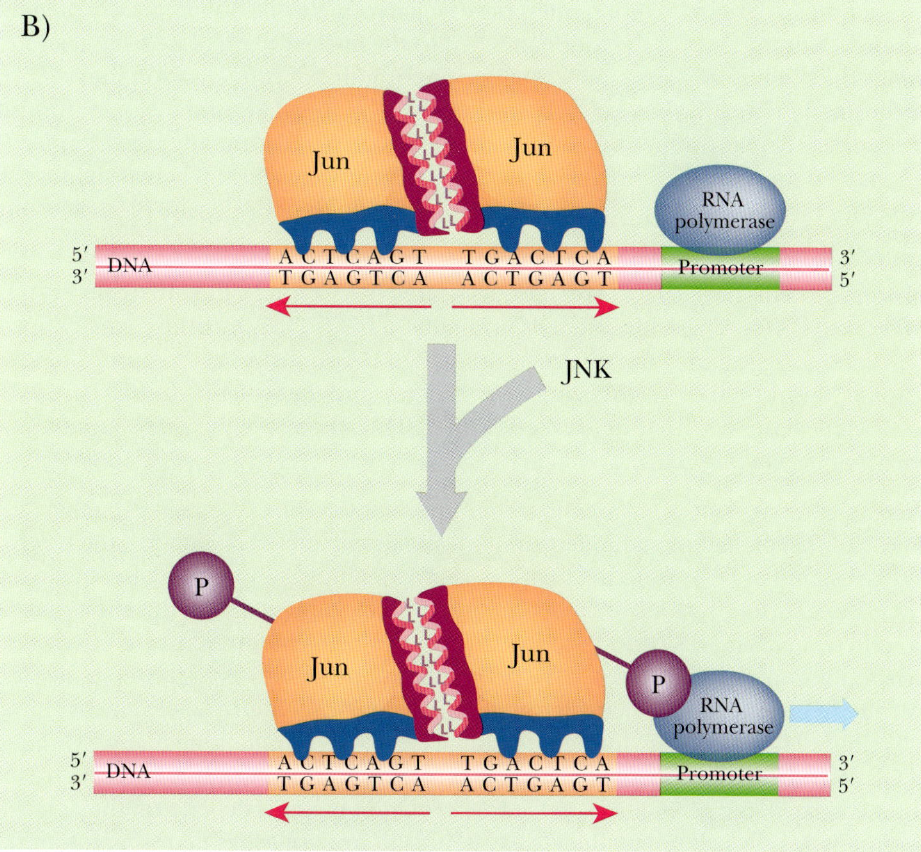

FIGURE 2.12 Eukaryotic Regulation of Transcription
(A) AP-1 is a eukaryotic transcription factor that consists of Fos and Jun. These two proteins interact through their leucine zippers. (B) To activate transcription, AP-1 must itself first be activated by phosphorylation by the kinase, JNK. Only then does Jun stimulate RNA polymerase II to transcribe the appropriate genes.

together to form dimers, plus a region of basic residues that contacts the DNA. (Leucine zippers are alpha helical segments where one side of the helix is mostly leucine. Two such leucine patches will associate by hydrophobic bonding and can bind two different proteins together.) Jun can either dimerize with itself or bind to Fos. In contrast, Fos can bind to Jun but cannot dimerize by itself. Fos and Jun also have activation domains that receive cellular signals that increase or attenuate their activity. Other protein partners may bind to Fos and Jun, and these change the recognition sequence. These other proteins, such as ATF, Maf, and Nr1, add to the complexity of AP-1 transcription control.

When AP-1 is stimulated, two different effects are involved. First, the cell makes more Fos and Jun proteins through increased expression of their genes. In addition, the proteins themselves become more stable and are not degraded as quickly. Second, the activity of Fos and Jun are stimulated by phosphorylation of their activation domain by **JNK (Jun amino-terminal kinase)**. Many other cellular signaling proteins can alter Jun and Fos activity, but JNK is the most potent. Phosphorylation of Jun and Fos triggers their interaction with the protein mediator complex and RNA polymerase II. It also affects other signal proteins and triggers other genes.

The eukaryotic transcription factor, AP-1, consists of two different proteins that work as a dimer. The family of proteins that form AP-1 are also controlled by posttranslational modifications such as phosphorylation.

DNA Structure Affects Access of Proteins to Promoter Regions

As noted earlier, the structure of eukaryotic DNA has a huge impact on the ability of regulatory proteins to bind. During interphase, eukaryotic chromosomes are uncondensed and looped, with the ends of the loops attached to the nuclear matrix (see Chapter 1). This helps the binding of transcription factors. The amount of condensation into heterochromatin is a critical factor. Eukaryotic DNA is wrapped around histones, forming nucleosomes. (These form the "beads on a string" structure of chromatin; see Chapter 1.) Loosely packed nucleosomes provide access for transcription factors, whereas tightly packed nucleosomes exclude all other proteins. Therefore, controlling the density of nucleosomes can regulate transcription initiation.

The histone proteins have tails that can be acetylated by enzymes called **histone acetyl transferases (HATs)**. These histone tails normally stabilize DNA by binding neighboring nucleosomes, so aggregating the nucleosomes. When HAT transfers acetyl groups to the tails, they can no longer bind to neighboring nucleosomes, and the structure loosens. In order to tighten the nucleosomes, **histone deacetylases (HDACs)** remove the acetyl groups, and the histones reaggregate (Fig. 2.13).

Another structural feature of eukaryotic DNA important to gene expression is methylation (see Fig. 2.13). In prokaryotes, differences in methylation patterns distinguish the newly synthesized strand of DNA from the template during replication. In eukaryotes, methylation is used to **silence** various regions of DNA and prevent their expression. Methylation of cytosine by two different enzymes occurs in the sequence CG or CNG. **Maintenance methylases** add methyl groups to newly synthesized DNA to give the same pattern as in the template strand. *De novo* **methylases** add new methyl groups, and of course, **demethylases** remove unwanted methyl groups. Many genes are located near stretches of DNA containing many CG sequences, called **CG islands**. If these are methylated, the nearby genes are not expressed, whereas if they are not methylated, the genes are expressed. Methylation patterns depend on the tissue. For example, muscle cells will not methylate CG islands in front of genes necessary for muscle function. The muscle-specific genes will have methylated CG islands in other tissues, though.

Silencing by DNA methylation can occur for one gene or for areas as large as an entire chromosome. Regions that become "silenced" first have methylated CG regions.

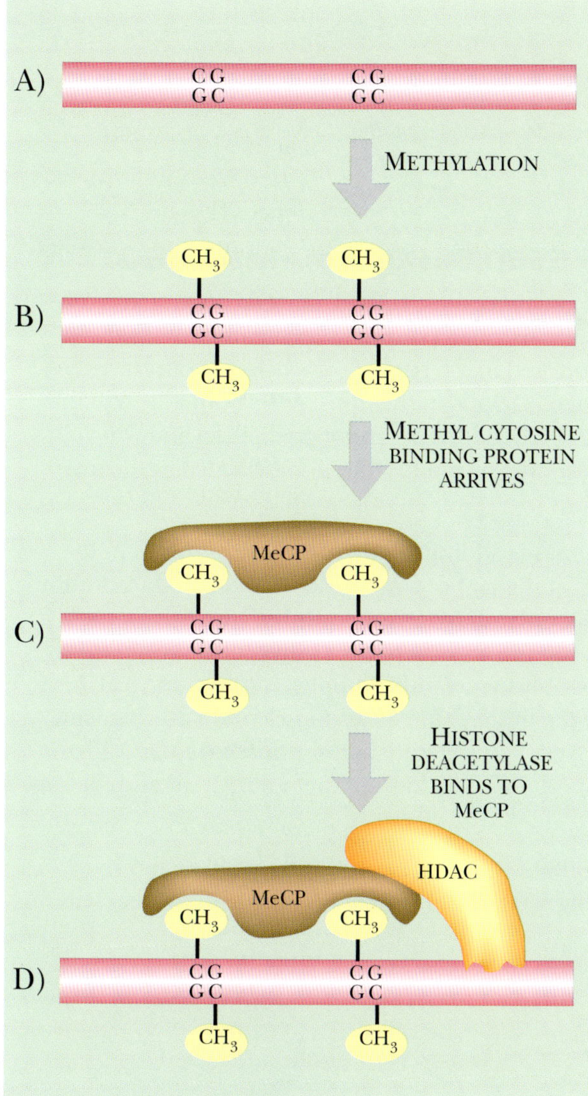

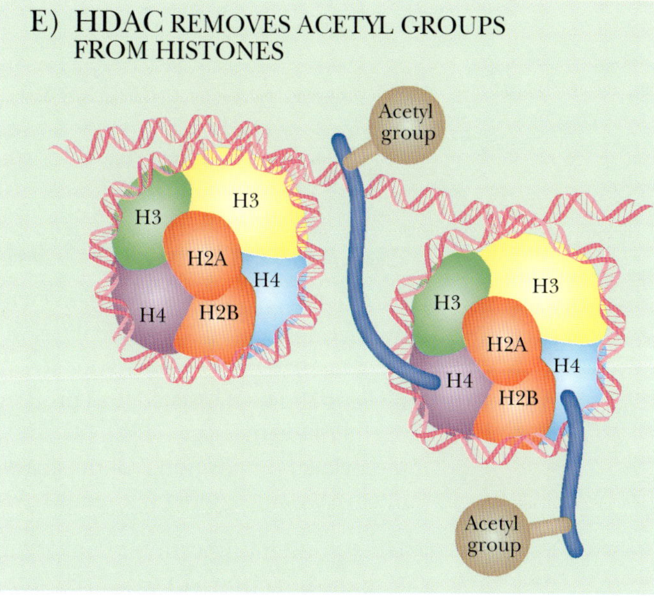

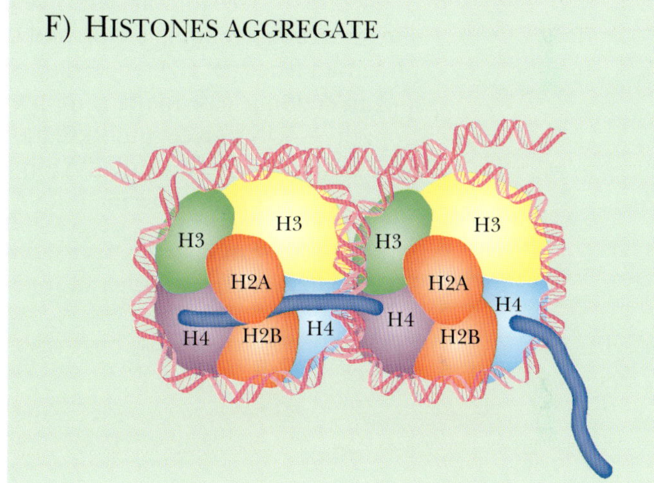

FIGURE 2.13 DNA Methylation Induces Gene Silencing
Gene expression in eukaryotes can be turned off by chromatin condensation. First, the area to be silenced is methylated. The methyl groups attract methyl cytosine binding protein, which in turn attracts histone deacetylases. Once HDAC removes the acetyl groups from the histone tails, the histones aggregate tightly. The closeness of histones excludes any DNA binding proteins and hence turns off gene expression in the area.

The methylated DNA attracts **methylcytosine binding proteins** that block binding of other DNA binding proteins. These proteins also recruit HDACs that deacetylate the histone tails, thus condensing the chromatin. The areas become heterochromatin, and this prevents any further gene expression.

Methylation patterns have a major impact during development, because gene expression here must be tightly controlled. Some genes remain methylated in the gamete, whereas other genes must be demethylated so they can become active. **Imprinting** occurs when a gene from the parent is methylated in the gamete and remains methylated in the new organism. This affects relatively few genes. In contrast, genes that are not imprinted change their methylation patterns during development. Imprinting can be different for male gametes and female gametes, setting the stage for sexual differences in gene expression. One special type of imprinting is **X-inactivation**, where the entire second X chromosome in females is silenced by methylation (except for a few loci).

The structural features of DNA itself lend control to gene expression. The density of chromatin packing can exclude transcription factors from binding to the enhancer regions.

Adding methyl ($-CH_3$) groups to the cytosine of CG areas controls expression of nearby genes. These groups can prevent binding of various transcription factors and have implications in setting up male/female differences during development.

EUKARYOTIC mRNA IS PROCESSED BEFORE MAKING PROTEIN

Bacterial mRNA may be translated without any processing. Indeed, bacteria often start translating their mRNA while it is still being transcribed (known as *coupled transcription/translation*). However, eukaryotic RNA is processed in a variety of ways before it can leave the nucleus and be translated into protein. First, eukaryotic mRNA must have a **cap** added to the 5′ end of the message (Fig. 2.14). The cap is a GTP added in reverse orientation and which is methylated on position 7 of the guanine base. Methyl groups may also be added to the first one or two nucleotides of the mRNA.

The second modification of eukaryotic mRNA is adding a long stretch of adenines to the 3′ end—the **poly(A) tail**. Three sequences at the end of a new mRNA mediate the addition of the tail: the recognition sequence for the **polyadenylation complex** (AAUAAA); the cut site for cleavage binding factor; and the recognition sequence for polyadenylation binding protein (a length of GU repeats). First, the polyadenylation complex binds to the AAUAAA, and an endonuclease in the complex cuts the mRNA after a CA dinucleotide downstream from the AAUAAA recognition sequence. Next poly(A) polymerase adds 100 to 200 adenine nucleotides. Finally, the poly(A) binding protein binds to both the poly(A) tail and the cap structure. This circularizes the mRNA.

A third modification made to eukaryotic mRNA is the removal of **introns**. Eukaryotic DNA contains many stretches of intervening sequence (introns) between regions that will

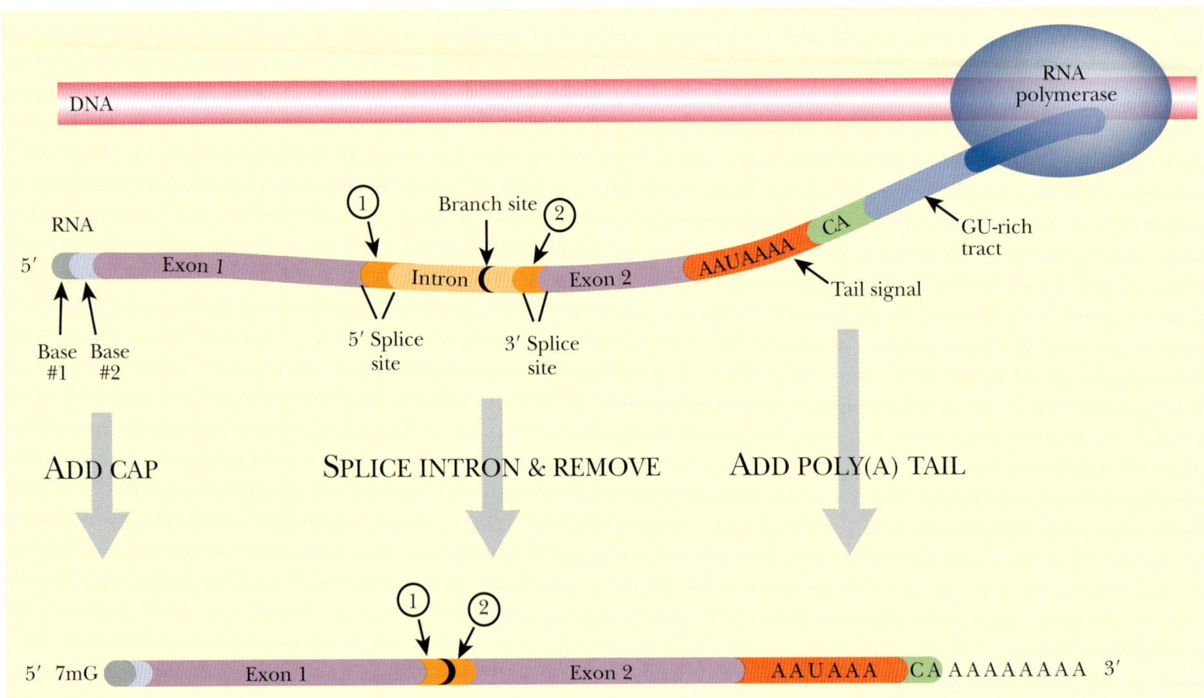

FIGURE 2.14 Processing Eukaryotic mRNA

Eukaryotic RNA is processed before exiting the nucleus for translation into protein. A guanine with a methyl group is added to the 5′ end of the message, a poly(A) tail is added to the 3′ end, and the introns are spliced out. These modifications stabilize the message and make it much shorter than the original RNA transcribed from the DNA.

ultimately code for a protein (**exons**). First the entire region is transcribed into an RNA molecule called the **primary transcript**. After capping and tailing, this is processed to remove the introns. The exons are spliced together to form the mRNA. Proteins called **splicing factors** recognize the exon/intron borders, cut the DNA, and join the neighboring exons.

Eukaryotic RNA is transcribed as a primary transcript, a cap is added at the 5′ end, the introns are removed, and a poly(A) tail is added. The mRNA is then transported from the nucleus to the endoplasmic reticulum for translation by ribosomes.

TRANSLATING THE GENETIC CODE INTO PROTEINS

The Genetic Code Is Read as Triplets or Codons

Messenger RNA provides the information a ribosome needs to make proteins. This process is known as translation because it involves translating information carried by nucleic acids to give the sequences of amino acids that make up proteins. Before the mechanism is discussed, the code used to assemble proteins must first be understood. Nucleic acids each have four different bases (the T in DNA is equivalent to U in RNA). However, proteins consist of 20 different amino acids. If each nucleotide corresponded to an amino acid, this would encode only four different amino acids. Two nucleotides give only 16 combinations, still not enough. To encode 20 amino acids, bases must be read in groups of three. This gives 64 different combinations—more than enough for all 20 amino acids (Fig. 2.15). Messenger RNA is therefore read in groups of three bases, known as **triplets** or **codons**. Each triplet of bases codes for one amino acid. Because there are more than 20 triplets, many are redundant so that multiple codons will be translated into the same amino acid. For example, valine is encoded by GUU, GUC, GUA, or GUG.

The genetic code listed in Fig. 2.15 is considered the **universal genetic code**. Not all organisms use precisely this code, although exceptions are rare. For example, UGA normally signals stop, but in *Mycoplasma*, UGA encodes tryptophan and, in the protozoan *Euplotes*, UGA encodes cysteine.

Small RNA molecules known as **transfer RNA (tRNA)** recognize the individual codons on mRNA and carry the corresponding amino acids. Although tRNA is synthesized as a single strand, it folds back on itself to form regions of double-stranded RNA. The final shape of tRNA is a folded "L" shape with the **anticodon** at one end and the **acceptor stem** at the other. The anticodon consists of three bases complementary to those of the corresponding codon and it therefore recognizes the codon by base pairing. The acceptor stem is where the amino acid is added to the free 3′ end of the tRNA (Fig. 2.16).

How does each specific tRNA carry the correct amino acid? A group of enzymes called **aminoacyl tRNA synthetases** attach the correct amino acid to the corresponding tRNA. These enzymes are very specific and recognize the correct tRNA by its sequence at the anticodon or elsewhere along the RNA structure. There is a specific aminoacyl tRNA synthetase for each amino acid.

The first base of the anticodon binds the third base of the codon in the mRNA. Because this nucleotide in tRNA is not constrained by neighboring nucleotides, it can **wobble** instead of forming a perfect double helix. This allows nonstandard base pairs to be created. For example, if the first anticodon base is G, it would normally pair with C in the third position of the codon. Because of wobble, G can also pair with U. Thus, the tRNA for histidine has the

FIGURE 2.15 The Genetic Code
The 64 codons found in mRNA are shown with their corresponding amino acids. As usual, bases are read from 5′ to 3′ so that the first base is at the 5′ end of the codon. Three codons (UAA, UAG, UGA) have no cognate amino acid but signal stop. AUG (encoding methionine) and, much less often, GUG (encoding valine) act as start codons. To locate a codon, find the first base in the vertical column on the left, the second base in the horizontal row at the top, and the third base in the vertical column on the right.

	2nd (middle) base				
1st base	U	C	A	G	3rd base
U	UUU Phe UUC Phe UUA Leu UUG Leu	UCU Ser UCC Ser UCA Ser UCG Ser	UAU Tyr UAC Tyr UAA stop UAG stop	UGU Cys UGC Cys UGA stop UGG Trp	U C A G
C	CUU Leu CUC Leu CUA Leu CUG Leu	CCU Pro CCC Pro CCA Pro CCG Pro	CAU His CAC His CAA Gln CAG Gln	CGU Arg CGC Arg CGA Arg CGG Arg	U C A G
A	AUU Ile AUC Ile AUA Ile AUG Met	ACU Thr ACC Thr ACA Thr ACG Thr	AAU Asn AAC Asn AAA Lys AAG Lys	AGU Ser AGC Ser AGA Arg AGG Arg	U C A G
G	GUU Val GUC Val GUA Val GUG Val	GCU Ala GCC Ala GCA Ala GCG Ala	GAU Asp GAC Asp GAA Glu GAG Glu	GGU Gly GGC Gly GGA Gly GGG Gly	U C A G

FIGURE 2.16
Structure of tRNA Allows Wobble in the Third Position
Transfer RNA recognizes the codons along mRNA and presents the correct amino acid for each codon. The first position of the anticodon on tRNA matches the third position of the codon.

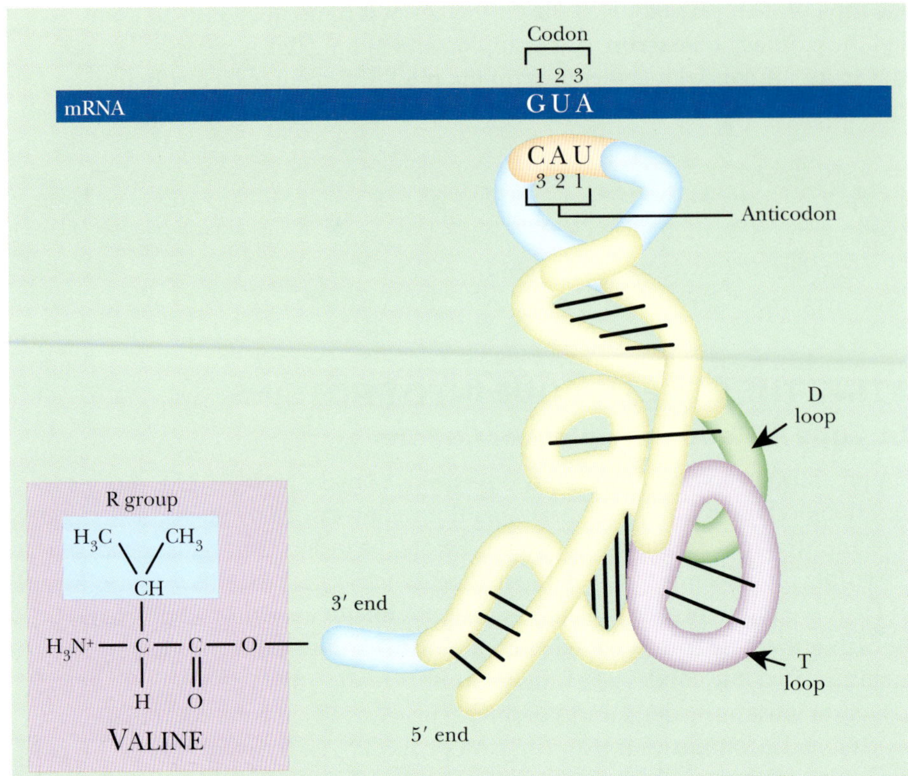

anticodon GUG and recognizes both CAC and CAU in the mRNA. Similarly, U in the first place in the anticodon can base-pair with A or G in the third position of the codon. Wobble explains how the same tRNA can read multiple codons all encoding the same amino acid.

During protein translation, each tRNA recognizes a specific three-nucleotide sequence and has the correct amino acid attached to the opposite end. A family of specific enzymes, aminoacyl tRNA synthetases, ensure that each tRNA has the correct amino acid.

Protein Synthesis Occurs at the Ribosome

The **ribosome** is a molecular machine that unites the mRNA with the appropriate tRNAs and then links the amino acids together into a chain. Prokaryotic ribosomes consist of two subunits called the 30S and 50S, which combine to form a functional 70S ribosome. A ribosome consists of several RNA molecules (**ribosomal RNA** or **rRNA**) and many proteins. The 30S subunit has a 16S rRNA plus 21 proteins; the 50S subunit has two rRNAs,

Box 2.1 Codon Bias

Several amino acids are encoded by multiple codons and have more than one corresponding tRNA. Thus valine is encoded by GUU, GUC, GUA, and GUG. One tRNA for valine recognizes GUU and GUC by wobble, but another tRNA is necessary for the other two codons. However, many organisms tend to use only one or two of the codons for amino acids with multiple codons—a phenomenon known as **codon bias**. Consequently, they make low amounts of tRNA for the rarely used codons. Furthermore, different organisms show different codon preferences. This becomes an issue when genes from one organism are expressed in another. Plants and animals often prefer different codons than bacteria for the same amino acids. When bacteria express plant or animal proteins, not enough tRNA is available for the nonpreferred codons, and the ribosomes stall and fall off, making protein yield very low. To remedy this problem, researchers may genetically engineer the genes so that abundant tRNAs recognize their codons (see Chapter 14). Alternatively, bacterial host strains may be engineered to express higher levels of the necessary tRNAs.

the 5S and 23S, plus 34 proteins. The larger subunit has three binding sites for tRNA, called A for acceptor, P for peptide, and E for exit, referring to the action occurring at each site. The 23S rRNA actually catalyzes the addition of amino acids to the growing polypeptide chain and is therefore a ribozyme. (Ribozymes are discussed in Chapter 5.)

In prokaryotes, various factors beside the ribosome are involved in protein synthesis (Fig. 2.17). First, a ribosome must assemble at the start site and begin protein synthesis at the correct start codon. The 5′ untranslated region of the mRNA (see above) has the signal for ribosome binding in front of the start codon. In prokaryotes, translation begins at the first AUG codon after the **Shine-Dalgarno sequence**, or ribosome binding site, which has the consensus sequence UAAGGAGG. The **anti-Shine-Dalgarno sequence** is found in the 16S rRNA of the smaller 30S subunit. So first, the small ribosomal subunit binds the Shine-Dalgarno sequence. A derivative of methionine, **N-formyl-methionine (fMet)**, and a special initiator tRNA **($tRNA_i$)** are used to initiate translation in prokaryotes. Only initiator tRNA charged with fMet (referred to as **$tRNA_i^{fMet}$**) can bind the small subunit of the ribosome.

Translation factors are proteins needed to recruit and assemble the components of the ribosome and translational complex. **Initiation factors** (IF1, IF2, and IF3) assemble the **30S initiation complex**, which is the 30S ribosomal subunit plus $tRNA_i^{fMet}$. The IF3 factor then leaves the complex, and the 50S ribosomal subunit binds, forming the **70S initiation complex** (see Fig. 2.17A).

Finally, polypeptide assembly can begin (see Fig. 2.17B). The $tRNA_i^{fmet}$ occupies the P-site on the ribosome. Another tRNA recognizes the next codon and enters the A-site, and a peptide bond is formed between the first and second amino acid by the **peptidyl transferase** activity of 23S rRNA. fMet releases its tRNA, which moves into the E-site. This allows the second tRNA to move into the P-site, and the cycle begins again. A third tRNA, complementary to the next codon, enters the A-site, a peptide bond forms between amino acids 2 and 3, and then the second tRNA moves into the E-site of the ribosome, and exits.

Adding successive amino acids is called **elongation** and requires **elongation factors**. EF-T is a pair of proteins (EF-Tu and EF-Ts) that uses GTP to catalyze the addition of a new tRNA into the A-site (EF-Tu) and then exchanges the GDP with GTP for the next cycle (EF-Ts). The movement of tRNA from the P-site to the E-site is called **translocation**, and the mRNA simultaneously moves one codon sideways relative to the ribosome. The E-site and A-site cannot be occupied at the same time, and the used tRNA must exit before the next tRNA enters. EF-G oversees the translocation step.

Amino acids are added to the growing chain and the process continues until the ribosome encounters a stop codon (UAA, UAG, or UAA). None of the tRNA recognizes the stop codon. Instead proteins known as **release factors** bind the stop codons (see Fig. 2.17C). RF1 and RF2 recognize the different stop codons and stimulate the 23S rRNA to split the bond between the last amino acid and its tRNA. The whole ribosome assembly falls off the mRNA and dissociates. Its components will be used again for translation of another mRNA. The new polypeptide chain folds to form its final structure. In prokaryotes, multiple ribosomes bind to the same mRNA to form a **polysome**. Because there is no nucleus, transcription and translation are often simultaneous. As partially made mRNA comes off the DNA, ribosomes bind and start synthesizing protein.

Translation in prokaryotes starts in the 5′ UTR of the mRNA message, where the ribosome scans for the first start codon. After an initiator methionine is added to the AUG, the ribosome catalyzes the addition of more amino acids. Ribosomes work with elongation factors and release factors to control the movement down the mRNA until the stop codon.

Ribosomes have three different sites of action. The A-site accepts the next tRNA with the correct anticodon and amino acid. The P-site holds the previous tRNA with amino acid. The E-site is occupied briefly after the amino acids are linked as the empty tRNA exits the ribosome.

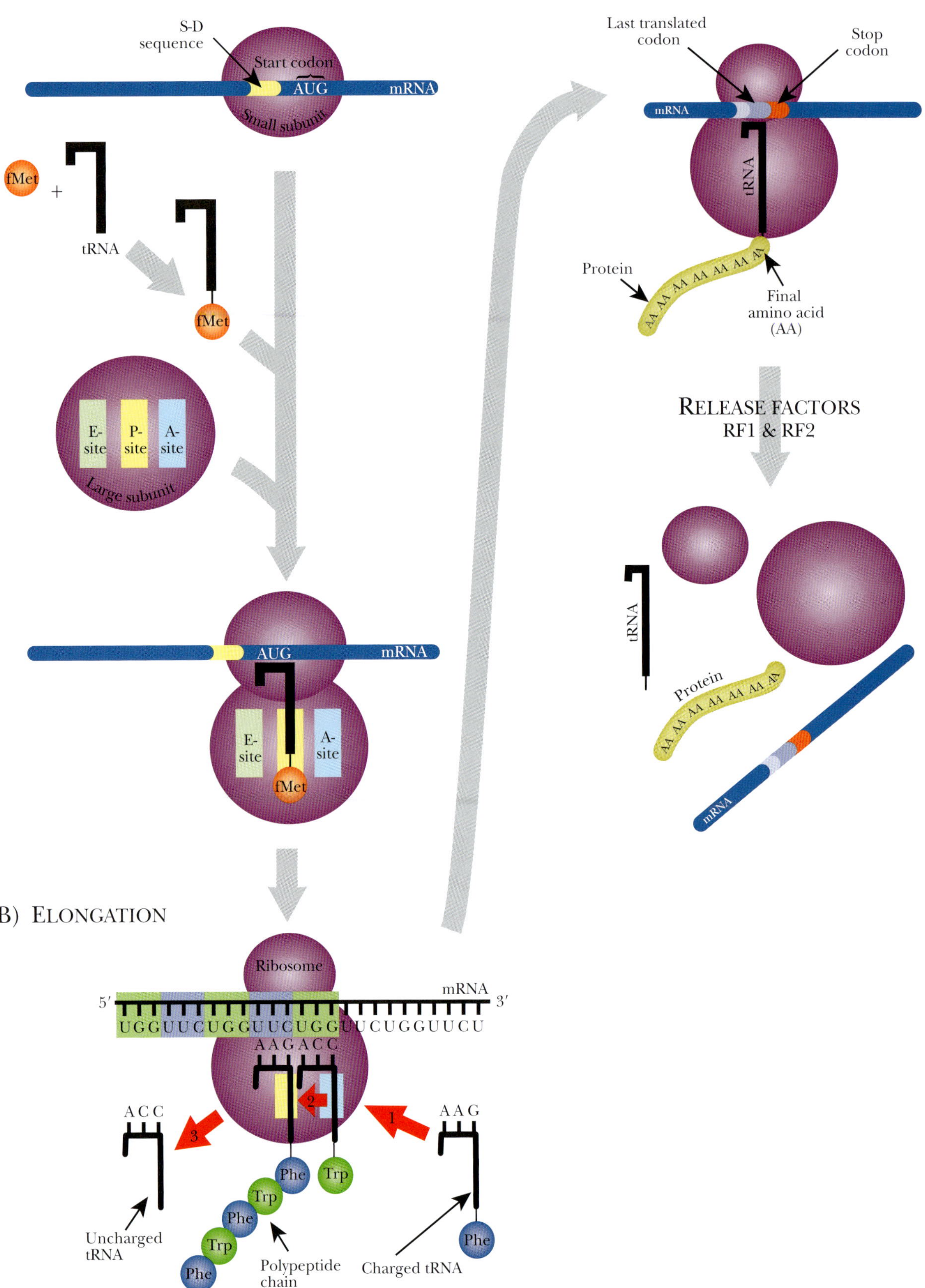

FIGURE 2.17 Translation in Prokaryotes
(A) Initiation of translation begins with the association of the small ribosome subunit with the Shine-Dalgarno sequence (S-D sequence) on the mRNA. Next, the initiator tRNA that reads AUG is charged with fMet. The charged initiator tRNA associates with the small ribosome subunit and finds the start codon. Assembly is helped by initiation factors (IF1, IF2, and IF3)—not shown. (B) During elongation peptide bonds are formed between the amino acids at the A-site and the P-site. The movement of the ribosome along the mRNA and addition of a new tRNA to the A-site are controlled by elongation factors (also not shown). (C) Termination requires release factors. The various components dissociate. The completed protein folds into its proper three-dimensional shape.

DIFFERENCES BETWEEN PROKARYOTIC AND EUKARYOTIC TRANSLATION

Translation in eukaryotes differs from prokaryotes in many ways (Fig. 2.18). First of all, mRNA is made in the nucleus, but translation occurs on the ribosomes in the cytoplasm. Therefore, there is no coupled transcription and translation in eukaryotes. Eukaryotic ribosomes have 60S and 40S subunits that combine to form an 80S ribosome, which is a little larger than bacterial ribosomes. Additionally, eukaryotes have more initiation factors than prokaryotes, and they assemble the initiation complex in a different order. Overall, more proteins are involved in eukaryotic translation to deal with the greater complexity of regulation.

Despite this, the binding of the mRNA is simpler in eukaryotes. Eukaryotic mRNA does not have a Shine-Dalgarno sequence. Instead, eukaryotic ribosomes recognize the 5′ cap structure and begin protein synthesis at the first AUG. Only one gene per mRNA is found (unlike bacteria, which often have polycistronic messages and whose ribosomes recognize separate Shine-Dalgarno sequences for each coding sequence). The first amino acid in each new polypeptide is methionine, as in bacteria. However, unlike in bacteria, this methionine is not modified with a formyl group. Finally, many eukaryotic proteins are modified after translation by addition of chemical groups. (Although bacteria do modify some proteins, this is much rarer and the variety of additions is much more limited.)

Prokaryotic mRNA has information for multiple proteins that are translated simultaneously by many different ribosomes. The mRNA has a ribosome binding site in front of each operon on the mRNA.

Eukaryotic mRNA has information for one protein. The ribosome recognizes the cap structure, scans until it finds the first AUG, and starts translating the message into protein.

MITOCHONDRIA AND CHLOROPLASTS SYNTHESIZE THEIR OWN PROTEINS

The mitochondria and chloroplasts found in eukaryotes have their own genome and make some of their own proteins. The **symbiotic theory** of organelle origin argues that these organelles were once free-living bacteria or blue-green algae (cyanobacteria) that formed a symbiotic relationship with a single-celled ancestral eukaryote. The bacteria supplied energy to the early eukaryote. Over time, the bacteria gave up many duplicate functions and came to rely on the host for precursor molecules. Eventually, the symbiotic mitochondria and chloroplasts lost the majority of their genes, yet today they still maintain a small version of their genome. These genomes have many genes associated with protein synthesis (Fig. 2.19). Organelle genes are often more closely related to bacterial genes than to eukaryotic (nuclear) genes. Moreover, the ribosomes in animal mitochondria are 28S and 39S in size, closer to the 30S and 50S subunits of bacteria. The ribosomal RNA of mitochondria and bacteria are also much more similar in sequence than either is to the rRNA encoded by the eukaryotic nucleus.

Mitochondria and chloroplasts have their own genome that includes many genes for transcription and translation. These may have been free-living bacteria that formed a symbiotic relationship with a unicellular eukaryote.

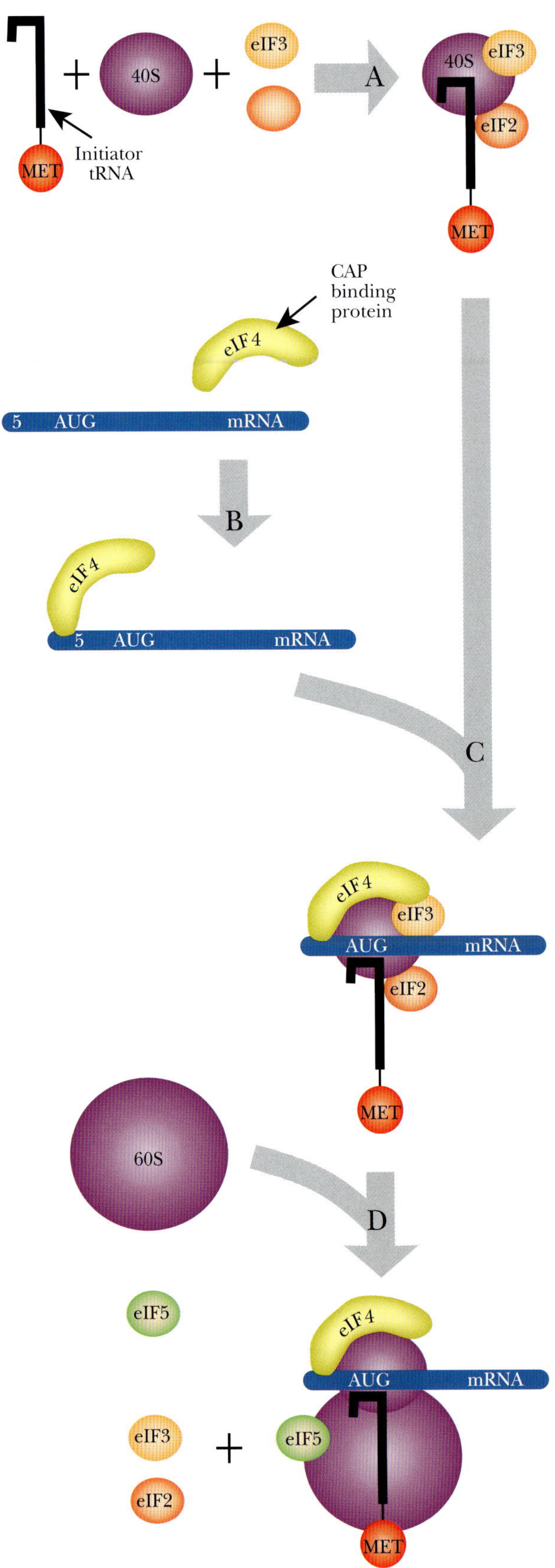

FIGURE 2.18 Translation in Eukaryotes

(A) Assembly of the small subunit plus initiator Met-tRNA involves the binding of factors eIF3 and eIF2. (B) The cap binding protein of eIF4 attaches to the mRNA before it joins the small subunit. (C) The mRNA binds to the small subunit via cap binding protein and the 40S initiation complex is assembled. (D) Assembly of the large subunit requires factor eIF5. After assembly, eIF2 and eIF3 depart.

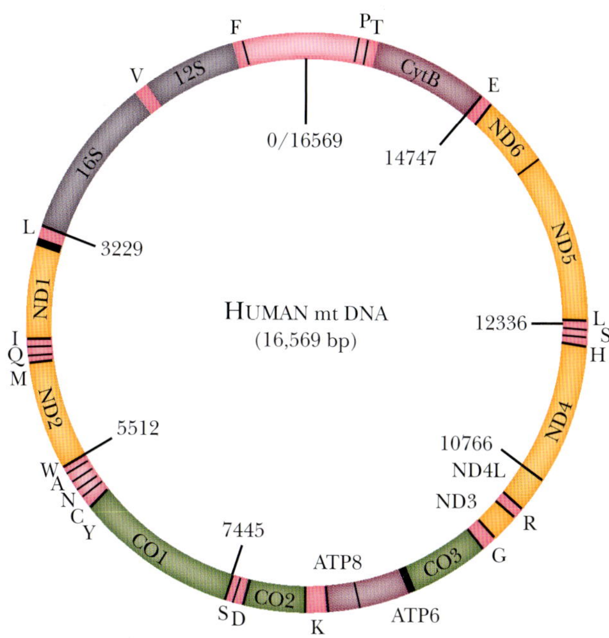

FIGURE 2.19 Human Mitochondrial DNA
The mitochondrial DNA of humans contains the genes for ribosomal RNA (16S and 12S), some transfer RNAs (single-letter amino acid codes mark these on the genome), and some proteins of the electron transport chain.

Summary

This chapter briefly explains the process of transcription and translation, highlighting the differences between eukaryotes and prokaryotes. Transcription occurs when RNA polymerase makes a complementary copy of the gene using ribose, phosphate, uracil, guanine, cytosine, and adenine. The complementary copy is called mRNA, and this form is used to translate into protein. The ribosome holds the mRNA so that two triplet codons starting at AUG are stable. Then a complementary tRNA that is holding the correct amino acid is held close to the mRNA by the ribosome. A second tRNA-amino acid complex moves next to the first, and the ribosome connects the two amino acids using its peptidyl transferase activity. The ribosome translocates to the next triplet codon on the mRNA and continues to link the amino acids to form a polypeptide. This basic mechanisms of transcription and translation are very similar in prokaryotes and eukaryotes.

The regulation of transcription and translation varies significantly between prokaryotes and eukaryotes. First, proteins called transcription factors control the expression of the correct gene at the correct time with the correct amount. In prokaryotes, the lactose operon demonstrates how activator proteins and repressor proteins work together so that lactose utilization genes are only expressed when lactose is the only sugar source for the bacteria. Prokaryotes have different sigma factors, an integral part of RNA polymerase, which specify the correct gene expression. In eukaryotes, many different transcription factors control gene expression by binding to the mediator complex. In eukaryotes, the condensation of DNA around histones prohibits many transcription factors from access to certain genes. Methylation of CG islands inhibits gene expression also. During translation, eukaryotes are actually less complex and express the mRNA transcript as a single message. In prokaryotes, the mRNA may contain multiple coding regions that are translated into proteins simultaneously as the transcript is made from the DNA.

End-of-Chapter Questions

1. Which of the following are important features for transcription?
   a. promoter
   b. RNA polymerase
   c. 5′ and 3′ UTRs
   d. ORF
   e. all of the above

2. For which of the following nitrogenous bases does DNA substitute thymine?
   a. uracil
   b. adenine
   c. guanine
   d. cytosine
   e. inosine

3. Which of the following is not necessary during Rho-independent termination of transcription?
   a. RNA polymerase
   b. Rho protein
   c. hairpin structure
   d. repeating A's in the DNA sequence
   e. All of the above are necessary.

4. Which of the following statements is not true about mRNA?
   a. Prokaryotic mRNA may contain multiple structural genes on the same transcript, known as polycistronic mRNA.
   b. Eukaryotes only transcribe one gene at a time on mRNA, called monocistronic mRNA.
   c. Eukaryotes are capable of having polycistronic mRNA; however, only the first cistron will be translated.
   d. Eukaryotes almost always produce polycistronic mRNA.
   e. The genes for metabolic pathways in bacteria are typically located close together and transcribed on one mRNA.

5. In what way is eukaryotic transcription more complex than prokaryotic transcription?
   a. Eukaryotes have three different RNA polymerases, whereas prokaryotes only have one RNA polymerase.
   b. Eukaryotic transcription initiation is much more complex than prokaryotic initiation because of the various transcription factors involved.
   c. Upstream elements are required for efficient transcription in eukaryotic cells, but these elements are not usually necessary in prokaryotes.
   d. Eukaryotic mRNA is made in the nucleus.
   e. All of the above statements outline ways that eukaryotic transcription is more complex.

6. Why is the *lac* operon of *E. coli* important to biotechnology research?
   a. IPTG is a cheaper additive than lactose to growing cultures.
   b. The *lac* operon is not used in biotechnology research.
   c. The inducers and regulators of the *lac* operon are used to control the expression of genes in model organisms.
   d. The *lac* operon controls the amount of lactose that *E. coli* metabolizes.
   e. All of the above.

7. What feature about eukaryotic transcription factors is useful to biotechnology research?
   a. They have two domains, both of which bind to DNA.
   b. They have two domains, both of which bind to separate proteins.
   c. They have two domains: one domain binds DNA and the other binds to some part of the transcription apparatus.
   d. They have only one domain that binds to RNA polymerase.
   e. They have two domains but neither domain can be engineered and are therefore not useful to biotechnology research.

8. Which of the following DNA structure modifications are used to regulate transcription?
   a. acetylation/Deacetylation of the histone tails
   b. methylation of specific bases in the DNA sequence
   c. imprinting, such as X-inactivation
   d. chromatin condensation
   e. All of the above are important modifications for transcription regulation.

9. Which of the following statements about eukaryotic mRNA processing is not correct?
   a. The mRNA transcript must be exported from the nucleus.
   b. A 5′ cap and a 3′ poly(A) tail must be added.
   c. The introns are removed.
   d. A 3′ cap and a 5′ poly(A) tail must be added.
   e. Exons are spliced together to form the mRNA transcript.

10. Which of the following statements about protein translation is not correct?
   a. The genetic code is read in triplets, also called codons.
   b. The enzyme, aminoacyl tRNA synthetase, is responsible for adding the amino acid to the tRNA.
   c. The anticodon of the tRNA must recognize the codon on the mRNA exactly.
   d. Because of the wobble effect, a tRNA for one amino acid often recognizes multiple codons in the mRNA.
   e. For the most part, the genetic code is considered universal.

11. Condon bias can be overcome by which scenario?
   a. genetically engineering host organisms to express rarer tRNAs
   b. Nothing can be done to overcome codon bias when expressing proteins.
   c. genetically engineering the gene so that the codons are recognized by more abundant tRNAs
   d. Genetically engineer the gene to remove the codons for rare tRNAs.
   e. Both A and C are suitable scenarios.

12. Choose the statement about translation that is not correct.
   a. The ribosome is comprised of multiple subunits containing both ribosomal RNA and proteins.
   b. The consensus sequence UAAGGAGG is called the Shine-Dalgarno sequence and is recognized by the ribosome.
   c. Translation requires three initiation factors, two elongation factors, and two release factors.
   d. Transcription and translation are coupled in eukaryotes.
   e. There are three sites (E, P, and A) on the ribosome that can be occupied by a tRNA.

(*Continued*)

13. Which of the following statements does not highlight a difference in eukaryotic and prokaryotic translation?
    a. The first methionine in eukaryotic translation contains a formyl group.
    b. In eukaryotes, mRNA is made in the nucleus but translated in the cytoplasm.
    c. Prokaryotes often couple transcription and translation, forming a polysome.
    d. Eukaryotic mRNA does not have a Shine-Dalgarno sequence, but prokaryotic mRNA does.
    e. Many eukaryotic proteins are chemically modified after translation, which is a much rarer phenomenon in prokaryotes.

14. Why do mitochondria and chloroplasts contain their own genes?
    a. They are free-living prokaryotes, able to survive outside of the host cell.
    b. They are thought to have once been free-living organisms similar to bacteria that formed a symbiotic relationship with a unicellular eukaryote.
    c. They do not contain their own genetic material.
    d. They contain genetic material but do not make their own proteins.
    e. None of the above is correct.

15. Which of the following is not the correct composition of each subunit of the ribosome?
    a. 30S = 16 rRNA + 21 proteins
    b. 50S = 5S + 23S + 34 proteins
    c. 30S = 5S + 23S + 21 proteins
    d. 70S = 30S + 50S + $tRNA_i^{fMet}$
    e. All of the above are correct.

Further Reading

Clark DP (2005). *Molecular Biology: Understanding the Genetic Revolution*. Elsevier Academic Press, San Diego, CA.

Wisdom R (1999). AP-1: One switch for many signals. *Exp Cell Res* **253**, 180–185.

CHAPTER 3

Recombinant DNA Technology

DNA ISOLATION AND PURIFICATION

Basic to all biotechnology research is the ability to manipulate DNA. First and foremost for recombinant DNA work, researchers need a method to isolate DNA from different organisms. Isolating DNA from bacteria is the easiest procedure because bacterial cells have little structure beyond the cell wall and cell membrane. Bacteria such as *E. coli* are the preferred organisms for manipulating any type of gene because of the ease at which DNA can be isolated. *E. coli* maintain both genomic and plasmid DNA within the cell. Genomic DNA is much larger than plasmid DNA, allowing the two different forms to be separated by size.

To release the DNA from a cell, the cell membrane must be destroyed. For bacteria, an enzyme called **lysozyme** digests the peptidoglycan, which is the main component of the cell wall. Next, a detergent bursts the cell membranes by disrupting the lipid bilayer. For other organisms, bursting the cells depends on their architecture. Tissue samples from animals and plants have to be ground up to release the intracellular components. Plant cells are mechanically sheared in a blender to break up the tough cell walls, and then the wall tissue is digested with enzymes that break the long polymers into monomers. DNA from the tail tip of a mouse is isolated after enzymes degrade the connective tissue. Every organism or tissue needs slight variations in the procedure for releasing intracellular components.

Once released, the intracellular components are separated from the remains of the outer structures by either centrifugation or chemical extraction. Centrifugation separates components according to size, because heavier or larger molecules sediment at a faster rate than smaller molecules. For example, after the cell wall has been digested, its fragments are smaller than the large DNA molecules. Centrifugation causes the DNA to form a pellet, but the soluble cell wall fragments stay in solution. Chemical extraction uses the properties of **phenol** to remove unwanted proteins from the DNA. Phenol is an acid that dissolves 60% to 70% of all living matter, especially proteins. Phenol is poorly water soluble, and when it is mixed with an aqueous sample of DNA and protein, the two phases separate, much like oil and water. The protein dissolves in the phenol layer and the nucleic acids in the aqueous layer. The two phases are separated by centrifugation, and the aqueous DNA layer is removed from the phenol.

Once the proteins are removed, the sample still contains RNA along with the DNA. Because this is also a nucleic acid, it is not soluble in phenol. Luckily, the enzyme **ribonuclease (RNase)** digests RNA into ribonucleotides. Ribonuclease treatment leaves a sample of DNA in a solution containing short pieces of RNA and ribonucleotides. When an equal volume of alcohol is added, the extremely large DNA falls out of the aqueous phase and is isolated by centrifugation. The smaller ribonucleotides stay soluble. The DNA is ready for use in various experiments.

> DNA can be isolated by first removing the cell wall and cell membrane components. Next, the proteins are removed by phenol, and finally, the RNA is removed by ribonuclease.

ELECTROPHORESIS SEPARATES DNA FRAGMENTS BY SIZE

Gel electrophoresis followed by staining with **ethidium bromide** is used to separate DNA fragments by size (Fig. 3.1). The gel of gel electrophoresis consists of **agarose**, a polysaccharide extracted from seaweed that behaves like gelatin. Agarose is a powder that dissolves in water only when heated. After the solution cools, the agarose hardens. For visualizing DNA, agarose is formed into a rectangular slab about 1/4 inch thick. Inserting a comb at one end of the slab before it hardens makes small wells or holes. After the gel solidifies, the comb is removed, leaving small wells at one end.

Gel electrophoresis uses electric current to separate DNA molecules by size. The agarose slab is immersed in a buffer-filled tank that has a positive electrode at one end and a negative

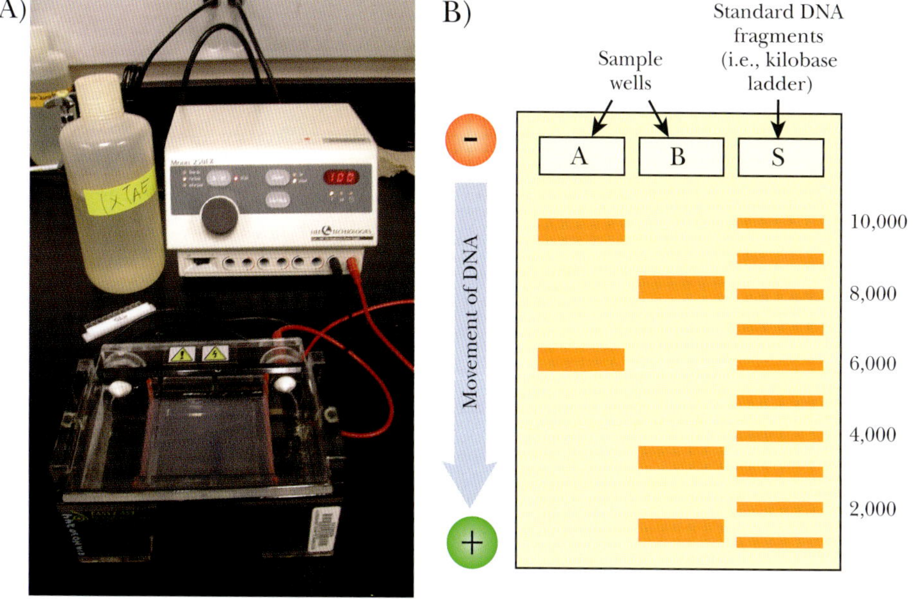

FIGURE 3.1
Electrophoresis of DNA
(A) Photo of electrophoresis supplies. Electrophoresis chamber holds an agarose gel in the center portion, and the rest of the tank is filled with buffer solution. The red and black leads are then attached to an electrical source. FisherBiotech Horizontal Electrophoresis Systems, Midigel System; Standard; 13 × 16-cm gel size; 800-mL buffer volume; Model No. FB-SB-1316. (B) Agarose gel separation of DNA. To visualize DNA, the agarose gel containing the separated DNA fragments is soaked in a solution of ethidium bromide, which intercalates between the bases of DNA. Under UV light, the DNA bands fluoresce a bright orange color. The size of the fragments can be calculated by comparing them to the standards on the right.

electrode at the other. DNA samples are loaded into the wells, and when an electrical field is applied, the DNA migrates through the gel. The phosphate backbone of DNA is negatively charged so it moves away from the negative electrode and toward the positive electrode. Polymerized agarose acts as a sieve with small holes between the tangled chains of agarose. The DNA must migrate through these gaps. Agarose separates the DNA by size because larger pieces of DNA are slowed down more by the agarose.

To visualize the DNA, the agarose gel is removed from the tank and immersed into a solution of ethidium bromide. This dye intercalates between the bases of DNA or RNA, although less dye binds to RNA because it is single-stranded. When the gel is exposed to ultraviolet light, it fluoresces bright orange. DNA molecules of the same size usually form a tight band, and the size can be determined by comparing to a set of **molecular weight standards** run in a different well. Because the standards are of known size, the experimental DNA fragment can be compared directly.

The size of DNA being examined affects what type of gel is used. The standard is agarose, but for very small pieces of DNA, from 50 to 1000 base pairs, polyacrylamide gels are used instead. These gels are able to resolve DNA fragments that vary by only one base pair and are essential to sequencing DNA (see Chapter 4). For very large DNA fragments (10 kilobases to 10 megabases), agarose is used, but the current is alternated at two different angles. **Pulsed field gel electrophoresis (PFGE)**, as this is called, allows very large pieces of DNA to migrate further than if the current only flows in one direction. Each change in direction loosens large pieces of DNA that are stuck inside agarose pores, letting them migrate further. Finally, gradient gel electrophoresis can be used to resolve fragments that are very close in size. A concentration gradient of acrylamide, buffer, or electrolyte can reduce compression (i.e., crowding of similar sized fragments) due to secondary structure and/or slow the smaller fragments at the lower end of the gel.

Fragments of DNA are separated by size using gel electrophoresis. A current causes the DNA fragments to move away from the negative electrode, and toward the positive. As the DNA travels through agarose, the larger fragments get stuck in the gel pores more than the smaller DNA fragments.

Pulsed field gel electrophoresis separates large pieces of DNA by alternating the electric current at right angles.

RESTRICTION ENZYMES CUT DNA; LIGASE JOINS DNA

The ability to isolate, separate, and visualize DNA fragments would be useless unless some method was available to cut the DNA into fragments of different sizes. In fact, naturally occurring **restriction enzymes** or **restriction endonucleases** are the key to making DNA fragments. These bacterial enzymes bind to specific recognition sites on DNA and cut the backbone of both strands. They evolved to protect bacteria from foreign DNA, such as from viral invaders. The enzymes do not cut their own cell's DNA because they are methylation sensitive, that is, if the recognition sequence is methylated, then the restriction enzyme cannot bind. Bacteria produce **modification enzymes** that recognize the same sequence as the corresponding restriction enzyme. These methylate each recognition site in the bacterial genome. Therefore, the bacteria can make the restriction enzyme without endangering their own DNA.

Restriction enzymes have been exploited to cut DNA at specific sites, since each restriction enzyme has a particular recognition sequence. Differences in cleavage site determine the type of restriction enzyme. **Type I restriction enzymes** cut the DNA strand 1000 or more base pairs from the recognition sequence. **Type II restriction enzymes** cut in the middle of the recognition sequence and are the most useful for genetic engineering. Type II restriction enzymes can either cut both strands of the double helix at the same point, leaving **blunt ends**, or they can cut at different sites on each strand leaving single-stranded ends, sometimes called **sticky ends** (Fig. 3.2). The recognition sequences of Type II restriction enzymes are usually inverted repeats, so that the enzyme cuts between the same bases on both strands. Since the repeats are inverted, the cuts may be staggered, thus generating single-stranded overhangs. Some commonly used restriction enzymes for biotechnology applications are listed in Table 3.1.

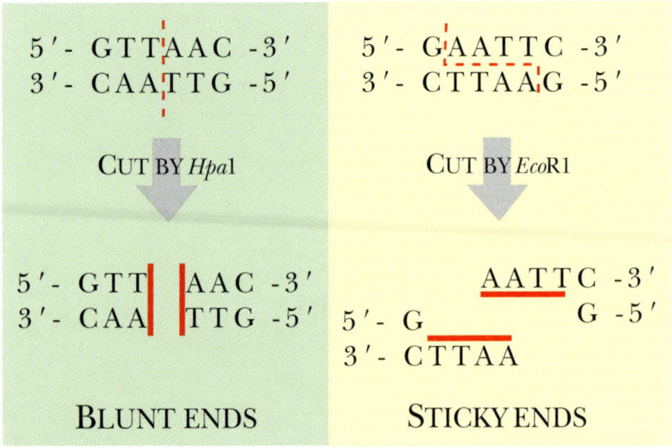

5′- GTTAAC -3′
3′- CAATTG -5′

CUT BY *Hpa*I

5′- GTT AAC -3′
3′- CAA TTG -5′

BLUNT ENDS

5′- GAATTC -3′
3′- CTTAAG -5′

CUT BY *Eco*R1

AATTC -3′
G -5′
5′- G
3′- CTTAA

STICKY ENDS

FIGURE 3.2 Type II Restriction Enzymes—Blunt versus Sticky Ends
*Hpa*I is a blunt end restriction enzyme, that is, it cuts both strands of DNA in exactly the same position. *Eco*RI is a sticky end restriction enzyme. The enzyme cuts between the G and A on both strands, which generates four base-pair overhangs on the ends of the DNA. Since these ends may base pair with complementary sequences, they are considered "sticky."

The number of base pairs in the recognition sequence determines the likelihood of cutting. Finding a particular sequence of four nucleotides is much more likely than finding a six base-pair recognition sequence. So to generate fewer, longer fragments, restriction enzymes with six or more base-pair recognition sequences are used. Conversely, four base-pair enzymes give more, shorter fragments from the same original segment of DNA.

When two different DNA samples are cut with the same sticky-end restriction enzyme, all the fragments will have identical overhangs. This allows DNA fragments from two sources (e.g., two different organisms) to be linked together (Fig. 3.3). Fragments are linked or **ligated** using **DNA ligase**, the same enzyme that ligates the Okazaki fragments during replication (see Chapter 4). The most common ligase used is actually from T4 bacteriophage. Ligase catalyzes linkage between the 3′-OH of one strand and the 5′-PO_4 of the other DNA strand. Ligase is much more efficient with overhanging sticky ends, but can also link blunt ends much more slowly. (Specific fragments for ligation are often isolated by agarose gel electrophoresis as described earlier.)

> Restriction enzymes are naturally occurring enzymes that recognize a particular DNA sequence and cut the phosphate backbone.
>
> When two pieces of DNA are cut by the same restriction enzyme, the two ends have compatible overhangs that can be reconnected by ligase.

Table 3.1 Table of Common Restriction Enzymes

Enzyme	Source Organism	Recognition Sequence
*Hpa*II	*Haemophilus parainfluenzae*	C/CGG GGC/C
*Mbo*I	*Moraxella bovis*	/GATC GATC/
*Nde*II	*Neisseria denitrificans*	/GATC GATC/
*Eco*RI	*Escherichia coli* RY13	G/AATTC CTTAA/G
*Eco*RII	*Escherichia coli* RY13	/CCWGG GGWCC/
*Eco*RV	*Escherichia coli* J62/pGL74	GAT/ATC CTA/TAG
*Bam*HI	*Bacillus amyloliquefaciens*	G/GATCC CCTAG/G
*Sau*I	*Staphylococcus aureus*	CC/TNAGG GGANT/CC
*Bgl*I	*Bacillus globigii*	GCCNNNN/NGGC CGGN/NNNNCCG
*Not*I	*Nocardia otitidis-caviarum*	GC/GGCCGC CGCCGG/CG
*Dra*II	*Deinococcus radiophilus*	RG/GNCCY YCCNG/GR

/, position where enzyme cuts.
N, any base; R, any purine; Y, any pyrimidine; W, A or T.

METHODS OF DETECTION FOR NUCLEIC ACIDS

Recombinant DNA methodologies require the ability to detect DNA. One of the easiest ways to detect the amount of DNA or RNA in solution is to measure the absorbance of ultraviolet light (Fig. 3.4). DNA absorbs ultraviolet light because of the ring structures in the bases. Single-stranded RNA and free nucleotides also absorb ultraviolet light. In fact, they absorb more light because their structures are looser. The amount of absorption is compared with a known set of standards and the concentration of DNA can be determined.

Box 3.1 Restriction Fragment Length Polymorphisms Identify Individuals

Restriction enzymes are useful for many different applications. Because the DNA sequence is different in each organism, the pattern of restriction sites will also be different. The source of isolated DNA can be identified by this pattern. If genomic DNA is isolated from one organism and cut with one particular restriction enzyme, a specific set of fragments can be separated and identified by electrophoresis. If DNA from a different organism is cut by the same restriction enzyme, a different set of fragments will be generated. This technique can be applied to DNA from two individuals from the same species. Although the DNA sequence differences will be small, restriction enzymes can be used to identify these differences. If the sequence difference falls in a restriction enzyme recognition site, it gives a **restriction fragment length polymorphism (RFLP)** (Fig. A). When the restriction enzyme patterns are compared, the number and size of one or two fragments will be affected for each base difference that affects a cut site.

(*Continued*)

Box 3.1 Restriction Fragment Length Polymorphisms Identify Individuals—cont'd

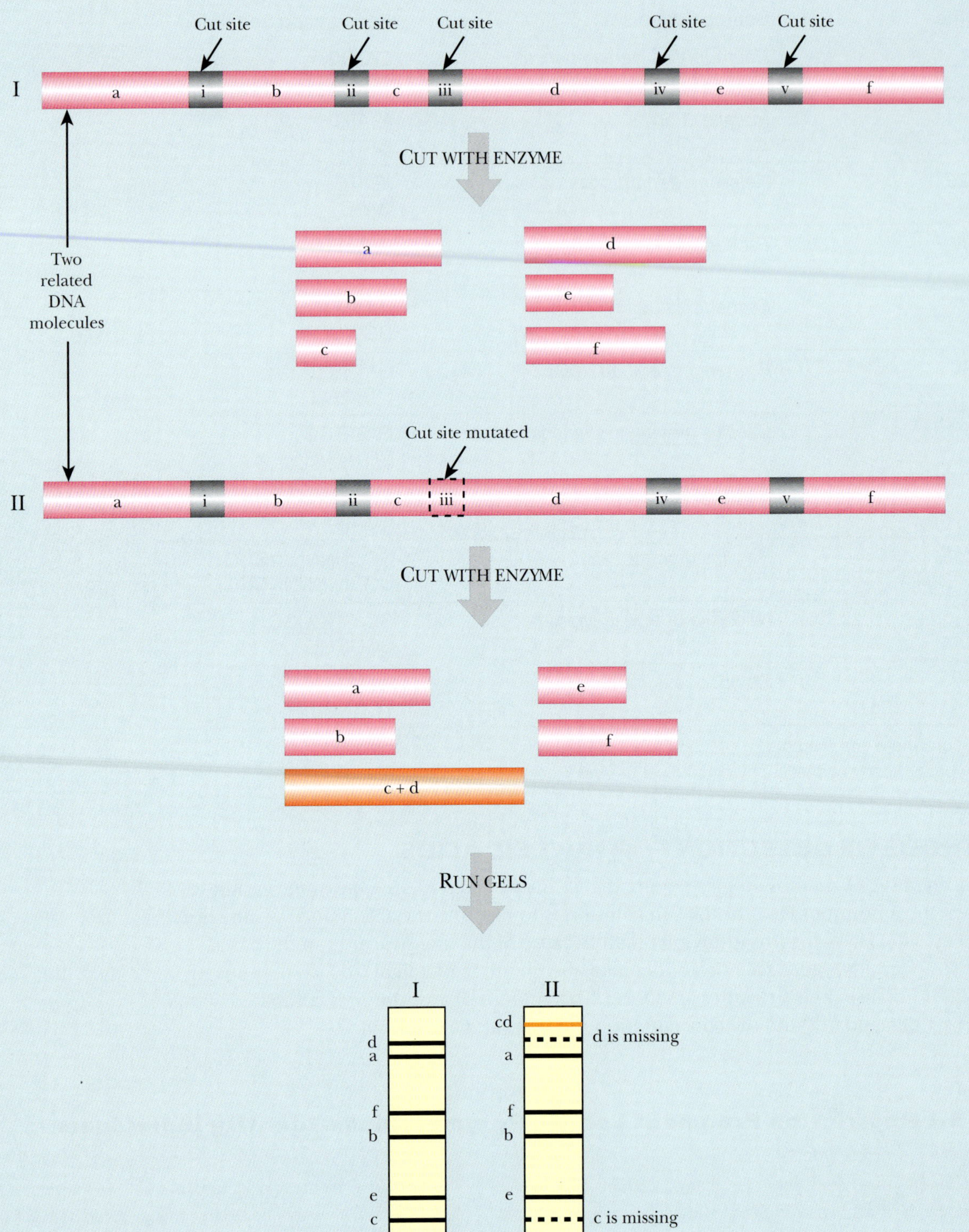

FIGURE A **RFLP Analysis**

DNA from related organisms shows small differences in sequence that cause changes in restriction sites. In the example shown, cutting a segment of DNA from the first organism yields six fragments of different sizes (labeled *a–f* on the gel). If the equivalent region of DNA from a related organism were digested with the same enzyme, a similar pattern would be expected. Here a single-nucleotide difference is present, which eliminates one of the restriction sites. Consequently, digesting this DNA produces only five fragments. Fragments *c* and *d* are no longer seen, but form a new band labeled *cd*.

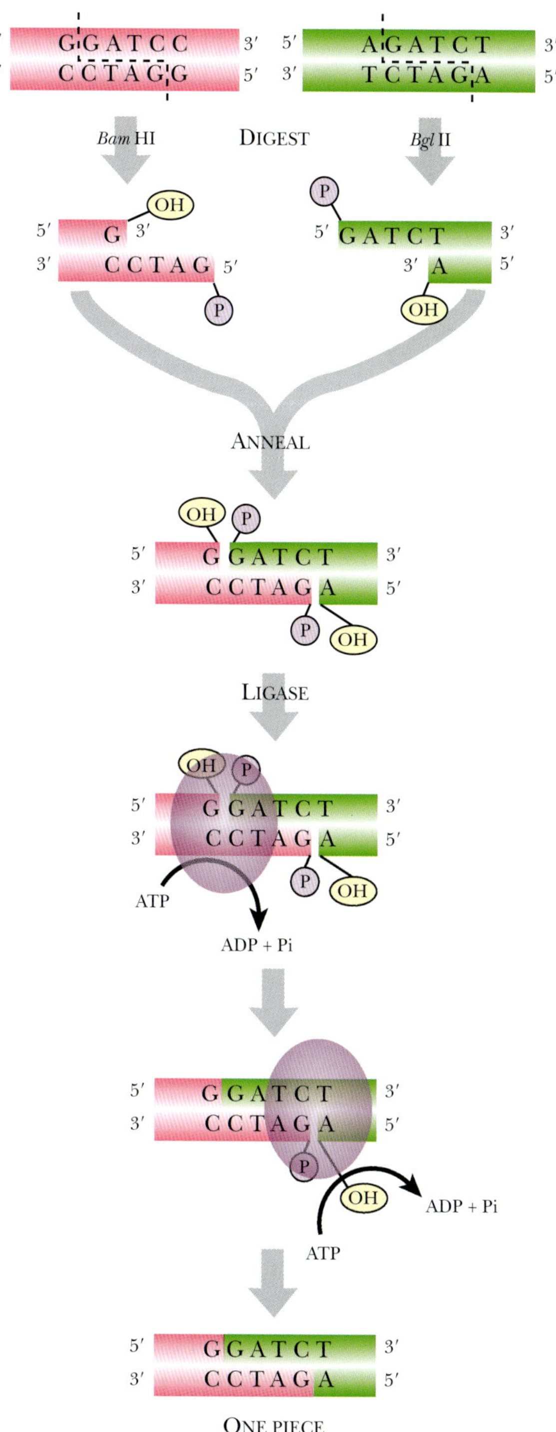

FIGURE 3.3 Compatible Overhangs Are Linked Using DNA Ligase
*Bam*HI and *Bgl*II generate the same overhanging or sticky ends: a 3′-CTAG-5′ overhang plus a 5′-GATC-3′ overhang. These are complementary and base pair by hydrogen bonding. The breaks in the DNA backbones are sealed by T4 DNA ligase, which hydrolyzes ATP to energize the reaction.

The concentration of DNA in a liquid can be determined by measuring the absorbance of UV light at 260 nm.

Radioactive Labeling of Nucleic Acids and Autoradiography

Ultraviolet light absorption is a general method for detecting DNA, but does not distinguish between different DNA molecules. DNA can also be detected with radioactive isotopes

FIGURE 3.4
Determining the Concentration of DNA
(A) All nucleic acids absorb UV light via the aromatic rings of the bases. Stacked nucleotides (on the left) absorb less UV than scattered bases (on the right) because of the ordered structure. (B) The concentration of DNA in solution is determined by measuring the absorbance of UV light at 260 nm. Graphing the absorbance versus concentration shows a linear relationship. The concentration of an unknown sample can be determined by measuring its absorbance at 260 nm, then extrapolating its concentration.

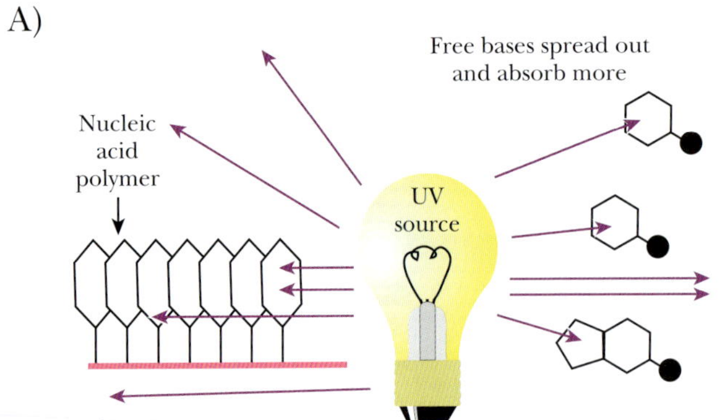

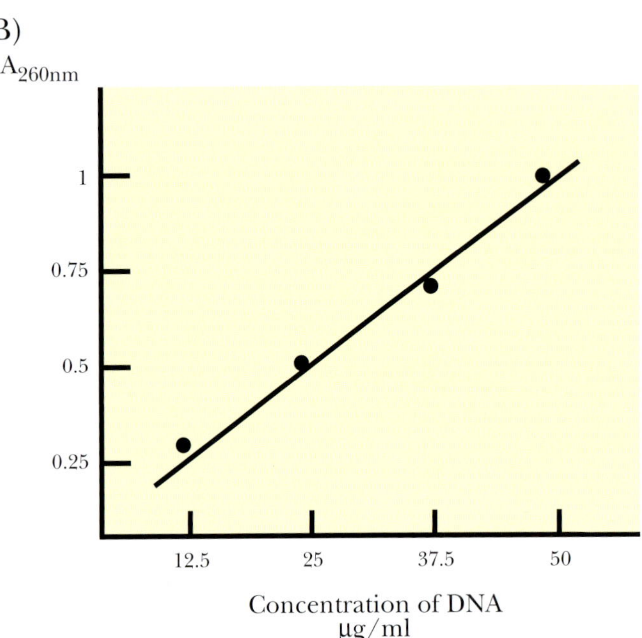

(Fig. 3.5). During replication, radioactive precursors such as ^{32}P in the form of a **phosphate group** and ^{35}S in the form of **phosphorothioate** can be incorporated. Because native DNA does not contain sulfur atoms, one of the oxygen atoms of a phosphate group is replaced with sulfur to make phosphorothioate. Most radioactive molecules used in laboratories are short lived. ^{32}P has a half-life of 14 days and ^{35}S has a half-life of 68 days, so the isotopes degrade fairly fast. Although radioactive DNA is invisible, photographic film will turn black when exposed to the radioactive DNA. Radioactively labeled DNA is considered **"hot,"** whereas unlabeled DNA is considered **"cold."**

The radioactive nucleotide precursors can be supplied to rapidly growing bacterial cultures. During replication, the radioactive precursor is incorporated into new DNA (see Chapter 4). The DNA is isolated from the bacteria and run on a gel. **Autoradiography** identifies the location of radioactively labeled DNA in the gel (Fig. 3.6). If the gel is thin, like most polyacrylamide gels, it is dried with heat and vacuum. If the gel is thick, like agarose gels, the DNA is transferred to a nylon membrane using capillary action (see Fig. 3.9, later). The dried gel or nylon membrane is placed next to photographic film. As the radioactive phosphate decays, the radiation turns the photographic film black. Only the areas next to radioactive DNA will have black spots or bands. The use of film detects where the hot DNA is on a gel, and the use of ethidium bromide shows where all of the DNA, hot or cold, is. These two methods allow distinguishing one DNA fragment from another.

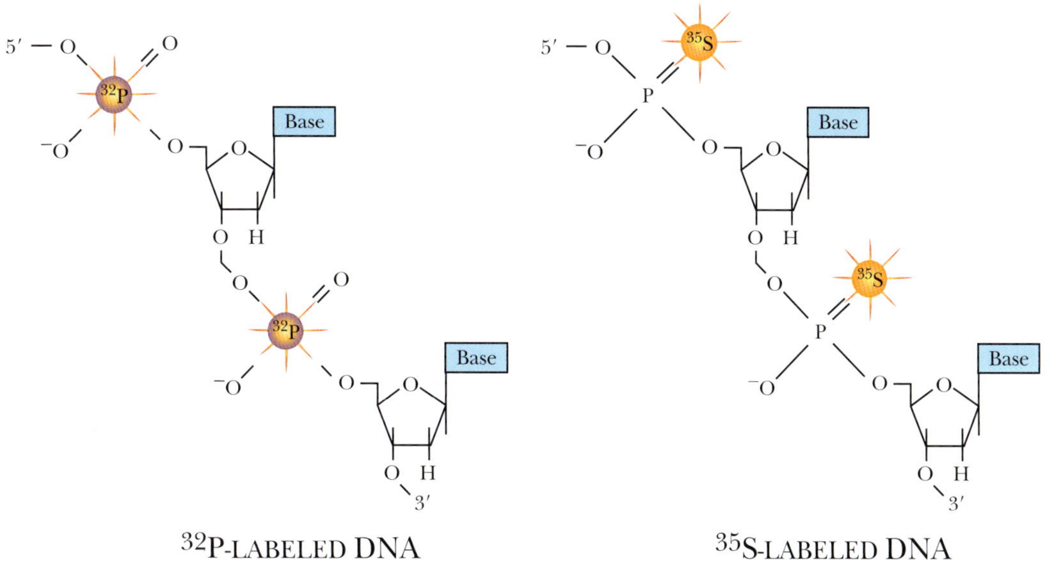

^{32}P-LABELED DNA

^{35}S-LABELED DNA

FIGURE 3.5 Radioactively Labeled DNA
DNA can be synthesized with radioactive precursor nucleotides. These nucleotides have ^{32}P (rather than nonradioactive ^{31}P phosphorus) or ^{35}S (replacing oxygen) in the phosphate backbone.

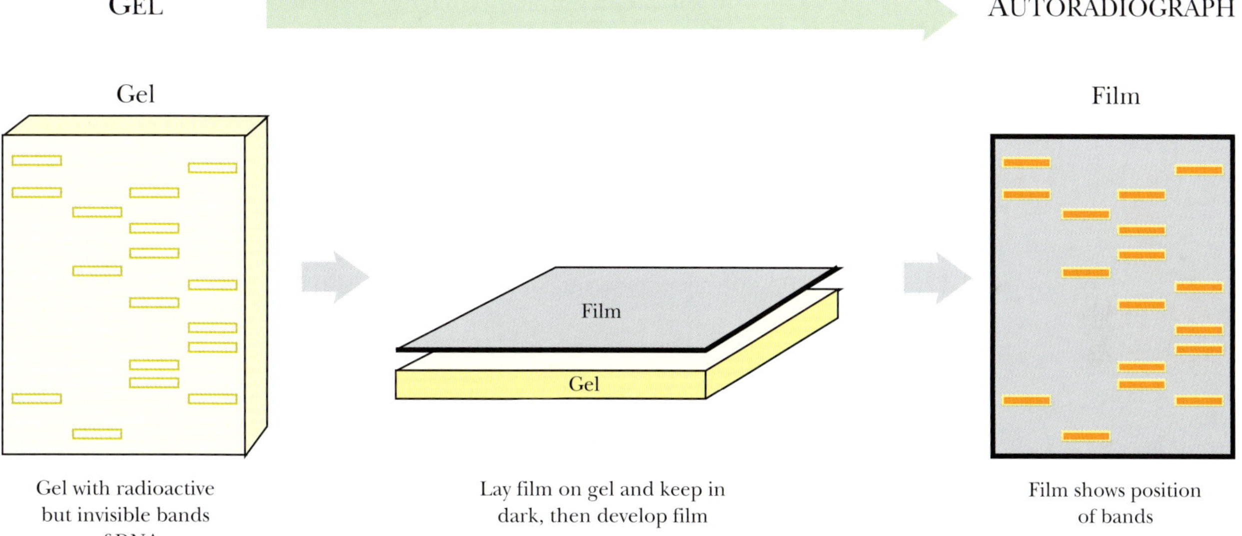

FIGURE 3.6 Autoradiography
A gel containing radioactive DNA (or RNA) is dried and a piece of photographic film is laid over the top. The two are loaded into a cassette case that prevents light from entering. After some time (hours to days), the film is developed and dark lines appear where the radioactive DNA was present.

Radioactive DNA can also be detected using **scintillation counting**. Here a small sample of the radioactive DNA is mixed with scintillation fluid. When the radioactive isotope decays, it emits a beta particle. Scintillation fluid emits a flash of light when excited by the beta particle. The scintillation counter detects light flashes with a photocell, counting them over a specified amount of time. Radioactive DNA concentrations can be determined by comparing to a set of known standards. Scintillation counting cannot detect the unlabeled or cold DNA, nor can it distinguish between multiple fragments of hot DNA, because it merely measures the total radioactivity in the sample.

Radioactive isotopes are incorporated into the DNA backbone during replication. Autoradiography or scintillation counting identifies the radioactive label.

Fluorescence Detection of Nucleic Acids

Autoradiography has its merits, but working with and disposing of radioactive waste is costly, both monetarily and environmentally. Using fluorescently tagged nucleotides was developed as a better method of DNA detection (Fig. 3.7). Fluorescent tags absorb light of one wavelength, which excites the atoms, increasing the energy state of the tag. This excited state releases a photon of light at a different (longer) wavelength and returns to the ground state. The emitted photon is detected with a **photodetector**. There are many different fluorescent tags, and each emits a different wavelength of light. Some photodetector systems are sensitive enough to distinguish between these different tags; therefore, if different bases have different fluorescent labels, the photodetector can determine which base is present. This is the basis for most modern DNA sequencing machines (see Chapter 4).

Fluorescently labeled nucleotides can be used to incorporate a fluorescent tag on DNA during replication.

Chemical Tagging with Biotin or Digoxigenin

Biotin is a vitamin and digoxigenin is a steroid from the foxglove plant. Using these two chemicals allows scientists to label DNA without radioactivity or costly photodetectors. Biotin or digoxigenin are chemically linked to uracil; therefore, DNA must be synthesized with the labeled uracil replacing thymine. The DNA is synthesized *in vitro* as described in Chapter 4. A single-stranded DNA template, DNA polymerase, a short DNA primer, and a mixture of dATP, dGTP, dCTP, plus dUTP linked to either biotin or digoxigenin are combined. DNA polymerase synthesizes the complementary strand to the template, incorporating biotin- or digoxigenin-linked uracil opposite all the adenines.

The labeled DNA is visualized in a two-step process (Fig. 3.8). First, for biotin, a molecule of **avidin** binds to the tag. For digoxigenin, a specific antibody binds to the tag. Both avidin and the digoxigenin antibody are conjugated to **alkaline phosphatase**, an enzyme that

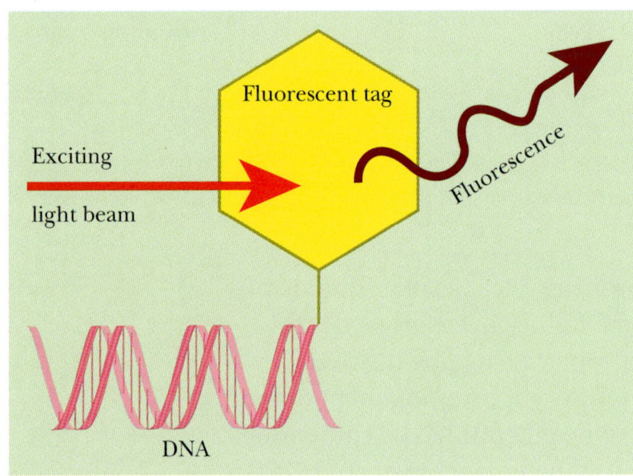

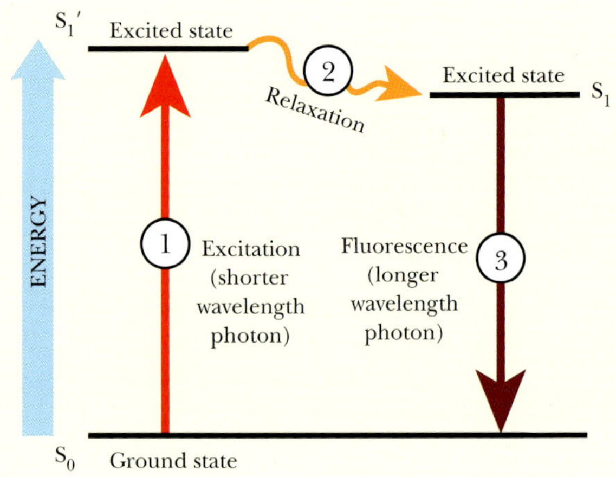

FIGURE 3.7 Fluorescent Labeling of DNA

(A) Fluorescent tagging of DNA. During synthesis, a nucleotide linked to a fluorescent tag is incorporated at the 3′ end of the DNA. A beam of light excites the fluorescent tag, which in turn, releases light of a longer wavelength. (B) Energy levels in fluorescence. The fluorescent molecule attached to the DNA has three different energy levels, S_0, S_1', and S_1. The S_0 or ground state is the state before exposure to light. When the fluorescent molecule is exposed to a light photon, the fluorescent tag absorbs the energy and enters the first excited state, S_1'. Between S_1' and S_1, the fluorescent tag relaxes slightly, but doesn't emit any light. Eventually the high-energy state releases its excess energy by emitting a longer wavelength photon. This release of fluorescence returns the molecule to the ground state.

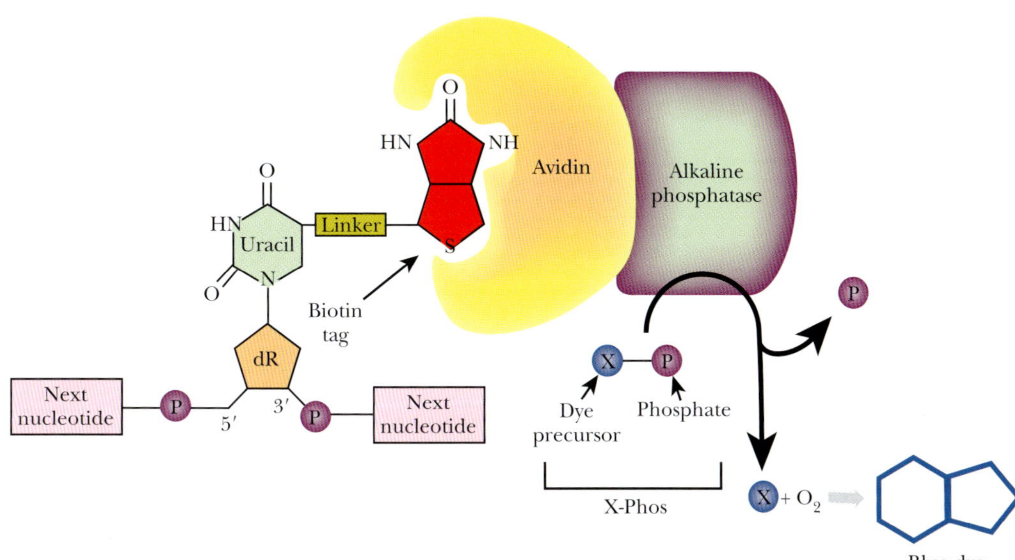

FIGURE 3.8 Labeling and Detecting DNA with Biotin
DNA can be synthesized *in vitro* with a uracil nucleotide linked with a biotin molecule. The biotin can be visualized by adding an avidin/alkaline phosphatase conjugate. The avidin half binds to biotin and the alkaline phosphatase half removes phosphates from different substrates. In this figure, alkaline phosphatase removes phosphate from X-Phos to form a blue dye.

removes phosphates from a variety of substrates. Several different chromogenic molecules act as substrates for alkaline phosphatase, but the most widely used one is **X-Phos**. Once alkaline phosphatase removes the phosphate group from X-Phos, the intermediate molecule reacts with oxygen and forms a blue precipitate. This blue color reveals the location of the labeled DNA. Another substrate of alkaline phosphatase is **Lumi-Phos**, which is chemiluminescent and emits visible light when the phosphate is removed. Much like autoradiography, when photographic film is placed over labeled DNA treated with Lumi-Phos, dark bands form wherever the Lumi-Phos glows.

Biotin and digoxigenin-labeled DNA is detected using either avidin or antibody to digoxigenin. Avidin and the antibody both are conjugated to alkaline phosphatase, which reacts with X-Phos to leave a blue precipitate or Lumi-Phos to emit visible light. Either detection method is used to identify, quantify, or locate the labeled DNA.

COMPLEMENTARY STRANDS MELT APART AND REANNEAL

The complementary antiparallel strands of DNA form an elegant molecule that is able to unzip or **melt** and come back together or **reanneal** (Fig. 3.9). The hydrogen bonds that hold the two halves together are relatively weak. Heating a sample of DNA will dissolve the hydrogen bonds, resulting in two complementary single strands. If the same sample of DNA is slowly cooled, the two strands will reanneal so that G matches with C and A matches with T, as before.

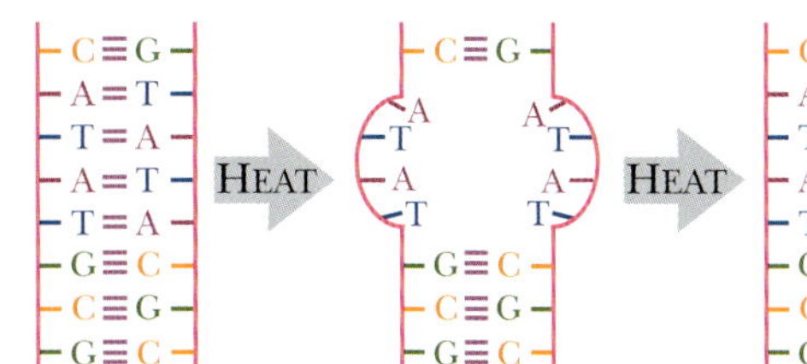

FIGURE 3.9 Heat Melts DNA; Cooling Reanneals DNA
Hydrogen bonds readily dissolve when heated, leaving the two strands intact, but separate. When the temperature returns to normal, the hydrogen bonds form again.

The proportion of G-C base pairs affects how much heat is required to melt a double helix of DNA. G-C base pairs have three hydrogen bonds to melt, whereas A-T base pairs have only two. Consequently, DNA with a higher percentage of GC will require more energy to melt than DNA with fewer GC base pairs. The **GC ratio** is defined as follows:

$$\frac{G + C}{A + G + C + T} \times 100\%$$

The ability to zip and unzip DNA is crucial to cellular function, and has also been exploited in biotechnology. Replication (see Chapter 4) and transcription (see Chapter 2) rely on strand separation to generate either new DNA or RNA strands, respectively. In molecular biology research, many techniques, from PCR to library screening, exploit the complementary nature of DNA strands.

The complementary strands of DNA are easily separated by heat, and spontaneously reanneal as the DNA mixture cools.

HYBRIDIZATION OF DNA OR RNA IN SOUTHERN AND NORTHERN BLOTS

If two different double helixes of DNA are melted, the single strands can be mixed together before cooling and reannealing. If the two original DNA molecules have similar sequences, a single strand from one may pair with the opposite strand from the other DNA molecule. This is known as **hybridization** and can be used to determine whether sequences in two separate samples of DNA or RNA are related. In hybridization experiments, the term **probe molecule** refers to a known DNA sequence or gene that is used to screen the experimental sample or **target DNA** for similar sequences.

Southern blots are used to determine how closely DNA from one source is related to a DNA sequence from another source. The technique involves forming hybrid DNA molecules by mixing DNA from the two sources. A Southern blot has two components, the probe sequence (e.g., a known gene of interest from one organism) and the target DNA (often from a different organism). A typical Southern blot begins by isolating the target DNA from one organism, digesting it with a restriction enzyme that gives fragments from about 500 to 10,000 base pairs in length, and separating these fragments by electrophoresis. The separated fragments will be double-stranded, but if the gel is incubated in a strong acid, the DNA separates into single strands. Using capillary action, the single strands can be transferred to a membrane as shown in Fig. 3.10. The DNA remains single-stranded once attached to the membrane.

FIGURE 3.10
Capillary Action Transfers DNA from Gel to Membrane
Single-stranded DNA from a gel will transfer to the membrane. The filter paper wicks buffer from the tank, through the gel and membrane, and into the paper towels. As the buffer liquid moves, the single-stranded DNA also travels from the gel and sticks to the membrane. The weight on top of the setup keeps the membrane and gel in contact and helps wick the liquid from the tank.

Next, the probe is prepared. First, the known sequence or gene must be isolated and labeled in some way (see earlier discussion). Identifying genes has become easier now that many genomes have been entirely sequenced. For example, a scientist can easily obtain a copy of a human gene for use as a probe to find similar genes in other organisms. Alternatively, using sequence data, a unique oligonucleotide probe can be designed that only recognizes the gene of interest (see Chapter 4). If an oligonucleotide has a common sequence, it will bind to many other sequences. Therefore, oligonucleotide probes must be long enough to have sequences that bind to only one (or very few) specific site(s) in the target genome. To prepare DNA probes for a Southern blot, they are labeled using radioactivity, biotin, or digoxigenin (see earlier discussion). Finally, the labeled DNA is denatured at high temperature to make it single-stranded. (Synthetic oligonucleotides do not require treatment, as they are already single-stranded.)

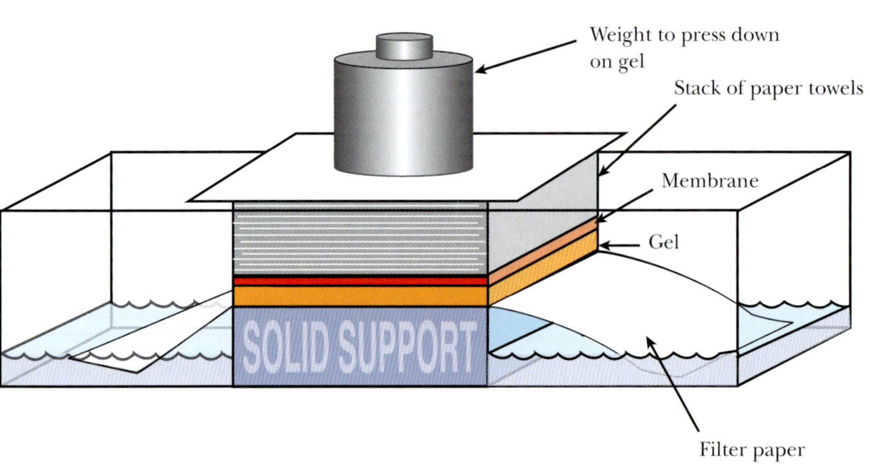

To perform the Southern blot, the single-stranded probe is incubated with the membrane carrying the single-stranded target DNA (Fig. 3.11). These are incubated at a temperature that allows

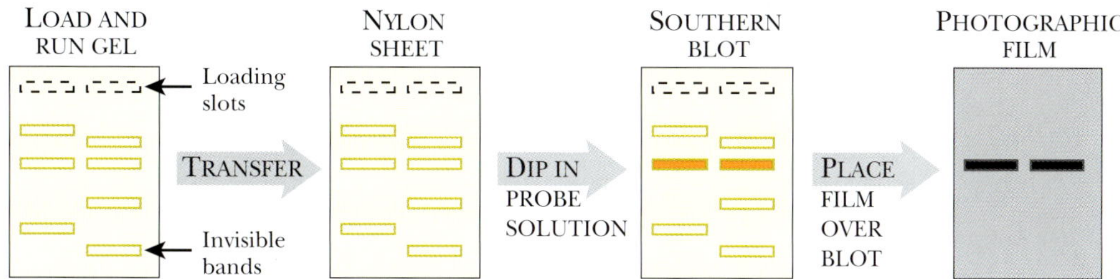

FIGURE 3.11 Hybrid DNA Molecules Can Detect Related Sequences in Southern Blots
Southern blotting requires the target DNA to be cut into smaller fragments and run on an agarose gel. The fragments are denatured chemically to give single strands, and then transferred to a nylon membrane. A radioactive probe (also single-stranded) is incubated with the membrane at a temperature that allows hybrids (with some mismatches) to form. When photographic film is placed over the top of the membrane, the location of the radioactive hybrid molecules is revealed.

hybrid DNA strands to form but allows a low amount of mismatching. The temperature, and hence the level of mismatching tolerated, can be varied depending on how specific a search is being run. If the probe is radioactive, then the membrane is exposed to photographic film. If the probe is labeled with biotin or digoxigenin, the membrane may be treated with chemiluminescent substrate to detect the labeled probe, and then exposed to photographic film. Dark bands on the film reveal the positions of fragments with similar sequence to the probe. Alternatively, biotin or digoxigenin labels may be visualized by treatment with a chromogenic substrate. In this case blue bands will form directly on the membrane at the position of the related sequences.

Northern blots are also based on nucleic acid hybridization. The difference is that RNA is the target in a Northern blot. The probe for a Northern blot is either a fragment of a gene or a unique oligonucleotide just as in a Southern blot. The target RNA is usually messenger RNA. In eukaryotes, screening mRNA is more efficient because genomic DNA has a lot of introns, which may interfere with probes binding to the correct sequence. Besides, mRNA is already single-stranded, so the agarose gel does not have to be treated with strong acid. Much like a Southern blot, Northern blots begin by separating mRNA by size using electrophoresis. The mRNA is transferred to a nylon membrane and incubated with a single-stranded labeled probe. As before, the probe can be labeled with biotin, digoxigenin, or radioactivity. The membrane is processed and exposed to film or chromogenic substrate.

A variation of these hybridization techniques is the **dot blot** (Fig. 3.12). Here the target sample is not separated by size. The DNA or mRNA target is simply attached to the nylon membrane as a small dot. As in Southern blots, the DNA sample must be made single-stranded before it is attached to the membrane. As before, the dot-blot membrane is allowed to hybridize with a labeled probe. The membranes are processed and exposed to film. If the dot of DNA or mRNA contains a sequence similar to the probe, the film will turn black in that area. Dot blots are a quick and easy way to determine if the target sample has a related sequence, before more detailed analysis by Southern or Northern blotting. Another advantage of dot blots is that multiple samples can be processed in a smaller amount of space.

DOT ssDNA OR mRNA
Different samples in each row
Dot different concentrations of sample
DIP IN PROBE SOLUTION
DOT BLOT
EXPOSE TO FILM
PHOTOGRAPHIC FILM

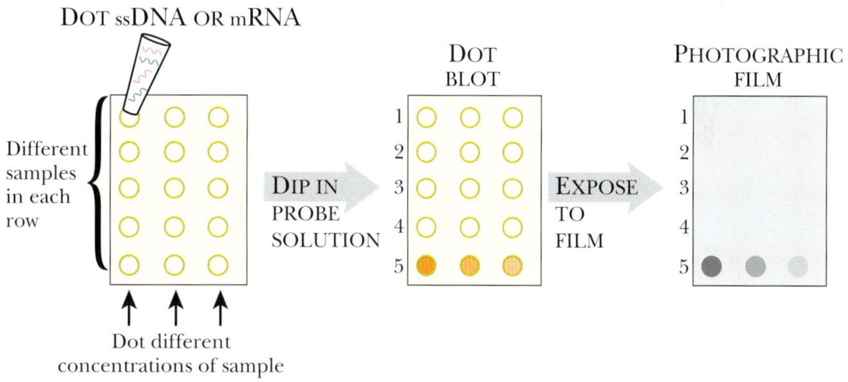

FIGURE 3.12 Dot Blot
Dot blots begin by spotting DNA or RNA samples onto a nylon membrane. Often, different concentrations of the sample are dotted side by side. The membrane is incubated with a radioactive probe and then exposed to photographic film. Samples that contain DNA or RNA complementary to the probe will leave a black spot on the film.

Southern blots form hybrid DNA molecules to determine if a sample of DNA has a homologous sequence to another DNA probe.

Northern blots determine if a sample of mRNA has a homologous sequence to a DNA probe. In large genomes, using mRNA is more efficient because all the introns are removed.

Box 3.2 Zoo Blot Compare Sequences among Different Species

Southern blots are used for a variety of experiments. If DNA from a variety of different organisms is examined for sequences similar to a human probe sequence, the Southern blot is called a **zoo blot**. The more closely related the organisms, the more likely a related sequence will be found. For example, if the human sequence were from a hemoglobin gene, the results might show a related sequence in chimpanzee, horse, and pig, but not in yeast or bacteria. The most interesting use for zoo blots is to test for noncoding versus coding DNA (Fig. B). Sequences of noncoding DNA evolve rapidly compared to coding sequences. Hence a probe that recognizes genomic DNA from many different related organisms is usually from a coding region. If the probe sequence is noncoding DNA, then very few matches will occur.

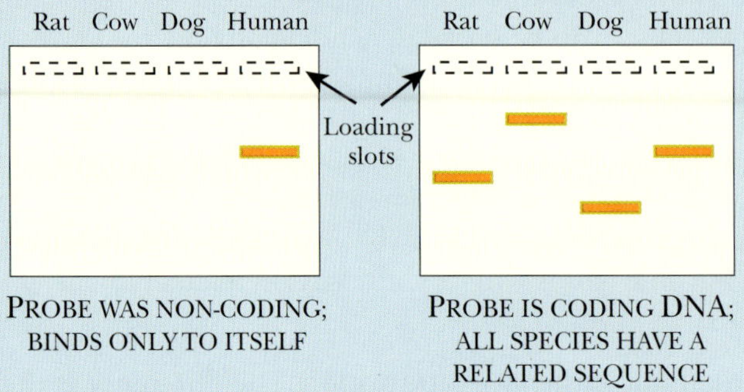

FIGURE B Zoo Blot
A specialized form of Southern blotting, called zoo blotting, is used to distinguish coding DNA from noncoding regions. The target DNA includes several samples of genomic DNA from different animals, hence the term "zoo." The probe is a segment of human DNA that may or may not be from a coding region. Since base sequences of noncoding DNA mutate and change rapidly, whereas coding sequences do not mutate as rapidly, a probe that recognizes genomic DNA from many different organisms is usually a coding region. On the left, the only hybrid seen was between the probe and the human DNA; therefore, the probe is probably noncoding. In the example on the right, the probe binds to related sequences in other animals; therefore, this probe is probably from a coding region.

FLUORESCENCE *IN SITU* HYBRIDIZATION (FISH)

The previously discussed hybridization techniques rely on purified DNA or RNA run in an agarose gel. In **fluorescence *in situ* hybridization (FISH)**, the probe is hybridized directly to DNA or RNA within the cell. As described earlier, the probe is a small segment of DNA that has been labeled with fluorescent tags in order to be visualized. The target DNA or RNA is located within the cell and requires some special processing. The target cells may be extremely thin sections of tissue from a particular organism. For example, when a person has a biopsy, a small piece of tissue is removed for analysis. This tissue is preserved and then cut into extremely thin sections to be analyzed under a microscope. These can be used to determine the presence of a gene with FISH. Another source of target cells for FISH is cultured mammalian or insect cells (see Chapter 1). Additionally, blood can be isolated and processed to isolate the white blood cells. (Note: Red blood cells do not contain a nucleus and therefore do not contain DNA.) Chromosomes from white blood cells can be isolated and dropped onto a glass slide. Blood sample DNA may be found both in interphase, with the chromosomes spread out, or in metaphase, with the chromosomes highly condensed.

Whether the target DNA is from blood, cells cultured in dishes, or actual tissue sections, the cells must be heated to make the DNA single-stranded. Samples where RNA is the target do not need to be heated. The fluorescently labeled probe hybridizes to complementary sequences in the DNA or RNA, and when the cells are illuminated at the appropriate wavelength, the probe location on the chromosome can be identified by fluorescence. Figure 3.13 shows FISH analysis of the *RUNX1* gene that is amplified in certain cases of human acute leukemia due to polysomy (that is, multiple copies) of chromosome 21.

FISH is a technique where a labeled probe is incubated with cells that have had their DNA denatured by heat. The probe hybridizes to its homologous sequence on the chromosome.

Box 3.3 Discovery of Recombinant DNA

In 1972, two researchers met at a conference in Hawaii to discuss plasmids, the small rings of extrachromosomal DNA found in bacteria. Herbert W. Boyer, PhD, was a faculty member at the University of California, San Diego, and he was studying restriction and modification enzymes. He had just presented his research on *Eco*RI. Stanley N. Cohen, MD, was a faculty member at Stanford, and he was interested in how plasmids could confer resistance to different antibiotics. His lab perfected laboratory transformation of *Escherichia coli* using calcium chloride to permeabilize the cells. After the talks ended, the two met over corned beef sandwiches and combined their two ideas.

They isolated different fragments of DNA from animals, other bacteria, and viruses and, using restriction enzymes, ligated the fragments into a small plasmid from *E. coli*. This was the first recombinant DNA to be made. Finally, they transformed the engineered plasmid back into *E. coli*. The cells expressed the normal plasmid genes as well as those inserted into the plasmid artificially. This sparked the revolution in genetic engineering, and since then every biotechnology lab has used some variation of their technique. Boyer and Cohen applied for a patent on recombinant DNA technology. In fact, Boyer cofounded Genentech with Robert Swanson, a venture capitalist. Genentech is one of the first biotechnology companies in the United States, and under Boyer and Swanson, the company produced human somatostatin in *E. coli*.

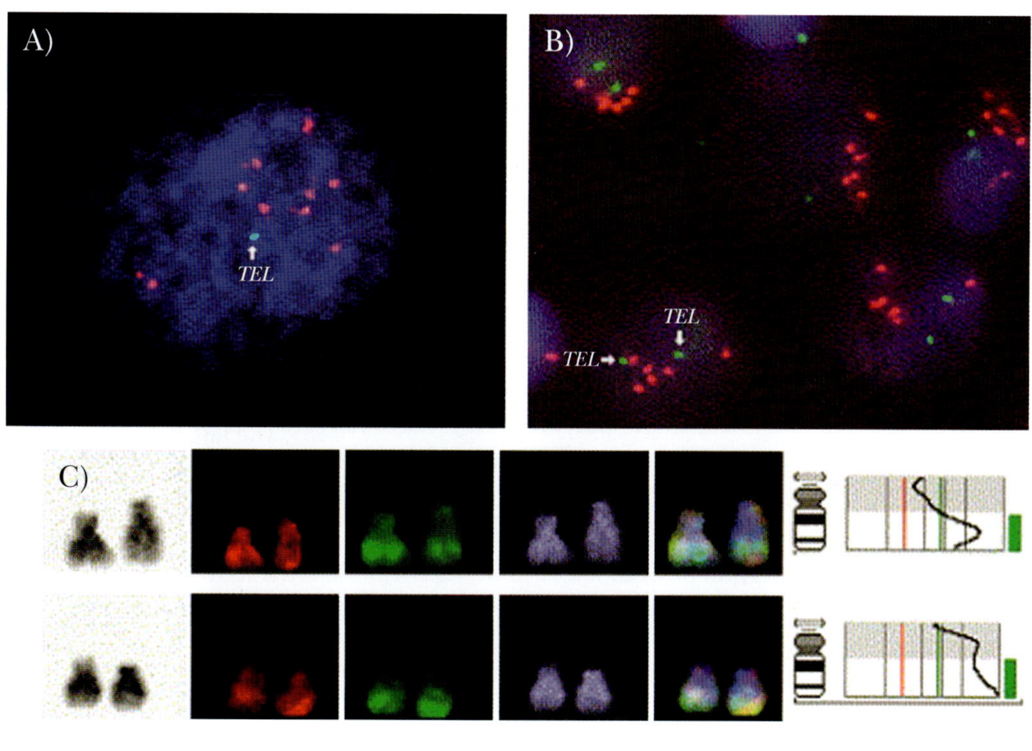

FIGURE 3.13
Gene Location on Chromosomes by FISH
(A) and (B) FISH of interphase nuclei with a dual-color DNA probe that shows *RUNX1* (= *AML1*) in red and *TEL* (telomerase) in green. (A) Patient 1 showed multiple *RUNX1* signals and lacked one *TEL* signal (arrow). (B) Patient 2 also showed extra copies of *RUNX1*, but had two normal *TEL* signals (arrows). (C) Partial GTG-banding karyotype and CGH profile of chromosomes 21 from both patients, showing that the amplification threshold is exceeded for the 21q22 region where *RUNX1* is located. From: Garcia-Casado *et al.* (2006). High-level amplification of the *RUNX1* gene in two cases of childhood acute lymphoblastic leukemia. *Cancer Genet Cytogenet* **170**, 171–174. Reprinted with permission.

GENERAL PROPERTIES OF CLONING VECTORS

Cloning vectors are specialized plasmids (or other genetic elements) that will hold any piece of foreign DNA for further study or manipulation. The numbers and types of plasmids available for cloning have grown. In addition, other DNA elements are now used, including viruses and artificial chromosomes. Once a fragment of DNA has been cloned and inserted into a suitable vector, large amounts of DNA can be manufactured, the sequence can be determined, and any genes in the fragment can be expressed in other organisms. Studying human genes in humans is virtually impossible because of the ethical ramifications. In contrast, studying a human gene expressed in bacteria provides useful information that can often be applied to humans. Modern biotechnology depends on the ability to express foreign genes in model organisms. Before discussing how a gene is cloned, the properties of vectors are considered.

Useful Traits for Cloning Vectors

Although many specialized vectors now exist, the following properties are convenient and found in most modern generalized cloning plasmids:

- Small size, making them easy to manipulate once they are isolated
- Easy to transfer from cell to cell (usually by transformation)
- Easy to isolate from the host organism
- Easy to detect and select
- Multiple copies helps in obtaining large amounts of DNA
- Clustered restriction sites (polylinker) to allow insertion of cloned DNA
- Method to detect presence of inserted DNA (e.g., alpha complementation)

Most bacterial plasmids satisfy the first three requirements. The next key trait of cloning vectors is an easy way to detect their presence in the host organism. Bacterial cloning plasmids often have antibiotic resistance genes that make bacteria resistant to particular antibiotics. When treated with the antibiotic, only bacteria with the plasmid-borne resistance gene will survive. Other bacteria will die. Other traits have been exploited to detect plasmids. Vectors derived from the yeast 2μ plasmid often carry genes for essential amino acids, such as leucine. The host strain of yeast is defective in the corresponding gene and unable to grow on media lacking the amino acid, unless the plasmid is present.

Plasmids vary in their **copy number**. Some plasmids exist in just one or a few copies in their host cells whereas others exist in multiple copies. Such multicopy plasmids are in general more useful as the amount of plasmid DNA is higher, making them easier to isolate and purify. The type of origin of replication controls the copy number, since this region on the plasmid determines how often DNA polymerase binds and induces replication.

Most cloning vectors have several unique restriction enzyme sites. Usually these sites are grouped in one location called the **polylinker** or **multiple cloning site (MCS**; Fig. 3.14). This allows researchers to open the cloning vector at one site without disrupting any of the vector's replication genes. Fragments of foreign DNA are digested with enzymes matching those in the polylinker. Ligase connects the vector and insert. Specific restriction enzyme sites can be added using PCR primers or synthetic DNA oligomers (see Chapter 4).

Some vectors have ways to detect whether or not they contain an insert. The simplest way to do this is **insertional inactivation** of an antibiotic gene (Fig. 3.15). Here, the vector has two different antibiotic resistance genes. The foreign DNA is inserted into one of the antibiotic-resistant genes. Thus the host bacteria will be resistant to one antibiotic and sensitive to the other.

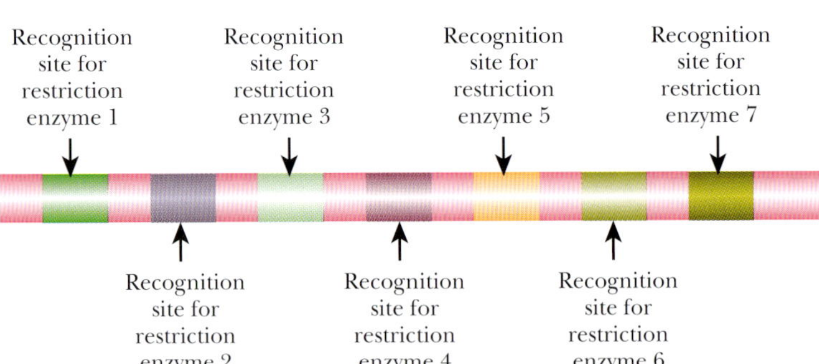

FIGURE 3.14 Typical Polylinker or Multiple Cloning Site
The restriction enzyme sites within the polylinker region are unique. This ensures that the plasmid is only cut once by each restriction enzyme.

Alternatively, **alpha complementation** may be used (see Fig. 3.15). The vector has a short portion of the β-galactosidase gene (the alpha fragment), and the bacterial chromosome has the rest of the gene. If both give rise to proteins, the subunits combine to form functional β-galactosidase. If DNA is inserted into the plasmid-borne gene segment, the encoded subunit is not made and β-galactosidase is not produced. When

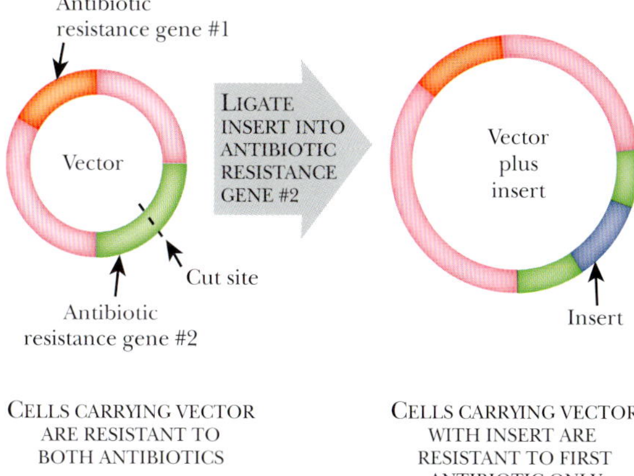

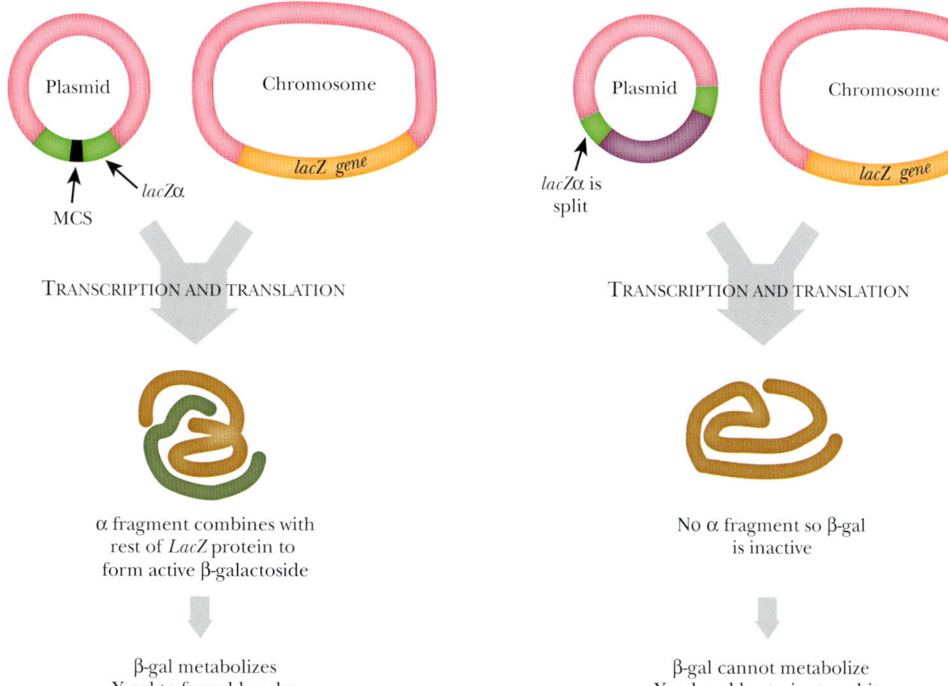

FIGURE 3.15
Detecting Inserts in Plasmids
(A) Insertional inactivation. Cells with an insert become sensitive to the second antibiotic. Cells without an insert remain resistant to the antibiotic. (B) Alpha complementation. Alpha complementation refers to the ability of β-galactosidase to be expressed as two protein fragments, which assemble to form a functional protein. In cells without an insert in the plasmid, β-galactosidase is active and splits X-gal to form a blue dye. In cells with an insert, the alpha fragment is not made and β-galactosidase is inactive. These cells remain white on media with X-gal.

β-galactosidase is expressed, the bacteria can degrade X-gal, which turns the bacterial colony blue. If a piece of DNA is inserted into the alpha fragment gene, the bacteria cannot split X-gal and stay white.

Once an appropriate vector has been chosen for the gene of interest or other insert, the two pieces are ligated into one **construct**. The term *construct* refers to any recombinant DNA molecule that has been assembled by genetic engineering. If both the vector and insert are cut with the same restriction enzyme, the two pieces have complementary ends and require only ligase to link them. Tricks are used to make two pieces of DNA with unrelated ends compatible. Sometimes, short oligonucleotides are synthesized and added onto the ends of the insert to make them compatible with the vector. These short oligonucleotides are called **linkers**, and they add one or a few new restriction enzyme sites to the ends of a segment of DNA.

Cloning vectors have multiple cloning sites with many unique restriction enzyme sites, they have genes for antibiotic resistance that make the bacterial cell able to grow with the antibiotic present, and they have a way to detect when a foreign piece of DNA is present such as alpha complementation.

SPECIFIC TYPES OF CLONING VECTORS

Because *E. coli* is the main host organism used for manipulating DNA, most vectors are based on plasmids or viruses that can survive in *E. coli* or similar bacteria. Most vectors have bacterial origins of replication and antibiotic resistance genes. The polylinker or multiple cloning site is usually placed between prokaryotic promoter and terminator sequences. Some vectors may also supply the ribosome binding site, so that any inserted coding sequence will be expressed as a protein. Many other features are present in specialized cloning vectors. The following discussion will introduce some of the different categories of vectors with their essential features (Fig. 3.16).

Many vectors are based on the **2μ circle** of yeast. The native version of the 2μ circle has been modified in a variety of ways for use as a cloning vector. A **shuttle vector** contains origins of replication for two organisms plus any other sequences necessary to survive in either organism (see Fig. 3.16B). Shuttle vectors that are based on the 2μ plasmid have the components needed for survival in yeast and bacteria, plus antibiotic resistance and a polylinker. The *Cen* sequence is a eukaryotic **centromere (Cen) sequence** that keeps the plasmid in the correct location during mitosis and meiosis in yeast. Because yeast cells are eukaryotic and also have such a thick cell wall, most antibiotics do not kill yeast. Therefore, a different strategy is used to detect the presence of plasmids in yeast. A gene for synthesis of an amino acid, such as leucine, allows strains of yeast that require leucine to grow.

Bacteriophage vectors are viral genomes that have been modified so that large pieces of nonviral DNA can be packaged in the virus particle. Lambda bacteriophages have linear genomes with two cohesive ends—***cos* sequences (lambda cohesive ends)**. These are 12-base overlapping sticky ends. When inside the virus coat, the cohesive ends are coated with protein to prevent them from annealing. After lambda attaches to *E. coli*, it inserts just the linear DNA. The proteins that protect the cohesive ends are lost, and the genome circularizes with the help of DNA ligase. The circular form is the **replicative form (RF)**, and it replicates by the rolling circle mechanism (see Chapter 4). Expression of various lambda genes produces the proteins that assemble into protein coats. Each coat is packaged with one genome, and after many of these are assembled, the *E. coli* host explodes, releasing the new bacteriophage to infect other cells.

The lambda bacteriophage is a widely used cloning vector (see Fig. 3.16C). The middle segment of the lambda genome has been deleted and a polylinker has been added. An insert of 37 to 52 kb can be ligated into the polylinker and packaged into viral particles. In order to work with the bacteriophage DNA without killing the entire *E. coli* culture, one or more genes necessary for packaging are deleted. When the researcher wants to form fully packaged bacteriophages, coat proteins from **helper virus** can be added (Fig. 3.17). The helper viruses do not contain foreign DNA, but supply the missing genes for the coat proteins. Because coat proteins self-assemble *in vitro*, helper lysates are mixed with recombinant lambda DNA and complete virus particles containing DNA are produced. This is known as *in vitro* packaging.

Cosmid vectors can hold pieces of DNA up to 45 kb in length (see Fig. 3.16D). These are highly modified lambda vectors with all the sequences between the *cos* sites removed and replaced with the insert. The DNA of interest is ligated between the two *cos* sites using restriction enzymes and ligase. This construct is packaged into a lambda particle

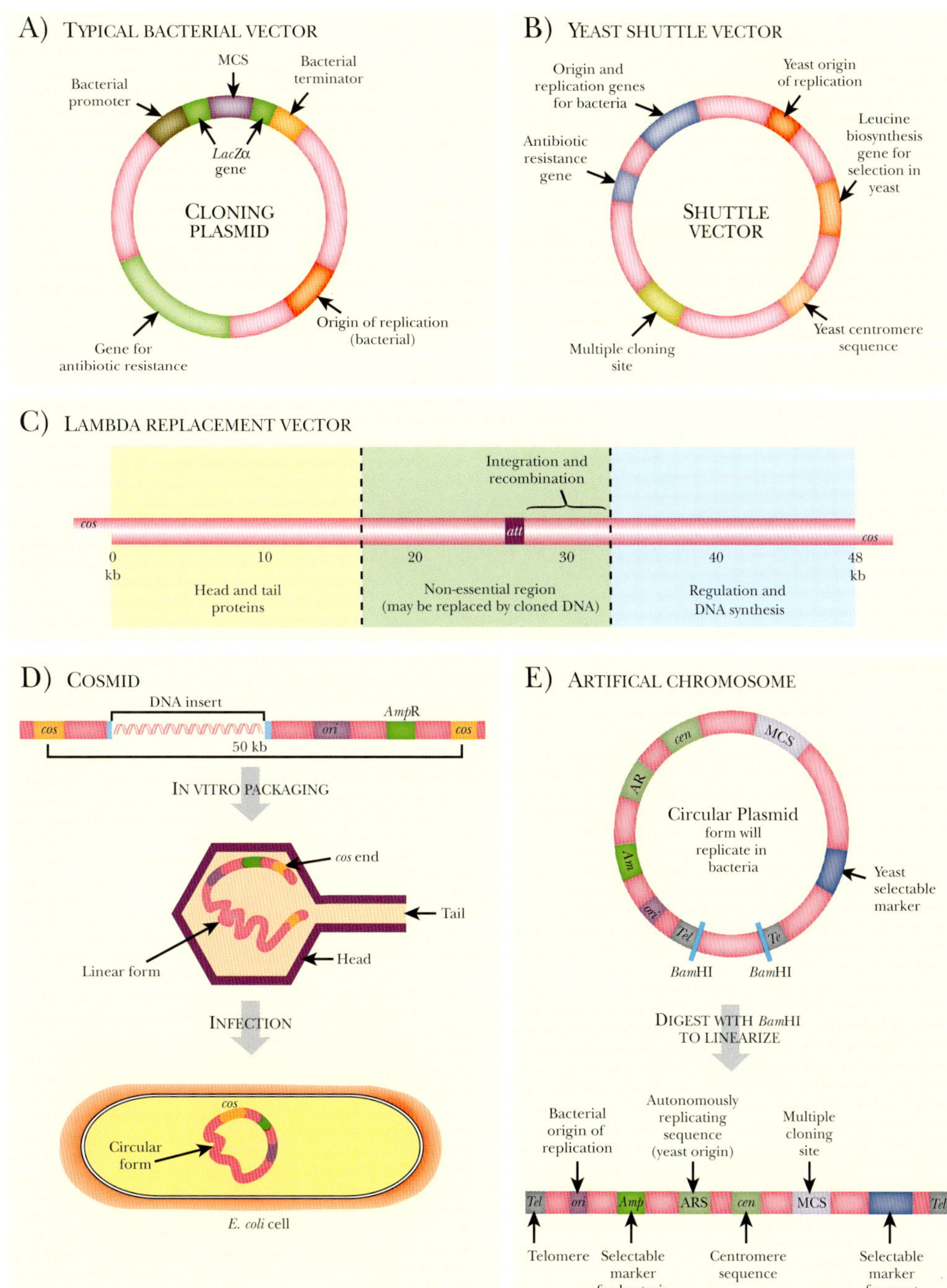

FIGURE 3.16 Various Cloning Vectors

(A) Typical bacterial cloning vector. This vector has bacterial sequences to initiate replication and transcription. In addition, it has a multiple cloning site embedded within the *lacZ* α gene so that the insert can be identified by alpha-complementation. The antibiotic resistance gene allows the researcher to identify any *E. coli* cells that have the plasmid. (B) Yeast shuttle vector. This vector can survive in either bacteria or yeast because it has both yeast and bacterial origin of replication, a yeast centromere, and selectable markers for yeast and bacteria. As with most cloning vectors, there is a polylinker. (C) Lambda replacement vectors. Because lambda phage is easy to grow and manipulate, its genome has been modified to accept foreign DNA inserts. The region of the genome shown in green is nonessential for lambda growth and packaging. This region can be replaced with large inserts of foreign DNA (up to about 23 kb). (D) Cosmids. Cosmids are small multicopy plasmids that carry *cos* sites. They are linearized and cut so that each half has a *cos* site (not shown). Next, foreign DNA is inserted to relink the two halves of the cosmid DNA. This construct is packaged into lambda virus heads and used to infect *E. coli.* (E) Artificial chromosomes. Yeast artificial chromosomes have two forms, a circular form for growing in bacteria and a linear form for growing in yeast. The circular form is maintained like any other plasmid in bacteria, but the linear form must have telomere sequences to be maintained in yeast. The linear form can hold up to 2000 kb of cloned DNA and is very useful for genomics research.

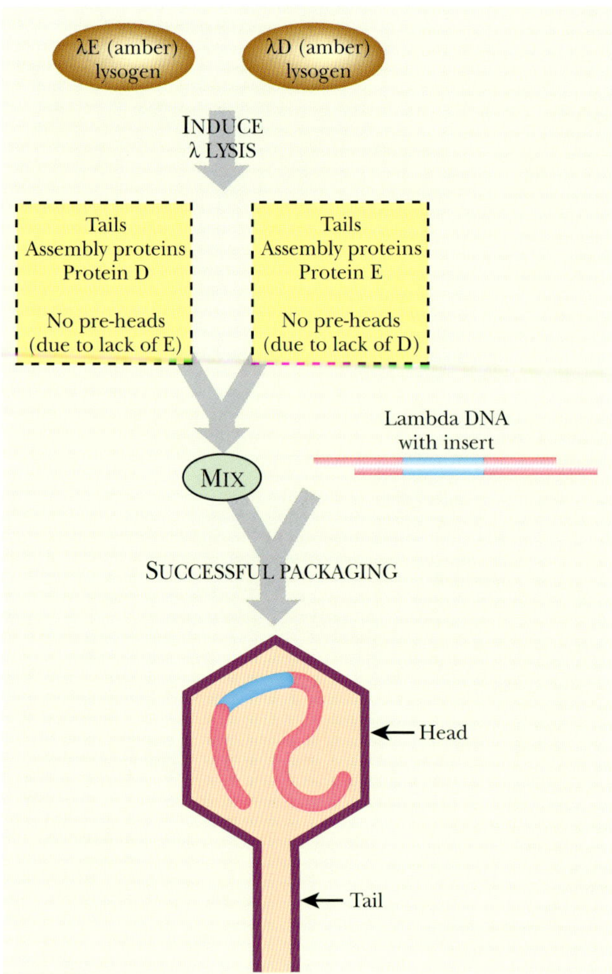

FIGURE 3.17 *In Vitro* Packaging
A lambda cloning vector containing cloned DNA must be packaged in a phage head before it can infect *E. coli*. First, one culture of *E. coli* cells is infected with a mutant lambda that lacks the gene for one of the head proteins called E. A different culture of *E. coli* is infected with a different mutant, which lacks the phage head protein D. The two cultures are induced to lyse, which releases the tails, assembly proteins, and head proteins, but no complete heads because of the missing proteins. When these are mixed with a lambda replacement vector, the three spontaneously form complete viral particles containing DNA. These are then used to infect *E. coli*.

produced by helper phage as shown earlier (see Fig. 3.17), and then these are used to infect *E. coli*.

Artificial chromosomes hold the largest pieces of DNA (see Fig. 3.16E). These include yeast artificial chromosomes (YACs), bacterial artificial chromosomes (BACs), and P1 bacteriophage artificial chromosomes (PACs). They are used to contain lengths of DNA from 150 kb to 2000 kb. YACs hold the largest amount of DNA, up to about 2000 kb. YACs have yeast centromeres and yeast telomeres for maintenance in yeast. BACs can be circularized and grown in bacteria; therefore, they have a bacterial origin of replication and antibiotic resistance genes. The flexibility of artificial chromosomes makes them most useful for sequencing entire genomes, especially those of higher organisms with vast amounts of DNA.

Many different cloning vectors are available to biotechnology research. The smaller genes are studied using bacterial plasmids or shuttle vectors, whereas the larger genes are studied in bacteriophage vectors, cosmids, and artificial chromosomes.

Shuttle vectors have sequences that enable them to survive in two different organisms such as yeast and bacteria.

Bacteriophage vectors have critical genes removed so that the bacteriophage cannot destroy the host cell by producing phage particles. Adding a helper phage restores this activity.

GETTING CLONED GENES INTO BACTERIA BY TRANSFORMATION

Once the gene of interest is cloned into a vector, the construct can be put back into a bacterial cell through a process called **transformation** (Fig. 3.18). Here the "naked" DNA that was constructed in the laboratory is mixed with **competent** *E. coli* cells. To make the cells competent, that is, able to take up naked DNA, the cell wall must be temporarily opened up. In the laboratory, *E. coli* cells are treated with high concentrations of calcium ions on ice, and then shocked at a higher temperature for a few minutes. Most of the cells die during the treatment, but some survive and take up the DNA. Another method to make *E. coli* cells competent is to expose them to a high-voltage shock. **Electroporation** opens the cell wall, allowing the DNA to enter. This method is much faster and more versatile. Electroporation is used for other types of bacteria as well as yeast.

When a mixture of different clones is transformed into bacteria as in a gene library (see later discussion), cells that take up more than one construct usually lose one of them. For example, if genes A and B are both cloned into the same kind of vector and both cloned genes get into the same bacterial cell, the bacteria will lose one plasmid and keep the other. This is due to **plasmid incompatibility**, which prevents one bacterial cell from harboring two of the same type of plasmid. Incompatibility stems from conflicts between two plasmids with identical or related origins of replication. Only one is allowed to replicate in any given cell. If a researcher wants a cell to have two cloned genes, then two different types of plasmids could be used, or both genes could be put onto the same plasmid.

CONSTRUCTING A LIBRARY OF GENES

Gene libraries are used to find new genes, to sequence entire genomes, and to compare genes from different organisms (Fig. 3.19). Gene libraries are made when the entire DNA from one particular organism is digested into fragments using restriction enzymes, and then each of the fragments is cloned into a vector and transformed into an appropriate host.

The basic steps used to construct a library are:

1. Isolate the chromosomal DNA from an organism, such as *E. coli*, yeast, or humans.
2. Digest the DNA with one or two different restriction enzymes.
3. Linearize a suitable cloning vector with the same restriction enzyme(s).
4. Mix the cut chromosome fragments with the linearized vector and ligate.

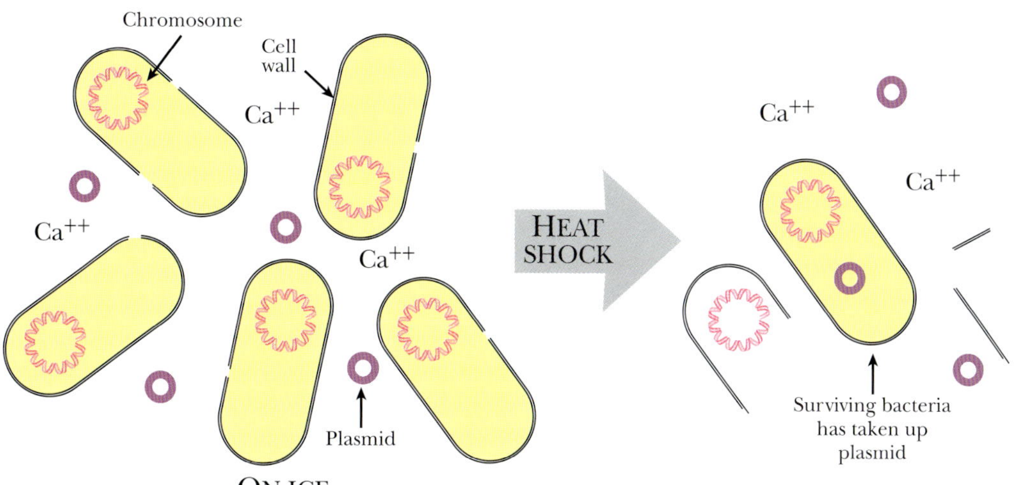

FIGURE 3.18
Transformation
Bacterial cells are able to take up recombinant plasmids by incubation in calcium on ice. After a brief heat shock, some of the bacteria take up the plasmid.

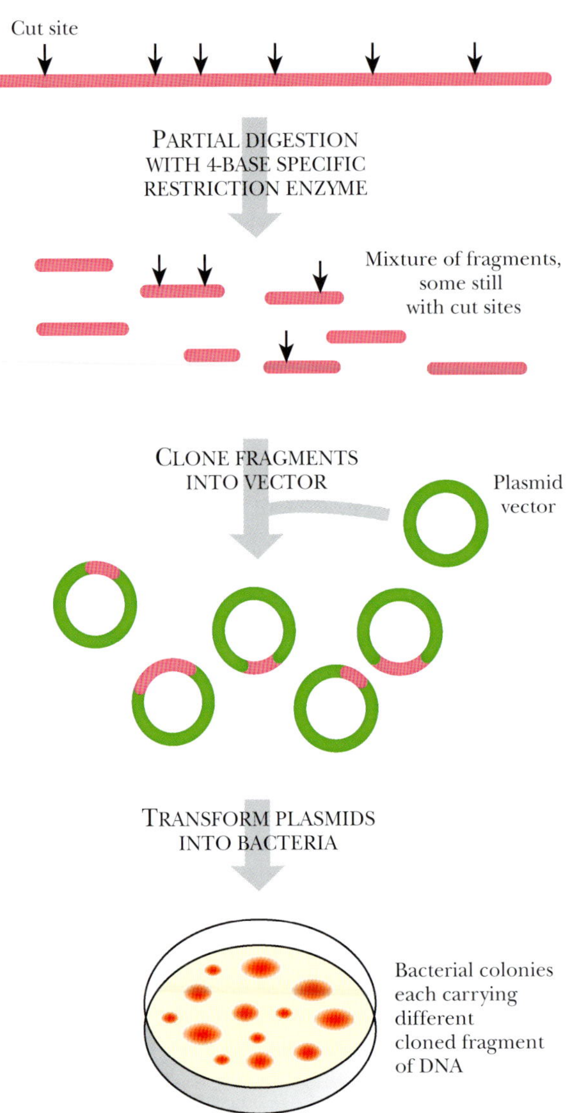

FIGURE 3.19 Creating a DNA Library
Genomic DNA from the chosen organism is first partially digested with a restriction enzyme that recognizes a four base-pair sequence. Partial digestions are preferred because some of the restriction enzyme sites are not cut, and larger fragments are generated. If every recognition site were cut by the restriction enzyme, then the genomic DNA would not contain many whole genes. The genomic fragments are cloned into an appropriate vector, and transformed and maintained in bacteria.

5. Transform this mixture into *E. coli*.
6. Isolate large numbers of *E. coli* transformants.

The type of restriction enzyme affects the type of library. Because restriction sites are not evenly spaced in the genome, some inserts will be large and others small. Using a restriction enzyme that recognizes only four base pairs will give a mixture of mostly small fragments, whereas a restriction enzyme that has a six or eight base-pair recognition sequence will generate larger fragments. (Note that finding a particular four base-pair sequence in a genome is more likely than finding a six base-pair sequence.) Even if an enzyme that recognizes a four base-pair recognition sequence is used to digest the entire genome, there may still be segments that are too large to be cloned. Conversely, clustered restriction sites will cause some genes will be cut into several pieces. To avoid this, partial digestion is often used. The enzyme is allowed to cut the DNA for only a short time, and many of the restriction enzyme sites are not cut, leaving larger pieces for the library. In addition, it is usual to construct another library using a different restriction enzyme.

Gene libraries are used for many purposes because they contain almost all the genes of a particular organism.

SCREENING THE LIBRARY OF GENES BY HYBRIDIZATION

Once the library is assembled, researchers often want to identify a particular gene or segment of DNA within the library. Sometimes the gene of interest is similar to one from another organism. Sometimes the gene of interest contains a particular sequence. For example, many enzymes use ATP to provide energy. Enzymes that bind ATP share a common signature sequence whether they come from bacteria or humans. This sequence can be used to find other enzymes that bind ATP. Such common sequence motifs may also suggest that a protein will bind various cofactors, other proteins, and DNA, to name a few examples.

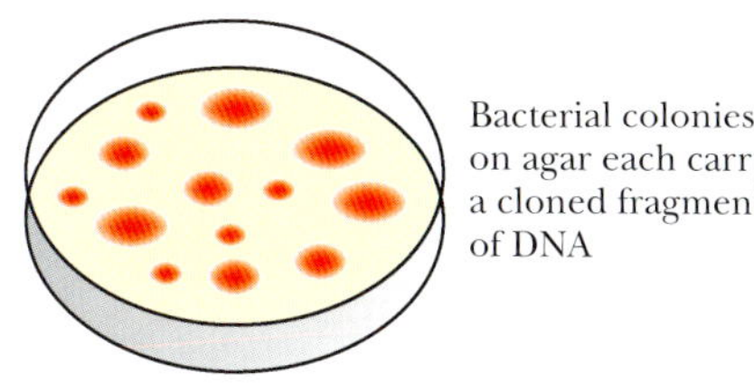

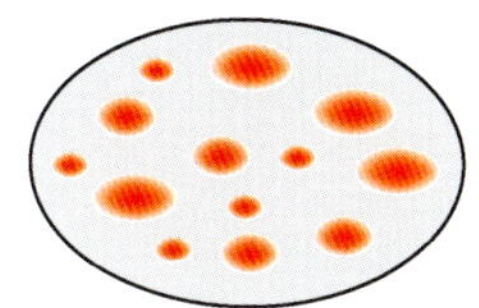

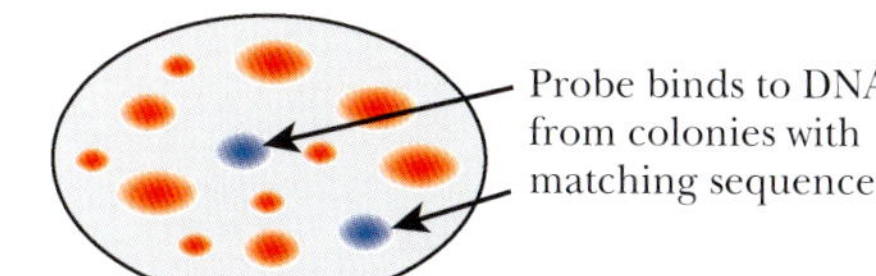

FIGURE 3.20
Screening a Library with DNA Probe
First, bacterial colonies containing the library inserts are grown and plated on large, shallow agar-filled dishes. Many different colonies are plated so that every cloned piece of DNA is present. These colonies are transferred to nylon filters and lysed open. The cell remains are washed away, while the genomic and plasmid DNA sticks to the nylon. The sequences are made single-stranded by incubating the filters in a strong base. When these are incubated with a radioactive single-stranded probe at the appropriate temperature, the probe hybridizes to any matching sequences.

Screening DNA libraries by hybridization requires preparing the library DNA and preparing the labeled probe. A gene library is stored as a bacterial culture of *E. coli* cells, each having a plasmid with a different insert. The culture is grown up, diluted, and plated onto many different agar plates so that the colonies are spaced apart from one another. The colonies are transferred to a nylon filter and the DNA from each colony is released from the cells by lysing them with detergent. The cellular components are rinsed from the filters. The DNA sticks to the nylon membrane and is then denatured to form single strands (Fig. 3.20).

If a scientist is looking for a particular gene in the target organism, the probe for the library may be the corresponding gene from a related organism. The probe is usually just a segment of the gene, because a smaller piece is easier to manipulate. The probe DNA can be synthesized and labeled either with radioactivity or with chemiluminescence. The probe is heated to make it single-stranded and mixed with the library DNA on the nylon filters. The probe hybridizes with matching sequences in the library. The level of match needed for binding can be adjusted by incubating at various temperatures. The higher the temperature, the more stringent, that is, the more closely matched the sequences must be. The lower the temperature, the less stringent. If the probe is labeled with radioactivity, photographic film will turn black where the probe and library DNA hybridized. The black spot is aligned with the original bacterial colony. Usually the most likely colony plus its neighbors are selected, grown, plated, and rescreened with the same probe to ensure that a single transformant is isolated. Then the DNA from this isolate can be analyzed by sequencing (see Chapter 4).

Screening a library has two parts. First, the library clones growing in *E. coli* are attached to nylon filters, and the cellular components washed away and then denatured to form single-stranded DNA pieces. Second, a probe is labeled with radioactivity, is heated to melt the helix into single strands, and finally added to the nylon membranes where it hybridizes to its matching sequence.

EUKARYOTIC EXPRESSION LIBRARIES

In **expression libraries**, the vector has sequences required for transcription and translation of the insert. This means that the insert DNA is expressed as RNA, and this may be translated into a protein. An expression library, in essence, generates a protein from every cloned insert, whether it is a real gene or not. When studying eukaryotic DNA, expression libraries are constructed using **complementary DNA (cDNA)** to help ensure the insert is truly a gene. Eukaryotic DNA, especially in higher plants and animals, is largely noncoding, with coding regions spaced far apart. Even genes are interrupted with noncoding introns. cDNA is a double-stranded DNA copy of mRNA made by using reverse transcriptase. Reverse transcriptase was first identified in retroviruses (see Chapter 1). It is used in eukaryotic research to eliminate the introns and generate a version of a gene consisting solely of an uninterrupted coding sequence.

In contrast, bacteria have very little noncoding DNA and their genes are not interrupted by introns; therefore, genomic DNA can be used directly in expression libraries.

Eukaryotic DNA is first made into cDNA in order to construct an expression library (Fig. 3.21). To make cDNA, the messenger RNA is isolated from the organism of interest by

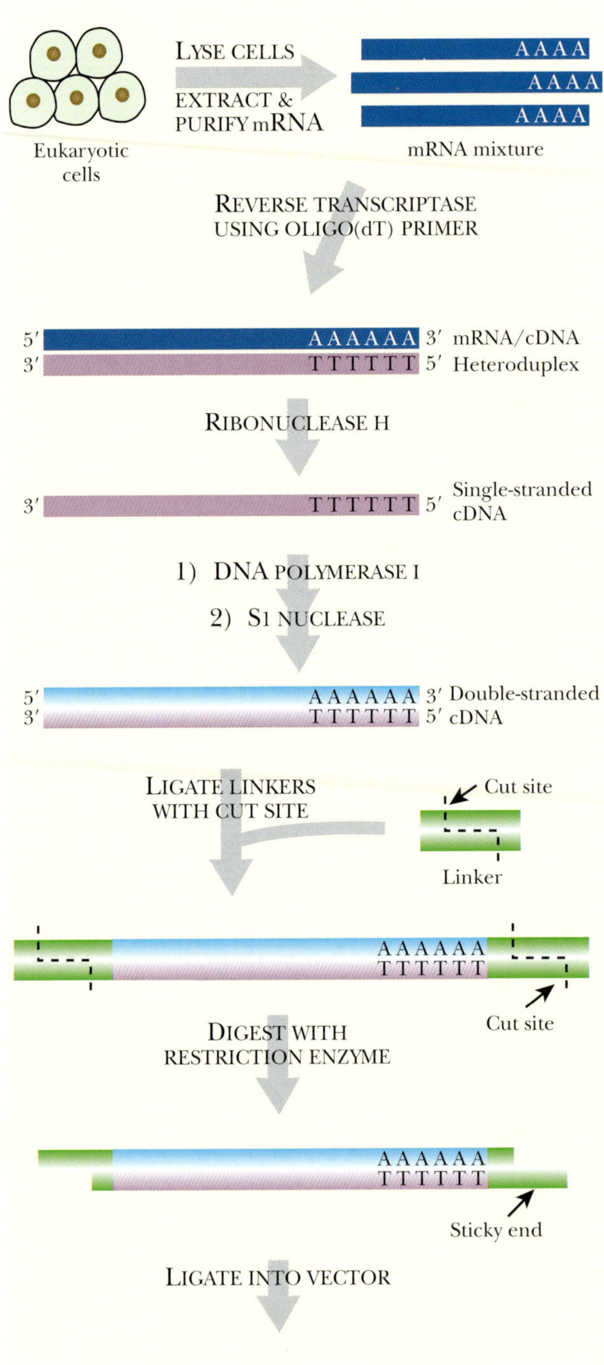

FIGURE 3.21 Making a cDNA Library from Eukaryotic mRNA
First, eukaryotic cells are lysed and the mRNA is purified. Next, reverse transcriptase plus primers containing oligo(dT) stretches are added. The oligo(dT) hybridizes to the adenine in the mRNA poly(A) tail and acts as a primer for reverse transcriptase. This enzyme makes the complementary DNA strand, forming an mRNA/cDNA hybrid. The mRNA strand is digested with ribonuclease H, and DNA polymerase I is added to synthesize the opposite DNA strand, thus creating double-stranded cDNA. Next, S1 nuclease is added to trim off any single-stranded ends, and linkers are added to the ends of the dsDNA. The linkers have convenient restriction enzyme sites for cloning into an expression vector.

binding to a column containing poly(T) (i.e., a DNA strand consisting of repeated thymines). This isolates only mRNA because poly(T) anneals to the poly(A) tail of eukaryotic mRNA. The mRNA is made double-stranded using reverse transcriptase, which synthesizes a DNA complement to mRNA. The hybrid mRNA:DNA molecule is converted to double-stranded DNA using RNase H and DNA polymerase I. RNase H cuts nicks into the mRNA backbone, and DNA polymerase displaces the mRNA strand and synthesizes DNA (see Fig. 3.21).

The cDNA is then ligated into an **expression vector** with sequences that initiate transcription and translation of the insert. In some cases, the insert will have its own translation start site (e.g., a full-length cDNA). If, as often occurs, the insert does not contain a translation start, then the reading frame becomes an issue. Because the genetic code is triplet, each insert can be translated in three different reading frames. A protein may be produced for all three reading frames, but only one frame will actually produce the correct protein. To ensure obtaining inserts with the correct reading frame, each cDNA is cloned in all three reading frames by using linkers with several different restriction sites. The number of transformants to screen is therefore increased greatly.

The cloned genes are transformed into bacteria, which express the foreign DNA. The bacteria are grown on agar and the colonies are then transferred to a nylon membrane and lysed. The proteins released are attached to the nylon membranes and are screened in various ways. Most often, an antibody to the protein of interest is used (see Chapter 6). This recognizes the protein and can be identified using a secondary antibody that is conjugated to a detection system. Usually alkaline phosphatase is conjugated to the secondary antibody. The whole complex can be identified because alkaline phosphatase cleaves X-Phos, leaving a blue color where the bacterial colony expressed the right protein (Fig. 3.22). *E. coli* cannot perform most of the posttranslational modifications that eukaryotic proteins often undergo. Therefore, the proteins are not always in their native form. Nonetheless, appropriate antibodies can detect most proteins of interest.

Complementary DNA or cDNA is constructed by isolating mRNA and making a DNA copy with reverse transcriptase.

Expression libraries express the foreign DNA insert as a protein because expression vectors contain sequences for both transcription and translation. The protein of interest is identified by incubating the library with an antibody to the protein of interest.

FEATURES OF EXPRESSION VECTORS

Because foreign proteins, especially if made in large amounts, can be toxic to *E. coli*, the promoter used to express the foreign gene is critical. If too much foreign protein is made, the host cell may die. To control protein production, expression vectors have promoters with on/off switches; therefore, the host cell is grown up first and the foreign protein is expressed later. One commonly used promoter is a mutant version of the *lac* promoter (Fig. 3.23). This *lacUV* promoter drives a very high level of transcription, but only under induced conditions. It has the following elements: a binding site for RNA polymerase, a binding site for the LacI repressor protein, and a transcription start site. The vector has strong transcription stop sites downstream of the polylinker region. The vector also has the gene for LacI so that high levels of repressor protein are made, thus keeping the cloned genes repressed. Like all vectors, there is an origin of replication and antibiotic resistance gene for selection in bacteria. When a gene library is cloned behind this promoter, the genes are not expressed due to high levels of LacI repressor. When an inducer, such as IPTG, is added, LacI is released from the DNA and RNA polymerase transcribes the foreign, cloned, genes.

Another common promoter in expression vectors is the lambda left promoter, or P_L. It has a binding site for the lambda repressor. The gene of interest or library fragment is not expressed unless the repressor is removed. Rather than using its natural inducer, a mutant

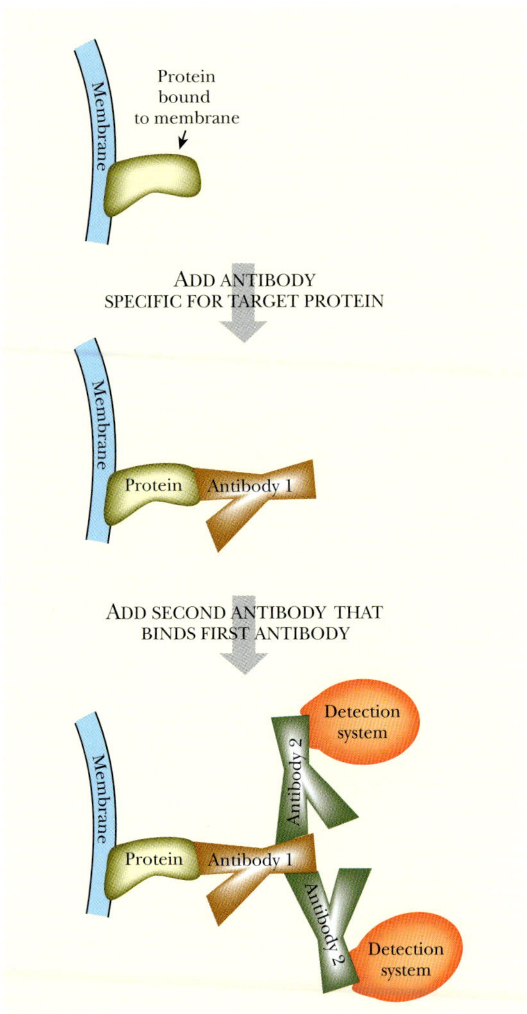

FIGURE 3.22 Immunological Screening of an Expression Library
Bacteria expressing foreign genes are grown on an agar plate, transferred to a membrane, and lysed. Released proteins are bound to the membrane. This figure shows only one attached protein, although in reality many different proteins are present. These include both expressed library clones and bacterial proteins. The membrane is incubated with a primary antibody that binds only the protein of interest. To detect this protein:antibody complex, a second antibody with a detection system such as alkaline phosphatase is added. The bacterial colony expressing the protein of interest will turn blue when X-Phos is added. This allows the vector with the correct insert to be isolated.

FIGURE 3.23
Expression Vectors Have Tightly Regulated Promoters
An expression vector contains sequences upstream of the cloned gene that control transcription and translation of the cloned gene. The expression vector shown uses the *lacUV* promoter, which is very strong, but inducible. To stimulate transcription, the artificial inducer, IPTG, is added. IPTG binds to the LacI repressor protein, which then detaches from the DNA. This allows RNA polymerase to transcribe the gene. Before IPTG is added, the LacI repressor prevents expression of the cloned gene.

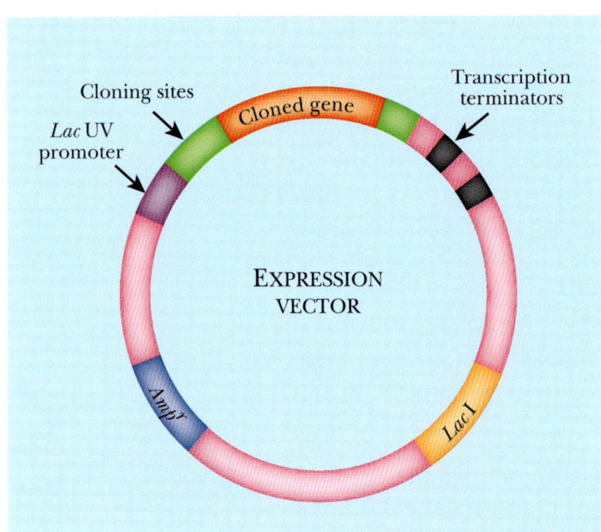

version of the repressor has been isolated that releases its binding site at high temperatures. So when the culture is shifted to 42 °C, the repressor falls off the DNA and RNA polymerase transcribes the cloned genes.

Another expression system uses a promoter whose RNA polymerase binding site only recognizes RNA polymerase from the bacteriophage T7. Bacterial RNA polymerase will not transcribe the gene of interest. This system is designed to work only in bacteria that have the gene for T7 RNA polymerase integrated into the chromosome and under the control of an inducible promoter.

Some expression vectors contain a small segment of DNA that encodes a protein tag. These are primarily used when the gene of interest is already cloned, rather than for screening libraries. The gene of interest must be cloned in frame with the DNA for the protein tag. The tag can be of many varieties, but **6HIS, Myc,** and **FLAG® tag** are three popular forms (Fig. 3.24). 6HIS is a stretch of six histidine residues put at the beginning or end of the protein of interest. The histidines bind strongly to nickel. This allows the tagged protein to be isolated by binding to a column with nickel attached. Myc and FLAG are epitopes that allow the expressed protein to be purified by binding to the corresponding antibody. The antibodies may be attached to a column, used for a Western blot, or seen *in vivo* by staining the cells with fluorescently tagged versions of the Myc or FLAG antibodies. (The histidine tag can also be recognized with a specific antibody, if desired.)

The most important feature of expression vectors is a tightly controlled promoter region. The proteins of the expression library are expressed only under certain conditions, such as presence of an inducer, removal of a repressor, or change in temperature.

Small tags can be fused into the protein of interest using expression vectors. These tags allow the protein of interest to be isolated and purified.

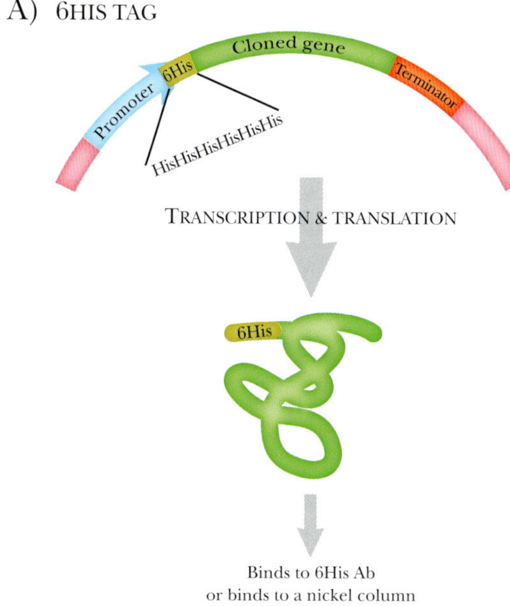

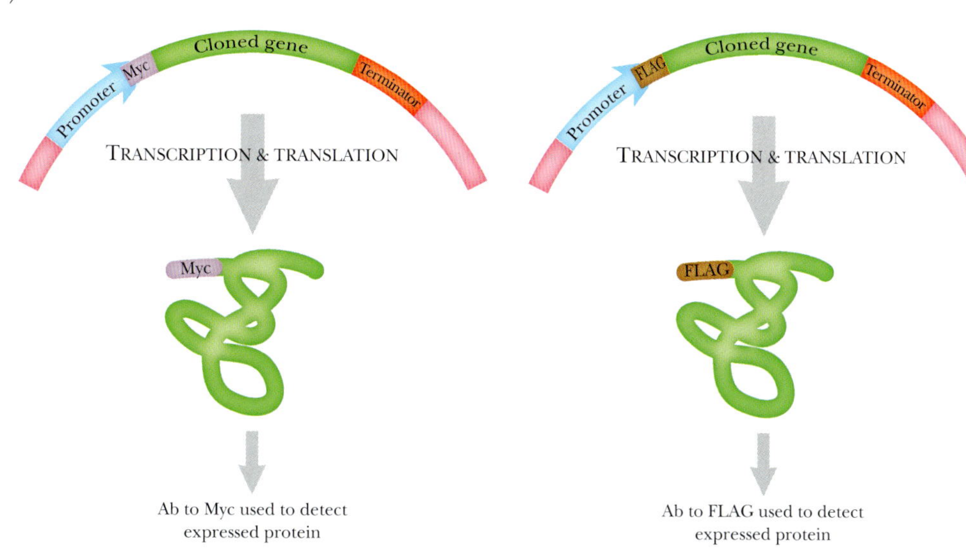

FIGURE 3.24 Using Tags to Isolate Proteins
Some expression vectors have DNA sequences that code for short protein tags. The 6HIS tag (A) codes for six histidine residues. When fused in-frame with the coding sequence for the cloned gene, the tag is fused to the protein. The 6HIS tag specifically binds to nickel ions; therefore, binding to a nickel ion column isolates 6HIS-tagged proteins. Other tags, such as Myc or FLAG (B), are specific antibody epitopes that work in a similar manner. Myc-tagged or FLAG-tagged proteins can be isolated or identified by binding to antibodies to Myc or FLAG, respectively.

SUBTRACTIVE HYBRIDIZATION

Subtractive hybridization is a screening method that allows researchers to find genes that are "missing." For example, the gene responsible for a hereditary defect may be totally deleted in one particular victim. A healthy person will have the complete gene. Therefore, the DNA of the two people will be identical, except for an extra segment in the person without the disease. The DNA from the person with the deletion is isolated and cut with one restriction enzyme (Fig. 3.25). The DNA from the healthy person is isolated and cut with a second, different restriction enzyme. An excess amount of mutant DNA is mixed with healthy DNA, and then heating denatures the mixture. Slowly cooling the mixture allows hybrid molecules to form from the normal and mutant DNA. If two mutant fragments anneal, the dsDNA fragment will have sites for restriction enzyme 1 at each end. If two healthy fragments hybridize, the ends will have sites for restriction enzyme 2. If healthy DNA hybridizes with mutant DNA, the two will have nonmatching ends that cannot be cut by either restriction enzyme. All regions of the DNA will be able to form mutant:normal hybrids except for the

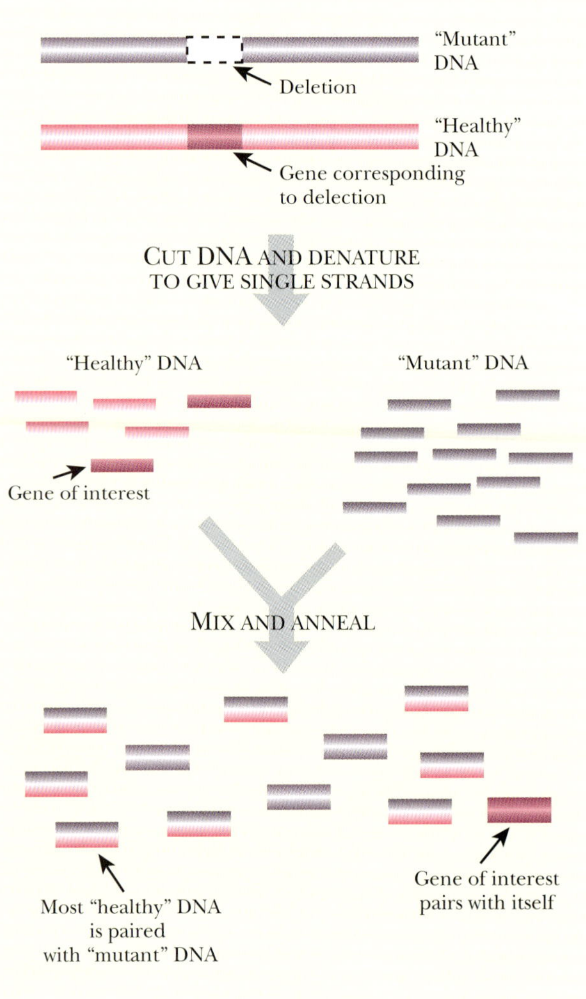

FIGURE 3.25 Cloning by Subtractive Hybridization
The key to subtractive hybridization is to hybridize all the wild-type or "healthy" DNA fragments (pink) with an excess of mutant DNA (purple). In this example, the mutant DNA is digested with restriction enzyme 1 and the wild-type DNA is digested with enzyme 2. Next, the two DNA samples are mixed, denatured, and allowed to anneal. The DNA will hybridize to form mutant/mutant strands (flanked by sites for restriction enzyme 1); mutant/healthy strands (flanked by incompatible ends); or healthy/healthy strands (flanked by sites for enzyme 2). The healthy/healthy hybrids will be rare and should correspond to the region of the deletion. They are easily separated from the other hybrids by cloning into a vector cut with restriction enzyme 2.

region that is missing in the mutant DNA. This region of DNA can only self-hybridize, and the dsDNA formed will have sites for restriction enzyme 2 at the end. These segments can be cloned into a vector that has a corresponding restriction site. Overall, DNA that does not encode the gene of interest is excluded or subtracted by hybridization.

Subtractive hybridization is also used to compare gene expression under two different conditions (Fig. 3.26). For example, a researcher can compare the genes expressed by *E. coli* at low temperature versus those expressed in heat shock. First, separate cultures of bacteria are grown in both conditions and mRNA is isolated from both cultures. The mRNA from the bacteria kept at low temperature is made into cDNA to provide complementary sequences for hybridization. The cDNA is bound to a filter, denatured to give ssDNA, and incubated with the mRNA from the bacteria grown under the experimental conditions (i.e., heat shock). The mRNA that is present under both conditions will hybridize to the cDNA on the filter. However, mRNA that is expressed only under the experimental condition will not find a complementary sequence and will be left in solution. These unique mRNAs are then made into cDNA, cloned into a vector, and sequenced to identify genes that increase in expression under hot conditions.

Subtractive hybridization can be used to identify a gene that causes a disease or a set of genes that are expressed under certain conditions.

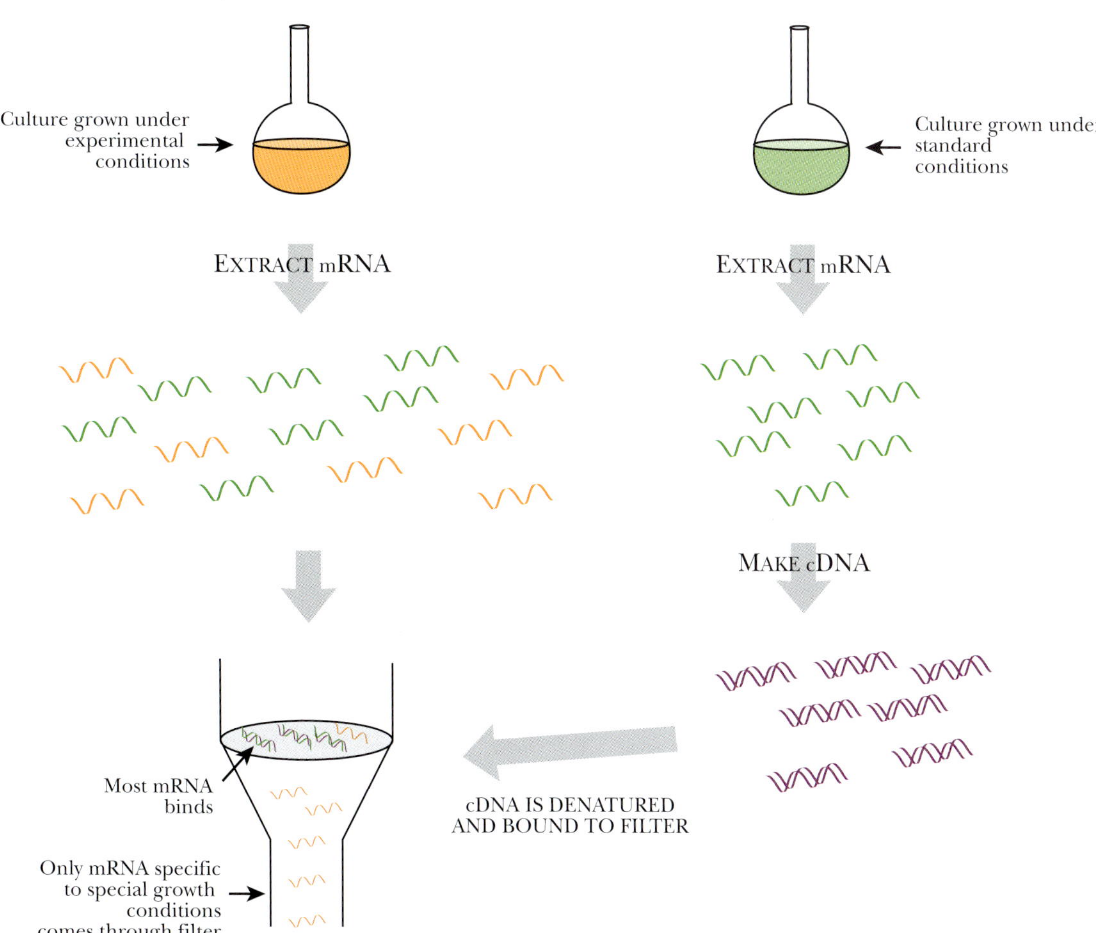

FIGURE 3.26 Subtractive Hybridization Captures mRNA Expressed under Specific Conditions
Two different cultures are grown, one under standard conditions (green) and the other under experimental conditions (orange). The mRNA is isolated from each culture, and one sample is converted to cDNA (purple). The cDNA is bound to the filter and denatured to make it single-stranded. The experimental mRNA (orange) is passed through the filter. Only mRNA that is unique to the experimental conditions will fail to find a partner to hybridize. Consequently, this mRNA will come through the filter. Conversely, any mRNA that is present under both growth conditions will be trapped on the filter.

Summary

Recombinant DNA technology is the basis for almost all biotechnology research. Understanding these techniques is tantamount to understanding the rest of the textbook. First, DNA must be isolated from the organism in order to identify novel genes, to recover new recombinant vectors, or to purify a new library clone. The DNA is isolated from the cellular components using enzymes followed by centrifugation, RNA digestion with RNase, and precipitation with ethanol. Each organism will require special adaptations of this basic process in order to remove the cellular and extracellular components.

The purified DNA can be manipulated in many different ways. This chapter describes how restriction enzymes can cut the phosphate backbone of the DNA into smaller fragments, then how these fragments can be visualized by gel electrophoresis. The chapter then describes various methods to label specific DNA pieces so only one particular piece can be visualized. Hybridization is a key technique for FISH, Southern blots, Northern blots, and dot blots, as well as the screening of a genomic library for a particular sequence.

The chapter also outlines the key characteristics of vectors, including plasmids, bacteriophage vectors, cosmids, and artificial chromosomes. These extrachromosomal genetic elements vary in their uses but are very important to getting a foreign gene expressed in a host organism. Vectors require a region that is convenient to adding a foreign piece of DNA such as a multicloning site, they need a gene for selection, and they need some easy way to identify whether or not the vector contains the foreign piece of DNA.

Finally, the chapter describes how to construct a library of genes from an organism. A genomic library simply contains all the DNA of the organism of interest, cut into smaller fragments, and cloned into a vector. The library recombinant clones are then returned to a host cell. In addition, the expression library uses mRNA rather than genomic DNA, and these foreign pieces of DNA are then turned into proteins in the host. Libraries are easily screened for particular sequences of interest by hybridization or antibodies to a protein of interest.

End-of-Chapter Questions

1. Which of the following statements about DNA isolation from *E. coli* is not correct?
   - **a.** Chemical extraction using phenol removes proteins from the DNA.
   - **b.** RNA is removed from the sample by RNase treatment.
   - **c.** Detergent is used to break apart plant cells to extract DNA.
   - **d.** Lysozyme digests peptidoglycan in the bacterial cell wall.
   - **e.** Centrifugation separates cellular components based on size.
2. Which of the following is important for gel electrophoresis to work?
   - **a.** Negatively charged nucleic acids to migrate through the gel.
   - **b.** Ethidium bromide to provide a means to visualize the DNA in the gel.
   - **c.** Agarose or polyacrylamide to separate the DNA based on size.
   - **d.** Known molecular weight standards.
   - **e.** All of the above are important for gel electrophoresis.

3. How are restriction enzymes and ligase used in biotechnology?
   a. Restriction enzymes cut DNA at specific locations, producing ends that can be ligated back together with ligase.
   b. Only restriction enzymes that produce blunt ends after cutting DNA can be ligated with ligase.
   c. Only restriction enzymes that produce sticky ends on the DNA can be ligated with ligase.
   d. Restriction enzymes can both cut DNA at specific sites and ligate them back together.
   e. Restriction enzymes randomly cut DNA, and the cut fragments can be ligated back together with ligase.
4. Which of the following is an appropriate method for detecting nucleic acids?
   a. Measuring absorbance at 260 nm.
   b. Autoradiography of radiolabeled nucleic acids.
   c. Chemiluminescence of DNA labeled with biotin or digoxigenin.
   d. Measuring the light emitted after excitation by fluorescent-labeled nucleic acids on a photodetector.
   e. All of the above are appropriate methods for detecting nucleic acids.
5. Why does the GC content of a particular DNA molecule affect the melting of the two strands?
   a. The G and C bond only requires two hydrogen bonds, thus requiring a lower temperature to "melt" the DNA.
   b. Because G and C base-pairing requires three hydrogen bonds and a higher temperature is required to "melt" the DNA.
   c. The percentage of As and Ts in the molecule is more important to melting temperature than the percentage of Gs and Cs.
   d. The nucleotide content of a DNA molecule is not important to know for biotechnology and molecular biology research.
   e. None of the above.
6. What is the difference between Southern and Northern hybridizations?
   a. Southern blots hybridize a DNA probe to a digested DNA sample but Northern blots hybridize a DNA probe to, usually, mRNA.
   b. Southern blots use an RNA probe to hybridize to DNA but Northern blots use an RNA probe to hybridize to RNA.
   c. Southern blots determine if a particular gene is being expressed but Northern blots determine the homology between mRNA and a DNA probe.
   d. Southern blots determine the homology between mRNA and a DNA probe but Northern blots determine if a particular gene is being expressed.
   e. Southern and Northern blots are essentially the same technique performed in different hemispheres of the world.

(Continued)

7. What might be a use for fluorescence *in situ* hybridization (FISH)?
   a. For identification of a specific gene in a DNA extraction by hybridization to a DNA probe.
   b. For identification of a specific gene by hybridization to a DNA probe within live cells that have had their DNA denatured by heat.
   c. For identification of an mRNA within an RNA extraction by hybridization to a DNA probe.
   d. For identification of both mRNA and DNA in cellular extracts using an RNA probe.
   e. None of the above.

8. Which of the following are useful traits of cloning vectors?
   a. An antibiotic resistance gene on the plasmid for selection of cells containing the plasmid.
   b. A site that contains unique, clustered restriction enzyme sequences for cloning foreign DNA.
   c. A high copy number plasmid so that large amounts of DNA can be obtained.
   d. Alpha complementation to determine if the foreign DNA was inserted into the cloning site.
   e. All of the above are useful traits.

9. Which of the following vectors holds the largest pieces of DNA?
   a. plasmids
   b. bacteriophage
   c. YACs
   d. PACs
   e. cosmids

10. Besides a high voltage shock, what is another method to make *E. coli* competent to take up “naked” DNA?
   a. high concentrations of calcium ions followed by high temperature
   b. high concentrations of calcium ions and several hours on ice
   c. large amounts of DNA added directly to a bacterial culture growing at 37 °C
   d. high concentrations of minerals followed by high temperature
   e. A high voltage shock is the only way to make *E. coli* competent.

11. Why are gene libraries constructed?
   a. To find new genes.
   b. To sequence whole genomes.
   c. To compare genes to other organisms.
   d. To create a “bank” of all the genes in an organism.
   e. All of the above.

12. Which of the following statements about gene libraries is correct?
   a. Genes in a library can be compared to genes from other organisms by hybridization with a probe.
   b. A gene library is only necessary to maintain known genes.
   c. Every gene in the library must be sequenced first in order to compare genes in the library to genes from other organisms.
   d. Gene libraries are only created for eukaryotic organisms.
   e. Gene libraries can only be created in prokaryotes.

13. Why must reverse transcriptase be used to create a eukaryotic expression library?

- **a.** Reverse transcriptase is only used to create prokaryotic expression libraries.
- **b.** Reverse transcriptase creates cDNA from mRNA in prokaryotes.
- **c.** Reverse transcriptase ensures the gene is in the correct orientation within the expression vector to create protein.
- **d.** Reverse transcriptase creates cDNA from mRNA because genes in eukaryotes have large numbers of non-coding regions.
- **e.** No other enzymes are used to create expression libraries except restriction enzymes.

14. Which of the following are common features of expression vectors?

- **a.** Small segments of DNA that encode tags for protein purification.
- **b.** Transcriptional start and stop sites.
- **c.** A tightly controlled promoter than can only be induced under certain circumstances.
- **d.** Antibiotic resistance gene.
- **e.** All of the above are common features of expression vectors.

15. How is subtractive hybridization useful?

- **a.** To eliminate genes from a gene library.
- **b.** To create expression libraries based on genes that are currently being expressed.
- **c.** To identify and construct new probes for Southern and Northern hybridizations.
- **d.** To identify sets of genes that are only expressed under certain conditions.
- **e.** All of the above are useful traits of subtractive hybridization.

Further Reading

Clark DP (2005). *Molecular Biology: Understanding the Genetic Revolution*. Elsevier Academic Press, San Diego, CA.

CHAPTER 4

DNA Synthesis *in Vivo* and *in Vitro*

INTRODUCTION

Replication copies the entire set of genomic DNA, so that the cell can divide in two. During replication, the entire genome must be uncoiled and copied exactly. This elegant process occurs extremely fast in *E. coli*, where DNA polymerase copies about 1000 nucleotides per second. Although the process is slower in eukaryotes, DNA polymerase still copies 50 nucleotides per second. Many biotechnology applications use the principles and ideas behind replication; therefore, this chapter first introduces the basics of DNA replication as it occurs in the cell. We then review some of the most widely used techniques in genetic engineering and biotechnology that rely on DNA polymerase, including chemical synthesis of DNA, polymerase chain reaction, and DNA sequencing.

REPLICATION OF DNA

In order to maintain the integrity of an organism, the entire genome must be replicated identically. Even for gene creatures such as plasmids, viruses, or transposons, replication is critical for their survival. The complementary two-stranded structure of DNA is the key to understanding how it is duplicated during cell division. The double-stranded helix unwinds, and the hydrogen bonds holding the bases together melt apart to form two single strands. This Y-shaped region of DNA is the **replication fork** (Fig. 4.1). Replication starts at a specific site called an **origin of replication (ori)** on the chromosome. The origin is called *oriC* on the *E. coli* chromosome and covers about 245 base pairs of DNA. The origin has mostly AT base pairs, which require less energy to break than GC pairs.

Once the replication fork is established, a large assembly of enzymes and factors assembles to synthesize the complementary strands of DNA (see Fig. 4.1). **DNA polymerase** starts synthesizing the complementary strand on one side of the fork by adding complementary bases in a 5′ to 3′ direction. This strand is synthesized continuously because there is always a free 3′-OH group. This strand is called the **leading strand**. Because DNA polymerase synthesizes only in a 5′ to 3′ direction, the other strand, called the **lagging strand**, is synthesized as small fragments called **Okazaki fragments**. The lagging strand fragments are ligated together by an enzyme called **DNA ligase**. Ligase links the 3′-OH and the 5′-PO_4 of neighboring nucleotides, forming a phosphodiester bond. The final step is to add methyl (CH_3) groups along the new strand (see later discussion). The original double-stranded helix is now two identical double-stranded helices, each containing one strand from the original molecule and one new strand. This is why the process is called **semiconservative replication**.

In replication, DNA polymerase synthesizes the leading strand as one continuous piece, and the lagging strand as small Okazaki fragments. Each copy has one strand from the original helix and one new strand.

Uncoiling the DNA

Because DNA is condensed into supercoils in order to fit inside the cell, several different enzymes are needed to open and relax the DNA before replication can start (Fig. 4.2). **DNA helicase** and **DNA gyrase** attach near the replication fork and untwist the strands of DNA. DNA gyrase removes the supercoiling, and helicase unwinds the double helix by dissolving the hydrogen bonds between the paired bases. The two strands are kept apart by **single-stranded binding protein**, which coats the single-stranded regions. This prevents the two strands from reannealing, so that other enzymes can gain access to the origin and begin replication.

As DNA polymerase travels along the DNA, more positive supercoils are added ahead of the replication fork. Because the bacterial chromosome is negatively supercoiled, initially the new positive supercoils relax the DNA. After about 5% of the genome has been replicated, though, the positive supercoils begin to accumulate and need to be removed. DNA gyrase cancels the positive supercoils by adding negative supercoils. When circular chromosomes

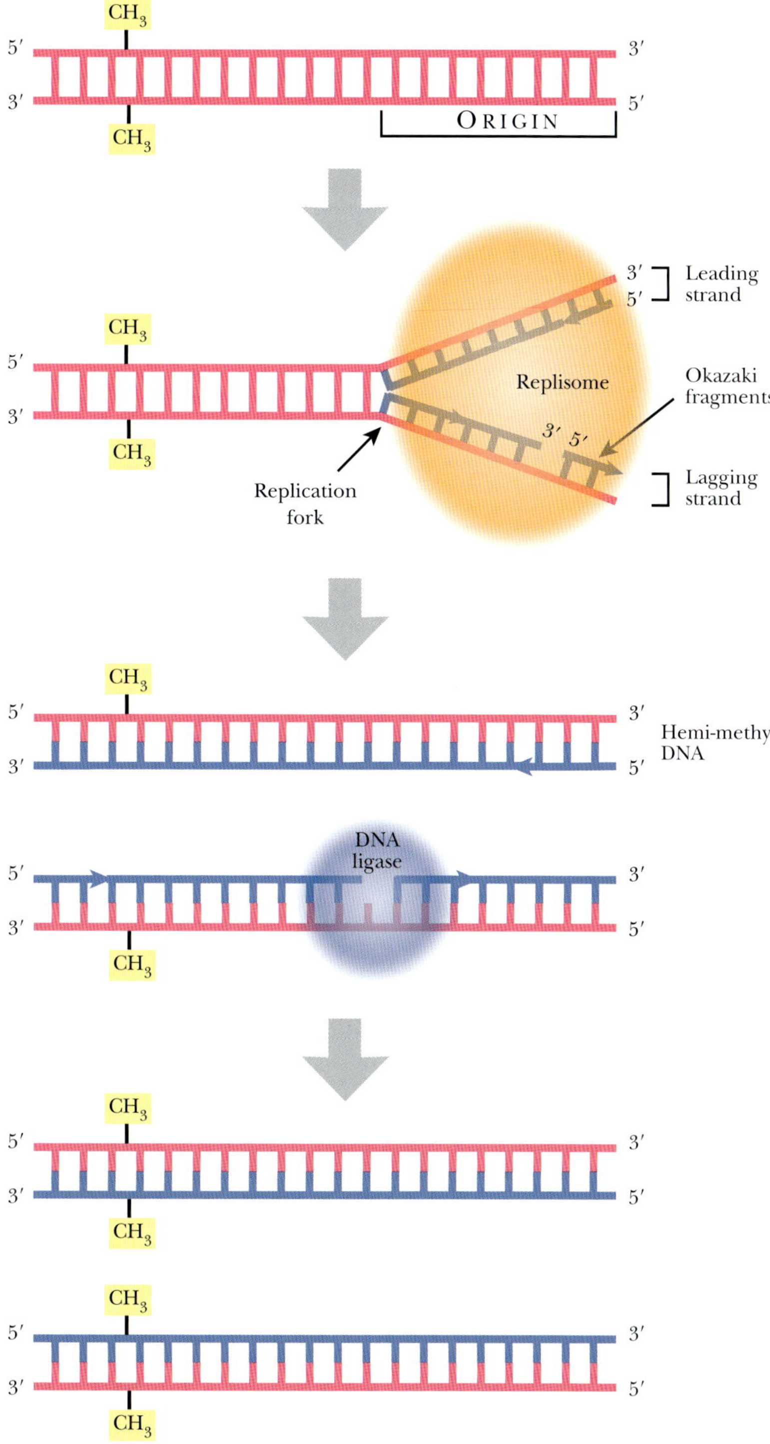

FIGURE 4.1
Replication
Replication enzymes open the double-helix around the origin to make it single-stranded. DNA polymerase adds complementary nucleotides to each side in a 5′ to 3′ direction; therefore, one strand is synthesized continuously (leading strand) and the other strand (lagging strand) is synthesized in short pieces called Okazaki fragments. DNA ligase seals any nicks or breaks in the phosphate backbone. Finally, methylases add methyl groups to the newly synthesized strands.

are replicated, the two daughter copies may become **catenated**, or connected like two links of a chain. Topoisomerase IV releases catenated daughter strands by introducing double-stranded nicks into one chromosome. The second copy can then pass through the first, giving two separated molecules.

DNA helicase, DNA gyrase, and topoisomerase IV untwist and untangle the supercoiled DNA during replication.

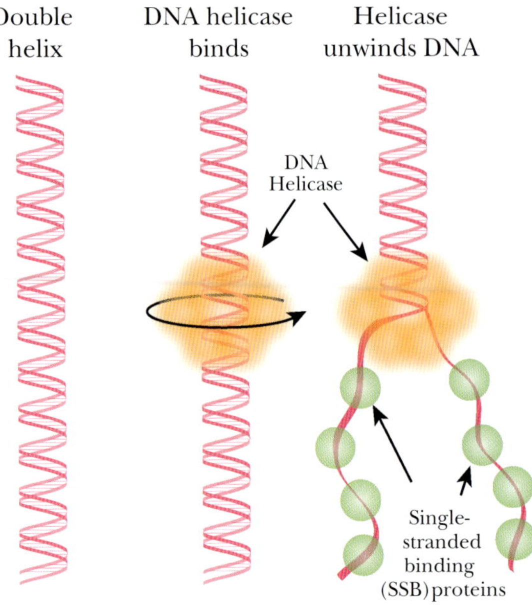

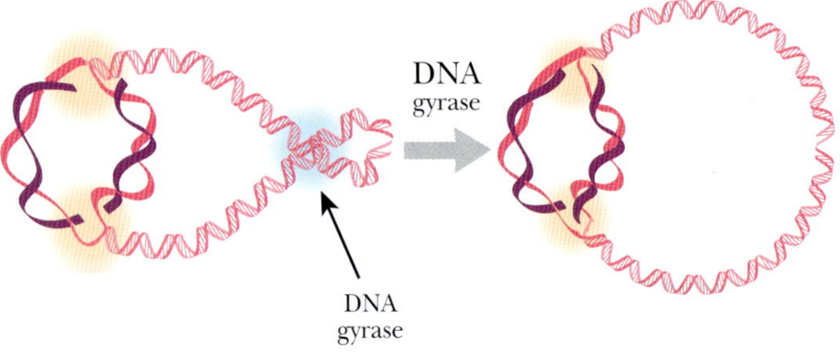

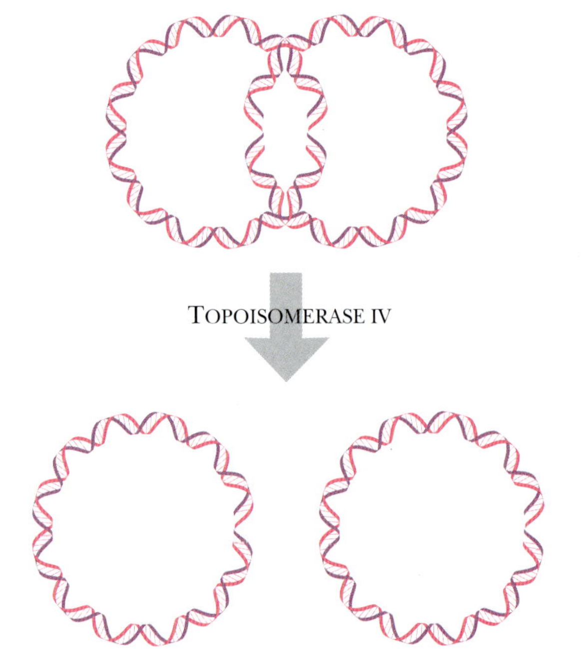

FIGURE 4.2
Untwisting DNA
(A) Opening the origin for replication. Helicase binds to the DNA and breaks the hydrogen bonds holding the two strands together. Then SSB binds to the free strands to keep them from reannealing. (B) Removing supercoils. For the replication enzymes to proceed along the entire chromosome, the supercoils must be removed by DNA gyrase. (C) Untangling chromosomes. Sometimes after replication of circular genomes is complete, the two rings are catenated or linked together like links in a chain. Topoisomerase IV untangles the two chromosomes so they can partition into the daughter cells.

Priming DNA Synthesis

Once the origin is opened up, the **replisome** is assembled there. The replisome contains the enzymes for synthesizing the leading and lagging strands. The replisome also contains helicase and SSB protein, which were described earlier. Unlike RNA polymerase, DNA polymerase needs a preexisting 3′-OH to add bases to. Consequently it cannot initiate new strands of nucleic acid but can only elongate. Therefore, an 11 to 12 base-pair length of RNA (an RNA primer) is made. DNA polymerase then makes DNA starting from this RNA primer. At the origin, a protein called **PriA** displaces the SSB proteins so a special RNA polymerase, called **primase** (DnaG), can enter and synthesize short RNA primers using ribonucleotides. Primase makes a single primer at the origin of the leading strand and makes multiple primers for the lagging strand. Two molecules of DNA polymerase III bind to the primers on the leading and lagging strands and synthesize new DNA from the 3′ hydroxyls (Fig. 4.3).

Primase, a special RNA polymerase, works with PriA to displace the SSB proteins and synthesize a short RNA primer at the origin. DNA polymerase then starts synthesis of the new DNA strand using the 3′-OH of the RNA primer. This occurs at multiple locations on the lagging strand.

Structure and Function of DNA Polymerase

DNA polymerase III (PolIII) is the major form of DNA polymerase used to replicate bacterial chromosomes and consists of multiple protein subunits (Fig. 4.4). The **sliding clamp** is a donut-shaped protein consisting of a dimer of DnaN protein. Two clamps encircle the two single strands of DNA at the replication fork. A cluster of accessory proteins, the **clamp loader complex**, loads the clamps onto DNA strands. The two sliding clamps bind two **core enzymes**, one for each strand of DNA. The core enzyme consists of three subunits, DnaE (α subunit), which links the nucleotides together, DnaQ (ε subunit), which proofreads the new strand, and HolE (θ subunit), whose role is uncertain. As the α subunit adds new nucleotides, the ε subunit recognizes any distortions and removes any mismatched bases. A correct nucleotide is then added. Bacterial DNA polymerase III can add up to 1000 bases per second, which is an extraordinarily fast rate of enzyme activity.

The multiple subunits of DNA polymerase III work together to synthesize a new strand of DNA. The core has two essential subunits: α subunit links the nucleotides, and ε subunit ensures that they are accurate.

Synthesizing the Lagging Strand

After the new lagging strand of DNA has been made, it has many segments of RNA derived from multiple RNA primers, as well as multiple breaks or **nicks** along the backbone that need to be sealed (Fig. 4.5). One model is that RNaseH removes the RNA primers from the lagging strand. Then DNA polymerase I fills in the regions of single-stranded DNA. Alternatively, DNA polymerase I can also excise the RNA primers itself by identifying the nicks in the DNA backbone. It then removes and replaces about 10 nucleotides downstream of each nick. Finally, the DNA fragments of the lagging strand are linked together with a ligation reaction by DNA ligase. DNA polymerase I and DNA ligase are both very important enzymes in molecular biology, and are used extensively in biotechnology.

Because the lagging strand is synthesized in small pieces, either DNA polymerase I or RNaseH must excise the many regions of RNA primer and replace the areas with DNA. DNA ligase must then close the nicks in the sugar/phosphate backbone of the new DNA strand.

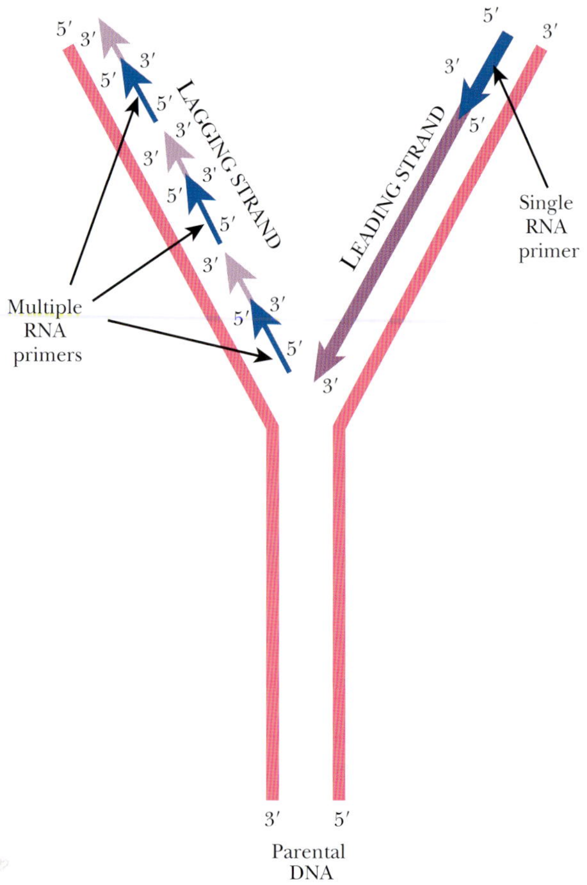

FIGURE 4.3 Strand Initiation Requires an RNA Primer
DNA polymerase cannot synthesize new DNA without a preexisting 3′-OH. Thus, DNA replication requires an RNA primer to initiate strand formation. One RNA primer is needed for the leading strand, and multiple primers are needed for the lagging strand to be synthesized.

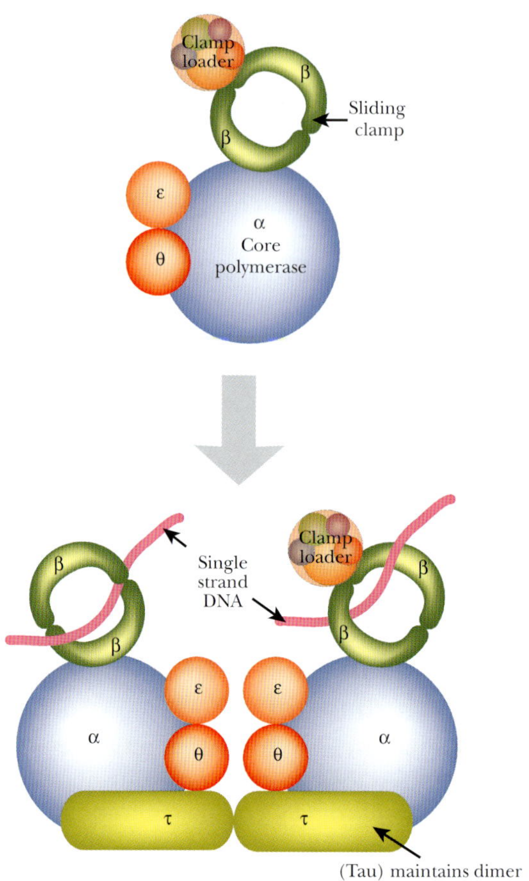

FIGURE 4.4 DNA Polymerase III—Assembly of Subunits
A single core subunit is shown in the upper part of the figure, and its assembly into a dimeric unit is shown in the lower part. The dimeric subunit contains only one clamp loader complex, which is associated with the lagging strand synthetic unit. The two core proteins are bound together by the tau subunit.

Repairing Mistakes after Replication

After replication is complete, the **mismatch repair system** corrects mistakes made by DNA polymerase. If the wrong base is inserted and DNA polymerase does not correct the error itself, there will be a small bulge in the helix at that location. Identifying which of the two bases is correct is critical. The cell assumes that the base on the new strand is wrong and the original parental base is correct. The mismatch repair system of *E. coli* (MutSHL) deciphers which strand is the original by monitoring methylation. Immediately after replication, the DNA is **hemimethylated**, that is, the old strand still has methyl groups attached to various bases but the new strand has not been methylated yet (see Fig. 4.1). Some methyl groups protect against restriction enzymes produced by the bacteria (see Chapter 3); others mark the parental strand of DNA. Two different *E. coli* enzymes add the latter type of methyl groups: **DNA adenine methylase (Dam)** adds a methyl group to the adenine in GATC, and **DNA cytosine methylase (Dcm)** adds a methyl group to the cytosine in CCAGG or CCTGG. These enzymes methylate the new strand after replication, but they are slow. This allows mismatch repair to find and fix any mistakes first.

Three genes of *E. coli* are responsible for mismatch repair, *mutS,mutL,* and *mutH* (Fig. 4.6). MutS protein recognizes the bulge or distortion in the sequence. MutH finds the nearest GATC site and nicks the nonmethylated strand—that is, the newly made strand. MutL holds the MutS plus mismatch and the MutH plus GATC site together (these may be far apart on the DNA helix). Finally, the DNA on the new strand is degraded and replaced with the correct sequence by DNA polymerase III.

In *E. coli*, mismatch repair proteins (MutSHL) identify a mistake in replication, excise the new nucleotides around the mistake, and recruit DNA polymerase III to the single-stranded region to make the new strand without a mistake.

COMPARING REPLICATION IN GENE CREATURES, PROKARYOTES, AND EUKARYOTES

Although the basic mechanism for replication is the same for most organisms, the timing, direction, and sites for initiation and termination are variable. The major differences in replication occur mainly because of the special challenges posed by circular and linear genomes. Normal DNA replication occurs bidirectionally in prokaryotes and eukaryotes, whether the genome is linear or circular. Two replication forks travel in opposite directions, unwinding the DNA helix as they go. In bacteria such as *E. coli*, there is only one origin, *oriC*, and replication occurs in both directions around the circular chromosome until it meets at the other side, the terminus, *terC*. Halfway through this process, the chromosome looks like the Greek letter θ; therefore, this process is often called **theta-replication** (Fig. 4.7). The single circular chromosome then becomes two. Theta replication is also used by many plasmids, such as the F plasmid of *E. coli*, when growing and dividing asexually (as opposed to transferring its genome to another cell via conjugation).

Some plasmids and many viruses replicate their genomes by a process called **rolling circle replication** (Fig. 4.8). At the origin of replication, one strand of the DNA is nicked and unrolled. The intact strand thus rolls relative to its partner (hence "rolling circle"). DNA is synthesized from the origin using the circular strand as a template. As DNA polymerase circles the template strand, the new strand of DNA is base-paired to the circular template. Meanwhile the other parental strand is dangling free. This dangling strand is removed, ligated to form another circle, and finally a second strand is synthesized. This results in two rings of plasmid or viral DNA, each with one strand from the original molecule and one newly synthesized strand.

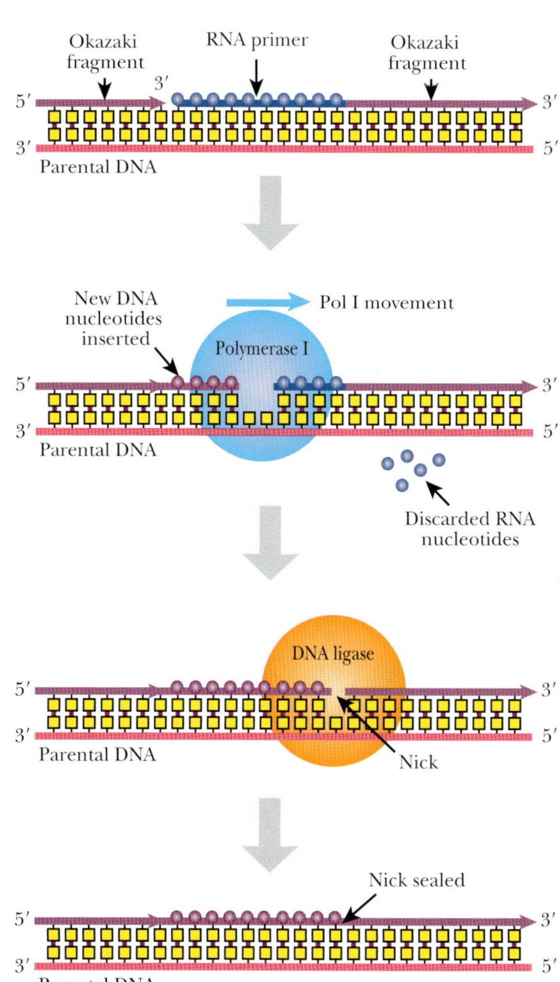

FIGURE 4.5
Joining the Okazaki Fragments
When first made, the lagging strand is composed of alternating Okazaki fragments and RNA primers. Next, DNA polymerase I binds to the primer region, and as it moves forward, it degrades the RNA and replaces it with DNA. Finally, DNA ligase seals the nick in the phosphate backbone.

Some viral genomes use rolling circle replication, but continue to make more and more copies of the original circular template. They continue rolling around the circle, synthesizing more and more copies that are all dangling as a long single strand. The long strand of new DNA may be made double-stranded or left single-stranded (depending on the type of virus). Finally the dangling strand is chopped into genome-sized units and packaged into viral particles. Some viruses circularize these copies before packaging; others simply leave the genomes linear.

Long linear DNA molecules such as human chromosomes pose several problems for replication. The ends pose a particularly difficult problem because the RNA primer is synthesized at the very end of the chromosome. When the RNA primer is removed by an exonuclease (MF1), there is no upstream 3′-OH for addition of new nucleotides to fill the gap. (In eukaryotes, there is no equivalent to the dual-function DNA polymerase I. A separate

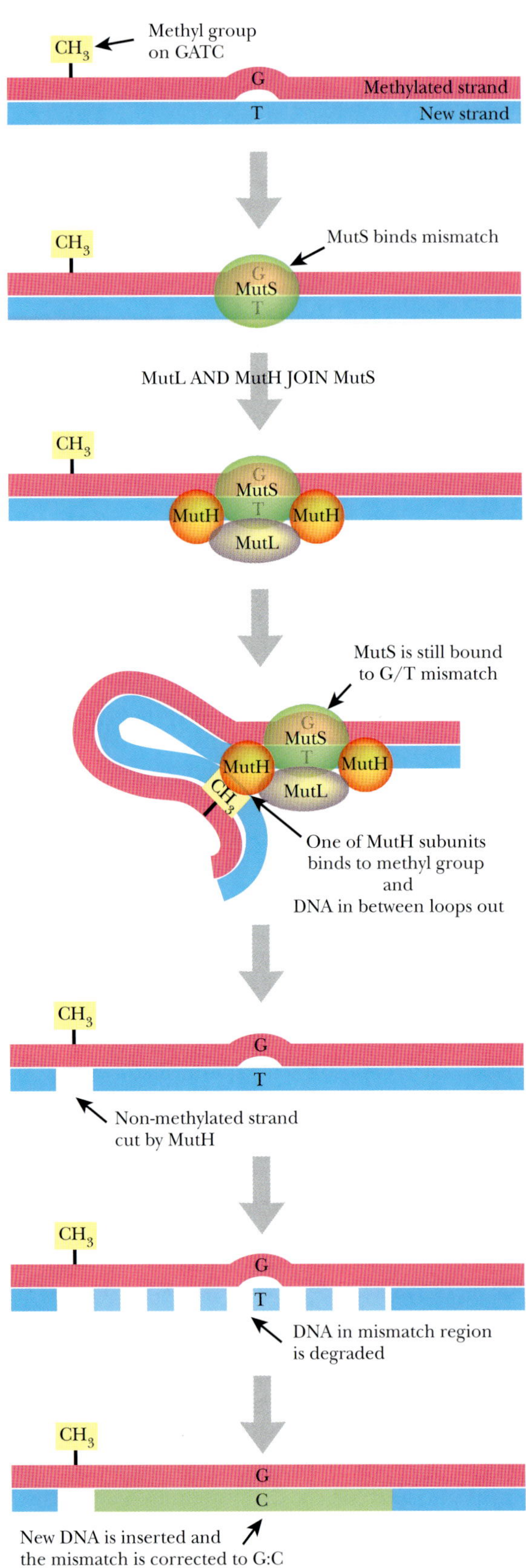

FIGURE 4.6 Mismatch Repair Occurs after Replication

MutS recognizes a mismatch shortly after DNA replication. MutS recruits MutL and two MutH proteins to the mismatch. MutH locates the nearest GATC of the new strand by identifying the methyl group attached to the "mother" strand. MutH cleaves the nonmethylated strand and the DNA between the cut and the mismatch is degraded. The region is replaced and the mismatch is corrected.

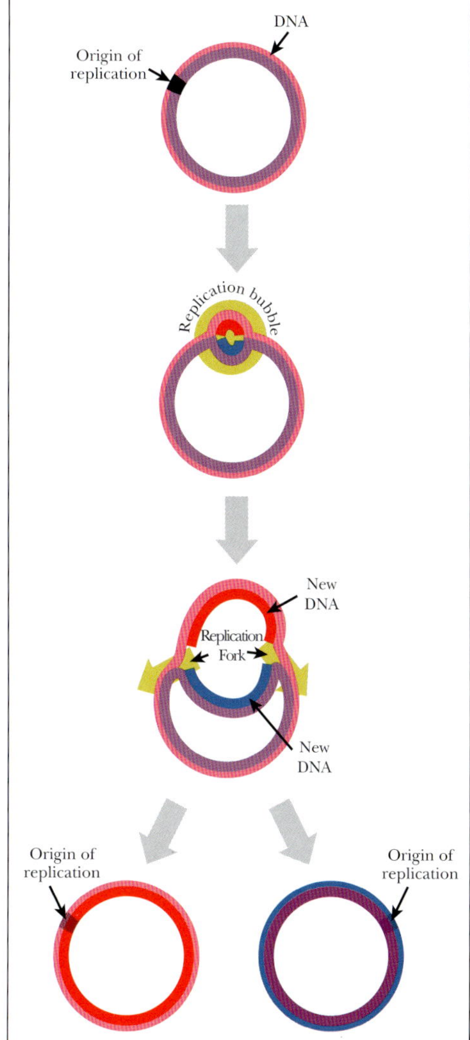

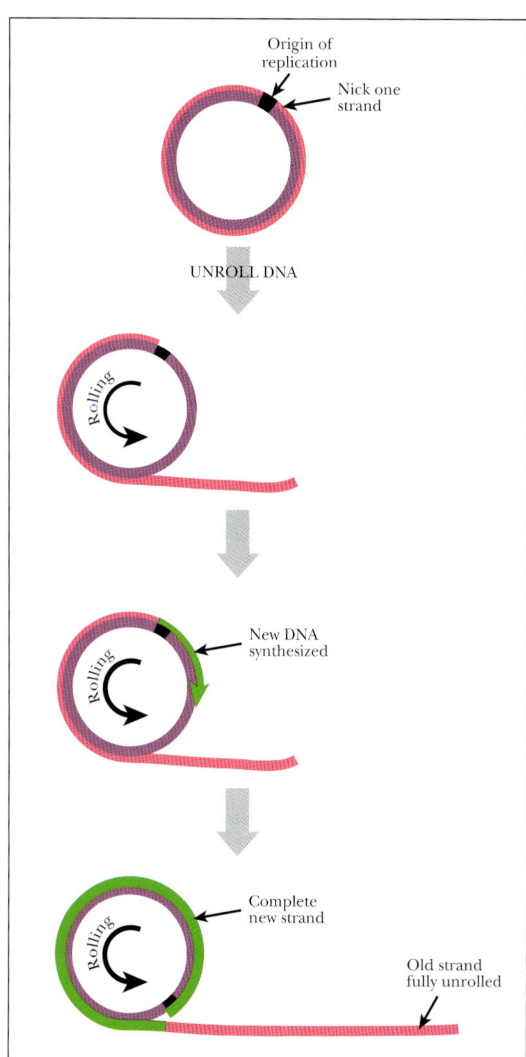

FIGURE 4.7 (left) Theta Replication
In circular genomes or plasmids, replication enzymes recognize the origin of replication, unwind the DNA, and start synthesis of two new strands of DNA, one in each direction. The net result is a replication bubble that makes the chromosome or plasmid look similar to the Greek letter theta (θ). The two replication forks keep moving around the circle until they meet on the opposite side.

FIGURE 4.8 (right) Rolling Circle Replication
During rolling circle replication, one strand of the plasmid or viral DNA is nicked, and the broken strand (pink) separates from the circular strand (purple). The gap left by the separation is filled in with new DNA starting at the origin of replication (green strand). The newly synthesized DNA keeps displacing the linear strand until the circular strand is completely replicated. The linear single-stranded piece is fully "unrolled" in the process.

exonuclease, MF1, removes the RNA primers, and DNA polymerase δ fills in the gaps.) Over successive rounds of replication, the ends of linear chromosomes get shorter and shorter. Special structures called **telomeres** are found at the tips of each linear chromosome and prevent chromosome shortening from affecting important genes. Telomeres have multiple tandem repeats of a short sequence (TTAGGG in humans). The enzyme **telomerase** can regenerate the telomere by using an RNA template to synthesize new repeats. This only happens in some cells; in others, the telomeres shorten every time the cell replicates its DNA. One theory regards telomere shortening as a molecular clock, aging the cell, and eventually triggering suicide (see Chapter 20).

The length of linear chromosomes also poses a problem. The time it takes to synthesize an entire human chromosome would be too long if replication began at only one origin. To solve this issue, there are multiple origins, each initiating new strands in both directions. These are elongated until they meet the new strands from the other direction.

The cellular structure of eukaryotes also poses some problems for replication. (In bacteria, the chromosome simply replicates, the two copies move to each end of the cell, and a new wall forms in the middle. There are no nuclear membranes or organelles to divide; there is just one chromosome plus, perhaps, some plasmids.) In eukaryotes, the cell has a specific **cell cycle**, with four different phases and replication occurs at specific points (Fig. 4.9). During G_1, the

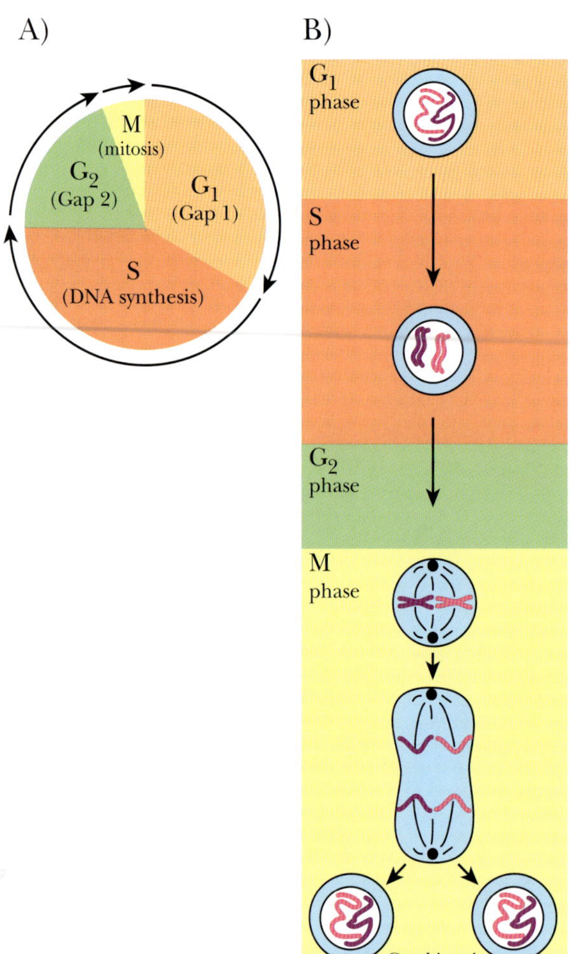

cell rests for a period before DNA synthesis begins. This period varies, lasting about 25 minutes for yeast. The next phase is S or synthesis, during which the entire genome is replicated. This is usually the longest phase, lasting about 40 minutes in yeast. The third phase is G_2 and is another resting phase before the cell undergoes mitosis, in the M phase. During mitosis, cells divide their walls and membranes into two separate cells, partitioning the new chromosomes and other cellular components into each half. The signal that triggers cell division depends on many factors, including environment, size, and age. Most of these are still not understood.

Eukaryotic mitosis is a dynamic process with much movement and repositioning of cellular components. First, the nuclear membrane must be dissolved before the chromosomes can separate. After replication, the two sets of chromosomes are partitioned to separate sides of the cell. The chromosomes attach to long fibers making up the **spindle** via special sequences called **centromeres**. They slide along the spindle fibers until they reach separate ends of the cell. A new cell membrane separating the two halves is then synthesized. Other cellular components including mitochondria, endoplasmic reticulum, lysosomes, and so forth are split between the two daughter cells. Finally, a new nuclear membrane must be assembled around the chromosomes of each new daughter cell. The dynamics of this process are still being investigated, and new proteins and molecules are still being discovered that mediate different parts of mitosis in eukaryotic cells.

FIGURE 4.9
Eukaryotic Cell Cycle
DNA replication occurs during the S phase of the cell cycle but the chromosomes are actually separated later, during mitosis or M phase. The S and M phases are separated by G_1 and G_2.

Bacteria and viruses use either theta replication or rolling circle replication to create new genomes. Eukaryotic cells have chromosomes with multiple origins of replication. Telomeres protect the ends of the chromosomes because each round of replication shortens the DNA. Replication in eukaryotes only occurs at a specific point during the cell cycle.

IN VITRO DNA SYNTHESIS

Making DNA in the laboratory relies on the same basic principles outlined for replication (Fig. 4.10). DNA replication needs the following "reagents": enzymes to open up the double helix, an RNA primer with a 3′-hydroxyl for DNA polymerase to extend, and a pool of nucleotide precursors, plus DNA polymerase to catalyze the addition of new nucleotides.

To perform DNA replication in the laboratory, a few modifications are made. First, the enzymes that open and unwind the template DNA are not used. Instead, double-stranded DNA is converted to single-stranded DNA using heat or a strong base to disrupt the hydrogen bonds that hold the two strands together. Alternatively, template DNA can be made by using a virus that packages its DNA in single-stranded form. For example, M13 is a bacteriophage that infects *E. coli*, amplifies its genome using rolling circle replication, and packages the single-stranded DNA in viral particles that are released without lysing open the *E. coli* cell. If template DNA is cloned into the M13 genome, then the template will also be manufactured as in a single-stranded form. This DNA can be isolated directly from the viral particles.

In laboratory synthesis of DNA, an RNA primer is not used because RNA is very unstable and degrades easily. Instead, a short single-stranded oligonucleotide of DNA is used as a primer.

(As long as the primer has a free 3′-hydroxyl, DNA polymerase will extend either RNA or DNA.) The primers are synthesized chemically (see later discussion) and mixed with the single-stranded template DNA. The oligonucleotide primer has a sequence complementary to a short region on the DNA template. Therefore, at least some sequence information must be known for the template. If the sequence of the template DNA is unknown, it may be cloned into a vector, and the primer is then designed to match sequences of the vector (such as the polylinker region) that are close to the inserted DNA.

Finally, purified DNA polymerase plus a pool of nucleotides (dATP, dCTP, dGTP, and dTTP) is added to the primer and template. The primer anneals to its complementary sequence, and DNA polymerase elongates the primer, creating a new strand of DNA complementary to the template DNA.

In vitro replication requires a single-stranded piece of template DNA, a primer, nucleotide precursors, and DNA polymerase.

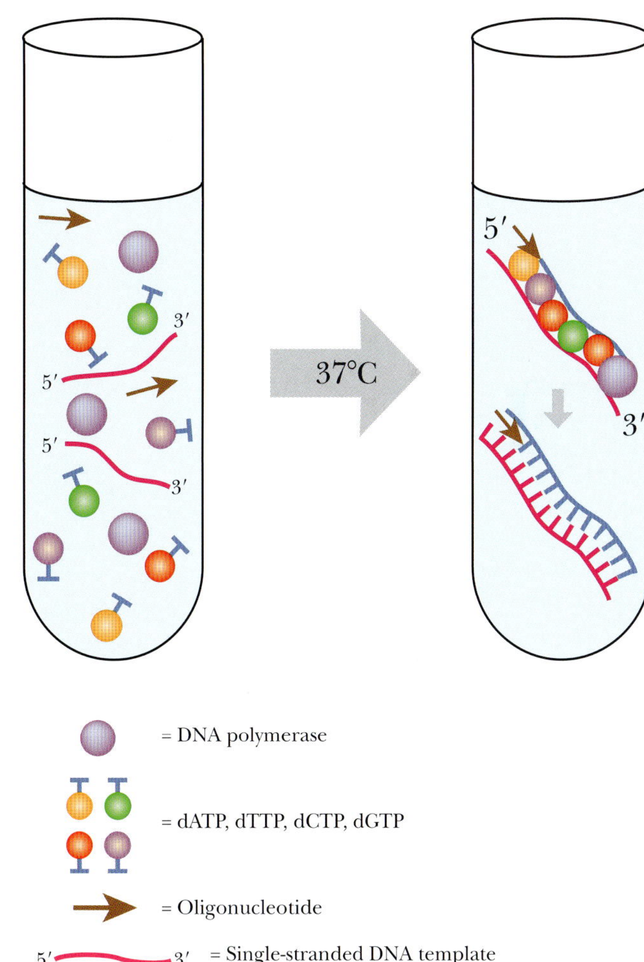

FIGURE 4.10 *In Vitro* DNA Synthesis
DNA synthesis in the laboratory uses single-stranded template DNA, plus DNA polymerase, an oligonucleotide primer, and nucleotide precursors. After incubating all the components at the appropriate temperature, double-stranded DNA is made.

CHEMICAL SYNTHESIS OF DNA

Making DNA chemically rather than biologically was one of the first new technologies to be applied by the biotechnology industry. The ability to make short synthetic stretches of DNA is crucial to using DNA replication in laboratory techniques. DNA polymerase cannot synthesize DNA without a free 3′-OH end to elongate. Therefore, to use DNA polymerase *in vitro*, the researcher must supply a short primer. Such primers are used to sequence DNA (see later discussion), to amplify DNA with PCR (see later discussion), and even to find genes in library screening (see Chapter 3). So a short review of how primers are synthesized is included here.

Research into chemical synthesis of DNA began shortly after Watson and Crick published their research on the crystal structure of DNA. H. Gobind Khorana at the University of Chicago was an early pioneer in the study of **oligonucleotide** synthesis. Technically, oligonucleotides are any piece of DNA less than 20 nucleotides in length, but today, *oligonucleotide* denotes a short piece of DNA that is chemically synthesized. In 1970, Khorana's lab synthesized an active tRNA molecule of 72 nucleotides (Agarwal *et al.*, 1970). The chemistry he used was inefficient and cumbersome, but some of his ideas are still used in current oligonucleotide synthesis. Today chemical synthesis is done with an automated **DNA synthesizer** that creates DNA by sequentially adding one nucleotide after another in the sequence specified.

Unlike *in vivo* DNA synthesis, artificial synthesis is done in the 3′ to 5′ direction. The first step is attaching the first nucleotide to a porous glass bead made of **controlled pore glass (CPG)**. The first nucleotide is not attached directly, but linked to the bead via a spacer molecule that bonds to the 3′-OH of the nucleotide (Fig. 4.11). Many beads are held in a column so that reagents can be washed through and removed easily. (Using CPG is one improvement over

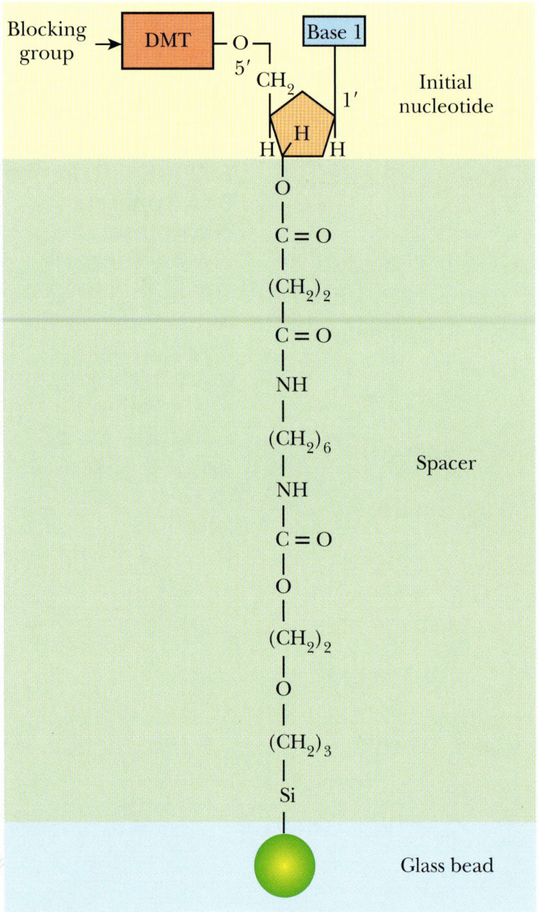

FIGURE 4.11
Addition of a Spacer Molecule and First Base to the CPG
The first nucleotide is linked to a glass bead via a spacer molecule attached to its 3′-OH group.

Khorana's technology. He used polymer beads to couple the reaction, but found that the polymer swelled as the reagents passed through the column, which inhibited synthesis. CPG is superior because it does not swell.)

When linking the spacer to the nucleotide 3′-OH, a chemical blocking group is attached to the 5′-OH. Thus the 3′-OH is the only available reactive group. Khorana's early synthesis was revolutionary in this respect because he chose the **dimethyloxytrityl (DMT) group**, which is still used as a blocking group in today's synthesizers. DMT has a strong orange color and is easily removed from the 5′-OH so that another nucleotide can be linked to the first. In practice, the CPG–spacer–first nucleotide is washed, and then the DMT group is removed by mild acid such as trichloroacetic acid (TCA). The 5′-OH is now ready to accept the next nucleotide. The efficiency of removing DMT is critical. If DMT is not removed completely, many of the potential oligonucleotides will fail to elongate. The orange color reveals the efficiency of removal and is easily measured optically.

Each nucleotide is added as its **phosphoramidite**, which consists of a blocking group protecting a 3′-phosphite group (Fig. 4.12). (One problem with early oligonucleotide synthesis technology was branching. Rather than the incoming nucleotide adding to the 5′ end, it sometimes attached to the phosphate linking two nucleotides.) Nowadays every added nucleotide has a di-isopropylamine group attached to a methylated 3′ phosphite group. This structure protects against unwanted branching. It is also stable and can be stored for long periods. Before adding another nucleoside, the 3′ phosphite group is activated by tetrazole. The next nucleotide is then added, and reacts with the phosphite to form a dinucleotide (Fig. 4.13).

If the terminal nucleotide of a growing chain fails to react with an incoming nucleotide, the chain must be capped off to prevent generation of an incorrect sequence by later reactions. The 5′-OH of all unreacted nucleotides is acetylated with acetic anhydride and dimethylaminopyridine. This terminates the chain, so that no other nucleoside phosphoramidites can be added.

Now the column has CPG–spacer–first nucleoside–phosphite–second nucleoside–DMT. Phosphites are used because they react much faster, but they are unstable. Adding iodine oxidizes the phosphite triester into the normal phosphodiester, which is more stable under acidic conditions (Fig. 4.14).

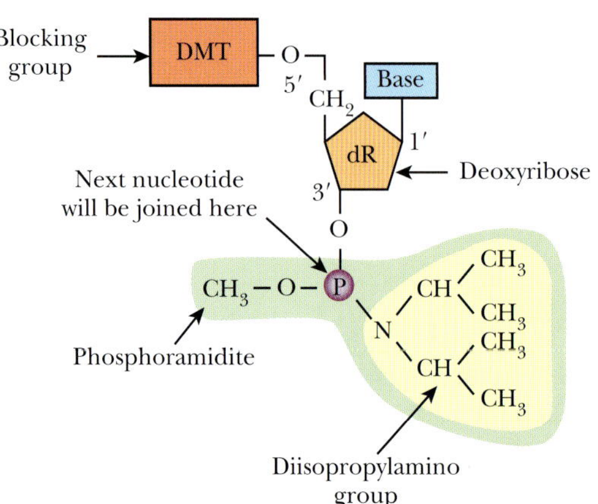

FIGURE 4.12
Nucleoside Phosphoramidites Are Used for Chemical Synthesis of DNA
Nucleotides are modified to ensure that the correct group reacts with the next reagent. Each nucleotide has a DMT group blocking its 5′-OH. The 3′-OH is activated by a phosphoramidite group, which is originally also protected by di-isopropylamine.

The column can now be prepared to add the third nucleotide. The DMT is removed with TCA, and the third phosphoramidite is added. The chains are capped so that any dinucleotides that didn't react with the third phosphoramidite are prevented from adding any more nucleosides. Finally, the phosphite

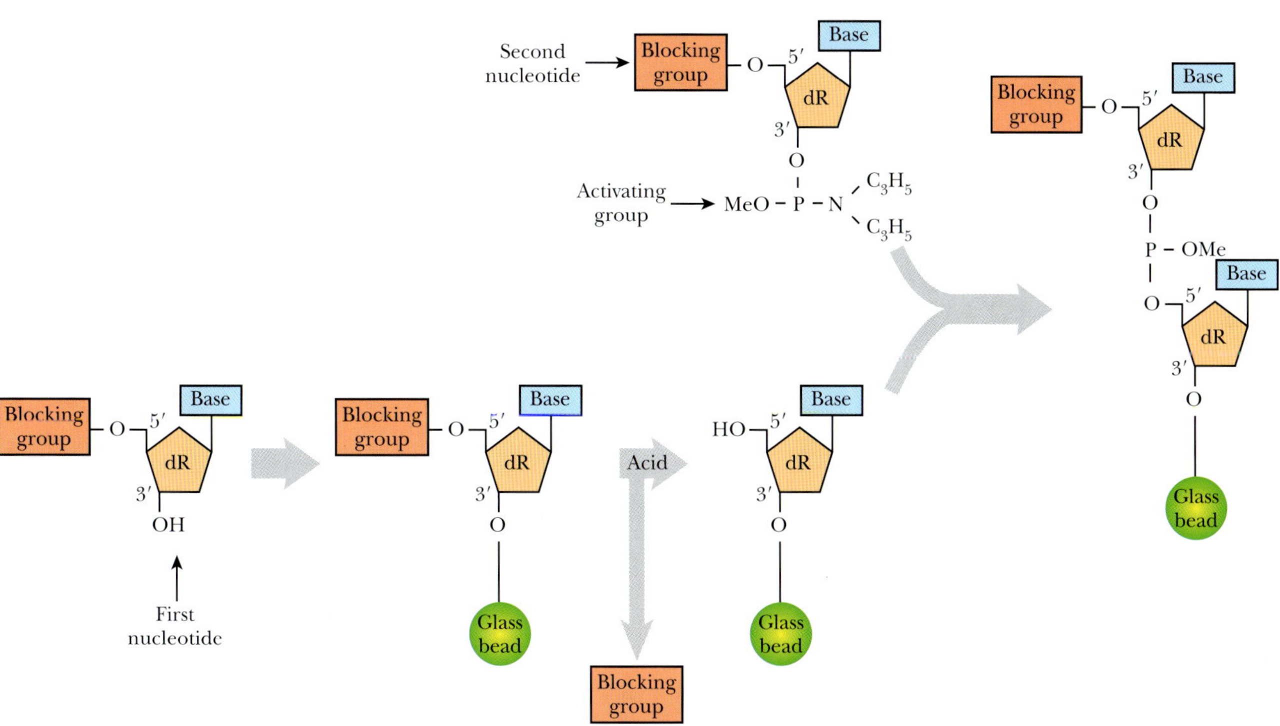

FIGURE 4.13 Adding the Second Nucleotide

During chemical synthesis of DNA, nucleotides are added in a 3′ to 5′ direction (the opposite of *in vivo* DNA synthesis). Therefore, the 3′-OH of an incoming nucleotide must be activated, but the 5′-OH must be blocked (see top nucleotide). For nucleotides already attached to the bead, the opposite must be done. Here the blocking group on the 5′-OH of nucleotide 1 is removed by treatment with a mild acid. When the second nucleotide is added, it reacts to form a dinucleotide.

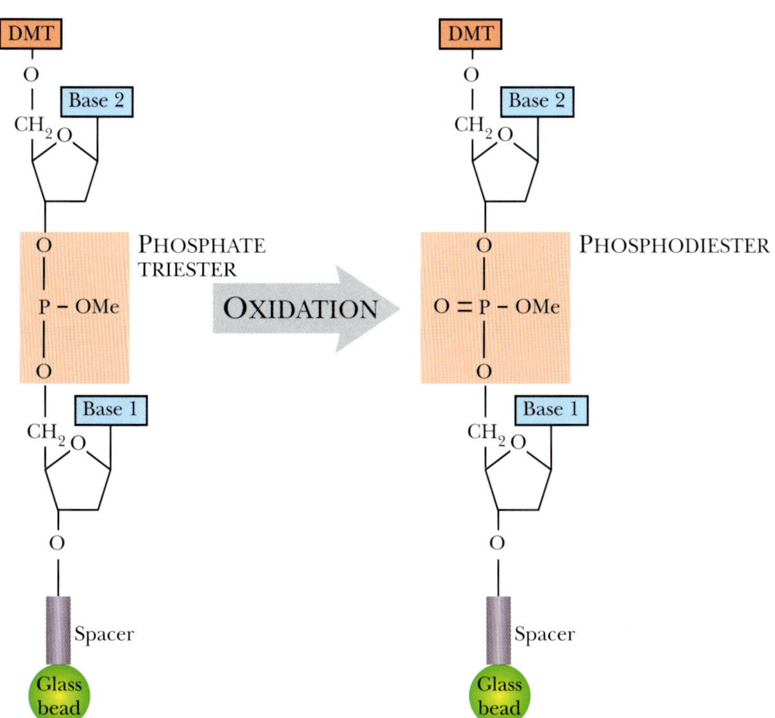

FIGURE 4.14 Oxidation Converts Phosphite Triester into a Phosphodiester

The phosphite triester is oxidized to a phosphodiester by adding iodine. This stabilizes the dinucleotide for further additions.

triester is oxidized to phosphodiester. This process continually repeats until all the desired nucleotides are added and the final oligonucleotide has the correct sequence (Fig. 4.15).

After the final phosphoramidite nucleoside is added, the oligonucleotide still has DMT protecting the 5′-OH, methyl groups attached to the phosphates, and amino-protecting groups on the bases. (Amino groups would react with the reagents during synthesis; therefore, chemical groups are added to protect the bases before they are added to the column.) All three types of protective groups must be removed before the oligonucleotides are released from the column. Finally, to make the oligonucleotides biologically active, the 5′-OH must be phosphorylated. Usually, kinase from bacteriophage T4 is used to transfer a phosphate group from ATP to the 5′ end of the oligonucleotides. The newly synthesized oligonucleotide is now ready for use.

Chemical synthesis of DNA occurs by successively adding nucleosides to the previous base attached to beads. As each base is added, the blocking groups must be modified to accept the next base. The final step of synthesis is adding a phosphate group to the 5′ end.

Box 4.1 Khorana, Nirenberg, and Holley

Har Gobind Khorana, Marshall W. Nirenberg, and Robert W. Holley are pioneers in the field of molecular biology. The three scientists received the Nobel Prize in Physiology or Medicine in 1968 for their combined efforts in identifying which triplet codons coded for which amino acid. Khorana originally began chemical synthesis of DNA in order to help elucidate the role of different enzymes. He wanted to understand the mode of action for nucleases and phosphodiesterases, but without being able to chemically synthesize a defined nucleic acid, the work on enzymes would be very difficult.

Khorana's lab determined ways to synthesize dinucleotide, trinucleotide, and tetranucleotide sequences using chemical synthesis. Rather than using single nucleotide additions, his lab focused on synthesizing nucleotides in blocks. His ability to chemically synthesize blocks of DNA was the backbone experiment, but many other discoveries were instrumental in determining the amino acid codes.

Matthaei and Nirenberg (1961) experimentally determined that polyuridylate (poly(U)) mixed with a bacterial cell-free amino acid incorporating system created polyphenylalanine. This experiment determined that the codon UUU encoded for the amino acid phenylalanine. During this time, Robert Holley was working on tRNA. He specifically identified the structure of the tRNA for alanine by purifying tRNA-alanine from yeast, fragmenting the tRNA into pieces with nucleases, and logically piecing together the size of the fragments and the sites at which the enzymes were recognized. Other important discoveries included the purification of DNA polymerase and RNA polymerase.

These experiments were woven into an elegant method of determining which triplet nucleotide sequence encoded which amino acid. First, Khorana's groups began synthesizing dinucleotide, trinucleotide, and tetranucleotide double-stranded DNA fragments. For example, one of these fragments had the following structure:

5' TCTCTC 3'
3' AGAGAG 5'

Arthur Kornberg had previously won the Nobel Prize for his discovery and purification of DNA polymerase I. Khorana's group mixed their short synthesized DNA with pure DNA polymerase to create long polydeoxynucleotides with a known sequence. Next, the DNA pieces were mixed with RNA polymerase to create long polyribonucleotides of known sequence. These were mixed with the cell-free system devised by Matthaei and Nirenberg, which made polypeptides. The dinucleotide example just shown resulted in a polypeptide of repeating serine and leucine. The experiment demonstrated that TCT or CTC encoded serine or leucine. There was no way to determine definitively which codon matched which amino acid, so more experiments were needed.

The final important contributions to make the final assignments were using purified tRNAs labeled with ^{14}C. Nirenberg and Leder (1964) mixed Khorana's synthetic trinucleotides and mixed them with the labeled tRNAs and ribosomes. (Note: The isolation of pure tRNA was not possible without Robert W. Holley's work.) They looked for binding of the labeled tRNA to the trinucleotide sequence. These experiments provided clear answers to many of the trinucleotide sequences, but many times the results were not very clear. It was the combination of these experiments with Khorana's work that determined the direct genetic code.

CHEMICAL SYNTHESIS OF COMPLETE GENES

As mentioned earlier, at each nucleoside addition in chemical synthesis, a proportion of oligonucleotides do not react with the next base, and these are capped with an acetyl group. The efficiency for nucleoside addition is critical, because if each step has low efficiency, the number of full-length oligonucleotides will decrease exponentially. For example, if the efficiency is 50% at each round, only half of the oligonucleotides add the second base, one-fourth would add the third base, one-eighth would get four bases; one-sixteenth would get the fifth base, and so on. Even if the final product were merely 10 bases in length, poor coupling would yield minuscule amounts of full-length product. It is critical for DNA synthesizers to have about 98% efficiency in each round, and then truncated products are the minority of the final sample. With high efficiencies, it is possible to synthesize longer segments of DNA. At 98% efficiency, an oligonucleotide that is 100 nucleotides long would give about 10% final yield. If the desired oligonucleotide is separated from the truncated products by electrophoresis (see Chapter 3), it is possible to get plenty of full-length products.

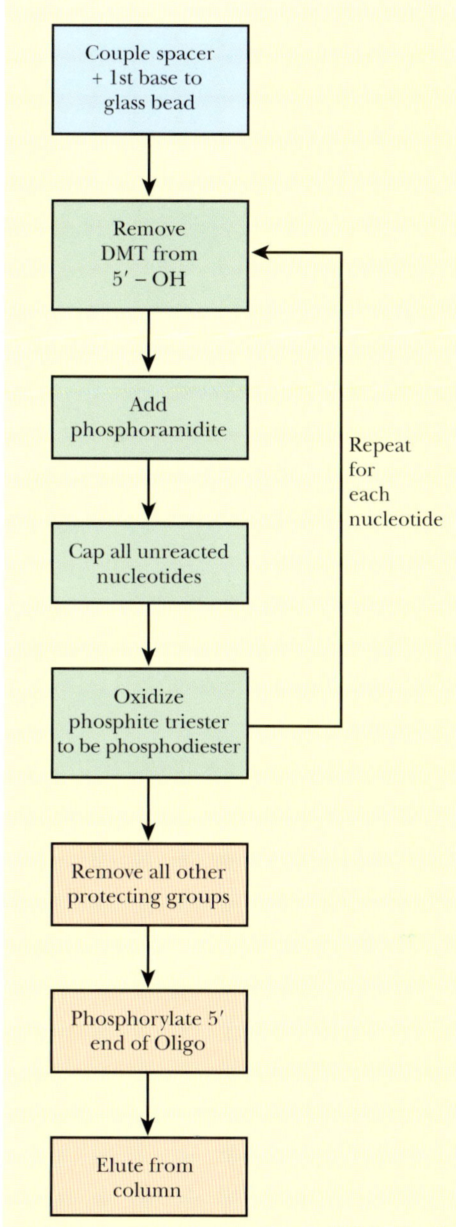

FIGURE 4.15
Flow Chart of Oligonucleotide Synthesis
Oligonucleotide synthesis has many steps that are repeated. The first nucleotide is coupled to a bead with a spacer molecule. Next the 5′-DMT is removed, and activated phosphoramidite nucleotide is added to the 5′ end of the first nucleotide. All the first nucleotides that were not linked to a second nucleotide are capped to prevent any further extension. Next, the phosphite triester is converted to a phosphodiester. These steps (in green) are repeated for the entire length of the oligonucleotide. Once the oligonucleotide has the appropriate length, the steps in tan are performed on the entire molecule.

Complete genes can be synthesized by linking smaller oligonucleotides together (Fig. 4.16). If the complete sequence of a gene is known, then long oligonucleotides can be synthesized identical to that sequence. The efficiency of the DNA synthesizer usually limits the length of each segment to about 100 bases; therefore, the gene segments are made with overlapping ends. Because oligonucleotides are single-stranded, both strands of the gene must be synthesized and annealed to each other, and then the segments are linked using ligase. Another strategy for assembly is to create strands that overlap only partially, and then use DNA polymerase I to fill in the large single-stranded gaps.

DNA can be synthesized in long segments provided each base is added efficiently. These long segments can be linked into one complete gene.

IN VITRO SYNTHESIS OF DNA CAN DETERMINE THE SEQUENCE OF BASES

Being able to quickly and easily determine the sequence of any gene has been the driving force for the recent advances made in biotechnology. Before sequencing became commonplace, identifying the gene responsible for a particular trait or disease was challenging. Frederick Sanger developed a method for sequencing a gene *in vitro* in 1974. He was interested in the amino acid sequence of insulin and decided to deduce the

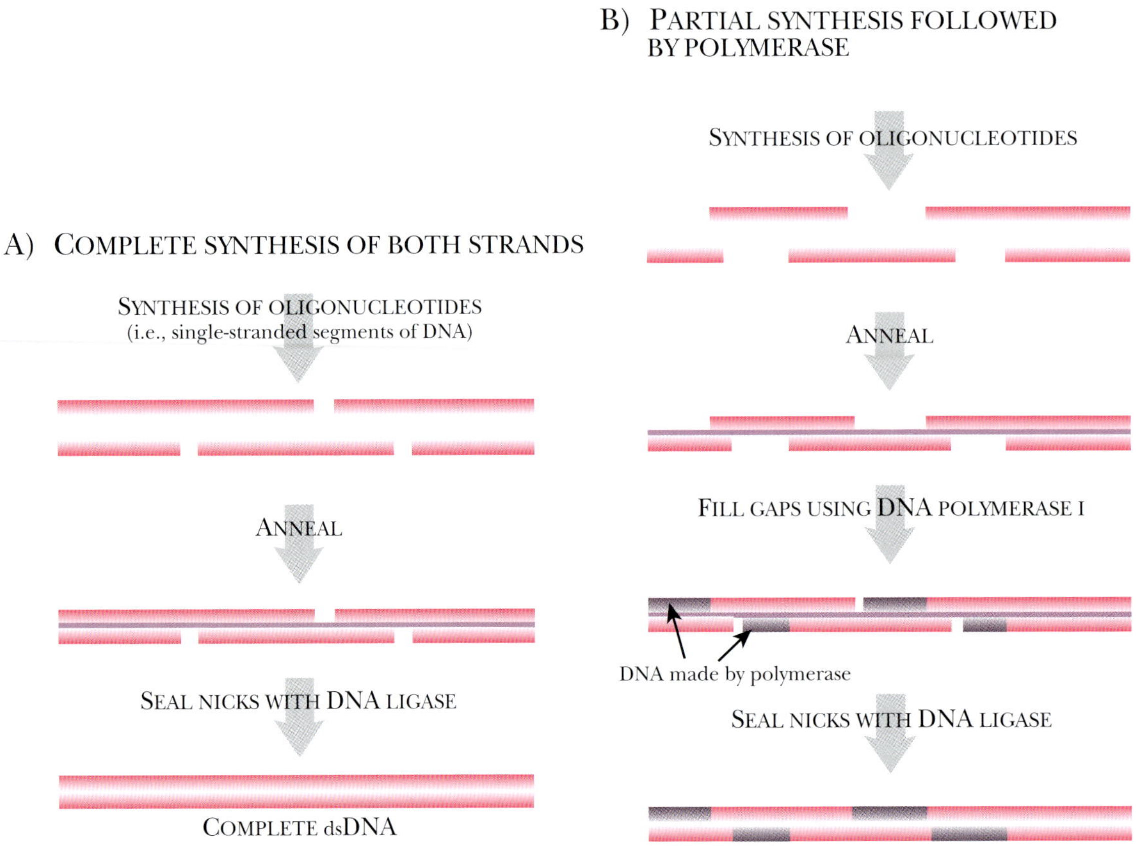

FIGURE 4.16 Synthesis and Assembly of a Gene

(A) Complete synthesis of both strands. Small genes can be chemically synthesized by making overlapping oligonucleotides. The complete sequence of the gene, both coding and noncoding strands, is made from small oligonucleotides that anneal to each other, forming a double-stranded piece of DNA with nicks along the phosphate backbone. The nicks are then sealed by DNA ligase. (B) Partial synthesis followed by polymerase. To manufacture larger stretches of DNA, oligonucleotides are synthesized so that a small portion of each oligonucleotide overlaps with the next. The entire sequence is manufactured, but gaps exist in both the coding and noncoding strands. These gaps are filled using DNA polymerase I and the remaining nicks are sealed with DNA ligase.

sequence of the protein from the nucleotide sequence. He invented the **chain termination sequencing** method, whose basic principles are still used today (Fig. 4.17). Much like DNA replication, chain termination sequencing requires a primer, DNA polymerase, a single-stranded DNA template, and deoxynucleotides. During *in vitro* sequencing reactions, these components are mixed and DNA polymerase makes many copies of the original template. The trick needed to deduce the sequence is to stop synthesis of some DNA chains at each base pair. Consequently, the fragments generated will differ in size by one base pair. These are then separated by gel electrophoresis. If the final base pair for each fragment is known, the sequence may be directly read from the gel (reading from bottom to top).

But how do we know what the final base is for each fragment on the sequencing ladder? DNA polymerase synthesizes a new strand of DNA based on the template sequence. The chain consists of deoxynucleotides, each with a hydroxyl group at the 3′ position on the deoxyribose ring. DNA polymerase adds the next nucleotide by linking its 5′-phosphate to the 3′-hydroxyl of the previous nucleotide. If a nucleotide lacks the 3′-hydroxyl, no further nucleotides can be added and the chain is terminated (Fig. 4.18). During a sequencing reaction, a certain percentage of nucleotides with no 3′-hydroxyl, called **dideoxynucleotides**, are mixed with the normal deoxynucleotides. Historically, the dideoxynucleotides were labeled with ^{32}P-phosphate, which is thus incorporated into

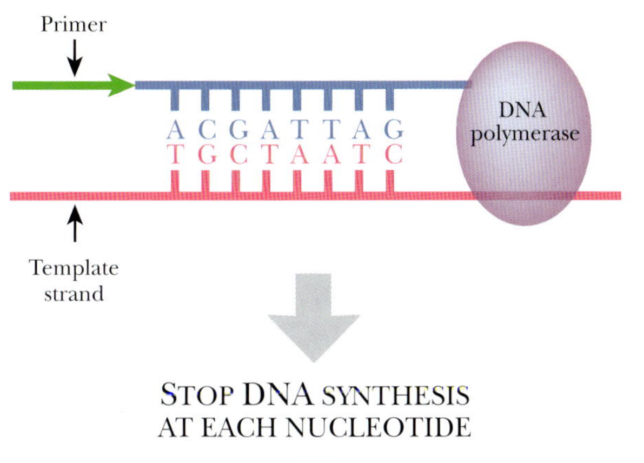

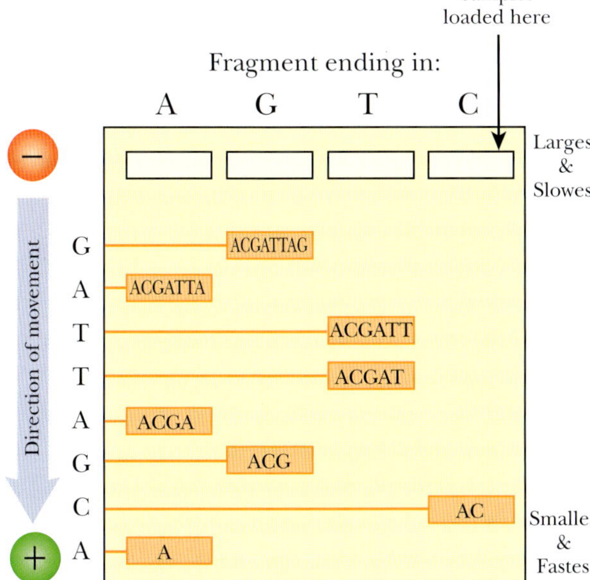

FIGURE 4.17 Chain Termination Method of Sequencing
During chain termination, DNA polymerase synthesizes many different strands of DNA from the single-stranded template. DNA polymerase will stop at each nucleotide, such that strands of all possible lengths are made. They are separated by size using electrophoresis. The smallest fragments are at the bottom and represent the primer plus only the first one or two nucleotides of the template DNA. Longer fragments contain the primer plus longer stretches corresponding to the template DNA.

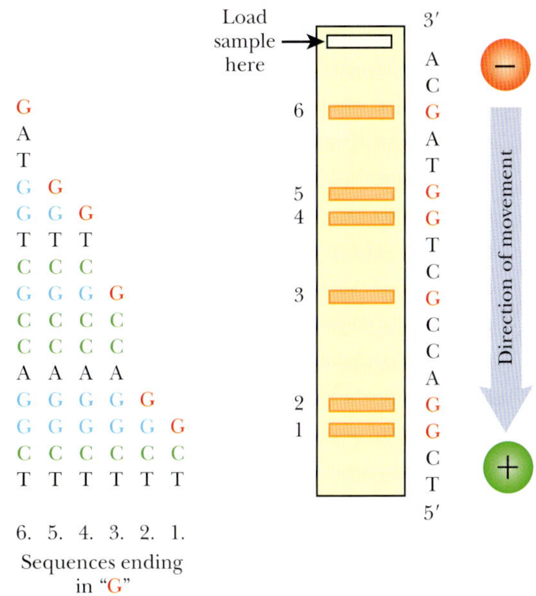

FIGURE 4.18 Chain Termination by Dideoxynucleotides
(A) During the sequencing reaction, DNA polymerase makes multiple copies of the original sequence. Sequencing reaction mixtures contain dideoxynucleotides that terminate growing DNA chains. The example here shows the G reaction, which includes triphosphates of both deoxyguanosine (dG) and dideoxyguanosine (ddG). Whenever ddG is incorporated (shown in red), it causes termination of the growing chain. If dG (blue) is incorporated, the chain will continue to grow. (Note: The normal dCTP is radioactively labeled, so that each fragment will be detected by autoradiography.) (B) When the sequencing reaction containing the ddG is run on a polyacrylamide gel, the fragments are separated by size. Each band on the autoradiogram directly represents a G in the original sequence.

each strand at the end. To get a stronger signal, one of the four normal deoxynucleotides is now radioactively labeled with ^{32}P-phosphate. The radioactive label on the fragments is detected by autoradiography of the sequencing gel (see Chapter 3). To identify the end nucleotide, four separate reactions are run for each template, each reaction getting one of the four possible dideoxynucleotides. For example, reaction 1 contains all the normal deoxynucleotides plus a fraction of dideoxyguanine; therefore, all fragments synthesized in

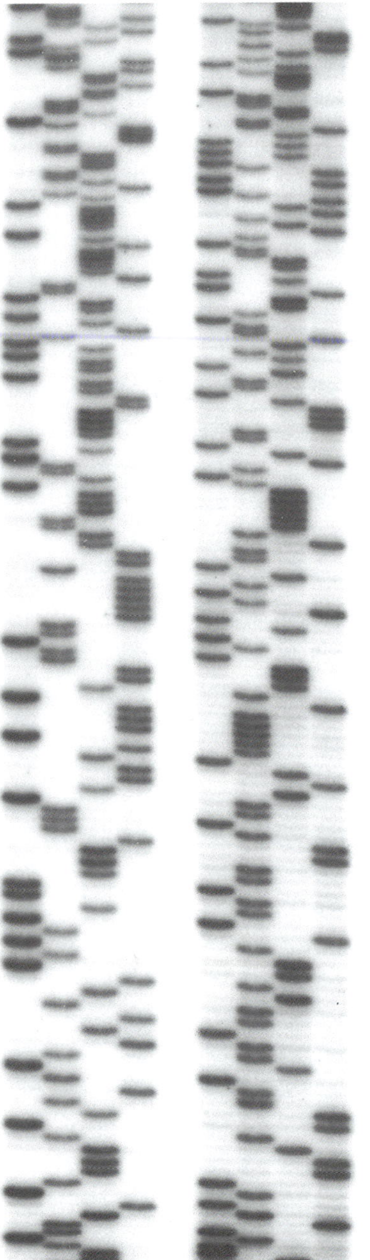

FIGURE 4.19
Sequencing Gel Autoradiogram
The sequencing of two different DNA templates is shown in this figure. The two sequences each consist of four lanes that represent the four different bases. The sequence is read from the bottom of the gel toward the top. This gel was run by Kiswar Alam in the author's laboratory.

this mixture will end at a guanine in the sequence. Reaction 2 is similar, but contains dideoxycytosine as the terminating nucleotide and hence stops the chains at each of the cytosines. Each reaction has identical template DNA and identical primer so that each segment starts at the same location. Only the stopping points are different. Such reactions typically have a maximum length of about 300 nucleotides.

The fragments are relatively small for DNA and vary in length by only one nucleotide; therefore, they must be separated by size using polyacrylamide gel electrophoresis (see Chapter 3). The principle is the same as for agarose gel electrophoresis, but polyacrylamide has smaller pores, and so smaller fragments can be separated. Each reaction is run in a different lane, side by side to each other. The fragments are then visualized by autoradiography. When the sequencing gel is dried and exposed to photographic film, the dark bands represent each of the terminated fragments. The sequence is actually read from the bottom of the gel to the top, because the fragments terminated closest to the primer are smaller (hence run faster) than the ones further from the primer. The bands appear as a ladder, each separated by one nucleotide; therefore, each band is a different nucleotide on the sequence (Fig. 4.19).

The nature of the template DNA affects whether or not readable sequence can be obtained. Single-stranded template DNA provides the best results and can be made by cloning the template DNA into bacteriophage M13, which has single-stranded DNA. When the recombinant virus infects its host, *E. coli*, rolling circle replication and viral packaging create thousands of single-stranded copies of the template for sequencing. Alternatively, the sequence of interest is cloned into a typical plasmid vector (see Chapter 3), and the double-stranded template is denatured either with heat or chemicals to make it single-stranded.

Natural DNA polymerase has been modified from its original form to enhance *in vitro* sequencing. Native DNA polymerase I has a domain that excises any mismatched or defective bases and replaces them. This activity would remove dideoxynucleotides rather than incorporate them. This repair domain is easily removed from purified DNA polymerase by protease digestion without affecting the other activities of DNA polymerase. The resulting polymerase is called Klenow polymerase. Further modifications of DNA polymerase have streamlined it for sequencing. For example, sequencing polymerase has been engineered to increase processivity, that is, the polymerase is less likely to fall off the template and can make longer strands.

The order of nucleotides on a strand of DNA can be determined using *in vitro* replication.

Chain-terminating dideoxynucleotides are the key to determining the sequence. When these are incorporated into an *in vitro* replication reaction, DNA polymerase cannot add any more nucleotides and the synthesis reaction ends. If only one type of dideoxynucleotide is present, the resulting strands will vary in length, each size corresponding to the position of the dideoxynucleotide relative to the primer.

POLYMERASE CHAIN REACTION USES *IN VITRO* SYNTHESIS TO AMPLIFY SMALL AMOUNTS OF DNA

The **polymerase chain reaction (PCR)** amplifies small samples of DNA into large amounts, much as a photocopier makes many copies of one sheet of paper. The DNA is amplified using the principles of replication, that is, the DNA is replicated over and over by

DNA polymerase until a large amount is manufactured. Kary Mullis invented this technique while working at Cetus in 1987. He eventually won the Nobel Prize in Chemistry for PCR because of its huge impact on biology and science. PCR is used in forensic medicine to identify victims or criminals by amplifying the minuscule amounts of DNA left at a crime scene (see Chapter 24); PCR can identify infectious diseases such as AIDS before symptoms emerge (see Chapter 22); PCR can amplify specific segments of genes without the need for cloning the segment first; in fact, PCR is now used in all aspects of the biological sciences.

Just as the photocopier needs more paper, ink, and a mechanism to copy, PCR requires specific reagents. The sample to be copied is called the template DNA, and this is often a known sequence or gene. The template DNA is double-stranded, and extremely small quantities are sufficient. PCR can amplify a particular gene segment directly from one *Caenorhabditis elegans* or even a single cell. The second reagent needed for PCR is a pair of primers, which have sequences complementary to the ends of the template DNA. The DNA primers are oligonucleotides about eight to 20 nucleotides long. One primer anneals to the 5′ end of the sense strand, and the other anneals to the 3′ end of the antisense strand of the target sequence. Because the primers are specific, the template can be mixed with other DNA sequences. The third reagent is a supply of nucleoside triphosphates, and the fourth is *Taq* **DNA polymerase** from *Thermus aquaticus*, which actually makes the copies.

The basic mechanism of PCR relies on heat denaturation of the template, annealing of the primers, and making a complementary copy using DNA polymerase, all being steps found in DNA replication. The three steps are repeated over and over until one template strand generates millions of identical copies. An amount of DNA too small to be seen can be exponentially copied so that it can be cloned into a vector, or visualized on an agarose gel (see Chapter 3). The process requires changing the temperature in a cyclic manner. Changing temperatures is accomplished by a **thermocycler**, a machine designed to change the temperature of its heat block rapidly so that each cycle can be completed in minutes. The temperature cycles between 94 °C to denature the template; 50–60 °C to anneal the primer (depending on the length and sequence of the primer); and 70 °C for *Taq* polymerase to make new DNA. Before thermocyclers were developed, PCR was accomplished by moving the mixture among three different water baths at different temperatures every few minutes, which was very tedious.

In principle, the PCR cycle resembles DNA replication with a few modifications (Fig. 4.20). Like other *in vitro* DNA synthesis reactions, the double-stranded template is denatured with high heat rather than enzymes. Then the temperature is lowered so that the primers anneal to their binding sites. The primers are made so that each binds to opposite strands of the template, one at the beginning and one at the end of the gene. Then DNA polymerase elongates both primers and converts both single template strands to double-stranded DNA. (Note: During sequencing, only one primer is used and only one strand of the template is replicated, but during PCR both strands are copied.) *Taq* polymerase is the most widely used polymerase for PCR because it is very stable at high temperatures and does not denature at the high temperatures needed to separate the strands of the template DNA. *Taq* polymerase comes from *Thermus aquaticus*, a bacterium that grows in the thermal vents in the ocean. After the first replication cycle, the whole process is repeated. The two DNA strands are denatured at high heat, and then the temperature drops to allow the primers to anneal to their target sequences. *Taq* polymerase synthesizes the next four strands, and now there are four double-stranded copies of the target sequence. Early in the process, some longer strands are generated; however, eventually only the segment flanked by the two primers is amplified. Ultimately the template strands and early PCR products become the minority. The shorter products become the majority.

The primers are key to the process of PCR. If the primers do not anneal in the correct location, if the span between the primers is too large, or if the primers form hairpin regions rather than annealing to the target, then *Taq* polymerase will not be able to amplify the segment. Also, if both primers anneal to the same strand, the reaction will not work. If the template has a

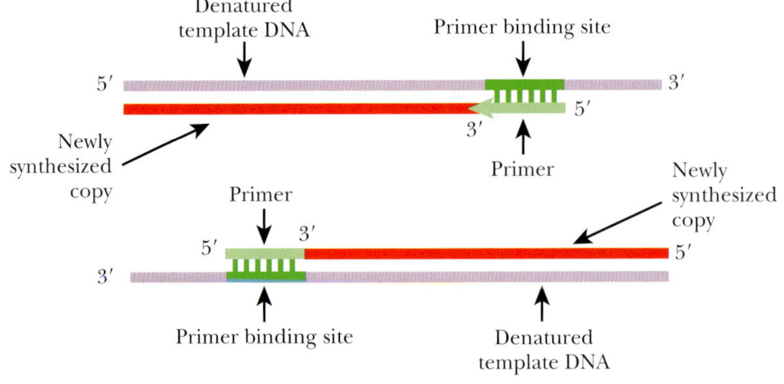

REPEAT

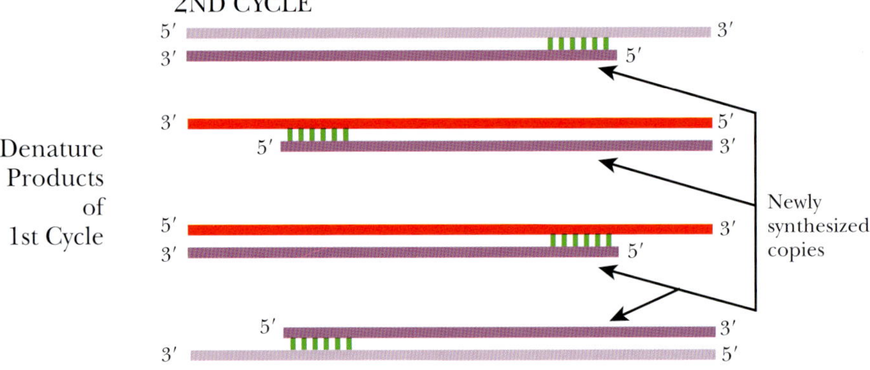

REPEAT

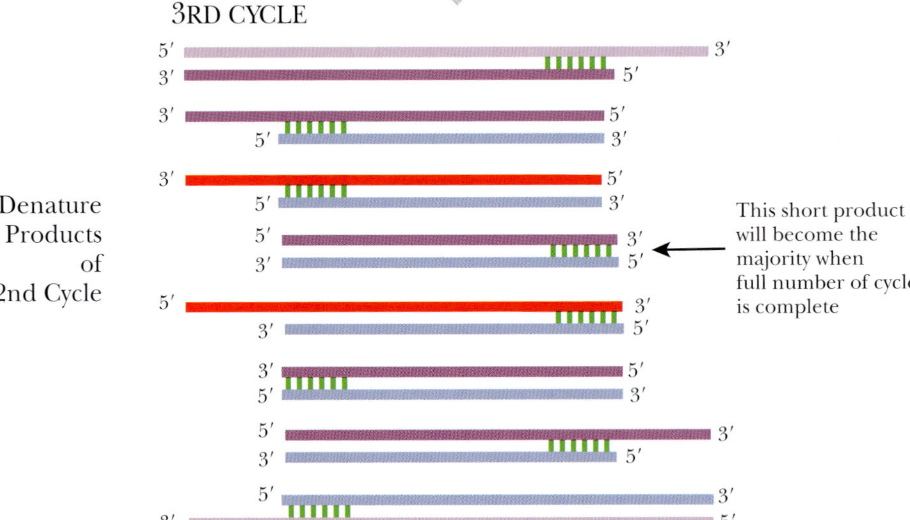

FIGURE 4.20 PCR, the First Three Cycles

In the first cycle, double-stranded template DNA (light purple) is denatured, complementary primers are annealed to the primer binding sites, and a new copy of the template is generated by Taq polymerase (red). In the second cycle, the two double-stranded products from the first cycle are denatured to form four single-stranded templates. The same set of primers anneals to the four template strands, and *Taq* polymerase makes each of the four double-stranded (dark purple). In the third cycle, the four double-stranded products from the second cycle are denatured, the primers anneal, and the four products from the second cycle become eight. Each subsequent round of denaturation, primer annealing, and extension doubles the number of copies, turning a small amount of template into a large amount of PCR product.

known sequence, primers are synthesized based on the sequences upstream and downstream of the region to be amplified. Modifications exist that allow researchers to analyze unknown sequences by PCR (see later discussion).

PCR is another process that uses DNA polymerase in an *in vitro* sequencing reaction. Here, a double-stranded template is replicated to make two copies. Each of these products is replicated to make four, and the process continues exponentially.

AUTOMATED DNA CYCLE SEQUENCING COMBINES PCR AND SEQUENCING

Automated DNA sequencing uses a PCR-type reaction to sequence DNA. This is an improved way to sequence DNA because of its speed and because it can be analyzed by computer rather than a person. In PCR sequencing, or **cycle sequencing**, the template DNA with unknown sequence is amplified by *Taq* polymerase rather than Klenow polymerase. As for DNA polymerase I, this *Taq* DNA polymerase was modified to remove its proofreading ability. Cycle sequencing reaction mixtures includes all four deoxynucleotides, all four dideoxynucleotides, a single primer, template DNA, and *Taq* polymerase. Unlike regular sequencing, each dideoxynucleotide is labeled with a fluorescent tag of a different color; therefore, all four dideoxynucleotides are in the same reaction mix.

The samples are amplified in a thermocycler. First, the template DNA is denatured at a high temperature, then the temperature is lowered to anneal the primer, finally the temperature is raised to 70 °C, the optimal temperature for *Taq* polymerase to make DNA copies of the template. During polymerization, dideoxynucleotides are incorporated and cause chain termination, just as with standard sequencing. The ratio of dideoxynucleotides to deoxynucleotides is adjusted to ensure that some fragments stop at each G, A, T, or C. After *Taq* polymerase makes thousands of copies of the template, each stopping at a different nucleotide, the entire mixture is run in one lane of a sequencing gel (Fig. 4.21). Bands of four different colors are seen, corresponding to the four fluorescently labeled dideoxynucleotides and hence the four bases.

Cycle sequencing has some advantage over regular sequencing. During regular sequencing reactions, the temperatures are lower; therefore, the template DNA is more likely to fold back on itself into different secondary structures. This is especially true for double-stranded DNA templates, which can reanneal before Klenow polymerase has a chance to work. During cycle sequencing, each round brings the temperature to 95 °C, which destroys any secondary structures or double-stranded regions. Another advantage of cycle sequencing is to control primer hybridization. Some primers do not work well with regular sequencing reactions because they bind to closely related sequences. During cycle sequencing, the primer annealing temperature is controlled and can be set quite high in order to combat nonspecific binding. Finally, cycle sequencing requires much less template DNA than regular sequencing; therefore, sequencing can be done from smaller samples.

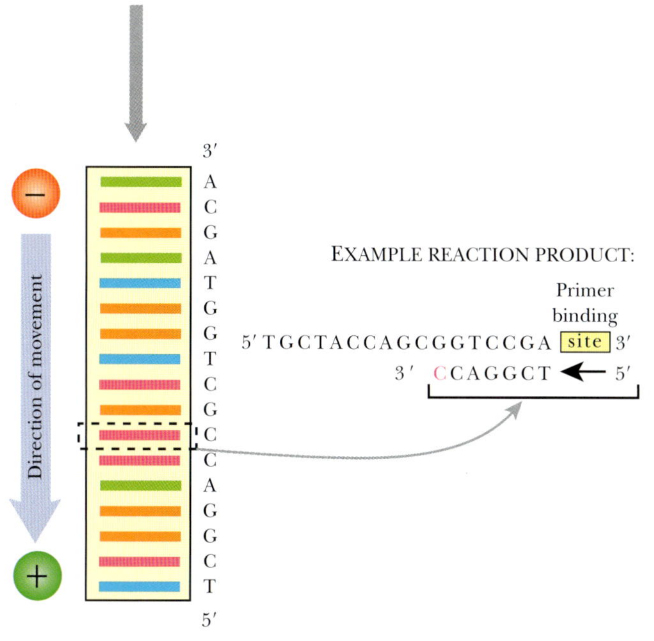

FIGURE 4.21 Cycle Sequencing
During cycle sequencing, the reaction contains template DNA, primer, *Taq* polymerase, deoxynucleotides, and dideoxynucleotides. Each of the different dideoxynucleotides has a different fluorescent label attached. The automated sequencer detects the color and compiles the sequence data.

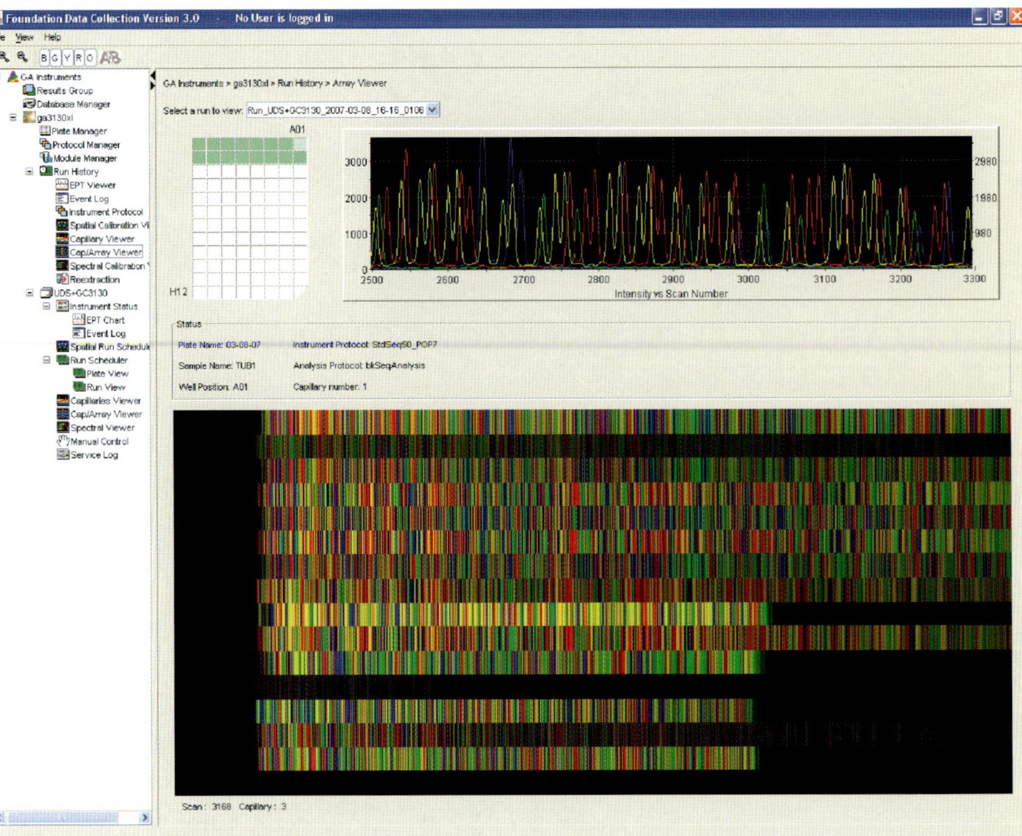

FIGURE 4.22 Data from an Automated Sequencer
Raw fluorescent data from an automated sequencer (Applied Biosystems 3130XL Genetic Analyzer) that was detected and recorded by the CCD camera from 16 different samples. The fluorescent peaks for individual bases are shown at the top for one of the channels. Courtesy of Brewster F. Kingham, Sequencing & Genotyping Center, University of Delaware.

Another advance in sequencing has been the detection system. Traditionally, researchers have used radioactively labeled nucleotides, followed by autoradiography to detect the bands. A major improvement has been to use fluorescently labeled dideoxynucleotides; therefore, the terminal nucleotide of each DNA strand will fluoresce. Different fluorescent tags can be used for each of the four dideoxynucleotides, so that the whole reaction can be done in one tube. **Automatic DNA sequencers** detect each of the fluorescent tags and record the sequence of bases (Fig. 4.22). Much like regular sequencing, the different fragments are run on a gel, but instead of four lanes, the entire mix is all in the same lane. Some automatic DNA sequencers can read up to 96 samples side-by-side on each gel. At the bottom of each lane is a fluorescent activator, which emits light to excite the fluorescent dyes. On the other side is the detector, which reads the wavelength of light that the fluorescent dye emits. As each fragment passes the detector, it measures the wavelength and records the data as a peak on a graph. For each fluorescent dye, a peak is recorded and assigned to the appropriate base. An attached computer records and compiles the data into the DNA sequence.

Automated sequencing has a large startup cost because the sequence analyzer is quite expensive, but they run multiple samples at one time, and thus the cost per sample is quite low. Many universities and companies have a centralized facility that does the sequencing for all the researchers. In fact, sequencing has become so automated that many researchers just send their template DNA and primers to a company that specializes in sequencing.

Cycle sequencing combines PCR technology with standard sequencing methodology. Each PCR reaction is the same, except a controlled amount of fluorescently labeled dideoxynucleotides is included. *Taq* polymerase stops adding nucleotides when a dideoxynucleotide is incorporated. The fluorescent tag is used to identify the ending base of each fragment using an automated sequencer.

MODIFICATIONS OF BASIC PCR

Many different permutations of PCR have been devised since Kary Mullis developed the basic procedure. All rely on the same basic PCR reaction, which takes a small amount of DNA and amplifies it by *in vitro* replication. Many of these variant protocols are essential tools for recombinant DNA research.

Several strategies allow amplifying a DNA segment by PCR even if its sequence is unknown. For example, the unknown sequence may be cloned into a vector (whose sequence is known). The primers are then designed to anneal to the regions of the vector just outside the insert.

In another scenario, the sequence of an encoded protein is used to generate PCR primers. Remember that most amino acids are encoded by more than one codon. Thus, during translation of a gene, one or more codons are used for the same amino acid. Therefore, if a protein sequence is converted backwards into nucleotide sequence, the sequence is not unique. For example, two different codons exist for histidine and glutamine, and four codons exist for serine. Consequently, the nucleotide sequence encoding the amino acid sequence histidine-glutamine-valine can be one of 16 different combinations (Fig. 4.23).

RELATED PROTEIN SEQUENCE: His-Gln-Val

His codons: CAT, CAC
Gln codons: CAA, CAG
Val codons: GTT, GTG, GTC, GTA

POSSIBLE DNA SEQUENCE:

CATCAAGTT	CACCAAGTT
CATCAAGTG	CACCAAGTG
CATCAAGTC	CACCAAGTC
CATCAAGTA	CACCAAGTA
CATCAGGTT	CACCAGGTT
CATCAGGTG	CACCAGGTG
CATCAGGTC	CACCAGGTC
CATCAGGTA	CACCAGGTA

FIGURE 4.23
Degenerate DNA Primers
A protein sequence of three amino acids can be encoded by many different DNA sequences. Degenerate primers are a mixture of all these possible sequences. Therefore the unknown DNA sequence for His-Gln-Val will be matched by at least one primer sequence, and the DNA will then be amplified by PCR.

If primers are made that depend on protein sequence they will be **degenerate primers** and they will have a mixture of two or three different bases at the wobble positions in the triplet codon. During oligonucleotide synthesis, more than one phosphoramidite nucleotide can be added to the column at a particular step. Some of the primers will have one of the nucleotides, whereas other primers will have the other nucleotide. If many different wobble bases are added, a population of primers are created, each with a slightly different sequence. Within this population, some will bind to the target DNA perfectly, some will bind with only a few mismatches, and some won't bind at all. Of course, the annealing temperature for degenerate primers is adjusted to allow for some mismatches.

Inverse PCR is a trick used when sequence information is known only on one side of the target region (Fig. 4.24). First, a restriction enzyme is chosen that does not cut within the stretch of known DNA. The length of the recognition sequence should be six or more base pairs in order to generate reasonably long DNA segments for amplification by PCR. The target DNA is then cut with this restriction enzyme to yield a piece of DNA that has compatible sticky ends, one upstream of the known sequence, and one downstream. The two ends are ligated to form a circle. The PCR primers are designed to recognize the end regions of the known sequence. Each primer binds to a different strand of the circular DNA and they both point "outwards" into the unknown DNA. PCR then amplifies the unknown DNA to give linear molecules with short stretches of known DNA at the ends, and the restriction enzyme site in the middle.

Degenerate primers are designed based on amino acid sequences and contain different nucleotides at the wobble position.

Inverse PCR sequences DNA near a known sequence by finding a restriction enzyme recognition sequence away in the unknown region, cutting out this template, and amplifying the entire piece with *Taq* polymerase.

RANDOMLY AMPLIFIED POLYMORPHIC DNA

Another PCR application, called **randomly amplified polymorphic DNA (RAPD)**, allows the researcher to compare the genetic relatedness of two DNA samples (Fig. 4.25). First, two DNA samples are isolated from two different organisms. Thus there are two samples of target DNA, which are compared by using the same set of primers. As before,

STEP 1: MAKING THE TEMPLATE

Left side Known sequence Right side

Recognition site for restriction enzyme

CUT WITH RESTRICTION ENZYME; LIGATE ENDS

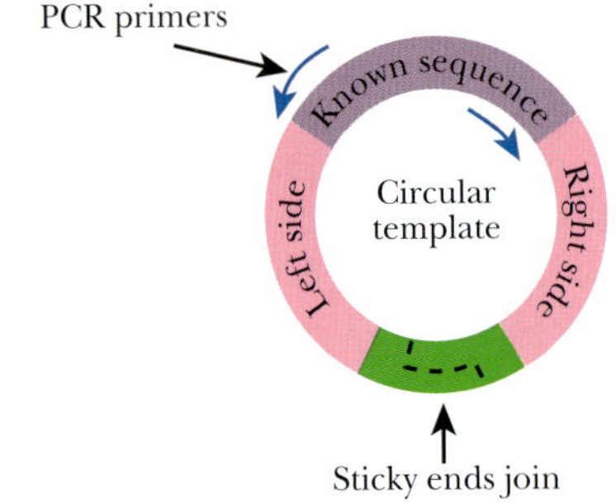

STEP 2: RUN PCR REACTION

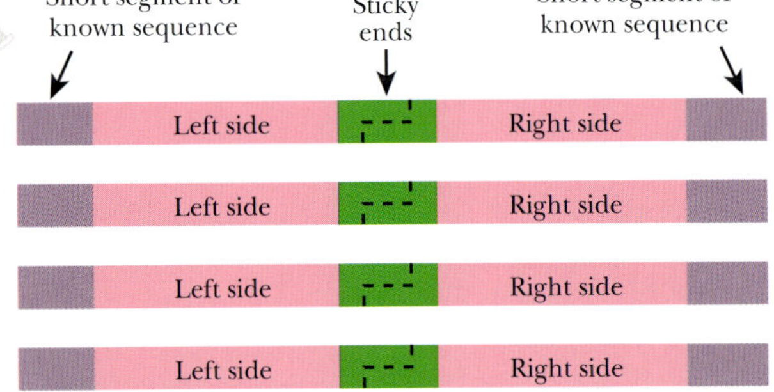

FIGURE 4.24 Inverse PCR
Inverse PCR allows unknown sequences to be amplified by PCR provided that they are located near a known sequence. The DNA is cut with a restriction enzyme that cuts upstream and downstream of the known region but not within it. The linear piece of DNA is circularized and then amplified with primers that anneal in the known region. The PCR products have the unknown DNA from the left and right of the known sequence. These can be cloned and sequenced.

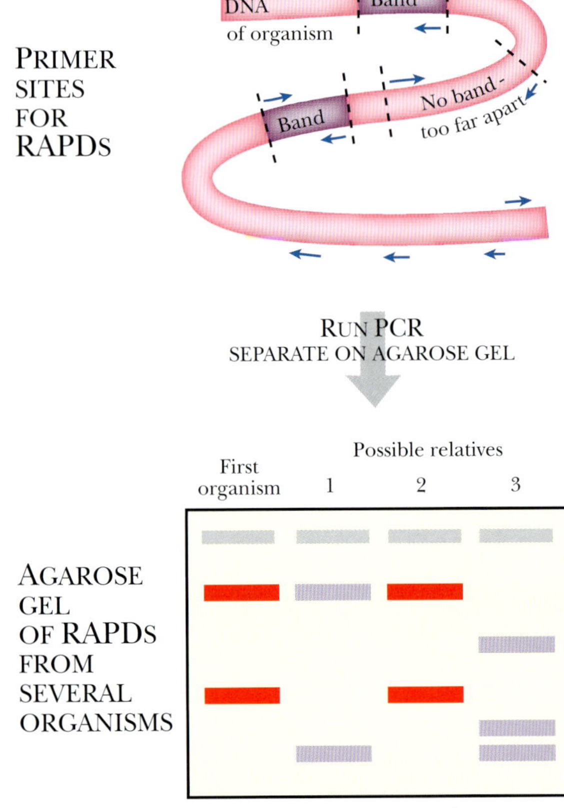

FIGURE 4.25 Randomly Amplified Polymorphic DNA
The first step of RAPD analysis is to design primers that will bind to genomic DNA at random sites that are neither too rare nor too common. In this example, the primers are sufficiently long to bind the genomic DNA at a dozen places. For PCR to be successful, two primers must anneal at sites facing each other but on opposite strands. In addition, these paired primer sites must be close enough to allow synthesis of a PCR fragment. In this example, there are three pairs, but only two of these are close enough to make a PCR product. Consequently, this primer design will result in two PCR products as seen in the first lane of the gel (marked "First organism"). The same primers are used to amplify genomic DNA from other organisms that are suspected of being related. In this example, suspect 2 shows the same banding pattern as the first organism and is presumably related. The other two suspects do not match and so are not related.

the sequence information is unknown, but rather than using degenerate primers, primers with a randomly chosen sequence are made. The length of a primer determines how often it will bind within the target DNA. If a particular primer were, say, 5 bases long, it would bind once on average every 4^5 bases = $4 \times 4 \times 4 \times 4 \times 4 = 1024$ bases. If the target DNA were a sample from a large genome, such a primer would bind far too many times. In practice longer primers of around 10 bases are often suitable. The random primer is mixed with nucleotides, *Taq* polymerase, and each of the target DNA samples as for normal PCR reactions. In order to amplify any target DNA fragments, two of the random primers must bind to the target DNA, on opposite strands, usually within a few thousand bases. The results of the two PCR reactions are compared using gel electrophoresis. The number and size of PCR products will vary for the two samples. If two organisms are very closely related, then their DNA will be close in sequence. Hence, the PCR products will be very similar with only one or two different fragments. If the pattern of PCR products is totally different, then the two organisms are not related. Comparing RAPDs from two organisms can thus give an estimate of relatedness.

Two closely related organisms have similar DNA sequences, and therefore, when the same primers are used to amplify the genomic pieces of DNA, the same pattern of PCR products will be obtained.

REVERSE TRANSCRIPTASE PCR

Reverse transcriptase PCR (RT-PCR) uses the enzyme reverse transcriptase to make a cDNA copy of mRNA from an organism, and then uses PCR to amplify the cDNA (Fig. 4.26). The advantage of this technique is evident when trying to use PCR to amplify a gene from eukaryotic DNA. Eukaryotes have introns, some extremely long, which interrupt the coding segments. After transcription, the primary RNA transcript is processed to remove all the introns, hence becoming mRNA. Using mRNA as the source of the target DNA relies on the cell removing the introns. In practice, RT-PCR has two steps. First, reverse transcriptase recognizes the 3′ end of primers containing repeated thymines and synthesizes a DNA strand that is complementary to the mRNA. (The thymines base-pair with the poly(A) tail of mRNA.) Then the RNA strand is replaced with another DNA strand, leaving a double-stranded DNA (i.e., the cDNA). Next, the cDNA is amplified using a normal PCR reaction containing appropriate primers (one usually recognizes the poly(A) tail), *Taq* polymerase, and nucleotides.

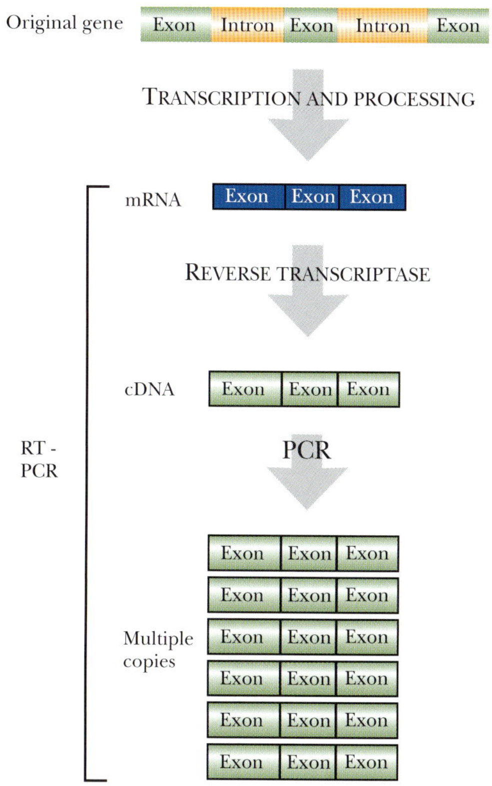

FIGURE 4.26 Reverse Transcriptase PCR
RT-PCR is a two-step procedure that involves making a cDNA copy of the mRNA, and then using PCR to amplify the cDNA.

RT-PCR uses reverse transcriptase to convert mRNA into double-stranded DNA, and then the gene without any introns can be amplified by regular PCR.

PCR IN GENETIC ENGINEERING

PCR allows the scientist to clone genes or segments of genes for identification and analysis. PCR also allows the scientist to manipulate a gene that has already been identified. Various modified PCR techniques allow scientists to hybridize two separate genes or genes segments into one, delete or invert regions of DNA, and alter single nucleotides to change the gene and its encoded protein in a more subtle way.

PCR can make cloning a foreign piece of DNA easier. Special PCR primers can generate new restriction enzyme sites at the ends of the target sequence (Fig. 4.27). The primer is synthesized so that its 5′ end has the desired restriction enzyme site, and the 3′ end has sequence complementary to the target. Obviously, the 5′ end of the

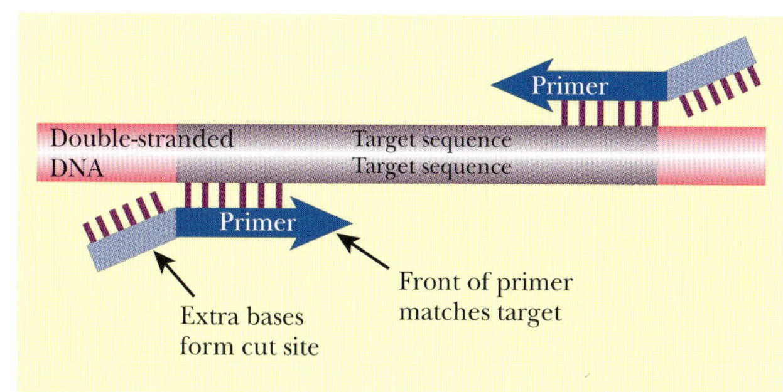

FIGURE 4.27 Incorporation of Artificial Restriction Enzyme Sites
Primers for PCR can be designed to have nonhomologous regions at the 5′ end that contain the recognition sequence for a particular restriction enzyme. After PCR, the amplified product has the restriction enzyme sites at both ends. If the PCR product is digested with the restriction enzyme, this generates sticky ends that are compatible with a chosen vector.

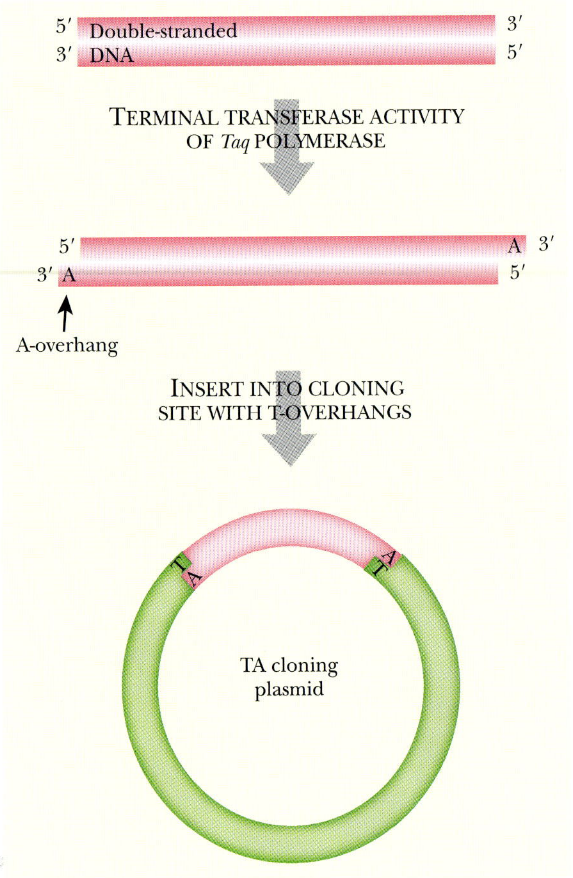

primer does not bind to the target DNA, but as long as the 3′ end has enough matches to the target, then the primer will still anneal. *Taq* polymerase primes synthesis from the 3′ end; therefore, the enzyme is not bothered by mismatched 5′ sequences. The resulting PCR product can easily be digested with the corresponding restriction enzyme and ligated into the appropriate vector.

Rather than incorporating restriction enzyme sites into the ends of the PCR product, **TA cloning** will clone any PCR product directly (Fig. 4.28). *Taq* polymerase has terminal transferase activity that generates a single adenine overhang on the ends of the PCR products it makes. Special vectors containing a single thymine overhang have been developed, and simply mixing the PCR product with the TA cloning vector plus DNA ligase clones the PCR product into the vector without any special modifications.

PCR can be used to manipulate cloned genes also. Two different gene segments can be hybridized into one using **overlap PCR** (Fig. 4.29). Here, PCR amplification occurs with three primers: one is complementary to the beginning of the first gene segment, one is complementary to the end of the second gene segment, and a third is half complementary to the end of gene segment 1 and half complementary to the beginning of gene segment 2. During PCR, the two gene segments become fused into one by a mechanism that is hard to visualize, but probably involves looping of some of the early PCR products.

FIGURE 4.28
TA Cloning of PCR Products
When *Taq* polymerase amplifies a piece of DNA during PCR, the terminal transferase activity adds an extra adenine at the 3′ ends. The TA cloning vector was designed so that when linearized, it has a single 5′-thymine overhang. The PCR product can be ligated into this vector without the need for special restriction enzyme sites.

PCR can be used to create large deletions or insertions into a gene (Fig. 4.30). Once again the design of the PCR primers is key to the construction. For example, primers to generate insertions have two regions: the first half is homologous to the sequence around the insertion point; the second half has sequences complementary to the insert sequence. For example, suppose an antibiotic resistance gene such as *npt* (confers resistance to neomycin) is to be inserted into a cloning vector. The primers would have their 5′ ends complementary to the sequence flanking the insertion point on the vector and their 3′ ends complementary to the

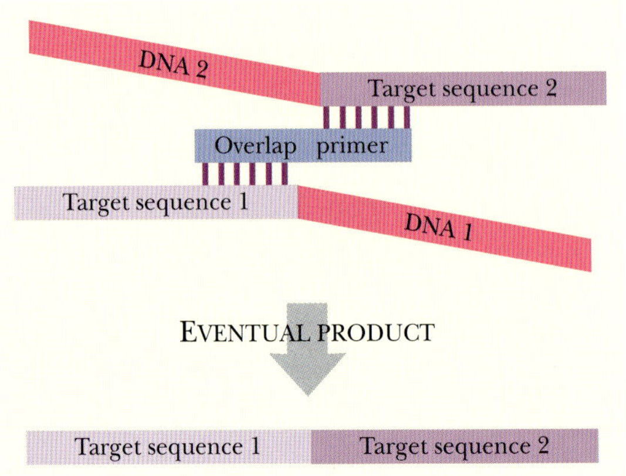

FIGURE 4.29 Overlap PCR
Overlapping primers can be used to link two different gene segments. In this scheme, the overlapping primer has one end with sequences complementary to target sequence 1, and the other half similar to target sequence 2. The PCR reaction will create a product with these two regions linked together.

ends of the *npt* gene. First the primers are used to amplify the *npt* gene and give a product with sequences homologous to the vector flanking both ends. Next, the PCR product is transformed into bacteria harboring the vector. The *npt* gene recombines with the insertion point by homologous recombination, resulting in insertion of the *npt* gene into the vector.

The insertion point(s) will determine whether the antibiotic cassette causes just an insertion or both an insertion plus a deletion. If the two PCR primers recognize separate homologous recombination sites, then the incoming PCR segment will recombine at these two sites. Homologous recombination then results in the *npt* gene replacing a piece of the vector rather than merely inserting at one particular location.

PCR can also generate nucleotide changes in a gene by **directed mutagenesis** (Fig. 4.31). Usually only one or a few adjacent nucleotides are changed. First, a mutagenic PCR primer is

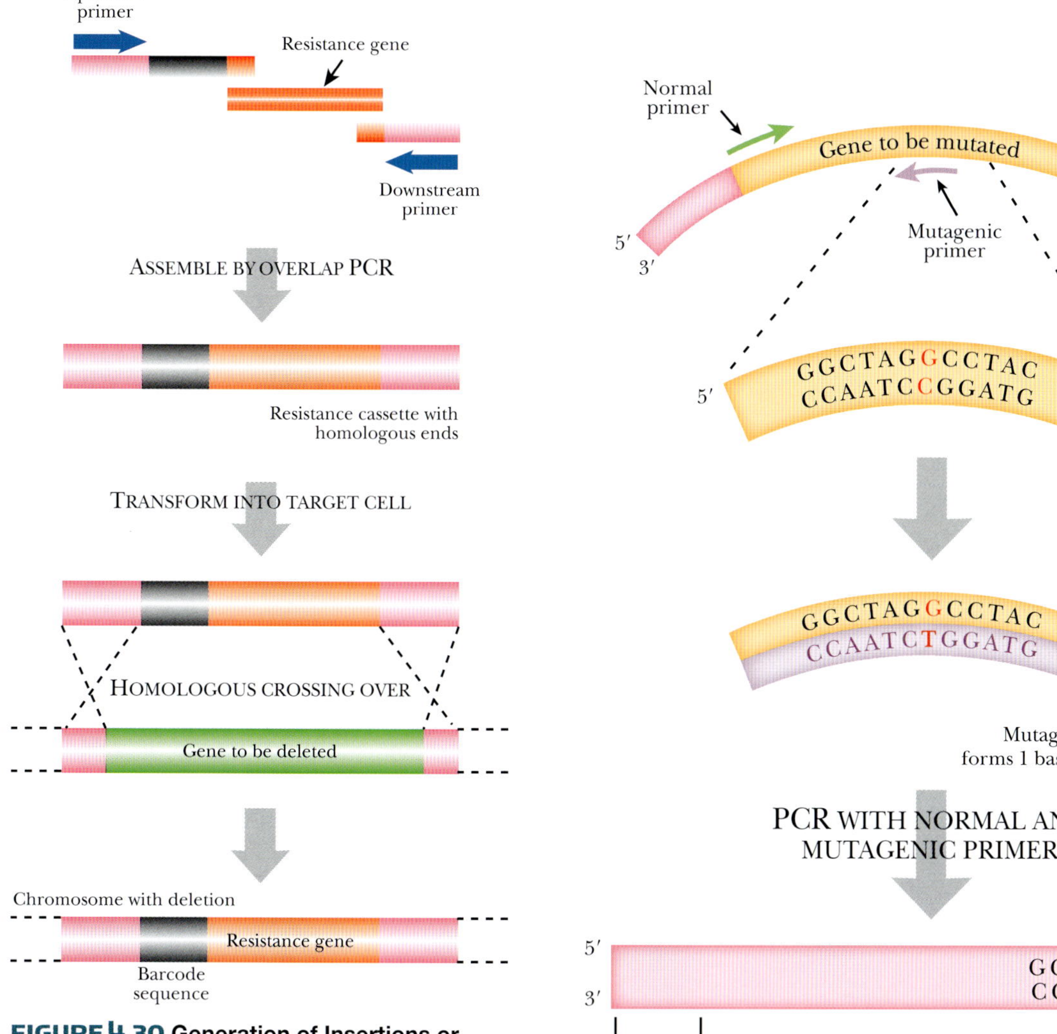

FIGURE 4.30 Generation of Insertions or Deletions by PCR
In the first step, a specifically targeted cassette is constructed by PCR. This contains both a suitable marker gene and upstream and downstream sequences homologous to the target site. The engineered cassette is transformed into the host cell and homologous crossing over occurs. Recombinants are selected by the antibiotic resistance carried on the cassette. The barcode sequence is a unique marker used to identify the cassette.

FIGURE 4.31 Direct Mutagenesis Using PCR
The gene to be mutated is cloned and the entire sequence is known. To alter one specific nucleotide, normal and mutagenic primers are combined in a PCR reaction. The mutagenic primer will have a mismatch in the middle, but the remaining sequences will be complementary. The PCR product will incorporate the sequence of the mutagenic primer.

synthesized that has nucleotide mismatches in the middle region of the primer. The primer will anneal to the target site with the mismatch in the center. The primer needs to have enough matching nucleotides on both sides of the mismatch so that binding is stable during the PCR reaction. The mutagenic primer is paired with a normal primer. The PCR reaction then amplifies the target DNA incorporating the changes at the end with the mutagenic primer. These changes may be relatively subtle, but if the right nucleotides are changed, then a critical amino acid may be changed. One amino acid change can alter the entire function of a protein. Such an approach is often used to assess the importance of particular amino acids within a protein.

PCR products can have added restriction enzymes site by designing the primers, or they can have added adenines by *Taq* polymerase. These traits allow the PCR product to be cloned into a vector.

PCR can be used to delete, insert, and even fuse different gene segments.

PCR can be used to make small changes in nucleotide sequences by directed mutagenesis.

Summary

Chapter 4 outlines the process of DNA replication. First, in order to replicate the DNA, DNA gyrase and DNA helicase relax the coiling in the DNA. The relaxed DNA is open and ready for the replisome to assemble at the origin. The replisome adds single-stranded binding protein along the open DNA, which keeps the DNA stable. Then PriA prepares an RNA primer at the origin to provide a 3′-OH group for DNA polymerase to attach the complementary bases during replication. DNA polymerase only makes new DNA in a 5′ to 3′ direction, so on the leading strand, the whole strand is made in one piece. Because the lagging strand is antiparallel, DNA polymerase has to make the strand in smaller segments called Okazaki fragments.

In vitro DNA synthesis can be made by purified DNA polymerase or by chemically linking nucleotides. In reactions done with chemical reagents, the DNA is single-stranded and is short because the process is not very efficient. Chemical synthesis of DNA is primarily used for making short primers or oligonucleotides.

In vitro DNA synthesis by DNA polymerase is very versatile. Here, DNA polymerase of different types can be used to determine the sequence of a cloned gene and amplify a piece of DNA from a few copies to millions using PCR. In sequencing reactions, the final base of each of the copies made by DNA polymerase is determined with one of two methods. The original sequencing method used a separate reaction for each of dideoxynucleotides that stop DNA polymerase and radioactively labeled nucleotides to determine the final base. In the cycle sequencing method, one reaction tube uses four different fluorescent labels on each of the dideoxynucleotides in order to identify the final base. PCR can also be used to assess how closely related two organisms are by RAPD analysis. Modifications of PCR include using degenerate primers and inverse PCR to amplify unknown regions of DNA. RT-PCR uses mRNA rather than DNA because mRNA is already missing noncoding intron regions. Additionally, PCR can be used to clone copies of genomic DNA into a vector using TA cloning or by adding novel restriction enzyme sites at the end of the PCR product. Finally, PCR can mutate template DNA by inserting or deleting regions, linking two separate regions together, or by mutating single nucleotides.

End-of-Chapter Questions

1. Which of the following enzymes aid in uncoiling DNA?
   a. DNA gyrase
   b. DNA helicase
   c. topoisomerase IV
   d. single-stranded binding protein
   e. all of the above
2. Why is an RNA primer necessary during replication?
   a. DNA polymerase III requires a 3′-OH to elongate DNA.
   b. An RNA primer is not needed for elongation.
   c. DNA polymerase requires a 5′-phosphate before it can elongate the DNA.
   d. A DNA primer is needed for replication instead of an RNA primer.
   e. An RNA primer is only needed once the DNA has been elongated and DNA polymerase is trying to fill in the gaps.
3. What are the functions of the two essential subunits of DNA polymerase III?
   a. Both subunits synthesize the lagging strand only.
   b. One subunit links nucleotides and the other ensures accuracy.
   c. They both function as a clamp to hold the complex to the DNA.
   d. The subunits function to break apart the bonds in the DNA strand.
   e. One subunit removes the RNA primer and the other synthesizes DNA.
4. Which of the following statements about mismatch repair is incorrect?
   a. MutSHL excise the mismatched nucleotides from the DNA.
   b. Mismatch repair proteins identify a mistake in DNA replication.
   c. The mismatch proteins recruit DNA polymerase III to synthesize new DNA after the proteins have excised the mismatched nucleotides.
   d. MutSHL can synthesize new DNA after a mismatch has been excised.
   e. MutSHL monitors the methylation state of the DNA to determine which strand contains the correct base when there is a mismatch.
5. Which of the following statements is incorrect regarding DNA replication?
   a. Rolling circle and theta replication are common for prokaryotes and viruses.
   b. Each round of replication for linear chromosomes, such as in eukaryotes, shortens the length of the chromosome.
   c. Prokaryotic chromosomes have multiple origins of replication.
   d. Eukaryotic replication only occurs during the S-phase of the cell cycle.
   e. Eukaryotic chromosomes have multiple origins of replication.
6. During *in vitro* DNA replication, which of the following components is not required?
   a. single-stranded DNA
   b. a primer containing a 3′-OH
   c. DNA helicase to separate the strands
   d. DNA polymerase to catalyze the reaction
   e. nucleotide precursors

(*Continued*)

7. Which of the following is not a step in the chemical synthesis of DNA?
   a. The 3′ phosphate group is added using phosphorylase.
   b. The addition of a blocking compound to protect the 3′ phosphite from reacting improperly.
   c. The 5′-OH is phosphorylated by bacteriophage T4 kinase.
   d. The addition of acetic anhydride and dimethylaminopyridine to cap the 5′-OH group of unreacted nucleotides.
   e. The amino groups on the bases are modified by other chemical groups to prevent the bases from reacting during the elongation process.

8. During chemical synthesis of DNA, a portion of the nucleotides does not react. How can the efficiency of such reactions be increased?
   a. The unreacted nucleosides are not acetylated so that more can be added in subsequent reactions.
   b. The efficiency of the reaction is not critical. Instead, the quality of the final product is more important than the quantity.
   c. The desired oligonucleotide can be separated from the truncated oligos by electrophoresis.
   d. Oligonucleotides should be made using DNA polymerase III instead of *in vitro* chemical synthesis.
   e. The reaction times can be increased to allow the reaction to be more efficient.

9. Which of the following components terminates the chain in a sequencing reaction?
   a. dideoxynucleotides
   b. Klenow polymerase
   c. DNA polymerase III
   d. deoxynucleotides
   e. DNA primers

10. Which of the following statements about PCR is incorrect?
   a. The DNA template is denatured using helicase.
   b. PCR is used to obtain millions of copies of a specific region of DNA.
   c. A thermostable DNA polymerase is used because of the high temperatures required in PCR.
   d. Template DNA, a set of primers, deoxynucleotides, a thermostable DNA polymerase, and a thermocycler are the important components in PCR.
   e Primers are needed because DNA polymerase cannot initiate synthesis, but can only elongate from an existing 3′-OH.

11. Which of the following is not an advantage of automated cycle sequencing over the chain termination method of sequencing?
   a. The reactions in an automated sequencer can be performed faster.
   b. The reactions performed in an automated sequencer can be read by a computer rather than a human.
   c. Higher temperatures are used during cycle sequencing, which prevent secondary structures from forming in the DNA and early termination of the reaction.
   d. In cycle sequencing, nonspecific interactions by the primer can be controlled by raising the annealing temperature.
   e. All of the above are advantages of cycle sequencing.

12. Which of the following statements about degenerate primers is not correct?
- **a.** Degenerate primers have a mixture of two or three bases at the wobble position in the codon.
- **b.** Because of the nature of degenerate primers, the annealing temperature during PCR using these primers must be lowered to account for the mismatches.
- **c.** Degenerate primers are often designed by working backwards from a known amino acid sequence.
- **d.** Degenerate primers are used even when the sequence of DNA is known.
- **e.** Within a population of degenerate primers, some will bind perfectly, some will bind with mismatches, and others will not bind.

13. Which of the following techniques would allow a researcher to determine the genetic relatedness between two samples of DNA?
- **a.** inverse PCR
- **b.** reverse transcriptase PCR
- **c.** TA cloning
- **d.** overlap PCR
- **e.** randomly amplified polymorphic DNA

14. Why would a researcher want to use RT-PCR?
- **a.** RT-PCR is used to compare two different samples of DNA for relatedness.
- **b.** RT-PCR creates an mRNA molecule from a known DNA sequence.
- **c.** RT-PCR generates a protein sequence from mRNA.
- **d.** RT-PCR generates a DNA molecule without the noncoding introns from eukaryotic mRNA.
- **e.** All of the above are applications for RT-PCR.

15. Which of the following is an application for PCR?
- **a.** site-directed mutagenesis
- **b.** creation of insertions, deletions, and fusions of different gene segments
- **c.** amplification of specific segments of DNA
- **d.** for cloning into vectors
- **e.** all of the above

Further Reading

Agarwal KL, Büchi H, Caruthers MH, Gupta N, Khorana HG, Kleppe K, Kumar A, Ohtsuka E, Rajbhandary UL, Van De Sande JH, Sgaramella V, Weber H, Yamada T (1970). Total synthesis of the gene for an alanine transfer ribonucleic acid from yeast. *Nature* **227**, 27–34.

Clark DP (2005). *Molecular Biology: Understanding the Genetic Revolution*. Elsevier Academic Press, San Diego, CA.

Glick BR, Pasternak JJ (2003). *Molecular Biology: Principles and Applications of Recombinant DNA* 3rd ed. ASM Press, Washington, DC.

Kresge N, Simoni RD, Hill RL (2005). Arthur Kornberg's discovery of DNA polymerase I. *J Biol Chem* **280**, 46.

Matthaei JH, Nirenberg MW (1961). Characteristics and stabilization of DNase-sensitive protein synthesis in *E. coli* extracts. *Proc Natl Acad Sci USA* **47**, 1580–1588.

Nirenberg MW, Leder P (1964). RNA codewords and protein synthesis. *Science* **145**, 1399.

CHAPTER 5

RNA-Based Technologies

RNA plays a multifaceted role in biology that is adaptable for many different applications in biotechnology. The most widely understood role of RNA is as messenger RNA. For many years, RNA was simply thought of as this intermediary with an important albeit limited function. It is now known that important gene regulation occurs at the level of RNA. Rather than simply controlling gene expression from the genome, organisms produce RNA that controls the mRNA translation into protein. In some organisms, **antisense RNA** is one way by which RNA controls protein translation. Antisense RNA binds to the complementary mRNA and blocks translation. From this discovery came the potential use of antisense RNA to block or attenuate synthesis of proteins that cause various diseases. A second method of RNA-regulated gene expression is **RNA interference (RNAi)**. Here small noncoding RNAs identify specific mRNAs and trigger their degradation. This fortuitous finding opened the door to a specific technique for controlling protein translation. Since RNA interference was discovered in 1993, its application has become widespread. The third novel role for RNA was the discovery that some RNA sequences can catalyze enzyme reactions themselves. **Ribozymes**, as they are called, are found in many different organisms, catalyzing cleavage and ligation of various substrates. This chapter focuses on the roles of antisense RNA, RNA interference, and ribozymes in biology and biotechnology.

ANTISENSE RNA MODULATES mRNA EXPRESSION

Antisense refers to the orientation of complementary strands during transcription. The two complementary strands of DNA are referred to as *sense* (= coding or plus) and *antisense* (= noncoding or minus; see Chapter 2). Transcription occurs using the antisense strand as template, resulting in an mRNA that is identical in sequence to the sense strand (except for the replacement of uracil for thymine). Antisense RNA is synthesized using the sense strand as template; therefore, it has a sequence complementary to mRNA (Fig. 5.1).

Antisense RNA is sometimes made in normal cells of many different organisms, including humans. Artificial antisense RNA is also made for manipulating gene expression in laboratory settings. When a cell has both the mRNA (i.e., the sense strand of RNA) plus a complementary antisense copy, the two single strands anneal to form double-stranded RNA. The duplex can either inhibit protein translation by blocking the ribosome binding site, or inhibit mRNA splicing by blocking a splice site (Fig. 5.2A). When antisense sequences are made in the laboratory, they are usually synthesized as DNA as this is more stable than RNA (see Chapter 4). In this case, the DNA:RNA duplex is digested by RNase H (see Fig. 5.2B). RNase H is a cellular enzyme that normally functions during replication. It recognizes and cleaves the RNA backbone of an DNA:RNA duplex, targeting the antisense DNA:mRNA duplex for further degradation. RNase H recognizes a 7-base-pair heteroduplex, so the region of homology between the antisense DNA and target mRNA need not be very long.

Antisense RNA sequences are complementary to a target mRNA. Antisense RNA forms double-stranded regions that block either protein translation or splicing of introns.

Antisense Controls a Variety of Biological Phenomena

Naturally occurring antisense genes have been found that control a variety of different processes. One example of natural antisense control is found in *Neurospora*. This fungus follows a strict schedule, based on circadian rhythms, and forms hyphae only at specific times during the day. Many mutants have been identified that do not follow such a strict standard of time. The genes affected by these mutations are regulators in the circadian rhythm of *Neurospora*. One of the first circadian rhythm genes identified was *frequency* (*frq*).

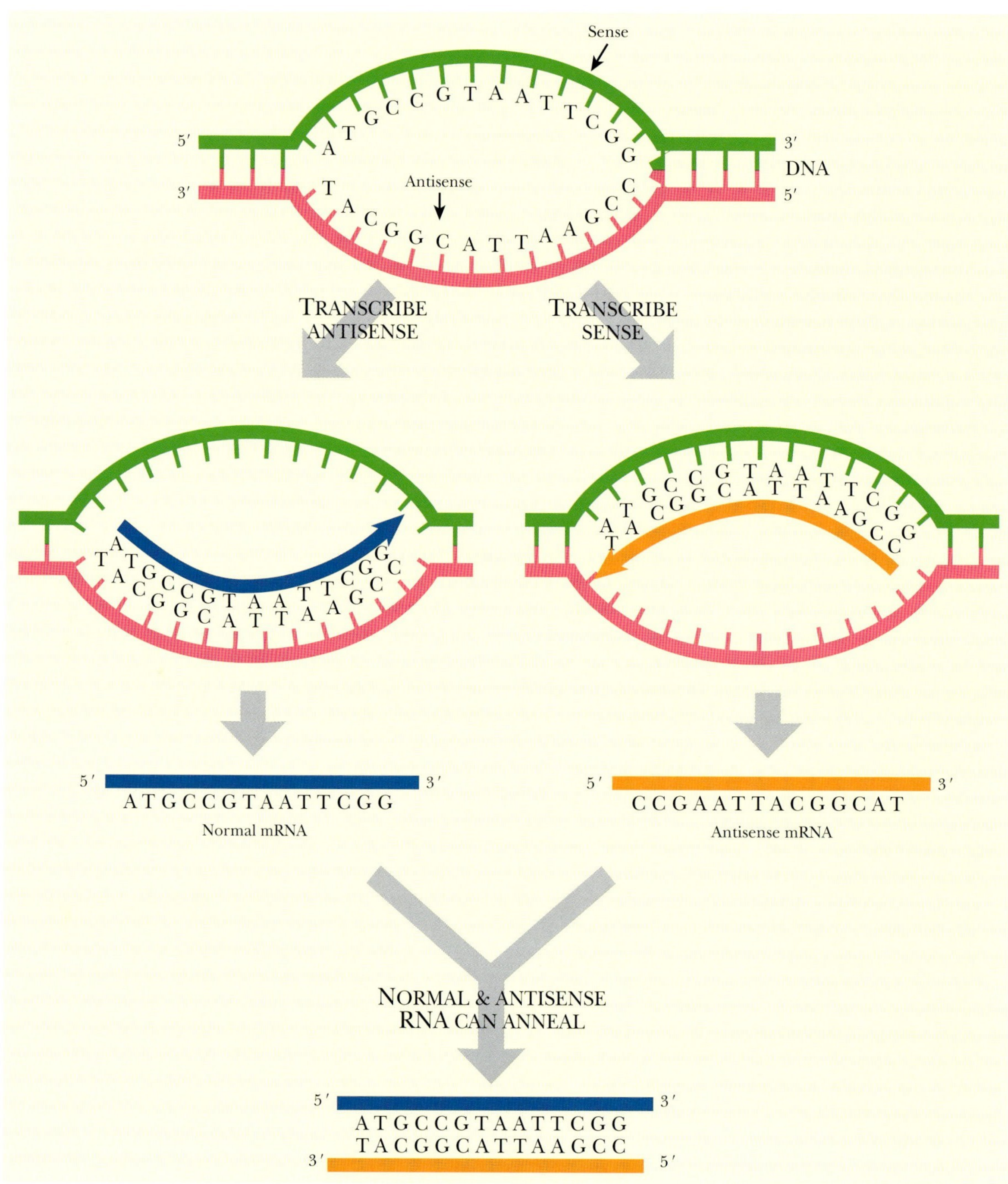

FIGURE 5.1 Antisense RNA Is Complementary to Messenger RNA
Transcription from both strands of DNA creates two different RNA molecules, on the left, the messenger RNA, and on the right, antisense RNA. These two have complementary sequences and can form double-stranded RNA.

Mutations in this gene change how often the fungus forms hyphae. The amount of *frq* mRNA fluctuates, with highest levels during the day and lowest at night. Interestingly, antisense *frq* RNA also cycles, but in reverse, with the lowest levels during the day and the highest during the night. Although the exact mechanism is uncertain, *Neurospora* that do not produce antisense *frq* RNA have disrupted circadian rhythms. In addition, both the antisense and sense mRNAs are induced by light and therefore respond directly to the environment to maintain the correct circadian rhythm.

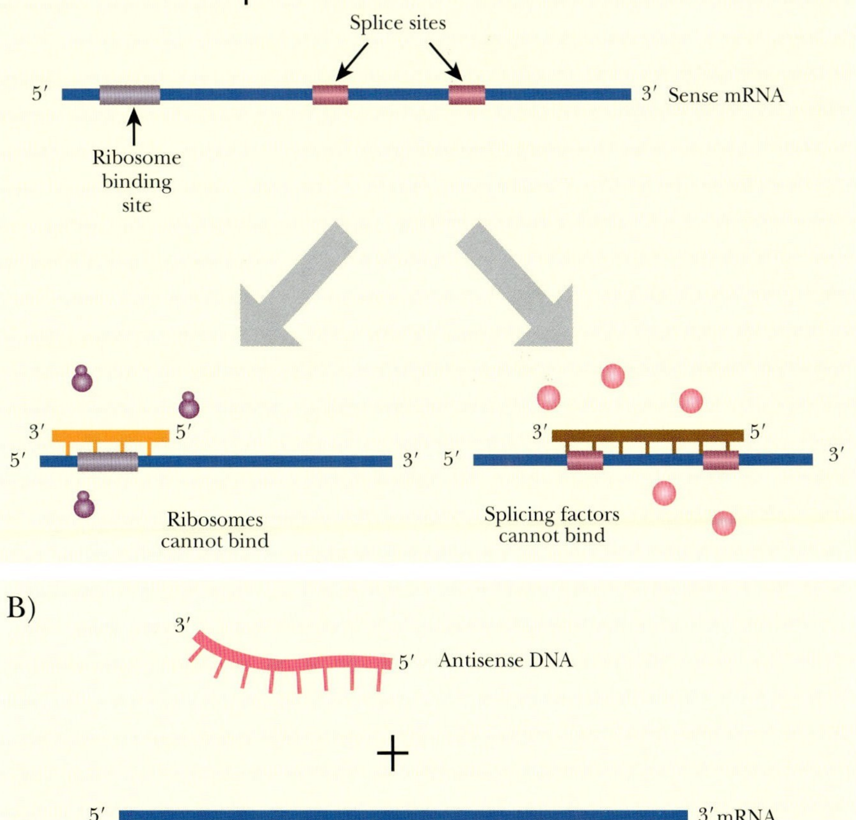

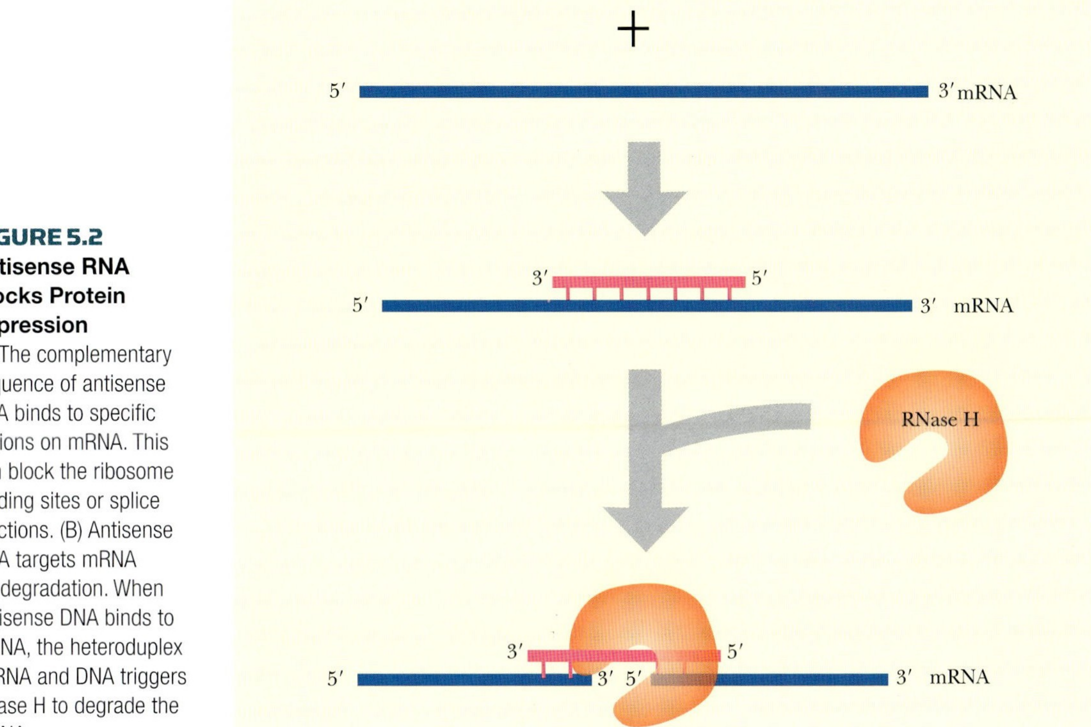

FIGURE 5.2
Antisense RNA Blocks Protein Expression
(A) The complementary sequence of antisense RNA binds to specific regions on mRNA. This can block the ribosome binding sites or splice junctions. (B) Antisense DNA targets mRNA for degradation. When antisense DNA binds to mRNA, the heteroduplex of RNA and DNA triggers RNase H to degrade the mRNA.

Using antisense to regulate gene expression is so widespread in nature that scientists became curious how many potential antisense/sense partners exist in various genomes. Computer algorithms have been devised to search for sequences that could function as antisense. In the human genome, there are a predicted 1600 different partners. The most interesting finding in the antisense field is the realization that small noncoding regulatory RNAs called **microRNAs (miRNAs)** inhibit gene expression through an antisense mechanism (see later discussion). Using computer searches, an additional 250 potential microRNAs have been identified in humans, but because these are only about 20 nucleotides long, identifying them conclusively by computer is very difficult.

EXAMPLES OF NATURAL ANTISENSE CONTROL

- Control of ColE1 Plasmid Replication: RNAI and RNAII mRNA are sense/antisense partners that prevent DNA polymerase from initiating plasmid replication. The amount of antisense RNAI controls how often replication is initiated.
- Control of X-Chromosome Inactivation: Antisense RNA binds to the *Xist* gene to prevent its expression and thus prevents X chromosome inactivation.
- Control of Circadian Rhythm in *Neurospora*: The time of day controls when the fungus forms hyphae by regulating the antisense and sense mRNA for the *frq* gene. The antisense and sense mRNAs are controlled by light and dark cycles.
- Iron Metabolism in Bacteria: FatB/RNAα are sense/antisense partners that control regulation of iron uptake in the fish pathogen *Vibrio anguillarum*. When iron is plentiful and iron scavenging is not necessary, higher amounts of RNAα prevent *fatA* and *fatB* expression. When iron is scarce, the bacteria need to get iron from the environment. RNAα mRNA is degraded and *fatA* and *fatB* are expressed so *Vibrio* can ingest iron.
- Control of HIV-1 Gene Expression: Antisense *env* mRNA binds to the Rev Response Element (RRE) on *env* mRNA. When antisense blocks the RRE, Env protein is not produced. When antisense *env* mRNA is absent, Env protein is produced.
- Control of RNA Editing: Antisense/sense loops are formed between complementary exon and intron sequences of the gene for the glutamate-gated ion channel in human brain. These loops are recognized by dsRNA-specific adenosine deaminase (DRADA), which converts adenosine to inosine by deamination. This alters the sequence of the final mRNA and hence of the protein. This reduces the permeability of the ion channel.
- Alternate Splicing of Thyroid Hormone Receptor mRNA: Antisense RNA transcribed from the thyroid hormone locus inhibits splicing of the *c-erbAα* gene. Two alternately spliced transcripts give the authentic thyroid hormone receptor and a decoy receptor that does not bind thyroid hormone. These two forms modulate cellular responses to thyroid hormone.
- Developmental Control of Basic Fibroblast Growth Factor (bFGF): Basic FGF controls various developmental and repair processes, including mesoderm induction, neurite outgrowth, differentiation, and angiogenesis (blood vessel development). Antisense bFGF mRNA binds to cellular bFGF mRNA during development of *Xenopus laevis* (African clawed frog), targeting the bFGF mRNA for degradation.
- Control of Eukaryotic Transcription Factors: The transcription factor hypoxia-induced factor (HIF-1) is a basic helix-loop-helix dimeric protein that turns on genes associated with oxygen and glucose metabolism, including glucose transporters 1 and 3 and enzymes of the glycolytic pathway. Antisense mRNA to the α subunit mRNA controls the expression of the transcription factor. The level of antisense RNA is modulated by the amount of oxygen in the environment.

Organisms have antisense genes and microRNAs that bind to a target mRNA and prevent its translation into protein. These modulate a large number of systems, including hyphae formation in *Neurospora*, development, replication, and many more.

ANTISENSE OLIGONUCLEOTIDES

In the laboratory, antisense RNA can be made by two different methods (Fig. 5.3). The easiest method is to chemically synthesize oligonucleotides that are complementary to the target gene. The oligonucleotides are then injected or transformed into the target cell (see later discussion). Alternatively, the gene of interest can be cloned in the opposite orientation, so that transcription gives antisense RNA. The vector carrying the anti-gene is then transformed into the target organism.

Chemically synthesized antisense oligonucleotides are traditionally made of DNA rather than RNA for two reasons. DNA is more stable in the laboratory, and DNA synthesis is an established and automated procedure. In the cell, DNA oligonucleotides are still very susceptible to degradation by endonucleases; therefore, various chemical modifications are added to increase intracellular stability. The most common modification is to replace one of the nonbridging oxygens in the phosphate groups with sulfur (Fig. 5.4) to make a **phosphorothioate oligonucleotide**. This makes the phosphorus a chiral center; one diastereomer is resistant to nuclease degradation but the other is still sensitive, leaving about half of the antisense molecules functional inside the cell. This modification does not affect the solubility of the oligonucleotides or their susceptibility to RNase H degradation. These types of antisense oligonucleotides have been developed to inhibit cancers such as melanoma and some lung cancers. The most common side effect with phosphorothioate oligonucleotides is nonspecific interactions, especially with proteins that interact with sulfur-containing molecules.

Two other modifications have fewer nonspecific interactions than phosphorothioate oligonucleotides. Adding an *O*-alkyl group to the 2′-OH of the ribose makes the oligonucleotide resistant to DNase and RNase H degradation (see Fig. 5.4). Inserting an amine into the ribose ring, thus changing the five-carbon ribose into a morpholino ring, creates **morpholino-antisense oligonucleotides** (see Fig. 5.4). In addition to the morpholino ring, a second amine replaces the nonbridging oxygen to create a **phosphorodiamidate** linkage. This amine neutralizes the charged phosphodiester of typical oligonucleotides. The loss of charge affects their uptake into cells, but alternative methods have been developed to get these antisense molecules into the cells (see later discussion). Both types of modified oligonucleotides are resistant to RNase H. Therefore, they do not promote degradation of an mRNA:DNA hybrid target. Consequently, their use is restricted to blocking splicing sites in the pre-mRNA transcript or to block ribosome binding sites.

The most different modified oligonucleotides are **peptide nucleic acids (PNAs)**, which have the standard nucleic acid bases attached to a polypeptide backbone (normally found in proteins) rather than a sugar-phosphate backbone (see Fig. 5.4). The polypeptide backbone has been modified so that the RNA bases are spaced at the same distance as the typical oligonucleotide. The spacing is critical to function because the bases of a PNA must match the bases in the target RNA. This molecule is also uncharged and works through non-RNase H dependent mechanisms as with morpholino-antisense oligonucleotides. Antisense PNA has been developed to inhibit translation of the HIV viral transcript, *gag-pol*, and to block translation of two cancer genes, *Ha-ras* and *bcl-2*.

As noted earlier, antisense oligonucleotides that are resistant to RNase H must be made to target splice sites and/or ribosome binding sites in order to block the target mRNA. In some cases these sequences are not well characterized in the target mRNA, so modified antisense oligonucleotides become useless. Making mixed or chimeric antisense oligonucleotides can restore the targeting of RNase H to the mRNA and allow the researcher to use the modified structures to prevent degradation (see later discussion). In these chimeric antisense oligonucleotides, the core has a short (~7) base-pair span of phosphorothioate linkages, which are RNase H sensitive, flanked on each side by sequences consisting of one of the RNase H–resistant modifications (Fig. 5.5). The flanking regions contain 2′-*O*-methyl groups, morpholino structures, or even PNA. These chimeric molecules can target any accessible regions of the mRNA, not just splice sites or ribosome binding sites.

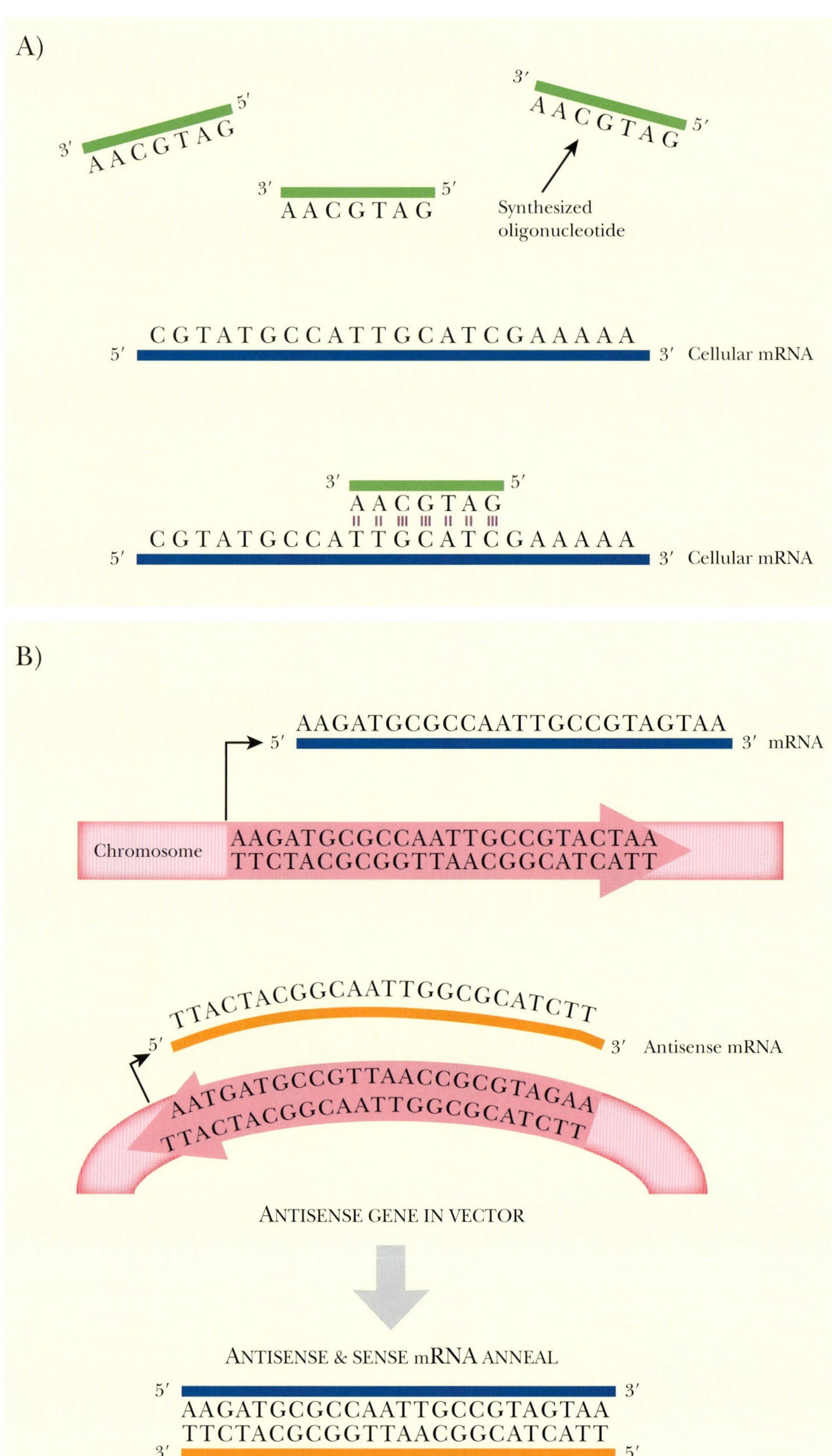

FIGURE 5.3 Making Antisense in the Laboratory
(A) Antisense oligonucleotides. Small oligonucleotides are synthesized chemically and injected into a cell to block mRNA translation. (B) Antisense genes. Genes are cloned in inverted orientation so that the sense strand is transcribed. This yields antisense RNA that anneals to the normal mRNA, preventing its expression.

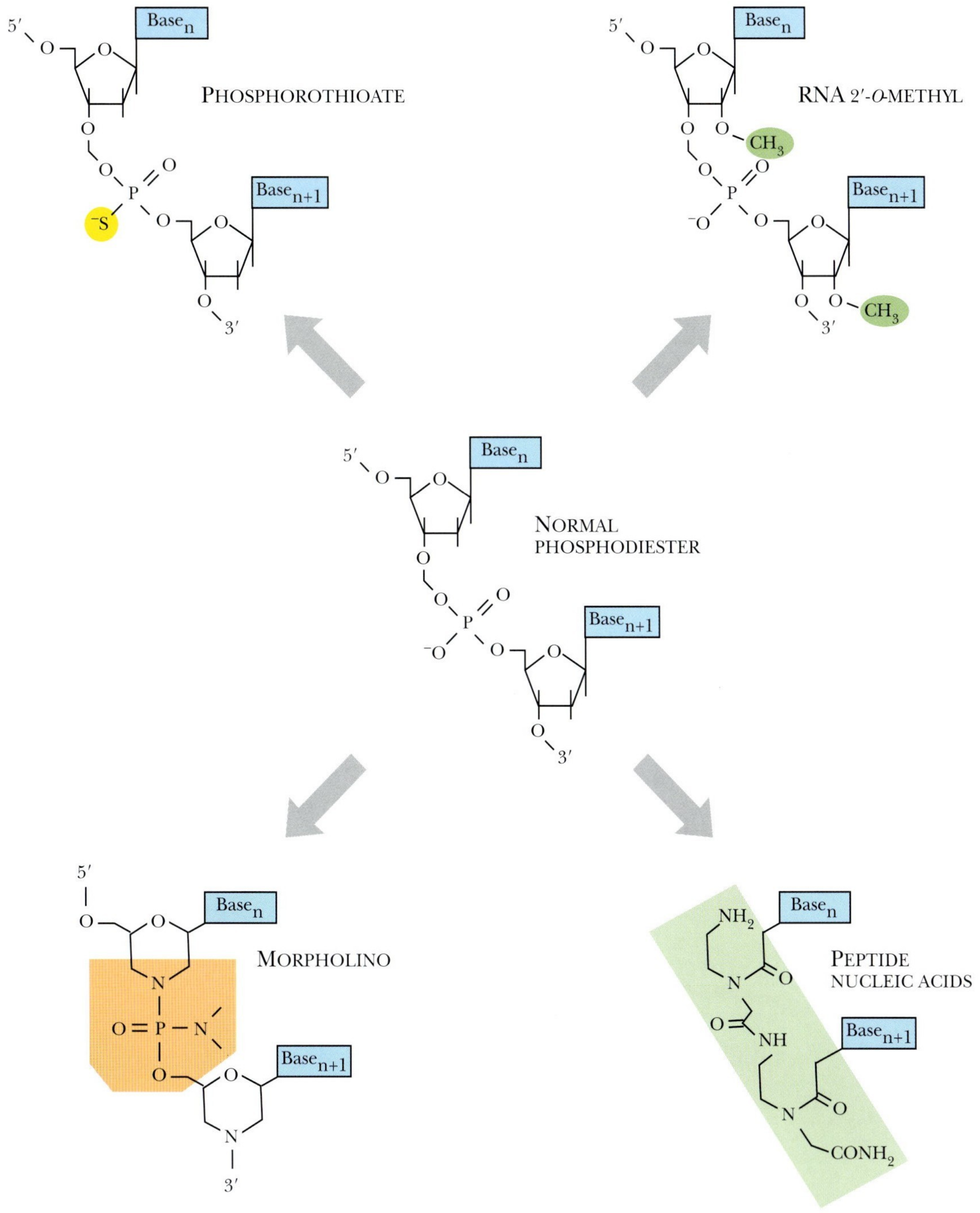

PHOSPHOROTHIOATE

RNA 2′-O-METHYL

NORMAL PHOSPHODIESTER

MORPHOLINO

PEPTIDE NUCLEIC ACIDS

FIGURE 5.4 Modifications to Oligonucleotides
Replacing the nonbridging oxygen with sulfur (*upper left*) increases oligonucleotide resistance to nuclease degradation. Adding an *O*-alkyl group to the 2′-OH on the ribose (*upper right*) makes the oligonucleotide resistant to nuclease degradation and also to RNase H. Morpholino-antisense oligonucleotides and peptide nucleic acids are two more substantial changes in the standard oligonucleotide structure (*lower left and lower right*). Both are resistant to RNase H degradation. The RNase H resistant oligonucleotides are used to target splice junctions or ribosome binding sites in order to prevent translation of their target mRNA.

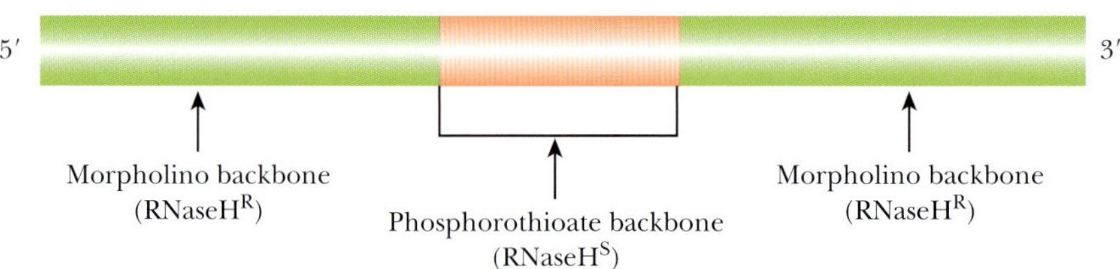

FIGURE 5.5 Chimeric Oligonucleotides Use RNase H Sensitive Cores to Degrade Target mRNA
Chimeric oligonucleotides are made using different chemistries. The core region maintains RNase H sensitivity, whereas the outer regions are RNase H resistant. When the oligonucleotide hybridizes to target molecules in the cell, RNase H will digest the hybrid of oligonucleotide plus mRNA only where the central domain forms a heteroduplex. RNase H will not digest any nonspecific complexes between the chimeric oligonucleotide ends and the wrong mRNA.

To stabilize antisense oligonucleotides and guard these from degradation in the cell, various modifications to the antisense oligonucleotide structure can be incorporated.

When RNase H does not recognize the antisense oligonucleotide, the oligonucleotide must bind to the ribosome binding site or splice junctions to inhibit protein translation.

Chimeric antisense oligonucleotides include a core sequence of about seven nucleotides that is recognized by RNase H, and then RNase H degrades the oligonucleotide and its target mRNA.

Antisense Oligonucleotides May Alter Splicing

Alternative splicing is widely used to control gene expression in eukaryotes. Some diseases are caused by aberrant patterns of splicing. For example, beta-thalassemia is a blood disorder where red blood cells cannot carry enough oxygen because of defective hemoglobin. Some cases of the disorder arise from aberrant splicing of the beta-globin pre-mRNA. Antisense morpholino-oligonucleotides have been designed that target the mutant splice site between exons two and three in the beta-globin gene. The antisense oligonucleotide corrects the splicing pattern, and restores the correct protein in red blood cells taken from patients with beta-thalassemia (Fig. 5.6A).

The *Bcl-x* gene in humans produces two different proteins because of alternate splicing. Bcl-xL, the longer protein, includes a segment of coding material between exons 1 and 2. Bcl-xS, the shorter protein, lacks this segment. Bcl-xL protein promotes cell growth by blocking apoptosis, whereas Bcl-xS opposes this and causes cells to die via apoptosis (see Chapter 20). In some cancers, the long form is overexpressed. Thus, devising a method to inhibit Bcl-xL expression and enhance Bcl-xS expression could induce these cancer cells to undergo apoptosis and die. Antisense 2′-*O*-methyl-oligonucleotides to the splice site in the pre-mRNA prevent the longer form from being made. The Bcl-xS protein then counteracts the cancerous growth by promoting apoptosis (see Fig. 5.6B). Those cancer cells that were "fixed" by antisense therapy also became more sensitive to chemotherapeutic agents because of the restoration of apoptosis.

Antisense oligonucleotides are used to correct the genetic mistakes of beta-thalassemia and some cancers.

Problems Associated with Antisense Oligonucleotides

There are two major problems associated with antisense technology. The most important is the design of the antisense oligonucleotide itself. Although mRNA is usually depicted as a linear molecule, mRNA can fold back on itself, forming normal Watson-Crick base pairs between different regions. In addition, some regions form loops and turns, creating a

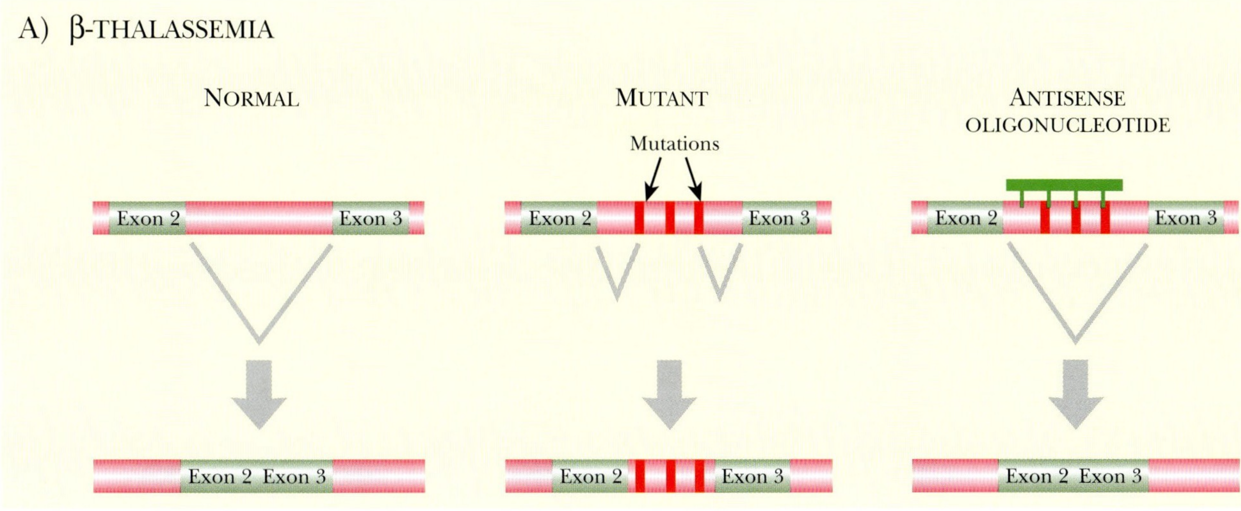

FIGURE 5.6 Antisense Oligonucleotides Correct Splicing Errors
(A) Beta-thalassemia is a blood disorder where extra splice sites are found between exon 2 and exon 3 (*middle*). Antisense oligonucleotides that block the extra splicing junctions restore the original structure of the gene during splicing (*right*). (B) The *Bcl-x* gene makes two different proteins through alternate splicing. Bcl-xS is made in a normal defective cell and promotes cell death via apoptosis (*left*). Some cancerous cells do not produce Bcl-xS. Instead, the longer form, Bcl-xL, is produced via alternate splicing, and it protects the cancerous cell from apoptosis (*middle*). To resensitize the cancerous cell to apoptosis, antisense oligonucleotides that block the splicing junction for Bcl-xL restore Bcl-xS protein production and, ultimately, sensitivity to apoptosis (*right*).

three-dimensional structure. Because some target regions are annealed to other regions via base pairing, and other target regions may be hidden deep within the three-dimensional structure, finding a region of the target mRNA accessible to a laboratory-designed antisense oligonucleotide can be difficult.

The second major challenge with antisense technology is nonspecific interactions with nontarget molecules. As mentioned earlier, phosphorothioate oligonucleotides have a propensity to bind cellular proteins that interact with sulfur-containing compounds, essentially pulling the oligonucleotides away from the target mRNA. In addition, small stretches of an oligonucleotide may base-pair with nonspecific mRNA sequences. Normally, this would not pose any problem, because the heteroduplex region is short and unstable. However, if the oligonucleotides are sensitive to RNase H, this endonuclease may degrade the wrong mRNA. The extent of homology between RNA and DNA only needs to be about seven base pairs for RNase H recognition, so when a short unstable heteroduplex forms, it can be sufficient to activate RNase H. In order for the antisense oligonucleotide to specifically target only one mRNA, the region of homology needs to be about 20 bases in length to be specific. Using chimeric antisense oligonucleotides is one method to overcome this problem, because RNase H can cleave only the core region of the chimera (Fig. 5.7). Even if the ends of the chimeric antisense oligonucleotide find a seven-base-pair region of homology with a nontarget mRNA, RNase H cannot cleave these nonspecific targets.

Antisense oligonucleotides may have a difficult time finding their target binding site if the target mRNA has extensive secondary structure.

Chimeric antisense nucleotides are used to prevent nontarget mRNA degradation.

EXPRESSION OF ANTISENSE RNA CONSTRUCTS

Rather than using antisense oligonucleotides, antisense RNA can be transcribed from a vector that has been inserted into the cell. (See Chapters 14 and 15 for more details on inserting foreign DNA into plant and animal cells.) First, the target gene is cloned in reverse orientation so that antisense RNA is produced instead of sense mRNA (see Fig. 5.3B). This method is believed to inactivate the cellular target mRNA by forming a heteroduplex of sense/antisense RNA. Heteroduplex formation relies on both RNAs to unfold. If either RNA has a very stable secondary or tertiary structure, then the construct may not work inside the cell.

The advantage of internal synthesis of antisense RNA is that the antisense expression can be controlled. If the antisense gene is cloned behind an inducible promoter, then the antisense RNA is not made until the gene is induced by specific signals or conditions. This may be useful to allow organ-specific expression of an antisense gene. Another advantage is that the antisense RNA may be continuously expressed internally over a long-term period. This avoids the inconvenience and expense of constant administration of external antisense oligonucleotides.

A target gene can be cloned in the inverse direction to create an antisense vector. When this is expressed in the cell, the antisense mRNA and endogenous mRNA bind, thus preventing expression into protein.

DELIVERY OF ANTISENSE THERAPIES

Getting antisense oligonucleotides into cells requires special techniques, because they do not cross cell membranes easily enough on their own to be effective. Getting the antisense oligonucleotide into the correct cellular location poses another obstacle to devising methods of antisense therapy. Although the uptake of oligonucleotides occurs by an unknown mechanism, the process is active. That is, uptake depends on temperature, the concentration of the oligonucleotide, and the type of cell. It is thought that oligonucleotides are taken up by endocytosis and pinocytosis. (Pinocytosis is similar to endocytosis, but the droplets that enter the cell are much smaller.) There is also a possibility that oligonucleotides enter via a membrane-bound receptor at very low concentrations, whereas at higher concentrations, they enter via pinocytosis and endocytosis.

A common method to deliver oligonucleotides to cells is to encapsulate the oligonucleotides in **liposomes** (Fig. 5.8A). Liposomes are small vesicles made of bilayers of phospholipids and cholesterol. Whether the liposome is neutral or positively charged depends on the type of phospholipid used to manufacture them. The oligonucleotides ride on the exterior if the liposome is positively charged or are encapsulated inside the aqueous interior. Positively charged liposomes are drawn to the cell surface because it is negatively charged, and the entire

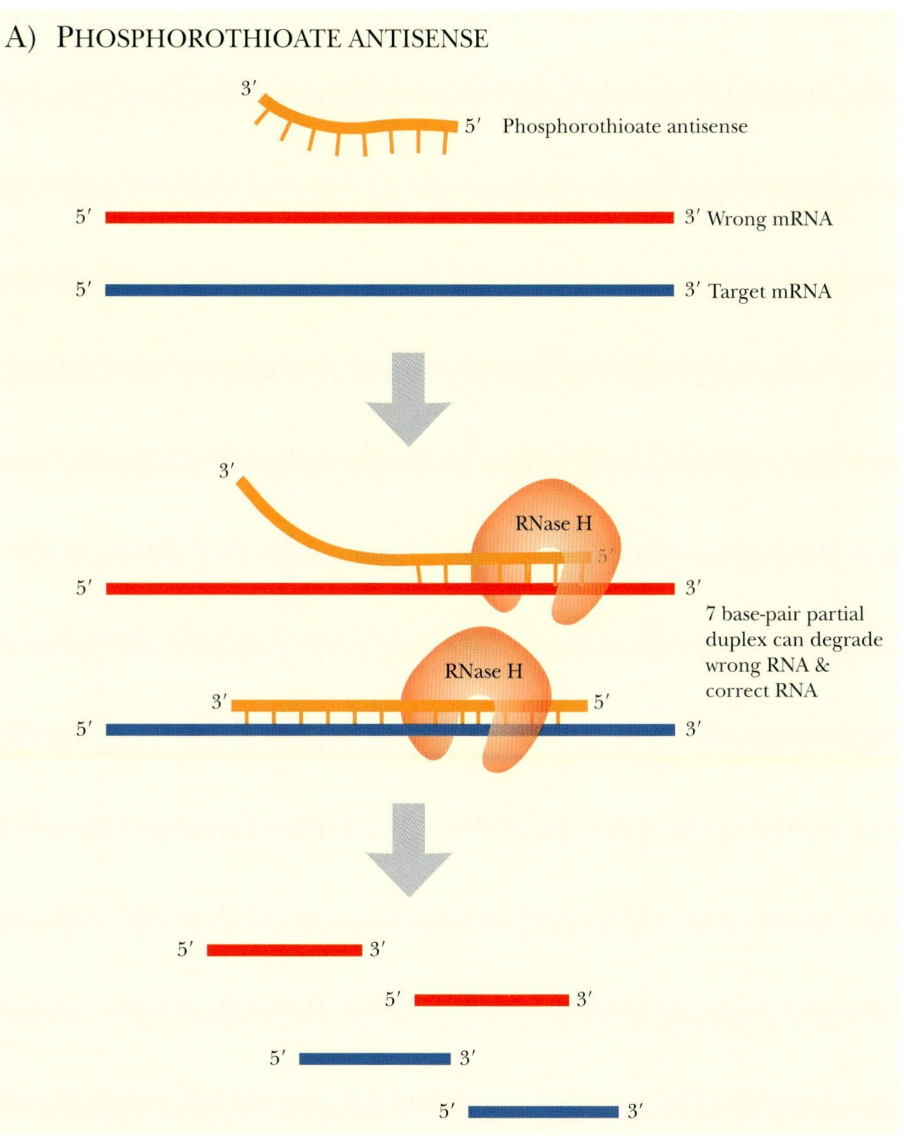

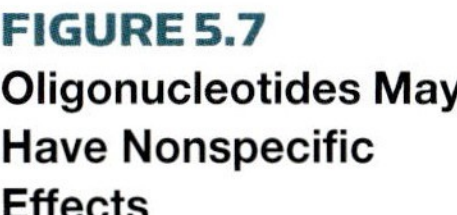

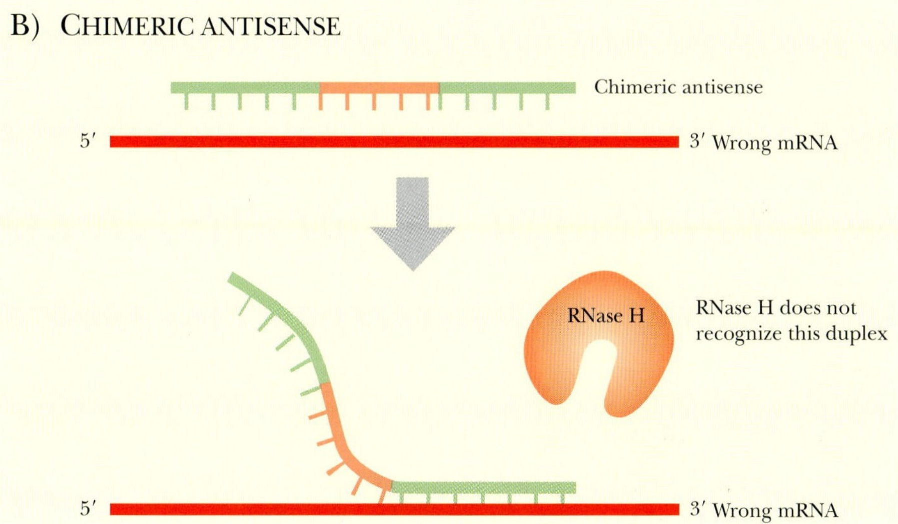

FIGURE 5.7
Oligonucleotides May Have Nonspecific Effects
(A) Misdirected RNase H cleavage can occur if the oligonucleotide recognizes a short seven base-pair region of homology.
(B) Chimeric oligonucleotides prevent nonspecific RNase H degradation. When the RNase H resistant ends of the chimeric oligonucleotide bind to the wrong mRNA, the structure of the oligonucleotide prevents RNase H from recognizing and cutting the heteroduplex.

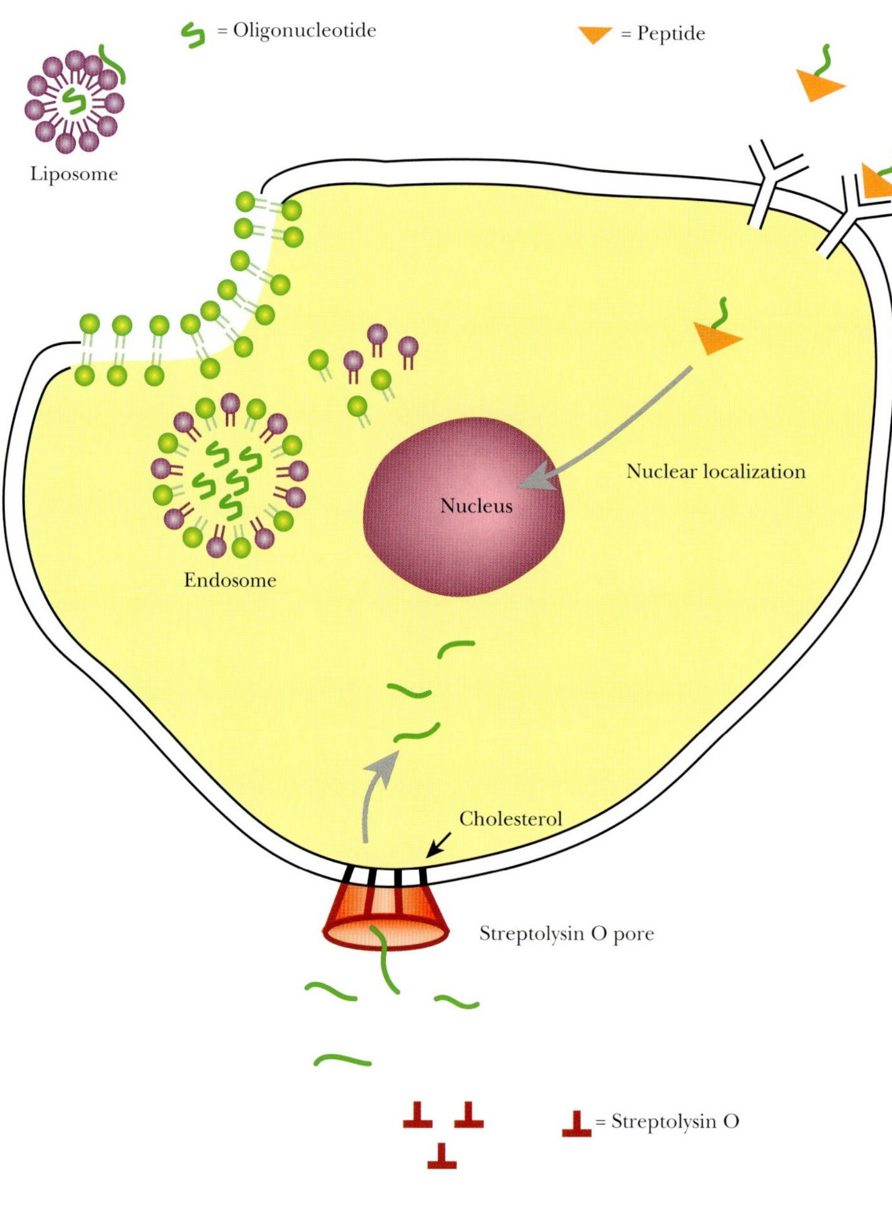

FIGURE 5.8
Methods of Antisense Oligonucleotide Uptake by Cells
(A) Liposomes are spherical structures made of lipids and cholesterol. Oligonucleotides are either encapsulated in the central core or ride on the exterior surface of the liposome. The complexes enter the cell via endocytosis and are released into the cytoplasm. (B) Basic peptides are naturally occurring proteins that normally enter the nucleus of target cells. The oligonucleotide can be fused to these peptides and ride into the nucleus with the basic peptide. (C) Streptolysin O, a toxin from *Streptococci* bacteria, aggregates at the membrane to form a porelike structure. The oligonucleotides can pass into the cell through the pore.

liposome, oligonucleotides and all, is engulfed by endocytosis. Some liposomes contain "helper" molecules that make the endosomal membrane unstable and release the liposome directly into the cytoplasm. Other delivery "vehicles" are cationic polymers, which include poly-L-lysine and polyethyleneimine. These operate via electrostatic interactions as discussed earlier, but they are toxic when taken into the cell; therefore, they are not used very often.

When the endosomal pathway is used for uptake, as with liposomes, there is a good chance that the antisense oligonucleotide will be degraded or not released to the cytoplasm. To alleviate this problem, antisense oligonucleotides may be delivered to the cell attached to **basic peptides** (see Fig. 5.8B). These include the Tat protein of HIV-1, the N-terminal segment of HA2 subunit of influenza virus agglutinin protein, and Antennapedia peptide from *Drosophila* (which normally acts as a transcription factor). These peptides are able to enter the cell nucleus. By attaching these to oligonucleotides, the antisense oligonucleotides are taken directly into the nucleus.

Other methods to get oligonucleotides into the cells require chemically or manually disrupting the membrane (see Fig. 5.8). Membrane pores can be generated by **streptolysin O** permeabilization (see Fig. 5.8C) or electroporation (see Chapter 3). Streptolysin O is a toxin from *Streptococci* bacteria that aggregates after binding to cholesterol in the membrane, forming a pore. The oligonucleotide passes through the pore and enters the cytoplasm directly. Antisense oligonucleotides can also be microinjected directly into each cell, but this method cannot be used for treating patients and is only useful for small-scale experiments on cultured cells (Fig. 5.9). Another mechanical method is called **scrape-loading** (see Fig. 5.9). Here, adherent cultured cells are gently scraped off the dish while the oligonucleotide bathes the cells. Removal of the cells probably creates small openings that allow the oligonucleotides to enter the cytoplasm.

Antisense oligonucleotides enter the target cell by endocytosis of oligonucleotide-filled liposomes, by riding on basic peptides that normally enter the nucleus, by passing through pores created by streptolysin O, or by mechanical shearing.

RNA INTERFERENCE USES ANTISENSE SEQUENCES TO INHIBIT GENE EXPRESSION

RNA interference (RNAi) is a recently discovered pathway for endogenous gene regulation where short double-stranded RNA (dsRNA) segments trigger an enzyme complex to degrade a target mRNA. In essence, the short dsRNA pieces decrease target protein expression by degrading its complementary mRNA. RNAi was discovered in a variety of different organisms, including plants, fungi, mammals, flies, and worms, and consequently has multiple names. Different organisms have variations of the same basic response. Mutations in the enzymes responsible for RNAi affect a wide range of cellular processes. Some affect development of the organism; others affect the ability to fend off viruses, particularly RNA viruses. In still other cases, mutations affecting RNAi increase transposon movement, suggesting that RNAi may also prevent transposon jumping. All these processes rely on regulating mRNA translation or mRNA degradation.

RNAi is divided into two different phases, the initiation phase and the effector phase. Initiation begins by the formation of dsRNA. The dsRNA can arise from three main sources. First, externally infecting RNA viruses replicate through a dsRNA intermediate, which can trigger RNAi. One theory suggests that the RNAi mechanism may have evolved to combat these infecting viruses. Next, the organism's own genomic DNA contains sequences that code for microRNAs, which are specific mediators of RNAi (see later discussion). Finally, dsRNA can be produced from aberrant transcription of a genetically engineered gene (Fig. 5.10A). During the next step of initiation, dsRNA binds to an endonuclease called **Dicer**, which cuts the dsRNA into small fragments about 21 to 23 nucleotides in length called **short interfering RNAs** (**siRNAs**; see Fig. 5.10B). Dicer is a dsRNA-dependent RNA endonuclease that belongs to the RNase III family. The siRNAs have a two-nucleotide overhang on the 3′ ends (characteristic of RNase III–type enzymes). The 5′ ends are phosphorylated by a kinase associated with Dicer, making the siRNAs competent for the next phase.

In the effector phase, Dicer transfers the siRNA to a ribonucleoprotein complex called the **RNA-induced silencing complex (RISC)**. RISC is activated by the siRNAs and uses an RNA helicase to unwind the double-stranded fragments, making single strands. The antisense single-stranded siRNA is then kept as a guide to find complementary sequences in the cytoplasm. When RISC binds complementary sequences, these are first cleaved by an endonuclease within the RISC complex and are then further degraded by exonucleases in the cytosol. This destroys all of the mRNA that is complementary to the siRNA. Both Dicer

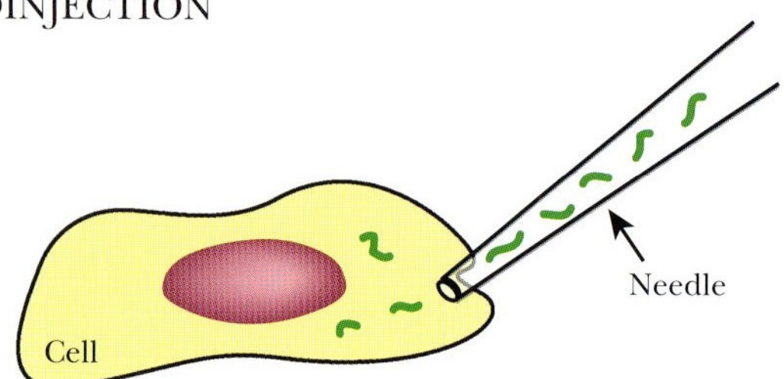

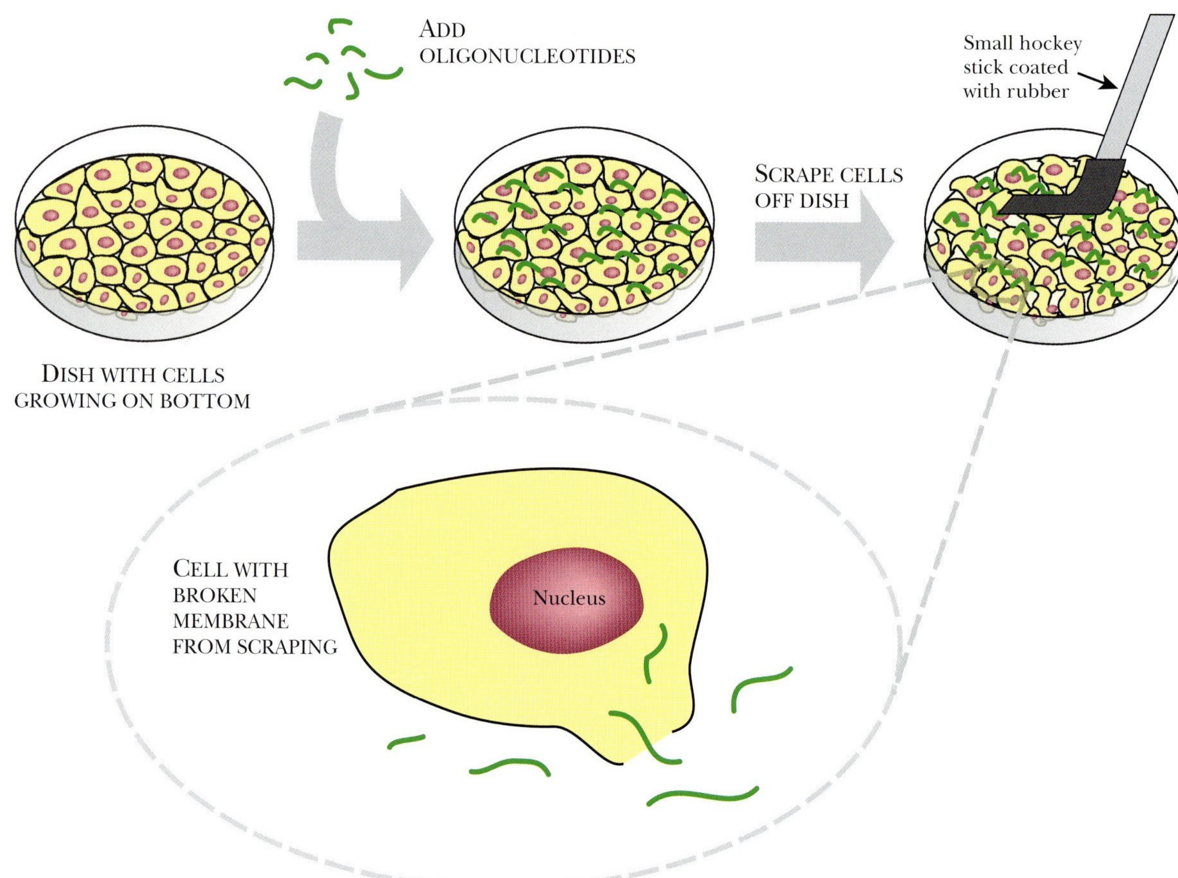

FIGURE 5.9 Microinjection and Scrape-Loading
(A) Oligonucleotides can be injected directly into a cell using a very fine needle. Microinjection can be done on individual cells grown in culture. (B) Scrape-loading is a mechanical method of getting oligonucleotides into cultured cells. As the cells are scraped off the bottom of the culture dish, the membranes break open, allowing the oligonucleotides to enter. When the cell membrane reseals, the oligonucleotides are trapped within the cells.

and the RISC complex are dependent on ATP for energy. The antisense siRNA specifies which mRNA is targeted, ensuring that no nonspecific mRNAs are degraded.

RNAi does not require many molecules of siRNA. In fact, as few as 50 copies of siRNA may destroy the entire cellular content of target mRNA. The ability to target so many mRNA molecules with so few siRNA copies relies on amplification by the enzyme **RNA-dependent RNA polymerase (RdRP)**, which creates dsRNA. RdRP uses the cleaved target mRNA as template to synthesize more dsRNA. Dicer recognizes the new dsRNA and cleaves it into more siRNA, thus greatly increasing the number of siRNA molecules (Fig. 5.11).

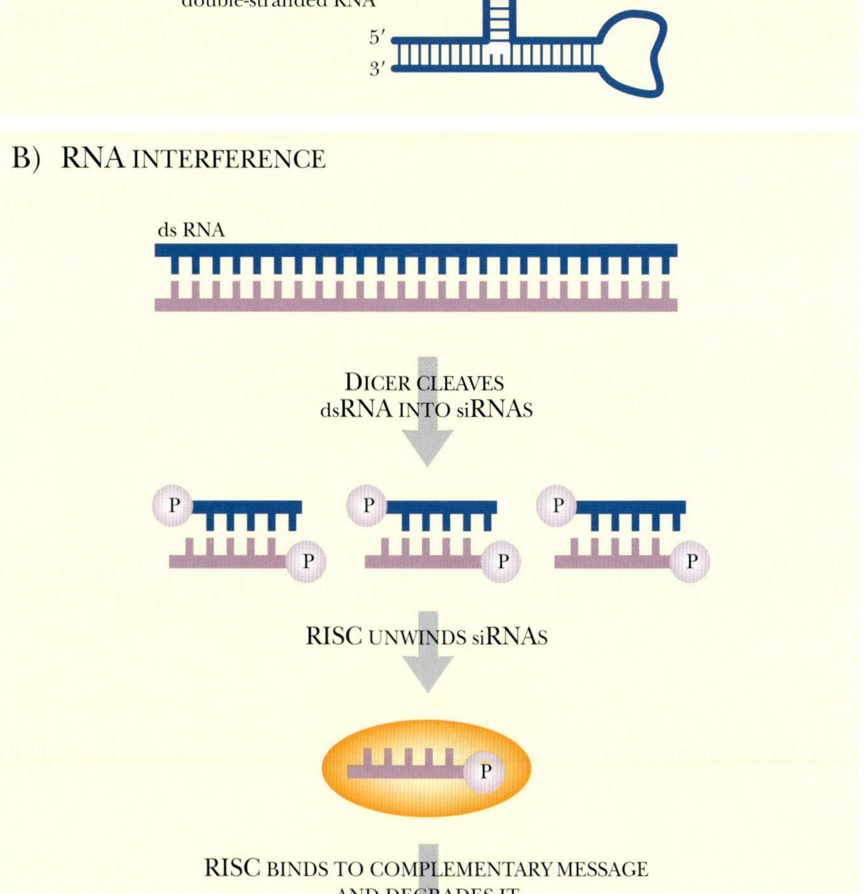

FIGURE 5.10
Cellular Mechanism of RNAi
(A) Double-stranded RNA triggers RNA interference. dsRNA is produced by RNA viruses during infections, microRNA encoded by the genome, or overexpression of transgenes. (B) RNA interference degrades all the RNA that is complementary to segments of double-stranded RNA. First, Dicer recognizes dsRNA and cuts it into pieces of 21–23 nucleotides. A kinase phosphorylates the 5′ end of each piece. Next, RISC unwinds the siRNAs and uses one strand to search out complementary mRNA, which is degraded by associated enzymes.

The final aspect of RNAi is its ability to modulate DNA expression by converting copies of the target gene into heterochromatin (Fig. 5.12). The siRNA can direct the heterochromatin-forming enzymes and proteins to the target gene location. Once the open, expressed DNA conformation is converted into heterochromatin, no more mRNA is produced. Therefore, RNAi can repress gene expression permanently.

RNAi has two phases: The initiation phase forms double-stranded RNA approximately 21 to 23 nucleotides long called siRNA, and the effector phase makes the double-stranded siRNA into a single-stranded template that searches out complementary mRNA and destroys them.

RNAi IN PLANTS AND FUNGI

RNAi was actually first observed in plants, where it was named **posttranscriptional gene silencing (PTGS)**. The phenomenon was first noted when some early transgenic experiments in plants gave strange results. When an extra copy of a gene was inserted to increase production of a particular protein, both the inserted gene (i.e., the transgene) and the resident gene copy were silenced. The result was a plant that made less of the target protein rather than more. For example, in 1990, researchers inserted a gene to make petunia flowers a darker purple. Instead, the plant made white flowers. Both the transgene and the endogenous gene were suppressed, leaving the flower without any pigment. A similar phenomenon was seen in *Neurospora*, where it was called **quelling**. After the discovery of RNAi in *Caenorhabditis elegans*, it was recognized that RNAi, PTGS, and quelling all operate via the same mechanism. Initially, none of these processes affected the level of transcription. There was plenty of mRNA produced from the transgene. After some time, the mRNA for the transgene was found in two fragments, suggesting an endonuclease cleaved it in two. Later, the target mRNA was found in smaller and smaller fragments, suggesting that exonucleases were digesting the large mRNA segments. Finally, the genes were converted into heterochromatin.

How does an extra copy of a gene induce a system that is triggered by dsRNA? One theory is that overproduction of certain mRNAs triggers RdRP to make dsRNA from the excess. This dsRNA activates Dicer to create siRNAs that quench mRNA, both from the transgene and from any closely related endogenous gene. Alternatively, when certain transgenes are expressed, some regions of the mRNA may fold back on themselves to form **hairpins**. These double-stranded segments may also activate Dicer. Genetic analysis of the model plant, *Arabidopsis*, has shown that the RdRP encoded by the SDE1 gene is necessary for transgene silencing but is not needed for antiviral RNAi. (In the latter case, the virus RNA polymerase would make dsRNA and the plant RdRP enzyme would therefore not be necessary.) This favors the first model for transgene-triggered silencing.

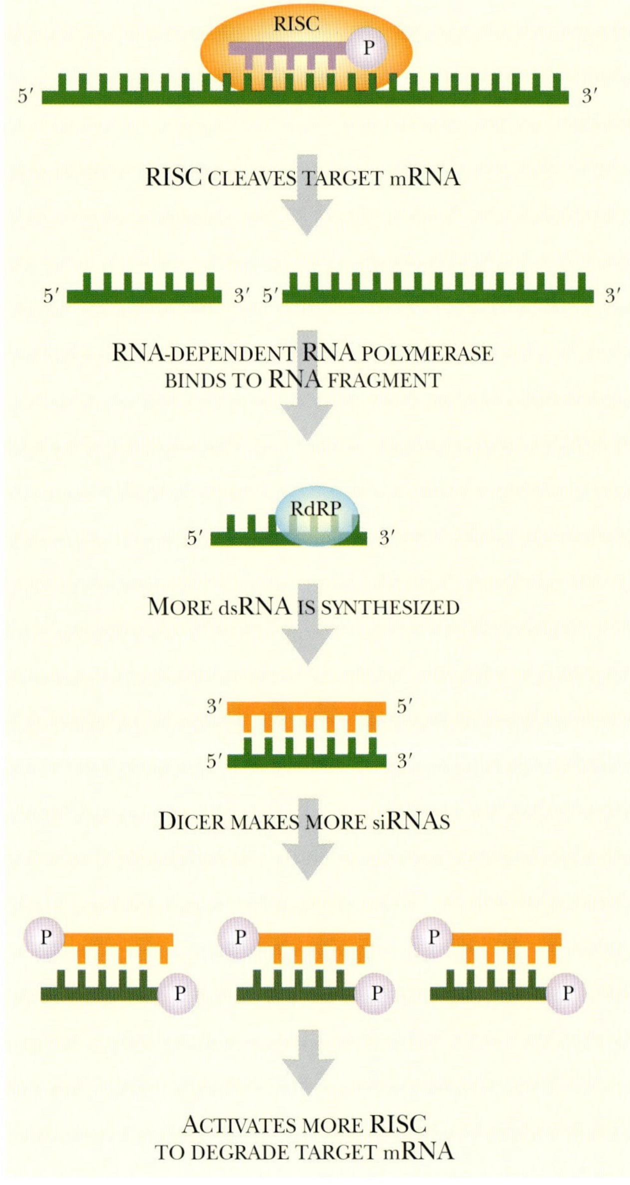

FIGURE 5.11
Amplification of RNAi
After RISC-associated enzymes cleave the complementary mRNA in a cell, another enzyme, RNA-dependent RNA polymerase, binds to some of these fragments. RdRP synthesizes complementary strands, making more double-stranded RNA. Dicer recognizes these fragments and creates more siRNA.

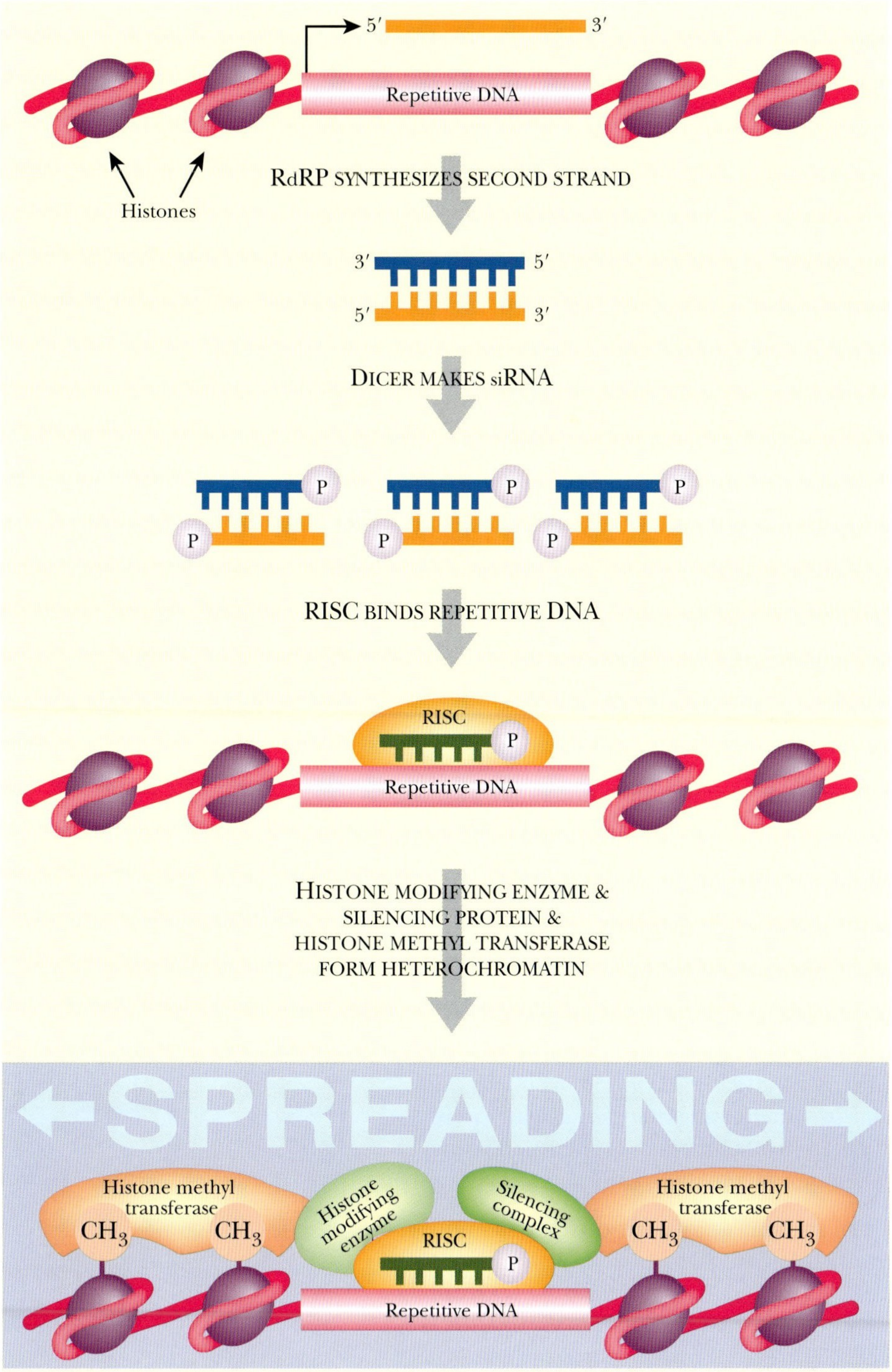

FIGURE 5.12
Heterochromatin Formation by RNAi
The RISC complex containing single-stranded siRNA can also recognize and bind to complementary DNA sequences. When RISC associates with a repetitive DNA element, various histone modifying enzymes and silencing complexes are activated to turn that region of DNA into heterochromatin. Once this is done, the region is no longer transcribed into mRNA.

The most interesting trait of PTGS is the ability of silencing to propagate from one part of the plant to the next. Plants can be grafted, that is, a leaf or stem can be attached to a different plant. If the graft has a transgene silenced by PTGS, the scion (piece of grafted plant) will then silence the corresponding endogenous gene. The effect of RNAi travels through the vascular system of the plant, and affects regions without the transgene. RNAi in *C. elegans* also has the ability to spread, not only from tissue to tissue, but also from parent to progeny. It does not appear that mammalian systems have the ability to spread the RNAi signal.

The ability to spread may not rely solely on siRNA. In plants, the potyviruses produce an inhibitor of RNAi called **helper component proteinase (Hc-Pro)**. This protein blocks the accumulation of siRNA. Despite this, the RNAi signal still spreads to other parts of the plant and triggers methylation of DNA, thus turning it into heterochromatin. Other viral genes that inhibit different steps of the RNAi process will, it is hoped, illuminate the mechanism of spread.

Other terms used to describe variants of RNAi are **transcriptional gene silencing, co-suppression**, and **virus-induced silencing**. Virus-induced silencing occurs when the viral genome has a double-stranded RNA intermediate, which triggers Dicer and RISC. Co-suppression is an early name for PTGS. Transcriptional gene silencing refers to the silencing of gene expression by converting the gene into heterochromatin. A comprehensive term, **GENE impedance (GENEi)**, has been proposed to encompass all these phenomena but is rarely used.

Early experiments in making transgenic plants identified a phenomenon now called RNAi. RNAi is also known as posttranscriptional gene silencing (PTGS), quelling, transcriptional gene silencing, co-suppression, and virus-induced silencing.

MicroRNAS ARE ANTISENSE RNAs THAT MODULATE GENE EXPRESSION

The development from embryo to adult of the worm *C. elegans* requires RNAi to turn off genes at appropriate times. In this case, RNAi is not triggered by intrusion of external sequences such as transgenes or viruses. During development, small noncoding RNA molecules known as **microRNAs (miRNA)** are transcribed from the worm's own genome. These miRNAs regulate gene expression by blocking the translation of developmentally appropriate target mRNA. MicroRNAs, first identified in *C. elegans*, have now been identified in various plants and animals, including humans. RNAi induced by miRNAs is similar to the mechanism described above. The mRNA targets are identified by antisense, that is, the miRNA has sequences that are complementary to the part of the target mRNA. Some miRNAs bind to the target mRNA and block the initiation of translation. In other cases, the miRNA binds to the 3′ UTR region of the mRNA.

MicroRNAs are transcribed as longer precursor molecules, **pre-microRNAs**, of approximately 70 nucleotides in length. In *Drosophila*, miRNAs are transcribed as polycistronic messages that are first cleaved by an endonuclease called Drosha. In plants, the pre-microRNAs can be even longer, up to 300 nucleotides. Dicer cleaves the plant pre-microRNA into segments of approximately 20 nucleotides. After cleavage of the precursor, the released miRNA forms a stem-loop structure by complementary base pairing. Dicer recognizes the stem-loop and cleaves the loop structure, thus separating the strands. The RISC complex then unravels the two strands. The miRNA found in animals such as *C. elegans* can tolerate a few mismatched base pairs within the binding domain. In animals, the antisense miRNA strand blocks translation of the target mRNA (Fig. 5.13), which is not degraded. In contrast, in plants, microRNA must have perfect matches and relies on RISC-mediated recognition and cleavage to degrade the target mRNA.

MicroRNAs (miRNAs) modulate expression of various genes during development in many organisms. These are first translated as pre-miRNAs from the organism's own genome, processed into 21 to 23 nucleotide pieces, and then RISC separates the strands and destroys the complementary target mRNA.

APPLICATIONS OF RNAi FOR STUDYING GENE EXPRESSION

By inhibiting translation, RNAi provides the ability to remove a particular protein from an organism, without the need for genetic modification. RNAi thus provides a powerful tool

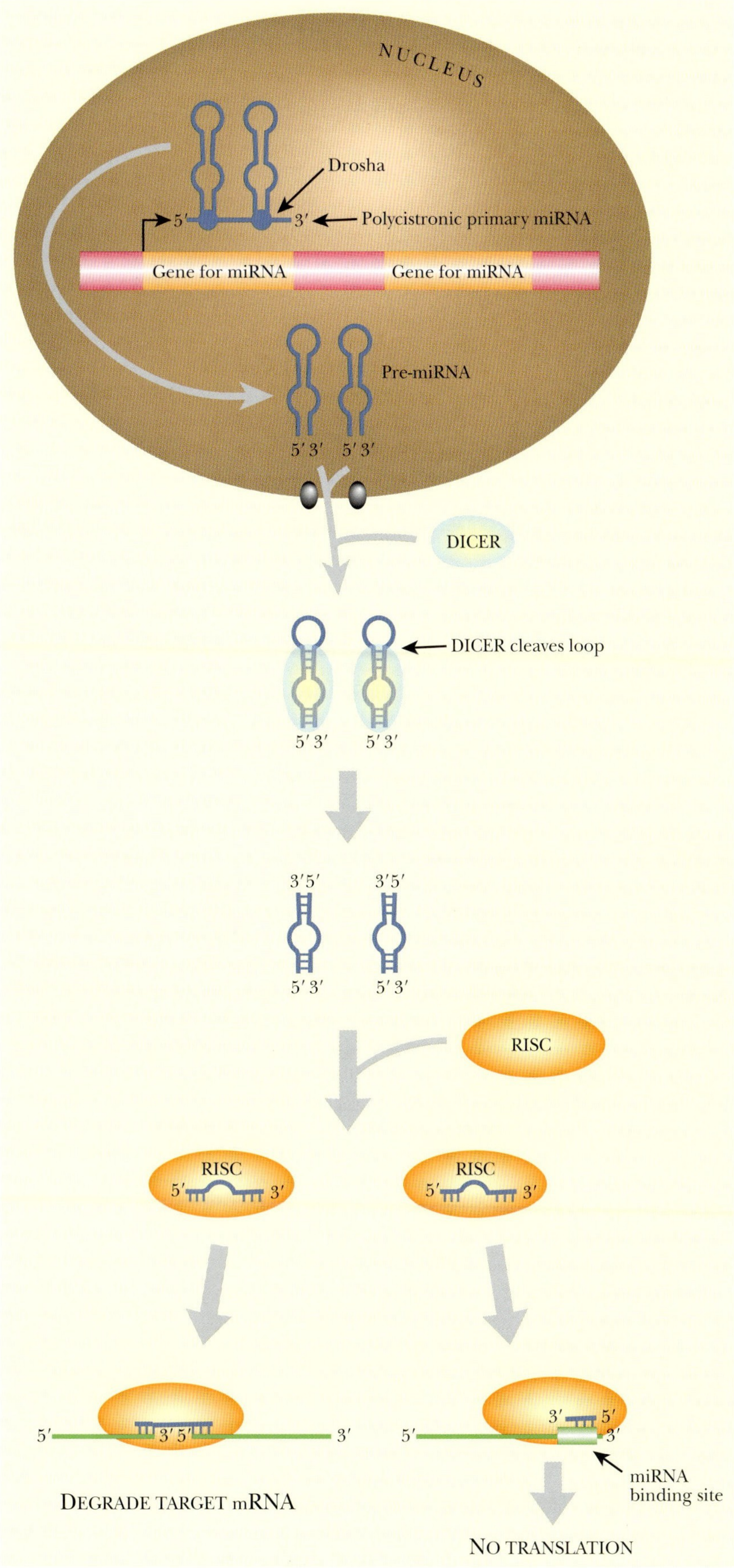

FIGURE 5.13
Pathway of miRNA Inactivation of Gene Translation in *Drosophila*
Drosophila has several genes for miRNAs that control mRNA expression during development. The genes are polycistronic, and an enzyme called Drosha cuts each of the hairpin structures to form pre-miRNAs. These exit the nucleus through nuclear pores. In the cytoplasm Dicer recognizes them and cleaves the loop end. The resulting double-stranded miRNAs are recognized by RISC, separated into single strands, and used to block translation or degrade complementary mRNA.

to study the roles that particular proteins play in development. The application of RNAi to studying *C. elegans* is especially well understood. Three different methods have been devised to get dsRNA constructs into *C. elegans*, and stimulate RNAi to block expression of a target gene (Fig. 5.14). The little worm has an uncanny ability to take up exogenous DNA or RNA. Worms can be fed *Escherichia coli* bacteria expressing the dsRNA of interest, and the worm takes up the dsRNA into its cells. Dicer then cleaves the dsRNA into small siRNAs that activate RISC to block the target mRNA. Another method to deliver dsRNA is simply bathing the worms in a solution of dsRNA. The exogenous dsRNA is absorbed into the worm, where it activates RNAi. A third method to induce RNAi is to inject worm eggs with dsRNA. The worm develops with the dsRNA inside, and RNAi is activated in all the cells. Bathing worms in dsRNA or feeding them dsRNA-expressing bacteria may be incomplete—some cells are not penetrated—yet because the signal can spread from one cell to the next, the method works. Surprisingly, the RNAi effect can pass from parent to offspring. The progeny of a worm that has a silenced gene will also silence the same gene.

Delivering dsRNA to *Drosophila* is not quite as easy. The flies do not take up external dsRNA by eating it or absorbing it from solution. *Drosophila* requires the dsRNA to be microinjected directly into a developing egg. The dsRNA is absorbed into the cells as the embryo develops. The advantage of this method is the ability to knock out the protein of interest through all stages of development. Obviously, if the protein is essential for development, activating RNAi too early can stop development and kill the fly. Luckily, *Drosophila* has a feature not found

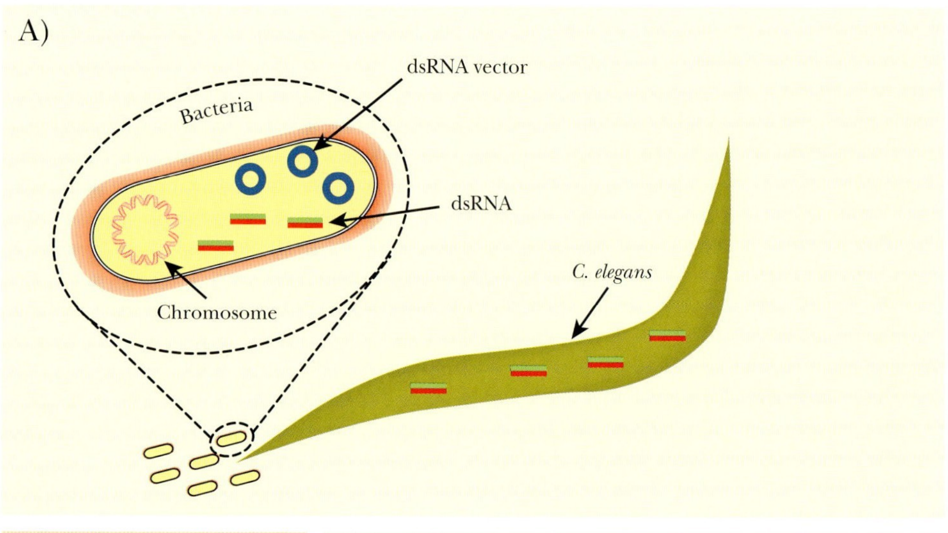

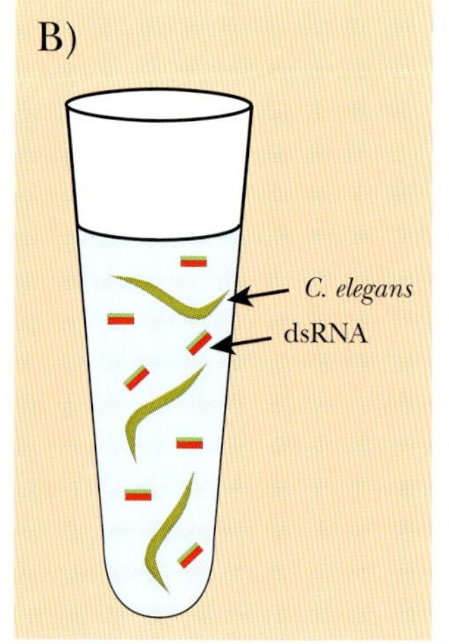

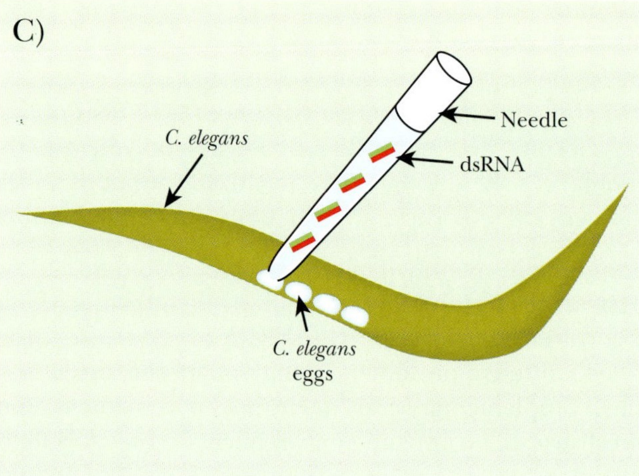

FIGURE 5.14
Delivering dsRNA to *C. elegans*
(A) *C. elegans* can absorb dsRNA expressed in bacteria that they eat.
(B) *C. elegans* can absorb dsRNA by swimming in a solution containing dsRNA.
(C) Injecting dsRNA into an egg will trigger gene silencing in the developing worm.

in *C. elegans*. Fly cells can be cultured *in vitro* in nutrient medium. Because dsRNA can be transfected into the cultured cells, these cell lines can be used to study protein function using RNAi. Therefore, if RNAi kills the embryo during development, the corresponding protein can still be examined using cultured cell lines.

C. elegans can take up dsRNA pieces and activate RNAi by ingesting transgenic bacteria that are expressing dsRNA, by bathing in a solution of the pure dsRNA, or by having the dsRNA injected into the eggs.

Drosophila is also used as model organism to study protein function with RNAi by microinjecting dsRNA into *Drosophila* eggs or cultured cell lines.

RNAi FOR STUDYING MAMMALIAN GENES

An important application for RNAi is testing the individual roles of human proteins, which can reveal new targets for curing diseases. Until recently, using RNAi to assess the function of mammalian proteins was not possible. The application of dsRNA to cultured mammalian cells or whole mice induces a potent antiviral response. Interferon is produced, which triggered the cells to degrade all RNA transcripts and shut down protein synthesis. Thus, the methods used in *C. elegans* and *Drosophila* killed mammalian cells, and no alternatives were known until recently.

Recognizing that short interfering RNA (siRNA) pieces mediated RNAi was the key to its application in mammalian systems. Instead of using long dsRNA as in *C. elegans* and *Drosophila*, exposing mammalian cells to dsRNA shorter than 30 nucleotides activated the mammalian counterparts to Dicer and RISC. This in turn abolished expression of the target mRNA. Such short dsRNA segments thus act as endogenously produced siRNA. As described earlier, siRNAs made by Dicer are short double-stranded RNA about 21 to 23 base pairs in length. In addition, the siRNAs have a two-base 3′ overhang that is more stable when it consists of two uracils. To study a particular target mRNA in mammalian cells, chemically synthesized siRNAs with these characteristics are designed to have the complementary sequence to the target. Much like antisense oligonucleotides, these may have modifications to make them more stable, such as methyl groups added to the 2′-OH of the ribose. The most important and the most difficult aspect of *in vitro* siRNA construction is determining an effective sequence. The sequence on the target mRNA must be accessible to the siRNA, which may be challenging because of RNA secondary structure. Many suitable siRNA are designed and synthesized, and each is tested for activating RNAi. These siRNA are delivered to mammalian cells much like antisense oligonucleotides, including transfection, liposomes, and microinjection.

Rather than chemically synthesizing siRNA, the target mRNA can be mixed with purified Dicer to cleave the target mRNA into siRNA pieces (Fig. 5.15). Although the reaction occurs in a tube, the purified Dicer can generate multiple siRNAs as it would *in vivo*. In order to create the target mRNA, the chosen target gene is amplified with PCR using PCR primers that also contain a promoter sequence for T7 RNA polymerase. The double-stranded DNA is then converted into dsRNA by T7 RNA polymerase. The RNA pieces are allowed to anneal spontaneously, and then they are mixed with purified Dicer, which digests it into multiple different siRNAs. These are then transfected into mammalian cells.

Additionally, mammalian cells can be induced to activate RNAi by expressing **short hairpin RNAs (shRNAs)** that mimic the structure of microRNAs. A gene for the shRNA is constructed in a vector. The shRNA can be transcribed as two complementary strands by two different promoters facing in opposite directions, or simply made as one transcript with complementary ends interrupted by a sequence that forms a loop (Fig. 5.16). In both constructs, the complementary sequences come together to form a double-stranded RNA region. The most common promoter for expressing shRNA *in vivo* is the promoter for mammalian RNA polymerase III. (Note: This enzyme normally transcribes small noncoding RNAs.) RNA polymerase III starts transcription at a specific sequence and stops transcription when it encounters four to five consecutive thymidines. In addition, RNA polymerase III does not activate the enzymes that cap and add a poly(A) tail onto the transcript. This polymerase is precisely what is necessary to create an shRNA that mimics one found in eukaryotic cells.

Rather than designing the shRNA constructs from scratch, another strategy uses preexisting microRNA. First, the stem portion is replaced with sequences that match the target mRNA. The newly designed microRNA will trigger RNAi for the target mRNA rather than its endogenous target.

Vectors that express shRNAs have some advantages over adding siRNA. When siRNA is added to mammalian cultures, the effect is temporary. When the siRNA is gone, the effect ends (unless heterochromatin formation was triggered, as seen in some organisms). When an shRNA vector is transfected into a mammalian cell, the effect is more sustained. The vector continues to produce shRNA for a considerable time. In addition, expression of the shRNA may be controlled using promoters that are inducible or tissue-specific. In a clinical setting, the ability to deliver the siRNA only to those tissues that need it is crucial, and using a vector system can accomplish this.

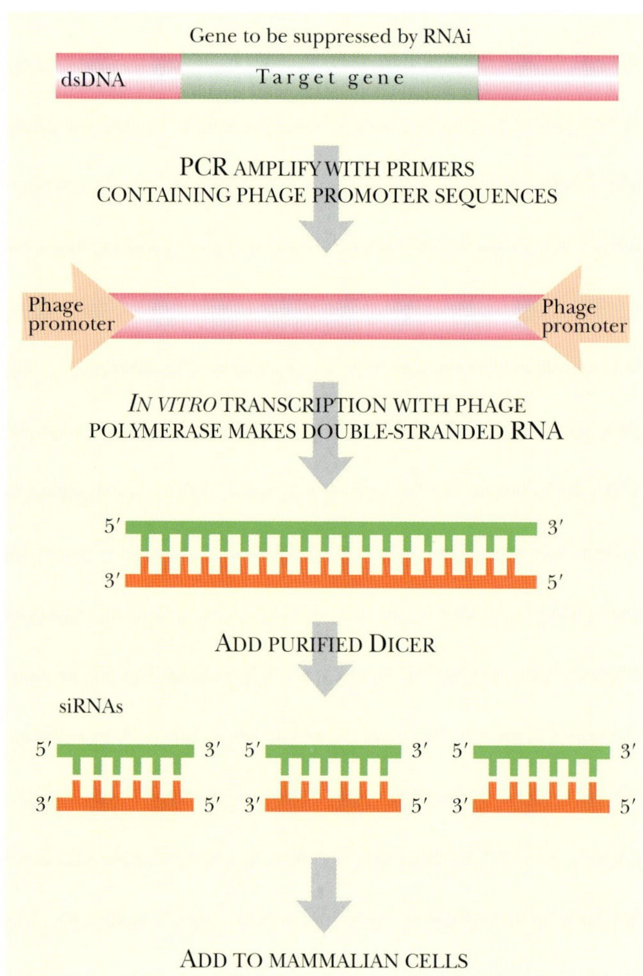

FIGURE 5.15
***In Vitro* Treatment with Dicer Generates siRNAs**
The key to making siRNAs *in vitro* is cloning the target gene so that both the sense and antisense strands are expressed into mRNA. The two strands anneal spontaneously, and when purified Dicer is added, small siRNAs are produced.

RNAi can be triggered in mammalian cells using chemically synthesized siRNA, creating an shRNA that degrades the target mRNA, or by modifying existing shRNA to recognize a different target mRNA.

FUNCTIONAL SCREENING WITH RNAi LIBRARIES

RNAi has been used to study the entire genome in model organisms. An **RNAi library** containing about 86% of the predicted genes in the genome of *C. elegans* has been constructed in *E. coli*. An RNAi library expresses each of the genes as dsRNA. *E. coli* carrying any particular gene can be used to feed *C. elegans*, which triggers RNAi for the gene of interest and allows the effect of removing each individual protein from the worm to be assayed. Using this library, more than 900 genes have been identified whose suppression kills the embryo or causes gross developmental defects. Most of the proteins encoded by the RNAi library target mRNA that had no known function before. These libraries demonstrate how RNAi can link a particular gene to a specific cellular process. Similar experiments in *Drosophila* have begun. An RNAi library of about 7200 conserved genes (about 91% of the predicted genes in the genome) has been constructed. So far the library has been used to identify proteins associated with cell growth and viability in *Drosophila* cell lines, but the library can be applied to many different processes.

In order to construct an RNAi library, the genes from the organism of interest are isolated as cDNA clones (Fig. 5.17). The cDNA clones are then amplified using PCR (see Chapter 4).

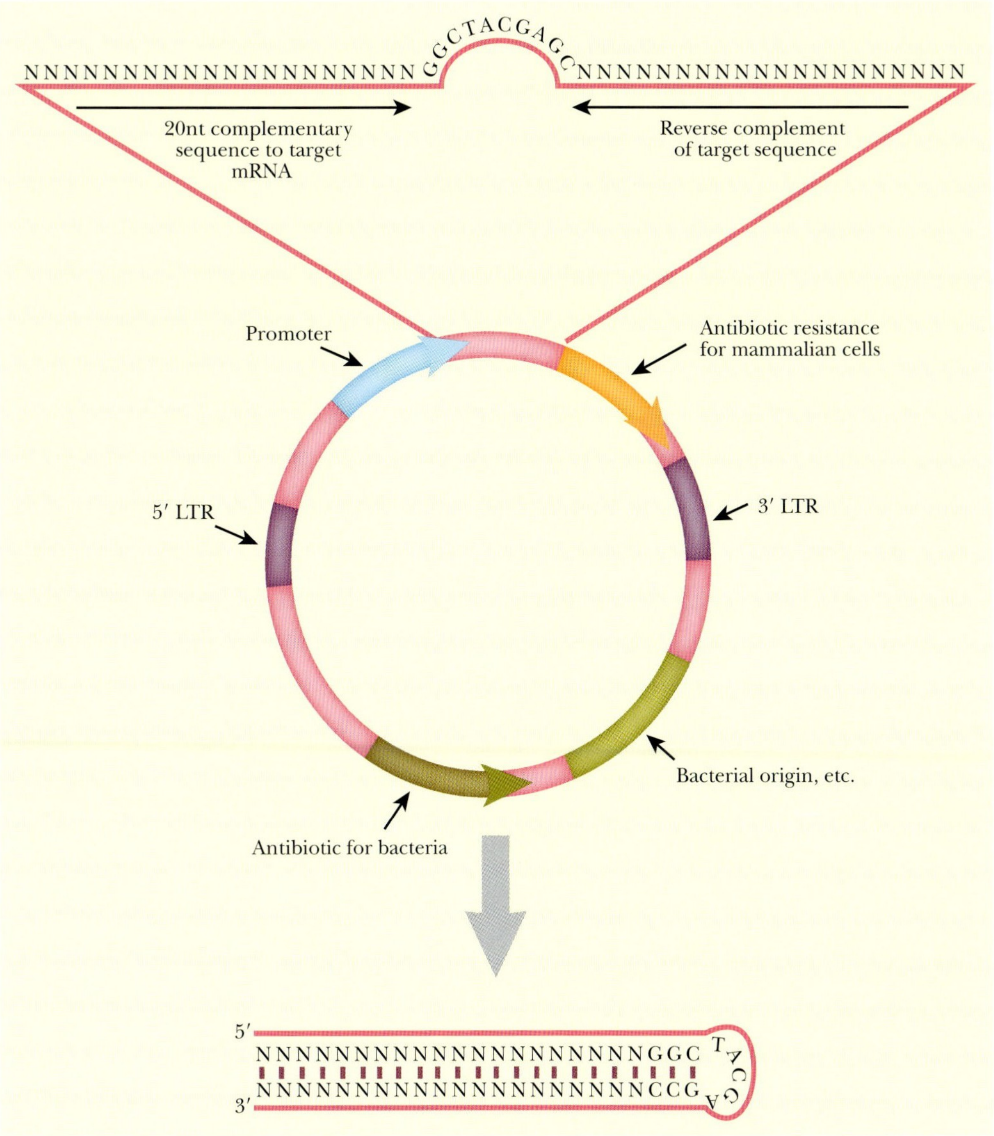

FIGURE 5.16
Design of shRNA Expression Vectors to Activate RNAi
Vectors can be designed to express different shRNA molecules. This retrovirus-based vector has two complementary sequences about 20 nucleotides in length that form a stem, separated by a loop region. When the vector is transformed into a cell, the shRNA is transcribed and activates gene silencing.

FIGURE 5.17
Constructing an RNAi Library
Each clone in the library must have two different promoters flanking the coding region. When the clone is transcribed into mRNA, both an antisense and a sense transcript will be produced and the two strands will come together to form double-stranded RNA.

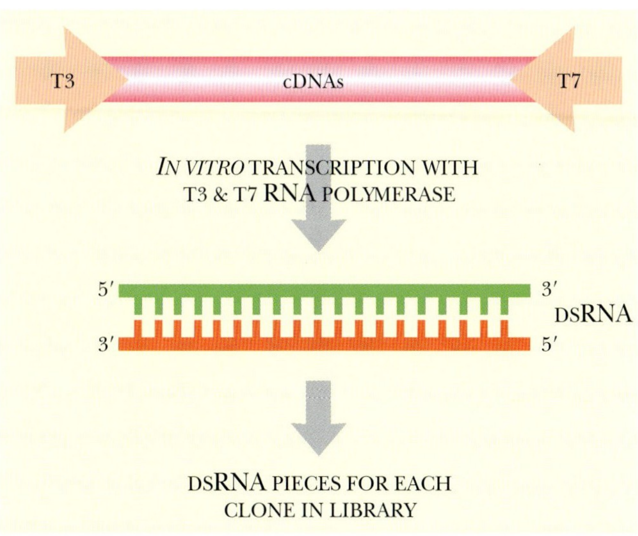

Because the genomes concerned have been sequenced, PCR primers that are specific for each gene are designed with special features. They are designed to add two different promoters at the ends of each gene. For example, the 3′ end would have a T7 polymerase promoter, and the 5′ end a T3 polymerase promoter. The PCR products are then cloned into a suitable vector. When the vector is present in a cell with both T7 and T3 RNA polymerase, both an antisense and a sense transcript are transcribed for each of the clones. The two strands spontaneously anneal to form dsRNA, which then activates RNAi.

For mammalian systems, RNAi libraries must be constructed with siRNAs or shRNAs rather than full-length dsRNA, because the full-length dsRNAs are toxic. These RNAi libraries are made by two different methods (Fig. 5.18). In the first method, short synthetic siRNAs are chemically synthesized for each of the

Box 5.1 Does RNAi Protect Mammalian Cells from Viral Infections?

It is well established that plants use RNAi to protect themselves from viruses. It has been shown that virus-derived siRNAs are made when the plant is infected with either DNA or RNA viruses. Plants that have mutations in various components of RNAi become more susceptible to viral diseases. Plants have a system to spread the RNAi signal to uninfected parts of the plant, thus protecting the neighboring tissue. Finally, plant viruses have genes/proteins that suppress the RNAi pathway.

In mammalian cells, there is only some evidence that RNAi protects the organism from viruses. When a mammal is infected with a virus, the innate and adaptive immune responses are potent and protect the organism from many different assaults. First, cellular proteins such as the toll-like receptor, protein kinase R, and retinoic acid-inducible gene I are activated by the entry of the virus. These turn on the transcription and translation of many different genes, but most notably, type I interferons and nonspecific RNases. These work in unison to fight the infection. In addition, mammalian systems may not rely on RNAi for fighting viruses, because they only have one gene for a Dicer homolog. In contrast, *Arabidopsis* has four different Dicer homologs. Mammalian cells do not spread their RNAi signal to uninfected tissues as do the plant systems.

Some recent evidence does suggest that RNAi functions to prevent virus destruction in mammalian cells. First, some proteins from mammalian specific viruses target RNAi proteins. For example, NS1 from the influenza virus binds to siRNA *in vitro* and suppresses RNA silencing when expressed in plants. Another viral protein, tat from HIV, has been shown to inhibit purified Dicer *in vitro*. Although the evidence so far is weak, the experimental work that uses siRNA and plasmid-encoded shRNA to block viral infections in mammalian cells is the most convincing. Many studies have found that administering siRNA or shRNA to animal models has reduced virus replication and protected the organism from lethal infections. So whether or not mammalian cells use RNAi as a primary form of defense against viruses, activating this system is protective against viral assaults.

genes that will be represented in the library. All of the siRNAs are pooled and cloned into the appropriate vector. The vector must have the promoters and terminators needed to express the siRNA inserts as double-stranded RNA. The promoters are usually two RNA polymerase promoters that transcribe the sense and antisense strands of the insert. As described earlier, these spontaneously anneal to make siRNA. The vector must also contain other appropriate sequences, such as an origin of replication for mammalian and bacterial cells, and antibiotic resistance genes for selecting cells carrying the vector.

A second approach to constructing a mammalian RNAi library involves the use of restriction enzymes to generate small shRNA fragments. **Restriction enzyme-generated siRNA (REGS)** libraries start with either single or multiple genes that are going to be represented in the library (see Fig. 5.18B). Whether a large or small sample of genes is used, the technique begins with double-stranded cDNAs. The library is cut with restriction enzymes that cut often and leave identical overhangs. This generates many different small fragments for each gene. Because each of the fragments is a different size, they must be trimmed to the same size. For this, the fragments are linked to a 3′ hairpin loop, which has a site for the restriction enzyme *Mme*I. This cuts precisely 20 nucleotides from the 3′ end of its recognition sequence, thus making each fragment the same length.

To make many siRNA copies of these random constructs, they are converted into double-stranded loops. First, a 5′ hairpin is ligated onto the end cut by *Mme*I. A viral DNA polymerase (from ϕ29) then converts the loop into dsDNA. This polymerase can separate the strands in double-stranded regions and so synthesizes a second strand for the entire circle. Since ϕ29 DNA polymerase displaces dsDNA, it continues to circle around and around making a long concatemer for each loop, as in rolling circle replication. The concatemer is digested with *Bgl*II and *Mly*I, to generate two complementary gene fragments connected by the 3′ loop. The 3′ loop is shortened by digestion with *Bam*HI, and then the two fragments are ligated together. The final product is cloned into a vector that will transcribe the insert as a single strand of RNA. Finally, the two complementary sequences base-pair to form a short hairpin that is processed into siRNA by Dicer.

The RNAi libraries can be analyzed either with multiwell plates or **live cell microarrays** (Fig. 5.19). The multiwell plates use a **barcode** sequence on the shRNA in order to identify which clone corresponds to which gene. A barcode or zipcode sequence is approximately 20 nucleotides long and is unique to each clone. Each clone can be identified by this sequence rather than sequencing the entire gene insert. To analyze the library, it is transformed into a large number of cells that are inoculated into the wells of a multiwell plate. A defined number of cells go into each well, so that only one siRNA is found in each well. The cells are then analyzed for any noticeable symptoms. For those showing effects, the siRNA is identified by the zipcode sequence. Another method for screening a siRNA or shRNA library is to spot each clone onto slides in a microarray (see Chapter 8). Live cells are then added and take up the siRNA or shRNA at these locations. The change in cell phenotype over a spot corresponds to a particular siRNA.

If each protein in a genome is studied for detectable phenotypes via RNAi, a **phenotype catalogue** can be constructed. For example, many genes/proteins involved with cell membrane division will arrest cells with a bridge structure during cell division. An unknown protein that causes the same phenotype in an RNAi screen would potentially belong to the same family of proteins. These proteins are said to have the same **phenotypic signature**. Databases have already been devised for *C. elegans* (http://nematode.bio.nyu.edu:8001) and *Drosophila* (http://www.flyRNAi.org) where this type of information is compiled to share with other scientists.

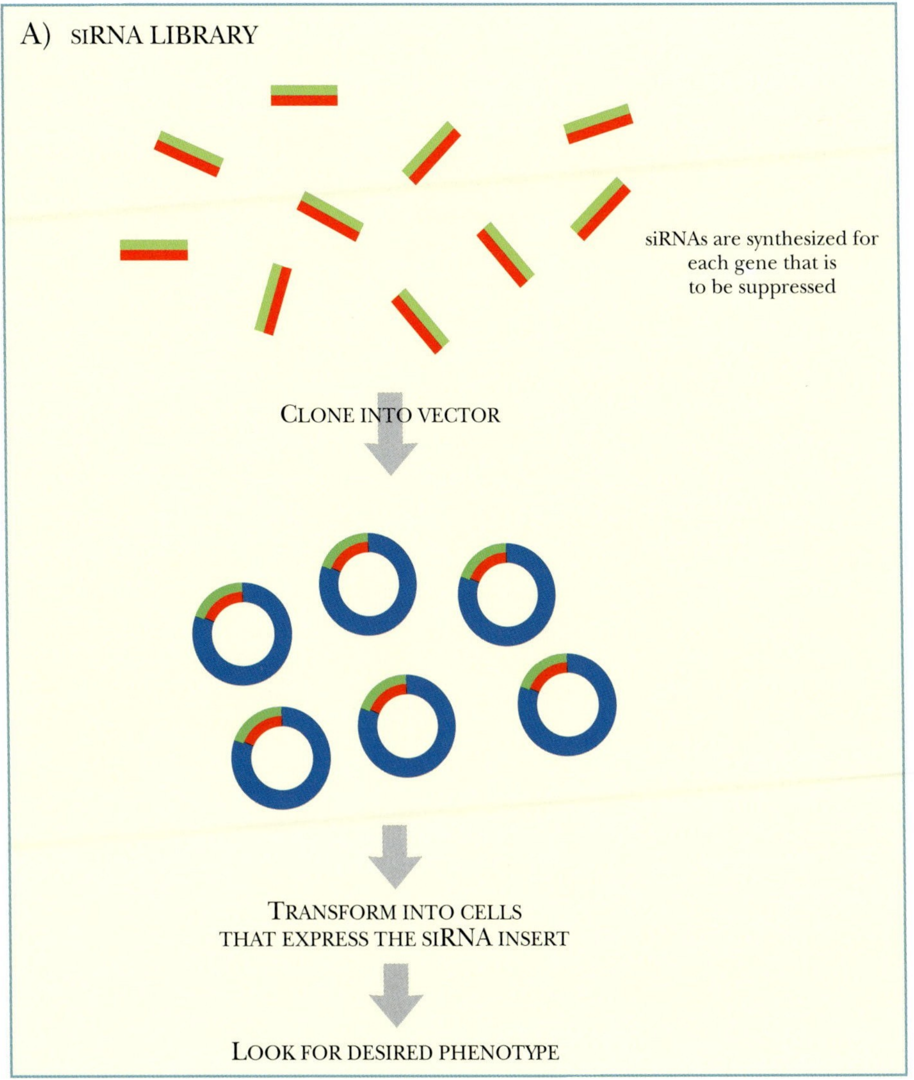

FIGURE 5.18 Making siRNA or shRNA Libraries
(A) Chemically synthesized oligonucleotides are cloned into a vector that has promoters to express both strands into mRNA. When transcribed, the two strands pair together to form siRNA.

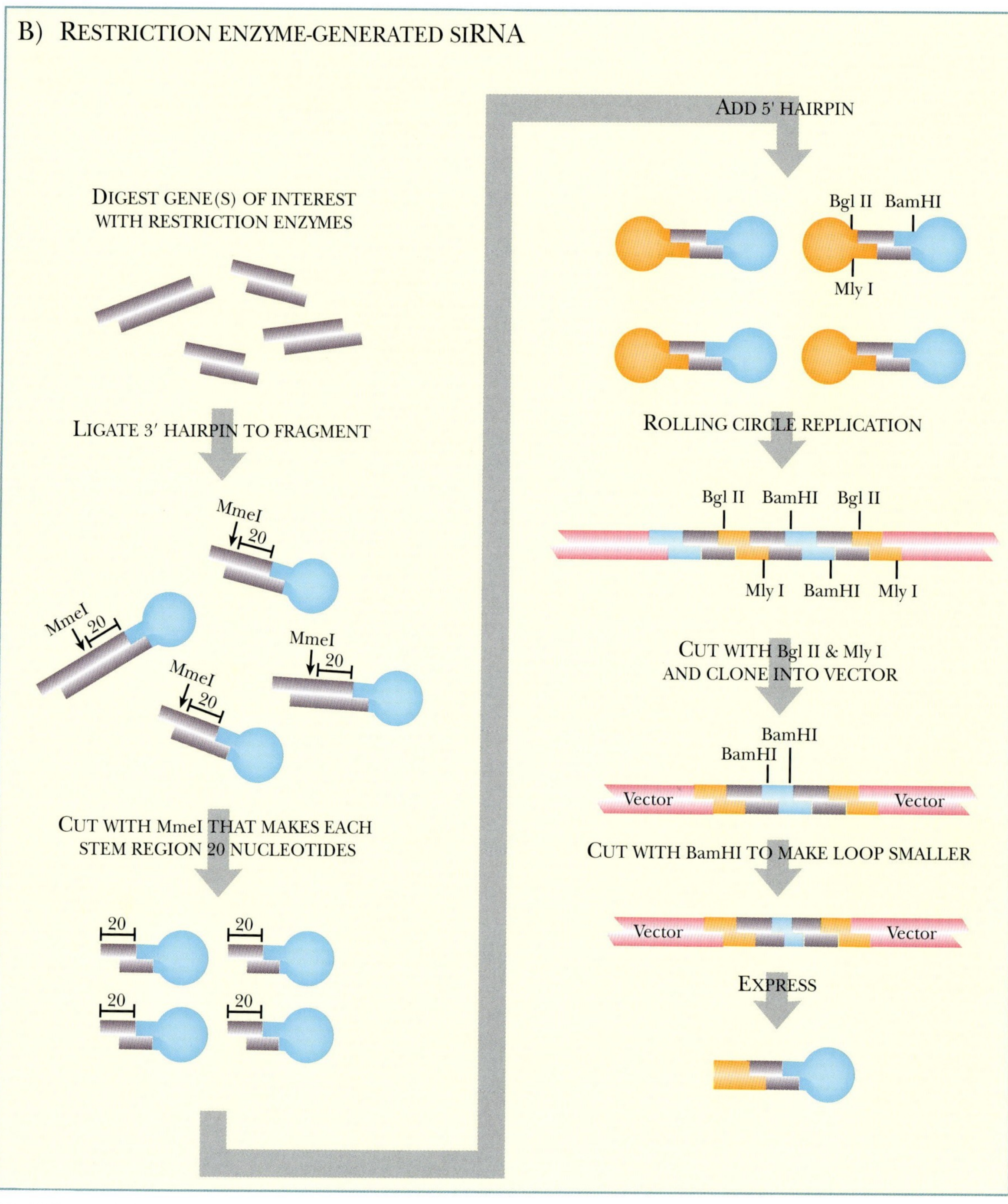

FIGURE 5.18, cont'd

(B) Restriction enzyme-generated siRNA (REGS) creates a library of short hairpin RNA (shRNA). First the genes of interest are randomly digested with restriction enzymes. Next, a 3′ hairpin is ligated to the pieces. *MmeI* then cuts the fragments to uniform lengths. *MmeI* is a type I restriction enzyme that has its recognition sequence in the 3′ hairpin and its cutting site 20 nucleotides into the restriction enzyme fragment. The fragments are next ligated to a 5′ hairpin and subjected to rolling-circle replication to create a short hairpin coding region. The replicated linear fragment has repeating segments of 5′ hairpin (orange), cloned fragment (gray), 3′ hairpin (blue), cloned fragment (gray), 5′ hairpin (orange), etc. Next, the repeating fragment is cut with restriction enzymes that only recognize the 5′ hairpin (orange). The resulting piece has two complementary sequences (orange and gray) that are separated by a loop (blue). When this is cloned and expressed into mRNA, the two complementary sequences (orange and gray) anneal to form a short hairpin.

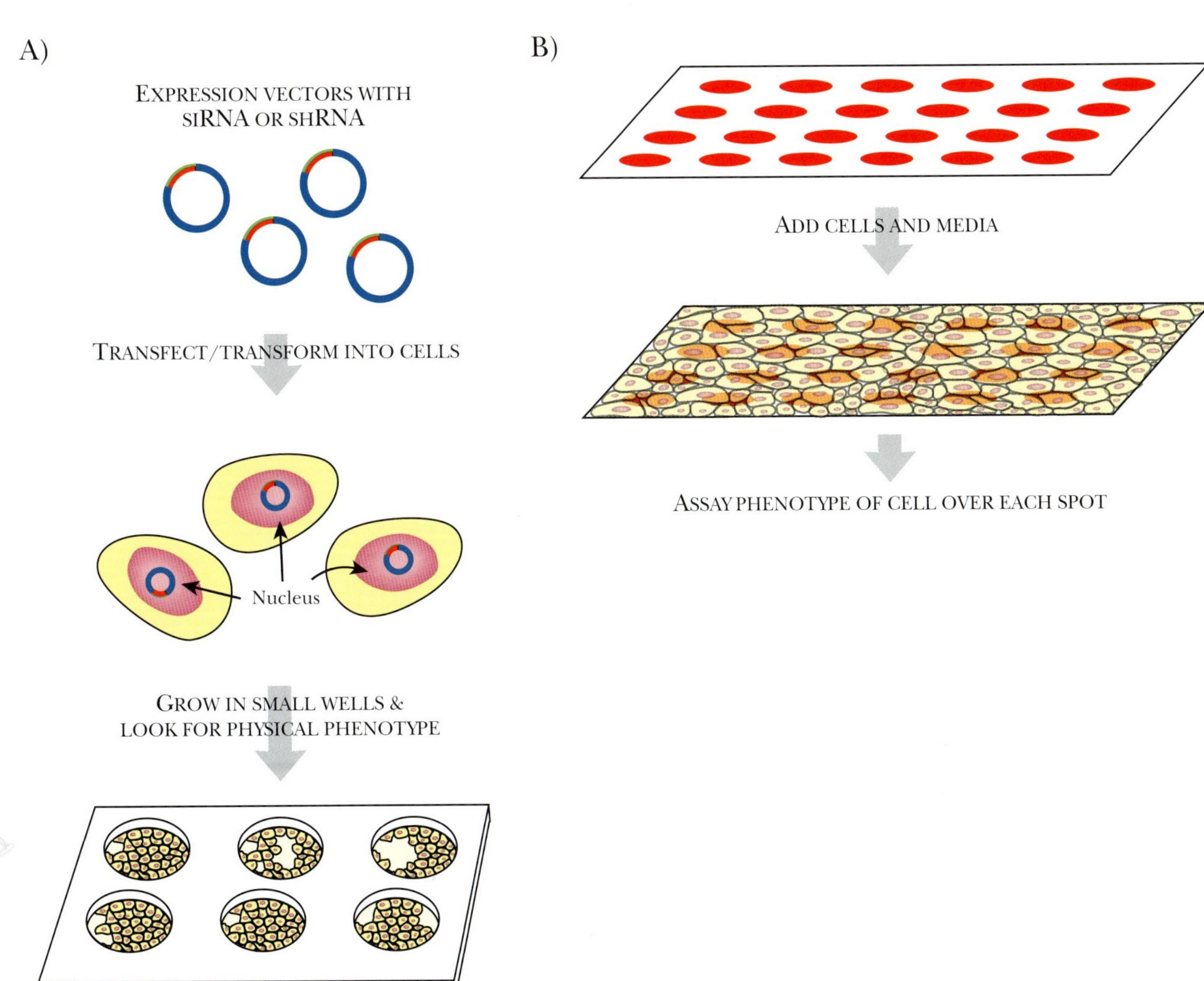

FIGURE 5.19 Multiwell Plate Assays and Live Cell Microarrays

(A) Short-interfering RNA and short-hairpin RNA libraries can assess the function of mammalian proteins. If the library is transfected into mammalian cells, each cell can be assessed for different physical phenotypes, such as loss of adherence, mitotic arrest, or changed cell shape. The shRNA or siRNA that causes an interesting phenotype can be isolated and sequenced to identify the protein that was suppressed. (B) Rather than transforming cells, the siRNA or shRNA can be spotted onto microscope slides. As cells grow and divide on the slide, the siRNAs are taken up by the cells and initiate RNAi. The cells over each spot are assessed for a particular phenotype.

RNAi libraries are used to identify the function of unknown protein from an organism by degrading all the cellular mRNA for that particular protein. Each library clone targets one protein.

RNAi libraries are generated using chemically synthesized siRNA sequences cloned into a vector that converts the insert into dsRNA.

Restriction enzymes can also be used to create artificial short-hairpin RNAs. These can be assembled into an RNAi library.

Mammalian cells can be screened for mutations induced by an RNAi library clone using a live cell microarray or by using a multiwell-plate assay.

RIBOZYMES CATALYZE CLEAVAGE AND LIGATION REACTIONS

Ribozymes are RNA molecules that bind to specific targets and catalyze enzymatic reactions. Some ribozymes consist of RNA associated with proteins, but the RNA catalyzes the actual reaction. Some ribozymes work like allosteric enzymes, that is, binding an effector molecule alters the ribozyme structure so that the ribozyme becomes competent to cleave its substrate. Ribozymes are naturally occurring, but biotechnology research has started to exploit their unique characteristics for medical and industrial applications.

There are eight known classes of ribozymes at present, with the distinct possibility that many more will be identified. Ribozymes are classified into large or small. The large ones range in size from several hundred nucleotides to 3000 nucleotides in length. Large ribozymes were the first identified, and the first of these were the group I introns of *Tetrahymena*. These are intron sequences found in pre-mRNA that are able to self-splice. They do not use splicing factors such as U1, U2, U4/U6 snRNA (aka snurps). Group I introns are common in fungal and plant mitochondria, in nuclear rRNA genes, in chloroplast DNA, in bacteriophage and eukaryotic viruses, and in the tRNA of chloroplasts and eubacteria. The important aspect of intron self-cleavage is the RNA structure. RNA is a linear polymer, but because of base-pairing between different regions, RNA also has a secondary structure. Multiple stem-loop structures fold into different configurations leading to a three-dimensional structure much like a protein (Fig. 5.20). The example shown is the second group I intron within the *orf142* gene of bacteriophage Twort, which infects *Staphylococcus aureus*. The three-dimensional structure of group I introns brings the two exons close together, facilitating removal of the intron between them (Fig. 5.21).

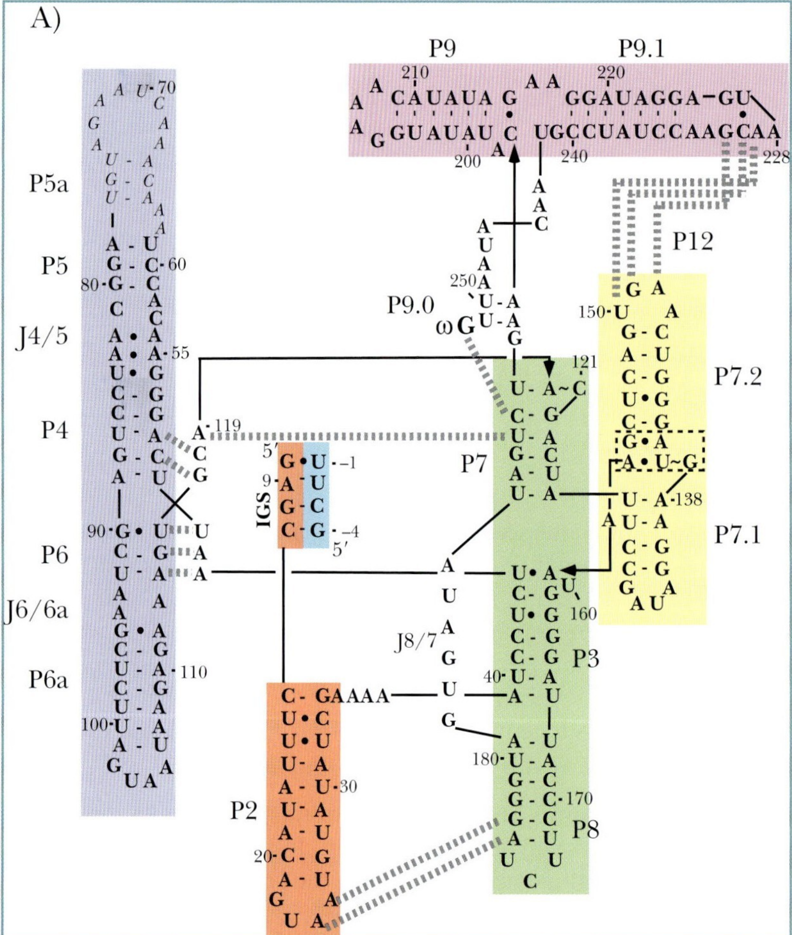

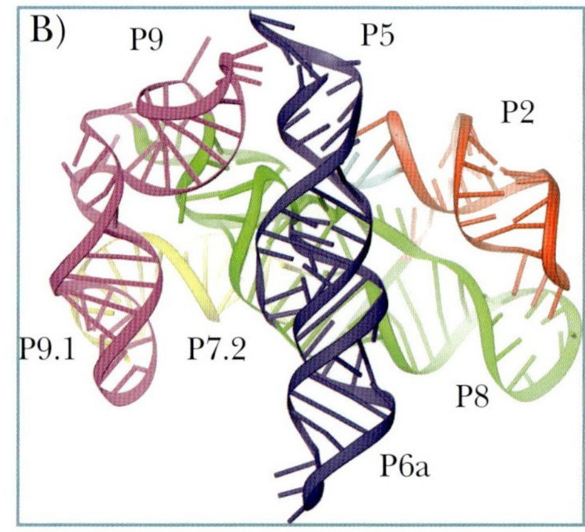

FIGURE 5.20 Structure of the Twort Ribozyme

(A) Primary and secondary structure of the wild-type intron. The P1-P2 domain is highlighted in red, the P3-P7 region is green, the P4-P6 domain is blue, the P9-P9.1 region is purple, the P7.1-P7.2 subdomain is yellow, and the product oligonucleotide is cyan. Dashed lines indicate key tertiary structure contacts. Nucleotides in italics (P5a region) are disordered in the crystal. IGS, internal guide sequence. (B) Ribbon diagram colored as in (A). The backbone ribbon is drawn through the phosphate positions in the backbone. From: Golden, Kim, and Chase (2004). Crystal structure of a phage Twort group I ribozyme-product complex. Reprinted by permission from: Macmillan Publishers Ltd. *Nat Struct Mol Biol* **12**, 82–89, copyright 2005.

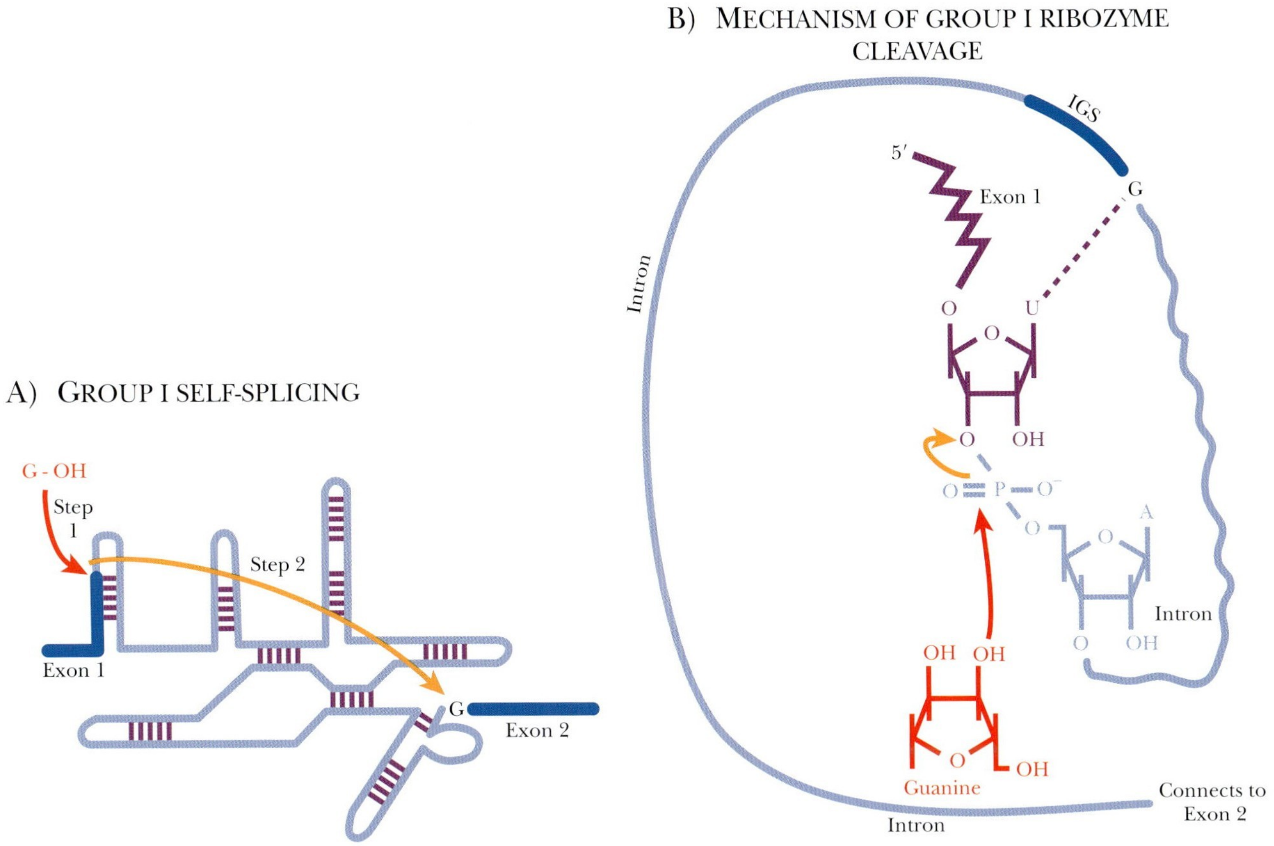

A) Group I self-splicing

B) Mechanism of group I ribozyme cleavage

FIGURE 5.21 Mechanism of Group I Self-Splicing Reaction
(A) The secondary structure of group I introns shows multiple hairpins that mediate the cleavage reaction. In step I, a free guanosine (Red, G-OH) mediates the cleavage of the exon 1–intron boundary. In step 2, the free end of exon 1 cleaves and ligates to exon 2. (B) Mechanism of group I ribozyme cleavage. First the exon sequences are brought near the catalytic core via the internal guide sequence (IGS). Exon 1 has an important uridine (U) that forms a U=G base pair with the IGS (dotted line). The other end of the ribozyme has a binding site for the nucleophile, a free guanosine (Red), which initiates intron removal by attacking the end of exon 1 with the 3′-OH of its ribose. The free 3′-OH on the exon than reacts with the splice site on exon 2. The intron is spliced out and the two exons are united (not shown). Although it appears that this reaction requires energy, the actual number of bonds stays the same, and no net energy is needed.

Group II introns are also self-splicing sequences found within genes. They are less common than group I introns, being found only in fungal and plant mitochondria, in plant chloroplasts, in algae, in eubacteria, and in chloroplasts of *Euglena gracilis*. These introns do not self-splice *in vitro* and require far from physiological conditions to work. The three-dimensional structure of the intron creates these abnormal conditions *in vivo*, affecting the microenvironment to create the correct ionic concentrations. The 3D structure of group II introns brings the two exons together, facilitating intron removal and exon ligation (Fig. 5.22).

Another naturally occurring large ribozyme is RNase P from bacteria. This is an RNA-protein complex, but the RNA component is the catalytic entity. RNase P cleaves the 5′ end of pre-tRNA molecules. RNase P can act on multiple substrates, unlike the group I and group II introns that naturally act only on themselves.

Ribozymes are naturally occurring RNAs that can facilitate an enzymatic reaction. Group I and group II introns are two types of ribozymes that are able to cleave the phosphate backbone and release themselves from the mRNA molecule without any cellular enzymes.

SMALL NATURALLY OCCURRING RIBOZYMES

In contrast to the large ribozymes, small ribozymes are only about 200 to 500 nucleotides long. Small ribozymes include the hammerhead and hairpin ribozymes, the hepatitis delta virus ribozyme, and the Varkud satellite ribozyme. These are often found in viroids, virusoids, and satellite viruses, which are **subviral agents**. Viroids are self-replicating pathogens of plants that are merely naked single strands of RNA with no protein coat. Satellite viruses are small RNA molecules that require a helper virus for either replication or capsid formation. Their genomes may encode proteins. Virusoids are even less functional and are often considered a subtype of satellite virus. Virusoids are single strands of circular RNA that encode no proteins. They rely on helper viruses for both replication and a protein coat.

The **hammerhead ribozyme** is a small catalytic RNA that can catalyze a self-cleavage reaction. Hammerhead ribozymes take part in the replication of some viroids and satellite RNAs (Fig. 5.23). These both exist as single-stranded RNA genomes that form rodlike structures that are resistant to cellular ribonucleases. During viroid replication, the positive RNA strand is replicated by the host cell RNA polymerase, resulting in a long concatemer of negative-strand genomes. RNA polymerase then uses this as template to make a positive strand. The long RNA is cut into individual unit genomes by the hammerhead motif. Hammerhead ribozymes first cleave the ribose phosphate backbone of RNA, and then ligate the linear unit genomes into circular genomes.

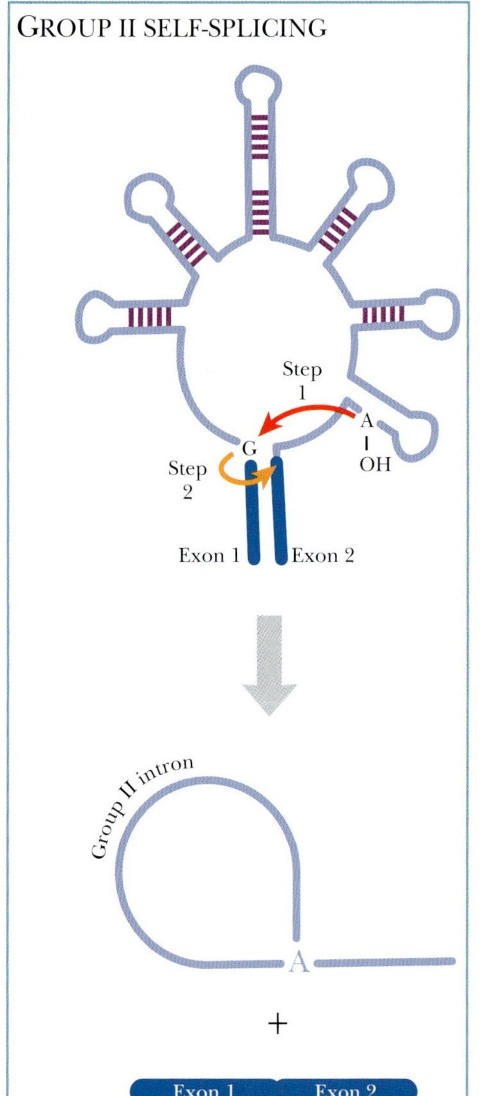

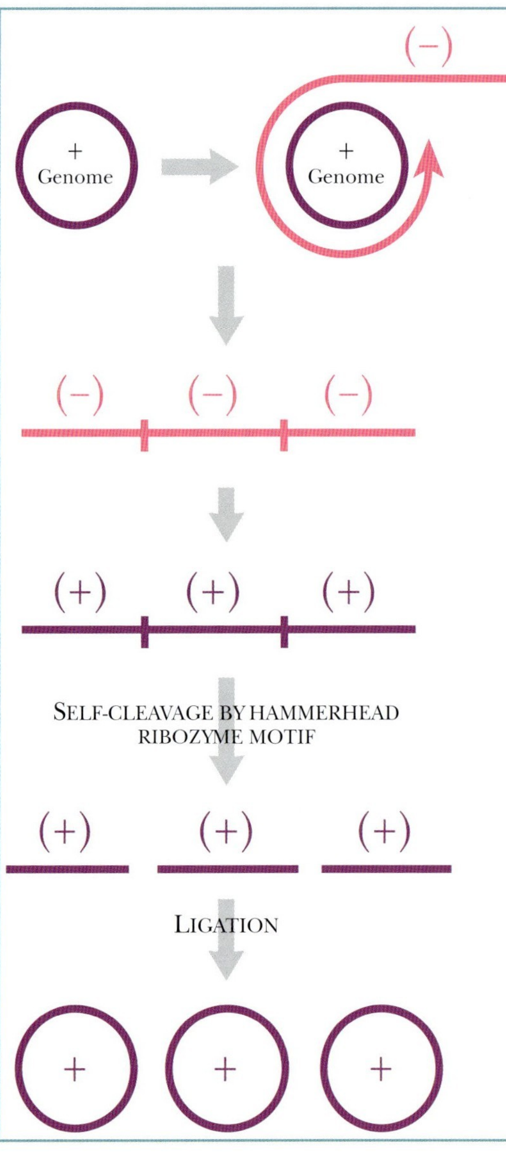

FIGURE 5.22 (left) Group II Intron Splicing Reactions
The secondary and tertiary structure of group II introns brings the two exons together, but the reaction mechanism does not require an external nucleophile. Instead, the 2′-OH of an internal conserved adenine acts as a nucleophile, attacking the 5′ splice site and cleaving the phosphate backbone. The 3′ OH of the 5′ splice site attacks the 3′ splice site, resulting in two ligated exons and a free intron. The intron forms a lariat structure.

FIGURE 5.23 (right) Life Cycle of Viroids
Viroids are single-stranded circular RNA genomes with no protein coat, but the ability to self-replicate. First, the plus-stranded genome is converted into a concatemer of negative-stranded genomes with rolling-circle replication. RNA polymerase converts the negative-stranded genomes into plus-stranded genomes, which are separated and ligated into circular genomes. The hammerhead ribozyme, embedded within the viroid genome, catalyzes the separation and ligation into a circle.

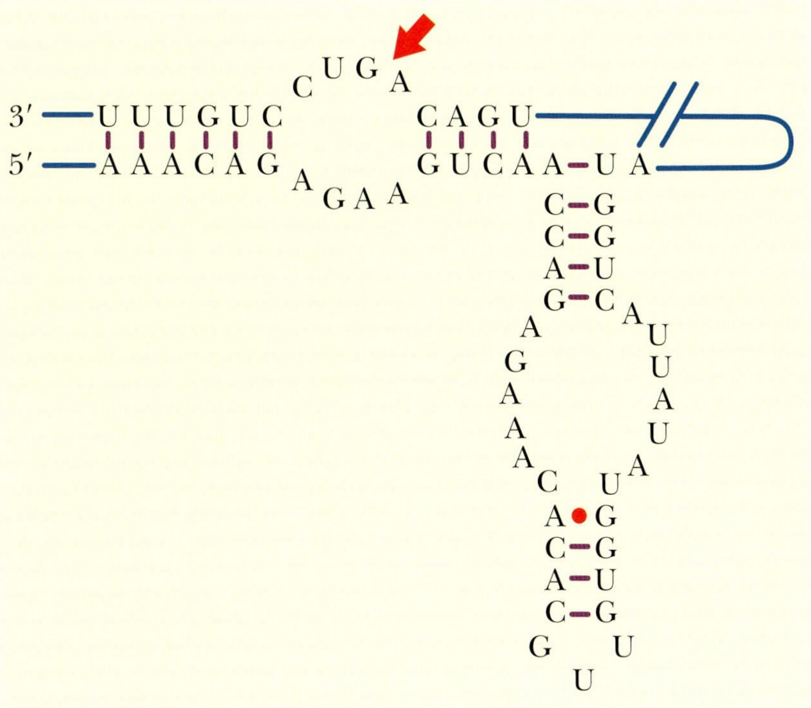

Another small ribozyme is the **hairpin ribozyme** (Fig. 5.24). This is found in pathogenic plant satellite viruses such as tobacco ring spot virus. The hairpin ribozyme from tobacco ring spot virus was originally called the "paperclip" ribozyme, and this may be a better description of the structure. *In vivo*, hairpin ribozymes cleave the linear concatemers of single-stranded RNA genomes, much like hammerhead ribozymes, and then ligate the linear segments into circular genomes.

FIGURE 5.24
Secondary Structure of Hairpin Ribozyme
The minus strand of the tobacco ring spot virus genome is shown with the cleavage site indicated by the red arrow.

Two other small ribozymes are the **Varkud satellite ribozyme** from *Neurospora* and the **hepatitis delta virus (HDV)** of humans. Both work by similar reaction mechanisms to initiate self-cleavage and ligation reactions. The Varkud satellite ribozyme helps replicate the small Varkud plasmid found within the mitochondria of *Neurospora*. The hepatitis delta virus is a viroid-like satellite virus of the hepatitis B virus. Hepatitis B infects the liver and can cause liver scarring, liver failure, and sometimes death. In patients with hepatitis B, the presence of HDV amplifies the symptoms, causing a very severe and often fatal case of the disease. The satellite virus, HDV, has a single-stranded RNA genome. It can be found in both positive and negative forms in liver cells. Both the genomic and antigenomic sequences, as the positive and negative forms are called, have regions that fold into an active ribozyme, which can catalyze RNA cleavage and ligation. Unlike plant viroids, HDV also has an open reading frame that encodes a protein called the *delta antigen*. Delta antigen plus coat proteins from hepatitis B virus are required to package HDV into small spherical particles. These can be spread from cell to cell, and from person to person via bodily fluids such as saliva and semen.

Small naturally occurring ribozymes are found in small subviral agents such as viroids and satellite viruses. These have common motifs that catalyze RNA cleavage.

ENGINEERING RIBOZYMES FOR MEDICAL AND BIOTECHNOLOGY APPLICATIONS

Engineered ribozymes are used to suppress the expression of genes, especially those that promote growth of cancers or those from pathogenic viruses such as human immunodeficiency virus (HIV). To engineer such a molecule, a ribozyme catalytic core is combined with a region that recognizes the target gene. Often the target gene is recognized through antisense technology, thus combining the two strategies (Fig. 5.25). First, an appropriate region on the target mRNA is identified. The region must be relatively free of secondary structure and have no protein binding sites. Next, antisense RNA is synthesized to match the target mRNA. The antisense sequence is split, so that the 5′ half is in front of the ribozyme catalytic core, and the 3′ half is behind. When this chimeric ribozyme is mixed with target mRNA, the antisense regions base-pair with the target and the ribozyme cleaves the target mRNA. The two halves of the target mRNA are further degraded by other enzymes.

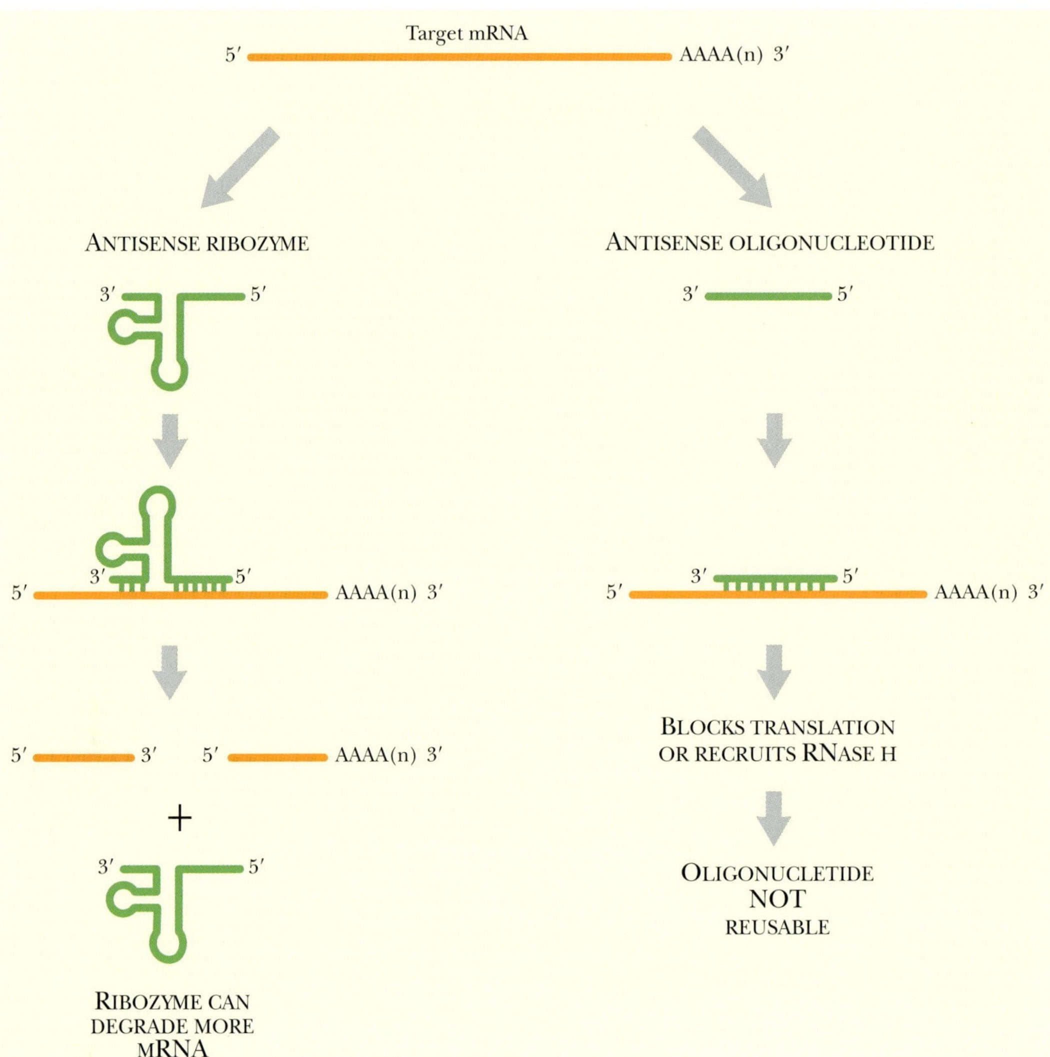

FIGURE 5.25 Antisense Construct with Ribozyme Inactivates Target mRNA
The chimeric antisense ribozyme not only has the ability to bind to a specific target mRNA, but also cleaves the target mRNA. A traditional antisense oligonucleotide must rely on recruiting RNase H to digest the target mRNA. However, RNase H also degrades the antisense oligonucleotide, which cannot therefore be reused.

The engineered ribozyme can attack many target mRNA molecules because it is an enzyme, not merely an inhibitor. The addition of a ribozyme is much better than using antisense inhibition alone, because antisense constructs are degraded along with the target mRNA.

The ribozyme catalytic core used for the constructs just described is usually from either hairpin or hammerhead ribozymes (Fig. 5.26). Altering the ribozymes from group I introns, from group II introns, or from RNase P is difficult because of their large size and complex secondary and tertiary structure. Small ribozymes have a natural division between their catalytic centers and the sequences that specify their target. Thus, it is easy to manipulate the type of target the ribozyme will cleave. Hammerhead ribozymes have a lower propensity for ligation and are often used preferentially over hairpin ribozymes.

> Adding a ribozyme motif such as the hammerhead or hairpin region from the small ribozymes can make antisense oligonucleotides more stable as these are not degraded. These constructs cut the target mRNA without the use of RNase H.

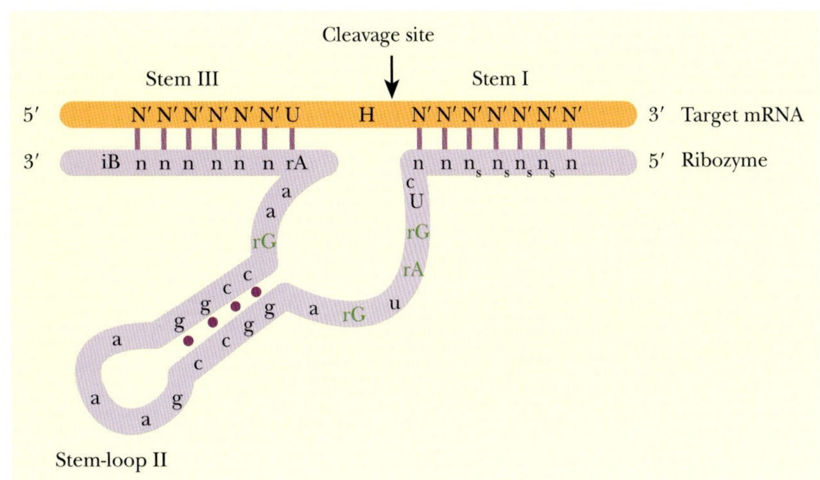

FIGURE 5.26
Nuclease Resistant Ribozyme Bound to a Target mRNA
The ribozyme consists of 2′-*O*-methyl nucleotides and phosphorothioate linkages. At the 3′-end, a 3′-3′ deoxyabasic sugar (iB) is added. All three modifications prevent nuclease degradation of the ribozyme. The five green nucleotides (rA or rG) form the catalytic core and cut the target mRNA at the cleavage site. The H at the cleavage site represents an A, C, or U.

RNA SELEX IDENTIFIES NEW BINDING PARTNERS FOR RIBOZYMES

Natural ribozymes normally act on only one specific substrate. Moreover, natural ribozymes are often degraded and so do not normally process more than one substrate molecule. An exception is RNase P, which can catalyze multiple cleavages of different tRNA molecules. One goal of biotechnology is to increase the number of substrates for the known ribozymes.

To identify new potential substrates for existing ribozymes, a procedure called **RNA SELEX** can be used. SELEX (Systematic Evolution of Ligands by EXponential enrichment) isolates new substrates for existing enzymes from a large (10^{15}) population of random-sequence oligonucleotides (Fig. 5.27). First, the mixture of random oligonucleotides is chemically synthesized as single-stranded DNA. To make the RNA, the random oligonucleotides are converted into double-stranded DNA using a 5′ primer and Klenow polymerase. The 5′ primer contains the promoter sequence for T7 RNA polymerase, which is added to the pool of dsDNA to make multiple single-stranded RNA copies. The ribozyme of interest is then mixed with this large pool of ssRNA oligonucleotides, and those RNA molecules that bind to the ribozyme are isolated. In order to facilitate isolation, the ribozyme can be immobilized on beads or linked to biotin. The binding sequences can be directly identified, or they can be pooled and put through repeated cycles of selection, thus eliminating those that bound nonspecifically. In order to identify the new binding substrate, the RNA must be released from the ribozyme. It is then converted into cDNA using a 3′ primer and reverse transcriptase. Because the actual number of specific binding molecules is low, these are amplified using PCR before sequencing.

The use of SELEX extends beyond ribozymes. It can be applied to drug design and delivery. The process can be applied to finding DNA binding substrates for different enzymes. In **DNA SELEX**, the initial pool of random-sequence oligonucleotides is not converted to mRNA with RNA polymerase. Instead, the oligonucleotides are used directly in substrate binding and selection.

New substrates for a known ribozyme are found by incubating the pure ribozyme with a large pool of random RNA sequences. Any RNA sequence that binds to the ribozyme is a potential substrate.

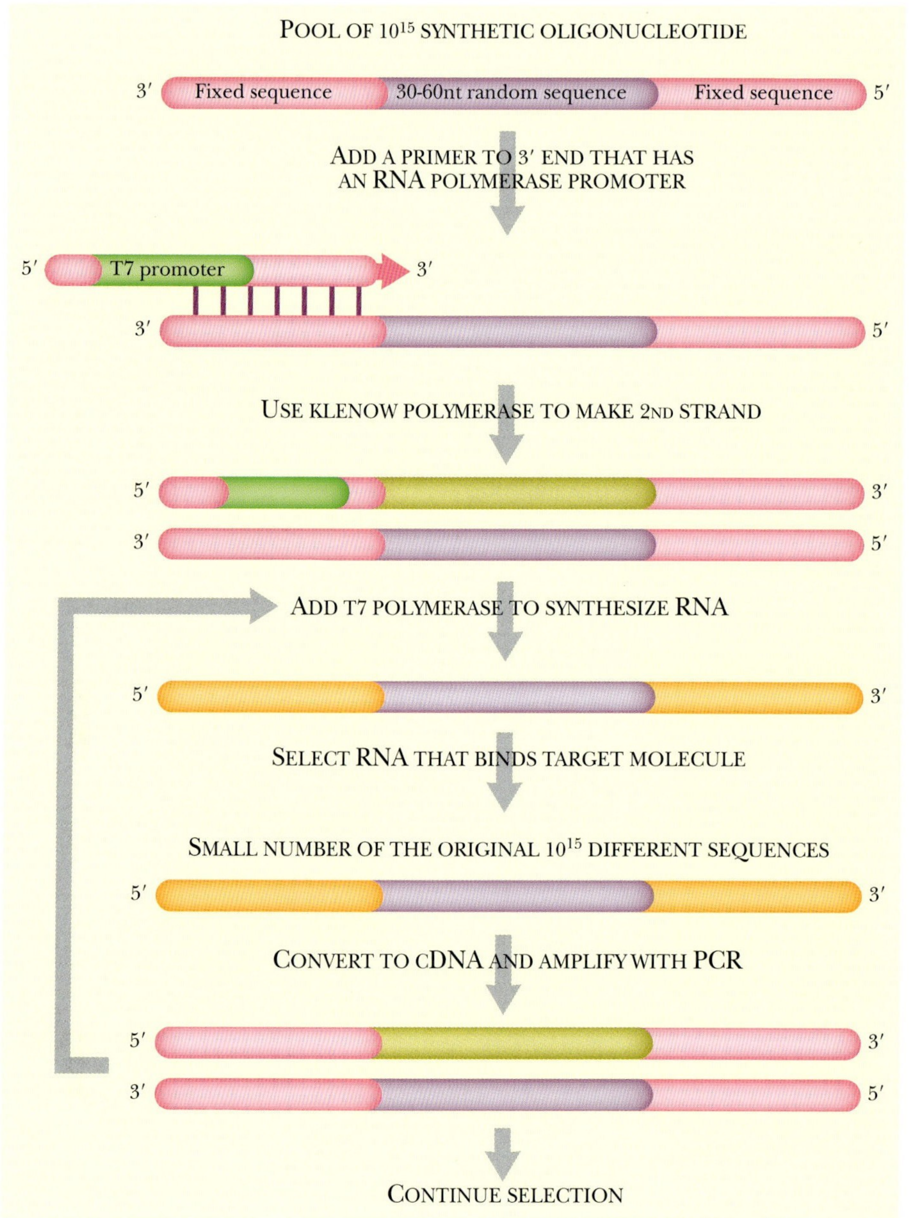

FIGURE 5.27 RNA SELEX Identifies New Ribozyme Substrates
The key to RNA SELEX is to create a very large pool of random RNA sequences. First, DNA oligonucleotides are chemically synthesized to create a large pool of random sequences. These are converted into double-stranded DNA with a primer and Klenow polymerase. The primer adds an RNA polymerase binding site. The dsDNA oligonucleotides are transcribed into RNA with RNA polymerase. This large pool of random RNA is then screened for binding to the ribozyme. Any RNA molecules that bind are kept, and the rest are discarded. Those that bind are converted to cDNA and amplified with PCR. The selection process can be repeated numerous times to enrich for RNA sequences that bind.

IN VITRO EVOLUTION AND *IN VITRO* SELECTION OF RIBOZYMES

It is also possible to generate new ribozymes with novel enzymatic capabilities from large pools of random RNA sequences. This is worthwhile because small natural ribozymes are mostly limited to cleavage and ligation reactions. Using *in vitro* selection allows new ribozyme reactions to be identified from random nucleotide sequences (Fig. 5.28).

For example, a ribozyme that catalyzes the ligation of a particular sequence can be identified. This approach begins by synthesizing a set of random oligonucleotide sequences. However, these represent the pool of potential ribozymes rather than substrates as seen in RNA SELEX. Each random sequence is flanked by two known sequences. The 5′ end sequence is one substrate for the desired ligation reaction. The 5′ end also has a terminal triphosphate to energize ligation. The 3′ end has a sequence domain that binds a chosen effector molecule. This allows the enzyme reaction to be regulated. In addition, knowing the sequence at the two ends allows amplification of the total construct by PCR.

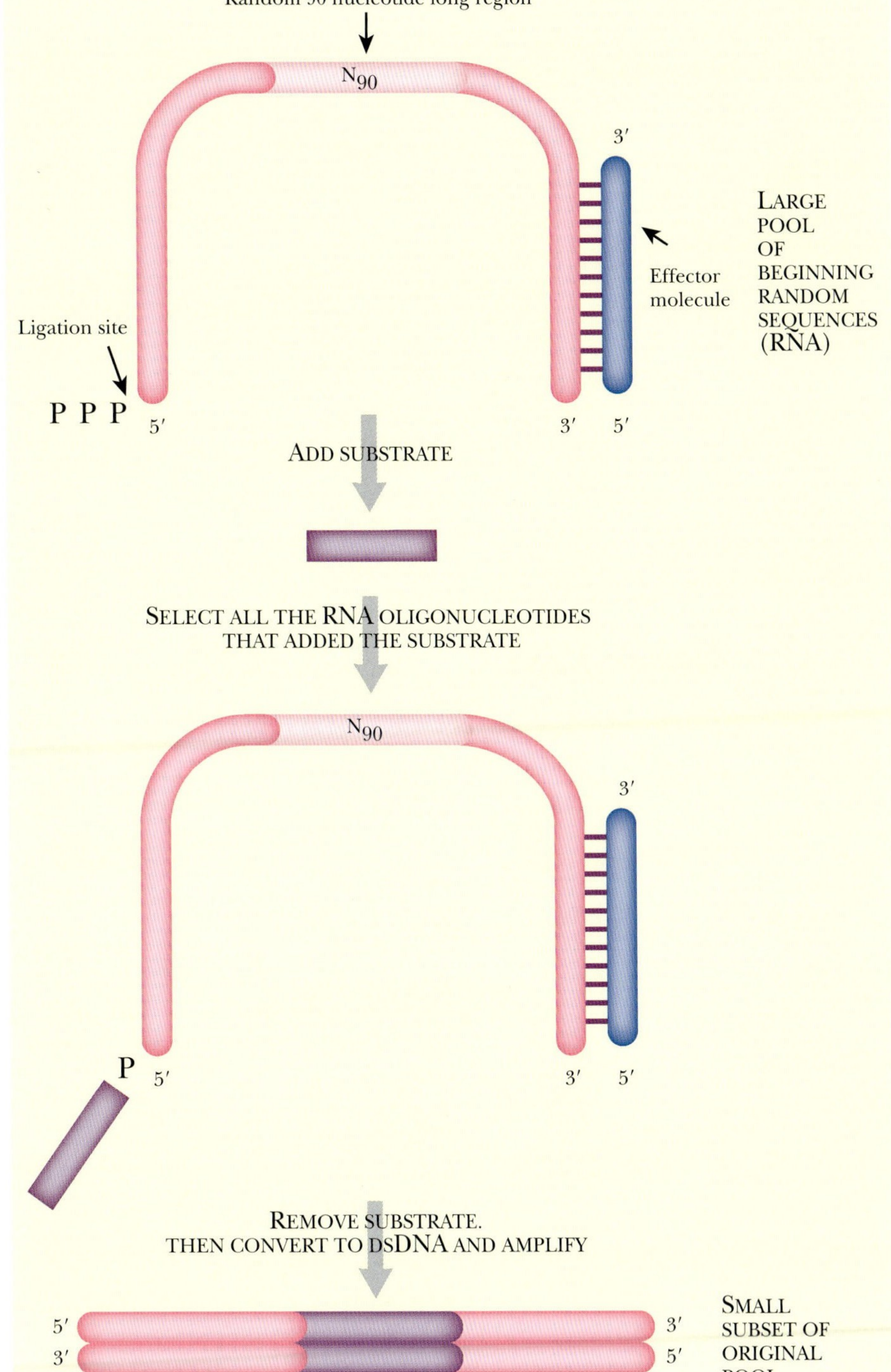

FIGURE 5.28
***In Vitro* Selection of Ribozyme Ligation**
The pink molecule represents the large pool of random RNA sequences. At the 5′ end, there is a substrate sequence with a terminal triphosphate. At the 3′ end, a blue effector molecule is bound to facilitate ligation. The second substrate (purple) is then incubated with the random pool of oligonucleotides. If any of the random sequences catalyze the ligation of the substrates, the resulting species will be larger and may be separated out by gel electrophoresis. The ligated oligonucleotide is isolated from the gel, amplified by PCR, and finally sequenced.

The other substrate for ligation is mixed with the potential ribozymes and incubated in conditions that favor ligation. If one of the random potential ribozyme sequences ligates the substrate to its 5′ end, the resulting RNA molecule (i.e., ribozyme plus ligation product) will run more slowly on an agarose gel. The slower molecules are isolated from the gel. The ribozyme suspect is then converted to DNA with reverse transcriptase. Finally, the DNA is amplified with PCR using primers that match the 5′ end and 3′ end of the original RNA constructs.

In vitro evolution enhances *in vitro* selection by adding a mutagenesis step after each cycle of selection (Fig. 5.29). This method begins with a pool of random oligonucleotides, as before.

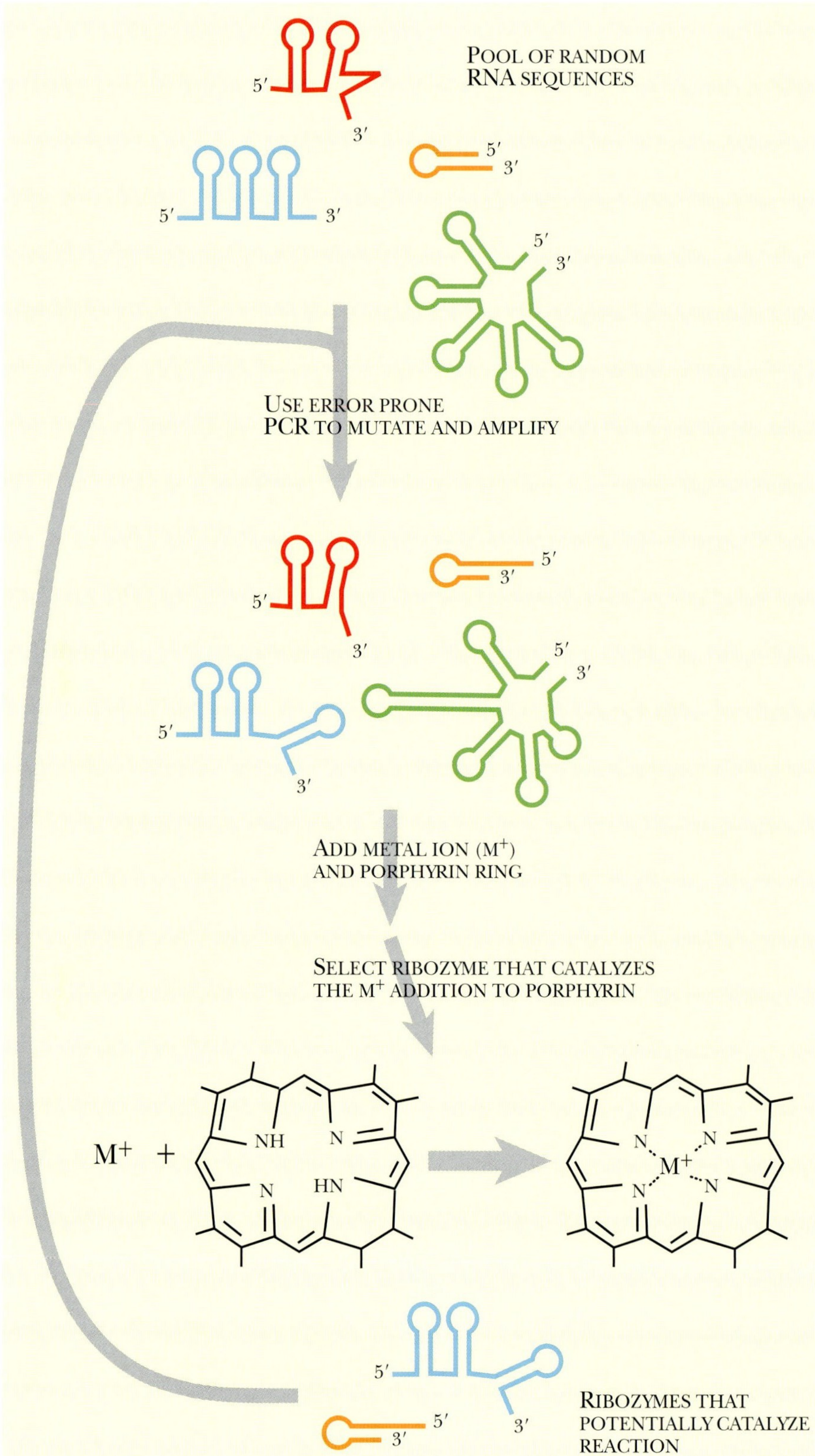

FIGURE 5.29
***In Vitro* Evolution of Ribozymes**
In vitro evolution tries to find an RNA sequence that works as a ribozyme. In this example, the researcher is looking for a ribozyme that catalyzes the addition of a metal ion (M^+) to a porphyrin ring. The pool of random RNA sequences is created and amplified with error-prone PCR to increase the odds of finding one or two sequences that catalyze the reaction. Each successive round of selection and mutation improves any ribozymes that are found.

These sequences may be of any length, provided they are long enough to carry out the desired reaction. The pool of random sequences is then mutagenized. The most efficient method is to use error-prone PCR (see Chapter 4) to amplify the initial pool of sequences. The pool both becomes larger and gains even more different sequences. The next step selects for the specific sequence that carries out the desired reaction. For example, artificial ribozymes have been evolved to add metal ions to mesoporphyrin IX (see Fig. 5.29). The mutagenesis and selection steps can be repeated over and over to improve the ribozyme. Once an efficient ribozyme is obtained, the sequence is determined after converting the RNA into cDNA.

Artificial ribozymes have been made to carry out nucleophilic attacks at various centers, including phosphoryl, carbonyl, and alkyl halides. There is also an artificial ribozyme that can isomerize a 10-member ring structure. In each of these cases, the initial pools of oligonucleotides or ribozymes were selected for the ability to carry out the specific reaction. In both *in vitro* selection and *in vitro* evolution, the key to success is the selection step. It must be stringent enough that most of the nonfunctional RNA molecules are eliminated, but not so stringent that ribozymes with weak activity are eliminated too early.

In vitro selection can also generate new ribozymes by mixing random sequences that represent potential ribozymes with a specific substrate.

Adding a mutagenesis step to the *in vitro* selection procedure allows the ribozyme to "evolve" into a better enzyme.

SYNTHETIC RIBOZYMES USED IN MEDICINE

Ribozymes are beginning to be used in medical applications. Researchers studying AIDS have derived a hammerhead ribozyme that can inhibit HIV replication. This engineered ribozyme was in clinical trials as of 2006. It is administered by expressing the ribozyme gene in a viral vector. The vector is transfected into peripheral blood T lymphocytes from HIV-infected patients. It is hoped that the expressed ribozyme will cleave the RNA version of the HIV genome, thus preventing replication of the HIV virus.

Another ribozyme has been developed to cleave an RNA virus, hepatitis C virus (HCV). HCV is the leading cause of chronic hepatitis, and no vaccine is available. Various engineered ribozymes have been identified that can efficiently cleave HCV RNA, but these studies are still *in vitro*. The engineered ribozymes have worked efficiently in cell culture where liver cells from infected individuals have been harvested and grown in dishes, but they have not yet been tested directly in patients.

The clinical use of ribozymes has many of the same obstacles as for any new drug. Each new ribozyme must be delivered to the correct location and expressed in cells that are diseased. Each ribozyme must be stable and resistant to degradation. In this regard, many engineered ribozymes contain modified bases, which prevent degradation by cellular endonucleases. Finally, the ribozyme must not have any deleterious side effects. High specificity to their target provides ribozymes with more potential than many preexisting therapies. For example, chemotherapy of cancer patients kills any rapidly dividing cells, not just the cancerous cells. This is why chemotherapy patients lose their hair. Ribozymes recognize one specific target mRNA; therefore, ribozyme treatments may avoid side effects seen in chemotherapy treatments.

Ribozymes that treat illness or cancer have much promise because the ribozyme will precisely affect only the actual defect.

ALLOSTERIC DEOXYRIBOZYMES CATALYZE SPECIFIC REACTIONS

Because some RNA has catalytic properties, researchers investigated whether DNA has catalytic potential. Although no natural catalytic DNA molecules are known, DNA nonetheless has the ability to catalyze various reactions in a manner similar to RNA-based ribozymes. Indeed, *in vitro* selection was used to create a deoxyribozyme that can split thymine dimers caused by UV radiation of DNA. Different organisms have various mechanisms to deal with these dimers. For example, excision repair removes the damaged strand and replaces it with new DNA. Another mechanism involves **photolyase** enzymes, which are activated by light. These enzymes recognize and repair thymine dimers in response to blue light.

In order to identify a DNA sequence that could accomplish the photolyase reaction, *in vitro* selection was carried out on a pool of random DNA oligonucleotide sequences. The random sequences were first linked to a substrate that consisted of two DNA oligonucleotides joined via a thymine dimer. If a random oligonucleotide split the thymine dimer after exposure to blue light, then the overall length of the DNA construct would be smaller. The smaller species were isolated by gel electrophoresis. This was successful, and a specific deoxyribozyme (UV1C) that could catalyze a photolyase reaction was identified (Fig. 5.30).

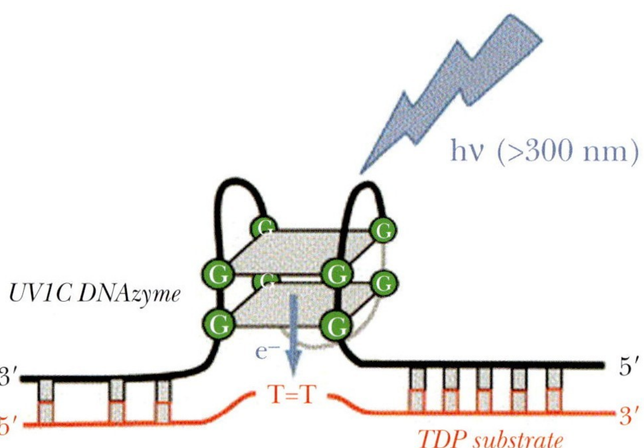

FIGURE 5.30
Deoxyribozyme That Repairs Thymine Dimers
A model for the deoxyribozyme UV1C–substrate complex. Light energy is absorbed by the guanine quadruplex. The thymine dimer is thought to lie close to the guanine cluster within the folded deoxyribozyme. This allows electron flow from the excited guanines to the thymine dimer. From: Chinnapen and Sen (2007). Towards elucidation of the mechanism of UV1C, a deoxyribozyme with photolyase activity. *J Mol Biol* **365**, 1326–1336. Reprinted with permission.

> Deoxyribozymes are DNA sequences that catalyze an enzymatic reaction.

RIBOSWITCHES ARE CONTROLLED BY EFFECTOR MOLECULES

Transcription is controlled primarily by protein factors. Nonetheless, in prokaryotes, conserved sequences have been identified that control gene expression at the RNA level. These sequences are an integral part of the messenger RNA molecules that they control and are called **riboswitches**. Unlike miRNAs or siRNAs, which work via base pairing, riboswitches bind small effector molecules, such as nutrients or cAMP. The riboswitches work by alternating between two different RNA secondary structures. In most cases, effector binding terminates mRNA transcription prematurely or prevents mRNA translation.

Riboswitches are found in several genes for biosynthetic enzymes. In *E. coli*, the thiamine riboswitch is controlled by thiamine pyrophosphate, a vitamin. When the vitamin is abundant, it binds to the TH1 box (i.e., a riboswitch) close to the 5′ end of the mRNA, and transcription of the mRNA is aborted. When the vitamin is absent, the mRNA is transcribed and translated to give enzymes that make more thiamine. Similar control occurs for riboflavin biosynthesis in *Bacillus subtilis*. The vitamin itself binds to the riboswitch domain of the mRNA and controls whether or not the mRNA is expressed.

Riboswitches normally work by changing the stem and loop structure of the mRNA transcript. In **attenuation riboswitches**, the effector molecule binds to the mRNA as it is being transcribed. If the effector binds, changes in structure create a terminator loop, which causes the transcriptional machinery to fall off prematurely. The incomplete mRNA is degraded. When the effector is in short supply, then the mRNA is transcribed to completion (Fig. 5.31A). Alternatively, some riboswitches work through translational inhibition. Here the riboswitch controls whether or not protein translation occurs by sequestering the Shine-Dalgarno sequence. When the effector molecule is abundant, its binding changes the stem-loop structure so that the Shine-Dalgarno sequence is not accessible to the ribosomes (see Fig. 5.31B).

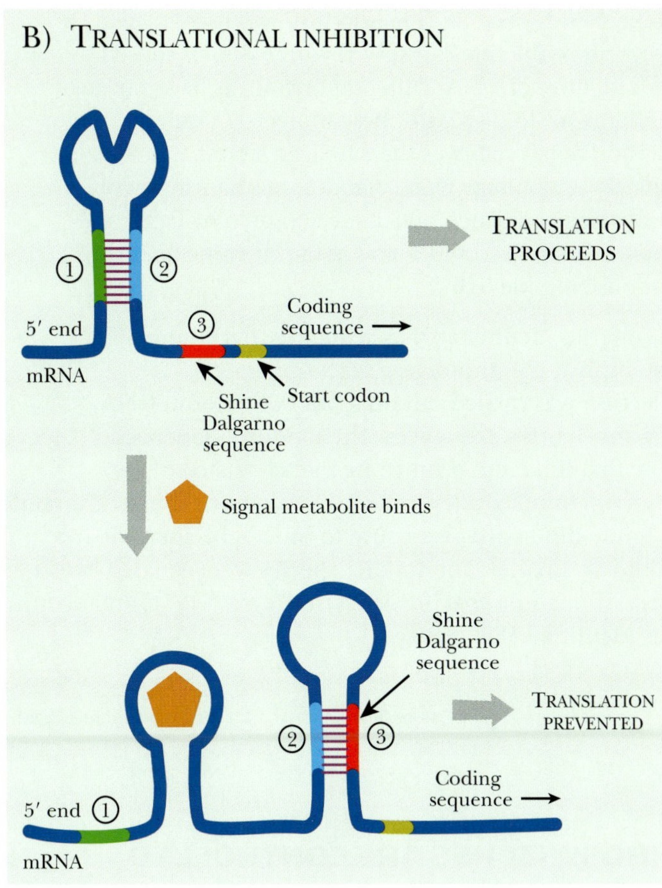

FIGURE 5.31 Riboswitches Control mRNA Expression
Riboswitches alternate between two stem and loop structures depending on the presence or absence of the signal metabolite. (A) In the attenuation mechanism, the presence of the signal metabolite results in formation of the terminator structure and transcription is aborted. (B) In the translational inhibition mechanism, the presence of the metabolite results in sequestration of the Shine-Dalgarno sequence, which prevents translation of the mRNA.

Recently, a novel riboswitch was identified in *Bacillus subtilis* that controls the expression of a biosynthetic gene (*glmS*) for a cell wall component (Fig. 5.32). As for other riboswitches, a product of the biosynthetic pathway controls whether or not the mRNA is expressed. However, instead of hiding the Shine-Dalgarno sequence or creating a terminator loop, the change in RNA secondary structure creates a self-cleaving ribozyme. The *glmS* gene of *B. subtilis* codes for the enzyme glutamine fructose 6-phosphate amidotransferase, which converts fructose 6-phosphate plus glutamine into glucosamine 6-phosphate (GlcN6P). This is further converted into a component of the cell wall, UDP-GlcNAc. When this is abundant, it binds to *glmS* mRNA, altering the secondary structure. The new structure functions as a ribozyme that cuts the mRNA, preventing any further translation.

Other riboswitches have been identified that respond directly to thermal stress. For example, the *rpoH* gene of *E. coli* is involved in the heat shock response. In addition to other forms of regulation, the mRNA contains a thermosensor domain, which controls the amount of translation. At normal temperatures, the thermosensor has a stem-loop structure that prevents translation. When the heat increases, the stem-loop structure falls apart and translation can occur.

Riboswitches are mRNA sequences that bind directly to effector molecules to control the expression of the mRNA into protein.

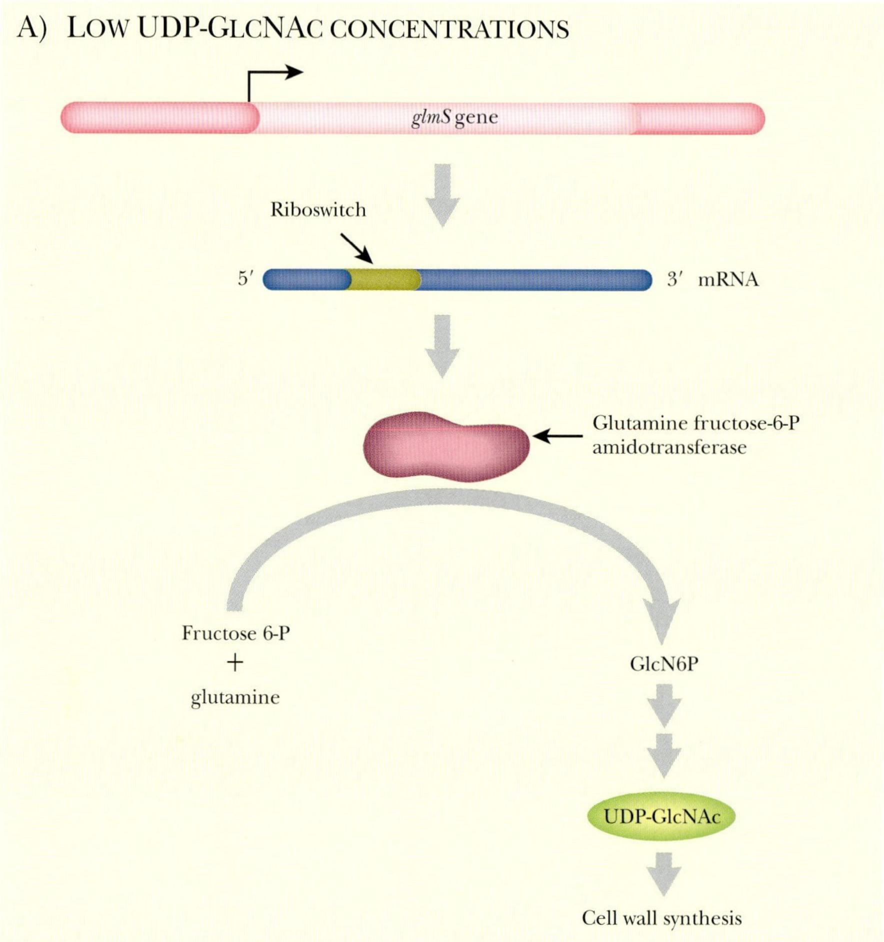

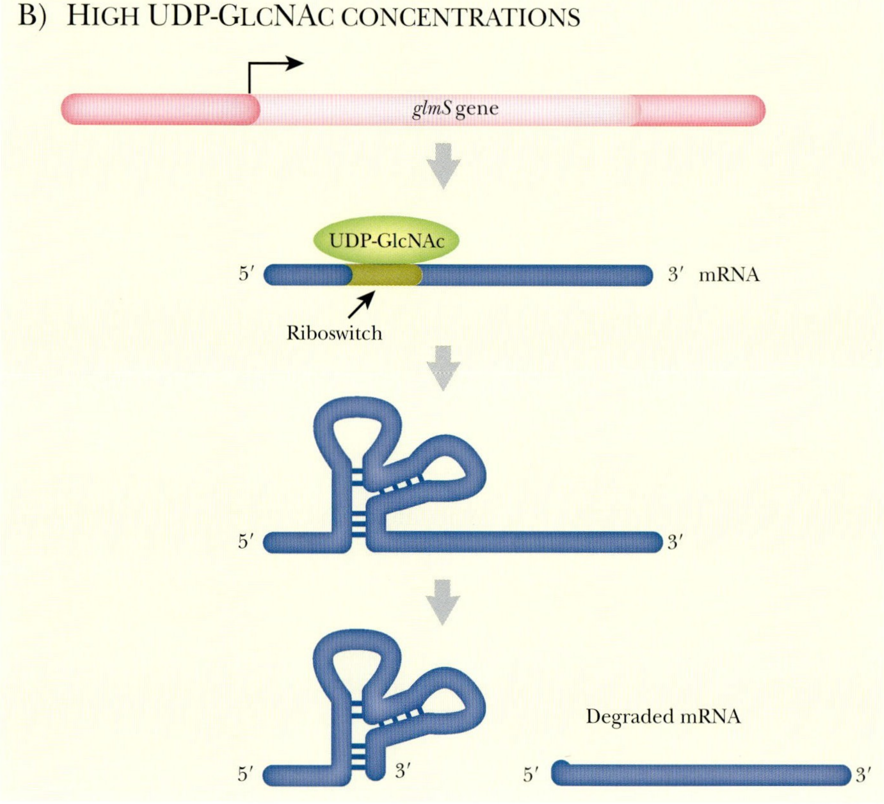

FIGURE 5.32
Ribozyme Riboswitch of *B. subtilis GlmS* Gene
(A) Cell wall synthesis occurs during growth conditions. When the cell is growing, levels of UDP-GlcNAc are low and are quickly converted into cell wall components. (B) If the cell is not growing, UDP-GlcNAc is not incorporated into the cell wall and accumulates. The excess UDP-GlcNAc binds to a riboswitch on the *glmS* gene. Once bound, it activates the self-cleaving ribozyme to degrade the mRNA, and halts the production of glutamine fructose 6-phosphate amidotransferase.

ENGINEERING ALLOSTERIC RIBOSWITCHES AND RIBOZYMES

Artificial or modified ribozymes have enormous potential in medicine and biotechnology. The ability to control the activity of a ribozyme would be very advantageous. If a ribozyme were engineered to cleave mRNA that causes rampant growth of cancer cells, controlling its action could prevent cancer from spreading. Additionally, ribozymes could be engineered into genes used for gene therapy. By controlling the ribozyme, the clinician could modulate when and where the gene is expressed. Such control could be exerted by building riboswitches into the ribozymes and then controlling self-destruction by the presence or absence of a small effector molecule, as happens naturally in the *glmS* gene described earlier.

In order to engineer a ribozyme to cleave only in the presence of a certain effector molecule, a combination of **modular design** and *in vitro* selection is used. Modular design takes various domains from different ribozymes and merges them to create a new molecule. For example, the catalytic core of a particular hammerhead ribozyme can be genetically linked to the binding domain of another, changing the binding specificity of the original ribozyme (Fig. 5.33A).

Artificial allosteric riboswitches have been selected by combining the ribozyme catalytic core with a pool of many different random sequences (see Fig. 5.33B). Some of the random sequences will, it is hoped, have the ability to bind the chosen effector and thus represent a pool of possible riboswitches. Some of the combinations will catalyze self-cleavage or substrate cleavage without regulation, and these must be eliminated. If the ribozyme construct cleaves itself, the products will move faster during electrophoresis. Therefore, the pool of possible riboswitch/ribozymes is electrophoresed, and the slower moving, uncleaved RNAs are isolated from the gel. Next, the uncleaved ribozymes are mixed with the chosen effector and incubated under cleavage-promoting conditions. This step is the positive selection step, and any ribozyme that undergoes cleavage in the presence of the effector is isolated. As before, the ribozymes are separated by gel electrophoresis, but this time the cleaved (shorter and faster) molecules are isolated. Cloning and sequencing of the isolated ribozyme constructs determines the sequence of the riboswitch domain.

Some effectors that researchers have used to control riboswitches include cyclic GMP, cyclic AMP, and cyclic CMP. Allosteric ribozymes have been artificially created that respond not only to small organic molecules such as cyclic AMP, but also to oligonucleotides, proteins, and even metal ions.

Ribozymes can be created with riboswitches. The riboswitch controls the ribozyme so that it is only active when the effector molecule is present.

Summary

Although RNA was only thought of as an intermediary molecule, it is now recognized that many different processes control the expression of RNA into protein. In the first part of this chapter, antisense genes and oligonucleotides are described. These complementary RNA sequences match the mRNA, bind to the sequence, and inhibit protein translation. The double-stranded RNA cannot be converted to proteins because the ribosomes are unable to bind. In addition, double-stranded RNA triggers RNase H to degrade the duplex, which obliterates the target mRNA. Biotechnology has looked to this technology in order to suppress genes that cause disease. Making changes to the phosphate-sugar backbone is essential to making these constructed antisense oligonucleotides more stable. The oligonucleotides are

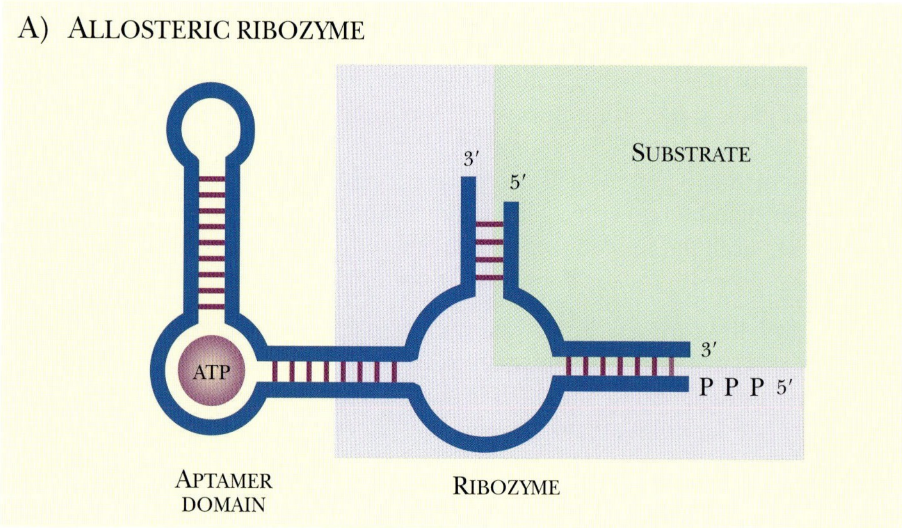

FIGURE 5.33
Designing Allosteric Ribozymes
(A) Modular design of a ribozyme. The ribozyme has three different domains joined together. The substrate domain (light green background) base-pairs with the ribosyme domain (light purple background), and the aptamer domain binds the allosteric effector (ATP in this example). (B) *In vitro* selection scheme to identify ribozymes that are active only when bound to an effector (i.e., are allosteric). First, all ribozymes that catalyze substrate cleavage without an effector are removed. If the substrate is cleaved without the effector, the ribozyme will move faster during electrophoresis. Only the uncleaved ribozyme/substrate band is isolated from the gel. Next, the ribozymes are mixed with an effector molecule. This time, the ribozymes that cleave the substrate are isolated. Repeated cycles of isolation will identify a ribozyme that works only with the effector.

delivered to cells by a variety of methods, including attaching them to liposomes and basic peptides. These oligonucleotides can pass through pores created by streptolysin O and, *in vitro*, by mechanical shearing of the membranes. In contrast to oligonucleotides, a vector containing the target gene in the opposite orientation creates an entire antisense gene. This can be expressed in cells to determine the function of an unknown gene product.

The next part of this chapter describes RNAi or RNA interference. This is a cellular process where double-stranded RNA activates Dicer to cleave the RNA into 21- to 23-nucleotide pieces called siRNA, for short interfering RNA. These short segments are taken into another enzyme complex called RISC. Here the double-stranded siRNAs are unwound and the single-stranded template is used to find similar sequences in the cytoplasm. When RISC finds these complementary sequences, it cuts the target, thus preventing protein translation. This cellular process is used during embryonic development of many different organisms, except the original double-stranded RNA derives from the genome in a structure called a microRNA. Biotechnology has begun using this cellular process to destroy expressed cellular proteins with an unknown function. RNAi is used to remove one protein from the cell, and how that affects the cellular functioning is then determined. This provides insight into cellular proteins that do not have any known function.

Ribozymes are RNA sequences that catalyze enzymatic reactions. Naturally occurring ribozymes are classified as large or small based on their size. The small ribozymes have compact motifs that catalyze the reaction and therefore are useful for designing ribozymes that destroy cellular mRNA that causes disease. For example, the hammerhead motif can be engineered into an antisense oligonucleotide that recognizes the mRNA from a disease-causing virus. The antisense oligonucleotide will bind the mRNA, and the hammerhead motif will cut the target mRNA. Creating large pools of random RNA sequences can be used to either find new substrates for an existing ribozyme or find new sequences that are ribozymes. These can be "evolved" into better substrates or ribozymes by adding a step of mutagenesis before the selection. In addition to RNA ribozymes, DNA has also been shown to have catalytic power.

The final part of this chapter discusses riboswitches, which are RNA sequences that alter shape after binding an effector molecule like cAMP. These are used in biotechnology to control the expression of various genetic constructs.

End-of-Chapter Questions

1. Which of the following statements about antisense RNA is true?
   - **a.** Antisense RNA binds to form double-stranded regions on RNA to either block translation or intron splicing.
   - **b.** Antisense RNA is transcribed using the sense strand of DNA as a template.
   - **c.** The sequence of antisense RNA is complementary to mRNA.
   - **d.** Antisense RNA is made naturally in cells and also artificially in the laboratory.
   - **e.** All of the above statements about antisense RNA are true.
2. Which biological function is not controlled by antisense RNA?
   - **a.** iron metabolism in bacteria
   - **b.** the circadian rhythm of *Neurospora*
   - **c.** replication of prokaryotic genomic DNA
   - **d.** replication of ColE1 plasmid
   - **e.** developmental control of basic fibroblast growth factor
3. Which of the following is a modification of antisense oligonucleotide structure to increase intracellular stability?
   - **a.** insertion of an amine into the ribose ring to create a morpholino structure
   - **b.** attachment of nucleic acid bases to a peptide backbone instead of a sugar-phosphate backbone
   - **c.** replacement of one of the oxygen atoms in the phosphate group with a sulfur atom to inhibit nuclease degradation in some molecules
   - **d.** addition of an O-alkyl group to the 2′-OH of the ribose group to make the molecule resistant to nuclease degradation
   - **e.** all of the above

4. How can antisense RNA be expressed within a cell?
   a. The target gene can be cloned inversely into a vector and under the control of an inducible promoter.
   b. The antisense RNA cannot be expressed within a cell and instead must be delivered via liposomes.
   c. Antisense RNA can be expressed within cells, but this is unfavorable because of the high degree of non-specific interactions.
   d. No system has been designed to express antisense RNA within a cell.
   e. None of the above is correct.

5. Which of the following terms describes when gene regulation occurs by short is dsRNA molecules triggering an enzymatic reaction that degrades the mRNA of a target gene?
   a. post-transcriptional gene silencing
   b. quelling
   c. co-suppression
   d. RNA interference
   e. all of the above

6. Which statement about RNAi is not correct?
   a. RNAi was first discovered in plants.
   b. RNAi has two phases: initiation and effector.
   c. During the initiation phase of RNAi, a protein called Dicer cuts dsRNA into small fragments called siRNAs.
   d. Non-specific interactions between the antisense siRNA and mRNA often cause mRNAs to be degraded that should not have been.
   e. The RNA-induced silencing complex has both helicase and endonuclease activities.

7. Which of the following is not a method for delivering dsRNA for RNAi into *Drosophila* and *C. elegans* cells?
   a. ingestion of transgenic bacteria that express dsRNA
   b. cDNA library clone
   c. injection of dsRNA into eggs
   d. bathing in a solution of pure dsRNA
   e. injection of dsRNA into cell culture lines

8. How can RNAi be triggered in mammalian cells?
   a. transfection of siRNA
   b. chemically synthesized siRNA
   c. degradation of target mRNA through shRNA creation
   d. modification of an existing shRNA to recognize a different mRNA
   e. all of the above

9. What information has been obtained through the creation of RNAi libraries?
   a. the function of unknown proteins by degrading all of the mRNA for that protein
   b. the mechanism by which *E. coli* delivers dsRNA to *C. elegans*
   c. the mechanism by which heterochromatic formation occurs after some RNAi
   d. all of the above
   e. none of the above

(*Continued*)

10. What is a ribozyme?
    a. an enzyme that cuts ribosomes
    b. an RNA molecule that binds to specific targets and catalyzes reactions
    c. an enzyme that catalyzes the degradation of dsRNA
    d. an RNA molecule that catalyzes the degradation of ribonucleases
    e. none of the above

11. Which of the following is a large ribozyme?
    a. hairpin ribozyme
    b. hammerhead ribozyme
    c. Twort ribozyme
    d. hepatitis delta virus
    e. Varkud satellite ribozyme

12. What process is used to identify possible ribozyme substrates?
    a. DNA SELEX
    b. DNA BLAST
    c. RISC
    d. RNA SELEX
    e. GENEi

13. What property must a ribozyme possess in order to be used in clinical medicine?
    a. stability and resistance to degradation
    b. no deleterious side effects to the host
    c. expression within a diseased cell only
    d. be able to be delivered to the correct location
    e. all of the above

14. What is a riboswitch?
    a. an mRNA sequence that binds directly to an effector molecule to control the translation of the mRNA into protein
    b. an enzyme that converts ribozymes into deoxyribozymes
    c. the effector molecule responsible for translational control of a particular mRNA
    d. an RNA molecule that switches between being translated into protein or being a ribozyme
    e. none of the above

15. What is an example of an effector molecule for riboswitches?
    a. some cyclic mononucleotides
    b. oligonucleotides
    c. metal ions
    d. some proteins
    e. all of the above

Further Reading

Allen E, Xie Z, Gustafson AM, Carrington JC (2005). MicroRNA-directed phasing during *trans*-acting siRNA biogenesis in plants. *Cell* **121**, 207–221.

Breaker RR (2002). Engineered allosteric ribozymes as biosensor components. *Curr Opin Biotechnol* **13**, 31–39.

Chen Q, Crosa JH (1996). Antisense RNA, Fur, iron and the regulation of iron transport genes in *Vibrio anguillarum*. *J Biol Chem* **271**, 18885–18891.

Chinnapen DJ-F, Sen D (2003). A deoxyribozyme that harnesses light to repair thymine dimers in DNA. *Proc Natl Acad Sci USA* **101**, 65–69.

Chinnapen DJ-F, Sen D (2007). Towards elucidation of the mechanism of UV1C, a deoxyribozyme with photolyase activity. *J Mol Biol* **365**, 1326–1336.

Clark DP (2005). *Molecular Biology: Understanding the Genetic Revolution*. Elsevier Academic Press, San Diego, CA.

Crosthwaite SK (2004). Circadian clocks and natural antisense RNA. *FEBS Lett* **567**, 49–54.

Dias N, Stein CA (2002). Antisense oligonucleotides: Basic concepts and mechanisms. *Mol Cancer Ther* **1**, 347–355.

Elgin SC, Grewel SI (2003). Heterochromatin: Silence is golden. *Curr Biol* **13**, R895–R898.

Fitzwater T, Polisky B (1996). A SELEX primer. *Methods Enzymol* **267**, 275–301.

Golden BL, Kim H, Chase E (2004). Crystal structure of a phage Twort group I ribozyme-product complex. *Nat Struct Mol Biol* **12**, 82–89.

Khan AU, Lal SK (2003). Ribozymes: A modern tool in medicine. *J Biomed Sci* **10**, 457–467.

Knee R, Murphy PR (1997). Regulation of gene expression by natural antisense RNA transcripts. *Neurochem Int* **31**, 379–392.

Lai EC (2003). MicroRNAs: Runts of the genome assert themselves. *Curr Biol* **13**, R925–R936.

Lochmann D, Jauk E, Zimmer A (2004). Drug delivery of oligonucleotides by peptides. *Eur J Pharm Biopharm* **58**, 237–251.

Maeda I, Kohara Y, Yamamoto M, Sugimoto A (2001). Large-scale analysis of gene function in *Caenorhabditis elegans* by high-throughput RNAi. *Curr Biol* **11**, 171–176.

Preall JB, Sontheimer EJ (2005). RNAi: RISC gets loaded. *Cell* **123**, 543–545.

Rossignol F, de Laplanche E, Mounier R, Bonnefont J, Cayre A, Godinot C, Simonnet H, Clottes E (2004). Natural antisense transcripts of HIF-1α are conserved in rodents. *Gene* **339**, 121–130.

Sazani P, Vacek MM, Kole R (2002). Short-term and long-term modulation of gene expression by antisense therapeutics. *Curr Opin Biotechnol* **13**, 468–472.

Sen G, Wehrman TS, Myers JW, Blau HM (2004). Restriction enzyme-generated siRNA (REGS) vectors and libraries. *Nat Genet* **36**, 183–189.

Tanner NK (1998). Ribozymes: The characteristics and properties of catalytic RNAs. *FEMS Micro Rev* **23**, 257–275.

Tsang J, Joyce GF (1996). *In vitro* evolution of randomized ribozymes. *Methods Enzymol* **267**, 410–426.

van Rij RP, Andino R (2006). The silent treatment: RNAi as a defense against virus infection in mammals. *Trends Biotechnol* **24**, 186–193.

Vanhée-Brossollet C, Vanquero C (1998). Do natural antisense transcripts make sense in eukaryotes? *Gene* **211**, 1–9.

Vaucheret H, Béclin C, Fagard M (2001). Post-transcriptional gene silencing in plants. *J Cell Sci* **114**, 3083–3091.

Warashina M, Takagi Y, Stec WJ, Taira K (2000). Differences among mechanisms of ribozyme-catalyzed reactions. *Curr Opin Biotechnol* **11**, 354–362.

CHAPTER 6

Immune Technology

ANTIBODY STRUCTURE AND FUNCTION

The world is full of infectious microorganisms, all looking for a suitable host to infect. Bacteria, viruses, and protozoans are constantly attempting to gain entry into our tissues. If nothing was done about these attempts at invasion, no human could survive. Fortunately, cells of the immune system patrol the internal tissues, the bloodstream, and the body surfaces, both outside and inside, protecting the entire organism from attack. Any foreign macromolecules that are not recognized as being "self" will be regarded as signs of an intrusion and will trigger an immune response. Invading microorganisms have their own distinctive proteins, which differ in sequence and therefore in 3-D structure from those of the host animal. In particular, those molecules exposed on the surfaces of invading microorganisms will attract the attention of the immune system. These foreign molecules are referred to as **antigens**, and the immune system molecules that recognize and bind to them are known as **antibodies** (Fig. 6.1).

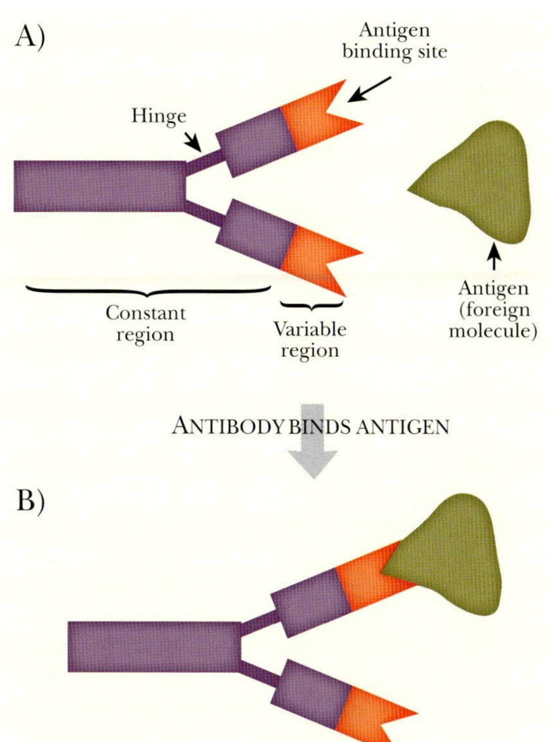

FIGURE 6.1
Foreign Antigens Are Recognized by Antibodies
(A) Antibodies are Y-shaped molecules produced by the immune system in vertebrates. These bind to specific proteins, or antigens, of any invading pathogen.

Although we often speak as if the body makes antibodies in response to invasion by a foreign antigen, this is rather misleading. In fact, long before infection, the immune system generates billions of different antibodies. Each **B cell** of the immune system has the capability to make just one of these. The immune system keeps a few B cells on standby for each of its colossal repertoire of antibodies. This happens before encountering the antigens and without knowing which antibodies will actually be needed later. If enough different antibodies are available, at least one or two should match an antigen of any conceivable shape, even if the body has never come into contact with it before.

Eventually, a foreign antigen appears (Fig. 6.2). Among the billions of predesigned antibodies, at least one or two will fit the antigen reasonably well. Those B cells that make antibodies that recognize the antigen now divide rapidly and go into mass production. Thus the antigen determines which antibody is amplified and produced. Once a matching antibody has bound invading antigens, the immune system brings other mechanisms into play to destroy the invaders.

Sometime later there is a stage of refinement during which those antibodies that bound to the invading antigen are modified by mutation to fit the antigen better. In addition, the immune system keeps a record of antibodies that are actually used. If the same invader ever returns, the corresponding antibodies can be rushed into action, faster and in greater numbers than before. Vaccines exploit this capacity by stimulating the immune system to store the antibodies that recognize and destroy a pathogenic virus such as smallpox. Yet the vaccines cause no disease symptoms themselves (see later discussion).

The immune system keeps a repertoire of B cells that are poised to make antibodies to invading pathogens. Each B cell makes a different antibody. When one of these B cell antibodies is needed, the B cell starts dividing so that many clones are available to attack the pathogen. Some of these clones are refined by mutation to make a more specific antibody.

ANTIBODIES, ANTIGENS, AND EPITOPES

The term *antigen* refers to any foreign molecule that provokes a response by the immune system. In practice, most antigens are proteins made by invading bacteria or viruses. In

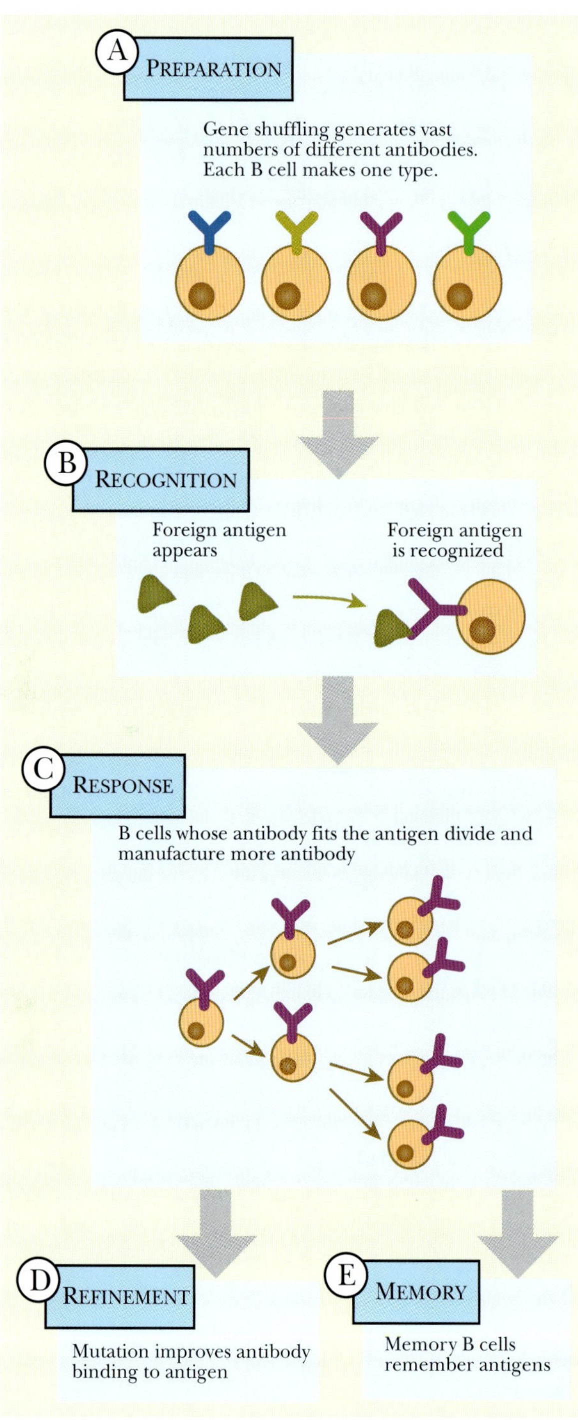

FIGURE 6.2 Pre-Designed Antibodies Are Ready for Foreign Antigens
Long before an attack by a pathogen, an army of B cells produces a large repertoire of antibodies (A). When one of the antibodies binds to an antigen (B), that particular B cell starts dividing and expanding (C). The majority of the B cells refine the antibody so that the antigen/antibody complex binds more tightly, and they fight the pathogens (D). A small subset of B cells become memory cells that never die, waiting for another attack by the same pathogen (E).

particular, glycoproteins, which carry carbohydrate residues, and lipoproteins, which carry lipid residues, generate strong immune responses, that is, they are highly antigenic. Other macromolecules can also work as antigens. Polysaccharides are often found as surface components of infiltrating germs and may act as antigens. Even DNA can be antigenic under certain circumstances.

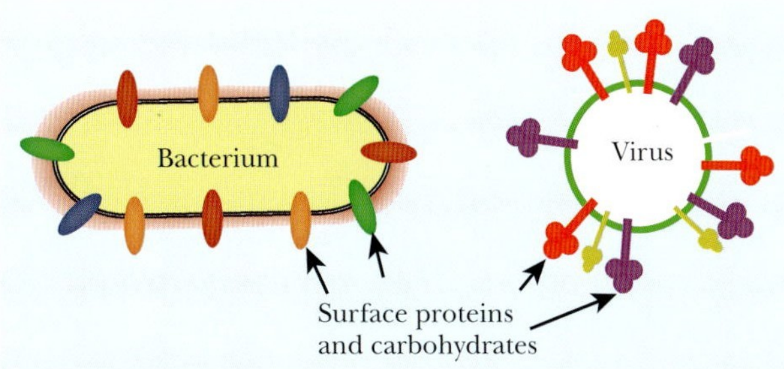

FIGURE 6.3
Surface Antigens of Microorganisms
The surfaces of bacteria and viruses are coated with glycoprotein and lipoproteins that are recognized by antibodies in the host organism.

Not surprisingly, the antigens exposed on the surface of an alien microorganism will usually be detected first by the immune system (Fig. 6.3). Later in infection, especially after the cells of some invaders have been disrupted by the immune system, molecules from the interior of the infectious agent may be liberated and also act as antigens.

The immune system mediates immunity to various infectious agents through **specific immunity** or **acquired immunity**. Acquired immunity can be subdivided into **humoral immunity** and **cell-mediated immunity**. Humoral immunity is mediated by antibodies, or **immunoglobulins**, in the blood plasma, which are produced by B cells. Cell-mediated immunity is mediated by antigen-specific cells called **T lymphocytes**, which are divided into T_H or T helper cells, and T_C or T cytotoxic cells. Antibodies generally bind to whole proteins, whereas T cell receptors bind to fragments of protein. When an antibody binds to a protein, it recognizes a relatively small area on the surface of the protein, such as dimples or projections sticking out from the surface. Such recognition sites are known as **epitopes** (Fig. 6.4). Because intact proteins are large molecules, they may have several epitopes on their surfaces. Consequently, several different antibodies may be able to bind the same protein, although not usually simultaneously, because antibodies are themselves large molecules.

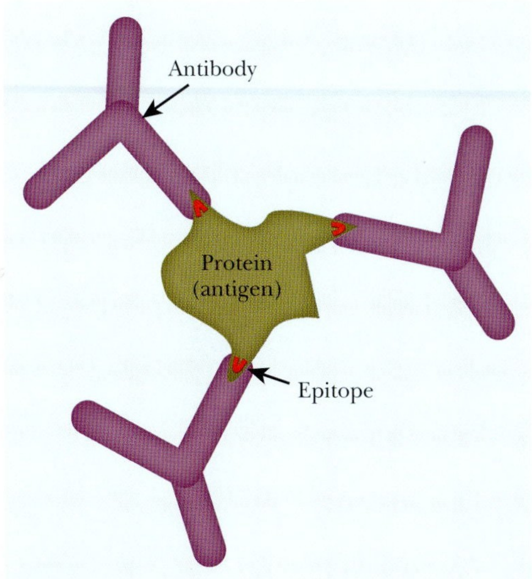

FIGURE 6.4
Antibodies Bind to Epitopes on an Antigen
Antibodies only recognize a small ridge or dimple on the surface of a protein. The region of the antigen that binds to the antibody is called an epitope.

T cells work in the same manner but recognize only antigens expressed on the surface of other body cells, particularly macrophages, cells infected with a virus, or antibody-making B cells, as opposed to the microorganism itself. T cells recognize these other cells via cell surface receptor proteins called the **class I** and **class II major histocompatibility complexes (class I** and **class II MHCs).** Class I MHCs activate T_H cells, and class II MHCs activate T_C cells. MHC receptors are encoded by a family of genes that are different for every person. These may be used to distinguish people, and must be matched in organ transplantation to prevent rejection. Another name for the MHC receptors are **human leukocyte antigens (HLAs)**.

Acquired immunity is divided into two halves. Humoral immunity is mediated by antibodies in blood plasma, which are produced by B cells. The second part is cellular immunity, which is mediated by T cells.

Antibodies recognize epitopes or specific regions of the invading pathogen.

T cells recognize cell surface receptors called class I and class II major histocompatibility complexes that are expressed on the surface of body cells that become infected with an invading pathogen.

THE GREAT DIVERSITY OF ANTIBODIES

Antibodies are proteins that recognize and bind to alien molecules. Because there are an almost infinite variety of possible alien molecules, a correspondingly vast number of different antibody molecules are needed. The amino acids making up protein molecules can certainly be arranged to give an almost infinite number of different sequences and, therefore, of different shapes. However, this leads to a major genetic problem. If a separate gene encoded each antibody protein, this would require a gigantic number of genes and a vast amount of DNA. Even if all the DNA in the mammalian genome was coding DNA, it could encode only a few million antibodies, which is far too few.

The answer is that the immune system generates a vast array of different protein sequences from a relatively small number of genes by shuffling gene segments. Instead of storing complete genes for each antibody, the immune system assembles antibody genes from a collection of shorter DNA segments. Shuffling and joining these partial genes allows the generation of an immense variety of antibodies. In Fig. 6.5 this idea is illustrated using three alternative front ends and three rear ends. Combining them in all possible ways gives nine different genes. Suppose that we had 10 different front, middle, and end segments. These could be combined to give $10 \times 10 \times 10 = 1000$ variants. This is closer to what really happens with antibody genes. Note that generating 1000 different proteins in this way would require only 30 segments of coding DNA to be stored on the chromosomes.

The remarkable genetic economy of the immune system stands in bizarre contrast to the phenomenon of junk DNA. In mammals, typically 95% or more of the DNA may be noncoding. In contrast, the immune system is a fascinating example of how massive genetic diversity can be generated by shuffling relatively few segments of genetic information. Animals can make billions of possible antibodies from only a few thousand gene segments.

The detailed genetics of antibody diversity is a complex issue and is described in textbooks on immunology. The rest of this chapter discusses those aspects of immunology of importance to biotechnology. These include antibody structure, the bioengineering of antibodies, biotechnological techniques that use antibodies, and finally vaccines. The chapter ends with techniques used to identify and produce new vaccines, which is a huge biotechnology business.

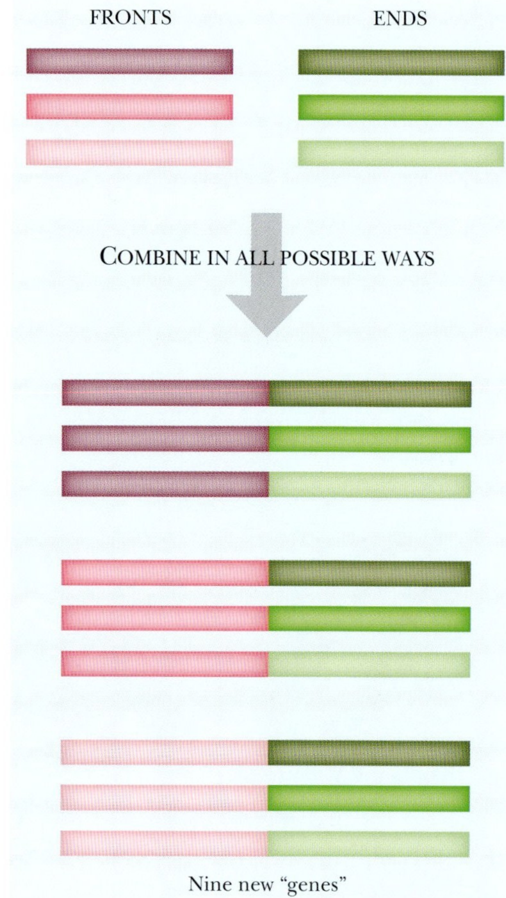

FIGURE 6.5 Modular Gene Assembly
Linking different segments of genes creates exponential numbers of unique combinations.

Antibodies are very diverse in structure so that all the pathogens can be recognized.

Antibodies are produced by shuffling gene segments rather than having one gene code for each different antibody.

ANTIBODY STRUCTURE

Each antibody consists of four protein subunits, two **light chains** and two **heavy chains**, arranged in a Y-shape (Fig. 6.6). Disulfide bonds between cysteine amino acid residues hold the chains together. Each of the light and heavy chains consists of a **constant region** and a **variable region**. The constant region is the same for all chains of the same class. The variable region binds to the target molecule, the antigen. There are millions of different variable regions, which are generated by genetic shuffling.

Breaking an antibody at the "hinge" where the heavy chains bend yields three chunks, two identical **Fab fragments** and one **Fc fragment** (Fig. 6.7). Fab, meaning "fragment, antigen binding," consists of one light chain plus half of a heavy chain. Fc, meaning "fragment, crystallizable," contains the lower halves of both heavy chains. Other components of the immune system often recognize and bind to the Fc region of an antibody (see later discussion).

Antibodies have a Y-shaped structure. The hinge or bend region divides the two Fab fragments from the Fc fragment. There are two light chains and two heavy chains.

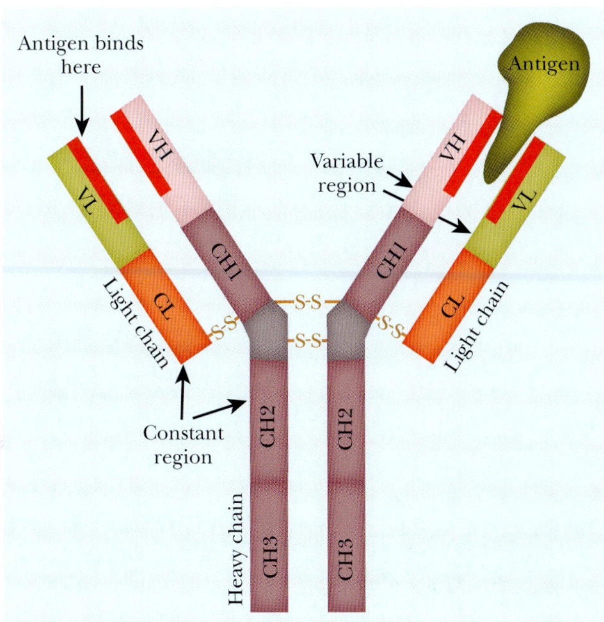

FIGURE 6.6 Structure of an Antibody
Y-shaped antibodies consist of two light chains and two heavy chains. Each consists of segments: CH1, CH2, and CH3 are heavy-chain constant regions; CL is the light-chain constant region; VH is the heavy-chain variable region; and VL is the light-chain variable region. Antigens bind to the variable regions.

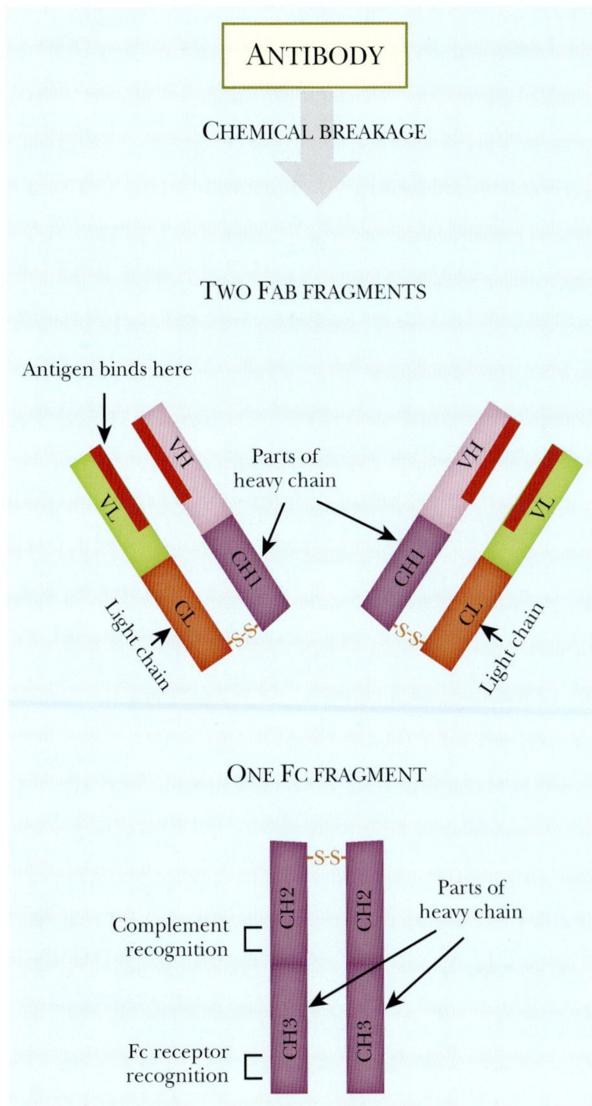

FIGURE 6.7 Fab Fragments and Fc Fragment of an Antibody
Antibodies can be spit into two Fab fragments and one Fc fragment by breaking the molecule at the hinge region.

STRUCTURE AND FUNCTION OF IMMUNOGLOBULINS

Depending on the type of heavy chain, antibodies are categorized into different classes, and assume different roles in the immune system (see Table 6.1). The most abundant and typical antibody has a gamma heavy chain and is called **immunoglobulin G (IgG)**. IgG has four different subclasses, but as a whole, IgG is found mainly in blood serum. About 75% of the serum antibodies are IgG, and these are critical to stimulate immune cells to engulf invading pathogens. IgG is the only antibody able to transfer across the placenta during pregnancy. The second most common antibody in serum is secretory IgA. This antibody is also found in mucosal secretions as well as colostrum and breast milk. It is extremely important in fighting respiratory and gastrointestinal infections, especially in infants, where gastrointestinal illnesses are particularly deadly. The third most common is IgM, which is usually found as a pentamer. The unusual structure of IgM provides multiple binding sites for antigens (10 in IgM versus two in IgG). This structure makes IgM good for clumping microorganisms, and then stimulating immune cells to digest the entire complex. IgD is found at low levels, and its role is still uncertain. IgE is the least common antibody in serum and is primarily found

Table 6.1 Different Types and Functions of Human Antibodies

Antibody	Subtype	Light Chain	Heavy Chain	Structure
IgA	IgA_1 IgA_2	κ or λ	α_1 α_2	J chain; Secretory piece; Secretory; Monomer
IgE	none	κ or λ	ε	Extra domain
IgD	none	κ or λ	δ	Tail piece
IgM	none	κ or λ	μ	Extra domain; Tail piece; J chain; Monomer; Pentamer
IgG	IgG_1	κ or λ	γ_1	IgG_1, IgG_2, IgG_4
	IgG_2	κ or λ	γ_2	
	IgG_3	κ or λ	γ_3	IgG_3
	IgG_4	κ or λ	γ_4	

Note: Light chains are depicted in light blue and heavy chains are purple.

attached to mast cells. IgE is the antibody that stimulates allergic responses by releasing the histamines that cause all the common symptoms of allergies, including runny noses, sneezing, and coughing.

MONOCLONAL ANTIBODIES FOR CLINICAL USE

There are many clinical uses for antibodies. They are used in diagnostic procedures (including the ELISA—see later discussion), for pregnancy testing, and to detect the presence of proteins characteristic of particular disease-causing agents. In the future, they may be used to specifically kill cancer cells or destroy viruses. Such uses need relatively large amounts of a pure antibody that specifically recognizes a single antigen. Even if an experimental animal is inoculated with a purified single antigen, its blood serum will contain a mixture of antibodies to that antigen. Remember that a single antigen has multiple epitopes, and thus antibodies will vary in both specificity and affinity. Nowadays, such a mixture is referred to as **polyclonal antibody** because it results from antibody production by many different clones of B cells, which all recognized the same antigen. Such a mixture is of little use either for a specific, accurate assay or for other techniques in biotechnology.

Somehow, a single line of B cells all making one particular antibody must be isolated and grown in culture. Such a pure antibody made by a single line of cells is known as a **monoclonal antibody**. Unfortunately, B cells live for only a few days and survive poorly outside the body. The solution to this problem is to use cancer cells. **Myelomas** are naturally occurring cancers derived from B cells and which therefore express immunoglobulin genes. Like many tumor cells, myeloma cells will continue to grow and divide in culture forever if given proper nutrients. To make monoclonal antibodies, the relatively delicate B cell, which is making the required antibody, is fused to a myeloma cell (Fig. 6.8). (To avoid confusion, a myeloma that has lost the ability to make its own antibody is used.) The resulting hybrid is called a **hybridoma**. In principle it can live forever in culture and will make the desired antibody.

In practice, an animal, such as a mouse, is injected with the antigen against which antibodies are needed. When antibody production has reached its peak, a sample of antibody-secreting B cells is removed from the animal. These are fused to immortal myeloma cells to give a mixture of many different hybridoma cells. The tedious part comes next. Many individual hybridoma cell lines must be screened to find one that recognizes the target antigen. Once found, the hybridoma is grown in culture to give large amounts of the monoclonal antibody.

Monoclonal antibodies recognize only one epitope on the antigen and derive from one single B cell.

Fusing antigen-stimulated B cells from a mouse spleen with a myeloma cell line produces an immortalized hybridoma. Each of the cells can be grown *in vitro* and evaluated for its affinity to the original antigen to make a monoclonal antibody.

HUMANIZATION OF MONOCLONAL ANTIBODIES

Monoclonal antibodies could theoretically be used as magic bullets to kill human cancer cells by aiming them at specific molecules appearing only on the surface of cancer cells. Ironically, the main problem is that the human immune system regards antibodies from mice or other animals as foreign molecules themselves, and so attempts to destroy them!

One approach that may partly solve this problem is using genetic engineering to make **humanized** monoclonals (Fig. 6.9A). This takes advantage of the fact that only the variable or V-region of the antibody recognizes the antigen. The constant or C-region may therefore be replaced. The first-generation hybridoma is isolated and cultured, generally using mouse B cells. Then the DNA encoding the mouse monoclonal antibody is isolated and cloned. The DNA for the constant region of the mouse antibody is replaced with the corresponding human DNA sequence. The V region is left alone. The human/mouse hybrid gene is then put back into a second mouse myeloma cell for production of antibody in culture. Although not fully human, the hybrid is less mouselike and provokes much less reaction from the human immune system.

Further humanization can be accomplished by altering those parts of the V region that are not directly involved in binding the antigen. A closer look at the V-region of each chain shows that most of the variation is restricted to three short segments that form loops on the surface of the antibody, thus forming the antigen binding site (see Fig. 6.9B). These are known as hypervariable regions or as **complementarity determining regions (CDRs)**. Overall, each antigen binding site consists of six CDRs, three from the light chain and three from the heavy chain. Full humanization of an antibody involves cutting out the coding regions for these six CDRs from the original antibody and splicing them into the genes for human light and heavy chains.

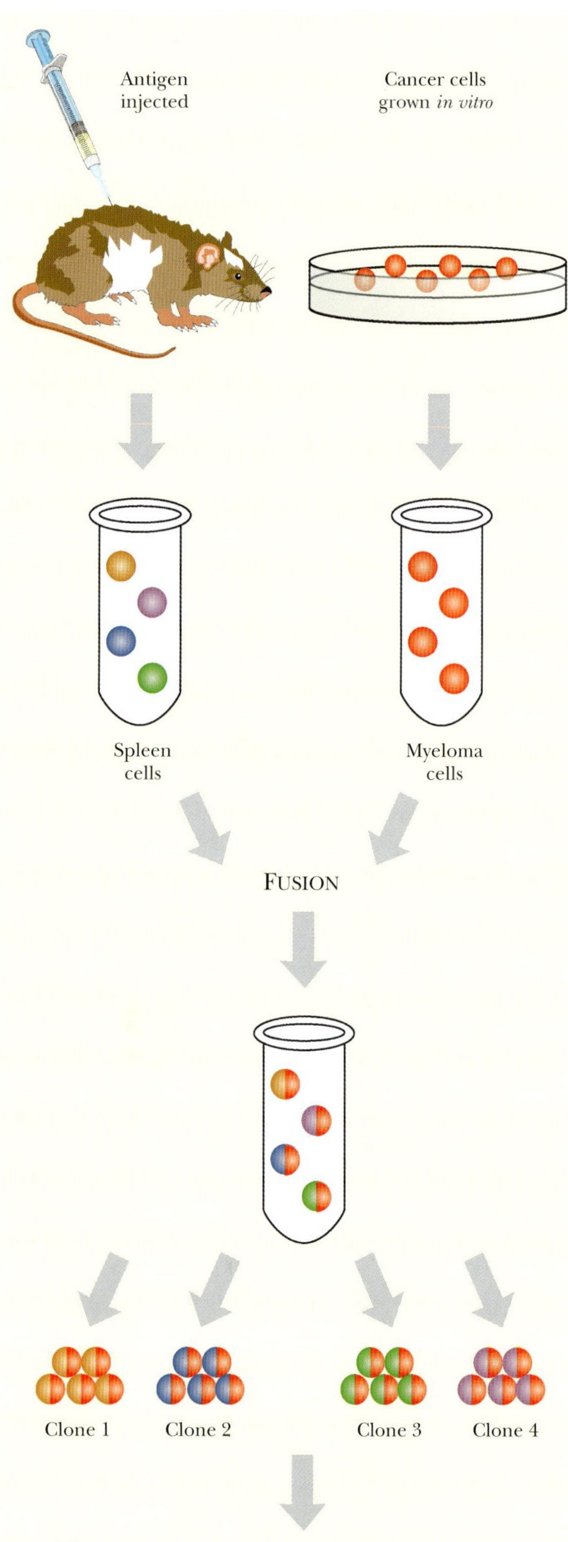

FIGURE 6.8 Principle of the Hybridoma
Monoclonal antibodies derive from a single antibody-producing B cell. The antigen is first injected into a mouse to provoke an immune response. The spleen is harvested because it harbors many activated B cells. The spleen cells are short-lived in culture, so they are fused to immortal myeloma cells. The hybridoma cells are cultured and isolated so each hybrid is separate from the other. Each hybrid clone can then be screened for the best antibody to the target protein.

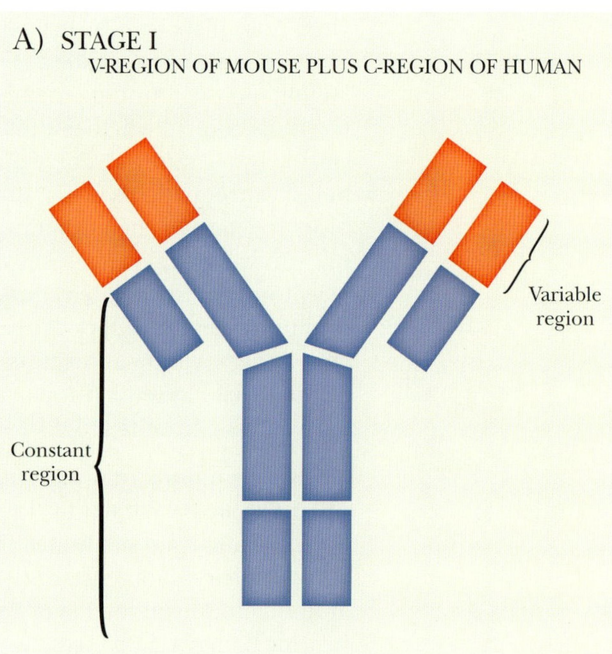

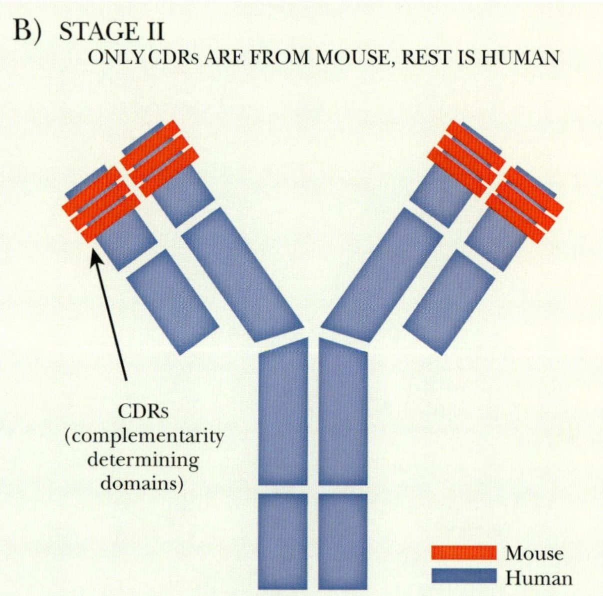

FIGURE 6.9 Humanization of Monoclonal Antibodies
Antibodies from a mouse can be altered to become more like a human antibody. (A) The entire constant region of the heavy and light chain can be replaced with constant regions from a human. (B) Antibodies have six CDRs that determine the actual antigen binding site. The entire antibody except the CDR region can be replaced with human sequence.

Removing the constant regions of a mouse antibody and replacing these with human constant regions makes humanized antibodies. Human cells do not reject these antibodies.

HUMANIZED ANTIBODIES IN CLINICAL APPLICATIONS

There are currently many different humanized monoclonal antibodies in development to treat a variety of conditions. The first humanized monoclonal antibody approved for clinical use, trastuzumab **(Herceptin)**, is for the treatment of breast cancer. The Federal Drug Administration (FDA) approved this therapeutic agent in 1998. Herceptin recognizes a cell surface receptor called HER2. This receptor is part of a larger family, including HER3, HER4, and the founding member, the epidermal growth factor receptor (EGFR). These receptors control whether a cell proliferates, differentiates, or undergoes programmed suicide by signaling a variety of intracellular proteins that modulate gene expression. In breast cancer patients, when the HER2 receptor is overproduced, the breast cancer is much more resistant to chemotherapy. Excess receptor is thus a good indicator that the patient will not survive as long. Herceptin binds to the extracellular domain of HER2, preventing the receptor from being internalized. This prevents the cancer cell from dividing and induces the immune system to attack the cell (Fig. 6.10). When Herceptin is used in combination with chemotherapy to treat breast cancer, patients survive much longer. The main point to keep in mind is that Herceptin binds one specific protein; therefore, the particular breast cancer must have excess amounts of HER2 in order for the treatment to be effective.

One of the most common problems in today's health care system is that hospital patients often acquire bacterial infections from just being in the hospital. **Nosocomial infections**, as they are called, are implicated in thousands of deaths and cost billions of dollars. Even worse, some infections are resistant to all or most antibiotics. The development of a humanized antibody to one of the major bacterial agents will hopefully provide another method for treating patients. *Staphylococcus aureus* is a bacterium that causes serious infections. Strains have been identified that are methicillin resistant, called MRSA (methicillin-resistant *S. aureus*), and other strains are vancomycin resistant, VRSA. These two antibiotics are the most effective available for *S. aureus*, so when these fail, the infection can become deadly. Humanized monoclonal antibodies against *S. aureus* are being tested for their effectiveness. This monoclonal antibody binds to the **ClfA** protein (clumping factor A), which is found on the cell surface of *S. aureus*. The ClfA protein is responsible for the bacteria adhering to

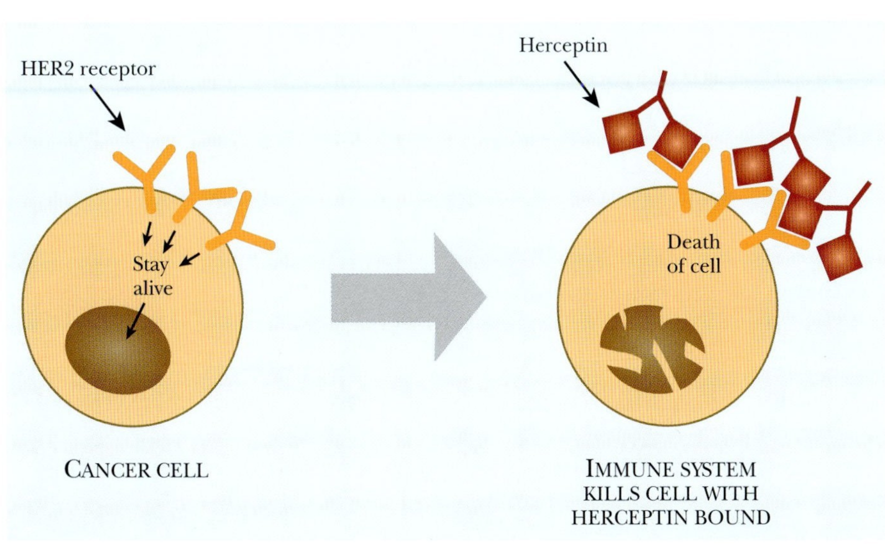

FIGURE 6.10 Herceptin Helps Kill Cancer Cells with HER2
Herceptin is a humanized monoclonal antibody that recognizes the HER2 receptor on breast cancer cells. When the antibody binds to the receptor, the immune system helps destroy the cancer cell, and the cancer cell becomes more sensitive to chemotherapeutic treatments.

fibrinogen, a protein found at the surface of the host cell. When the antibody binds ClfA, the bacteria cannot adhere to the cell surface, and therefore, cannot invade and cause damage (Fig. 6.11). The humanized version of the antibody has been tested in rabbits. The rabbits were infected with *S. aureus* and then treated with the antibody. When two doses of the antibody were given, 100% of the rabbits had no bacteria in their blood for 96 hours. Only about two thirds of the blood cultures were negative for *S. aureus* when the rabbits were treated with just vancomycin. Clinical trials for the antibody began in March 2004 and are ongoing.

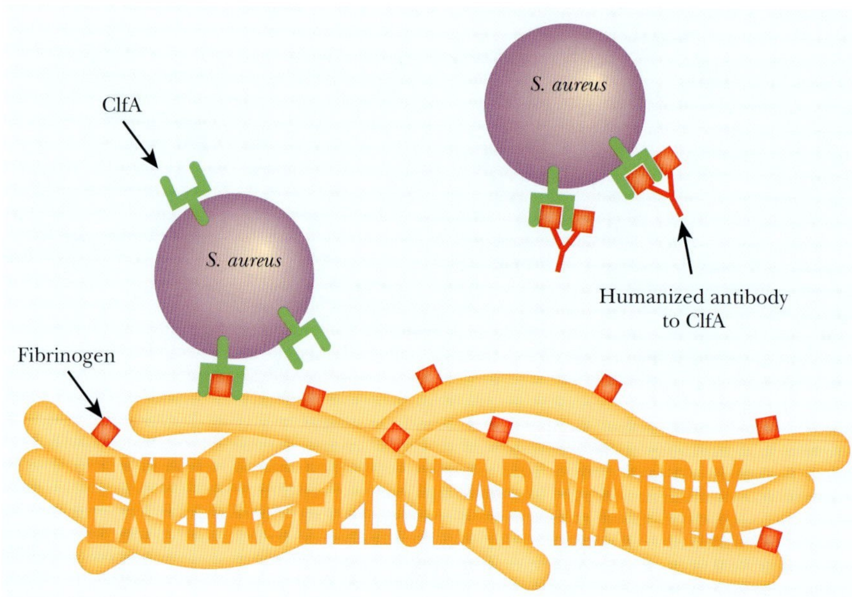

FIGURE 6.11 Humanized Antibodies to *S. aureus* Prevent Colonization
Antibodies to the cell surface protein, ClfA, prevent the bacteria from binding to the extracellular matrix protein, fibrinogen. If *S. aureus* cannot bind to the extracellular matrix, the bacteria cannot colonize and therefore do not cause infections.

> Monoclonal antibodies to HER2 (human epidermal growth factor receptor type 2) inhibit breast cancer cells from growing and are used as a treatment for breast cancer patients.
>
> Humanized antibodies to ClfA prevent *S. aureus* from binding and causing bacterial infections in rabbits. Current clinical trials are determining their efficacy in preventing *S. aureus* infections in humans.

ANTIBODY ENGINEERING

Natural antibodies consist of an antigen binding site joined to an effector region that is responsible for activating complement and or binding to immune cells. From a biotechnological viewpoint, the incredibly high specificity with which antibodies bind to a target protein is useful for a variety of purposes. Consequently, antibody engineering uses the antigen binding region of the antibody. These are manipulated and are attached to other molecular fragments.

To separate an antigen binding site from the rest of the antibody, gene segments encoding portions of antibody chains are subcloned and expressed in bacterial cells. Bacterial signal sequences are added to the N terminus of the partial antibody chains, which results in export of the chains into the periplasmic space. Here the VH and VL domains fold up correctly and form their disulfide bonds. The antibody fragments used include Fab, Fv, and **single-chain Fv (scFv)** (Fig. 6.12). In a Fab fragment, an interchain disulfide bond holds the two chains together. However, the Fv fragment lacks this region of the antibody chains and thus is less stable. This led to development of the single-chain Fv fragment in which the VH and VL domains are linked together by a short peptide chain, usually 15 to 20 amino acids long. This is introduced at the genetic level so that a single artificial gene expresses the whole structure (VH-linker-VL or VL-linker-VH). A tag sequence (such as a His6-tag or FLAG-tag—see Chapter 9) is often added to the end to allow detection and purification. Such an scFv fragment is quite small, about 25,000 in molecular weight.

Such scFv fragments are attached to various other molecules by genetic engineering. The role of the scFv fragment is to recognize some target molecule, perhaps a protein expressed only on the surface of a virus-infected cell or a cancer cell. A variety of toxins, cytokines, or enzymes may be attached to the other end of the scFv fragment, to provide the active portion of the final recombinant antibody. In principle, this approach provides a way of delivering a therapeutic agent in an extremely specific manner. At present the clinical applications of engineered antibodies are under experimental investigation.

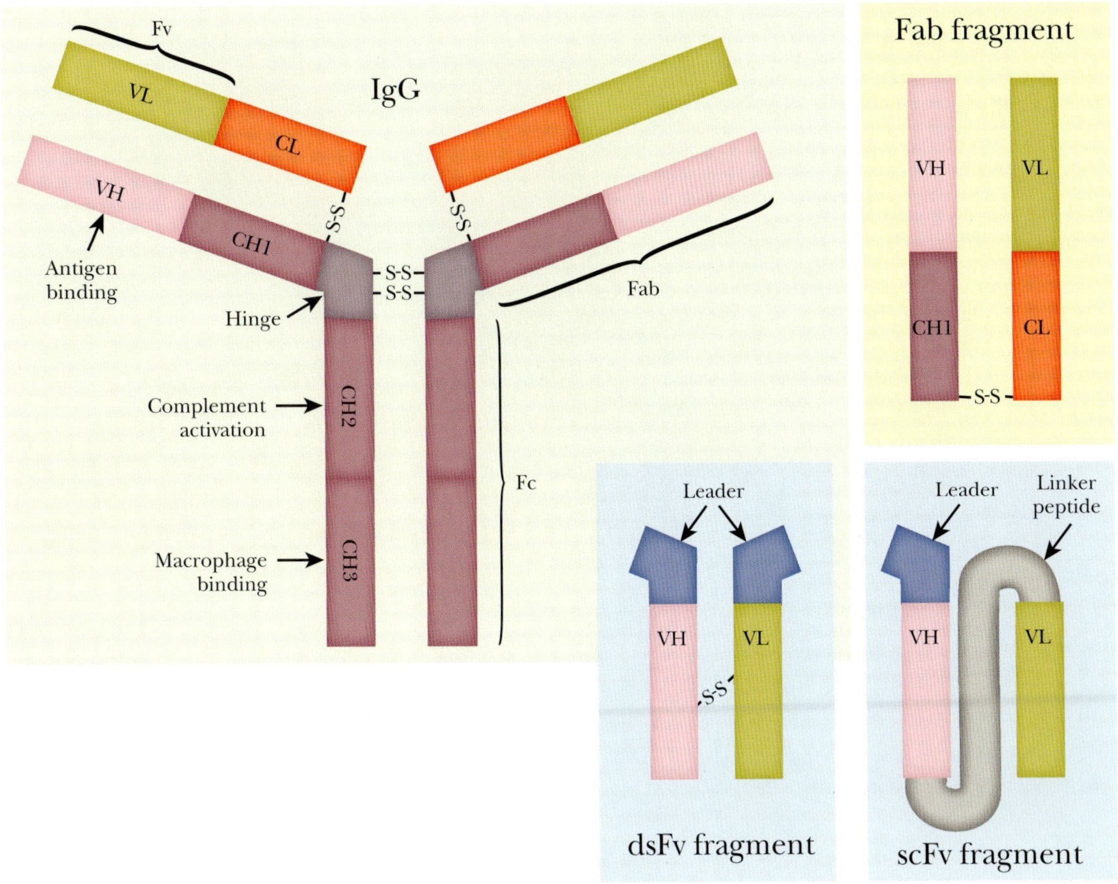

FIGURE 6.12 Fab and Fv Antibody Fragments
Fab fragments are produced by protease digestion of the hinge region. A disulfide bond holds the heavy and light chains together. To make an antibody fragment without any constant region, the genes for the VH domain and the VL domain are expressed on a bacterial plasmid. This structure is unstable because of a lack of disulfide bonds. Therefore, disulfide bonds are engineered into the two halves (dsFv fragment), or a linker is added to hold the VH and VL domains together (scFv fragment).

The antigen binding regions used in antibody engineering may be derived from characterized monoclonal antibodies. Alternatively, a library of DNA segments encoding V-regions may be obtained from a pool of B cells obtained from an animal or human blood sample. Such a library should in theory contain V-regions capable of recognizing any target molecule. Using a human source avoids the necessity for the complex humanization procedures described earlier. However, in this case it is necessary to screen the V-region library for an antibody fragment that binds to the desired target molecule. This may be done by the phage display procedure outlined in Chapter 9. The library of V-region constructs is expressed on the surface of the phage, and the target molecule is attached to some solid support and used to screen out those phages carrying the required antibody V region.

Single-chain Fv antibody binding domains are linked to various toxins, cytokines, or enzymes to create recombinant antibodies. These can be used to precisely deliver the toxin, cytokine, or enzyme to the antigen that the scFv recognizes *in vivo*.

DIABODIES AND BISPECIFIC ANTIBODY CONSTRUCTS

A variety of engineered antibody constructs are presently being investigated. A **diabody** consists of two single-chain Fv (scFv) fragments assembled together. Shortening the linker from 15 amino acids to five drives dimerization of two scFv chains. This no longer allows intrachain assembly of the linked VH and VL regions. The dimer consists of two scFv fragments arranged in a crisscross manner (Fig. 6.13). The resulting diabody has two antigen binding sites pointing in opposite directions. If two different scFv fragments are used, the result is a bispecific diabody that will bind to two different target proteins simultaneously. Note that formation of such a bispecific diabody requires that VH-A be linked to VL-B and VH-B to VL-A. It is of course possible to engineer both sets of VH and

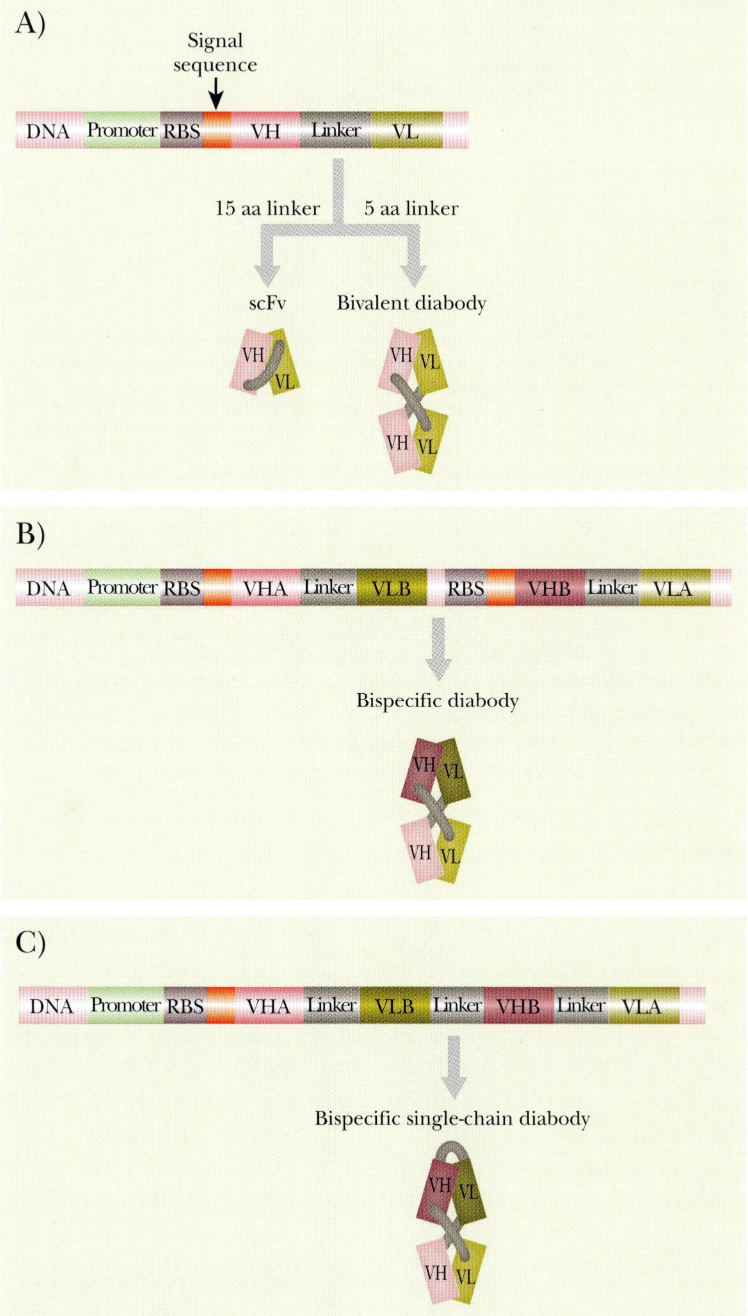

FIGURE 6.13 Engineered Diabody Constructs

(A) Engineering a diabody construct begins by genetically fusing the variable domains of the heavy and light chain (VH and VL) with a linker. The long linker allows a single polypeptide to form into a single antibody binding domain. The short linker allows two polypeptides to complex into a diabody with two antibody binding domains. The construct is expressed in bacteria using a bacterial promoter and RBS (ribosome binding site). The signal sequence tells the bacteria to secrete the engineered protein. (B) Instead of identical Fv units, two different Fv chains can be coexpressed in the bacterial cell. The two different Fv chains will unite into a diabody with two different antibody binding domains, a different one on each side. (C) Bispecific antibodies can be made as one single transcript with a linker between VHA and VLB, a linker between the two halves, and finally a linker between VHB and VLA.

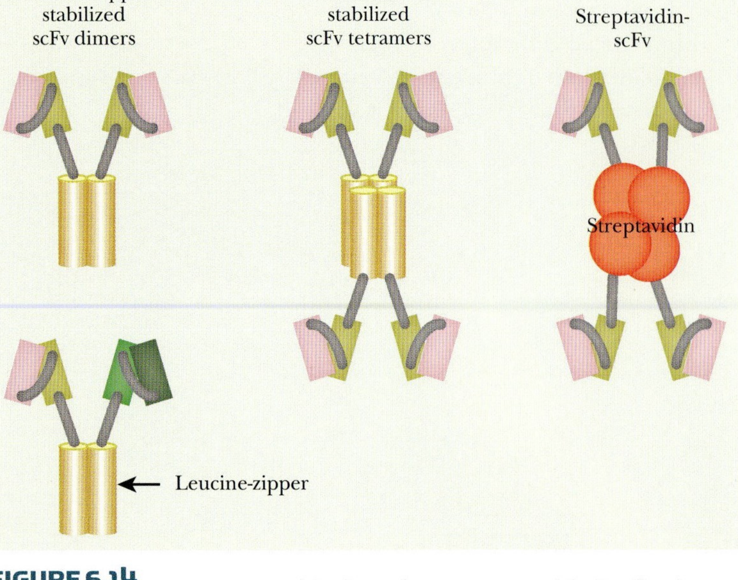

FIGURE 6.14
Engineered Bispecific Antibody Constructs
Instead of genetic linkers to hold diabodies, various proteins can also hold scFv fragments together. Proteins with a leucine zipper domain dimerize; therefore, when scFv genes are genetically fused to these, the scFv domains come together as dimers. Proteins such as streptavidin or proteins with four helix bundle domains can be genetically fused to scFv domains. When expressed, there are four scFv domains on the outside, providing four different antibody binding sites.

VL regions onto a single polypeptide chain encoded by a single recombinant gene, as shown in Fig. 6.13. Bispecific diabodies have a variety of potential uses in therapy, because they may be used to bring together any two other molecules; for example, they might be used to target toxins to cancer cells.

Another way to construct an engineered bispecific antibody is to connect the two different scFv fragments to other proteins that bind together (Fig. 6.14). Two popular choices are **streptavidin** and leucine zippers. Streptavidin is a small biotin binding protein from the bacterium *Streptococcus*. It forms tetramers, so it allows up to four antibody fragments to be assembled together. Furthermore, binding to a biotin column can purify the final constructs. Leucine zipper regions are used by many transcription factors that form dimers (see Chapter 2). Often, such proteins form mixed dimers when their leucine zippers recognize each other and bind together. Leucine zipper regions from two different transcription factors that associate (e.g., the Fos and Jun proteins) may therefore be used to assemble two different scFv fragments.

Linking two scFv fragments together with either polypeptide linker regions or proteins (e.g., streptavidin or leucine zipper proteins) creates divalent antibodies, that is, each side of the antibody will recognize a different antigen. These constructs are useful to bring two different proteins in close proximity in the cell.

ELISA ASSAY

The **enzyme-linked immunosorbent assay (ELISA)** is widely used to detect and estimate the concentration of a protein in a sample. The protein to be detected is regarded as the antigen. Therefore, the first step is to make an antibody specific for the target protein. A detection system is then attached to the rear of the antibody. Usually this consists of an enzyme that generates a colored product from a colorless substrate. Alkaline phosphatase, which converts X-Phos to a blue dye (see Chapter 8), is a common choice. The samples to be assayed are immobilized on the surface of a membrane or in the wells of a microtiter dish (Fig. 6.15). The antibody plus detection system is added and allowed to bind. The membrane or microtiter dish is then rinsed to remove any unbound antibody. The substrate is added, and the intensity of color produced indicates the level of target protein.

A variety of modifications of the ELISA exist. Often binding and detection are done in two stages, using two different antibodies. The first antibody is specific for the target protein. The second antibody recognizes the first antibody and carries the detection system. For example, antibodies could be raised in rabbits to a series of target proteins. The second antibody, which recognizes rabbit antibodies, could be produced in sheep. These are called *secondary antibodies* and are often described as, for example, sheep anti-rabbit. The secondary antibody has the detection system, and because it will recognize any antibody made in a rabbit, it does not have to be reengineered for each different target protein. This allows the use of the same final antibody detection system in each assay even if different primary antibodies are used to identify different proteins.

Antibodies are used in ELISA assays to determine the relative concentration of the target protein or antigen in a sample.

Primary antibodies are the antibody that recognizes the target protein or antigen. Secondary antibodies recognize the primary antibody and often carry a detection system. Secondary antibodies are made to recognize any antibody that is made in sheep, cow, rabbit, goat, or mouse.

THE ELISA AS A DIAGNOSTIC TOOL

The ELISA is used in many different fields. Diagnostic kits that rely on the ELISA are produced for clinical diagnosis of human disease, dairy and poultry diseases, and even for plant diseases. The diagnostic kits are so simple that most require no laboratory equipment and take as little as 5 minutes. ELISA kits can be used to detect a particular plant disease by crushing a leaf and smearing the leaf tissue on the antibody. When the disease-specific antigen reacts with the antibody, the antibody spot turns blue. In clinical applications, ELISA kits can detect the presence of minute amounts of pathogenic viruses or bacteria, even before the pathogen has a chance to cause major damage. Clinical ELISA kits detect various disease markers. In certain diseases, characteristic proteins mark the start of disease progression long before the patient exhibits any symptoms. Detecting such markers can help diagnose and treat a problem before the disease causes serious damage.

ELISA diagnostic testing is even available for you to try at home. Home pregnancy kits are a simple, over-the-counter ELISA assay for **human chorionic gonadotropin (hGC)**. This is a protein produced by the placenta and secreted into the bloodstream and urine of pregnant women. The actual pregnancy test has four important features (Fig. 6.16). First, the entire test is on a piece of paper that wicks the urine from one end to the other. This paper has three regions: first, a region where anti-hCG antibody is loosely attached to the paper strip; second, a region called the pregnancy window; and finally, a control window. As the urine wicks up the paper strip, any hCG present is bound by the anti-hCG antibody. If the woman is pregnant, the anti-hCG/hCG complex moves up the paper strip. If the woman is not pregnant, the anti-hCG antibody moves up the paper strip alone. (Even if the woman is pregnant, there is excess anti-hCG, and so unbound anti-hCG antibody is always found.) If the woman is pregnant, the anti-hCG/hCG complex reaches the pregnancy window where it binds to

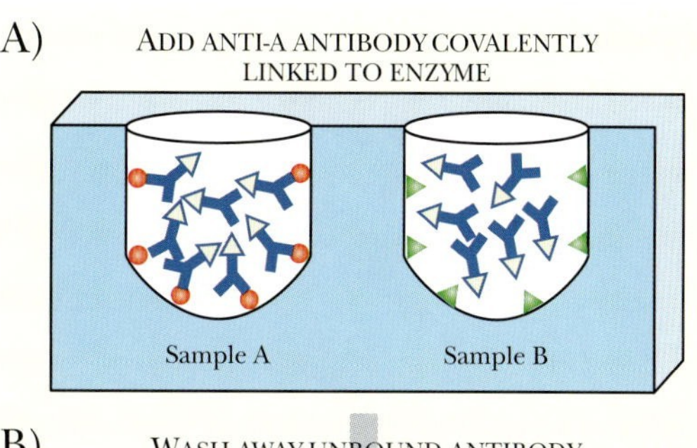

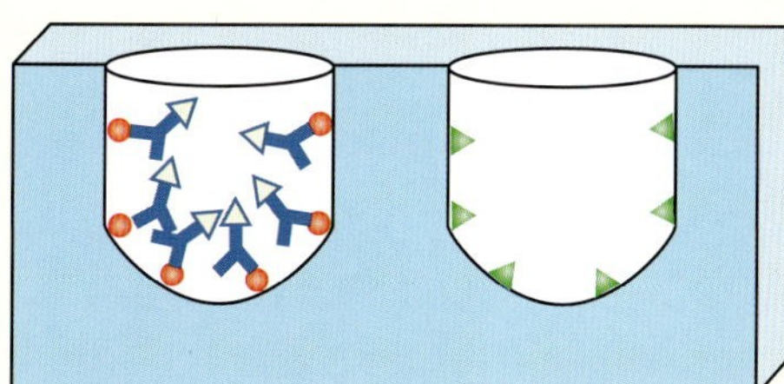

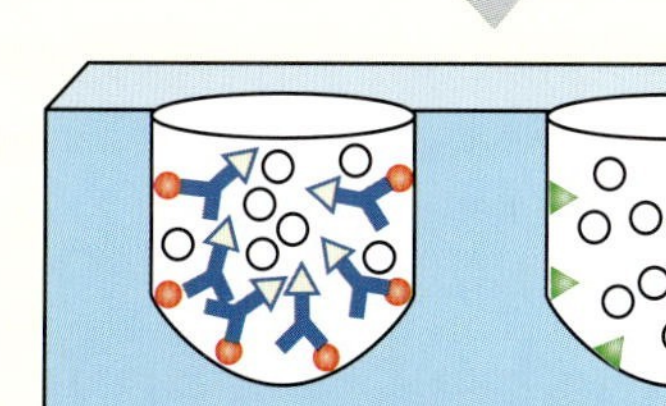

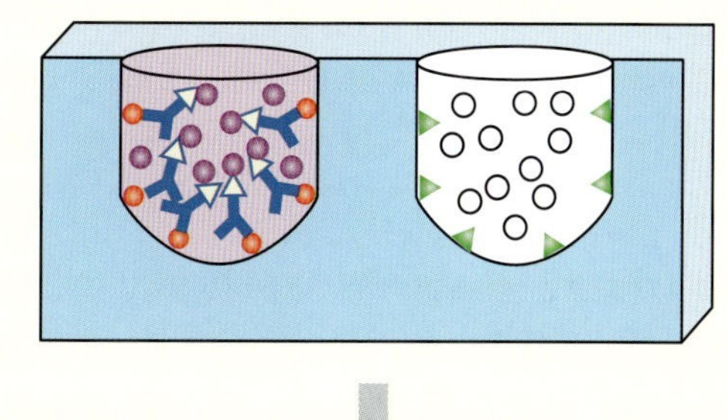

FIGURE 6.15 Principle of the ELISA

ELISA detects and quantifies the amount of a particular protein bound to the well of a microtiter dish. Anti-A antibody is linked to an enzyme such as alkaline phosphatase. The antibody recognizes only the orange protein, and not the green protein (A). After the antibody binds to its target, the unbound antibody is washed from the dish (B). A colorimetric substrate of alkaline phosphatase is added to each well (C), and wherever there is antibody, the substrate is cleaved to form its colorful product (D). The amount of color is proportional to the amount of protein.

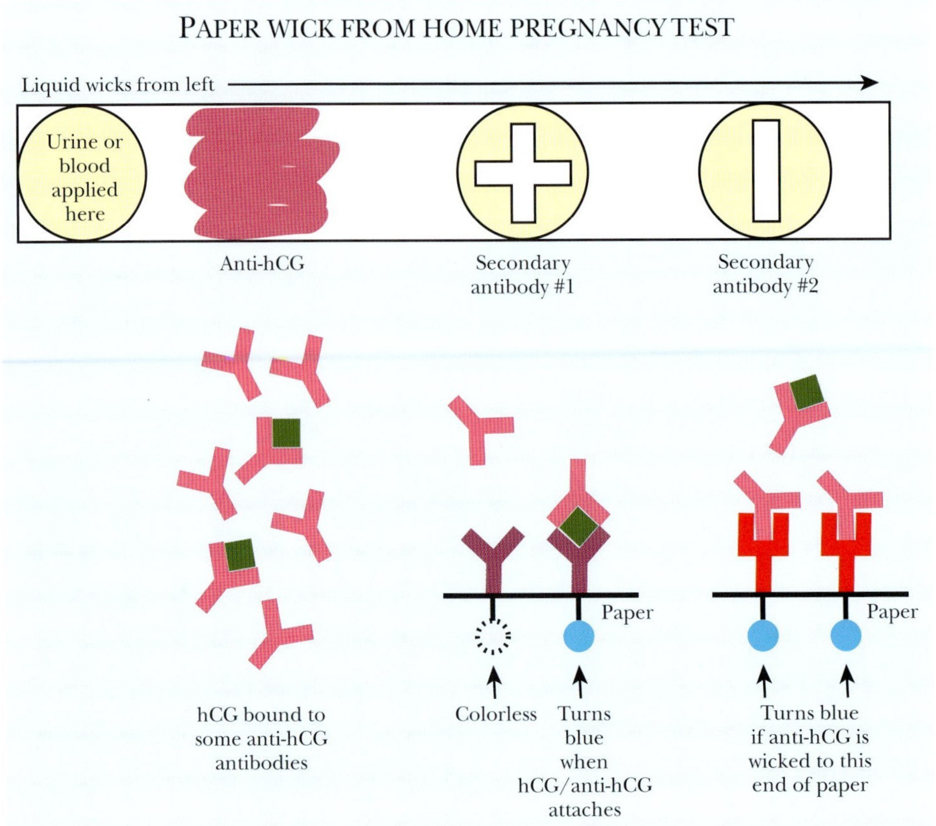

FIGURE 6.16 Home Pregnancy Tests Are an ELISA Diagnostic Tool
The pregnancy test shown here has four important areas along the paper wick. The urine or blood is applied on the far left and wicks to the right. The anti-hCG antibodies loosely attached to the paper are next. If the urine has hGC this binds to its antibody and travels along the paper as a complex. In the next area, a secondary antibody that only recognizes the hCG–primary antibody complex is firmly attached in a plus pattern. When the hCG complex binds to the secondary antibody, the detection system turns blue. The final spot is a different secondary antibody that recognizes the primary antibody without any hCG. This is a positive control, to ensure that the antibody was released and wicked up the paper with the urine.

secondary antibody 1. This is attached to the paper in the shape of a plus sign and cannot move. The secondary antibody has a color detection system attached to it. When the anti-hCG/hCG complex binds to the secondary antibody it triggers color release and a plus sign forms. The control window contains secondary antibody 2. This recognizes only anti-hCG antibody that is not bound to hCG, so its color is activated whether or not the woman is pregnant.

ELISA is a powerful diagnostic tool because antibodies can be made to almost any protein. For pregnancy tests, any hCG in the urine binds to antibodies to hCG, which in turn bind to immobilized secondary antibody to form the plus sign.

VISUALIZING CELL COMPONENTS USING ANTIBODIES

Antibodies can be used to visualize the location of specific proteins within the cell. **Immunocytochemistry** refers to the visualization of specific antigens in cultured cells, whereas **immunohistochemistry** refers to their visualization in prepared tissue sections. In either technique, the first step is to prepare the cells. They must be treated to maintain their cellular architecture, so that the cells appear much as they would if still alive. Usually, the cells are treated with crosslinking agents such as formaldehyde or with denaturants like acetone or methanol. In immunohistochemistry, tissue samples can be frozen and then sliced into small thin sections (about 4 mm), providing a two-dimensional view of the cell. Another option is to embed the tissue sample in paraffin wax. Here the cells are first dehydrated in a series of alcohol solutions, and then treated with the wax. The tissue is then sectioned into thin two-dimensional slices as for frozen tissues.

Preserved cells then need to be permeabilized so that the antibody can enter. Once a single, thin layer of prepared cells or tissue sections is readied, the cells are treated to make the antigen more accessible to the antibody. If in wax, the tissue sections are dewaxed and rehydrated in solution. Fixed tissue sections can be irradiated with microwaves, which break the crosslinks induced by the fixative, or the samples can be heated under pressure. After permeabilization, the primary antibody finds its antigen within the sample and binds. A secondary antibody contains the detection system. The secondary antibody binds to the primary antibody/antigen complex, and then the appropriate reagents are added to visualize the location of the complex. (In some cases, a single antibody, with an attached detection system, is used.)

The antibody is detected using enzymes or by fluorescent labels. A common enzyme-mediated detection system is alkaline phosphatase, as with the ELISA—see Fig. 6.15. Fluorescently labeled antibodies must be excited with UV light, on which the fluorescent label emits light at a longer wavelength. Samples are directly visualized with a microscope attached to a UV light source (Fig. 6.17). Fluorescent antibodies tend to bleach out when exposed to excess UV; therefore, the microscope is attached to a camera to record the data as a digital image.

Immunocytochemistry and immunohistochemistry use a primary antibody to a specific cellular target protein to visualize its location within the cell. The primary antibody is visualized by adding a secondary antibody with a detection system.

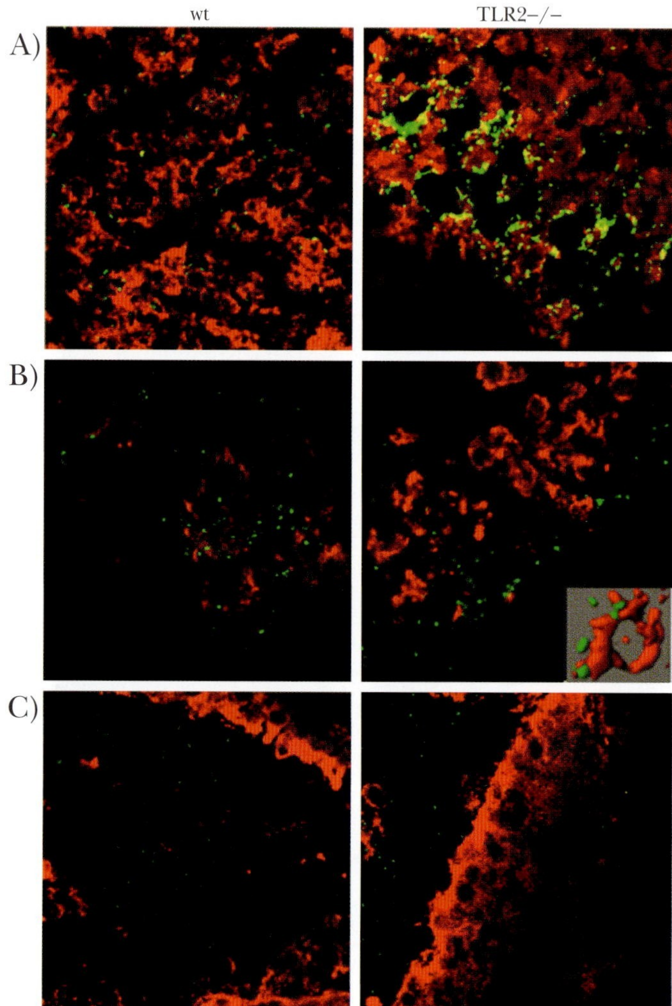

FIGURE 6.17 Fluorescent Antibody Staining

Colocalization of *Streptococcus pneumoniae* and membrane antigens in mouse brain sections as seen by confocal microscopy at 63-fold magnification. The bacteria are expressing GFP and appear green. All three antigens were visualized separately by a red fluorescent stain. Yellow regions indicate colocalization of bacteria with antigens. At left, wild-type mice and at right mice lacking Toll-like receptor 2. (A) GLT1v-stained plexus choroideus epithelial cells. (B) Third ventricle with infiltrating Gr1-stained granulocytes (insert: 3D picture of bacteria taken up by granulocytes). (C) Third ventricle and GFAP-stained astrocytes. From: Echchannaoui *et al.* (2005). Regulation of *Streptococcus pneumoniae* distribution by Toll-like receptor 2 *in vivo*. *Immunobiology* **210**, 229–236. Reprinted with permission.

FLUORESCENCE ACTIVATED CELL SORTING

As explained in the last section, fluorescent antibodies are used to find the location of intracellular proteins. Fluorescent antibodies are also able to bind to surface antigens. Many cells of the immune system have specific antigens on their surface that distinguish them from others. Each immune cell can have over 100,000 antigen molecules on their surface. These surface antigens characterize the different types of immune cells and are systematically named by assigning them a **cluster of differentiation (CD) antigen** number. The antigens were mostly identified before their physiological function was known. For example, CD4 antigens are associated with T-helper cells, and CD8 with killer T cells. Monoclonal antibodies are available to label many CD antigens, especially the most common.

Fluorescence activated cell sorting (FACS) involves the mechanical separation of a mixture of cells into different tubes based on their surface antigens (Fig. 6.18). Because the antibody attaches to the outside of the cell, the cell does not have to be prepared like above. In this example, helper T cells and killer T cells can be separated from other white blood cells based on the presence of CD4 or CD8 surface antigens. First, the cell suspension is labeled with monoclonal antibodies to the surface antigens of interest. In this example, antibodies to CD4 and CD8 are used. Both antibodies have fluorescent labels that are different for the two antibodies. The labeled cell suspension is loaded into a charging electrode. Drops of liquid containing only one cell are released to the bottom, and the fluorescence detector notes whether or not the drop of liquid is labeled for CD4 or CD8 by the color of its fluorescence. If the drop has an antigen, an electrical charger pulls or pushes the droplet to the right or left, separating the two antigens into separate tubes. If the drop has no antigen in it, it gets no electrical charge and goes into a third tube. Usually two different antibodies are used, but some of the newer FACS machines can sort up to

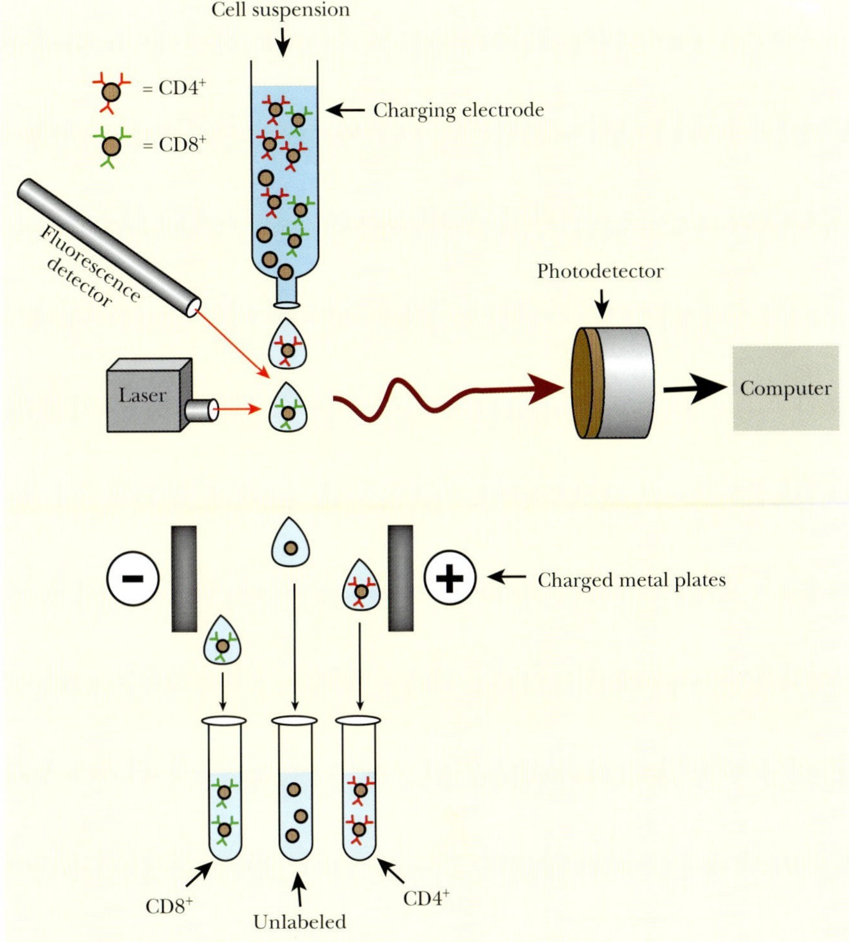

FIGURE 6.18 FACS Separates $CD4^+$ and $CD8^+$ Cells
FACS machines can separate fluorescently labeled cells into different compartments. A mixture of $CD4^+$, $CD8^+$, and unlabeled cells is separated based on their fluorescence. When the fluorescence detector notes green, the charged metal plates pull that drop to the left or minus plate, allowing those cells to collect into the left tube. If no fluorescence is detected, the drop stays neutral and is collected in the middle tube. If the drop fluoresces red, the charged plates pull the drop to the plus side, and it collects in the right tube.

12 different fluorescently labeled antibodies and can sort up to 300,000 cells per minute.

Flow cytometry is a related technique to analyze fluorescently labeled cells. As with FACS, cells are labeled with monoclonal antibodies to cell-surface antigens. The antibodies are conjugated to a variety of different fluorescent labels, and each antibody is detected based on its fluorescence. The cells are loaded into a charging electrode and released in small droplets. During flow cytometry, the cells are not sorted and saved; instead the sample of cells is measured and discarded. As the cells pass the detector, the computer records the fluorescence and plots the number of cells with each of the fluorescent labels. These are plotted with a small dot representing each of the cells (Fig. 6.19).

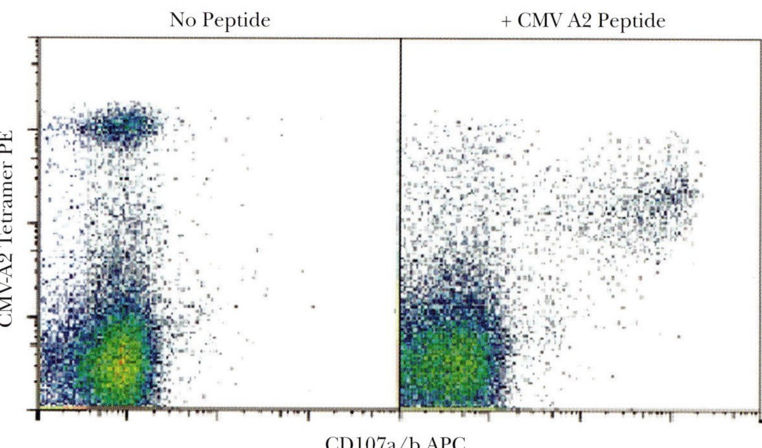

FIGURE 6.19
Example of Flow Cytometry Data
Expression of proteins on the surface of white blood cells. White blood cells were stained with an MHC-class I tetramer and anti-CD107 labeled with APC (allophycocyanin, blue) and treated as follows: (*Left panel*) Stained with anti-CD3 labeled with phycoerythrin (red) and anti-CD8 labeled with peridinin chlorophyll protein (green), and analyzed without further incubation. (*Right panel*) Stained after stimulation with the cognate peptide (NLVPMVATV). From: Betts *et al.* (2003). Sensitive and viable identification of antigen-specific CD8^{+} T cells by a flow cytometric assay for degranulation. *J Immunol Methods* 281, 65–78. Reprinted with permission.

FACS and flow cytometry use monoclonal antibodies to surface antigens. The FACS machine can sort the cells into individual samples, and flow cytometry simply records the fluorescent label and plots the data on a graph.

IMMUNE MEMORY AND VACCINATION

Individuals who survive an infection normally become immune to that particular disease, although not to other diseases. This is because the immune system "remembers" foreign antigens, a process called **immune memory**. Next time the same antigen appears, it triggers a far swifter and more aggressive response than before. Consequently, the invading microorganisms will usually be overwhelmed before they cause noticeable illness.

Immune memory is due to specialized B cells called **memory cells**. As discussed earlier, virgin B cells are triggered to divide if they encounter an antigen that matches their own individual antibody. Most of the new B cells are specialized for antibody synthesis, and they live only a few days. However, a few active B cells become memory cells, and instead of making antibodies, they simply wait. If one day the antigen they recognize appears again, most of the memory cells switch over very rapidly to antibody production.

Vaccination takes advantage of immune memory. **Vaccines** consist of various derivatives of infectious agents that no longer cause disease but are still antigenic, that is, they induce an immune response. For example, bacteria killed by heating are sometimes used. The antigens on the dead bacteria stimulate B-cell division. Some of the B cells form memory cells so, later, when living germs corresponding to the vaccine attack the vaccinated person, the immune system is prepared. The makers of vaccines are constantly trying to find different ways to stimulate the immune system, without causing disease.

CREATING A VACCINE

Because vaccines are such a huge part of the biotechnology industry, and such an important part of our health care system, much research and money are invested in finding new and improved vaccines. Many vaccines are administered to young babies; thus, ensuring the safety and effectiveness of vaccines is critical. Many different methods of developing a vaccine exist.

Most vaccines are simply the disease agent, killed with high heat or denatured chemically. Heat or chemical treatment inactivates the virus or bacterium so it cannot cause disease. Yet enough of the original structure exists to stimulate immunity. When the live agent infects the vaccinated person, memory B cells are activated and the disease is suppressed. Such whole vaccines elicit the best immune response, but many diseases cannot be isolated or cultured to make whole vaccines. Other times, the cost of culturing the pathogen is prohibitive. Moreover, growing live viruses is a dangerous job, with potential exposure of lab workers. With these limitations in mind, many different strategies have been developed to make improved vaccines.

Attenuated vaccines are still-living pathogens that no longer express the toxin or proteins that cause the disease symptoms (Fig. 6.20). Sometimes, viruses or bacteria are genetically engineered to remove the genes that cause disease. Other attenuated vaccines are related but nonpathogenic strains of the infectious agent (see Box 6.1). Making attenuated virus does not pose the same risks as for live virus. However, much research is needed to identify those genes that cause disease. Another disadvantage is that an attenuated virus might revert to the pathogenic version, especially if the attenuated virus was mutated in only one disease-causing gene.

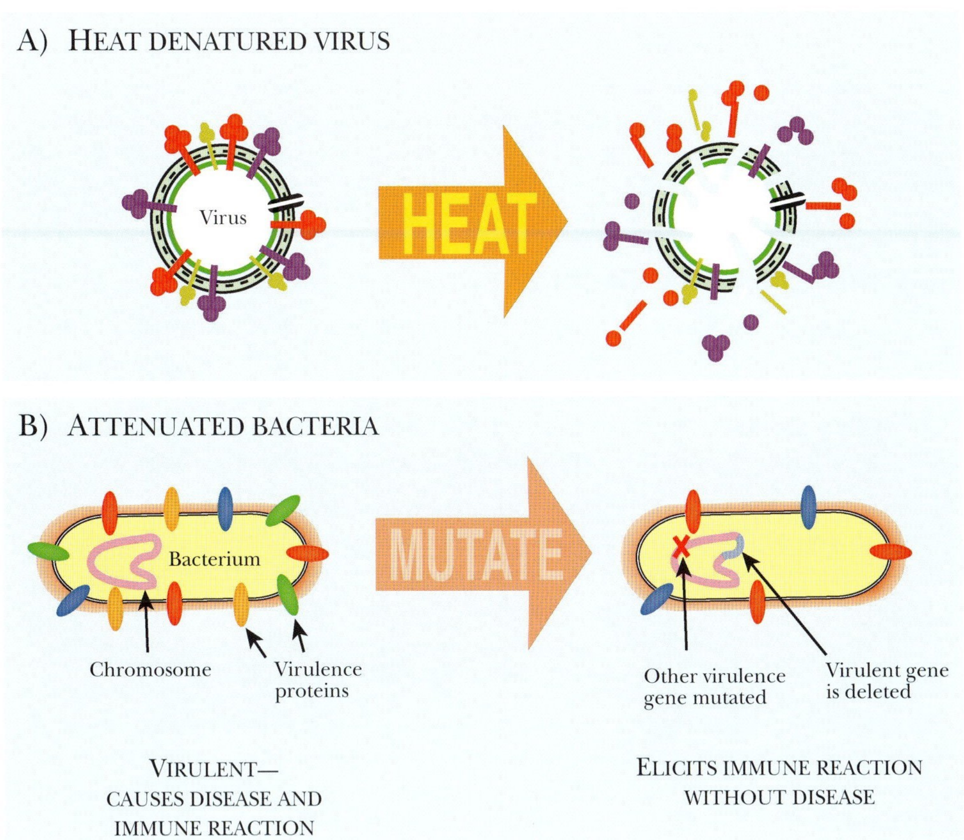

FIGURE 6.20
Whole Vaccines Include Killed or Attenuated Pathogens
(A) High heat or chemical treatment kills pathogens, but leaves enough antigens intact to elicit an immune response. Once exposed to a dead virus or bacterium, memory B cells are established and prevent the live pathogen from making the person sick. (B) Attenuated viruses or bacteria have been mutated or genetically engineered to remove the genes that cause illness. The immune system generates antibodies to kill the attenuated pathogen and establishes memory B cells that prevent future attack.

Box 6.1 Cowpox and Smallpox

Infection with cowpox produces only mild disease but gives immunity to the frequently fatal smallpox. In medieval times, a substantial proportion of the population caught smallpox. About 20% to 30% of those infected died, and the survivors ended up with ugly pockmarks on their faces—hence the name *smallpox*. Milkmaids rarely suffered from smallpox because most had already caught cowpox from their cows. Consequently, milkmaids were seldom pockmarked and gained a reputation for beauty due to their unblemished skin. This led to Edward Jenner's classic experiments in which he inoculated children with cowpox and demonstrated that this protected against infection with smallpox. The term *vaccination* is derived from *vacca*, the Latin for "cow."

Subunit vaccines are vaccines against one component or protein of the disease agent, rather than the whole disease (Fig. 6.21). Subunit vaccines are available only because of recombinant DNA technology. The first step in creating a subunit vaccine is identifying a potential protein or part of a protein that elicits a good immune response. Interior proteins from the pathogen would not trigger the immune system when the real virus challenged it. Because of this, most subunit vaccines are made from proteins found on the outer surface of the virus or bacterium. Experiments must be done to evaluate the protein chosen for the subunit vaccine. Once a suitable protein is identified, it is expressed in cultured mammalian cells, eggs, or some other easily maintained system. The target protein is isolated from other proteins and used to immunize mice to test its effectiveness. After extensive testing in animals, the purified protein is used as a vaccine.

Sometime subunit vaccines fail, perhaps because the protein does not form the correct structure when expressed in mammalian cells or eggs. In these cases, **peptide vaccines** are created. These vaccines just use a small region of the protein. Since such peptides are small, they are conjugated to a **carrier** or **adjuvant** (Fig. 6.22). These range from live nonpathogenic viruses or bacteria, to other proteins that stimulate a better immune response than the vaccine protein itself.

Killed pathogens, attenuated pathogens, single proteins, or epitopes from a disease-causing pathogen are used as vaccines. These are isolated and injected into people to elicit their immune response without causing the disease.

MAKING VECTOR VACCINES USING HOMOLOGOUS RECOMBINATION

Another method of displaying a foreign antigen for use as a vaccine is the **vector vaccine**. Here genetic engineering is used to express a disease-causing antigen on the surface of a nonpathogenic virus or bacterium. When this infects a person, it induces immunity both to the nonpathogenic microorganism and to the attached antigen. For example, vaccinia virus is a nonpathogenic relative of the smallpox virus. Using vaccinia virus is so effective that smallpox was eradicated. If vaccinia virus were engineered to express an antigen from another deadly virus, the person vaccinated would gain immunity to smallpox and the other virus at the same time. Indeed, multiple genes could be inserted, conferring resistant to multiple diseases. The benefit of using vaccinia virus is that it is very potent and stimulates development of both B cells and T cells. In contrast, many other vaccines, particularly subunit vaccines, stimulate only a B-cell response.

Inserting genes into the vaccinia genome is awkward because the genome has very few restriction enzyme sites. However, the vaccinia genome sequence is known. This allows genes to be added to the genome using **homologous recombination** (Fig. 6.23). In homologous recombination, two segments of similar or homologous DNA align, and one strand of each DNA helix is broken and exchanged to form a **crossover**. A single crossover creates a hybrid molecule; if two crossovers occur close together, entire regions of DNA are exchanged. During homologous recombination in vaccinia, a region of single-stranded DNA is generated from a double-stranded break in the incoming new gene. The single-stranded region invades the double helix of the vaccinia genome to form a triple helix. One of the strands from vaccinia then is free to hybridize to the single-stranded homologous region on the incoming gene. If this occurs on both sides, the foreign gene is inserted into the vaccinia genome.

Changing a harmless virus or bacteria so that it expresses a protein from a disease-causing pathogen on its surface can trick the immune system into making antibodies to the disease-causing pathogen.

FIGURE 6.21 (LEFT) Subunit Vaccines Rely on a Single Antigen
A single antigenic protein from a pathogen is isolated and its gene is cloned into an expression vector. The gene is expressed in cultured mammalian cells (such as Chinese hamster ovary [CHO] cells), isolated, purified, and used as a vaccine.

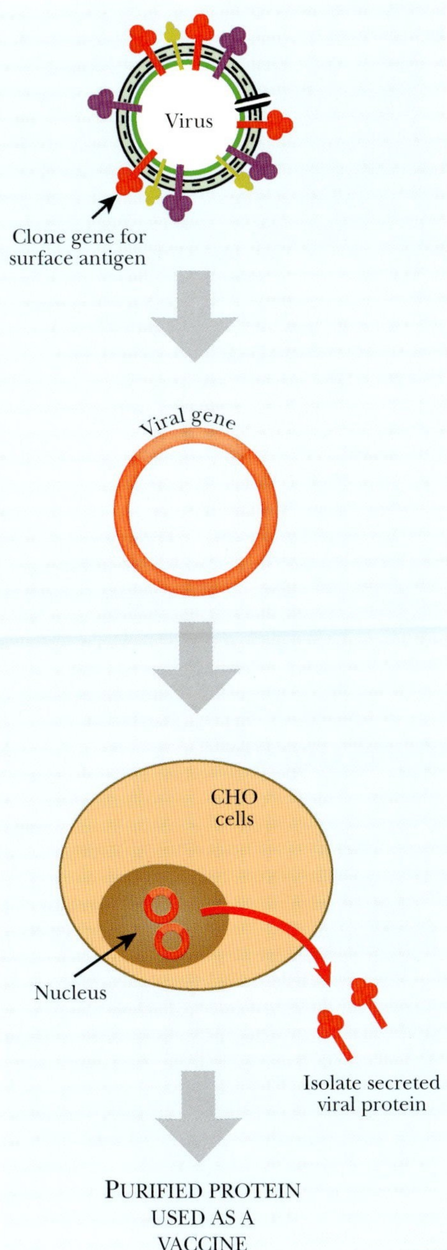

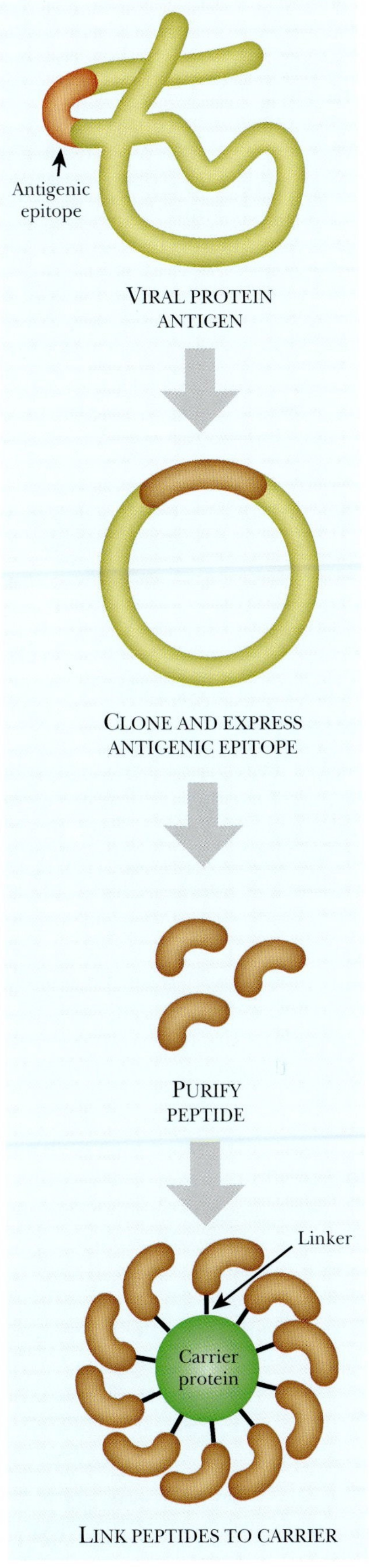

FIGURE 6.22 (RIGHT) Peptide Vaccines Are Conjugated to Carrier Proteins
Peptide vaccines are small regions of an antigenic protein from a pathogen. The peptide is often an epitope that elicits a strong immune response. Because the peptide is small, multiple peptides are conjugated to a carrier protein to prevent degradation, and to stimulate the immune system.

Antigenic epitope

VIRAL PROTEIN ANTIGEN

CLONE AND EXPRESS ANTIGENIC EPITOPE

PURIFY PEPTIDE

Linker

Carrier protein

LINK PEPTIDES TO CARRIER

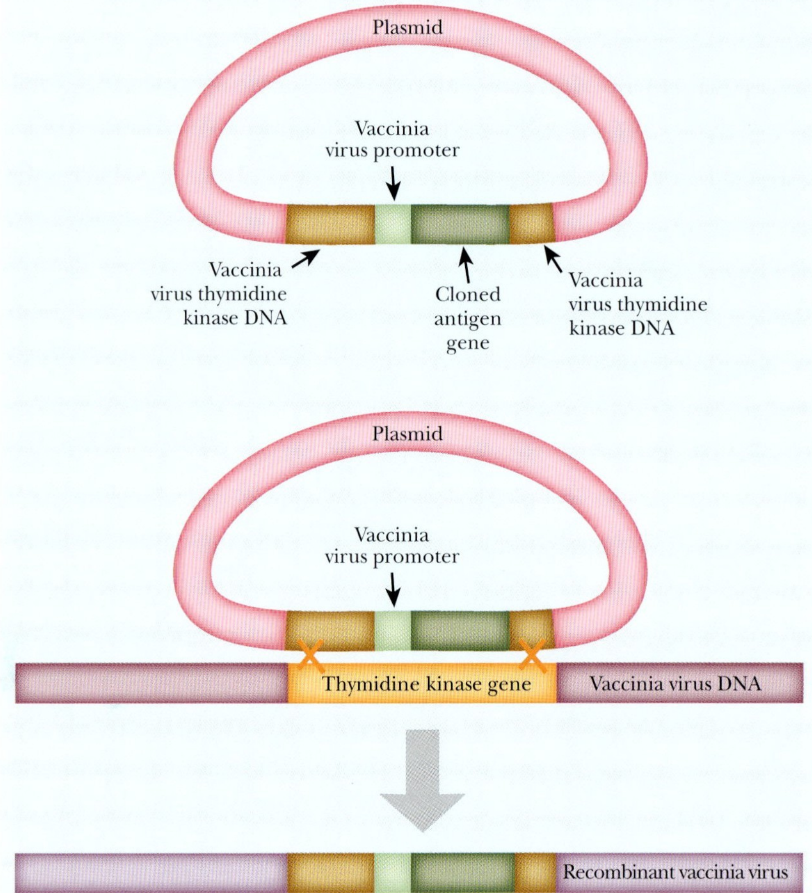

FIGURE 6.23 Homologous Recombination Adds New Genes to the Vaccinia Genome
The plasmid contains two regions homologous to the virus thymidine kinase gene and flanking the cloned antigen gene. When the plasmid aligns with the vaccinia genome, the regions of homology elicit a recombination event. The recombinant vaccinia will gain the cloned antigen gene and lose the gene for thymidine kinase.

REVERSE VACCINOLOGY

Many genomes from infectious agents have now been sequenced. **Reverse vaccinology** takes advantage of this information to find new antigens for use in immunization (Fig. 6.24). The primary research begins with cloning each of the genes from the infectious organism into an expression library. All the proteins are expressed, isolated, purified, and then screened in mice for immune response. Each protein is tested for stimulating the immune system and for its ability to protect the mice from the actual infectious agent. The proteins that elicit the best response can either be combined into a subunit vaccine or used as separate vaccines.

Reverse vaccinology has been used to create a vaccine for *Neisseria meningitidis* serogroup B, which is a major cause of meningitis in children. Attenuated bacteria were not effective as vaccines, and until the sequencing of the *N. meningitidis* genome, no vaccine was available. A library of 350 different *N. meningitidis* proteins was expressed in and purified from *E. coli*. These were individually screened for their presence on the surface of the bacteria using ELISA and FACS. Surface proteins were then screened for immune effectiveness. Of the 350 tested proteins, only 29 became potential candidates and are currently being developed into a vaccine. Without the ability to sequence genomes, vaccine development was often impossible, but now new and emerging diseases can be studied to find potential vaccines.

Reverse vaccinology uses the expressed genomic sequences to find new potential vaccines. Normal vaccines are created using the pathogenic organism. The term *reverse* refers to the use of expressed DNA over the purified proteins from the organism itself.

IDENTIFYING NEW ANTIGENS FOR VACCINES

Another approach to creating vaccines is to identify bacterial pathogen genes that are expressed when the pathogen enters the host. These usually encode proteins that are different from surface antigens. They encompass a variety of adaptations the pathogen makes in order to live within the host organism. Typically bacteria that enter animals are engulfed by phagocytes and digested by the enzymes within the lysosome. Some pathogens are engulfed as usual, but avoid digestion. Modifications required to live intracellularly include changes in nutrition and metabolism and mechanisms to protect against host attacks. Many different types of genes are needed for this switch, and the products of these genes are potential antigens for vaccine development.

EXPRESSION LIBRARY OF GENES FROM INFECTIOUS ORGANISM

ISOLATE PROTEINS

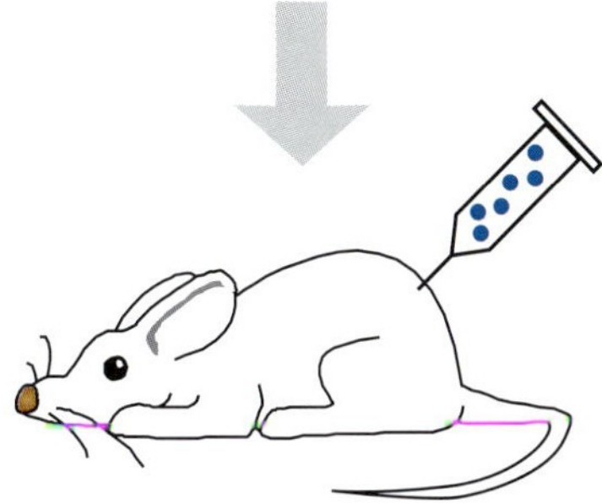

CHECK EACH PROTEIN FOR IMMUNE RESPONSE IN MOUSE

FIGURE 6.24
Reverse Vaccinology
Reverse vaccinology uses the genes identified in the genome of pathogenic agents. First, the genes are cloned into expression vectors and expressed to give proteins. Each protein is tested in mice for an immune response.

Traditionally, identifying genes that are expressed only in host cells relies on gene fusions. Suspected genes are genetically fused to a reporter such as β-galactosidase, luciferase, or green fluorescent protein (GFP), or to epitope tags such as FLAG or myc (see Chapter 9). The fusion gene is introduced into the pathogenic organism, which is then allowed to infect the host. The amount of reporter gene expression correlates with the expression level of the target gene. For example, if the target gene is linked to GFP, the amount of fluorescence is monitored with fluorescence microscopy or fluorescence-based flow cytometry. If the fluorescence increases with host cell invasion, then the target gene is a potential vaccine candidate because it may be important for bacterial pathogenesis.

Individual gene fusions are fine for suspected genes, but screening for novel genes with this method would be tedious. Instead, **differential fluorescence induction (DFI)** uses a combination of GFP fusions and FACS sorting (see earlier discussion) to identify novel genes involved with host invasion (Fig. 6.25). First, a library of genes from the pathogenic organism is genetically linked to GFP. The library is transformed into bacterial cells where the gene fusions are expressed to give GFP. The bacteria are then given a specific stimulus related to host invasion. For example, when phagocytes engulf them, bacteria leave a neutral environment (pH 7) and enter a compartment that is acidic (pH 4). To determine if pH change induces gene expression, the bacteria with the fusion library are shifted to an acidic environment. They are then sorted using FACS to collect those with high GFP expression. If the novel gene fused to GFP is truly induced by low pH, its GFP levels should drop when it is shifted back to neutral pH. Therefore the cells with high GFP expression are shifted to pH 7 and resorted, but this time bacteria with low levels of GFP are collected. The smaller pool of bacteria are again stimulated with low pH and sorted, collecting those with high GFP expression. This sorting scheme eliminates genes that are constitutively expressed plus those that are not induced by low pH. The remaining genes are acid-induced genes that adapt the organism to living within the host. These may then be evaluated as antigens for vaccine development.

Another method to identify new antigens for vaccine development is ***in vivo* induced antigen technology (IVIAT**; Fig. 6.26). This method takes serum from patients who have been infected with a particular disease to which a vaccine is needed. The serum is a rich source of antibodies against the chosen disease agent. The serum is then mixed

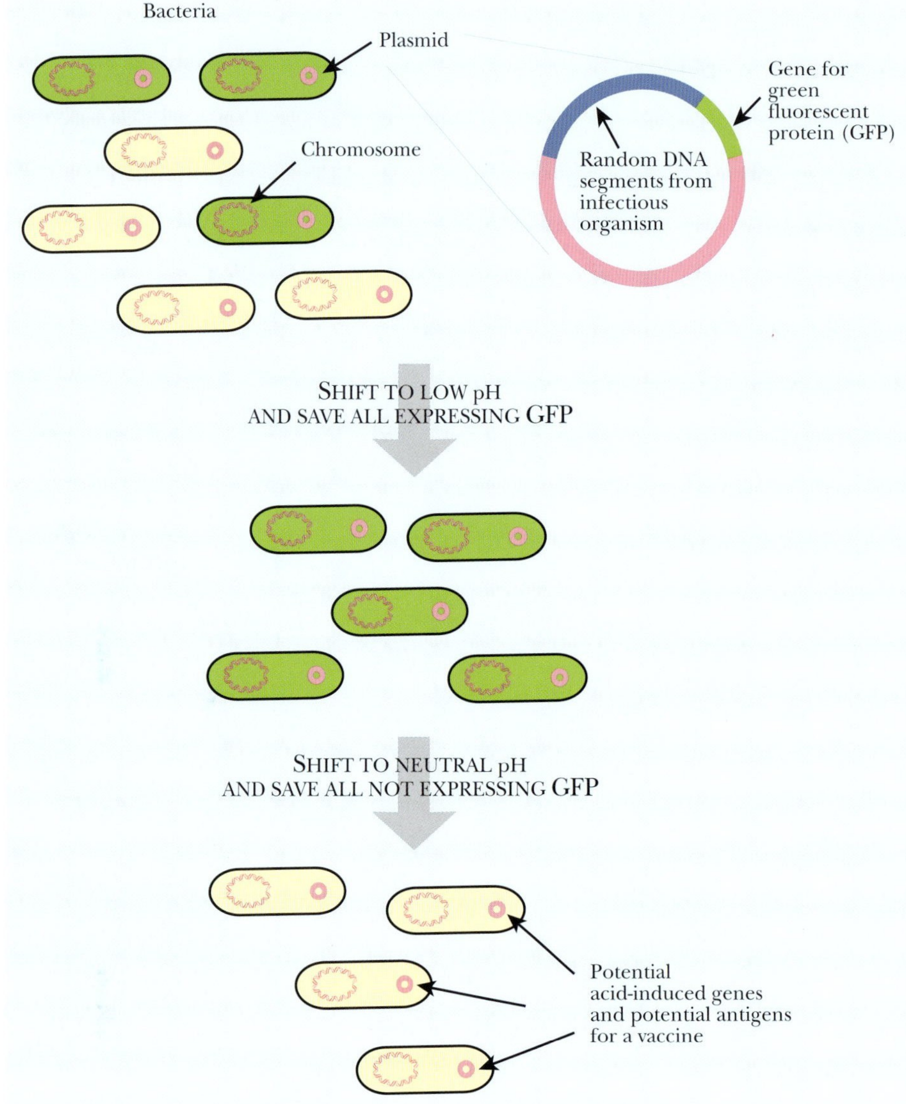

FIGURE 6.25 Differential Fluorescence Induction (DFI) First, genes from the pathogen of interest are cloned in frame with the GFP gene. The fusion proteins are then expressed in bacteria. First the entire bacterial population is exposed to low pH. Those bacteria that are expressing GFP are isolated. These clones either express the GFP protein constitutively or were induced by the low pH. To isolate the clones that are expressed only at low pH, the green cells are shifted to neutral pH, and this time, only the colorless cells are kept. Repeating this procedure will ensure a pure set of genes that are induced only under low pH.

with a sample of the disease-causing microorganism. This removes antibodies that bind to proteins expressed by the microorganism while outside the host. This leaves a pool of antibodies against proteins that are expressed only during infection. To identify the proteins corresponding to these antibodies, a genomic expression library is constructed containing all the genes from the microorganism. The library is expressed in *E. coli* and is probed by the remaining antibodies. When an antibody matches a library clone, the gene insert is sequenced to identify the protein antigen. This method directly identifies protein antigens that stimulated antibody production during a genuine infection; therefore, antigens identified by this method are likely vaccine candidates.

Pathogens must change their metabolism when changing from a free-living organism to the environment within their host. The proteins that help the pathogen adapt to this switch are potential proteins to which a vaccine could be made.

DFI and IVIAT are two techniques to identify proteins that allow pathogens to live within an organism. DFI fuses the potential proteins to fluorescent tags and selects the clones that are only expressed inside the organism. IVIAT uses serum from patients who have been infected with the pathogen to find the antibodies that bind intracellular pathogenic proteins.

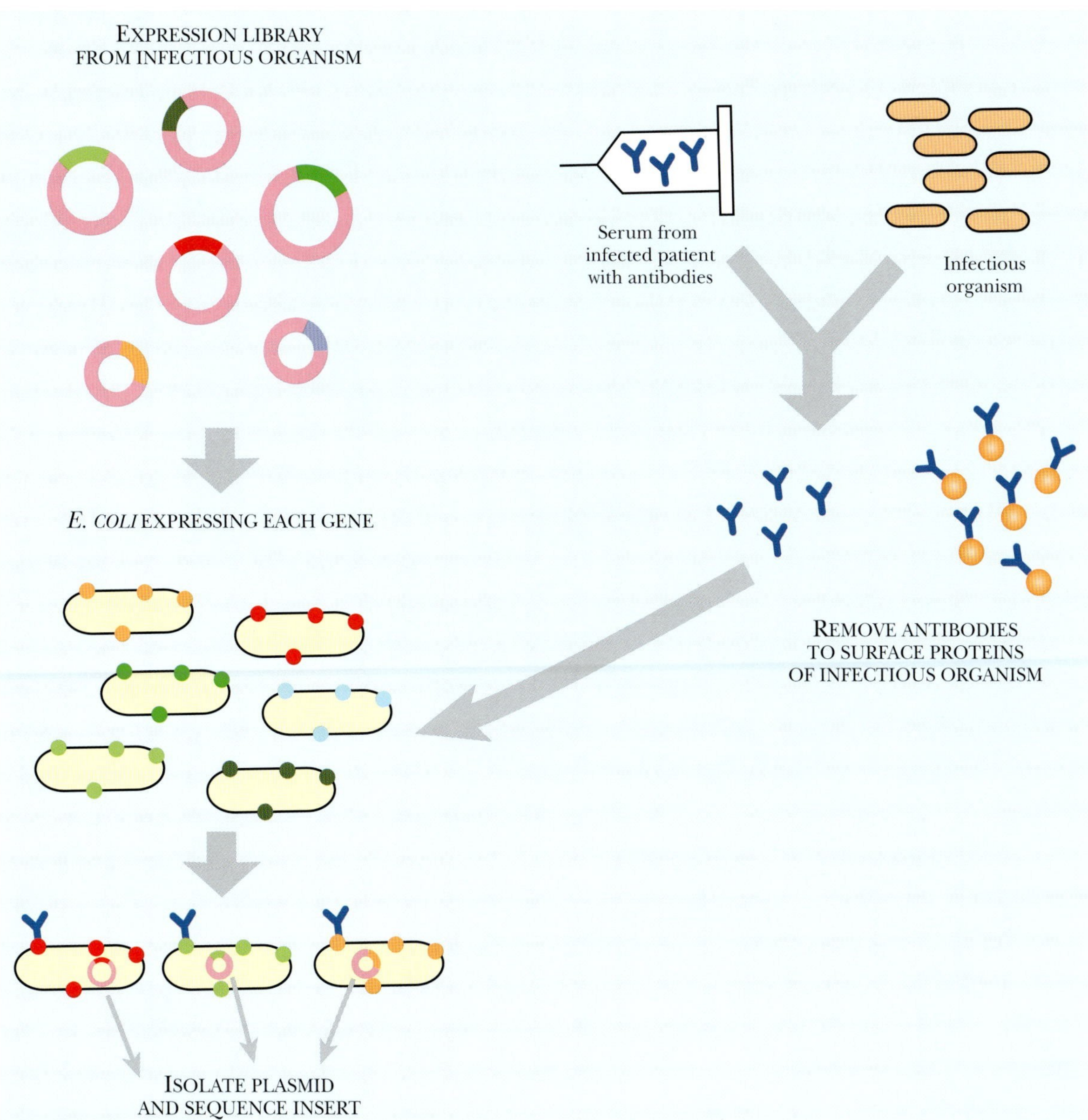

FIGURE 6.26 *In Vivo* Induced Antigen Technology (IVIAT)

Finding novel antigens to make a new vaccine relies on identifying proteins that elicit an immune response. IVIAT identifies antigens directly from patients who have been exposed to the pathogenic organism. First, an expression library is established that includes each of the genes from the pathogen of interest. Next, serum from infected patients is collected and preabsorbed to the infectious organism (grown in culture) to remove the antibodies that recognize surface proteins. The remaining antibodies are used to screen the expression library. When an antibody recognizes a cloned protein, the specific DNA clone is sequenced to identify the gene product.

DNA VACCINES BYPASS THE NEED TO PURIFY ANTIGENS

The principle of the **DNA vaccine** is to just administer DNA that encodes appropriate antigens instead of providing whole microorganisms or even purified proteins. Naked DNA vaccines consist of plasmids carrying the gene for the antigen under control of a strong promoter. The intermediate early promoter from cytomegalovirus is often used because of its strong expression. The DNA is then injected directly into muscle tissue. The foreign genes are expressed for a few weeks and the encoded protein is made in amounts sufficient to trigger an immune response. The immune response is localized to the chosen muscle, which helps avoid side effects. In addition, purified DNA is much cheaper to prepare than purified protein and can be stored dry at room temperature, avoiding the need for refrigeration. The best method of delivering DNA is attaching it to a microparticle with a cationic surface (Fig. 6.27) because the surface binds to the negatively charged phosphate backbone. After the DNA-coated microparticle enters the cells, the DNA is slowly released from the bead and is then converted into protein. The slow release of DNA elicits a better immune response than a large direct dose of DNA. The immune system has to create more and more antibodies to the proteins.

One problem with DNA vaccines is that certain DNA sequences may induce an immune response directly. In particular, some DNA sequence motifs found in bacterial DNA may elicit strong immune responses, which in turn may cause the body to target its own DNA, thus generating an autoimmune response.

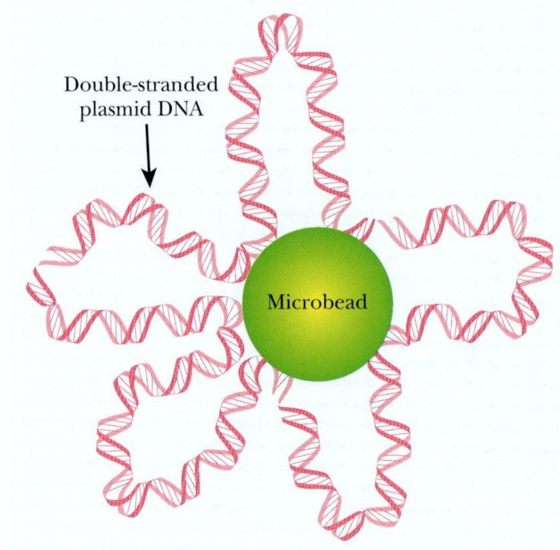

FIGURE 6.27
DNA-Coated Microbeads
Microbeads are coated with plasmid DNA encoding an antigen gene and injected into a patient. Once inside the cells, the plasmid DNA is slowly released and the protein antigen is expressed over a period of time. The expressed protein elicits an immune response without causing disease, thereby vaccinating the person against future exposures to the pathogen.

Rather than injecting a protein, some vaccines are simply DNA of a gene that will elicit an immune response. After the DNA enters the cell, it is converted into protein, which elicits the immune response, providing immunity.

EDIBLE VACCINES

Many vaccines are susceptible to heat, and degrade when not refrigerated. In developed countries, this is not an issue, but in developing countries, proper storage is hard to find. In addition, needles and qualified personnel are needed to administer injected vaccines. An alternative to injection is to use **oral vaccines**. These are taken by mouth in liquid or pill form. Of course, the antigen that is delivered orally must not be degraded by digestive enzymes and must still stimulate the immune system. One example is the oral polio vaccine, which contains live attenuated polio virus, whereas the injected polio vaccine contains inactivated virus. The advantage of the oral vaccine is that the attenuated viruses colonize the intestine and stimulate the immune system the same way that the virulent form of polio would. The disadvantage is the possibility that the live attenuated virus may convert back to a virulent form and the recipient would get polio. The estimate for this is 1 virulent dose in 2.5 million. Where polio itself is very rare, this risk is too great. Most children in the United States now receive the inactivated form of polio vaccine.

Another method of creating heat-stable, low-cost vaccines is to express the antigens in plants and then eat the plant. The benefits of **edible vaccines** include being able to "manufacture" the vaccine in large quantities cheaply. The patient has to eat a certain portion of plant tissue to acquire immunity. Distributing the vaccine in developing countries is easy, and storage is the same as for standard crops. Recent advances in expressing foreign proteins in plants (see Chapter 14) have facilitated the development of edible vaccines. Genetically engineered potatoes containing a hepatitis B vaccine have currently entered human trials. The volunteers ate finely chopped chunks of raw potato expressing a surface protein from hepatitis B. Sixty percent of those who ate the vaccine had more antibodies against hepatitis B. All participants had previously received the traditional vaccine, so the potato vaccine simply boosted immunity. The main drawback of using vaccines in a food source is the possibility of the vaccine vegetables being confused with normal vegetables and used as food.

Instead of food-based vaccines, researchers are now developing **heat-stable oral vaccines**. Instead of crops like corn and potato, other edible plants are being developed to express the vaccine. One potential plant is *Nicotiana benthamiana*, a relative of tobacco that is edible, but is not used for food. The plant is easy to make transgenic, and the vaccine antigens are expressed in the leaf tissue. In order to keep the antigens in the leaf tissue, the recombinant protein is produced only in the chloroplast. Because chloroplasts are only inherited maternally, pollen from the transgenic plant does not contain the transgene. (If pollen did contain recombinant genes, it could be carried to fields of normal tobacco and affect the genetics of normal plants.) Once the vaccine-expressing plant has grown, the leafy part is harvested, washed, ground up, and freeze dried. The powder is then loaded into gelatin capsules. This type of vaccine is easy to transport and distribute, the freeze-dried leaves are heat stable, and the pills are easily administered. Of course, the effectiveness and potential side effects need to be assessed through a series of clinical trials.

Edible vaccines are either live attenuated virus like the oral polio vaccine, or an antigenic protein that is expressed in a food.

Box 6.2 Vaccine Safety

In the United States, infants receive vaccines for many different illnesses, including diphtheria, tetanus, pertussis (whooping cough), measles, mumps, rubella, chickenpox, polio, and hepatitis A and B. All these vaccines are given to children before they enter school. The list is long, but many of the vaccines are combined into one shot. Paradoxically, the effectiveness of vaccines has made many question their use. Many argue that vaccines are not needed because so few people actually get these diseases. It is easy to forget that the reason why very few people get diphtheria or measles is that so many are vaccinated. In 1980, about 4 million people contracted measles, but only about 10% of the world population had received measles vaccine. In 2002, about 500,000 cases of measles were recorded in the world, but about 70% of the world population had received the vaccine for measles. In the United States, about 216 cases of measles were reported from 2001 to 2003, so if you are not vaccinated, the likelihood of contracting measles is very slim. However, the more people who opt not to vaccinate their children, the more cases of the disease there will be, and those who remain unvaccinated will gradually be at increased risk.

Other vaccines have been eliminated from the childhood immunization schedule because the diseases have been eradicated. For example, so many people across the world were vaccinated against smallpox that the disease was not seen at all for years. Now, smallpox vaccine is no longer given to the entire population. The only smallpox that exists is kept in two different labs, one in the United States and one in Russia. After the 2001 attacks on the World Trade Center, much focus has been given to the possibility that smallpox might be used as a biowarfare agent (see Chapter 23). Some people are calling for the reinstatement of smallpox vaccination to prevent this from occurring.

Other vaccines have the opposite issue: Even with widespread vaccination for pertussis, the number of cases of whooping cough is on the rise. In 2004, the state of Wisconsin alone reported more than 5000 cases of whooping cough. Other states also reported increases. In 2002, the Centers for Disease Control reported 9771 cases in the entire United States—the highest number since 1964. Wisconsin had about half the number of cases. Many different theories exist that try to explain the increase in whooping cough. Some attribute the use of a more sensitive test to diagnose whooping cough, and others suggest this may be a natural cycle of *B. pertussis* pathogenicity. Others attribute the increase to waning immunity. Once the last booster shot is received at age 5, the immunity to whooping cough wanes after about 10 years. At the time of writing, no adult booster shots are available, and thus adults are relatively unprotected.

Vaccines cause some adverse side effects. In most cases, vaccines cause a local reaction, pain and swelling at the injection site. Other possible side effects are systemic, perhaps a fever or a mild form of the disease, as is the case with the flu shot. Some vaccines can cause allergic reactions because of impurities in the vaccine. Some vaccines are made in eggs and traces of egg proteins may remain in the vaccine. People with allergies to eggs often still tolerate the vaccine, but some may have an allergic reaction. Another potential allergenic component is gelatin.

Other safety concerns about vaccines are based on the preservatives. Until 1999, the most common preservative was thimerosal, a mercury-containing compound. Thimerosal can cause allergic reactions in some children and has also been thought to cause autism. Since 1999, many manufacturers have completely removed thimerosal from their vaccines, yet the number of cases of autism has not dropped. Yet because there are anecdotal cases of autism developing in children after receiving vaccines containing thimerosal, many manufacturers have switched to other types of preservatives. Conceivably a certain genetic background makes some children more susceptible to mercury than others.

Summary

The immune system has two different components, humoral immunity and cell-mediated immunity. Humoral immunity includes the production of antibodies by B cells that are found in the serum and other bodily fluids. The antibodies have a general Y shape that consists of two heavy chains and two light chains. The hinge region of the Y divides the molecule into the Fc (constant) and Fab (variable) regions. Cell-mediated immunity involves the activation of T cells, a subset of white blood cells. The T cells become active when a pathogen invades a cell, and the cell starts presenting fragments of the pathogen on the cell surface major histocompatibility complexes. In both arms of the immune reaction, the antibody or T cell only recognizes small epitopes or distinct regions of the pathogenic proteins. The immune system can make many different antibodies to one protein because only these small areas are recognized.

In the laboratory, antibodies can be made to specific proteins by injecting an animal such as a mouse or rabbit with a pure sample of the protein. To make monoclonal antibodies, mouse B cells are fused to immortal myeloma cells to make hybridomas. Each B-cell fusion makes an antibody to one specific epitope of the protein. Polyclonal antibodies, on the other hand, include all the antibodies to the protein, that is, the antibodies recognize multiple epitopes. Antibodies are used in ELISA, where the amount of the target protein in a mixture can be estimated by the amount of antibody that binds. In immunohistochemistry and immunocytochemistry, an antibody to the target protein is used to localize its position within the cell. Antibodies are also used to sort samples of cells by FACS and are used to count a specific type of cell in flow cytometry.

Vaccines stimulate our immune systems to form antibodies and memory B cells without causing the disease for which the vaccine is providing protection. Vaccines are live attenuated viruses, inactivated or dead viruses, a subunit of the virus, or simply a peptide from a viral protein. The vaccine could also be made from a related but harmless virus or bacteria that is expressing a protein from the pathogenic virus or bacteria. Reverse vaccines and DNA vaccines are created from genomic DNA sequences that are expressed into protein. Reverse vaccines are made in a laboratory, whereas DNA vaccines are injected directly into the muscular tissue as DNA. Also, some vaccines are also made by expressing pathogenic proteins in edible crops. A person can receive resistance to the pathogen by simply ingesting these plants. New antigenic proteins are the key to making a good vaccine. DFI and IVIAT are two methods to identify potential antigenic proteins from the pathogenic organism.

End-of-Chapter Questions

1. What are antigens and antibodies?
   - **a.** Antigens are foreign bodies and antibodies are immune system components that recognize antigens.
   - **b.** Antibodies are foreign bodies and antigens recognize them and work to destroy them.
   - **c.** Antigens are produced by B cells in response to antibody accumulation.
   - **d.** Antigens are foreign bodies and antibodies are a specific cell type from the immune system.
   - **e.** none of the above

(Continued)

2. Which of the following is an accurate description of B and T cells?
   a. B cells recognize antigens expressed on the surface of other cells and T cells produce antibodies.
   b. B cells are components of the cell-mediated immunity and T cells comprise the humoral immunity.
   c. Major histocompatibility complexes are associated with B cells whereas T cells produce antibodies.
   d. B cells produce antibodies and T cells recognize antigens expressed on the surface of other cells.
   e. none of the above

3. How are the variants of antibodies produced?
   a. Each variant is encoded on one gene.
   b. by post-translational modification of the antibodies
   c. by shuffling a small number of gene segments around
   d. by splicing the transcript into various configurations
   e. all of the above

4. Which of the following statements about antibodies is not correct?
   a. Antibodies consist of two light chains and two heavy chains.
   b. Polyclonal antibodies are derived from hybridomas.
   c. Antibodies are classified into classes and have distinct roles in the immune system.
   d. One particular antibody made from a clonal B cell is called a monoclonal antibody.
   e. Monoclonal antibodies are made by fusing B cells to myelomas, culturing the hybridomas, and screening for appropriate antigen recognition.

5. Which of the following statements about humanized antibodies is correct?
   a. Humanized antibodies to the ClfA protein of *S. aureus* may provide a way to eliminate the antibiotic-resistant pathogen in patients with nosocomial infections.
   b. Herceptin has been effective in treating some patients with breast cancer.
   c. Humanized monoclonal antibodies are created by removing the constant regions of mouse antibodies and replacing them with human constant regions.
   d. Full humanization of an antibody involves removing the hypervariable regions and splicing them into the heavy and light chains of human antibodies.
   e. All of the above are correct.

6. How is the creation of recombinant antibodies useful to researchers?
   a. Recombinant antibodies can be used to precisely deliver toxins, cytokines, and enzymes directly to the antigen.
   b. The production of recombinant antibodies is strictly theoretical and probably will serve no purpose to biotechnology research.
   c. Recombinant antibodies allow for more efficient production and isolation of the scFv.
   d. Recombinant antibodies can deliver toxins, cytokines, and enzymes but are disseminated throughout the organism.
   e. none of the above

7. Why is an ELISA used?
   a. to quantify the amount of a specific protein or antigen in a sample
   b. to quantify the amount of DNA in a sample
   c. to determine the amount of antibody within a sample
   d. to dilute out antibody from serum in a microtiter plate
   e. none of the above

8. Which of the following is an example of how ELISA is used?
   a. home pregnancy test
   b. detection of pathogenic organisms
   c. detection of plant diseases
   d. detection of dairy and poultry diseases
   e. all of the above

9. In which application are fluorescent antibodies used?
   a. immunocytochemistry
   b. flow cytometry
   c. immunohistochemistry
   d. fluorescence activated cell sorting
   e. all of the above

10. Which of the following statements about immunity is not true?
   a. Vaccines use a live infectious agent that is still capable of producing disease in order to elicit an immune response.
   b. The immune system remembers foreign antigens through memory B cells.
   c. Vaccines consist of an antigen from an infectious agent that induces an immune response.
   d. Immunity to a fatal disease can often be triggered by infection with a closely related infectious agent, as in the cases of cowpox and smallpox.
   e. Antibody-producing B cells normally live only a few days but memory cells survive for a long time.

11. How are vaccines made so that they do not cause disease?
   a. killing the infectious agent with heat or denaturing the infectious agent with chemicals
   b. using a component or protein of the infectious agent instead of the organism itself
   c. genetically engineering the infectious agent to remove the genes that cause disease
   d. using a related but non-pathogenic strain of an infectious agent
   e. all of the above

12. What is reverse vaccinology?
   a. the removing of B cells from a person's body, exposing them to an infectious agent *in vitro*, and then returning them to the body
   b. the use of expressed genes from an expression library to find proteins that elicit an immune response in mice to create new vaccine candidates
   c. the vaccination of a person with a related, but non-pathogenic, strain to elicit an immune response
   d. the vaccination of a person after he or she has already been exposed to the pathogen
   e. none of the above

(Continued)

13. What is critical to finding novel antigens for vaccine development?
- **a.** the growth of live infectious agents to create whole vaccines
- **b.** the engineering of genes to attenuate infectious agents
- **c.** the identification of proteins that elicit an immune response
- **d.** the identification of the immune system components unique to specific infectious agents
- **e.** none of the above

14. Which of the following statements about edible vaccines is not true?
- **a.** In developing countries, proper storage and availability of needles and personnel to administer the vaccine are limiting factors in vaccinating the population.
- **b.** Edible vaccines must not be destroyed by the digestive system and must still elicit an immune response.
- **c.** A problem with using edible vaccines is the possibility that vaccine vegetables could be mistaken for normal vegetables used as food.
- **d.** Edible vaccines are usually too expensive to be manufactured in large quantities.
- **e.** All of the above are true.

15. Which of the following is not a risk associated with vaccines?
- **a.** adverse side effects
- **b.** allergic reactions
- **c.** preservatives containing mercury
- **d.** induction of autoimmunity in some individuals
- **e.** all of the above are potential risks associated with vaccination

Further Reading

Betts MR, Brenchley JM, Price DA, De Rosa SC, Douek DC, Roederer M, Koup RA (2003). Sensitive and viable identification of antigen-specific CD8+ T cells by a flow cytometric assay for degranulation. *J Immunol Methods* **281**, 65–78.

Clark DP (2005). *Molecular Biology: Understanding the Genetic Revolution*. Elsevier Academic Press, San Diego, CA.

Clark M (2000). Antibody humanization: A case of the "Emperor's new clothes"? *Immunol Today* **8**, 397–402.

Elgert KD (1996). *Immunology: Understanding the Immune System*. Wiley-Liss, New York.

Fischer OM, Streit S, Hart S, Ullrich A (2003). Beyond Herceptin and Gleevec. *Curr Opin Chem Biol* **7**, 490–495.

Glick BR, Pasternak JJ (2003). *Molecular Biotechnology: Principles and Applications of Recombinant DNA*, 3rd ed. ASM Press, Washington, DC.

Handfield M, Brady LJ, Progulske-Fox A, Hillman JD (2000). IVIAT: A novel method to identify microbial genes expressed specifically during human infections. *Trends Microbiol* **8**, 336–339.

Patti JM (2004). A humanized monoclonal antibody targeting *Staphylococcus aureus*. *Vaccine* **228**, S39–S43.

Scarselli M, Giuliana MM, Adu-Bobie J, Pizza M, Rappuoli R (2005). The impact of genomics on vaccine design. *Trends Biotechnol* **23**, 84–91.

Valdivia RH, Falkow S (1997). Probing bacterial gene expression within host cells. *Trends Microbiol* **5**, 360–363.

CHAPTER 7

Nanobiotechnology

INTRODUCTION

In 1959, Richard Feynman was the first scientist to suggest that devices and materials could someday be fabricated to atomic specifications: "The principles of physics, as far as I can see, do not speak against the possibility of maneuvering things atom by atom."

Molecular biology originated largely from the study of microorganisms. One micrometer is one millionth of a meter, and cells of *Escherichia coli*, the geneticist's favorite bacterium, are roughly 1 micrometer (= "micron") in length. A nanometer is one thousandth of a micrometer = 10^{-9} meters (Fig. 7.1). The terms *micro-* and **nano-** are both from Greek. *Mikros* means "small." More imaginative is *nanos*, a little old man or dwarf. *Pico-* comes from Spanish where it means a small quantity, or *beak* (from Latin *beccus*, "beak," ultimately of Celtic origin). Prefixes for even smaller quantities are shown in Table 7.1. As far as length is concerned, these are applicable only to subatomic dimensions. Nevertheless, when dealing with masses and volumes on the nanoscale we may find femtograms and zeptoliters.

Recently, science has advanced into the area of **nanotechnology**. As the name indicates, the impetus has come from pursuing practical applications, especially in the fields of electronics and materials science, rather than a quest for theoretical knowledge. Nanotechnology involves the individual manipulation of single molecules or even atoms. Building components atom-by-atom or molecule-by-molecule in order to create materials with novel or vastly improved properties was perhaps the original goal of nanotechnologists. However, the field has expanded in a rather ill-defined way and tends to include any structures so tiny that their study or manipulation was impossible or impractical until recently. At the nanoscale, quantum effects emerge and materials often behave strangely, compared to their bulk properties.

The internal components of biological cells are on the same scale as those studied by nanotechnology. As a consequence, nanotechnologists have looked to cell biology for useful structures, processes, and information. Cellular organelles such as ribosomes may be regarded as programmable "nanomachines" or "nano-assemblers." Thus nanotechnology is spilling over into molecular biology. Much of "nanobiotechnology" is in fact molecular biology viewed from the perspective of materials science and described in novel terminology.

All chemical reactions operate at a molecular level. What distinguishes true nanotechnology is that single molecules or nanostructures are assembled following specific instructions. A ribosome does not merely polymerize amino acids into a chain. It takes specified amino acids, one at a time, according to information provided, and links them in a specific order. Thus the critical properties of a nanoassembler include the ability not only to assemble structures at the molecular level but to do so in a specific and controlled manner.

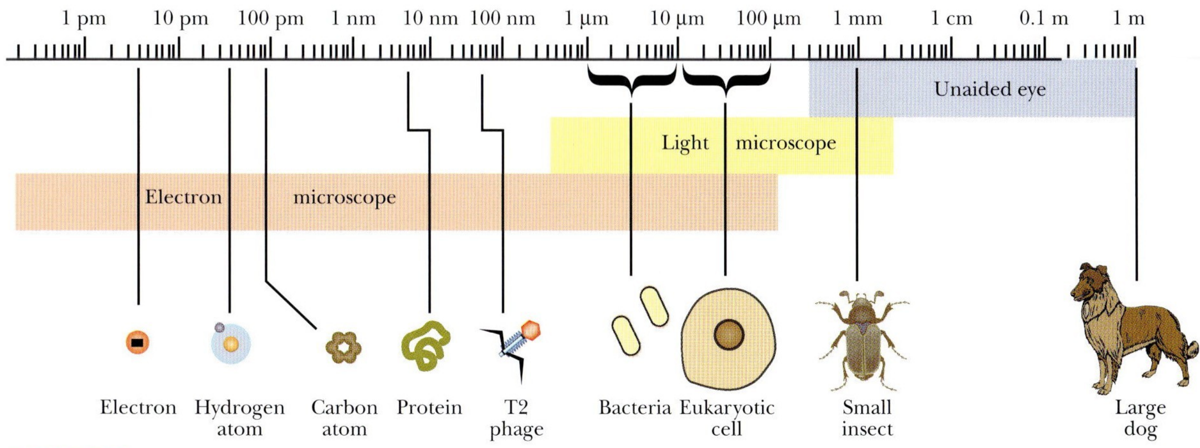

FIGURE 7.1 Size Comparisons
The objects range in size from 1 meter to 1 picometer.

Table 7.1 Prefixes and Sizes

Length Unit	Meters	Examples
5.9 terameters		mean distance from sun to Pluto
Terameter	10^{12}	
150 gigameters		distance to the sun
Gigameter	10^{9}	
380 megameters		distance to the moon
6.3 megameters		radius of the earth
3.2 megameters		length of Great Wall of China
Megameter	10^{6}	
Kilometer	10^{3}	
30 meters		blue whale
Meter	1	large dog
Millimeter	10^{-3}	small insect
Micrometer	10^{-6}	bacterial cell
500 nanometers		wavelength of visible light
100 nanometers		size of typical virus
3.4 nanometers		one turn of DNA double helix
Nanometer	10^{-9}	molecules
350 picometers		molecular diameter of water
260 picometers		atomic spacing in solid copper
77 picometers		atomic radius of carbon (= resolution limit of atomic force microscope as of 2004)
32 picometers		atomic radius of hydrogen
Ångstrom		= 100 picometers = 10^{-10} meter
2.4 picometers		wavelength of electron
Picometer	10^{-12}	
Femtometer	10^{-15}	radius of atomic nucleus
Attometer	10^{-18}	radius of proton
Zeptometer	10^{-21}	
Yoctometer	10^{-24}	radius of neutrino

The main practical objectives of nanobiotechnology are using biological components to achieve nanoscale tasks. Some of these tasks are nonbiological and have applications in such areas as electronics and computing, whereas others are applicable to biology or medicine. The purpose of this chapter is to show, by selected examples, how biological approaches can contribute to nanoscience.

Many internal components of biological cells are in the nanoscale range. As nanotechnology advances it is developing many links with biotechnology and genetic engineering.

VISUALIZATION AT THE NANOSCALE

In order to manipulate matter on an atomic scale, we need to see individual atoms and molecules. Although individual molecules have been visualized with the electron microscope, it was the development of scanning probe microscopes that opened up the field of nanotechnology. These instruments all rely on a miniature probe that scans across the surface under investigation.

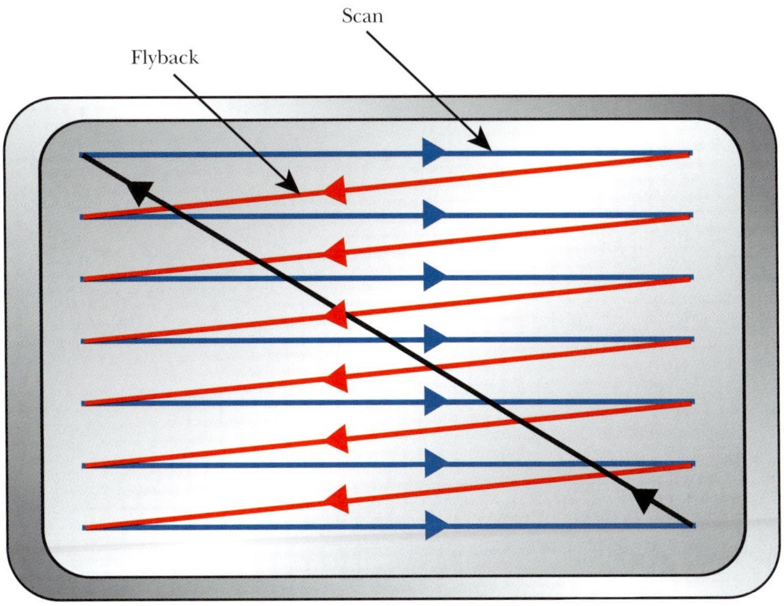

All scanning probe microscopes work by measuring some property, such as electrical resistance, magnetism, temperature, or light absorption, with a tip positioned extremely close to the sample. The microscope **raster-scans** the probe over the sample (Fig. 7.2) while measuring the property of interest. The data are displayed as a raster image similar to that on a television screen. Unlike traditional microscopes, scanned-probe systems do not use lenses, so the size of the probe rather than diffraction limits their resolution. Some of these instruments can be used to alter samples as well as visualize them.

FIGURE 7.2
Principle of Raster Scanning
In raster scanning, the probe moves to and fro across the target region. The probe scans only while moving in one direction ("scan"). When the probe travels in the reverse direction, it moves more rapidly without making contact ("flyback").

The first of these instruments was the **scanning tunneling microscope (STM)**, which was developed by Gerd Binnig and Heinrich Rohrer at IBM (see following section). They received the Nobel Prize in 1986. The STM sends electrons, that is, an electric current, through the sample and so measures electrical resistance. The **atomic force microscope (AFM)** is especially useful in biology and measures the force between the probe tip and the sample.

Visualization of individual molecules or even atoms is possible using scanning probe microscopes.

SCANNING TUNNELING MICROSCOPY

When a metal tip comes close to a conducting surface, electrons can tunnel from one to the other, in either direction. The probability of tunneling depends exponentially on the distance apart. Surface contours can be mapped by keeping the current constant and measuring the height of the tip above the surface. This allows resolution of individual atoms on the surface being studied. This is the principle of the scanning tunneling microscope (Fig. 7.3).

Atoms may also be moved using the STM. In 1989, in perhaps the most famous experiment in nanotechnology, D. M. Eigler and E. K. Schweizer fabricated the IBM logo by arranging 35 xenon atoms on a nickel surface. They chose nickel because the valleys between rows of nickel atoms are deep enough to hold xenon atoms in place, yet small enough to allow the xenon atoms to be pulled over the surface. To move xenon atoms, the STM tip was placed above a xenon atom, using imaging mode. Next, scanning mode was turned off and the tip lowered until the tunneling current increased severalfold ("fabrication mode"). The xenon atom was attracted to the STM tip and was dragged by moving the tip horizontally. The atom was deposited at its new location by reducing the tunneling current. Since then several diagrams have been made in the same way. Carbon monoxide man is shown in Fig. 7.4.

From a biological perspective, the weakness of STM is that it requires a conducting surface, in practice generally a metal layer of some sort. The atomic force microscope (see following

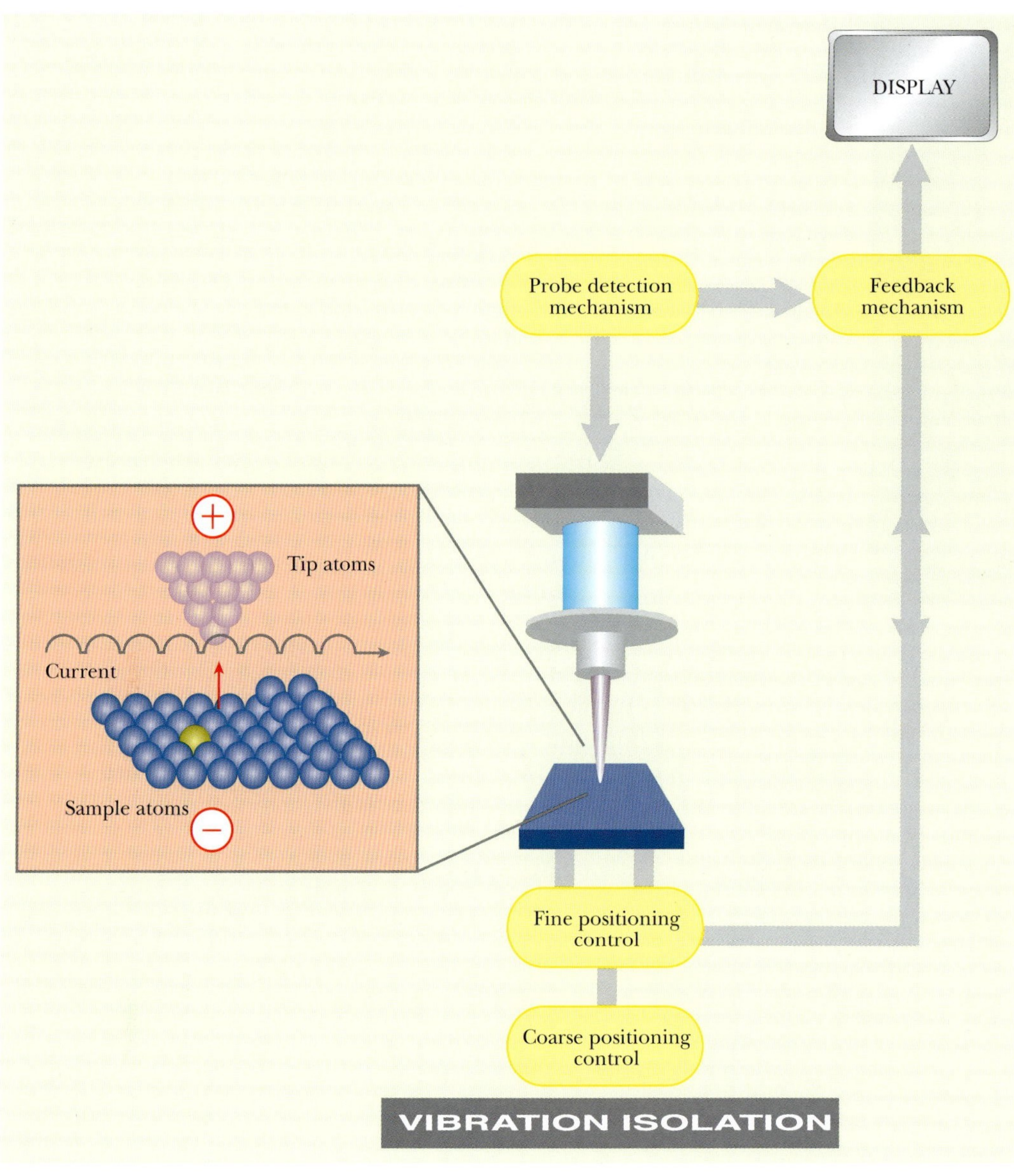

FIGURE 7.3
Principle of Scanning Tunneling Microscope
The probe tip and surface atoms of the sample are shown in the inset.

section) has the advantage of not needing conductive material and has therefore been more widely applied in biology.

> The scanning tunneling microscope can be used to detect or move individual atoms on a conducting surface.

ATOMIC FORCE MICROSCOPY

Visualization at the nanoscale is often performed using atomic force microscopy. As the name indicates, this operates by measuring force, not by using a stream of particles such as photons (as in light microscopy) or electrons (as in electron microscopy).

Physicists sometimes compare the operation of an AFM to an old-fashioned record player, which uses a needle to scrape the surface of a record. Perhaps to a biologist, the difference between a light microscope and AFM is like the difference between reading text with the eyes and feeling Braille.

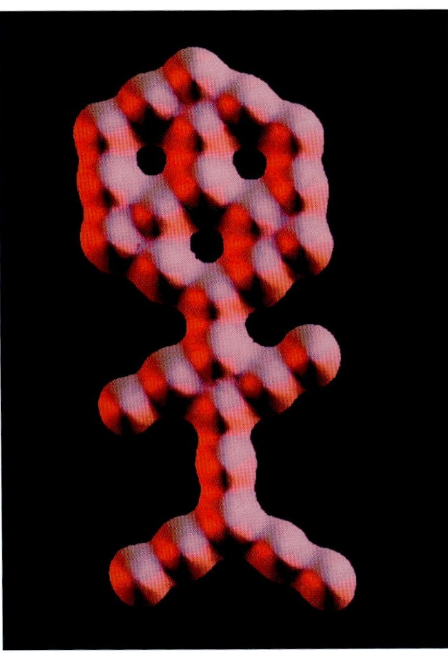

FIGURE 7.4 Carbon Monoxide Man by Zeppenfeld
The atoms were arranged by STM. The medium is carbon monoxide on platinum. Courtesy of International Business Machines Corporation. © 1995, IBM.

The atomic force microscope was invented in 1985 by Gerd Binnig, Calvin Quate, and Christof Gerber. The AFM uses a sharp probe that moves over the surface of the sample and which bends in response to the force between the tip and the sample. The movement of the probe performs a raster scan and the resulting topographical image is displayed onscreen.

During scanning, the movement of the tip or sample is performed by an extremely precise positioning device and is made from **piezoelectric ceramics**. (These are materials that change shape in response to an applied voltage.) It usually takes the form of a tube scanner that is capable of sub-Ångstrom resolution in all three directions.

The AFM probe is a tip on the end of a cantilever. As the cantilever bends because of the force on the tip, its displacement is monitored by a laser, as shown in Fig. 7.5. The beam from the laser is reflected onto a split photodiode. The difference between the A and B signals measures the changes in the bending of the cantilever. For small displacements, the displacement is proportional to the force applied. Hence the force between the tip and the sample can be derived.

The distance between tip and sample is adjusted so that it lies in the repulsive region of the intermolecular force curve; that is, the AFM probe is repelled by its molecular interaction with the surface. The repulsion gives a measure of surface topography, and this is what is generally displayed, with color coding indicating relative height. It is possible to scan a surface for topography and then raise the AFM probe and rescan to detect electrostatic or magnetic forces. These can then be plotted for comparison with the topography.

As with STM, it is possible to use AFM to move single atoms, although this was only achieved in 2003. Researchers at Osaka University in Japan removed a single silicon atom from a surface and then replaced it.

Using AFM, it is possible to visualize polymeric biological molecules such as DNA or cellulose and even to see the individual monomers and, at high resolution, even the atoms of which they are composed.

> The atomic force microscope can detect atoms or molecules by scanning a surface for shape or electromagnetic properties.

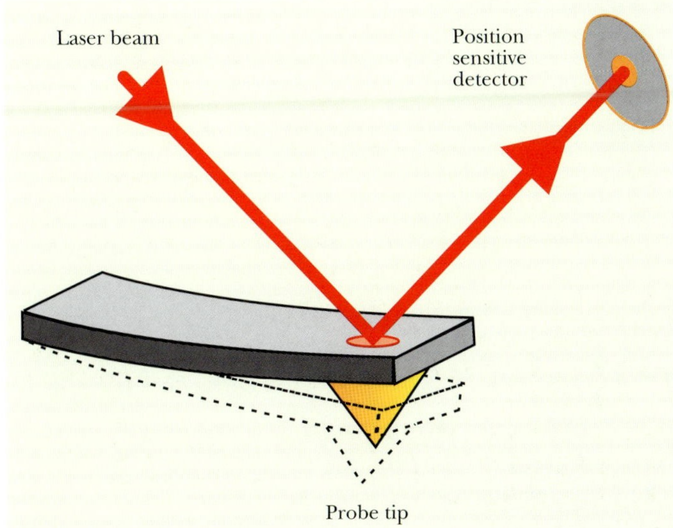

FIGURE 7.5 The Atomic Force Microscope (AFM)
The deflection of the tip of the probe by the surface is monitored by a laser.

VIRUS DETECTION VIA AFM

The AFM can be used to monitor for the presence of virus particles, using a device known as a "ViriChip." A silicon chip is coated with antibodies specific for the virus of interest. In practice, several different antibodies are applied to separate regions of the chip, thus allowing multiple viruses to be monitored simultaneously. A single microliter of liquid from the sample to be analyzed is applied to the chip, and any viruses that are recognized by the antibodies are bound to the chip. The chip surface is then scanned by AFM to detect the presence of virus particles, which are identified by their location on the chip (Fig. 7.6).

The device has been successfully tested with a variety of viruses, including human pathogens, but is presently still under development. In principle, it

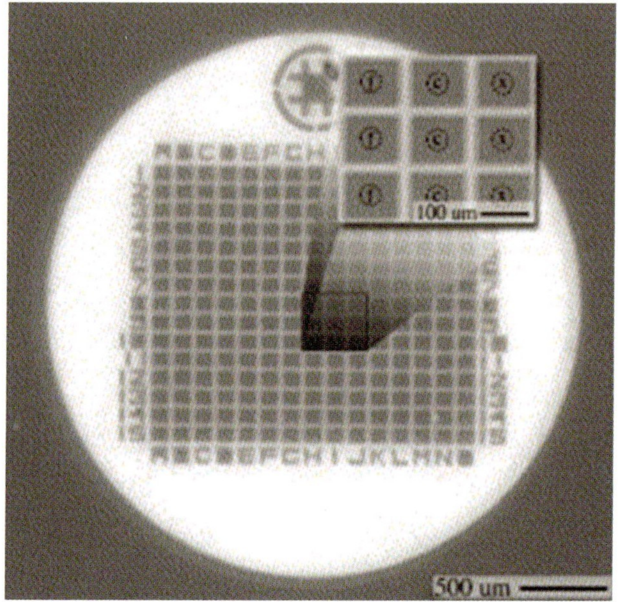

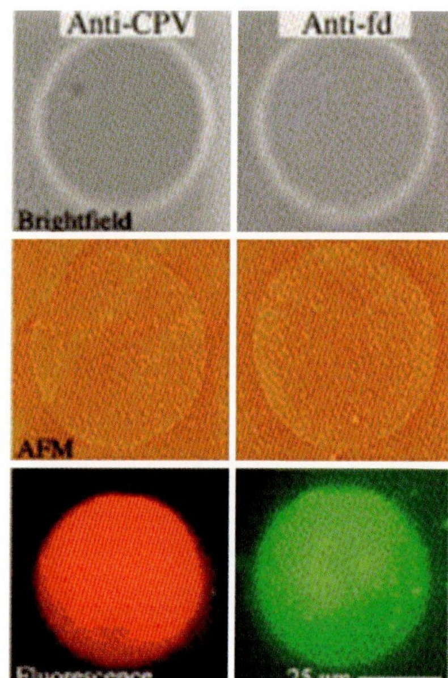

FIGURE 7.6
Principle of the ViriChip
(*Left panel*) An optical micrograph of the ViriChip platform. The inset shows the arrangement of antibody spots in each three by ten array (3 × 3 shown). (*Right panel*) Higher resolution images of two antibody spots (anti-fd and anti-CPV) obtained by brightfield (*top*), AFM (*middle*, Z = 15 nm), and fluorescence (*bottom*) methods. The fluorescence staining used mouse anti-CPV and rabbit anti-fd followed by anti-mouse IgG (labeled with Alexa 594—red) and anti-rabbit IgG (labeled with Cy2—green). Each image field is 60 micrometers square. From: Nettikadan *et al.* (2003). Virus particle detection by solid phase immunocapture and atomic force microscopy. *Biochem Biophys Res Commun* **311**, 540–545. Reprinted with permission.

should be possible to monitor for thousands of different viruses on a single chip. Eventually patients may be tested for multiple viruses using a single drop of blood.

Atomic force microscopy can detect and identify individual viruses.

WEIGHING SINGLE BACTERIA AND VIRUS PARTICLES

It has been known for many years that bacteria are on the order of 1000 nanometers in size and 1 picogram in weight. However, in addition to detecting microorganisms via nanotechnology it is now possible to weigh them individually.

The oscillation frequency of a diving board depends on the mass applied. Scaling down, it is possible to construct a cantilever of micrometer dimensions (approximately 6 microns long by 0.5 micron wide with an end platform about 1 micron square). The oscillation frequency can be measured by using a laser and observing the altered light reflection. Addition of single bacterial cells or even virus particles changes the oscillation frequency of the cantilever. The mass of single cells or virus particles has been measured this way in the laboratory of Harold Craighead at Cornell University (Fig. 7.7).

To hold the bacteria or viruses in place, the cantilever is coated with an antibody that recognizes the microorganism to be weighed. A single cell of *Escherichia coli* was 1430 × 730 nanometers in size and weighed 665 femtograms (665×10^{-15} grams). Viruses (weighing around 1 femtogram) can be detected by reducing the size of the cantilever and enclosing it in a vacuum. By mid-2005, this technique had been refined to weigh a single macromolecule—a double-stranded DNA of approximately 1500 base pairs (roughly the size of a typical coding sequence). Future developments should allow measurements of small proteins and other molecules in the zeptogram range (10^{-21} grams).

Laser monitoring of the oscillation of a nanoscale cantilever allows single bacteria or viruses to be individually weighed.

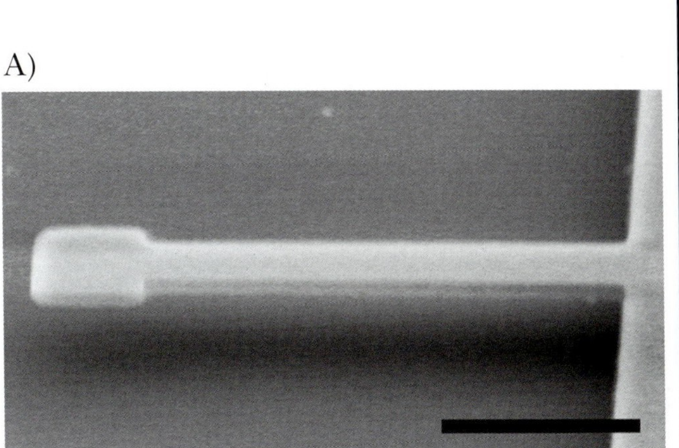

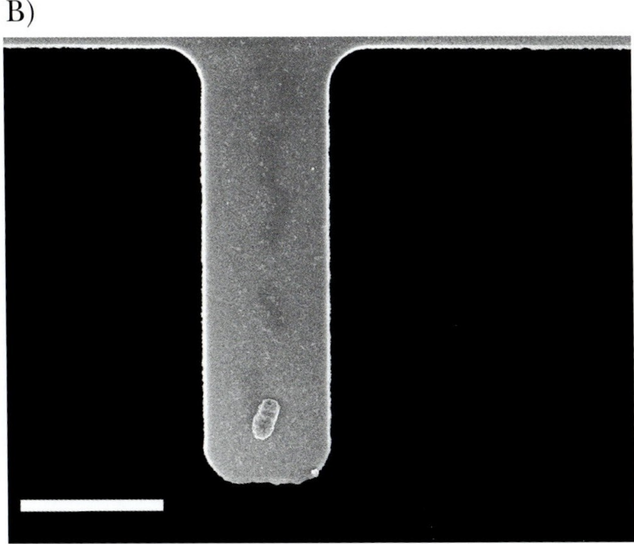

FIGURE 7.7 Weighing a Single Bacterium
(A) Scanning electron micrograph of cantilever oscillator with length 6 microns, width 0.5 microns, and a 1 micron by 1 micron paddle. Scale bar corresponds to 2 microns. Ilic, B., et al., "Virus detection using nanoelectromechanical devices," Appl. Phys. Lett. (2004), Vol. 95, pgs. 2604–2607. Copyright 2004. Reprinted with permission from the American Institute of Physics. (B) Scanning EM of a single *E. coli* cell attached to the cantilever by antibody. Courtesy of Craighead group, Cornell University. Ilic, B., et al, "Single cell detection with micromechanical oscillators," J. Vac. Sci. Technol. B (2001), Vol. 19, pgs. 2825–2828. Copyright 2001. Reprinted with permission from the American Institute of Physics.

NANOPARTICLES AND THEIR USES

Nanotechnology began with advances in viewing and measuring the incredibly small. It then moved on to building structures at the nanoscale. Simple nanostructures are now being used for a variety of analytical purposes, and a second generation is being developed for clinical use.

As their name indicates, **nanoparticles** are particles of submicron scale—in practice, from 100 nm down to 5 nm in size. They are usually spherical, but rods, plates, and other shapes are sometimes used. They may be solid or hollow and are composed of a variety of materials, often in several discrete layers with separate functions. Typically, there is a central functional layer, a protective layer, and an outer layer allowing interaction with the biological world.

FIGURE 7.8 Typical Layered Structure of Nanoparticles
Several layers surround the physically active core. Chemical groups are often added to the exterior to allow attachment of biological molecules.

The central functional layer usually displays some useful optical or magnetic behavior. Most popular is fluorescence. The protective layer shields the functional layer from chemical damage by air, water, or cell components and conversely shields the cell from any toxic properties of the chemicals composing the functional layer. The outer layer(s) allow nanoparticles to be "biocompatible." This generally involves two aspects, water solubility and specific recognition. For biological use, nanoparticles are often made water soluble by adding a hydrophilic outer layer. In addition, chemical groups must be present on the exterior to allow specific attachment to other molecules or structures (Fig. 7.8).

Nanoparticles have a variety of uses in the biological arena:

(a) Fluorescent labeling and optical coding
(b) Detection of pathogenic microorganisms and/or specific proteins
(c) Purification and manipulation of biological components
(d) Delivery of pharmaceuticals and/or genes
(e) Tumor destruction by chemical or thermal means
(f) Contrast enhancement in magnetic resonance imaging (MRI)

Nanoparticles are now widely used in a range of biological procedures. These include both analytical and clinical applications.

NANOPARTICLES FOR LABELING

Consider luminescent CdSe nanorods as an example of nanoparticles used for labeling (Fig. 7.9). These nanorods can be used as fluorescent labels for molecular biology because they absorb light from the UV to around 550 nm and emit strongly at 590 nm. They were made—appropriately enough—in the lab of Thomas Nann, in Freiberg, Germany.

These nanorods measure approximately 3 nm in width by 10 to 20 nm in length. A core of luminous cadmium selenide (CdSe) is surrounded by a shell of ZnS (zinc sulfide, wurtzite) to protect the core against oxidation. Outside this is a layer of silica, which allows coupling of phosphonates or amines to the exterior of the nanorod. These hydrophilic groups make the nanorods water soluble. These outer chemical groups also allow attachment of the nanorods to proteins.

The scaffold inside eukaryotic cells is built from cylindrical protein structures known as microtubules. These are often disassembled into monomers (known as tubulin) and

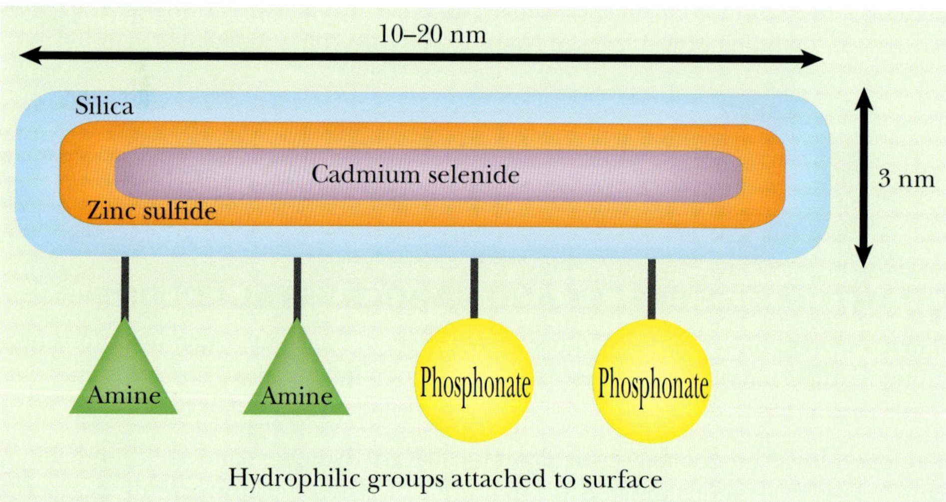

FIGURE 7.9 CdSe Nanorods
Luminescent CdSe nanorods are encased in protective layers of zinc sulfide and of silica. Hydrophilic chemical groups on the outside allow proteins or other biological molecules to be attached.

Box 7.1 Trendy Terminology

Nanoparticles are referred to by a variety of nanoterms, depending on their shape and structure. The meanings of nanorod, nanocrystal, nanoshell, nanotube, nanowire, and so forth, should be obvious enough. And, despite what you might think, quantum dots are not a new brand of frozen snack but an alternative name for fluorescent nanocrystals, small enough to show quantum confinement and used in biological labeling.

reassembled in different locations. **Nanorods** can be used to follow this remodeling by attaching them to the tubulin monomers. On addition of guanosine triphosphate (GTP), assembly of microtubules is stimulated and the fluorescent nanorods can be seen aggregating into linear structures.

Why use a complex multilayered nanostructure instead of a simple fluorescent dye?

(a) Although nanocrystals have narrow emission peaks, they have broad absorption peaks (rather than narrow ones like typical dyes). Consequently they do not bleach during excitation and can therefore be used for continuous long-term irradiation and monitoring.
(b) Nanocrystals have high brightness—the product of molar absorptivity and quantum yield. (**Molar absorptivity** is the absorbance of a one molar solution of pure solute at a given wavelength; the higher it is, the more light is absorbed. The **quantum yield** is the ratio of photons absorbed to photons emitted during fluorescence.)
(c) The emission maximum of a nanocrystal depends on the size and so can be set to any desired wavelength by making crystals of the appropriate size (see later discussion).

Nanoparticles can also be targeted to specific tissues, such as cancer cells, by adding appropriate antibodies or receptor proteins to the nanoparticle surface. Fluorescent nanoparticles are often known as quantum dots and are now commercially available for a wide range of biological labeling. Although fluorescent dyes can be attached to other molecules, nanoparticles are more versatile in this regard. Quantum dots can be used to label DNA molecules as well as proteins. Thus labeling of PCR primers with quantum dots results in fluorescently labeled PCR products—a variant referred to as *quantum dot PCR*.

A variety of materials have been used to give better contrast enhancement in MRI. Nanoparticles containing a variety of materials are beginning to see increasing use in this area. For example, superparamagnetic iron oxide (SPIO) nanoparticles act as good MRI contrast agents. Their magnetic properties vary with particle size. Larger particles, of greater than 300 nm, are used for bowel, liver, and spleen. Smaller particles, of 20 to 40 nm, have shown higher diagnostic accuracy for detecting early tumors in lymph nodes than conventional materials.

Fluorescent nanoparticles are widely used in biological labeling. They last longer than traditional fluorescent dyes and are often brighter.

QUANTUM SIZE EFFECT AND NANOCRYSTAL COLORS

When materials are subdivided into sufficiently small fragments, quantum effects begin to influence their physical properties. The fluorescent nanoparticles discussed earlier are in fact **semiconductors** that are small enough to show such quantum effects.

Semiconductors are substances that conduct electricity under some conditions but not others. In *N-type* semiconductors (as in normal electric wires) the current consists of negatively charged electrons. In *P-type* semiconductors the current consists of **holes**. A hole is the absence of an electron from an atom. Although not physical particles, holes can move from atom to atom. Electrons and holes may combine and cancel out, a process that releases energy. Conversely, energy absorbed by certain semiconductors may generate an electron-hole pair whose two components may then move off in different directions.

Nanoparticle labels can be made with different emission wavelengths, covering the UV, visible spectrum, and near infrared. Emission wavelengths obviously vary depending on the semiconductor material. However, in addition, the quantum size effect (Fig. 7.10) allows the same semiconductor to emit at different wavelengths, depending on the size of the nanoparticle. The smaller the nanoparticle the shorter the wavelength (i.e., the higher the energy).

Fluorescent nanoparticles may be regarded as miniaturized light-emitting diodes (LEDs). These are semiconductors that work by absorbing energy (either electrical or light) and creating electron-hole pairs. When the electrons and holes recombine, light is emitted. For bulk material, the energy, and hence the wavelength, of the emitted light depends on the chemical composition of the semiconductor. However, at nanoscale dimensions, quantum effects become significant. If the physical size of the semiconductor is smaller than the natural radius (the **Bohr radius**) of the electron-hole pair, extra energy is needed to confine the electron-hole pair. This is referred to as **quantum confinement** and occurs with nanocrystals of around 20 nm or less. The smaller the semiconductor crystal, the more energy is needed and the more energetic (shorter in wavelength) is the light released.

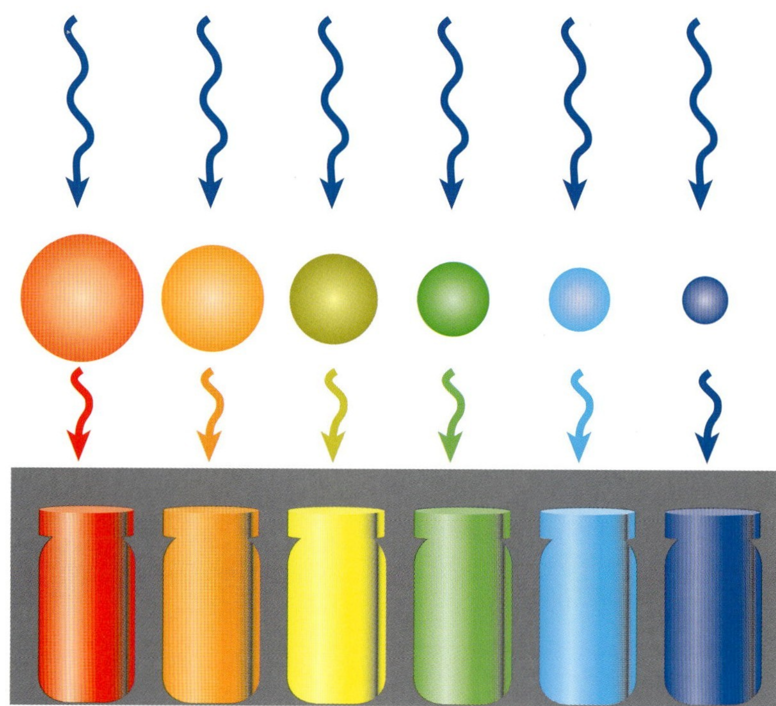

FIGURE 7.10
Quantum Size Effect
Nanocrystals of different sizes absorb UV light and re-emit the energy. The wavelength of emission depends on the size of the nanocrystal. The smaller the crystal the more energetic the emission. From: Riegler, Nann (2004). Application of luminescent nanocrystals as labels for biological molecules. *Anal Bioanal Chem* **379**(7–8), 913–919. With kind permission from Springer Science and Business Media.

The emission wavelength of a fluorescent nanoparticle depends on its size and may therefore be easily modified by the experimenter.

NANOPARTICLES FOR DELIVERY OF DRUGS, DNA, OR RNA

Because nanoparticles can be targeted to specific tissues, they can be used to deliver a variety of biologically active molecules, including both pharmaceuticals and genetic engineering constructs.

Large polymeric molecules such as DNA may themselves be compacted to form nanoparticles of around 50 to 200 nm in size. This involves addition of positively charged molecules (e.g., cationic lipids, polylysine) to neutralize the negative charge of the phosphate groups of the nucleic acid backbone. Other molecules may be added to promote selectivity for certain cells or tissues.

Alternatively, hollow nanoparticles **(nanoshells)** may obviously be used to carry other, smaller molecules. Such nanoshells must be made from biocompatible materials such as polyethyleneimine (PEI) or chitosan. The latter alternative seems popular at present, because it is both naturally derived and biodegradable. Chitin is a beta-1,4-linked polymer of *N*-acetyl-D-glucosamine. It is found in the cell walls of insects and fungi and among biopolymers is second only in natural abundance to cellulose. Chitosan is derived from chitin by removing most of the acetyl groups by alkali treatment.

An interesting approach that combines two trendy technologies is using nanoshells to carry siRNA (short interfering RNA). Delivery of siRNA triggers RNA interference, which results in the destruction of target mRNA (see Chapter 5). The siRNA may be targeted against mRNA from genes expressed preferentially in cancer cells or genes characteristic of certain viruses.

Hollow nanoparticles may be used to deliver DNA, RNA, or proteins.

NANOPARTICLES IN CANCER THERAPY

It is possible to destroy tumor cells by a variety of toxic chemicals or localized heating. In both cases a major issue is delivering the fatal reagent to the cancer cells and avoiding nearby healthy tissue. When using toxic chemical reagents, the reagent must be not only delivered specifically to the target cells but also prevented from diffusing out of the cancer cells. Both related objectives may be achieved by using hollow nanoparticles to carry the reagent. Nanoparticles may be targeted to tumors by adding specific receptors or reactive groups to the outside of the nanoparticles. These are chosen to recognize proteins that are solely or predominantly displayed on the surface of cancer cells. It is hoped that such nanoparticles will be safe to give by mouth.

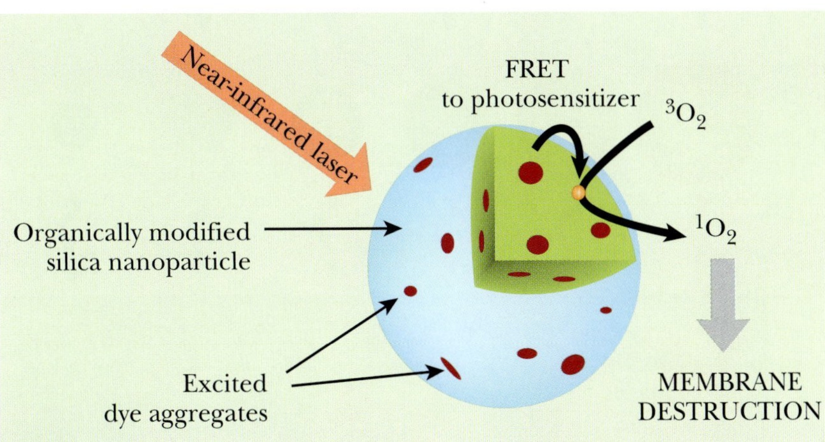

FIGURE 7.11
Nanoparticle for Singlet Oxygen Release
The near-infrared laser excites the dyes attached to the nanoparticle. Energy transfer to photosensitizers by FRET (fluorescent resonance energy transfer) results in conversion of normal (triplet) oxygen to singlet oxygen.

Diffusion is more difficult to deal with, but may be limited to some extent by designing nanoparticles for slow release of the reagent. A clever alternative is to produce the toxic agent inside the nanoparticle after it has entered the cancer cell. Photodynamic cancer therapy involves generating singlet oxygen by using a laser to irradiate a photosensitive dye. The singlet oxygen is highly reactive and in particular destroys biological membranes via oxidation of lipids. After diffusing out of the nanoparticle, the toxic oxygen reacts so fast that it never leaves the cancer cell (Fig. 7.11).

Nanoparticles may also be used to kill cancer cells by localized heating. In one approach nanoparticles with a magnetic core are used. An alternating magnetic field is used to supply energy and heats the nanoparticle to a temperature lethal to mammalian cells. Another approach uses metal nanoshells. These consist of a core, often silica, surrounded by a thin metal layer, such as gold. Varying the size of the core and thickness of the metal layer allows such nanoparticles to be tuned to absorb from any region of the spectrum from UV through the visible to the IR. Because living tissue absorbs least in the near infrared, the nanoparticles are designed to absorb radiant energy in this region of the spectrum. This results in external near infrared being specifically absorbed and heating the surrounding tissue.

> Nanoparticles may be used to kill cancer cells by localized heating or local generation of a toxic product.

Box 7.2 The Nanobacteria—Nanotechnology Meets Nanomythology

Cryptozoology is the study of "undiscovered" creatures such as the Loch Ness monster or Bigfoot. However, students of microbiology no longer need to feel left out. The new field of nanocryptobiology is here. It is perhaps not surprising that some investigators claimed to have discovered "nanobacteria." These were supposedly 100-fold smaller than typical bacteria, yet capable of growth and replication. They were proposed as causative agents in the formation of kidney stones and then linked to heart disease and cancer. Unfortunately, "nanobacteria" are too small to contain ribosomes or chromosomes, and it has become clear that they are mineral artifacts. Their supposed replication was due to the fact that certain minerals can act as nuclei for further crystallization. It scarcely needs adding that "fossilized nanobacteria" have also been seen in meteorites from Mars and have been claimed as evidence for life on Mars. However, similar mineral structures have been found in both lunar meteorites and terrestrial rocks.

ASSEMBLY OF NANOCRYSTALS BY MICROORGANISMS

It has been known for many years that bacteria may accumulate a variety of metallic elements and may modify them chemically, usually by oxidation or reduction. For example, many bacteria accumulate anions of selenium or tellurium and reduce them to elemental selenium or tellurium, which is then deposited as a precipitate either on the cell surface or internally. Certain species of the bacterium *Pseudomonas* that live in metal-contaminated areas and the fungus *Verticillium* can both generate silver nanocrystals.

Recently, it has been found that when *Escherichia coli* is exposed to cadmium chloride and sodium sulfide, it precipitates cadmium sulfide as particles in the 2- to 5-nm size range. In other words, bacteria can "biosynthesize" semiconductor nanocrystals.

Rather more sophisticated is the use of phage display to select peptides capable of organizing semiconductor nanowires. As described in Chapter 9, phage display is a technique that allows the selection of peptides that bind any chosen target molecule. In brief, stretches of DNA encoding a library of peptide sequences are engineered into the gene for a bacteriophage coat protein. The extra sequences are attached at either the C terminus or N terminus, where they do not disrupt normal functioning of the coat protein. When the hybrid protein is assembled into the phage capsid, the inserted peptides are displayed on the outside of the phage particle. The library of phages is then screened against a target molecule. Those phages that bind the target are kept.

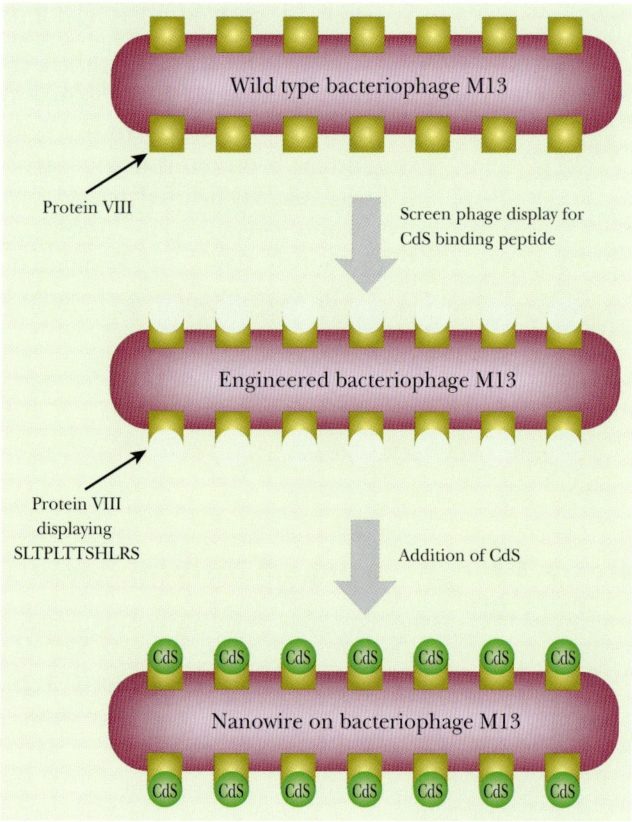

FIGURE 7.12
Nanowire Assembly on Bacteriophage
Phage display yielded engineered versions of the M13 coat protein (protein VIII) with inserted peptides. Some of these are capable of binding CdS. In the presence of CdS crystals a nanowire forms along the surface of the bacteriophage.

Phage display libraries have been screened to find peptides capable of binding ZnS or CdS nanocrystals. Protein VIII of bacteriophage M13 was used for peptide insertion. For example, ZnS was bound by the peptide VISNHAGSSRRL and CdS on the peptide SLTPLTTSHLRS. Because the bacteriophage capsid contains many copies of the coat protein, the displayed peptide is also present in many copies. Consequently an array of nanocrystals forms on the phage surface. Because M13 is a filamentous phage, the result is a semiconductor nanowire (Fig. 7.12).

Nanocrystals and nanowires may be assembled using unmodified bacteria or sophisticated phage display techniques.

NANOTUBES

Carbon **nanotubes** are cylinders made of pure carbon with diameters of 1 to 50 nanometers. However, they may be up to approximately 10 micrometers long. Pure elemental carbon exists as diamond or graphite. In diamond, each carbon is covalently linked to four others forming a 3-D tetrahedral lattice that is extremely strong. In contrast, graphite consists of flat sheets of carbon atoms that form a hexagonal pattern. In the sheets of graphite, each carbon atom is covalently bonded to three neighbors and the sheets can slide sideways over each other, because there are no covalent linkages between atoms in different sheets.

To form a nanotube, a single sheet of graphite is rolled into a cylinder. The sheets may be rolled up straight or at an angle to the carbon lattice and may be of various diameters. Depending on the diameter and the torsion, the nanotube may act as a metallic conductor or a semiconductor. Not surprisingly, nanotubes are now finding many uses in electronics, a topic beyond the scope of this book.

In biotechnology, nanotubes are just beginning to find applications. The critical issue is attaching other biomolecules, such as enzymes, hormone receptors, or antibodies, to the nanotube surface. One idea is to build biosensors where interaction of, say, an antibody with its target antigen will change the electrical behavior of the nanotube. Hence it should become possible to generate an electrical signal on detecting hormones, pathogens, pollutants, and so forth.

A major problem in attaching proteins is that the surface of carbon nanotubes is hydrophobic. One way to attach biomolecules is to first modify the surface by adding nonionic detergents, such as Triton X100. The hydrophobic portion of the detergent binds to the nanotube surface and the hydrophilic region can be used for binding of proteins. Alternatively, chemical reagents may be used that react with the carbon surface of the nanotube and generate side chains carrying reactive functional groups. Proteins can then be linked covalently by reaction with these (Fig. 7.13). Creation of devices by combining biological molecules with nanotubes is still in its infancy, but progress in this area is likely to be rapid.

Hollow nanotubes may be fabricated to carry a variety of biologically useful side chains.

Box 7.3 Magnetosomes: Natural Bacterial Magnetic Nanoparticles

Naturally occurring magnetic nanoparticles are made by magnetotactic bacteria, such as *Magnetospirillum*. These microorganisms can detect magnetic fields and orient themselves in response. They contain **magnetosomes**, consisting of nanosized crystals of magnetic iron oxide (magnetite, Fe_3O_4) or, less often, iron sulfide (greigite, Fe_3S_4) inside an envelope of protein. The magnetosomes are aligned in chains along the cell axis. Synthesis of the protein shell and mineralization of the magnetic core are under genetic control. At least in some cases, the genes responsible for the magnetosome are clustered on the bacterial chromosome.

It is possible to attach other molecules to the outside of magnetosomes by genetically modifying proteins of the magnetosome envelope (Fig. A). The gene for the Mms16 protein of *Magnetospirillum magneticum* has been fused to the genes for luciferase and the dopamine receptor in the lab of Dr. Tadashi Matsunaga of the Tokyo University of Agriculture and Technology. The fused proteins are displayed on the surface of the magnetosomes. After disrupting the bacterial cells, the magnetosomes carrying the attached proteins can be purified by magnetic separation. This should allow easier analysis of membrane-bound receptors, such as those from the human nervous system, in a simplified system.

A) Foreign protein (hydrophilic)

Anchor protein (transmembrane)

B) Anchor protein (anchoring membrane)

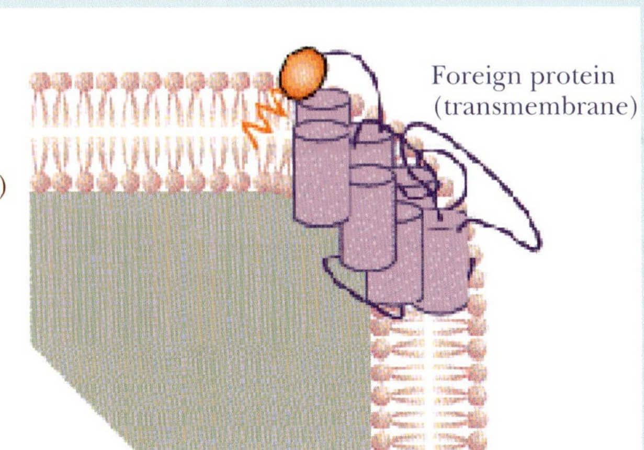

FIGURE A Protein Display on Magnetosome Membrane
(A) Display of hydrophilic protein using MagA as an anchor. (B) Display of transmembrane protein using Mms16 as an anchor. From Matsunaga and Okamura (2003). Genes and proteins involved in bacterial magnetic particle formation. *Trends Microbiol* **11**, 536–542. Reprinted with permission.

ANTIBACTERIAL NANOCARPETS

Nanocarpets are formed by stacking a large number of nanotubes together, with their cylindrical axes aligned vertically. Nanocarpets capable of changing color and of killing bacteria have been assembled from specially designed lipids that spontaneously assemble into a variety of nanostructures depending on the conditions. In water, nanotubes are formed. Partial rehydration of dried nanotubes generates a side-by-side array—the nanocarpet.

The lipid consists of a long hydrocarbon chain (25 carbons) with a diacetylenic group in the middle of the chain. The individual nanotubes are about 100 nm in diameter by 1000 nm in length. The walls of the nanotubes consist of five bilayers of the lipid. Both the separate lipid molecules and the assembled nanocarpet kill bacteria. Like other long-chain amino compounds, they act as detergent molecules and disrupt the cell membrane. Consequently, the nanocarpet provides a surface lethal to bacteria. This property could be very useful if nanocarpets are used in biomedical applications.

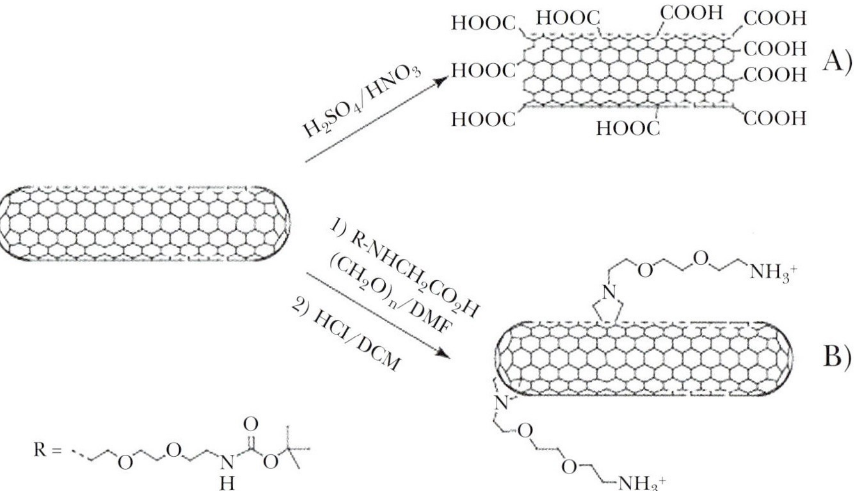

FIGURE 7.13
Attaching Organic Functional Groups to Nanotubes
(A) Carbon nanotubes can be treated with acids to purify them and generate carboxylic groups.
(B) Alternatively, they may react with amino acid derivatives and aldehydes to add more complex hydrophilic groups to the external surface. From: Bianco, Kostarelos, and Prato (2005). Applications of carbon nanotubes in drug delivery. *Curr Opin Chem Biol* **9**, 674–679. Reprinted with permission.

Diacetylenic compounds have the interesting ability to change color. The nanocarpet starts out white, but if exposed to ultraviolet light, it turns deep blue. UV irradiation causes crosslinks to form by reaction between acetylenic groups on neighboring molecules. This polymerization stabilizes the nanocarpet. Blue nanocarpets change color on exposure to a variety of reagents. Detergents and acids change them from blue to red or yellow, and the presence of bacteria, such as *E. coli*, gives red and pink shades. Eventually such materials may be used both as biosensors and for protection against bacterial contamination.

Nanotubes may be assembled to create surfaces (nanocarpets) that are antibacterial or act as biosensors.

DETECTION OF VIRUSES BY NANOWIRES

Nanowires are what their name suggests. They have nanoscale diameters but may be several microns long. They may be metallic and act as electrical conductors or they may be made from semiconductor materials.

Biosensors can be made using silicon semiconductor nanowires. These may be coated with antibodies that bind to a specific virus. Binding of the virus to the antibody triggers a change in conductance of the nanowire. For a p-type silicon nanowire, the conductance decreases when the surface charge on the virus particle is positive and, conversely, increases if the virus surface is negative. Single viruses may be detected by this approach (Fig. 7.14). It is also possible to attach single-stranded DNA to the nanowire. In this case, conductance changes are triggered by binding of the complementary single strand. Possible future applications include both clinical testing and sensors for monitoring food, water, and air for public health and/or biodefense.

Nanowire sensors are capable of detecting specific individual viruses. Binding of a virus particle changes the conductance of the nanowire.

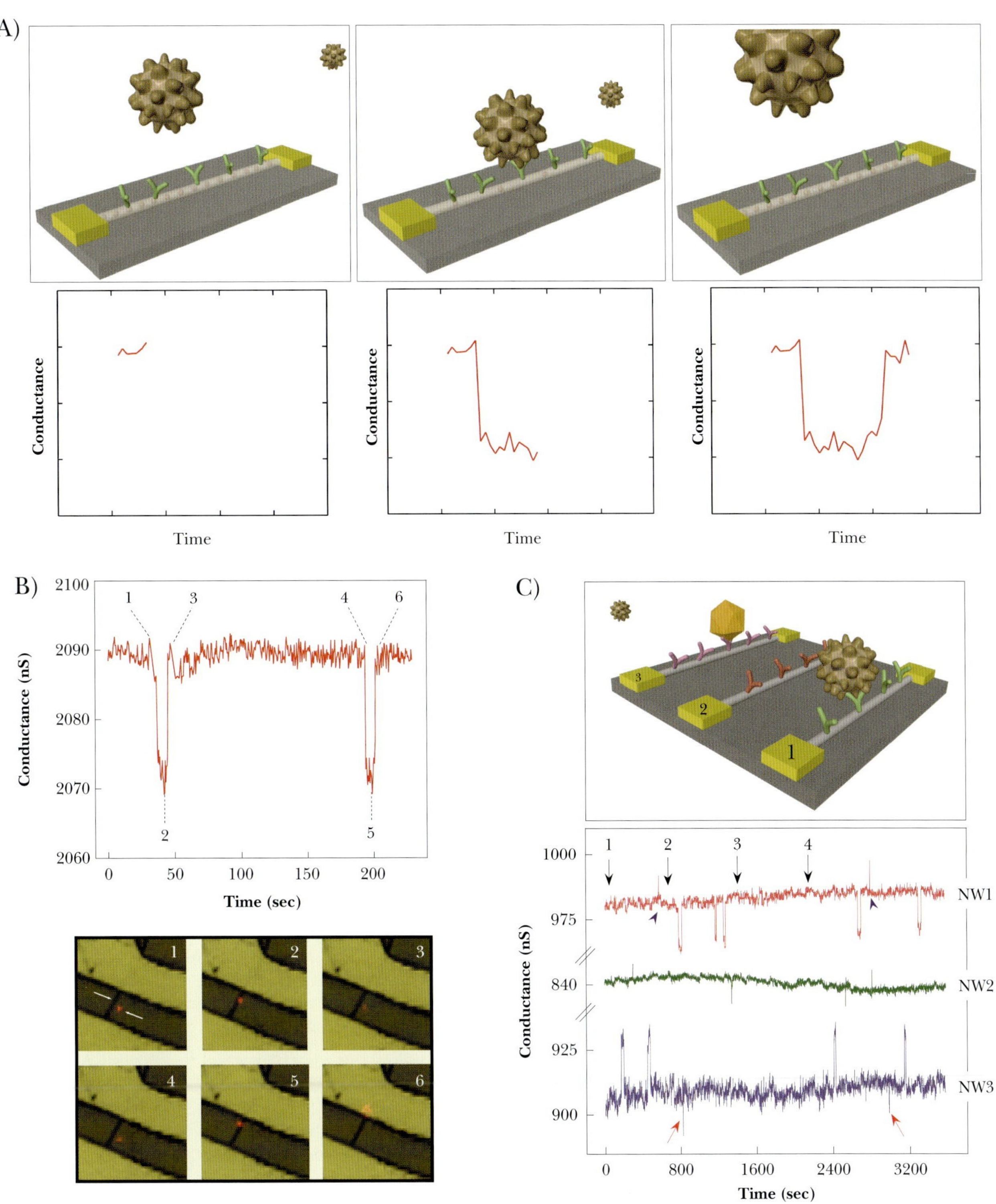

FIGURE 7.14 Nanowire Biosensors

(A) A single virus particle binding to and detaching from the surface of a SiNW nanowire coated with antibody receptors. The corresponding changes in conductance are shown for each step. (B) Conductance and optical data on addition of influenza A virus. (C) Schematic of multiplexed single virus detection. Conductance versus time is recorded simultaneously for three channels specific for different viruses. NW1 responds to influenza A and NW3 to adenovirus. The NW2 channel was not used here. Black arrows 1–4 correspond to addition of adenovirus, influenza A, pure buffer, and a 1:1 mixture of adenovirus plus influenza A. Red and blue arrows indicate conductance changes due to the diffusion of viral particles past the nanowire without specific binding for influenza and adenovirus, respectively. Courtesy of Charles M. Lieber, Harvard University, Cambridge, MA.

ION CHANNEL NANOSENSORS

Somewhat more complex than nanotubes and nanowires are **nanoscale ion channels** that are assembled into membranes. These channels are designed so that they can be controlled to permit the movement of ions under only certain conditions. The ion flow generates an electrical current that is detected, amplified, and displayed by appropriate electronic apparatus.

Ion channels can be used as biosensors by attaching a binding site for the target molecule at the entry to the channel. Attached antibodies are often used for the binding sites as described in Chapter 23 (see Fig. 23.16). The simplest arrangement results in the channel being open in the absence of the target molecule and shut when it is detected. A drop in ion flow therefore signals detection of the target molecule.

At present, such ion channels are being developed using modified biological components. The ion channel itself can be made using the peptide antibiotic gramicidin A (made by the bacterium *Bacillus brevis*). This transports monovalent cations, especially protons and sodium ions. Natural gramicidin spans half of a standard biological membrane. A short-lived channel is formed when two gramicidin molecules line up, as shown in Fig. 7.15. Permanent channels may be made by covalently linking two gramicidin molecules together. Up to 10^7 ions/second flow through a single gramicidin channel. This gives a picoampere current that is easily measured. An alternative is to monitor a change in the pH due to movement of H^+ ions. This may be done by using an optical sensor and a fluorescent pH indicator.

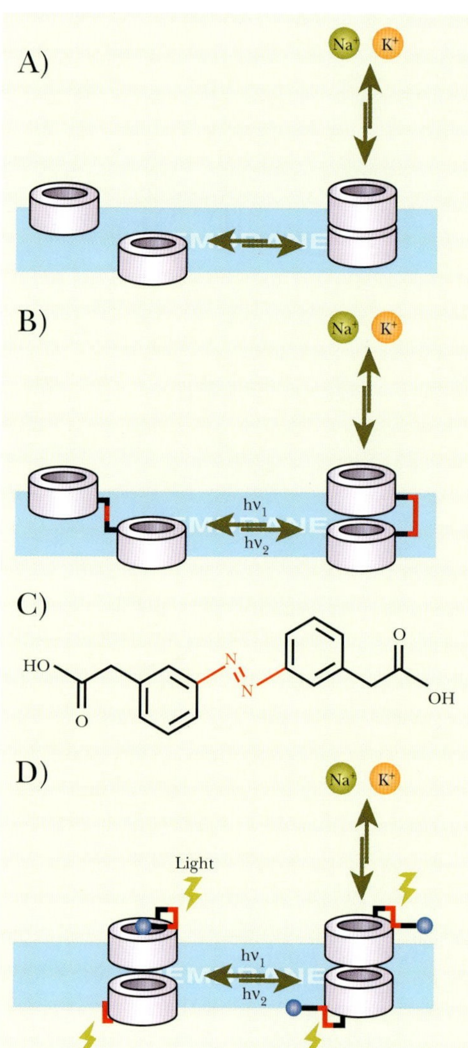

FIGURE 7.15
Natural and Modified Gramicidin Ion Channels
Gramicidin forms a transchannel for Na^+ and K^+ ions. (A) Natural gramicidin channels are formed when two gramicidin molecules align within a membrane. (B) Two gramicidin monomers joined by a photosensitive linker (C). Absorption of light changes the conformation of the N=N bond (red) of the linker from *cis* to *trans* and opens or closes the channel. (D) Channels are opened or closed by blocking groups (blue circles) attached by photosensitive linkers to each gramicidin monomer.

The channels are made responsive by attaching an appropriate ligand molecule to the front end of the gramicidin, so that it projects outward from the membrane surface. This ligand is chosen to bind the target molecule and may be an antigen or other small molecule that is recognized by an antibody or protein receptor. It is also possible to attach a single-stranded segment of DNA that will recognize and bind the complementary sequence. Thus biosensors may be designed to respond to the presence of a variety of biological molecules.

The membrane itself may be a lipid bilayer made using natural membrane lipids. Typical phospholipids span half a membrane (i.e., one monolayer), and the two monolayers can therefore slide relative to one another. The membrane may be stabilized by including lipids that span the whole membrane. These may be found naturally in certain archaebacteria or they may be synthesized artificially. Lipid bilayers are relatively fragile and in practice must be assembled on some solid support. Building a long-lasting and stable membrane structure has so far proven difficult, and practical ion channel sensors are still under development.

> Ion channel sensors operate by opening or closing the channel in response to binding a specific molecule. They may be used to detect a variety of target molecules.

NANOENGINEERING OF DNA

In "classical" genetic engineering, the sequence of DNA is deliberately altered in order to generate new combinations of genetic information. Even when major rearrangements are

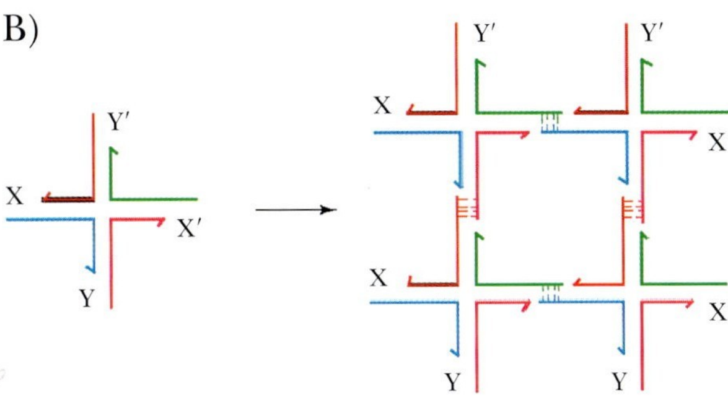

FIGURE 7.16
Branched DNA from Four Single Strands
(A) A branched DNA molecule with four arms. Four different color-coded strands combine to produce four arms (I, II, III, and IV). The branch point of this molecule is fixed.
(B) Formation of a two-dimensional lattice from a four-arm junction with sticky ends. X and Y are sticky ends, and X′ and Y′ are their complements. Four monomers are complexed in parallel orientation to yield the lattice structure. DNA ligase can seal the nicks left in the lattice. From: Seeman (1999). DNA engineering and its application to nanotechnology. *Trends Biotechnol* **17**, 437. Reprinted with permission.

made, in order to function as genetic information, the DNA must remain as a base-paired double-stranded helix with an overall linear structure.

In nanoengineering, the objective is to make structures using DNA merely as a structural element, rather than to manipulate genetic information. DNA is attractive because the double helix is a convenient structural module. Moreover, its natural base-pairing properties can be used to link separate DNA molecules together. However, a critical requirement for assembling 3D structures is branched DNA. Although branched structures do form in biological situations (especially the Holliday junction involved in crossing over during recombination), they are not permanent or stable.

Cross-shaped DNA can be generated by mixing four carefully designed single strands with different sequences. Each strand base-pairs with two of the other strands over half its length (Fig. 7.16). If sticky ends are included in the initial strands, it is possible to link the crosses together into a two-dimensional matrix. The nicks can be sealed by DNA ligase if desired. The principles used in branching can be extended to three dimensions, and it is possible to build cubical DNA lattices.

The DNA double helix is about 2 nm wide with a helical pitch of about 3.5 nm. Hence it can be used to build nanoscale frameworks. These can be used for the assembly of other components, such as metallic nanowires or nanocircuits. Note that while DNA is flexible over longer distances, it is relatively rigid over nanoscale lengths, up to about 50 nm.

Such cross-shaped DNA molecules (and the 3D counterparts) have the drawback that the junctions are flexible and do not maintain rigid 90-degree angles. Rigid DNA components have been made by using double crossover (DX) DNA molecules. Two isomers of antiparallel DX DNA exist, with an odd (DAO) or even (DAE) number of half-turns between the crossover points (Fig. 7.17). (Double crossover molecules with parallel strands also exist, but behave poorly from a structural viewpoint.) DAE or DAO units can be assembled into a rigid array by providing appropriate sticky ends. It is possible to replace the central, short DNA strand of the DAE structure with a longer protruding strand of DNA (DAE+J). This allows assembly of branched structures.

What is the purpose of building arrays and 3D structures from DNA? Perhaps the most plausible use suggested so far is to use the DNA as the framework for assembling nanoscale electronic circuits. So far, normal, unbranched DNA has been used as a scaffold to create linear metallic nanowires. Various metals (gold, silver, copper, palladium, platinum) have been used to coat the DNA, and diameters range from 100 nm down to 3 nm. Eventually, it should be possible to put together these two approaches and build circuits from metal-covered 3D DNA structures.

DNA may be viewed solely as a structural molecule. Three-dimensional frameworks may be built from DNA whose sequence is designed to generate branched structures. Such DNA structures may be used as nanoscale scaffolds for metallic nanowires and circuits.

DNA MECHANICAL NANODEVICES

A rather more futuristic use for 3-D DNA structures is as frameworks for mechanical nanodevices. The essential components are some sort of moving parts. Several experimental prototype "DNA machines" have been designed or constructed that illustrate the concept. These all involve reversible changes in conformation of a DNA structure driven by changes in base-pairing. Such changes may be caused either by changing the physical conditions (heat, salt, etc.) or by adding segments of single-stranded DNA (ssDNA) that base-pair to some region of the DNA machine, as illustrated in Fig. 7.18. If ssDNA is used, then another single strand, complementary to the first, is added to convert the machine back to its original conformation. The result is a mechanical cycle that could in principle be used to perform some sort of task. The two ssDNA molecules may be regarded as "fuel," and the final waste product is a double-stranded DNA consisting of the two paired ssDNA fuel elements. (Note that this scheme does not involve breaking covalent chemical bonds. It is thus not an enzymatic reaction and is distinct from the use of DNA as a deoxyribozyme as described in Chapter 5.)

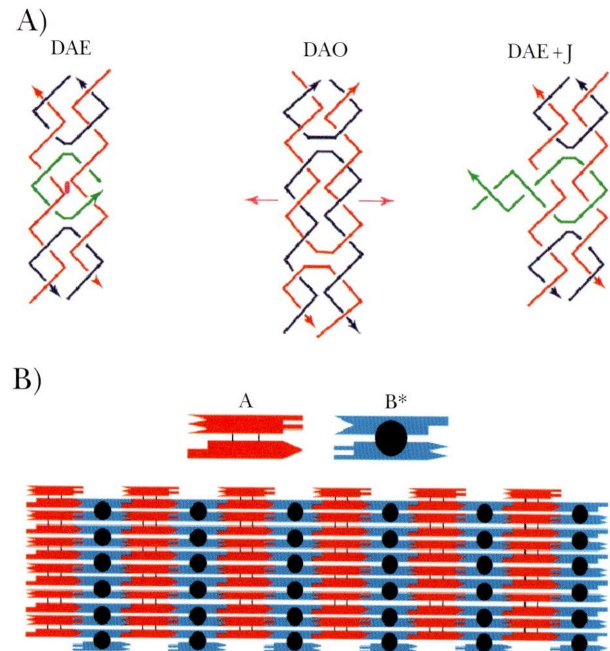

FIGURE 7.17 Rigid DNA Nanomodules
Arrays may be assembled from double-crossover (DX) molecules. (A) DAE and DAO are two anti-parallel DX isomers. DAE+J is a DAE molecule in which an extra junction replaces the nick in the green strand of DAE. (B) Two-dimensional array derived from DX molecules. Complementary sticky ends are depicted by complementary geometrical shapes. A is a conventional DX molecule but B* is a DX+J molecule with a vertically protruding DNA hairpin (black circle). From: Seeman (1999). DNA engineering and its application to nanotechnology. *Trends Biotechnol* **17**, 437. Reprinted with permission.

DNA has been proposed as a framework for nanomachines. Proof-of-concept prototypes have been constructed.

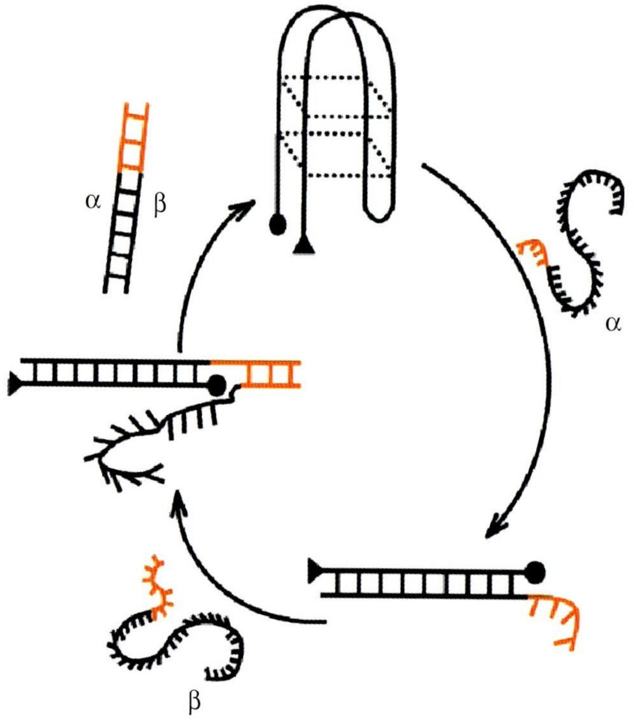

FIGURE 7.18 Prototype DNA Machine
A DNA nanomotor designed by J. J. Li and W. Tan. Successive addition of the complementary DNA strands labeled alpha and beta causes a change in conformation. The DNA nanomotor alternates between a folded quadruplex structure and a double-stranded structure. The nanomotor expands and contracts in a wormlike motion. From: Ito and Fukusaki (2004). DNA as a nanomaterial. *J Mol Catalysis B: Enzymatic* **28**, 155–166. Reprinted with permission.

CONTROLLED DENATURATION OF DNA BY GOLD NANOPARTICLES

DNA hybridization is widely used to detect target sequences, both in the laboratory and in clinical diagnosis. Before hybridization can occur, the DNA double helix must be denatured into single strands. This is accomplished by the heating of bulk DNA. However, newly emerging nanotechnology may allow specific individual DNA molecules to be dissociated when required.

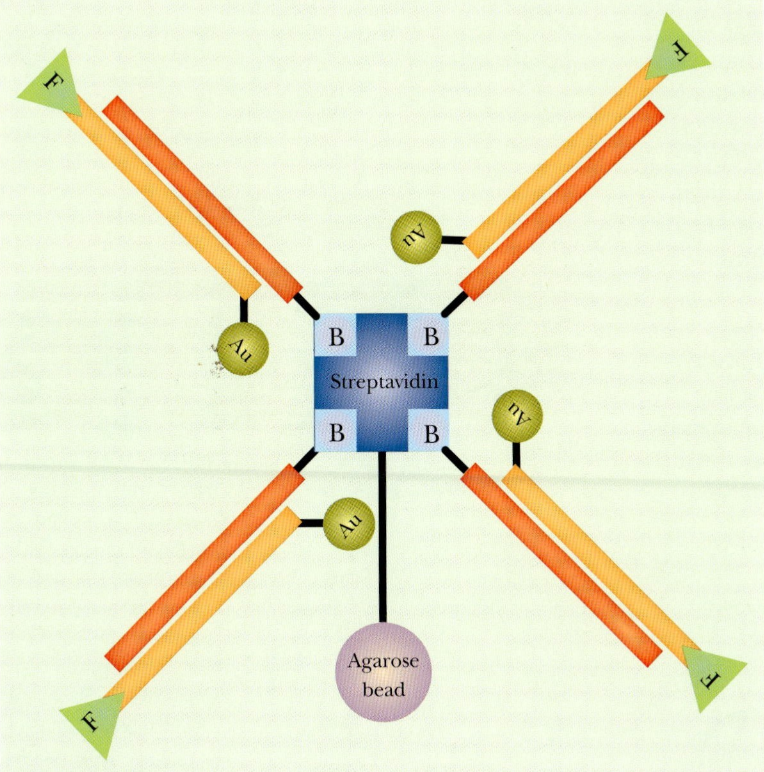

FIGURE 7.19
Controlling DNA Denaturation by Gold Nanoparticles
One strand of DNA has a gold nanoparticle (Au) attached to one end and a fluorescent dye (F) at the other end. The complementary strand has a biotin tag (B) at one end. The biotin is bound by streptavidin and therefore binds the DNA strand to an agarose bead. When the gold absorbs energy, it melts the two strands of DNA. The DNA strand with the fluorescent dye is released into the supernatant and its fluorescence is monitored.

Nanoparticles of about 1.4 nm and containing fewer than 100 atoms of gold are attached to double-stranded DNA. When the structure is exposed to radio waves (generated by an alternating magnetic field), the gold acts as an antenna. It absorbs energy and heats the DNA molecule to which it is attached. This melts the DNA double helix and converts it to single strands. Heating extends over a zone of about 10 nm so surrounding molecules are unaffected. The heat is dissipated in less than 50 picoseconds, so the DNA may be rapidly switched between the double- and single-stranded states by turning the magnetic field on and off. The procedure may be applied to dsDNA made of two separate single strands (Fig. 7.19) or to stem-and-loop structures formed by folding from a single strand of DNA.

Practical applications are several years away. However, because radio waves penetrate living tissue very effectively, it may eventually be possible to control the behavior of individual DNA molecules from outside an organism. Metal antennas of different materials or sizes could be used to tune different DNA molecules to radio waves of different frequencies.

> Attachment of a metallic antenna allows DNA to be melted into single strands by radio waves. It might eventually be possible to control the behavior of DNA from outside an organism.

CONTROLLED CHANGE OF PROTEIN SHAPE BY DNA

Allosteric proteins change shape in response to the binding of signal molecules (allosteric effectors) at a specific site. The essence of allosteric control is that the shape change is transmitted through the protein and affects the conformation of distant regions of the protein. In allosteric enzymes, binding of an allosteric effector at a distant site alters the conformation of the active site and may change its affinity for the substrate. In this way, some enzymes are switched on and off in response to signal molecules. For example, phosphofructokinase is switched on by the buildup of AMP, which signals that energy is in short supply. The response increases flow into the glycolytic pathway. Similarly, many DNA binding proteins, such as repressors and activators, also change shape on binding small signal molecules.

It is possible to change the shape of a protein artificially by mechanical force. This has been demonstrated by attaching a single-stranded 60-base segment of DNA between the two poles of a protein. Attaching the DNA requires chemical "handles." These are engineered into the

A)

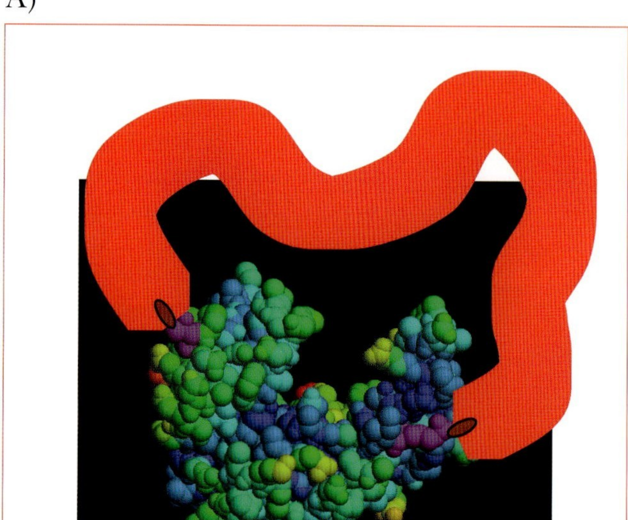

B)

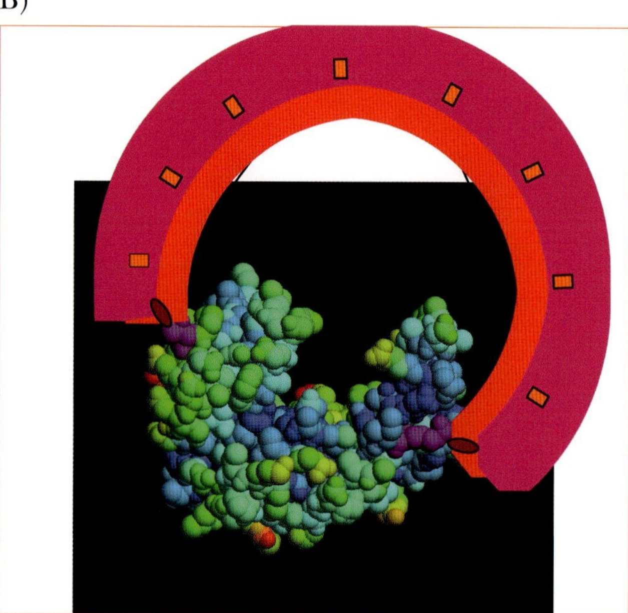

FIGURE 7.20 Controlling Protein Shape by DNA

Protein-ssDNA chimera for guanylate kinase from *Mycobacterium tuberculosis* (PDB structure 1S4Q). The purple attachment points for the molecular spring correspond to mutations Thr75 → Cys and Arg171 → Cys. (A) Unstretched—a single strand of DNA is attached to the protein. (B) The protein is stretched by addition of the complementary DNA strand (purple). Courtesy of Giovanni Zocchi.

target protein by replacing amino acids at appropriate positions with cysteine. The reactive SH group is then used to chemically attach the DNA.

Double-helical DNA is much more rigid than ssDNA. Consequently, the addition of the complementary strand generates tension as it binds and creates a double helix. This approach has so far been demonstrated in the laboratory of Giovanni Zocchi at UCLA with maltose binding protein and an enzyme, guanylate kinase. When maltose binding protein is stretched, its binding site for maltose opens wider than optimum and the affinity for the sugar decreases. For guanylate kinase (Fig. 7.20), applying tension decreases enzyme activity by lowering affinity for substrate binding. In this case releasing the tension by adding DNase to digest the DNA switches the enzyme on again.

Potential applications are a long way in the future. However, it is possible to imagine biosensors that detect DNA sequences based on this mechanism. In addition, it might be possible to externally control enzymes or other proteins by adding appropriate ssDNA (or, of course, RNA).

The shape of a protein may be changed artificially by applying force. This may be demonstrated by attaching DNA strands to the protein. When single-stranded DNA pairs with its complementary strand, this generates tension and stretches the protein.

BIOMOLECULAR MOTORS

A major aim of nanotechnology is to develop molecular-scale machinery that can carry out the programmed synthesis (or rearrangement) of single molecules (or even atoms), or other similar nanoscale tasks. The term (nano)**assembler** refers to a nanomachine that can build nanoscale structures, molecule by molecule or atom by atom. And the term (nano)**replicator** refers to a nanomachine able to build copies of itself when provided with raw materials and energy. This, of course, sounds remarkably like a living cell. Indeed, the organelles of living cells may be regarded as nanomachines

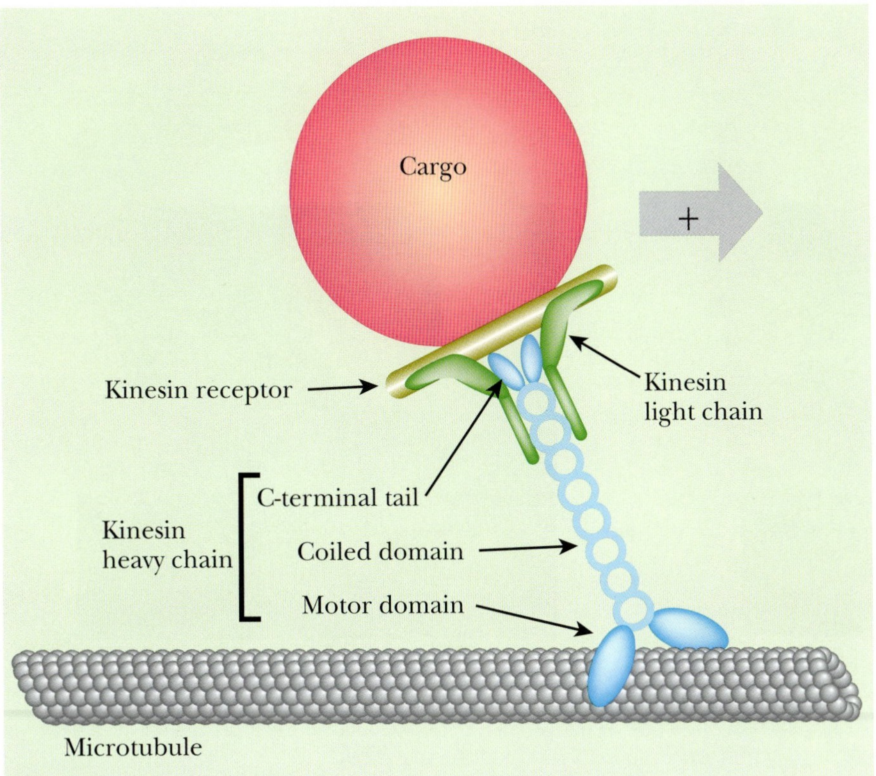

FIGURE 7.21 Kinesin Linear Motor on Microtubules
Kinesin consists of light and heavy chains. The light chains binds to kinesin receptors on vesicles that are to be transported. The heavy chains each include motor domains that use ATP as energy to move kinesin plus the attached cargo along the surface of a microtubule.

and have provided both inspiration and components for nanotechnologists.

To operate, nanomachines will need energy, which will be provided by "molecular motors." At present such devices are still in development. It has been suggested that biological structures might be used for this purpose. Examples include the ATP synthase, the flagellar motor of bacterial cells, various enzymes that move along DNA or RNA, and assorted motor proteins of eukaryotic cells. Several of these systems are presently being investigated in the hope of making usable nanodevices that can be coupled to nanomachines to provide energy and/or moving parts.

The ATP synthase is a rotary motor whose natural role is to generate ATP. It is embedded in the mitochondrial membrane and uses energy from the proton motive force. The ATP synthase takes three steps to complete each rotation, and at each step it makes an ATP. For use in nanotechnology, the F1 subunit would be detached from the membrane and run in reverse (i.e., it would be given ATP as fuel and, from a biological perspective, rotate backwards).

Kinesin and dynein are motor proteins that use ATP as energy to move along the microtubules of eukaryotic cells. They therefore act as linear step motors (Fig. 7.21). Their natural role is to transport material. Kinesin moves cargo from the center to the periphery of the cell, whereas dynein carries cargo from the periphery to the center. Kinesin takes steps of 8 nanometers and can move at 100 steps per second (approximately 3 mm/hour!). Each step consumes one ATP for energy. The microtubules they use as tracks are protein cylinders with an outside diameter of 30 nm.

Box 7.4 From Merely Micro to Truly Nano: Lab-in-a-Cell

Microfluidics (sometimes known as "lab-on-a-chip") refers to the manipulation of liquid samples at the scale of micrometers. Microfluidic devices are available today and are used to process large numbers of small samples. Applications include DNA or protein analysis of blood samples. The volumes involved are usually in the microliter range, though some microfluidics devices can use volumes less than 1 microliter, that is, nanoliter volumes. You might think that this entitles them to be regarded as nanotechnology, but remember that the dimensions of volume are the cubes of linear measure. Thus a cube with sides of 1 micrometer (10^{-6} m) has a volume of 10^{-18} cubic meters or 10^{-15} liters (one femtoliter). A nanoliter (10^{-9} liters) is the volume of a cube with sides of 100 micrometers. So handling nanoliters is *not* nanotechnology!

Future prospects are scaling down liquid sample processing to true nanoscale—"lab-in-a-cell." This would involve a microchip platform that uses modified single cells as analytical devices. This is still in the conceptual stage, but given the rapid progress in nanotechnology, it may not be so far in the future.

Proteins that interconvert chemical and mechanical energy have been suggested as possible molecular motors to power future nanomachines.

Summary

Many techniques from nanotechnology are now being applied to biological systems. Individual bacteria and viruses can be detected and weighed. Nanoparticles are already in use for biological labeling and various other analytical purposes and are being developed for clinical use. More complex nanodevices made from protein or DNA components are being assembled and their properties investigated.

End-of-Chapter Questions

1. What is nanotechnology?
   - **a.** the individual manipulation of molecules and atoms to create materials with novel or improved properties
   - **b.** the creation of new terms to describe very small, almost unimaginable, particles in physics
   - **c.** the term used to describe the size of cellular components
   - **d.** the transition of molecular biology into the physical sciences
   - **e.** none of the above
2. Which property is measured with a scanning probe microscope?
   - **a.** magnetism
   - **b.** electric resistance
   - **c.** light absorption
   - **d.** temperature
   - **e.** all of the above
3. What is considered a weakness of scanning tunneling microscopy (STM)?
   - **a.** the inability to move and arrange atoms to create a design
   - **b.** the possibility of destroying the surface with the metal tip on the microscope
   - **c.** the requirement for a conducting surface to work properly
   - **d.** the inability to apply this technology to biology
   - **e.** all of the above
4. What is an atomic force microscope?
   - **a.** The AFM detects the force between molecular bonds in an object.
   - **b.** The AFM detects atoms or molecules by scanning the surface.
   - **c.** The AFM uses photons to predict the structures present on any surface.
   - **d.** The AFM detects atoms or molecules on a conducting surface.
   - **e.** none of the above

(Continued)

5. Which principle is utilized to weigh a single bacterial cell or virus particle?
   - **a.** Oscillation frequency is dependent upon the mass applied.
   - **b.** It is impossible to weigh a single cell or particle.
   - **c.** Oscillation frequency affects the amount of light reflection.
   - **d.** Scanning electron microscopy can identify the length and width of a cell, which can further be converted to mass.
   - **e.** none of the above

6. What is a potential use of nanoparticles in the field of biology?
   - **a.** delivery of pharmaceuticals or genetic material
   - **b.** tumor destruction
   - **c.** fluorescent labeling
   - **d.** detection of microorganisms or proteins
   - **e.** all of the above

7. What is an advantage to using a complex, multilayered nanocrystal over a fluorescent dye?
   - **a.** They do not bleach during excitation because they have broad absorption peaks.
   - **b.** Nanocystals are often brighter than fluorescent dyes.
   - **c.** The emission maximum of nanocrystals can be controlled by adjusting the size of the crystal.
   - **d.** Nanostructures are longer-lived than fluorescent dyes.
   - **e.** All of the above are advantages.

8. Why is chitin the most popular material to construct nanoshells?
   - **a.** Chitin is easy to synthesize.
   - **b.** Chitin has properties that enable it to bind strongly to DNA, RNA, and other small molecules.
   - **c.** Chitin is stable and easy to store at room temperature.
   - **d.** Chitin is naturally derived and biodegradable.
   - **e.** Chitin is easier to manipulate than the alternative for the creation of nanoshells.

9. How can nanoparticles be used to treat cancer?
   - **a.** Nanotubes can create pores in the cancer cells, thus leaking out the cellular components and killing the cell.
   - **b.** Some nanoparticles can bind to specific enzymes in cancer cell metabolism to block reactions.
   - **c.** Nanoparticles can be designed to absorb radiant energy in the IR spectrum, which produces heat that destroys only the cancer cells because living tissue does not absorb IR energy.
   - **d.** Nanoparticles can recruit immune system components directly to the cancer cells.
   - **e.** All of the above are uses.

10. What characteristic of bacteriophage M13 makes it ideal for synthesizing nanowires?
- **a.** M13 accumulates certain nanoparticle building blocks in high concentrations.
- **b.** M13 phage is easily manipulated in the laboratory to secrete peptides that nanowires can be assembled upon.
- **c.** Nanowires can be constructed directly on M13 capsid proteins without any further modifications.
- **d.** M13 is filamentous.
- **e.** Nanowires are usually created in bacterial systems, not viral systems.

11. What purpose could nanotubes serve in biotechnology?
- **a.** as a metallic conductor or semiconductor
- **b.** for the creation of components of electronic equipment
- **c.** for attachment of biomolecules, including enzymes, hormone receptors, and antibodies
- **d.** for the detection of a specific molecule in a sample, such as blood
- **e.** all of the above

12. Why do nanocarpets have antibacterial activity?
- **a.** The long-chain amino compounds in the nanocarpet act like a detergent and disrupt the cell membrane.
- **b.** The nanocarpet tubes act as spears and shear the bacterial cells.
- **c.** The nanocarpet binds to bacterial cells and blocks the uptake of nutrients.
- **d.** The nanocarpet immobilizes the bacterial cell so that the cells can be targeted by treatment with UV light.
- **e.** Nanocarpets can act as biosensors so that people can treat the area with antibacterial agents.

13. Which of the following is a structure that can be created by nanoengineering of DNA?
- **a.** cubical structures
- **b.** nanoscale scaffolds for circuits and nanowires
- **c.** frameworks for mechanical nanodevices
- **d.** cross-shaped DNA to create 2D matrices
- **e.** all of the above

14. How might the behavior of individual DNA molecules be controlled from outside the body?
- **a.** exposure to UV light
- **b.** attachment of a metallic antenna, allowing DNA to be melted with radio waves
- **c.** addition of fluorescent tags
- **d.** using an electrical current to align the DNA molecules
- **e.** none of the above

15. Which cellular component is considered to be a nano(assembler)?
- **a.** chromatin
- **b.** lipids
- **c.** ribosomes
- **d.** DNA
- **e.** mRNA

Further Reading

Bogunia-Kubik K, Sugisaka M (2002). From molecular biology to nanotechnology and nanomedicine. *BioSystems* **65**, 123–138.

Bianco A, Kostarelos K, Prato M (2005). Applications of carbon nanotubes in drug delivery. *Curr Opin Chem Biol* **9**, 674–679.

Choi B, Zocchi G, Wu Y, Chan S, Jeanne Perry L (2005). Allosteric control through mechanical tension. *Phys Rev Lett* **95**, 78–102.

Hamad-Schifferli K, Schwartz JJ, Santos AT, Zhang S, Jacobson JM (2002). Remote electronic control of DNA hybridization through inductive coupling to an attached metal nanocrystal antenna. *Nature* **415**, 152–155.

Hess H, Bachand GD, Vogel V (2004). Powering nanodevices with biomolecular motors. *Chem Eur J* **10**, 2110–2116.

Ilic B, Yang Y, Craighead HG (2004). Virus detection using nanoelectromechanical devices. *Appl Phys Lett* **85**, 27.

Li JJ, Tan W (2002). A single DNA molecule nanomotor. *Nano Lett* **2**, 315–318.

Liu D, Park SH, Reif JH, LaBean TH (2004). DNA nanotubes self-assembled from triple-crossover tiles as templates for conductive nanowires. *Proc Natl Acad Sci USA* **101**, 717–722.

Matsunaga T, Okamura Y (2003). Genes and proteins involved in bacterial magnetic particle formation. *Trends Microbiol* **11**, 536–541.

Niemeyer CM, Mirkin CA (2004). *Nanobiotechnology: Concepts, Applications and Perspectives*. Wiley-VCH.

Papazoglou ES, Parthasarathy A (2007). *BioNanotechnology (Synthesis Lectures on Biomedical Engineering)*. Morgan and Claypool Publishers, San Rafael, CA.

Patolsky F, Zheng G, Hayden O, Lakadamyali M, Zhuang X, Lieber CM (2004). Electrical detection of single viruses. *Proc Natl Acad Sci USA* **101**, 14017–14022.

Riegler J, Nann T (2004). Application of luminescent nanocrystals as labels for biological molecules. *Anal Bioanal Chem* **379**, 913–919.

Zhang S (2003). Fabrication of novel biomaterials through molecular self-assembly. *Nat Biotechnol* **21**, 1171–1178.

CHAPTER 8

Genomics and Gene Expression

INTRODUCTION

Sequencing the entire human genome was a daunting task conceived by an initiative from the Department of Energy in 1986. The goal was to have a high-quality reference set of sequence information from each human chromosome. The initiative was strengthened when the National Institutes of Health (NIH) joined the effort in 1990. During the 1990s, many other collaborators around the world joined the effort. Finally, in June 2000, the first working draft of the human genome was announced. The sequence was refined, and a final high-quality sequence was finished in April 2003. The Human Genome Project has also involved sequencing entire genomes of model organisms such as mouse and *Drosophila*, enhancing computational methods for sequence data analysis, comparing the function of genes among different organisms, and studying human variation.

The availability of genome sequences has revolutionized many areas of biology from the construction of evolutionary trees to the design and testing of pharmaceuticals. Genomics has also affected gene expression studies by providing methods to analyze thousands of genes simultaneously instead of just examining individual genes.

GENETIC MAPPING TECHNIQUES

Genomes are sequenced by making libraries of genomic DNA segments and then sequencing each of the segments. These stretches must then be compiled into the final sequence. To structure the sequence data into a draft genome, the Human Genome Project started by compiling a working genome map. **Genome maps** provide various landmarks for use when putting together sequence data. There are two different categories for genome maps, **genetic maps** and **physical maps**. Genetic maps are based on the relative order of genetic markers, but the actual distance between the markers is hard to determine. Physical maps are more precise and give the distance between markers in base pairs.

Traditional genetic maps are based on the recombination frequency between genes. In eukaryotic cells, recombination occurs between homologous pairs of chromosomes during meiosis. If two genes are close together on the same chromosome, recombination between them will be rare. If the two genes are located far apart on the same chromosome, recombination is relatively frequent. Early genetic maps were based on measuring recombination frequencies between genes.

Genetic maps are based on landmarks called **genetic markers**. Many different types of markers can be used. The order of these markers is determined by how often the two markers are found in offspring. The most useful markers are genes, but often in large genomes, genuine genes are too few and far between to give a good map. Genes that encode specific traits are wonderful markers in organisms such as *Drosophila* because mating can be controlled and directed. After mating many different flies, the number of flies with both markers can be determined. The more often the markers appear together in the offspring, the closer these are in the genome. In humans, deliberate mating experiments are unethical. Moreover, the human genome contains only a few percent of coding DNA; thus, using real genes does not produce enough points on the map. A sparse map makes it difficult to order the sequences obtained in the genome sequencing project. Therefore, other markers, including physical markers, are also used on genomic maps.

An example of a physical marker is the RFLP, or restriction fragment length polymorphism (see Chapter 3). RFLPs are commonly used because of the ease of identification. For small genomes such as yeast, monitoring the frequency of recombination between two RFLP markers is easy. Diploid yeast cells undergo meiosis and form four haploid cells called a tetrad. Each of these haploid cells can be isolated, grown into many identical clones, and examined individually. Thus each RFLP marker can be followed easily from one generation to the next. In humans, following such markers is more challenging, but studies on groups of closely related people, such as large families or small cultures like the Amish, have allowed some RFLPs to be followed in this manner (Fig. 8.1).

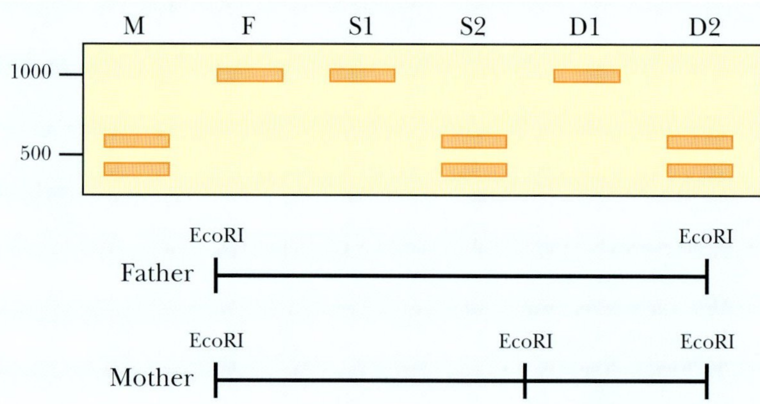

FIGURE 8.1 RFLPs of Family Members
The mother (M) and father (F) of this family have a difference in the sequence of their DNA. In the mother the difference adds a restriction enzyme site for *Eco*RI. The children (S1, S2, D1, and D2) have inherited one or the other fragment from their parents.

Another marker used for making genetic maps is the **VNTR**, or **variable number tandem repeat** (Fig. 8.2). These sequence anomalies occur naturally in the genome and consist of tandem repeats of 9 to 80 base pairs in length. The number of repeats differs from one person to the next; therefore, these can be used as specific markers on a genetic map. They can also be used to identify individuals in forensic medicine or paternity testing. Some repeats are found in many different locations throughout the genome and cannot be used for making genetic maps, but other repeat sequences are found only in one unique location.

A third type of marker is the **microsatellite polymorphism**, which is also a tandem repeat. However, unlike VNTRs, microsatellite polymorphisms are repeats of 2 to 5 base pairs in length, and usually consist of cytosine and adenosine.

A fourth type of genetic marker used in mapping is the **SNP** (pronounced "snip"), or **single nucleotide polymorphism** (Fig. 8.3). SNPs are individual substitutions of a single nucleotide that do not affect the length of the DNA sequence. These changes can be found within genes, in regulatory regions, or in noncoding DNA. When found within the coding regions of genes SNPs may alter the amino acid sequence of the protein. This in turn may affect protein function. If a SNP correlates with a genetic disease, identifying that SNP may diagnose the disease before symptoms appear. When a SNP falls within a restriction enzyme site, it coincides with an RFLP.

Determining the order of various markers makes genetic maps. Some markers include RFLPs, SNPs, VNTRs, and microsatellite polymorphisms.

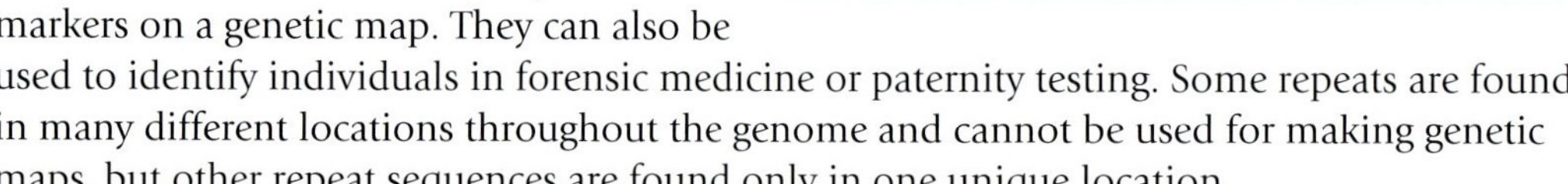

FIGURE 8.2 Tandem Repeat of 12 Base Pairs
This individual has only four repeats of this 12 base-pair sequence. Other people may have more or fewer repeats.

5′ AAGGTAT 3′ to 5′ AAGCTAT 3′

FIGURE 8.3 SNP
The same DNA segment from two different individuals has a single nucleotide difference, that is, a SNP. Such changes are common when comparing DNA sequences between individual people.

PHYSICAL MAPS USE SEQUENCE DATA

Genetic markers such as SNPs, VNTRs, RFLPs, and microsatellites are useful, but for large genomes like the human genome, these still do not provide enough markers. The map builders needed other types of markers, such as **sequence tagged sites (STSs)** (Fig. 8.4). These are simply short sequences of 100–500 base pairs that are unique and can be detected by PCR. A specialized type of STS is the **expressed sequence tag (EST)**, so called because it was identified in a cDNA library. This means that the EST is expressed as mRNA. These small pieces of sequence data are just portions of larger genes; therefore, many different ESTs may be found for one single gene.

Mapping physical markers resembles linkage analysis for genes in the sense that the closer they are, the more likely they will remain together. However, linkage for physical markers

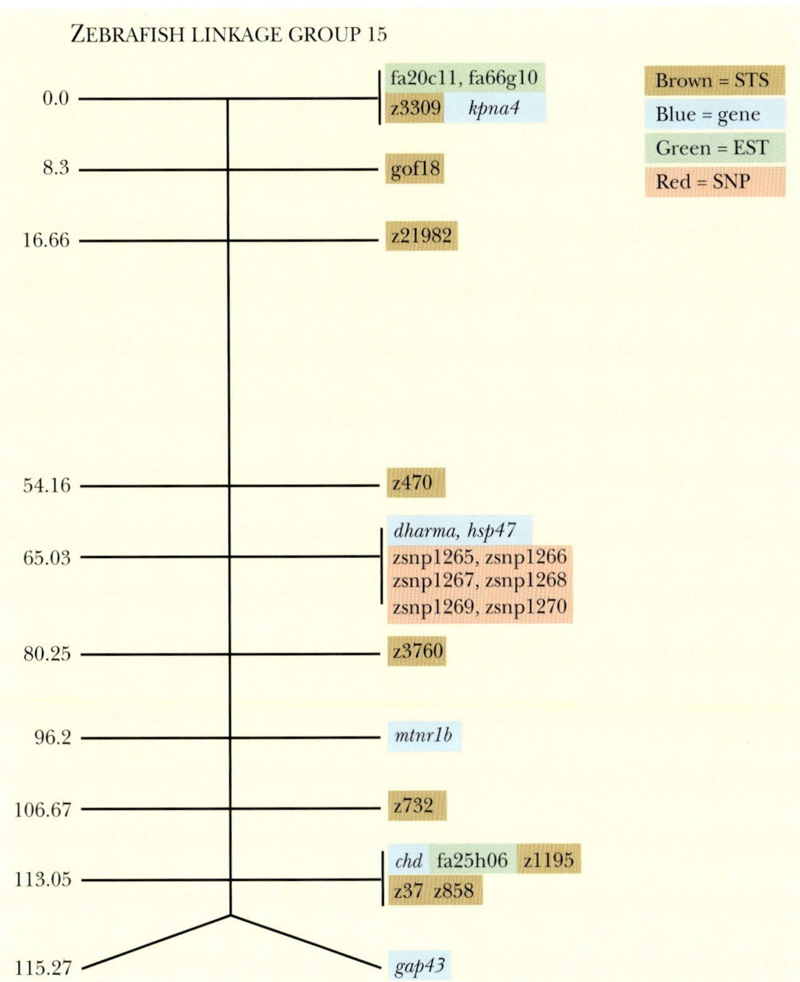

FIGURE 8.4 STS and EST Markers on Zebrafish Linkage Map
The relative positions of various markers are shown on the zebrafish map. The markers include STSs and ESTs that were identified and mapped relative to one another. In addition, the positions of real genes and SNPs are shown relative to the others. The linkages were established using meiotic recombination frequencies and are presented in centimorgans (cM). The GAT linkage group 15 map information for this figure was retrieved from the Zebrafish Information Network (ZFIN), the Zebrafish International Resource Center, University of Oregon, Eugene, OR 97403-5274; http://zfin.org/.

is determined by restriction enzyme digestion (Fig. 8.5). Either entire genomes or single large clones from a library are digested with a variety of different restriction enzymes. Each enzyme will digest the DNA into different sized fragments, which are then probed for several different STS or EST markers. If two markers are close together, they will often be found on the same restriction fragment, but if they are far apart, they will be on different fragments. The fragment sizes are determined, and from this the approximate distance between two markers can be estimated.

One library clone will often overlap with another when many different clones from one genome are compared. The overlap region has identical sequence and, therefore, identical restriction enzyme sites, identical STS markers, and identical ESTs. **Contig maps** compare the overlapping regions, thus ordering the clones into a longer linear segment, known as a **contig** (Fig. 8.6). As more and more clones are analyzed, the contig gets longer and longer, producing large segments of chromosomes.

Physical maps are based on actual distance in base pairs. Markers such as ESTs and STSs are used to determine these distances.

Contig mapping determines regions that overlap from one library clone to another clone. Sequences of overlapping library clones are continuous and provide information for physical maps.

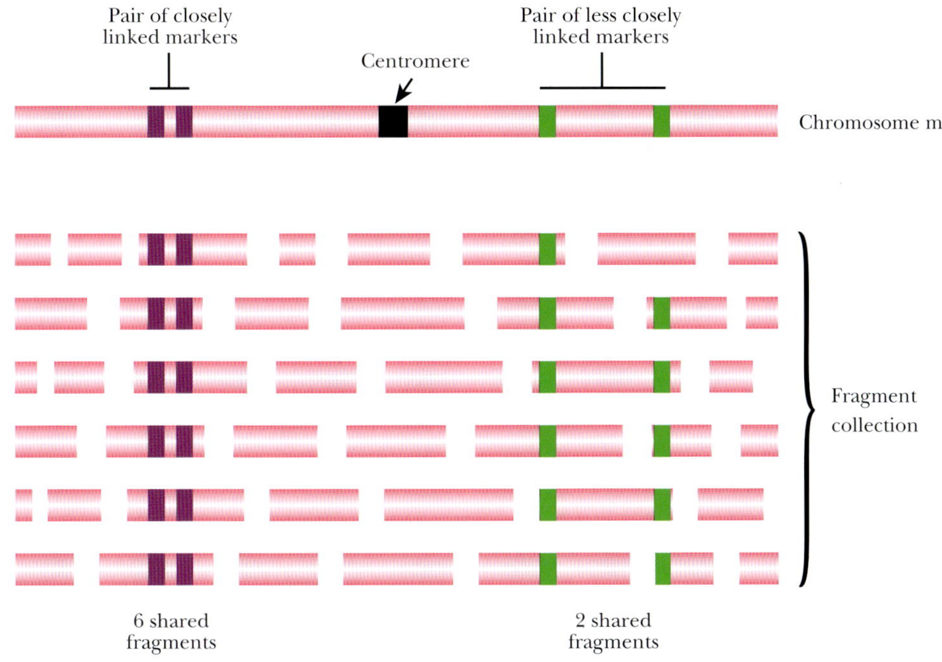

FIGURE 8.5 STS Mapping Using Restriction Enzyme Digests
STS mapping is shown for four STS sites on a single chromosome. A variety of restriction enzyme digests are performed to cut the chromosome into many different sized fragments. The number of times two STS sequences are found on the same fragment reveals the proximity of the two markers. In this example, the two purple STSs are found on the same fragment six times and must be close to each other on the chromosome. The two green STSs are found on the same fragment only two times and are therefore farther apart. The purple and green STSs are never found on the same fragment; therefore, they must be far apart.

RADIATION HYBRID AND CYTOGENETIC MAPPING

Library clones can sometimes be unreliable because large cloned segments may actually consist of two fragments of DNA, from different parts of the genome, inserted into the same vector. **Radiation hybrid mapping** overcomes these limitations by examining STSs or ESTs on original chromosomal fragments (Fig. 8.7). To generate these, cultured human cells are treated with x-rays or γ-rays to fragment the chromosomes. The radiation dosage controls how often the chromosome breaks, and thus the average length of the fragments. The human cells possess a marker enzyme that allows them to grow on selective media. After irradiation, the human cells are fused to cultured hamster cells using polyethylene glycol or Sendai virus. The hamster cells do not have the selective marker. Consequently, only those hamster cells that fuse with human cells survive. The fragments of human chromosomes become part of the hamster nucleus, and the individual hybrid cell lines can be examined by STS or EST mapping. Because the average fragment length is known, these maps determine relative distance between two markers.

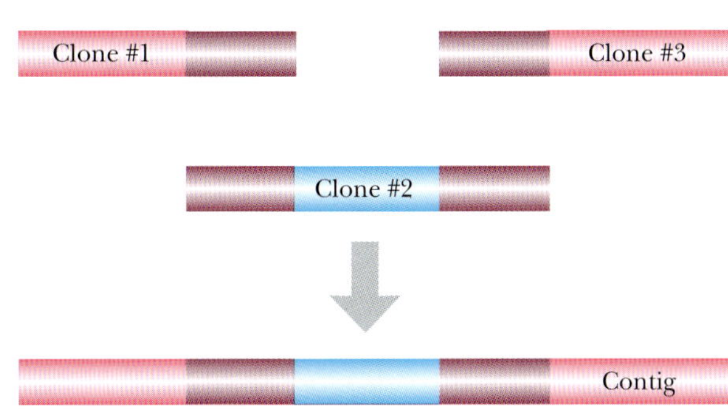

FIGURE 8.6 Contig Mapping
Small clones have regions that overlap with each other. Ordering the small clones into one sequence forms a contig map.

Cytogenetic mapping is another physical technique that uses original chromosomes. When chromosomes are placed on microscope slides and stained, they form banding patterns that are visible under a light microscope. This **cytogenetic map** shows where a gene or marker lies relative to the stained bands (Fig. 8.8). Cytogenetic maps are very low resolution compared with the other mapping techniques, yet they are useful to compare gene locations on a large scale.

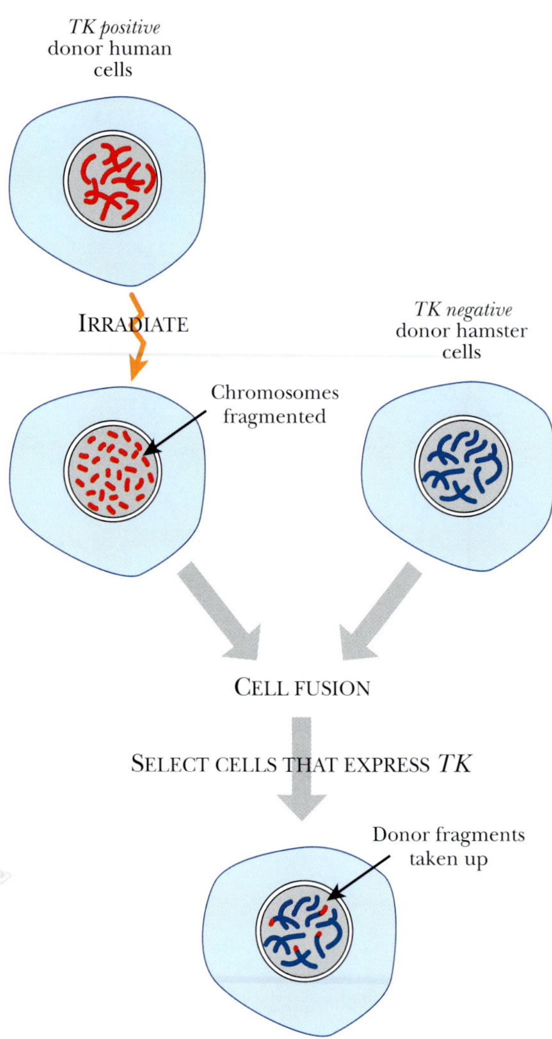

FIGURE 8.7 Radiation Hybrid Mapping
To determine how close STSs and ESTs are to each other, many large chromosome fragments must be analyzed. Radiation hybrid mapping allows large human chromosome fragments to be inserted into hamster cells. First, the human chromosomes, which carry the thymidine kinase gene (TK^+), are fragmented by irradiation. The human cells are then fused with hamster cells, which are TK^-. Such hybrid cells should express thymidine kinase and will grow on selective medium. Random loss of human chromosome fragments occurs during this process; therefore, each radiation hybrid cell line has a different set of human chromosome fragments, which can be screened for the STSs and ESTs.

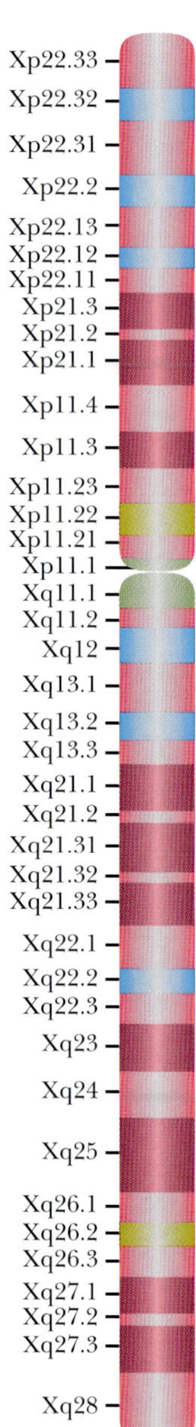

FIGURE 8.8 Banding Pattern of X Chromosome
Staining condensed mitotic chromosomes with various DNA-binding dyes forms a distinct banding pattern. The location of a gene or marker can be determined relative to the bands. For example, a gene located at Xp21.1 is on chromosome X, on the p arm, and on band number 21.1.

Radiation hybrid mapping orders the physical markers such as STSs and ESTs along larger regions of human chromosomes.

Human chromosomes have distinct banding patterns that are used to order the physical markers.

SEQUENCING ENTIRE GENOMES

Sequencing the entire genome from one organism can be accomplished in different ways. **Chromosome walking** allows the researcher to identify and sequence one clone and then, using those data, to find overlapping clones (Fig. 8.9). After those are identified and sequenced, more overlapping clones are identified. The process goes in order either up or down the chromosome, compiling the sequence piece by piece. Usually the first clone is located relative to a particular marker, such as an STS or RFLP. Chromosome walking is often used to characterize genes responsible for a particular disease. Analysis of DNA from people with the disease may have revealed a particular RFLP that is always present in those with the disease, but absent in unaffected people. This RFLP can be identified in a library clone. Then chromosome walking both upstream and downstream of the RFLP will, it is hoped, provide the whole gene sequence.

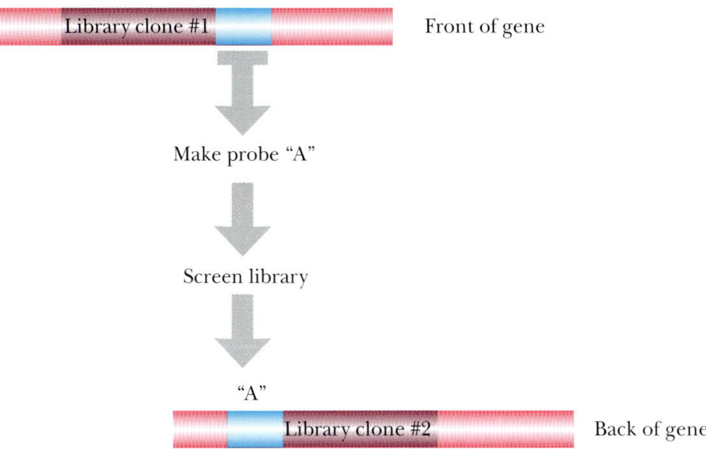

FIGURE 8.9 Chromosome Walking
Researchers identify the downstream and upstream regions of a gene using chromosome walking. In this example, the end of library clone 1 is converted into a probe. The probe is used to screen a library, and a second clone is identified. The two clones overlap and are linked to form a complete gene.

Although chromosome walking is a powerful tool to identify genes, its use for sequencing an entire genome is too arduous. Instead, **shotgun sequencing** is used to assemble sequence data from an entire genome (Fig. 8.10). Here genomic libraries are constructed and random

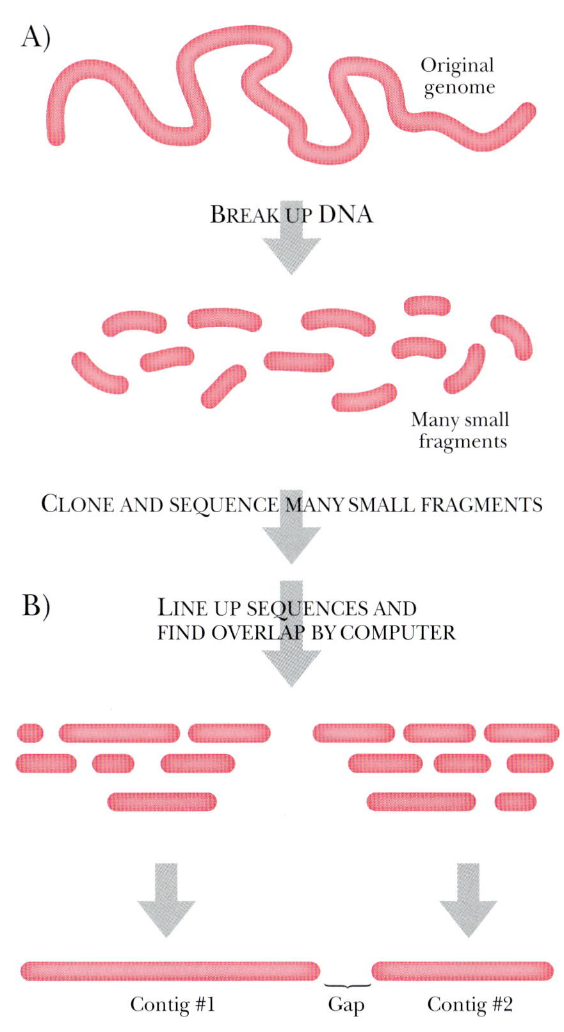

FIGURE 8.10 Shotgun Sequencing
The first step in shotgun sequencing an entire genome is to digest the genome into many small fragments that are cloned and sequenced individually. Computers analyze the sequence data for overlapping regions and assemble the sequences into several large contigs. Because some regions of the genome are unstable when cloned, some gaps may remain even after this procedure is repeated several times.

clones are sequenced. A computer compiles the sequence information, identifying the overlapping regions between clones, and ordering the clones into a complete sequence. This procedure is repeated to eliminate as many gaps as possible. This method was used to sequence *Haemophilus influenzae*, the first cellular genome to be sequenced. This genome has 1.8 Mb and took less than 3 months to complete using shotgun sequencing.

Large genomes can be sequenced by chromosome walking, where one end of a library clone is used to find a different clone with downstream regions.

Shotgun sequencing was the technique used to sequence the human genome. Each library clone is randomly sequenced, and computer analysis ordered the clones by identifying overlapping regions.

RACE FOR THE HUMAN GENOME

The Human Genome Project is a federally funded program that set out to sequence the entire 3 billion base-pair human genome. The project started in 1990 and was to be completed in 15 years. At the start of the project, sequencing technology was in its infancy. Chain termination sequencing was becoming refined, but sequencing more than 1000 base pairs of DNA a day was challenging. Within 10 years, the rate of sequencing was substantially greater, and the cost per base had decreased from $10 in 1990 to about 50 cents per base sequenced in the late 1990s. In 1998, Celera Genomics, led by Craig Venter, decided to sequence the entire human genome faster and cheaper. Celera Genomics proved its point by sequencing the entire 180-Mb genome of the fruit fly, *Drosophila*, between May and December of 1999. Celera used the shotgun sequencing approach, which most researchers thought would not work for such large numbers of data. Celera Genomics then continued with the shotgun approach. The official Human Genome Project was only 85% complete when Celera announced that it had sequenced 99% of the human genome. In June 2000, a joint announcement from Celera Genomics and the Human Genome Project presented the first working draft of the entire sequence of the human genome.

The Human Genome Project started with a good game plan for sequencing the genome. The project first mapped large fragments of human DNA (such as YACs and BACs) to their respective chromosomal locations. Then the fragments of DNA were sequenced. As described earlier, mapping is time consuming but necessary to order the sequence data. At the time of startup, computers were unable to order more sequence data than found in large chromosomal fragments. During the 1990s, computing power increased so rapidly that Craig Venter decided the mapping steps were unnecessary. He sequenced many small fragments of DNA and entered the data into the computer. The computer assembled the information into a working draft. Venter was able to beat his competition largely because of the increase in computer power.

The human genome was sequenced used shotgun sequencing and was done much sooner and at a lower cost than was projected.

GAPS REMAIN IN THE HUMAN GENOME

Although the sequence of the genome is considered complete, there are still gaps. The gaps fall in highly condensed regions of highly repetitive **heterochromatin** that is difficult to sequence. Three features characterize heterochromatin: hypoacetylation (i.e., lack of acetyl groups on the histones); methylation of histone H3 on a specific lysine; and methylation on CpG or CpNpG sequence motifs. Heterochromatin is not transcribed and comes in two forms, **facultative heterochromatin** and **constitutive heterochromatin** (Fig. 8.11). The constitutive form is found around the centromeres and telomeres of the chromosome and does not change from one generation to the next.

Facultative heterochromatin, on the other hand, is found in other regions of the chromosomes, and its presence is cell-specific. Once a specific region of a chromosome becomes heterochromatin, all of its descendent cells will maintain this pattern. However, neighboring cells may not have heterochromatin in the same regions, leading to **position effect variegation (PEV)**. The theory is that heterochromatin forms in defense against invading retrotransposons and viruses, and also due to gene silencing during RNAi (see Chapter 5). When a gene is near a region of heterochromatin, the gene is no longer transcribed and becomes dormant.

The amount of methylation on lysine-9 in histone H3 determines whether or not heterochromatin is considered facultative or constitutive.

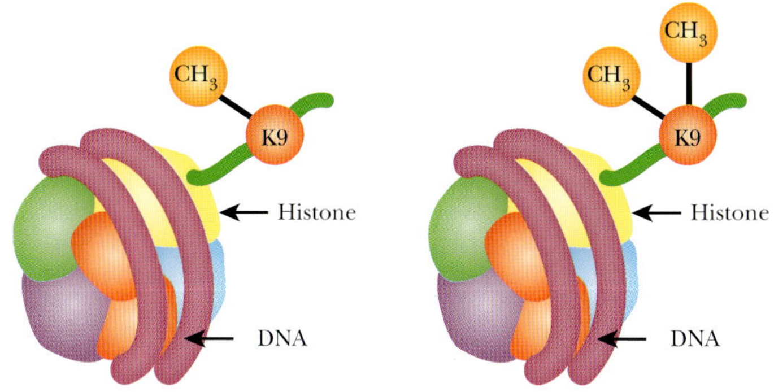

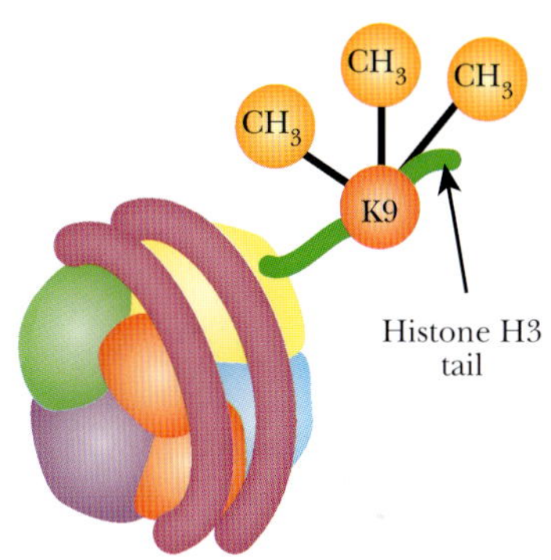

FIGURE 8.11
Facultative versus Constitutive Heterochromatin
The amount of methylation on lysine-9 in histone H3 determines whether or not heterochromatin is considered facultative or constitutive.

> Heterochromatin is highly condensed DNA that is found in specific sites throughout the genome. The physical nature of heterochromatin makes it difficult to sequence. The constitutive heterochromatin near the centromere is consistent throughout the body, whereas facultative heterochromatin varies from person to person and organ to organ.

SURVEY OF THE HUMAN GENOME

The sequence of the human genome is 2.9×10^9 base pairs (2.9 Gbp or gigabase pairs) in length. If the sequence were typed onto paper, at about 3000 letters per page, it would fill 1 million pages of text. This extraordinary amount of information is encoded by the sequence of just four bases, cytosine, adenosine, guanine, and thymine. Most people expected the human genome sequence to reveal the actual number of genes found in human beings. In reality the massive amount of sequence needs sophisticated interpretation in order to determine how many genes it contains. The best estimates so far predict only 25,000 genes, but the number may be more or less. Of the identified genes, we only know the function of around 50%. More than 40% of the predicted human proteins are similar in structure to proteins in organisms such as fruit flies or worms.

The genome sizes and estimated gene numbers are given for several organisms in Table 8.1. Note that plants presently hold the record for gene numbers, although they have less total DNA than higher animals. Among the animals, the ciliate protozoan *Paramecium* has less DNA but more genes than any multicellular animal sequenced so far. The largest bacterial genomes have more genes than the smaller eukaryotic genomes. An example is *Streptomyces*, famous as the source of many antibiotics. The smallest bacterial genomes, such as *Mycoplasma*, have fewer than 500 genes, although because these bacteria are parasitic they rely on the eukaryotes they infect for many metabolites.

In eukaryotic DNA, genes encompass thousands or even millions of base pairs, most of which are actually introns that are spliced out of the mRNA transcript. For example, the gene for dystrophin (defective in Duchenne's muscular dystrophy) is 2.4 million base pairs long, and some of its introns are 100,000 base pairs or more in length. In contrast, the coding sequence, consisting of multiple exons, is about 3000 base pairs. In such situations, it is not always easy to find coding sequences among all the noncoding DNA. On the one hand, this may result in genes (or individual exons) being completely missed. On the other hand, widely separated exons that are in reality parts of a single coding sequence may be interpreted as separate genes.

Another confounding factor in determining the number of genes is the presence of **pseudogenes**. These are duplicated copies of real genes that are defective and no longer expressed. They may be found next to the original, or they may be far away, on different chromosomes. Determining whether or not a "gene" is a pseudogene or genuine may be difficult with sequence data alone. Often the expression of a particular region of

Table 8.1 Various Genomes and Number of Estimated Genes

Organism	Genome Size (Megabase pairs)	Estimated Genes (Protein encoding)
Plants		
Black poplar (*P. trichocarpa*)	500	45,000
Rice (*Oryza sativa*)	390	38,000
Mustard weed (*Arabidopsis thaliana*)	125	26,000
Animals		
Paramecium tetraaurelia	72	40,000
Human (*Homo sapiens*)	2,900	25,000
Mouse (*Mus musculus*)	2,500	25,000
Roundworm (*Caenorhabditis elegans*)	97	19,000
Fruit fly (*Drosophila melanogaster*)	180	13,600
Fungi		
Aspergillus nidulans	30	9,500
Yeast (*Saccharomyces cerevisiae*)	13	5,800
Bacteria		
Streptomyces coelicolor	8.7	7,800
Escherichia coli	4.6	4,300
Mycoplasma genitalium	0.58	470

DNA must be confirmed by finding corresponding mRNA transcripts. DNA microarrays are a popular approach to confirming whether or not a gene is expressed (see later discussion).

The number of genes also hinges on how we define a gene. In addition to the 25,000 protein-encoding genes, there are a thousand or more genes that encode nontranslated RNA. The ribosomal RNA and transfer RNA genes are the most familiar of these. However, a variety of other small RNA molecules are involved in splicing of mRNA and in the regulation of gene expression. Other sequences of DNA may not even be transcribed, yet are nonetheless important. Should these also be regarded as genes?

The number of genes in an organism depends on the definition of *gene* and the distinction between real gene and pseudogene. The absolute number of genes in any sequence is approximate.

NONCODING COMPONENTS OF THE HUMAN GENOME

Many noncoding DNA elements have been identified in the genome of humans and other organisms. Many scientists have referred to these regions as **junk DNA** because they are not expressed, either as proteins or RNA, nor do they comprise regulatory regions associated with genes. However, some of this DNA does have a functional role. Tables 8.2 and 8.3 list the major components of the human genome and what percentage of each is present.

One common type of noncoding DNA is the introns between genes. Most introns have no function, but there are examples of whole genes being found within introns of a different gene. Introns may also contain binding sites for transcription factors and therefore play a role in gene regulation.

Table 8.2 Components of the Eukaryotic Genome

Unique sequences	
Protein-encoding genes—comprising upstream regulatory region, exons and introns	
Genes encoding nontranslated RNA (snRNA, snoRNA, 7SL RNA, telomerase RNA, Xist RNA, a variety of small regulatory RNAs)	
Nonrepetitive intragenic noncoding DNA	
Interspersed Repetitive DNA	
Pseudogenes	
Short Interspersed Elements (SINEs)	
Alu element (300 bp)	~1,000,000 copies
MIR families (average ~130 bp) (mammalian-wide interspersed repeat)	~400,000 copies
Long Interspersed Elements (LINEs)	
LINE-1 family (average ~800 bp)	~200,000–500,000 copies
LINE-2 family (average ~250 bp)	~270,000 copies
Retrovirus-like elements (500–1300 bp)	~250,000 copies
DNA transposons (variable; average ~250 bp)	~200,000 copies
Tandem Repetitive DNA	
Ribosomal RNA genes	5 clusters of about 50 tandem repeats on 5 different chromosomes

(Continued)

Table 8.2 Components of the Eukaryotic Genome—cont'd

Transfer RNA genes	multiple copies plus several pseudogenes
Telomere sequences	several kb of a 6-bp tandem repeat
Mini-satellites (= VNTRs)	blocks of 0.1 to 20 kbp of short tandem repeats (5–50 bp), most close to telomeres
Centromere sequence (alpha-satellite DNA)	171-bp repeat, binds centromere proteins
Satellite DNA	blocks of 100 kbp or longer of tandem repeats of 20 to 200 bp, most close to centromeres
Mega-satellite DNA	blocks of 100 kbp or longer of tandem repeats of 1 to 5 kbp, various locations

Numbers of copies given is for the human genome.

Table 8.3 Percentages of Components of Human Genome

Genetically active	
25%	transcribed into RNA
24%	introns (spliced out)
1%	exons (translated into protein)
Retro-elements	
13%	SINEs
20%	LINEs
8%	defunct retroviruses
Other DNA	
4%	DNA-based transposons
3%	microsatellites and VNTRs
5%	repetitive DNA at telomeres and centromeres
22%	nonrepetitive intergenic DNA

Another feature of the genome is moderately repetitive sequences. Ribosomal RNA genes are found in great numbers because many ribosomes are needed. These genes are considered moderately repetitive elements, but of the coding variety. Noncoding repetitive elements include **LINE**, for **long interspersed element**, which is found in 200,000 to 500,000 copies (Fig. 8.12). These are retrovirus-like elements that contain genes inside long terminal repeats (LTRs) similar to retroviruses. They are autonomous, that is, the LINE is able to make copies of itself and insert new copies into other sites in the genome. However, most copies of LINEs are defective; only a few are still mobile and functional. There are many different types of LINEs, the most common by far in mammals being L1.

LINE retro-elements are transcribed into mRNA using the internal promoter. The mRNA goes to the cytoplasm, where ribosomes translate it, giving the proteins. One of these proteins, the reverse transcriptase/endonuclease dual function protein, then binds to the mRNA to form a **ribonucleoprotein** (RNP). This is transported back into the nucleus where the endonuclease portion nicks the DNA to generate a 3′-OH. Then the reverse transcriptase

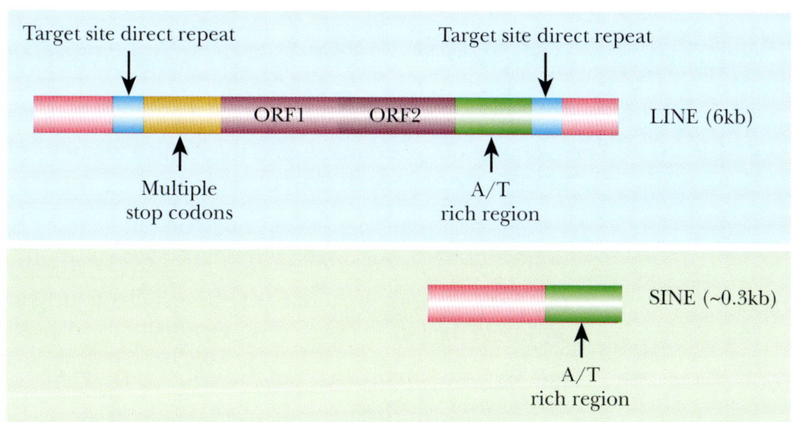

FIGURE 8.12 General Structure of LINE and SINE
LINE elements contain an internal promoter for RNA polymerase II and two open reading frames that encode proteins. The first protein has an unknown function. The second is a bifunctional protein with reverse transcriptase and DNA endonuclease domains. SINE elements usually contain only an internal promoter for RNA polymerase III, and some sort of tRNA stem-loop structure, followed by a poly(A) tail.

portion makes a DNA copy of the LINE that inserts into a new location in the genome. Cellular repair enzymes fill in the gaps to create duplicated sequences flanking both sides of the new LINE. When a LINE moves to a new location, it may disrupt an essential gene, which would prove fatal to the cell. Control of LINE movement is critical. Too much movement is disruptive and might destroy both the host cell and the LINEs it contains. Conversely, too little movement, and the LINE will fail to reproduce effectively. In humans, many LINEs are found in gene-poor, A/T-rich regions of the genome, suggesting that some mechanism exists to keep these elements from disrupting cellular function.

In addition to moderately repetitive sequences, the human genome is filled with highly repetitive DNA. The **SINEs** or **short interspersed elements** (see Fig. 8.12) are retro-elements like the LINEs. The most common type of SINE is called the **Alu element** because an *Alu* restriction enzyme site falls within the element. In the human genome, about 300,000 to 500,000 Alu elements exist. SINE elements are not able to move to new locations in the genome without help from the LINE reverse transcriptase and endonuclease protein. Unlike LINEs, SINEs are found in gene-rich regions of the human genome, but they are shorter and often inert, so their presence does not usually interfere with gene function.

Another type of highly repetitive element found in the human genome is the **minisatellite** or VNTR. These were used in mapping the human genome, and are scattered around the entire genome (see earlier discussion).

Junk DNA or noncoding genomic DNA has many different types of elements, including LINE, SINE, and satellite DNA.

LINE sequences are transcribed into RNA and converted into proteins. One of the proteins reassociates with the mRNA. The RNP complex reenters the nucleus, where it integrates itself into a new location of the genome.

BIOINFORMATICS AND COMPUTER ANALYSIS

As noted before, the use of computers has revolutionized the way in which genetic information has been gathered and analyzed. The term **bioinformatics** has been coined to describe the emerging scientific discipline of using computers to handle biological information. It encompasses a large number of fields (Table 8.4). Bioinformatics includes the

Table 8.4 Fields of Study Related to Bioinformatics

Field	Description
Computational biology	evolutionary, population, and theoretical biology; statistical models for biological phenomena
Medical informatics	the use of computers to improve communication, understanding, and management of medical information
Cheminformatics	the combination of chemical synthesis, biological screening, and data mining to guide drug discovery and development
Genomics	analysis and comparison of the entire genetic complement of one or more species
Proteomics	the global study of proteins
Pharmacogenetics	using genomic/bioinformatic methods to identify individual differences in response to drugs
Pharmacogenomics	applying genomics to the identification of drug targets

storage, retrieval, and analysis of data about biomolecules. By far the greatest achievement of the bioinformatics revolution has been the sequencing of the human genome. The term *bioinformatics* is now used to include analyses associated with DNA microarrays (see later discussion) and assessing the function of genomes.

Because bioinformatics is so widely used, it is important to make genomic data available to researchers. The data from the Human Genome Project are available on the Internet through the National Center of Biotechnology website (http://www.ncbi.nlm.nih.gov/). Some other websites that present sequence data are listed next. At the NCBI home page you can explore the human genome many different ways. Using Entrez Gene, a specific gene can be identified by name. The record for each gene contains the gene name and description, its location, a graphical representation of the introns and exons for all the protein isoforms that are known, and a summary of all the information known about the gene. Additionally, the various domains within the protein, such as actin binding sites, are listed with links to explain the domain and its function. Finally, genes and/or regions of DNA from other organisms that are homologous to the gene are shown. The page also contains links to research papers on the gene's function.

Some Bioinformatics Websites:

- GenBank and linked databases
  - http://www.ncbi.nlm.nih.gov/Entrez/
  - http://www.ncbi.nlm.nih.gov/mapview/
  - http://www.ncbi.nlm.nih.gov/genome/guide/human/
- Institute for Genomics Research (TIGR)
  - http://www.tigr.org/tdb
- Genome Database (GDB) (human genome)
  - http://www.gdb.org
- European Bioinformatics Institute (including EMBL and Swissprot)
  - http://www.ebi.ac.uk/
- Flybase (*Drosophila* genome)
  - http://flybase.bio.indiana.edu:82
- RCSB Protein Data Bank
  - http://www.rcsb.org/pdb/
- PIR Protein Information Resource (PIR)
  - http://www-nbrf.georgetown.edu/pir/searchdb.html

The program Map Viewer (http://www.ncbi.nlm.nih.gov/mapview/) is used to browse the human genome without any particular gene in mind. For example, individual chromosomes can be explored via a graphical interface that allows you to zoom in and out of various regions. Another **genome browser** can be found at http://www.ensembl.org.

The amount of information generated by the Human Genome Project is tremendous; therefore, understanding this information without the use of computers is too difficult. **Data mining** refers to the use of computer programs to search and interpret the data. Many bioinformatics researchers develop programs that search the genomic data banks and sift, sort, and filter the raw sequence data. Data mining programs often process information using the following steps:

1. Selection of the data of interest.
2. Preprocessing or "data cleansing." Unnecessary information is removed to avoid slowing or clogging the analysis.
3. Transformation of the data into a format convenient for analysis.
4. Extraction of patterns and relationships from the data.
5. Interpretation and evaluation.

These programs can be designed to search for related sequences, determine areas of coding and noncoding DNA by looking at codon bias, or search for known consensus sequences, just to name a few applications. Searching for related sequences or **similarity searches** allows researchers to identify a potential function for a gene. If a gene of unknown function from humans is very similar to a characterized gene from flies, the two encoded proteins may have similar functions. This type of research is called **comparative genomics**. More than one gene can be compared. For example, entire pathways are often similar in different species. For example, human insulin attaches to a receptor on the cell surface and controls gene transcription via several intracellular proteins. Remarkably, very similar insulin signaling proteins are found in the roundworm *Caenorhabditis elegans*.

Difficulties may arise if scientists only use comparative genomics to study a new protein. Sometimes sequences that are similar have radically different roles as similar proteins may evolve new functions. Thus sequence similarity does not always imply functional similarity. Finally, the databases themselves are not perfect and may contain mistakes that are misleading. Comparative genomics must be complemented with other studies to reliably assign a role to a novel protein.

Other programs determine coding and noncoding areas of the genome by looking at **codon bias**. Identifying coding regions is critical for finding genes and can be accomplished by looking at the wobble position (third base) of the codon. Although a particular amino acid is often encoded by multiple codons, some codons are used preferentially. This codon bias varies from one organism to another. Most of the tRNAs for a particular amino acid will recognize the favored codon(s). For example, *Escherichia coli* genes preferentially use CGA, CGU, CGC, and CGG for the amino acid arginine, but rarely use AGA or AGG. Very few tRNAs for arginine are produced that recognize the AGA and AGG codons. Codon bias will thus be evident in regions that encode proteins, but in noncoding regions, the wobble position will not maintain this bias. Thus a potential gene in *E. coli* would contain relatively few AGA and AGG codons.

Finally, programs that identify consensus sequences allow researchers to find various signatures or motifs associated with particular functions. For example, a site that binds to ATP has specific amino acids in specific locations. These sequences in an unknown protein may help identify one of its functions. Other motifs include actin binding domains, which indicate that the protein binds to the cytoskeleton, or protease cleavage sites that suggest the protein is subject to intracellular modification by proteases. Any potential motif in the sequence must be confirmed experimentally. For example, a protein with an ATP binding site signature must be shown to bind ATP experimentally. Thus, sequence analysis provides a basis for further experiments.

Bioinformatics is the study of biological information using computers.

Data mining uses computer algorithms to study, sort, and compile information from genomic databases.

Information about genes can be obtained by comparing sequences from different organisms.

MEDICINE AND GENOMICS

One of the greatest applications for human genomics data is in disease and diagnoses. Medical applications of genomics are abundant and the remaining chapters of this book cover some of these. **Gene testing** is the most common present application. Once genes have been associated with particular diseases, people can be screened for genetic mutations within the gene. Such tests can diagnose diseases such as muscular dystrophy, cystic fibrosis, sickle cell anemia, and Huntington's disease because these are strictly inherited disorders. In diseases with an environmental component, genetic testing offers information that may change how a person lives his or her life. Perhaps those with a genetic predisposition to colon cancer will have more screenings, earlier than usual, and perhaps alter their diet to minimize the chance of cancer developing. Other applications include gene therapy (see Chapter 17).

Genomics has a wide-reaching effect on biotechnology and medicine.

DNA ACCUMULATES MUTATIONS OVER TIME

The Human Genome Project has opened the doors to improved analyses for many areas, including evolutionary biology. The sequence features of the human genome arose over millions of years, as **mutations**, or alterations of the genetic material, occurred and were passed on to successive generations. During the course of human history many different events have molded and sculpted our genetic history and resulted in our current genetic state. Each individual has undergone some sort of genetic recombination and/or mutation to become unique in physical and emotional constitution. Many genetic mutations happen throughout all the cells of our bodies. Most of the defective cells die and undergo apoptosis (see Chapter 20). When a mutation occurs in the **somatic cells**, the children or offspring do not inherit the mutation, only when a mutation occurs in the **germline** or sex cells are the mutations passed on to the next generation.

There are many different types of mutations that cause genetic diversity (Table 8.5). The most common are **base substitutions**, where one nucleotide is exchanged for another. When a purine base is replaced by another purine, or a pyrimidine is replaced by another pyrimidine, this is called a **transition**. If a pyrimidine is exchanged for a purine, or vice versa, this is a **transversion**. These mutations create the SNPs used to make genomic maps. Because different human individuals have variations every 1000 to 2000 bases, there are on average 2.5 million SNPs over the whole genome.

SNPs, or single base substitutions, can fall anywhere in the genome, in either coding or noncoding DNA. When the SNP falls within a gene, it may alter protein sequence and function. When the base substitution alters one amino acid in a protein, the mutation is called a **missense mutation**. Some missense substitutions have little effect on protein structure or function because one amino acid is replaced by another with similar properties. This is known as a **conservative substitution.** An example would be replacing threonine by serine, as these vary only slightly in size but not in chemistry (both have an -OH group). **Radical replacements**, on the other hand, can alter the protein function or structure because they involve replacing amino acids with others that have a different chemistry. For example, aspartic acid or serine are often involved in hydrogen bonding, and when either

Table 8.5 Types of Mutations

Base Changes	Normal	Mutant
Transitions	GAACGT	GAGCGT
Transversions	GAACGT	GATCGT
Missense mutation	GAA CGT Glu Arg	GAT CGT Asp Arg
Conservative substitution	ACT CGT Thr Arg	TCT CGT Ser Arg
Radical replacement	GAT CGT Asp Arg	GCT CGT Ala Arg
Nonsense mutation	GAA CGT Glu Arg	TAA CGT Stop
Insertions	GAACGT	GAAACGT
Deletions	GAACGT	GACGT

is replaced by a neutral amino acid like valine, the protein structure may become unstable. Sometimes missense substitutions create **conditional mutations**, where the protein will work under certain conditions but not others. A common conditional mutation is a **temperature-sensitive mutation**, where the mutation does not alter the protein function at the permissive temperature, but the protein is defective at the restrictive temperature. When base substitutions change a codon for an amino acid into a stop codon, this results in a truncated version of the original protein. These are **nonsense mutations**.

Besides base substitutions, mutations may result in the **insertion** or **deletion** of one or more bases. As for single base substitutions, location is the key to what effect the mutation will have. If the deletion or insertion of a few bases falls within a gene, it may alter the reading frame of the protein, which will create random polypeptide after the mutation. Often the altered reading frame creates a stop codon, which truncates the protein. Large deletions may of course completely remove all or part of a gene. Larger segments of DNA can undergo alterations due to **inversions**, **translocations**, and **duplications**. Inversions occur when DNA segments become inverted relative to the original sequence. Translocations occur when DNA segments are moved to new locations. Duplications are when the DNA segment is copied and then moved, resulting in two identical regions.

In some bacteria, reversible inversions are deliberately used to cause **phase variation**. Here, an enzyme called **invertase** (strictly, **DNA invertase**) inverts a segment of DNA involved in gene regulation. This alters the phenotype of the bacteria. For example, *Salmonella* and *E. coli* turn different surface antigen proteins on or off by inverting a promoter segment. In one orientation, the gene is expressed and the corresponding antigen appears on the cell surface. In the other orientation, the promoter is backward relative to the gene, which is consequently not expressed. This allows the bacteria to vary their appearance to the human immune system, making it easier to infect us.

To a first approximation, mutations occur randomly throughout the genome. However, **mutation hot spots** are regions where mutations occur at much higher frequencies. Mutations often occur at methylated sites because methylated cytosine often loses an amino group, turning it into thymine. DNA polymerase can also induce mutations during DNA replication. Occasionally the proofreading ability of the polymerase fails and single wrong bases are incorporated. More often, DNA polymerase undergoes **strand slippage**, when a segment of DNA is highly repetitive. The result is either a duplication or deletion, depending on the orientation of the slippage.

The rates at which mutations occur help in understanding how mutations have affected the course of evolution. The rate of mutation is low and depends on the organism and even the particular gene being considered. Nonetheless, over long periods of time, many mutations will occur. As Table 8.6 suggests, the rate of mutation is much lower in genomes that are larger. In *E. coli*, mutations occur at 5.4×10^{-7} per 1000 base pairs per generation, but in humans, mutations occur more than 10 times more slowly, at only 5.0×10^{-8} per 1000 base pairs. However, when the mutation rate is corrected for effective genome size (i.e., coding capacity rather than total DNA), it is approximately the same for most organisms. This suggests that some mechanism must actively control the mutation rate.

Mutations occur in all organisms at random places in the genome with an approximately similar rate.
Mutations can be simple base substitutions, inversions, deletions, or insertions. The length of insertions and deletions is variable.

Table 8.6 Mutation Rates in DNA Genomes

		Mutation Rate per Generation		
Organism	Genome Size (Kilobases)	Per kb	Per Genome (Uncorrected)	Per Effective Genome
Bacteriophage M13	6.4	7.2×10^{-4}	0.005	0.005
Bacteriophage Lambda	49	7.7×10^{-5}	0.004	0.004
Escherichia coli	4,600	5.4×10^{-7}	0.003	0.003
Saccharomyces cerevisiae	12,000	2.2×10^{-7}	0.003	0.003
Caenorhabditis elegans	80,000	2.3×10^{-7}	0.018	0.004
Drosophila	170,000	3.4×10^{-7}	0.058	0.005
Human	3,200,000	5.0×10^{-8}	0.16	0.004

GENETIC EVOLUTION

Genome sequencing has expanded the field of **molecular phylogenetics**, which is the study of evolutionary relatedness using DNA and protein sequences. Comparing sequences from different organisms shows the number of changes that have occurred over millions of years. All cellular organisms, including bacteria, plants, and animals, have ribosomal RNA. These sequences can be compared and the differences can be used to determine the relatedness of different organisms. This system is less subjective than using physical characters for **taxonomy**. The **cladistic** approach assumes that any two organisms ultimately derive from the same common ancestor (if we go far enough back) and that at some point **bifurcation**, or separation into two **clades**, occurred in their line of descent. The difference between the two organisms indicates how long ago the split occurred. Taxonomy may be based on visible characteristics—that is, the phenotype. This approach works well, at least to a first approximation, in organisms with plenty of obvious features, such as mammals and plants. But in organisms such as bacteria, the method falls apart. However, molecular phylogenetics has opened the door to making family trees for every organism.

When using molecular data to study relatedness, it is essential that the sequences be correct and truly have come from the organisms under study. This can be complicated in the human genome because some sequences have been derived from other organisms, such as viruses or bacteria. This problem applies to all organisms, to some extent. For example, many bacterial genomes contain inserted bacteriophage genomes. Another important point is to ensure that sequences being compared are truly homologous, that is, they have all descended from one shared ancestral sequence. When gene sequences are compared, they are **aligned**, so that the regions of highest similarity correspond (Fig. 8.13).

This type of alignment can determine the relatedness of two or more proteins or genes. The relatedness can be represented graphically by drawing **phylogenetic trees**. The tree has various features: a root, nodes, and branches (Fig. 8.14). The root represents the overall common ancestor, and the branching indicates the bifurcations or separations that occurred during evolution.

```
Human alpha 1       MVLSPADKTNVKAAWGKVGAHAGEYGAEALERMFLSFPTTKTYFPHFDLS  50
Human alpha 2       MVLSPADKTNVKAAWGKVGAHAGEYGAEALERMFLSFPTTKTYFPHFDLS  50
Rat alpha 1         MVLSADDKTNIKNCWGKIGGHGGEYGEEALQRMFAAFPTTKTYFSHIDVS  50
Mouse alpha 1       MVLSGEDKSNIKAAWGKIGGHGAEYGAEALERMFASFPTTKTYFPHFDVS  50
Chicken alpha-A     MVLSAADKNNVKGIFTKIAGHAEEYGAETLERMFTTYPPTKTYFPHFDLS  50
                    ****  **.*:*   : *:..*. *** *:*:***  ::*.*****.*:*:*
```

```
Human alpha 1       HGSAQVKGHGKKVADALTNAVAHVDDMPNALSALSDLHAHKLRVDPVNFK  100
Human alpha 2       HGSAQVKGHGKKVADALTNAVAHVDDMPNALSALSDLHAHKLRVDPVNFK  100
Rat alpha 1         PGSAQVKAHGKKVADALAKAADHVEDLPGALSTLSDLHAHKLRVDPVNFK  100
Mouse alpha 1       HGSAQVKGHGKKVADALASAAGHLDDLPGALSALSDLHAHKLRVDPVNFK  100
Chicken alpha-A     HGSAQIKGHGKKVVAALIEAANHIDDIAGTLSKLSDLHAHKLRVDPVNFK  100
                     ****:*.*****. **  .*. *::*:..:** *****************
```

```
Human alpha 1       LLSHCLLVTLAAHLPAEFTPAVHASLDKFLASVSTVLTSKYR  142
Human alpha 2       LLSHCLLVTLAAHLPAEFTPAVHASLDKFLASVSTVLTSKYR  142
Rat alpha 1         FLSHCLLVTLACHHPGDFTPAMHASLDKFLASVSTVLTSKYR  142
Mouse alpha 1       LLSHCLLVTLASHHPADFTPAVHASLDKFLASVSTVLTSKYR  142
Chicken alpha-A     LLGQCFLVVVAIHHPAALTPEVHASLDKFLCAVGTVLTAKYR  142
                    :*.:*:**.:* * *.  :**  :********.:*.****:***
```

FIGURE 8.13
Alignment of Related Hemoglobin Sequences
The hemoglobin sequences were aligned using ClustalW (http://www.ebi.ac.uk/clustalw/). Amino acids marked with * are identical in all sequences; those marked with : and . are not identical but are conserved in the type of amino acid.

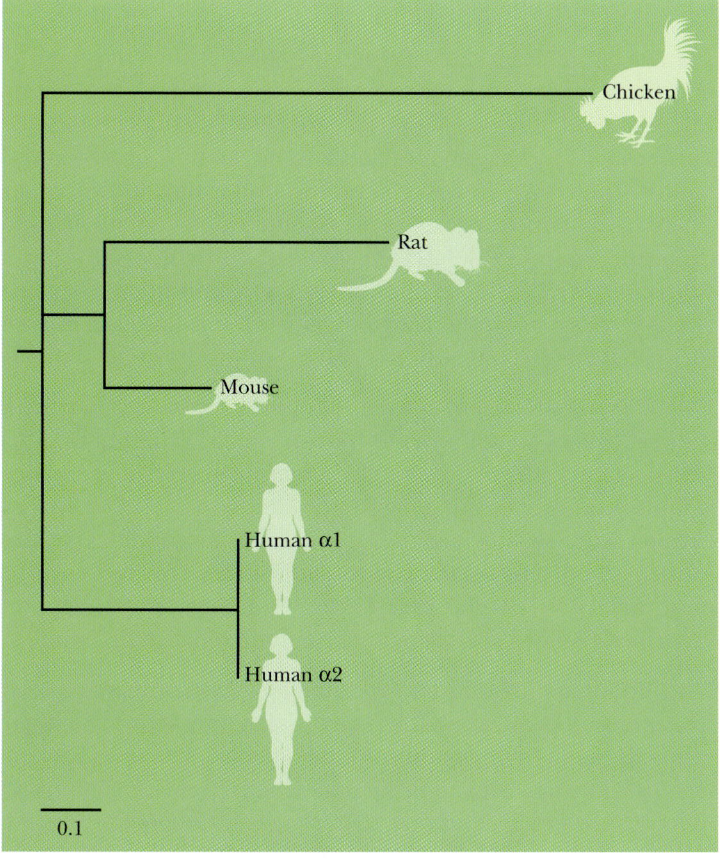

FIGURE 8.14
Phylogenetic Tree of Hemoglobin
The amino acid sequences of chicken globin-A, rat hemoglobin alpha 1 chain, mouse hemoglobin alpha 1 chain, and the alpha 1 and alpha 2 chains from humans were compared. The length of lines represents the number of sequence differences, the longer the line, the more changes in sequence. The differences—were analyzed with ClustalW and the tree was drawn using Phylodendron (http://iubio.bio.indiana.edu/treeapp/).

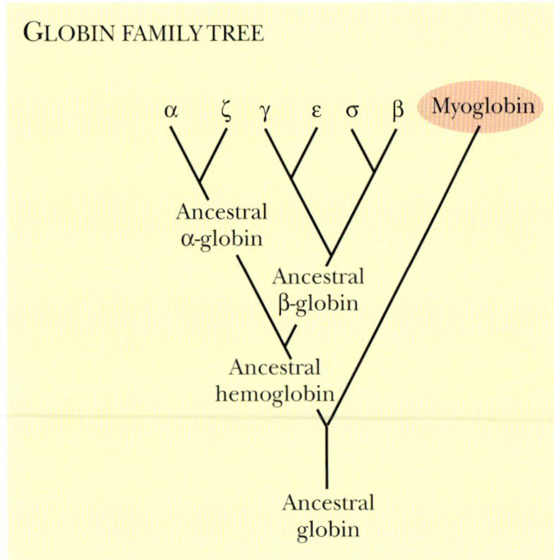

FIGURE 8.15 Globin Family of Genes
Over the course of evolution, a variety of gene duplication and divergence events gave rise to a family of closely related genes. The first ancestral globin gene was duplicated, giving hemoglobin and myoglobin. After another duplication, the hemoglobin gene diverged into the ancestral alpha-globin and ancestral beta-globin genes. Continued duplication and divergence created the entire family of globin genes.

Individual nodes represent common ancestors between two subgroups of organisms. Branches represent clades, that is, groups of organisms with a common ancestor. The length of the branches indicates the number of sequence changes, so if the branches are short, the two organisms bifurcated relatively recently, and if the branches are long, the bifurcation occurred long ago.

Based on alignments, genes have been grouped into **families**, groups of closely related genes that arose by successive duplication and divergence. **Gene superfamilies** occur when the functions of the various genes have steadily diverged until some are hard to recognize. For example, the transporter superfamily encompasses many proteins that transport molecules across biological membranes. This superfamily has members that transport sugars into bacteria, transport water into human cells, and even export antibiotics out of bacteria. They are found in almost all organisms. Another gene superfamily is the globin family (Fig. 8.15). The family includes myoglobin and hemoglobin from different organisms. These proteins all carry oxygen bound to iron, but myoglobin is specific to muscle cells whereas hemoglobin is specific to blood. The theory is that early in evolution one gene for an ancestral globin existed. At some point this gene was duplicated and the copies diverged so that one was specialized for blood and the other for muscle. Hemoglobin itself also diverged later into different forms, each used at various stages of development.

New genes may be generated one at a time, but in addition, whole chromosomes or genomes may be duplicated. In some organisms, particularly plants, genome duplications are relatively stable and have occurred quite often. An example is the modern wheat plant. Its ancestor was a typical diploid, but modern wheat used to make flour is tetraploid. The wheat used to make pasta, durum wheat, is hexaploid and is derived from three different ancestral plants. These varieties arose by natural mutation and were exploited because of the higher protein content and better yield.

The rate of mutation can vary greatly between different genes. Although the human genome undergoes a steady average rate of mutation, individual genes may mutate at different rates. Essential proteins evolve or mutate more slowly than average. Conversely, the less critical a gene is for survival, the more mutations can be tolerated and the protein evolves more rapidly. Thus, the gene for cytochrome *c*, an essential component in the electron transport chain, has incorporated only 6.7 changes per 100 amino acids in 100 million years. In contrast, fibrinopeptides, which are involved in blood clotting, have had 91 mutations per 100 amino acids in 100 million years. As noted earlier, ribosomal RNA is useful to establish family trees for distantly related organisms. It is found in every organism and is essential to survival; therefore, it is slow to evolve.

What happens if a scientist wants to classify organisms that are closely related? Essential gene sequences do not provide enough genetic variation to differentiate such organisms. Nonessential genes may help, but sometimes even these are too close. In such cases, the wobble position of coding regions or even noncoding regions may be used. As noted in Chapter 2, the wobble position is the third nucleotide of a codon. The same amino acid is often encoded by several codons, which vary only in this third base. Alterations at this position usually have no net effect on protein function or structure and may occur between very closely related species or between individuals of the same species.

Mitochondrial or chloroplast genomes are also compared in order to determine the relatedness of organisms. These accumulate mutations at a higher rate than the nuclear genomes in the same organisms. The organelle genomes vary particularly in the noncoding regions. One drawback to using organelle genomes is that mitochondria and chloroplasts are inherited maternally and thus trace the evolutionary lineage only on the maternal side.

Molecular phylogenetics uses genomic sequences of different organisms in order to determine their evolutionary relatedness.

Essential proteins have fewer mutations over time. Less essential proteins have more mutations over time.

FROM PHARMACOLOGY TO PHARMACOGENETICS

Another field that has undergone many changes due to the Human Genome Project is pharmacology, the development of drugs to treat disease. Drug development has traditionally been a hit-or-miss matter, with drug discovery often a by-product of other research. Penicillin is one of the 20th century's greatest discoveries, but was found by accident. Alexander Fleming was growing *Staphylococcus* bacteria and left his plates while on vacation. When he came back, mold had contaminated the plates, and miraculously the staphylococci were not growing close to the mold. The mold was evidently secreting something that stopped bacterial growth.

Even Viagra was discovered by accident. Scientists were trying to develop heart medications when they noticed the "side effect." One factor that makes drug development costs so high is that many of the paths chosen lead to dead ends.

Another problem of drug development is **adverse drug reaction (ADR)**. Adverse reactions may happen in some patients while others respond well and are cured by the same drug. Most drugs are developed with the average patient in mind, yet there is often a subset of people who react badly to the drug. For example, many people are allergic to penicillin or other types of antibiotics. Adverse drug reactions are a major cause of hospitalizations and death. The differences in reaction to certain drugs often depend on the person's genetic makeup.

Pharmacogenomics is the study of all the genes that determine drug response in humans. More specifically, **pharmacogenetics** is the study of inherited differences in drug metabolism and response. The goal of these fields of study is to reduce the number of ADRs by determining the genetic makeup of the patient before offering a specific drug. The key to making a "genetic" diagnosis is the use of SNPs (see earlier discussion). Single changes in coding regions can often be correlated with adverse drug reactions. For example, if a certain subpopulation of people does not respond to a drug, then their DNA can be examined for a specific SNP that is absent in patients who do respond. Before the drug is given to any new patients, DNA from a blood sample can be tested for the presence of that SNP. Testing for SNPs can be done by microarray analysis (see later discussion) in the doctor's office, thus reducing the number of office or hospital visits.

SNP analysis is also used to screen for hereditary defects. Specifically, SNPs can be identified using a technique called Zipcode analysis (Fig. 8.16). Here many different SNPs can be examined simultaneously. First, PCR is used to amplify the region containing each different SNP being investigated. The PCR fragments could be sequenced in full, but because SNPs differ by only one base, single base extension analysis is done instead. For this, a primer is designed to anneal just one base pair away from the SNP location. This primer also carries a "zipcode" region that is used to identify this specific SNP, and each SNP has a different zipcode. After the Zipcode primer anneals to the PCR fragments, DNA polymerase plus fluorescently labeled dideoxynucleotides are added. This results in a single base being added to the primer. (Note that dideoxynucleotides block chain elongation, and so only one base can be added.) Each base is labeled with a different fluorescent dye, allowing it to be identified. Next, beads linked to complementary zipcode (cZipcode) sequences are added to grab the zipcoded primers. The trapped Zipcode primer with the labeled nucleotide has a different color based on which base was incorporated. The different colors can be sorted and counted by FACS or fluorescent activated cell sorting (see Chapter 6).

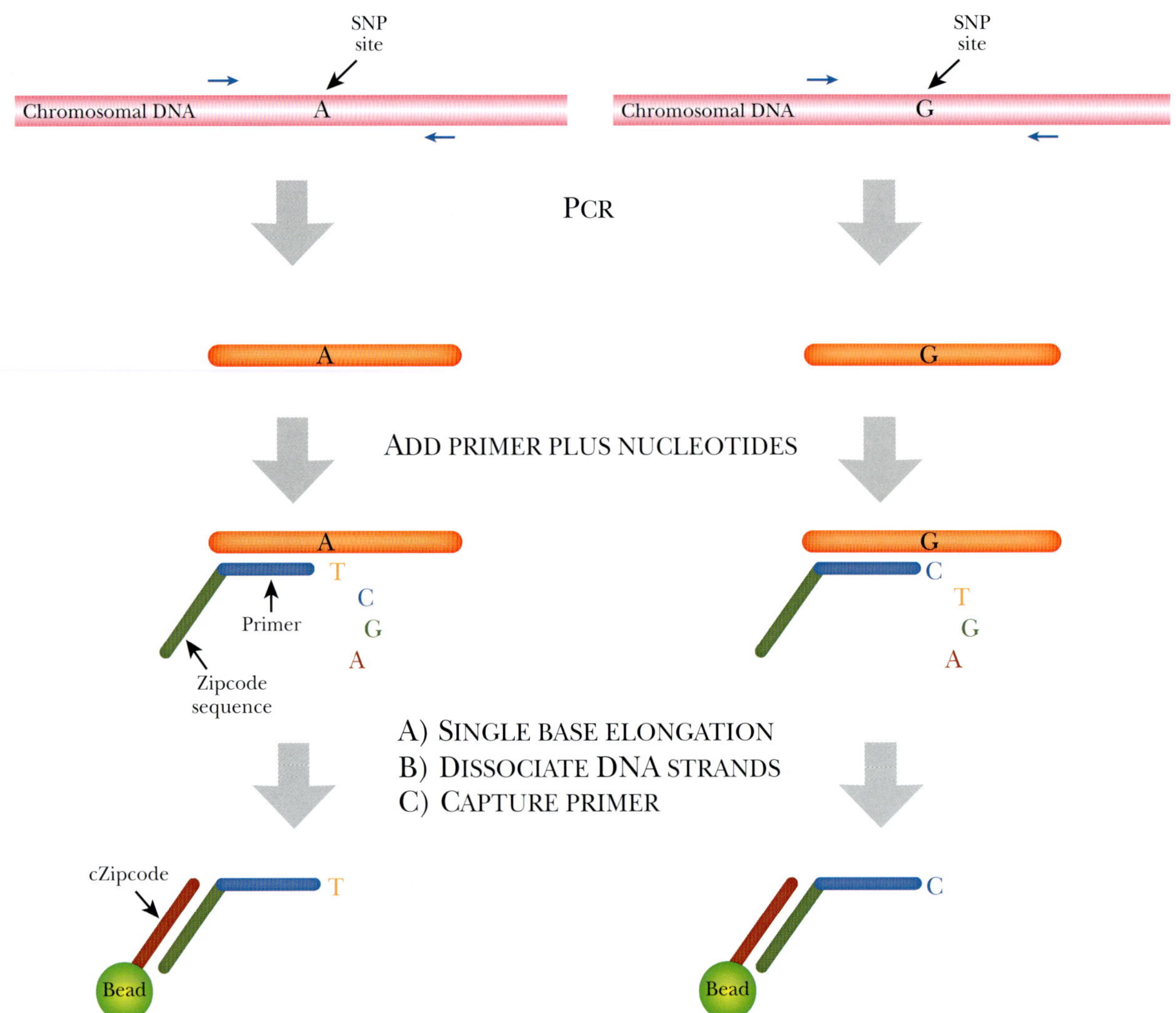

FIGURE 8.16 Zipcode Analysis and Single Base Extension of SNPs
A segment of DNA that includes a SNP site is generated by PCR (only a single strand of DNA is shown here for simplicity). Single base extension is performed with a primer that binds one base in front of the SNP. Person I has an A at the SNP site and therefore ddT is incorporated. In person II, a G at the same position results in incorporation of ddC. The bases are labeled with different fluorescent dyes. Use of dideoxynucleotides prevents addition of further bases. The elongated primer is then trapped by binding its Zipcode sequence to the complementary cZipcode, which is attached to a bead or other solid support for easy isolation.

One of the spin-offs from the Human Genome Project is the Pharmacogenetics and Pharmacogenomics Knowledge Base (PharmGKB; http://www.pharmgkb.org/). This records genes and mutations that affect drug response. Consider asthma, a condition where people overreact to inhaled irritants by cutting airflow in and out of the lungs. The muscle cells around the bronchial tubes constrict, decreasing airflow. Albuterol is a drug used to open the airways in people with asthma. This drug opens the bronchial tubes by relaxing the muscle cells. Albuterol affects the beta2-adrenergic receptor, and mutations in this receptor alter the efficacy of albuterol. A single nucleotide change that replaces glycine at position 16 with arginine gives a receptor protein with a better response to albuterol. Whether a patient has this SNP will determine whether or not albuterol will be effective.

Another key area of pharmacogenetics concerns the cytochrome P450 family of enzymes. These play a role in the oxidative degradation of many foreign molecules, including many pharmaceuticals. The CYP2D6 isoenzyme oxidizes drugs of the tricyclic antidepressant class, and different alleles of this enzyme affect how well a person metabolizes these drugs. Much as for albuterol, identifying which allele a patient has will prevent overdosages or adverse reactions. As time goes on, more medical treatments will be designed for the individual rather than the average person.

> Pharmacogenetics is the study of inherited differences in drug metabolism and response.
> Some SNPs affect how a person metabolizes a certain drug. By determining what SNP correlates with what drug sensitivity, new patients can be screened and possibly avoid adverse drug reactions (ADRs).

GENE EXPRESSION AND MICROARRAYS

As noted earlier, one of the major issues in determining the correct number of genes is deciding whether or not a sequence is really a gene. Measuring whether a gene is transcribed into mRNA is the first step to deciding whether or not a presumed gene is real. Gene expression was traditionally done on a single gene basis, but now, **functional genomics** studies gene expression using the entire genome. Functional genomics encompasses the global study of all the RNA transcribed from the genome—the **transcriptome**, all the proteins encoded by the genome—the **proteome** (see Chapter 9); and all the metabolic pathways in the organism—the **metabolome** (see Chapter 9).

Because the entire human genome contains only around 25,000 different genes, using microarray technology to study gene expression is feasible. **DNA microarrays**, or **DNA chips**, contain thousands of different unique sequences bound to a solid support, such as a glass slide. Microarrays are based on the principle of hybridization between a "probe" and target molecules in the experimental sample. However, in a microarray the probes are attached to the solid support and the experimental sample is in solution. The microarray represents the genome of the organism being tested and consists of sequences corresponding to each gene in the organism.

To monitor gene expression, RNA extracted from a cell sample is tested against a microarray. The experimental RNA sample is usually fluorescently tagged. Hybridization of the mRNA from the experimental sample to the DNA probes on the solid support indicates whether or not a gene is expressed and to what degree. The level of fluorescence at each point on the array correlates with the level of the corresponding mRNA in the sample.

Microarrays can be used to hybridize to RNA isolated from cells grown under a variety of experimental conditions—for example, heat shock, cancer, or other disease states. The same array can be hybridized to two or even more samples of RNA (control versus experimental), thus comparing gene expression. For this comparison, each RNA sample is labeled with a different fluorescent dye, for example, red and green. If a particular RNA is present in only one of the samples, the corresponding spot on the microarray will be red or green (Fig. 8.17), whereas, if that particular RNA is present in both experimental and control samples, the spot will be yellow (i.e., red plus green). Modern arrays can accommodate thousands or millions of different probes, allowing the entire genome for most organisms to be examined at once. Some arrays are clustered so that all the genes involved in protein synthesis are together, or the genes involved with heat shock are clustered. The computer reads the color and fluorescence intensity for each of the spots and carries out an analysis.

The results can provide a global view of gene expression in different experimental conditions. This powerful technique

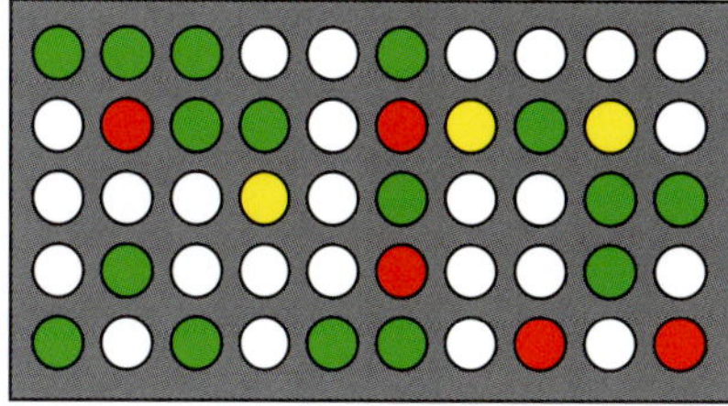

FIGURE 8.17 DNA Chip Showing Detection of mRNA by Fluorescent Dyes DNA chips can monitor many different mRNAs at one time. Each spot on the grid has a different DNA sequence attached. To determine which genes are expressed under which conditions, mRNA is isolated and each sample is labeled with a different fluorescent dye. If two different dyes are used (as shown here), the same chip can be used for both. It is then visualized in three different ways: one shows only the red dye, another only the green, and the third merges the two images so overlapping spots look yellow.

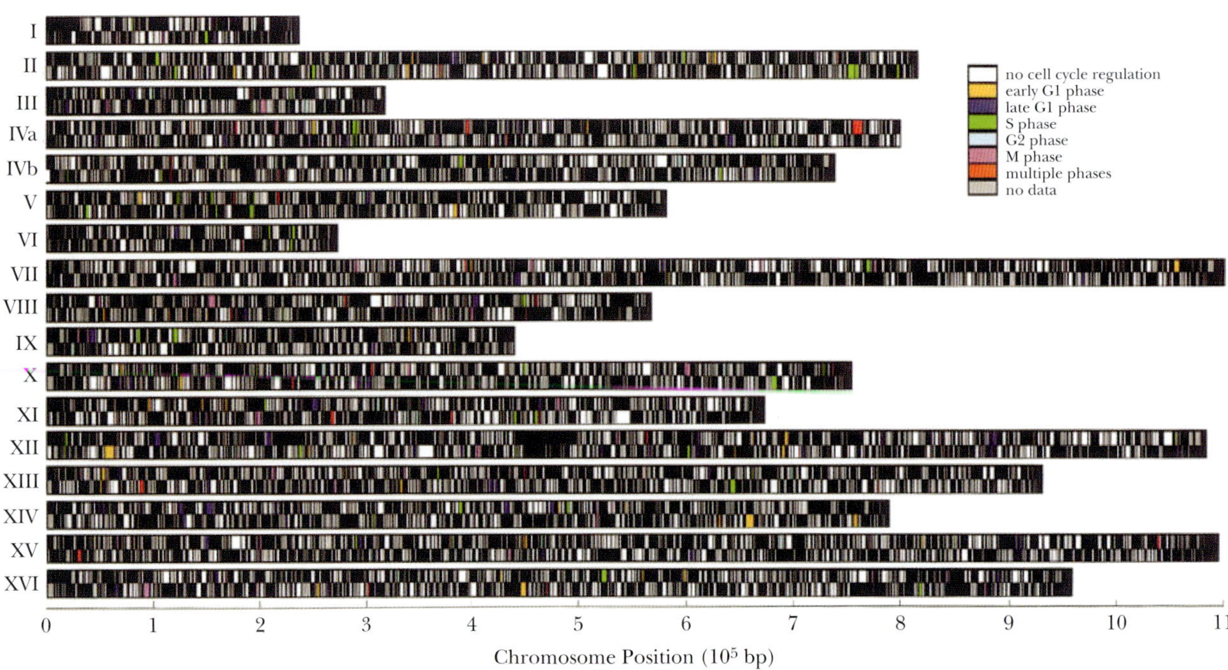

FIGURE 8.18 Gene Expression during Yeast Cell Cycle
Color coding indicates the time of the cell cycle for maximum gene expression. More than 800 different genes that respond to changes in the cell cycle were monitored on the 16 yeast chromosomes. From: Cho *et al.* (1998). A genome-wide transcriptional analysis of the mitotic cell cycle. *Mol Cell* **2**, 65–73. Reprinted with permission.

allows the researcher to analyze thousands to millions of sequences at one time. For example, slides can be made with every named gene from the yeast genome. These genes may be analyzed for expression at different stages of the cell cycle. Thus, a culture of yeast can be synchronized and arrested at different stages of the cell cycle by adding α factor or by using mutant yeast that freeze at particular stages of mitosis. The gene expression patterns for each stage are compared and compiled (Fig. 8.18).

Functional genomics encompasses the global study of all the RNA transcribed from the genome.

DNA microarrays, or DNA chips, contain thousands of different unique sequences bound to a solid support, such as a glass slide. When fluorescently labeled RNA is incubated on top of the microarray, complementary sequences hybridize. The amount of fluorescence corresponds to the amount of RNA that is bound to the DNA microarray.

MAKING DNA MICROARRAYS

There are two major types of DNA microarrays: one type contains cDNA fragments 600 to 2400 nucleotides in length, and the other uses oligonucleotides of 20 to 50 nucleotides in length. Each type of microarray is manufactured differently. When making a cDNA microarray, each of the different probes must be chosen independently and made by PCR or traditional cloning. Then all the DNA probes are spotted onto the slide. When making an oligonucleotide array, the oligonucleotide is synthesized directly on the slide.

cDNA Microarrays

The first step in making a cDNA microarray is to determine the numbers and types of probes to attach to the solid support. Since entire genomes have been sequenced for a variety of organisms, identifying potential genes is relatively easy. During the sequencing of these genomes, many cloned segments of DNA containing all or part of various genes were generated. Researchers can either obtain these clones or amplify genes from a sample of DNA using PCR. Each PCR product must be purified before attachment to the glass slide, so that all the extra nucleotides, *Taq* polymerase, and salts are removed and only pure DNA attaches to the slide. Pure cDNA samples can be used directly.

The next step is to create the chip using a microarray robot. A purified sample of each DNA is put into small wells arranged in a grid in microtiter plates. The size of the grid depends on the number of probes. If every predicted human gene is present once, approximately 25,000 different wells are needed. In practice, probes for each gene are attached more than once, in different areas of the chip, to provide several readings for each gene. A grid of pens or quills is dipped into the wells, one pen for each well, using a robotic arm. The pen tips are then touched to a glass slide, where a small drop of DNA is left behind. The robotic arm continues to manufacture spotted slides until the DNA in the well is used up. The use of a microarray robot makes each chip cheap and easy to produce. Finally, the DNA is crosslinked to the glass slide with ultraviolet light, which causes thymines in the DNA to crosslink to the glass. Figure 8.19 shows DNA on a microarray grid visualized by an atomic force microscope.

FIGURE 8.19 AFM of DNA on Microarray
A region of a yeast microarray after hybridization. The DNA is clearly deposited in sufficient density to permit many strand-to-strand interactions. The width of the figure represents a scanned distance of 2 micrometers. Reprinted by permission from Macmillan Publishers Ltd.: Duggan DJ, Bittner M, Chen Y, Meltzer P, Trent JM. ***Nature Genetics*** 12, 82–89, copyright 1999.

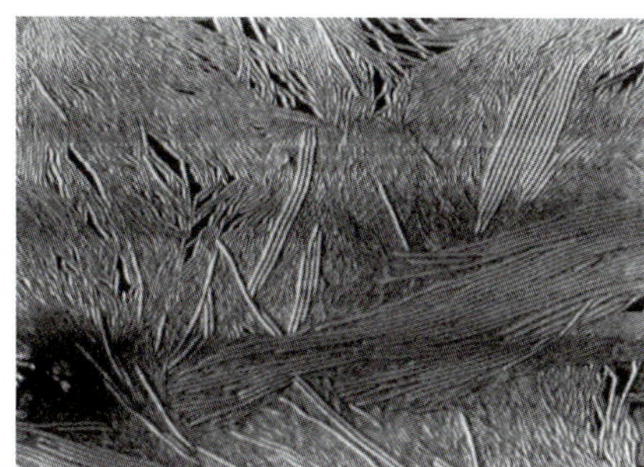

Newer technology has been developed to decrease the variation in size of spotted samples in microarrays. In newer cDNA microarrays, the samples are spotted onto a glass slide using inkjet printer technology. The cDNA samples are sucked into separate chambers of the inkjet printer head, and then spotted onto the glass slide much as ink is spotted onto paper in a printer. Inkjet technology prevents variations in size and quantity of cDNA in the sample spots. Special adaptors have been developed to prevent the inkjet sample channels from mixing, thus preventing cross-contamination.

Oligonucleotide Microarrays

Oligonucleotides are traditionally synthesized chemically on beads of controlled pore glass (CPG; see Chapter 4). Therefore, it is not too much of a logical leap to synthesize many different oligonucleotides side-by-side on a glass slide. The main differences between synthesizing single nucleotides on beads versus making arrays on glass slides is that the array has thousands of different oligonucleotides, and each must be synthesized in its proper location with a unique sequence. To accomplish this, **photolithography** and solid-phase DNA synthesis are combined (Fig. 8.20). Photolithography is a process used in making integrated circuits, where a mask makes a specific pattern of light on a solid surface. The light activates the surface it reaches, while the remaining surface remains inactivated.

A glass slide is first covered with a spacer that ends in a reactive group. This is then covered with a photosensitive blocking group that can be removed by light. In each synthetic cycle,

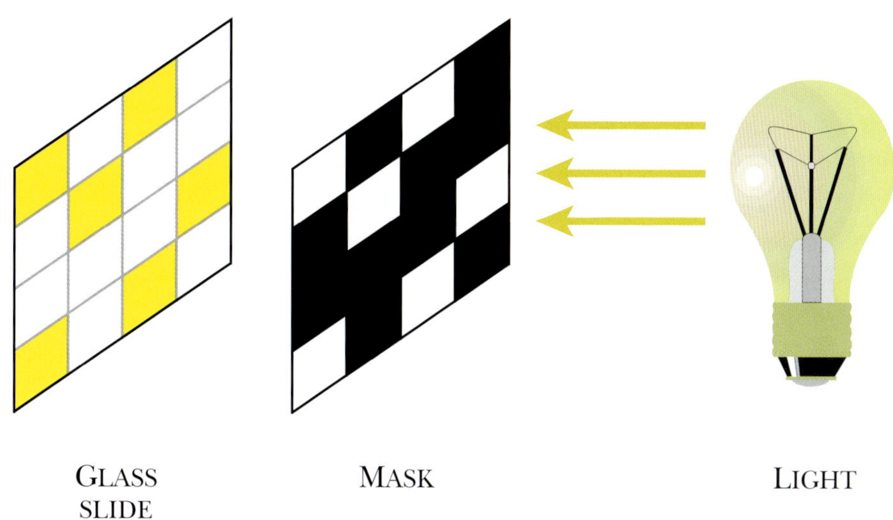

FIGURE 8.20 Photolithography
Light passing through the mask makes a particular pattern on the glass slide. If the slide is coated with a light-activated substance, only the regions that are illuminated will be activated for the addition of another nucleotide.

those sites where a particular nucleotide will be attached are illuminated to remove the blocking group. Each of the four nucleotides is added in turn. At each addition, a mask is aligned with the glass slide. Light passes through holes in the mask and activates the ends of those growing oligonucleotide chains that it illuminates. Much as in traditional chemical synthesis, each nucleotide has its 5′-OH protected. Thus after each addition, the end of the growing chain is blocked again. These protective groups are light activated, so at each step, a new mask is aligned with the slide, and light deprotects the appropriate nucleotides. The entire process continues for each nucleotide at each position on the glass slide. Making the masks is the key to this technology (Fig. 8.21).

DNA microarrays are made with cDNA or oligonucleotides.

Microarrays are created by spotting pure samples of cDNA clones onto the glass in a small spot. The DNA is cross-linked to the glass with UV light.

Oligonucleotide arrays are created by synthesizing the DNA directly on the glass slide.

HYBRIDIZATION ON DNA MICROARRAYS

Hybridization on a microarray is similar to the hybridization of DNA during other hybridization experiments, such as Southern blots, Northern blots, or dot blots. All these techniques rely on the complementary nature of double-stranded DNA. When two complementary strands of DNA are near each other, the bases match up with their complement, that is, thymine with adenine, and guanine with cytosine. On a DNA microarray, hybridization is affected by the same parameters as in these other techniques.

How the DNA is attached to the slide can affect how well the probe DNA and target DNA hybridize, especially for oligonucleotide microarrays (Fig. 8.22). The short length of oligonucleotides requires that the entire piece be accessible to hybridize. The length of the spacer between the oligonucleotides and the glass slide optimizes hybridization. An oligonucleotide attached with a short spacer has many of its initial nucleotides too close to the glass and inaccessible to incoming RNA or DNA. Oligonucleotides with longer spacers may fold back and tangle up; therefore, again the sequence is inaccessible for hybridization. Oligonucleotides attached with medium-sized spacers are far enough from the glass, but not so far as to get tangled. Thus medium-sized spacers give the best accessibility for hybridization.

Hybridization of two lengths of DNA (or RNA with DNA) requires certain sequence features. One important property is the relative number of A:T base pairs versus G:C base pairs. Because G:C base pairs have three hydrogen bonds holding them together, it takes more energy to dissolve the bonds. A:T base pairs have only two hydrogen bonds and require less energy. Thus more GC base pairs give stronger hybridization. If the sequence has many A:T base pairs, the duplex may form slowly and be less stable. Another important consideration is secondary structure. If the probe sequence can form a hairpin structure, it will hybridize poorly with the target. If the probe has several mismatches relative to the target, the duplex may not form efficiently. All these issues must be addressed when making an oligonucleotide microarray. Computer programs are available that identify suitable regions of genes with sequences that will produce effective probes.

cDNA arrays are less prone to the problems seen in oligonucleotide arrays. cDNAs are double-stranded, so secondary structures such as hairpins are less likely to be a problem. During a hybridization reaction, cDNA arrays must be denatured either with heat or chemicals, making the probes single-stranded. Then the single-stranded RNA samples are allowed to hybridize on the slide under conditions that promote duplex RNA:cDNA without any mismatches.

Oligonucleotide microarrays must have a sufficient spacer and little secondary structure in order to hybridize with the samples.

1)

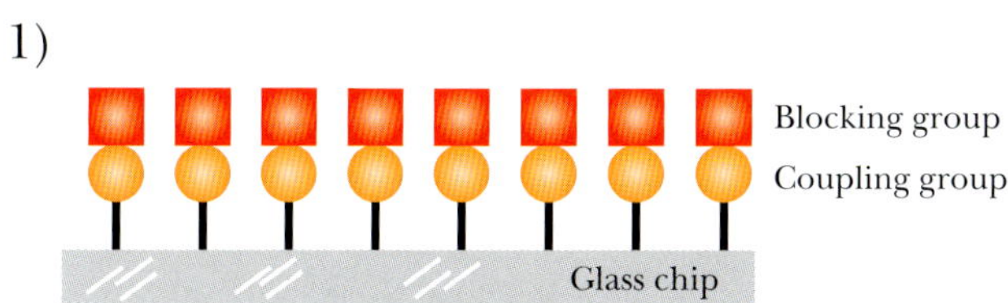

LIGHT TREATMENT

2)

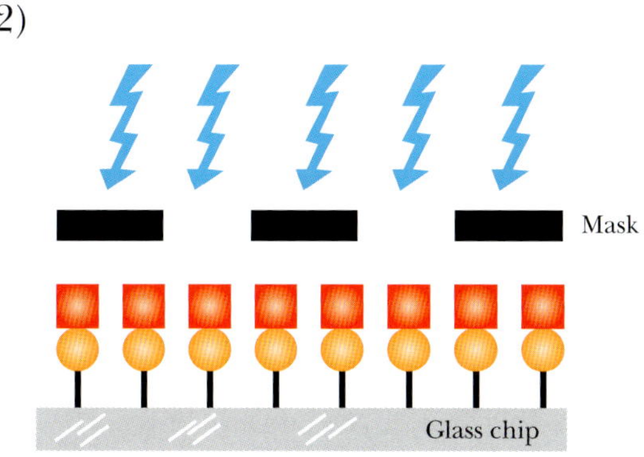

COUPLING GROUPS REVEALED

3)

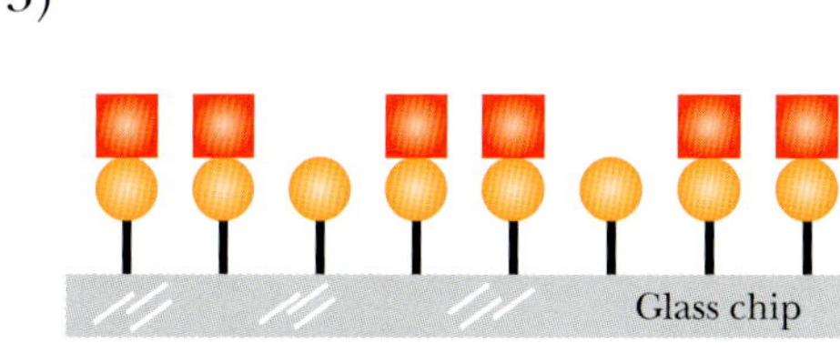

COUPLE WITH ACTIVATED AND BLOCKED G

4)

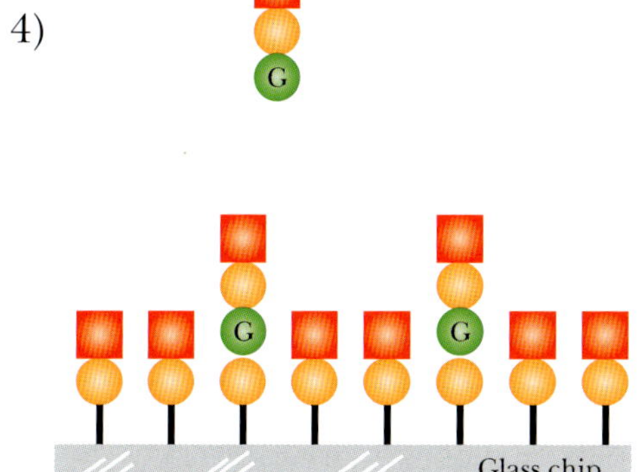

LIGHT TREATMENT

5)

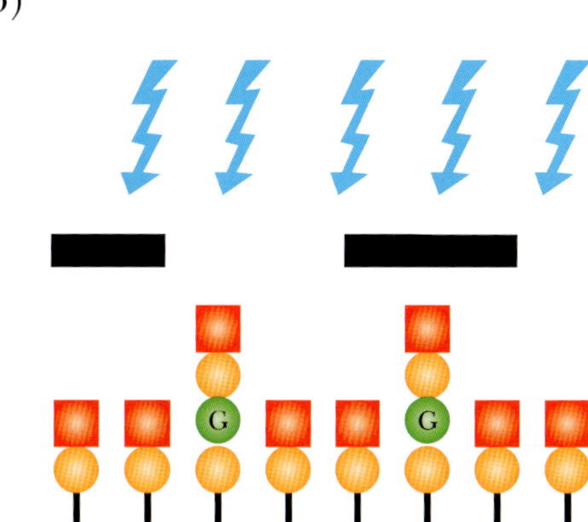

COUPLING GROUPS REVEALED

6)

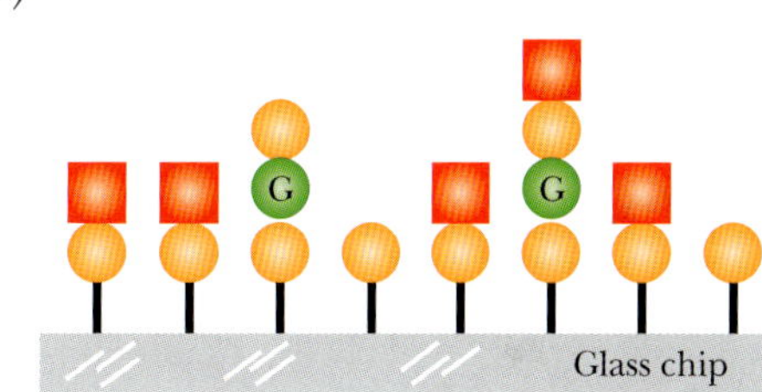

COUPLE WITH ACTIVATED AND BLOCKED T

7)

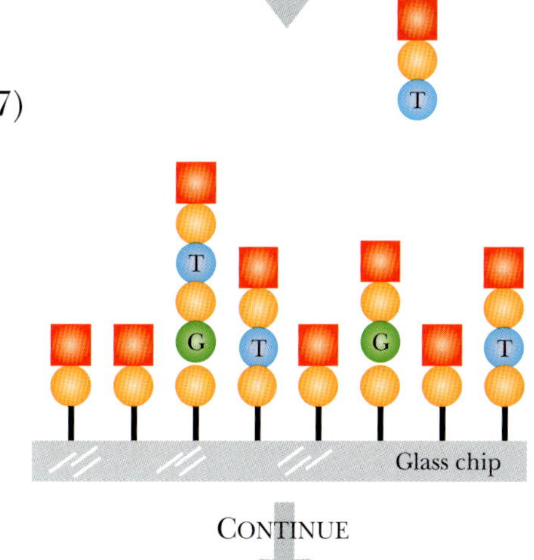

CONTINUE

FIGURE 8.21 On-Chip Synthesis of Oligonucleotides

Arrays may be created by chemically synthesizing oligonucleotides directly on the chip. First, spacers with reactive groups are linked to the glass chip and blocked. Then each of the four nucleotides is added in turn (in this example, G is added first, then T). A mask covers the areas that should not be activated during any particular reaction. Light activates all the groups not covered with the mask, and a nucleotide is added to these. The cycle is repeated with the next nucleotide.

MONITORING GENE EXPRESSION USING WHOLE-GENOME TILING ARRAYS

Whole-genome tiling arrays (WGAs) are oligonucleotide microarrays that cover the entire genome. The first entire genome to be represented by a whole-genome array was from *Arabidopsis*. A gene chip was designed to have 25-mer oligonucleotides that overlapped each other and covered the entire sequence of the genome. Complementary oligonucleotide sequences were tiled back to back along each entire chromosome and ordered so that the array could be conveniently analyzed for gene expression (Fig. 8.23).

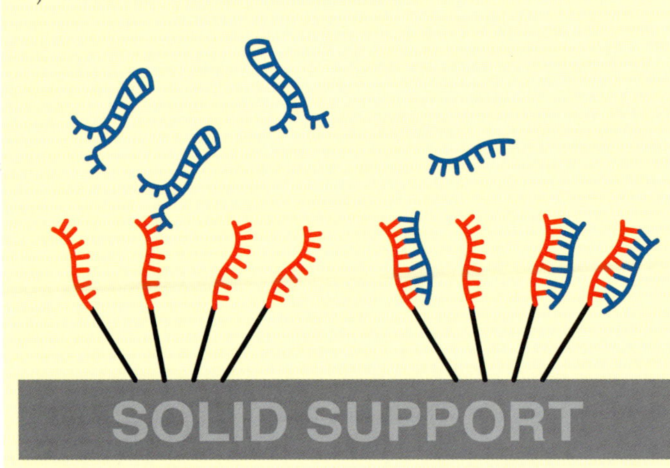

FIGURE 8.22
Length of Spacers and Target Molecules Affect Hybridization on Microarrays
(A) When the spacer between the glass slide and oligonucleotide is too short, the oligonucleotides are condensed and not accessible to hybridize. If the spacer region is too long, the oligonucleotides and spacers tangle and fold, preventing optimal hybridization. (B) When the target for hybridization is too long, the target sequences may form hairpins with themselves rather than bind to the array oligonucleotides.

For the human genome, tiling arrays have been made to cover the entire sequences of chromosomes 21 and 22. These also use 25-mer oligonucleotides, but rather than being overlapped, the oligonucleotides were spaced 35 base pairs apart along the sequence. These are therefore strictly only "quasi-whole-genome arrays." Compared to arrays that include only known genes, tiling arrays have the potential to identify novel regions that are transcribed, whether these encode unknown protein-encoding genes or nontranslated RNA. The RNA extracted from many different cell lines and tissues has been used to monitor gene expression, assess differences in splicing patterns, find new genes, and find RNA-binding protein target sequences.

The most interesting finding from studying human chromosomes 21 and 22 is that much larger portions of these chromosomes are transcribed into mRNA than previously predicted from computer analyses of exon regions. About 90% of the transcribed regions occurred outside the known exons. The majority of the transcribed regions generated noncoding RNA, mostly of less than 75 base pairs in length. This suggests that noncoding RNA may have a much greater role in human biology than previously thought. These arrays have also identified new exons that were previously unknown. In addition, these arrays can identify novel alternatively spliced proteins. The WGAs for chromosomes 21 and 22 have also been used to compare the level of expression of exons within the same gene. About 80% of the genes had exons with varied levels of expression, implying most genes have some sort of alternate splicing.

Another potential use for whole genome arrays is to analyze results of **chromatin immunoprecipitation (ChIP)**. ChIP begins by crosslinking all the various transcription factors to chromatin, essentially freezing them in place. Next the chromatin is sheared into smaller fragments, and the DNA/transcription factor complexes are isolated. Affinity purification isolates one particular transcription factor from all the others (e.g., antibodies to the transcription factor Jun isolates all the Jun/DNA complexes from this mixture). Finally the DNA sequences that are bound to the chosen transcription factor are identified using WGA. The entire procedure, including the analysis on a gene chip, is called **ChIP-chip**. This type of analysis can precisely identify transcription factor binding sites on a variety of genes. Curiously, binding locations for NF-κB, for example, have been found within both coding and noncoding regions, such as introns or the 3′ ends of genes. These surprising findings suggest that transcription factors may also function outside of the traditional upstream promoter region.

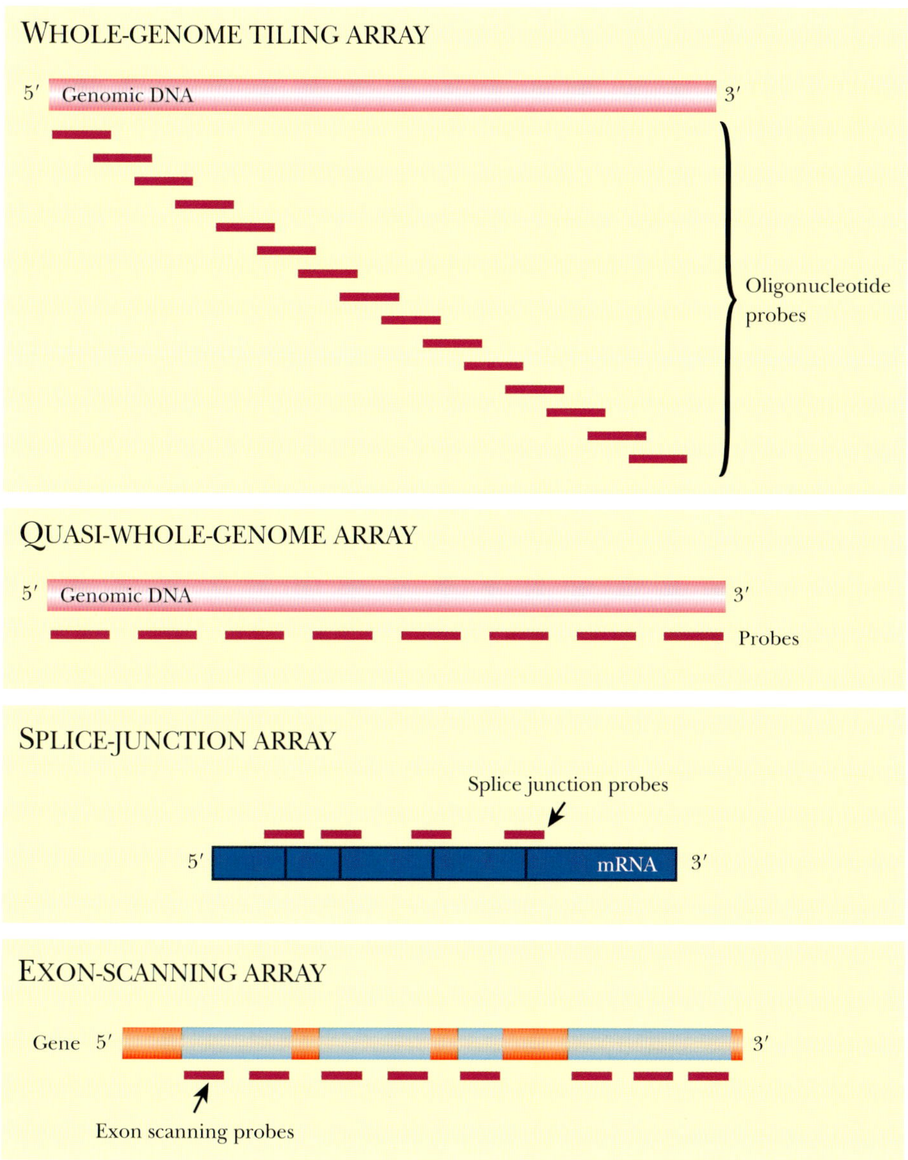

FIGURE 8.23 Whole-Genome Array Designs
Whole-genome arrays (WGAs) contain oligonucleotide probes that cover the entire genome in an overlapping set. In quasi-whole-genome arrays, the probes are spaced equal distances apart through the genome. The probes thus cover the entire genome, except for the gaps between the probes. Splice junction arrays have only probes that span the upstream and downstream regions of known splice junctions in mRNA. Exon-scanning arrays have probes derived from exon sequences only.

Another use for WGA is to identify regions of the genome that are methylated. Methylation prevents the inappropriate expression of various genes, especially those used only during development of young organisms, or those genes from transposons or viruses that could be detrimental. Cancerous cells have methylation patterns much different from those of normal cells, suggesting that this type of regulation is critical to proper growth control of normal cells. In order to identify the methylated regions, genomic DNA is first treated with **sodium bisulfite**, which deaminates nonmethylated cytosine to uracil, yet does not affect methylated cytosine. The treated DNA is then hybridized to a WGA. Those regions with nonmethylated cytosine no longer hybridize to the array because the cytosines have been converted to uracil (which pairs with A, not G). Those regions of the genome that are methylated still hybridize well because methylated cytosine and guanine form a stable base pair.

Of course, finding genetic variations and polymorphisms is critical to genome analysis, and whole-genome arrays offer a nonbiased method to analyze samples. In fact, a WGA that has the reference sequence for the human genome can be used to identify and catalogue all different types of polymorphisms, including SNPs, VNTRs, and repetitive elements. In fact, an overlapping WGA made to the entire reference sequence of the human genome spaced at a single base pair could be used to effectively resequence the entire genome with ease and speed.

Whole-genome arrays are oligonucleotide arrays that have sequences which cover the entire genome. These can be used to identify transcription factor binding sites, regions of methylation, SNPs, VNTRs, repetitive elements, and so forth.

MONITORING GENE EXPRESSION OF SINGLE GENES

Although microarrays provide a global view of gene expression, the results do not provide much specific information on individual genes. Once candidate genes have been identified by microarrays, these are analyzed individually. To determine the exact location and/or level of expression, the gene of interest is genetically linked to a reporter gene (see later discussion), creating a **gene fusion**. The regulatory region of the gene of interest is isolated first. This segment is normally found upstream of the gene of interest and includes sites for transcription factors to bind, plus various enhancer elements. The coding sequence of the gene of interest is replaced with the reporter gene, so that the regulatory elements now control the reporter gene rather than the original gene of interest.

Reporter genes usually encode enzymes whose activity is easy to assay. One of the most widely used reporters is the ***lacZ* gene** from *E. coli*, which encodes the enzyme **β-galactosidase** (Fig. 8.24). This enzyme splits disaccharide sugar molecules into their monomers, but also cleaves various artificial substrates. When the substrate ONPG is cleaved, one of the cleavage products forms a visible yellow dye. When X-Gal is cleaved by β-galactosidase, one of the products reacts with oxygen to form a blue dye.

The ***phoA* gene** is another reporter gene that encodes **alkaline phosphatase**, which removes phosphate groups from many different substrates (Fig. 8.25). Artificial substrates are designed so that when the phosphate is removed, they either change color or fluoresce.

Another popular reporter gene is **luciferase**, which emits a pulse of visible light when the correct substrate, **luciferin**, is supplied (Fig. 8.26). Luciferase is an enzyme encoded by the ***lux* gene** in bacteria or the ***luc* gene** in fireflies. The two luciferases are not related and have different enzyme mechanisms. Both genes work well as reporter genes and have been cloned onto vectors that work in a variety of different organisms. Detecting the light emitted by luciferase is difficult because of its low levels and requires special equipment such as a luminometer or scintillation counter.

Another popular reporter protein is **green fluorescent protein** or **GFP**, which is not an enzyme (Fig. 8.27). This protein has natural fluorescence that does not require any cofactors or substrates. Additionally, the fluorescence is active in living tissues, so that when the protein is expressed the organism gains a green fluorescence. This is especially noticeable when the organism is transparent like zebrafish or *Caenorhabditis elegans*. GFP is excited by long-wavelength UV light of 395 nm and then emits light at the green wavelength of 510 nm. The original protein is from the jellyfish *Aequorea victoria* and is encoded by the gene *gfp*. Many new variants of GFP have been developed that emit light at different wavelengths, including red, blue, and yellow. The main advantage of using GFP as reporter is the ability to see expression in living tissues.

Other techniques are useful to confirm data obtained with gene expression microarrays. Differential display PCR (see Chapter 4) is useful to compare mRNA expression patterns

I

GALACTOSE β(1,4) GLUCOSE
= LACTOSE

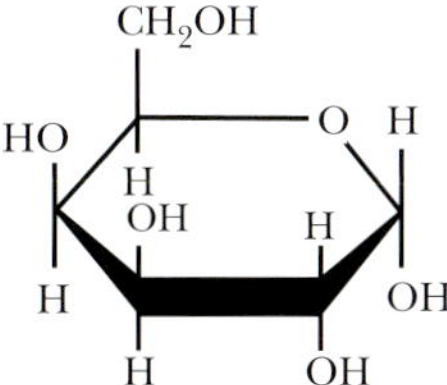

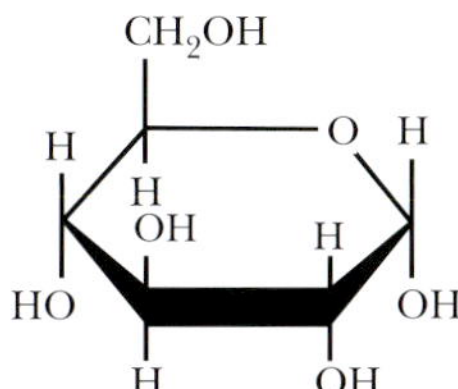

II

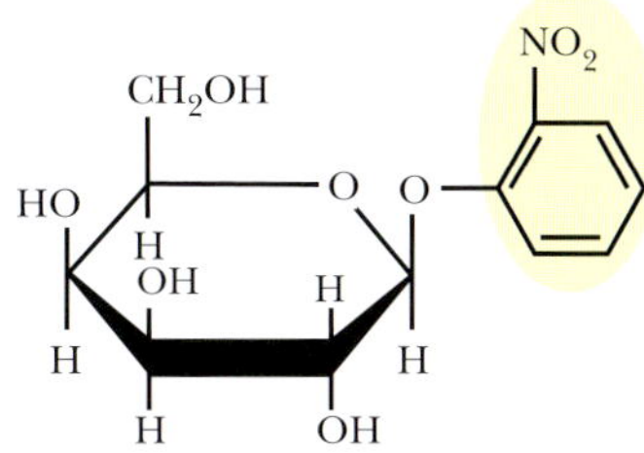

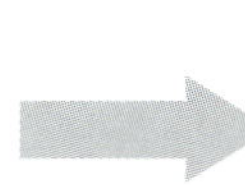

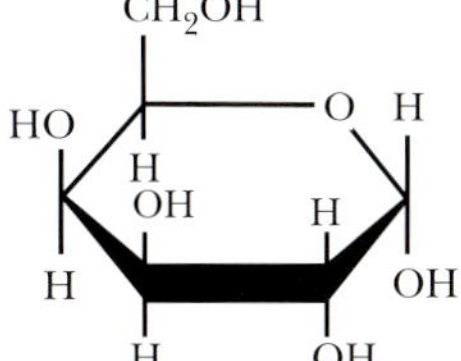

OH
NO2
o-NITROPHENOL
bright yellow

III

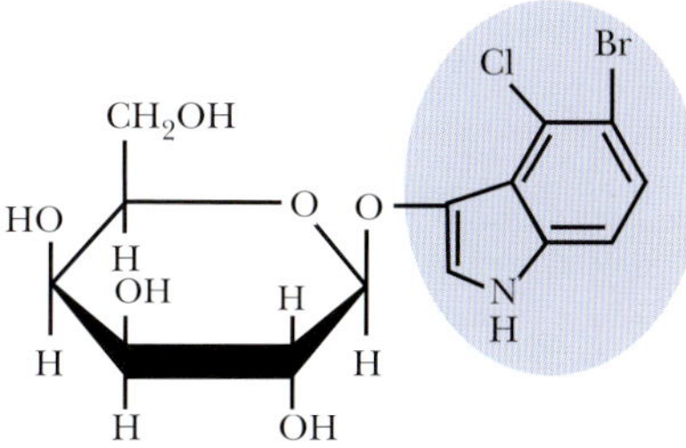

5-BROMO-4-CHLORO-3-
INDOLYL GALACTOSIDE
= X-GAL

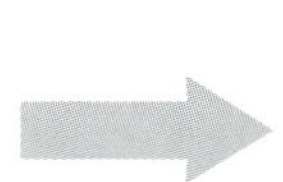

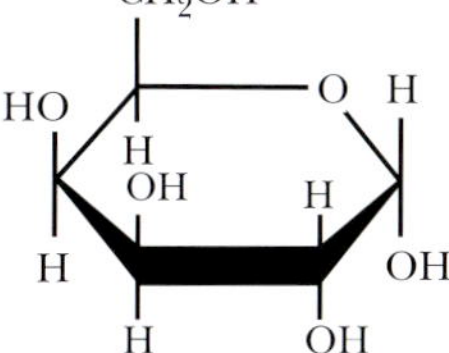

D-GALACTOSE

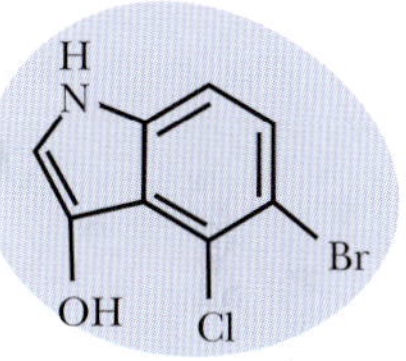

5-BROMO-4-CHLORO-
3-INDOXYL
unstable

SPONTANEOUSLY
REACTS WITH
OXYGEN IN AIR

INDIGO-TYPE DYE
dark blue and insoluble

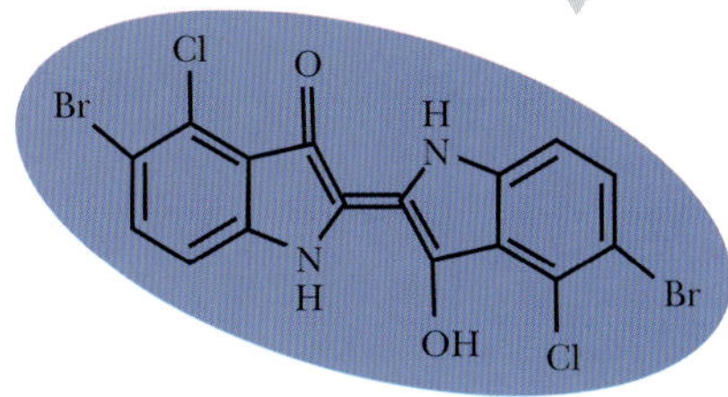

FIGURE 8.24 β-Galactosidase Has Multiple Substrates

The enzyme β-galactosidase normally cleaves lactose into two monosaccharides, glucose and galactose. β-Galactosidase also cleaves artificial substrates, such as ONPG and X-Gal, releasing groups that form visible dyes. ONPG releases the bright yellow substance *o*-nitrophenol, whereas X-Gal releases an unstable group that reacts with oxygen to form a blue indigo dye.

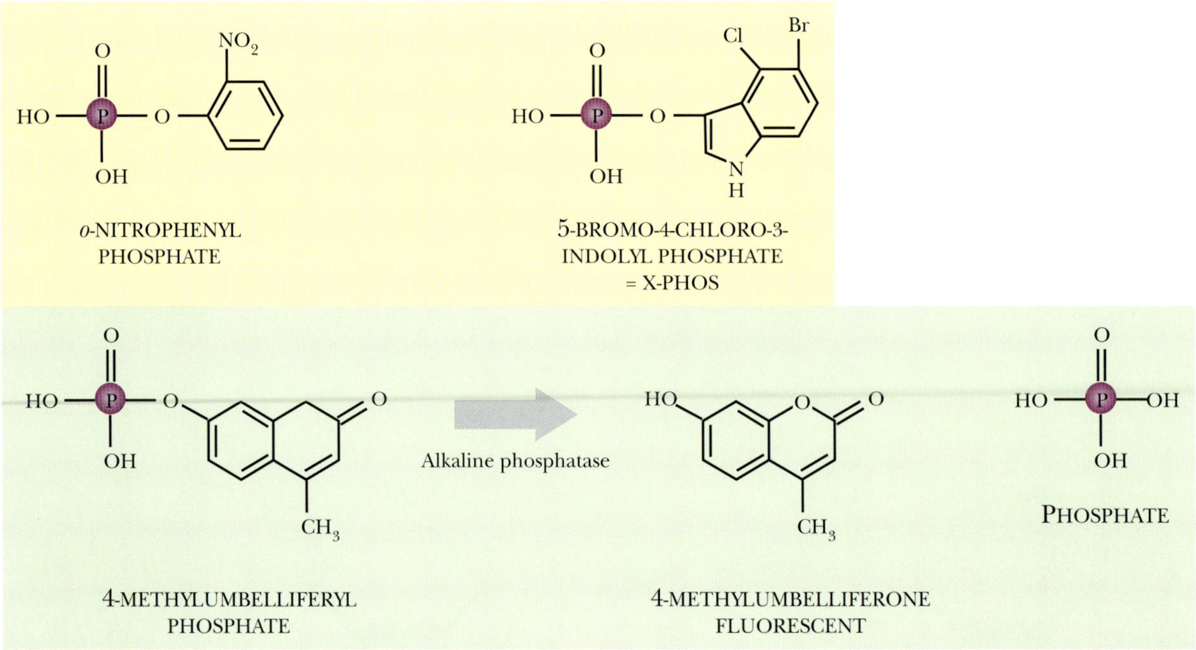

FIGURE 8.25 Substrates Used by Alkaline Phosphatase
Alkaline phosphatase removes phosphate groups from various substrates. When the phosphate group is removed from *o*-nitrophenyl phosphate, a yellow dye is released. When the phosphate is removed from X-Phos, further reaction with oxygen produces an insoluble blue dye. When the phosphate is removed from 4-methylumbelliferyl phosphate, this releases a fluorescent molecule.

Bacterial Luciferase:

$FMNH_2 + O_2 + R\text{-}CHO \rightarrow R\text{-}COOH + FMN + H_2O + \text{light}$

Firefly Luciferase:

$\text{Luciferin} + O_2 + ATP \rightarrow \text{oxidized luciferin} + CO_2 + H_2O + AMP + PPi + \text{light}$

FIGURE 8.26 The Luciferase Reaction Emits Light from Luciferin
Luciferase from bacteria uses a long-chain aldehyde, oxygen, and the reduced form of the cofactor FMN (flavin mononucleotide) as its luciferin. Firefly luciferase uses ATP, oxygen, and firefly luciferin to produce light.

from different tissue samples or experimental conditions. Subtractive hybridization can also identify genes that are expressed in one experimental condition and not elsewhere (see Chapter 3). Finally, Northern blot (see Chapter 3) analysis can monitor expression levels of mRNA that vary in different experimental conditions.

Fusing regulatory sequences from the gene of interest to a reporter gene approximates the actual expression pattern of the gene.

Reporter genes such as β-galactosidase, alkaline phosphatase, and luciferase are enzymes that cleave their substrates to form a visible dye or light. Green fluorescent protein has luminescent properties that allow it to absorb one wavelength of light and emit a longer wavelength.

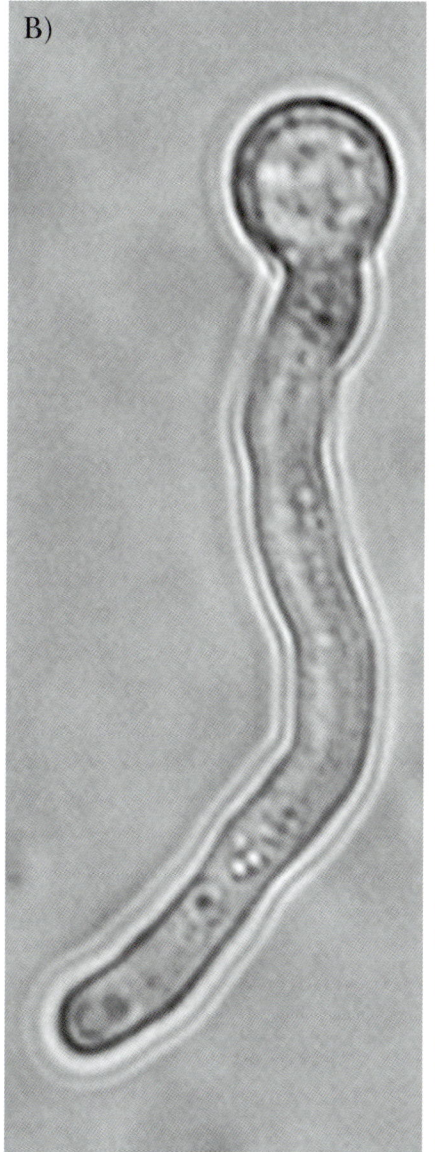

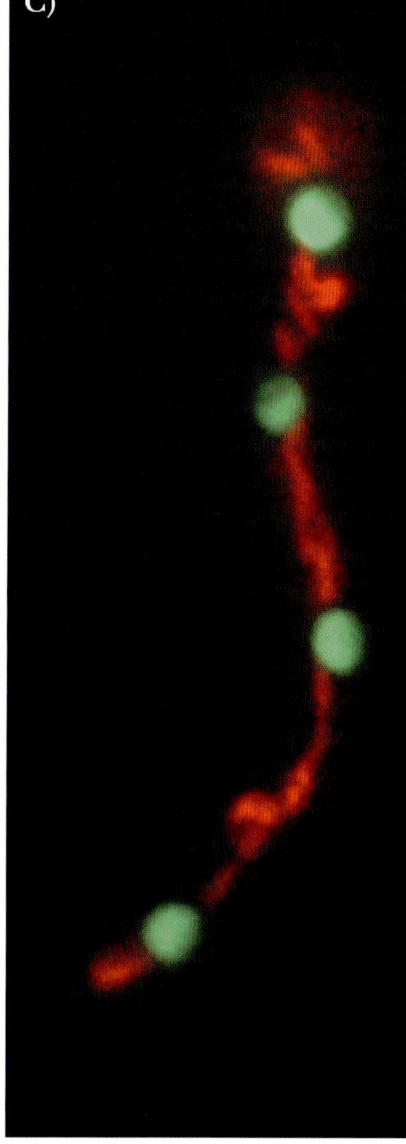

FIGURE 8.27
Transgenic Organisms with Green Fluorescent Protein
The gene for GFP has been integrated into the genome of animals, plants, and fungi. After exposure to long-wavelength UV, the organisms emit green light. (A) Transgenic mice with GFP among normal mice from the same litter. The *gfp* gene was injected into fertilized egg cells to create these mice. GFP is produced in all cells and tissues except the hair. Credit: Eye of Science, Photo Researchers, Inc. (B) Phase contrast and (C) fluorescent emission of germlings of the fungus *Aspergillus nidulans*. Original GFP was used to label the mitochondria and a red GFP variant (DsRed) for the nucleus. From: Toews *et al.* (2004). *Curr Genet* **45**, 383–389.

Box 8.1 Endomesoderm Specification in Sea Urchin Embryos

Using a variety of techniques together can elucidate the network of gene regulation controlling the formation of an entire embryo. In the sea urchin *Strongylocentrotus purpuratus*, the embryo undergoes a specific set of spatial and temporal events that control development of the endomesoderm. The precursor cells of the blastula start the process and continue to develop and divide into the adult sea urchin. The control of the development is due both to altering gene expression and varying protein-protein interactions. Perturbing the function of these genes with various techniques, such as morpholino antisense mRNA, mRNA overexpression, and two-hybrid analysis in the sea urchin, together with methods to confirm location such as whole-mount *in situ* hybridization, has allowed a network of gene functions to be proposed (Fig. A). Arrows show each gene that exerts its influence on other genes or proteins. Such networks can be constructed and further tested to refine the model. The latest version of the endomesodermal network is at http://sugp.caltech.edu/endomes/.

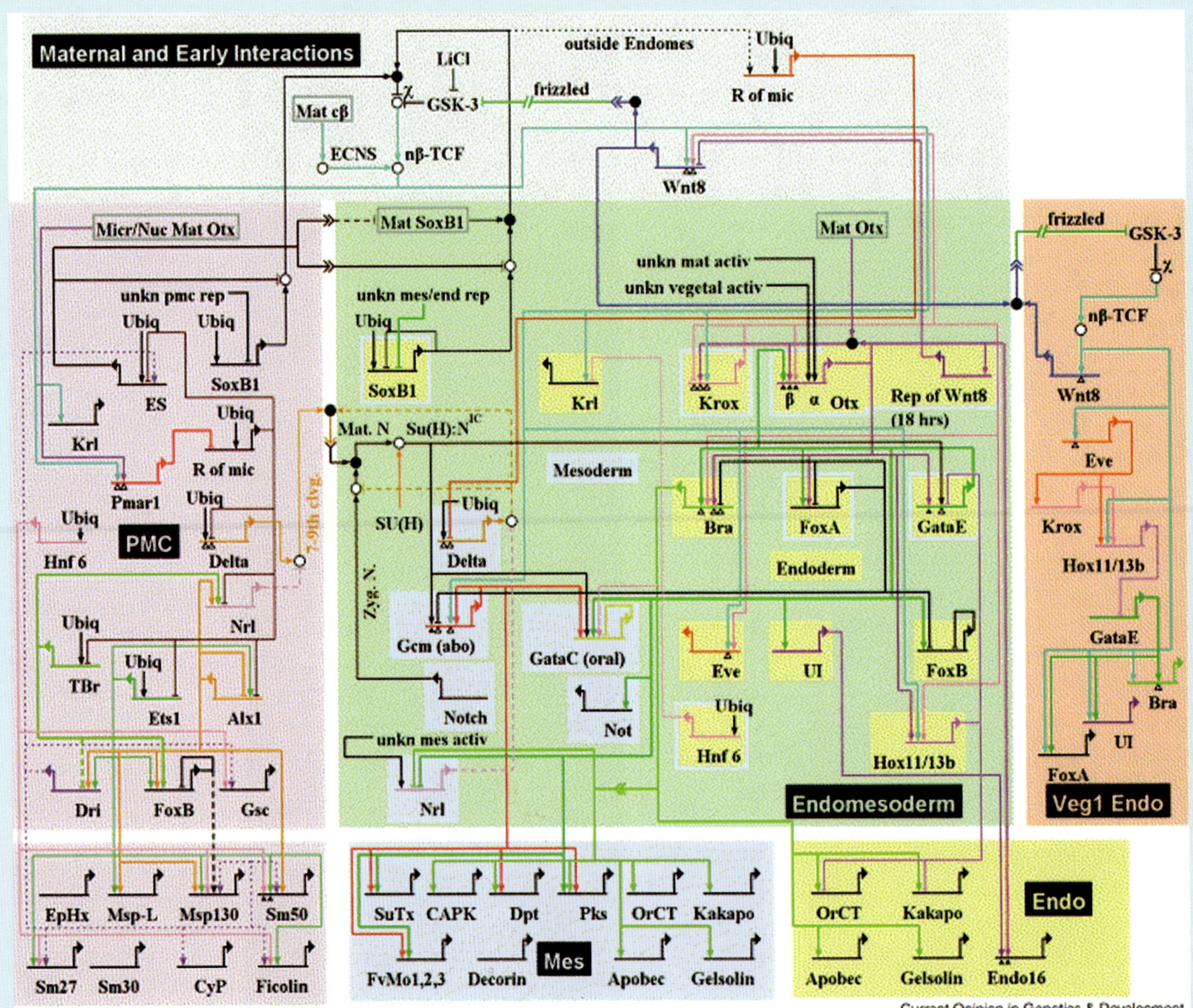

FIGURE A **Genome View of Endomesodermal Gene Regulatory Network**
The gene regulatory network is divided into spatial domains. Each gene is depicted as a short horizontal line from which extends a bent arrow indicating transcription. Genes are indicated by the names of the proteins they encode. From: Oliveri and Davidson (2004). Gene regulatory network controlling embryonic specification in the sea urchin. *Curr Opin Genet Dev* **14**, 351–360. Reprinted with permission.

Summary

Genomics is the study of all nucleotide sequences for an organism of interest, including genes, pseudogenes, noncoding regions, and regulatory regions. In human genomics, identifying the sequence of the entire set of chromosomes was a major achievement. The human genome sequence was determined by first forming DNA libraries, sequencing each of the clones, and finally ordering the sequences using computer algorithms, genetic maps, and physical maps. Without the great advances in computing, the Human Genome Project would have taken much longer and cost more money. Data mining on the information has identified many potential protein coding regions, regulatory elements, and different types of repetitive elements. The human genome is predicted to contain about 25,000 genes, SINES, LINEs, and tandem repeats such as telomeres, centromeres, and satellite DNA. Computer analysis of these complex sequences is called *bioinformatics.*

Genomics has changed many different fields of study, including medicine, evolutionary biology, and pharmacology. Understanding and identifying new genes related to diseases has changed the way new diseases are treated and diagnosed. Much of the textbook is devoted to these advances. The study of genomics focuses on mutations in the genome, by identifying single nucleotide polymorphisms, methylation patterns, or differences in tandem repeats. Mutations include single nucleotide changes, inversions, deletions, and insertions. Pharmacologists hope to correlate these differences with drug sensitivity, thus preventing adverse drug reactions. In evolutionary biology, physical features have always determined the relatedness of two organisms. Since the genomes of many organisms have been sequenced, the genetic code for highly conserved genes is used to determine relatedness. Over time, mutations accumulate within every genome. The more essential genes change slowly over time, whereas less essential genes incorporate more changes.

Genomics also encompasses gene expression, which is done on a global scale using genomic microarrays. These arrays have DNA from the genome, either a pure cDNA or a synthesized oligonucleotide, linked to a glass slide. The fluorescently labeled mRNA sample of interest is then hybridized to the microarray. When an mRNA hybridizes to the immobilized cDNA or oligonucleotide, that region will fluoresce. The amount of fluorescence correlates to the amount of mRNA in the sample. These microarrays are very flexible and can be designed to cover the whole genome as in whole-genome arrays, or they can be designed to a subset of genes. In single gene analysis, specific regulatory regions defined by genomics are linked to a variety of different reporter genes, including β-galactosidase, alkaline phosphatase, luciferase, and green fluorescent protein. These studies replace the actual gene of interest with the reporter gene, but leave the regulatory regions upstream or downstream. The amount of reporter gene product is a direct measure of the strength of expression for the regulatory region under study.

End-of-Chapter Questions

1. Which of the following is utilized in genomic research?
   - **a.** microsatellite polymorphism
   - **b.** restriction fragment length polymorphism
   - **c.** single nucleotide polymorphism
   - **d.** variable number tandem repeat
   - **e.** all of the above

(*Continued*)

2. What is contig mapping?
   a. determination of regions that overlap from one clone to the next in a library
   b. the distance in base pairs between two markers
   c. the use of landmarks in the genes to put together sequencing data
   d. the relative order of specific markers in a genome
   e. the mapping that determines if a library sequence is from one continuous gene or two gene segments cloned into one vector

3. Which method was used to sequence the human genome?
   a. cytogenetic mapping
   b. shotgun sequencing
   c. chromosome walking
   d. radiation hybrid mapping
   e. All of the above were used in combination to complete the project.

4. Which organism has the most genes?
   a. *H. sapiens*
   b. *D. melanogaster*
   c. *O. sativa*
   d. *P. trichocarpa*
   e. *A. thaliana*

5. What is a gene?
   a. a segment of DNA that encodes a protein
   b. a segment of DNA that encodes nontranslated RNA
   c. sequences of DNA that are not transcribed
   d. a segment of DNA that is transcribed
   e. all of the above

6. Which of the following is considered a field of study related to bioinformatics?
   a. proteomics
   b. computational biology
   c. genomics
   d. cheminformatics
   e. all of the above

7. How is data mining useful to biotechnology research?
   a. It allows researchers to determine sequence similarity, which usually translates into functional similarity.
   b. Data mining allows researchers to use computers to study, sort, and compile the vast amounts of raw data generated through bioinformatics.
   c. Data mining is the act of gathering the raw data from research projects such as sequencing into one central location.
   d. Data mining usually provides too much information, which only slows down the research project and is therefore not very useful.
   e. none of the above

8. Which type of mutation is the most common?
   a. insertion of one or more bases
   b. deletion of one of more bases
   c. base substitutions
   d. inversion of DNA segments
   e. duplications of DNA segments

9. Which of the following statements about mutations is not true?
   a. Mutations occur in all organisms at the same rate.
   b. DNA polymerase never produces mutations during replication because of the proofreading ability of this enzyme.
   c. Mutations often occur at methylated cytosine residues.
   d. Mutations such as duplications or deletions occur due to repetitive sequences causing strand slippage.
   e. When comparing mutation rates to coding capacity, mutation rates are usually the same for most organisms, which suggests a mechanism to control the rate.

10. Which one of the following is often used to establish family trees for organisms because it is present in all organisms and does not accumulate mutations quickly?
   a. rRNA
   b. fibrinopeptides
   c. hemoglobin
   d. chloroplasts
   e. mitochondrial DNA

11. Which of the following statements about DNA microarrays is not correct?
   a. Fluorescently labeled mRNA from the organism hybridizes to the DNA on the glass slide.
   b. DNA microarrays contain thousands of DNA segments on a support, such as a glass slide.
   c. Hybridization to a DNA microarray can only occur once.
   d. The amount of fluorescence correlates with the amount of mRNA in the sample.
   e. The data obtained from DNA microarrays represents a global view of gene expression, even under particular growth conditions.

12. What is the term used to describe the process of synthesizing oligonucleotides directly on the glass slide?
   a. photosynthesis
   b. photolithography
   c. light-activated oligosynthesis
   d. on-chip oligosynthesis
   e. protected oligosynthesis

13. Which of the following statements highlights the issues surrounding oligonucleotide microarrays?
   a. A duplex may not properly form if the mRNA probe has several mismatches compared to the oligonucleotide sequence.
   b. The ability to hybridize to the oligonucleotides will be decreased if the probe is able to form a stem-loop structure.
   c. The A:T content of the oligonucleotide may affect the stability of the duplex.
   d. Depending on the size of the spacer, incoming probes may not be able to hybridize if the spacer is too small or the oligonucleotide may fold back on itself if the spacer is too long.
   e. All of the above are issues surrounding oligonucleotide microarrays.

(Continued)

14. What can whole-genome arrays identify?
  a. regions on the DNA that are methylated
  b. transcription factor binding sites
  c. various polymorphisms
  d. repetitive elements
  e. all of the above

15. Which one of the following fusion proteins does not require some kind of chemical substrate to observe activity?
  a. luciferase
  b. alkaline phosphatase
  c. green fluorescent protein
  d. β-galactosidase
  e. all of the above

Further Reading

Brown PO, Botstein D (1999). Exploring the new world of the genome with DNA microarrays. *Nat Genet* **21**, 33–37.

Brown TA (1999). *Genomes.* Wiley-Liss, New York.

Chenna R, Sugawara H, Koike T, Lopez R, Gibson TJ, Higgins DG, Thompson JD (2003). Multiple sequence alignment with the Clustal series of programs. *Nucleic Acids Res* **31**, 3497–3500. (See also www.ebi.ac.uk/clustalw/.)

Cho RJ, Campbell MJ, Winzeler EA, Steinmetz L, Conway A, Wodicka L, Wolfsberg TG, Gabrielian AE, Landsman D, Lockhart DJ, Davis RW (1998). A genome-wide transcriptional analysis of the mitotic cell cycle. *Mol Cell* **2**, 65–73.

Clark DP (2005). *Molecular Biology: Understanding the Genetic Revolution.* Elsevier Academic Press, San Diego, CA.

Davidson EH, Rast JP, Oliveri P, et al. (2002). A provisional regulatory gene network for specification of endomesoderm in the sea urchin embryo. *Dev Biol* **246**, 162–190.

Duggan DJ, Bittner M, Chen Y, Meltzer P, Trent JM (1999). Expression profiling using cDNA microarrays. *Nat Genet* **21**, 10–12.

Grewel SIS, Rice JC (2004). Regulation of heterochromatin by histone methylation and small RNAs. *Curr Opin Cell Biol* **16**, 230–238.

Khambata-Ford S, Liu Y, Gleason C, Dickson M, Altman RB, Batzoglou S, Myers RM (2003). Identification of promoter regions in the human genome by using a retroviral plasmid library-based functional reporter gene assay. *Genome Research* **13**, 1765–1774.

Lima JJ, Thomason DB, Mohamed MH, Eberle LV, Self TH, Johnson JA (1999). Impact of genetic polymorphisms of the beta2-adrenergic receptor on albuterol bronchodilator pharmacodynamics. *Clin Pharmacol Ther* **65**, 519–525.

Lipshutz RJ, Fodor SPA, Gingeras TR, Lockhart DJ (1999). High density synthetic oligonucleotide arrays. *Nat Genet* **21**, 20–24.

Lodish H, Berk A, Zipursky LS, Matsudaira P, Baltimore D, Darnell J (2000). *Molecular Cell Biology.* 4th ed. WH Freeman, New York.

Madden SL, Wang CJ, Landes G (2000). Serial analysis of gene expression: From gene discovery to target identification. *DDT* **5**, 415–425.

Mockler TC, Ecker JR (2005). Applications of DNA tiling arrays for whole-genome analysis. *Genomics* **85**, 1–15.

Oliveri P, Davidson EH (2004). Gene regulatory network controlling embryonic specification in the sea urchin. *Curr Opin Genet Dev* **14**, 351–360.

Southern E, Mir K, Shchepinov M (1999). Molecular interactions on microarrays. *Nat Genet* **21**, 5–9.

Toews MW, Warmbold J, Konzack S, Rischitor P, Veith D, Vienken K, Vinuesa C, Wei H, Fischer R (2004). Establishment of mRFP1 as a fluorescent marker in *Aspergillus nidulans* and construction of expression vectors for high-throughput protein tagging using recombination *in vitro* (GATEWAY). *Curr Genet* **45**, 383–389.

Weiner AM (2002). SINEs and LINEs: The art of biting the hand that feeds you. *Curr Opin Cell Biol* **14**, 343–350.

Zebrafish Information Network (ZFIN). The Zebrafish International Resource Center, University of Oregon, Eugene, OR 97403-5274; available at: http://zfin.org/; retrieved May 8, 2005.

CHAPTER 9

Proteomics

Introduction
Gel Electrophoresis of Proteins
Western Blotting of Proteins
High-Pressure Liquid Chromatography Separates Protein Mixtures
Digestion of Proteins by Proteases
Mass Spectrometry for Protein Identification
Peptide Sequencing Using Mass Spectrometry
Protein Quantification Using Mass Spectrometry
Protein Tagging Systems
Phage Display Library Screening
Protein Interactions: The Yeast Two-Hybrid System
Protein Interactions by Co-immunoprecipitation
Protein Arrays
Metabolomics

INTRODUCTION

Today we have the genome sequences for humans as well as many other animals, plants, fungi, and bacteria. All these data have given scientists a global view of the various genes in humans and others. However, genes are only the first step to understanding how an organism works. Genes are transcribed into mRNA and then translated into protein. So in order to truly understand gene function, the gene product or protein must be characterized also—hence the advent of **proteomics**. Proteomics refers to the global analysis of proteins, and the **proteome** refers to the entire protein complement of an organism. The **translatome**, or the complement of proteins under specific circumstances, also falls into this field of study. Note that the translatome is dynamic and changes when environmental conditions change.

The relationships among the genome, proteome, and translatome are not linear. The genome of a species is the most stable, but differences do exist between one person and the next, and between one generation and the next. The proteome correlates highly with the genome because proteins are the products of the majority of genes. However, some genes encode nontranslated RNA, and so do not contribute to the proteome. In addition, some genes, especially in higher eukaryotes, may give rise to multiple proteins because of alternative splicing. In contrast, the translatome is highly dynamic, changing from minute to minute depending on many different stimuli.

The genome ultimately dictates the changes in the translatome and proteome, but genomic changes do not always affect the translatome or proteome. Sometimes, for example, mRNA transcripts are made, but never translated into protein. MicroRNAs and siRNAs control the expression of many different proteins at the translational level. The rate of mRNA degradation and translation will have a huge impact on how much protein is actually made. Thus, although some genes give rise to a lot of mRNA, very little protein is made, because the transcripts are very unstable.

The translatome and proteome are also affected by modifications that occur after translation. For example, the function of many proteins is altered by addition or removal of various groups, such as phosphate, acetyl, AMP, or ADP-ribose. Also, many proteins, especially in eukaryotes, are altered by chemical modification of amino acid residues. Proteins also undergo proteolytic cleavage. Hence, the composition of the translatome is affected by the rate of protein degradation, and protein stability has a major influence. Finally, some proteins themselves may affect the expression of other proteins via assorted regulatory effects. All of these factors affect the protein makeup of the cell.

Proteomics is the study of the protein complement of an organism.

GEL ELECTROPHORESIS OF PROTEINS

Studying proteins requires the ability to isolate and identify the proteins in a particular sample. The first step is to separate them. Much as electrophoresis on agarose gels is used to separate DNA by size, so **polyacrylamide gel electrophoresis (PAGE)** is used to separate proteins by size (Fig. 9.1). Polyacrylamide has smaller pores than agarose and is thus suitable for proteins, because these are generally smaller than DNA molecules. When an electric field is applied to a sample of proteins, those with a smaller size are able to circumnavigate the pores of the acrylamide more easily and will migrate away from the negative pole faster than larger proteins.

Unlike DNA, most proteins do not have a net negative charge; indeed, some have a positive charge. Therefore, protein samples are treated by boiling with **sodium dodecyl sulfate (SDS)**. This unfolds the polypeptide chains and the entire strand of amino acids is coated with negatively charged SDS. The amount of SDS, and therefore the amount of charge, correlates with the length (i.e., molecular weight) of the protein. As in DNA electrophoresis,

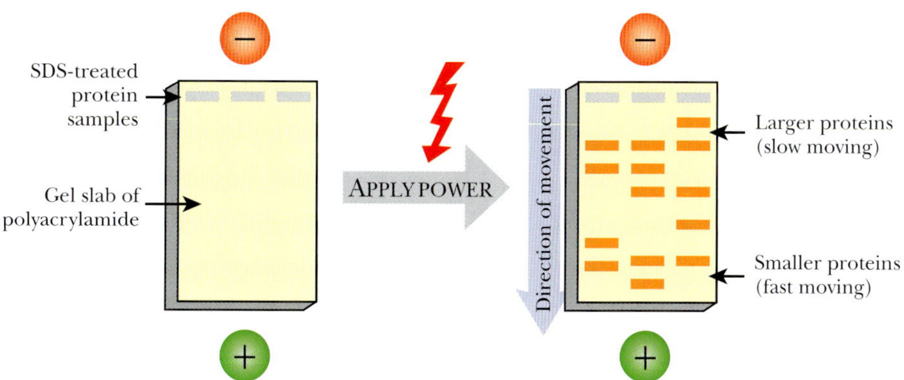

FIGURE 9.1 SDS Polyacrylamide Gel Electrophoresis
Once a protein has been boiled in SDS, it gets a net negative charge. The total charge correlates to the size of protein. After being loaded into a sample well of a polyacrylamide gel, the proteins migrate away from the negative cathode and toward the positive anode. The sieving action of the gel allows the smaller proteins to move faster than the larger. The distance traveled in a given time is proportional to the log of the molecular weight.

the protein sample is loaded into a well, and an electric current is applied so that the proteins migrate through the acrylamide gel. Finally, the separated proteins are visualized using either **Coomassie Blue**, a dark blue dye, or the more sensitive **silver stain**, both of which bind tightly to all proteins.

Thousands of proteins are found in cells, and many of these proteins are similar in size. However, resolving individual proteins into single bands on an acrylamide gel requires that the sample have only a few proteins, and these must be of different sizes. If the sample is from an entire cell, then the sample has so many proteins that individual bands will smear together. Using **two-dimensional PAGE (2D-PAGE)** can help alleviate this problem by first separating the proteins by their native charge in one dimension, then separating by size in a second dimension (Fig. 9.2). **Isoelectric focusing** is the term for separating proteins by their native charge. All proteins have an inherent natural charge due to the side chains of their amino acid residues. The total number of positively and negatively charged amino acids determines the natural charge on a protein. For separation by charge, the sample is loaded into the top of a gel with a pH gradient. When an electric field is applied, the proteins move along the pH gradient until their charge is neutralized. This step is usually done on a tube-shaped gel. After running, the gel is removed from its tube. The second dimension of 2D-PAGE is separation by size. The tube gel containing the separated proteins is treated with SDS in order to denature the proteins and coat them with negative charges as in regular PAGE. The tube gel is laid along the top of a polyacrylamide slab gel, and the proteins are separated by size. The gel is stained as described earlier.

2D-PAGE can resolve many proteins. Early studies on *E. coli* used 2D-PAGE to characterize all the proteins present under different conditions, about 1000 different

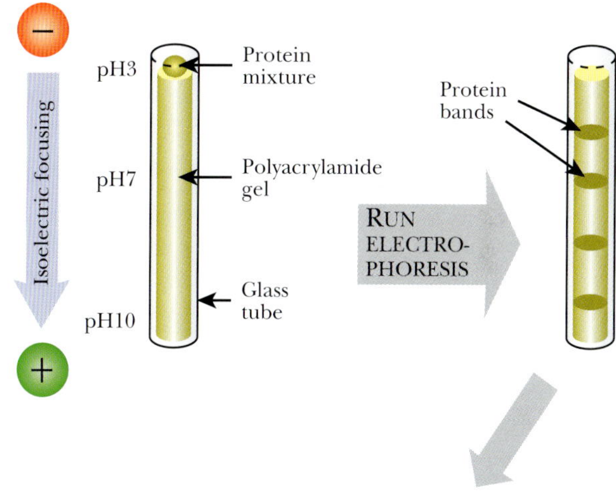

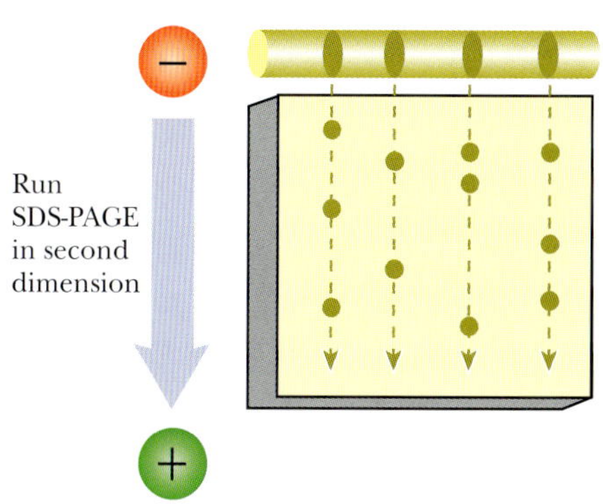

FIGURE 9.2 Two-Dimensional Polyacrylamide Gel Electrophoresis
First, a sample containing large numbers of proteins is separated by their natural charge. The proteins are electrophoresed through a tube of polyacrylamide gel with a pH gradient. The proteins move in the gradient until their charge is neutralized. The tube gel is then removed, treated with SDS, and placed on top of a slab gel. The proteins are then separated according to size in the second dimension as described before.

proteins. Larger 2D-PAGE gels have been developed to study larger proteomes and can separate more than 10,000 different proteins into single spots. Identifying each spot on a 2D gel is a major task. Scientist interested in studying the proteome or translatome can stain the proteins and cut out each spot of interest. The protein trapped in each piece of gel can be digested into peptide fragments and identified by mass spectrometry (see later discussion). Another key feature of 2D-PAGE is the ability to quantify the relative amounts of different proteins. The size of the spot indicates the relative abundance of that protein. This can be quantified by scanning with a laser. Computer analysis then determines the density of the spot and relative abundance.

Two different samples can be run together on the same gel to compare the proteins present under different conditions (Fig. 9.3). Each sample is labeled with a different fluorescent dye. Fluorescent labels such as methyl-Cy5 and propyl-Cy3 will fluoresce different colors, but do not affect protein separation in 2D-PAGE. After both samples have been run on the same gel, the dyes are activated by light. Where sample 1 has a protein not found in sample 2, the spot will fluoresce red, and when sample 2 has a protein not found in sample 1, the spot will be green. When both samples have the same amount of the same protein, the spot will be yellow. The unique proteins can be isolated, digested, and identified using mass spectroscopy (see later discussion).

Although 2D-PAGE has been widely used, it does have some disadvantages. Certain classes of proteins are underrepresented on a gel because they do not migrate through acrylamide. Extremely large proteins often cannot enter into the gel matrix, whereas very small proteins may travel off the end of the gel. Hydrophobic proteins may travel through the gel, but their dynamics is altered because of their hydrophobic surfaces. These proteins often run at different positions than expected based on molecular weight. Proteins that are scarce within the cell, such as transcription factors, are barely visible, if at all, with even the most sensitive dyes. Abundant proteins, on the other hand, can distort the location of nearby proteins within the gel because the gel becomes saturated at that location. Another issue with 2D-PAGE is that proteins must be isolated from the acrylamide for analysis by mass spectrometry (see later discussion).

> Polyacrylamide gel electrophoresis separates a mixture of proteins based on their size. The proteins must first be coated with SDS, which gives the proteins a net negative charge. Applied electrical current moves the protein away from the negative pole and toward the positive pole.
>
> Two-dimensional PAGE first separates proteins based on their inherent charge using isoelectric focusing. The proteins are then separated as in SDS-PAGE.

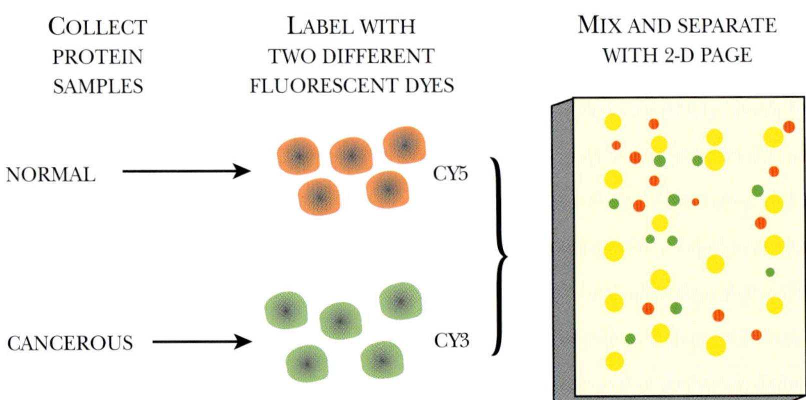

FIGURE 9.3 Two-Color 2D-Gel

Proteins from two different conditions (e.g., normal and cancerous) can be compared directly on the same gel by labeling each with a different fluorescent dye. When the gel is visualized to see the dyes, the proteins found only in normal tissue form red spots, the proteins found only in cancerous tissue form green spots, and proteins found in both normal and cancerous tissue look yellow.

WESTERN BLOTTING OF PROTEINS

Often researchers will use **Western blotting** to identify proteins. Western blots rely on having an antibody to the protein (see Chapter 6). Antibodies are extremely specific and will bind only to one target protein.

The first step is to separate the proteins by size by either standard SDS-PAGE or 2D-PAGE. The proteins are then transferred from the gel to a type of paper membrane made of **nitrocellulose**. Other types of membranes are made from nylon and are stronger. Either way, the membrane must have a positive charge so that the negatively charged proteins will stick to its surface. The proteins are moved from the gel to the membrane with an electric current as shown in Fig. 9.4.

After the proteins are attached to the nitrocellulose membrane, the **primary antibody** is used to find the location of the target protein. However, many areas of the membrane will not have any protein bound, because the corresponding area of the protein gel was empty. These blank areas are positively charged and can bind nonspecifically to the antibody. Therefore, these sites must be blocked. Often, the membranes are soaked in reconstituted nonfat dry milk. The milk proteins mask the unused sites on the membrane and will not bind to the antibody. Next, the antibody is added to the membrane in a solution. The antibody will bind only to its target protein, and nowhere else on the membrane, because it recognizes only one specific epitope of its target protein (see Chapter 6).

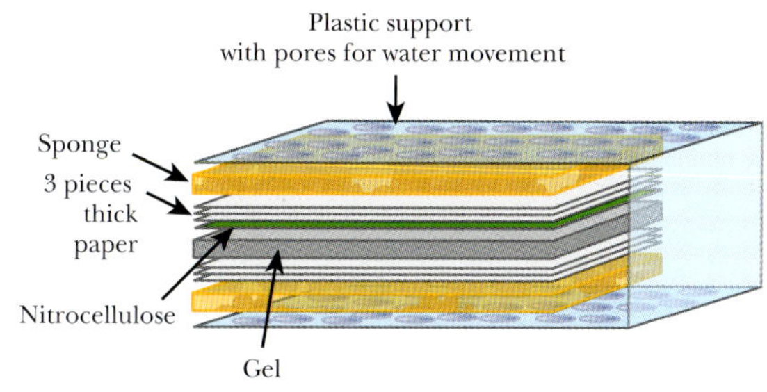

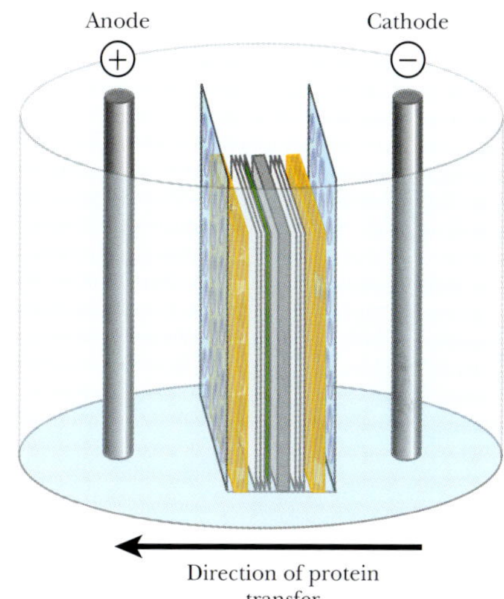

FIGURE 9.4 Electrophoretic Transfer of Proteins from Gel to Nitrocellulose
A "sandwich" is assembled to keep the gel in close contact with a nitrocellulose membrane while in a large tank of buffer. The sandwich consists of the gel (gray) and nitrocellulose (green) between layers of thick paper and a sponge (yellow). The entire stack is squeezed between two solid supports so that none of the layers can move. The sandwich is then transferred to a large tank filled with buffer to conduct the current. As in SDS-PAGE, the proteins are repelled by the negatively charged cathode and attracted to the positively charged anode. As the proteins move out of the gel, they travel into the nitrocellulose, where they adhere.

The next step is to visualize the location of the primary antibody, thus revealing the location of the target protein (Fig. 9.5). To achieve this, a **secondary antibody** is added. This antibody recognizes the stem of the primary antibody without affecting its binding to the protein. The secondary antibody has a tag or label on its stem that is easily detected. Often the tag is the enzyme alkaline phosphatase, which removes phosphate from various substrates (see Chapter 8). If the antibody complex is incubated with a chromogenic substrate such as X-Phos, the alkaline phosphatase removes the phosphate from the X-Phos. The remaining indolyl group reacts with oxygen to form a blue precipitate in the location of the target protein. If a chemiluminescent substrate is used, the nitrocellulose membrane is placed next to a piece of photographic film. The light pulses turn the film black where the secondary antibody/primary antibody/target protein complex is located on the membrane. Western blots are used extensively both to prove that a protein is being expressed and to estimate its level, because the intensity of the spot directly correlates with the amount of protein.

Western blots identify the location of a specific protein after it has been separated by SDS-PAGE. First, a primary antibody that recognizes one epitope binds to the protein of interest. The location of the primary antibody is visualized by adding a secondary antibody conjugated to a detection system.

HIGH-PRESSURE LIQUID CHROMATOGRAPHY SEPARATES PROTEIN MIXTURES

Chromatography is a general term for many separation techniques, where a sample of molecules, the **analyte**, is dissolved in a **mobile phase** and then forced through a **stationary phase**. For 2D-PAGE, the mobile phase is the buffer and the stationary phase is the gels. In **high-pressure liquid chromatography (HPLC)** the sample is dissolved in a mobile phase

Nitrocellulose membrane with attached proteins

ADD NON-SPECIFIC MILK PROTEINS AND PRIMARY ANTIBODY

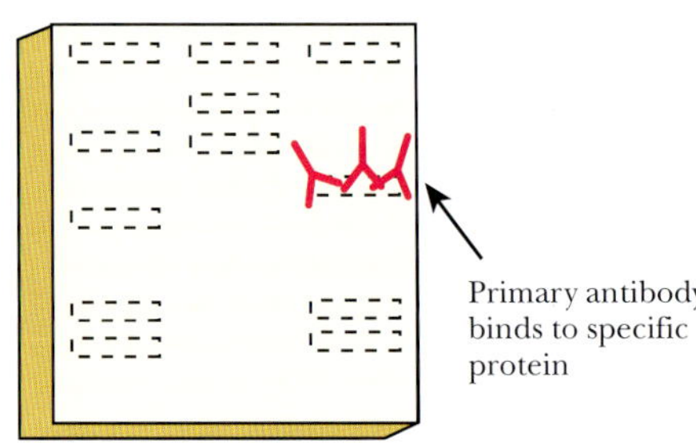

ADD SECONDARY ANTIBODY CONJUGATED TO ALKALINE PHOSPHATASE

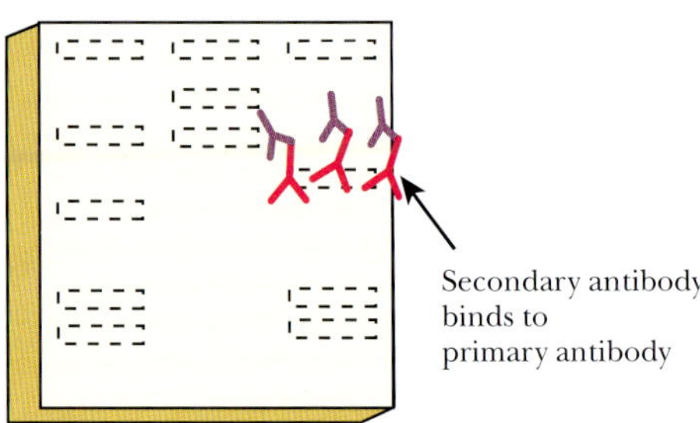

ADD X-PHOS TO DETECT LOCATION OF SECONDARY ANTIBODY

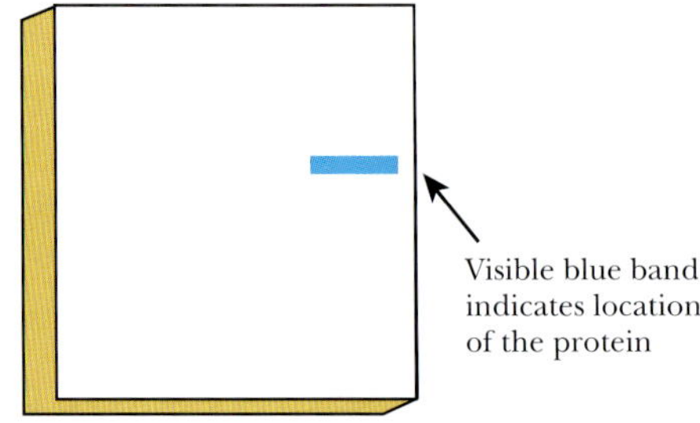

FIGURE 9.5 Western Blot

After a mixture of proteins has adhered to the nitrocellulose membrane, one specific protein can be detected using an antibody. The antibody is added with a solution of milk proteins and incubated with the membrane. The milk proteins block those regions of the nitrocellulose that do not have any proteins attached. The primary antibody attaches only to the target protein. A secondary antibody with attached alkaline phosphatase binds specifically to the primary antibody. This allows the position of the target protein to be visualized. In this example, the alkaline phosphate reacts with the substrate X-Phos to form a blue dye (see Chapter 8).

and separated based on one specific characteristic by passing over a stationary phase (Fig. 9.6). In HPLC, the mobile phase is forced through a **chromatography column**, that is, a narrow tube packed with the stationary phase, under high pressure. As the mobile phase travels through the column, the mixture separates and different fractions are collected at the column exit. The mixture is forced through the column by constantly adding more mobile phase, in a process called **elution**. As the mobile phase exits the column, a detector emits a response to molecules in the eluting sample and draws a peak on the chromatogram.

HPLC has many applications including separation, identification, purification, and quantification of proteins or other analytes. Preparative HPLC is used to isolate and purify one specific protein from a mixture. Using HPLC to identify a compound requires a specific detection method. For example, if the target protein carries a fluorescent label, then a fluorescence detector would be used. One application of HPLC, quantitative HPLC, can be used to determine the amount of target protein by comparing it to a set of standard proteins with known amounts. This allows measurement of changes in level of a specific protein under different conditions. A major benefit that HPLC offers to proteomics researchers is that the separated proteins are already in a liquid state, making further analysis easier.

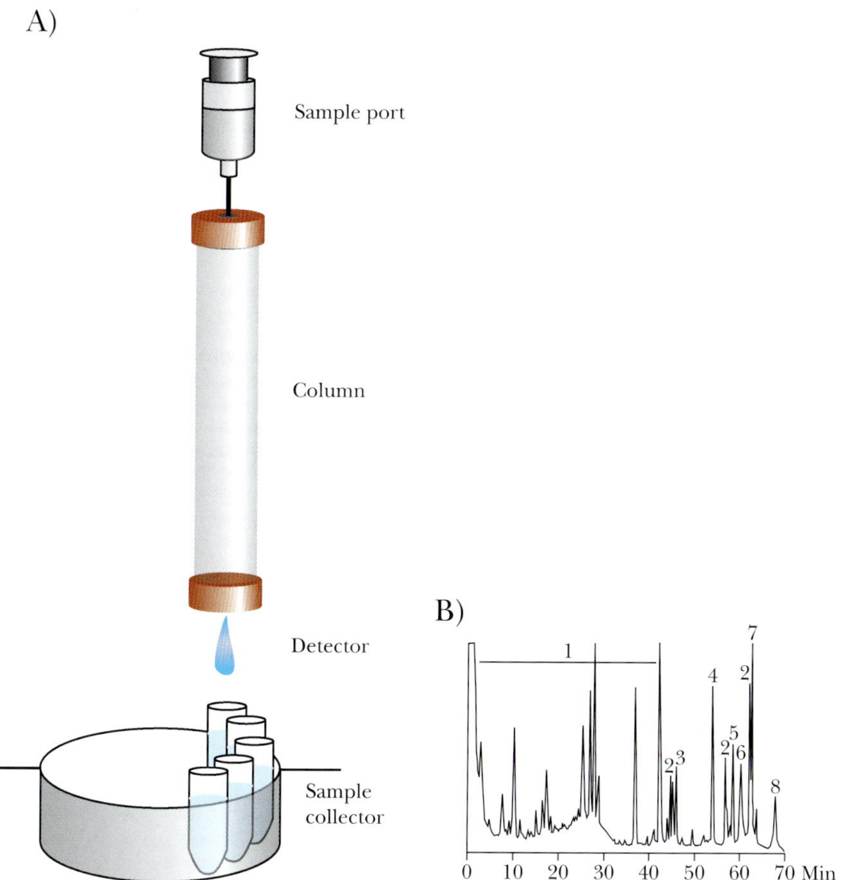

FIGURE 9.6 High-Pressure Liquid Chromatography
(A) The mobile phase has a protein sample dissolved in a solvent, the simplest being water. This is loaded into a syringe and injected into the sample port. The sample flows through the column under high pressure, and the components of the mixture travel at different rates based on the column type. The mobile phase fractions exit the column, and are collected into tubes. (B) A mixture of dyes is separated using a reverse phase column (Grom™ Sil 120 ODS-5 ST). The dyes were detected using UV light at 254 nm, and eleuted in the following order: 1. Tartrazine, 2. Amaranth, 3. Indigo Carmine, 4. New Coccine, 5. Sunset Yellow, 6. Fast green, 7. Brilliant Blue, 8. Erythosine, 9. Acid Red, 10. Phloxine, 11. Rosé Bengale. Reprinted with permission from Grace Davison Discovery Sciences. http://www.discoversciences.com.

Although HPLC seems simple in theory, the actual process of separating the mixture into its components is complex. Each mixture has different chemistries, and so many different solid phases are used to separate them. Even before it is loaded into the column, the mixture can be manipulated to remove certain components or change their chemistry. For example, treating with a phosphatase will remove phosphate groups from proteins. Such manipulations can increase the efficiency with which the protein of interest is isolated from the mixture.

HPLC is very adaptable because of the availability of different types of stationary phase materials. **Size exclusion chromatography** columns contain porous beads that separate mixtures of proteins by size. Large molecules do not enter the pores of the beads and travel through the column quickly, while smaller compounds are delayed by entering the beads. Many different pore sizes are available for mixtures of different size ranges. **Reverse phase HPLC** uses columns packed with hydrophobic alkyl chains attached to silica-based material. The column binds and delays hydrophobic molecules while hydrophilic molecules elute faster.

Ion-exchange HPLC uses a stationary phase with charged functional groups that bind oppositely charged molecules in the sample. Such molecules remain in the column after the sample has passed through. To elute them, the mobile phase is changed. For example, if the pH of the mobile phase is adjusted, the net charge on many proteins will be altered and they will be released from the column. Other stationary phases form hydrogen bonds with the analyte and separate based on overall polarity. For **affinity** HPLC, the stationary phase contains a molecule that specifically binds the target protein, for example, an antibody. When a mixture passes over the stationary phase, only the target protein is bound, and other proteins pass through. Changing the mobile phase so as to disrupt the interaction releases the protein of interest.

As molecules exit the column, they must be detected. Many different detectors exist. These usually respond by plotting peaks as molecules pass by. **Refractive index detectors** monitor whether the exiting mobile phase refracts any light by shining a light beam through it. Compounds present in exiting fractions will scatter light, and a photodetector records this as a positive signal. The amount of scatter affects the height of the peak and the length of time of the scatter determines the width of the peak. **Ultraviolet detectors** have a UV light source and a detector to determine when the passing mobile phase absorbs the UV light. Such detectors may monitor one or more wavelengths depending on the substance being examined. **Fluorescence detectors** detect compounds that fluoresce, that is, absorb and re-emit light at different wavelengths; **radiochemical detectors** detect radioactively labeled compounds; and **electrochemical detectors** measure compounds that undergo oxidation or reduction reactions. An approach that is increasingly used for proteomics is detection by mass spectrometry. This combination allows proteins separated by HPLC to be fed directly into a mass spectrometer for identification.

A critical aspect of HPLC is getting a good separation between the different proteins in the sample, that is, good **resolution**. Each peak that comes off the column should be symmetrical and as narrow as possible. For high resolution, many experimental conditions can be adjusted. The most obvious is changing the stationary phase. Sometimes just changing the particle size of the stationary phase improves separation. An alternative is to adjust the composition of the mobile phase. Temperature also affects many separations and may need to be controlled.

An analyte is a sample of molecules in a liquid. This is the mobile phase that moves through the column.

The stationary phase is the actual column packed with different materials. The properties of these materials determine what proteins are retained in the column, and what proteins are expelled from the column.

Different detectors are at the end of the column to determine when the protein exits. These can identify proteins by refractive index, ultraviolet waves, fluorescence, radioactivity, or electrochemistry.

DIGESTION OF PROTEINS BY PROTEASES

Proteases (also known as **proteinases** or **peptidases**) hydrolyze the peptide bond between amino acid residues in a polypeptide chain. Proteases may be specific and limited to one or more sites within a protein, or they may be nonspecific, digesting proteins into individual amino acids. The ability to digest a protein at specific points is critical to mass spectrometry (see later discussion) and many other protein experiments. For example, proteases are used to cleave fusion proteins or remove single amino acids for protein sequencing.

Proteases are found in all organisms and are involved in all areas of metabolism. During programmed cell death, proteases digest cellular components for recycling (see Chapter 20). Plants deploy proteases to protect themselves from fungal or bacterial invaders. In the biotechnology industry, proteases have many uses. For example, they are included as additives in detergents to digest proteins in ketchup, blood, or grass stains.

Proteases are classified by three criteria: the reaction catalyzed, the chemical nature of the catalytic site, and their evolutionary relationships. **Endopeptidases** cleave the target protein internally. **Exopeptidases** remove single amino acids from the amino- or carboxy-terminal ends of a protein. Exopeptidases are divided into **carboxypeptidases** or **aminopeptidases** depending on whether they digest proteins from the carboxy- or amino-terminus, respectively. Proteases are also divided based on their catalytic site architecture. **Serine proteases** have a serine in their active site that covalently attaches to one of the protein fragments as an enzymatic intermediate (Fig. 9.7). This class includes the chymotrypsin family (chymotrypsin, trypsin, and elastase), and the subtilisin family. **Cysteine proteases** have a similar mechanism, but use cysteine rather than serine. These include the plant proteases (papain, from papaya, and bromelain, from pineapple) as well as mammalian proteases such as calpains. **Aspartic proteases** have two essential aspartic acid residues that are close together in the active site, although far apart in the protease sequence. This family includes the digestive enzymes pepsin and chymosin. **Metalloproteases** use metal ion cofactors to facilitate protein digestion and include thermolysin. Finally, the fifth class of proteases, **threonine proteases**, has an active-site threonine.

Researchers who study proteases have not escaped the "-omics" culture, and sometimes the term **degradome** is used for the complete set of proteases expressed at one specific time by a cell, tissue, or organism. Understanding the degradome of an organism relies on much the same techniques as used for proteomics, although modifications are made to look only at proteases rather than at the entire proteome. For example, protease chips contain only antibodies to known proteases rather than to all proteins (see later discussion).

Many different proteases exist in nature, and these are useful for a variety of applications in biotechnology. Specific proteases recognize specific amino acids and cut the peptide bond in a specific location.

MASS SPECTROMETRY FOR PROTEIN IDENTIFICATION

Mass spectrometry is a technique to determine the mass of molecules. In mass spectrometry, a molecule is fragmented into different ions whose masses are accurately measured. The ions generate a spectrum of peaks that is unique and therefore determines the identity of the original molecule. The molecule 2-pentanol is used as an example in Fig. 9.8. Here an electron beam aimed at the sample fragments the 2-pentanol into different ions: the molecular ion gains an electron, m-1 loses hydrogen from the alcohol group, m-15 loses a methyl group, m-17 loses the alcohol group, m/e = 45 loses the alkyl chain.

These ions are accelerated into a vacuum tube by an ion accelerating array. The ions travel through the tube at different speeds due to a magnetic field that causes the ions to follow

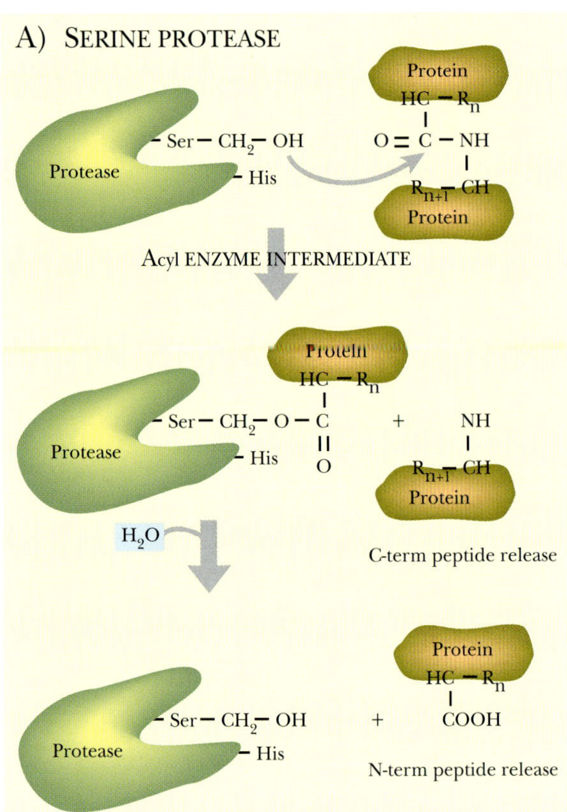

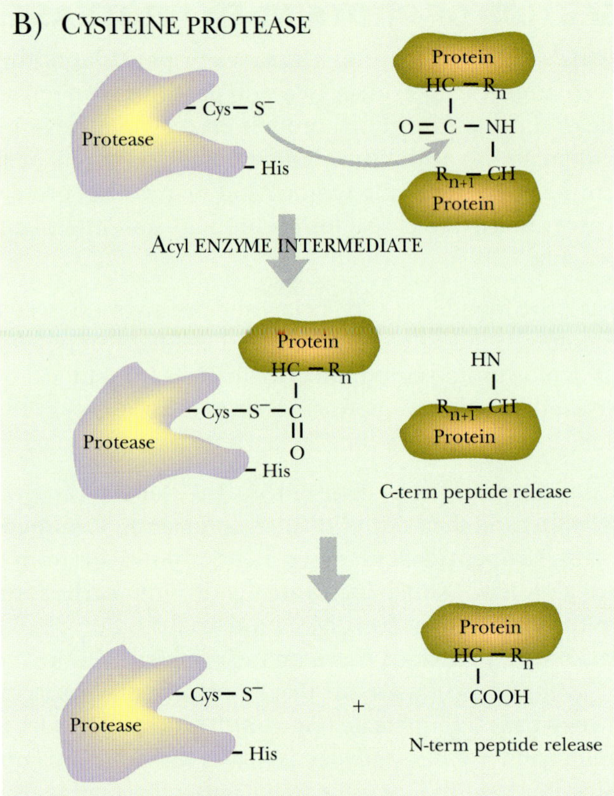

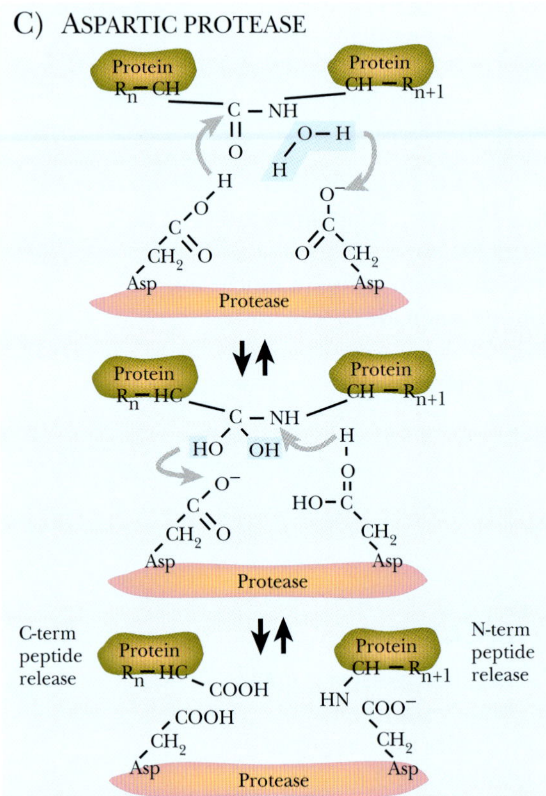

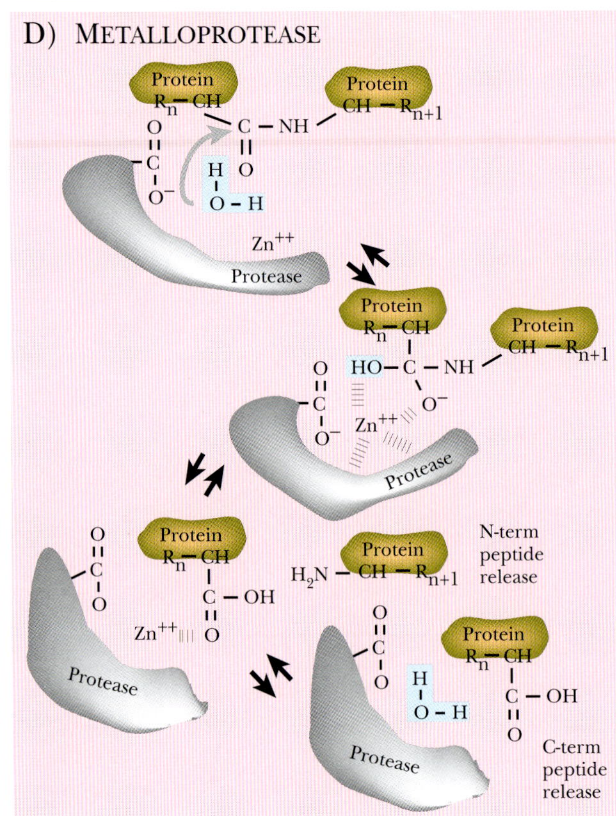

FIGURE 9.7 Mechanism of the Four Classes of Endopeptidase

(A) Serine proteases cleave the peptide bond of a protein by the formation of an acyl enzyme intermediate. The active-site serine forms a temporary bond with the amino-terminal half of the digested protein. (B) Cysteine proteases are similar serine proteases but use cysteine at the active site. (C) Aspartic proteases have two active-site aspartic acids that coordinate hydrolysis of the peptide bond in the target protein. (D) Metalloproteases hydrolyze a target protein using a metal ion such as Zn^{2+}.

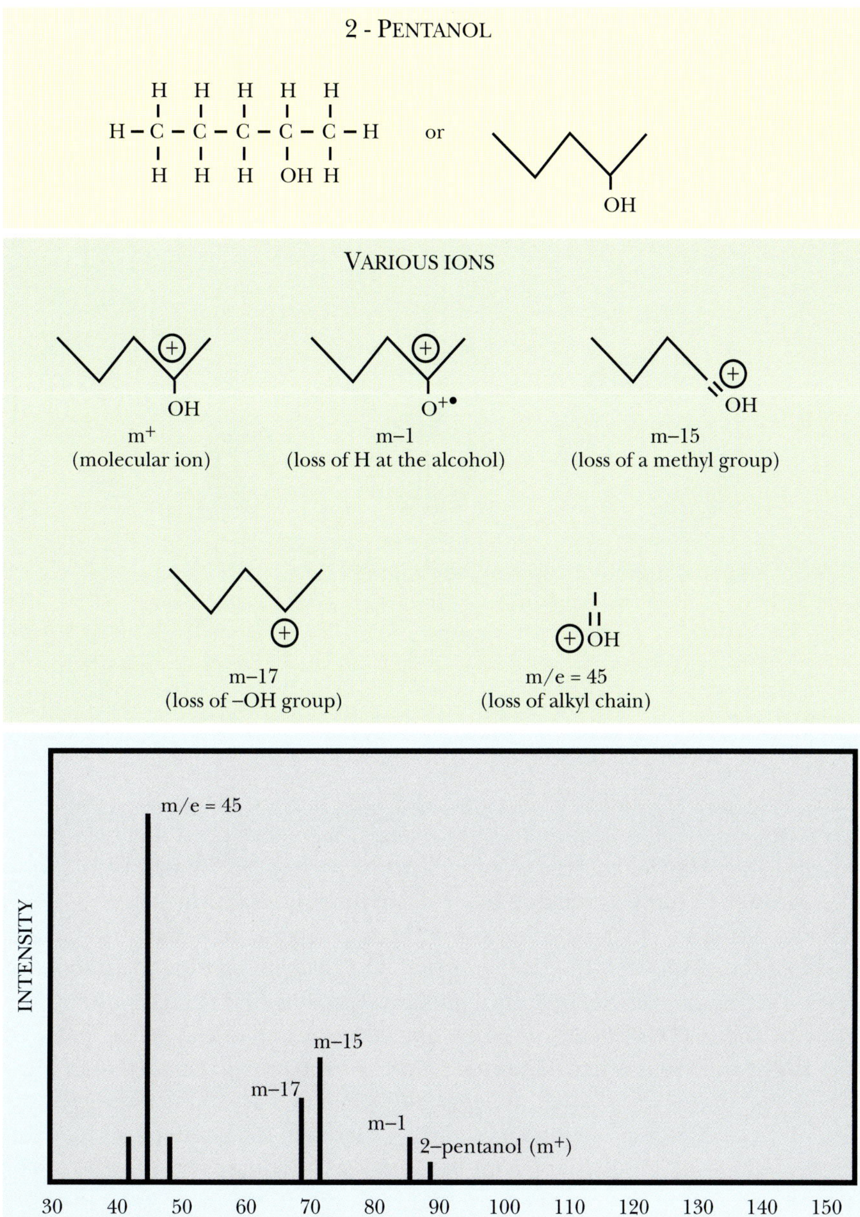

FIGURE 9.8 Basic Mass Spectrometry for 2-Pentanol
Every substance can be fragmented into multiple ions. This example shows the molecular structure of all the ions from 2-pentanol. A mass spectrometer separates these ions by size and graphs the results. The spectrum is always the same for each substance, and so an unknown substance can be identified by comparing its spectrum with a database of known substances.

a curved path within the tube (Fig. 9.9). The curves eliminate ions that are too small or too big. Ions that are too small gain so much momentum from the magnetic field that they collide with the wall. Those that are too big are not deflected by the magnetic field and also collide with the wall. Ions in the right size range are deflected by the magnetic field around both curves to hit the collector, where they are recorded as peaks in the mass spectrum. The **base peak** is the most intense, and other peaks are measured relative to this. The time ions take from the accelerator to the collector correlates directly to the size of the ion. Each peak is plotted based on the mass/charge ratio (m/z). The losses of mass (such as m-17 or m-15) refer to the loss of specific groups from the parent molecule and are most informative to the structure of the sample molecule because each such group has a characteristic mass.

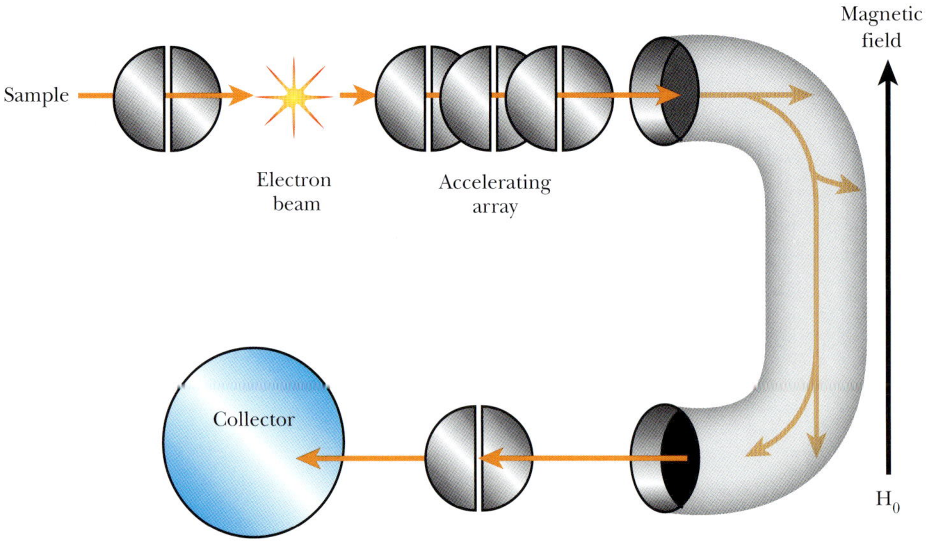

FIGURE 9.9 **Schematic Diagram of a Mass Spectrometry Tube**
The sample travels through a narrow slit and then passes through a beam of electrons that disrupts it into a mixture of ion fragments. The accelerating array moves the fragments into the C-shaped tube. This is surrounded by a strong magnetic field that prevents ions that are too small or too large from exiting the tube. A collector detects the exiting fragments and measures the time it took for them to travel the tube. The computer then converts the time of travel into size and charge information and plots this as a mass spectrum (not shown).

Until recently, very large molecules such as proteins were beyond the range of mass spectrometry. Two different ionization techniques have been developed that have made proteins manageable. The first technique embeds proteins in a solid matrix before ionization, and is called **MALDI**, or **matrix-assisted laser desorption-ionization** (Fig. 9.10). Here the proteins are embedded in a material such as 4-methoxycinnamic acid that absorbs laser light. The matrix absorbs and transfers the laser energy to the proteins, causing them to release different ions. The ions are accelerated through a vacuum tube by a charged grid. At the far end, the **time-of-flight (TOF)** detector records the intensity and calculates the mass. In between is a **flight tube** that is free of electric fields. The ions are accelerated with the same kinetic energy, and when they reach the flight tube, the lighter ions move faster than the heavier ions. The time-of-flight is proportional to the square root of mass to charge ratio (m/z). MALDI is able to handle ions up to 100,000 daltons.

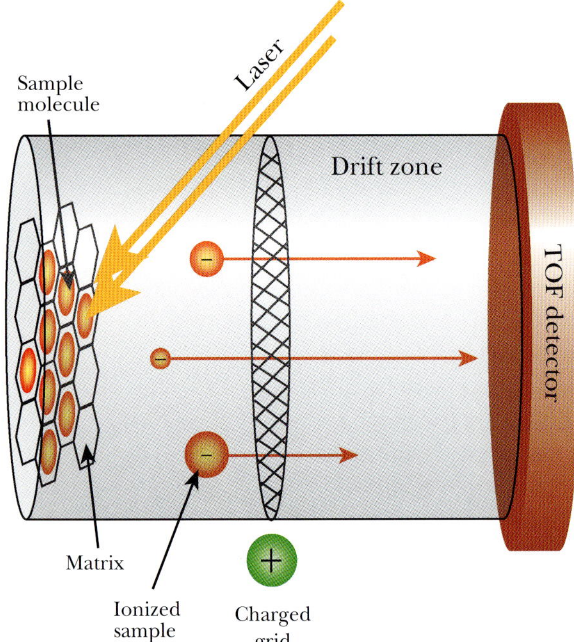

FIGURE 9.10
MALDI/TOF Mass Spectrometer
Mass spectrometry can be used to determine the molecular weight of proteins. The proteins are crystallized in a solid matrix and exposed to a laser, which releases ions from the proteins. These travel along a vacuum tube, passing through a charged grid, which helps separate the ions by size and charge. The time it takes for ions to reach the detector is proportional to the square root of their mass to charge ratio (m/z). The molecular weight of the protein can be determined from these data.

The other method for preparing ions from proteins is **electrospray ionization (ESI**; Fig. 9.11). Here the protein is dissolved in liquid and very small droplets are released from a narrow capillary tube. The droplets enter the electrostatic field, where a heated gas, such as hydrogen, causes the solvent to evaporate and the droplets to break up. This causes the protein to release ions into the vacuum tube, where they are accelerated by the electric field. The detector at the far end varies based on the sample being studied. A TOF detector may be used, as described earlier. Other detectors use quadrupole ion traps or Fourier transform ion cyclotron resonance to determine the mass of the ions. Quadrupole ion traps capture the ions in an electric field. The ions are then ejected into the detector by a second electric field. The electric field controls what size ions can pass to the detector, and varying the field allows

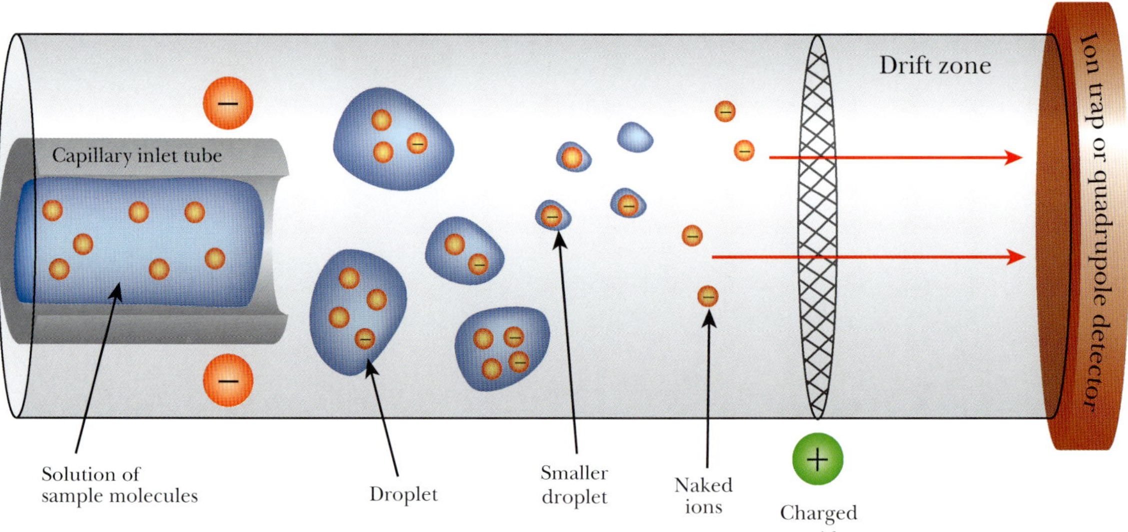

FIGURE 9.11 Electrospray Ionization (ESI) Mass Spectrometer

ESI mass spectrometry uses a liquid sample of the protein held in a capillary tube. After exposure to a strong electrostatic field, small droplets are released from the end of the capillary tube. A flow of heated gas within the drift zone evaporates the solvent and releases small charged ions. The charged ions vary in size and charge and the pattern of ions produced is unique to each protein. The ions are separated by size using a charged grid to either impede or promote the flow toward the detector.

different-sized ions to be detected. Combination detectors exist that use both TOF and quadrupole ion traps. The advantage that ESI has over MALDI is that proteins isolated from HPLC (see earlier discussion) require no special preparation and can be used directly. The disadvantage of ESI is that masses of about 5000 are the maximum.

The use of mass spectroscopy will become more prevalent as techniques improve. For example, **surface-enhanced laser desorption-ionization (SELDI)** mass spectroscopy takes liquid samples and ionizes the proteins that adhere to a treated metal bar. The technique shows great promise for bodily fluids such as blood and, it is hoped, will help in identifying a particular protein profile for different disease states. Perhaps one day patients will be diagnosed with cancer long before any symptoms are detected. A change in the protein profile in their blood could denote that cancer cells are forming. MALDI and ESI are sensitive enough to detect changes in proteins due to phosphorylation, glycosylation, and so forth. The technique can identify which amino acid is modified because only one specific ion is altered.

Mass spectroscopy ionizes a sample into smaller parts and measures the time it takes for these ions to reach the detector. The amount of time correlates with the size of the ion.

Proteins can be ionized for mass spectroscopy after they are either embedded in a matrix for MALDI or in liquid as for ESI. Both techniques can identify the protein by its pattern of fragmentation, and the techniques then can identify any modifications such as phosphorylation and glycosylation.

PEPTIDE SEQUENCING USING MASS SPECTROMETRY

Determining the peptide sequence of a short peptide is readily achieved using current mass spectroscopy techniques (Fig. 9.12). Determining the entire sequence of a protein is too complex at this point, but may someday be feasible. To determine the sequence, a pure sample of the protein must be obtained either by cutting a spot from a two-dimensional gel or by HPLC purification. The protein is then digested into fragments using a protease such as trypsin, which cuts proteins on the carboxy-terminal side of arginine and lysine. Cutting a protein into peptides helps reduce undesirable characteristics of the entire protein.

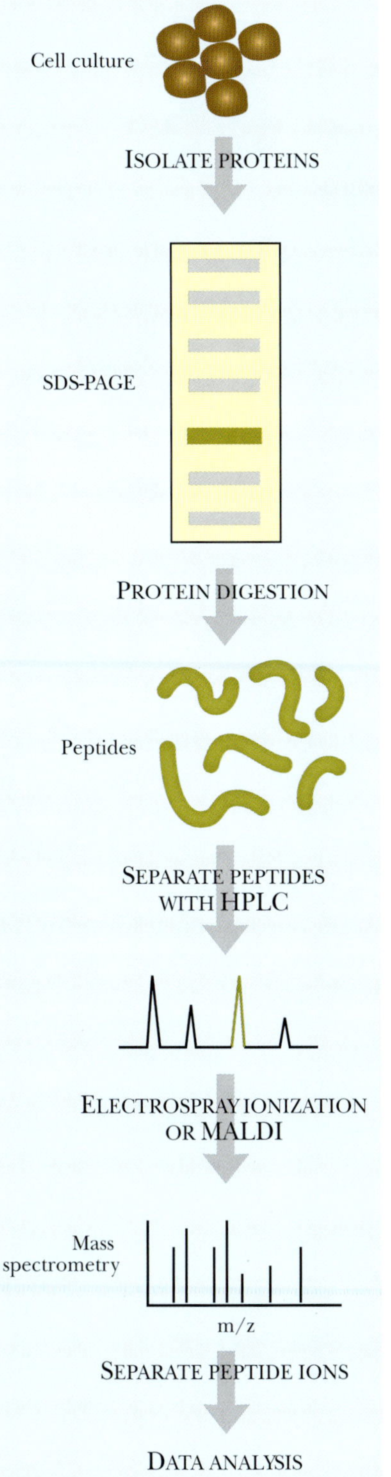

FIGURE 9.12 Mass Spectrometry of Peptides
Because mass spectrometry is so sensitive, the use of large whole proteins is limited. Instead, peptide fragments are generated by protease digestion. The peptides are easily separated with HPLC, and then specific peptides are subjected to mass spectrometry.

For example, membrane proteins are hydrophobic and stick together, and digesting these into peptide fragments destroys this characteristic. Solubility issues can also often be resolved by digesting a protein into peptides. Determining the sequence of these peptides will yield the sequence of the original protein.

The most common method of determining peptide sequences begins by separating the peptides using an HPLC column directly attached to the mass spectrometer. The column is a microscale capillary in order to keep the sample volume as small as possible. The mobile phase is an organic solvent that elutes the peptides in order of their hydrophobicity.

From the HPLC capillary tube, the samples enter the mass spectrometer chamber. Each peptide is ionized into multiple fragments (Fig. 9.13). For peptides, common ions include a doubly protonated form $(M + 2H)^{2+}$, where M is the mass of the peptide and H^+ is the mass of a proton. The ion peaks are plotted versus the mass to charge ratio or m/z. For the doubly protonated peptide ion, this would be the mass of the ion divided by 2. For example, if the original peptide was 1232.55 daltons, the double protonated ion would have a mass of 1232.55 daltons + (2×1.0073) for each added hydrogen. The peak would appear at 617.2828. (Note: The mass to charge ratio is where the peak is plotted, that is, the mass for this ion is 1234.5646 and the charge is +2. The peak appears at m/z or 1234.5646/2.) When the mass spectrometer separates peptide ions, the first step is to determine the charge state of the ion. Usually a cluster of peaks occurs for each peptide ion. If peaks are 1 dalton apart, the charge state of the peptide is 1. If the peaks are 0.5 dalton apart, the charge state is 2.

To determine the peptide sequence, two rounds of mass spectroscopy are used. This is called **tandem mass spectroscopy** because one ion is produced in the first round of mass spectroscopy, then that ion is fragmented by collision with a gas such as hydrogen, argon, or helium. As before, the ion fragments are separated based on their mass to charge ratio. Each peak usually varies by one amino acid, and the size difference between the peaks determines

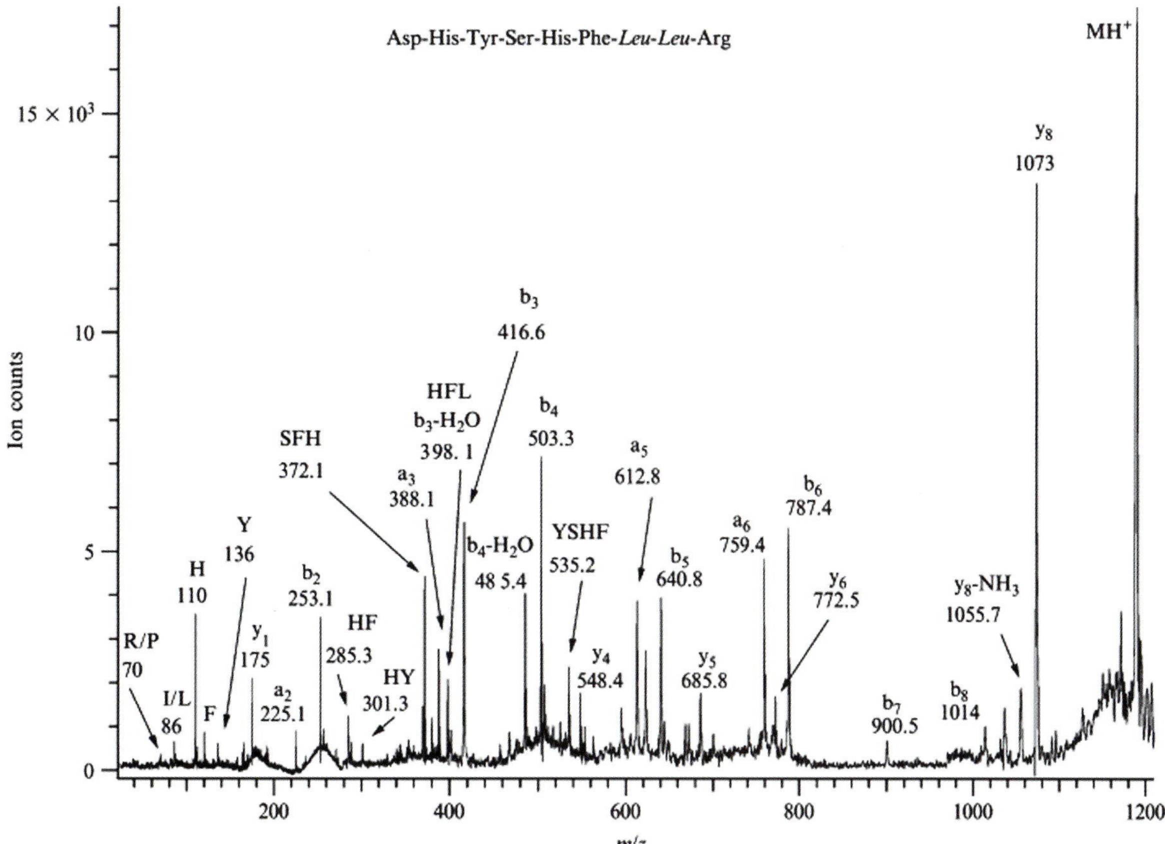

FIGURE 9.13 Mass Spectroscopy Trace of Peptide Fragments

Post-source decay spectrum of a tryptic peptide (m/z 1187.6) from the 50- kDa subunit of DNA polymerase from *Schizosaccharomyces pombe*. The spectrum was acquired on a Voyager mass spectrometer (Applied Biosystems). From: Medzihradszky KF (2005). Peptide sequence analysis. *Methods Enzymol* **402**, 209–244. Reprinted with permission.

the amino acid sequence. Sometimes the spectrum obtained for a peptide ion is ambiguous, so databases of peptide ion spectra are used for comparison.

Each amino acid degrades into predicted ions in mass spectroscopy. The amino acid sequence for an unknown peptide can be deduced based on the pattern of ions produced in comparison with the known patterns.

PROTEIN QUANTIFICATION USING MASS SPECTROMETRY

Mass spectrometry can also be used to quantify a particular peptide from a protein, which directly correlates to the amount of protein (Fig. 9.14). To compare the relative amounts of one protein, samples of cells are grown under two experimental conditions. One sample is grown in the presence of an amino acid tagged with a heavy isotope. The heavy isotope is usually deuterium, ^{2}H, but could also be ^{13}C or ^{15}N. These isotopes increase the mass of all proteins in that particular sample. The cells from each condition are lysed, and the proteins isolated. The two samples are mixed and analyzed using HPLC coupled to ESI mass

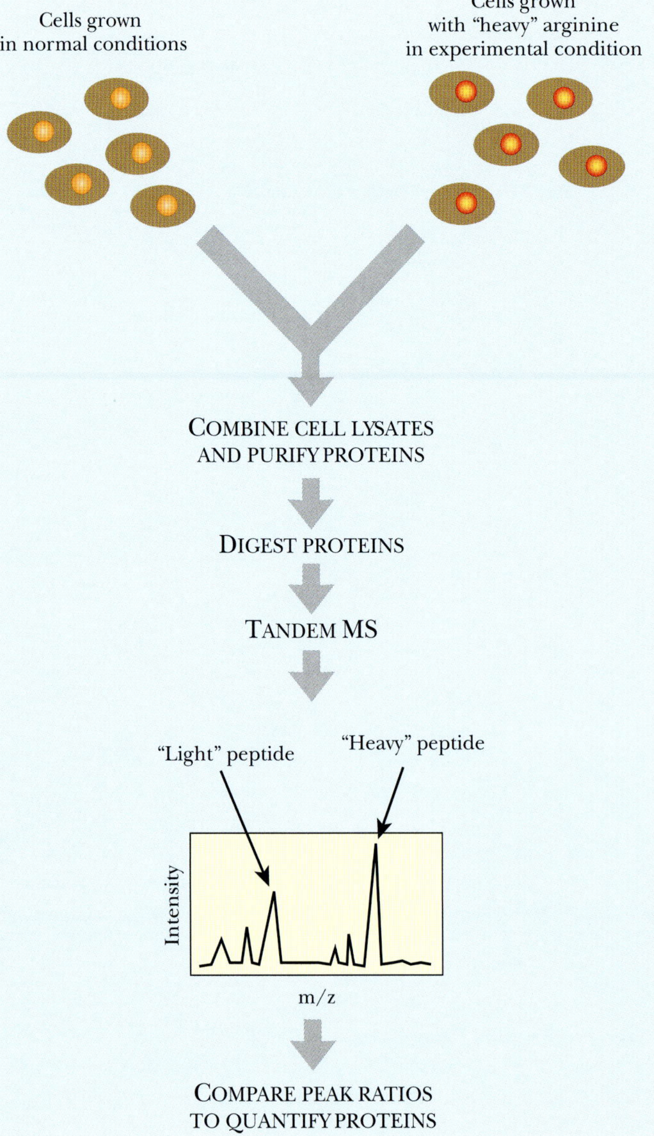

FIGURE 9.14
Quantification of Peptides Using Mass Spectrometry
Mass spectrometry can compare the amount of a particular protein in samples from two different conditions. Cells in one condition are grown with a "heavy" amino acid, such as arginine, which is incorporated during protein synthesis. The proteins are digested with proteases into small peptide fragments. These are ionized by ESI mass spectrometry. The analysis gives pairs of light and heavy peaks for each peptide. Peak sizes correlate to the amount of peptide, and hence protein. The "heavy" peak is more abundant in this example; thus this protein is more abundant under the experimental conditions.

spectroscopy. Each individual peptide will now have two peaks, one from the normal and one from the heavy sample. The ratio of the two peaks will determine the relative change in level of the protein of interest between the samples.

Changing experimental conditions affects the amount of a particular protein. This change can be determined using mass spectroscopy by adding heavy isotopes to one of the samples.

Box 9.1 Proteomic Pattern Diagnostics

Because mass spectra are very sensitive to slight variations in protein samples, they can potentially be used to diagnose disease. Potential applications include determining whether a serum protein sample from a patient is diseased or normal. To standardize such measurements, profiles for disease and normal states must first be determined. Blood serum samples from healthy patients and patients with ovarian cancer, for example, can be collected and subjected to mass spectroscopy. Computers can analyze the data for each of the spectra, but because the patterns are so complex, only subsets of the spectra are actually analyzed. First the computer compiles each

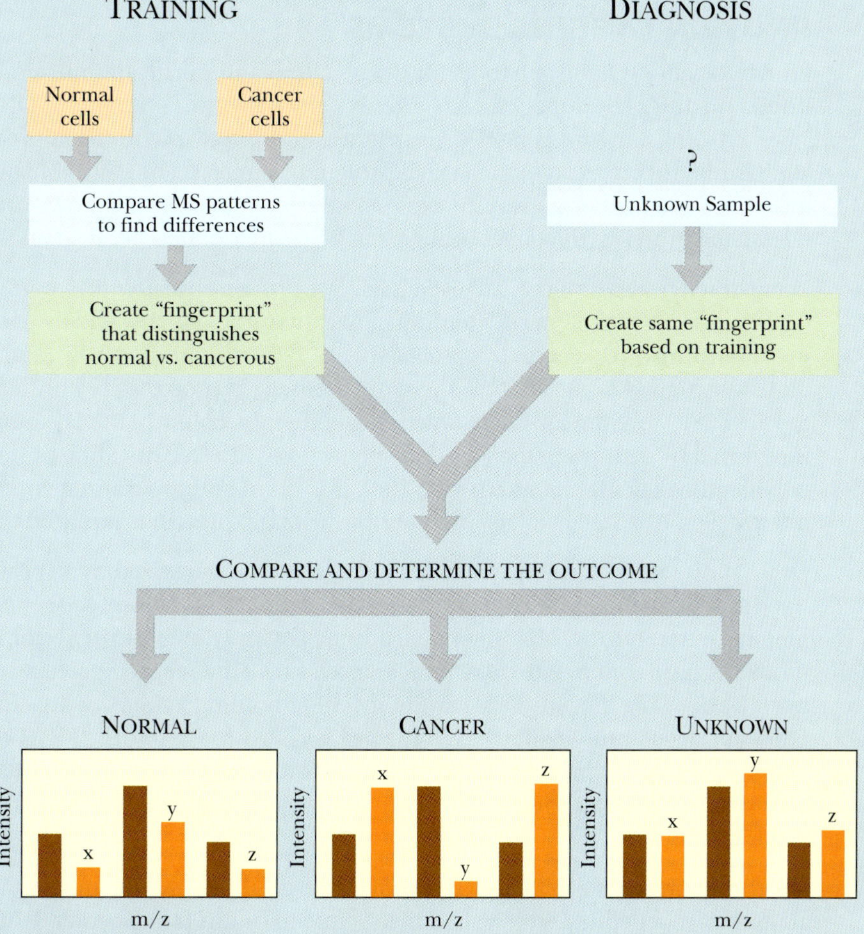

FIGURE A Proteomic Pattern Diagnostics
Computer training involves letting a computer define subsets of mass spectra data that differs between two conditions. For example, mass spectra from both cancerous and normal cells are analyzed to find proteins that show differences only in the cancerous condition. To ensure reliable results, a large number of samples must be analyzed. Once the computer has generated a "fingerprint" for cancer, unknown samples can be analyzed for cancerous, normal, or unknown fingerprints. In this example X, Y, and Z are the only proteins that change in cancerous states; X and Z increase while Y decreases in amount. In the unknown, X, Y, and Z all increase; therefore, it does not fit the true cancer profile.

(Continued)

Box 9.1 Proteomic Pattern Diagnostics—cont'd

spectrum into different "data packets" that contain only 5 to 20 different proteins. These data packets are then compared to find those that are different in the normal and cancerous samples. The others are ignored and the best data packets are kept. This form of computer training is based on artificial intelligence in a process called **selective pressure**. Continued refinement is accomplished by restricting the computer analysis to the differences between normal and cancer spectra. After the computer has been trained, new patients can be screened for ovarian cancer by mass spectroscopy on a sample of blood serum. The computer then determines whether the serum from the new patient has a spectrum that is normal, cancerous, or unknown (Fig. A).

The major drawback of this type of analysis is that the identity of the protein peaks is unknown. Great care must be taken for the peaks to truly represent tumor markers and not other differences. For example, different profiles may be due to age, sex, level of exercise, menopausal state, nutritional habits, drug use, and so on. Profiles may also be affected by how the sample has been stored or handled, because extended freeze/thaw cycles can affect proteins. Other factors are making sure the mass spectrometer works consistently for all samples, and that the patient samples are collected and processed identically. Once these factors are addressed, pattern diagnostics may become a powerful method for detecting various diseases long before any overt symptoms appear.

PROTEIN TAGGING SYSTEMS

Protein tagging systems are tools for the isolation and purification of single target proteins from a mixture. The target protein is genetically fused to a segment of DNA that codes for a "tag," creating a hybrid gene. This is inserted into a vector with the appropriate promoters and terminators to express the tagged protein of interest. The gene construct is transformed into a suitable host organism for expression. When the cells are grown and disrupted to release the proteins, the target protein can be easily isolated because of its tag.

Many different types of tags are used to isolate proteins because the chemistry and size of the tag may affect the protein of interest in a negative way. The first widely used tag, called the **polyhistidine** or **His6 tag**, consists of six histidine residues in a row (Fig. 9.15). Histidine binds very tightly to nickel ions; therefore, His-tagged proteins are purified on a column to which Ni^{2+} ions are attached. Once attached to the column, the His6-tagged protein is removed by disrupting the Ni^{2+}–His interaction with free histidine or imidazole. The polyhistidine tag may be attached to the carboxy- or amino-terminal end of the protein of interest. Because the His6 tag is very short, the target protein is rarely affected by adding it.

Other short tags for proteins include FLAG, which is recognized by a specific antibody. FLAG has the peptide sequence AspTyrLysAspAspAspAspLys. As before, the gene for the target protein plus a short DNA segment encoding FLAG is cloned into a vector, and the hybrid protein is produced in either bacteria or a cell line. FLAG-tagged proteins can be isolated from a cell lysate using the anti-FLAG antibody either bound to beads or attached to a column. Only the tagged protein attaches to the beads/column. Finally, the FLAG-tagged protein is separated from the antibody by adding free FLAG peptide. The short peptide is present in surplus and competes for antibody with the tagged protein, which is therefore eluted from the beads or column.

Another short tag is the "Strep" tag, provided by a short DNA segment that encodes a 10-amino-acid peptide with a similar structure to biotin (see Chapter 3). The biotin-like peptide binds tightly to avidin and streptavidin, so Strep-tagged proteins are isolated by binding to streptavidin-coated beads or a streptavidin column.

Besides short tags, longer tags that consist of entire proteins are used for some applications. Three popular tags include **protein A** from *Staphylococcus*, **glutathione-S-transferase (GST)** from *Schistosoma japonicum*, and **maltose-binding protein (MBP)** from *Escherichia coli*. Just like the short tags, the genes for these longer tags are genetically fused to the target gene. The hybrid gene constructs are expressed by using appropriate transcriptional promoters and terminators. Once the host cells express the hybrid gene, the fusion protein is isolated by

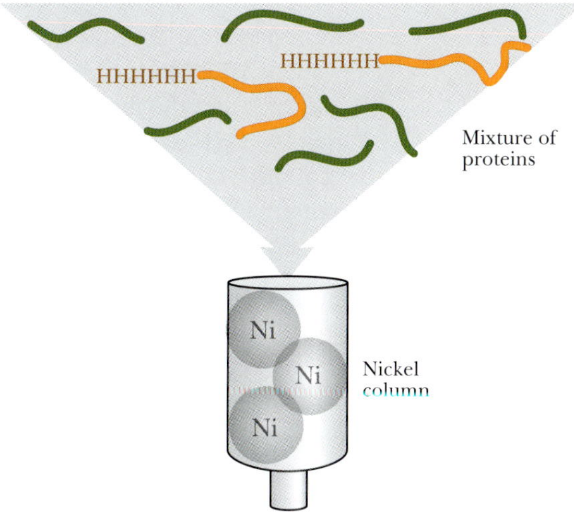

PROTEINS POURED THROUGH COLUMN

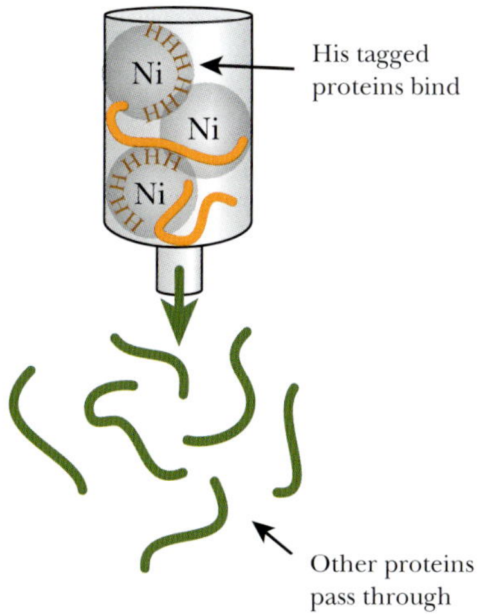

ELUTE WITH HISTIDINE OR IMIDAZOLE

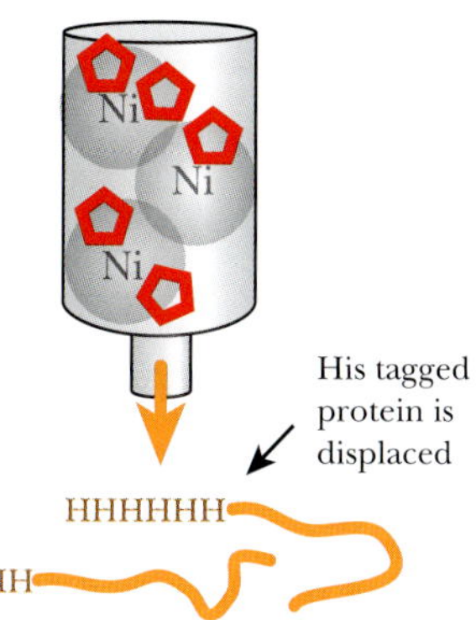

FIGURE 9.15 Nickel Purification of His6 Tagged Protein
To isolate a pure sample of one specific protein, the gene for the protein is genetically linked to a coding region for six histidine residues. The expressed fusion protein can be isolated from a mixture of proteins because of the chemistry of histidines. The histidines bind to nickel-coated beads, and the remaining untagged proteins pass through the column. The histidine-tagged protein is then eluted by passing free histidine or imidazole over the column.

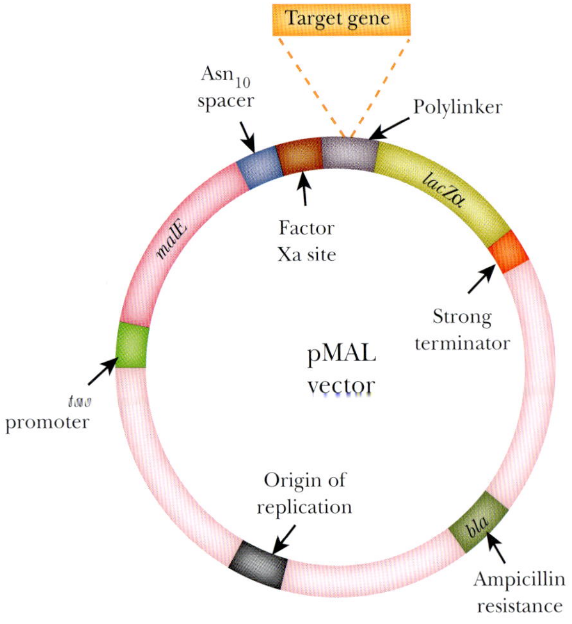

purifying the protein tag. A specific antibody is available to bind to protein A, and lowering the pH releases the fusion protein. MBP binds to maltose (attached to beads or a column), and the fusion protein is released by adding free maltose. GST binds to its substrate, glutathione (on beads or a column), and free glutathione is used to release the hybrid protein. Once the fusion protein has been isolated, it must be cleaved to separate the target protein from the tag protein.

A useful feature of the longer tags is the presence of protease cleavage site between the target protein and the tagging protein. The vector for pMAL (New England Biolabs, Inc., Ipswich, MA) has the gene for MBP, followed by a spacer region, a recognition site for factor Xa, then the polylinker region for cloning the target gene (Fig. 9.16). Factor Xa is a specific protease used in the blood clotting system, and inserting its recognition sequence allows the MBP portion of the hybrid protein to be cleaved from the target protein. After the hybrid protein is eluted from the purification column with maltose, the original protein is isolated by protease treatment. This is extremely useful when pure, native protein is needed for analysis.

FIGURE 9.16
Maltose-Binding Protein Fusion Vector
Vectors such as pMAL have polylinker regions for cloning a target gene in frame with the gene for a tag protein such as MBP (the *malE* gene). The fusion protein is easily isolated because MBP binds to maltose columns. The fusion protein also has a binding site for the protease, factor Xa. When the fusion protein is bound to the column, factor Xa will release the target protein and leave behind the MBP domain.

An even easier way to obtain a pure sample of native protein is the self-cleavable **intein** tag. The approach is based on inteins, self-splicing intervening segments found in some proteins. Inteins are the protein equivalent to introns in RNA. The intein removes itself from its host protein via a branched intermediate (Fig. 9.17).

The Intein Mediated Purification with Affinity Chitin-binding Tag (IMPACT) system from New England Biolabs uses a modified intein from the *VMA1* gene of *Saccharomyces cerevisiae*. Intein cleavage is used to release the target protein after purification of the fusion protein (Fig. 9.18). This yeast intein originally cleaved both its N-terminus and its C-terminus, but it has been modified so that it only cleaves its N-terminus. The chitin-binding tag of this system is the small chitin-binding domain (CBD) from the chitinase A1 gene of *Bacillus circulans*. (Chitin is the substance that forms the exoskeleton of insects.) The vector has a polylinker or cloning site for the target gene followed by the DNA segment encoding the intein, followed by the CBD. The fusion protein is expressed, and cell lysates containing the hybrid protein are isolated. When the lysate passes through a chitin column, the hybrid protein binds to the column via the CBD, and the remaining cellular proteins elute. The column is then incubated at 4 °C with dithiothreitol (DTT), a thiol reagent that activates the intein to cleave its N-terminus. Thus the target protein is released from the column, leaving behind the intein and CBD regions.

Protein tags are either short peptides or entire genes fused onto a protein of interest. Protein tags provide a means to isolate the protein of interest from the rest of the cellular proteins.

PHAGE DISPLAY LIBRARY SCREENING

A **phage display** library is a collection of bacteriophage particles that have segments of foreign proteins protruding from their surface (Fig. 9.19). Normal bacteriophages have outer coats made of proteins. The outer coat of M13 bacteriophage has about 2500 copies of the major coat protein (gene VIII protein) and about five copies of the minor coat protein (gene III protein). One particular phage display system fuses gene III protein to the foreign protein, so that M13 now has about five copies of the foreign protein on its surface. M13 is convenient because it does not lyse the bacteria it infects; the viral particles are simply secreted through the bacterial cell envelope.

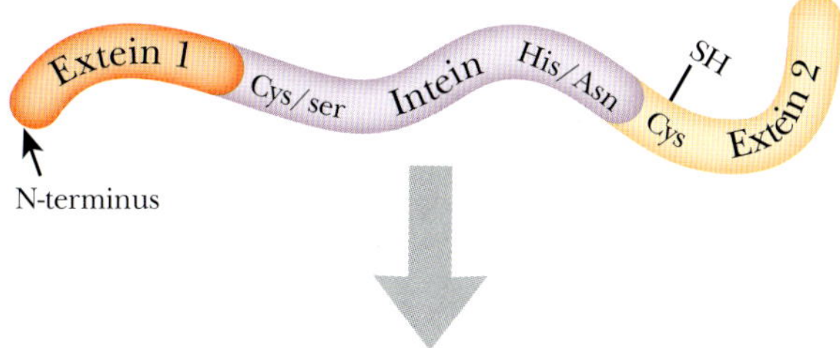

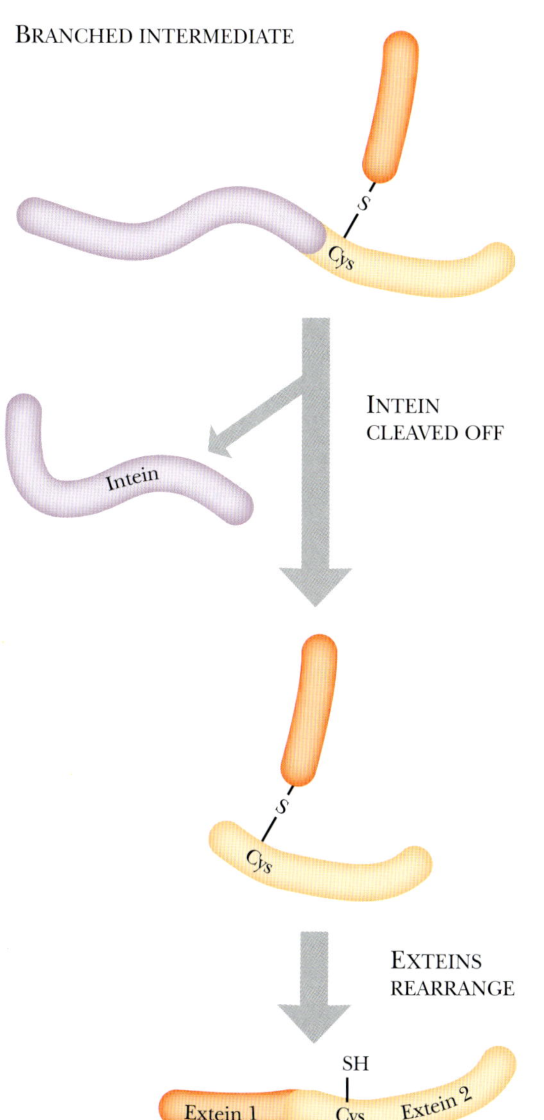

FIGURE 9.17 Mechanism of Intein Removal
The intervening intein segment splices itself out in two stages. The intein has a Cys or Ser at the boundary with extein 1 and a basic amino acid at its boundary with extein 2. The downstream extein (2) has a Cys residue at the splice junction. Extein 1 is cut loose and attached to the sulfur side chain of the cysteine at the splice junction. This forms a temporary branched intermediate. Next the intein is cut off and discarded and the two exteins are joined to form the final protein.

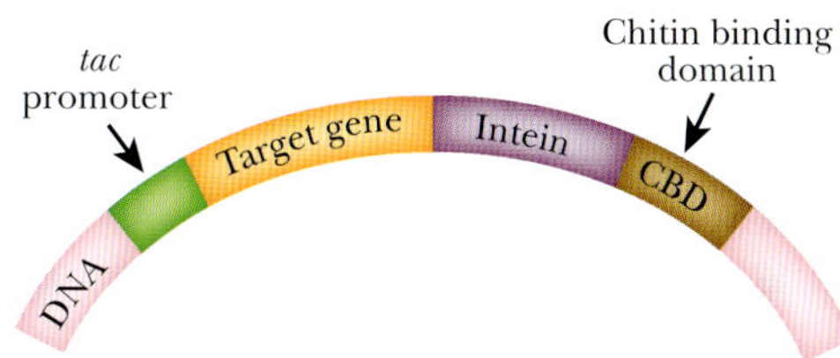

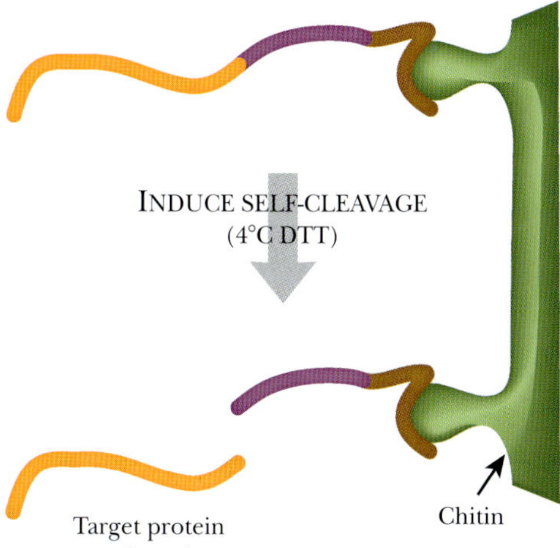

FIGURE 9.18 Intein-Mediated Purification System
Inteins that can self-cleave at their amino-terminus allow specific proteins to be purified and cleaved from a fusion protein in one step. First, the fusion protein is purified by passing over a column made of chitin. (Note: The CBD or chitin binding domain recognizes and binds the chitin molecule.) The column is incubated with DTT in a refrigerator. The intein cleaves itself at its amino terminus and releases the target protein.

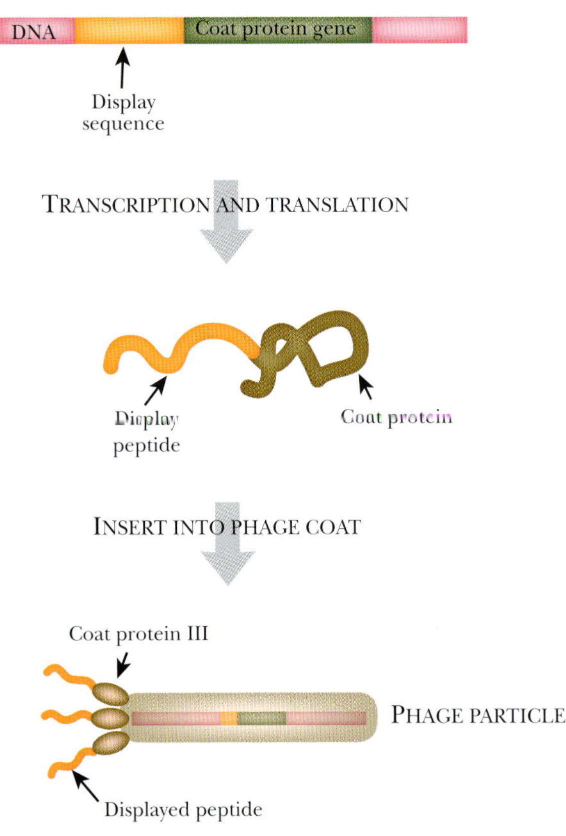

FIGURE 9.19
Principle of Phage Display
In order to display a peptide on the surface of a bacteriophage, the DNA sequence encoding the peptide must be fused to the gene for a bacteriophage coat protein. In this example, the chosen coat protein is encoded by gene III of phage M13. The N-terminal portion of gene III protein is on the outside of the phage particle, whereas the C-terminus is inside. Therefore, the peptide must be fused in frame at the N-terminus to be displayed on the outside of the phage.

In order for the phage to display a foreign protein, the gene for that protein must be fused to gene III to produce a hybrid protein. The gene of interest must be in frame with gene III for proper expression. The M13 genome can accommodate extra DNA because the bacteriophage particles are filamentous and are made longer to package a larger genome. The M13 genome containing the gene of interest is transformed into *E. coli*, where the bacteriophage DNA directs the synthesis of new particles containing the protein of interest in the coat.

Bacteriophage can also display small segments of genes. Random oligonucleotides generated by PCR can be cloned and fused to gene III of M13. Each random oligonucleotide will encode a different peptide. These clones are transformed into the bacteriophage, and each transformant will display its foreign peptide fused to gene III protein.

The collection of displayed peptides can be useful in many ways. They can be screened by **biopanning** to find a particular target, perhaps a specific protein binding domain, or a specific peptide that binds an antibody (Fig. 9.20). In biopanning, the library of phages displaying the foreign peptides is incubated with the target, such as an antibody, bound to a bead or membrane. All the recombinant phages that bind to that antibody adhere to the solid support, and the others are washed away. All the bound phages are eluted and incubated with *E. coli* to replicate the phage. The procedure is usually repeated in order to enrich for peptides that bind specifically, because some nonspecific binding could occur. Once a phage with a useful peptide is identified, the clone is sequenced to determine the structure of the peptide.

Full-length protein libraries can also be studied using phage display, but pose some extra problems. Coding sequences for full-length proteins must be cloned in frame with both the signal sequence at the N-terminal end and gene III at the C-terminal end. (The signal sequence is required to direct the hybrid protein to the viral coat.) Ensuring the correct reading frame is reasonable for one or two genes, but for an entire library, there is too much room for error. Besides, the possible creation of a stop codon at either fusion junction would prevent the hybrid protein from being expressed. The solution is to use T7 bacteriophage for libraries of full-length proteins. T7 has a coat protein whose C-terminal tail is exposed to the outside. To be expressed on the bacteriophage surface, the protein library must therefore be fused to the C-terminus of the coat protein. This requires only one fusion junction. Furthermore, even if library sequences are cloned out of frame or contain stop codons, the coat protein itself is unaffected and still assembled, although the attached library proteins will be defective.

Being able to express full-length proteins for biopanning is very useful to proteomics researchers. To identify a protein that binds to a particular cell surface receptor, a phage display library can be biopanned for receptor binding. Another example is finding RNA binding proteins. Here, RNA is anchored to a solid support and the phage display library is incubated with this RNA "bait." The phages that stick to the RNA bait are isolated and enriched by repeating the procedure. Each isolated clone can then be sequenced to identify which proteins bind RNA.

Phage display is a technique where foreign proteins or peptides are fused to a coat protein on the surface of the phage. The phage then displays these for analysis.

Biopanning identifies binding partners for a protein of interest. The protein of interest is incubated with phage display library. When a phage binds to the protein of interest, it is isolated and the sequence for the displayed peptide is determined.

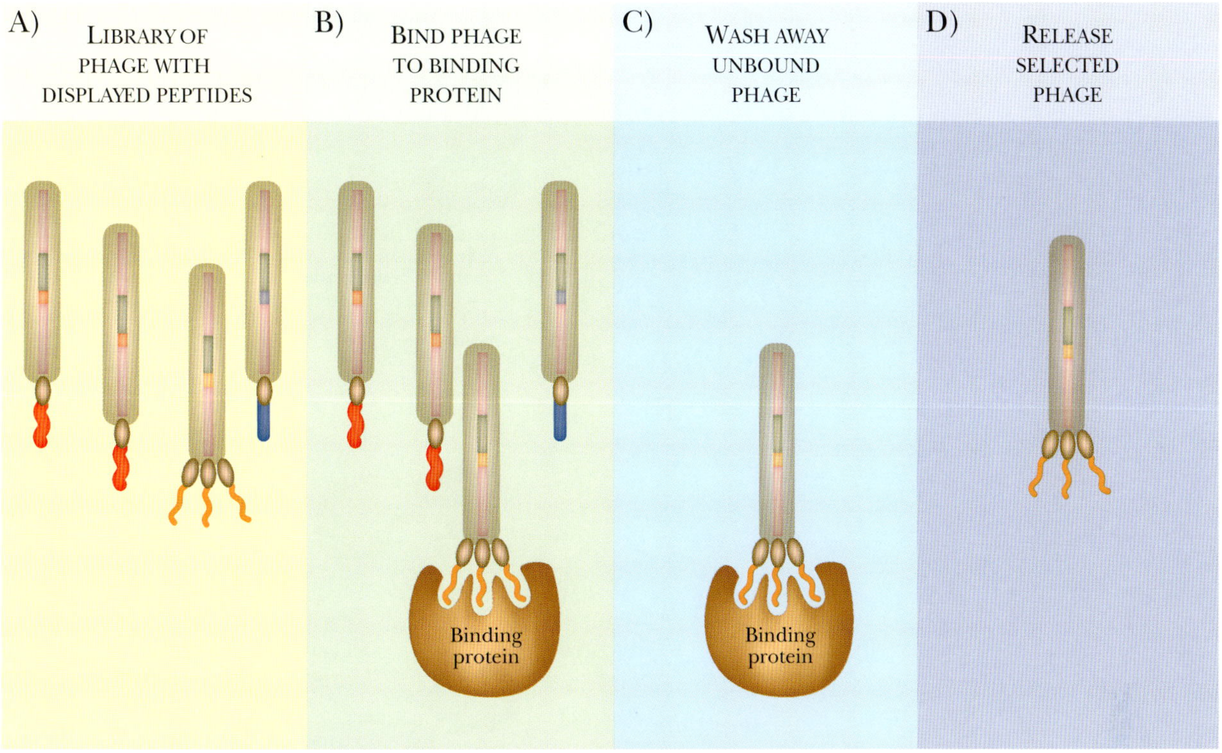

A) LIBRARY OF PHAGE WITH DISPLAYED PEPTIDES
B) BIND PHAGE TO BINDING PROTEIN
C) WASH AWAY UNBOUND PHAGE
D) RELEASE SELECTED PHAGE

Binding protein
Binding protein

FIGURE 9.20 Biopanning of Phage Display
Biopanning is used to isolate peptides that bind to a specific target protein, which is usually attached to a solid support such as a membrane or bead. The phage display library (A) is attached to the binding protein (B). Those phages that bind to the target protein will be retained (C) but the others are washed away. The phage that does recognize the binding protein can be released, isolated, and purified.

PROTEIN INTERACTIONS: THE YEAST TWO-HYBRID SYSTEM

In addition to protein function and expression, proteomics attempts to find relevant protein interactions. For those who like "-omics" terminology, the total of all protein-protein interactions is called the **protein interactome**. For example, hormones usually bind to receptors that pass the signal on. Often this involves a protein relay where one protein activates another, which in turn activates yet another. To understand hormone function, researchers must identify all the proteins in the signal cascade. Phage display is one way to identify interactions, but the displayed proteins may not fold correctly or specific cofactors may be missing when mammalian proteins are expressed in bacteria.

An approach to overcoming these difficulties is the yeast **two-hybrid system**, where the binding of two proteins activates a reporter gene. The binding of a transcriptional activator protein, GAL4, to the promoter region of the reporter gene activates transcription and translation. GAL4 contains two domains needed to turn on the reporter gene. The DNA binding domain (DBD) recognizes the promoter element and positions the second domain, the activation domain (AD) next to RNA polymerase, where it activates transcription. These two domains can be expressed as separate proteins, but cannot activate the reporter gene unless they are brought together (Fig. 9.21).

In the two-hybrid system, the two domains are each fused to different proteins by creating hybrid genes. The **bait** is the DBD genetically fused to the protein of interest, and the **prey** is the AD fused to proteins that are being screened for interaction with the bait. When the bait and prey bind, the DBD and AD activate transcription of the reporter gene.

Two vectors are needed to perform two-hybrid analysis (Fig. 9.22). The first vector has a multiple cloning site for the bait protein 3′ of the GAL4-DBD; therefore, the fusion protein has the Bait protein as its C-terminal domain. The second vector has a multiple cloning site

A)

B)

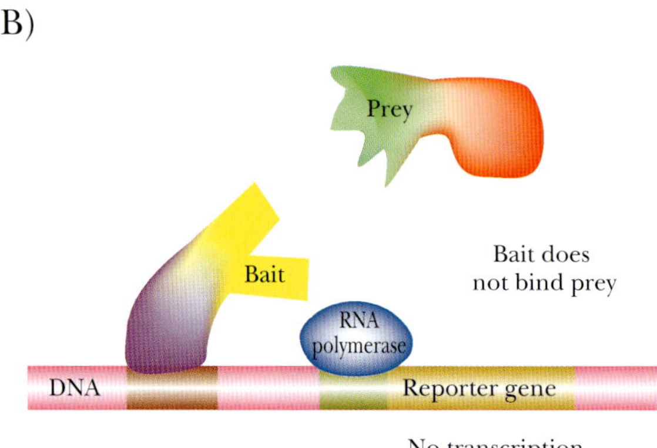

C)

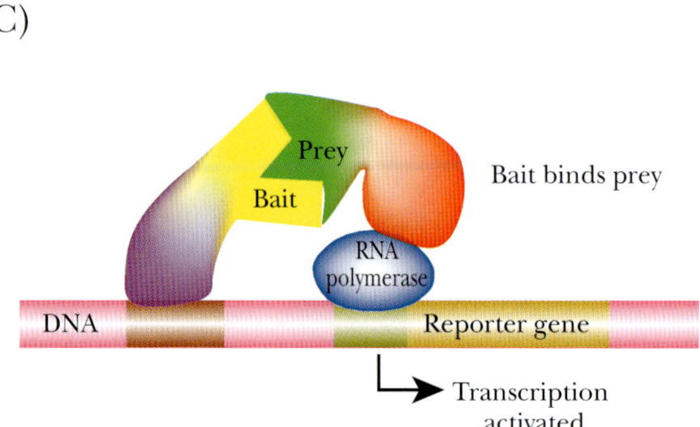

FIGURE 9.21 Principle of Two-Hybrid

Analysis (A) Yeast transcription factors have two domains: the DBD (purple) recognizes regulatory sites on DNA, and the AD (red) activates RNA polymerase to start transcription of the reporter gene. For two-hybrid analysis, two proteins (Bait and Prey) are fused separately to the DBD and AD domains of the transcription factor. The Bait protein is joined to the DBD and the Prey protein to the AD. (B) The Bait protein and Prey protein do not interact and the reporter gene is not turned on. (C) The Bait binds the Prey, bringing the transcription factor halves together. The complex activates RNA polymerase and the reporter gene is expressed.

for the Prey protein 5′ of GAL4-AD and the fusion protein has the Prey as its N-terminal domain. Both plasmids must be expressed in the same yeast cell. If the bait and prey proteins interact, the reporter gene is turned on.

The reporter genes must be engineered to be under control of the GAL4 recognition sequence. Common reporter genes include *HIS3*, which encodes an enzyme in the histidine pathway and whose expression allows yeast cells to grow on media lacking histidine, or *URA3*, which allows growth without uracil. These reporter systems require yeast host cells that are defective in the corresponding genes. However, they do allow direct selection of positive isolates.

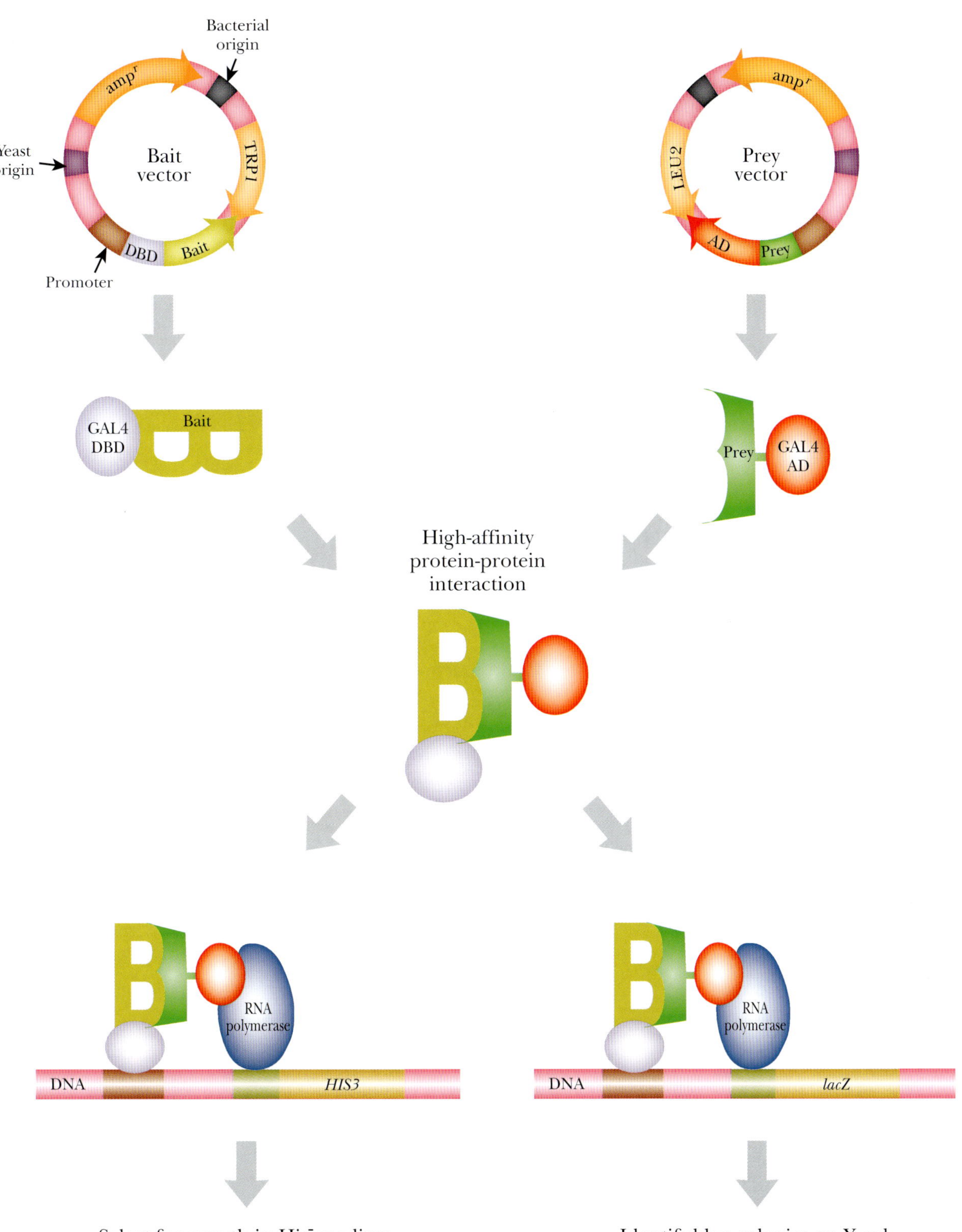

FIGURE 9.22 Vectors for Two-Hybrid Analysis

Two different vectors are necessary for two-hybrid analysis. The bait vector has coding regions for the DBD and for the Bait protein. The Prey vector has coding regions for the AD and for the Prey protein. These two different constructs are expressed in the same yeast cell. If the Bait and Prey interact, the reporter gene is expressed. Two reporter systems are shown here. The *His3* gene allows yeast to grow on histidine-free media. The *lacZ* gene encodes β-galactosidase, which cleaves X-gal, forming a blue color.

Another reporter used is *lacZ* from *E. coli,* which encodes β-galactosidase. Both bacteria and yeast that express *lacZ* turn blue when grown with X-Gal. β-galactosidase cleaves X-Gal, releasing a blue product. The reporter genes are usually integrated into the yeast genome, rather than being carried on a separate vector.

The yeast two-hybrid system has been used to identify all the protein interactions in the yeast proteome by mass screening with mating (Fig. 9.23). Yeast has about 6000 different proteins, and each of these has been cloned into both vectors via PCR. This way, each protein can be used as both bait and prey. All the bait vectors were transformed into haploid yeast of one

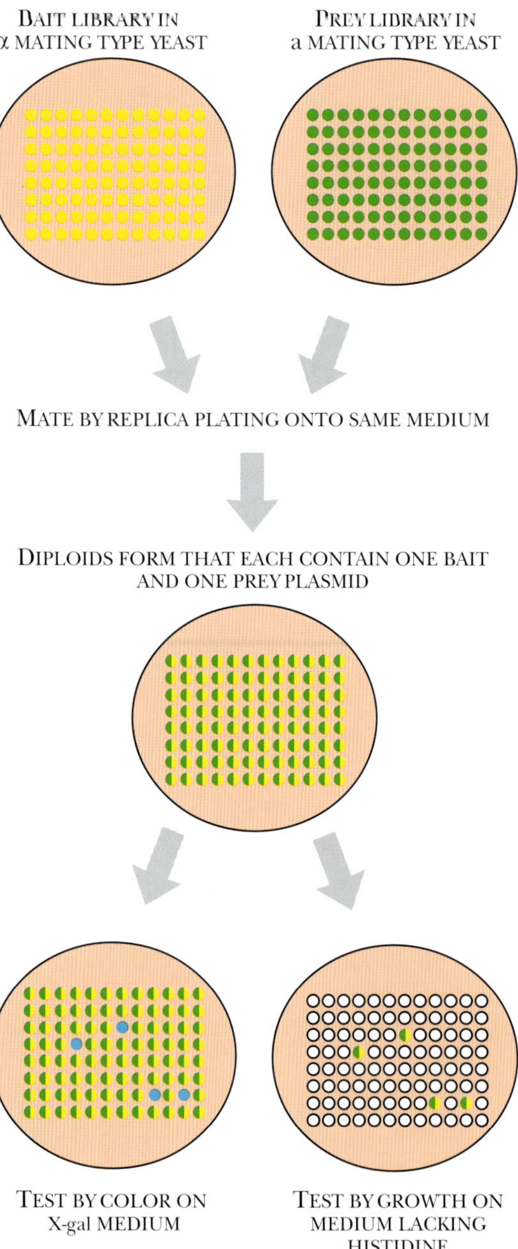

FIGURE 9.23 Two-Hybrid Analysis: Mass Screening by Mating

To identify all possible protein interactions using the two-hybrid system, haploid α yeast are transformed with the Bait library, and haploid **a** yeast are transformed with the Prey library. When the two yeast types are mated with each other, the diploid cells will each contain a single bait fusion protein and a single prey fusion protein. If the two proteins interact, they activate the reporter gene, which allows the yeast to grow on media lacking histidine (yeast *His3* gene) or turn the cells blue when growing on X-gal medium (*lacZ* from *E. coli*). This process can be done for all 6000 predicted yeast proteins using automated techniques.

mating type, and the prey vectors into the other mating type. Haploid cells carrying bait are fused to haploid cells with prey and the resulting diploid cells are screened for reporter gene activity. This analysis thus examined 6000 × 6000 combinations for protein interaction.

The yeast two-hybrid system has some limitations. Because transcription factors must be in the nucleus to work, the target proteins must also function in the nucleus. For some proteins, entering the nucleus may cause the protein to misfold. For other proteins, the nucleus does not have the proper cofactors and the protein may be unstable. Large proteins may not be expressed well, or may be toxic to the yeast, leading to false negative results.

Yeast two-hybrid analysis finds proteins that bind together. Cellular proteins are linked to the AD of GAL4 or the DBD of GAL4. When two cellular proteins bring the AD and DBD together, the completed GAL4 binds to the reporter gene promoter. Reporter gene products allow the yeast to grow on histidine-free media or turn blue on media that contain X-Gal.

PROTEIN INTERACTIONS BY CO-IMMUNOPRECIPITATION

Co-immunoprecipitation is a technique to examine protein interactions in the cytoplasm rather than the nucleus (Fig. 9.24). Here, the target protein is expressed in cultured mammalian cells, which are lysed to release the cytoplasmic contents. The target protein is precipitated from the lysate with an antibody. Other proteins that are associated with the target protein remain associated with the antibody-protein complex. (If no antibody exists for the target protein, a small tag [such as FLAG or His6; see earlier discussion] can be engineered onto the protein.) Protein A from *Staphylococcus*, in turn, binds the antibodies. The protein A is attached to beads before it is added to the cell lysate. This generates very large target protein/antibody/protein A/bead complexes, which are gently isolated from the rest of the cellular proteins by

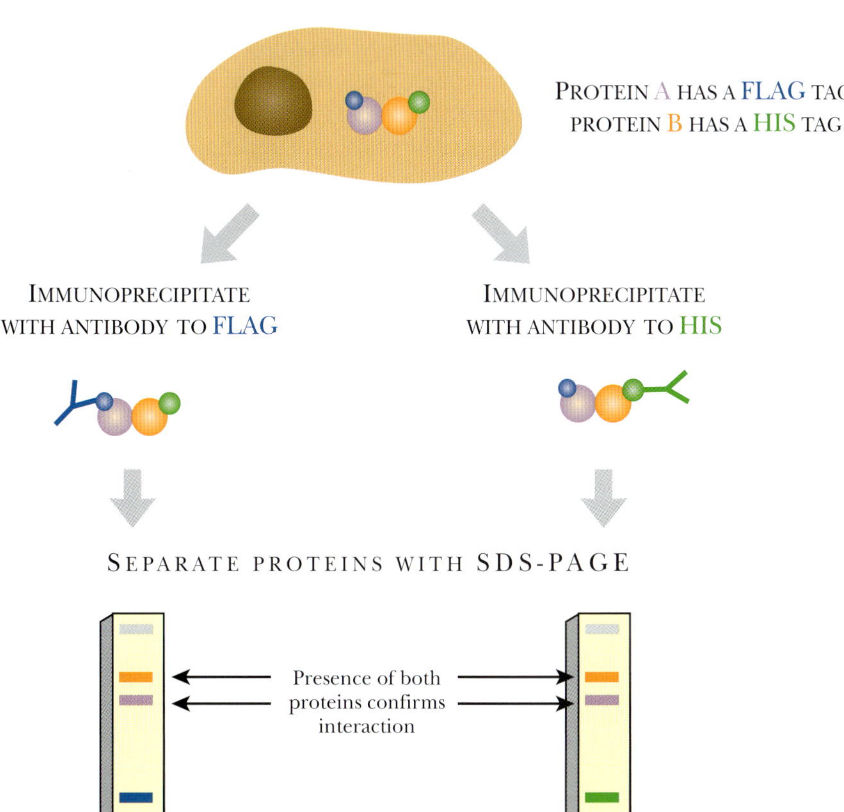

FIGURE 9.24 Co-immunoprecipitation
To determine whether protein A and B interact within the cytoplasm, each protein is fused to a different tag for easy isolation. Each fusion protein is expressed in mammalian cells, which are then lysed to release the cell proteins. The cells must be lysed gently to avoid disrupting the protein interactions. The fusion proteins are isolated using the tag sequence. Each tagged protein and all its associated proteins are isolated independently. For example, on the left, Flag-tagged protein A is isolated with an antibody to the Flag sequence, and on the right, His6-tagged protein B is isolated with an antibody to the His6 sequence. The protein complexes are separated by SDS-PAGE. This example shows the two tagged proteins A and B interacting.

centrifugation. The complexes are separated by size with SDS-PAGE. The gel should show the target protein, the antibody, protein A, and other bands that represent interacting proteins. These can be identified with protein sequencing and/or mass spectrometry.

Co-immunoprecipitation is often used to confirm the results from yeast two-hybrid analysis, especially for mammalian proteins. Many two-hybrid experiments reveal novel uncharacterized proteins. To confirm the interaction, both proteins are tagged for easy isolation. Adding a tag is much easier than generating a specific antibody to each new protein. For example, protein A is tagged with FLAG, while protein B is tagged with His6. (Note: This protein A is *not* the protein A from *Staphylococcus*.) Each vector construct is transformed into a mammalian cell line, and each protein is expressed. The cell lysate is harvested and divided into two samples. The protein A complexes are isolated from the first sample, whereas the protein B complexes are isolated from the second sample. Each of the complexes is isolated with protein A–coated beads. The different proteins from each sample are separated by SDS-PAGE. If the two proteins interact, both proteins will be found in both samples.

Co-immunoprecipitation determines whether two proteins bind together in the cytoplasm.

PROTEIN ARRAYS

Protein-detecting arrays may be divided into those that use antibodies and those based on using tags. In the **ELISA assay** (see Chapter 6), antibodies to specific proteins are attached to a solid support, such as a microtiter plate or glass slide. The protein sample is then added and if the target protein is present, it binds its complementary antibody. Bound proteins are detected by adding a labeled second antibody.

Another antibody-based protein-detecting array is the **antigen capture immunoassay** (Fig. 9.25). Much like the ELISA, this method uses antibodies to various proteins bound to a solid surface. The experimental protein sample is isolated and labeled with a fluorescent dye. If two conditions are being compared, proteins from sample 1 can be labeled with Cy3, which fluoresces green, and proteins from sample 2 can be labeled with Cy5, which fluoresces red. The samples are added to the antibody array, and complementary proteins bind to their cognate antibodies. If both sample 1 and 2 have identical proteins that bind the same antibody, the spot will fluoresce yellow. If sample 1 has a protein that is missing in sample 2, then the spot will be green. Conversely, if sample 2 has a protein missing from sample 1, the spot will be red. This method is good for comparing protein expression profiles for two different conditions.

In the third method, the **direct immunoassay** or **reverse-phase array**, the proteins of the experimental sample are bound to the solid support (Fig. 9.26). The proteins are then probed with a specific labeled antibody. Both presence and amount of protein can be monitored. For example, proteins from different patients with prostate cancer can be isolated and spotted onto glass slides. Each sample can be examined for specific protein markers or the presence of different cancer proteins. The levels of certain proteins may be related to the stages of prostate cancer. This immunoassay helps researchers to decipher these correlations.

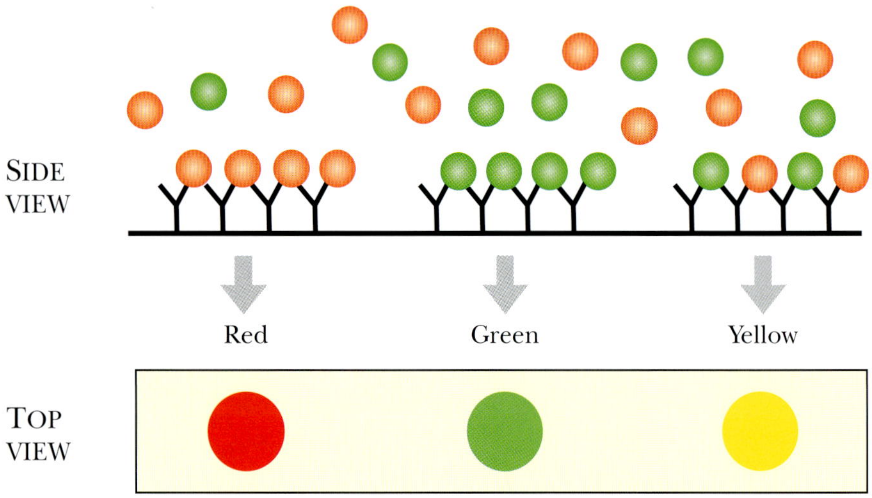

FIGURE 9.25 Ideal Results for Antigen Capture Immunoassay
A variety of different antibodies are fused to different regions of a solid surface. Each spot has a different antibody. If the antibody recognizes only proteins labeled with Cy5, the region will fluoresce red (*left*). If the antibody recognizes only proteins labeled with Cy3, the region will fluoresce green (*middle*). If the antibody recognizes proteins in both conditions, the spot will fluoresce yellow (*right*).

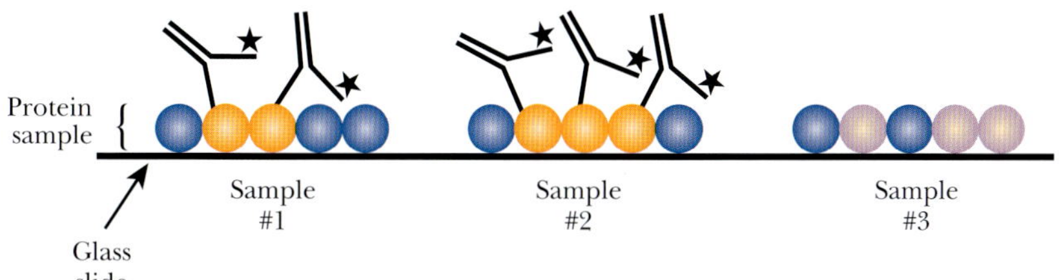

FIGURE 9.26 Direct Immunoassay
The direct immunoassay binds the protein samples to different regions on a solid support. Each spot has a different protein sample. Next, an antibody labeled with a detection system is added. The antibody binds only to its target protein. In this example, the antibody recognizes only a protein in patient sample 1 and 2.

The main problem with immuno-based arrays is the antibody. Many antibodies cross-react with other cellular proteins, which generates false positives. In addition, binding proteins to solid supports may not be truly representative of intracellular conditions. The proteins are not purified or separated; therefore, samples contain very diverse proteins. Some proteins will bind faster and better than others. Also, proteins of low abundance may not compete for binding sites. Another problem is that many proteins are found in complexes, so other proteins in the complex may mask the antibody binding site.

Rather than using antibodies, protein interaction arrays use a fusion tag to bind the protein to a solid support (Fig. 9.27). The use of protein arrays to determine protein interactions and protein function is a natural extension of yeast two-hybrid assays and co-immunoprecipitation. Protein arrays can assess thousands of proteins at one time, making this a powerful technique for studying the proteome. Protein arrays are often used in yeast because its proteome contains only about 6000 proteins. Libraries have been constructed in which each protein is fused to a His6 or GST tag. The proteins are then attached by the tags to a solid support such as a glass slide coated with nickel or glutathione. To build the array, each protein is isolated individually and spotted onto the glass slide. The tagged proteins bind to the slide and other cellular components are washed away. Each spot has only one unique tagged protein.

Once the array is assembled, the proteins can be assessed for a particular function. In the laboratory of Michael Snyder at Yale University, the yeast proteome has been screened for proteins that bind **calmodulin** (a small Ca^{2+} binding protein) or phospholipids (Fig. 9.28). Both calmodulin and phospholipid were tagged with biotin and incubated with a slide coated with each of the yeast proteins bound to the slide via

FIGURE 9.27
Protein Interaction Microarray—Principle
To assemble a protein microarray, a library of His6-tagged proteins is incubated with a nickel-coated glass slide. The proteins adhere to the slide wherever nickel ions are present.

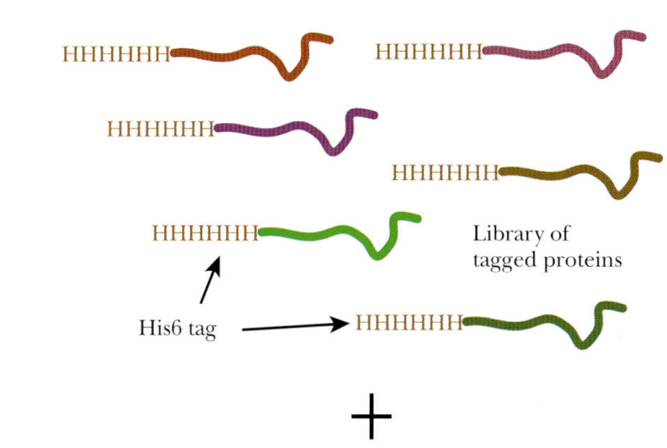

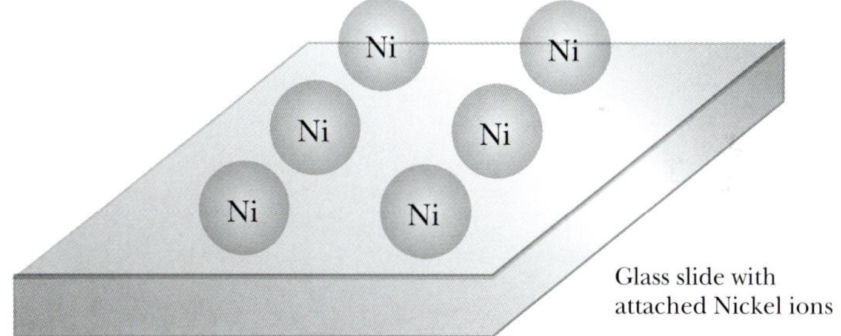

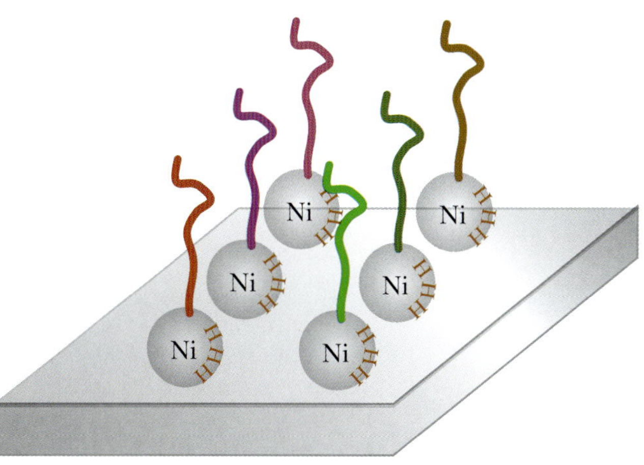

Proteins to be screened
His6 tag
Ni Ni Ni Ni Ni Ni
Slide

PHOSPHOLIPID TAGGED WITH BIOTIN

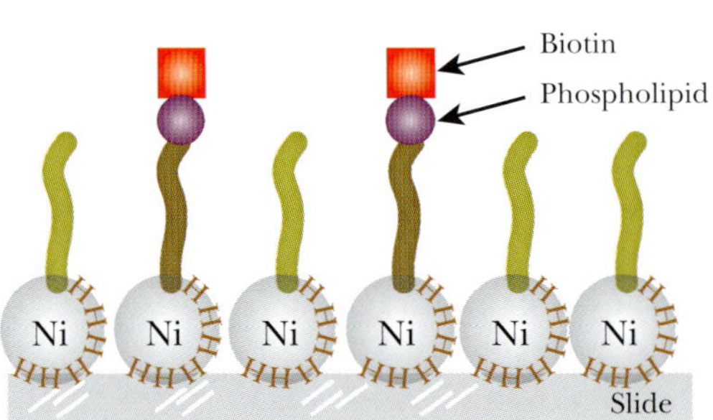

AVIDIN WITH CY3 FLUORESCENT LABEL

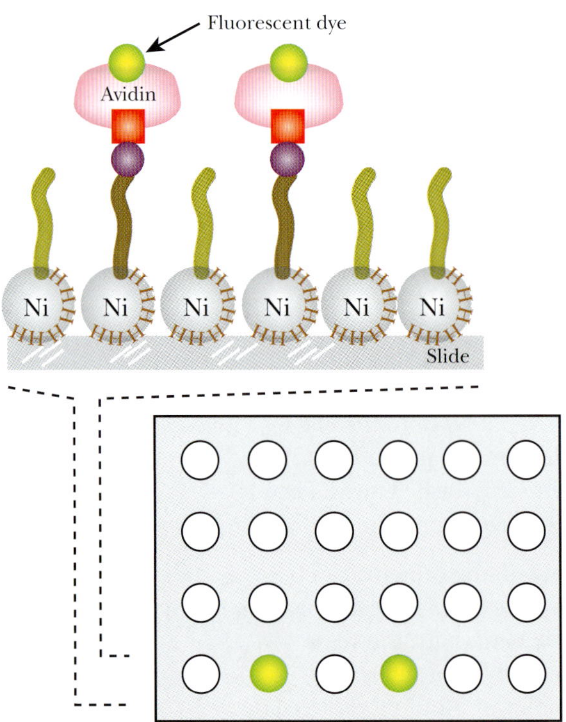

FIGURE 9.28 Screening Protein Arrays Using Biotin/Streptavidin
Protein microarrays can be screened to find proteins that bind to phospholipids. The protein microarray is incubated with phospholipid bound to biotin. Then the bound phospholipid is visualized by adding streptavidin conjugated to a fluorescent dye. Spots that fluoresce represent specific proteins that bind phospholipids.

His6-nickel interactions. The biotin-labeled calmodulin or phospholipid was then visualized by incubating the slide with Cy3-labeled streptavidin. (Streptavidin binds very strongly to biotin.) The results identified 39 different calmodulin binding proteins (only six had been identified previously), and 150 different phospholipid binding proteins.

Antigen capture immunoassay links different antibodies to a solid support. This array is incubated with a sample of proteins labeled with fluorescent dye. When an antibody binds its cellular protein, then that spot will fluoresce.

Reverse-phase arrays fuse the protein samples to the solid support, where each spot has many different cellular proteins. The support is incubated with labeled antibody. The detection system on the antibody determines which protein sample interacted with the antibody.

METABOLOMICS

As methods for identifying small molecules become more accurate and sensitive, metabolic research has become more global. The **metabolome** consists of all the small molecules and metabolic intermediates within a system, such as a cell or whole organism, at one particular time. Understanding the metabolome is complex because small metabolites affect many other components of a cell. Metabolites flow in a complex network and form many different transient complexes. The network of metabolites may be compared to city streets. At each corner, a decision on which route to take must be made, and such decisions continue until the person reaches the final destination. Each metabolite molecule follows a specific pathway, often with several potential branches, and at each junction, a decision is made before moving on to the next step. Characterizing the metabolome under particular conditions is known as **metabolic fingerprinting**.

Several techniques that involve separating and/or identifying small metabolites have made metabolomics possible. Nuclear magnetic resonance (NMR) of extracts from cells grown with ^{13}C-glucose has allowed simultaneous measurement of multiple metabolic intermediates. Metabolites have also been identified by thin layer chromatography after growth in ^{14}C-glucose. These methods are not very sensitive, and some metabolites may not be separated or identified.

Mass spectroscopy offers the best way to analyze whole metabolomes. The technique can identify many different metabolites (even novel ones) and is extremely sensitive. Mass spectroscopy can determine the exact molecular formula for a compound, so every metabolite can be identified. Even if isomers exist, the fragmentation pattern will be different although the molecular formula is the same.

The use of mass spectroscopy is often combined with other methods to simplify analysis. Different types of chromatography can be used to separate the metabolites before analysis by mass spectroscopy. Obviously, using HPLC will separate the complex cellular extract into different fractions, which can then be analyzed by mass spectroscopy.

Metabolomics is especially valuable in studying plants, because metabolites affect the pigments, scents, flavors, and nutrient content. These are all commercially important traits, and using mass spectroscopy to analyze these metabolites will aid in developing better-tasting and fresher produce. For example, in strawberries, there are 7000 metabolites that can be identified by mass spectroscopy (Fig. 9.29). Comparing white and red strawberries has identified which of the metabolite peaks in the mass spectrum corresponds to the intermediates in pigment synthesis.

The metabolome consists of the entire complement of small molecules and metabolites within a system, such as a cell or whole organism.

Because the metabolome is dynamic, a metabolic fingerprint is the entire complement of small metabolites at one particular point in time.

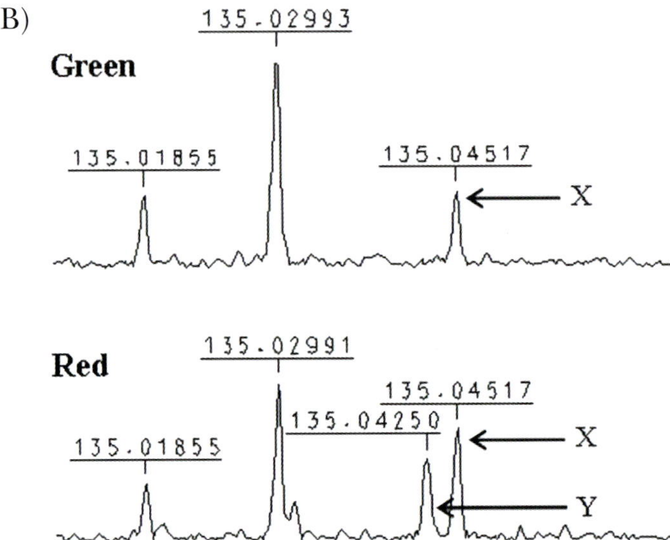

FIGURE 9.29 Metabolome Analysis of Strawberry
Nontargeted metabolic analysis in strawberry. (A) Four consecutive stages of strawberry fruit development (G, green; W, white; T, turning; R, red) were subjected to metabolic analysis using Fourier transform mass spectrometry (FTMS). Similar fruit samples were used earlier to perform gene expression analysis using cDNA microarrays. (B) An example of high-resolution (>100,000) separation of very close mass peaks in data obtained from the analysis of green and red stages of fruit development. Peaks marked with an X have the same mass, while peak Y is different by a mere 3 ppm. Courtesy of Phenomenome Discoveries Inc., Saskatoon, Canada.

Summary

Proteomics is the study of the protein complement for an organism. Because proteins change in response to many conditions, the proteome is dynamic, adapting to new challenges and environments. The term *translatome* refers to the proteome at one particular point in time. In order to study proteins, SDS-PAGE separates proteins by size. These proteins are trapped inside a gel, but can be transferred to nitrocellulose for Western blot analysis. In the Western blot, a protein of interest can be visualized by adding labeled antibodies. The Western blot can be used to determine the relative abundance of the protein of interest. Another useful tool to separate proteins is HPLC. This method keeps the proteins in a liquid, and the column materials separate the proteins of interest from the mixture. These columns vary greatly, making HPLC a key method to separating protein mixtures.

Many proteins are difficult to isolate because of their chemistry or size. Proteases are enzymes that break the peptide bond. There are a huge variety of proteases, some with specific binding/cutting sites, and other with nonspecific effects. Using proteases to digest a protein makes them more manageable for research.

Mass spectroscopy is a relatively new tool for the study of proteins. The method breaks molecules into their ions and records their mass to charge ratio, which is calculated by the time it takes for the ions to travel through the flight tube. Until the advent of MALDI and ESI, proteins were too large and complex for mass spectroscopy. MALDI and ESI are two new methods for preparing the protein for ionization. ESI is particularly useful because the proteins are ionized from a liquid solvent. This method can be linked with HPLC.

Once the protein of interest has been identified by HPLC, mass spectroscopy, or SDS-PAGE, the gene sequence can be found. Once cloned, the gene then can be expressed into protein under the control of a regulated promoter. In order to identify the protein of interest from the remaining cellular proteins, researchers use genetic fusions to either short tags or full proteins. A function for the protein of interest is often hard to determine, and being able to express the protein and isolate the protein via these tags is key to studying their function.

Sometimes finding new protein binding partners can further the understanding of the protein's function. In yeast two-hybrid analysis, the protein of interest is genetically fused to one half of a transcription factor, GAL4. Potential binding partners are genetically fused to the other half of GAL4. When the protein of interest binds to a different protein, then the transcription factor functions to turn on the reporter gene. This changes the yeast physiology, marking the cell. The gene for the binding partner can be isolated from the yeast, sequenced, and possibly identified. In a related experiment, co-immunoprecipitation finds binding partners for a protein of interest. This technique works in the cytoplasm and does not require reporter genes.

Finally, this chapter discusses global protein analysis by using protein arrays. Like the microarrays discussed in the previous chapter, the protein array has a different antibody or different proteins in each of the spots on a solid support. These can be studied with fluorescently labeled samples to identify changes in protein expression in different conditions.

End-of-Chapter Questions

1. Why is SDS used in the electrophoresis of proteins?
   - **a.** SDS coats the protein with a negative charge so that the sample can run through the gel.
   - **b.** SDS is a specific protease that digests large proteins in the sample.
   - **c.** SDS allows the coomassie blue stain to bind to the proteins in the gel so that they may be visualized.
   - **d.** SDS adds more molecular weight to each sample so that the proteins do not run off the end of the gel.
   - **e.** none of the above
2. What is an issue with using 2D-PAGE?
   - **a.** Hydrophobic proteins may not run as expected due to the hydrophobic surfaces.
   - **b.** Highly expressed proteins may cover up proteins that are not as abundant but running in the gel nearby.
   - **c.** Some proteins may not migrate through polyacrylamide and therefore not be represented on the gel.
   - **d.** Rare cellular proteins are hard to visualize with Coomassie blue protein stain.
   - **e.** All of the above are issues with 2D-PAGE.

(Continued)

3. Which one of the following is not used during Western blotting?
   - a. secondary antibody with a conjugated detection system
   - b. agarose gel electrophoresis
   - c. non-fat dry milk
   - d. primary antibody that recognizes the protein
   - e. nitrocellulose membrane

4. Which of the following statements about HPLC is not correct?
   - a. There are two phases to HPLC: mobile and stationary.
   - b. Separation, identification, and purification of proteins are just a few of the applications for HPLC.
   - c. The downside to HPLC is that it is not very adaptable due to the availability of stationary phase material.
   - d. Adjusting the experimental conditions, changing the particle size of the stationary phase, and controlling temperature are factors that affect resolution.
   - e. All of the above statements are true.

5. Which of the following is not an example of protease activity?
   - a. Some proteases cleave the phosphodiester bond between nucleic acid residues.
   - b. Some proteases cleave within a protein sequence and other proteases snip off residues from either end.
   - c. Some proteases contain serine, cysteine, threonine, or aspartic acid residues within their active sites.
   - d. Proteases hydrolyze the peptide bond between amino acid residues.
   - e. Metalloproteases contain metal ion cofactors within their active site.

6. Which of the following statement about mass spectroscopy is incorrect?
   - a. MS ionizes the sample and then measures the time it takes for the ions to reach the detector.
   - b. SELDI-MS has great potential for analyzing protein profiles of body fluids and may, in the future, be used to identify diseases before symptoms appear.
   - c. Glycosylation and phosphorylation of proteins can be identified using ESI or MALDI.
   - d. ESI is able to handle much larger ions than MALDI.
   - e. The time-of-flight for ions is directly correlated with the mass of the ion in mass spectroscopy.

7. Which of the following statements is not true about sequencing peptides with mass spectroscopy?
   - a. The entire protein can be sequenced all at once using mass spectroscopy.
   - b. Some purified proteins must be digested with proteases to eliminate undesirable characteristics such as hydrophobicity and solubility.
   - c. Two rounds of mass spectroscopy are used to determine the sequence.
   - d. In order to determine the sequence, a pure sample of protein is obtained through 2D-PAGE or HPLC.
   - e. A database of protein ion spectra is used to compare the peaks of the unknown peptide to determine the sequence.

8. Which of the following is used to quantify proteins with mass spectroscopy?
   a. ^{2}H
   b. ^{33}P
   c. ^{35}S
   d. ^{32}P
   e. ^{125}I
9. Why are protein tags useful?
   a. Protein tags are exactly the same thing as reporter fusions and perform similar functions.
   b. Tags allow the protein to be isolated and purified from other cellular proteins.
   c. Tags allow the protein to be quantitated.
   d. Protein tags enable the protein to which they are fused to perform their function more readily.
   e. None of the above.
10. How is biopanning useful to proteomics research?
   a. To express large amounts of protein on the cell surface of yeast.
   b. To screen expression libraries in *E. coli*.
   c. To alter the cell membrane structures of cells by expressing foreign proteins on the cell surface.
   d. To isolate specific peptides that bind to a specific target protein.
   e. all of the above
11. Which of the following is needed to perform yeast two-hybrid assays?
   a. Two vectors are needed to express the bait and prey proteins.
   b. A reporter gene under the control of the GAL4 recognition sequence.
   c. The DBD of a transcription factor genetically fused to the protein of interest, also called the bait.
   d. The AD domain of a transcription factor genetically fused to proteins that are being screened for interactions with bait.
   e. All of the above are needed.
12. For what is co-immunoprecipitation used?
   a. to determine if a protein-of-interest binds to a specific DNA sequence
   b. to examine protein-protein interactions in the nucleus instead of in the cytoplasm
   c. to examine protein-protein interactions in the cytoplasm instead of the nucleus
   d. to allow a protein to be expressed in mammalian cell culture
   e. none of the above

(Continued)

13. What is a problem associated with immuno-based arrays?
- **a.** Proteins that are bound to solid supports may not be representative of intracellular conditions.
- **b.** The antibody may cross-react with other cellular proteins, producing a false positive.
- **c.** Low concentrations of some proteins may not be able to compete for active sites compared to those that are in abundance.
- **d.** Proteins that are often found in complexes may have the antibody binding site masked by the other proteins in the complex.
- **e.** All of the above are problems associated with immuno-based arrays.

14. Which of the following has been extensively studied using protein interaction arrays?
- **a.** proteins in yeast that bind calmodulin or phospholipids
- **b.** proteins in yeast that bind to glutathione-*S*-transferase
- **c.** proteins that are able to bind to biotin and streptavidin
- **d.** proteins that are able to bind to various cofactors present in the sample
- **e.** none of the above

15. Which of the following methods is the best way to analyze a metabolome?
- **a.** high-pressure liquid chromatography
- **b.** mass spectroscopy
- **c.** nuclear magnetic resonance
- **d.** thin layer chromatography
- **e.** ELISA

Further Reading

Clark DP (2005). *Molecular Biology: Understanding the Genetic Revolution*. Elsevier Academic Press, San Diego, CA.

Hughes TR, Robinson MD, Mitsakakis N, Johnston M (2004). The promise of functional genomics: Completing the encyclopedia of a cell. *Curr Opin Microbiol* **7**, 546–554.

López-Otin C, Overall CM (2002). Protease degradomics: A new challenge for proteomics. *Mol Cell Biol* **3**, 509–519.

MacBeath G (2002). Protein microarrays and proteomics. *Nat Genet* **32**, 526–532.

Medzihradszky KF (2005). Peptide sequence analysis. *Methods Enzymol* **402**, 209–244.

Steen H, Mann M (2004). The ABC's (and XYZ's) of peptide sequencing. *Nat Rev Mol Cell Biol* **5**, 699–711.

Zhang H, Yan W, Aebersold R (2004). Chemical probes and tandem mass spectrometry: A strategy for the quantitative analysis of proteomes and subproteomes. *Curr Opin Systems Biol* **8**, 66–75.

Recombinant Proteins

Proteins and Recombinant DNA Technology

Expression of Eukaryotic Proteins in Bacteria

Translation Expression Vectors

Codon Usage Effects

Avoiding Toxic Effects of Protein Overproduction

Increasing Protein Stability

Improving Protein Secretion

Protein Fusion Expression Vectors

Expression of Proteins by Eukaryotic Cells

Expression of Proteins by Yeast

Expression of Proteins by Insect Cells

Protein Glycosylation

Expression of Proteins by Mammalian Cells

Expression of Multiple Subunits in Mammalian Cells

Comparing Expression Systems

PROTEINS AND RECOMBINANT DNA TECHNOLOGY

Proteomics has opened the door to identify more and more clinically relevant proteins. Once identified, these proteins need to be studied in detail, including expression of the protein in model organisms by using recombinant DNA techniques (see Chapter 3). Some proteins will become therapeutic agents and large amounts of purified protein will be required.

Once a gene has been cloned, the protein it encodes can be produced in large amounts with relative ease. Some examples of such **recombinant proteins** are given in Table 10.1. Smaller, nonprotein molecules, which seem simpler to an organic chemist, would need half a dozen proteins (enzymes) working in series to synthesize them. Thus, paradoxically, proteins, despite being macromolecules, have been more susceptible to genetic engineering than simpler products such as antibiotics. Pathway engineering to produce small organic molecules will be discussed in the following chapter (Chapter 11).

Expressing a gene for large-scale production brings extra problems compared to a laboratory setting. The more copies of a gene that a cell contains, the higher the level of the gene product. Thus cloning a gene onto a high-copy-number plasmid will usually give higher yields of a gene product. However, high-copy plasmids are often unstable, especially in the dense cultures used in industrial situations. Although the presence of antibiotic resistance genes on most plasmids provides a method to maintain the plasmid in culture, antibiotics are expensive, especially on an industrial scale. One solution to prevent plasmid loss is to integrate the foreign gene into the chromosome of the host cell. This, however, decreases the copy number of the cloned gene to one. Attempts have been made to insert multiple copies of cloned genes in tandem arrays. However, the presence of multiple copies results in instability due to recombination between homologous sequences of DNA.

Recombinant proteins are clinically relevant proteins produced in large scale. The gene for the protein of interest is cloned into a vector and expressed into protein in a model organism.

Table 10.1 Proteins Produced by Recombinant Technology

Protein	Function
Erythropoietin	Promotes red blood cell formation in the treatment of anemia
Factor VIII	Helping blood clots form in hemophiliacs
Filgrastim and sargramostim (blood-cell-stimulating bone marrow factors)	Boosting white blood cell counts after radiation therapy or transplantation
Insulin	Treatment of diabetes
Interferon (alpha)	Treatment of hepatitis B and C, genital warts, certain leukemias and other cancers
Interferon (beta)	Treatment of multiple sclerosis
Interferon (gamma)	Treatment of chronic granulomatous disease
Interleukin-2	Killing tumor cells
Somatotropin	Treatment of growth hormone deficiency
Tissue plasminogen activator (t-PA)	Dissolving blood clots to prevent heart attacks and lessen their severity

EXPRESSION OF EUKARYOTIC PROTEINS IN BACTERIA

In Chapter 3, Recombinant DNA Technology, we discussed the basics of cloning genes onto a variety of vectors. Successful expression of cloned genes depends on several factors. Obviously, bacterial genes will usually be expressed when carried on cloning vectors in bacterial host cells, provided that they are next to a suitable bacterial promoter. Special plasmids known as **expression vectors** are often used to enhance gene expression. As noted in Chapter 3, these provide a strong promoter to drive expression of the cloned gene. Expression vectors also contain genes for antibiotic resistance to allow selection of the vector and therefore the recombinant protein. In addition, they must have an origin of replication appropriate to the host.

The expression of eukaryotic proteins is more problematic. Although eukaryotic cells can be used to express eukaryotic proteins, bacteria are simpler to grow and manipulate genetically. Therefore it is often desirable to express eukaryotic proteins in bacteria (Fig. 10.1). Because eukaryotic promoters do not work in bacterial cells, it is necessary to provide a bacterial promoter. In addition, bacteria cannot process introns; therefore it is standard procedure to clone the cDNA version of eukaryotic genes, which lacks the introns and consists solely of uninterrupted coding sequence. In fact, the cDNA version of eukaryotic genes is generally used, even for expression in eukaryotic cells, not only to avoid possible processing problems, but also because the amount of DNA is much smaller and consequently easier to handle.

Even if a cloned gene is transcribed at a high level, production of the encoded protein may be limited at the stage of protein synthesis. Different mRNA molecules are translated with differing efficiencies. Several factors are involved:

(a) Strength of the ribosome binding site–ribosome interaction
(b) mRNA stability and/or secondary structure
(c) Codon usage

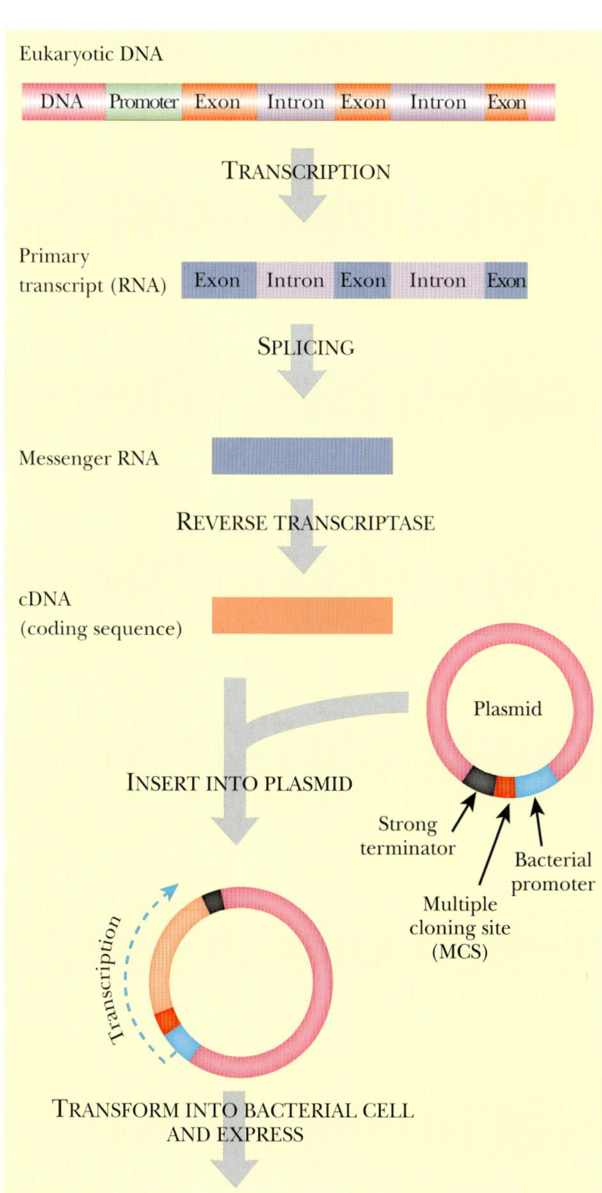

FIGURE 10.1 Expression of Eukaryotic Gene in Bacteria—Overview Eukaryotic genes must be adapted for expression in bacteria. First, the mRNA from the gene of interest is converted to cDNA to provide uninterrupted coding DNA. The cDNA is cloned between a bacterial promoter and a bacterial terminator so the bacterial transcription and translation machinery express the coding sequence.

In addition to standard expression vectors, more sophisticated vectors exist to optimize these other aspects of protein production. In this chapter we will discuss the use of translation vectors and fusion vectors to increase the synthesis of a recombinant protein from a cloned gene.

Expression vectors are used to make eukaryotic proteins in bacteria. The vector has the ribosome binding site, terminator sequences, and a strong regulated promoter. The eukaryotic gene is a cDNA copy of the mRNA.

TRANSLATION EXPRESSION VECTORS

As discussed in Chapter 2, bacterial ribosomes bind mRNA by recognizing the ribosome binding site (RBS) (also known as the Shine-Dalgarno sequence). The RBS base pairs with the sequence AUUCCUCC on the 16S rRNA of the small subunit of the ribosome. The closer the RBS is to the consensus sequence (i.e., UAAGGAGG), the stronger the association. Generally, this leads to more efficient initiation of translation. In addition, for optimal translation, the RBS must be located at the correct distance from the start codon, AUG.

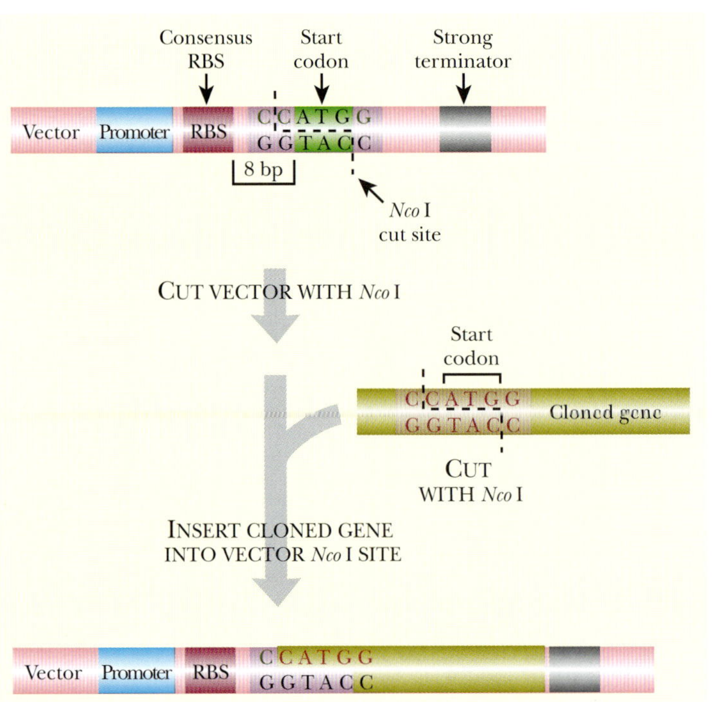

FIGURE 10.2
Translational Expression Vector
The recognition site for *NcoI* is C/CATGG, which has an ATG in the middle. The gene of interest and the translation vector each have an *NcoI* site, allowing the gene of interest to be cloned exactly into the correct site for highest protein expression. The vector also has a consensus RBS spaced 8 base pairs from the ATG, which provides an optimum binding site for ribosomes. The terminator sequence(s) is also strong to prevent transcriptional run-on.

Expression vectors are designed to optimize gene expression at the level of transcription (see Chapter 3). However, it is also possible to design **translational expression vectors** to maximize the initiation of translation. These vectors possess a consensus RBS plus an ATG start codon located an optimum distance (8 bp) downstream of the RBS. The cloned gene is inserted into a cloning site that overlaps the start codon. The restriction enzyme *NcoI* is very convenient because its recognition site (C/CATGG) includes ATG. Therefore it is possible to insert a cloned gene so that its ATG coincides exactly with the ATG of the translational expression vector (Fig. 10.2). The gene to be expressed is cut with *NcoI* at its 5′-end and with another convenient restriction enzyme at its 3′-end. If necessary, an artificial *NcoI* site may be introduced into the gene by site-directed mutagenesis or by PCR.

Translational expression vectors also possess a convenient selective marker, usually resistance to ampicillin or some other antibiotic, and a strong, regulated transcription promoter. Downstream of the cloning site are two or three strong terminator sequences, to prevent transcription continuing into plasmid genes.

Although translational expression vectors can optimize the initiation of translation, they do not control the other factors listed earlier, which are properties of the sequence of each individual gene. The secondary structure of the mRNA may have significant effects on the level of translation. If the mRNA folds up so that the RBS and/or start codon are blocked, then translation will be hindered. In particular, the sequence of the first few codons of the coding sequence should be checked for possible base pairing with the region around the RBS. If necessary, bases in the third (redundant) position of each codon may be changed to eliminate such base pairing. Getting active and abundant translation is the key to preventing mRNA instability. Any mRNA that is not being actively translated becomes subject to degradation, which decreases the protein yield.

Translational expression vectors provide the optimal ribosomal binding site. They also have strong regulated promoters and multiple terminator sequences for controlled transcription.

CODON USAGE EFFECTS

When genes from one organism are expressed in different host cells the problem of **codon usage** arises. As explained in Chapter 2, the genetic code is redundant in the sense that several different codons can encode the same amino acid. So even though a protein has a fixed amino acid sequence, there is still considerable choice over which codons to use. In practice, different organisms favor different codons for the same amino acid. For example, the amino acid lysine is encoded by AAA or AAG. In *E. coli*, AAA is used 75% of the time and AAG only 25%. In contrast, *Rhodobacter* does the exact opposite and uses AAG 75% of the time, even though *E. coli* and *Rhodobacter* are both gram-negative bacteria.

Codons are read by transfer RNA (see Chapter 2). When a cell uses a particular codon only rarely, it has lower levels of the tRNA that reads the rare codon. So if a gene with many AAA codons is put into a cell that rarely uses AAA for lysine, the corresponding tRNA may be in such short supply that protein synthesis slows down (Fig. 10.3).

Consequently, to optimize protein production, codon usage must be considered. Although it takes a lot of work, genes may be altered in DNA sequence so as to change many of the bases in the third position of redundant codons. This is done by artificially synthesizing the DNA sequence of the whole engineered gene. For a whole gene, individual segments of DNA that overlap are synthesized and then assembled as discussed in Chapter 4. Genes that have been codon-optimized for new host organisms may show a 10-fold increase in the level of protein produced, due to more rapid elongation of the polypeptide chain by the ribosome. This approach has been used successfully for several of the insect toxins, originally from *Bacillus thuringiensis*, which have been cloned into transgenic plants (see Chapter 14 for details).

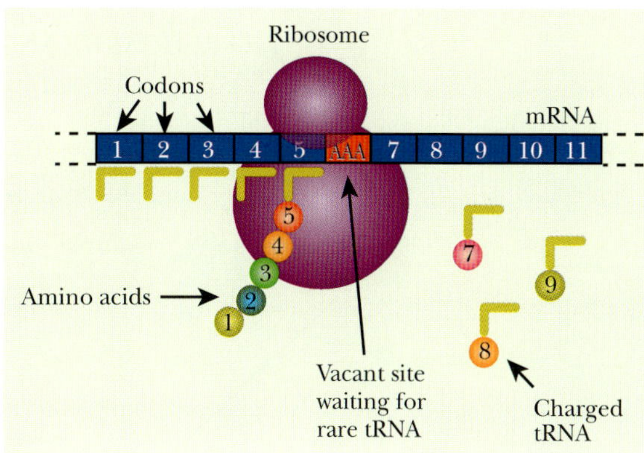

FIGURE 10.3 Codon Usage Affects Rate of Translation
Bacteria prefer one codon for a particular amino acid to other redundant codons. In this example, the ribosome is stalled because it is waiting for lysine tRNA with a UUU anticodon. *Escherichia coli* does not use this codon very often and there is a limited supply of this tRNA.

Another less labor-intensive approach to the codon issue is to supply the rare tRNAs. The rarest codons are not used because the corresponding tRNAs are in short supply. For example, the rarest codons in *E. coli* are AGG and AGA, which both encode arginine and only occur at frequencies of about 0.14% and 0.21%, respectively. The *argU* gene encodes the tRNA that reads both the AGA and AGG codons. If this gene is supplied on a plasmid, the host *E. coli* no longer has a tRNA shortage, and the recombinant proteins can be made. Plasmids carrying all the genes for rare codon tRNAs are available commercially. These plasmids have a p15A origin of replication, which makes them compatible with most expression vectors as these usually have a ColE1 origin of replication (see Chapter 3 for plasmid incompatibility).

Each organism has certain preferred codons for an amino acid. When the sequence for the recombinant protein encodes a rarely used tRNA, the protein production slows. The addition of the rare tRNA or changing the rare codon wobble position solves this problem.

AVOIDING TOXIC EFFECTS OF PROTEIN OVERPRODUCTION

Although higher yields are usually desirable, overproduction of a recombinant protein may harm the host cell. In bacteria, when too much protein is manufactured too fast, the surplus forms **inclusion bodies**. These are dense crystals of misfolded and nonfunctional protein (see below). Thus expression systems for recombinant proteins have features to control when and how much protein the host cell makes. Two common expression systems are the pET and pBAD systems for *E. coli*. These have control mechanisms to switch recombinant protein production on or off, and the pBAD system can also modulate the amount of protein produced.

The pET vectors have a hybrid T7/*lac* promoter, multiple cloning site, and T7 terminator. Transcription from the hybrid T7/*lac* promoter requires T7 RNA polymerase. The host *E. coli* has the gene for T7 RNA polymerase engineered into the chromosome, but the *lac* operon repressor, LacI, represses the gene. When IPTG is added, this induces release of the LacI protein and expression of the gene. T7 RNA polymerase is then made and binds to the T7/*lac* hybrid promoter, and the cloned protein of interest is manufactured (Fig. 10.4).

The pBAD system is based on the arabinose operon. This is induced by adding arabinose, which binds to the AraC regulatory protein. Activated AraC exits the O_2 site and binds to the I site (Fig. 10.5). This activates transcription of the cloned gene. The pBAD system is modulated by the amount of arabinose added to the culture. If a lot of recombinant protein is needed, then more arabinose is added. However, if the recombinant protein is toxic to the host cells, then less arabinose is added, and less recombinant protein is made. This effect is based on

A) NO RECOMBINANT PROTEIN EXPRESSION

OFF
T7 Promoter
lac operon
ATG/MCS
T7 Terminator
T7 RNA poly gene
lac operator
lac promoter
pET
lac I
ori
Antibiotic resistance gene
Lac repressor protein

B) RECOMBINANT PROTEIN EXPRESSION

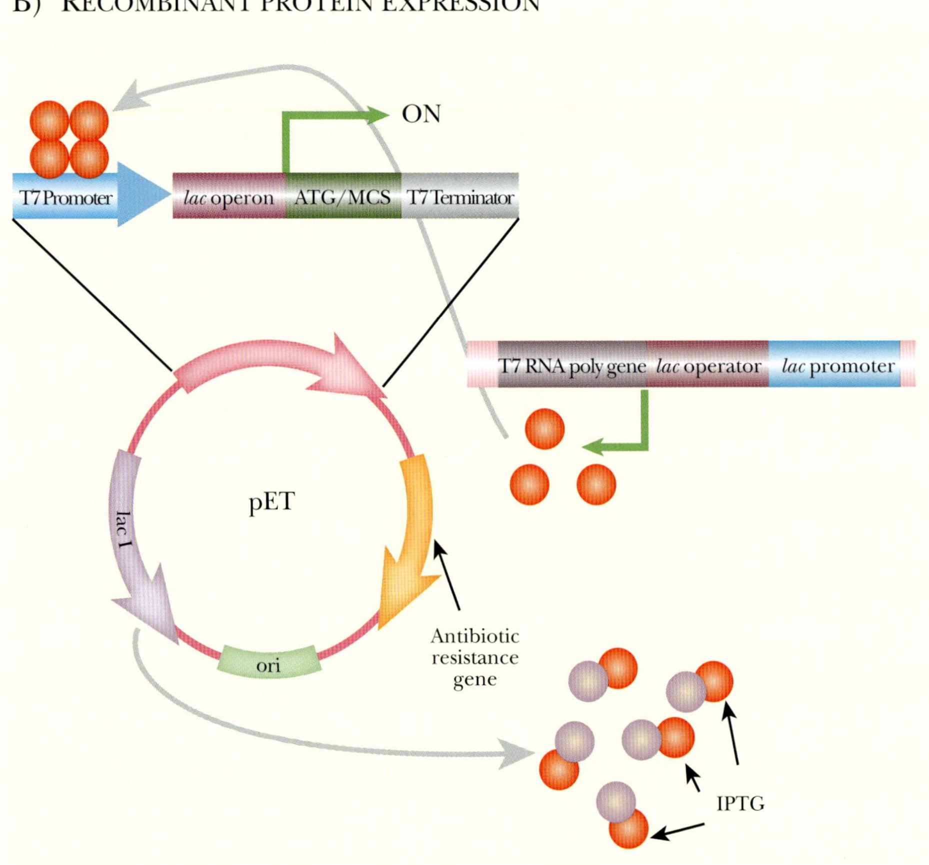

FIGURE 10.4 pET Protein Expression System
(A) Recombinant proteins are not expressed in the cell until induced. The pET plasmid has the gene for LacI protein, which represses both the gene for T7 RNA polymerase on the bacterial chromosome and the recombinant protein gene on the pET plasmid. (B) When the inducer IPTG is added, it causes release of LacI from both promoters. The gene for T7 RNA polymerase is then expressed. This polymerase then transcribes the gene for the recombinant protein.

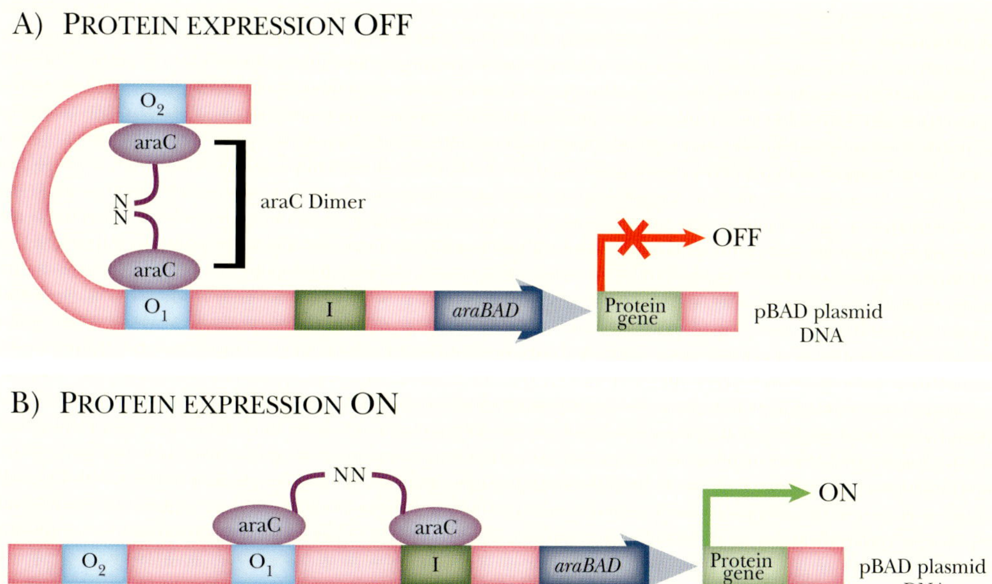

FIGURE 10.5 pBAD Expression System
(A) Recombinant proteins are not expressed when the AraC protein dimer binds to the O_1 and O_2 regulatory regions on the pBAD plasmid. (B) When arabinose is present, the sugar induces AraC to switch conformation, and it now binds to the O_1 and I sites. This conformation stimulates transcription of the recombinant protein.

the culture, not the cell. Each cell manufactures the recombinant protein in an "all-or-none" fashion. Low levels of arabinose will not activate every cell in the culture, whereas higher levels of arabinose do activate every cell.

The pET and pBAD expression vectors are used to control protein production by controlling the gene expression with inducers and repressors.

INCREASING PROTEIN STABILITY

Different proteins vary greatly in stability. The lifetime of a protein inside a cell depends mainly on how fast the protein is degraded by proteolytic enzymes or proteases (see Chapter 9). Apart from the overall 3D structure of the protein, there are two specific factors that greatly affect the rate of protein degradation. These are believed to act by altering the recognition of proteins by the degradation machinery.

(a) The identity of the N-terminal amino acid greatly affects the half-life of proteins. Although all polypeptide chains are made with methionine as the initial amino acid, many proteins are processed later. Therefore the N-terminal amino acid of mature proteins varies greatly. The N-terminal rule for stability is shown in Table 10.2.

(b) The presence of certain internal sequences greatly destabilizes proteins. **PEST sequences** are regions of 10 to 60 amino acids that are rich in P (proline), E (glutamate), S (serine), and T (threonine). The PEST sequences create domains whose structures are recognized by proteolytic enzymes.

Alteration of the N-terminal sequence of a protein by genetic engineering is relatively simple. A small segment of artificially synthesized DNA encoding a new N-terminal region can be incorporated by standard methods, e.g., PCR (see Chapter 4). The effectiveness of this approach has been demonstrated experimentally for β-galactosidase. Changing an internal PEST sequence without disrupting protein function is a vastly more complex

Table 10.2 N-Terminal Rule for Protein Stability

N-Terminal Residue	Approximate Half-Life (Minutes)
Met, Gly, Ala, Ser, Thr, Val	120
Ile, Glu, Tyr	30
Gln, Pro	10
Leu, Phe, Asp, Lys, Arg	2–3

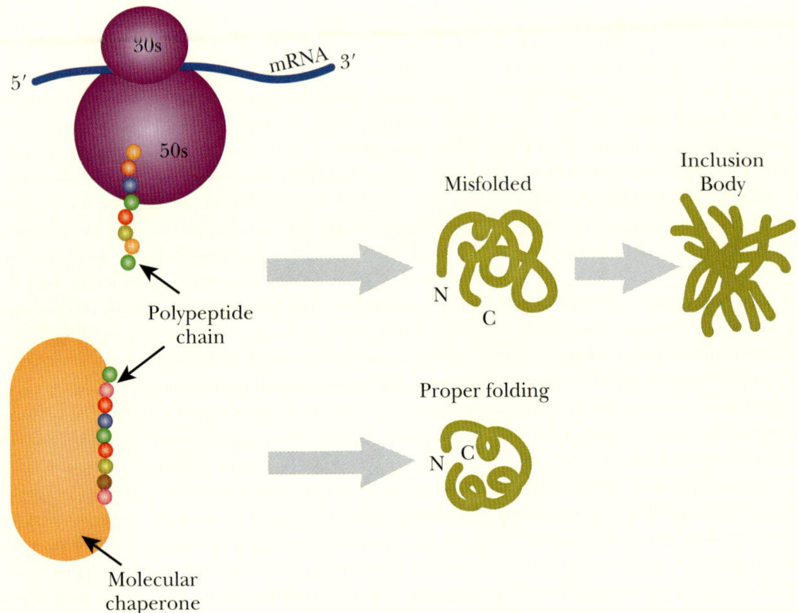

FIGURE 10.6
Molecular Chaperones Help Prevent Inclusion Body Formation
If recombinant proteins fold aberrantly, they aggregate and form inclusion bodies. If, during translation, molecular chaperone proteins hold the recombinant protein, the new polypeptide will fold properly and become fully functional.

undertaking, even though in theory it could lead to greater stability and therefore greater yields.

In some instances, proteins, which would be stable if folded correctly, fold aberrantly during synthesis. The misfolded proteins form inclusion bodies, which are dense particles visible with optical microscopy. Inclusion bodies are usually formed when recombinant proteins are poorly soluble or produced too fast. One method to alleviate the aggregation is to provide **molecular chaperones** that help fold the recombinant protein correctly. Molecular chaperones attach themselves to polypeptides while they are being translated and help keep the protein unfolded until translation is complete. Then the protein can fold into its correct shape (Fig. 10.6).

To enhance recombinant protein production, the N-terminal amino acid can be altered, the PEST sequences can be removed, or the recombinant protein can be coexpressed with a molecular chaperone that helps the protein fold correctly.

IMPROVING PROTEIN SECRETION

When a bacterial cell synthesizes a recombinant protein, it may end up in the cytoplasm, the periplasmic space between the inner and outer membranes, or be exported out of the cell into the culture medium. Secretion across the inner, cytoplasmic, membrane is directed by the presence of a hydrophobic signal sequence at the N-terminal end of the newly synthesized protein. The signal sequence is cut off after export by signal peptidase (also known as leader peptidase). Although bacteria such as *E. coli* export few proteins, special secretion systems do exist that allow export of proteins across both membranes into the culture medium.

It is obviously easier to purify a secreted protein than one that remains in the cytoplasm with the majority of the bacterial cell's own proteins. On the other hand, many proteins are unstable when outside the cellular environment and/or exposed to air. Generally, such proteins are not used in biotechnology since they are awkward to purify or to use. Most proteins and enzymes chosen for practical use are relatively stable outside the cell. A standard goal is to arrange for them to be secreted in order to help isolation and purification. Several approaches exist for this:

(a) A signal sequence is engineered into the cloned gene. Consequently, the recombinant protein will have an N-terminal signal sequence when newly synthesized. This will direct its export to the periplasm by the **general secretory system**. The bacterial cells are then harvested and treated so as to permeabilize or remove the outer membrane, thus releasing the recombinant protein.
One major problem is that large amounts of a recombinant protein may overload the secretory machinery and aggregate into inclusion bodies. Substantial amounts of the recombinant protein may accumulate inside the cell with the signal sequence still attached. Secretion may sometimes be increased dramatically by providing elevated levels of the secretory system proteins (Sec proteins). Extra copies of the sec genes can be cloned and expressed on plasmids to increase the protein amounts. Mutant *E. coli* strains also have been developed that facilitate secretion.

(b) The recombinant protein may be fused to a bacterial protein that is normally exported (see the following section on fusion vectors). The maltose binding protein of *E. coli* is efficiently exported to the periplasmic space and is a favorite carrier for recombinant proteins. The recombinant protein is later released from the carrier protein by protease cleavage. This technique is especially useful for relatively short peptides, including many hormones and growth factors, which are often unstable alone.

(c) The gene of interest can be expressed in gram-positive bacteria, such as *Bacillus*, which lack an outer membrane. Consequently, exported proteins go out into the culture medium. Although this is convenient, the genetics of *Bacillus* is still far behind that of *E. coli*. Animal cells are another alternative as they have only a single cytoplasmic membrane, and therefore secrete proteins directly into the medium.

(d) Secretion across both membranes of gram-negative bacteria such as *E. coli* may be achieved by specialized export systems (Fig. 10.7). Most of these export systems are used naturally for the secretion of toxins by pathogenic strains of bacteria. Thus the **type I secretory system** spans both inner and outer membranes and is used to secrete hemolysin by some *E. coli* strains that cause urinary infections. The **type II secretory system** spans the outer membrane only. Export into the medium therefore

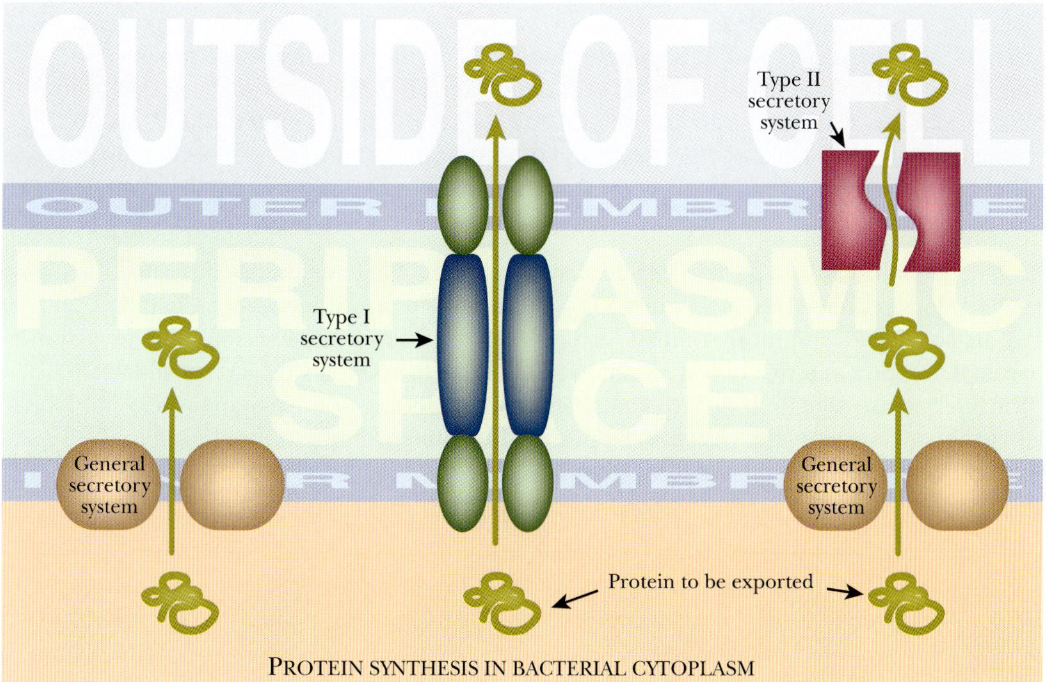

FIGURE 10.7
Secretion across Both Membranes
Protein secretion in *E. coli* involves a general secretory system that recognizes a signal sequence and exports those particular proteins. The protein is only transported to the periplasmic space. Type I secretory systems transport the protein from the cytoplasm, through the periplasmic space, to the outside of the cell. Type II secretory systems take proteins already in the periplasmic space and transport them across the outer membrane.

requires the general secretory system (i.e., the Sec system) to export the protein across the inner membrane first. Normally, this two-step process is used by *Pseudomonas* to secrete exotoxin A.

Strains of *E. coli* engineered to deploy these alternative export systems are beginning to be used. In these cases it is necessary to modify the protein to be exported so that it is recognized by the chosen export system. Engineering the gene with the appropriate signal sequences allows the protein to be recognized and exported.

Recombinant proteins are usually exported out of the cell into the culture medium. The use of bacterial secretory systems facilitates the export.

PROTEIN FUSION EXPRESSION VECTORS

Joining the coding sequences of two proteins together in frame makes a **protein fusion** (see Chapter 9). Consequently, a single, longer polypeptide is made during translation. If the first (i.e., N-terminal) protein is normally secreted, then the fusion protein will be secreted, too. Thus it is possible to achieve export of a recombinant protein by joining it to a protein that the cell normally exports.

Protein fusions also help address the issues of stability and purification. Many eukaryotic proteins are unstable inside the bacterial cell. This is especially true of growth factors, hormones, and regulatory peptides, which are often too short to fold into stable 3D configurations. Attaching them to the C terminus of a stable bacterial protein protects them from degradation. If the carrier protein is carefully chosen, purification may be greatly facilitated.

A **protein fusion vector** is a plasmid that allows the gene of interest to be fused to the gene for a suitable carrier protein. This is chosen so that it is an exported protein that is easy to purify. In addition, it should have a ribosomal binding site close to consensus and be translated efficiently. One of the best examples is the **MalE protein** of *E. coli*. This is a protein that is normally exported into the periplasmic space, where it transports maltose from the outer membrane to the inner membrane. Binding it to an amylose resin purifies MalE protein.

The gene of interest is cloned downstream and in frame with the MalE coding sequence. Between the coding sequences is a protease cleavage site, which allows the fused proteins to be cleaved apart after synthesis and purification. A strong promoter is also provided to maximize protein production. After protein expression, all the proteins are harvested from the periplasmic space. These are passed over an amylose column to bind the MalE fusion. The protein of interest is released from MalE and thus from the column by addition of the protease. Finally the protease is removed from the protein sample by another column that specifically binds protease.

A variety of other fusion protein systems have been used. We have already discussed the use of peptide tags, such as His6, FLAG, or GST to aid in protein purification—see Chapter 9. Here we are concerned with more sophisticated protein fusions that allow secretion of the fusion protein. Another effective fusion/secretion system uses the periplasmic enzyme β-lactamase. The β-lactamase domain binds to borate. This allows binding of the fusion protein (and the β-lactamase fragment after cleavage) by **phenyl-boronate** resins and the elution of the fusion protein from the resin with soluble borates.

Sophisticated fusion protein vectors provide strong, regulated promoters, optimal RBSs, and a fusion gene for a secreted protein. The recombinant protein is fused to the secreted protein to facilitate the export of the recombinant protein. There is a protease digestion site at the junction that releases the recombinant protein from the fusion protein during purification.

EXPRESSION OF PROTEINS BY EUKARYOTIC CELLS

Although bacterial cells have successfully expressed many eukaryotic proteins, there are cases where it is best to express eukaryotic proteins using eukaryotic cells. Some eukaryotic proteins are unstable or inactive after being made by bacterial cells. This is especially true of proteins that require posttranslational modification.

A variety of eukaryotic modifications may occur after the polypeptide chain has been made (Fig. 10.8). These include:

- **(a)** Chemical modifications that form novel amino acids in the polypeptide chain.
- **(b)** Formation of disulfide bonds between correct cysteine partners (e.g., the assembly of insulin, Chapter 19).
- **(c)** Glycosylation, that is, the addition of sugar residues at specific locations on the protein. Many cell surface proteins are glycosylated and will not assemble correctly into membranes or function properly if lacking their glycosyl components.
- **(d)** Addition of a variety of extra groups, such as fatty acid chains, acetyl groups, phosphate groups, sulfate groups.
- **(e)** Cleavage of precursor proteins. This may occur in several stages, as illustrated by insulin (see Chapter 19). Cleavage may be involved with secretion, correct folding, and/or activation of proteins.

The enzymes required for modification and processing are normally absent from bacterial cells, making it necessary to express eukaryotic proteins in eukaryotic cells. Related processing enzymes are often present in a range of higher organisms; thus it is rarely absolutely necessary to express a protein in its original organism. Here we are concerned with protein production in cultured cells. However, as discussed in Chapters 14 and 15, it is now possible to engineer whole transgenic animals or plants to produce recombinant proteins. A further advantage of expressing eukaryotic proteins in eukaryotic cells is that contamination with bacterial components is avoided. Despite purification, bacterial components that are toxic or promote immune reactions or cause fever may be present in traces if bacteria are used for production.

As discussed in Chapter 3, shuttle vectors are designed to move genes between different groups of organisms. Because genetic engineering is more difficult for eukaryotes, most expression vectors for eukaryotic cells are in fact shuttle vectors. Such vectors allow genetic engineering to be carried out in bacteria, usually *E. coli*, and allow transfer to other organisms for gene expression. We will consider the use of yeasts, insect cells, and mammalian cells for expression of recombinant proteins.

Many proteins require special modifications that are only available in eukaryotic cells.

Box 10.1 Selecting Mammalian Cell Lines That Produce Large Amounts of Recombinant Protein

The main issue with expressing recombinant proteins in mammalian cell culture is finding one cell that has high expression of the recombinant protein. The single cell must then be grown to a large culture so that the recombinant protein is identical throughout the culture. If more than one mammalian cell is grown into a large culture, the cells may be producing slightly different variations of the recombinant protein. Many techniques are used to isolate a single mammalian cell with high production of recombinant protein.

Limiting Dilution Cloning: Cells are transfected with the recombinant protein vector, and then suspended in a solution. The solution is diluted and small amount of the suspension is added to small wells of a dish at a density of less than one cell per well. The cells are allowed to grow, and small amounts of the media in each well are screened for the amount of recombinant protein (usually by ELISA; see Chapter 6). High producers are isolated and grown into the large cultures for production.

(Continued on page 317)

A) NOVEL AMINO ACIDS

PYRROLYSINE

LYSINE

CYSTEINE

SELENOCYSTEINE

HISTIDINE

DIPHTHAMIDE

B) DISULFIDE BOND

HS — CH_2 — C — COOH (H; NH_2)

CYSTEINE

HOOCCHCH_2 — S — S — CH_2CHCOOH (NH_2; NH_2)

CYSTINE

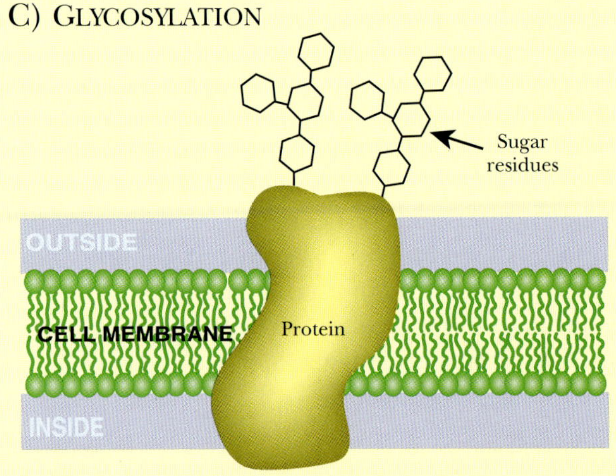

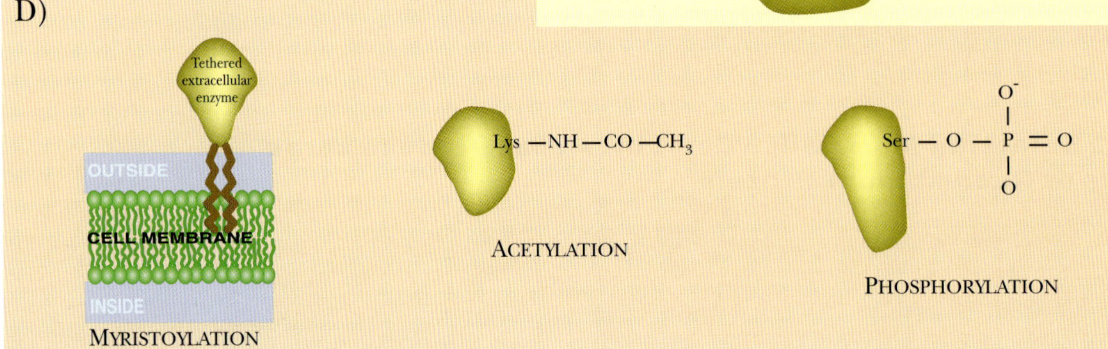

FIGURE 10.8 Protein Modifications in Eukaryotes

(A) Novel amino acids may be inserted during translation (e.g., pyrrolysine, selenocysteine) or made after translation by modifying other amino acids (e.g., diphthamide from histidine). (B) Disulfide bonds link two cysteine molecules together by their side chains. (C) Glycosylation is the addition of sugar residues to the surface of proteins. (D) Myristoylation, acetylation, and phosphorylation all add different groups to amino acids within a protein. These added groups alter the function of a protein. Myristoylation adds fatty acid chains that tether a protein to the cell membrane. Acetylation and phosphorylation often activate or deactivate proteins.

Box 10.1 Selecting Mammalian Cell Lines That Produce Large Amounts of Recombinant Protein—cont'd

Flow Cytometry and Cell Sorting: Cells are transfected with the vector for the recombinant protein and a reporter gene such as a membrane protein, green fluorescent protein, or dihydrofolate reductase. The transfected cells are sorted based on the expression of the reporter gene using flow cytometry or FACS (see Chapter 6). The amount of reporter gene correlates with the amount of recombinant protein in most cases. This method is indirect and only assesses the amount of reporter gene rather than recombinant protein.

Gel Microdrop Technology: Cells are transfected with the vector for the recombinant protein. The cells are then suspended in biotinylated agarose, that is, agarose polymers with biotin attached at points along the chain. The secreted recombinant protein is captured by these biotin side chains by using avidin and a biotinylated capture antibody. The amount of recombinant protein is determined with a fluorescently labeled antibody. This method actually assesses the amount of recombinant protein rather than a coexpressed reporter gene. This method also immobilizes the cell in a gel, making it easy to remove and grow into large cultures.

Automated Systems for Cell Selection: Many automated systems have been developed to identify a single mammalian cell that produces large amounts of recombinant protein. Automated systems are very efficient and can screen large numbers of transfected cells. One system, laser-enabled analysis and processing (LEAP), ablates all the low-producing cells of the original transfected cells. The cells are first immobilized in a matrix, and then a fluorescently labeled antibody to the recombinant protein is added. The brightest cell is marked, and a laser kills the remaining cells. Another system, automated colony pickers, works by a similar method. The transfected cells are immobilized and labeled with a fluorescent antibody to the recombinant protein. The brightest cells are chosen, but rather than killing the rest of the cells, an automated pipetting system removes the bright cells and places them in a new dish.

EXPRESSION OF PROTEINS BY YEAST

As already discussed, brewer's yeast, *Saccharomyces cerevisiae*, has been used as a model single-celled eukaryote in molecular biology (Chapter 1). The yeast genome has been sequenced and many genes have been characterized. From the viewpoint of biotechnology, several other factors favor the use of yeast for production of recombinant proteins:

(a) Yeast can be grown easily on both a small scale or in large bioreactors.
(b) After many years of use in brewing and baking, yeast is accepted as a safe organism. It can therefore be used to produce pharmaceuticals for use in humans without needing extra government approval.
(c) Yeast normally secretes very few proteins. Consequently, if it is engineered to release a recombinant protein into the culture medium, this can be purified relatively easily.
(d) DNA may be transformed into yeast cells either after degrading the cell wall chemically or enzymatically, or, more usually, by electroporation (see Chapter 3).
(e) A naturally occurring plasmid, the 2-micron circle, is available as a starting point for developing cloning plasmids.
(f) A variety of yeast promoters have been characterized that are suitable for driving expression of cloned genes.
(g) Although only a primitive single-celled organism, yeast nonetheless carries out many of the posttranslational modifications typical of eukaryotic cells, such as addition of sugar residues (glycosylation). However, yeast only glycosylates proteins that are secreted.

Recombinant proteins may be engineered for secretion by yeast (Fig. 10.9). Adding a signal peptide upstream of the coding region flags the protein for secretion. The signal sequence used is taken from the gene for mating factor α, a protein that is normally secreted (see Chapter 1). Yeast signal peptidase recognizes the sequence Lys-Arg and cleaves off the signal peptide once the protein has crossed the cell membrane. It is therefore necessary to position the two codons for Lys-Arg immediately upstream of the coding sequence for the recombinant protein to ensure that the engineered protein has a correct N-terminal sequence.

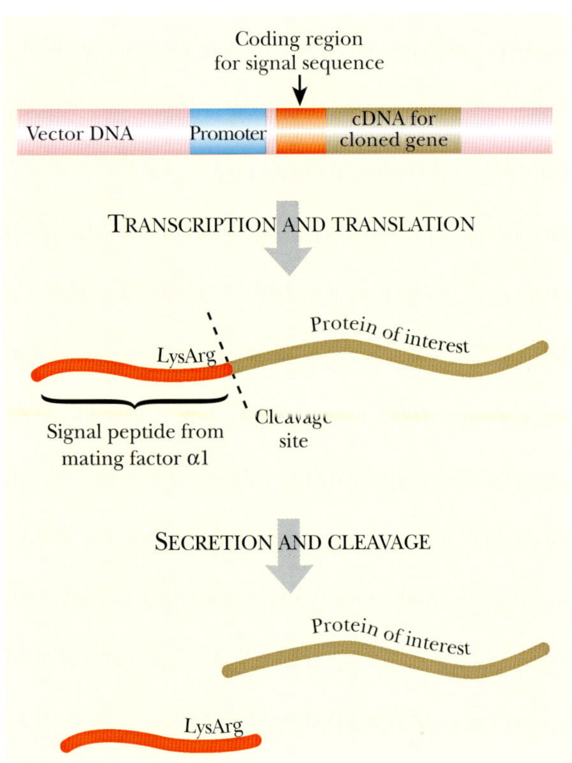

FIGURE 10.9
Engineering Proteins for Secretion by Yeast
To make yeast cells secrete the protein of interest, a small signal sequence is fused just upstream of the coding region. The signal sequence ends with a Lys-Arg, which is recognized by signal peptidase. Cleavage occurs after the Arg during secretion.

A variety of vectors exist for use with yeast. These may be classified into three main classes:

(a) Plasmid vectors are used most often. These are typically yeast shuttle vectors that can replicate in *E. coli* as well as yeast. The yeast replication system is normally derived from the yeast 2-micron circle (see Chapter 1).
(b) Vectors that integrate into the yeast chromosomes. Because plasmids may be lost, especially in large-scale cultures, this approach has the advantage of stability. The disadvantage is that only a single copy of the cloned gene is present, unless tandem repeats are used. These, however, are unstable because of crossing over. In practice this approach is rarely used.
(c) Yeast artificial chromosomes (see Chapter 3). These are used for cloning and analysis of large regions of eukaryotic genomes. However, they are not convenient for use as expression vectors.

Assorted proteins for human use are now produced using *Saccharomyces cerevisiae.* These include insulin, factor XIIIa (for blood coagulation), several growth factors, and several virus proteins (from human immunodeficiency virus, hepatitis B, hepatitis C, etc.) used as vaccines or in diagnostics.

Although many recombinant proteins have been expressed successfully in yeast, yields are often low. The problems include:

(a) Loss of expression plasmids during growth in large bioreactors.
(b) Proteins supposed to be secreted are often retained in the space between membrane and wall rather than exiting into the culture medium.
(c) Glycosylation of proteins is often excessive and the final protein may have too many sugar residues attached to function correctly.

Yeasts provide many advantages to producing recombinant proteins because they are true eukaryotes, have plasmid vectors, grow fast and easily, and finally because they have well-known characteristics.

EXPRESSION OF PROTEINS BY INSECT CELLS

Mammalian cells are relatively delicate and have complex nutritional requirements. This makes them both difficult and expensive to grow in culture compared to bacteria or yeasts. However, cultured insect cells are relatively robust and can be grown in simpler media than mammalian cells. Consequently, insect-based expression systems have been developed. These have the additional advantage of providing posttranslational modifications that are very similar to those found in mammalian cells.

The vectors used in cultured insect cells are almost all derived from a family of viruses, the **baculoviruses**, which infect only insects (and related invertebrates such as arachnids and crustaceans). Baculoviruses are unusual in forming packages of virus particles, known as **polyhedrons**. Some infected cells release single virus particles that can infect neighboring cells within the same insect. But when the host insect is dead or dying, packages of virus particles embedded in a protein matrix are released instead. The matrix protein is known as **polyhedrin**, and the polyhedron structure protects the virus particles while they are outside the host organism in the environment. When swallowed by another insect, the polyhedrin is dissolved by digestion and the polyhedron falls apart. This releases individual virus particles that can infect the cells of the new host insect.

The polyhedrin gene has an extremely strong promoter, and late during infection, the polyhedrin protein is made in massive amounts. Because polyhedrin is not actually needed for virus infection of cultured insect cells, the polyhedrin promoter may be used to express recombinant proteins. The polyhedrin coding sequence is removed and replaced by cDNA encoding the protein to be expressed.

There are many different types of baculoviruses, and the one most often used is **multiple nuclear polyhedrosis virus (MNPV)**. This infects many insects and replicates well in many cultured insect cell lines. A popular cell line used to propagate this baculovirus is from the fall armyworm (*Spodoptera frugiperda*). Yields of polyhedrin—and therefore of a recombinant protein using the polyhedrin promoter—are especially high in this cell line.

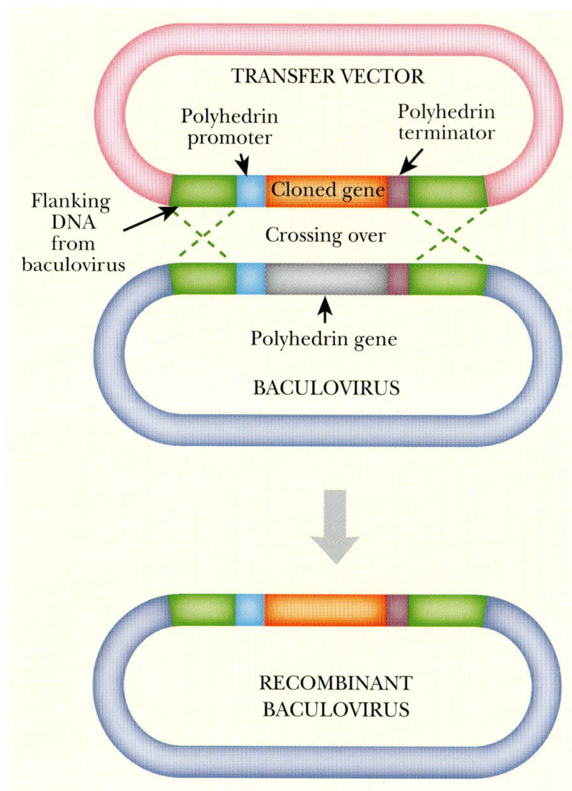

FIGURE 10.10
Baculovirus Expression Vector
To express a gene in insect cells, the gene must be inserted into a baculovirus genome. First, the gene of interest is cloned into a transfer vector containing the baculovirus polyhedrin gene promoter followed by a multiple cloning site and the polyhedrin terminator. This is done in *E. coli*. The construct is transfected into insect cells along with the normal baculovirus genome. A double cross-over between the polyhedrin promoter and terminator replaces the polyhedrin gene with the gene of interest.

Because the expression vector is a virus, construction is carried out in two stages (Fig. 10.10). In the first stage, a "transfer" vector is used to carry the cDNA version of the cloned gene. The transfer vector is an *E. coli* plasmid that carries a segment of MNPV DNA, which initially included the polyhedrin gene and flanking sequences. The polyhedrin coding sequence was then replaced with a multiple cloning site. When the cloned gene is inserted, it is under control of the polyhedrin promoter. Construction up to this point is done in *E. coli*. In the second stage, the segment containing the cloned gene is recombined onto the baculovirus, thus replacing the polyhedrin gene. To achieve this, insect cells are transfected with both the transfer vector and with MNPV virus DNA. A double-crossover event generates the required recombinant baculovirus.

Because this procedure sometimes gives undesirable results, other insect vectors have been constructed. One possibility is a shuttle vector that replicates as a plasmid in *E. coli* and as a virus in insect cells. Such baculovirus-plasmid hybrids are referred to as **bacmids** (Fig. 10.11). They consist of an almost entire baculovirus genome into which a segment of DNA from an *E. coli* plasmid has been inserted. This region carries a bacterial origin of replication, a selective marker, and a multiple cloning site. The inserted segment replaces the polyhedrin gene of the baculovirus. The cloned gene is inserted into the MCS giving a recombinant bacmid. Bacmid DNA is then purified from *E. coli* and transfected into insect cells, where it replicates as a virus.

The recombinant baculovirus may then be used to infect cultured insect cells and the desired protein is produced over several days. Several hundred eukaryotic proteins have been successfully made using the baculovirus/insect cell system. Of these, the vast majority were correctly processed and modified.

FIGURE 10.11
Bacmid Shuttle Vector
Bacmids are primarily baculovirus genomes with the addition of a bacterial origin of replication, a multiple cloning site, and an antibiotic resistance gene. These sequences allow the bacmid to survive in *E. coli* but still infect insect cells.

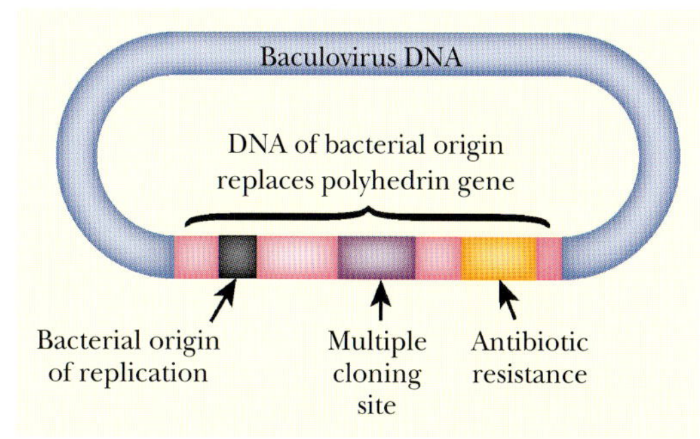

Insect cells are easy to grow for making recombinant protein. Insect cells are infected with a virus genome that has the recombinant protein gene. Rather than making viral particles, the infected insect cells make recombinant protein.

PROTEIN GLYCOSYLATION

Many proteins of higher organisms are glycosylated, that is, they have short chains of sugar derivatives added after translation. Glycosylation is required for proper biological function of many proteins.

One advantage of expressing proteins in insect cells is that they do possess a glycosylation pathway. The pathway for posttranslational glycosylation of asparagine residues is shared by mammals and insects up to the addition of mannose (Fig. 10.12). However, the pathways diverge beyond this. Nonetheless, partial glycosylation is better than none. Furthermore, insect cell lines that have been engineered to express the full mammalian glycosylation pathway are now available.

Insect cells glycosylate recombinant proteins.

EXPRESSION OF PROTEINS BY MAMMALIAN CELLS

Some cloned animal genes may ultimately need to be expressed in cultured mammalian cells. To facilitate this, mammalian shuttle vectors exist that allow movement of a cloned gene from bacteria to mammalian cells (Fig. 10.13). As usual, such vectors contain a bacterial

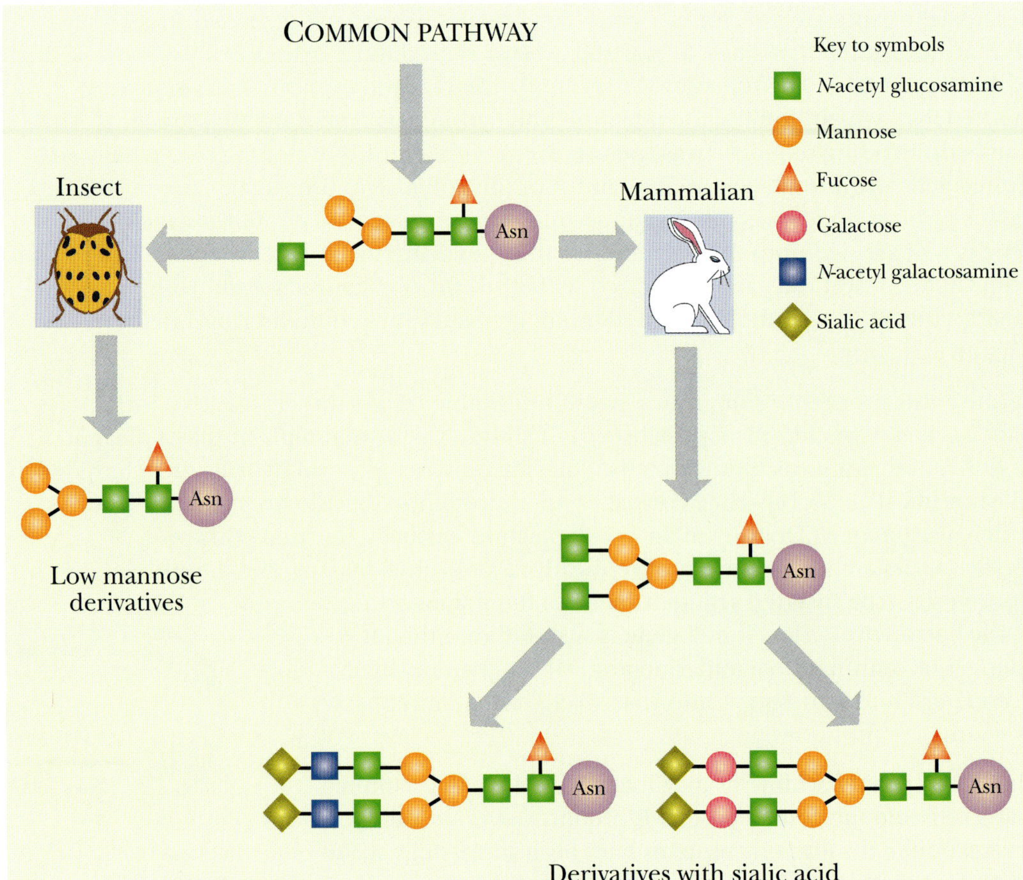

FIGURE 10.12
Protein Glycosylation Pathways
Mammals and insects share several steps in the pathway for posttranslational glycosylation of asparagine residues in proteins. However, the final modifications differ as shown. In particular, mammals add sialic acid residues to the ends of the glycosyl chains, whereas insects do not.

origin of replication and an antibiotic resistance gene allowing selection in bacteria. These vectors must also possess an origin of replication that works in mammalian cells. Usually this is taken from a virus that infects animal cells, such as SV40 (simian virus 40). Viral promoters are often used because they are strong, producing copious amounts of protein. Alternatively, promoters from mammalian genes that are expressed at high levels (e.g., the genes for metallothionein, somatotropin, or actin) may be used. The multiple cloning sites lie downstream of the strong promoter. Since animal genes are normally cloned as the cDNA, the vector must also provide a polyadenylation signal (i.e., tail signal) at the 3′-end of the inserted gene.

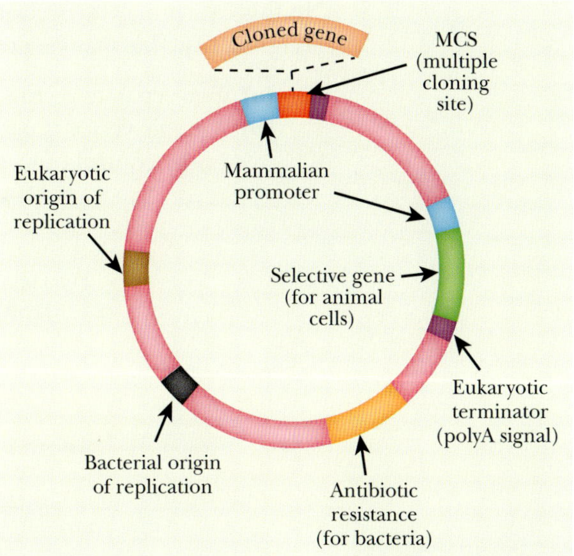

FIGURE 10.13 Mammalian Shuttle Vector
Mammalian shuttle vectors contain an origin of replication and an antibiotic resistance gene for growth in bacteria. The vector has a multiple cloning site between a strong promoter and a polyadenylation signal as well as a eukaryotic origin of replication and a eukaryotic selection gene, such as *npt*.

Several selective markers may be used with animal cells. Very few antibiotics kill animal cells. However, the antibiotic **genetecin**, also known as **G-418**, kills animal cells by blocking protein synthesis. G-418 is related to antibiotics such as neomycin and kanamycin that are used against bacteria. The *npt* (NeoR) gene, encoding neomycin phosphotransferase, adds a phosphate group to neomycin, kanamycin, and G-418, which inactivates these antibiotics. Consequently, *npt* can be used to select animal cells using G-418 or bacterial cells using neomycin or kanamycin.

Because of the lack of antibiotics for mammalian cells, mutant eukaryotic cells lacking a particular enzyme are sometimes used for selection. The plasmid then carries a functional copy of the missing gene. This rather inconvenient approach has been used in both yeast and animal cells. In yeast, genes for amino acid biosynthesis have been used (see Chapter 1). In animal cells, the ***DHFR* gene**, encoding **dihydrofolate reductase**, is sometimes used. This enzyme is required for synthesis of the essential cofactor folic acid and is inhibited by **methotrexate**. Mammalian cells lacking the *DHFR* gene are used, and a functional copy of the *DHFR* gene is provided on the shuttle vector. Treatment with methotrexate inhibits DHFR and hence selects for high-level expression of the DHFR gene on the vector. Methotrexate levels can be gradually increased, which selects for a corresponding increase in copy number of the vector. (The chromosomal *DHFR* genes must be absent to avoid selecting chromosomal duplications rather than the vector.)

An alternative approach is to use a metabolic gene as a dominant selective marker. The enzyme glutamine synthetase protects cells against the toxic analog, **methionine sulfoximine**. The resistance level depends on the copy number of the glutamine synthetase gene. Therefore, multicopy plasmids carrying the glutamine synthetase gene can be selected even in cells with functional chromosomal copies of this gene.

Mammalian shuttle vectors have features for producing recombinant proteins in mammalian cell cultures.
Because antibiotics do not harm mammalian cells, a different method of keeping the mammalian shuttle vector in the cell must be used. Some vectors have the *DHFR* gene, which is needed to protect mammalian cells from methotrexate. Other vectors use the glutamine synthetase gene, which protects the cell from methionine sulfoximine.

EXPRESSION OF MULTIPLE SUBUNITS IN MAMMALIAN CELLS

To further complicate recombinant protein expression, many mammalian proteins have multiple subunits that must assemble inside a mammalian cell. Thus, manufacturing the separate subunits in separate cultures and mixing them later fail to yield active protein.

In this case it may be necessary to synthesize more than one polypeptide in the same cell (Fig. 10.14). This may be done in three main ways:

(a) Two separate vectors are used, each carrying the gene for one of the subunits.
(b) A single vector is used that carries two separate genes for the two subunits, each under control of its own promoter.

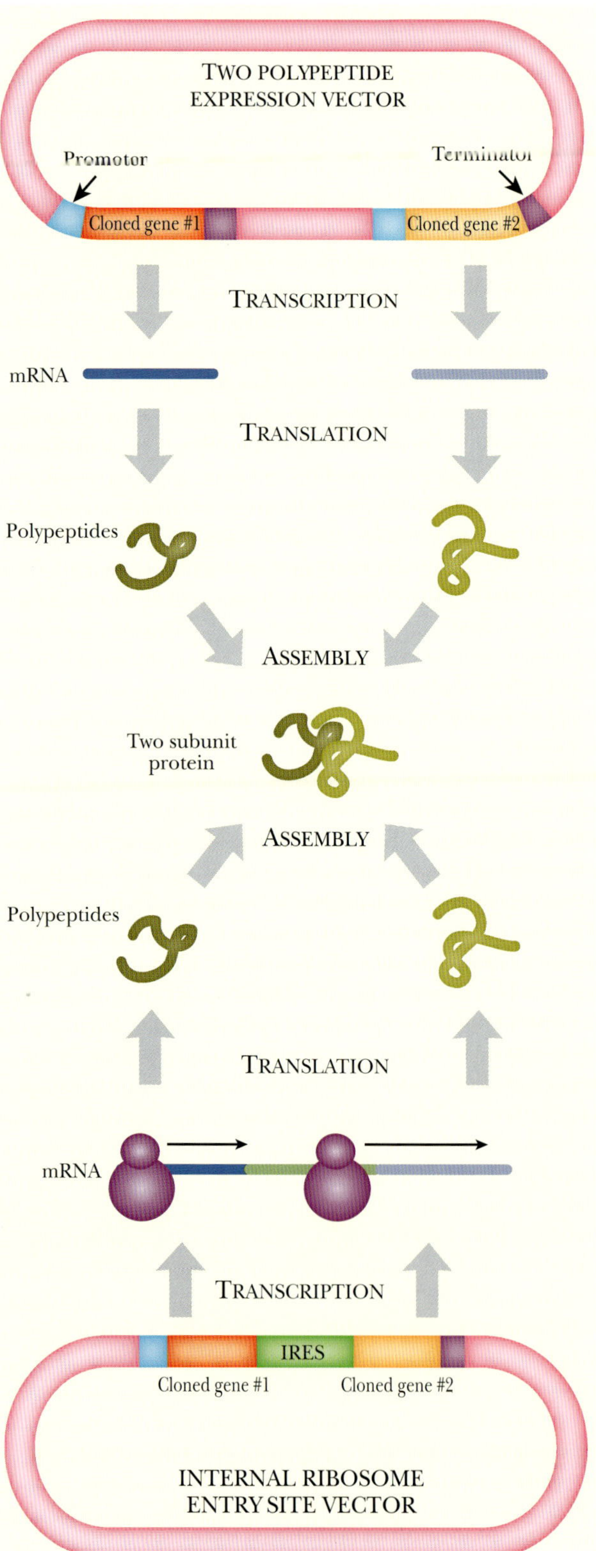

FIGURE 10.14
Expression of Multiple Polypeptides in the Same Cell
Some proteins have two subunits that assemble properly only when inside a mammalian cell. These can be expressed on one shuttle vector with two different promoters (blue), two multiple cloning sites, and two polyadenylation sites (purple). After transfection into a mammalian cell, both proteins are expressed. The two protein subunits assemble using the necessary cellular components. Alternatively, the vector may carry the two genes separated by an internal ribosome entry site (IRES), which allows ribosomes to bind to the mRNA upstream of both genes. As before, the two proteins are made and associate.

(c) A single vector is used that carries an artificial "operon" in which the two genes are expressed using the same promoter. Transcription gives a single polycistronic mRNA carrying both genes. As discussed in Chapter 2, eukaryotic cells normally only translate the first open reading frame in an mRNA. However, certain animal viruses have sequences known as **internal ribosomal entry sites (IRESs)** that allow the translation of multiple coding sequences on the same message. Note that the mRNA is read twice by two different ribosomes that bind in two different places (the normal 5′-end and the IRES). This is different from the translation of polycistronic mRNA in bacteria where a single ribosome proceeds along the mRNA and translates all the open reading frames in the operon.

If the recombinant protein has multiple subunits, each subunit must be produced separately. Sometimes each subunit is expressed from separate vectors, sometimes one vector contains separate genes for each of the subunits, and finally, some vectors have the subunits expressed as an operon.

COMPARING EXPRESSION SYSTEMS

Each expression system has its own unique strengths and weaknesses when it comes to characteristics such as cost, speed, and ability to glycosylate or fold proteins, as well as government regulations. Bacterial cells are by far the cheapest and fastest method for recombinant protein expression, but they do not glycosylate proteins or fold mammalian proteins correctly. Cultured mammalian cells glycosylate and fold proteins properly, but they are expensive to maintain. Therefore, determining what system to use to express recombinant proteins depends on the protein and its particular idiosyncrasies (Fig. 10.15).

FIGURE 10.15 Comparison of Recombinant Protein Expression Systems
Each protein expression system falls on a continuum of worst to best for characteristics such as speed, cost, glycosylation, folding, and government regulations. Transgenic animals (rabbit) and transgenic plants (plants) are discussed in Chapters 14 and 15. The other symbols include mammalian cultured cells, insect cell culture, yeast, and bacteria.

No expression system is perfect, and each recombinant protein must be evaluated for which system will work the best.

Summary

Recombinant proteins are clinically relevant proteins produced in large scale. To make recombinant proteins, their gene is isolated and cloned into an expression vector. Most recombinant proteins are from humans but are expressed in model organisms such as bacteria, yeast, or insect cells. Human genes are very complex, containing large introns; therefore, the recombinant proteins are often cloned from mRNA by converting the mRNA into cDNA. Because the cDNA is purely coding sequence without any regulatory regions, the expression vectors must provide the promoter, ribosome binding site, and terminator sequences.

Many issues affect the expression of the recombinant proteins. Because each organism prefers to use one or two codons for an amino acid, the model organism has lower concentrations of certain tRNAs. If the gene for the recombinant protein uses the rare codon, the low amount of tRNA becomes rate limiting for expression. Adding the rare tRNA or changing the coding sequence for the gene of interest can resolve this issue. When the recombinant protein is produced, too much protein can overwhelm the model organism, resulting in protein aggregation into inclusion bodies. Using regulated expression vectors can control the rate at which recombinant protein is produced. In addition to changing the codons, the gene for the recombinant protein can be altered to make the protein more stable. Changing the N-terminal amino acid, removing the PEST sequences, or coexpressing a molecular chaperone are all strategies for increasing the protein stability. Another issue with making recombinant proteins is getting the protein outside the cell. It is easier to isolate and purify the recombinant proteins if they are exported outside the cell. Many strategies that facilitate recombinant protein export employ cellular secretory proteins.

Many recombinant proteins require protein modifications that are available only in eukaryotic cells; therefore, the chapter discusses some features for yeast expression systems, insect cell culture expression systems, and mammalian cell culture systems.

End-of-Chapter Questions

1. Which of the following is a clinically relevant protein produced by recombinant technology?
   - **a.** insulin
   - **b.** interferon
   - **c.** factor VIII
   - **d.** tissue plasminogen
   - **e.** all of the above
2. Which of the following is associated with translation efficiency for the mRNA of genes cloned for protein expression?
   - **a.** mRNA stability
   - **b.** strength of the ribosomal binding site
   - **c.** codon usage
   - **d.** mRNA secondary structure
   - **e.** all of the above

3. Which of the following is not a feature of translational expression vectors?
   - a. a selectable marker (i.e., antibiotic resistance)
   - b. terminator sequences
   - c. the native promoter for the gene that is to be cloned into the vector
   - d. consensus RBS plus the ATG initiation codon
   - e. a strong, but regulated, transcriptional promoter

4. Which method to overcome codon usage problems is the least labor-intensive?
   - a. Supply the genes for the rare tRNAs on a separate plasmid.
   - b. Codon-optimize the gene of interest, usually in the third position of the codon.
   - c. Express the protein in the organism from which the gene originated.
   - d. Grow the host cells that are expressing the protein longer so that they have more time to make the protein containing rare codons.
   - e. all of the above

5. In expression systems, which of the following forms when too much protein is manufactured too quickly?
   - a. activated AraC
   - b. inclusion bodies
   - c. T7 RNA polymerase
   - d. functional protein
   - e. inactivated Lac repressor

6. How can recombinant protein production be enhanced?
   - a. alteration of the C-terminal sequence
   - b. inclusion of PEST sequences
   - c. addition of a detergent to help prevent inclusion body formation
   - d. co-expression of a molecular chaperone
   - e. all of the above

7. What can be done to ensure a recombinant protein is excreted outside the cell?
   - a. addition of a signal sequence
   - b. using specialized export systems
   - c. fusion to a protein that is normally exported
   - d. expressing the protein in an alternative system, such as gram-positive bacteria
   - e. all of the above

8. What could be used to increase the stability and purification of eukaryotic proteins from bacterial cells?
   - a. a peptide tag
   - b. a protein fusion
   - c. a protease site
   - d. PEST sequence
   - e. a signal sequence

(Continued)

9. Why would it be necessary to express eukaryotic proteins in eukaryotic cells rather than in bacterial cells?
- **a.** Many eukaryotic proteins are modified and bacteria usually do not have the enzymes needed for these modifications.
- **b.** Codon usage between eukaryotes and prokaryotes is different.
- **c.** Molecular chaperones are more readily available in eukaryotes.
- **d.** Protein production in eukaryotic cells never produces inclusion bodies, which sometimes form in bacterial cells due to overexpression.
- **e.** none of the above

10. Which of the following is not a problem associated with low yields of proteins expressed in yeast?
- **a.** Proteins are not getting secreted into the medium.
- **b.** Yeast have no known naturally occurring plasmids from which to construct expression vectors.
- **c.** Expression plasmids are lost in large batch cultures of yeast.
- **d.** Too much modification of the protein occurs (i.e., glycosylation).
- **e.** None of the above is a problem associated with low yield.

11. What is a bacmid?
- **a.** a virus used to infect bacterial cells
- **b.** a recombinant protein from bacteria that is expressed in insect cells
- **c.** a shuttle vector that replicates as a plasmid in *E. coli* and a virus in insect cells
- **d.** a general term for bacterial expression vectors
- **e.** none of the above

12. Are insect cells fully able to glycosylate foreign proteins?
- **a.** yes
- **b.** no
- **c.** partially, only up to the addition of mannose
- **d.** uncertain, has never been experimentally tested
- **e.** none of the above

13. Which of the following is not used as a selectable marker to maintain mammalian shuttle vectors within mammalian cells?
- **a.** genetecin
- **b.** fungicides
- **c.** *DHFR* gene
- **d.** glutamine synthetase gene
- **e.** all of the above

14. What function does an internal ribosomal entry site serve?
- **a.** as a way to translate polycistronic mRNA in eukaryotes that carries the genes for multiple subunits of a protein
- **b.** as a protease cleavage site in between two proteins expressed on one transcript
- **c.** as a way to control the expression of two different proteins
- **d.** as a way to differentially express two proteins on the same transcript
- **e.** none of the above

15. Which factor affects which protein expression system to utilize?
- **a.** cost
- **b.** ability to glycosylate or fold proteins
- **c.** speed
- **d.** government regulations
- **e.** all of the above

Further Reading

Browne SM, Al-Rubeai M (2007). Selection methods for high-producing mammalian cell lines. *Trends Biotechnol* **25**, 425–432.

Clark DP (2005). *Molecular Biology: Understanding the Genetic Revolution*. Elsevier Academic Press, San Diego, CA.

Sørensen HP, Mortensen KK (2005). Advanced genetic strategies for recombinant protein expression in *Escherichia coli*. *J Biotechnol* **115**, 113–128.

Protein Engineering

INTRODUCTION

A variety of enzymes have been in industrial use since before genetic engineering appeared. However, merely a couple of dozen enzymes account for over 90% of total industrial enzyme use. Some common examples are listed in Table 11.1. These proteins are used under relatively harsh conditions and are exposed to oxidizing conditions not found inside living cells. Consequently, these particular proteins are unusually robust and stable and are not at all representative of typical enzymes in this respect. It is notable that most of them are **hydrolases** that degrade either carbohydrate polymers or proteins.

Increasing the range of industrial enzymes has three facets. First, modern biology has identified many novel enzyme-catalyzed reactions that may be of industrial use. Second, as discussed in the previous chapter, it is now possible to produce desired proteins in large amounts because of gene cloning and expression systems. Third, the sequence of the protein itself may be altered by genetic engineering to improve its properties. This is known as **protein engineering** and is the subject of the present chapter.

Methods for manipulating DNA sequences and for expressing the encoded proteins have been discussed in previous chapters. Therefore we shall omit these details here. Rather we will emphasize the possibilities for altering the biological properties of proteins. In practice, most protein engineering has so far been concerned with making more stable variants of useful enzymes. The objective here is to engineer proteins so that they may be used under industrial conditions without being denatured and losing activity. However, it is also possible to alter proteins to change the specificity of their enzyme activities or even to create totally new enzyme activities. Ultimately, it may be possible to design proteins from basic principles. This will require the ability to predict three-dimensional protein structure from the polypeptide sequence.

Many enzymes are used, often under harsh conditions, by the biotechnology industry. There is a growing need for both tougher and novel enzymes.

Table 11.1 Important Proteins Used Industrially

Protein	Function
Amylases	Hydrolysis of starch for brewing
Lactase	Hydrolysis of lactose in milk processing
Invertase	Hydrolysis of sucrose
Cellulase	Hydrolysis of cellulose from plant materials
Glucose isomerase	Conversion of glucose to fructose for high-fructose syrups
Pectinase	Hydrolysis of pectins to clarify fruit juices, etc.
Proteases (ficin, bromelain, papain)	Hydrolysis of proteins for meat tenderizing and clarification of fruit juices
Rennet	Protease used in cheese making
Glucose oxidase	Antioxidant in processed foods
Catalase	Antioxidant in processed foods
Lipases	Lipid hydrolysis in preparing cheese and other foods

ENGINEERING DISULFIDE BONDS

The formation of disulfide bonds is a major factor in maintaining the 3D structure of many proteins. Disulfide bonds are especially important for those proteins found outside the cell in oxidizing environments. In practice, most enzymes used industrially will be exposed to such oxidizing conditions, and therefore disulfide bonds are particularly relevant.

Introduction of extra disulfide bonds is a relatively straightforward way to increase the stability of proteins. The first step is to simply introduce two cysteine residues into the polypeptide chain. Then, under oxidizing conditions, these will form a disulfide bond provided that the polypeptide chain folds so as to bring the two cysteines into close contact. Obviously, for this approach to work, the tertiary structure of the protein must be known so that the cysteines can be inserted in appropriate positions (Fig. 11.1). In general, the longer the loop of amino acids between the two cysteines, the greater the increase in stability. Note however, that formation of a disulfide linkage can create a strained conformation if the two cysteines are not properly aligned. This may result in a decrease in stability of the protein.

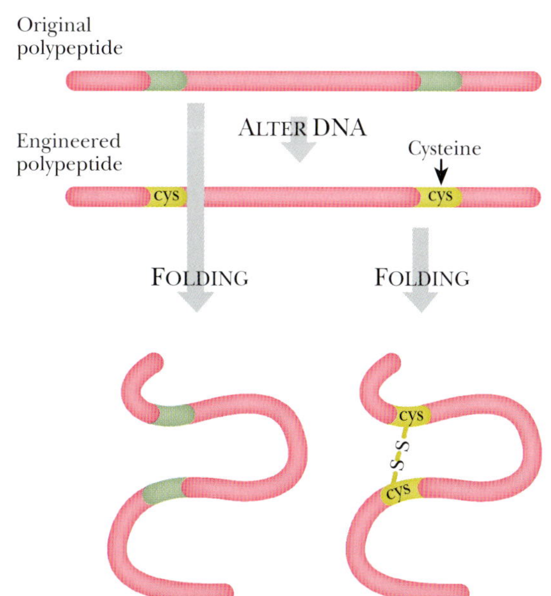

FIGURE 11.1
Introduction of Disulfide Bonds
A disulfide bond can be added to a protein by changing two amino acids into cysteines by site-directed mutagenesis. When the engineered protein is put under oxidizing conditions, the two cysteines form a disulfide bond, holding the protein together at that site.

This approach has been demonstrated using the enzyme lysozyme from bacterial virus T4. The structure of this enzyme has been solved by x-ray crystallography. The polypeptide chain of 164 amino acids folds into two domains and has two cysteines, neither involved in disulfide bond formation in the wild-type protein. One of these, Cys54, was first mutated to Thr to avoid formation of incorrect disulfides. The other, Cys97, was retained for use in disulfide formation. Extensive analysis of possible locations for disulfides was carried out. Those disulfides that might impair other stabilizing interactions in the protein were eliminated. This left three possible disulfide bonds that should theoretically promote stability, located between positions 3 and 97, 9 and 164, and 21 and 142 (Fig. 11.2).

To test these experimentally, five amino acids (Ile3, Ile9, Thr21, Thr142, and Leu164) were converted to Cys in various combinations. Engineered proteins with each individual disulfide as well as proteins with two or three disulfide linkages were tested for stability and for enzyme activity (Table 11.2). Stability was measured by thermal denaturation; the **melting temperature, Tm**, is the temperature at which 50% of the protein is denatured. Although all three disulfides increased stability of T4 lysozyme, those engineered proteins with the 21–142 disulfide had lost their enzyme activity. The precise reason is uncertain, but presumably some structural distortion has altered the active site. Nonetheless, the engineered protein with both of the two other disulfides showed a massive increase in stability and retained almost all of its enzyme activity. Overall, this approach is highly effective.

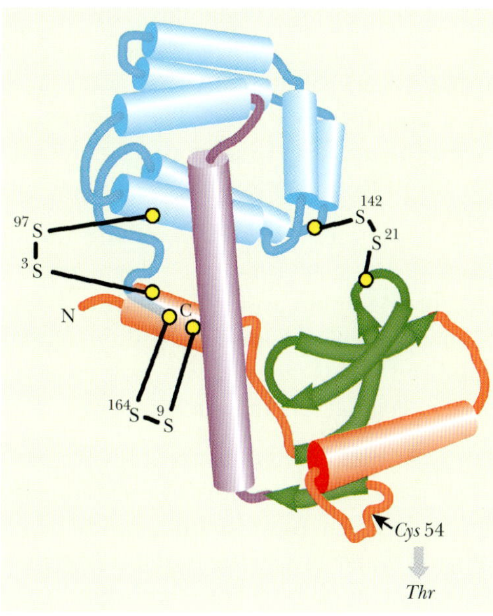

FIGURE 11.2
Disulfide Engineering of T4 Lysozyme
T4 lysozyme has two domains. The N-terminal region is shown in green and red, and the C-terminal region is in blue. These are linked by an alpha helix (purple). Disulfide bonds were added at three locations in T4 lysozyme to increase the stability. The first disulfide was between positions 9 and 164. This links the first alpha helix at the N terminus with the C-terminal tail. The second disulfide is between positions 2 and 97, which links the N- and C-terminal domains. Finally, the third disulfide links position 21 in the N-terminal domain with 142 in the C-terminal domain. The figure depicts alpha helices as barrel-shaped and beta-sheets as green arrows.

Introduction of extra disulfide bonds, by altering the coding sequence for an enzyme, often provides major increases in protein stability.

Table 11.2 Disulfide Stabilization of T4 Lysozyme

Protein	Disulfide Bonds Present			Stability as Tm	Activity (%)
	3–97	9–164	21–142		
Original	–	–	–	41.9	100
1	+	–	–	46.7	96
2	–	+	–	48.3	106
3	–	–	+	52.9	0
4	+	+	–	57.6	95
5	–	+	+	58.9	0
6	+	+	+	65.5	0

IMPROVING STABILITY IN OTHER WAYS

Although the introduction of disulfide bonds is the simplest and most effective way to increase protein stability, a variety of other alterations may also help. Because of the effects of entropy, the greater the number of possible unfolded conformations, the more likely a protein is to unfold. Decreasing the number of possible unfolded conformations therefore promotes stability. This may be done in several ways, in addition to introducing extra disulfide linkages as described earlier. Glycine, whose R-group is just a hydrogen atom, has more conformational freedom than any other amino acid residue. In contrast, proline, with its rigid ring, has the least conformational freedom. Therefore replacing glycine with any other amino acid or increasing the number of proline residues in a polypeptide chain will increase stability. Such replacements must avoid altering the structure of the protein, especially in critical regions. When tested experimentally, such changes do contribute small increases in stability.

Because hydrophobic residues tend to exclude water, these residues tend to cluster in the center of proteins and avoid the outer surface. If cavities exist in the hydrophobic core, filling them should increase protein stability. This may be done by replacing small hydrophobic residues, which are already in or near the core, with larger ones. For example, changing Ala to Val or Leu to Phe will achieve this. However, most proteins have hydrophobic cores that are already fairly stable and have few cavities. Furthermore, inserting larger hydrophobic amino acids to fill these cavities often causes twisting of their side chains into unfavorable conformations. This cancels out any gains of stability from packing the hydrophobic core more completely.

Because of its asymmetrical structure, the alpha helix is actually a dipole with a slight positive charge at its N-terminal end and a slight negative charge at its C-terminal end. The presence of amino acid residues with the corresponding opposite charge close to the ends of an alpha helix promotes stability. In natural proteins the majority of alpha helixes are stabilized in this manner. However, in cases where such stabilizing residues are absent, protein engineering may create them.

Asparagine and glutamine residues are relatively unstable. High temperature or extremes of pH convert these amides to their corresponding acids, aspartic acid and glutamic acid. The replacement of the neutral amide by the negatively charged carboxyl may damage the structure or activity of the protein. This may be avoided by engineering proteins to replace Asn or Gln by an uncharged hydrophilic residue of comparable size, such as Thr.

A variety of alterations to the amino acid sequence may yield proteins with moderate increases in stability.

CHANGING BINDING SITE SPECIFICITY

In addition to altering the overall stability of a protein, it is possible to deliberately change the active site. The most straightforward alterations to make are those that change the binding specificity for the substrate or a cofactor, but do not disrupt the enzyme mechanism. Changing the specificity for a cofactor or substrate may be useful, either to make the product of the enzyme reaction less costly or to change it chemically.

This principle has been demonstrated with several enzymes that use the cofactors **NAD** or **NADP** to carry out dehydrogenation reactions. Both cofactors carry reducing equivalents and both react by the same mechanism. Although a few enzymes can use either NAD or NADP, most use one or the other. Generally, NAD is used by **dehydrogenases** in degradative pathways, and the respiratory chain oxidizes the resulting NADH. In contrast, biosynthetic enzymes use NADP. Structurally they differ only in NADP having an extra phosphate group attached to the ribose ring (Fig. 11.3). This gives NADP an extra negative charge and, not surprisingly, enzymes that prefer NADP have somewhat larger binding pockets with positively charged amino acid residues at the bottom. Enzymes that favor NADH often have a negatively charged amino acid residue in the corresponding position.

Several enzymes that use NAD or NADP have been engineered to change their preference. For example, the **lactate dehydrogenase (LDH)** of most bacteria uses reduced NAD, not NADP, to convert pyruvate to lactate. A conserved aspartate provides the negative charge at the bottom of the cofactor binding pocket that excludes NADP. If this is changed to a neutral residue, such as serine, the enzyme becomes able to use both NADP and NAD. If, in addition, a nearby hydrophobic residue in the cofactor pocket is replaced by a positively charged amino acid (such as lysine or arginine), the enzyme now prefers NADP to NAD (Fig. 11.4).

The specificity of LDH for its substrate can be altered in a similar way. The natural substrate lactate is a three-carbon hydroxyacid. It is possible to alter several residues surrounding the substrate binding site without impairing

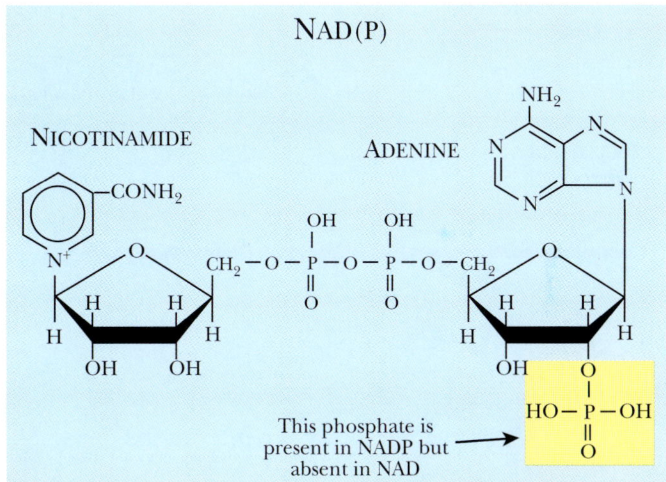

FIGURE 11.3 Difference in Structure between NAD and NADP
NAD (nicotinamide adenine dinucleotide) differs from NADP by one single phosphate (yellow).

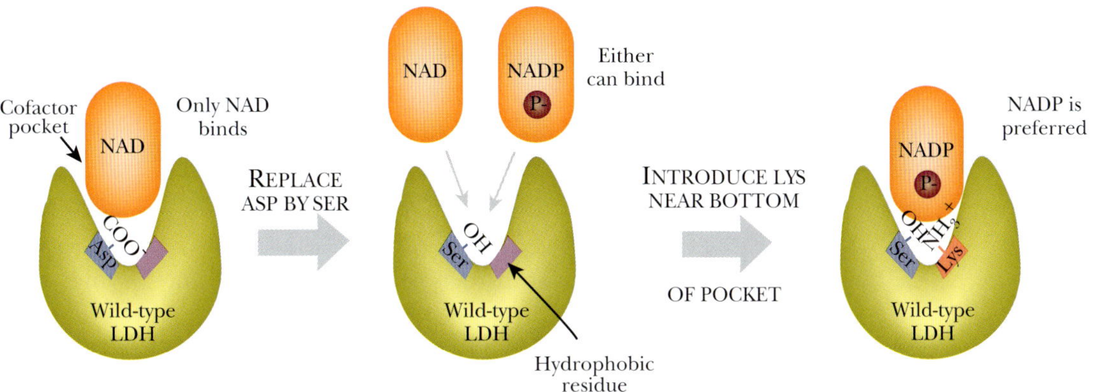

FIGURE 11.4 Changing Cofactor Preference of Lactate Dehydrogenase
Lactate dehydrogenase (LDH) preferentially binds NAD because the binding pocket has an aspartic acid. The negatively charged carboxyl repels the negatively charged phosphate of NADP. Changing the aspartic acid to serine allows either NAD or NADP bind to LDH. Adding a positively charged lysine makes the pocket more attractive to the NADP.

the enzyme reaction mechanism. By replacing a pair of alanines with glycines, the binding site can be made larger. By replacing hydrophilic residues (Lys, Gln) with hydrophobic ones (Val, Met), the site becomes more hydrophobic. Alteration of multiple residues gives an engineered LDH that accommodates five or six carbon analogs of lactate and uses them as substrates.

The substrate specificity of an enzyme may be changed by altering amino acids in and around the active site.

STRUCTURAL SCAFFOLDS

Relatively few of the amino acid residues in a protein are actually involved in the active site. Most of the protein provides the 3D platform or scaffold needed to correctly position the active site residues. Quite often the scaffold is much larger than really necessary. For example, the β-galactosidase of *Escherichia coli* (LacZ protein) has approximately 1000 amino acids, whereas most simple hydrolytic enzymes have only 200 to 300. Presumably it should be possible to redesign a functional β-galactosidase that is only 25% to 30% the size of LacZ protein. From an industrial viewpoint, such a smaller protein would obviously be more efficient.

The concept of using only the active site region of useful proteins has already been applied to antibody engineering as discussed in Chapter 6. In addition, the technique of phage display (see Chapter 9) has been used to select relatively short peptides for binding to a variety of target molecules. Once such a binding domain has been selected, it can be grafted on to another protein. Such techniques will allow the engineering of smaller proteins whose biosynthesis consumes less energy and material (Fig. 11.5).

Many proteins are larger than necessary from the industrial viewpoint. Removing extra sequences and simplifying the protein may increase industrial efficiency.

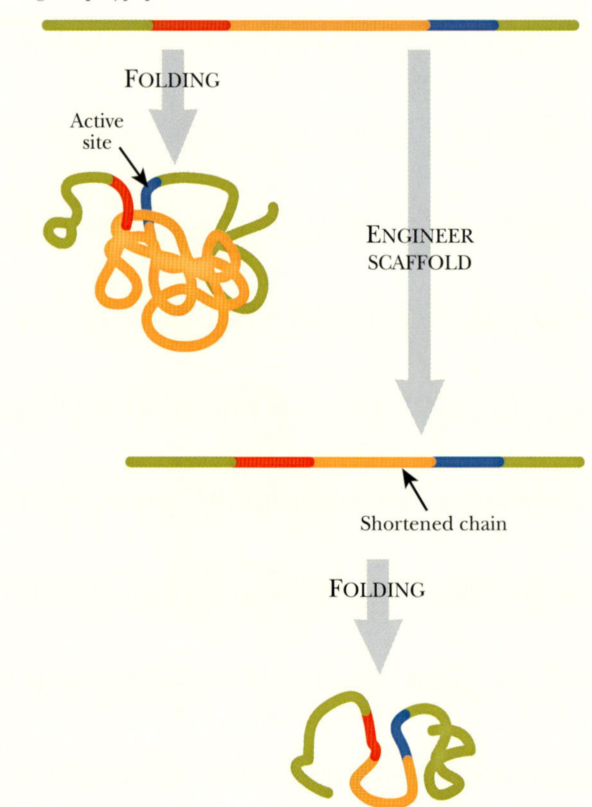

FIGURE 11.5
Scaffold Minimization
Proteins often have large structural domains (orange) outside the active site (blue/red). These domains can be engineered to make them smaller (which often enhances the proteins' level of expression) or to make isolation and purification easier.

DIRECTED EVOLUTION

The protein engineering techniques described to this point require knowledge of protein structure plus detailed knowledge of active-site function. Very few enzymes have been studied this intensively. **Directed evolution** is a powerful technique to alter the function of an enzyme without the need for exhaustive structural and functional data. Directed evolution can be used to change substrate specificity, either changing the enzyme to recognize a totally different substrate, or making subtle changes where the substrate is slightly different. The main premise of directed evolution is the random mutagenesis of the gene of interest, followed by a selection scheme for the new desired function. As already described in Chapter 5, new ribozymes are isolated using a similar principle.

Directed evolution screens for new enzyme activities by constructing a library of different enzymes derived from the same original protein. Each protein in the library is slightly different because of random mutagenesis. Random mutants may be generated over the entire length of the gene sequence. Alternatively, certain target amino acids can be replaced by random amino acids. The third main method for generating mutants relies on recombination (homologous or nonhomologous).

These mutant genes are then screened for the new, desired enzyme activity after insertion into a suitable expression vector and host cell.

Random mutagenesis usually starts with a gene whose function is close to that desired. The gene is randomly mutated throughout the entire sequence using **error-prone PCR** (epPCR). Different methods exist to induce errors during PCR amplification. The most straightforward is to add $MnCl_2$ to the PCR reaction. *Taq* polymerase has a fairly high rate of incorporating the wrong nucleotide, and $MnCl_2$ stabilizes the mismatched bases. The error will be copied in subsequent rounds of amplification. Adding nucleotide analogs such as 8-oxo-dGTP and dITP, which form mismatches on the opposite strand, can also enhance the PCR error rate. These analogs in combination with $MnCl_2$ can induce a wide variety of different mutations along the length of a gene. Some random mutations that occur outside the active site may cause subtle changes with profound effects on substrate recognition and enzyme function.

Target mutagenesis is much more focused and requires some knowledge of the enzyme structure, including the active site. For example, **tyrosyl-tRNA synthetase** from *Methanococcus jannaschii* was mutated through directed evolution. The gene for this enzyme had been sequenced, but no structural data were available. By comparing the sequence with that of another tyrosyl-tRNA synthetase, whose structure had been solved, the researchers identified amino acids potentially involved in tyrosine recognition. These residues were then randomly mutagenized. Altering residues with known functions in substrate recognition is more likely to have potent effects.

The third method to form new enzymes via directed evolution involves various schemes for recombining different domains. These are based on homologous or nonhomologous sequences and encompass a variety of different protocols, including DNA shuffling and combinatorial protein libraries. Some of these are discussed in detail later.

Proteins may be altered by directed evolution, which consists of random mutagenesis of the coding sequence, followed by biological selection for improved or novel properties.

ADDING NEW FUNCTIONAL GROUPS USING NONNATURAL AMINO ACIDS

Many nonnatural amino acids have different functional groups that are useful in protein engineering. For example, adding *p*-benzoyl-L-phenylalanine (*p*Bpa) into one position of glutathione-S-transferase (GST) adds a crosslinking group that can be activated by UV irradiation. When GST is modified in this way, UV irradiation creates a covalently linked homodimer (Fig. 11.6).

Incorporating a nonnatural amino acid into a protein can be done on a small scale by isolating the tRNA for a particular amino acid and attaching the nonnatural amino acid. The charged tRNA is then added to an *in vitro* protein translation system, which incorporates the nonnatural amino acid into the growing polypeptide chain. The chemical method is too costly and time consuming for large-scale use. For large-scale incorporation, *E. coli* may be modified to insert the nonnatural amino acid *in vivo*.

Inserting a nonnatural amino acid during *in vivo* protein synthesis requires a mutant **aminoacyl-tRNA synthetase** that charges a tRNA with the nonnatural amino acid. The laboratory of Peter G. Schultz, at the Scripps Research Institute, has developed an *E. coli* strain that incorporates *p*Bpa at a specific amber codon. Directed evolution was used to mutate a tyrosyl-tRNA synthetase from *M. jannaschii*. The enzyme from *M. jannaschii* was used because it does not recognize any endogenous *E. coli* tRNA. Consequently, it needs the gene for its specific partner tRNA to be provided as well. In addition, the partner tRNA was altered so that it recognizes the amber stop codon instead of its original natural

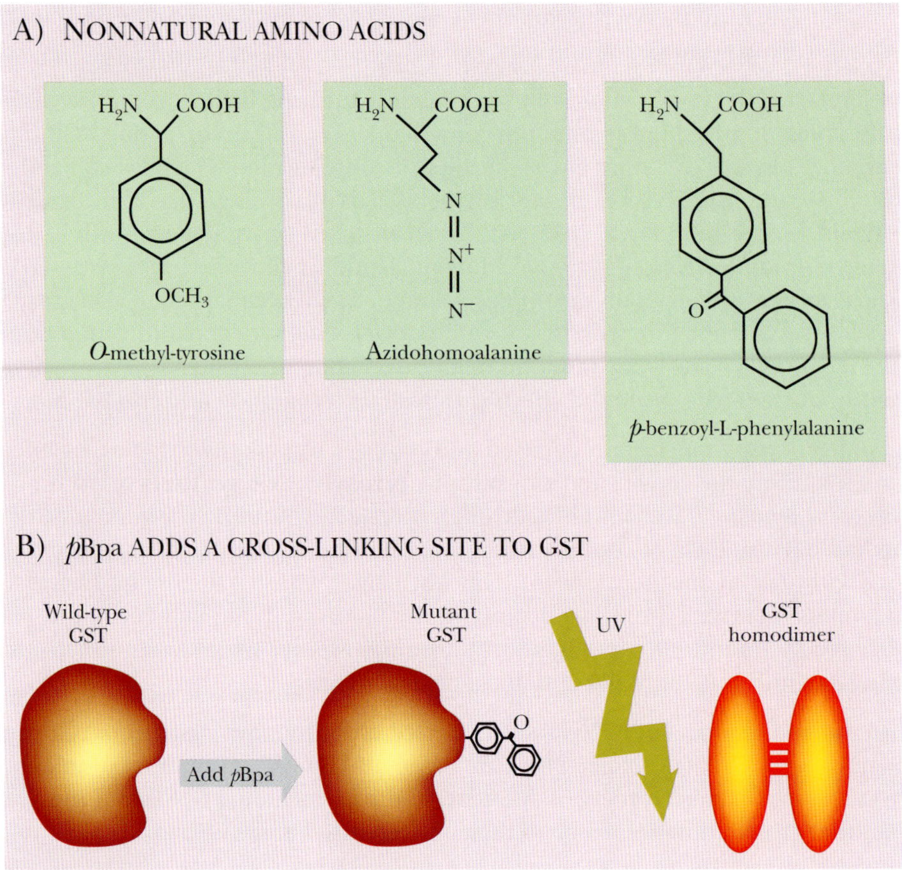

FIGURE 11.6 Adding New Functional Groups to Proteins
(A) Nonnatural amino acids add new functional groups. These can be incorporated into a protein during translation. (B) The nonnatural amino acid, *p*Bpa, crosslinks the GST mutant protein to form a homodimer.

codon (i.e., it is an amber mutant tRNA). The result is that when the mutant tyrosyl-tRNA synthetase is expressed in *E. coli,* it inserts whichever amino acid it is charged with at amber stop codons.

To alter the tyrosyl-tRNA synthetase, a library of mutant enzymes was generated by random mutation of each amino acid residue involved in recognition of the amino acid substrate (originally tyrosine). The library of mutant tRNA synthetase genes was transformed into *E. coli* that possess a gene for the partner tRNA, and a gene for chloramphenicol resistance with an amber codon in the middle. The *E. coli* were grown in the presence of *p*Bpa and chloramphenicol. If the mutant tRNA synthetase was able to insert an amino acid at the amber stop codon, then the chloramphenicol resistance gene (CAT) was expressed, and the cells lived. Otherwise, the cells died. This was the positive selection (Fig. 11.7).

This positive selection does not exclude mutant tRNA synthetases that charge the amber tRNA with a natural amino acid, so a negative selection scheme was used next. The plasmids carrying the mutant tRNA synthetases were isolated and transformed into a different *E. coli* strain. This *E. coli* had a toxin gene with an amber suppressor mutation plus the amber tRNA. Here the mutants were grown without any *p*Bpa. If the mutant tRNA synthetase could charge the amber tRNA with a natural amino acid, the toxin would be made and the *E. coli* would die. This eliminated mutant tRNA synthetases that used natural amino acids. The selection scheme was repeated numerous times, and finally a specific mutant tRNA synthetase was isolated (Fig. 11.7) that recognized the amber tRNA and linked *p*Bpa to it.

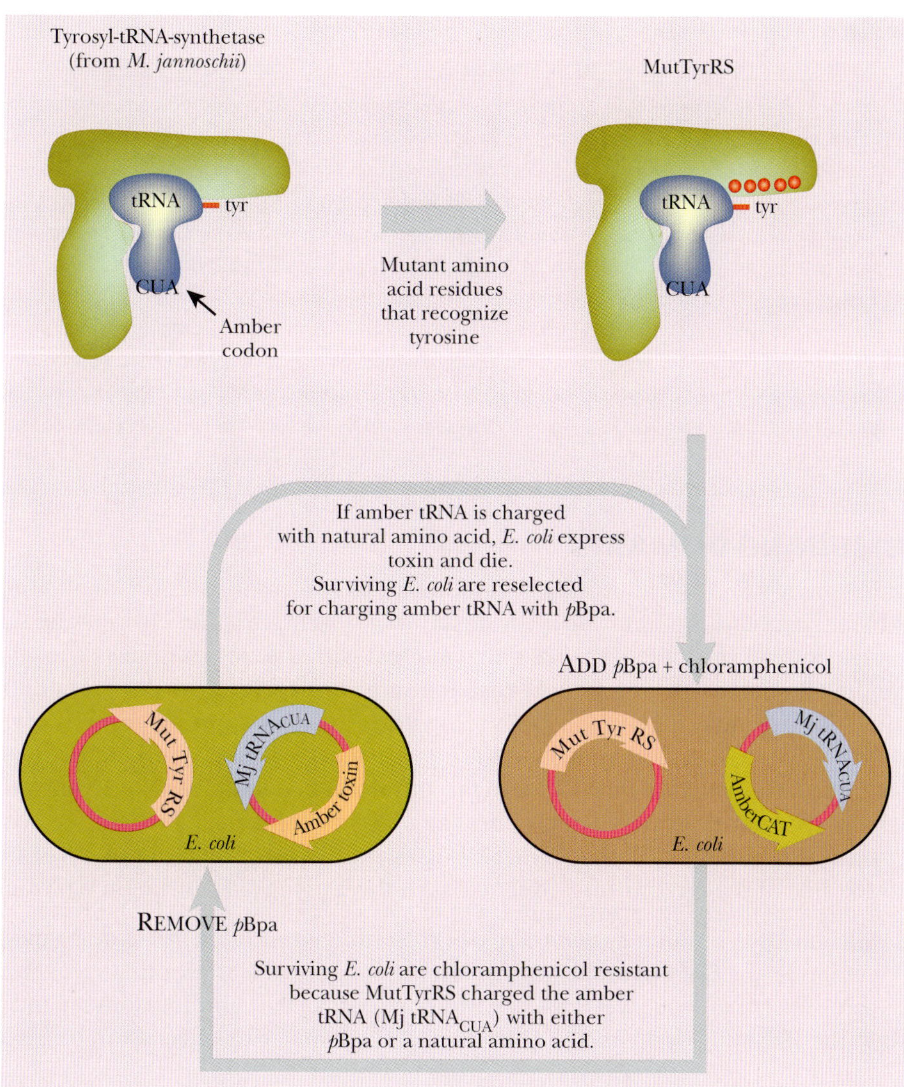

FIGURE 11.7 Positive and Negative Selections for Mutant tRNA Synthetase

Tyrosyl-tRNA synthetase normally attaches tyrosine to the tRNA for the CUA amber codon. The amino acids that recognize tyrosine were randomly mutagenized to form a library of different tRNA synthetases that still recognize the same tRNA, but might attach different amino acids. Next, these library clones (MutTyrRS) were expressed in a cell containing another plasmid carrying genes for the amber tRNA (Mj $tRNA_{CUA}$) and for chloramphenicol acetyl transferase (CAT). In the middle of the CAT gene is an amber codon (amberCAT). These *E. coli* are grown with *p*Bpa and chloramphenicol. MutTyrRS must charge the Mj $tRNA_{CUA}$ with *p*Bpa or another amino acid to express CAT, which protects the *E. coli* from chloramphenicol. The library clones that survive this selection are expressed in a different *E. coli* host (*left*). This strain has the gene for the amber tRNA (Mj $tRNA_{CUA}$) plus a toxin gene with an amber codon. The toxin protein is made if the MutTyrRS charges the amber tRNA with any amino acid (no *p*Bpa is present). This eliminates clones that charge the amber tRNA with an endogenous amino acid. The positive and negative selection schemes are alternated to find the best mutant tRNA synthetase.

This altered tRNA synthetase can be used to insert the nonnatural amino acid *p*Bpa into other proteins, such as the GST mutant described earlier. Thus it can be used to engineer crosslinking agents at any point in any target protein. To achieve this, the researcher must insert an amber codon into the gene that encodes the protein of interest and express the mutant gene in an *E. coli* host cell that expresses the altered tRNA synthetase.

A variety of schemes have been devised to insert nonnaturally occurring amino acids into proteins. The most sophisticated schemes consist of changing the coding properties of selected codons.

Box 11.1 Glycoengineering

In addition to the polypeptide chain, many proteins, especially in higher animals, have attached sugars that are crucial for activity. When such glycoproteins are manufactured via genetic engineering, they must be correctly modified, as discussed in Chapter 10. However, just as changing the amino acid sequence may generate proteins with novel properties, so altering the glycosylation pattern of a glycoprotein may result in useful alterations in protein behavior.

One widely used glycoprotein that is produced by genetic engineering is recombinant human erythropoietin, marketed as Epogen by the Amgen Corporation. Erythropoietin stimulates the production of red blood cells (erythrocytes). It is used to treat anemia and reduce the need for blood transfusions.

Natural erythropoietin has four attached oligosaccharides. When erythropoietin was engineered to have extra N-linked glycosylation sites (Asn-Xxx-Ser/Thr) its activity was changed in a paradoxical manner. Although its affinity for its receptor decreased, its *in vivo* activity increased. This was due to a longer half-life after administration. In practice the longer half-life outweighed the lowered receptor affinity, and the resulting protein had significantly increased clinical activity. This reengineered variant is marketed as Aranesp. It seems likely that similar glycoengineering could be applied to other glycoprotein hormones.

RECOMBINING DOMAINS

A related approach to directed evolution is to deliberately recombine functional domains from different proteins. An example is the creation of novel restriction enzymes by linking the cleavage domain from the restriction enzyme *Fok*I with different sequence-specific DNA binding domains. *Fok*I is a **type II restriction enzyme** with distinct N-terminal and C-terminal domains that function in DNA recognition and DNA cutting, respectively. By itself the endonuclease domain cuts DNA nonspecifically. However, when the nuclease domain is attached to a DNA-binding domain, this domain determines the sequence specificity of the hybrid protein. The two domains may be joined via a sequence encoding a linker peptide such as $(GlyGlyGlyGlySer)_3$ (Fig. 11.8). Cleavage of the DNA occurs several bases to the side of the recognition sequence, as in the native FokI restriction enzyme.

Several different DNA binding domains have been combined with the nuclease domain of FokI. For example, the **Gal4 protein** of yeast is a transcriptional activator that recognizes a 17 base-pain consensus sequence. Gal4 has two domains, a DNA binding domain and a transcription activating domain. The N-terminal 147 amino acids of Gal4 can be fused

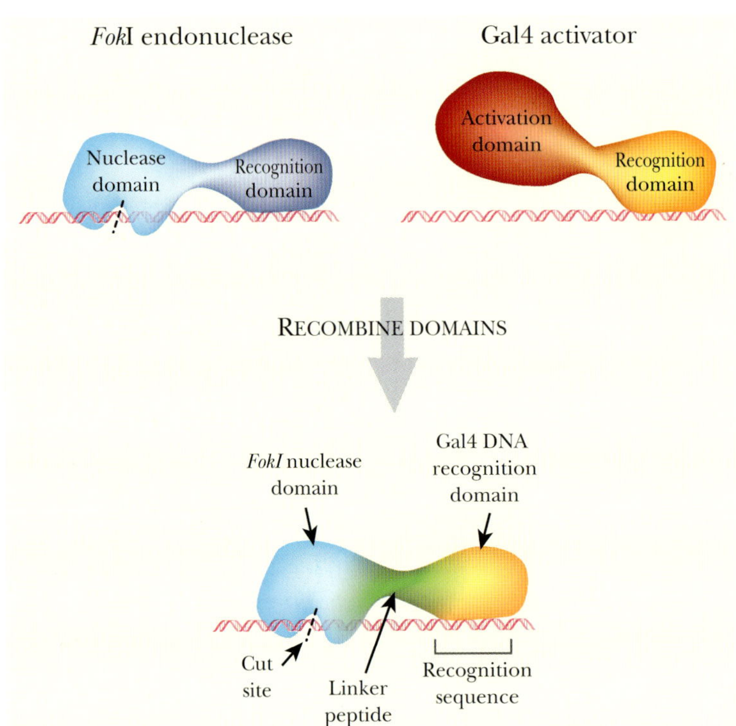

FIGURE 11.8 Recombining Domains to Create a Novel Endonuclease The *Fok*I endonuclease has separate nuclease and sequence recognition domains. Using genetic engineering, the recognition domain of *Fok*I can be replaced with a Gal4 recognition domain, which binds to a different DNA sequence. The two domains are joined with an artificial linker peptide. The new hybrid enzyme now cuts DNA at different locations from the original *Fok*I protein.

to the nuclease domain of *Fok*I, giving a hybrid protein that binds to the Gal4 consensus sequence and cleaves the DNA at that location.

Zinc finger domains have also been joined to the nuclease domain of *Fok*I. The zinc finger is a common DNA binding motif found in many regulatory proteins. The zinc finger consists of 25 to 30 amino acids arranged around a Zn ion, which is held in place by binding to conserved cysteines and histidines. Each zinc finger motif binds three base pairs, and a zinc finger domain may possess several motifs. In this example, domains approximately 90 amino acids long and comprising three zinc finger motifs, and which therefore specifically recognize nine base DNA sequences, were connected to the *Fok*I nuclease domain (Fig. 11.9).

Both the amino acid sequence and the corresponding DNA recognition specificity are now known for a wide range of zinc fingers. This information allows zinc finger domains to be designed to read any chosen DNA sequence. DNA segments that encode zinc finger domains may be derived either from naturally occurring DNA binding proteins or by artificial synthesis, because individual zinc finger motifs are about 25 amino acids (75 nucleotides) in length. In principle, engineered proteins can now be provided with a zinc finger domain that will enable them to bind specifically to any chosen DNA sequence.

Many proteins consist of domains that possess individual functions. Proteins with novel combinations of properties may be generated by recombining domains from different proteins.

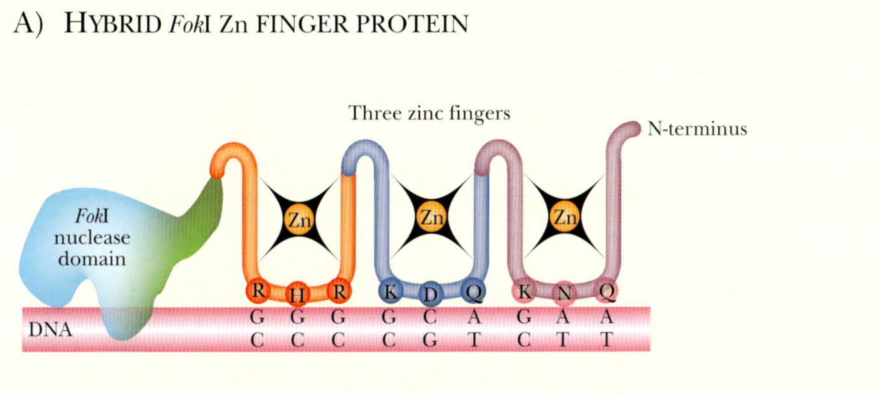

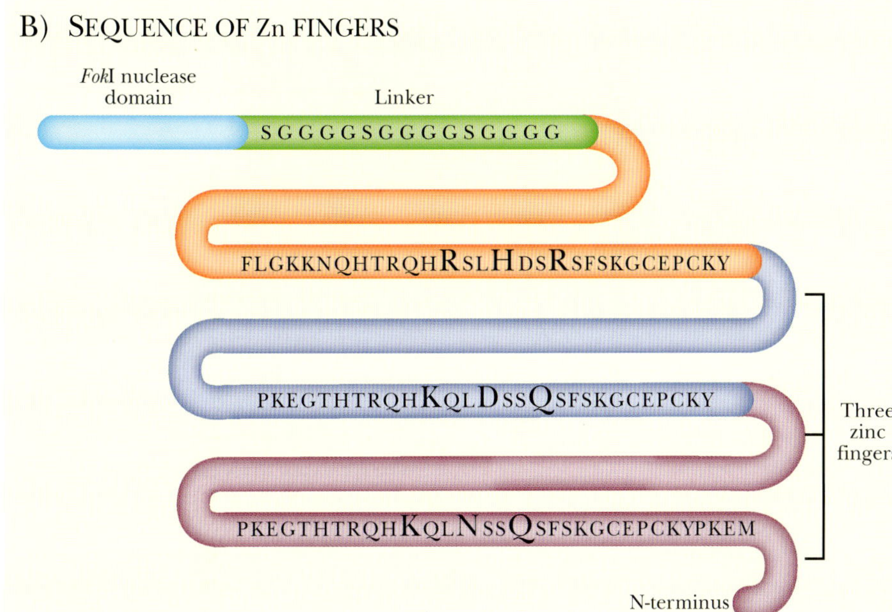

FIGURE 11.9
Assembly of Zinc Finger Domains
(A) The nuclease domain of *Fok*I can be linked to a zinc finger domain containing three zinc finger motifs. Zinc fingers recognize three nucleotides each; therefore, any 9-base-pair recognition sequence can in principle be linked to the nuclease domain. (B) The sequence of the hybrid between the *Fok*I nuclease and the zinc finger domain. The letters represent the amino acid sequence. The amino acids in large letters recognize and bind the DNA sequence.

DNA SHUFFLING

Natural selection works on new sequences generated both by mutation and recombination. **DNA shuffling** is a method of artificial evolution that includes the creation of novel mutations as well as recombination. The gene to be improved is cut into random segments around 100 to 300 base pairs long. The segments are then reassembled by using a suitable DNA polymerase with overlapping segments or by using some version of overlap PCR (see Chapter 4). This recombines segments from different copies of the same gene (Fig. 11.10).

Mutations may be introduced in several ways, including the standard mutagenesis procedures already described. In addition, the DNA segments may be generated using error-prone PCR (see earlier discussion) instead of by using restriction enzymes. Alternatively, mutations may be introduced during the reassembly procedure itself by using a DNA polymerase that has impaired proofreading capability. The result is a large number of copies of the gene, each with several mutations scattered at random throughout its sequence. The final shuffled and mutated gene copies must then be expressed and screened for altered properties of the encoded protein.

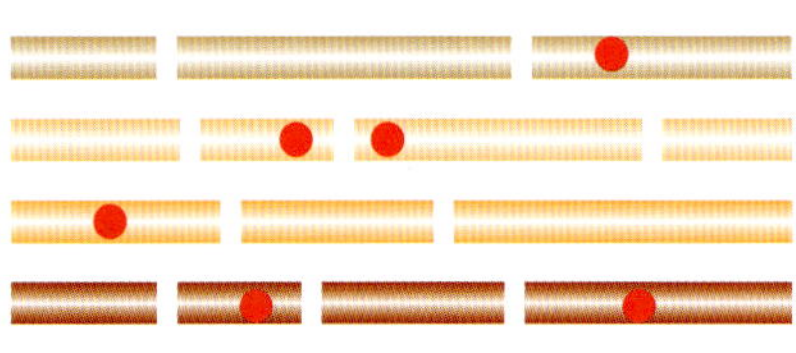

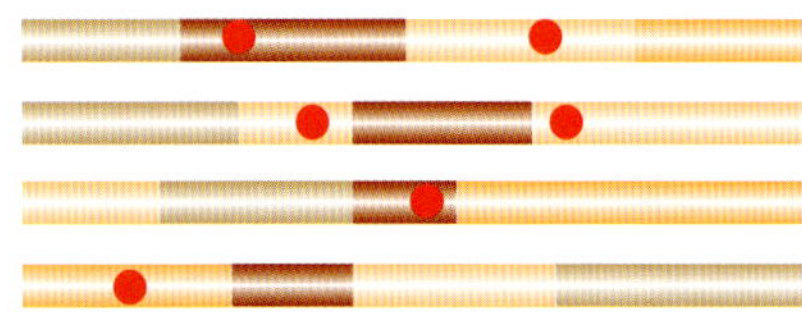

FIGURE 11.10 DNA Shuffling for a Single Gene

Introducing point mutations and shuffling gene segments can generate a better version of a protein. First, many copies of the original gene are generated with random mutations. The genes are then cut into random segments. Last, the fragments are reassembled using overlap PCR. The new constructs must be assessed for enhanced protein function.

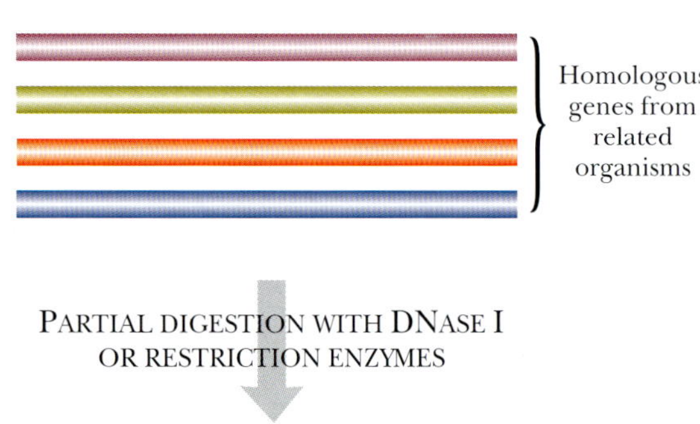

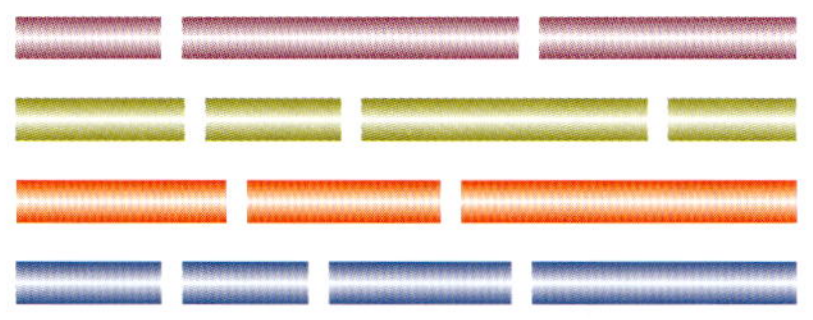

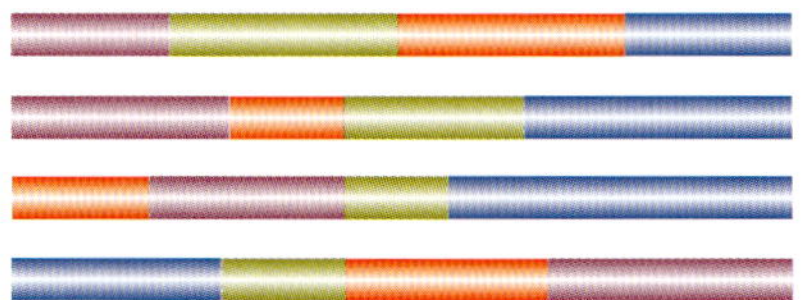

FIGURE 11.11 DNA Shuffling for Multiple Related Genes

Shuffling segments from related genes can also enhance the function of a particular protein. The original set of related genes are digested into small fragments and reassembled using PCR. The new combinations are tested for a change in function.

A more powerful variant of DNA shuffling is to start with several closely related (i.e., homologous) versions of the same gene from different organisms. The genes are cut at random with appropriate restriction enzymes and the segments mixed before reassembly. The result is a mixture of genes that have recombined different segments from different original genes (Fig. 11.11). Note that the reassembled segments keep their original natural order. For example, several related β-lactamases from different enteric bacteria have been shuffled. The shuffled genes were cloned onto a plasmid vector and transformed into host bacteria. The bacteria were then screened for resistance to selected β-lactam antibiotics. This approach yielded improved β-lactamases that degraded certain penicillins and cephalosporins more rapidly and so made their host cells up to 500-fold more resistant to these β-lactam antibiotics.

In DNA shuffling, the coding sequence for a protein is rearranged in the hope of generating novel or improved activities. Mutations may also be introduced during the procedure to provide more variation.

COMBINATORIAL PROTEIN LIBRARIES

So far we have discussed ways to modify a useful protein that already exists. Another approach to protein engineering is to generate large numbers of different protein sequences and then screen them for some useful enzyme activity or other chemical property. (Screening is often done by phage display or related techniques as described in Chapter 9.) Rather than merely generating large numbers of random polypeptides, **combinatorial screening** usually uses pre-made modules of some sort to create a **random shuffling library**. For example, protein motifs known to provide a binding site for metal ions or metabolites might be combined with segments known to form structures such as an alpha helix.

In a common approach, DNA modules of around 75 base pairs (i.e., 25 codons) are made by chemical DNA synthesis. Several modules are then assembled to give a new artificial gene. The modules are usually joined by PCR using overlapping primers (Chapter 4). Modules may be joined in a chosen order or in a randomized manner. For proper expression of the assembled sequence, the front and rear modules are normally specified to provide suitable promoter and terminator sequences. The intervening modules may then be randomly shuffled to generate more possible variation (Fig. 11.12).

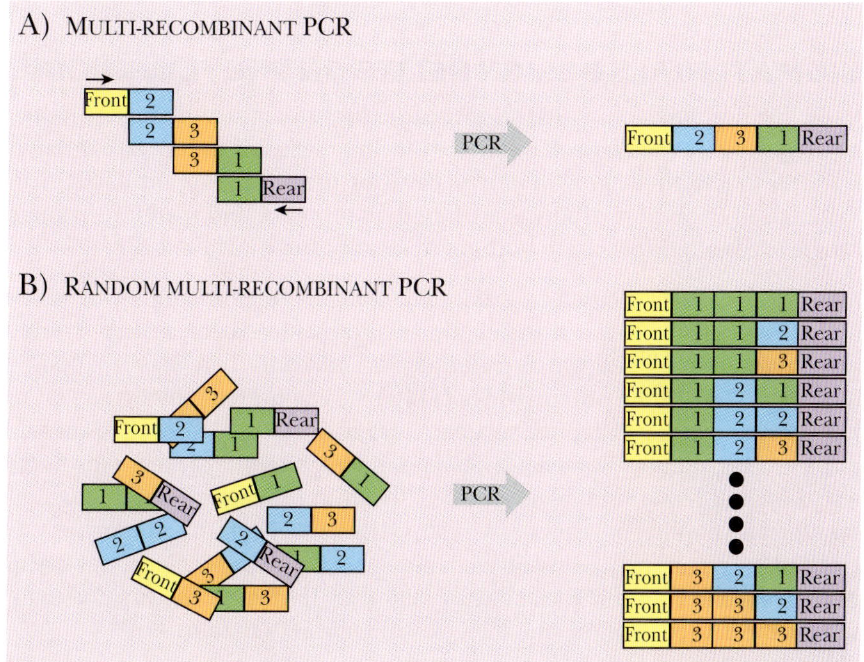

FIGURE 11.12 Generation of Random Shuffling Library
To create a library of new proteins, different modules can be randomly joined together. The first module (yellow) has sequences for the promoter; therefore, this is always added at the front. Similarly, the last module (purple) has the terminator sequences. Using overlapping PCR primers, the modules are joined together in a particular order (part A), or randomly (part B). Because random assembly creates many different combinations, this method creates a library of new proteins that can be screened for a particular function or set of functions.

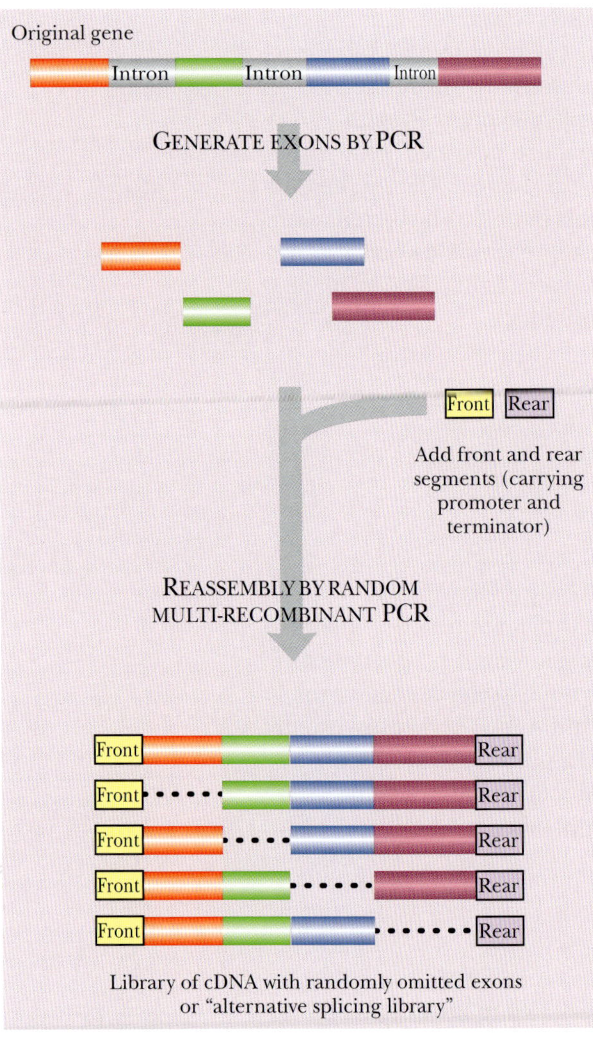

FIGURE 11.13
Generation of Alternative Splicing Library
The exons from an original gene can be recombined such that one exon is missing in each novel construct. The new genes are then screened for new or altered function.

The resulting protein library is then screened for some activity associated with the modules used. For example, if modules for binding an organic metabolite and for binding Fe ions were included, then the library of products might be screened for an enzymatic activity that oxidizes the metabolite via iron-mediated catalysis.

A related approach is based on the idea that the exons of eukaryotic genes encode modular segments of proteins, such as binding sites and structural motifs. Although it is by no means always true, in many cases exon boundaries do correspond to the ends of structural domains within the encoded proteins. Consequently, it is believed that at least some eukaryotic genes have evolved by the natural shuffling of exons.

In the exon shuffling approach a **combinatorial library** is generated from an already existing eukaryotic gene. Each of the exons of the eukaryotic gene is generated by a separate PCR reaction. The segments are then mixed and reassembled by overlap PCR. Two variants exist, depending on the design of the overlap primers for the PCR assembly. Reassembling the exons in random order generates a random splicing library, much as described earlier. Less radically, an **alternative splicing library** retains the order of the exons but includes or excludes any particular exon at random (Fig. 11.13). The final products are prescreened to obtain sequences long enough to encode most of the original exons, as this obviously greatly increases the chances of a functional protein.

Another shuffling technique to generate novel proteins is to use premade protein modules. The coding sequences for the modules are combined in various combinations by a PCR-based approach.

BIOMATERIALS DESIGN RELIES ON PROTEIN ENGINEERING

In the medical field, biomaterials are crucial for reconstructive surgery, tissue engineering, and regenerative medicine. Biomaterials include vascular grafts and cartilaginous tissue scaffolds that facilitate growth of new tissue by providing support and structure. The materials used in these products are based on proteins, and therefore, protein engineering can be used to improve them both mechanically and biochemically.

Many biomaterials are based on extracellular matrix proteins that provide support and structure *in vivo*. For example, collagen and elastin are proteins found in cartilage. *In vivo*, these proteins are secreted from cells known as chondrocytes and form a hard elastic support that cushions our joints. **Elastin-like polypeptides (ELPs)** are engineered proteins similar to native elastin. ELPs possess a repeated peptide sequence such as $(VPGZG)_x$, where Z is any amino acid except proline. If the amino acid repeat is changed, the physical properties of the final material will also change. If lysine residues are inserted, then two ELP strands can be crosslinked. Varying the location and number of lysines can create various types of films. Alternatively, UV-responsive crosslinking groups may be engineered into the ELP peptide. The peptides stay as soluble strands until exposed to UV light. The ability to control gel formation allows a doctor to inject the liquid form at the desired location, and then crosslink the ELPs to form a gel.

Besides supplying support, these materials can promote healing and tissue regeneration by attracting cells to the area. Adding different protein binding domains to the repeated peptide can promote cell migration and adherence. For example, the cell membrane receptor integrin recognizes the extracellular matrix protein fibronectin. If the integrin-binding domain of fibronectin is alternated with the ELP repeat, then integrin-expressing cells will recognize and migrate into the ELP substance. Another example is the peptide Val-Ala-Pro-Gly, which is recognized by a membrane-bound receptor on vascular smooth muscle cells. When this peptide is alternated with the ELP sequence, the resulting material promotes the movement and growth of vascular smooth muscle cells only. Another method to induce cellular migration and growth is to encapsulate various growth factors in the ELP. For example, if vascular endothelial growth factor or platelet-derived growth factor are mixed with ELPs, then blood vessels are induced to form within the matrix.

Biomaterials for medical use, especially extracellular proteins used in reconstructive surgery, may be improved by protein engineering.

ENGINEERED BINDING PROTEINS

Making drugs specific for a particular organ can eliminate many unwanted side effects. One way of achieving this is to attach the drug to a reagent that recognizes proteins specific to particular tissues. As noted in Chapter 6, antibodies are the most widely used reagents for binding specific target proteins. However, antibodies require disulfide crosslinks to function,

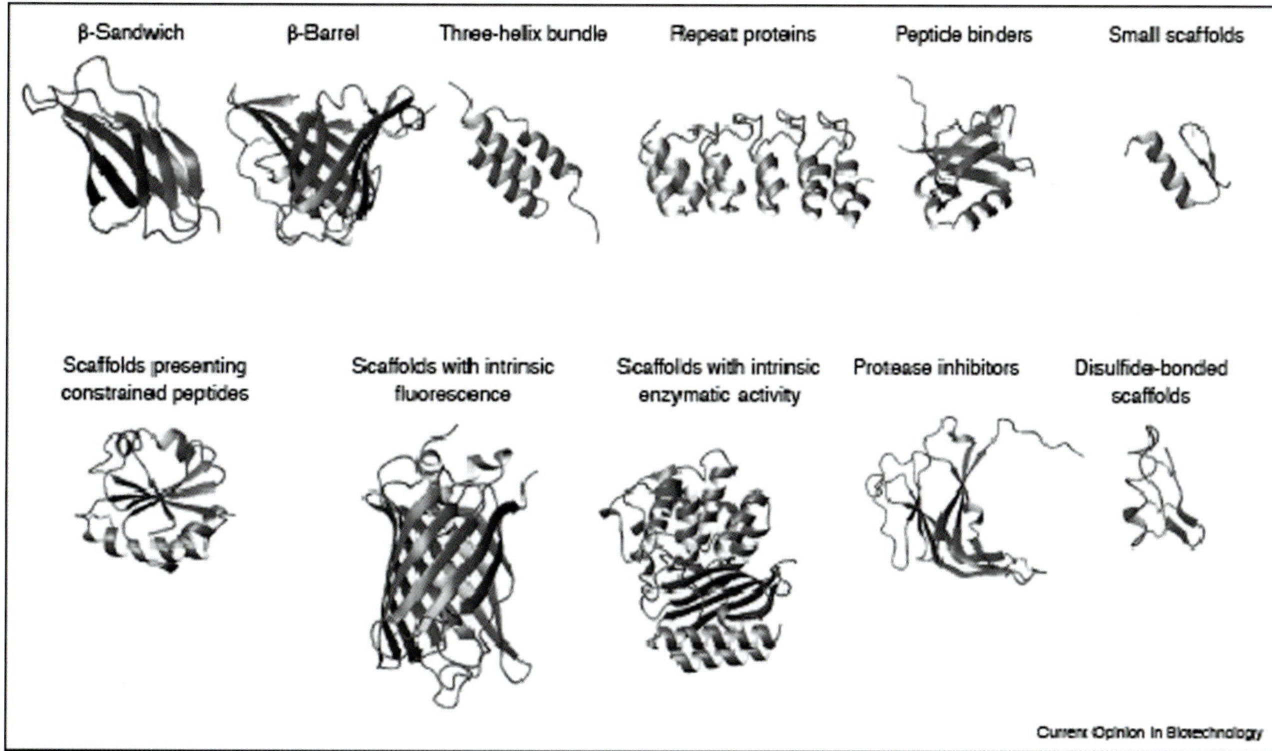

FIGURE 11.14 Structural Domains Involved in Protein-Protein Binding

Some types of protein backbones used as scaffolds for protein-binding agents. The proteins used (with their Protein Data Bank ID numbers) are beta sandwich (1FNA, fibronectin); beta barrel (A chain of 1BBP, lipocalin); thee-helix bundle (1Q2N, SpA domain); repeat proteins (1MJO, AR protein); peptide binders (chain A of 1KWA, PDZ domain); small scaffolds (chain F of 1MEY, zinc-finger protein); scaffolds presenting constrained peptides (chain A of 2TrX, thioredoxin A); proteins with intrinsic fluorescence (chain A of 1GFL, GFP) or intrinsic enzyme activity (1M40, beta-lactamase); protease inhibitors (1ECY, ecotin); and disulfide-bonded scaffolds (chain A of 1CMR, scorpion toxin). Cysteine residues and disulfide bonds are depicted in yellow. From: Binz and Plückthun (2005). Engineered proteins as specific binding reagents. *Curr Opin Biotechnol* **16**, 459–469. Reprinted with permission.

and these are often hard to maintain during large-scale manufacture. Some researchers have therefore been seeking alternatives to antibodies.

Most research into finding nonantibody binding partners has focused on certain protein structural domains. Because many different proteins have been crystallized and their structures have been determined, many different binding domains can be compared. The binding domains of one family of proteins share the same general structure, such as a β-barrel or three-helix bundle (Fig. 11.14). To generate novel binding domains, a binding protein with a known structure is chosen and the amino acid residues associated with binding are identified. The binding protein is modified by mutation of these residues and then screened for new binding partners. It is hoped that the targeted directed evolution approach will find new, more easily isolated proteins for targeting drugs to specific target cells within our bodies.

Creating binding proteins for a variety of uses is still in the experimental stages. However, it may eventually prove possible to replace cumbersome antibodies with smaller and more stable proteins.

Summary

Many enzymes are used industrially, especially in food processing. Protein engineering uses genetic technology both for improving enzymes already in use and to create proteins with novel properties. Methods range from introducing single specific changes into protein sequences to screening large numbers of semirandom polypeptides for new activities.

End-of-Chapter Questions

1. What is protein engineering?
   - **a.** alteration of a protein through genetic engineering to improve the protein's properties
   - **b.** modification of the amino acid residues after a protein has been translated
   - **c.** engineering a protein to be expressed in an artificial expression system
   - **d.** production of large quantities of proteins for use in industrial settings
   - **e.** none of the above

2. Which alteration to the protein sequence provides a major increase in protein stability?
   - **a.** glycosylation of amino acid residues
   - **b.** acetylation of amino acid residues
   - **c.** oxidation of the protein
   - **d.** introduction of cysteine residues for disulfide bond formation
   - **e.** reduction of the protein

3. Which of the following concepts increases protein stability?
   - **a.** decreasing the number of possible unfolded conformations
   - **b.** decreasing the number of hydrophobic interactions
   - **c.** increasing the number of hydrophilic interactions
   - **d.** decreasing the number of disulfide bonds
   - **e.** none of the above

4. What is a benefit to changing the active site or cofactor binding site of an enzyme?
   a. to prevent a reaction from taking place
   b. to produce the enzyme's product more cost-effectively
   c. to make the enzyme more stable
   d. to ensure that only the native substrate or cofactor are able to bind to the enzyme
   e. none of the above
5. Why might it be sensible to simplify an industrially relevant enzyme?
   a. to decrease the possibility of misfolding in the protein
   b. to increase the amount of protein being made
   c. to increase efficiency of the enzyme
   d. to decrease the amount of amino acid supplements to the growth medium for the growing cells
   e. none of the above
6. What term describes the process of creating random mutations in a gene and then selecting for improved function or altered specificity of the resulting protein product?
   a. indirect evolution
   b. protein mutagenesis
   c. translational evolution
   d. directed evolution
   e. instant evolution
7. Which of the following is required to insert a nonnatural amino acid during *in vivo* protein synthesis, but is not required for *in vitro* protein synthesis?
   a. a mutant aminoacyl-tRNA synthetase
   b. a tRNA charged with the nonnatural amino acid
   c. a natural aminoacyl-tRNA synthetase
   d. amino acids
   e. none of the above
8. Which enzyme identified in the book had significantly increased clinical activity after engineering more glycosylation sites into the protein?
   a. pectinase
   b. erythropoietin
   c. calmodulin
   d. insulin
   e. glucose isomerase
9. Which of the following is a way to create novel enzymes?
   a. glycoengineering
   b. using nonnatural amino acids to add new functional groups to amino acids within binding sites
   c. recombining functional domains from different enzymes
   d. DNA shuffling
   e. all of the above

(Continued)

10. How were improved β-lactamases made?
   a. by glycoengineering the N-terminal sequence of β-lactamase
   b. through the addition of nonnatural amino acids within the active site of the enzyme
   c. by DNA shuffling several related β-lactamase genes from different enteric bacteria
   d. by recombining the functional domains of different β-lactamases
   e. none of the above

11. With regards to exons, how is a combinatorial library constructed?
   a. The exons of a eukaryotic gene are created by PCR and then reassembled randomly.
   b. The order of the exons from a eukaryotic gene is maintained but the exclusion or inclusion of a particular exon is random.
   c. The exons are digested with restriction enzymes and then religated.
   d. The exons are assembled and screened by phage display.
   e. none of the above

12. With regards to exons, how is an alternative splicing library created?
   a. The exons of a eukaryotic gene are created by PCR and then reassembled randomly.
   b. The order of the exons from a eukaryotic gene is maintained but the exclusion or inclusion of a particular exon is random.
   c. The exons are digested with restriction enzymes and then religated.
   d. The exons are assembled and screened by phage display.
   e. none of the above

13. What are ELPs?
   a. engineered proteins that are similar to collagen
   b. engineered proteins that contain regenerated tissues
   c. engineered proteins that adhere to cell surface receptors
   d. engineered proteins that are similar to elastin
   e. none of the above

14. Which of the following is a structural domain associated with protein-protein binding?
   a. β-barrel
   b. three-helix bundle
   c. beta sandwich
   d. peptide binders
   e. all of the above

15. Which of the following statements about protein engineering is not correct?
   a. Protein engineering uses genetic technology to improve already useful industrial enzymes.
   b. Protein engineering has the capacity to create enzymes with novel functions.
   c. Protein engineering can improve biomaterials both biochemically and mechanically.
   d. The methods that exist to engineer protein sequences are often not useful.
   e. Enzymes can be made more robust and industrially stable through protein engineering.

Further Reading

Baltz RH (2006). Molecular engineering approaches to peptide, polyketide and other antibiotics. *Nat Biotechnol* **24**, 1533–1540.

Binz HK, Plückthun A (2005). Engineered proteins as specific binding reagents. *Curr Opin Biotechnol* **16**, 459–469.

Bloom JD, Meyer MM, Meinhold P, Otey CR, MacMillan D, Arnold FH (2005). Evolving strategies for enzyme engineering. *Curr Opin Struct Biol* **15**, 447–452.

Branden C, Tooze J (1998). *Introduction to Protein Structure,* 2nd ed. Garland Publishing, New York.

Chaparro-Riggers JF, Polizzi KM, Bommarius AS (2007). Better library design: Data-driven protein engineering. *Biotechnol J* **2**, 180–191.

Chin JW, Martin AB, King DS, Wang L, Schultz PG (2002). Addition of a photocrosslinking amino acid to the genetic code of *Escherichia coli*. *Proc Natl Acad Sci USA* **99**, 11020–11024.

Hohsaka T, Sisido M (2002). Incorporation of non-natural amino acids into proteins. *Curr Opin Chem Biol* **6**, 809–815.

Kaur J, Sharma R (2006). Directed evolution: An approach to engineer enzymes. *Crit Rev Biotechnol* **26**, 165–199.

Kittendorf JD, Sherman DH (2006). Developing tools for engineering hybrid polyketide synthetic pathways. *Curr Opin Biotechnol* **17**, 597–605.

Leisola M, Turunen O (2007). Protein engineering: Opportunities and challenges. *Appl Microbiol Biotechnol* **75**, 1225–1232.

Lesk AM 2004. *Introduction to Protein Science: Architecture, Function, and Genomics*. Oxford University Press, New York.

Maskarinec SA, Tirrell DA (2005). Protein engineering approaches to biomaterials design. *Curr Opin Biotechnol* **16**, 422–426.

Qian Z, Fields CJ, Yu Y, Lutz S (2007). Recent progress in engineering alpha/beta hydrolase-fold family members. *Biotechnol J* **2**, 192–200.

Wang L, Brock A, Herberich B, Schultz PG (2001). Expanding the genetic code of *Escherichia coli*. *Science* **292**, 498–500.

Woycechowsky KJ, Vamvaca K, Hilvert D (2007). Novel enzymes through design and evolution. *Adv Enzymol Relat Areas Mol Biol* **75**, 241–294.

Environmental Biotechnology

INTRODUCTION

The environment around us has always been a source of new products and stimulated our imagination in developing new technologies. Our species is very successful at harnessing the environment for our benefit. We are also good at destroying or harming the environment for immediate gains. The readily visible world has been charted and mapped, but areas under the ocean and in the deep recesses of jungles are still unknown. In fact, many parts of the visible world still harbor unknown life forms invisible to the naked eye, including bacteria and viruses. These are found in the air, water, and land. Many have unique metabolisms, and some have abilities never seen before. Many can live in extreme environments, once thought too hot or too dry for life to exist.

Estimates predict that about 10^{31} to 10^{32} viral particles are present in the biosphere, an order of magnitude more than host cells. The **virosphere**, as it is sometimes known, is probably one of the biggest sources of novel genes. At the time of writing, only about 0.1% to 1% of microorganisms have been cultured. Even the majority of those found growing at moderate temperatures in soil or other normal habitats have not been cultured. In addition to DNA inside life forms, there is much free DNA in the environment that might also be a source of new genes. The field of environmental biotechnology has revolutionized the study of these previously hidden life forms and DNA. What kinds of secrets do they harbor? What kinds of new enzymes and proteins can be identified?

Molecular biology techniques are now being applied directly to the environment to investigate the uncultured viruses and bacteria. PCR is routinely used to amplify random sequences from many environmental samples in the hope of identifying new genes. After PCR, the DNA is sequenced (see Chapter 4 for PCR and sequencing). Then bioinformatics reveals whether or not the sequence (or a close relative) has already been identified or if it is completely novel (see Chapter 8). Microarrays are also being created to compare the numbers and types of organisms present in different environments (see Chapter 8). Almost every recombinant DNA methodology discussed in the first half of this book can be applied to environmental samples.

Environmental biotechnology is divided into different areas. These include direct studies of the environment, research with a focus on applications to the environment, or research that applies information from the environment to other venues. This chapter focuses on direct analyses of the environment and the natural biochemical processes that are present, whereas upcoming chapters cover research with environmental applications or results from environmental research with practical applications. Surveying different environments may identify new life forms, new metabolic pathways, or novel individual genes. Genomics techniques have revolutionized this field, and it is rapidly expanding.

The environment is filled with invisible life forms such as viruses, bacteria, and other gene creatures.

IDENTIFYING NEW GENES WITH METAGENOMICS

Metagenomics is the study of the genomes of whole communities of microscopic life forms. Approaches include shotgun DNA sequencing, PCR, RT-PCR, and other genetic methods. Metagenomic research sometimes allows us to identify microorganisms, viruses, or free DNA that exist in the natural environment by identifying genes or DNA sequences from the organisms. Metagenomics applies the knowledge that all creatures contain nucleic acids that encode various protein products; therefore, organisms do not have to be cultured, but can be identified by a particular gene sequence, protein, or metabolite. The term *meta-*, meaning more comprehensive, is also used in *meta-analysis*, which is the process of statistically combining separate analyses. Metagenomics is the same as genomics in its approach. The difference between genomics and metagenomics is the nature of the sample.

Genomics focuses on one organism, whereas metagenomics deals with a mixture of DNA from multiple organisms, "gene creatures" (i.e., viruses, viroids, plasmids, etc.), and/or free DNA. Most microorganisms have never been cultured or previously identified. Using metagenomics, researchers investigate, catalogue, and analyze the current microbial diversity. They identify new proteins, enzymes, and biochemical pathways. They also hope to provide insight into the properties and functions of the new organisms. The knowledge garnered from metagenomics has the potential to affect how we use the environment to our benefit or harm.

Metagenomics has been used to identify new beneficial genes from the environment, such as novel antibiotics, enzymes that biodegrade pollutants, and enzymes that make novel products (Table 12.1). Historically, studying microbes in the environment has identified many useful products. In the early 1900s, Selman Waksman was studying actinomycetes in soil when he discovered the antibiotic streptomycin. Similarly, metagenomics research has identified (by accident) another antibiotic, called turbomycin. The researchers were looking for hemolysin-related genes in the soil by screening a metagenomic library (see later discussion). Hemolysin is a bacterial toxin that punctures holes in susceptible cell membranes, allowing the cellular contents to leak out and the cell then dies. Hemolysin lyses red blood cells and creates a clear zone around a bacterial colony growing on blood agar plates. Some *E. coli* clones from the library had dark red or orange colors. Further investigation of these clones found two novel antibiotics, turbomycin A and B (Fig. 12.1). Novel biodegradation pathways have also been found in bacteria that inhabit contaminated sites. Enzymes that can reduce the toxic effects of oil- and petroleum-based contaminants are found in bacteria that utilize the pollutants as an energy source. Even bacteria that thrive in environments contaminated with radioactivity have been identified.

Table 12.1 Genes and Proteins Identified via Gene Mining

Gene	Environmental Sample
Esterases	Urania deep-sea hypersaline anoxic basins in the Mediterranean Sea
Lipases	Soil outside University of Göttingen, Germany; soil from the Madison, Wisconsin, USA, Agricultural Research Station
Amylases	Soil outside University of Göttingen, Germany; soil from the Madison, Wisconsin, USA, Agricultural Research Station
Chitinases	Estuarine and coastal seawater collected in Delaware Bay, USA
Antibiotics	
Turbomycin A and B	Soil from the Madison, Wisconsin, USA, Agricultural Station
Polyketide synthases	Soil from an arable field in La Cote Saint Andre, Isere, France
Vitamin biosynthesis	Avidin-enriched forest soil from outside Göttingen, Germany; sandy soil from a beach near Kavalla, Greece; volcanic soil from Mt. Hood, Oregon, USA
4-Hydroxybutyrate dehydrogenase	Soils from sugar beet field near Göttingen, meadow near Northeim, and the River Nieme valley, Germany

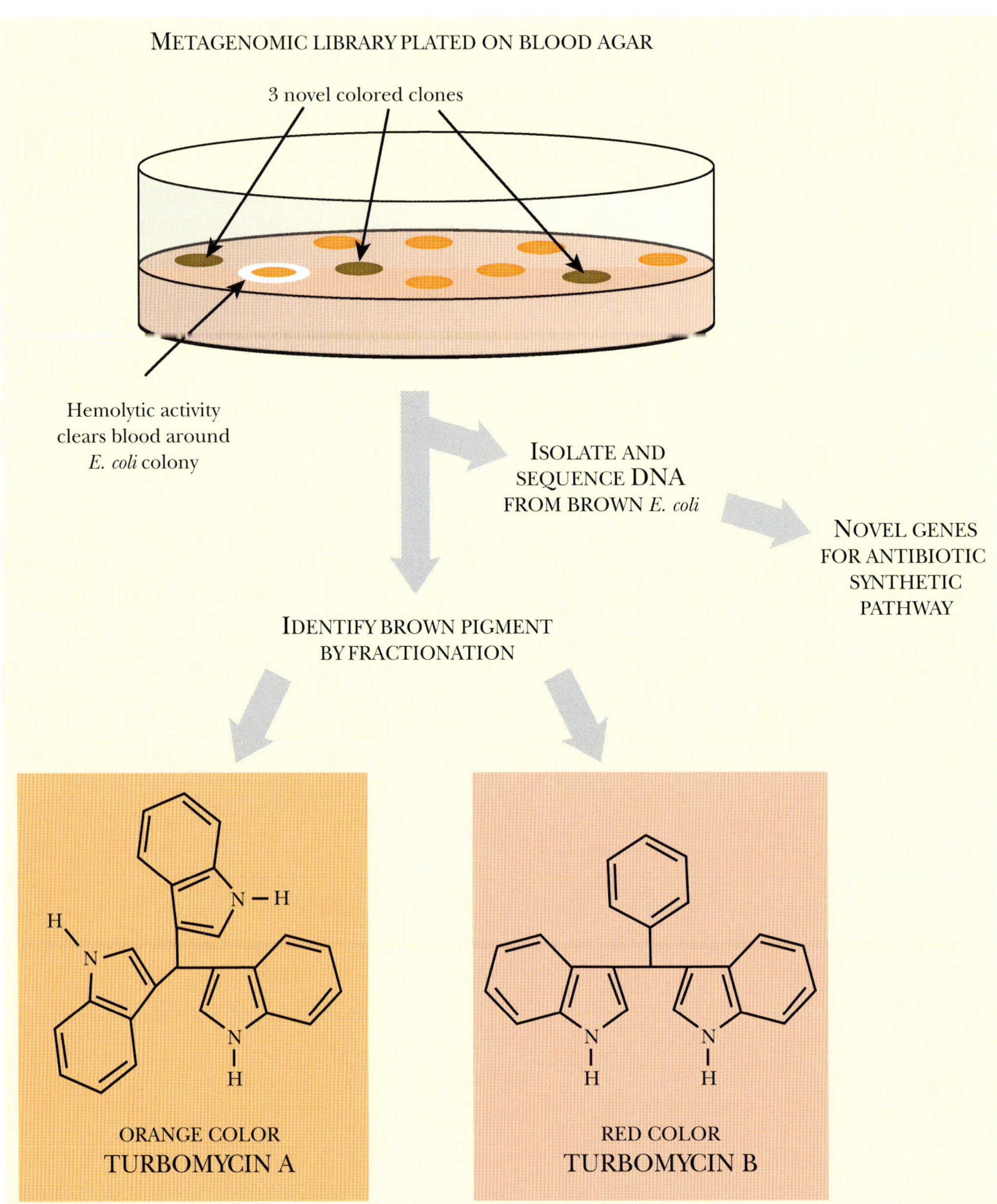

FIGURE 12.1 Novel Antibiotics Discovered in a Metagenomic Library
While screening a metagenomic library for hemolytic activity, three *E. coli* colonies were identified that had a dark brown color. The colored pigment was secreted and found to be a mixture of two different pigments (red and orange). The pure pigments have strong antimicrobial activity against both gram-positive and gram-negative bacteria. Sequencing the cloned insert identified the genes responsible for synthesizing the antibiotics. Adapted from Gillespie *et al.* (2002). *Appl Environ Microbiol* **68**(9), 4301–4306.

Metagenomics is the study of the genomes of whole communities of microscopic life forms. Many new proteins have been identified from examining DNA collected from an environment.

CULTURE ENRICHMENT FOR ENVIRONMENTAL SAMPLES

Various methods are used to enhance the starting material for metagenomics research, because a metagenomic library is only as good as its contents. Enrichment strategies include stable isotope probing (SIP), BrdU enrichment, and suppressive subtraction hybridization.

Stable isotope probing (SIP) was originally developed to trace single carbon compounds during their metabolism by cultured methylotrophs (bacteria specialized for growth on single-carbon compounds). Labeled precursor carbons were traced into fatty acids during bacterial growth. The method was adapted to the environmental samples used to create metagenomic libraries. Here, an environmental sample of water or soil is first mixed with a precursor such as methanol, phenol, carbonate, or ammonia, that has been labeled with a stable isotope such as ^{15}N, ^{13}C, or ^{18}O (Fig. 12.2). If the organisms in the sample metabolize the precursor substrate, the stable isotope is incorporated into their genome. Then, when the DNA from the sample is isolated and separated by centrifugation, the genomes that incorporated the labeled substrate will be "heavier" and can be separated from the other DNA in the sample. The heavier DNA will

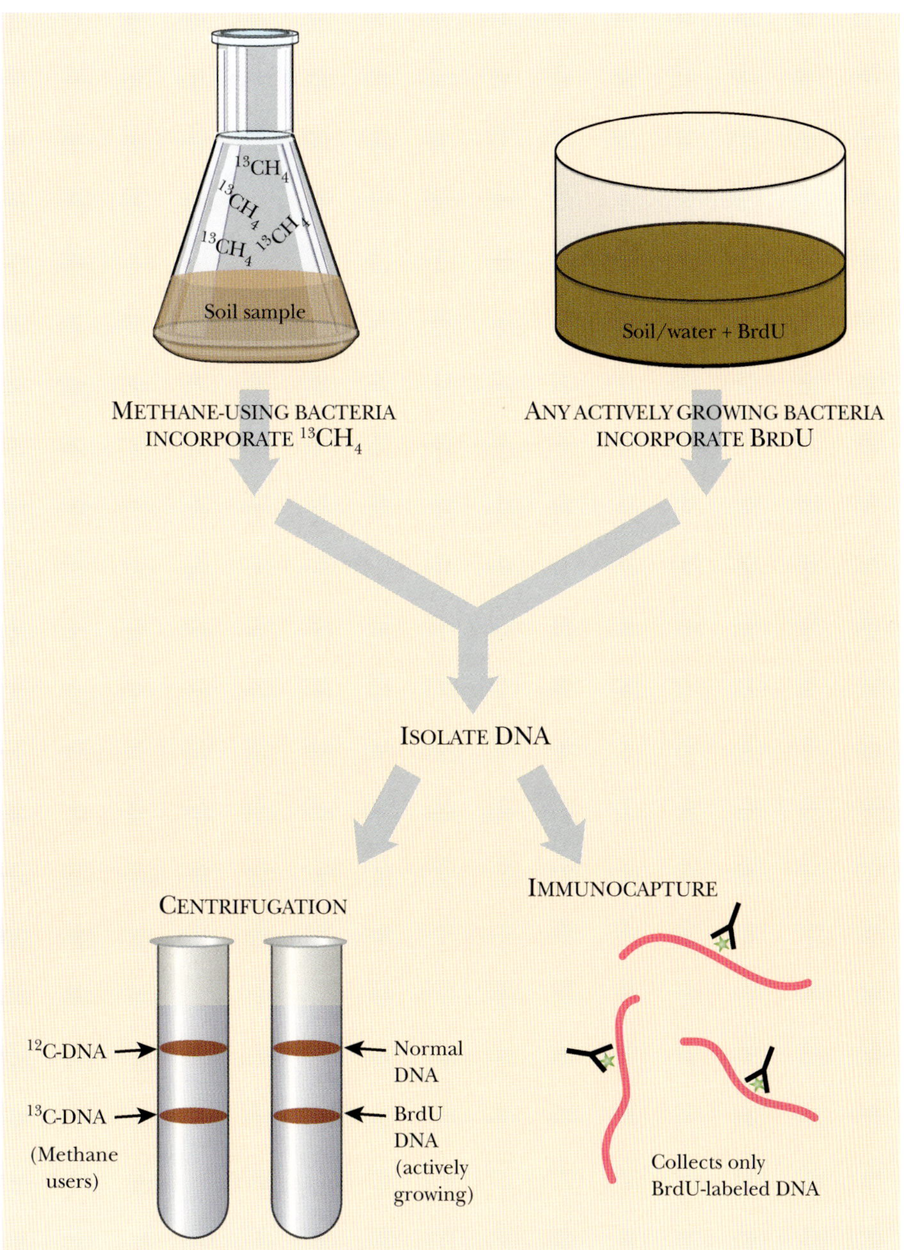

FIGURE 12.2 Stable Isotope Probing and BrdU Enrichment
Samples from the environment can be enriched for particular bacterial populations by incubating them with a labeled precursor such as ^{13}C-methane (*left*) or BrdU (*right*). After the DNA is isolated from the sample, the labeled DNA can be separated from the rest of the soil DNA using centrifugation or immunocapture.

migrate further in a cesium chloride gradient during centrifugation. As described later, the DNA can either be used directly or cloned into vectors to make a metagenomic library. This technique is particularly useful to find new organisms that can degrade contaminants, such as phenol in the example given.

Adding **5-bromo-2-deoxyuridine (BrdU)** is a related technique for enriching for the DNA of active bacteria in an environmental sample. Rather than entering only a metabolic subset of bacteria, BrdU is incorporated into the DNA and RNA of any actively growing bacteria or viruses. Note that bacteria and viruses that are dormant or dead, as well as free DNA, will not be labeled by this method. As before, the soil or water is isolated and incubated with BrdU. Any bacteria that are actively growing will take up the nucleotide analog and incorporate it into its DNA. Next, the BrdU-labeled DNA is isolated either with antibodies to BrdU or by density gradient centrifugation (see Fig. 12.2).

RNA-SIP focuses on isolating RNA from the environmental sample (rather than DNA). **Small subunit ribosomal RNA (SSU rRNA)**, that is, the 16S rRNA of bacteria or the 18S rRNA of eukaryotes, is an excellent biomarker because it is essential to all cellular life, it is very abundant within a cell, it is variable among different species, and there is an enormous database of different SSU rRNA sequences making identification relatively easy. In RNA SIP the SSU rRNA in the environmental sample is labeled. As described earlier, ^{13}C-labeled precursors are supplied to the environmental sample. These are incorporated into SSU rRNA independently of cell division because ribosomal RNA is produced in any cell that is making proteins, not just cells undergoing replication. This technique provides information on bacteria that are dormant as well as those that are more active. Much as before, the RNA is isolated and separated on a gradient by centrifugation. The rRNA bands tend to aggregate together during centrifugation. Therefore, each fraction must be repeatedly separated from the others. The final SSU rRNA fraction may still contain some contaminating nonlabeled rRNAs, so the fraction must be evaluated with care.

RNA-SIP can be used to identify a variety of microorganisms in environmental samples. For example, water from an aerobic industrial wastewater plant was evaluated for phenol-degrading microorganisms. The water was incubated with ^{13}C-labeled phenol, and the SSU rRNA was isolated by centrifugation. The rRNAs were isolated and amplified with RT-PCR followed by denaturing gradient gel electrophoresis. The bands were subjected to mass spectrometry to identify which rRNA sequence was most abundant. Interestingly, an organism belonging to the genus *Thauera* in the β-Proteobacteria was abundant, even though this organism was most usually found in denitrifying conditions. It was previously thought that pseudomonads were degrading the phenol.

Another culture-enrichment technique, **suppressive subtraction hybridization (SSH)**, takes advantage of the genetic differences between samples from two different areas. During standard subtractive hybridization, two different samples are hybridized and the mRNA that is the same is removed, leaving only mRNA that is different between the two samples (see Chapter 3). SSH works by the same principle. First, two different conditions must be established. For example, one soil sample from a polluted site could be compared with nearby soil that is not contaminated. The two soil samples will differ in their content of microorganisms, and those microorganisms enriched in the contaminated site could potentially metabolize the contaminant.

When DNA from each sample is isolated, the contaminated soil is considered the **tester** DNA and the "normal" soil is the **driver** sample. The tester sample is divided into two, and two different linkers are added to the ends of the DNA to form tester A and tester B. Tester A (with linker A), tester B (with linker B), and driver DNA are all mixed, denatured to make them single-stranded, and then rehybridized. The driver DNA is in excess to the testers, which ensures that DNA fragments from bacteria outside the contamination site outnumber those from the contaminated site. The driver DNA will anneal to all the common DNA fragments, making these double-stranded and with only one strand connected to the linker. All the tester DNA that is unique and not found in the uncontaminated soil will be free to hybridize with itself, forming A:A, B:B, or A:B hybrids. PCR primers are added to the hybridization mix; one primer recognizes linker A and the other primer is for linker B. As shown in Fig. 12.3, PCR will amplify only those hybrids that are tester:tester. Furthermore, because the A:A and B:B hybrids have inverted linkers, these hybrids will form a "panlike" structure during annealing and will not be amplified by PCR. Thus only A:B hybrids are amplified, and these represent unique sequences found only in the contaminated site.

Stable isotope probing is adding heavy isotopes to an environmental sample, which distinguishes the actively growing bacteria or viruses from the remaining organisms.

RNA-SIP is similar to SIP, but the heavy isotope is incorporated into the SSU rRNA. These sequences do not require the organism to be undergoing cell division, but they must be metabolically active.

Suppressive subtraction hybridization (SSH) compares the microorganisms present in two different samples by removing the identical DNA and evaluating the unique sequences.

SEQUENCE-DEPENDENT TECHNIQUES FOR METAGENOMICS

Historically, sequence-based metagenomics techniques were performed directly on samples from the environment after culturing. In 1985, Pace and colleagues directly sequenced the 5S and 16S rRNA gene sequences from the environment without culturing. This was technically difficult at the time. After PCR techniques became more prevalent in the early 1990s, the analyses of specific groups of organisms became simpler. Researchers used PCR primers specific to 16S rRNA sequences to identify the different organisms within the environmental sample. Yet these techniques rely on some prior knowledge of the sequences. Moreover, this method identifies a new organism only by its 16S rRNA sequence. It does not reveal the physiology or genetics of the organism.

Making and using **metagenomic libraries** is a way to identify the entire genetic complement of newly discovered life forms, without culturing them. From the genome sequences the physiology and function of the organism can be determined. DNA (or RNA) from the environment is isolated (and in some cases may be enriched as described earlier). The DNA, which is a mixture of fragments from multiple genomes, is then made into a library. Such libraries are like traditional genomic libraries, except that the starting DNA has sequences from many different organisms rather than one.

Basically, the total DNA isolated from an environmental sample is cloned into a vector plasmid, and constructs are maintained in *Escherichia coli*. Of course, all the same library permutations apply as for a single genome library. Large DNA fragments can be cloned into BACs (bacterial artificial chromosomes), and small pieces can be cloned into plasmids. The environmental DNA can be cloned into various expression vectors to determine if any specific enzyme or protein is present in the library. Environmental mRNA can also be isolated and converted to cDNA using reverse transcriptase before cloning. Such a cDNA library will represent genes that are being actively expressed in the environment. This approach circumvents the cloning of noncoding DNA. Of course, the same caveat exists, that is, the library is only as good as its original DNA or mRNA. If these are contaminated or sheared into small pieces, the library is worthless. In addition, some DNA from the environment may not be represented in the library because it cannot be isolated.

Metagenomic libraries can be sequenced to identify new genes or can be screened for new protein functions (Fig. 12.4). Many environmental biologists want to develop or find new enzymes that degrade pollutants or contaminants. Screening the library with a sequence

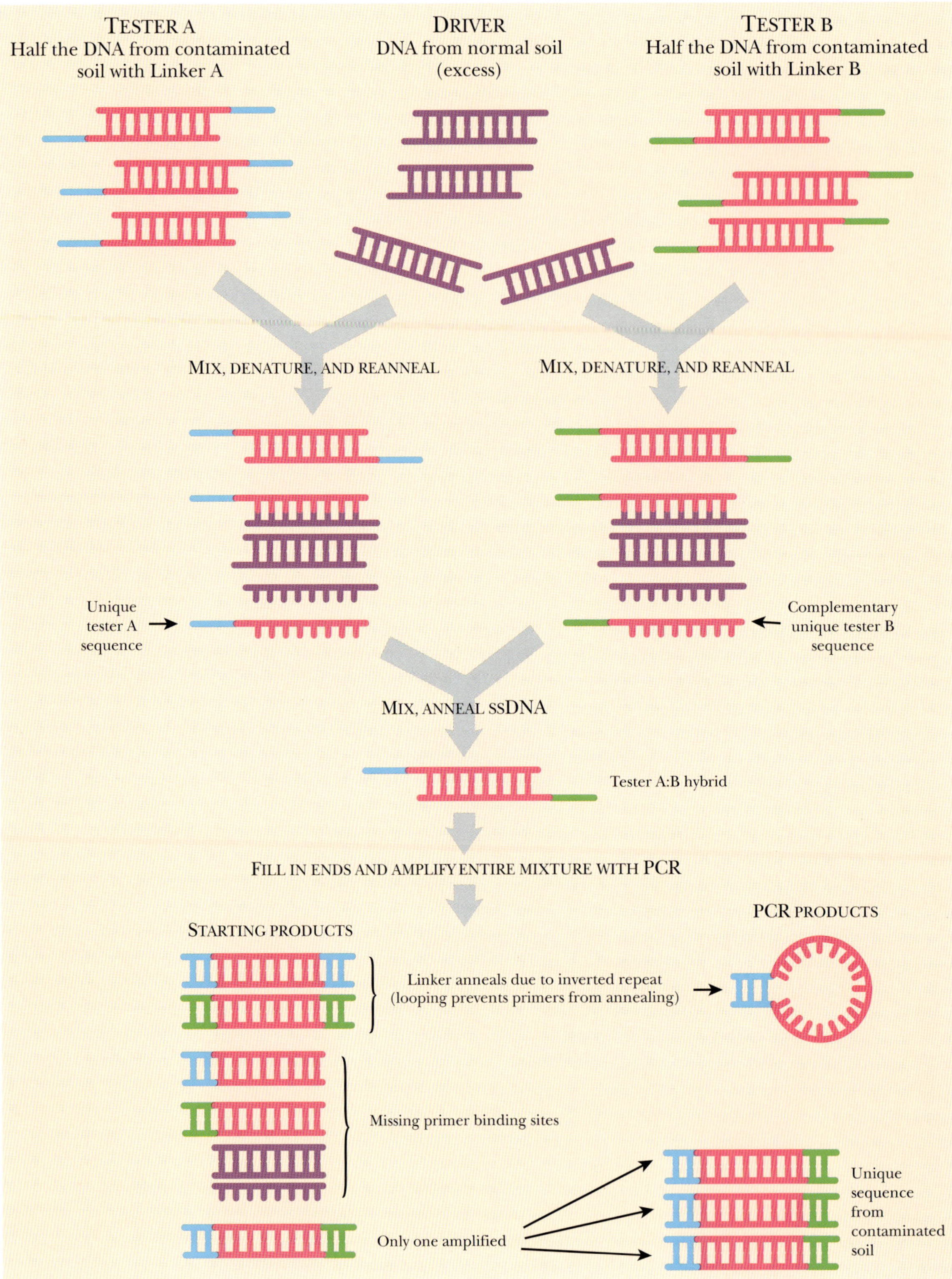

FIGURE 12.3 Suppressive Subtractive Hybridization
SSH begins by dividing the experimental sample in half and adding two different linkers to each half. Then each of these is mixed with an excessive amount of normal DNA, which will create an abundance of tester:driver hetero-hybrids (pink/purple strands). Each sample will still have some single-stranded pieces that did not find a complementary strand, including some normal or driver DNA and some tester DNA with a linker. The two pools are then mixed together (they are not denatured this time). The single-stranded pieces will anneal from tester A and tester B to form a hybrid molecule with two different linkers on each side. The entire pool is prepared for PCR by filling in the single-stranded regions, and then primers are added to the two different linkers. PCR only amplifies the tester A:tester B hetero-hybrids because the other hybrids either are missing a primer binding site or self-anneal via the linker to form a loop structure.

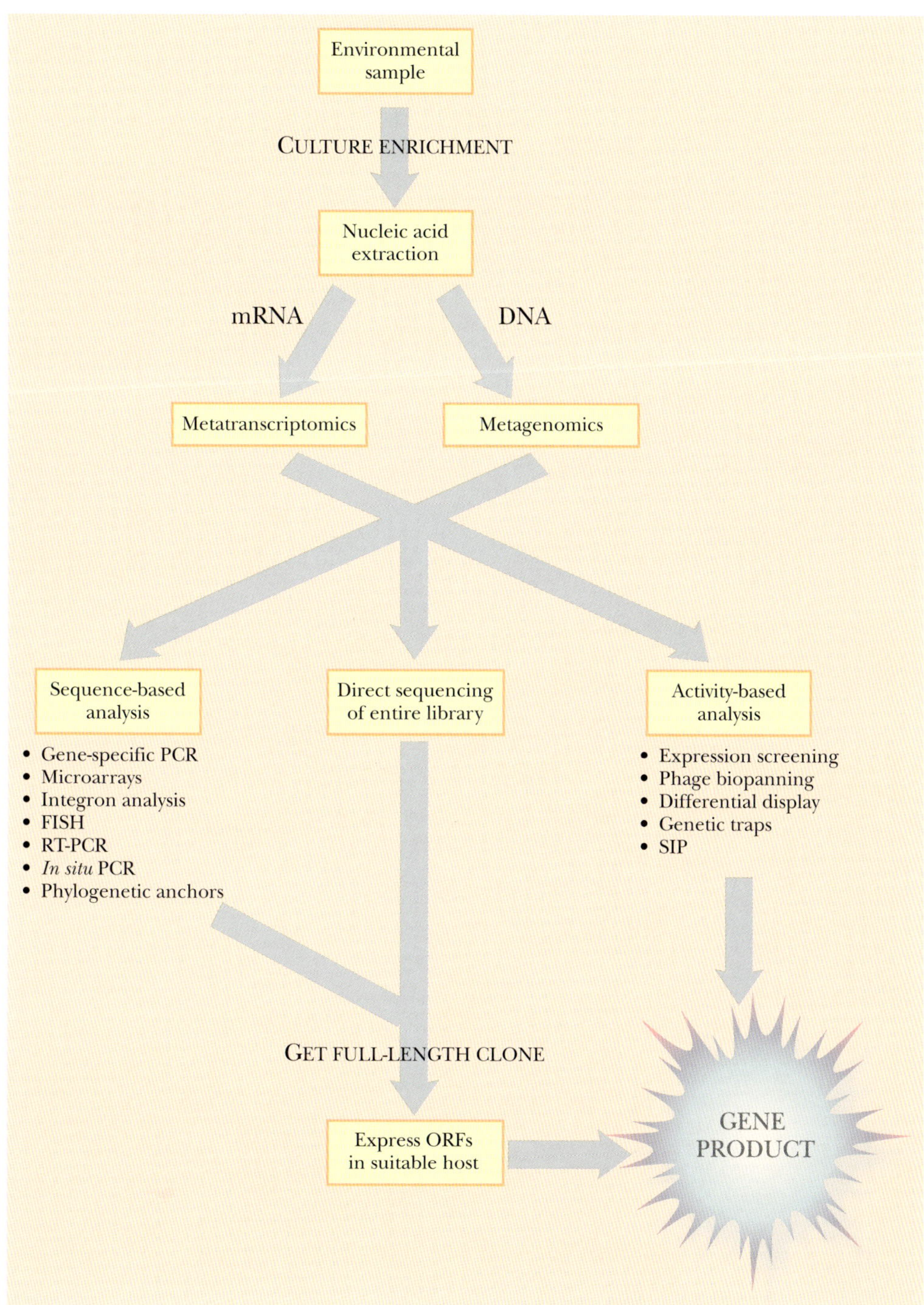

FIGURE 12.4
Techniques to Study Environmental Samples
Many different techniques can be applied to DNA or mRNA isolated from the environment. These approaches all converge on discovering the gene product and identifying its function in the environment.

containing conserved domains of known enzymes can find new enzymes. Along the same lines, PCR primers to conserved domains of known enzymes can amplify never-before-seen genes from the library. For example, aromatic oxygenases facilitate the degradation of aromatic hydrocarbons found in oil and coal. These contaminants can be degraded by a variety of different organisms. The genes and pathways involved are key targets for pathway engineering (see Chapter 13). Using PCR primers to conserved regions of known aromatic oxygenases amplifies the genes for novel (but related) oxygenases found in the environment.

Some sequence-based analyses for metagenomic libraries include microarrays (see Chapter 8), FISH (see Chapter 3), RT-PCR (see Chapter 4), sequencing using phylogenetic anchors, and integron analysis (see later discussion). Sequencing with phylogenetic anchors begins by identifying the sequence of a known gene. Often a marker gene such as 16S rRNA gene is identified first, and then the regions upstream or downstream of the marker are sequenced.

For example, a 16S rRNA sequence from seawater classified a particular genomic fragment to the γ-Proteobacteria. Adjacent to the 16S rRNA gene was a gene similar to bacteriorhodopsin, a transmembrane proton pump that responds to light. The genes for bacteriorhodopsin were originally thought to exist only in the Archaea, but this analysis revealed that other inhabitants of the ocean had similar genes.

Microarrays can also be used to identify the types and numbers of different organisms in an environmental sample. First, a microarray of unique sequences from known organisms is created, and then fluorescently labeled environmental DNA is hybridized to the array. The results can confirm whether or not a particular bacterium, virus, or gene creature is present, and the relative abundance can be determined by the intensity of fluorescence. FISH and RT-PCR can yield similar information, but the environmental DNA is analyzed with a only few probes or primers, respectively. The results are much more direct and focus on identifying known organisms.

Integron analysis identifies open reading frames that are used by integrons and can identify more novel unknown genes than many previous techniques (Fig. 12.5). **Integrons**

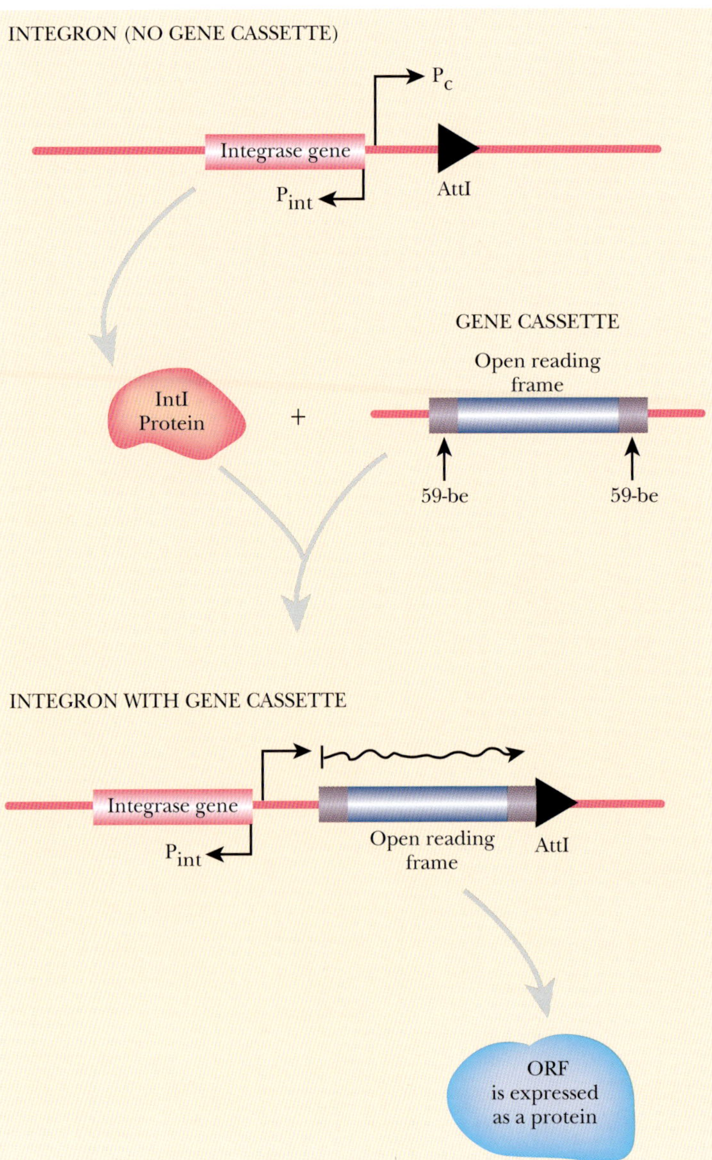

FIGURE 12.5
Integron Analysis
Integrons are genetic elements that capture and express gene cassettes. The integron has three main components: an integrase gene under the control of its own promoter (P_{int}), an *attI* site for integration of the gene cassette, and a promoter to express the gene cassette (P_c). When integrase is expressed, it searches the genome for gene cassettes. Integrase then excises the gene cassette and integrates it into the integron at the *attI* site. Using PCR primers to the 59-be sites allows isolation of any open reading frames in gene cassettes and potentially identifies new genes.

are related to transposons (see Chapter 1), but are particularly important for the spread of genes for antibiotic resistance and other properties that give the host a growth advantage in a particular environment. Integrons are genetic elements that contain a site (*attI*) to integrate a segment of DNA known as a **gene cassette**, a promoter to express the gene cassette (P_c), and a gene for **integrase (*intI*)**, the enzyme that recombines the gene cassette into the integron. Gene cassettes are segments of DNA with one or two open reading frames (ORFs) that lack promoters and are flanked by **59-base elements** (**59-be**, also known as *attC* sites). When the integrase recognizes the 59-be sites, it excises the gene cassette and integrates it downstream of the P_c promoter. This allows the open reading frame to be expressed into protein. The 59-be sequence may vary in length, but must contain a conserved, seven-nucleotide sequence.

The gene cassettes are the interesting part of the scenario and the key to integron analysis. Gene cassettes were first identified because many encode antibiotic resistance genes, but they may encode any type of gene. Screening metagenomic libraries using PCR primers that recognize the 59-be elements amplifies these open reading frames. This approach has identified novel genes related to DNA glycosylases, phosphotransferases, and methyl transferases. Additionally, new antibiotic resistance genes may be identified. Integron analysis is also useful to study bacterial evolution and gene transfer because these elements can pass from bacterium to bacterium during conjugation.

Metagenomic libraries identify the entire genetic complement of newly discovered life forms, without the need to culture the organism.

Environmental DNA can be analyzed using a variety of methods, such as FISH, microarrays, sequencing, and RT-PCR.

Searching the regions adjacent to known genes identifies novel genes with many diverse functions. This is called sequencing with phylogenetic anchors.

Integrons are like transposons, but have the ability to capture genes from different organisms and move them to others. Integron analysis of a metagenomic library uses PCR to amplify the sequence between the 59-be sites, which identifies the genes captured by the integron.

Box 12.1 Sequencing the Sargasso Sea

In 2004, *Science* magazine published one of the largest scale metagenomic analyses (*Science* 304, 66–74). Craig Venter (who led his company to sequence the human genome) took on a large-scale effort to sequence DNA isolated from the Sargasso Sea. This is part of the North Atlantic Ocean that has a relatively high salt content and is thus thought to have less biodiversity. He took samples of water from a 5-foot depth at various points along a journey on his sailboat. The water was passed through a series of three successively finer filters and then frozen. At intervals the filters were sent to Maryland for DNA sequencing. The DNA was extracted and sheared into small fragments by forcing it through a small opening at high pressure. Much like their work on the human genome, Venter's research group used shotgun sequencing and then allowed the computer to assemble overlapping stretches of sequence. Analysis revealed that the Sargasso Sea has a more diverse population than expected, with about 1800 different species of microbes, 150 new bacterial and Archaea species, and more than 1.2 million new genes. Interestingly, more than 700 different bacteriorhodopsin genes were identified. Bacteriorhodopsin is used by bacteria to harvest light for energy and is closely related to rhodopsin, a protein that detects light in mammalian eyes. The sequences from this exploration are found online at the National Center for Biotechnology Information (NCBI; http://www.ncbi.nlm.nih.gov) and are free to the public.

FUNCTION- OR ACTIVITY-BASED EVALUATION OF THE ENVIRONMENT

Besides these sequence-based approaches, metagenomic libraries can be screened for various functions (Fig. 12.6). Functional approaches include expression screening with various genetic traps and phage biopanning. Even stable isotope probing (see earlier discussion) could be categorized as a function-based approach if the labeled substrate is a specific metabolite that enriches the culture based on metabolic function. Screening a metagenomic library by sequencing has its limits. The function of many genes from exotic organisms cannot be identified by their sequence. In addition, using known genes to screen for new members of a gene family might miss an entire novel class of genes. For example, if a researcher were trying to find enzymes similar to bacteriorhodopsin, primers similar to known bacteriorhodopsins would be used to screen the library. Some genes would be identified, but others would be missed if their sequences were too divergent. Screening this library for light-driven proton pumping would identify any enzyme that carried out this function, regardless of its sequence.

Expression screening depends on the choice of an expression vector. Proteins are expressed from the metagenomic DNA fragments when they are cloned into a vector with transcriptional and translational start and stop sequences. Then an easy assay for the target function must be devised. For example, the library clones might be plated on a particular toxic pollutant. If any library inserts encoded an enzyme that metabolized the pollutant, that particular library clone would grow. The DNA insert is then isolated and identified by sequencing.

Another functional screen involves fusing the metagenomic DNA in frame with a promoterless gene for green fluorescent protein (GFP). The genes for many enzymes are turned on by their own substrates. Therefore, if this type of gene is cloned in front of promoterless GFP, the GFP gene will be regulated by the same substrate. If the substrate of interest is included in the growth medium for the library, any clone with genes that are activated by the substrate will also produce GFP. These cells will fluoresce green. FACS (see Chapter 6) provides a quick and easy way to isolate the fluorescent clones.

The main barrier to function-based analyses is successful gene expression. Getting a library host such as *E. coli* to express foreign genes is hit or miss, because some may be toxic to the host, some may require other factors for expression, and others may have very low activity. Another problem is simply volume. The number of potential clones that have any chosen gene of interest is usually low, and excessive numbers of clones must be screened in order to identify just one or two genes. For example, the lipases identified from German soil (see Table 12.1) were only found in 1 of 730,000 different metagenomic clones. In another example, only two novel Na^+/H^+ antiporters were found after screening 1,480,000 clones. This is why culture enrichment strategies are an important aspect of creating metagenomic libraries (see earlier discussion).

Metagenomic phage biopanning uses basically the same method as phage display (see Chapter 9). Cloned DNA inserts are expressed as fusion proteins with a phage coat protein and displayed on the phage surface. Because the displayed proteins carry only segments of foreign protein, the problems associated with heterologous expression are lessened. The cloned DNA is from any organisms found within the environmental sample. Thus the expressed protein segments could be any part of an enzyme, a membrane protein, etc.; therefore, the success with a phage display library relies on the screening method. For example, phage biopanning can identify binding partners for a particular pollutant, metabolite, or even another protein. If the target molecule is immobilized on a bead, then any phage carrying a protein segment that binds the target will stick to the bead. The phage can then be isolated, and the DNA insert sequenced to identify the sequence responsible.

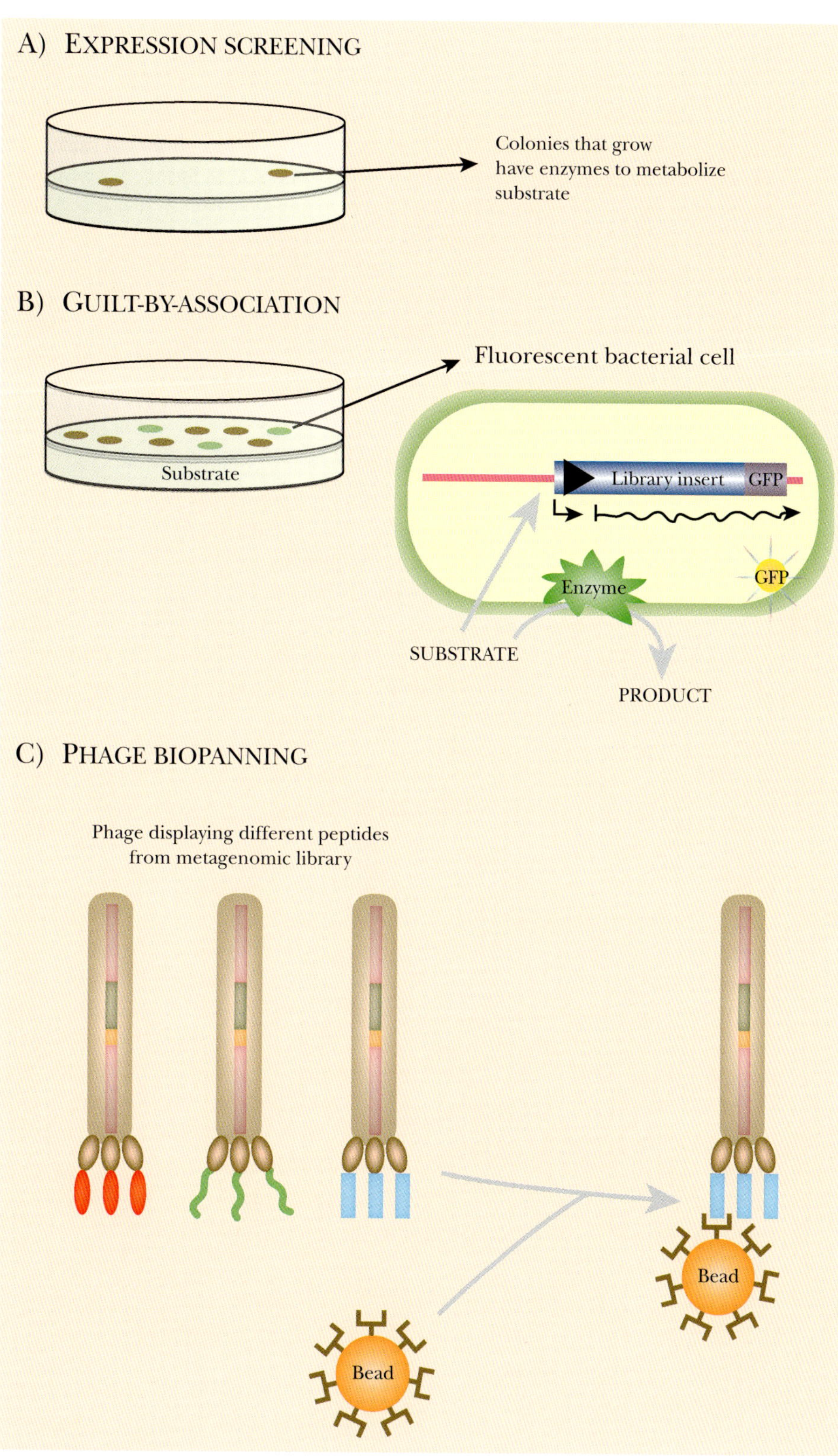

FIGURE 12.6 Function-Based Approaches for Metagenomics

(A) Expression screening identifies those library clones that can grow by metabolizing a particular substrate. In order to suppress the growth of the other library clones, the substrate must be the only source of carbon, nitrogen, or sulfur in a minimal medium. (B) Guilt-by-association identifies expressed genes that may metabolize a particular substrate. The premise is that if the library insert is responsible for metabolizing the substrate, the natural promoter for the genes responsible is also present. If this promoter is active, it can express a reporter gene such as *gfp*, encoding green fluorescent protein. Positive cells can be isolated from the others by FACS sorting (see Chapter 6). (C) Phage biopanning identifies protein binding partners. Peptides from a metagenomic library are fused to a coat protein from a bacteriophage and expressed on its extracellular surface. When the phage is mixed with the binding partner of interest, immobilized on a bead, those metagenomic peptides that bind to the protein of interest will be attached to the bead. These are easily isolated from the rest of the phage clones.

Screening the metagenomic expression library for a particular function such as growth on a pollutant can identify novel genes that metabolize that pollutant.

In expression screening, the metagenomic library is an expression library, that is, the vector contains the transcription and translation initiation and terminator sequences. When the library clone is grown with a single substrate, only those clones that can metabolize the substrate actually multiply.

Guilt-by-association relies on the library insert to carry the natural promoters that are regulated by the substrate. When the substrate activates its natural promoter, the library vector sequences contain a reporter gene that is expressed.

Metagenomic phage biopanning can alleviate problems associated with expressing so many diverse genes from a metagenomic library. This method identifies new binding partners for a particular protein or substrate.

ECOLOGY AND METAGENOMICS

As noted in the introduction, metagenomics research analyzes bacteria, viruses, and even simple gene creatures found within an environmental sample. The results obtained from metagenomic research have major potential for many different applications, including the study of ecology. Metagenomics techniques have been used to identify the entire genome sequence of symbiotic organisms. For example, *Buchnera* are symbiotic bacteria that live within aphids. These bacteria produce amino acids essential to the aphid, and in return, the aphids provide carbon and energy sources to the bacteria. The relationship is so intertwined that neither organism can live without the other. The bacteria have lost so many of their original functions that they are almost organelles. Because there is no possible way to culture the bacteria outside the aphid, a traditional genomic library cannot be established. Instead, a metagenomic library containing both aphid and *Buchnera* DNA was constructed and sequenced. Only when both genomes were examined was the true level of dependence deciphered.

The same scenario was used to sequence the entire genome of the bacteria that coexist within deep-sea tube worms. Tube worms live near thermal vents that are rich in sulfide and reach temperatures of 400 °C. The worms lack mouths and digestive tracts and rely completely on symbiotic members of the Proteobacteria to provide nutrition. The bacteria live within a specialized structure called a trophosome where they oxidize hydrogen sulfide to make energy. The energy is used to manufacture amino acids that feed the worm. In return, the worm collects hydrogen sulfide, oxygen, and carbon dioxide and transports these to the bacteria. The metagenomic library contained both worm and bacterial genomes, but yielded information about the bacteria previously unknown. For instance, the bacterial genome had genes for flagella, suggesting that the bacteria may also have a motile phase. Indeed, other observations suggest that the bacteria move through the seawater to colonize juvenile worms.

Metagenomics can also help in understanding microbial competition and communication. This research may have far-reaching applications to all environments, whether they are within the digestive tract of humans or in the deep-sea vents in the oceans. Functional metagenomics can identify small molecules important to microbial survival, such as antibiotics. Metagenomic libraries can be assessed for antimicrobial activity using functional assays to identify new antibiotics. Additionally, sequence-based analysis of metagenomic libraries can identify synthases that make novel polyketides (antibiotics related to erythromycin and rifamycin; see Chapter 13 for polyketide synthesis). Other functional metagenomic screens have been used to identify quorum-sensing molecules. These are indicators of bacterial population density. Because many bacteria only infect eukaryotic cells or make toxins when they are present in sufficient numbers, interference with quorum sensing provides a new approach to antibacterial therapy. Thus this area is of direct clinical importance. New quorum-sensing molecules were identified by coexpression of metagenomic clones with the reporter GFP. When clones express a quorum-sensing molecule, this activates the expression of GFP, making the bacteria fluorescent. The quorum-sensing metagenomic clone can then be isolated with FACS or microscopy, and then sequenced to determine the identity of the genes involved.

Identifying new genes using metagenomics has much promise for understanding the world around us and potentially for solving some of our health problems.

NATURAL ATTENUATION OF POLLUTANTS

Metagenomics can also be applied to biogeochemical research. Perhaps identifying how bacteria affect the environment will help us figure out ways to maintain our species. Understanding how bacteria can live in extreme environments can reveal useful biochemical processes for biotechnology. Most important, finding out how bacteria cope with contaminated sites may provide useful enzymes for cleaning up our pollution.

Bioremediation is one avenue in which biotechnology has made rapid advances. Many different humanmade compounds have contaminated the environment around us, through everyday use, accidental spillage, or intentional dumping. Many environmental biotechnologists are working on "biological" means of cleaning the environment. In fact, releasing an organism that can degrade a pollutant would provide a very easy, low-cost way of cleaning up a polluted site.

Naturally occurring microorganisms often have the ability to degrade humanmade pollutants. For example, *Rhodococcus* has a highly diverse repertoire of pathways to degrade pollutants, such as short- and long-chain alkanes, aromatic molecules (both halogenated and nitro-substituted), and heterocyclic and polycyclic aromatic compounds, including quinolone, pyridine, thiocarbamate, *s*-triazine herbicides, 2-mercaptobenzothiazole (a rubber vulcanization accelerator), benzothiophene, dibenzothiphene, MTBE (see later discussion), and the related ethyl *tert*-butyl ether (ETBE). *Rhodococcus* has several features that contribute to its ability to degrade so many compounds. First, it has a range of different enzymes that degrade toxic compounds, including cytochrome P450 enzymes. These are very efficient and versatile in oxidation pathways and catalyze a variety of reactions, including epoxidation (Fig. 12.7). Other enzymes that catalyze key degradation steps include monooxygenases and dioxygenases, which help degrade aromatic compounds (see Chapter 13, Pathway Engineering).

Furthermore, several strains of *Rhodococcus* can survive in solvents such as ethanol, butanol, dodecane, and toluene, which would kill many other bacteria. The oil-degrading strains actually adhere to oil droplets! *Rhodococcus* species are found in all types of environments, including nuclear waste sediments, tropical soil, Arctic soil, and sites in Europe, Japan, and the United States. Genetically speaking, *Rhodococcus* also has unique attributes that are advantageous in biodegradation. The genome of *Rhodococcus* sp. strain RHA1 has 9.7 Mb of DNA, including one chromosome and three large linear plasmids. The plasmids may be critical because they are important for gene transfer and recombination events. The genes for the catabolic enzymes are often found in clusters, flanked by inverted repeats, suggesting that they are acquired and passed from one strain to another by recombination. Such horizontal gene transfer can also transfer these catabolic regions to other bacteria, including *Pseudomonas* and *Mycobacterium*. Chapter 13, Pathway Engineering, describes some plasmids that encode pollutant-degrading enzymes.

Different pollutants are degraded in various ways. Sometimes a single naturally occurring organism can completely degrade a pollutant. Other pollutants require more than one

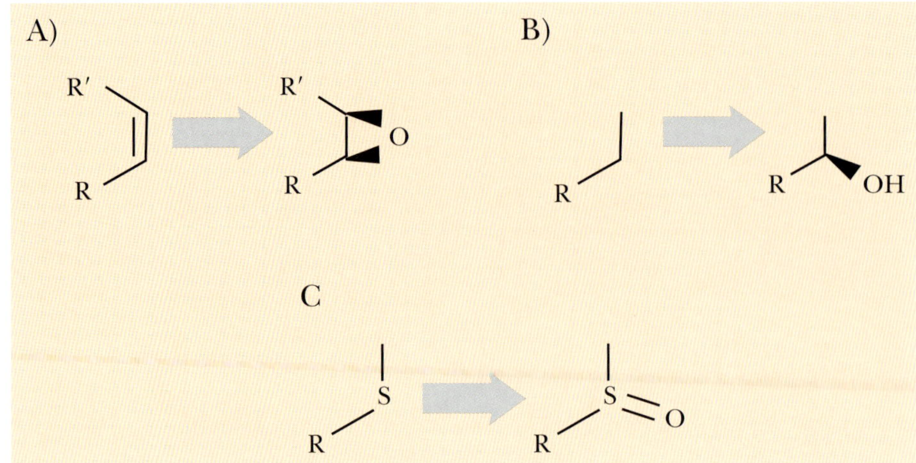

FIGURE 12.7
Oxygenase Reactions Catalyzed by *Rhodococcus*
Rhodococcus can catalyze various reactions, including epoxidation by cytochrome P450 enzymes (A), hydroxylation of alkyl groups by alkane monooxygenases (B), and sulfoxidation of sulfide to sulfoxide (C).

type of bacterium to achieve complete degradation. Some pollutants are degraded very slowly. Heavy metals cannot be chemically degraded, so they pose more of a challenge than organic molecules. Most environmental biotechnologists look for microorganisms that sequester the heavy metal in a solid phase. The conversion of such heavy metals as uranium from an aqueous phase to a solid phase can clean drinking water supplies. Thus, certain anaerobic microorganisms can reduce uranium(VI) to uranium(IV) by utilizing the metal as a terminal electron acceptor. This converts the uranium from a soluble to an insoluble form. In one uranium-contaminated site (Old Rifle, Colorado, USA), one study actually injected acetate into the ground water. Acetate is an electron donor that stimulates metal-reducing bacteria to sequester uranium into the solid phase. Within 50 days, some contaminated wells had uranium concentrations lower than the regulated level. Although these results are promising, over time the acetate dissipated and the soluble uranium levels increased again. More research is necessary to find a permanent means to keep uranium out of ground water.

Another recalcitrant compound that can contaminate ground water, **methyl tert-butyl ether (MTBE)**, oxygenates gasoline so that it burns more efficiently. Many cases of MTBE contaminating groundwater have been reported. Finding a natural method to clean these sites has great applicability. The United States Environmental Protection Agency has classified MTBE as a possible human carcinogen, and drinking water must contain less than 20 to 40 μg/L. One MTBE-contaminated site in South Carolina, United States, had a large plume of MTBE-contaminated gasoline leaking from an underground storage tank at a gas station. The plume ended at a drainage ditch. The concentration of MTBE in the water was low, and in the 2-meter gap between the anaerobic and aerobic zones, the MTBE was metabolized by naturally occurring microorganisms. This led to studies that determined that MTBE could be degraded by bacteria such as *Methylobium petroleophilum* PM1 in areas that transition from anaerobic (anoxic) to aerobic (oxic). In fact, if anaerobic regions of the MTBE plume in South Carolina were injected with a compound that released oxygen, the concentration of MTBE decreased from 20 mg/L to 2 mg/L, suggesting that **biostimulation** may be a good approach to clean up contamination. Biostimulation is the release of nutrients, oxidants, or electron donors into the environment to stimulate naturally occurring microorganisms to degrade a contaminant. In other areas of MTBE contamination, the site may have to undergo **bioaugmentation**, that is, specific microorganisms plus their energy sources may need to be added to the site. Such microorganisms may be naturally occurring, a mixture of different organisms, or even genetically modified (see Chapter 13, Pathway Engineering).

Bioremediation uses biologicals, such as bacteria, to clean up a polluted area.

Different pollutants are degraded in various ways. Sometimes a single naturally occurring organism can completely degrade a pollutant. Other pollutants require more than one type of bacterium to achieve complete degradation.

Biostimulation is the release of nutrients, oxidants, or electron donors into the environment to stimulate naturally occurring microorganisms to degrade a contaminant.

Bioaugmentation is adding specific microorganisms plus their energy sources to decontaminate a polluted area.

Summary

The environment has a huge biodiversity of bacteria, viruses, and gene creatures. The majority of these species are unknown and have never been cultured in a lab. The unique niches in which these organisms live make culturing the organism almost impossible. The use of metagenomics foregoes the culturing of individual organisms, and focuses on just sequencing the DNA from the environment.

Since the diversity is so great in a single sample, ways to enrich the sample for certain functions or organisms is critical. Stable isotope probing is a method where precursor molecules with stable isotopes are added to the environmental sample. All the active bacteria and viruses take in the precursors and incorporate the isotope into their cells. The labeled organisms can then be isolated from the unlabeled ones. Another method, suppressive subtraction hybridization (SSH), takes advantage of the genetic differences between samples from two different areas. This technique removes all the sequences that are identical between the two areas. The leftover sequences represent potential sequences that are unique to one region.

Metagenomic libraries are much like traditional libraries, but contain sequences from many different organisms. These libraries can be screened for new genes using all the traditional methods, such as RT-PCR, shotgun sequencing, probing the library for homologous sequences by hybridization, and microarray analysis. Integron analysis is a method to identify new genes that are flanked by 59-be sequences. These sequences flank integrons, which are transposon-like gene creatures that capture genes from one organism and move them into different organisms.

Metagenomic libraries can also be screened for functions, such as finding new genes that metabolize a specific substrate, that carry out a specific enzyme reaction, or that simply bind to a known protein. Two techniques that identify proteins which metabolize a particular substrate are expression screening and guilt-by-association. For expression screening, the metagenomic library is made with an expression vector, which contains the transcription and translation initiation and terminator sequences. In contrast, guilt-by-association relies on the library insert to carry the natural promoters that are regulated by the substrate. When the substrate activates its natural promoter, the library vector sequences contain a reporter gene that is expressed. Finally, new proteins from a metagenomic library can be identified by phage biopanning, where a known protein is used to find novel binding partner.

Ecology examines the relationship between organisms and their environment, and the application of metagenomics to these studies has changed the field of research. Metagenomic sequencing projects such as the one in the Sargasso Sea have identified many more organisms than previously thought existed. Symbiotic relationships can be studied in detail, because the genome of the symbiont can be isolated and identified. Extreme environments are hard to duplicate in the laboratory; therefore, organisms from these areas were impossible to study. Now, the genomic sequences from these organisms can be used to elucidate their metabolisms and special characteristics, which will, it is hoped, find new ways to combat human diseases.

Finally, the chapter discusses biological methods of removing pollutants, called *bioremediation*. Many naturally occurring bacteria, such as *Rhodococcus*, are able to metabolize a variety of different pollutants. Biostimulation increases the growth and metabolism of the naturally occurring bacteria to remove the pollutant by giving them the necessary food or precursor substrate. In other cases, bioaugmentation adds both the specific microorganism and their energy source to the contamination site. These techniques have been used to clean MTBE from spill sites.

End-of-Chapter Questions

1. According to the book, about what percentage of microorganisms have been cultured?
   - **a.** <0.1%
   - **b.** 0.1%–1%
   - **c.** 1%–10%
   - **d.** 10%–100%
   - **e.** unknown percentage
2. To date, what has metagenomics research been able to identify?
   - **a.** turbomycin A and B
   - **b.** enzymes that can reduce oil contaminants in the soil
   - **c.** bacteria that live in environments contaminated with radioactivity
   - **d.** novel biochemical pathways
   - **e.** all of the above
3. Why must samples be enriched?
   - **a.** The diversity in one sample is too great and enriching selects for a specific organism or function.
   - **b.** Enriching the sample allows all of the organisms to grow in culture.
   - **c.** Enriching the sample prevents organisms from releasing toxins.
   - **d.** The samples do not need to be enriched to identify the organisms growing.
   - **e.** none of the above
4. Which method is better suited for the identification of actively growing bacteria or viruses?
   - **a.** BrdU-enrichment
   - **b.** suppressive subtraction hybridization
   - **c.** stable isotope probing
   - **d.** RT-PCR
   - **e.** DNA sequencing
5. Why is small subunit ribosomal RNA (SSU rRNA) an excellent biomarker for RNA-SIP?
   - **a.** abundant within the cell
   - **b.** essential to all life
   - **c.** database exists containing SSU rRNA sequences
   - **d.** variable among different species
   - **e.** all of the above

6. What does suppressive subtractive hybridization do?
   a. hybridizes sample mRNA to known mRNA sequences
   b. identifies the sequences from two areas that are identical, and then uses the leftover sequences to identify unique sequences to one area
   c. suppresses the unknown sequences by binding to all of the known sequences
   d. identifies the sequences of one sample that are already known
   e. none of the above
7. Which of the following is not a potential downfall regarding metagenomic libraries?
   a. contaminated DNA or mRNA samples
   b. DNA that is difficult to isolate and therefore underrepresented in the sample and library
   c. sheared DNA
   d. lack of a cultured organism
   e. none of the above
8. Why are integrons important?
   a. Integrons can spread genes for antibiotic resistance or for a growth advantage under certain environmental conditions.
   b. Integrons encode the industrially relevant enzyme called *integrase.*
   c. Integrons are important for overexpressing certain proteins.
   d. Integrons only contain noncoding DNA.
   e. none of the above
9. What genes have been identified using integron analysis?
   a. phosphotransferases
   b. DNA glycosylases
   c. new antibiotic resistance genes
   d. methyltransferases
   e. all of the above
10. Which of the following was not identified by Venter from the Sargasso Sea?
    a. 1.2 million new genes
    b. 1800 different species of microbes
    c. 700 different bacteriorhodopsin genes
    d. 150 new bacterial/archaea species
    e. all of the above
11. What is a potential problem associated with function-based analyses of genes?
    a. The foreign protein has a toxic effect to the host organism.
    b. Successful expression of the gene of interest in the host organism may fail.
    c. Other cofactors are needed for proper activity that the host organism cannot provide.
    d. The number of clones to screen for the gene of interest is usually too many.
    e. All of the above are problems associated with this type of analysis.

(*Continued*)

12. What is the general term used to describe the degradation of pollutants using a biological approach?
    a. biostimulation
    b. bioremediation
    c. biodegradation
    d. bioprocessing
    e. bioaugmentation

13. Which genera of microorganisms have the most diverse pathways for bioremediation?
    a. *Mycobacterium*
    b. *Pseudomonas*
    c. *Rhodococcus*
    d. *Escherichia*
    e. *Methylobium*

14. What is biostimulation?
    a. the release of nutrients, oxidants, or electron donors into the environment to stimulate microorganisms to degrade a pollutant
    b. the addition of bacteria to a specific contaminated site for bioremediation
    c. the research term used to describe how bacteria can sequester certain heavy metals from a contaminated site
    d. the act of decreasing the amount of microorganisms in a contaminated site
    e. none of the above

15. Which pollutant has been cleaned from contaminated sites using either biostimulation or bioaugmentation?
    a. MTBE
    b. acetate
    c. aromatics
    d. ETBE
    e. alkanes

Further Reading

Bamford DH, Grimes JM, Stuart DI (2005). What does structure tell us about virus evolution? *Curr Opin Struct Biol* **15**, 655–663.

Clark DP (2005). *Molecular Biology: Understanding the Genetic Revolution*. Elsevier Academic Press, San Diego, CA.

Cowan D, Meyer Q, Stafford W, Muyanga S, Cameron R, Wittwer P (2005). Metagenomics gene discovery: Past, present and future. *Trends Biotechnol* **23**, 321–329.

Daniel R (2004). The soil metagenome—a rich resource for the discovery of novel natural products. *Curr Opin Biotechnol* **15**, 199–204.

Fu X, Huang Y, Deng S, Zhou R, Yang G, Ni X, Li W, Shi S (2005). Construction of a SSH library of *Aegiceras corniculatum* under salt stress and expression analysis of four transcripts. *Plant Sci* **169**, 147–154.

Galperin MY, Baker AJM (2004). Environmental biotechnology. From biofouling to bioremediation: The good, the bad and the vague. *Curr Opin Biotechnol* **15**, 167–169.

Galvão TC, Mohn WW, de Lorenzo V (2005). Exploring the microbial biodegradation and biotransformation gene pool. *Trends Biotechnol* **23**, 497–506.

Gillespie DE, Brady SF, Bettermann AD, Cianciotto NP, Liles MR, Rondon MR, Clardy J, Goodman RM, Handelsman J (2002). Isolation of antibiotics turbomycin A and B from a metagenomic library of soil microbial DNA. *Appl Environ Microbiol* **68**, 4301–4306.

Handelsman J (2004). Metagenomics: Application of genomics to uncultured microorganisms. *Microbiol Mol Biol Rev* **68**, 669–685.

Kröger S, Law RJ (2005). Sensing the sea. *Trends Biotechnol* **23**, 250–256.

Larkin MJ, Kulakov LA, Allen CCR (2005). Biodegradation and *Rhodococcus*—masters of catabolic versatility. *Curr Opin Biotechnol* **16**, 282–290.

Lloyd JR, Renshaw JC (2005). Bioremediation of radioactive waste: Radionuclide-microbe interactions in laboratory and field-scale studies. *Curr Opin Biotechnol* **16**, 254–260.

McDonald IR, Radajewski S, Murrell JC (2005). Stable isotope probing of nucleic acids in methanotrophs and methylotrophs: A review. *Org Geochem* **36**, 779–787.

Parales RE, Haddock JD (2004). Biocatalytic degradation of pollutants. *Curr Opin Biotechnol* **15**, 374–379.

Rowe-Magnus DA, Mazel D (1999). Resistance gene capture. *Curr Opin Microbiol* **2**, 483–488.

Schloss PD, Handelsman J (2003). Biotechnological prospects from metagenomics. *Curr Opin Biotechnol* **14**, 303–310.

Scow KM, Hicks KA (2005). Natural attenuation and enhanced bioremediation of organic contaminants in groundwater. *Curr Opin Biotechnol* **16**, 246–253.

Streit WR, Daniel R, Jaeger K-E (2004). Prospecting for biocatalysts and drugs in the genomes of non-cultured microorganisms. *Curr Opin Biotechnol* **15**, 285–290.

van Beilen JB, Funhoff EG (2005). Expanding the alkane oxygenase toolbox: New enzymes and applications. *Curr Opin Biotechnol* **16**, 308–314.

Venter JC, Remington K, Heidelberg JF, Halpern AL, Rusch D, Eisen JA, Wu D, Paulsen I, Nelson KE, Nelson W, Fouts DE, Levy S, Knap AH, Lomas MW, Nealson K, White O, Peterson J, Hoffman J, Parsons R, Baden-Tillson H, Pfannkoch C, Rogers Y-H, Smith HO (2004). Environmental genome shotgun sequencing of the Sargasso Sea. *Science* **304**, 66–74.

Whiteley AS, Manefield M, Lueders T (2006). Unlocking the "microbial black box" using RNA-based stable isotope probing technologies. *Curr Opin Biotechnol* **17**, 67–71.

Wiatrowski HA, Barkay T (2005). Monitoring of microbial metal transformations in the environment. *Curr Opin Biotechnol* **16**, 261–268.

Zylstra G, Kukor JJ (2005). What is environmental biotechnology? *Curr Opin Biotechnol* **16**, 243–245.

CHAPTER 13

Pathway Engineering

INTRODUCTION

From the genetic viewpoint, the production of a small molecule such as ethanol may well be more complex than production of a protein such as somatotropin. Although proteins are macromolecules, single genes encode them, whereas small molecules must be made by biochemical pathways that require several steps, each catalyzed by a separate enzyme. Thus multiple genes are involved, together with their regulatory systems. **Pathway engineering** involves the assembly of a new or improved biochemical pathway, using genes from one or more organisms. Most efforts to date have been directed to modifications and improvements of existing pathways, rather than the assembly of completely new synthetic schemes. However, totally novel pathways will no doubt begin to appear over the next few years.

Pathway engineering may be applied both to degradative pathways and to biosynthesis. Engineered bacteria may be used to degrade agricultural waste, pollutants, including industrial chemicals, as well as excess herbicides, weed killers, and so forth, in a process called **bioremediation**. In addition, microorganisms are used to produce a variety of products including alcohol, solvents, food additives, dyes, and antibiotics. The most efficient pathways are those that convert otherwise useless material into useful products. We will start by considering one such scheme, alcohol fermentation. This process was developed long before modern science and is probably humankind's earliest venture into biotechnology.

Genetic engineering can be used to assemble novel or more efficient metabolic pathways. Both degradative and biosynthetic pathways may be engineered.

ETHANOL, ELEPHANTS, AND PATHWAY ENGINEERING

Humans weren't the first to appreciate alcohol. Elephants, monkeys, and other wild animals deliberately consume fruit that has naturally fermented, yielding alcohol. Indeed, elephants in both Africa and Asia may run amok after consuming fermented fruit. Occasionally, elephants will even raid local villages and knock over houses to "recover" fermented liquids from their human competitors! There is even a species of the fruit fly *Drosophila*, from the sherry-producing regions of Spain, that relies on sherry as its sole source of nutrition. These insects spend their lives circling around in the caves where sherry is processed and presumably do not need to fly straight. The earliest cultural remains from human alcohol consumption date to about 5000 B.C. Analysis of a yellowish residue found in Neolithic pottery from Iran showed that it was derived from wine.

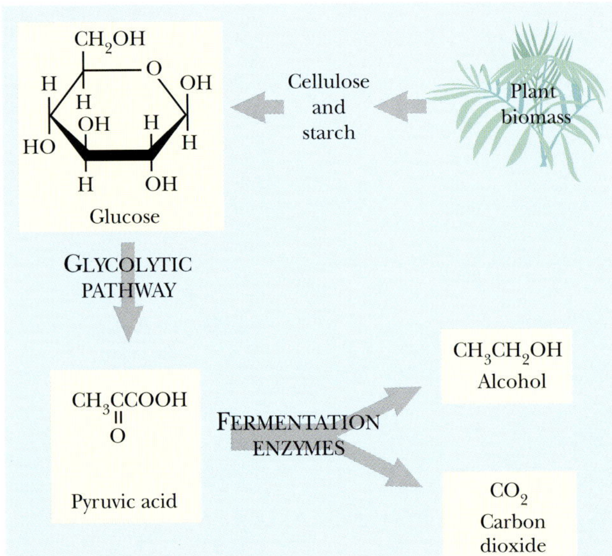

FIGURE 13.1 From Sugars to Ethanol Plant material contains polysaccharides such as starch and cellulose. Enzymes degrade the starch and release the glucose molecules. The glycolytic pathway then converts glucose to pyruvic acid, which is fermented into alcohol and carbon dioxide.

Alcohol is made from sugar (Fig. 13.1). Sugars are components of the carbohydrates making up much of the bulk of plant matter. So, in principle, alcohol can be made from almost any plant-derived material. Yeast is used to ferment sugars derived from grain or grapes, which produces an alcoholic liquid—the basis of beer or wine, respectively. Distillation is then used to make concentrated liquors such as whiskey or vodka. The lone exception to yeast is the use of a bacterium, *Zymomonas*, which ferments sugar from the sap of the agave plant to give a liquid known as pulque. Distillation converts this into tequila.

There is little need for genetic engineering in the area of alcoholic drinks. However, alcohol may be blended with gasoline to give "gasohol," which works well in most internal combustion engines. Thus conversion of waste biomass to fuel alcohol would not only get rid of large amounts of waste material but would also reduce gasoline consumption. If the United States converted the 100 million tons of waste paper it generates each year into fuel-grade alcohol, this could replace 15% of the gasoline used. Ethanol can also be made from corn, which is very economical because many acres of corn are grown each year in the United States and a large surplus is generated. Unlike wood pulp for paper, corn can be regenerated in 1 year.

The advantage of using *Zymomonas* and yeast is that they make only alcohol during fermentation, whereas most microorganisms generate mixtures of fermentation products. For example, *Escherichia coli* makes a mixture of ethanol, acetate, succinate, lactate, and formate. Although many fermentation products are potentially useful, purification is an expensive drawback. The problem with *Zymomonas* is that it lives entirely on glucose and lacks the enzymes to break down other sugars, let alone those needed to degrade carbohydrate polymers such as starch and cellulose. Yeast is almost as narrow in its growth requirements. *Zymomonas* grows faster than yeast and makes alcohol faster as well. On the other hand, yeasts are more alcohol resistant and are therefore capable of accumulating higher concentrations of ethanol in the medium before growth is halted.

Genetic engineering is being used to make improved strains of both yeast and *Zymomonas* that can use a wider range of sugars. In addition, genes for enzymes capable of breaking down starch, cellulose, or other plant polysaccharides can be inserted (see later discussion). Finally, these organisms can also be engineered for improved resistance to alcohol or for other properties that optimize growth and production under industrial conditions.

Xylose is a five-carbon sugar that is a major component of various polysaccharides (xylans) found in plant cell walls (see later discussion). Vast amounts of waste material from plants are available for possible biodegradation. Breakdown of the polysaccharide polymers would release large amounts of xylose. Consequently, it is worthwhile to develop strains of *Zymomonas* that efficiently ferment xylose to ethanol. This has been done in two stages. First the genes for metabolism of xylose itself must be introduced, because *Zymomonas* does not naturally use this sugar. The *xylA* and *xylB* genes encode the enzymes xylose isomerase and xylulose kinase, respectively, which convert xylose to xylulose and then to xylulose 5-phosphate. These two genes were placed on a shuttle vector that carries replication origins for both *E. coli*, in which the genetic engineering was done, and *Zymomonas* (Fig. 13.2).

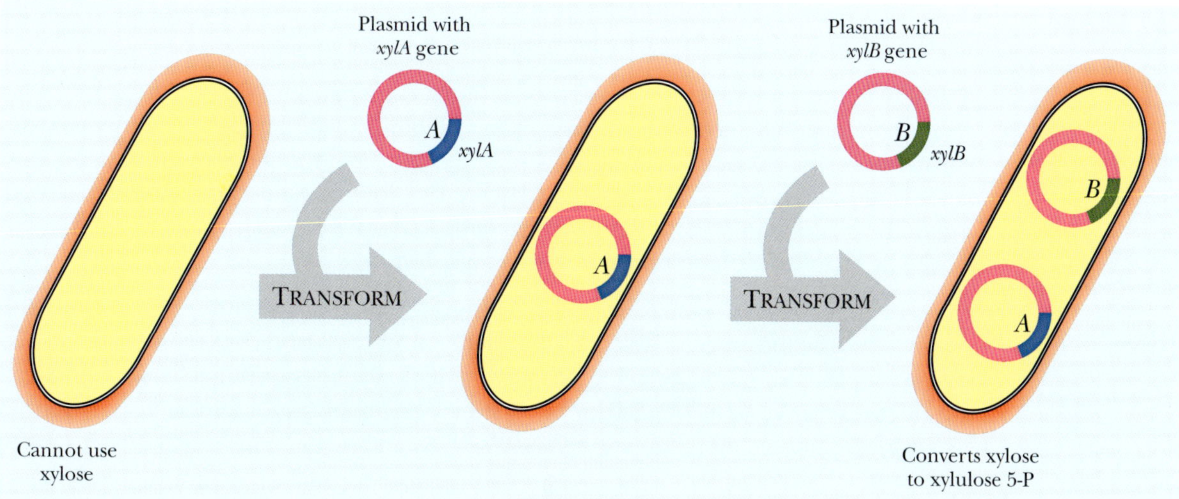

FIGURE 13.2 Pathway Engineering for Xylose Utilization—Genes
Xylose must be degraded by a specific set of reactions before its conversion to alcohol. Two genes are necessary for the initial xylose degradation, *xylA* and *xylB*. The XylA protein converts xylose to xylulose, and XylB phosphorylates this to form xylulose 5-phosphate. The two genes are carried on shuttle plasmids and transformed into bacteria such as *Zymomonas*.

The strain with just the extra *xylAB* genes grew poorly because it accumulated xylulose 5-phosphate as well as the phosphates of other pentose sugars, including ribose 5-phosphate. The genes for transketolase (*tktA*) and transaldolase (*tal*), two enzymes that convert pentose phosphates back into hexose phosphates, were then included on the plasmid, under control of a separate promoter. The resulting *Zymomonas* was then able to convert xylose first to xylulose 5-phosphate and then to fructose 6-phosphate and glyceraldehyde 3-phosphate. Finally, these central intermediates were fermented efficiently to ethanol (Fig. 13.3).

Conversion of sugars to alcohol is one of the oldest industrial processes. Pathway engineering may help convert waste plant-derived material, including paper, into fuel alcohol.

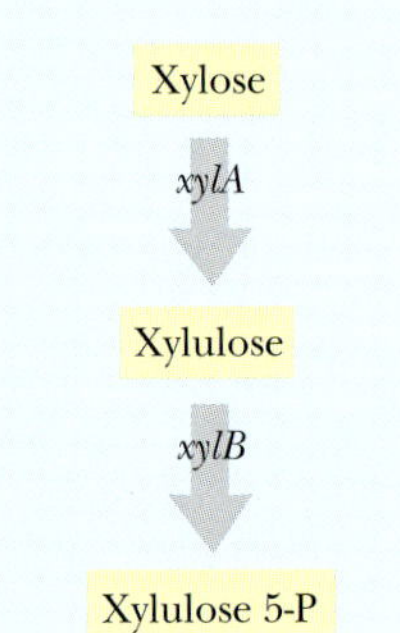

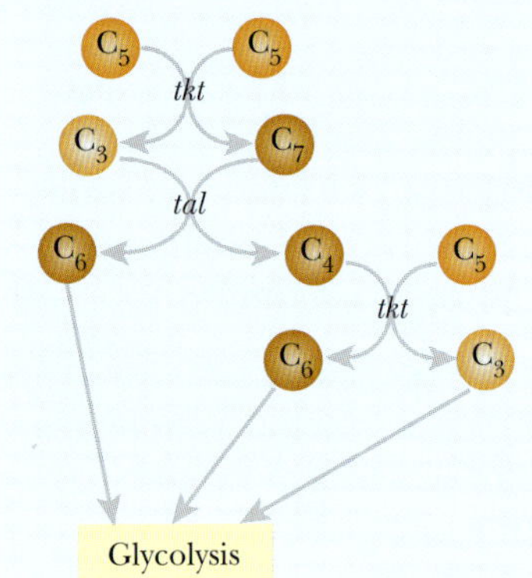

FIGURE 13.3
Pathway Engineering for Xylose Utilization—Reactions
(A) Xylose is converted into xylulose 5-phosphate using the *xylA* and *xylB* genes. (B) Transketolase (*tktA*) converts two five-carbon (C_5) sugar molecules into one three-carbon (C_3) and one seven-carbon sugar (C_7). Next, transaldolase (*tal*) converts these products into a four-carbon sugar and fructose 6-P, a six-carbon sugar. Fructose 6-P is degraded by glycolysis into ethanol. The four-carbon sugar (C_4) and another pentose 5-P (C_5) are converted by transketolase into a second six-carbon sugar (C_6) and a three-carbon sugar (C_3), which both feed into the glycolytic pathway to make ethanol.

DEGRADATION OF STARCH

Starch is a storage polysaccharide found in many plants. It is a polymer of glucose linked by α-1,4 bonds. Starch actually consists of a mixture of linear polymers—**amylose**—and branched polymers—**amylopectin**. The branches of amylopectin are due to α-1,6 bonds, and they occur approximately every 20 glucose residues along the polymer chain. Chain lengths vary from 100 to 500,000 glucose residues for amylose and up to 40 million for amylopectin. The proportions of linear and branched polymers also vary depending on the source of the starch. **Glycogen** is a storage polysaccharide of animals. It is essentially starch with a high proportion of branched polymers.

Starch is used in the food and brewing industries and is mostly converted to glucose by the use of the purified enzymes α-amylase and glucoamylase, rather than by microorganisms (Fig. 13.4). The α-amylase cleaves the linear regions of the starch chains at random, whereas the glucoamylase is needed for cutting the branches. The glucose may then be converted to fructose by the enzyme glucose isomerase or to alcohol by microbial fermentation. Because of the large size of the food and brewing industries, α-amylase and glucoamylase plus glucose isomerase account for over 25% of the cost of all enzymes used industrially. Many microorganisms make these enzymes. However, industrially, α-amylase is usually obtained from the bacterium *Bacillus amyloliquefaciens* and glucoamylase from the fungus *Aspergillus niger*.

A variety of improvements to the process of starch degradation might be made by genetic engineering. Recombinant organisms could be made that produce more enzyme. Furthermore, the enzymes themselves could be engineered for better thermal stability or higher rates of reaction as discussed in Chapter 11, Protein Engineering. Alternatively, it should be possible to construct microorganisms that carry out all or part of starch hydrolysis, as well as converting the resulting glucose to alcohol.

The gene for glucoamylase has been cloned from *Aspergillus niger* and inserted into a suitable yeast strain. The fungal gene was placed under control of a strong yeast promoter and carried on a yeast plasmid. The engineered yeast was able to degrade solubilized starch and ferment the glucose released to alcohol. Ultimately it may be possible to engineer yeast strains that also express (and secrete) high enough levels of α-amylase to completely convert raw starch to ethanol.

Starch is a widely available raw material. Higher enzyme levels and more stable enzymes would improve the breakdown of starch to sugars.

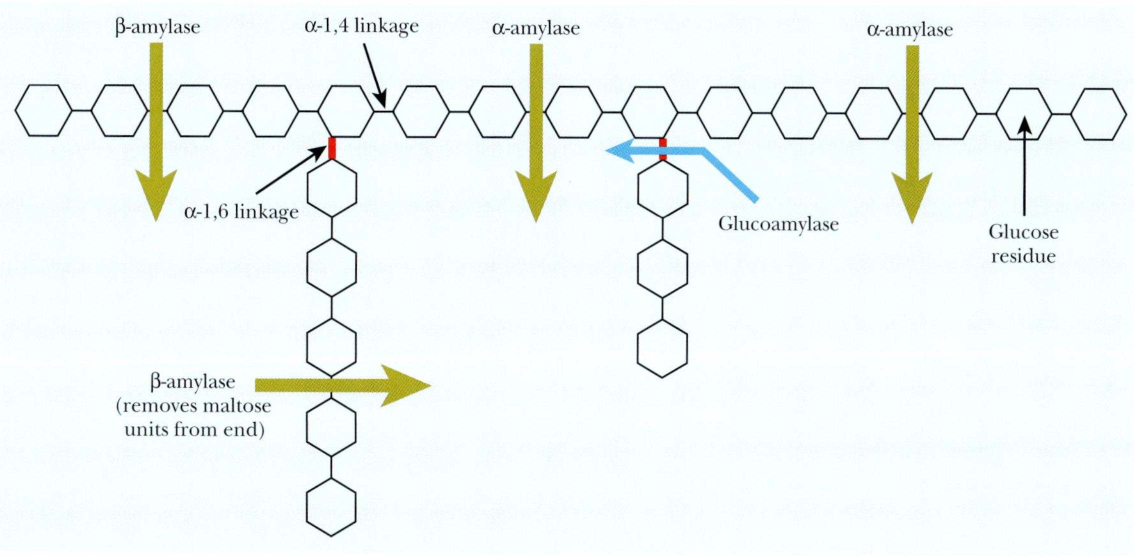

FIGURE 13.4 Enzymatic Breakdown of Starch by Amylases
Starch consists of long chains of glucose residues with other glucose chains branching off the main backbone. The main chain has glucose residues linked by alpha-1,4 linkages, and the side chain starts with an alpha-1,6 linkage. Amylase cuts between the glucose residues. Alpha-amylase cuts the alpha-1,4 linkages in the main chain, whereas beta-amylase cuts units of two glucose residues (maltose) from the ends of the chains. Glucoamylase cuts the alpha-1,6 linkage to remove the side branches from the main glucose chain.

DEGRADATION OF CELLULOSE

Plant cell walls contain a mixture of polysaccharides of high molecular weight. The major components are **cellulose, hemicellulose,** and **lignin**. Cellulose is a structural polymer of glucose residues joined by β-1,4 linkages. This contrasts with starch and glycogen which are storage materials also consisting solely of glucose, but with α-1,4 linkages. Hemicellulose is a mixture of shorter polymers consisting of a variety of sugars, especially mannose, galactose, xylose, and arabinose, in addition to glucose. Lignin differs from these other polymers in two major respects. First, it is not a polysaccharide but consists of aromatic residues (primarily phenylpropane rings). Second, lignin is crosslinked into an insoluble three-dimensional meshwork structure.

Vast amounts of material derived primarily from plant cell walls are available as agricultural waste products. In nature, fungi and soil bacteria degrade this material slowly. Engineering these pathways to enhance the rate of reaction could be very beneficial. Cellulose, with its simple composition and regular structure, is the easiest to degrade and lignin the most difficult. Paper (plus cardboard and related materials) accounts for the largest fraction of the trash of industrial nations. Because paper consists almost entirely of cellulose, this too may potentially be converted to glucose by cellulose-degrading microorganisms.

The polymer chains of cellulose are packed tightly side by side in a crystalline array with loosely packed, noncrystalline zones at intervals (Fig. 13.5). The challenge is to break down cellulose, yielding glucose that can be turned into alcohol or other products. Breakdown of cellulose requires several steps (Fig. 13.5), each catalyzed by a separate enzyme, as follows:

1. Endoglucanase snips open the polymer chains in the middle. This enzyme can only attack the polymer chains in the loosely packed "amorphous" zones.
2. Cellobiohydrolase cuts off sections with 10 or more glucose units from the free ends created by endoglucanase.

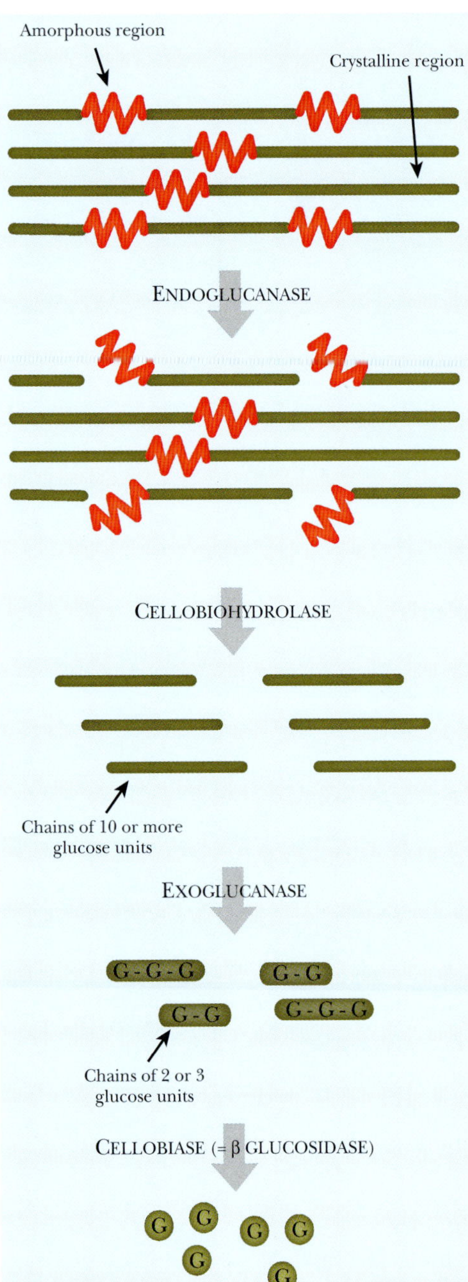

FIGURE 13.5
Cellulose Degradation
Cellulose is broken down into glucose molecules by a series of enzymatic reactions. Cellulose has both amorphous and crystalline regions. Endoglucanase digests within the amorphous areas to begin the degradation. Each of the subsequent enzymes shortens the glucose chains further. See text for details.

3. Exoglucanase chops off units of two or three glucose units from the exposed ends, which are called cellobiose and cellotriose, respectively.
4. β-Glucosidase (also known as cellobiase) converts cellobiose and cellotriose to glucose.

The genes for each of these four enzymes have been cloned from various microorganisms. Because cellulose is too big to enter the cell, the first three enzymes must be secreted and work outside. Cellobiose and cellotriose (respectively the β-1,4-linked dimer and trimer of glucose) are released from cellulose outside the cell and may then be transported inside. They are finally broken into individual glucose molecules, which may be fermented to alcohol. So far, assorted pilot projects have demonstrated the degradation of cellulose from waste paper to glucose by adding separate enzymes purified from different sources. Alternatively, a series of cellulose-degrading microorganisms, each chosen for high levels of one particular enzyme, are used. Finally, yeast or *Zymomonas* is added to convert the glucose to alcohol.

Such multistage procedures are not very efficient because each step has problems with efficiency. Cellobiose acts as a feedback inhibitor of cellulose degradation and, similarly, glucose inhibits hydrolysis of cellobiose. Therefore, it is crucial that the end products of cellulose breakdown should be rapidly removed to allow continuous degradation of the starting materials. In fact, cellobiose degradation appears to be the limiting step in many natural cellulose degraders, and these organisms can often be improved by cloning the gene for cellobiase, placing it under a strong promoter, and putting it back into the organism in question.

Overall, what is desirable is a complete recombinant organism, which possesses genes for all four enzymes, expresses these at high levels, and efficiently converts cellulose to alcohol on its own. In practice, genes for cellulose degradation from both bacteria and fungi are first cloned and expressed in *E. coli* for ease of manipulation. At present, genes for the individual stages have been isolated and characterized. Eventually, a cellulose degradation pathway may be assembled in either yeast or *Zymomonas* in order to convert waste cellulose materials to alcohol.

Cellulose is relatively difficult to metabolize, compared to sugars or starch. Assembling an effective pathway in an easily cultivated organism would be of great value in biodegrading waste paper and related material.

ICE-FORMING BACTERIA AND FROST

Perhaps the simplest "pathway" of all is the conversion of water to ice. This process may be "catalyzed" by proteins known as **ice nucleation factors**. On a microscopic scale, solidifying water forms crystals of ice. However, ice crystals need a microscopic nucleus or "seed" to form around. In the absence of structures allowing nucleation, water will supercool down to −8 °C without solidifying. Thus, ice nucleation factors are specialized proteins, mostly found in certain bacteria, which provide nuclei for crystallization.

Each year, frost causes more than a billion dollars in damage to crops in the United States alone. It is not the low temperature itself that does the damage. When water freezes to form ice, it expands, damaging plant tissues. The seeding of ice crystals on and within plants is mostly due to proteins on the surface of bacteria, especially *Pseudomonas syringae* and related species, which live on plants. The ice crystals that form damage the plant tissues and disrupt the vessels (xylem and phloem) that carry water and nutrients throughout the plant. If ice-nucleating bacteria are absent, ice fails to form and instead the water supercools, leaving the plants unharmed.

The best known ice-nucleation protein is encoded by the ***inaZ* gene** of *Pseudomonas syringae*. Like most bacteria, *E. coli* does not normally promote ice formation, although if it expresses a cloned *inaZ* gene it will gain ice-nucleating ability. Conversely, when the *inaZ* gene of *Pseudomonas syringae* is disrupted, ice-nucleating ability is lost (Fig. 13.6). The wild, "ice-plus" strains of *Pseudomonas syringae* can be displaced by spraying the "ice-minus" mutants onto crops that are at risk from frost damage. Subsequently, even if the temperature falls below freezing, very few ice crystals form and most of the plants are unharmed.

Bacteria that express ice-nucleation proteins cause major damage to crop plants. Disruption of the corresponding gene abolishes ice nucleation.

DEGRADATION OF AROMATIC RING COMPOUNDS

Many bacteria contain plasmids allowing the metabolism of materials they would otherwise be unable to use. Of major interest are a variety of pathways allowing the degradation of both linear and cyclic hydrocarbons and related compounds. Many industrial chemicals that are derived from petroleum, including various pesticides and herbicides, contain aromatic rings. These chemicals, together with oil residues or spills, are often responsible for significant pollution. Consequently, their biodegradation by bacteria is of considerable practical importance. Chemical compounds that possess significant biological activity, but are foreign to the environment and were introduced artificially by human industrial activity, are sometimes referred to as **xenobiotics**.

Many species of *Pseudomonas* and related bacteria contain plasmids conferring the ability to grow using aromatic hydrocarbons as an energy source. *Pseudomonas* is widely distributed in soil, water, etc., and is an obligate aerobe. Different pseudomonads can use a wide range of substrates including alkanes, mono- and polycyclic aromatics, heterocyclics, phenols, terpenes,

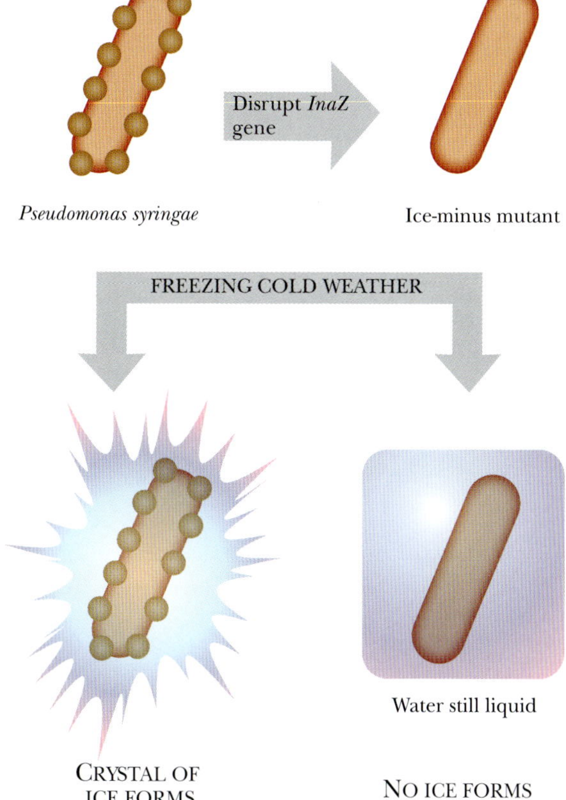

FIGURE 13.6
Disruption of *inaZ* Prevents Ice Nucleation
Cell surface proteins of *P. syringae* provide a nucleation point for ice. The *inaZ* gene encodes an ice-nucleating protein. Under freezing temperatures, wild-type *P. syringae* allow ice crystals to form, disrupting any plant tissues the bacteria are on or within. If the *inaZ* gene is disrupted, the *P. syringae* mutant will not nucleate any ice crystals, allowing the water to supercool.

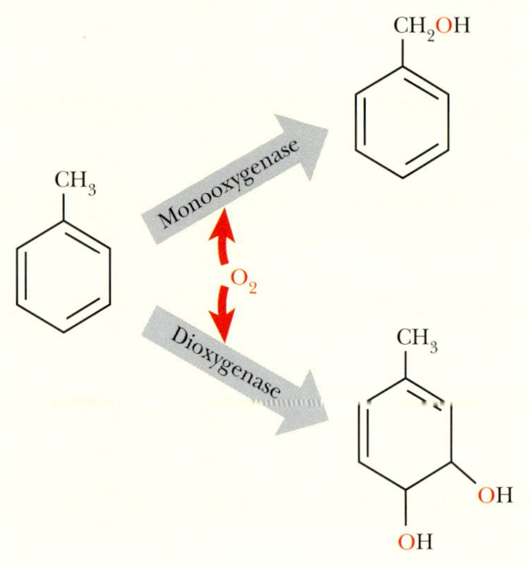

halogenated compounds, and so forth. Many of the degradation systems are of wide specificity and can handle many related substrates. The earliest pathways discovered were those for toluene/xylene and naphthalene. The plasmids were originally named after their substrates (e.g., the **TOL plasmid**, now known as pTOL, degrades toluene) and are mostly large and capable of self-transfer between different related bacterial strains.

Most of these pathways involve direct insertion of oxygen by **oxygenases** (Fig. 13.7) as a key step. Monooxygenases insert a single oxygen atom to give one -OH group, whereas **dioxygenases** insert two oxygen atoms and yield diols. When attacking aromatic compounds, the monooxygenases generally attack the side chains, whereas the dioxygenases attack the aromatic rings themselves.

FIGURE 13.7
Oxygen Insertion into Hydrocarbons
Monooxygenase and dioxygenase take molecular oxygen and insert one or two oxygen atoms, respectively, into aromatic compounds, thus creating hydroxyl groups. Toluene is shown as substrate in this example.

Many industrial chemicals as well as crude oil contain aromatic rings. Degradation of such rings is initiated by oxygenases, enzymes that insert molecular oxygen.

INDIGO AND RELATED NATURAL PIGMENTS

Enzymes that attack aromatic rings often have a wide substrate range. Often they may work on related compounds whose rings contain sulfur, oxygen, or nitrogen atoms, as well as on hydrocarbons. The **indole** ring system is similar to naphthalene except that indole contains a nitrogen in the ring. Consequently, naphthalene oxygenases also attack indole and its derivatives (Fig. 13.8). They convert indole to its diol, which oxidizes spontaneously in air to yield **indigo**, a bright blue pigment. In practice, most aromatic ring dioxygenases attack indole at least to some extent. This allows rapid color screening for the presence of ring dioxygenases, and the presumed aromatic pathways to which they belong.

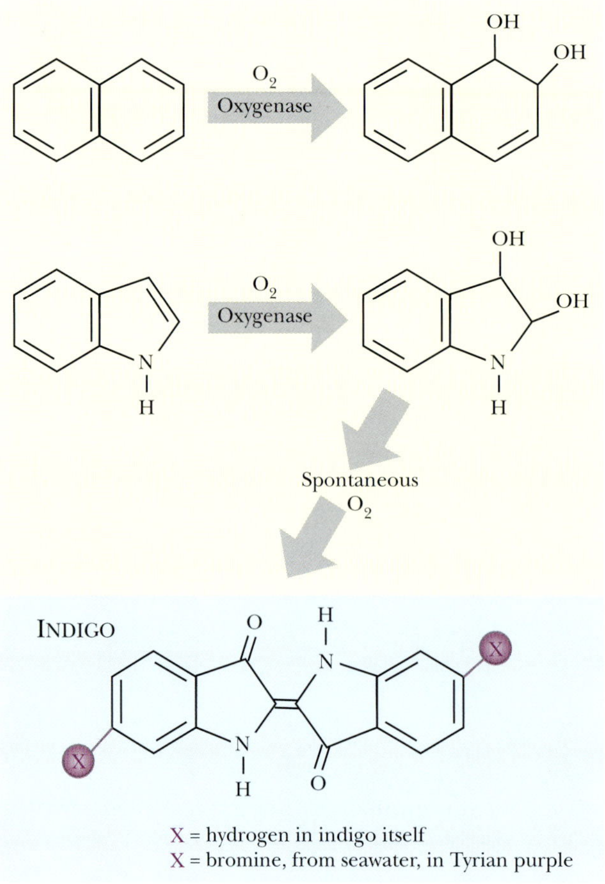

In the Bible, Moses referred to a special blue dye, hyacinthine purple, which the Israelites were to wear. The dye was (probably) a 50:50 mixture of Tyrian purple and indigo. The secret of this sacred dye was lost about 1400 years ago when its suppliers, the Phoenicians, were overrun by invading Arabs. Tyrian purple,

FIGURE 13.8
Similar Reactions of Naphthalene and Indole
When naphthalene (*top*) is attacked by an oxygenase, two hydroxyls are added to form its diol. The same oxygenase attacks the indole ring because this is similar to naphthalene. The indoxyl oxidizes spontaneously into indigo, which is blue. The indigo ring system can have alternative groups attached at the positions shown as X. If X is hydrogen, the molecule is indigo itself, but if X is bromine, the molecule becomes more purple than blue and is called Tyrian purple.

a reddish purple dye, is made by a sea snail called the spiny murex. A related snail, the banded murex, secretes a mixture of Tyrian purple plus indigo, that is, presumably the lost hyacinthine purple. Tyrian purple is very closely related to indigo (Fig. 13.8). It has two bromine atoms, extracted from seawater by the sea snail, on the indigo ring. Both dyes are secreted as colorless precursors that turn blue (or purple) by reacting with the oxygen in air. Indigo itself is used for dyeing wool and cotton blue. Blue jeans are made of cotton dyed with indigo.

In earlier times, indigo was extracted from plants, but nowadays it is chemically synthesized. Recently, genetically altered bacteria that can make indigo were discovered, largely by accident. The *nah* genes, carried on the NAH plasmid, encode the enzymes that break down naphthalene. When the *nah* system was originally analyzed, genes from the NAH plasmid were cloned into *E. coli* and some of the bacteria turned blue! These blue bacteria turned out to possess the genes for **naphthalene oxygenase**, the enzyme that carries out the first step in breaking down naphthalene. As discussed earlier, naphthalene oxygenase works very well against indole and converts it into indoxyl. Oxygen in the air converts indoxyl to indigo (see Fig. 13.8). *E. coli* itself provides the indole by degrading the amino acid tryptophan. Thus the engineered *E. coli* must be grown in rich medium containing protein hydrolysate or some other source of tryptophan in order to generate indigo (Fig. 13.9). Commercialization of such an engineered pathway would involve putting the recombinant bacteria with the naphthalene oxygenase gene onto a solid support in a bioreactor. Tryptophan would be added at one end and indigo would emerge from the other.

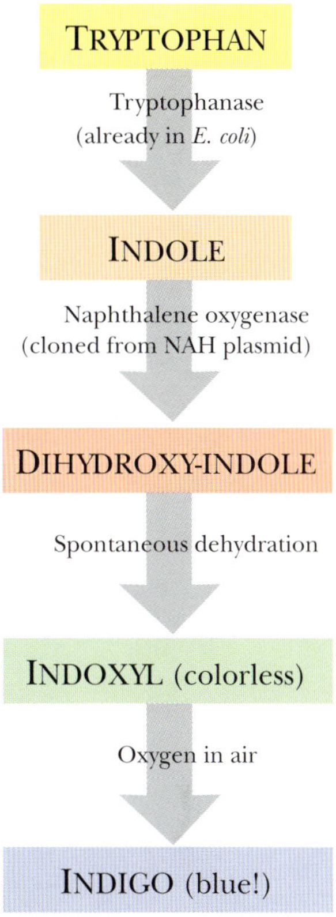

TRYPTOPHAN

Tryptophanase (already in *E. coli*)

INDOLE

Naphthalene oxygenase (cloned from NAH plasmid)

DIHYDROXY-INDOLE

Spontaneous dehydration

INDOXYL (colorless)

Oxygen in air

INDIGO (blue!)

FIGURE 13.9
Engineered Indigo Pathway
E. coli can be engineered to produce indigo by expressing naphthalene oxygenase from the NAH plasmid. First, *E. coli* converts tryptophan into indole using tryptophanase. Next, the indole is converted to dihydroxyindole by the cloned oxygenase, and this spontaneously dehydrates to indoxyl. Once exposed to the air, the indoxyl turns blue, forming indigo.

Indole is oxidized by naphthalene oxygenase. This key reaction allows the construction of a pathway to manufacture the dye indigo from the amino acid tryptophan.

THE TOLUENE/XYLENE PATHWAY

The toluene/xylene pathway is the best characterized of the aromatic degradation systems. It is carried on the pTOL and pXYL plasmids. (Historically, these two types of plasmids were isolated and named for the degradation of toluene or xylene. However, although the plasmids differ in other respects they have essentially identical sets of degradative genes.) The same pathway (Fig. 13.10) degrades toluene (R = H) and xylene (R = CH_3) as well as some derivatives. The "upper" pathway attacks the side chains and the "lower" pathway breaks open the aromatic ring.

Ring dioxygenases such as xylene or toluene oxygenase usually consist of three components. The first component (e.g., the XylX protein of the xylene pathway) transfers reducing equivalents from NADH or NADPH via an enzyme bound FAD to an Fe_2S_2 cluster. The second and third proteins (e.g., XylY and XylZ) carry more Fe_2S_2 clusters and consume molecular O_2. Together they form the terminal oxygenase.

The *xyl* genes are induced by toluene or benzyl alcohol. These inducers bind to the XylR protein, which activates the promoters for the upper pathway genes (the *xylCMABN* operon) as well as the *xylS* gene (Fig. 13.11). The upper pathway produces benzoate, which acts as the inducer for the lower pathway by binding to XylS protein. This then activates the lower pathway promoter and induces the *xylXYZLEGFJKIH* operon. (Promoters activated by the XylR protein require the alternative sigma factor RpoN [= NtrA], which is normally needed by genes for nitrogen metabolism. The significance of this is uncertain.)

The Tol/Xyl system can be engineered to accept substrates that are not originally used. For example, the Xyl pathway cannot normally degrade the aromatic hydrocarbon, 4-ethylbenzoate (4-EB). There are two problems. First, XylS activator protein does not recognize 4-EB. However, *xylS** mutants have been selected that make an altered XylS protein, which has gained the ability

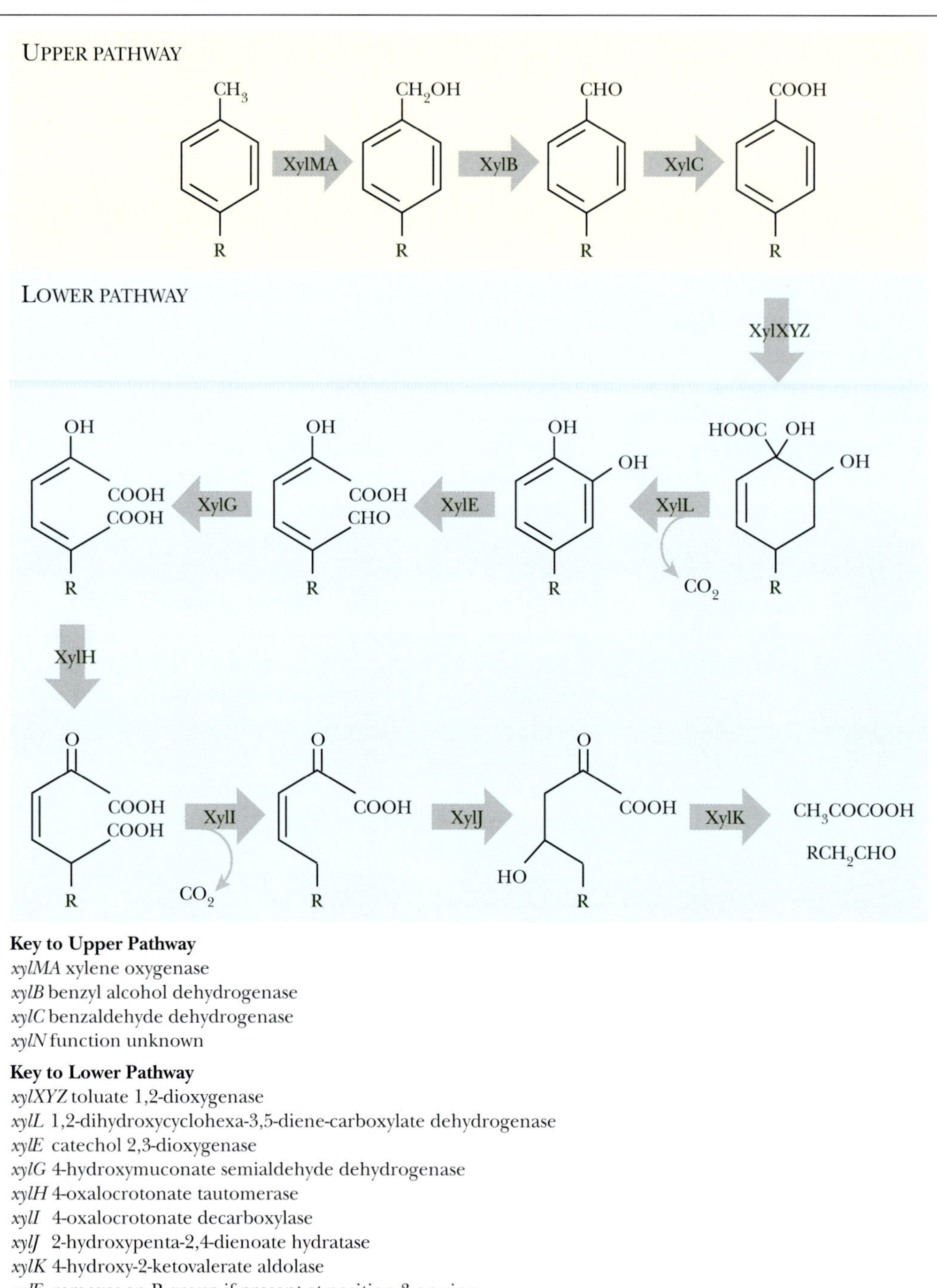

FIGURE 13.10 The Toluene/Xylene Pathway

Xylene/toluene degradations involve many enzymes that start by degrading the side chains, eventually breaking the ring apart. The pathway degrades toluene into acetaldehyde and pyruvate. Xylene is degraded into propionaldehyde and pyruvate.

to bind 4-EB and induces the upper pathway. This converts 4-EB to 4-ethylcatechol. The second problem is that the *xylE* product, catechol-2,3-dioxygenase, is inhibited by 4-ethylcatechol. However, a *xylE** mutant has been isolated that makes an altered enzyme which not only is resistant to 4-ethylcatechol, but can use it as a substrate. The *xylS* xylE** double mutant converts 4-ethylbenzoate completely to pyruvate plus acetaldehyde (Fig. 13.12). Such an improved system can easily be transferred to other host bacteria since it is already plasmid-borne.

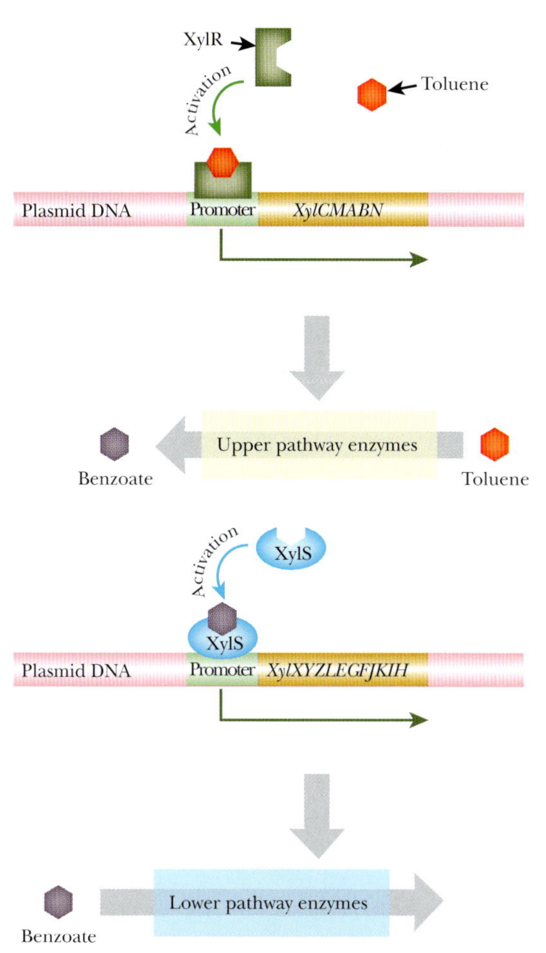

FIGURE 13.11 Regulation of the Toluene/Xylene Pathway
Toluene binds to the transcriptional activator, XylR, inducing XylR to bind the promoter for the upper pathway genes. The upper pathway genes convert toluene into benzoate, which binds to another transcriptional activator, XylS. The benzoate/XylS complex binds the promoter for the genes encoding the lower pathway enzymes. These break down benzoate into pyruvate, acetaldehyde, and carbon dioxide.

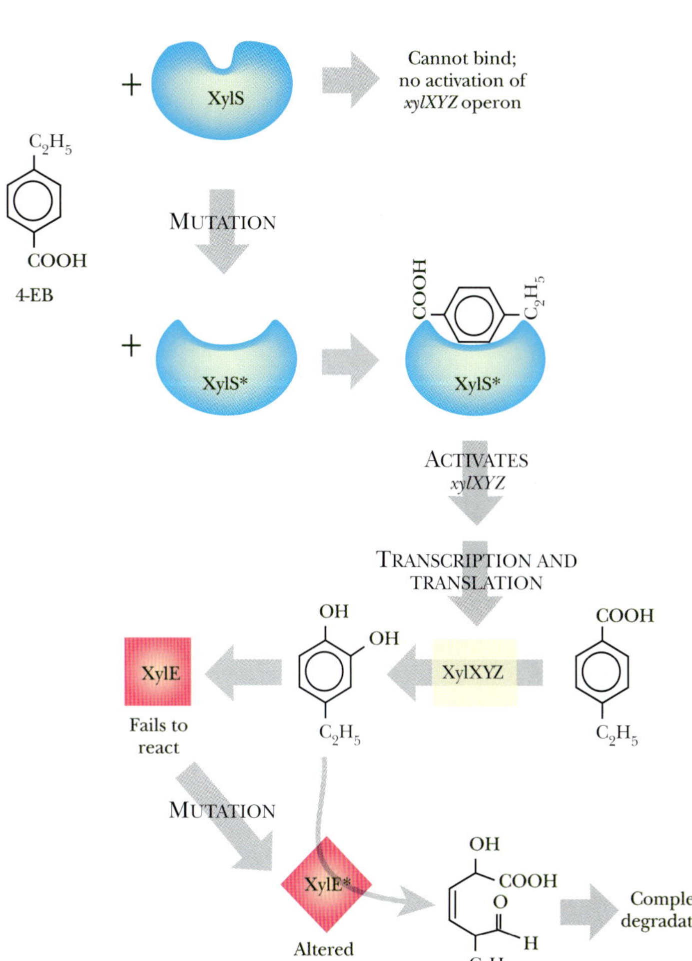

FIGURE 13.12 Engineering of the Toluene/Xylene Pathway
The enzymes that degrade toluene and xylene can also degrade other substances such as 4-ethylbenzoate (4-EB), but two proteins must be mutated to recognize the new substrate. To activate the upper pathway enzymes, the transcription factor XylS must be mutated so that 4-EB activates it. XylE normally breaks the ring structure of the diol, but cannot use the 4-EB diol. A mutant form of XylE will break the ring of the 4-EB diol. *E. coli* transformed a Tol/Xyl plasmid with both these mutant genes can effectively degrade 4-EB.

The toluene/xylene pathway has been engineered to degrade many other aromatic compounds. Full breakdown of these usually requires both mutant enzymes and mutant regulators with novel specificity.

REMOVAL OF HALOGEN, NITRO, AND SULFONATE GROUPS

A wide variety of aromatic compounds are found in the environment, some naturally, others due to human pollution. For almost all such compounds, bacteria can be isolated that degrade them (see Chapter 12 for *Rhodococcus*). Polychlorinated biphenyls (PCBs) are industrial pollutants. Other chlorinated aromatics include selective herbicides such as 2,4-D (2-4-dichlorophenoxyacetate). Many nitro and sulfonate derivatives are used in the pharmaceutical, dye, and detergent industries. The chlorine, nitro, and sulfonate groups may be removed during the dioxygenase reaction to release chloride, nitrite, or bisulfite (Fig. 13.13). Ring dioxygenases from certain bacteria will work on substituted rings, whereas chloro, nitro, or sulfonate groups inhibit others. About 10% of the organic pollution in the river Rhine is aromatic sulfonates from the German dye industry. Not surprisingly, many bacteria isolated from the Rhine possess dioxygenases that are good at knocking off sulfonate and nitro groups.

Many aromatic compounds carry chloro, nitro, or sulfonate groups. These can often be removed by bacterial dioxygenase enzymes.

FIGURE 13.13
Removal of Chloro and Nitro Groups by Dioxygenase
Dioxygenases can insert hydroxyl groups into aromatic rings that have chloro and nitro groups. Chloride and nitrite are then released.

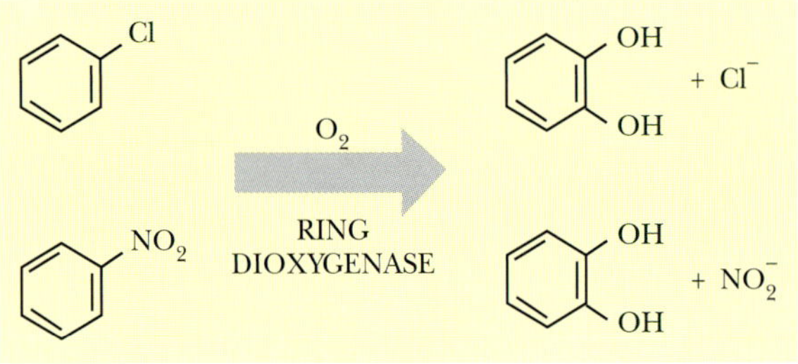

BIOREFINING OF FOSSIL FUELS

The growth of industrial civilization, in particular the use of fossil fuels for energy and the development of the organic chemical industry, have led to the pollution of the environment with a wide range of compounds of nonbiological origin. Many fossil fuel deposits, of both coal and oil, contain a high percentage of sulfur—up to 5% sulfur for many coals from Eastern Europe or the American Midwest. Burning high-sulfur coal releases large quantities of sulfur dioxide into the atmosphere, which leads to the formation of acid rain. Among the possible solutions to this problem is to develop bacteria capable of removing the offending sulfur compounds from the coal (or oil) before combustion.

Several naturally occurring sulfur bacteria such as *Thiobacillus* and *Sulfolobus* can convert pyrites (FeS_2) the major form of inorganic sulfur found in coal, into soluble sulfate that can be rinsed away. The crucial issue, therefore, is to remove the organic sulfur, especially that found in **thiophene** rings, which typically accounts for 70% or more of the organic sulfur (Fig. 13.14). Although compounds containing thiophene rings are almost never found in modern-day living organisms, they form a substantial part of the organic sulfur fraction of fossil fuels such as coal and oil. The major quinone of Archaebacteria such as *Sulfolobus* is caldariellaquinone, which contains a thiophene ring fused to a benzoquinone. Conceivably, the fused thiophenes of coal are the metabolic fossils of archaebacterial metabolism.

Dibenzothiophene (DBT) is a widely used model compound (Fig. 13.14) thought to be representative of the organic sulfur in coal and oil. Biodegradation of DBT and removal of sulfur involves several steps, a scheme known as the **4S pathway** (Fig. 13.15). Most bacteria capable of degrading thiophene derivatives show only partial breakdown. Many do not completely remove the sulfur from its organically bound form, and others only use DBT that has already been oxidized to the sulfone or sulfoxide level (see Fig. 13.15). Full desulfurization requires either finding a natural isolate that can carry out all the steps or the genetic assembly of the individual steps of the 4S pathway from different bacteria into a final engineered strain.

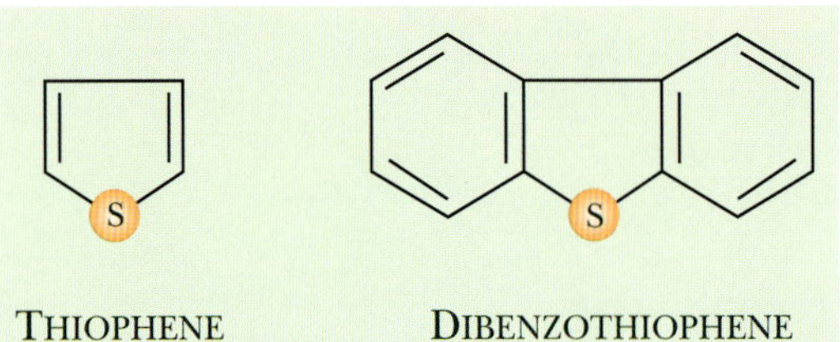

FIGURE 13.14
Thiophene and Dibenzothiophene
Thiophene and dibenzothiophene are the sulfur-containing molecules that account for much of the organic sulfur found in coal and fossil fuels.

Certain bacteria, especially certain species of *Rhodococcus,* do indeed completely desulfurize dibenzothiophene, as well as degrade related heterocyclics such as dibenzofuran and xanthones. The ***dszABC* operon** of *Rhodococcus* is found on a linear plasmid and encodes three enzymes responsible for the 4S pathway. In addition a flavin reductase encoded by the *dszD* gene is needed to supply reduced FMN (see Fig. 13.15). The *dszD* gene is not linked to the *dszABC* operon. The enzymes are as follows:

- Third step: DszA = dibenzothiophene sulfone oxygenase
- Fourth step: DszB = benzene sulfinate desulfinase
- First and second steps: DszC = dibenzothiophene oxygenase
- $FMNH_2$ supply: DszD = flavin reductase

However, there are several problems with these natural isolates:

(a) *Rhodococcus* is not well characterized genetically, so further modification is difficult.

(b) Desulfurizing coal or oil will require robust bacteria that can be grown easily to high density. Although not feeble, *Rhodococcus* is not especially convenient in this respect.

(c) The desulfurizing genes (*dszABC*) in *Rhodococcus* are used naturally to obtain sulfur (not to degrade benzothiophenes for carbon and energy). Consequently, they are expressed only at a low level because only low amounts of sulfur are required for bacterial growth. Moreover, other sulfur compounds, both inorganic and organic, repress the operon. Thus, the inorganic sulfur present in most high-sulfur coal and oil would repress the *dszABC* genes. Furthermore, pregrowth of the bacteria in organic material containing amino acids such as cysteine or methionine would also repress these genes.

The *dszABC* gene cluster has therefore been cloned from *Rhodococcus* and placed onto suitable plasmids for expression in *E. coli* and certain robust *Pseudomonas* strains. The *dszABC* genes have been placed under the control of strong promoters that may be induced as required and are not repressed by sulfur compounds (Fig. 13.16). The resulting strains desulfurize DBT better than the original *Rhodococcus*.

High-level operation of the 4S pathway requires a large flow of reducing equivalents. In cells carrying a cloned *dszABC* operon, reduction of FMN by flavin reductase becomes the limiting factor in removal of sulfur from dibenzothiophene. However, flavin reductases from several other bacteria work as well as or better than the *Rhodococcus* DszD enzyme. For example, the HpaC enzyme from *E. coli* has been cloned and expressed at high levels, and it greatly speeds desulfurization. As a further modification, the *hpaC* gene and the *dszABC* genes have been joined together to form a single operon under control of the ***tac* promoter**. (This is a hybrid promoter with the RBS from the *trp* promoter and the operator of the *lac* promoter. It is therefore a strong promoter that is induced by IPTG.) Thus the combined desulfurization module can be induced by IPTG when required.

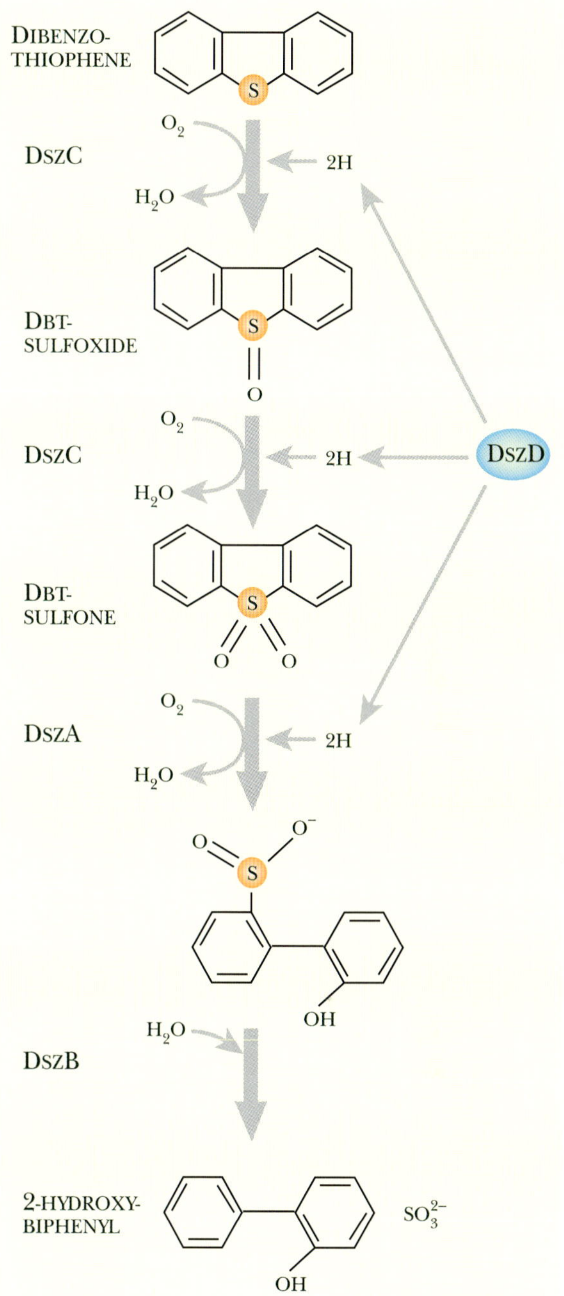

Despite adding oxygen to the sulfur of dibenzothiophene, the DszC enzyme is closely related to the ring dioxygenases that add two hydroxyl groups to aromatic rings. Thus phenanthrene dioxygenase hydroxylates phenanthrene (a three-ringed aromatic hydrocarbon) as well as converting DBT to its sulfone. Again, mutation of biphenyl dioxygenase by a gene shuffling approach (see Chapter 11) gives mutant enzymes capable of handling dibenzothiophene and related compounds. In these cases, the same enzyme hydroxylates the aromatic rings and also adds oxygen to the thiophene sulfur, giving the sulfoxide and then the sulfone. The DszC enzyme itself has also been mutated to

DIBENZO-THIOPHENE
DszC
O_2
H_2O
2H
DBT-SULFOXIDE
DszC
O_2
H_2O
2H
DszD
DBT-SULFONE
DszA
O_2
H_2O
2H
H_2O
DszB
2-HYDROXY-BIPHENYL
SO_3^{2-}

FIGURE 13.15 4S Pathway for Removal of Sulfur from Thiophene Rings
Dibenzothiophene is converted into a sulfoxide by the oxygenase DszC, aided by DszD, a flavin reductase, which supplies the reducing equivalents. The same two enzymes convert the sulfoxide into a sulfone. Next, DszA, another oxygenase, breaks the ring structure. Finally DszB releases the sulfur as sulfite.

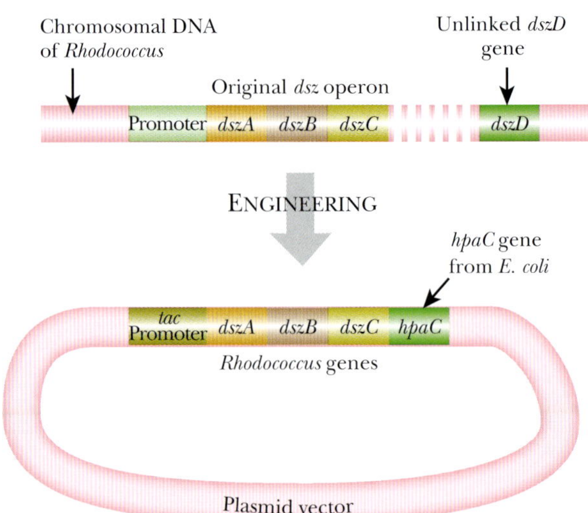

FIGURE 13.16
Engineering of Dsz System
The genes that encode the desulfurization pathway for dibenzothiophene have been cloned from *Rhodococcus* and moved onto a plasmid. The three enzymes DszC, DszA, and DszB need a flavin reductase to supply reducing equivalents. *Rhodococcus* uses the product of its *dszD* gene for this, but the *E. coli* flavin reductase (encoded by *hpaC*) works better. The *E. coli hpaC* gene is therefore overexpressed on the same plasmid. Transcription of the entire cluster of genes is controlled by the inducible *tac* promoter.

broaden its substrate range. For example, a Val261Phe mutation allows oxidation of methylbenzothiophene and alkyl thiophenes.

A major achievement of pathway engineering has been the assembly of a pathway to degrade the sulfur-containing thiophene rings often found in fossil fuels.

BIOSYNTHESIS OF MEDIUM-SIZED MOLECULES

Although we have covered the production of alcohol and of indigo, so far we have really been considering degradative pathways. The fermentation schemes that produce alcohol are designed to release energy from sugars, and indigo is an incidental by-product of the naphthalene degradation system. However, a variety of natural products are made industrially that rely on genuine biosynthetic pathways. For example, many amino acids are manufactured by microorganisms and used for a variety of purposes. Here we will discuss the synthesis of somewhat more sophisticated molecules. We will first look at selected sterols and antibiotics and then consider some examples of biopolymers. The objective is not to cover these pathways in detail but to provide examples of how genetic engineering can be used to improve these processes.

Although antibiotics and sterols are molecules of only intermediate complexity, they are the most difficult to tackle by genetic engineering. The reason is that their synthetic pathways may have 20 or more steps. Each step requires a separate enzyme, encoded by its own gene. In addition, many of these pathways are branched and/or interact with other metabolic pathways. Consequently, their regulation is often complex. Analyzing, cloning, and expressing all the genes that encode the enzymes and regulatory proteins for long and complex pathways requires a great deal of effort.

Biosynthesis of small molecules normally requires multiple enzymes acting sequentially. Consequently, manipulation of antibiotics, sterols, opiates, and so forth represents a major challenge.

Box 13.1 Integrated Circuits Approach to Pathway Engineering

Engineering new pathways can be approached in a more general way that focuses on the arrangement of genes and promoters rather than the actual gene products and metabolites. Scientists who focus on **genetic circuits** often use model genes rather than working with a particular pathway. This is because novel arrangements may develop new cellular behavior. For example, the arrangement of genes can make the final gene product very stable, the final product can oscillate at a particular frequency, or the product can be expressed only when a particular environment or input protein is present, thus making a **biosensory system**.

Box 13.1 Integrated Circuits Approach to Pathway Engineering—cont'd

An example of a biosensory system is where a particular pollutant (the input) causes a cell to fluoresce green (the output). The presence of the pollutant turns on the gene for green fluorescent protein. This requires a series of genes in a transcriptional cascade that controls whether or not GFP is made. A threshold of pollutant is needed before the cells fluoresce green. Manipulating the arrangement of control elements such as enhancers and repressors can make the cells highly sensitive to the pollutant or less sensitive.

Some motifs used to regulate gene expression are shown in Fig. A. The first is a simple on/off dosage compensation motif. In our example, the pollutant may bind to the repressor protein that controls the GFP gene. If the pollutant is present, the repressor is released from the GFP gene and the cells fluoresce. Another type of circuit is the feed forward motif, where one master gene regulates multiple pathways. A third motif involves regulatory feedback, where the final gene product represses the entire pathway. All of these motifs are found in nature and can be used in pathway engineering to control the production of the final protein.

Novel genetic circuits may be engineered into *E. coli*. For example, the system illustrated in Fig. B caused *E. coli* to produce a biofilm in response to DNA damage. The genetic circuit included the input, either UV light or mitomycin C, to induce DNA damage, a regulatory circuit of repressor genes, and an output module that had the biofilm gene (*traA*) controlled by a promoter that was responsive to the repressors. The cells produced a visible biofilm only when DNA damage activated the RecA protein. RecA protein prevents the regulatory circuit from repressing the output module, and therefore when RecA is activated, output—that is, biofilm—was made.

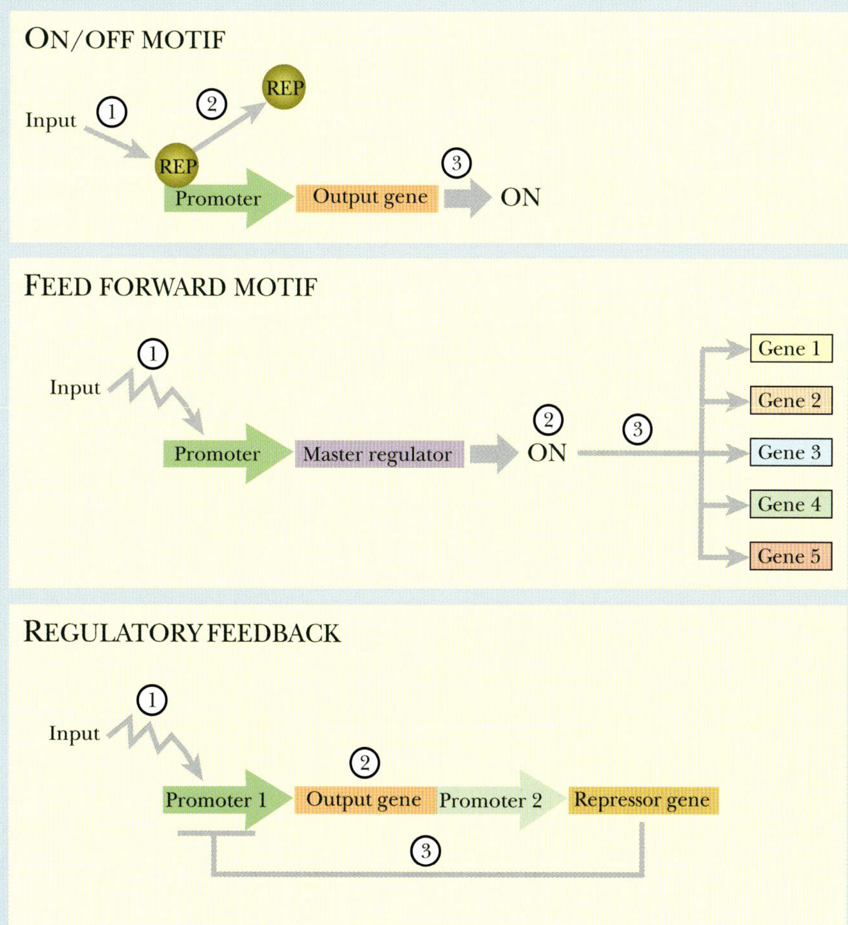

FIGURE A Schematic Representation of Different Genetic Circuits
A simple on/off motif (*top*) has a repressor protein that prevents the transcription and translation of the output gene. When the input signal is received, the repressor is released from the promoter and the output gene is made. In the feed forward motif (*middl*) a master regulator gene controls a variety of different genes (labeled 1–5). The input signal activates the promoter on the master regulator, which in turn, activates genes 1 through 5. In the regulatory feedback motif (*bottom*), two genes are linked in tandem with separate promoters. Promoter 2 is expressed continually to make repressor protein. This prevents the expression of promoter 1 unless an input signal is received. Then the repressor is released, and the output gene is expressed.

(*Continued*)

Box 13.1 Integrated Circuits Approach to Pathway Engineering—cont'd

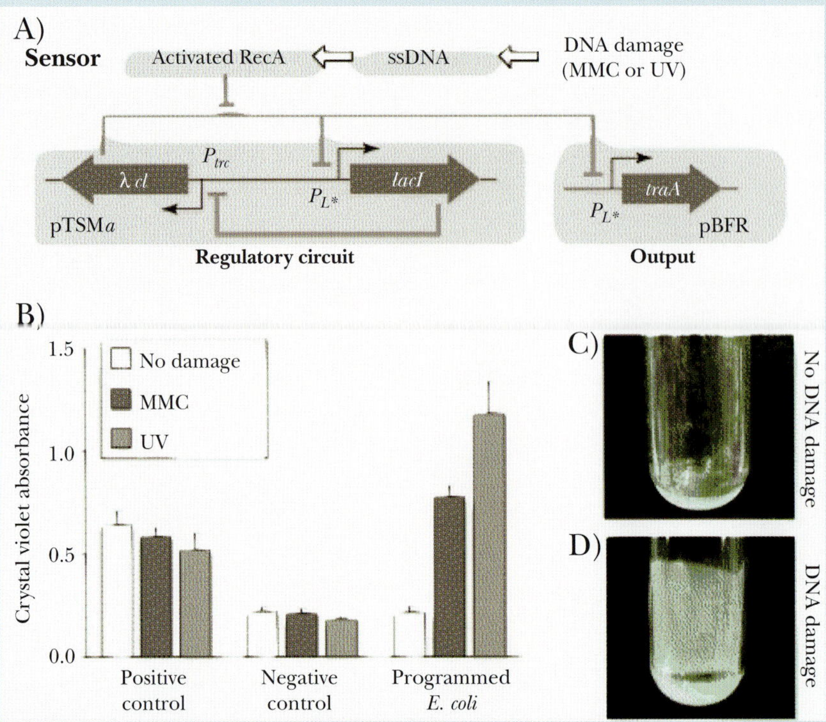

FIGURE B **Biofilm Produced by *E. coli* in Response to DNA Damage**

Novel input-output systems engineered to sense DNA damage and induce biofilm formation. (A) Genetic network connecting DNA damage (input) and biofilm formation (output). Repressor cI normally inhibits expression of *lacI* from promoter PL*. DNA damage causes activation of RecA, which cleaves cI, making it inactive. This allows expression of *lacI*, which inhibits cI expression. Without cI, the cell also produces TraA, which results in biofilm formation. (B) Crystal violet absorbance of control and engineered strains after exposure to the DNA-damaging agents mitomycin C (MMC) and UV radiation. Increased crystal violet absorbance indicates biofilm formation. (C) No biofilm formation is seen in a tube containing engineered cells that were not exposed to DNA-damaging agents. (D) Biofilm formation is seen in a tube containing engineered cells exposed to DNA-damaging agents. From: McDaniel and Weiss (2005). *Curr Opin Biotechnol* **16**, 476–483. Copyright 2005. National Academy of Sciences, USA.

STEROL SYNTHESIS AND MODIFICATION

The **sterol** nucleus consists of four fused rings: three six-membered rings and one five-membered ring (Fig. 13.17 shows examples). Plants, animals, and fungi make a wide variety of sterols. Most bacteria do not make sterols, perhaps because synthesis of sterols involves molecular oxygen and many bacteria can grow anaerobically. Nonetheless, many bacteria do possess enzymes that can modify the structures of sterols made by other organisms.

Sterols are used both as membrane components (e.g., cholesterol) and as hormones. (Steroid hormones activate gene expression via receptor proteins. Note that **steroids** are sterol derivatives having keto rather than hydroxyl groups.) Sterols made or modified by other organisms may exert major effects on animals by interfering with the roles of the natural steroid hormones. In addition, a variety of humanmade molecules also act as steroid mimics. Although steroid hormones regulate a wide variety of processes, the effects of environmental sterols on reproduction have been the most visible. Indeed, many of them act by binding to estrogen receptors. Consequently, these compounds are sometimes referred to as **xenoestrogens**.

Cholesterol and its analogs also have significant effects on human health. For example, the level of cholesterol in the blood has a major effect on the blocking of arteries by fatty deposits (atherosclerosis). Absorption of cholesterol from food is greatly inhibited by many plant sterols and especially by their hydrogenated derivatives, or **stanols**. The sterol nucleus has a double bond in the 5–6 position of the second ring (Fig. 13.17). When this is reduced to a single bond, the resulting stanol has very different biological properties. Eating plant material rich in such stanols reduces dietary intake of cholesterol and so has a positive effect on health. Consequently, plant sterols are sometimes hydrogenated to the corresponding stanols during food processing. This may be done chemically or by using microorganisms.

Many bacteria possess sterol-modifying enzymes that carry out a wide variety of reactions. Some are able to reduce the 5–6 double bond, thus converting sterols into stanols. In addition, some bacteria, especially species of *Eubacterium*, can reduce cholesterol itself to coprostanol, which is poorly absorbed by the intestine. Sterol-modifying enzymes are therefore of great potential use, both in food processing and possibly in breaking down xenoestrogens for decontamination of the environment. Several genes for sterol-modifying enzymes from *Eubacterium* have been cloned and expressed in *E. coli*, and the corresponding enzymes have been purified and characterized.

Pathway engineering may also be of use in manufacturing sterols for therapeutic use. Steroids such as cortisone and prednisone are used to treat inflammatory diseases (arthritis, colitis, allergies, skin problems). Other uses for steroids include contraception and treating syndromes due to hormone deficiency. Today these steroids are made starting with plant sterols and using a mixture of chemical and microbial steps. Ultimately, the role of pathway engineering will be to provide a single microorganism that can carry out all the modification steps required in the manufacture of a particular product.

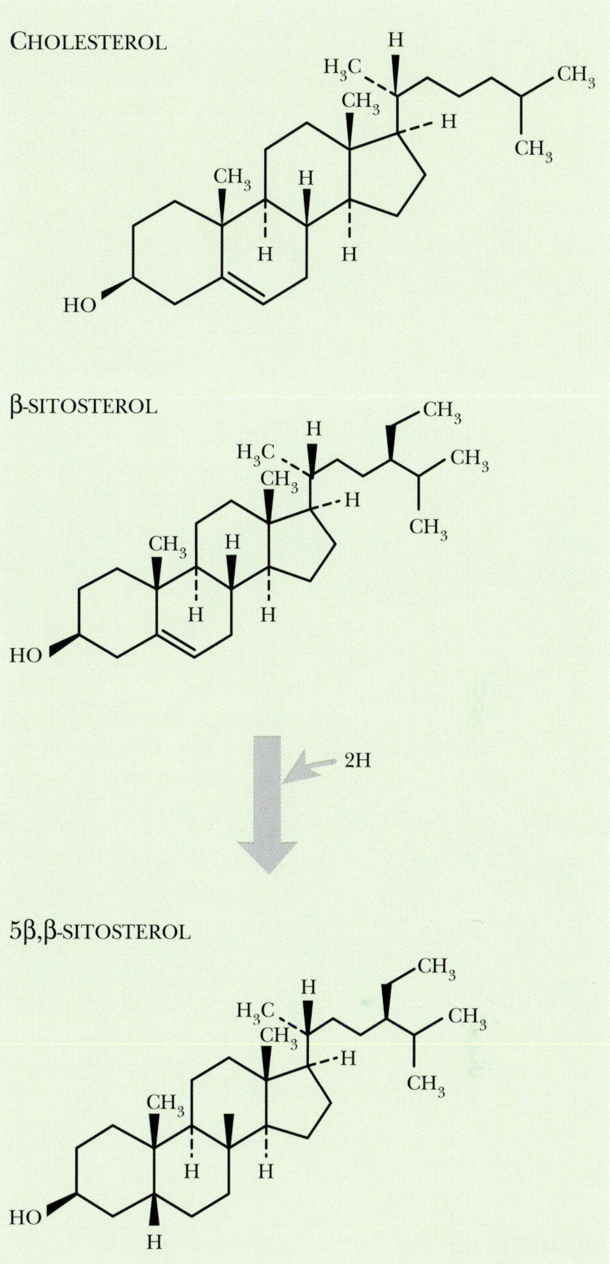

CHOLESTEROL

β-SITOSTEROL

2H

5β,β-SITOSTEROL

FIGURE 13.17
Sterols and Stanols
Cholesterol, an animal sterol, and β-sitosterol, a plant sterol, have similar structures. β-Sitosterol can be reduced to a stanol, 5-β,β-sitosterol, which can compete with cholesterol in dietary absorption, thus reducing cholesterol intake.

Pathway engineering is being used to modify sterols, especially those of plant origin, to yield a variety of products. Some are used as pharmaceuticals, others as food additives.

BIOSYNTHESIS OF β-LACTAM ANTIBIOTICS

Alexander Fleming discovered the mold that makes **penicillin** in the 1920s. The story is a classic case of chance favoring the prepared mind. Fleming left Petri dishes containing bacterial cultures lying around long enough to get moldy. He then noticed a clear zone, in which the bacteria had been killed and had disintegrated, around a blue mold of the *Penicillium* group. He found that the mold excreted a chemical toxic to bacteria but harmless to animals—penicillin. Fleming called the mold ***Penicillium notatum***. A related mold, *Penicillium chrysogenum*, makes a related antibiotic called **cephalosporin** C. Both antibiotics are members of the **β-lactam** family and are made by separate branches of the same biosynthetic pathway (Fig. 13.18).

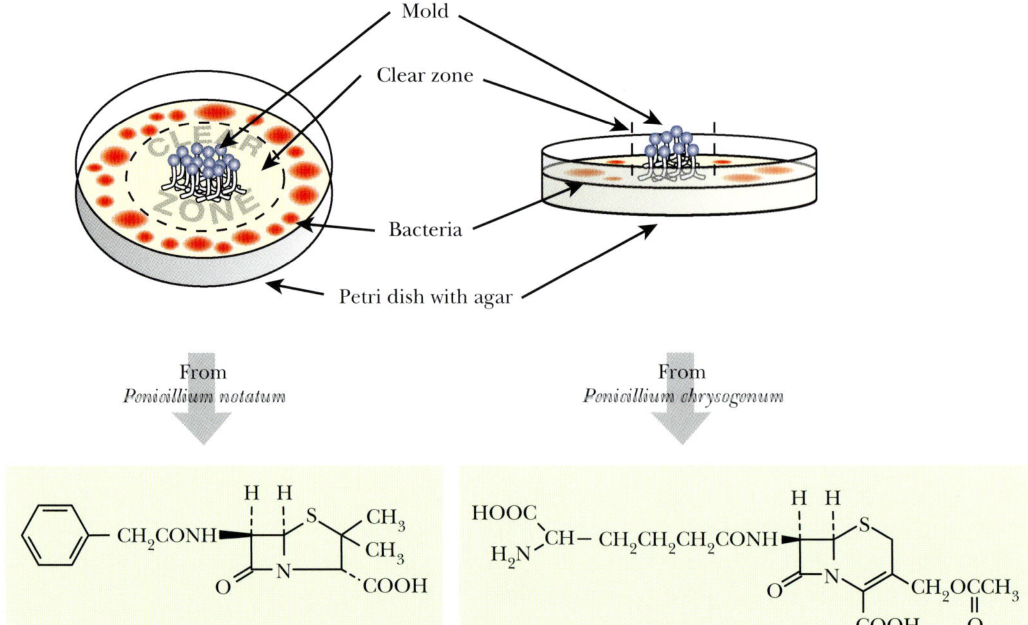

FIGURE 13.18 Biosynthesis of β-Lactam Antibiotics
When certain molds grow on agar originally covered with bacteria, a clear zone appears around the mold where no bacteria are able to grow. The clearing is due to release of antibiotics such as penicillin and cephalosporin C from the mold.

The original β-lactams made by molds can be altered chemically to give many different antibiotics. Although cephalosporin C itself has only feeble antibacterial activity, it is the starting point for a vast array of broad-spectrum antibiotics made by chemical modification. First, cephalosporin C must be converted to 7-ACA (7-aminocephalosporanic acid), which is not made by any known organism. Originally this step was done chemically and gave very low yields. Recently, a mold that makes cephalosporin C was engineered to convert this to 7-ACA. Two extra genes were inserted to create the extended pathway (Fig. 13.19). The gene for D-amino-acid oxidase was taken from a fungus (*Fusarium solani*) and the cephalosporin acylase gene from a bacterium (*Pseudomonas diminuta*).

The 7-ACA is used as base compound for a massive range of chemical modifications that provide antibiotics with different properties. Among these are variants that are resistant to bacterial β-lactamases and others that penetrate bacterial cell walls better as well as antibiotics with better pharmacological properties (e.g., superior absorption from the intestine).

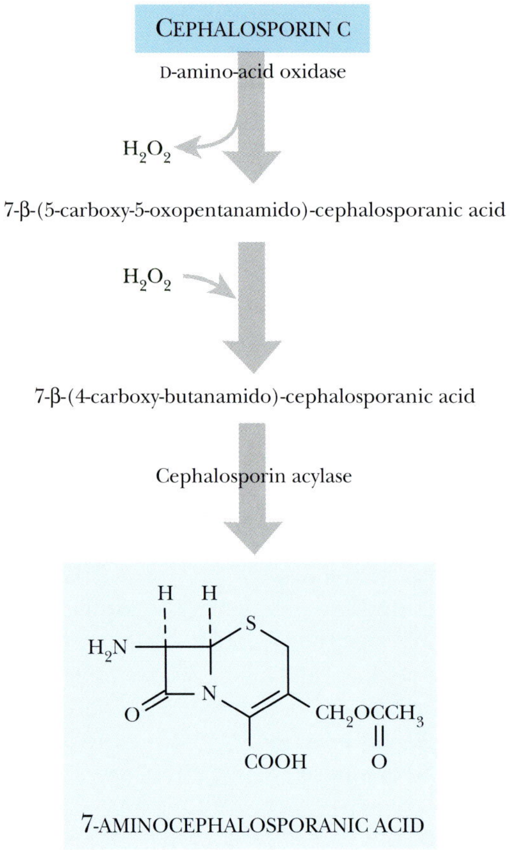

CEPHALOSPORIN C

D-amino-acid oxidase

H_2O_2

7-β-(5-carboxy-5-oxopentanamido)-cephalosporanic acid

H_2O_2

7-β-(4-carboxy-butanamido)-cephalosporanic acid

Cephalosporin acylase

7-AMINOCEPHALOSPORANIC ACID

FIGURE 13.19
Engineered Pathway to 7-ACA
To engineer new antibacterial compounds, cephalosporin C must be converted into 7-aminocephalosporanic acid (7-ACA).
The enzymes involved in this conversion are D-amino acid oxidase and cephalosporin acylase. The genes for these enzymes have been isolated, cloned, and expressed in different bacteria, as well as in molds producing cephalosporin C itself.

Most β-lactam antibiotics are the result of chemical modification. However, engineering the original β-lactam pathway of the *Penicillium* molds helps to increase the yield of some of these antibiotics.

POLYKETIDES AND POLYKETIDE ANTIBIOTICS

More and more disease-causing microorganisms are acquiring or mutating to antibiotic resistance. This is starting to cause problems in treating patients with certain infections. More novel antibiotics are becoming necessary simply to treat infections that were once susceptible to the earlier generation of antibiotics. Although new antibiotics are still to be found in the wild, and some may be made by modification of a base compound (e.g., 7-ACA as described earlier), other approaches are also useful.

Combinatorial biosynthesis manipulates genes for useful natural products to generate a wide range of possible variants. These can then be screened for different and/or improved activities. In practice, only certain rather flexible biochemical pathways are susceptible to this type of manipulation. The best example is the **polyketide** pathway (Fig. 13.20). Polyketides are linear polymeric molecules. Polyketide synthesis begins with a small initiator molecule that is elongated by addition of subunits of two, three, or four carbons. However, each subunit contributes two carbons to the growing polymer backbone and its other carbon atoms form branches (methyl or ethyl groups). As originally made, every other carbon in the polymer backbone carries a keto group—hence the name *polyketide*. However, some of the keto groups may be reduced to hydroxyl groups or to CH_2 groups during the polymer elongation process.

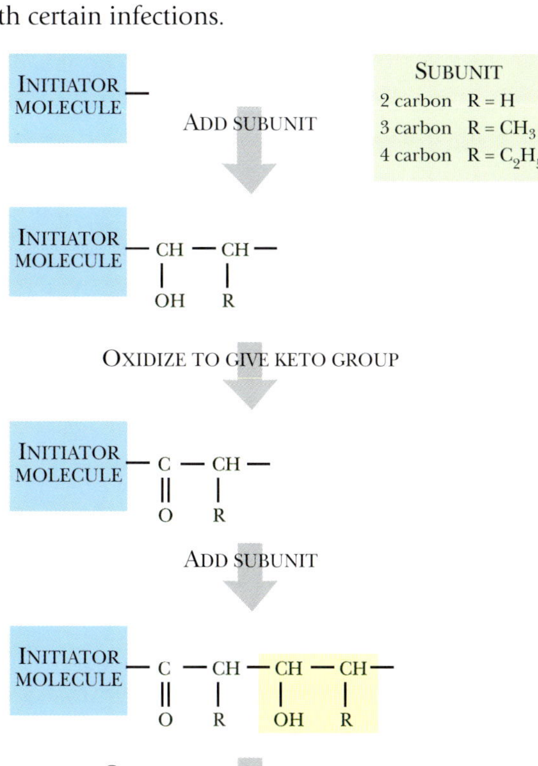

FIGURE 13.20
Principle of the Polyketide Pathway
Polyketides are synthesized from small subunits of two, three, or four carbons added onto an initiator molecule (blue). As each subunit is added, oxidation creates a keto group (C=O) next to the site of addition. Each new subunit extends the chain by two carbon atoms and any extra carbons project sideways to become side groups (labeled R on figure).

After synthesis, the polyketide chains form a variety of ring systems and are modified in various ways. Several families of antibiotics are polyketides, the best known being the **tetracyclines** and the **macrolides** (e.g., **erythromycin**). Certain bacteria make these antibiotics naturally, especially those of the actinomycete family. The genes for each antibiotic pathway are usually clustered together, which is convenient for cloning and manipulation.

The polyketide synthases consist of modules, each of which is responsible for addition and modification of a polyketide subunit. Each module contains activities for:

- Loading enzyme that carries the growing chain via a thiol ester
- Chain extension enzyme that selects correct incoming subunit (i.e., two-, three-, or four-carbon subunit) and links it to the growing chain
- Optional reductase that converts keto to hydroxyl
- Optional reductase that converts hydroxyl to methylene
- Transferase that passes growing polyketide to the next module

For example, the erythromycin system has six modules, each adding a 3C subunit, plus a loading domain that picks up the initiator molecule and a releasing domain. These modules are arranged on three separate polypeptide chains (Fig. 13.21). After release, the erythromycin precursor is circularized and modified by addition of two sugar derivatives.

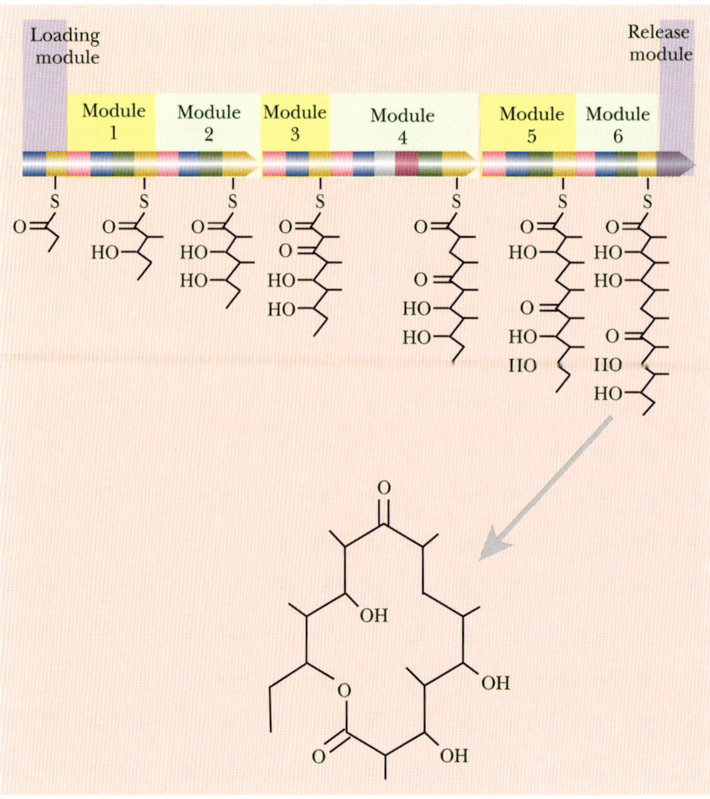

FIGURE 13.21 Modular Polyketide Pathway for Erythromycin
Polyketide synthase for erythromycin has a separate module for each subunit that is added to the polyketide chain. The loading module has a sulfhydryl for linking to the initiator molecule. Modules 1 and 2 each add a three-carbon subunit and reduce its keto group to a hydroxyl. Module 3 adds a three-carbon subunit. Each successive module adds a new subunit until the growing chain reaches the release module, which releases the finished chain. The polyketide chain then circularizes into a precursor that is further modified (not shown here) to give erythromycin.

Polyketide pathways, such as the one for erythromycin, may be engineered by two main methods. First, the polymerizing enzyme of module 1 is inactivated. This prevents elongation of the natural initiator molecule. An artificial analog is then added and joins the polyketide pathway at module 2. Chemical groups present on the initiator analog will then be present in the final polyketide product (Fig. 13.22).

An even greater variety of polyketides can be generated by genetically altering the modules of the polyketide synthase. For example, a module that converts the keto group to a hydroxyl could be altered to leave the keto unchanged or to reduce it further to a methylene (CH_2) group. Or a module that incorporates a two-carbon subunit could be altered to use a three- or four-carbon precursor, thus introducing methyl or ethyl branches along the polyketide chain. For genetic engineering, the polyketide synthase gene cluster is normally moved into *E. coli*. The synthase modules may be altered by three major approaches:

(a) Individual modules may be directly altered to inactivate or add enzymatic activities
(b) Modules may be replaced as a whole by other modules from the same pathway from the same original polyketide-producing organism
(c) Modules from the polyketide synthases of different polyketide-producing organisms may be combined

With six successive modules, the potential for introducing variety is immense. Engineered polyketide gene clusters may be reinserted into a natural polyketide producer, such as

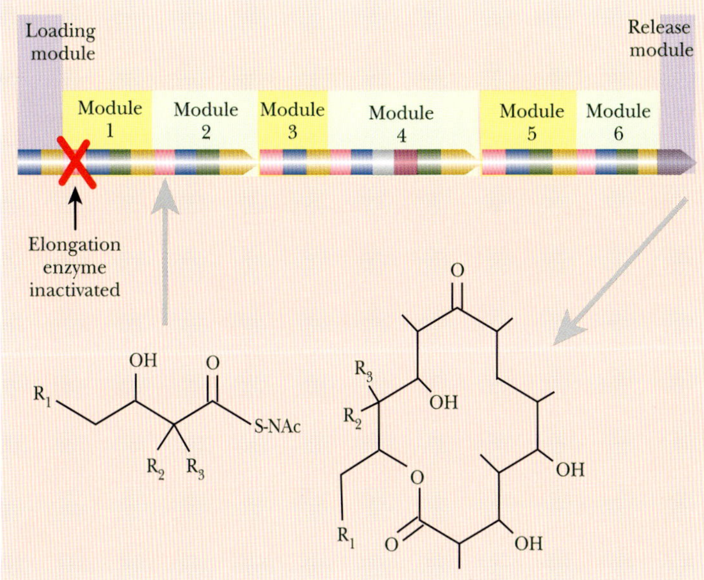

FIGURE 13.22 Engineering Polyketides by Changing the Initiator Molecule

Blocking the natural initiator module and loading an artificially synthesized precursor can create variations in the final structure. The rest of the polyketide synthase works the same as before; each module adds a two-, three-, or four-carbon subunit and may modify the keto group. A variety of final structures may be screened for enhanced antibacterial activity.

Streptomyces, or expressed in *E. coli*. A library of polyketides can then be generated and screened for useful properties.

> Large numbers of derivatives can be made by shuffling and modifying the enzyme modules of the polyketide pathway. This group of compounds contains several useful antibiotics such as tetracyclines and erythromycin.

BIOSYNTHETIC PLASTICS ARE ALSO BIODEGRADABLE

Plastics are polymers built from chains of monomer subunits, like proteins and nucleic acids. However, most plastics consist of the same monomer mindlessly repeated over and over again. Sometimes two or more closely related monomers may be mixed together and follow each other at random.

Certain bacteria make and store a group of related plastics known as **polyhydroxyalkanoates (PHAs)**. Their composition is shown in Fig. 13.23. The bacteria accumulate PHAs when they have surplus carbon and energy but are running low on other essential nutrients, such as nitrogen or phosphorus. When conditions improve, the PHA is broken down and used as a source of energy. The most commonly found PHA has four-carbon (hydroxybutyrate, HB) subunits and is therefore called **polyhydroxybutyrate (PHB)**. However, a plastic made by randomly mixing in 10% to 20% of five-carbon (hydroxyvalerate, HV) subunits has much better physical properties. Still other PHAs containing a proportion of subunits that are eight-carbon or longer give materials that are more rubbery.

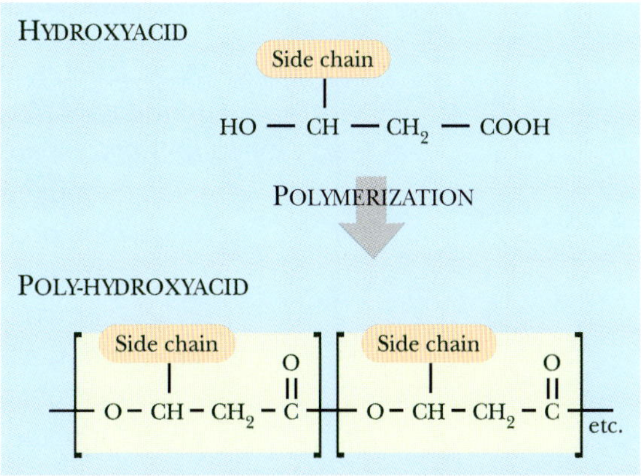

HYDROXYACID

Side chain

HO — CH — CH_2 — COOH

POLYMERIZATION

POLY-HYDROXYACID

[O − CH(Side chain) − CH_2 − C(=O)] [O − CH(Side chain) − CH_2 − C(=O)] etc.

FIGURE 13.23 Structure of Polyhydroxyalkanoate Plastics

Plastics are long chains of repeating subunits. Polyhydroxyalkanoates have repeated hydroxyacid subunits linked through their carboxyl and hydroxyl groups. The various possible side groups alter the final characteristics of the plastic.

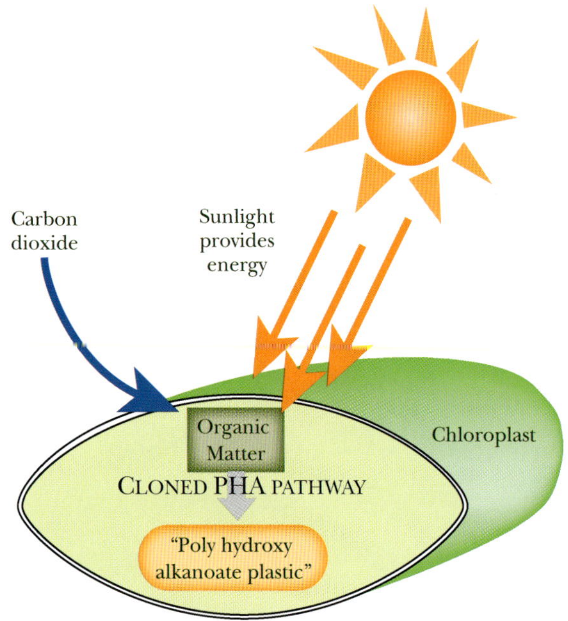

FIGURE 13.24
Expression of PHA Pathway in Plants
The PHA synthesis enzyme pathway has been cloned from *Alcaligenes eutrophus*. The genes were inserted into the chloroplast genome so that the enzymes are expressed only in the chloroplast. The enzymes use the newly synthesized organic matter from photosynthesis to create PHA.

A poly-HB/HV copolymer is manufactured by mutant bacteria of the species *Alcaligenes eutrophus* and is marketed by the Zeneca Corporation (United Kingdom) under the trade name of Biopol. It is more expensive than plastics made from oil, but it is completely biodegradable. Consequently, PHAs are restricted at present to specialized uses. For example, because they break down slowly inside the body to give natural, biochemical intermediates, they can be used for making slow-release capsules.

To make PHAs economically competitive, they will need to be produced cheaply and in bulk. One possible way to do this is to insert the genes for their synthesis into suitable crop plants. This is still in the experimental stage, but the genes for making PHA from *Alcaligenes eutrophus* have been successfully inserted into *Arabidopsis*, a plant widely used for genetic experiments. The engineered PHA genes were designed to be expressed inside the chloroplasts of the plant (Fig. 13.24). Because chloroplasts are the sites of photosynthesis, newly synthesized organic matter appears here first. Locating the PHA pathway in chloroplasts, rather than in the main compartment of the plant cells, gives a 100-fold increase in PHA yield. The next step will be to move the pathway into a genuine crop plant such as rapeseed or soybeans. Perhaps this will give new meaning to the phrase "plastic flowers"!

Polyhydroxybutyrate (PHB) is a plastic polymer that can be made by bacteria or engineered plants. Related plastics may be engineered for desirable properties and, as a bonus, are biodegradable.

Summary

Metabolic pathways may be created or modified by assembling genes from one or more organisms. Individual genes may themselves be subjected to engineering to produce regulatory proteins or enzymes with novel activities or binding specificities. In particular, pathway engineering has been successfully applied to the biodegradation of agricultural waste and industrial pollutants and to the biosynthesis of modified antibiotics.

End-of-Chapter Questions

1. What is the advantage of using either yeast or *Zymomonas* during fermentation?
   - **a.** *Zymomonas* and yeast generate a mixture of products, whereas other microorganisms only produce alcohol.
   - **b.** *Zymomonas* and yeast make only alcohol, but other microorganisms generate a mixture of products.
   - **c.** *Zymomonas* and yeast have very small growth requirements.
   - **d.** *Zymomonas* and yeast can only break down glucose and no other sugars or carbohydrates to produce alcohol.
   - **e.** none of the above

2. Which of the following best explains how pathway engineering is useful?
   a. the conversion of plant-derived waste from paper manufacturing into alcohol used for fuel
   b. the creation of better-tasting beers
   c. the genetic engineering of *E. coli* strains to grow only on one sugar source, such as glucose
   d. the genetic engineering of plants to decrease the use of cellulose, which is difficult to break down
   e. none of the above

3. How can the process of breaking down starch into sugar be improved?
   a. by beginning with glycogen rather than starch to generate sugar molecules
   b. by chemically treating the starch to break it down into sugars
   c. by identifying other enzymes through proteomics that can perform the same functions as the presently used enzymes
   d. by engineering an organism to produce higher levels and more stable enzymes
   e. all of the above

4. Which of the following enzymes is not involved in the breakdown of cellulose?
   a. cellulase
   b. endoglucanase
   c. cellobiohydrolase
   d. β-glucosidase
   e. exoglucanase

5. What are ice nucleation factors?
   a. Ice nucleation factors are proteins present within *E. coli* cells.
   b. Ice nucleation factors provide a way to prevent ice crystals from forming on plants and destroying plant tissue.
   c. Ice nucleation factors are specialized proteins in bacteria that provide a seed on which ice crystals are formed.
   d. When ice nucleation factors are present, water supercools instead of forming ice crystals.
   e. none of the above

6. Which enzyme degrades aromatic compounds by specifically adding a single oxygen atom?
   a. monooxygenase
   b. oxygenase
   c. dioxygenase
   d. arooxygenase
   e. all of the above

7. Which enzyme is capable of performing the first reaction in the production of indigo?
   a. β-galactosidase
   b. naphthalene hydrolase
   c. indigo oxygenase
   d. naphthalene oxygenase
   e. none of the above

(*Continued*)

8. Which of the following end products is common in the degradation pathway for toluene and xylene?
   - a. acetaldehyde
   - b. benzoate
   - c. pyruvate
   - d. propionaldehyde
   - e. none of the above

9. Which of the following groups is removed by bacterial dioxygenases?
   - a. nitro
   - b. chloro
   - c. sulfonate
   - d. all of the above
   - e. none of the above

10. What is the 4S pathway?
   - a. a set of reactions for the breakdown of DBT and removal of the sulfur
   - b. a set of reactions for the synthesis of sulfur-containing compounds
   - c. a pathway for the incorporation of FeS clusters into enzymes for stability
   - d. a pathway for the synthesis of the four nucleotides used in the genetic code
   - e. none of the above

11. What functions can sterol-modifying enzymes perform?
   - a. Sterol-modifying enzymes can be used to break down cholesterol into a product that is poorly absorbed in the intestine.
   - b. Sterol-modifying enzymes can break down xenoestrogens for decontamination of the environment.
   - c. Sterol-modifying enzymes can be used to create steroids, such as prednisone and cortisone.
   - d. Sterol-modifying enzymes can be used to treat hormone deficiencies or for contraception.
   - e. all of the above

12. What can be produced upon chemical modification of β-lactams?
   - a. β-lactam antibiotics that are resistant to β-lactamases
   - b. antibiotics that are more easily absorbed in the intestines
   - c. antibiotics that can penetrate bacterial cell walls more efficiently
   - d. increased yields of some β-lactam antibiotics
   - e. all of the above

13. Which of the following statements about polyketides is not true?
   - a. Tetracycline and erythromycin are families of antibiotics that are polyketides.
   - b. Polyketides have little potential of being modified.
   - c. Polyketide antibiotics are made naturally by *Actinomycetes*.
   - d. A variety of polyketides can be created by genetically modifying the modules of polyketide synthase.
   - e. Polyketide libraries can be generated and expressed, and then screened for useful properties.

14. Which of the following is not a property of polyketide synthase modules?

- **a.** a chain extension enzyme that selects the correct incoming subunit
- **b.** two optional reductases that convert a keto to hydroxyl and then hydroxyl to methylene
- **c.** a loading enzyme that carries the growing polyketide chain
- **d.** an oxidase that oxidizes the reduced groups on the growing chain
- **e.** all of the above are properties of polyketide synthase

15. How is the production of PHA being made more economical?

- **a.** by mass-marketing the product as an environmentally friendly alternative
- **b.** by engineering the PHA genes from *A. eutrophus* into the chloroplasts of *Arabidopsis*, and expression in plants
- **c.** by production in massive fermenters at Zeneca Corporation
- **d.** by expression of the PHA genes in all plant tissues, not just in the chloroplasts
- **e.** none of the above

Further Reading

Aldor IS, Keasling JD (2003). Process design for microbial plastic factories: Metabolic engineering of polyhydroxyalkanoates. *Curr Opin Biotechnol* **14**, 475–483.

Atkinson MR, Savageau MA, Myers JT, Ninfa AJ (2003). Development of genetic circuitry exhibiting toggle switch or oscillatory behavior in *Escherichia coli*. *Cell* **113**, 597–607.

Cochet N, Widehem P (2000). Ice crystallization by *Pseudomonas syringae*. *Appl Microbiol Biotechnol* **54**, 153–161.

Dien BS, Cotta MA, Jeffries TW (2003). Bacteria engineered for fuel ethanol production: Current status. *Appl Microbiol Biotechnol* **63**, 258–266.

Kern A, Tilley E, Hunter IS, Legisa M, Glieder A (2007). Engineering primary metabolic pathways of industrial microorganisms. *J Biotechnol* **129**, 6–29.

Lynd LR, van Zyl WH, McBride JE, Laser M (2005). Consolidated bioprocessing of cellulosic biomass: An update. *Curr Opin Biotechnol* **16**, 577–583.

McDaniel R, Weiss R (2005). Advances in synthetic biology: On the path from prototypes to applications. *Curr Opin Biotechnol* **16**, 476–483.

Meyer A, Pellaux R, Panke S (2007). Bioengineering novel *in vitro* metabolic pathways using synthetic biology. *Curr Opin Microbiol* **10**, 246–253.

Sandmann G, Römer S, Fraser PD (2006). Understanding carotenoid metabolism as a necessity for genetic engineering of crop plants. *Metab Eng* **8**, 291–302.

Sato F, Inui T, Takemura T (2007). Metabolic engineering in isoquinoline alkaloid biosynthesis. *Curr Pharm Biotechnol* **8**, 211–218.

Symons ZC, Bruce NC (2006). Bacterial pathways for degradation of nitroaromatics. *Nat Prod Rep* **23**, 845–850.

Thykaer J, Nielsen J (2003). Metabolic engineering of beta-lactam production. *Metab Eng* **5**, 56–69.

Urgun-Demirtas M, Stark B, Pagilla K (2006). Use of genetically engineered microorganisms (GEMs) for the bioremediation of contaminants. *Crit Rev Biotechnol* **26**, 145–164.

Yoshikuni Y, Keasling JD (2007). Pathway engineering by designed divergent evolution. *Curr Opin Chem Biol* **11**, 233–239.

Zaldivar J, Nielsen J, Olsson L (2001). Fuel ethanol production from lignocellulose: A challenge for metabolic engineering and process integration. *Appl Microbiol Biotechnol* **56**, 17–34.

Transgenic Plants and Plant Biotechnology

HISTORY OF PLANT BREEDING

For thousands of years humans have improved crop plants and domestic animals by selective breeding, mostly at a trial-and-error level. Over many years, animal and crop breeders have learned that improving their crops and animals has a biological basis. In fact, the father of genetics, Gregor Mendel, experimented with the common pea. He studied easily identified traits such as round versus wrinkled seeds or yellow versus green seeds. Mendel would take the pollen from one plant and put it on the stigma of another plant (Fig. 14.1), a procedure called **cross-pollination**. His experiments showed that plants have some traits that may dominate others. For example, a cross between a yellow-seeded pea plant and a green-seeded pea plant gave only seeds that were yellow. This work was published in 1865, but no one understood the importance of these findings until well after Mendel's death.

Many scientists around the world still use traditional breeding techniques to enhance crop yields, increase resistance to various pests or diseases, or increase the tolerance of a particular crop to heat, cold, drought, or wet conditions. Although simply crossing a high-yielding plant with another can produce offspring with even higher yields than either parent, the process is long and tedious. Many thousands of plants must be cross-pollinated to find the one offspring with higher yield. The crosses must be done by hand, that is, pollen must be taken from one plant and manually placed on another. In addition, the possibility of finding improved traits is limited by the amount of genetic diversity already present in the plants. Consequently, if the two plants that are crossed share many of the same genes, the amount of possible improvement is limited. If a plant has no genes for disease resistance, there is no way traditional cross-pollination will develop that trait. Therefore, scientists have searched for better ways to improve plants.

FIGURE 14.1 The Reproductive Organs of a Typical Plant
Pollen grains are the male reproductive cells of the plant. They are made in the anther (orange), the top portion of the stamen. The female reproductive cells, the ova, are sequestered in the ovary. Pollen reaches the ova via the stigma, which is attached to the ovary by the pistil.

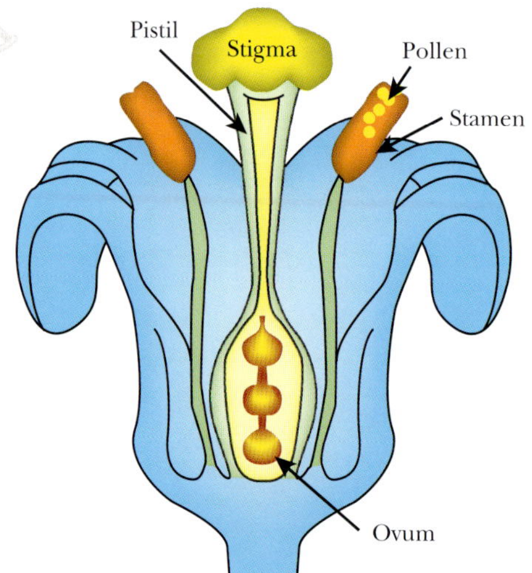

In the 1920s, scientists realized that mutations could be induced in seeds by using chemical mutagens or by exposure to X-rays or gamma rays. Although useful, the outcome of such treatments is even less predictable than traditional breeding methods. Nonetheless, **mutation breeding** has been successful, especially in the flower world. For example, new colors and more petals have been expressed in flowers such as tulips, snapdragons, roses, chrysanthemums, and many others. Mutation breeding has also been tried on vegetables, fruits, and crops. Some of the varieties of food we eat today were developed using this method. In short, a large number of seeds are exposed to the mutagen to generate various mutations in their DNA. The seeds are then planted and cultivated. However, the majority of seeds are killed by the treatment. After the viable seeds are grown, the fruit, flower, or grain of the plant is tested for improvements. If one plant is found with a desirable trait, then its progeny are tested for the trait. Novel traits are only useful if they are **heritable**, that is, passed from one generation to the next. Because only one original mutant plant would gain any particular desirable trait, this plant would need to be propagated a long time before any of the fruit, grain, or flower would be sold to market. It is important to realize that the actual fruits or flowers sold to the consumer were never exposed to the mutagen. Today, chemical mutagens are still used, but molecular biology techniques are being used to identify the actual gene associated with the desired phenotype (see later discussion).

Recently, the emergence of molecular biology has opened the door to a much more predictable way to enhance crops. Scientists have discovered ways to move genes from foreign sources into a specific plant, resulting in a **transgenic plant**. The foreign gene, or **transgene**, may confer specific resistance to an insect, or protect the plant against a specific herbicide, or enhance the vitamin content of the crop. The major difference between transgenic technology and traditional breeding is that a plant can be transformed with a gene from any source, including animals, bacteria, or viruses as well as other plants, whereas traditional cross-breeding methods move genes only between members of a particular genus of plants. Furthermore, the transgene has a known function and has been evaluated extensively before being inserted in the plant. During traditional breeding, the identity of genes responsible for improving the crop is rarely known.

Transgenic plants contain foreign genes, called transgenes, which have a desirable trait.

PLANT TISSUE CULTURE

One major advantage of plants is that they can often be regenerated from just a single cell, that is, each plant cell is **totipotent** and retains the ability to develop into any cell type of a mature plant (Fig. 14.2). There is no absolute separation of the germline from the somatic cells in plants (unlike animals). This unique feature of plants allows scientists to grow and manipulate plant cells in culture, then regenerate an entire plant from the cultured cells.

FIGURE 14.2 An Entire Plant Can Be Regenerated from a Single Cell
Small samples of tissue, or even single plant cells may be cultured *in vitro*. Under appropriate conditions, these may regenerate into complete plants.

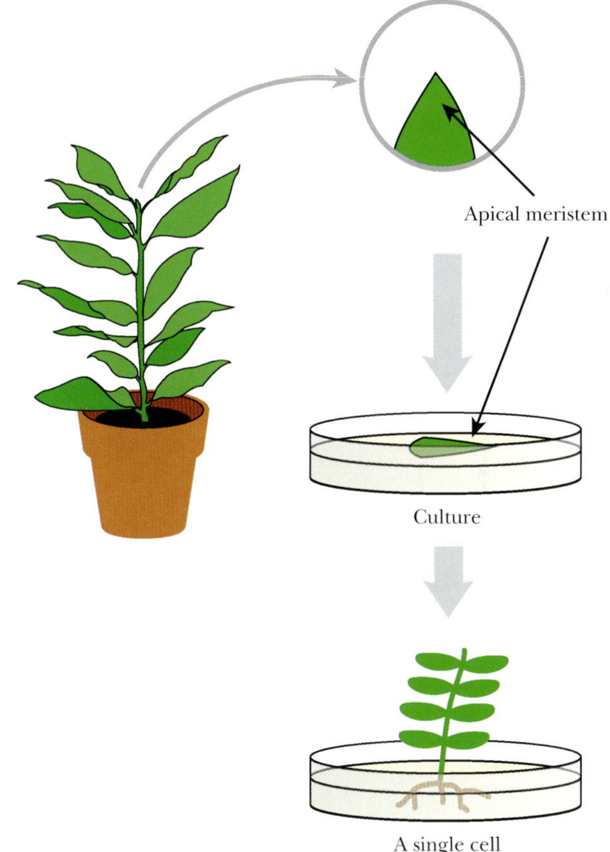

Plant tissue culture can be done on either a solid medium in a petri dish, called **callus culture**, or in liquid, called **suspension culture**. In both cases, a mass of tissue or cells, known as an **explant**, must be removed from the plant of interest. In callus culture, the tissue can be an immature embryo, a piece of the apical meristem (the region where new plant shoots develop), or a root tip. For liquid culture, cells must be dissociated from one another. Liquid culture usually uses **protoplasts** (plant cells from which the cell wall is removed), microspores (immature pollen cells), or macrospores (immature egg cells). The cells are then cultured with a mixture of nutrients and specific plant hormones that induce the undifferentiated cells to grow.

Different types of plants respond to different hormones. To culture wheat cells, for example, the explant is grown with 2,4-dichlorophenoxyacetic acid (2,4-D). This is an analog of the plant hormone auxin, which stimulates plant cells to dedifferentiate and grow. To culture tomato plants, the hormone cytokinin dedifferentiates the cells and induces cell division. In callus culture, undifferentiated cells form a crystalline white layer on top of the solid medium, called the **callus**. After about a month of growth, the mass of undifferentiated cells can be transferred to medium with a lower concentration of hormone or with a different hormone. Decreasing the amount of hormone allows some of the undifferentiated callus cells to develop into a plant shoot. In most cases the small shoots look like new blades of grass growing from the mass of cells. After another 30 days, the hormone is removed completely, which allows root hairs to start growing from some of the shoots. After another 30 days, small plants can be isolated and planted into soil. In liquid culture, hormones are also used to stimulate the growth of undifferentiated cells, but the shoot and root tissues grow simultaneously (Fig. 14.3).

Because plant tissue culture allows many plants to be produced from one source, the technique is useful for making clones of one particular plant. If a very rare plant is identified, it can be propagated using tissue culture. Only a small cutting is needed to generate many identical progeny. Certain special plant varieties that are hard to maintain by producing seed can be maintained for the long term in culture. Plant cell culture has also been used

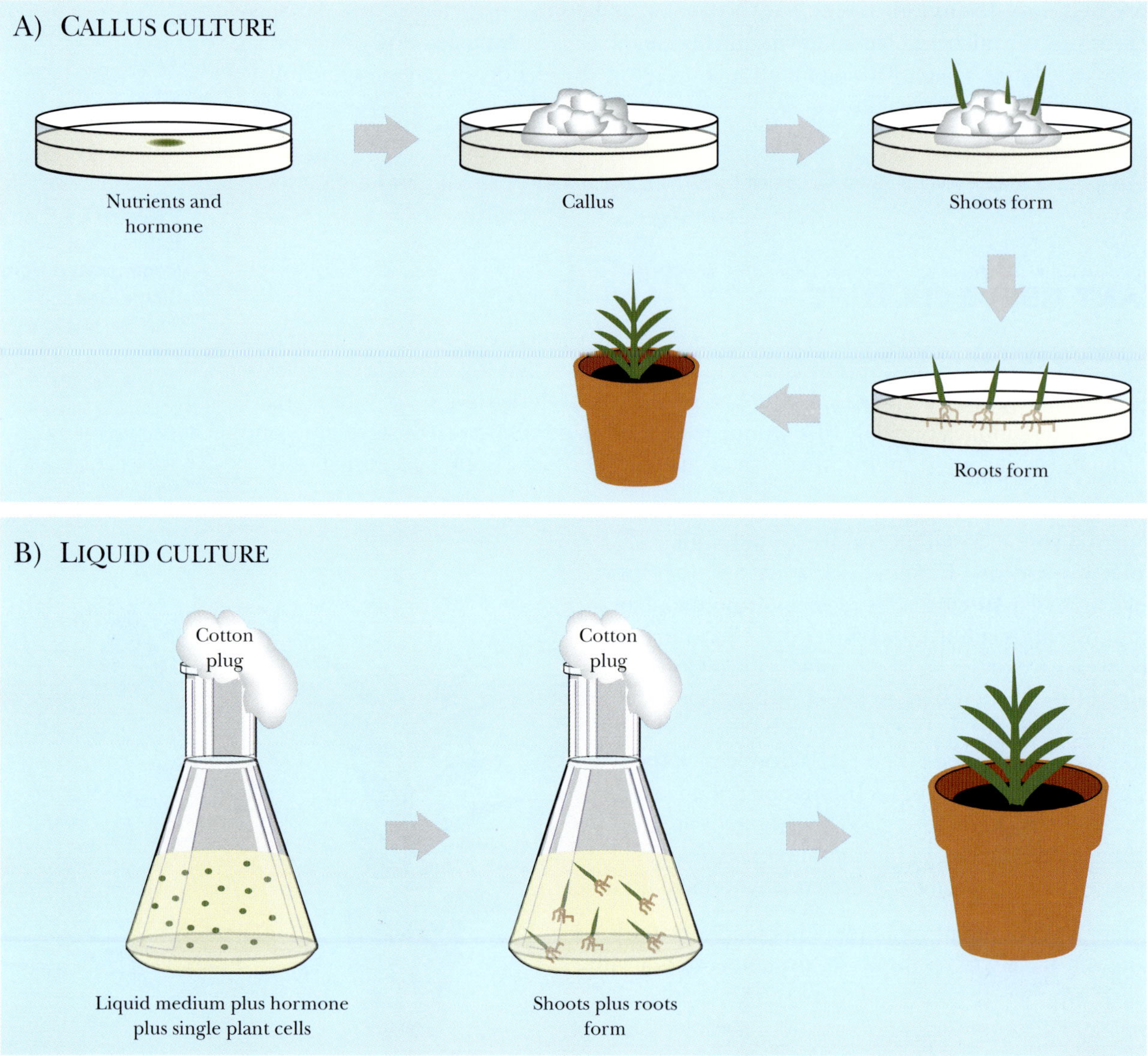

FIGURE 14.3 Callus or Liquid Culture of Plant Cells Can Regenerate Entire Plants
In callus culture a mass of undifferentiated cells grows on a solid surface. In liquid culture, separated single cells are grown. Both types of cultures can develop shoots and roots with appropriate manipulation of plant hormone levels.

for mutation breeding. Rather than using seeds, undifferentiated cells are exposed to the mutagen, and plants are regenerated from the mutagenized cells. Mutagens are more effective on the exposed cells of a callus rather than the protected cells within a seed.

Merely growing plants by tissue culture may induce mutations. The process of regenerating a plant from a single cell may cause three different types of alterations. Temporary physiological changes can occur in the regenerated plant. For example, when blueberry plants are regenerated via tissue culture, the plants are much shorter. These changes are not permanent, and after a few years growing in the field, the regenerated blueberries are no different from any other blueberries. Another alteration that may occur is an **epigenetic change**. This is an alteration that persists throughout the lifetime of the regenerated plant, but is not passed on to the next generation. Epigenetic changes are often due to alterations in DNA methylation (see Chapter 4). Finally, true **genetic changes** affect the regenerated plant and all its progeny. These may be due to changes in ploidy level, chromosome rearrangements, point mutations, activation of transposable elements, or changes in chloroplast or mitochondrial genomes. These types of changes are relatively common, but changes in the available nutrients and hormones during tissue culture can dramatically decrease the frequency of mutation.

Because plant cells are totipotent, regular plant tissue can be grown in culture. Callus culture and suspension culture cultivate plant cells from an undifferentiated state into small plants by altering the hormone concentrations.

Plant tissue culture can induce different types of alterations—some that are truly genetic, some that are epigenetic, and some that are physiological.

GENETIC ENGINEERING OF PLANTS

The first step in genetically engineering a plant is to identify a gene that will confer a specific desirable trait on the plant. In some ways, this is the hardest part of genetic engineering. The most desirable traits for a crop are increases in the amount of seed, grain, or other plant products. Increased resistance to disease or drought is also very useful. Finding the genes responsible is difficult, because multiple interacting genes usually control such traits. In addition, such genes may play other roles in plant physiology or development.

So far, most successful genetic engineering of plants has relied on inserting one or a few genes that supply simple, yet useful, properties. For example, resistance to the herbicide glyphosate is due to a single gene. Making a crop such as soybean resistant to glyphosate allows the farmer to kill the weeds in the field without harming the soybeans (see later discussion). Another desirable trait often due to a single gene is the production of toxins that kill harmful insects (see later discussion). Both these cases rely on transgenes derived from bacteria. As more research into plant physiology occurs, more genes can be identified that increase the value of a crop. For example, a two-gene pathway was engineered into rice to make it more resistant to drought (see later discussion).

Plants can also be engineered for novel products. Thus, golden rice expresses the biosynthetic pathway for vitamin A precursors. This rice was developed for people who rely on rice as the one main food in their diet. The addition of vitamin A precursors can prevent deficiencies that cause blindness or premature death in children in developing countries. Researchers have also engineered the human insulin gene for expression in *Arabidopsis* and safflower. At present, insulin is produced by engineered bacteria (see Chapter 19). However, plant-produced insulin is easier to isolate and purify in bulk and should cost much less. Further research will identify new genes and useful pathways that can be engineered into plants. Perhaps engineered plants will be used to clean up oil spills or other pollutants by growing them on contaminated soil.

Finding a gene with a specific useful trait for the plant is the first and most important step of plant genetic engineering.

GETTING GENES INTO PLANTS USING THE TI PLASMID

Plants suffer from tumors, though these are quite different from the cancers of animals. The most common cause is the Ti plasmid (tumor-inducing plasmid), which is carried by soil bacteria of the *Agrobacterium* group. Specifically, the Ti plasmid of *Agrobacterium tumefaciens* is an important tool for plant genetic engineering. The most important aspect of the infection is that a specific segment of the Ti plasmid DNA is transferred from the bacteria to the plant. Scientists have exploited this genetic transfer in order to get genes with desired properties into plant cells. *Agrobacterium* is unique in the ability to transfer a segment of its DNA from one kingdom to another. Most DNA transfers occur only between closely related organisms.

In nature, *Agrobacterium* is attracted to plants that have minor wounds by phenolic compounds such as acetosyringone, which are released at the wound (Fig. 14.4). These chemicals induce the bacteria to move and attach to the plant via a variety of cell surface receptors. The same inducers activate expression of the virulence genes on the Ti plasmid that are responsible for DNA transfer to the plant. This is under control of a two-component regulatory system (see Chapter 2). At the cell surface, the sensor, VirA, is autophosphorylated when it detects the plant phenolic compounds. Next, VirA transfers the phosphate to the DNA-binding protein, VirG, which activates transcription of the *vir* genes of the Ti plasmid. Two of the gene products (VirD1 and VirD2) clip the T-DNA borders to form a single-stranded immature T-complex. VirD2 then attaches to the 5′ end of the T-DNA, and bacterial helicases unwind the T-DNA from the plasmid. The single-stranded gap on the plasmid is repaired, and the T-DNA is coated with VirE2 protein to give a hollow cylindrical filament with a coiled structure. This is the mature form of T-DNA and traverses into the plant.

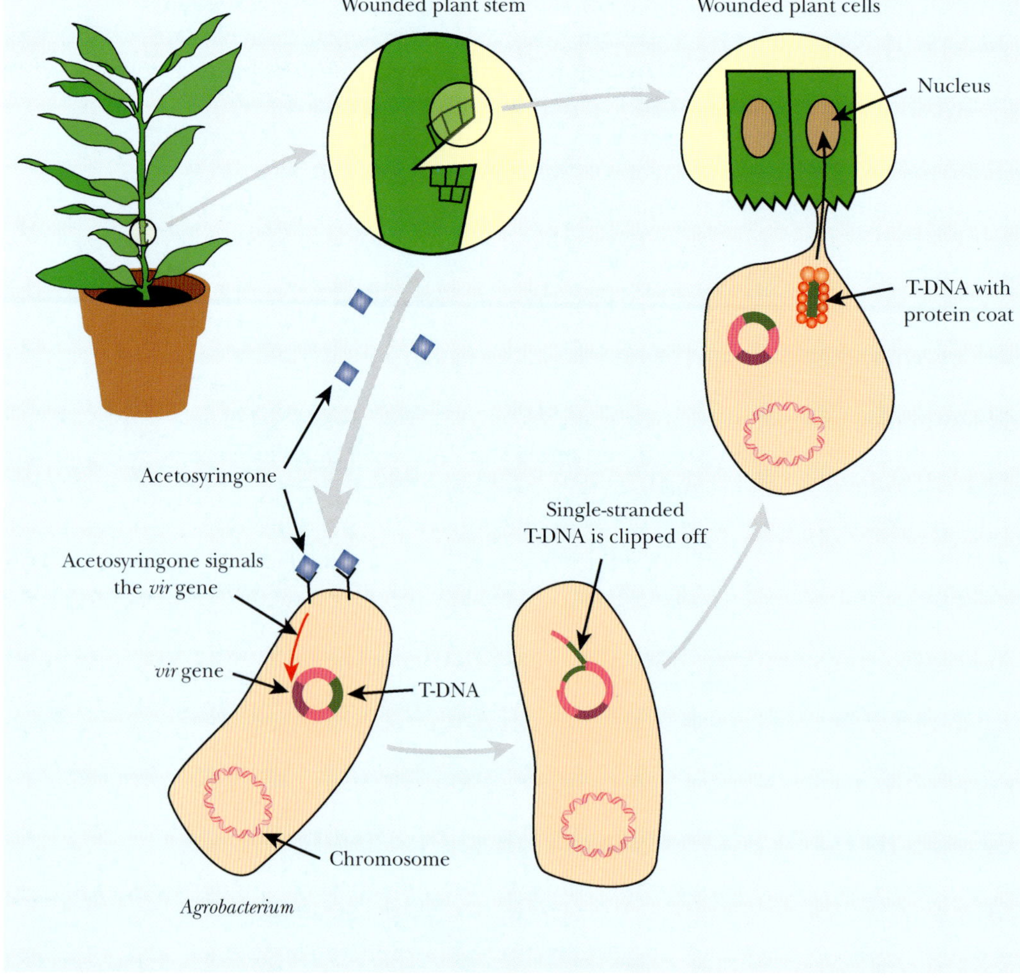

FIGURE 14.4 ***Agrobacterium*** **Transfers Plasmid DNA into Infected Plants**
Agrobacterium carrying a Ti plasmid is attracted by acetosyringone to a wounded plant stem. The Ti plasmid is cut by endonucleases to release single-stranded T-DNA, which is covered with protective proteins, and transported into the plant cell through a conjugation-like mechanism. The T-DNA enters the plant nucleus where it integrates into plant chromosomal DNA.

T-DNA is transferred to the plant in a process similar to bacterial conjugation. First, *Agrobacterium* forms a pilus. This rodlike structure forms a connection with the plant cell and opens a channel through which the T-DNA is actively transported into the plant cytoplasm. Both pilus and transport complex consist of proteins that are *vir* gene products. Once inside the plant cytoplasm, T-DNA is imported into the nucleus. Both VirE2 and VirD2 have nuclear localization signals that are recognized by plant cytosolic proteins. These proteins take the T-complex to the nucleus where it is actively transported through a nuclear pore. The single T-DNA strand is integrated directly into the plant genome and converted to a double-stranded form. The integration requires DNA ligase, polymerase, and chromatin remodeling proteins, which are all supplied by the plant.

Once they are part of the plant genome, the genes in the T-DNA are expressed. These genes have eukaryote-like promoters, transcriptional enhancers, and poly(A) sites and hence are expressed in the plant nucleus rather than in the original bacterium. The proteins they encode synthesize two plant hormones, auxin and cytokinin. Auxin makes plant cells grow bigger and cytokinin makes them divide. The infected plant cells begin to grow rapidly and without control, resulting in a tumor.

T-DNA also carries genes for synthesis of opines, which are a variety of different amino acid and sugar phosphate derivatives. The type of opine differentiates the various strains of *Agrobacterium*. Opines are made by plant cells that contain T-DNA but are used by the bacteria as carbon, nitrogen, and energy sources. Notice how the bacterium tricks the plant into using its resources to supply the bacteria with food. The Ti plasmid, which is still inside the *Agrobacterium*, carries genes that allow the bacteria to take up these opines and break them down for food. Note that other bacteria, which might be present by chance, cannot use opines because they do not possess the genes for uptake and metabolism. This ensures that the plant feeds only the bacteria with the Ti plasmid.

So how are Ti plasmids used to improve plants? First, the Ti plasmid is disarmed by cutting out the genes in the T-DNA for plant hormone and opine synthesis. Then, the transgene of interest, such as an insect toxin gene, is inserted into the T-DNA region of the Ti plasmid. The Ti plasmid is also streamlined by removing genes that are not involved in moving the T-DNA. These smaller plasmids are much easier to work with and can be manipulated in *Escherichia coli* rather than their original host, *Agrobacterium*. Now, when the T-DNA enters the plant cell and integrates into the chromosome, it will bring in the transgene instead of causing a tumor.

The transferred region of the plasmid must also have other elements in order for the transgene to function properly (Fig. 14.5). Expression of the transgene requires a promoter that works efficiently in plant cells. This may be one of two types. A **constitutive promoter** will turn the gene on in all the plant cells throughout development; thus every tissue, even the fruit or seed, will express the gene. A more refined approach is to use an **inducible promoter** that has an on/off switch. An example of this is the *cab* promoter from the gene encoding chlorophyll *a*/*b* binding protein. This promoter is turned on only when the plant is exposed to light; therefore, root tissues and tubers such as potatoes will not express the gene. Many different promoters may be used, but ideally, the promoter should turn on only in tissues that need transgene function. Another important component for the genetically modified T-DNA region is some sort of selectable marker. Including an herbicide or antibiotic resistance gene in the T-DNA region can be used to track whether the foreign DNA has been inserted into plant cells. The selectable marker may cause problems because it must be expressed constitutively throughout the plant. Many people worry that the protein product of the selectable marker could cause allergies or reactions if expressed in fruit, grain, or vegetables. However, systems exists that can remove this gene once the transgenic plant has been isolated (see later discussion).

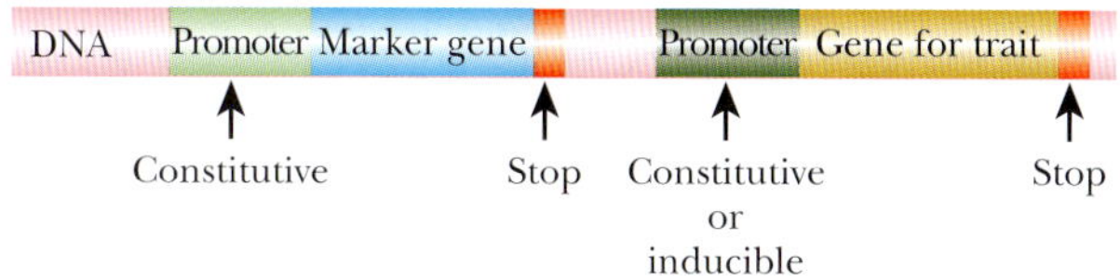

FIGURE 14.5
Essential Elements for Carrying a Transgene on Ti Plasmids
The T-DNA segment contains both a transgene and a selective marker or reporter gene. These have separate promoters and termination signals. The marker or reporter gene must be expressed all the time, whereas the transgene is often expressed only in certain tissues or under certain circumstances and usually has a promoter that can be induced by appropriate signals.

In practice, *Agrobacterium* is used to transfer genes of interest into plants using tissue culture. Either dissociated plant cells called *protoplasts* or a piece of callus are cultured with *Agrobacterium* harboring a Ti plasmid with modified T-DNA. After coculture, the plant cells are harvested and incubated with the herbicide or antibiotic used as the selectable marker. This kills all the cells that were not transformed with T-DNA or failed to express the genes on the T-DNA. The transformed cells can then be induced to produce shoot and root tissue by altering the hormone conditions in the medium as described earlier (Fig. 14.6). The small transgenic plants can then be screened for transgene expression levels (see later discussion).

Recently, a method for ***in planta Agrobacterium* transformation** was developed and has revolutionized the plant transformation world. *In planta* transformation is also known as the **floral dip** method. The method was developed using the model plant *Arabidopsis* but has been extended to other plants, such as wheat and maize. First, *Arabidopsis* plants are grown until flower buds begin to form. These buds are removed and allowed to regenerate for a few days. Once they begin to regenerate, the plants are dipped into a suspension of *Agrobacterium* containing a surfactant. The surfactant allows the *Agrobacterium* to adhere to the plant and transfer its T-DNA. Because the flower buds are just beginning to form, the T-DNA somehow becomes part of the germline through the ovarian tissue. The plant is allowed to finish growing and set seed. These seeds are harvested and grown in selective media to find those that have integrated and expressed T-DNA. Although the method gives a low percentage of transformants, so many seeds can be screened that the overall procedure works well.

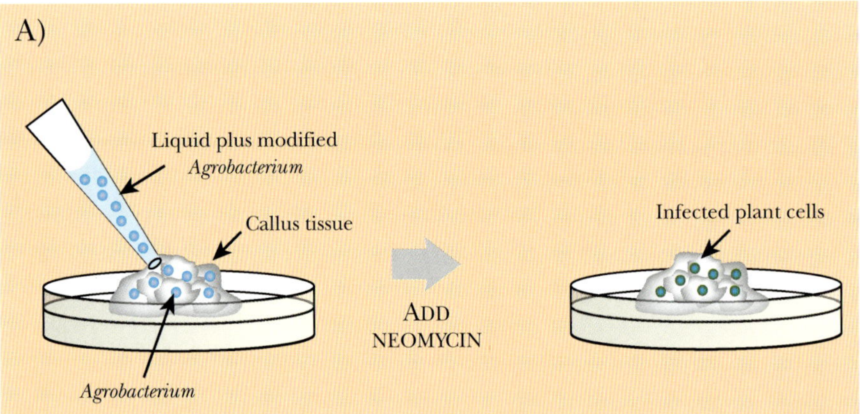

FIGURE 14.6
Transfer of Modified Ti Plasmid into a Plant
Agrobacterium carrying a Ti plasmid is added to plant tissue growing in culture. The T-DNA carries an antibiotic resistance gene (neomycin in this figure) to allow selection of successfully transformed plant cells. Both callus cultures (A) and liquid cultures (B) may be used in this procedure.

Agrobacterium are bacteria that infect wounded plants, inject a Ti plasmid into the plant cell, and transfer the T-DNA portion into the plant genome. The bacteria trick the plant into making its own sugar source, and then induce the formation of a tumor.

Transgenic plants are created by placing the foreign gene into the Ti plasmid of *Agrobacterium* and allowing the bacteria to transfer its T-DNA into the plant genome.

In planta Agrobacterium transformation or floral dip creates transgenic plants by integrating T-DNA into regenerating floral tissues.

PARTICLE BOMBARDMENT TECHNOLOGY

Another strategy for getting a gene of interest into plant tissue is to blast the DNA through the plant cell walls with a particle gun. Unlike Ti plasmid transfer by *Agrobacterium*, this technique works with all types of plants. The basic idea is that DNA is carried on microscopic metal particles. These are fired by a gun into plant tissue and penetrate the plant cell walls. This technique is rather nonspecific, yet it has been very successful in the plant world.

First, either a leaf disk (just a round piece of leaf tissue) or a piece of callus is isolated from the plant, placed on a dish, and put in a vacuum chamber. The DNA to be inserted (carrying the gene of interest, with any promoter and enhancer elements, plus selectable markers) is coated on microscopic gold or tungsten beads. Gold beads are preferable because tungsten can be toxic to some plants. The beads are placed at the end of a plastic bullet. One variant of the method uses a blast of air or helium to project the bullet toward the sample. In the first gene guns, actual firearm blanks were used to accelerate the bullet. Between the bullet and plant tissue is a plastic meshwork stop. When the bullet hits the stop, the DNA-coated beads are thrown forward through the meshwork and continue on through the vacuum chamber and into the plant tissue. An alternative method is to accelerate the beads by a strong electrical discharge. The high voltage vaporizes a water droplet, and the resulting shock wave propels a thin metal sheet covered with the particles at a mesh screen. The screen blocks the metal sheet but allows the DNA-coated particles to accelerate through into the plant tissue (Fig. 14.7). One advantage to this method is that the strength of the electrical discharge can be controlled; therefore, the amount of penetration into the tissue can be changed at will.

When the beads penetrate the tissue, some will actually enter the cytoplasm or nucleus of the leaf or callus cells. The DNA dissolves off the beads inside the cells and the DNA is free to recombine with the chromosomal DNA of the plant (Fig. 14.8). The leaf or callus tissue is then transferred to selection media where the cells that integrated the DNA carrying the selectable marker are able to grow, but other cells die. The transformed plants are regenerated using tissue culture techniques, and finally screened for the gene of interest.

FIGURE 14.7 Two Types of Particle Guns for DNA
In (A) a gene gun that operates via pressurized air is shown and in (B) a gene gun that operates via a high-voltage discharge. In both cases the stop plate halts the projectile and the microscopic metal particles carrying the DNA continue on to penetrate the plant tissue.

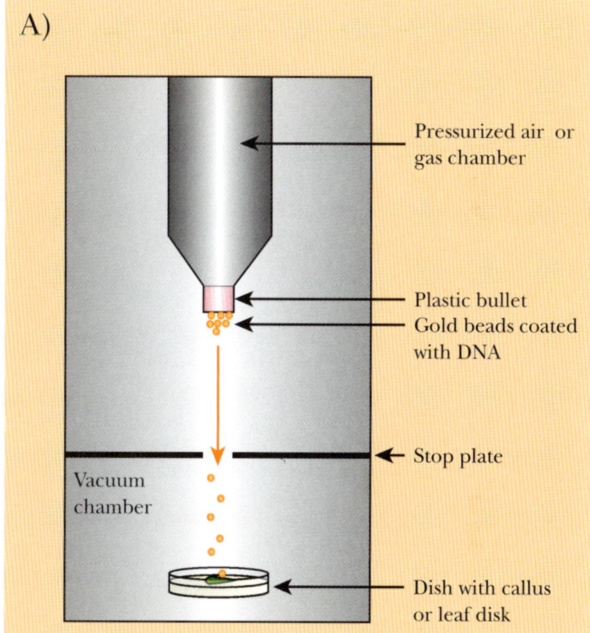

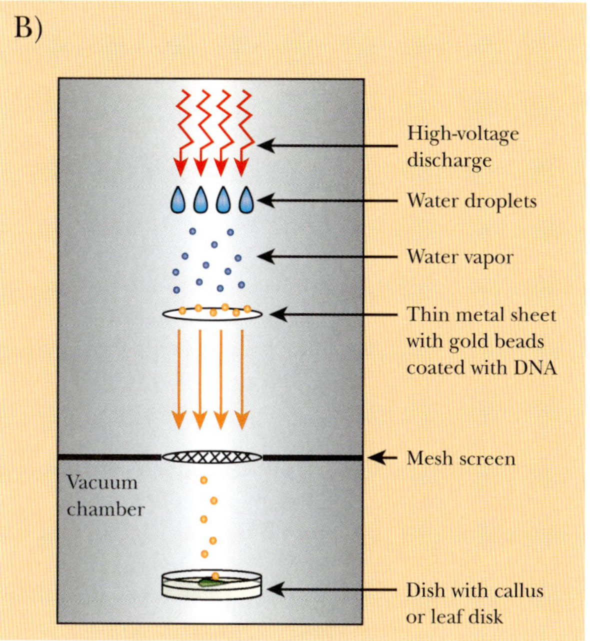

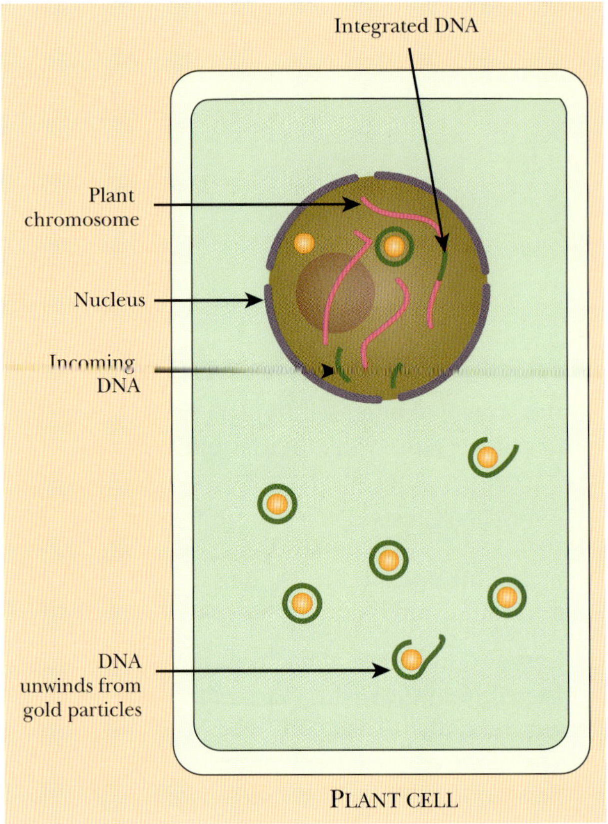

FIGURE 14.8 DNA Carried on Microscopic Gold Particles Can Integrate into Plant Chromosomes

After penetrating the cell, the DNA unwinds from around the gold carrier particle. Some of the DNA enters the nucleus and is successful in integrating into the plant chromosomes.

Particle guns have also been used on animal tissues (see Chapter 15, Transgenic Animals). In addition, scientists have modified the technology in order to transform DNA into the mitochondria of yeast, as well as the chloroplasts of *Chlamydomonas*, a small green alga.

Gold beads coated with transgene DNA can be accelerated into plant cells using a gun. The transgene DNA dissolves from the bead and integrates into the plant genome.

DETECTION OF INSERTED DNA

How do we know whether any DNA got into the target cells? Including a selectable marker or reporter gene on the same segment of DNA as the transgene provides a good indicator of whether the transgene is in the plant. One widely used reporter gene is ***npt***, which encodes **neomycin phosphotransferase**. This enzyme inactivates the antibiotic neomycin by attaching a phosphate group. Cells successfully incorporating DNA carrying the *npt* gene are no longer killed by neomycin. This allows direct selection of transformed cells as treatment with neomycin kills any cells that did not integrate the DNA.

A nonlethal diagnostic method is to include a reporter gene that codes for **luciferase**. This enzyme emits light when provided with its substrate, **luciferin** (Fig. 14.9A). Luciferase is found naturally in assorted luminous creatures, from fireflies to luminous squid. The genes encoding luciferase from eukaryotes and prokaryotes are referred to respectively as ***luc*** and ***lux***. Although both kinds of luciferase produce light, the chemical nature of their luciferins and mechanism of the reactions are different. If DNA carrying the eukaryotic *luc* gene is

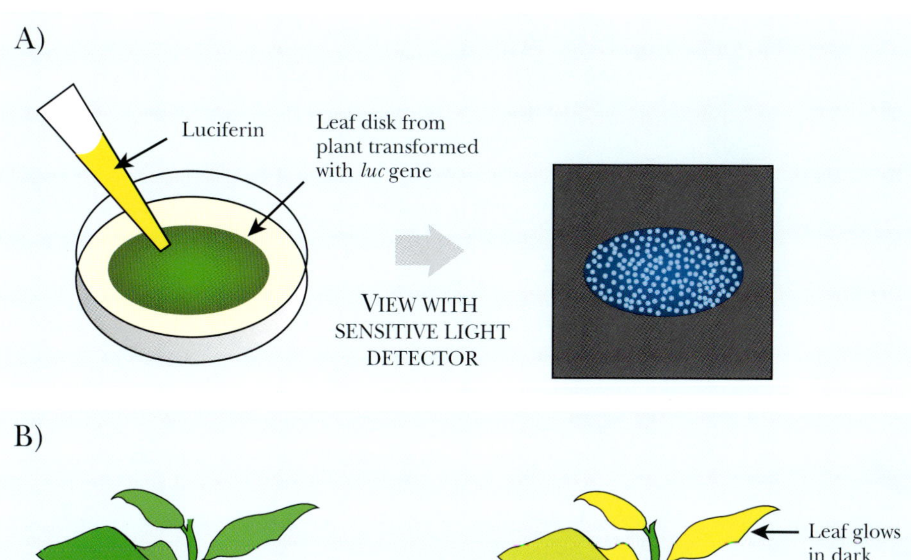

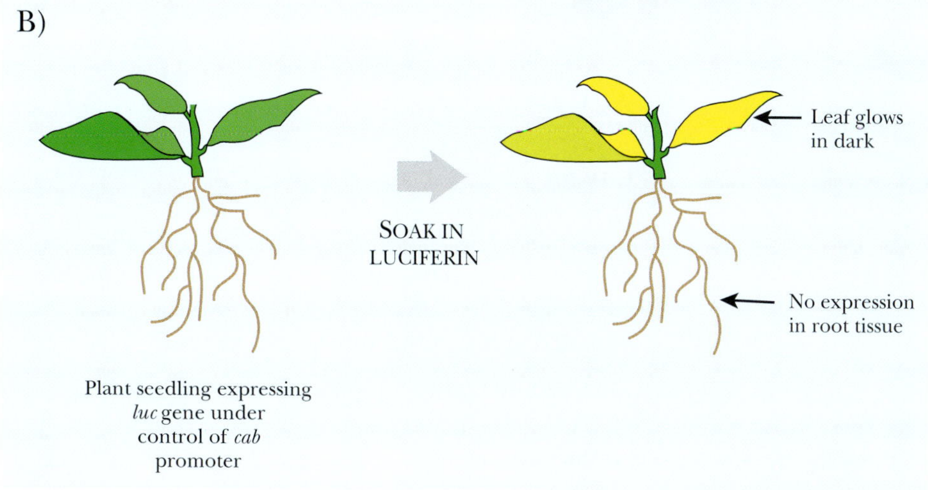

FIGURE 14.9 Luciferase as a Reporter in Plant Tissues

Plant tissue carrying the *luc* gene for firefly luciferase emits blue light when provided with the substrate luciferin. In (A) a leaf disk is viewed by a photocell detector. In (B) the *luc* gene in a seedling is expressed under control of an inducible promoter.

successfully incorporated into a target plant cell, light will be emitted when luciferin is added. In this case, the luciferin is oxidized using endogenous ATP and O_2. Although high-level expression of luciferase can be seen with the naked eye, usually the amount of light is small and must be detected with sensitive electronic apparatus such as a scintillation counter, a sensitive photocell detector, or a CCD camera.

This reporter gene also has another advantage. The luciferase protein is not stable for long in the plant, and so the amount of active protein correlates with the level of gene expression at any given time. Therefore, this reporter gene can be used to test the activity of specific promoters. For example, if the *cab* promoter controls the expression of the *luc* gene, then luciferase is only made when this promoter is turned on in the plant. Transgenic *Arabidopsis* plants containing this construct only emit light from luciferase in photosynthetic tissue that is exposed to light—the natural conditions that induce expression of the plant *cab* gene (see Fig. 14.9B). Other promoters can also be studied using the *luc* gene.

Once cells successfully expressing the reporter gene are identified, they are regenerated into plants using tissue culture techniques. The plants obtained are then screened for the gene of interest or transgene. Techniques such as PCR (see Chapter 4) can confirm the presence of the transgene. Further analyses such as Southern blots can pinpoint the relative chromosomal location of the inserted transgene. Several plant genomes have now been fully sequenced, so localizing transgenes has recently become much easier.

Different types of reporter genes are used to confirm the integration of the transgene. Neomycin phosphotransferase inactivates the antibiotic neomycin, making any cells with the transgene resistant to neomycin treatment. The luciferase gene makes the cell glow in ultraviolet light when the transgene is present.

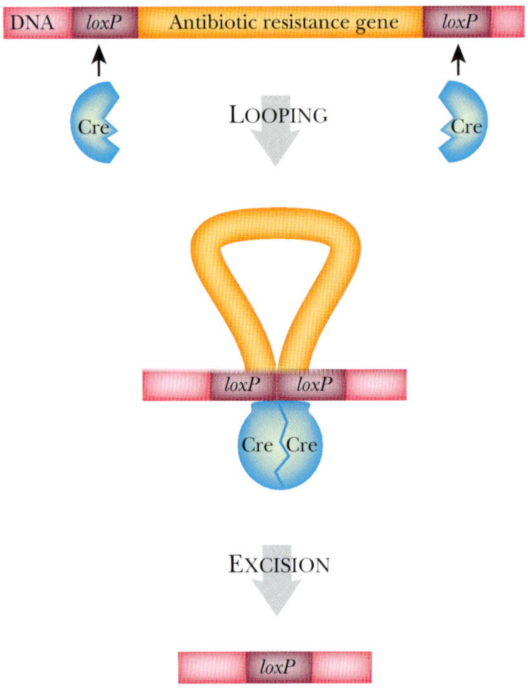

USING THE CRE/*loxP* SYSTEM

One of the main complaints about using transgenic technology in agriculture is the inclusion of the selectable marker or reporter gene. In particular, many consumers are concerned by the presence of a gene for antibiotic resistance in their food supply. However, genetic techniques are available that allow removal of the reporter or resistance gene, after integration of the incoming DNA has been checked.

Bacteriophage P1, which naturally infects *E. coli*, has a simple system for genetic recombination, the **Cre/*loxP*** system. Cre stands for "causes recombination." The **Cre** protein is a recombinase enzyme that recognizes a specific 34 base-pain DNA sequence, the ***loxP*** site. The Cre protein catalyzes recombination between two *loxP* sites (Fig. 14.10). By placing two *loxP* sites on either side of a segment of DNA, the enclosed region may be deleted by Cre recombination. To accomplish this, the *cre* gene is also included in the transgenic construct and is expressed when it is time to delete the unwanted DNA segment (see later discussion). This approach allows selectable marker genes to be removed from plant DNA after use. The segment of DNA that is recombined out of the chromosome has neither an origin of replication an nor any telomeres and will be either degraded or lost during mitosis and meiosis.

FIGURE 14.10
Cre/*loxP* System of Bacteriophage P1
The Cre protein binds to *loxP* recognition sites in the DNA. Two nearby *loxP* sites are brought together, and recombination between them eliminates the intervening DNA. A single *loxP* site remains in the target DNA molecule.

The *cre* gene can be added to the system by cross-pollination between two different plants, one carrying the transgene plus selectable marker flanked by two *loxP* sites and the other carrying the *cre* gene (inserted into a different location; Fig. 14.11). First the pollen from the plant with the *cre* gene is added to the stigma of the plant with the transgene. The resulting seeds are grown and checked for sensitivity to the selective agent (e.g., neomycin). If the Cre protein is present in the progeny, the selectable marker gene will be excised and lost during growth. This plant now has the transgene and the *cre* gene, but no longer has the gene for antibiotic resistance. If another cross is made between this transgenic plant and a wild-type plant, some of the progeny will have the transgene but lack the *cre* gene. Using a genetic scheme such as this will ensure that the final transgenic plant has only one extra gene, the transgene. This system has proven so easy and useful that every new variety of transgenic plant released to the public contains only the single transgene of interest.

By using the Cre/*loxP* system, the reporter genes can be removed from the genome. Cre is a recombinase that binds to *loxP* sites that are flanking the reporter gene and removes any DNA between the two sites.

PLANT BREEDING AND TESTING

Making a transgenic plant is a relatively small step; evaluating and testing the transformed plants is the most time consuming part of the whole process. The expression level of the transgene may vary considerably, depending on the number of integrated transgenes and their location. The term **event** refers to each independent case of transgene integration. For example, if one copy of the transgene inserted into chromosome 2 of the first transformant, this would be referred to as event 1. If, in the same experiment, a separate transformed plant received the same transgene, but integrated into chromosome 4, that would be a second event. The location of integration affects the expression of the transgene. If the transgene in event 1 integrated into a region of heterochromatin (see Chapter 2), the gene would probably be silenced and never be expressed at all, even if provided with a strong promoter. In contrast, if in event 2, the transgene integrated just downstream of a very active chromosomal gene, it would probably show high expression levels. The number of integrated transgenes can also vary. Often a single transformed plant will gain multiple copies of the same transgene.

The first issue to address is whether the transgene causes any harmful side effects to the plant. Does the transgene function as expected? Does the transgene affect the crop quality? Does the transgene affect the ecosystem? The answer to these questions depends on the individual transgene being used (see later discussion for specific examples).

If no harmful effects are found, then the transgene must be transferred from the experimental plant to one with a much higher yield. Most transgenic plants are made from old varieties that are good for work in laboratories, but do not make a lot of seeds per acre or are very susceptible to diseases. Furthermore, as discussed earlier, the regeneration of plants through tissue culture may itself cause mutations. In order to overcome these problems, the transgene is moved by traditional cross-breeding into high-yielding varieties that farmers are already using. First, the pollen from the plant with the transgene is harvested and put onto the corn silk or stigma of the high-yielding variety. The seeds from this cross are harvested and grown. This is the F1 generation, and the plants containing the transgene are selected. For example, if the transgene makes the plant resistant to an herbicide, the F1 generation is sprayed to kill the plants without the transgene. The pollen from the F1 plants that survive is **back-crossed** to the original high-yielding parent. The seeds are grown, plants with the transgene are selected, and the whole process is repeated about four or five times. This crossing scheme will ensure that about 98% of the genes in the final plant are from the high-yielding variety, and the remaining genes are from the original transgenic plant. Because it takes an entire summer for one generation of corn, soybeans, or cotton to grow, this backcrossing scheme can take many years to complete.

Once the transgene is back-crossed into a suitable variety, field tests are performed to determine how the transgene affects the growth, yield, disease resistance, and other important traits of the plant. These field tests must be done over many different locations so that soil type, terrain, rainfall, and other factors can be allowed for. The field tests may also take many years. Different amounts of rain from one year to the next can greatly affect crop yields. The plant breeder selects only the plants that consistently have the highest yield with the best disease resistance. The other plants are never grown again.

The other issue in releasing transgenic plants to the public is passing the tests of government regulatory agencies. These agencies regulate all stages of the transgenic construction process. An Institutional Committee for Biosafety regulates how the transgene is handled when making the transgenic plant, whether in *E. coli*, *Agrobacterium*, or the plant itself. These committees are usually associated with the university or company where the work is done, but they all follow guidelines from the National

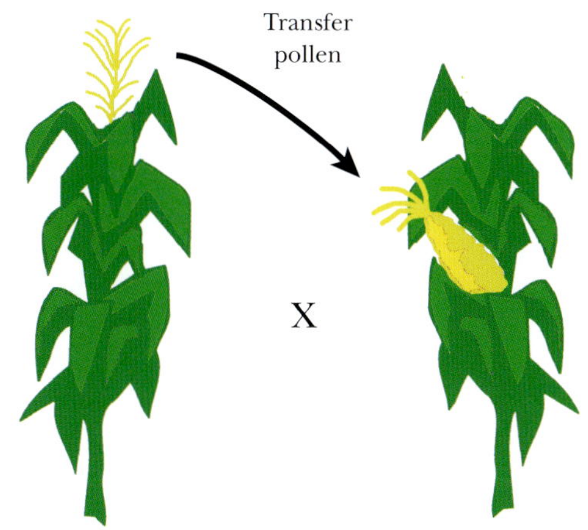

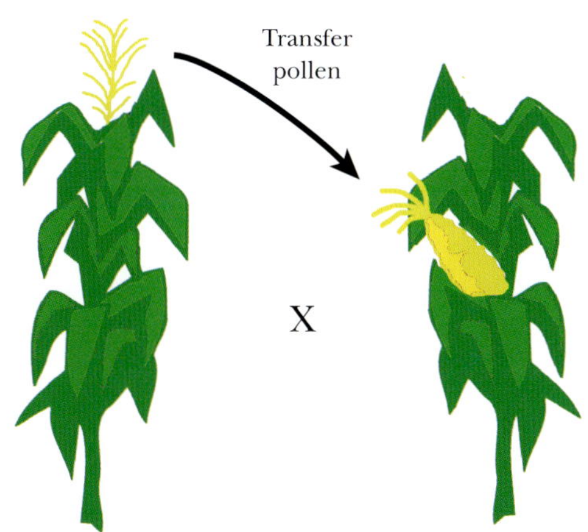

FIGURE 14.11 Cross-Pollination Scheme to Remove Unwanted Marker Genes
Pollen from a plant carrying the *cre* gene is transferred to a first-generation transgenic plant, still carrying a marker gene for antibiotic resistance. Expression of Cre protein in the progeny results in excision of the marker gene. Those plants that have lost the antibiotic resistance marker are finally crossed with wild-type plants. Normal mendelian assortment occurs and some progeny from this cross will have the *cre* gene; others will not. Those lacking *cre* are kept for further use.

Institutes of Health (NIH). The guidelines regulate the environment in which the transgenic plant may be grown (laboratory, greenhouse, etc.). In order to test the transgenic crop in the field, the Animal and Plant Health Inspection Service of the U.S. Department of Agriculture must be notified and must approve the plan. The scientist must provide extensive data on the transgene, its potential effect on the plant, the ecosystem, and any other crops similar or related to the transgenic plant.

Two other agencies must also approve the transgenic crop. If the transgenic plant gives a food product such as corn, the Food and Drug Administration (FDA) must do rigorous testing for possible allergies to the transgenic plant. The potential toxicity of the transgenic crop and whether or not the nutritional quality of the product is affected by the transgene are also tested. The Environmental Protection Agency also evaluates the transgenic crop for potential effects on the environment and on animals or insects that also inhabit the farmers' fields. These are just the beginning of the regulations since anything sold overseas must also satisfy the regulatory commissions from all the countries in which the product is sold. At this time, overseeing transgenic technology is a hot issue that is constantly changing. As transgenic technology becomes more common and more information becomes available, the regulatory issues will change and adapt to the type of crops being produced.

In order for a transgenic plant to be grown for commercial use, the transgene must be crossed into a different variety of the plant.

Transgenic plants are scrutinized by a variety of agencies before being used commercially.

TRANSGENIC PLANTS WITH HERBICIDE RESISTANCE

Herbicides cost the world's farmers more than $12 billion each year. Yet despite this, around 10% of the crop is lost due to weeds. One problem is that many of the chemicals used do not discriminate between crops and weeds. So they kill any plant they touch. One solution is to make the crop plants resistant to the herbicide by genetic engineering and then spray both weeds and crop together. The weeds are killed, but the crop survives.

One of the best herbicides on the market is **glyphosate**, sold by Monsanto under the trade name Round-Up. Glyphosate is environmentally friendly because it quickly

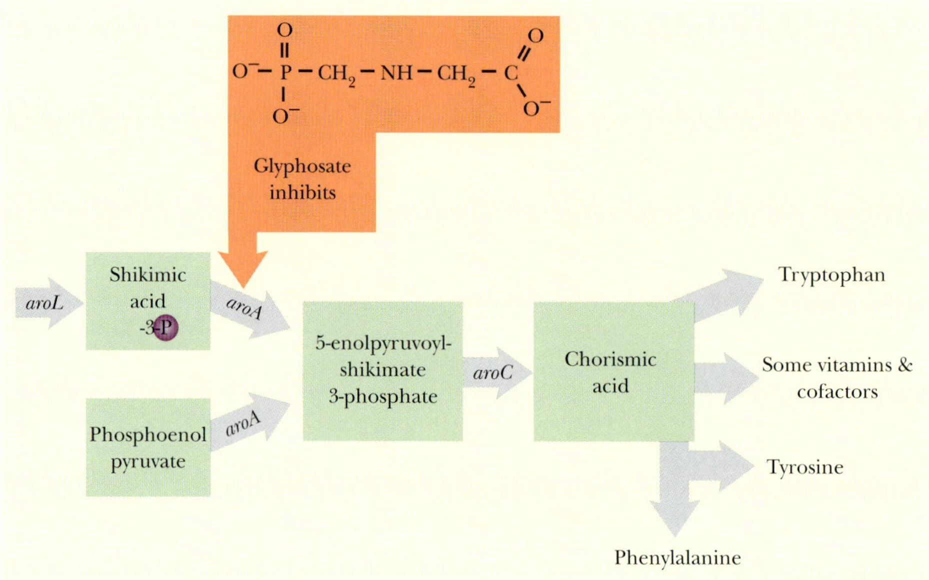

$$O^- - \overset{\overset{O}{\|}}{\underset{\underset{O^-}{|}}{P}} - CH_2 - NH - CH_2 - C(=O)O^-$$

Glyphosate inhibits

aroL → Shikimic acid -3-P → aroA → 5-enolpyruvoyl-shikimate 3-phosphate → aroC → Chorismic acid → Tryptophan; Some vitamins & cofactors; Tyrosine; Phenylalanine

Phosphoenol pyruvate → aroA → 5-enolpyruvoyl-shikimate 3-phosphate

FIGURE 14.12
Glyphosate Inhibits EPSPS in the Aromatic Pathway
The enzyme 5-enolpyruvoylshikimate-3-phosphate synthase (EPSPS) is the product of the *aroA* gene and makes 5-enolpyruvoylshikimate-3-phosphate, a precursor in the pathway to aromatic amino acids and cofactors. Glyphosate, an analog of phosphoenolpyruvate, inhibits EPSPS.

breaks down to nontoxic compounds in the soil. The glyphosate molecule is an amino acid phosphate derivative related to glycine. Glyphosate kills plants by blocking the synthetic pathway for the essential aromatic amino acids phenylalanine, tyrosine, and tryptophan. As shown in Fig. 14.12, glyphosate inhibits one particular enzyme in this pathway, **EPSPS** (5-enolpyruvoylshikimate-3-phosphate synthase), the product of the *aroA* gene, which is localized to the chloroplast. This enzyme is found naturally in all plants, fungi, and bacteria but is not present in animals; thus, the glyphosate target enzyme does not even exist in humans. Animals, including humans, must ingest aromatic amino acids because we cannot make them ourselves. When glyphosate is sprayed onto plants, it gets into the chloroplasts and binds to the EPSPS protein and blocks the pathway for aromatic amino acids. The plant essentially starves to death.

Because developing resistance directly in plants is difficult, scientists took a different route. EPSPS is also found in bacteria, so scientists turned there to find an EPSPS enzyme that was resistant to glyphosate, but still capable of synthesizing aromatic amino acids. Mutant strains of *Agrobacterium* that are resistant to glyphosate can be directly selected by plating onto medium containing glyphosate. Such mutants have EPSPS that is resistant to glyphosate but is still enzymatically active. This glyphosate-resistant version of the *aroA* gene was cloned (see Chapter 3) and modified for expression in plants. The bacterial promoter and terminator sequences were replaced with plant promoters and terminators. An antibiotic resistance gene was also added to the construct to allow selection. Finally, because EPSPS is localized to the chloroplast, DNA encoding a small **chloroplast transit peptide** was added to the front of the gene. The chloroplast transit peptide is present at the N-terminus of the protein, when it is first made, and targets it to the chloroplast. The transit peptide is cleaved off while crossing the chloroplast membranes, and only the functional enzyme enters the chloroplast (Fig. 14.13).

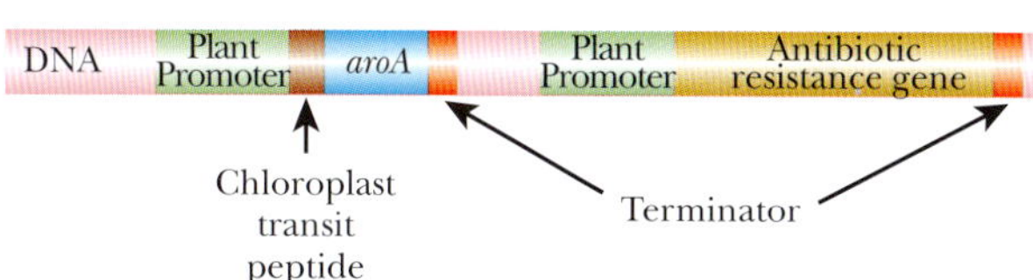

FIGURE 14.13
Expression of the *Agrobacterium aroA* Gene in Plants
The bacterial *aroA* gene must be placed under control of a promoter active in plants. Correct localization of the AroA protein (EPSPS) into the chloroplast requires addition of a chloroplast transit peptide at the N-terminus of the protein.

The glyphosate-resistant *aroA* gene from *Agrobacterium* has been transformed into several different crops, including soybean, cotton, and canola. Canola and cotton plants were engineered using the Ti-plasmid method of gene transfer. Soybean was modified using the gene gun approach. The resulting herbicide-resistant crops were easy to identify after regenerating the plants from tissue culture.

Comparison of the mutant bacterial *aroA* gene with the sensitive wild-type version revealed which amino acid changes were needed for glyphosate resistance. Because bacterial and plant *aroA* genes are homologous, equivalent changes should result in glyphosate resistance in plant *aroA* genes as well. Indeed, this information allowed the *aroA* gene from corn to be engineered by altering its DNA sequence *in vitro*. The altered corn *aroA* gene provided glyphosate resistance after being reinserted into corn plants by the gene gun.

Other herbicide tolerance genes have also been used to make transgenic crops, although these are not as widely used as the Round-Up–resistant transgenic plants. An example is resistance to sulfonylureas and imidazolinones, which inhibit an enzyme in the pathway that synthesizes the branched amino acids leucine, isoleucine, and valine. Plants resistant to these herbicides are quite common because a single amino acid substitution makes the enzyme resistant to the herbicide, but still able to synthesize the amino acids. Another example is resistance to glufosinate, an herbicide that blocks synthesis of glutamine. Glufosinate was originally discovered as an antibiotic produced by *Streptomyces*. Scientists therefore identified the enzyme that prevented *Streptomyces* from being harmed by its own antibiotic. This gene was cloned and transformed into crop plants.

Glyphosate inhibits EPSPS in the plant aromatic amino acid biosynthetic pathway. This is a good herbicide because humans do not have an enzyme like EPSPS.

To make transgenic plants resistant to glyphosate, the EPSPS gene must be altered to prevent the glyphosate binding. The mutant EPSPS was made in bacteria and modified to be expressed in plants.

TRANSGENIC PLANTS WITH INSECT RESISTANCE

Although weeds are a nuisance, even worse enemies of plants, and correspondingly more expensive to farmers, are:

1. Insects and roundworms
2. Fungal diseases (molds, blights, rusts, and rots)
3. Viral diseases of plants

It is possible to engineer plants for resistance to all of these, but we will consider just the insects here. Spraying crops with insecticides is a very costly and hazardous procedure. Insecticides are often more toxic to humans than are herbicides, because insecticides target species closer to our own. Many insect biochemical pathways are found not only in humans, but also in rodents or birds that may inhabit crop fields. Luckily, naturally occurring toxins exist that are lethal to insects but harmless to mammals. The prime example is the toxin from a soil bacterium called *Bacillus thuringiensis*. **Bt toxin**, as it is called, has been sprayed on crops to prevent insects such as the cotton bollworm and European corn borer from destroying cotton and corn, respectively. Damage from the European corn borer plus the cost of insecticides to control it cost farmers about $1 billion annually. Damage by the corn borer also makes corn plants susceptible to infection with a toxic fungus that can harm humans if ingested.

Bacteria of the genus *Bacillus* produce spores that contain a crystalline or **Cry protein**. When insects eat *Bacillus* spores, the Cry protein breaks down and releases **delta endotoxin** (i.e., the Bt toxin). This toxin binds to the intestinal lining of the insect and generates holes, which cripple the digestive system, and the insect dies (Fig. 14.14). Different species of *Bacillus* produce a family of different but related Cry proteins. These were originally classified according to insect susceptibility: CryI killed *Lepidoptera* (butterflies and moths), CryII killed both *Lepidoptera* and *Diptera* (flies), CryIII killed *Coleoptera* (beetles), and CryIV killed *Diptera* (but not *Lepidoptera*). However, as the number of known Cry variants increased, this classification became too simplistic, because sequence similarities did not always correspond to the spectrum of insecticidal activity. Cry proteins are now classified with Arabic numerals into some 20 subfamilies based on sequence relationships.

Instead of spraying insects with the toxin, scientists have used transgenic technology to insert the *cry* genes directly into plants. When a cloned toxin gene was inserted into tomato plants, it partly protected against tobacco hornworm. However, the plants made only low levels of the toxin because the toxin gene is from a bacterium and is designed to express well in bacteria, not plants.

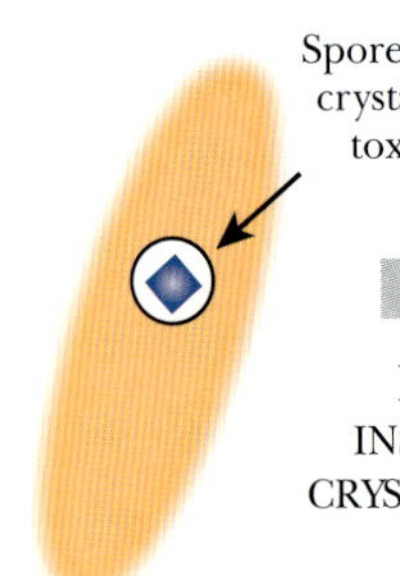

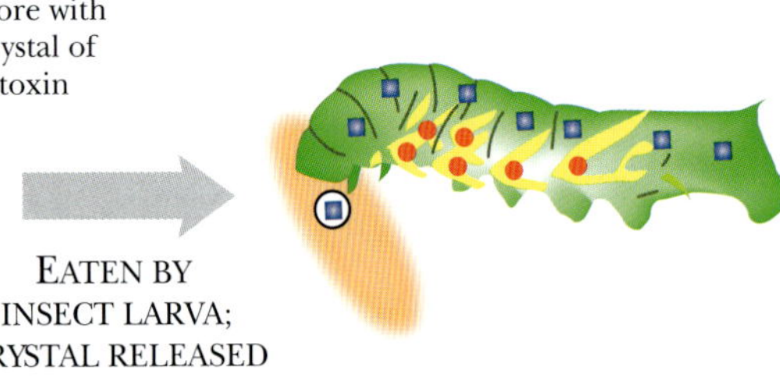

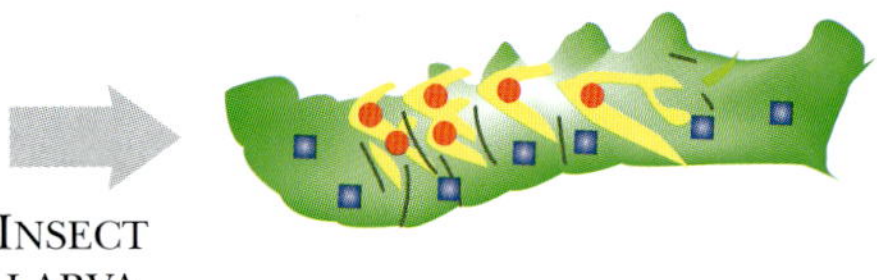

FIGURE 14.14
Insect Larvae Are Killed by Bt Toxin
Bacterial spores of *Bacillus* are found on food eaten by the caterpillar. The crystalline protein is released by digestion of the spore and its breakdown produces a toxin that kills the insect larvae.

Therefore the toxin gene was altered to enhance expression. The original toxin is a big protein that has 1156 amino acids. However, only the first 650 amino acids are necessary for the toxic effect; therefore, the protein was truncated by deleting the last half of the gene. The shorter, truncated protein requires less energy to produce. Next, the toxin gene was placed under the control of a promoter that gives constant high-level expression in plants. Certain promoters from plant viruses, such as cauliflower mosaic virus, satisfy these requirements and give 10-fold increases in toxin production.

When genes from one organism are expressed in a very different host cell, codon usage also becomes a problem. As explained in Chapter 2, the genetic code is redundant in the sense that several different codons can encode the same amino acid. So although a protein has a fixed amino acid sequence, there is considerable choice in which codons to use. Different organisms favor different codons for the same amino acid and have different levels of the corresponding tRNAs. If a gene uses codons for a rare tRNA, the supply of this may limit the rate of protein synthesis. In practice this is relevant only to genes expressed at high levels—exactly the situation here. Therefore the insect toxin gene was altered by changing many of the bases in the third position of redundant codons. Almost 20% of its bases were altered to make it more plantlike in codon usage. This did not alter the amino acids encoded and, therefore, the toxin protein itself was not affected by this procedure. However, the rate at which plant cells made the protein greatly increased and gave another 10-fold increase in toxin production.

Transgenic Bt crops such as cotton and corn have many advantages over spraying the fields with the toxin itself. One advantage is the toxin doesn't drift over other areas, which prevents cross-contamination. Using transgenic crops reduces the amount of insecticide needed. In 1998, about 450,000 kg less insecticide was used on the U.S. cotton crop alone. Yet only 45% of the cotton crop was actually transgenic.

Transgenic plants have been made to express the Cry protein from the soil bacterium *Bacillus thuringiensis*, which is a toxin that kills damaging insects such as the European corn borer and cotton bollworm.

To get good expression of the bacterial *cry* gene, the wobble position was altered to make the codon usage more compatible with plants. Unnecessary regions of the gene were removed so that the transgene would require less energy to be produced.

TREHALOSE IN TRANSGENIC PLANTS INCREASE STRESS TOLERANCE

Plants are highly adaptable and have many mechanisms to survive environmental stresses. Drought and high salinity of irrigation water are two stresses that pose problems in growing crop plants. In drought-tolerant plants, fungi, and bacteria, the sugar trehalose protects the organism during times of stress. **Trehalose** is a nonreducing storage carbohydrate that has the capacity to absorb or release water molecules. Two enzymes make trehalose: **Trehalose phosphate synthase** converts UDP-glucose and glucose 6-phosphate to trehalose 6-phosphate. Next, **trehalose 6-phosphate phosphatase** removes the phosphate to make trehalose (Fig. 14.15). Another enzyme, **trehalase**, degrades the trehalose molecule into two glucose molecules.

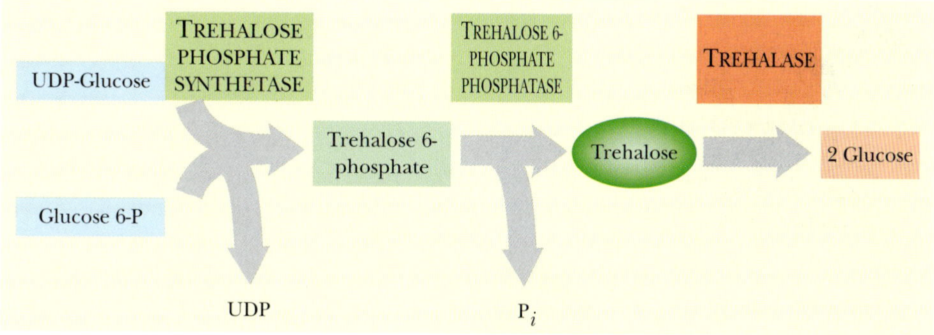

FIGURE 14.15
Trehalose Synthetic and Degradative Pathways
Two enzymatic reactions make trehalose. First, trehalose phosphate synthetase converts UDP-glucose plus glucose 6-phosphate into trehalose 6-phosphate. Next, trehalose-6-phosphate phosphatase removes the phosphate to make trehalose. Trehalose may be broken down into to glucose by trehalase.

To make rice more tolerant to high salinity or drought, a fusion gene encoding a protein with both trehalose phosphate synthase and trehalose 6-phosphate phosphatase activities was created and transformed into rice using the Ti plasmid of *Agrobacterium*. Two different constructs were tested that had the same fusion gene but different promoters. One construct had a stress-inducible promoter, and the other a light-inducible promoter. The insertion of these genes into rice made it much more tolerant to high salt concentrations and also to the lack of water. However, the same fusion gene with a constitutive promoter stunted plant growth. Thus, the expression pattern of transgenes can greatly change the outcome. These transgenic rice plants have not yet been used in the field, but show promise in finding ways to increase food production and quality in changing environments.

The biosynthetic enzymes that make trehalose were transformed into rice. The transgenic plants were more tolerant to high salt concentrations and drought.

FUNCTIONAL GENOMICS IN PLANTS

Because the complete DNA sequences of rice, poplar tree, and *Arabidopsis* are known, plant scientists are following functional genomics strategies. Rather than working on one specific gene, the entire genome can be screened. Most of this type of work is still done with *Arabidopsis*, but some has now moved into crop species such as rice, corn, and soybeans. Functional genomics in plants uses a variety of techniques, some of which have been discussed previously, but most rely on the removal or blockage of gene expression. Novel genes and metabolic pathways useful for understanding basic plant physiology are being analyzed in the hope of improving our current crops.

Box 14.1 Phytoremediation and Other Uses for Transgenic Plants

Making transgenic plants has many different applications important for agricultural gains. These include herbicide tolerance, pest resistance, drought tolerance, decreased use of fertilizers, increased yield, and increased nutritional value both to humans and animals. These uses are by far the most important for our survival and encompass most of the biotechnological applications of transgenic plants. Transgenic plants may have other applications outside of agriculture because they have such varied and unique qualities. Phytoremediation is one such use. Phytoremediation is the use of plants for environmental clean-up of soil and water contamination. This use of transgenic plants is very much in its infancy, but some strides have been made in the past few years.

There are two ways to clean soil contaminants with plants. The first, phytostabilization, simply provides ground cover for a contaminated site. The plant must provide good protection from wind and water erosion. Although most phytostabilization occurs with normal plants, the use of transgenics could increase the root system or enhance tolerance to the contaminant. The second method is called phytoextraction, where the plant actually assimilates the contaminant into its tissues. The plant would then be harvested and disposed of properly.

Some plants have natural systems to assimilate heavy metal ions, whereas other plants need modifications to take up toxins. One natural accumulator is the brake fern, *Pteris vittata*, which can accumulate up to 7500 μg/g of arsenic from a contaminated site. Some plants can concentrate 200 times more arsenic in their leaves than is found in the soil.

In other examples, transgenic plants are created to remove the toxic contaminant. *Arabidopsis*, tobacco, and poplar trees have had the *merA* and *merB* genes from bacteria added to their genome. These proteins remove mercury (Hg[II]) from organic mercury compounds and convert it to elemental mercury, which is volatile and escapes into the air. The actual transgenic plants are not as efficient at mercury removal from the soil, but the location of the MerA and MerB proteins may make it difficult to accumulate the mercury. Expressing MerA and MerB in the cell wall does enhance the ability of the transgenic plant to remove mercury.

Studying the genomics of natural accumulators of heavy metals will be very helpful in delineating the natural proteins and pathways that clean up toxic proteins. This type of knowledge will streamline the work in transgenic plants for phytoremediation.

Insertions are one method to find the function of new genes. Transposons or T-DNA insertions are two methods used to generate plant mutants. Here, instead of including a transgene, the T-DNA or transposon includes only a reporter gene. When the T-DNA or transposon integrates into the plant chromosome, the insertion may disrupt a plant gene. When the insertion knocks out the function of a plant gene, the resulting phenotype can be screened and assessed. By cloning the regions upstream and downstream of the insertion, the plant gene that corresponds to the phenotype can be identified.

Gene silencing is another method to identify the function of plant genes. As described in Chapter 5, gene silencing by RNA interference (RNAi) is a phenomenon that was originally described in plants. RNAi is triggered by double-stranded RNA, which is cut into short segments (siRNA, short-interfering RNA). The RISC enzyme complex uses siRNA to identify homologous RNA (in particular mRNA) and cut it up. This prevents mRNA from being expressed into protein. This is exploited in the laboratory by transforming a plant with small oligonucleotides that stimulate RISC to abolish the expression of a chosen gene. The plant can then be assessed for any visible phenotype associated with the gene knockdown.

Another method for generating gene knockouts is **fast neutron mutagenesis**. This uses **fast neutrons** to induce DNA deletions. Fast neutrons are created by nuclear processes such as nuclear fission, where free neutrons with a kinetic energy close to 1 MeV are generated. These neutrons cause deletions in exposed DNA. Therefore, seeds from the plant of interest exposed to fast neutrons acquire random mutations. The dose of fast neutrons and, consequently, the number of deletions per genome can be controlled. Seeds treated with fast neutrons, known as M1 seeds, are grown into plants. Each plant has a different deletion or set of deletions and a potentially different phenotype.

The seeds from each of these plants, called the M2 seeds, are collected. Most are saved as seed stock; the remaining M2 seeds are grown into plants. The DNA from the plants is isolated and collected into pools of varying sizes. For example, if the original pool had DNA from 100 plants, successively smaller pools of the 100 plants are made, down to just one or two plants per pool. These are usually screened by PCR to find specific genes with deletions. PCR primers are made to amplify a target gene from the largest pool of DNA. If a deletion was generated within the target gene in one of the plants, the PCR primers will amplify two bands, the wild-type gene plus a shorter segment from the deleted gene. The smaller DNA pools are then screened for the deletion until a specific M2 seed can be associated with the genetic deletion (Fig. 14.16).

Yet another method of creating plant mutations is called **TILLING (targeting-induced local lesions in genomes**; Fig. 14.17). First, point mutations are created in a collection of seeds by soaking them in a chemical mutagen, such as EMS (ethyl methane sulfonate), which induces G/C and A/T transitions in DNA. As before, the M1 seeds are grown into plants and the second-generation seeds (M2) are mainly saved as stocks. Some M2 seeds are grown and the DNA is harvested and pooled into a large megapool and successively smaller pools as described earlier. PCR primers are used to amplify selected regions of the DNA. The PCR primers carry fluorescent labels. Consequently, the PCR products are labeled with two different labels, one at either end.

The key to identifying point mutations (as opposed to deletions) is to create heteroduplexes of mutant and wild-type DNA. Therefore, the PCR products are denatured to single strands and then slowly cooled so that the DNA strands reanneal. During reannealing, some mutant strands will anneal with wild-type strands and the heteroduplex will have a mismatched nucleotide. The enzyme CEL-1 cleaves mismatched DNA. If the PCR product is cleaved by CEL-1, it will have only one fluorescent label, whereas uncleaved DNA (with no

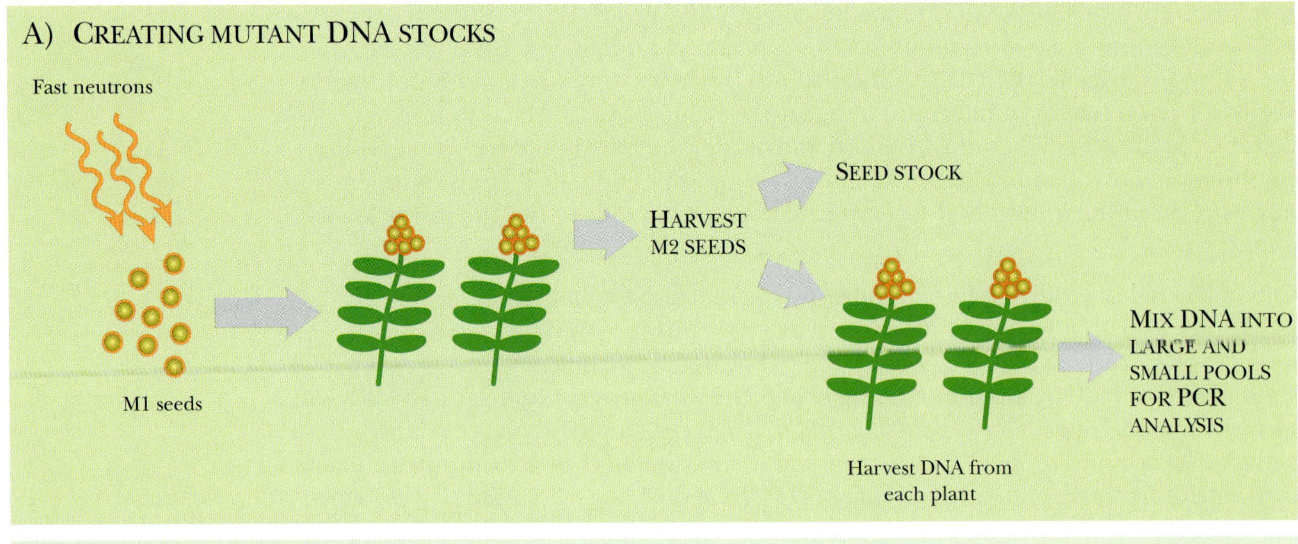

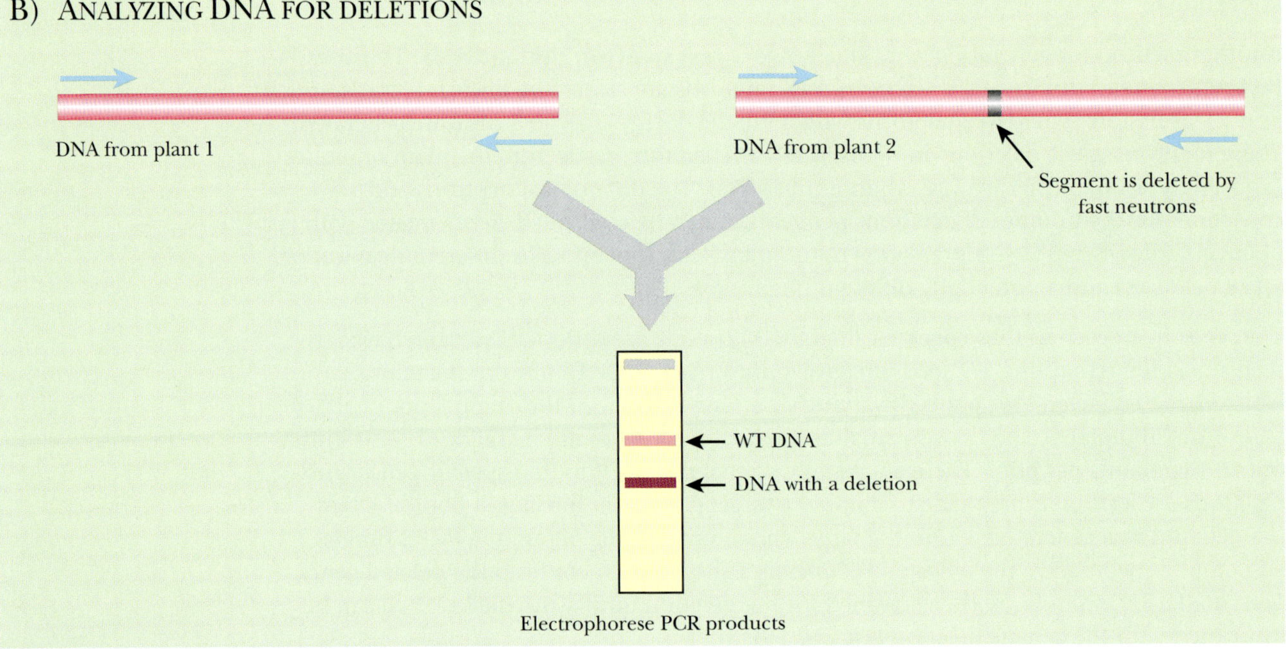

FIGURE 14.16 Identifying Fast Neutron Mutants with PCR

(A) M1 seeds are mutagenized by exposure to fast neutrons. The M2 seeds are grown and DNA harvested from each plant. The DNA is mixed to form large pools from many M2 seeds and successively smaller pools from fewer M2 seeds. (B) The seeds are analyzed for deletions using PCR. The primers recognize specific locations in the plant genome. If the DNA pool contains any deletions, the PCR primer will produce two bands, one from the wild-type (full-length) gene and one from the plant with the deletion.

mismatches) will have both fluorescent labels. When the PCR products are separated by gel electrophoresis, the digested mutant strands can be identified.

Disrupting random genes with transposon insertions can identify plant gene functions.

Gene silencing can also be used to identify the function of plant genes.

Fast neutron mutagenesis creates deletions in genomic DNA, which causes some type of new phenotype. PCR reactions are used to find the region of the genome that was deleted.

TILLING (targeting-induced local lesions in genomes) creates point mutations in the plant genome. When the mutation causes a visible mutant phenotype, the gene is identified using a mismatched hybrid PCR technique.

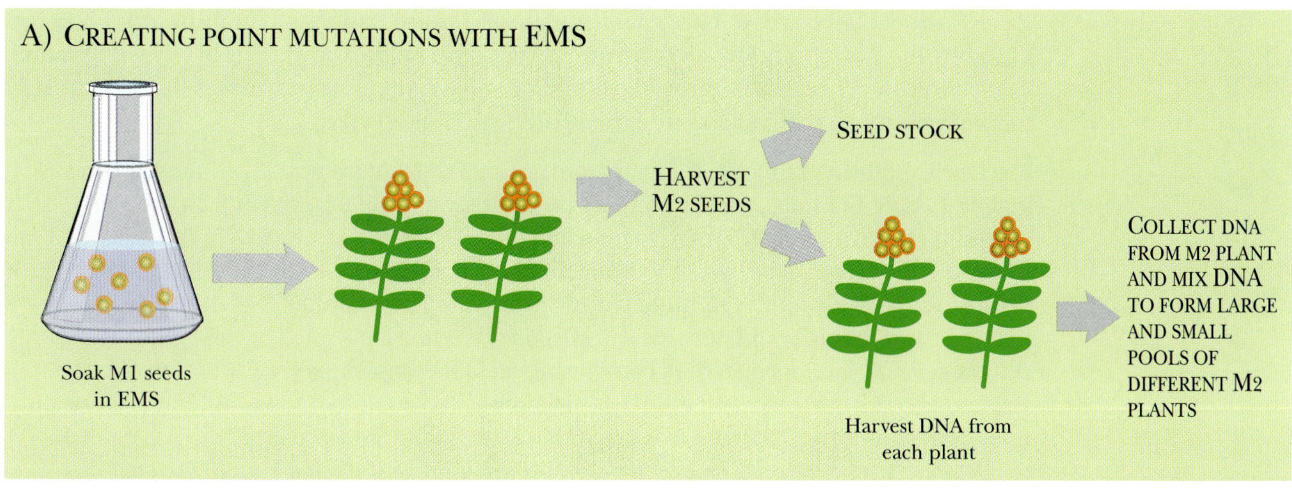

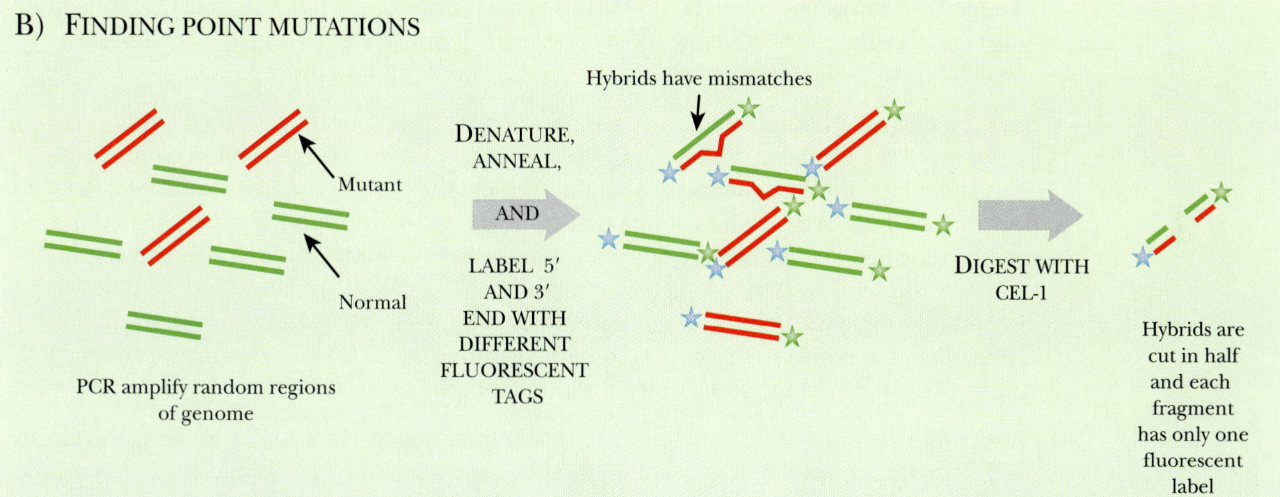

FIGURE 14.17 Identifying Point Mutations with TILLING

TILLING identifies point mutations in a library of plant DNA. (A) EMS, a chemical mutagen, induces point mutations in seeds. The M1 seeds are grown into plants and the M2 seeds are harvested. Most M2 seeds are stored as a stock, while the remaining M2 seeds are grown into plants. DNA is harvested from each plant and pooled. The larger pools contain DNAs from all the M2 plants, and the smaller pools contain DNA from one or two different M2 plants. (B) Point mutations are identified in the DNA pools using PCR to randomly amplify different areas of the plant genomes. Some PCR products will contain point mutations and others will be normal. These are denatured and annealed so that some of the normal and mutant strands form hybrids. The reannealed PCR products are labeled at each end with a different fluorescent tag. The PCR products are then digested with the enzyme CEL-1, which cuts only where the helix has a mismatch. This leaves any mutant:normal hybrids with a single fluorescent tag.

FOOD SAFETY ASSESSMENT AND STARLINK CORN

Many controversies exist over the use of genetically modified crops. The first point to make is that all crops have been genetically modified in some way. Even the oldest variety of edible corn bears no resemblance to its ancestor, teosinte. These changes have occurred through cross-pollination and selective breeding. Thus, using the term *transgenic crop* instead of *genetically modified crop* is more accurate.

A main concern with transgenic crops is the health issue. Does the transgenic crop include toxins or allergens, change the nutrient level of the food, or promote antibiotic resistance in humans or cattle? The spread of antibiotic resistance due to reporter genes or selective markers is no longer an issue because new transgenic varieties no longer contain these marker genes. The nutrient level of transgenic food is strictly controlled and evaluated before any transgenic crop is released to the public.

The allergenic potential of transgenic crops has caused much controversy. In 2000, an unapproved transgenic corn called **Starlink** was detected in taco shells found in the grocery store. Starlink corn has two transgenes. One makes it resistant to the European corn borer by encoding the toxin from *Bacillus thuringiensis* (see earlier discussion). This Bt transgene is the Cry9C isoform. The second transgene, from *Streptomyces hygroscopicus,* makes the corn resistant to a commonly used broad-spectrum herbicide.

The Cry9C isoform of Bt toxin is much more resistant to stomach acid. Also, after cooking and processing, Starlink corn had a higher concentration of Cry9C protein than expected, suggesting that this protein is more stable than other isoforms of Bt toxin. (In contrast, cooking, processing, and digestive enzymes readily break down the Cry1A isoform of Bt toxin.) Because the findings for Cry9C protein came from only one study, the EPA demanded more tests to ensure that it would not cause an allergic reaction if consumed by the public. The companies that developed Starlink corn pushed the EPA for some sort of approval. The EPA responded by giving split approval—that is, Starlink corn could be grown, as long as it was only used to feed livestock. What the EPA failed to realize is that after corn is grown, it is hauled to the nearest grain elevator. The corn is then mixed with all the other corn in the region and shipped to processing centers. So the company and farmers were following the EPA guidelines in good faith, but the next step in the process made it impossible to keep the Starlink corn separate from all the other varieties.

In September of 2000, a coalition of groups opposed to genetically modified foods announced they had detected traces of Starlink in taco shells. Further studies confirmed this and all the products were taken off the shelves. The Centers for Disease Control (CDC) examined all the people who complained of an allergic reaction to the contaminated taco shells. They first determined the type of antibody that the body would produce in response to Cry9C protein. Next they took blood samples and coded them. The blood samples were examined for the presence of Cry9C antibodies by both the CDC and an outside lab. Both concluded that none of the samples contained antibodies to Cry9C. This suggested that any allergic reactions were to some other component in the meals eaten.

After this, the company offered to buy back all remaining Starlink corn, providing the farmer with a premium price, so that no more food became contaminated. In addition, all Starlink seed was pulled from the market to prevent its future growth. In all, Starlink was on the market for only 2 years, 1999 and 2000. In 1999, the amount of Starlink grown in the United States represented only 0.4% of the corn crop and in 2000, 0.5%. Because this was such a small percentage of the overall crop, the Cry9C in the taco shells was massively diluted by other varieties of corn. Starlink is no longer grown anywhere in the world, and the EPA has revoked all approval.

Assaying allergic potential is critical to development of transgenic crops. In another example, soybean plants were transformed with a gene from the Brazil nut. The gene was intended to increase the methionine content of soybeans, which would improve them as cattle feed. Because many people are allergic to Brazil nuts, the FDA ordered tests for allergenicity by skin prick tests and immunoassays. This transgene was found to cause allergic reactions. The work was discontinued, and none of the transgenic plants were ever released to the public.

Transgenic plants are under great scrutiny by the public and government.

BT TOXIN AND BUTTERFLIES

Another issue with transgenic plants is the effect they may have on **nontarget organisms**, that is, any plants or animals inadvertently exposed to the transgenic crop or insecticide. An article in the journal *Nature* suggested that Monarch butterflies were killed by eating pollen from corn carrying the Bt gene. The study alarmed the science community as well as environmental advocacy groups. Monarch caterpillars spend the summer in parts of Canada and the Midwest, feeding exclusively on milkweed plants. In the fall, the adult butterflies migrate to specific areas in Mexico. In the spring, the adults migrate back to the Midwest to lay eggs and repeat the cycle. In this experiment, scientists at Cornell University dusted the leaves of milkweed plants with pollen from transgenic corn expressing the Bt toxin gene. This was compared to milkweed dusted with regular pollen and milkweed with no pollen. Caterpillars that ate the transgenic pollen had stunted growth, ate less, and died at a higher rate than the other two groups.

Much controversy surrounded this study. For example, it was never revealed how much pollen was actually put on the milkweed plants. The investigators simply put on as much pollen as they saw in cornfields. In addition, because caterpillars often refused to eat the Bt pollen–coated milkweed, some critics argued that a caterpillar would move to another milkweed plant in the natural environment. Environmental groups found this a landmark study, showing that transgenic crops could be as damaging to the environment as pesticides such as DDT. But the best result of the controversy was a surge of research on transgenic Bt plants and their effects on butterflies and other nontarget organisms.

The amount of pollen in cornfields and on milkweed plants that surround cornfields was measured. It takes 100 pollen grains per square centimeter for Monarch caterpillars to be affected by Bt pollen. This level of pollen would be found only on milkweed directly surrounding or within a cornfield and would never occur in neighboring open areas. At a distance of 2 meters from the cornfield edge, the pollen density dropped to 14 grains per square centimeter. In the natural environment, Monarch caterpillars simply do not eat these pollen-coated leaves, but move to a different part of the plant. In addition, rainfall or wind during pollen release greatly affects these numbers. One rainfall can remove more than half the pollen from the plants. Moreover, when given a choice, monarch butterflies lay their eggs on milkweed without pollen, further reducing the risk of harm.

Studies were also done to determine which types of Bt toxins affect the larvae of Monarch butterflies. Only certain types of Bt toxins were actually harmful, whereas other types were harmless. The only pollen that was consistently toxic was Cry1Ab, event #176, which made up about 1% of the entire corn crop in 2000. In 2001, this transgene event was no longer approved and was no longer grown in the United States. The Cry1Ab transgene of event #176 is under the control of a corn-pollen-specific promoter and a promoter from phosphoenolpyruvate carboxylase that ensures expression in all photosynthetic tissues. Other events that use the same toxin, Cry1Ab, were only toxic to Monarch caterpillars when at a density greater than 1000 pollen grains per square centimeter. The lower toxicity is due to use of different promoters. Other Cry toxins, Cry1F and Cry9C, showed no effect on any of the larval stages of the Monarch. These studies were done in a laboratory setting where the caterpillar had no choice but to eat milkweed contaminated with pollen.

Some studies have shown an adverse affect of transgenic corn pollen on the Monarch butterfly, but the studies are in the laboratory, which is much different from a cornfield. More studies in the actual environment are necessary to allow a true conclusion to be reached.

Summary

Transgenic crops are created to make plants resistant to different herbicides or insects. More complicated pathways, such as genes associated with vitamin biosynthesis or drought tolerance, have also been added to plants. In order to make transgenic plants, the plants must be cultured either as dissociated cells in liquid called suspension culture or in a layer of cells on top of nutrient media, called callus culture. Each method uses totipotent cells that retain the ability to develop into an entirely new plant. Normal plant cells are undifferentiated, and then the hormone concentrations are adjusted so that new plants form from these cells. Although useful, this method does introduce mutations.

Many transgenic plants are created using the Ti plasmid from *Agrobacterium*. These plasmids have a region, called the T-DNA, that normally integrates into the plant genome. In the natural state, the T-DNA has genes that subvert the plant into making food for the bacteria, and then induce the plant to grow a tumor. To use the T-DNA in making transgenic plants, the normal genes are removed and replaced with the gene of interest. The sequences that allow the T-DNA to insert itself into the genome are left. The gene of interest is often placed behind an inducible promoter, which basically has an on/off switch. And, in some cases, the wobble positions for each of the codons in the gene of interest are modified to be more compatible with plants (e.g., EPSPS gene of *Bacillus*).

Particle guns and floral dip are two methods used to get the transgenic *Agrobacterium* into the plant cell. In floral dip, flower buds are removed from a growing plant, and as the tissue starts to regenerate, the area is dipped into a solution containing the transgenic *Agrobacterium*. As the floral tissue begins to regenerate, the *Agrobacterium* infect the area and transfer the transgenic T-DNA into the cells. By chance, some of the T-DNA enters the developing egg cells, which develop into transgenic seeds. In the second method, the particle gun propels gold or tungsten beads coated with T-DNA at the leaf cells of a plant or a callus culture. As the beads penetrate the cell wall and enter the cytoplasm or nucleus, the T-DNA dissolves from the bead and integrates into the plant genome. Reporter genes such as neomycin transferase or luciferase are also transformed with the gene of interest, making identification of the transgenic plants easy.

Removing the reporter gene is important because the gene product may be allergenic to humans if the transgenic crop is ingested. Cre/*loxP* is a set of recombinase/DNA binding motifs that cause recombination in the P1 bacteriophage. The DNA sites are added around the reporter gene so the gene can be excised. The Cre recombinase is added to the plant by traditional cross-pollination.

Functional genomics in plants is another field of research that was developed with the advent of molecular biology. Scientists are constantly looking for genes that make plants healthier, genes that enhance crop yields, or simply genes that make them resistant to disease or adverse conditions. Random insertion of transposon or T-DNA into a plant genome disrupts all different types of genes. Assessing the physical changes associated with the insertions corresponds to the function of the disrupted gene. Gene silencing using RNAi also suppresses one gene at a time, and once again, the associated physical change defines the function of the gene. Finally, fast neutron mutagenesis and TILLING create mutations that are then associated with a physical change. These techniques are random, and the entire set of mutations is pooled and then successively smaller portions are assessed for the phenotype until one mutation is associated with the most interesting phenotype.

End-of-Chapter Questions

1. Which trait can be conferred onto a plant by a transgene?

- **a.** resistance to diseases
- **b.** enhance the nutritional value
- **c.** resistance to an insect
- **d.** protection from herbicides
- **e.** all of the above

2. What is the definition of totipotent?

- **a.** cells that retain the ability to develop into any cell type of a mature plant
- **b.** cells that cannot be reprogrammed to produce any type of cell in a mature plant
- **c.** plant cells that can naturally take up transgenes
- **d.** transgenic plant cells
- **e.** none of the above

3. What is the first and most important step in genetically engineering plants?
   a. growth of the plant cells in a culture system
   b. insertion of a transgene into the plant genome
   c. identification of a gene that will confer a specific useful trait onto the plant
   d. cloning the transgene into the appropriate plant shuttle vectors
   e. none of the above

4. Which one of the following statements about the Ti plasmid is true?
   a. The Ti plasmid originates from *Agrobacterium tumefaciens.*
   b. The Ti plasmid has been used by genetic engineers to transfer genes into plants to confer a particular trait.
   c. The Ti plasmid produces tumors on plant roots.
   d. The Ti plasmid carries genes for opine uptake and metabolism, as well as genes for virulence.
   e. All of the above.

5. What advantage does an inducible promoter have over a constitutive promoter to express transgenes in plants?
   a. Transgenes under the control of an inducible promoter are only expressed in the tissue that requires that particular trait.
   b. Constitutive promoters ensure that the transgene is turned on even when it is not necessary.
   c. An inducible promoter ensures that the transgene is always turned on, even when it is not necessary.
   d. Transgenes under the control of a constitutive promoter can produce toxins, even in parts of the plant that are consumed by animals.
   e. none of the above

6. What is DNA coated onto when transforming plant cells with a particle gun?
   a. silver
   b. aluminum
   c. helium
   d. gold
   e. calcium

7. Why is either the gene for luciferase or the gene for neomycin phosphotransferase included when transforming DNA into plant cells?
   a. The gene products, particularly luciferase, create interesting petal colors.
   b. These genes encode proteins that can be used to select for or identify cells that have been successfully transformed.
   c. The products of these genes can recruit *A. tumefaciens* to transfer T-DNA.
   d. These genes encode proteins that add nutritional value to the plant tissues.
   e. The products of these genes encode proteins that confer resistance to insects and diseases.

8. What is the significance of using the Cre/*loxP* system in plant biotechnology?
   a. This system prevents the transgene from recombining into the plant genome.
   b. This system creates a more efficient way to integrate useful genes into the plant chromosome.
   c. This system provides a way to remove the selectable marker or reporter gene from the transgenic plants.
   d. This system provides no added benefit to plant genetic engineering.
   e. none of the above

(Continued)

9. Which agency is responsible for regulation of all transgenic technology?
   a. FDA
   b. EPA
   c. USDA APHIS
   d. NIH
   e. all of the above
10. Which of the following statements about glyphosate is not true?
    a. Glyphosate is toxic to the environment according to the EPA.
    b. Glyphosate inhibits the EPSPS enzyme in the synthesis pathway for aromatic amino acids.
    c. Glyphosate is a particularly good herbicide because it is environmentally friendly and has no affect on humans since humans do not have the EPSPS enzyme.
    d. A mutant strain of *Agrobacterium* provided the *aroA* gene that conferred resistance to glyphosate in plants.
    e. All of the above are true.
11. Which of the following statements about Bt toxin is true?
    a. Bt toxin is produced by the soil bacterium *Bacillus thuringiensis.*
    b. Bt toxin kills insects like cotton bollworms and corn borers.
    c. Bt toxin is released by the Cry proteins of *Bacillus* spores that are ingested by insects.
    d. Transgenic crop plants expressing the Cry proteins have been very beneficial in decreasing the use of insecticides.
    e. All of the above are true.

12. What happened when the genes for drought and stress tolerance were placed under the control of a constitutive promoter and transformed into rice?
    a. The plant failed to produce rice.
    b. The plant grew faster and taller.
    c. The plant produced more rice than usual.
    d. The growth of the rice plant was stunted.
    e. nothing
13. Which of the following introduce mutations in plant genes?
    a. TILLING
    b. transposon insertions
    c. RNAi
    d. fast neutron mutagenesis
    e. all of the above
14. What product on the shelves at supermarkets was pulled because of contamination with a transgenic plant product?
    a. canned corn
    b. corn meal
    c. corn on the cob
    d. taco shells
    e. corn cereal

15. Which Cry protein, in high densities, was toxic to Monarch caterpillars?
- **a.** Cry1F Event #176
- **b.** Cry1Ab Event #176
- **c.** Cry9C Event #176
- **d.** Cry2A Event #176
- **e.** none of the above

Further Reading

Aljanabi S (2001). Genomics and plant breeding. *Biotechnol Annu Rev* **7**, 195–238.

Bent AF (2000). *Arabidopsis in planta* transformation. Uses, mechanisms, and prospects for transformation of other species. *Plant Physiol* **124**, 1540–1547.

Chua N-H, Tingey SV (2006). Plant biotechnology: Looking forward to the next ten years. *Curr Opin Biotechnol* **17**, 103–104.

Clark DP (2005). *Molecular Biology: Understanding the Genetic Revolution*. Elsevier Academic Press, San Diego, CA.

Denby K, Gehring C (2005). Engineering drought and salinity tolerance in plants: Lessons from genome-wide expression profiling in *Arabidopsis*. *Trends Biotechnol* **23**, 547–552.

Dodds JH, Roberts LW (1987). *Experiments in Plant Tissue Culture*, 2nd ed. Cambridge University Press, New York.

Garg AK, Kim J-K, Owens TG, Ranwala AP, Choi YD, Kochian LV, Wu RJ (2002). Trehalose accumulation in rice plants confers high tolerance levels to different abiotic stresses. *Proc Natl Acad Sci USA* **99**, 15898–15903.

Gatehouse AMR, Ferry N, Raemaekers RJM (2002). The case of the monarch butterfly: A verdict is returned. *Trends Genet* **18**, 249–251.

Hellmich RL, Siegfried BD, Sears MK, Stanley-Horn DE, Daniels MJ, Mattila HR, Spencer T, Bidne KG, Lewis LC (2001). Monarch larvae sensitivity to *Bacillus thuringiensis* purified proteins and pollen. *Proc Natl Acad Sci USA* **98**, 11925–11930.

Koornneef M, Dellaert LWM, van der Veen JH (1982). EMS- and radiation-induced mutation frequencies at individual loci in *Arabidopsis thaliana* (L.) Heynh. *Mut Res* **93**, 109–123.

Krämer U (2005). Phytoremediation: Novel approaches to cleaning up polluted soils. *Curr Opin Biotechnol* **16**, 133–141.

Leister, D (ed.)(2005). *Plant Functional Genomics.* Food Products Press, New York.

Li X, Song Y, Century K, Straight S, Ronald P, Dong X, Lassner M, Zhang Y (2001). A fast neutron deletion mutagenesis-based reverse genetics system for plants. *Plant J* **27**, 235–242.

Ma LQ, Komar KM, Tu C, Zhang W, Cai Y, Kennelley ED (2001). A fern that hyperaccumulates arsenic. *Nature* **409**, 579.

Nair RS, Fuchs RL, Schuette SA (2002). Current methods for assessing safety of genetically modified crops as exemplified by data on Roundup Ready soybeans. *Toxicol Pathol* **30**, 117–125.

Penna S (2003). Building stress tolerance through over-producing trehalose in transgenic plants. *Trends Plant Sci* **8**, 355–357.

Tzfira T, Citovsky V (2002). Partners-in-infection: Host proteins involved in the transformation of plant cells by *Agrobacterium*. *Trends Cell Biol* **12**, 121–129.

Zhu J, Oger PM, Schrammeijer B, Hooykaas PJJ, Farrand SK, Winans SC (2000). The bases of crown gall tumorigenesis. *J Bacteriol* **182**, 3885–3895.

Transgenic Animals

NEW AND IMPROVED ANIMALS

For thousands of years people have improved crop plants and domestic animals by selective breeding, mostly at a trial-and-error level. Woollier sheep and smarter sheep dogs have both been improved through many generations of selective breeding. Obviously, the more we know about genetics, the faster and more effectively we can improve our crops and livestock.

Today it is possible to alter plants, animals, and even humans by genetic engineering. Most early experiments in animal transgenics were done with mice, but many larger animals have now been engineered, including livestock such as sheep and goats, pets such as cats and dogs, and even monkeys. In a **transgenic** animal every cell carries new genetic information. In other words, novel genetic information is introduced into the germline, not merely into some somatic cells as in gene therapy (discussed in Chapter 17). Consequently, the novel genes in a transgenic animal are passed on to its descendants.

This novel genetic information generally consists of genes transferred from other organisms and so referred to as **transgenes**. They may be derived from animals of the same species, from distantly related animals, or even from unrelated organisms such as plants, fungi, or bacteria. The transgenes are themselves often engineered before being inserted into the host animal. The most frequent alteration is to place the transgene under control of a more convenient promoter. This may mean a stronger promoter or a promoter designed to express the transgene under specific conditions. These assorted manipulations have been dealt with in previous chapters. Here we will consider techniques to create transgenic animals and some applications of this technology.

Animals have been modified by selective breeding for thousands of years. Today, it is possible to insert foreign genes, thus creating transgenic animals with improved qualities.

CREATING TRANSGENIC ANIMALS

Once a suitable transgene is available the standard scenario for the creation of a transgenic animal by **nuclear microinjection** is as follows:

1. The transgene is injected into fertilized egg cells (Fig. 15.1). Just after fertilization, the egg contains its original female nucleus plus the male nucleus from the successful sperm. These two **pronuclei** will soon fuse together. Before this happens, the DNA is injected into the male pronucleus, which is larger and therefore a better target for microinjection. Nuclear microinjection requires specialized equipment and great skill. The success rate varies from 5% to 40% among various laboratories.
2. The egg is kept in culture during the first few divisions of embryonic development.
3. The engineered embryos are then implanted into the womb of a female animal, the **foster mother**. Here they develop into embryos and, if all goes well, into newborn animals.
4. Some of the baby animals will have the transgene stably integrated into their chromosomes. In others the process fails and the transgene is lost. Those that received the transgene and maintain it stably are called **founder animals**. A male and female are mated together to form a new line of animals carrying two copies of the transgene (Fig. 15.1). Note that the founder animals contain only a single copy of the transgene on one chromosome, and are heterozygous for the transgene.

When two such founder animals are bred together, 25% of the progeny will get two copies of the transgene and will be homozygous, 25% will get zero copies, and the remaining 50% will get one copy. Homozygous transgenic animals are most useful because if these are further interbred, all of their descendants will get two copies of the transgene.

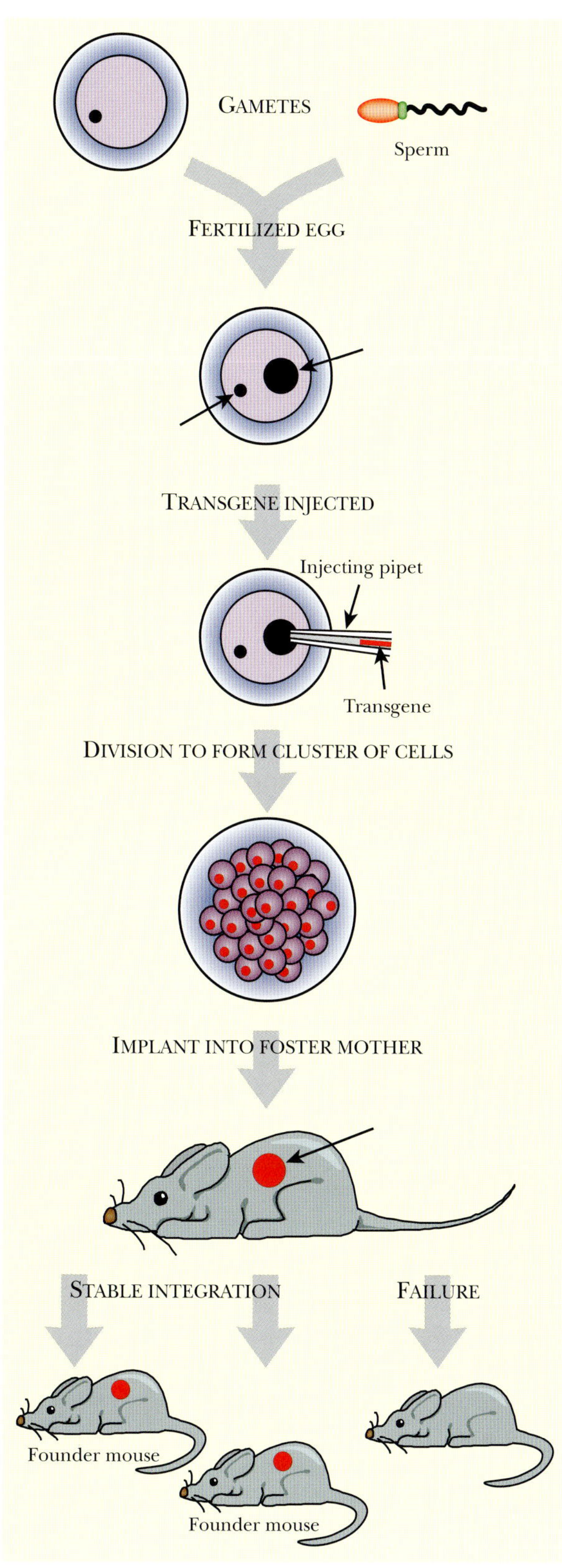

FIGURE 15.1 Creation of Transgenic Animals by Nuclear Injection
In vitro fertilization is used to start a transgenic animal. Harvested eggs and sperm are fertilized, and before the pronuclei fuse, the transgene is injected into the male pronucleus. The embryo continues to divide in culture and is then implanted into a mouse. The "foster mother" mouse has been treated with hormones so that she accepts the embryo and carries on with the pregnancy. The offspring are screened for stable integration of the transgene. Founder mice have one copy of the transgene.

There is some variability in the timing of integration of DNA injected into the male pronucleus. In some cases, the DNA integrates more or less immediately, so that all cells of the resulting animal will contain the transgene. Less often several cell divisions may happen before DNA integration occurs and the final result will be a **chimeric animal**, in which some cells contain the transgene and others do not. Sometimes multiple tandem copies of the transgene integrate into the same nucleus—such constructs are often unstable and the extra copies are often deleted out over successive generations. Integration of the incoming DNA into the host cell chromosomes occurs at random. Often, this is accompanied by rearrangements of the surrounding chromosomal DNA. This suggests that integration often occurs at the sites of spontaneous chromosomal breaks.

Transgenic animals are often generated by microinjection of DNA into the nucleus of a fertilized egg cell. Founder animals arising from such engineered eggs have one copy of the transgene. They must be bred to yield animals that are homozygous for the transgene.

LARGER MICE ILLUSTRATE TRANSGENIC TECHNOLOGY

The classic illustration of transgenic technology was the creation of larger mice by inserting the rat gene for growth hormone. Growth hormone, or **somatotropin**, consists of a single polypeptide encoded by a single gene. In 1982 the somatotropin gene of rats was cloned and inserted into fertilized mouse eggs. The eggs were then placed into foster mother mice that gave birth to the genetically engineered mice. These transgenic mice were larger (about twice normal size), although not as large as rats. This was the first case where a gene transferred from one animal to another not only was stably inherited, but also functioned more or less normally.

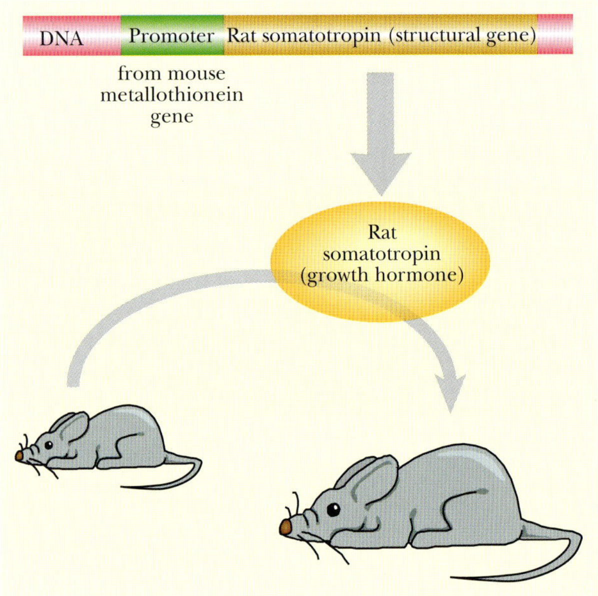

FIGURE 15.2 Large Transgenic Mice
A DNA construct containing the rat somatotropin gene under the control of the mouse **metallothionein promoter** was used to make a transgenic mouse. The transgene cause the mouse to grow to twice its normal body size.

To express the rat somatotropin gene, it was put under the control of the promoter from an unrelated mouse gene, **metallothionein**, which is normally expressed in the liver (Fig. 15.2). Instead of being made in the pituitary gland, the normal site for growth hormone, the rat somatotropin was mostly manufactured in the liver of the transgenic mice. Even though made in the "wrong" location, the hormone worked and made the mice larger. Human somatotropin has also been expressed in mice and also gives bigger mice.

Size does not depend solely on growth hormone. Defective growth hormone receptors also inhibit growth. African pygmies rarely grow taller than 4 feet, 10 inches, yet they have normal levels of human somatotropin. It appears that pygmies have defective growth hormone receptors. These normally bind the somatotropin circulating in the bloodstream and are necessary for the hormone to work on its target tissues. Dwarfism, among nonpygmies, may be due either to defects in production of somatotropin or to a shortage of receptors. **Recombinant human somatotropin (rHST)** is now used to treat the hormone-deficient type of dwarf. However, receptor-deficient dwarfs cannot yet be successfully treated.

In one of the first transgenic experiments the gene for growth hormone from larger animals was inserted into mice. It was expressed under control of the metallothionein promoter, allowing it to be induced by traces of zinc. The result was larger mice.

Box 15.1 Trendy Transgenic Mice

Genetically engineered mice have become all the rage and every so often hit the headlines:

Marathon Mouse can run about 1800 meters—more than a mile—before exhaustion. This is twice as far as a normal mouse can last. Marathon mouse has enhanced PPAR-delta—a regulator of several genes involved in burning fat and in muscle development.

Mighty Mouse was engineered to lack myostatin, a protein that slows muscle growth. The result is colossal muscle development. There is one known case of a human with a genetic defect leading to lack of myostatin. A German boy, born in Berlin in 2000, has muscles twice the size of other children his age.

Fierce Mouse has both copies of the *NR2E1* gene deleted and shows abnormal aggression (see Chapter 19 for more).

Smart Mouse is genetically modified to have improved learning and memory. It has extra copies of the *NR2B* gene encoding the NMDA receptor found in the synapse region of nerve cells in the brain.

And last but not least, the K14-Noggin mouse has an extra Noggin gene from chickens. This affects development of skin and external organs. In particular, the male is extra hairy and has a longer penis. As yet, this mouse, made in 2004, has no official nickname.

The improved athletic abilities of Marathon Mouse and Mighty Mouse have raised the issue of tampering with humans—so-called "gene-doping" in order to promote athletic prowess. See Chapter 25, Bioethics, for further discussion. As for K14-Noggin mouse, one can only imagine the implications!

RECOMBINANT PROTEIN PRODUCTION USING TRANSGENIC LIVESTOCK

The somatotropin gene from cows has been cloned and expressed in bacteria thus allowing the production of large amounts of the hormone, which is known as **rBST (recombinant bovine somatotropin)**. The rBST is used in the dairy industry to increase milk production. Unlike in mice, boosting an adult cow's somatotropin levels by injection of extra hormone results in increased milk production, rather than giant cows. Milk from treated cows is now widely marketed.

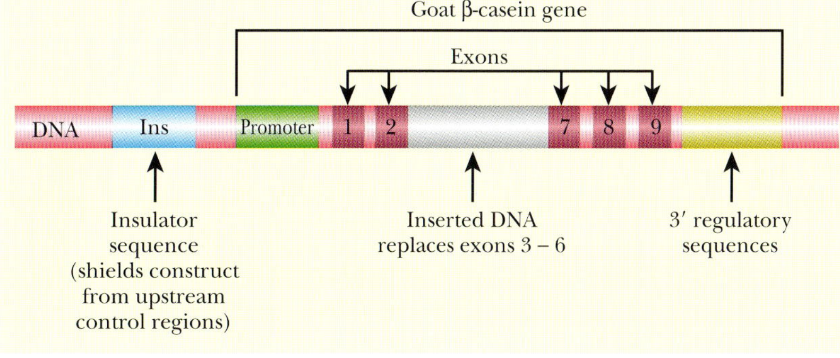

FIGURE 15.3 Milk Expression Construct for Transgenic Goats
In order to express a recombinant protein in goat milk, the gene of interest is inserted in place of the β-casein gene. The transgene will be expressed using the endogenous promoter and 3′ regulatory elements that restrict β-casein expression to goat milk. The construct also has insulator sequences that block other regulatory elements from affecting expression (see later discussion).

At present, bacteria, such as *Escherichia coli*, are cultured to make most recombinant proteins, such as human insulin or somatotropin. Such products are expensive and require a highly trained work force. However, using livestock to express these products may be cheaper. Dairy cows produce 10,000 quarts of milk each per year, and an industry to collect and process milk already exists. To take advantage of this, several recombinant proteins are now being produced in the milk from transgenic cows or other farm animals. To achieve this, cloned genes are placed under the control of a regulatory region that will allow gene expression only in the mammary gland. Consequently, the gene product will come out in the milk (Fig. 15.3). For small-scale production of proteins for clinical use, transgenic goats are often used. For example, transgenic goats have been made to produce **recombinant tissue plasminogen activator (rTPA)**, which is used for dissolving blood clots.

Recombinant proteins may be manufactured by expressing the corresponding genes in transgenic cattle or goats.

Box 15.2 Ecosensitive Swine Waste

Not only can genetics improve animals, it can also improve animal waste! From the University of Guelph in Canada comes **Enviropig™** (a trademark of Ontario Pork). These transgenic pigs have received the *appA* gene from *E. coli* under control of the PSP (parotid secretory protein) promoter. As a result they secrete in their saliva the enzyme phytase (or acid phosphatase; product of *appA*). This enzyme degrades "phytate" (inositol hexaphosphate, typically found in the outer layers of cereal grains)—which pigs cannot otherwise use.

Consequently, these ecosensitive swine no longer need phosphate supplements in their diet. More important, the phosphorus content of the manure is reduced by as much as 60%. This is significant because high phosphorus runoff into ponds and streams causes algal blooms that taint the water and rob it of oxygen, leading to the death of fish and other aquatic life.

KNOCKOUT MICE FOR MEDICAL RESEARCH

Transgenic animals, mostly mice, are of great value in the genetic analysis of inherited diseases and cancer. Here we are interested not so much in adding a cloned transgene as in discovering the function of genes already present. The general approach is to inactivate, or "knock out," the gene of interest and then ask what defect this causes.

To achieve this the target gene is first cloned. Then the gene is disrupted by inserting a **DNA cassette** into its coding sequence. (Most DNA cassettes also include an antibiotic resistance gene for easy detection.) The intruding DNA segment prevents the gene from making the correct protein product and thus abolishes its function. The inactive copy of the gene is then put back into the animal by following the procedure outlined earlier for transgene insertion. The incoming DNA, carrying the disrupted gene, sometimes replaces the original, functional, copy of the gene by homologous recombination. Founder mice are obtained that have one copy of the disrupted gene. By breeding them together, mice with both copies disrupted are obtained. Such mice are known as **knockout mice** and will completely lack gene function (Fig. 15.4). If the gene in question is essential, homozygous knockout mice may not survive, or may live only a short time.

Knockout mice have selected genes inactivated. They are widely used in medical research to investigate gene function.

ALTERNATIVE APPROACHES TO MAKING TRANSGENIC ANIMALS

Although nuclear injection was the first approach used to make transgenic animals, and is still the most widely used, there are several alternative procedures. As discussed in Chapter 17, engineered retroviruses have been used in gene therapy to introduce DNA into the chromosomes of animal cells. Retroviruses can infect the cells of early embryos, including embryonic stem cells (see later discussion). Hence, retrovirus vectors may introduce transgenes.

The advantages of using a retrovirus are that only a single copy of the retrovirus plus transgene is integrated into the genome. In addition, use of retroviruses does not require skill in microinjection. The retrovirus carrying the transgene construct is added to the fertilized egg and allowed to infect as normal. The egg is then transplanted into a pseudopregnant female mouse. The remainder of the procedure is as before.

Disadvantages are that virus DNA is introduced along with the transgene and that retroviruses can carry only limited amounts of DNA. Furthermore, founder animals made using retroviruses are always chimeras, because insertion of the virus does not occur exactly when the nuclei fuse. Consequently, retroviruses are rarely used in attempts to create fully transgenic animals. On the other hand, partially transgenic animals that have some sectors or tissues altered are useful because the transgenic tissues may be compared with normal tissue within the same animal. This can eliminate any doubts on whether the defects or changes caused by the transgene are merely due to differences among animals or truly due to transgene expression.

Embryonic stem cells may also be used to generate transgenic animals. **Stem cells** are the precursor cells to particular tissues of the body. Embryonic stem cells are derived from the **blastocyst**, a very early stage of the embryo, and retain the ability to develop into any body tissue, including the germline. Embryonic stem cells can be cultured and DNA can be introduced as for any cultured cell line. For successful creation of transgenic animals, embryonic stem cells must be maintained under conditions that avoid differentiation.

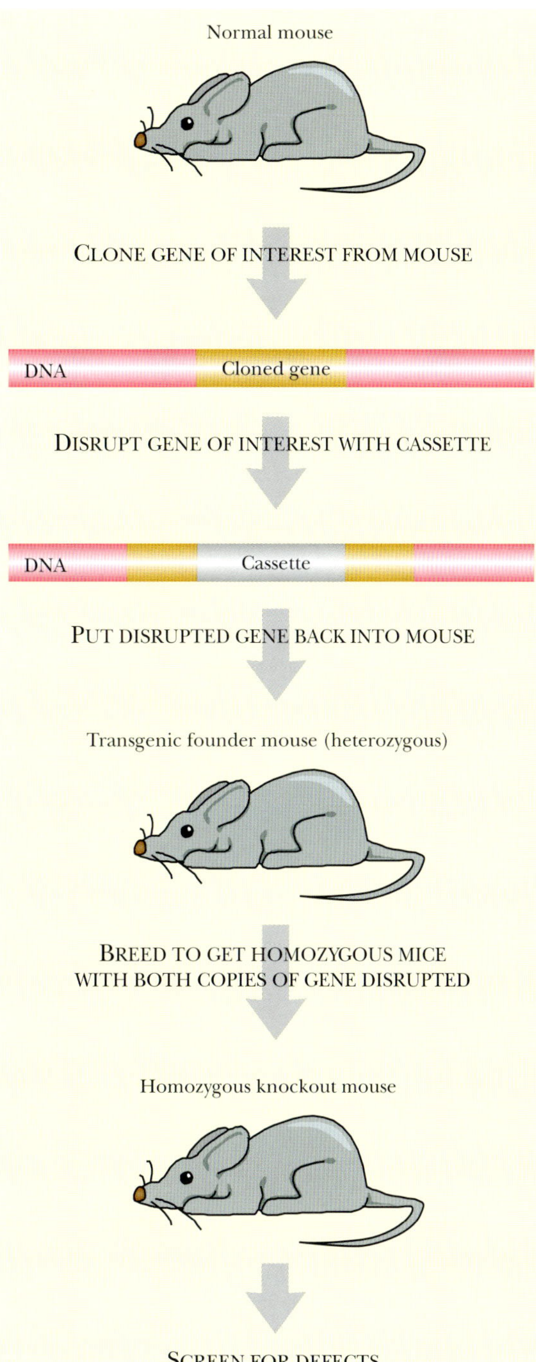

FIGURE 15.4
Knockout Mice
Like traditional transgenic mice, knockout mice are generated *in vitro*. The target gene is cloned and disrupted by inserting a DNA cassette. This work is usually done in bacteria. Once the construct is made, it is put back into a mouse by injection into the male pronucleus during fertilization (or by other methods outlined later). After the transgenic offspring are born, two heterozygotes are crossed to create a homozygous knockout mouse. These are then screened for defects due to inactivation of the target gene.

Engineered embryonic stem cells are then inserted into the central cavity of an early embryo at the blastocyst stage (Fig. 15.5). This creates a mixed embryo and results in an animal that is a genetic chimera consisting of some transgenic tissues and others that are normal.

If the host embryo and the embryonic stem cells are from different genetic lines with different fur colors, the result is an animal with a patchwork coat. This allows the transgenic sectors of the animal to be identified easily. This chimeric founder animal must then be mated with a wild-type animal. If the embryonic stem cells have contributed to the germline, then the coat color characteristic of this cell line will be transmitted to the offspring. In mice black (recessive) and **agouti** (dominant) coat colors are often used. The embryonic stem cells are usually taken from an agouti line, because the dominant fur color enables the transgenic cell lines to be tracked easily. Both the embryonic stem cells and the host embryo are usually taken from males because the resulting male chimeras can father many children when crossed with wild-type females.

Transgenic animals may be generated by several alternative methods. One approach is to use retrovirus vectors that insert into host chromosomes. Another is to engineer embryonic stem cells in culture before returning them to an early embryo.

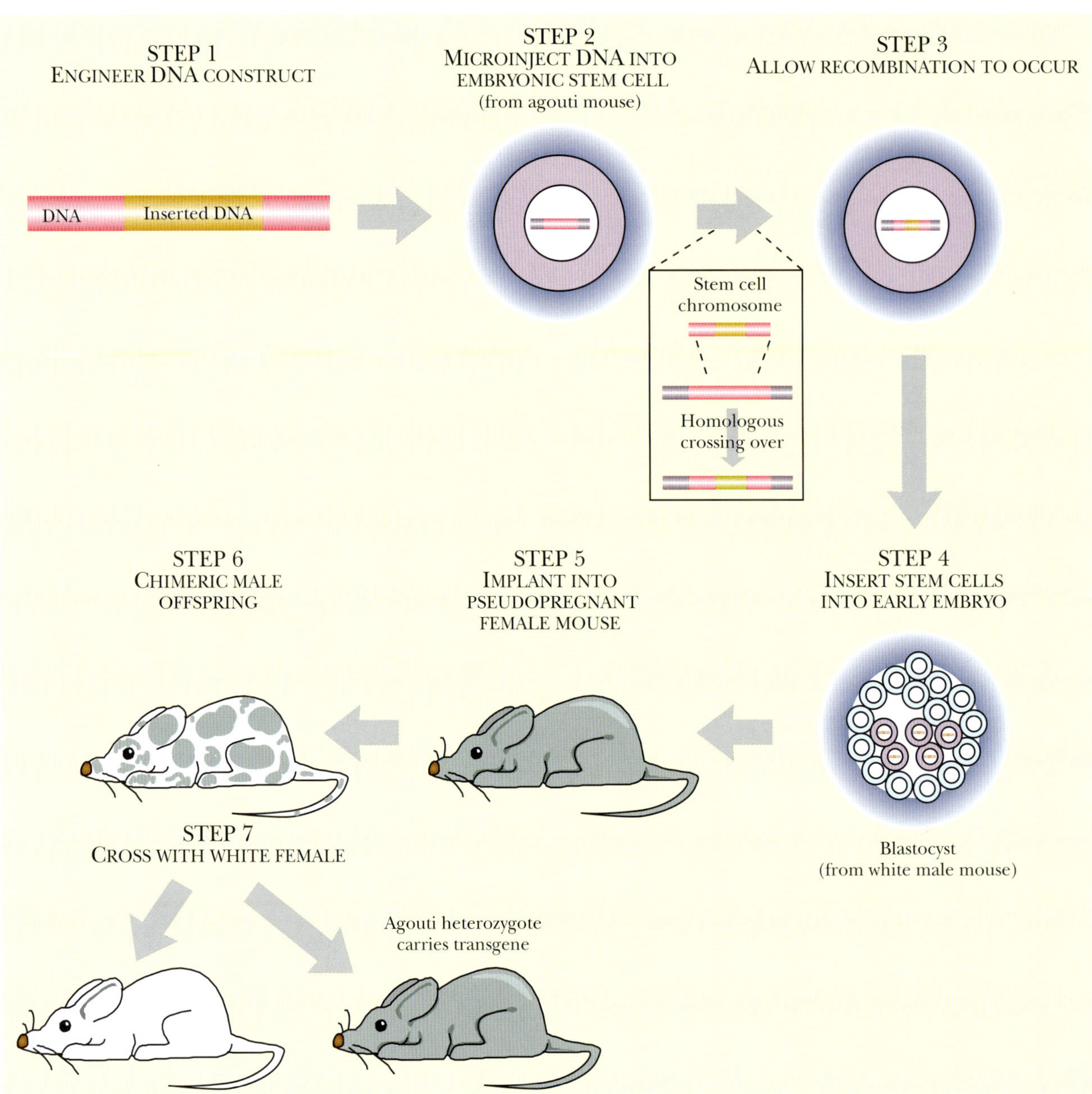

FIGURE 15.5 Use of Embryonic Stem Cells

To create a transgenic animal with embryonic stem cells, the transgene must first be inserted into these cells. The stem cells shown here are from an agouti mouse, that is, a mouse with grizzled brown fur. The stem cells are transformed with the transgene, which integrates by homologous recombination. Then the stem cells are injected into an early male embryo (blastocyst) from a white mouse. The embryo is put into a pseudopregnant female mouse. The offspring are chimeras because the majority of cells in the injected blastocyst are normal. The chimera will have a white coat with patches of brown derived from the injected stem cells. The chimera is crossed with a white female, and any fully brown (agouti) offspring will have the transgene incorporated into the germline.

LOCATION EFFECTS ON EXPRESSION OF THE TRANSGENE

Transgenic animals (or plants) carrying the same inserted transgene often differ considerably in expression. Both the level of expression and the pattern of expression in various tissues of the body may vary. Many of these effects are due to the location of the transgene. Expression of the inserted transgene will be affected by any nearby regulatory elements already present in the host animal chromosome. In particular, enhancer sequences work over considerable distances and will affect the expression of any transgenes integrated nearby. In addition, the physical state of the DNA is important. If the transgene is integrated into a region that consists largely of heterochromatin, the transgene will be expressed poorly or not at all. In such regions the DNA is tightly packed, often methylated, covered with nonacetylated histones, and consequently usually nontranscribed.

Such position effects have been confirmed experimentally by extracting transgene DNA from a transgenic animal in which the transgene was not expressed. This DNA was then used to construct another line of transgenic animals. If some of the new transgenic animals show proper expression of the transgene, this demonstrates that the gene itself is intact and its failure to express in the original host animal was due to its location (Fig. 15.6).

> The chromosomal location of a transgene can have major effects on its level of expression.

COMBATING LOCATION EFFECTS ON TRANSGENE EXPRESSION

Location effects may be avoided by targeting the transgene to a specific site (see later discussion). Alternatively, appropriate regulatory elements can be built in to the transgene construct itself:

(a) Dominant control elements. Some regulatory sequences control nearby genes or clusters of genes in a dominant manner. For example, the **locus control region (LCR)** in front of the β-globin gene cluster confers high expression (Fig. 15.7). Note that the LCR is distinct from the individual promoters and affects several clustered genes. LCR sequences dominate over any other nearby regulatory sequences and thus provide position-independent expression. Such LCR sequences may be placed in front of transgenes and, it is hoped, will confer high-level expression that is independent of chromosomal location.

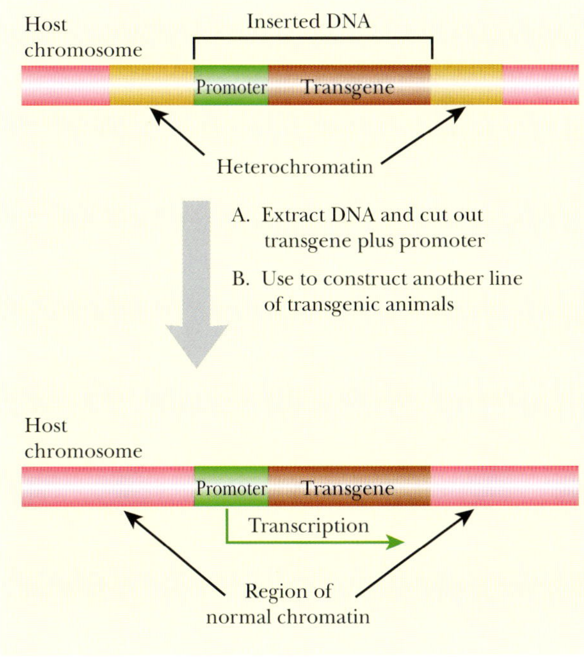

FIGURE 15.6 Failed Expression Due to Transgene Location
DNA carrying a transgene was inserted to generate a transgenic animal. In this instance, the DNA was inserted into a region of heterochromatin. Even though transgenic animals were obtained, the transgene was not expressed. The inserted DNA was removed and used to make another transgenic animal. The transgene was expressed in the second animal, showing that it was intact in the first transgenic animal. Because the location of integration was different, the earlier lack of expression must have been due to a position effect.

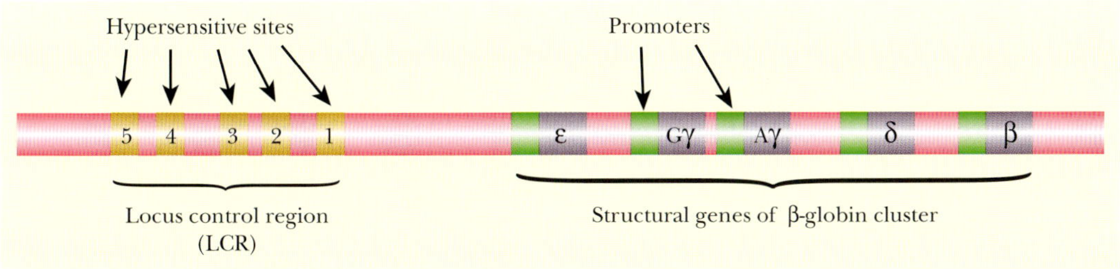

(b) **Insulator sequences** or boundary elements. These sequences block the activity of other regulatory elements. If a gene is flanked by two insulator sequences it is protected from

FIGURE 15.7 Locus Control Region (LCR)
The LCR of the β-globin gene cluster enhances expression of all five genes. This control region is outside the individual promoters. The LCR has five DNase I hypersensitive regions, which have multiple consensus sequences for transcription factor binding sites.

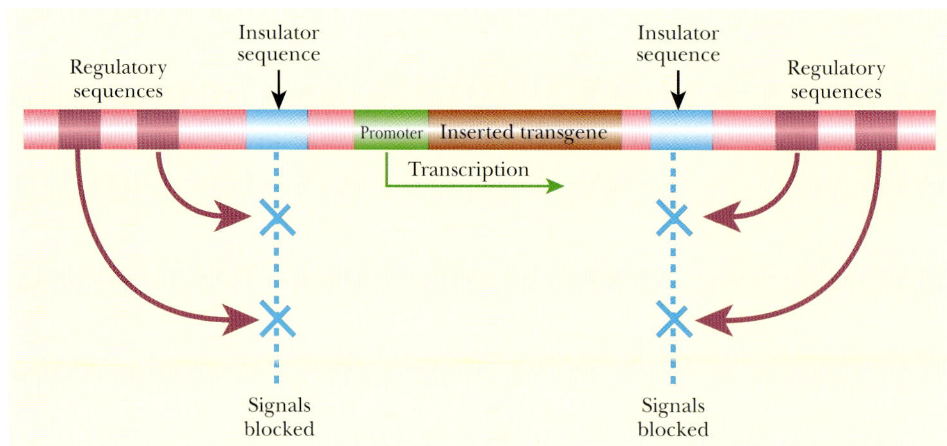

FIGURE 15.8 Protection of a Gene by Insulator Sequences Insulator sequences are placed flanking a transgene. These protect the transgene from regulatory sequences outside the insulator sequences.

the effects of any regulatory elements beyond the insulators (Fig. 15.8). Hence transgenes can be protected from position effects by including insulator sequences in the transgene construct. Transgenes flanked by insulators probably form independent loops of DNA from which heterochromatin is excluded.

(c) Use of natural transgenes. Most transgenes actually consist of cDNA and therefore differ from the original wild-type version of the gene in lacking introns. Furthermore, most transgenes are under control of viral or artificial promoters, which are shorter and more convenient to engineer than the natural promoters from the original gene. Nonetheless, full-length natural eukaryotic genes are often more resistant to position effects than the shortened engineered versions, especially if both upstream and downstream control elements are included.

Cloning and manipulating full-length animal genes is inconvenient because of the excessive lengths of DNA involved. Nonetheless, it is possible to carry such genes on **artificial chromosomes** (see Chapter 3). In a few cases, natural length transgenes carried on **yeast artificial chromosomes (YACs)** have been used to construct transgenic animals. YAC-based transgenes have been used in the study of long-range regulatory elements. They have also been used to introduce into mice full-length genes for humanized monoclonal antibodies (see Chapter 6).

Further engineering may be performed to protect transgenes from positional effects. This generally consists of inserting an appropriate regulatory sequence next to the transgene.

TARGETING THE TRANSGENE TO A SPECIFIC LOCATION

Targeting the incoming transgene to a particular location on the host chromosome requires homologous recombination, as opposed to the random integration that usually happens with injected DNA. Inserting a transgene in a specific location may be desirable for several reasons. First, chromosomal location often affects the expression of a transgene (see earlier discussion). Second, the transgene is not necessarily a novel gene. Sometimes the objective of genetic engineering is to replace the original version of a particular gene by an altered version. In this case it is obviously preferable to insert the incoming gene in the same location, and under the same regulation, as the gene it is replacing.

Gene targeting relies on homologous recombination, and special **targeting vectors** are designed to direct the integration. The DNA to be inserted is flanked by sequences that are homologous to those at the target location. Targeting vectors may be subdivided into those designed to insert novel DNA and those that replace DNA (Fig. 15.9). Targeting vectors are often linearized just before transforming the DNA into the cell because this promotes more efficient recombination. Integration of the required DNA segment may be selected by antibiotics or by some other positive selection.

Targeting vectors use homologous recombination to insert transgenes at specific chosen locations in the host genome.

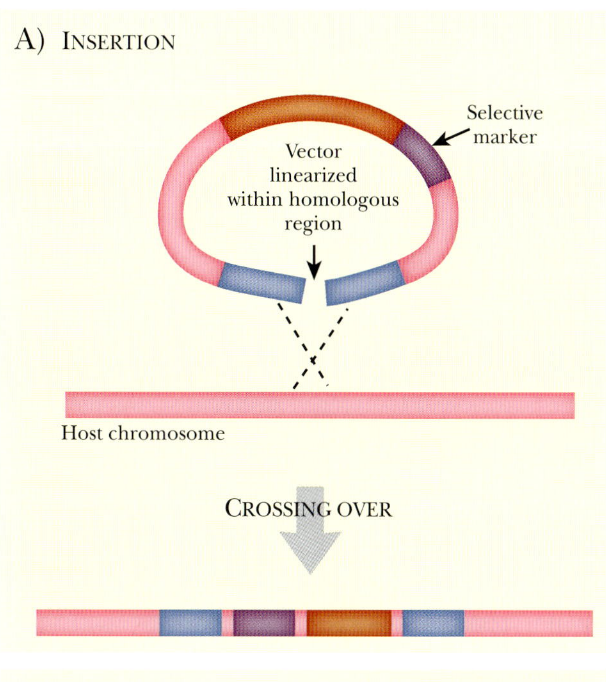

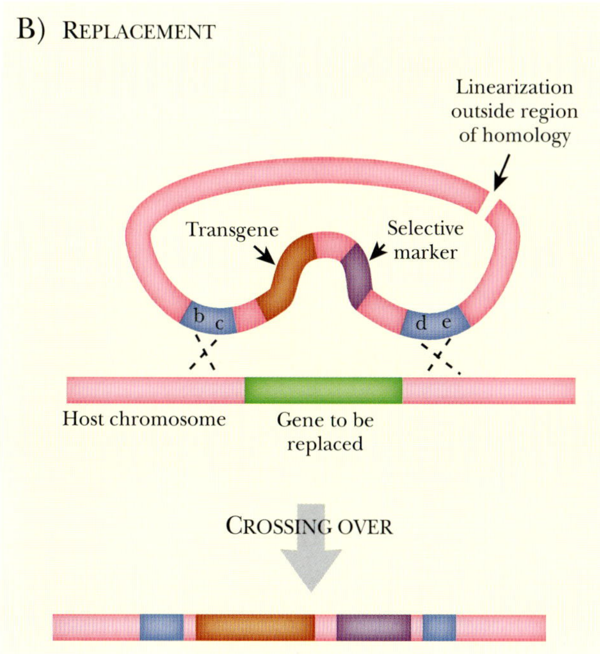

FIGURE 15.9 Targeting Vectors Rely on Homologous Recombination
(A) Targeting vectors can insert a transgene into a host chromosome at a specific location. The vector has sequences (blue) homologous to the insertion point on the host chromosome (pink). The linearized vector triggers a single crossover, thus integrating the transgene and selective marker into the host chromosome. (B) Some targeting vectors promote gene replacement. These vectors have two separate regions flanking the transgene that are homologous to the host. When the linearized vector enters the nucleus, homologous regions align, and crossovers occur on each side of the transgene. The host gene is replaced with the transgene and selective marker.

DELIBERATE CONTROL OF TRANSGENE EXPRESSION

There are many cases in which it would be helpful to control the expression of a transgene. For industrial production of a protein product, high-level expression of the transgene is usually preferred. But this is not always true—some proteins are toxic in high amounts, and gene expression needs to be kept low while establishing the line of transgenic animals. However, when transgenes are used for functional analysis, it is clearly convenient to be able to switch gene expression on and off as required. This is especially important for those genes that are normally expressed only in particular cell lines or at certain developmental stages. A variety of systems exist that allow the experimenter to control transgenes.

Inducible Endogenous Promoters

Early transgenic constructs often used natural promoters from the host animal (i.e., endogenous promoters) that respond to certain stimuli. For example, rat growth hormone was put under control of the mouse metallothionein promoter when inserted into mice. This promoter is induced by heavy metals such as lead, cadmium, and zinc. The least toxic of these, zinc, is actually used for induction, but even so there will be toxicity problems if continuous long-term induction is needed.

The heat shock promoter from the *Drosophila Hsp70* gene is another example of a natural promoter used to drive expression of transgenes. In this case, the promoter is inactive at room temperature; increasing the temperature to 37 °C is the inducing stimulus.

These simple inducible promoters have several drawbacks. First, there is often significant expression even in the absence of the inducing signal and the level of induction is often low—perhaps 10-fold or less. Second, there are often toxic side effects. These may be due directly to the inducing signal (e.g., zinc, high temperature) or they may be due to the induction of other, natural, genes that respond to the same signal. Third, the inducing signal may be taken up slowly and/or only penetrate some tissues of the organism. Some of these problems can be avoided by using foreign or artificially constructed promoters.

Recombinant Promoter Systems

Bacterial repressors have been used to control transgenes in animals and plants. Both the **LacI repressor** and the **TetR repressor** have been used in this manner. For use in transgenic animals the *lac* system must be modified as illustrated in Fig. 15.10. The LacI repressor is made by the *lacI* gene, which is modified by adding a Rous sarcoma virus (RSV) promoter to ensure expression in animal cells. Another animal virus promoter, from SV40 (simian virus 40), controls the transgene. In addition, a *lacO* operator site is inserted into this promoter. The LacI repressor binds to the operator site and prevents expression of the transgene via the SV40 promoter (see Fig. 15.10). When IPTG is added, the LacI protein binds the IPTG and is released from the operator site. Thus transgene expression is induced by IPTG.

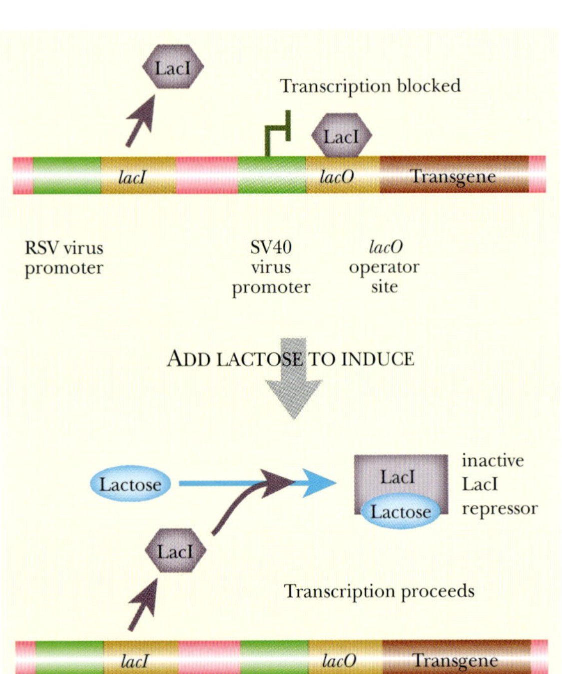

FIGURE 15.10
LacI Control of a Transgene
A Rous sarcoma virus (RSV) promoter drives transcription of the *lacI* gene. When the construct is expressed in a transgenic animal, LacI protein is produced. The LacI repressor binds to the operator site upstream of the transgene and blocks expression. When the inducer IPTG is added, it binds to LacI, which falls off the operator site. The transgene is then transcribed by the SV40 promoter.

The *tet* system operates in a similar manner to the *lac* system. The ***tet* operon** confers resistance to the antibiotic tetracycline. (It is usually found as part of a transposon, e.g., the Tn10 transposon of *E. coli*.) The TetR repressor protein binds to an operator site and so prevents expression of the tetracycline resistance genes. When tetracycline is present, it binds to the TetR protein, which is therefore released from the DNA. Consequently, the *tet* operon is induced by tetracycline. The *tet* system may be used in higher organisms like the *lac* system. The transgene has a *tetO* operator site inserted into its promoter and will then be induced by tetracycline.

Both the *lac* and *tet* systems work well in eukaryotic cells. They both have low background expression and a high ratio of induction. In particular, the *tet* system may show up to 500-fold induction. One problem is the need for constant high-level expression of LacI or TetR protein. These bacterial DNA binding proteins may become toxic to eukaryotic cells at high levels.

The *tet* system has been modified to avoid the need for high levels of TetR protein. Fusing TetR protein to eukaryotic transcription factors can give hybrid activator proteins. For example, the **VP16 activator** protein from herpes simplex virus (HSV) has been linked to TetR. When the hybrid TetR-VP16 protein binds to DNA it recognizes the ***tetO* operator** sequence (Fig. 15.11). However, when bound it causes activation (not repression) because the VP16 domain activates transcription. Thus the TetR recognition sequence now functions as a positive regulatory site. When tetracycline is added, the TetR-VP16 protein is released from the DNA and the gene is no longer expressed. Here, then, addition of tetracycline prevents gene expression. This system can give as much as 100,000-fold induction and background expression when the gene is turned off is extremely low.

A further improvement is the use of a mutant TetR protein whose DNA binding behavior has been inverted. Hence it binds to DNA in the presence of tetracycline and is released in its absence. Consequently, hybrid activator proteins made using this version of TetR activate the target gene in the presence of tetracycline. This is known as the reverse *tet* transactivator system (Fig. 15.12). Plasmid and virus vectors have been constructed that allow all components of this system, including the cloned transgene, to be inserted into the animal together.

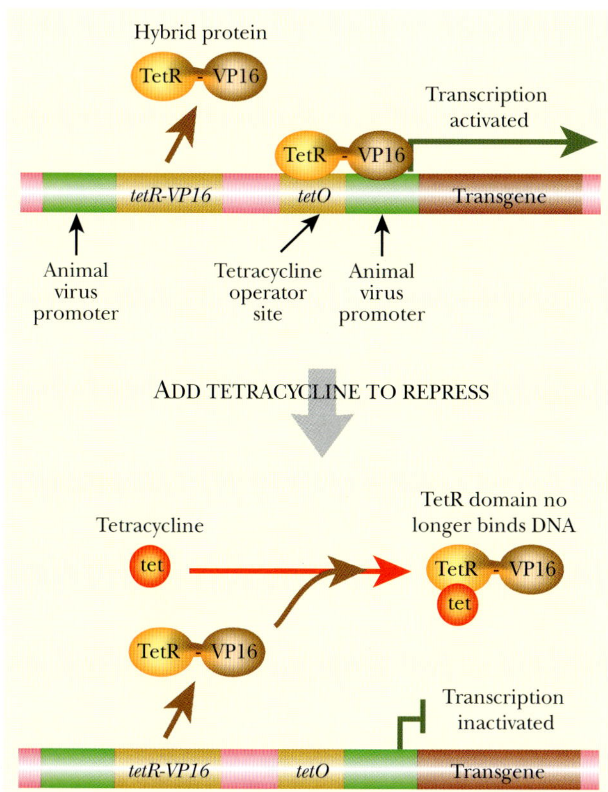

FIGURE 15.11 Hybrid TetR-VP16 Transactivator Systems
A hybrid protein made from TetR and VP16 controls transcription of a transgene. The hybrid protein binds to a TetR operator site, and the VP16 domain activates transcription. When tetracycline is added, the TetR domain binds this and is released from the *tetO* site. Therefore, tetracycline inhibits transgene expression.

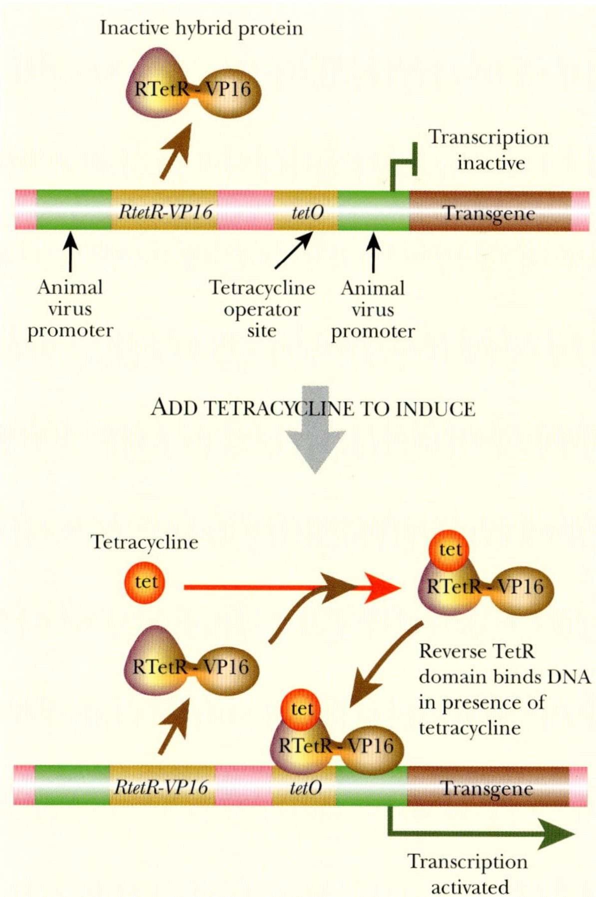

FIGURE 15.12 Reverse TetR-VP16 Transactivator System
The reverse TetR-VP16 hybrid protein only binds to *tetO* when tetracycline is present. So, in contrast to the previous case (see Fig. 15.11), the transgene is expressed only in the presence of tetracycline.

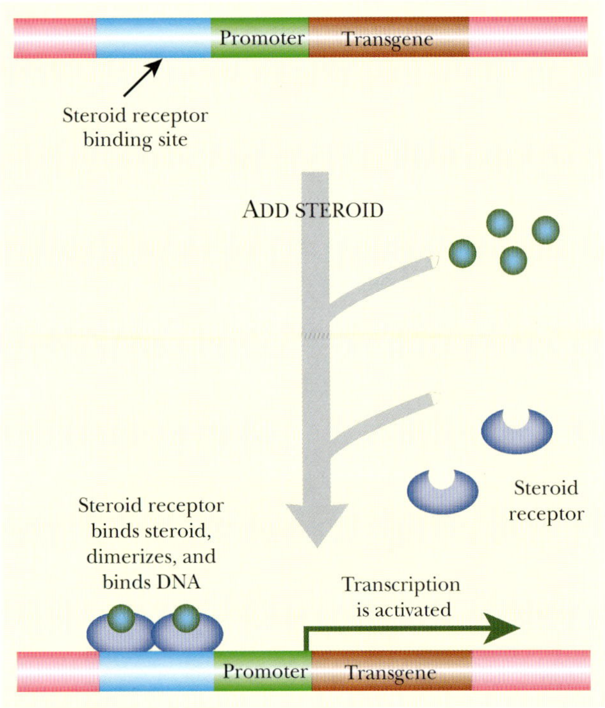

FIGURE 15.13
Steroid Hormone to Activate Transgene
Some transgenic animals have the transgene under the control of a steroid-controlled promoter. When the animal is treated with the steroid, the steroid diffuses across the cell membrane and binds to its receptor in the cytoplasm. The receptor-steroid complex then enters the nucleus, where it binds to the transgene promoter and turns on transcription.

Transgene Regulation via Steroid Receptors

A problem with molecules such as tetracycline is that they often penetrate different tissues unequally. Steroid hormones are lipophilic and consequently penetrate cell membranes rapidly. Once inside the cell, steroids bind to receptor proteins, which in turn bind directly to DNA and regulate gene expression. Steroids are also eliminated within a few hours, and consequently they are convenient molecules for inducing transgenes.

Here the issue is to avoid inducing other host genes that respond to steroids. One way to achieve this is to use a steroid not found naturally in the host animal. For example, the steroid **ecdysone** from insects may be used in transgenic mammals and, conversely, the mammalian **glucocorticoid hormones** may be used in transgenic insects or plants. Therefore the binding protein for the chosen hormone must also be supplied. A recognition sequence for this protein is placed in the upstream region of the target transgene (Fig. 15.13). The result is that the transgene is induced when the steroid is administered.

Altered and/or hybrid **steroid receptors** have been used for improved responses. One curious example is the use of a mutant version of the mammalian progesterone receptor. The mutant version can no longer bind progesterone, but still binds the antagonist **RU486** (active component of the abortion pill). When RU486 is used as inducer, the concentration needed is 100-fold less than required for its action in abortion. This receptor can be used without any interference with endogenous steroids, and treatment with RU486 does not induce abortion in the transgenic animal.

Transgenes may be artificially regulated by a variety of control systems. Engineered versions of the bacterial *lac* and *tet* regulators are widely used.

CONTROL BY SITE-SPECIFIC RECOMBINATION USING CRE OR FLP

Another way to control the expression of a transgene is via site-specific recombination. In this approach, segments of DNA are physically removed or inverted to achieve activation of the transgene. These DNA manipulations are done after the DNA carrying the transgene and associated sequences has been successfully incorporated into the germline chromosomes of the host animal.

Site-specific recombination involves recognition of short specific sequences by DNA binding proteins. Recombination then occurs between two of the recognition sequences. Some site-specific recombination systems require several proteins for recognition and crossing over. Others need only a single protein to bind the two recognition sequences and recombine them. These are obviously far more useful in genetic engineering.

Two such **recombinase** systems have been widely used: the **Cre recombinase** from bacterial virus P1 and the **Flp recombinase (flippase)** from the 2-micron plasmid of yeast. Both Cre and Flp recognize 34 base-pair sites (known as ***loxP*** and **FRT**, respectively) consisting of 13 base-pair inverted repeats flanking a central core of 8 base-pairs.

We have already described the use of the Cre/*loxP* system in plants (Chapter 14) to delete unwanted DNA segments after integration of incoming transgenic DNA. Similar manipulations can be carried out in animals using Cre/*loxP* or Flp/FRT. The specific removal of unwanted segments of DNA requires that they should be flanked by *loxP* or FRT sites (Fig. 15.14). (Segments flanked by two *loxP* sites are sometimes referred to as "floxed".) This approach may be used for a variety of purposes, some of which are summarized as follows:

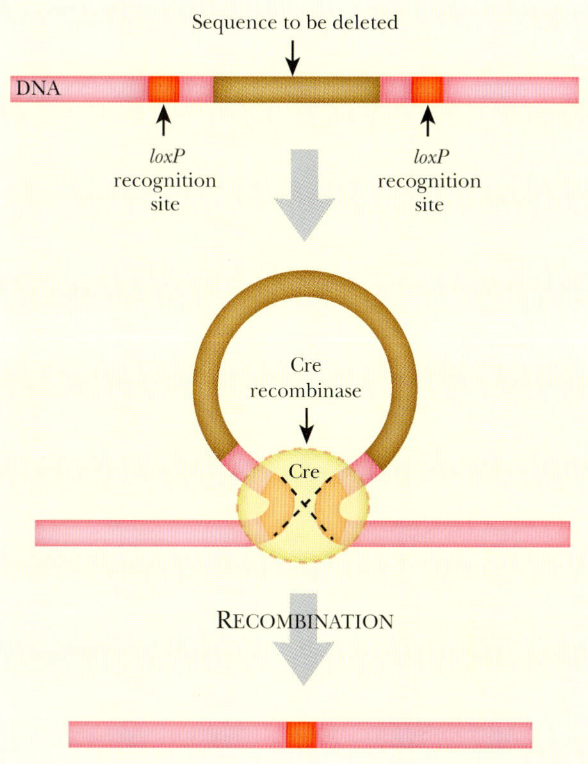

FIGURE 15.14 Site-Specific Deletions in Transgenic Animals Two *loxP* sites must be inserted into the DNA, flanking the sequence to be deleted. When Cre recombinase is activated, it recombines the two *loxP* sites and deletes the segment of DNA between them.

(a) Removal of selective markers. Once transgenic DNA has been successfully integrated, the antibiotic resistance gene used for selection and/or the reporter gene used for screening are no longer needed. If they are enclosed between *loxP* or FRT sites, they can be removed, leaving a transgenic organism with only the actual transgene (plus a single copy of the *loxP* or FRT site).

(b) Activation of transgene. Here the original transgenic construct is made with a blocking sequence between the promoter and the transgene. The blocking sequence is flanked by *loxP* or FRT sites. After integration of the incoming DNA, the blocking sequence is removed by Cre or Flp recombinase, thus activating the transgene.

(c) *In vivo* chromosome engineering. Large-scale deletions or rearrangements of eukaryotic chromosomes may be generated *in vivo* by using the Cre/*loxP* system. Two *loxP* sites are introduced by two rounds of DNA insertion at separate specific locations. Then the Cre recombinase is activated and deletions are generated.

(d) Creation of conditional knockout mutants. Transgenic constructs may be designed so that a specific gene can be deleted *in vivo*. Generally, two *loxP* sites are inserted into the introns flanking an essential exon of the target gene. On recombination, the exon will be deleted and the target gene will be inactivated. This allows investigation of genes whose knockout mutations are lethal at the embryonic stage. The animal can be allowed to grow into an adult before the recombinase is activated to generate the knockout.

When these recombinase systems are used, the recognition sequences are included in the transgenic constructs. Later, the recombinase itself is provided by one of three methods:

(a) The gene for recombinase may be carried on a plasmid that is transformed into the animal. The recombinase will be expressed transiently, assuming the plasmid does not integrate or survive over the long term.

(b) The recombinase gene may itself be part of the transgenic construct and be induced by some external stimulus.

(c) Two separate lines of transgenic organisms are used. The transgene plus recognition sites are present in one host line and a second line of transgenic organisms expresses the recombinase (Fig. 15.15). The two lines are then mated together and the deletions occur in their progeny.

This approach can save a lot of work. Instead of making separate transgenic constructs for each gene under each condition, two sets of transgenic animals, usually mice, are generated and are then crossed to investigate a wide range of genes and environmental conditions. One set of mouse lines has Cre under the control of a variety of promoters that are specific for different tissues or induced by different signals. The second set of mouse lines has a series of different target genes flanked by *loxP* sites. Thus the role of any particular target gene may be investigated under any of the available conditions by crossing the appropriate pair of strains.

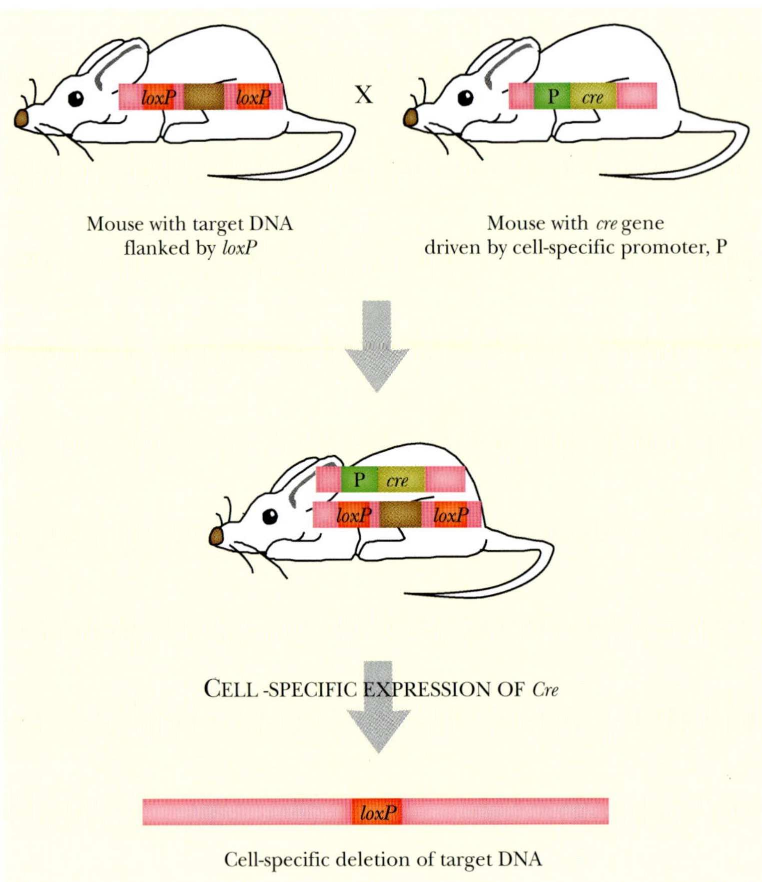

FIGURE 15.15
Two-Mouse Cre/Lox System
One mouse has the target DNA that is to be deleted flanked by two *loxP* sites. A second mouse has the gene for Cre recombinase under the control of a tissue-specific or inducible promoter. When these two mice mate, some of the offspring will receive a copy of both genetic constructs. When the Cre protein is induced, it will direct the deletion of the target DNA.

Transgene expression may be controlled by site-specific recombination. Rearrangements of the transgenic DNA that turn transgenes on or off are promoted by either the Cre or Flp recombinases.

TRANSGENIC INSECTS

Several insects can now be genetically modified. The fruit fly *Drosophila* has been investigated at the molecular level for a long time and, not surprisingly, methods exist for introducing novel genetic material into these flies.

P elements are transposons found in *Drosophila* and other insects, where they cause **hybrid dysgenesis**. In flies carrying a P element, the frequency of transposition is very low because of synthesis of a repressor protein encoded by the resident P element. When P-carrying males are crossed with P-negative females, the transposition frequency in the fertilized egg is very high for a brief period, due to lack of repressor. Random insertion of P elements then causes a high mutation rate and lowers the proportion of viable offspring, that is, hybrid dysgenesis.

P elements are flanked by perfect 31 base-pair inverted repeats. Any DNA sequence that is included between these inverted repeats will be transposed. Therefore engineered P elements can be used to introduce any sequence of DNA into a strain of fruit flies or other susceptible insects.

DNA may be microinjected into embryos of P-negative strains of *Drosophila*. In fruit flies the diploid nucleus resulting from fusion of the sperm and egg nuclei divides multiple times without cell division, resulting in a giant cell with many nuclei, known as a **syncytium**. Microinjection is normally done at this stage, and incoming DNA usually integrates into at least some of the nuclei that will give rise to the future germline cells (Fig. 15.16). These nuclei are clustered at one end of the fertilized egg—the *posterior end*. The nuclei then migrate to the outer membrane where a cleavage furrow forms around each nucleus. These furrows

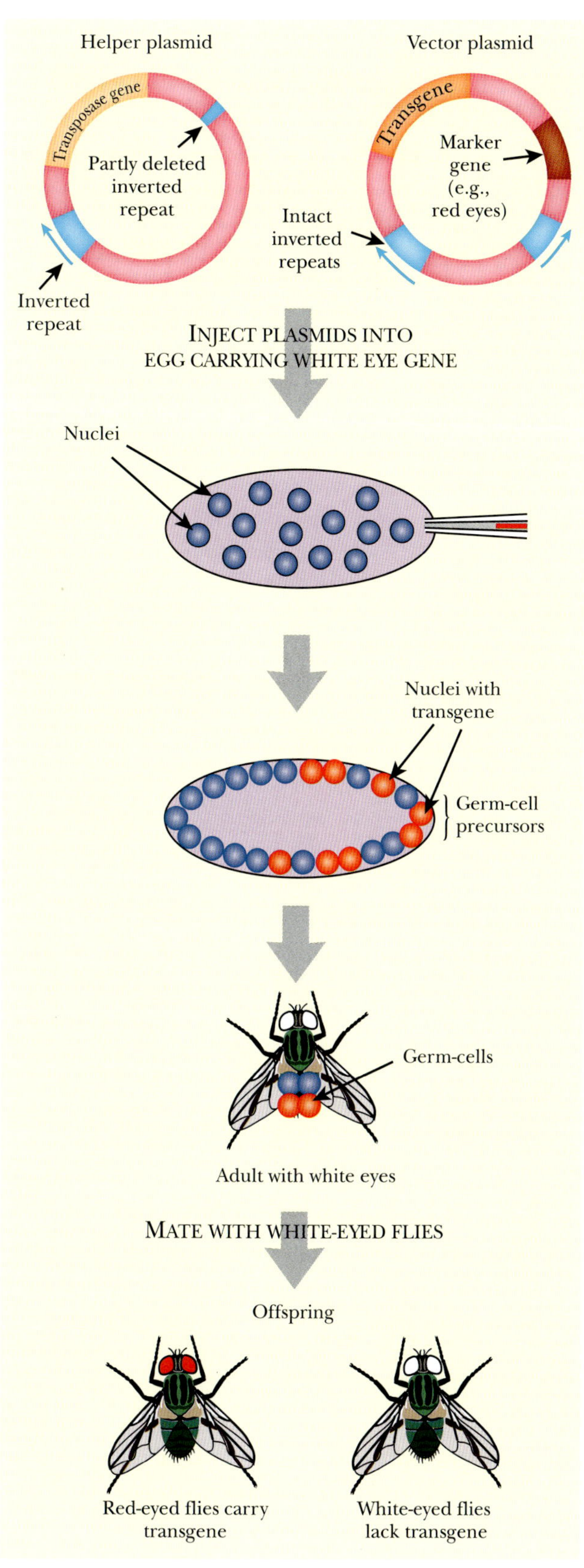

FIGURE 15.16 P Element Engineering in *Drosophila*

Two different plasmids are used to insert transgenes into *Drosophila*. The helper supplies the transposase. It carries an immobile P element with one of the inverted repeats deleted but with functional transposase. The second plasmid has the transgene plus marker (a gene for red eyes) flanked by the P element inverted repeats. Both vectors are injected into the posterior end of an egg, which has 2000 to 4000 nuclei within one membrane. The transposase is expressed, which results in random transposition of the transgene (plus marker) into various chromosomes in different nuclei. Hopefully, some insertions occur in germ cell nuclei. The egg is then allowed to form an adult (with white eyes in this case). This fly is then mated to another white-eyed adult. If insertion into the germline was successful, some offspring will express the marker gene and have red eyes.

expand to form individual cells for each nucleus. The center part remains undivided and acts as a yolk, providing nutrients to the developing larva.

The incoming P element is normally carried on a bacterial plasmid that was constructed in a bacterial host. The P element transposes into the *Drosophila* chromosomes, and the plasmid sequences are left behind. In practice, two P elements are often used. One, the helper, provides the transposase but cannot itself move because of defective 31 base-pair inverted repeats (see Fig. 15.16). The other P element, the vector, carries the desired transgene and has intact 31 base-pair inverted repeats, but lacks the transposase gene. Transposition of the vector depends on transposase made by the helper. Once the P element vector has inserted into a particular location on the insect chromosome, it cannot move in future cell generations, because it has no transposase of its own. Ideally, it will be inherited stably.

The presence of the P element is monitored by appropriate marker genes. Selectable markers used in flies include *neo* (neomycin resistance) and *adh* (alcohol dehydrogenase). Alternatively, eye color genes may be used to reveal the presence of a P element vector. Eye color cannot be positively selected; instead, flies are screened for changes in eye color. For example, flies defective in the ***rosy* gene** may be used as host. These flies have brown eyes because of lack of xanthine dehydrogenase, which is involved in synthesis of red eye pigment. If a wild-type copy of the *rosy* genes is included in the P element vector, it will restore the red eye color. If the offspring of a rosy$^{-/-}$ transgenic fly has red eyes, this implies that the transgene was inserted into the germline, and all the cells in the offspring will have the transgene.

Transposons known as P elements are widespread in *Drosophila* and other insects. They have been used to introduce transgenic DNA into insects.

GENETICALLY MODIFIED MOSQUITOES

The three diseases that kill most people are acquired immunodeficiency syndrome (AIDS), tuberculosis, and malaria. Each year, malaria infects some 300 to 500 million people and kills around 2 million, mostly African children. The ***Plasmodium*** parasite is transmitted by mosquitoes—as are several other major diseases including yellow fever, dengue fever, and filarial nematodes. At the moment malaria is spreading and mosquitoes that are resistant to insecticides such as DDT are emerging. The 260-Mb genome of the mosquito ***Anopheles gambiae***, which transmits malaria, has been fully sequenced and a variety of genetic markers is available. The genome of ***Aedes aegypti***, which transmits yellow fever, is about three times larger (800 Mb), and its sequencing is under way.

DNA can be inserted into germline cells of mosquitoes, as for flies. Several transposons have been used to insert DNA into the genomes of mosquitoes by an approach similar to the use of the P element in *Drosophila* described earlier. The ***piggyBac*** transposon from the cabbage looper (a butterfly) and the ***Minos*** transposon from *Drosophila hydei* are the most widely used. Eye color genes and green fluorescent protein (GFP) have been used as genetic markers. The transgenes are usually expressed from *Drosophila* promoters because these often work well in other insects.

One possible approach to controlling mosquito-borne diseases is to genetically engineer mosquitoes that are resistant to colonization by the disease agent. The noncarrier mosquitoes would then be released into the wild where they would displace the population of disease-transmitting mosquitoes. Several attempts have been made to engineer mosquitoes that will no longer carry malaria, or at least carry far fewer malarial parasites. So far, engineering has been done with species of malaria that attack birds or mice. It is hoped that similar approaches will work against human species of malaria and in the *Anopheles* mosquito, which carries them.

After taking in a meal of infected blood from a human or animal, the mosquito immune system attacks the incoming malarial parasites and does in fact kill a substantial proportion (Fig. 15.17). One approach to engineering mosquitoes therefore attempts to increase

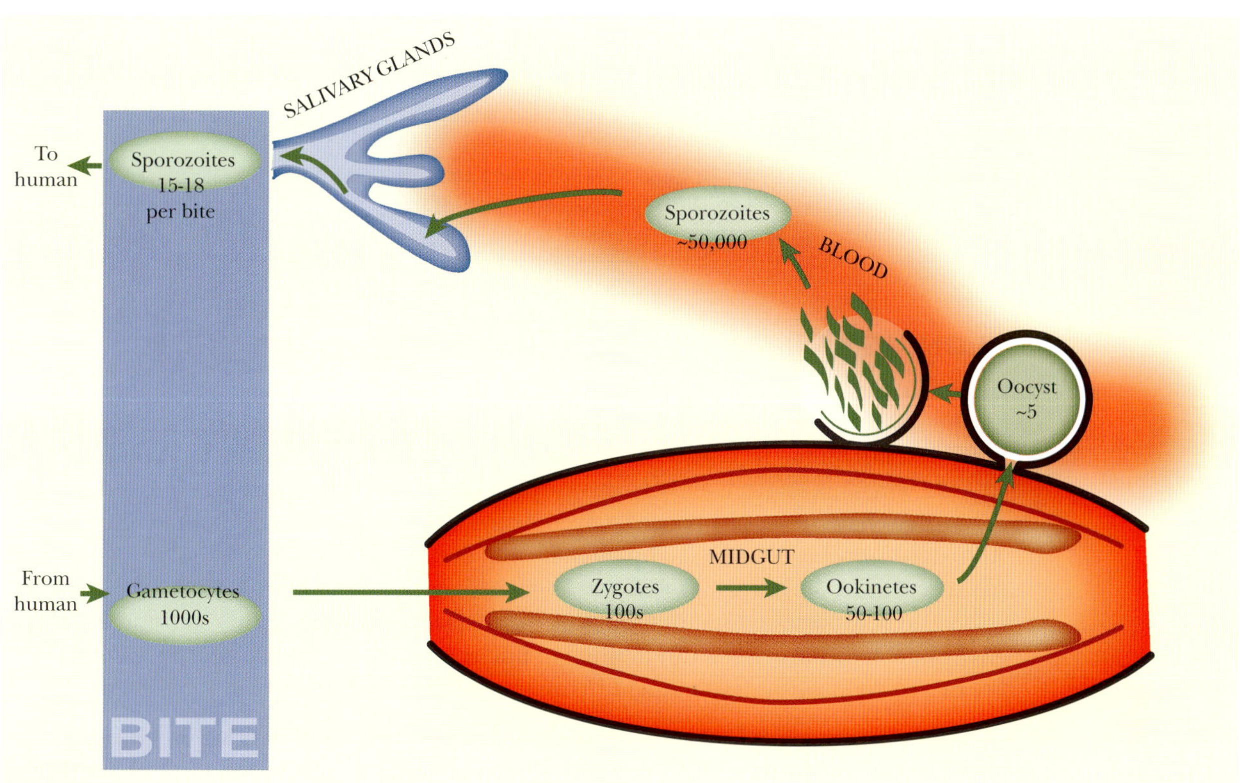

FIGURE 15.17 Development of *Plasmodium* in the Mosquito

When a mosquito bites a human with malaria, the human blood has thousands of *Plasmodium* gametocytes (*bottom of figure*). These travel into the midgut of the mosquito, where they produce hundreds of zygotes. About 50 to 100 of these develop into mobile ookinetes, which then migrate into the hemolymph or blood and give rise to about five oocysts. Each oocyst releases about 50,000 sporozoites that migrate to the mosquito salivary gland. When the mosquito bites another human, only about 15 to 18 sporozoites enter the bloodstream. *Plasmodium* life stages are shown in green ovals, and mosquito structures are labeled in uppercase.

expression of proteins such as **defensin A** that belong to the mosquitoes' own immune system. Proteins from other species have also been expressed in the mosquito midgut in attempts to block transmission. For example, transgenic mosquitoes that express bee venom **phospholipase** from a midgut-specific promoter destroy 80% to 90% of the incoming malarial parasites.

Another approach uses genetically engineered human antibodies. For example, artificial genes for **single-chain antibodies** or **scFv fragments** (see Chapter 6) to the **circum-sporozoite protein** of malaria have been constructed. (The sporozoite is the form of the parasite transferred from mosquito to mammal—see Fig. 15.17.) When expressed in the mosquito salivary glands, the antibody greatly reduced the numbers of malaria sporozoites.

After engineering a mosquito that no longer transmits malaria, the next problem is replacing the wild mosquito population with the engineered ones. One way to do this is to use a genetic suicide system consisting of two genes, A and B, which must both be inherited together for survival. Such a system is similar to those responsible for programmed cell death in bacteria (see Chapter 20). Engineered males with two copies each of gene A and gene B would be released.

Insects who inherit A and B together will survive, but those who get A without B or vice versa will die. Hybrids between engineered males and wild females will survive because they inherit one A and one B. However, in the next generation, when the hybrids mate with wild mosquitoes, some of the offspring will receive just gene A or gene B but not both and will therefore die. This generates a selective pressure that drives the A and B genes through the population, because the offspring of wild flies die more often than those of engineered or hybrid flies. Genes that kill the malaria parasite or make mosquitoes susceptible to insecticides would be linked to the suicide genes and would therefore spread through the population with them. Computer modeling suggests that modifying a mere 3% of the population is enough to spread the genes.

Engineered mosquitoes that no longer transmit malaria are being constructed. The ultimate aim is to replace wild populations of mosquitoes with noncarriers.

CLONING ANIMALS BY NUCLEAR TRANSPLANTATION

Although the cloning of Dolly the sheep in 1996 created a major furor in the media, from a scientific viewpoint it was a relatively small step in a developing technology. Cloning animals relies on the technique of **nuclear transplantation**. This actually dates back to 1952 when nuclei from early frog embryos were transplanted into eggs from which the nucleus had previously been removed. Some attempts gave rise to normal embryos. Although animal somatic cells differentiate and eventually become irreversibly committed to specialized roles, their nuclei nonetheless retain a complete genome. (A few exceptions, such as red blood cells, lose their nucleus.) Under some circumstances the cytoplasmic environment in the egg cell can reprogram nuclei from somatic cells. Not surprisingly, the earlier the stage of development in the nuclei, the easier they are to reprogram.

Nuclear transplantation can be used to generate a group of identical **cloned animals**. Several nuclei from the same donor are transplanted into a series of enucleated eggs. Since the 1980s, nuclear transfer in a variety of mammals has been performed successfully using nuclei from early embryos (morula or blastocyst stages). Fusing a somatic cell with an empty egg cell transfers the donor nucleus into a completely nondifferentiated cytoplasm. A brief electrical pulse fuses the two cell membranes into one embryo. In 1995, nuclei from cultured embryonic cells of sheep were successfully transplanted. Two lambs, Megan and Morag, were produced by this technique at the Roslin Institute in Edinburgh, Scotland. In 1996 the same research group produced Dolly by nuclear transplantation from an adult cell line—the epithelial layer of the mammary gland. Thus Dolly was the first mammal to be produced using a nucleus from a differentiated cell line.

Mammals may be cloned by nuclear transplantation in which the nucleus from a somatic cell is inserted into an egg cell whose own nucleus has been removed.

DOLLY THE CLONED SHEEP

Cells in an early embryo are **totipotent**; that is, they possess the ability to divide and give rise to any type of body cell (liver, spleen, brain, etc.). Later on, cells lose this ability. They become committed to generating a particular tissue such as the nervous system or the digestive tract. Most cells in an adult animal can either no longer divide or else only give rise to a particular, specialized type of cell. During development, different genes are expressed in different tissues and others are shut down. So while almost all adult cells retain a complete genome, they don't retain the ability to develop into new individuals.

The cloning of Dolly the sheep showed that it is possible to reset the clock of an adult cell to zero and start development again. In Dolly's case, the trick was to starve cultured udder cells from the donor animal so that both the cell and the DNA stopped dividing (i.e., the cells entered the **G_0 stage** of the cell cycle—see Chapter 4). What exactly happens to the DNA when the cell is starved is not known. However, there is probably some modification, including demethylation, which converts the DNA back to a form resembling that of an embryonic cell. When the resting G_0 nucleus is placed in an egg cell whose own nucleus has been removed, it starts dividing again. The egg is then transplanted into a female animal, where it will develop into an embryo. If all goes well, a baby will be born (Fig. 15.18).

Early in 1996, at the Roslin Institute in Scotland, the world's first cloned animal, Dolly the sheep, was born. The donor nucleus came from a mammary gland cell (also known as the udder) from a pregnant ewe. Since Dolly's birth, a variety of other animals, including cattle, pigs, goats, mice, and cats have been cloned (Table 15.1). Dolly herself has been mated and gave birth to a lamb of her own—named Bonnie—during Easter 1998.

Strictly speaking, Dolly is not a complete clone. In addition to the nucleus, which contains the majority of the genetic information, animal cells contain a few genes in their

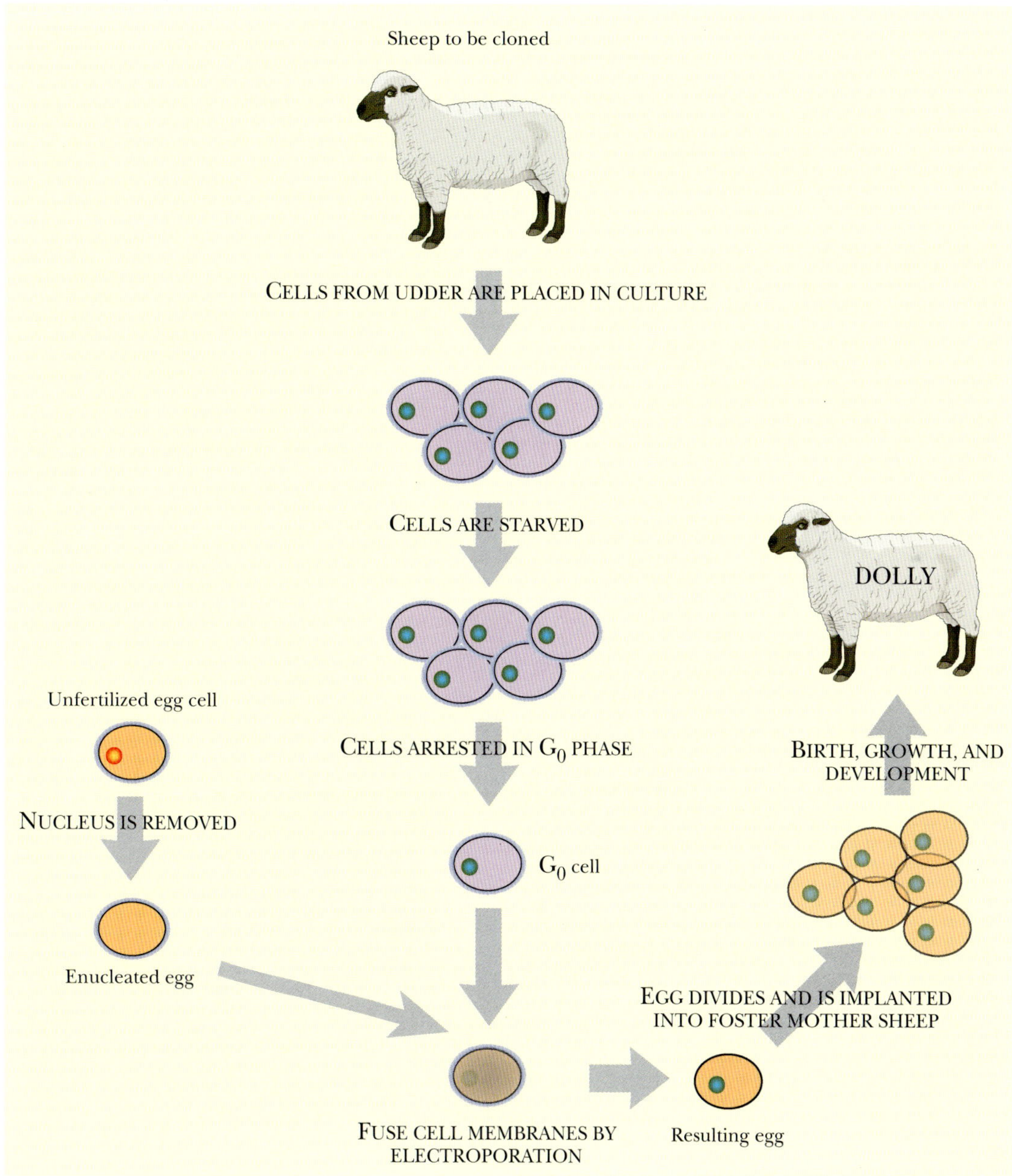

FIGURE 15.18 Scheme for Cloning Sheep

To clone a mammal such as a sheep, cells from the udder are isolated, grown in culture, and then starved in order to arrest them in G_0 of the cell cycle. Unfertilized egg cells from another sheep are also harvested and the nucleus is removed from them. An electrical stimulus fuses the G_0 udder cell with the enucleated egg, thus placing a somatic cell nucleus into an undifferentiated cytoplasm. The eggs that result are put back into a foster mother, and the offspring are screened for DNA identical to the donor sheep.

Table 15.1 Cloned Animals

Animal	Date	Name/Comments
Sheep	1996	Dolly
Mouse	1997	Cumulina (followed by 50 others!)
Cattle	1998	
Goat	1999	
Pig	2000	
Gaur	2000	Noah (endangered Asian ox; died of infection after 2 days)
Mouflon	2001	Endangered wild sheep (cloned from recently dead animal)
Cat	2001	CopyCat
Rabbit	2002	
Banteng	2003	Endangered Javanese cattle
Rat	2003	
Mule	2003	Idaho Gem
Horse	2003	
Deer	2003	
African wildcat	2003–4	Ditteaux (male), Madge and Caty (female)
Dog	2005	Snuppy

mitochondria. In Dolly's case, only the nuclear DNA was cloned. The mitochondrial DNA was provided by the egg cell that received the nucleus.

Following Dolly the cloned sheep, several mammals have now been cloned. However, the success rate is still low.

PRACTICAL REASONS FOR CLONING ANIMALS

Why clone sheep? Aside from showing that whole animals can be cloned, there are practical reasons. For millennia, humans have bred farm animals in attempts to improve them. Genetic duplication allows an improved animal to be widely distributed relatively quickly.

Box 15.3 Obituary for Dolly the Sheep

Dolly the sheep was put to sleep by lethal injection in February 2003, at an age of six and a half. The reason for euthanasia was a virus infection that gave rise to lung cancer. Dolly's breed of sheep, the Finn Dorset, often reaches 11 or 12 years. Like most cloned animals Dolly died somewhat prematurely. Dolly had telomeres that were 20% shorter than normal for a sheep her age, which might have contributed to premature aging. However, although Dolly had arthritis, no other signs of premature aging were found during the autopsy. Dolly was also overweight—perhaps as a result of getting extra snacks due to her celebrity status—and this may have been a factor in the development of her arthritis. Although it is possible that Dolly was less robust and more susceptible to infection as a result of being cloned, a significant number of other, normal, sheep die of similar virus infections, especially if kept inside in crowded conditions. Overall, it is impossible to draw definite conclusions on these matters based on a single animal—however much of a celebrity. Dolly's offspring, including Bonnie, born in 1998, and a later set of triplets, show no significant telomere shortening. However, these offspring were conceived naturally, and of course, half their DNA came from the fathers.

In addition, a flock or herd of genetically identical animals will give wool, milk, eggs, or meat of a more standardized quality. Conversely, genetically identical animals will all be susceptible to the same infections, and epidemics will spread faster and further.

Although cloning produces identical animals, the technique may paradoxically help protect genetic diversity. For example, in New Zealand, the last surviving cow of a rare breed was successfully cloned. Thus, cloning allows us to genetically rescue rare breeds of animals or endangered species but avoids mixing their genes with outsiders as would happen in crossbreeding. Table 15.1, which lists animals cloned to date, includes three rarities, the gaur, mouflon, and banteng.

The most important use of animal cloning is in combination with transgenics. Previously created transgenic animals may be cloned for speedier distribution of the product. However, it is also possible to introduce transgenic DNA during the cloning process. The Roslin Institute, where Dolly was born, has since cloned sheep carrying the gene for **human factor IX**. The transgene was inserted into the nuclear donor cells while in culture. A variety of other transgenic animals carrying pharmaceutically important proteins have also been cloned.

The reason why Dolly was cloned using a mammary gland cell should now be apparent. If a foreign protein is expressed in this tissue, it will be secreted into the milk and be easy to harvest commercially. Once a good transgenic cell line has been established in culture, nuclear transplantation can be used to generate several genetically identical animals for production purposes. Such transgenic cells can be stored over the long term by freezing in liquid nitrogen and have been referred to as *protoanimals*.

Cloned animals may be altered by transgenics, thus generating animals with useful properties more rapidly than traditional selective breeding.

IMPROVING LIVESTOCK BY PATHWAY ENGINEERING

It is possible to combine pathway engineering with cloning to produce improved livestock (as opposed to merely using transgenic animals as the source of a single useful protein). For example, the genes for the cysteine biosynthetic pathway are absent in mammals. Consequently mammals, including sheep, cannot make the sulfur-containing amino acid cysteine and must receive it in their diet. Sometimes cysteine is limiting for wool growth. Adding extra cysteine to the diet works poorly because of its uptake and degradation by microorganisms in the sheep's intestine.

Many bacteria do synthesize cysteine. Two steps are needed, starting from the amino acid serine plus inorganic sulfide (Fig. 15.19). In enteric bacteria these two key enzymes, **serine transacetylase** and **acetylserine sulfhydrylase**, are encoded by the *cysE* and *cysK* genes, respectively. These two bacterial genes have been cloned from *Salmonella* and placed under control of the mouse metallothionein promoter. The construct has been successfully integrated into transgenic mice, which expressed both enzymes. Furthermore, when these mice were deprived of the two sulfur-containing amino acids, cysteine and methionine, they remained healthy and made their own cysteine. The animals did need inorganic sulfide in their diet.

Although transgenic sheep carrying bacterial *cysE* and *cysK* genes have been made, high-level expression of these genes and synthesis of cysteine in the desired location (the epithelium of the rumen) have not yet been achieved. Several other schemes to improve livestock, including synthesis of other essential amino acids such as lysine and threonine, are in the early experimental stages.

One way to significantly improve livestock would be to engineer pathways to synthesize essential amino acids that must presently be provided in the diet. So far, partial success has been achieved using mice as model organisms.

A) CYSTEINE SYNTHESIS PATHWAY

Acetyl-CoA
Serine
Cys E
Serine transacetylase
O-Acetylserine
H_2S
Cys K
Acetylserine sulfhydrylase
Cysteine

B) VECTOR WITH CLONED *cys* GENES

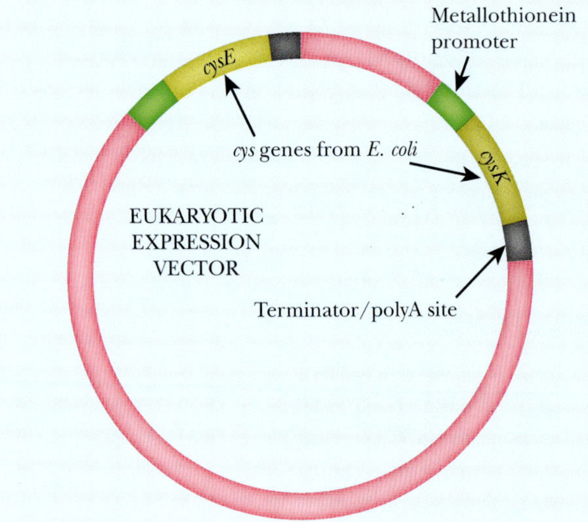

FIGURE 15.19
Biosynthesis of Cysteine from Serine plus Sulfide
(A) The cysteine biosynthetic pathway uses serine and acetyl-CoA. First, acetyl-CoA and serine are converted to *O*-acetylserine by serine transacetylase. Then acetylserine sulfhydrylase converts *O*-acetylserine to cysteine using hydrogen sulfide (H_2S). (B) The cysteine biosynthetic pathway has been cloned into a eukaryotic expression vector. The two *E. coli* genes are cloned behind the mammalian metallothionein promoter.

PROBLEMS AND ETHICS OF NUCLEAR TRANSPLANTATION

Cloning involves using genetic technology to make an exact genetic duplicate of an animal. Note that the new cloned animal starts out life as a single cell, develops into an embryo, and must proceed through childhood before becoming an adult. So although we can clone an animal, we do not get an instant full-grown duplicate.

So what about human cloning? The birth of Dolly the cloned sheep was followed by an outbreak of ethical discussion. Many critics insist that cloning humans is a threat to the sanctity of human life. In fact, human clones already exist. Identical twins, triplets, and so on are clones resulting from splitting of the same fertilized egg. Many ancient cultures regarded identical twins as supernatural in origin. Some cultures believed the twins were lucky; others thought they were evil and felt obliged to kill one of them.

Leaving aside moralistic arguments, there are major practical problems with cloning humans. The number of live clones born is just a few percent of the nuclei transplanted. Several lambs that died late in pregnancy or soon after birth were produced at the same time as Dolly, and some had developmental abnormalities. So cloning is fairly risky. Moreover, the chances of successful pregnancy, on implantation into a surrogate mother in humans, are estimated to be three- to 10-fold lower than in sheep. Thus the financial costs and effort to produce a cloned human would be vastly greater than for animals such as sheep. Outside of science fiction scenarios, why people would want to clone themselves is unclear. For one thing, the time scale of human development means that a cloned human duplicate would not be available for many years.

Human cloning is not only beset with moral issues, it also has major technical problems.

IMPRINTING AND DEVELOPMENTAL PROBLEMS IN CLONED ANIMALS

Even after the technological problems have been conquered, most attempts at cloning animals still fail. Successful nuclear transplantation involves reprogramming a nucleus from a differentiated cell. This is a complex process and the low success rate is probably due to failure to properly reprogram. Recent investigations implicate DNA methylation as the major factor.

Methylation patterns in cloned embryos are not identical to those in natural embryos. This is even true for most successfully cloned animals. However, Dolly and other cloned animals have given rise to genetically normal offspring, so cloning does not introduce permanent genetic alterations into an animal's germline.

Methylation generally shuts down eukaryotic genes that are not required in a particular tissue or at a particular stage in development. Methylation is also involved in **imprinting**, the regulatory mechanism that activates only the maternal or paternal copy of a gene (see Chapter 16 for more on imprinting). There are about 40 imprinted genes in humans. Usually, when the paternal copy is active fetal growth is promoted; whereas, when the maternal copy is active fetal growth is limited. Not surprisingly, incorrect imprinting patterns cause aberrant fetal growth and development as seen in many embryos derived from cloning.

Cloned mammals often display **large-offspring syndrome**. Their limbs, interior organs, and overall body size are abnormally large. The resulting animals are unhealthy. This syndrome is associated with incorrect imprinting of *IGF2R*, the gene for insulin growth factor 2 receptor. This gene normally shows maternal imprinting in mammals. In fetuses with large-offspring syndrome, the *IGF2R* gene showed altered methylation, and its expression was lower than normal. Curiously, the *IGF2R* gene is not imprinted in primates, so humans and monkeys should be less susceptible to large-offspring syndrome and might be expected to show less damage on cloning, at least from this particular cause.

Mammalian cloning suffers from a high failure rate. The major factor is probably the methylation pattern of DNA, which differs between natural embryos and those derived from cloning.

TRANSGENIC PEOPLE, PRIMATES, AND PETS

So far no primate has been successfully cloned, although transgenic rhesus monkeys have been generated. The first successful engineering of a transgenic primate resulted in the birth of ANDi, a rhesus monkey carrying the *gfp* gene, in late 2000. ANDi stands for "inserted DNA" (read backwards). A crippled retrovirus vector was used to deliver the gene for GFP to unfertilized eggs that were later fertilized *in vitro*. Treatment of 224 egg cells gave 20 embryos, five pregnancies, and eventually, three live male monkeys. Only one of these, ANDi, was transgenic and expressed GFP. ANDi does not fluoresce green because GFP levels are too low (also, rhesus monkeys have brown fur over much of their bodies).

There have been several exaggerated claims for primate cloning. Rhesus monkeys have been generated by splitting an embryo at the eight-cell stage into four genetically identical two-cell embryos, and this has been touted as "cloning." However, it is merely artificial twinning, rather than true nuclear transplantation as with Dolly. Nonetheless, it seems likely that by the time this book is published rhesus monkeys will have been cloned.

There seems no reason why cloning of monkeys, apes, and humans via nuclear transplantation should not be technically possible. One possible objective for human cloning is to obtain tissue for transplantation rather than to generate a new human individual. Reprogrammed human cells would be grown in culture to provide such material. This is known as **therapeutic cloning**, and developments in this area are likely to continue rapidly.

We will end this section on a cozy note. In November of 2001, the first pet, a kitten, named CC (for CopyCat; Fig. 15.20), was cloned at Texas A&M University. The objective of this cloning program is for people to have their favorite pets cloned—providing they have enough money, of course. One cannot help remembering that the ancient Egyptians mummified cats as well as people. CC is the only one of 87 implanted cloned embryos to survive. This is similar to the success rate for cloned sheep, mice, and so forth. Obviously, routine cloning of pets needs a higher success rate to be economically feasible.

Primates have not yet been successfully cloned; however, it has been possible to make transgenic rhesus monkeys using retrovirus vectors.

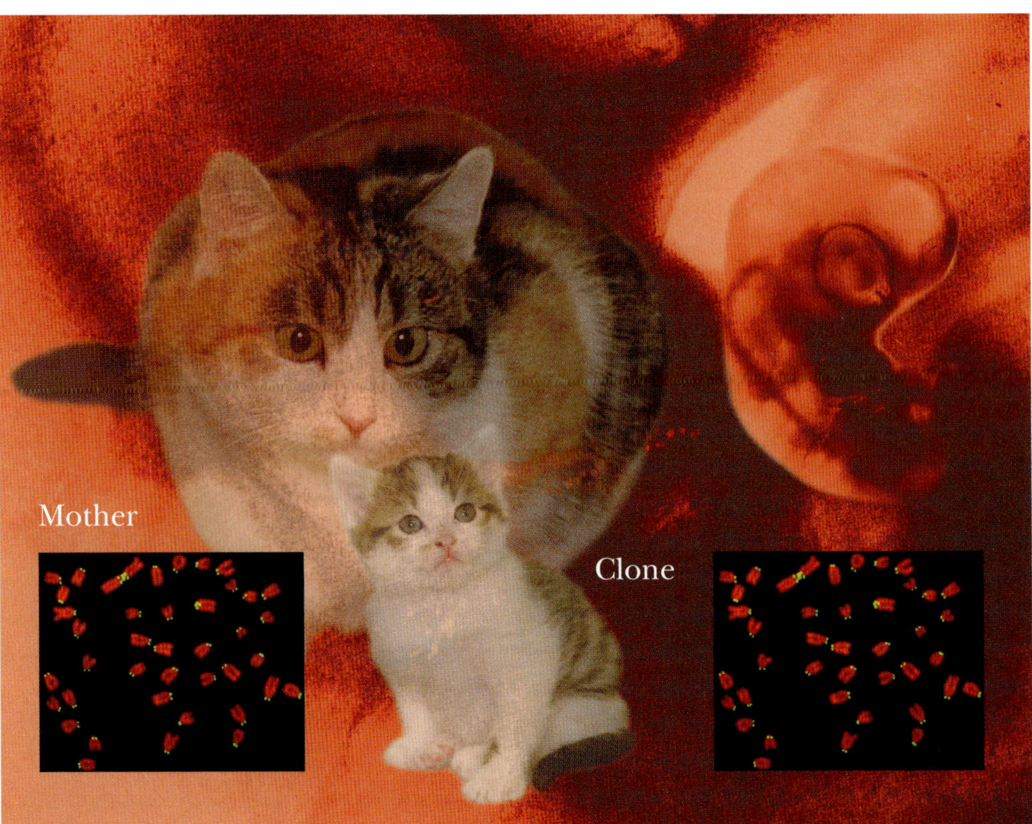

FIGURE 15.20 A Clone Is Not a Copy This digital collage shows CopyCat and his mother Rainbow. CC is a true clone generated by nuclear transplantation. Rainbow, a calico female is the nuclear donor. Both cats therefore have identical chromosomes. Although CC is a genuine clone, she does not look identical to the nuclear donor because the pattern of a cat's coat is partly due to randomized cell divisions and, in females, X-linked inactivation. Digital media by Hunter O'Reilly, 2002.

APPLICATIONS OF RNA TECHNOLOGY IN TRANSGENICS

As discussed earlier, transgenic technology at the DNA level allows us to add genes and to inactivate gene expression by insertion. It is also possible to manipulate genes at the RNA level by the use of **antisense, ribozymes,** or **RNA interference**. We have already discussed these topics (see Chapter 5); therefore, this section simply summarizes their application in transgenics.

It is possible to insert an antisense transgene when constructing a transgenic organism. The simplest way to construct an antisense gene is to invert the DNA sequence of the original gene. In higher organisms, the cDNA version of the gene is used as the starting point to avoid complications due to intervening sequences. Such an antisense gene will be transcribed to give antisense RNA, which will then bind to the mRNA of the corresponding sense gene (see Chapter 5). This results in inhibition of gene expression at the level of translation.

Insertion of antisense genes into mice has resulted in up to 95% inhibition of gene expression, although the effectiveness varies greatly. The antisense RNA may be full length or just a short segment corresponding to part of the sense gene. Short segments complementary to the 5′-untranslated region of the sense mRNA are often very effective at blocking translation. Obviously, antisense constructs can be placed under the control of inducible promoters. This allows antisense silencing of gene expression to be regulated by the experimenter.

It is also possible to insert transgenic constructs that express ribozymes. In such cases, the ribozyme is usually designed to cleave a specific mRNA. When the ribozyme is expressed it cleaves the mRNA of the target gene and so reduces gene expression. Occasional cases are known in both *Drosophila* and mice where transgenic expression of ribozymes has been successful in decreasing gene expression.

Gene expression may be decreased by insertion of genes encoding antisense RNA or encoding ribozymes targeted to cleave a specific mRNA.

APPLICATIONS OF RNA INTERFERENCE IN TRANSGENICS

RNA interference covers several related phenomena. The mechanisms are still only partly understood, but involve the formation of double-stranded RNA (dsRNA). This is not normally made by either prokaryotic or eukaryotic cells. Consequently it is regarded by eukaryotic cells as evidence of viral activity and therefore is degraded.

When both sense and antisense RNA are present in eukaryotic cells, the formation of dsRNA is followed by a massive drop in expression of the corresponding gene. This effect is due to degradation of the mRNA. Therefore the dsRNA must correspond to the exon sequences of the target gene. Only a few molecules of dsRNA are needed to trigger this effect. RNA interference is the favored procedure for gene inactivation in *Caenorhabditis elegans* and is also effective in other animals and plants.

RNA interference can be triggered by adding RNA directly or by expression of the corresponding transgenic DNA constructs. A construct with neighboring sense and antisense sequences will produce hairpin RNA (Fig. 15.21). Alternatively it is possible to use a single transgene flanked by two promoters, one pointing in each direction.

Homologous cosuppression is a type of RNAi characteristic of plants, though occasionally seen in other organisms. Here the presence of multiple copies of a transgene decreases the expression of related host genes (and of the transgenes themselves, too). Occasionally, single copies of transgenes that are highly homologous to an endogenous host gene may be sufficient to trigger cosuppression. The mechanism varies somewhat from case to case, but results in formation of dsRNA, which then triggers RNA interference. Transgenes carried by RNA viruses are especially effective at causing homologous cosuppression, although only if the virus is still capable of replication.

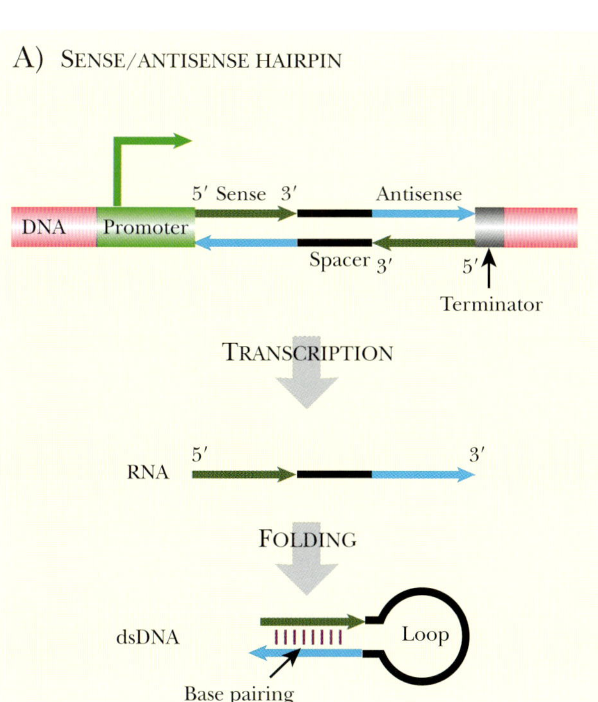

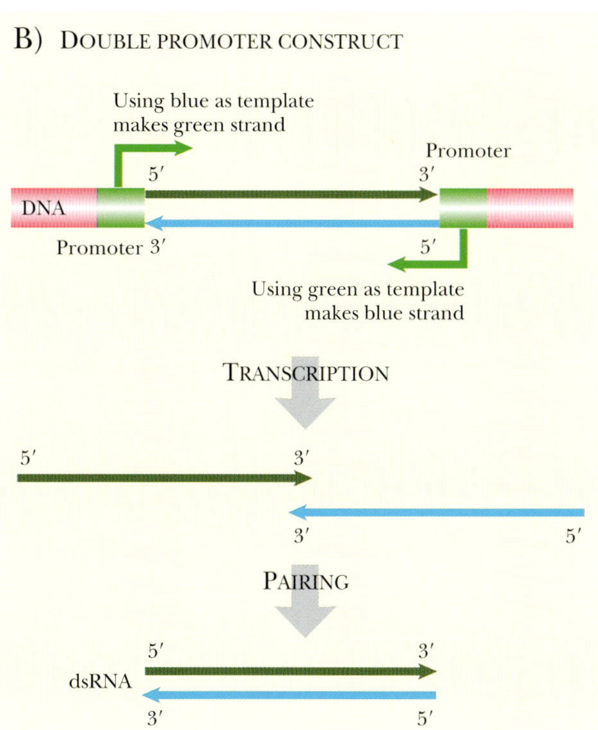

FIGURE 15.21 RNA Interference Constructs

RNA interference occurs when both the sense and antisense RNA of a gene are present and form dsRNA. Two constructs are shown that direct the synthesis of a dsRNA molecule. The first construct (A) has both sense and antisense regions that base pair. A spacer separates the sense and antisense regions and forms a loop at the end of the hairpin. The double promoter construct (B) has two promoters—one for the sense strand and the other for the antisense strand. The two resulting RNA molecules are complementary and form a dsRNA molecule. The presence of dsRNA triggers degradation of mRNA from the corresponding gene.

RNA interference may be used to knock down expression of chosen genes. A variety of RNAi constructs are available, some of which may be incorporated into the host genome and inherited.

NATURAL TRANSGENICS AND DNA INGESTION

Historically speaking, we are all transgenic. The human genome contains a significant number of genes of bacterial origin. These have presumably been picked up at various stages in evolutionary history from assorted bacteria. In addition, we carry quite a few genes that came originally from other higher organisms, some of which have been transmitted by retroviruses. Such movement of genes between organisms that are not direct descendants is generally known as lateral or horizontal gene transfer, but might just as well be called "natural transgenesis."

There are two major pathways for naturally acquiring genetic material from other organisms. One is by viral transduction and the other is by direct intake of DNA. In microorganisms DNA from the environment may be taken up by transformation. In the case of animals, DNA is constantly ingested along with other components of the diet. Certain protozoa such as *Paramecium* and some amoebas live by ingesting bacteria. It is thought that several genes involved in fermentative metabolism in *Entamoeba* are derived from such ingested bacteria.

Humans and other mammals, whether carnivorous or vegetarian, are constantly eating food that contains substantial amounts of DNA. Although it is generally assumed that ingested DNA is degraded into nucleotides in the intestinal tract, this is not entirely true. Recent findings suggest that a very small proportion of the DNA survives as fragments of moderate size (up to 1000 base pairs) and actually crosses the intestinal wall into the bloodstream of the animal in this form. At least transiently, DNA from food can be traced to several different organs and can also cross the placenta to fetuses and newborns.

DNA from the bacterial virus M13 and the gene for green fluorescent protein have been used as test molecules. More recently, naturally occurring DNA sequences for the plant-specific ribulose-1,5-bisphosphate carboxylase (Rubisco) gene have been tracked by PCR. All of these foreign DNA sequences have been found both in the bloodstream and in the cell nuclei of various tissues in animals that ingested the DNA. Sometimes the foreign DNA appears to be integrated into host chromosomal DNA. However, so far no evidence for expression of genes carried by ingested DNA has been found, nor has incorporation into the germline been seen for DNA eaten by mice. The overall likelihood of food-borne DNA infiltrating the genome and being expressed is unknown, although it seems low. Nonetheless, because vast numbers of animals have eaten DNA every day of their lives over millions of years, it seems likely that this must occur now and then. Furthermore, the fact that our genomes do include a small percentage of foreign genes argues that germline insertion does occur, albeit very rarely.

From a purely theoretical viewpoint, we should remember that living cells contain much more RNA than DNA. Thus the amount of RNA ingested in the food is at least 10-fold greater than the amount of DNA. Whether any of this RNA survives long enough to cross the intestinal wall and enter animal cells has so far not been investigated. Reverse transcriptase from the assortment of retroelements present in most animal genomes could in theory reverse transcribe such incoming RNA, so generating a DNA copy that might occasionally integrate into the host cell genome.

Short fragments of DNA in the diet can be taken up and incorporated into host chromosomes. Whether such DNA can be incorporated into the mammalian germline is unknown.

Summary

Genetic engineering combined with cloning has allowed the creation of an ever-growing range of genetically modified organisms. Some, with inactivated genes, are widely used in medical research. Transgenic animals with novel or improved capabilities may help improve livestock production. Controlling the expression of inserted transgenes is often problematic and may require assembly of highly sophisticated artificial control systems with components derived from several sources.

End-of-Chapter Questions

1. What process is used to create transgenic animals?
   - **a.** particle bombardment
   - **b.** nuclear microinjection
   - **c.** nuclear fusion
   - **d.** germ line transformation
   - **e.** none of the above
2. In the first transgenic animal experiment, which gene from rats was cloned into mice?
   - **a.** somatotropin
   - **b.** metallothionein
   - **c.** myostatin
   - **d.** plasminogen
   - **e.** PPAR-delta
3. According to the book, which protein is produced by transgenic goats?
   - **a.** recombinant tissue plasminogen activator
   - **b.** recombinant tissue fibrinogen activator
   - **c.** recombinant bovine somatotropin
   - **d.** recombinant human somatotropin
   - **e.** none of the above
4. Why are embryonic stem cells important?
   - **a.** They can be passed from one generation to the next.
   - **b.** These cells carry retroviral genes.
   - **c.** They can develop into any tissue in the body, including the germ line.
   - **d.** The cells are differentiated and can therefore be manipulated.
   - **e.** none of the above
5. How can location effects of transgenes in animals be avoided?
   - **a.** by placing LCR sequences in front of the transgene
   - **b.** by inclusion of insulator sequences on both ends of the transgene
   - **c.** by using natural transgenes instead of the cDNA version
   - **d.** by targeting the transgene to a specific location
   - **e.** all of the above
6. What is used by targeting vectors to insert transgenes at specific locations within the host genome?
   - **a.** homologous recombination
   - **b.** transfection
   - **c.** transduction
   - **d.** conjugation
   - **e.** all of the above

(Continued)

7. Which of the following systems can be used to control the expression of the transgene?
   a. a modified LacI repressor system
   b. the TetR recombinant promoter system
   c. use of a heat shock promoter from *Drosophila*
   d. the use of steroid receptors
   e. all of the above

8. Why are Cre/*loxP* or Flp/FRT used in transgenic animals?
   a. activation of a transgene by removing blocking sequences flanked by the *loxP* or FRT sites
   b. large-scale deletions and rearrangements of the chromosomes
   c. removal of selectable markers that are no longer needed
   d. creation of conditional knockout mutants
   e. all of the above

9. What can be engineered to introduce transgenes into *Drosophila*?
   a. Cre recombinase
   b. S elements
   c. P elements
   d. YACs
   e. none of the above

10. How can mosquito-borne diseases be controlled through biotechnology?
    a. by genetically engineering mosquitoes that are resistant to the disease
    b. by creating more efficient pesticides
    c. through the cloning, expression, and purification of novel therapeutic agents
    d. by engineering non-pathogenic malaria, dengue, and other mosquito-borne diseases
    e. none of the above

11. How might mammals be cloned?
    a. homologous recombination
    b. transfection with a retrovirus
    c. YACs
    d. nuclear transplantation
    e. BACs

12. Why might Dolly technically NOT be a true clone?
    a. Mitochondria contain DNA and these organelles were provided by the donor enucleated egg cell, not from the cell from which the nucleus was removed.
    b. Dolly contained 20% shorter telomeres.
    c. Dolly did not look like the donor of the nucleus.
    d. Dolly died prematurely, much earlier than her donor.
    e. none of the above

13. What is a reason for cloning animals?
    a. to give products of more standardized quality
    b. to insert a transgene into the nucleus during the cloning process to produce identical animals that express the foreign protein
    c. to be able to distribute improved animals more quickly

- **d.** to rescue rare breeds or endangered species of animals without crossbreeding
- **e.** all of the above

14. What is a goal of "therapeutic cloning"?
- **a.** to create a cloned human being
- **b.** to insert transgenes into humans via nuclear transplantation
- **c.** to obtain tissue for transplantation
- **d.** to obtain tissues that express foreign proteins, usually therapeutics
- **e.** none of the above

15. For which protein have foreign DNA sequences been found in the bloodstream and tissues of animals who ingested the DNA?
- **a.** Rubisco
- **b.** Nabisco
- **c.** reverse transcriptase
- **d.** prisco
- **e.** none of the above

Further Reading

Christophides GK (2005). Transgenic mosquitoes and malaria transmission. *Cell Microbiol* **7**, 325–333.

Houdebine LM (2007). Transgenic animal models in biomedical research. *Methods Mol Biol* **360**, 163–202.

Melo EO, Canavessi AM, Franco MM, Rumpf R (2007). Animal transgenesis: State of the art and applications. *J Appl Genet* **48**, 47–61.

Mitalipov SM, Wolf DP (2006). Nuclear transfer in nonhuman primates. *Methods Mol Biol* **348**, 151–168.

Trounson AO (2006). Future and applications of cloning. *Methods Mol Biol* **348**, 319–332.

Vilaboa N, Voellmy R (2006). Regulatable gene expression systems for gene therapy. *Curr Gene Ther* **6**, 421–438.

Wheeler MB (2007). Agricultural applications for transgenic livestock. *Trends Biotechnol* **25**, 204–210.

Whitelaw CB, Sang HM (2005). Disease-resistant genetically modified animals. *Rev Sci Tech* **24**, 275–283.

Inherited Defects

Introduction
Hereditary Defects in Higher Organisms
Hereditary Defects Due to Multiple Genes
Defects Due to Haploinsufficiency
Dominant Mutations May Be Positive or Negative
Deleterious Tandem Repeats and Dynamic Mutations
Defects in Imprinting and Methylation
Mitochondrial Defects
Identification, Location, and Cloning of Defective Genes
Cystic Fibrosis
Muscular Dystrophy
Genetic Screening and Counseling

INTRODUCTION

Genetic defects vary from trivial to life threatening. Although we tend to think of inherited conditions such as diabetes and muscular dystrophy as diseases, we often refer to cleft palates or color blindness as inherited defects. However, they are all the result of mutations in DNA, the genetic material. Not only are some diseases directly caused by mutations, but susceptibility to infectious disease and other damaging environmental factors, such as radiation, is also influenced by a variety of genes.

Precise rates of mutation are difficult to estimate, but for humans and apes, the mutation rate is around 5.0×10^{-8} per kilobase of DNA per generation. For rodents, the rate is some 10-fold less because fewer cell divisions are needed to form gametes from ancestral germ cells. Thus a considerable number of mutations are constantly accumulating in the germline of humans and other animals. Most of these have little or no effect, but a small percentage give rise to serious hereditary defects. A full listing of human hereditary defects, known as OMIM (Online Mendelian Inheritance in Man), is available on the Internet at http://www.ncbi.nlm.nih.gov/entrez/query.fcgi?db=OMIM.

Mutations in the human genome are responsible for a wide variety of inherited defects and diseases.

HEREDITARY DEFECTS IN HIGHER ORGANISMS

If the DNA of a single-celled organism is mutated, the mutation will be passed on to all of its descendants when it divides. The situation in multicelled creatures is more complex. In animals, the germline cells are reserved for reproductive purposes and give rise to the eggs and sperm in mature adults. The somatic cells forming the rest of the body are not passed on to the next generation (see Chapter 1). However, mutation of somatic cells is involved in cancer, which is dealt with separately in Chapter 18.

FIGURE 16.1
Inheritance of Recessive Mutations
A defective mutation usually occurs in only one copy of a gene. Therefore, most affected individuals will have one normal copy (A) and one mutated copy (a) of the gene. When two people, both carrying a recessive mutation in the same gene, have children, 25% of the children will inherit both mutant copies and exhibit the disease.

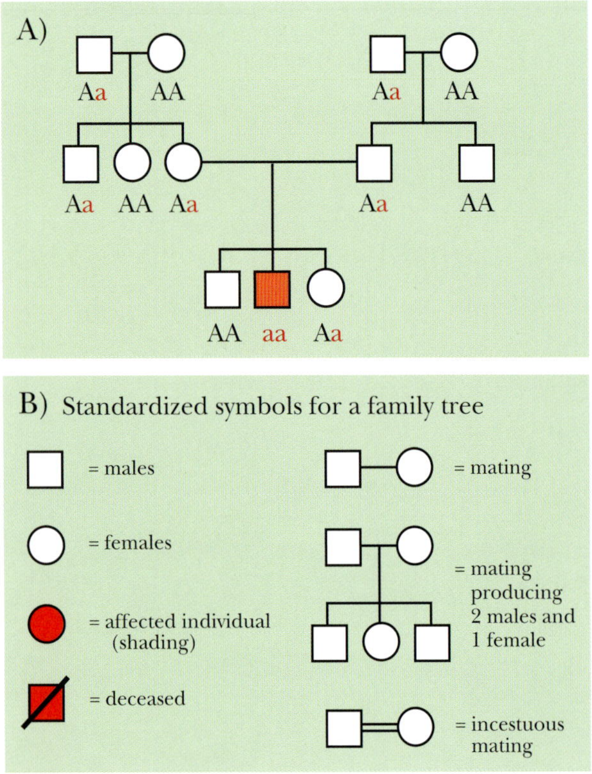

Higher organisms such as animals and plants are normally diploid and have two copies (i.e., two alleles) of each gene. Therefore, if one copy is damaged by a mutation, the other copy can compensate for the loss. Because most mutations are relatively rare, it is unlikely that both alleles of the same gene will carry mutations. Furthermore, most detrimental mutations are recessive to the wild type (Fig. 16.1 and Table 16.1). That is, a single functional allele is sufficient for normal growth and the defective copy has no noticeable effect on the phenotype.

Nonetheless, we humans all have quite a few mutations randomly scattered among our 25,000 genes. Obviously, close relatives tend to share many defects. Individuals share half of their genetic information with their brothers, sisters, father, and mother, although, of course, not the same half with each of them. Therefore, a child from the mating of close relatives (for example, as in brother/sister or father/daughter), has a much increased chance of getting two copies of the same defect, that is of being homozygous for the recessive allele (Fig. 16.2).

Mating between close relatives, even cousins, makes genetic disease more likely. This is because a rare recessive allele present in one ancestor may be passed down both sides of the family and two copies may end up

Table 16.1 Some Human Defects Due to a Single Gene

Disease	Frequency
Adenosine deaminase deficiency Lack of enzyme causes immune deficiency First defect approved for human gene therapy	Rare
Cystic fibrosis Defective ion transport indirectly affects mucous secretion in lungs	1/2000 (whites) Rare in Asians
Duchenne's muscular dystrophy Disintegration of muscle tissue Giant gene is a frequent target for mutations	1/3000 males (sex-linked)
Fragile X syndrome Common form of X-linked mental retardation	1/1500 males 1/3000 females (sex-linked)
Hemophilia Defect in blood coagulation	1/10,000 males (sex-linked)
Myotonic dystrophy Genetically dominant form of dystrophy	1/10,000
Phenylketonuria Mental retardation due to lack of enzyme Detected in newborns by urine analysis Special diet prevents symptoms	1/5000 (Western Europe) Rare elsewhere
Sickle cell anemia Defect in hemoglobin beta chain Heterozygotes are resistant to malaria but homozygotes are sick Common in Africa First molecular disease to be identified	1/400 (U.S. blacks)

in one particular child. In Europe, royal families insisted that their children marry other royalty; therefore, many marriages were between closely related individuals. This led to a high incidence of defects among the offspring, the best known being hemophilia. In marriages between unrelated people, recessive diseases will only affect the children if by chance an ancestor from each side of the family carried a copy of the defective allele. This is the underlying reason for the taboos against incest, which prevail in most societies, though the degree of forbidden relationships varies considerably between different cultures.

Many genetic defects are fatal to sperm cells, fertilized eggs, or early embryos so they are never seen in adults. Nevertheless, a wide range of genetic defects has been observed in living humans. Some of the most common defects due to a single gene are listed in Table 16.1. Unless otherwise noted, they are recessive and require both alleles to be defective for symptoms to appear. Large genes are bigger targets for random mutation. The genes responsible for cystic fibrosis, muscular dystrophy, and phenylketonuria are all abnormally large and all three genetic defects are relatively common.

FIGURE 16.2 Homozygous Recessive from Inbreeding
A brother and sister may both inherit a mutant gene (a) from their one parent (their father in this illustration), but will not be affected because they each have a normal gene from their mother. If the brother and sister have offspring, their child may inherit two copies of the defective gene and then exhibit the symptoms of the defect.

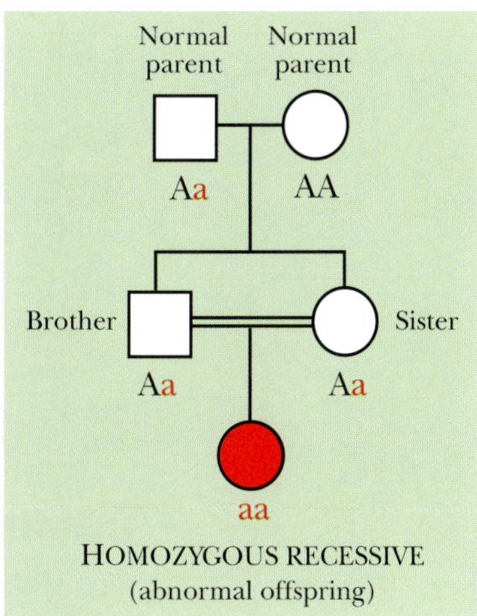

Many well-known hereditary defects are due to homozygous recessive mutations in a single gene.

HEREDITARY DEFECTS DUE TO MULTIPLE GENES

Several well-known hereditary defects are missing from Table 16.1 because they involve more than one gene. These may be subdivided into two types. Some multigene defects are due to the interacting effects of several individual genes. Some examples are cleft palate, spina bifida, certain cancers, and diabetes.

Other multigene defects are due to the presence of an extra copy of an entire chromosome. Although most errors involving whole chromosomes are lethal, a few are viable. The best known of these is **Down syndrome**, which causes mental retardation and is due to an extra copy of chromosome 21. The overall frequency of Down syndrome is about one in 800, but like most instances of getting a whole extra chromosome, it is due to an accident during cell division in the newly fertilized egg and it is not normally inherited. The relative chances of such a mishap increase with maternal age, but even so, most Down syndrome children are born to young women, simply because more younger women have babies. Extra sex chromosomes are found in about one of every 1000 people; there are three relatively common possibilities: XXY, XYY, and XXX. The XYY individuals are best known for their supposed tendencies toward violent crime, but the others show various abnormalities also.

Hereditary defects due to multiple interacting genes are difficult to analyze.

DEFECTS DUE TO HAPLOINSUFFICIENCY

Loss-of-function mutations are usually recessive and, in most cases, a defect in one of the two copies of a diploid gene has little effect. Only rarely is one functional copy of a gene insufficient. This situation is known as **haploinsufficiency**. Three main reasons explain most cases where gene dosage is important:

- **(a)** Some proteins may be needed in very high amounts in certain tissues. Thus a single functional gene may not allow sufficiently high levels of transcription and translation.
- **(b)** Some proteins interact with other proteins in strict ratios. Perturbing the amount of one component may have damaging effects.
- **(c)** Some regulatory networks respond in a quantitative manner. Therefore the absolute level of regulatory proteins that are involved may be critical for correct operation.

An example of the first case is the protein **elastin**, encoded by the *ELN* gene located on chromosome 7 in band 7q11. Elastin is found in the elastic tissues of skin, lung, and blood vessels. In people with a single defective *ELN* gene, most elastic tissues still function correctly. However, two copies of the *ELN* gene are needed to make a normal aorta, which is extremely elastic. People with one copy of *ELN* are unable to make sufficient elastin for this tissue, resulting in a narrowing of the aorta that sometimes requires surgery and is referred to as congenital supravalvular aortic stenosis (SVAS).

Usually one functional copy of a gene is sufficient. Occasionally two functional copies are required for optimal activity and a defect in one copy consequently results in a deficiency.

DOMINANT MUTATIONS MAY BE POSITIVE OR NEGATIVE

Occasionally a mutation causes a change of function or even a gain of function in the resulting gene product. In this case, a single mutant copy of the gene may cause significant phenotypic effects—that is, the mutation is dominant. Gain-of-function mutations are relatively rare in inherited disease but are frequent in the somatic mutations that cause cancer—see Chapter 18.

Gain-of-function mutations are much more specific than those causing loss of function. For example, **achondroplasia**, or short-limbed dwarfism, is due to mutation of a single copy of the *FGFR3* gene that encodes fibroblast growth factor (FGF) receptor #3. FGF receptors are signal-transducing proteins, that is, they receive signals at the cell membrane and transmit the information to the nucleus. Normally they are activated only when they receive a signal from outside the cell (i.e., bind FGF). Although most mutations inactivate the receptor, a few extremely rare mutations yield receptors that are active despite the absence of an external signal. Only mutations that cause the replacement of the amino acid glycine at position 380 with arginine (Gly380Arg) cause achondroplasia. Other mutations in this gene cause other symptoms.

Dominant negative mutations are those where the mutant protein loses its own function but, in addition, the defective protein interferes with the function of another protein. Thus a dominant negative mutation usually results from the presence of an altered, defective protein. Mutations in the same gene that merely abolish protein synthesis are usually recessive.

The simplest scenario for a dominant negative mutation is where the affected protein forms oligomers. If a mutation blocks protein synthesis in one copy, there will simply be a 50% reduction in the amount of protein. However, if the mutant allele produces an altered protein, this may bind to normal copies of the same protein and give inactive complexes. The final result may be that almost no active protein is available (Fig. 16.3). Many transcription factors bind to DNA as dimers and are susceptible to dominant negative effects.

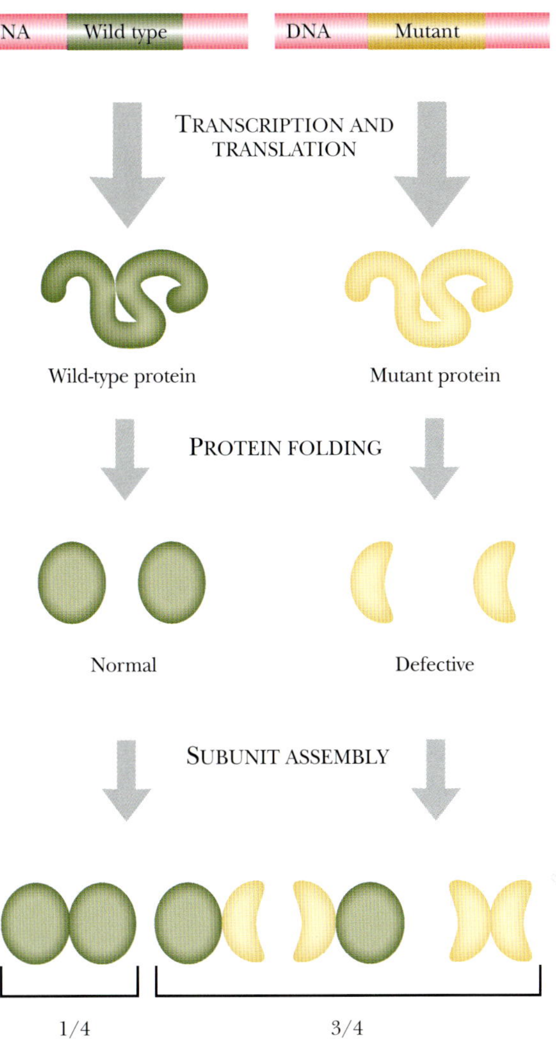

FIGURE 16.3 Dominant Negative Mutations
Dominant negative mutations occur when the defective copy of a gene interferes with the functional copy. For example, the defective protein may bind to and interfere with the normal protein. In this scenario, the proteins function as dimers. If the mutant protein is defective but still forms dimers, then three-fourths of the complexes will be defective.

Proteins that form ion channels are also liable to dominant negative effects. Cardiac arrhythmia (Romano-Ward syndrome) is due to a dominant mutation in the *KVLQT1* gene. This encodes a protein that assembles with others to form a transmembrane potassium channel. Certain defective KVLQT1 proteins can still bind as part of the assembly but block the operation of the channel. About 20% remaining activity is seen in this syndrome.

Hereditary defects due to dominant mutations are rare. They are often due to an altered protein that is not only defective itself but interferes with the function of other proteins.

DELETERIOUS TANDEM REPEATS AND DYNAMIC MUTATIONS

Several genetic disorders are known where the defect is due to a tandem repeat of three bases within a protein coding region. The three bases are usually CAG (in the coding strand), which encodes glutamine. The wild-type alleles contain several CAG repeats that give rise to a run of several glutamines within the encoded protein. The mutant alleles have increased numbers of these repeats and give rise to proteins with longer **polyglutamine tracts** (Fig. 16.4). Below a certain number (generally in the range 5–30), these repeats are relatively harmless and stable. Beyond this threshold, the mutations cause disease. In addition, the number of repeats is unstable and tends to expand each successive generation, sometimes to over 100. Hence these are sometimes referred to as **dynamic mutations**. Generally, the higher the number of tandem repeats, the more severe the pathogenic effects and the earlier the onset of disease.

A) EXPANSION OF CAG TRIPLETS IN DNA

NORMAL DNA REPLICATION

1 2 3
5′ CAG CAG CAG New strand 3′
3′ GTC GTC GTC GTC GTC GTC Template strand 5′
1 2 3 4 5 6

BACKWARD SLIPPAGE CAUSES DUPLICATION

CAG CAG
1 2
5′ CAG CAG New strand 3′
3′ GTC GTC GTC GTC GTC GTC Template strand 5′
1 2 3 4 5 6

1 2 3 4 5 6 7 8
5′ CAG CAG CAG CAG CAG CAG CAG CAG New strand 3′

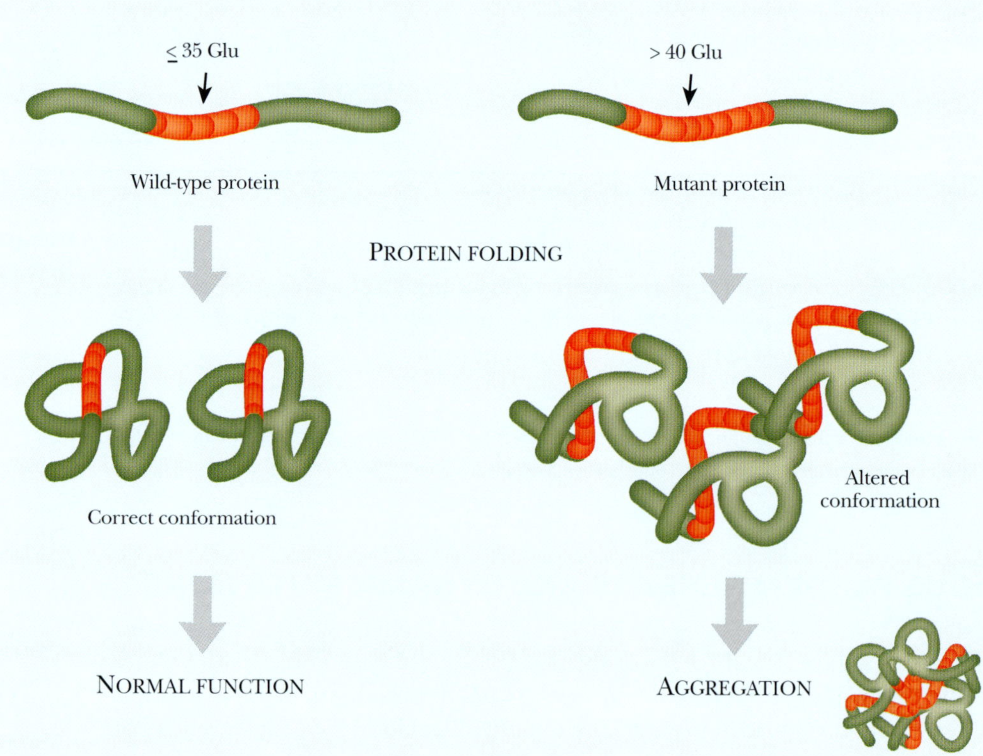

FIGURE 16.4 Tandem CAG Repeats and Polyglutamine Tracts

(A) Long stretches of CAG in DNA can cause errors during replication. If a germline cell has a stretch of six CAG repeats, DNA polymerase may sometimes slip during replication, causing the new strand to loop out. During meiosis, one of the gametes will have eight CAG repeats rather than six. (B) During protein synthesis CAG encodes glutamine and the number of CAG repeats can affect protein structure. Normally, the protein has fewer than 30 glutamines and forms the correct conformation. If there are more than 40 CAG repeats, the extra glutamines cause the protein to misfold, which in turn can cause the protein to aggregate into clumps.

The known polyglutamine/CAG repeat defects all cause neurodegenerative diseases with late onset. Almost all are autosomal dominants. In all cases, the extended polyglutamine tracts cause the proteins to clump into aggregates that ultimately kill nerve cells. Different CAG repeat diseases affect different proteins that are expressed at different levels in different nervous tissues. Consequently, the clinical symptoms vary. Perhaps the best known is **Huntington's disease**, located on chromosome 4 at 4p16. Transport of vesicles in nerve cells appears to be affected and results in loss of control of the limbs, impaired cognition, and dementia.

Unstable expanding tandem repeats may also occur outside coding sequences. In some cases these are harmless. In other cases they may affect the expression of nearby genes and have devastating results. An example is **fragile X syndrome**. The name refers to a fragile site within the long arm of the X chromosome, whose effects can be seen under the microscope. The tandem repeats may cause the chromosome to fragment into two parts or the two regions may remain held together by a thin thread of material. Fragile X affects about 1 in 4500 males, and causes mental retardation. The syndrome is less common and much less severe in females. Although differences in imprinting are also involved (see later discussion), this difference is largely due to females having two X chromosomes, whereas males have only one.

Fragile X syndrome is due to tandem repeats of CGG located within the 5′-untranslated region of the *FMR1* gene in band Xq27 on the affected X chromosome (Fig. 16.5). In the wild type there are around 30 copies of CGG. As with the CAG repeats described earlier, the number of repeats is unstable and tends to expand each successive generation. If the number of repeats expands to 50 to 200, no symptoms are observed, but such carriers often have children with more than 200 repeats, who do show symptoms. The fully expanded version of the fragile X allele may have up to 1300 CGG repeats. Although the CGG repeats are not translated into protein, they act as CG islands and tend to be methylated. This abolishes transcription of the *FMR1* gene, which regulates the synthesis of proteins in the synaptic junction of nerve cells. The defect causes mental retardation plus other symptoms. Another fragile X site (fragile X site E or FRAXE) is located in the *FMR2* gene about 600 kb distal of the *FMR1* gene (fragile X site A or FRAXA).

WILD TYPE

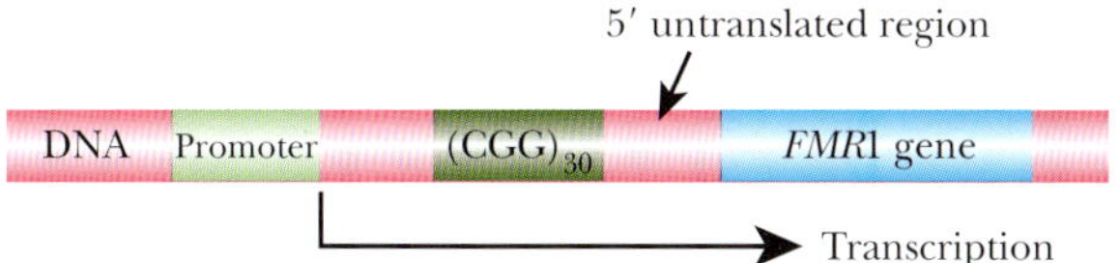

FRAXA MUTATION

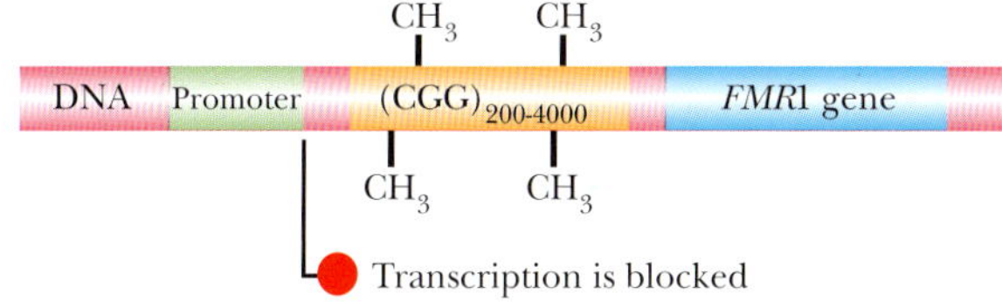

FIGURE 16.5
Fragile X
Between the promoter and gene for *FMR1* is a series of about 30 CGG repeats. These are usually present in the mRNA transcript, but are not translated into protein. During replication, DNA polymerase errors cause the number of repeats to expand. When more than 200 repeats occur, the CG sites become highly methylated. RNA polymerase is unable to transcribe the gene, and the protein is not produced.

Certain mutations consist of multiple tandem repeats of three bases (that is, of a whole codon). The severity depends on the number of repeats. Furthermore, the repeat number tends to increase every generation, giving dynamic mutations.

Box 16.1 RNA Gain of Function in Tandem Repeats

Some triplet repeat diseases have complex symptoms that cannot fully be explained by alterations in protein structure or in gene expression. Instead it appears that an altered RNA molecule, carrying the triplet repeat sequence, may directly cause harmful effects. This unusual situation is referred to as RNA gain of function. An example is myotonic dystrophy, which is caused by an unstable CTG repeat. This is located in the 3′-untranslated region of the *DMPK* gene on chromosome 19q13. The repeat sequence is present in the transcribed RNA (as a CUG repeat) but not in the final protein. This repeat hinders processing of the mRNA and its transport into the cytoplasm. Consequently, there is a deficiency in the amount of the DMPK (dystrophia myotonica protein kinase) protein. However, knockout mice with defective *DMPK* genes show only some of the symptoms of myotonic dystrophy.

(Continued)

Box 16.1 RNA Gain of Function in Tandem Repeats—cont'd

These further symptoms are due to direct action of the RNA with the CUG repeat (Fig. A). In RNA a long series of tandem CUG repeats will form a stable hairpin with a single long stem. This RNA is retained in the nucleus where it binds to several proteins, in particular the double-stranded RNA-dependent protein kinase, PKR. This enzyme is part of the antiviral defense system and is activated by binding to dsRNA. Activated PKR phosphorylates several target proteins, including translation factor eIF2α. This is intended to shut down protein synthesis in virus-infected cells. A variety of other damaging effects occur as a result of sequestration of other nuclear RNA-binding proteins and interference with their normal roles in RNA processing and splicing.

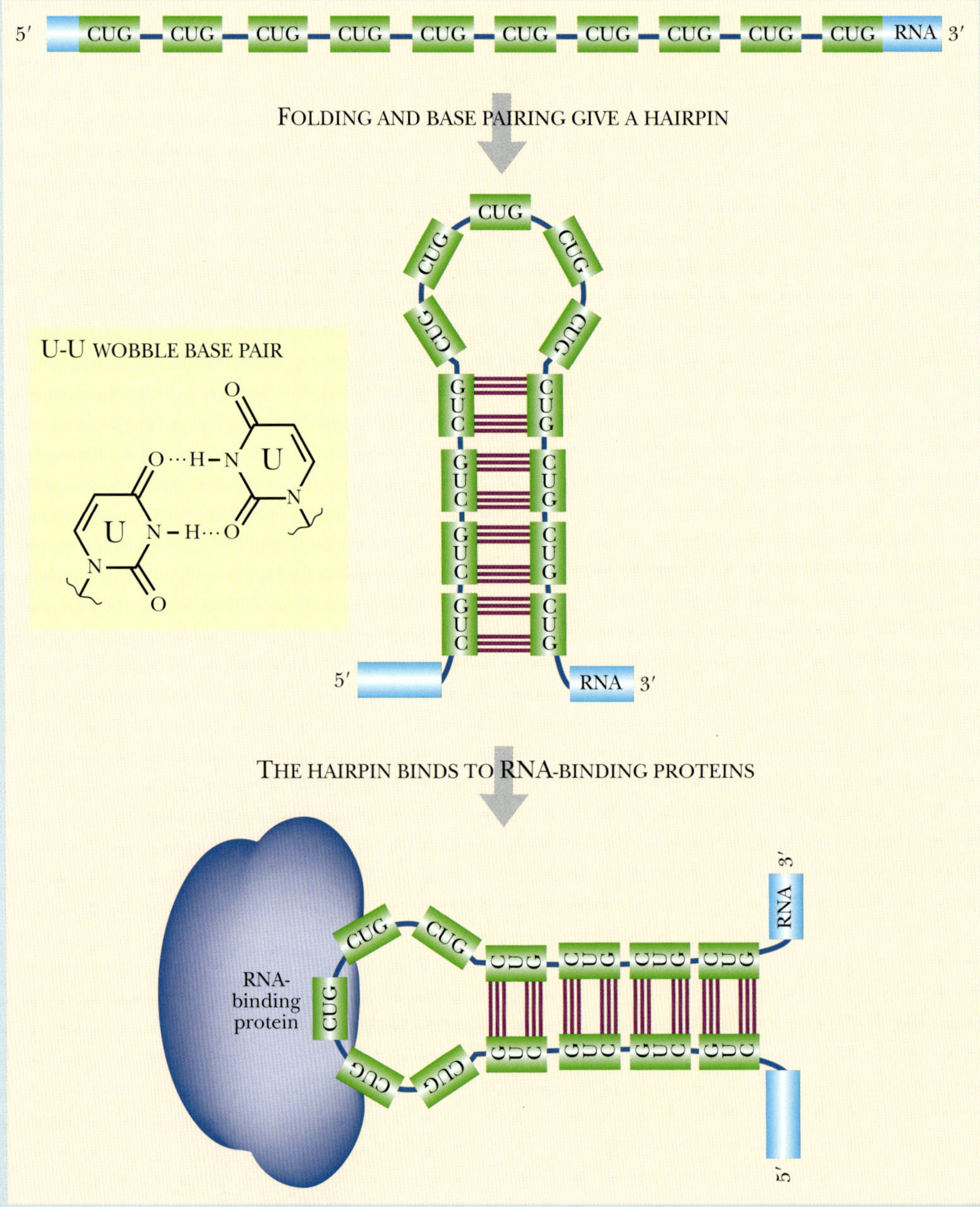

FIGURE A RNA Gain of Function
Tandem CUG repeats form stable hairpins in RNA that may bind tightly to a variety of RNA-binding proteins. The structure includes noncanonical U-U base pairs (*inset*).

DEFECTS IN IMPRINTING AND METHYLATION

Genetic imprinting is due to the effects of methylation on gene expression. Typically, one copy of a gene is methylated to prevent its expression. Imprinting occurs when methylation patterns present in the gametes survive to affect gene expression in the new organism. For a few genes, methylation patterns differ in the male and female gametes. Consequently, whether a particular copy of a gene is expressed may depend on whether it was inherited from the father or the mother. Such inherited changes that are not due to alterations in the DNA sequence are known as **epigenetic**.

Prader-Willi syndrome and **Angelman's syndrome** are due to defects in neighboring genes on chromosome 15 at 15q11-q13 that are subject to imprinting. Whether this region came from the sperm or the egg determines its pattern of methylation and therefore its pattern of gene expression (Fig. 16.6). About 75% of cases of Prader-Willi and Angelman's syndromes are due to deletions. Less often they result from point mutations, imprinting errors, or inheriting both copies of chromosome 15 from one parent. Prader-Willi syndrome develops when the paternal copy of 15q11-q13 is deleted or mutated and the maternal copy of this region is methylated. Conversely, Angelman's syndrome results from loss of function of genes on the maternally derived chromosome and methylation of the paternal chromosome. The imprinting patterns turn off the functional copy of these genes, thus causing the defects. Both syndromes cause mental retardation, but differ in other associated symptoms. For example, Prader-Willi syndrome also causes gross obesity, and Angelman's syndrome causes growth retardation and hyperactivity.

> A few genes have methylation patterns that differ depending on whether they came from the male or female parent. Complex syndromes may result when one copy of such a gene is defective and the other is silenced by methylations.

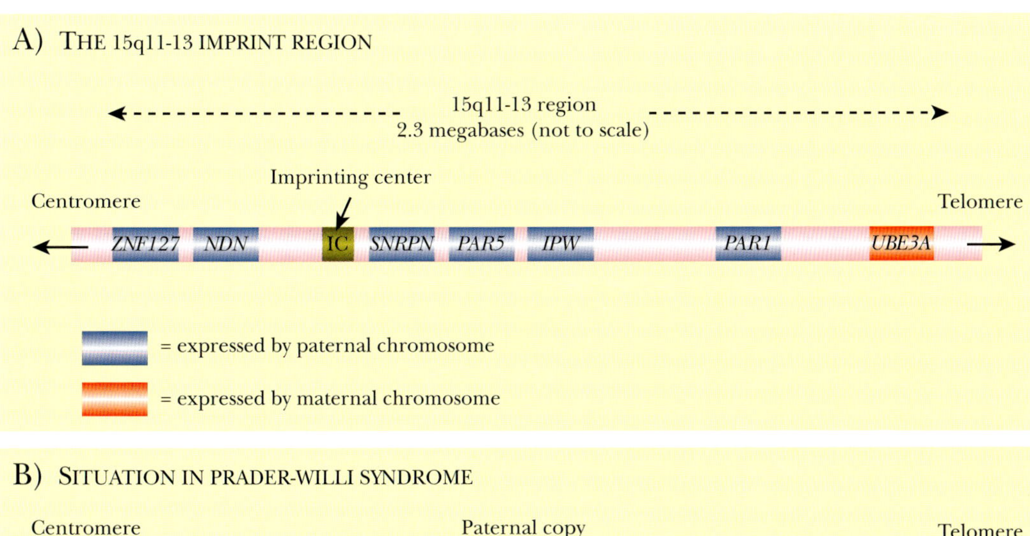

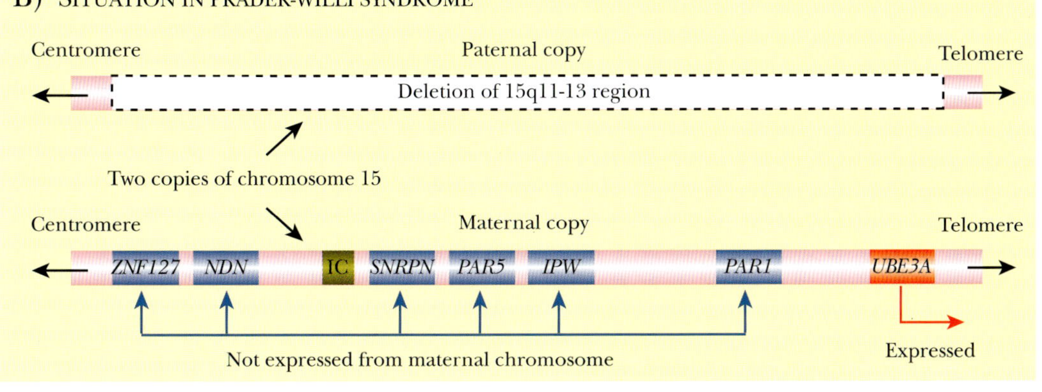

FIGURE 16.6
Prader-Willi and Angelman's Syndromes
(A) Chromosome 15 has a region between 15q11 and 15q13 that is imprinted. The paternal chromosome will express the genes shown in blue, and turn off the genes in red due to imprinting. The maternal chromosome expresses the genes shown in red and methylates the genes in blue. In a normal diploid individual, a single copy of each gene is expressed either from the paternal or maternal chromosome. (B) In Prader-Willi syndrome, the paternal copy of chromosome 15 is deleted between region 15q11 and 15q13. Even though the other copy of chromosome 15 has the genes shown in blue, they are turned off because of imprinting, resulting in overall loss of function.

MITOCHONDRIAL DEFECTS

The mitochondrial genome is a tiny fraction of an animal cell's genetic information. In terms of coding DNA there are about 100 mb in the nuclear genome and only 15.4 kb in human mitochondrial DNA (mtDNA). Consequently, the vast majority of mutations affecting coding DNA occur in nuclear DNA. Furthermore, there are thousands of copies of mitochondrial DNA in most cells, so a mutation in one gene on one copy has little effect on the whole organism. Because each original mutation must occur on a single molecule of mtDNA, the chances of its spreading in the mitochondrial population and becoming fixed seem highly unlikely. Despite these factors, the proportion of hereditary defects affecting mtDNA is surprisingly large.

One contributing factor is that mitochondrial DNA has a much higher mutation rate than nuclear DNA. Oxidation damage to DNA, due to reactive oxygen species generated by the respiratory chain, is much higher in mitochondria. In addition, there are fewer repair systems in mitochondria and the mtDNA is not protected by histones. Moreover, mitochondria go through many more divisions than the cells that contain them. (Although animal mtDNA is evolving very fast, the situation in plants is quite different. Plant mtDNA is much larger [150 kbp to 2.5 Mbp], contains introns, and evolves relatively slowly.)

Sperm cells contribute nuclear DNA during fertilization but do not donate their mitochondria to the zygote. Consequently, all the mitochondria in a particular individual are derived from ancestors in the female egg cell or **oocyte**, which has approximately 100,000 copies of mtDNA. Despite this, the rate at which mutations in mtDNA become established in the cell is some 10-fold greater than for nuclear DNA. Apparently, there is a bottleneck during the replication of mtDNA. When primordial germ cells develop into oocytes, only a few of their mitochondria divide and give rise to progeny (Fig. 16.7). This allows mutations to spread through the mitochondrial population in a specific cell line.

Disorders due to mitochondrial mutations are maternally inherited. In addition, the population of mitochondria within an individual may all carry the mutation, called **homoplasmy**, or there may be a mixture of mutant and normal mitochondria, called **heteroplasmy**. The severity of the defect will depend to some extent on the proportion of defective mitochondria, and this may vary among individuals with the same mutation.

Mitochondria are highly specialized in generating energy via respiration, and the majority of mitochondrial genes affect the respiratory chain (Fig. 16.8). Consequently, almost all genetic defects in mtDNA affect respiration and result in lowered energy production. Because different tissues of the body differ greatly in energy requirements, mitochondrial defects affect different tissues to very different extents. Brain and muscle are especially susceptible because of their high energy consumption.

Neuropathy, ataxia, and retinitis pigmentosa (NARP) syndrome is due to base changes at position 8993 of the mtDNA molecule. These affect the *ATP6* gene, encoding a component of the **ATP synthetase**. This enzyme normally couples energy from the respiratory chain to the synthesis of ATP. The result is a lack of ATP to energize the cell. Nerve and muscle cells are most affected, and muscular weakness, dementia, seizures, and sensory malfunction are among the results. Symptoms occur when 70% to 90% of the mitochondria are affected. If more than 90% are affected, death in infancy is the usual outcome.

Myoclonus epilepsy and ragged-red fibers (MERRF) syndrome is due to a defect in the mitochondrial lysine tRNA gene. Most cases are due to the point mutation A8344G (A to G at position 8344 of the mtDNA molecule). Defective tRNA results in decreased synthesis of several mitochondrial proteins and hence lowered respiration. Transient seizures, loss of muscle coordination, and loss of muscle cells result. The term *ragged-red fibers* refers to clumps of defective mitochondria that aggregate in muscle cells and were originally visualized by staining with a red dye. Symptoms are seen if 90% or more of the mitochondria in nerve and muscle cells are affected.

Mutations in the mitochondrial genome also cause hereditary defects, mostly affecting energy generation. Unlike nuclear genes, mitochondrial mutations are inherited only via the maternal line.

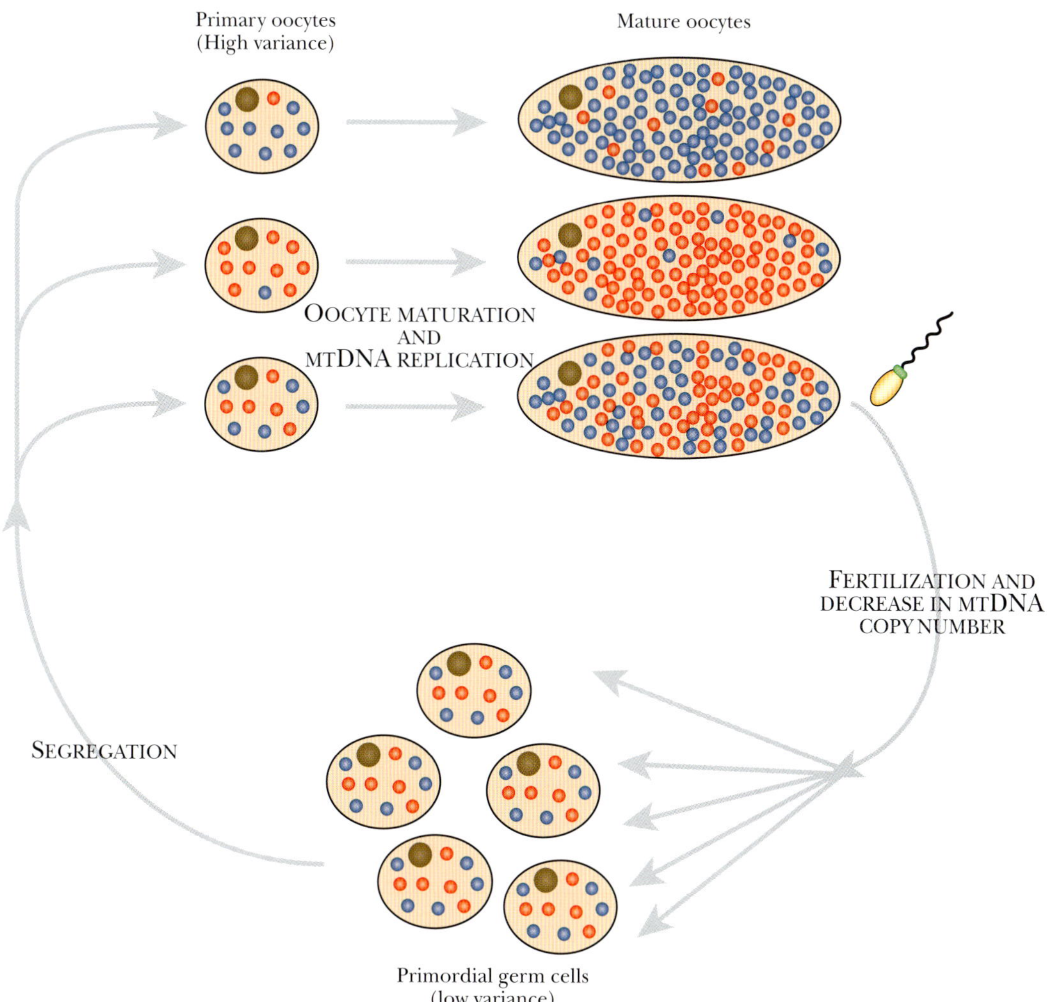

FIGURE 16.7 Mitochondrial DNA Bottleneck

In primordial germ cells (*bottom*), about 50% of the mitochondria have a mutation (red). As they develop into primary oocytes, only a few mitochondria are passed on. Therefore, some oocytes will have just a few affected mitochondria (*top oocyte*), some will have a majority of defective mitochondria (*middle*), and others will have an equal proportion (*bottom*). As the primary oocytes develop into mature eggs, the number of mitochondria increases, but the ratio of mutated to normal remains constant in each cell line.

IDENTIFICATION, LOCATION, AND CLONING OF DEFECTIVE GENES

The nature of mutations and many of the techniques used in analyzing them have already been discussed in previous chapters. Furthermore, the human genome has now been fully sequenced and in principle the DNA sequence is available for all of the genes responsible for observed hereditary defects. In practice, connecting a particular set of symptoms with a specific gene is not always so straightforward. Here we will outline some general approaches to identifying the genes responsible for hereditary defects and finding their location on the human chromosomes. Confirmation of identity normally involves cloning and further genetic analysis. After this general discussion we will consider two examples in more detail—cystic fibrosis and Duchenne's muscular dystrophy.

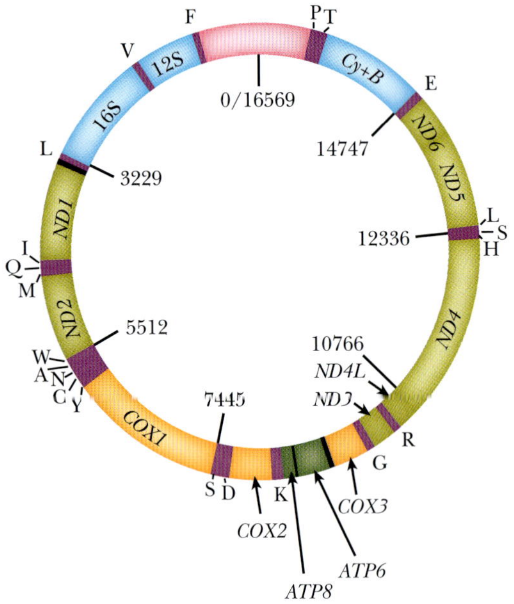

One way to identify genes responsible for hereditary defects is to analyze the symptoms and then make an informed guess as to what kinds of proteins are likely to be involved. Possible candidate genes are then chosen from the list of characterized genes and investigated further. This approach is therefore sometimes called **candidate cloning**. Because relatively few human genes have been characterized, this method is rarely successful. However, the recent vast increase in genomics and proteomics research is revealing the functions of many mammalian genes, so this method may become more valuable in the near future.

An improved variant of this approach comes from using model organisms, in particular mice. The vast majority of human genes have homologs in mice. Moreover, unlike humans, mice may be directly used for genetic experimentation. As described in Chapter 15, Transgenic Animals, it is possible to make mice in which both copies of any chosen gene have been artificially inactivated. Such **knockout mice** are then examined for symptoms. Several major programs are now in progress to systematically make mice with knockout mutations affecting every one of the 25,000 or so mammalian genes. Eventually, this information should allow many human genes to be matched with possible symptoms.

FIGURE 16.8 Map of Human Mitochondrial DNA
Mitochondrial DNA has genes that encode tRNA, rRNA, and proteins used in energy production. The tRNA genes are indicated by single-letter amino acid symbols. The two rRNA genes are 16S and 12S. The remaining genes are all for proteins in complexes I, III, IV, and V of the respiratory chain.

Functional cloning begins with a known protein that is suspected of involvement in a hereditary disorder. The amino acid sequence of the protein is determined. Nowadays this would most likely be done by mass spectrometry of peptide fragments generated from the protein by protease digestion. The protein sequence is then used to deduce the coding sequence of the gene and an oligonucleotide probe is synthesized. The probe may be used to screen a cDNA library by hybridization (see Chapter 3). Alternatively, the probe can be linked to a solid support and used to pull out a specific mRNA molecule from a pool of cellular mRNA. In the latter case, a single specific cDNA is made from the purified mRNA. The complete cDNA is then sequenced to confirm that the DNA matches the original protein.

The gene must then be localized to a specific region of a particular chromosome. This may be done by screening a set of radiation hybrid cells or by hybridization using a DNA probe with a fluorescent label (i.e., by FISH—see Chapter 3). Cloned DNA from the target region, carried on a vector capable of carrying large inserts, such as a cosmid or YAC, is then screened to narrow down the location.

Positional cloning is used when the nature of the gene product is unknown. In this case the disease gene must be mapped at least approximately by a genetic approach before further DNA-based screening can proceed. The easiest cases are those in which there is a major chromosomal abnormality, such as a deletion, inversion, or translocation that may be visualized under the light microscope. This may localize the defect to a specific band on a particular chromosome. Alternatively, linkage studies on individuals from families afflicted by the inherited defect may locate the damaged gene close to other genetic markers. These other markers may be known genes, but more often they will be RFLPs, VNTRs, or other sequence polymorphisms.

Such genetic mapping can localize a gene to around 1000 kb. This length of DNA may contain anywhere from 10 to 50 genes, depending on how crowded that region of the genome is. DNA from the suspect region is then cloned, as described earlier for functional cloning. However, in the case of positional cloning we have no previously identified protein that can be used to check for the corresponding gene. Therefore, the hereditary defect must be identified at the DNA level. The suspect DNA may be scanned for the presence of functional genes by a variety of approaches:

(a) The presence of open reading frames indicates a possible coding sequence. Note that in higher organisms, the coding sequence will typically be fragmented into several exons separated by noncoding introns. These introns may be very long and frequently account for more of the overall length of the gene than the exons.

(b) CpG (or CG) islands are often found upstream of the transcribed regions in vertebrate DNA. These are GC-rich regions that are often methylated for regulatory purposes. They may be identified by the presence of multiple cut sites for restriction enzymes whose recognition sequences consist solely of C and G (e.g., *Hpa*II cuts at C/CGG).

(c) Coding DNA tends to evolve more slowly than noncoding DNA. Consequently, coding DNA from one animal will often hybridize to DNA from a range of related organisms while noncoding DNA does not. Zoo blots are often used to identify coding DNA.

(d) Messenger RNA extracted from those tissues most severely affected by a genetic disease should contain significant levels of mRNA derived from the gene responsible for the defect. Hybridization can be used to see if candidate DNA sequences match those in the mRNA pool. (This assumes that the gene in question is expressed at a reasonably high level. This will usually be true for genes encoding structural proteins and enzymes but not for those encoding regulatory proteins. Note also that the mRNA should be isolated from a healthy person because the defective gene might not be transcribed in patients suffering from the defect.)

(e) Ultimately, sequencing of DNA from healthy and affected individuals should show a difference—if the suspected gene is truly responsible for the hereditary defect.

A variety of approaches are used to clone genes responsible for human hereditary defects. Some depend on the known location of the gene; others depend on gene function. Information from model organisms with equivalent defects has often been very useful.

CYSTIC FIBROSIS

About one in 2000 white children suffer from **cystic fibrosis**. This condition is due to homozygous recessive mutations. In other words two defective copies of the gene, one from each parent, must be inherited for the child to suffer from the disease. Humans with a single defective allele are carriers but do not show symptoms, as a single wild-type version of the gene is sufficient for normal health.

The protein encoded by the cystic fibrosis gene is referred to as the **CFTR protein** (for cystic fibrosis transporter) and is found in the cell membrane where it acts as a channel for chloride ions. In healthy people, this channel can be opened or shut as needed by the cell. The CFTR protein consists of a series of membrane-spanning segments with a central control module (Fig. 16.9). If the control module has a phosphate group attached, the channel is open, and when the phosphate group is removed, it shuts.

In cystic fibrosis patients, control of the chloride channel is defective. This, in turn, affects a variety of other processes. The most harmful is that the mucus that lines and protects the lungs is abnormally thick. Lack of chloride ions leads to lack of sufficient water accompanying the mucus. This not only causes obstructions but also allows the growth of harmful microorganisms. Cells lining the airways of the lungs are killed and replaced with fibrous scar tissue—hence the name of the disease. Eventually the patient succumbs to respiratory failure.

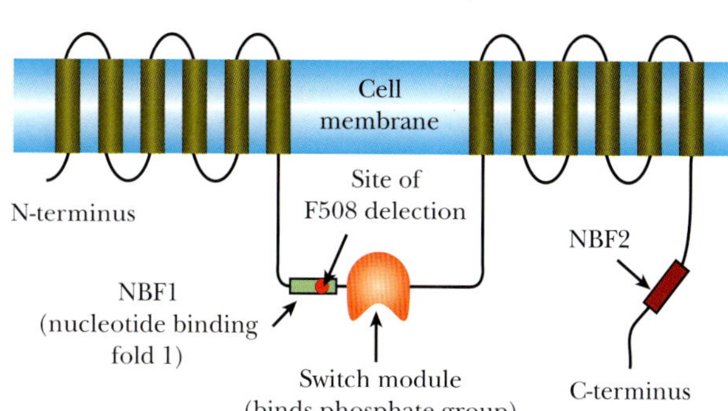

FIGURE 16.9 Cystic Fibrosis Protein Transports Chloride The CFTR protein has 12 hydrophobic segments that span the cell membrane. Small loops connect the 12 transmembrane segments, except for segments 6 and 7, which are linked by a large protein loop. In the three-dimensional configuration the 12 transmembrane segments organize to form a channel, and the large protein loop regulates whether the channel is open or shut.

The general location of the cystic fibrosis gene (to within 2 million base pairs) was found by screening large numbers of genetic markers (RFLPs) in members of families with a cystic fibrosis patient. The gene was found on the longer (q) arm of chromosome 7 between bands q21 and q31. After subcloning this region of the chromosome in large segments, the cystic fibrosis gene was found to occupy 250,000 base pairs and have 24 exons, encoding a protein of 1480 amino acids (Fig. 16.10). Because 1480 amino acids need only 4440 base pairs to encode them, this means that scarcely 2% of the cystic fibrosis gene is actually coding DNA. The rest consists of introns.

About 70% of the cystic fibrosis victims in North America share the same genetic defect. They all have a small deletion of three bases that code for the amino acid phenylalanine at position 508 of the normal protein. In Denmark, 90% of cystic fibrosis cases are due to this △F508 deletion (F = phenylalanine in single-letter code), whereas in the Middle East it accounts for only 30%. The other cystic fibrosis cases are the result of more than 500 different mutations, which makes genetic screening difficult.

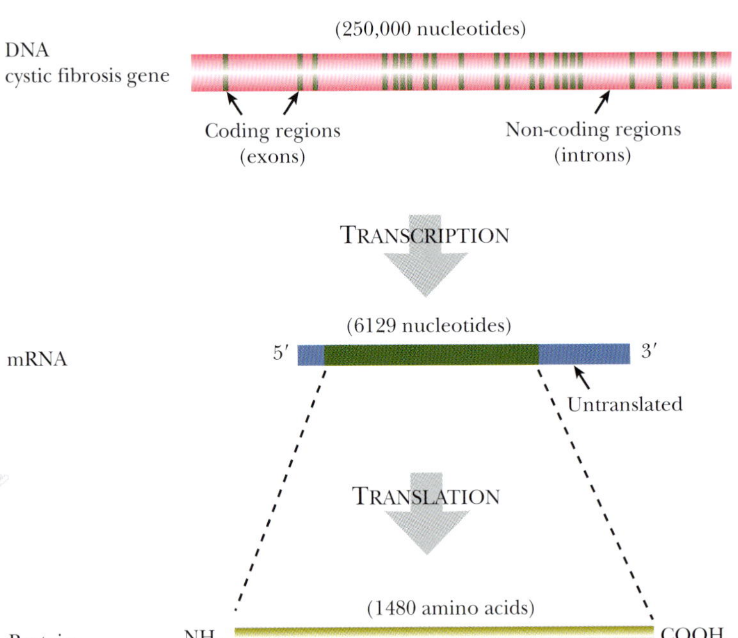

FIGURE 16.10
Cystic Fibrosis Gene
The cystic fibrosis gene covers an extremely large region of DNA (250,000 nucleotides) and has 24 exons. After transcription and splicing, the mRNA is only 6129 nucleotides; thus the majority of the gene consists of introns.

The distribution and relatively high frequency of the △F508 deletion suggest that it originated in Western Europe, perhaps around 2000 years ago, and has been positively selected. Although two defective copies of the *CFTR* gene are highly detrimental, heterozygotes with one △F508 allele and one normal allele show only moderate reduction of ion flow and water movement. This does not cause cystic fibrosis but is sufficient to significantly reduce the loss of fluids during diarrhea. Such heterozygotes are more resistant to the effects of enteric infections such as typhoid and cholera that cause dehydration via extensive diarrhea.

The primary defect in cystic fibrosis is in a chloride ion transport protein. Secondary effects include buildup of mucus in the lungs and bacterial infection. Heterozygotes, with only one defective CFTR gene, show resistance to infectious diseases that cause dehydration.

MUSCULAR DYSTROPHY

There are several forms of **muscular dystrophy**. These diseases result in the wasting away of muscle tissue and cause premature death, usually in the late teens or early 20s. There is no known cure. The most common form, **Duchenne's muscular dystrophy**, is due to defects in the protein **dystrophin**, which plays a role in attaching the internal muscle fibrils to the membranes of muscle cells (Fig. 16.11).

The ***DMD* gene**, responsible for this disease, is sex-linked and is located in the Xp21 band, close to the middle of the short (p) arm of the X chromosome (Fig. 16.12). Because the Y chromosome is shorter than the X chromosome, many genes that are present on the X chromosome do not have a corresponding partner on the Y chromosome. Therefore, females have two copies of the *DMD* gene, whereas males only have a single copy.

If the single copy of a sex-linked gene present in a male is defective, there is no backup copy and severe symptoms may result. In contrast, females with just one defective copy will usually have no symptoms, but they will be carriers and half of their male children will get the disease. The result is a pattern of inheritance in which only the male members of a family suffer from

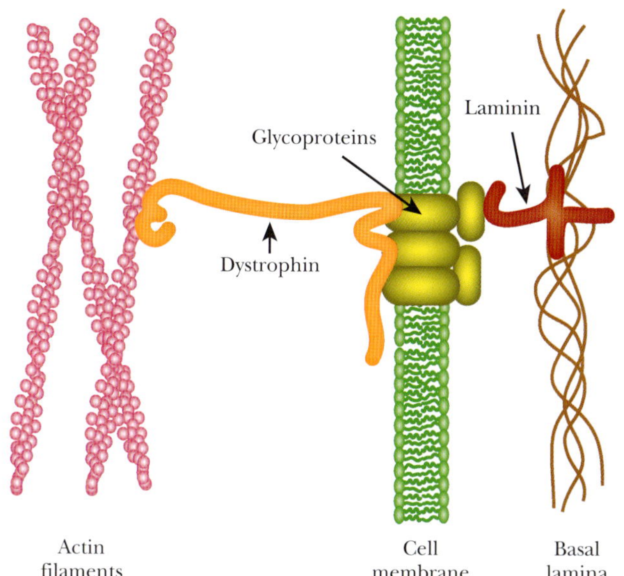

FIGURE 16.11 Dystrophin Attaches Actin to the Cell Membrane

The protein dystrophin anchors the long actin filaments to the membrane of muscle cells via glycoprotein interactions. The same glycoproteins anchor the cell membrane to the basal lamina in the extracellular regions.

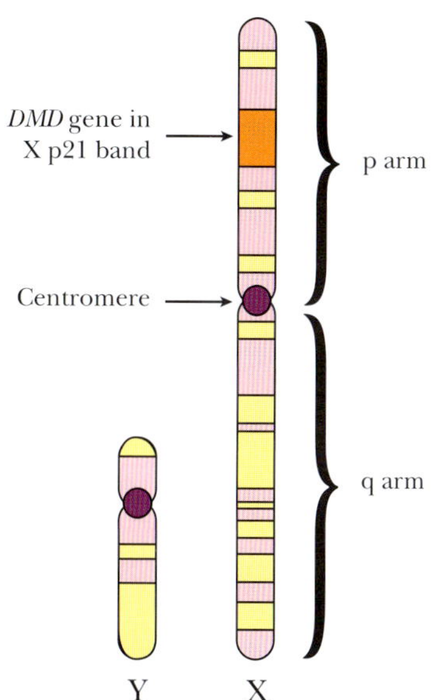

FIGURE 16.12 The *DMD* Gene Is Sex-Linked

DMD is located on the short arm of the X chromosome. Because males have only one X chromosome, Duchenne muscular dystrophy affects males only.

the disease. Figure 16.13 shows a family with several occurrences of an X-linked recessive disease. About two-thirds of Duchenne's muscular dystrophy patients inherit the disease from their mothers and the other third get it as the result of new mutations, which arise at a frequency of approximately 1 in 10,000 gametes.

The *DMD* gene is even odder than the cystic fibrosis gene. It takes up about 1.5% of the length of the X chromosome and is longer than the entire genomes of some bacteria. The *DMD* gene has roughly 75 exons and more than 2 million base pairs of DNA, of which less than 1% are used to encode the protein. Despite this the encoded protein, dystrophin, is gigantic. It has 4000 amino acids, so it is roughly 10 times as large as an average protein. Such huge coding sequences are bigger targets for detrimental mutations, which explains why the mutation rate for the *DMD* gene is more than 10 times greater than for typical human genes. In most victims the defect is due to alterations in just one or a few bases of the *DMD* gene. However, about 10% of the victims have a deletion of DNA, which includes all or part of the *DMD* gene.

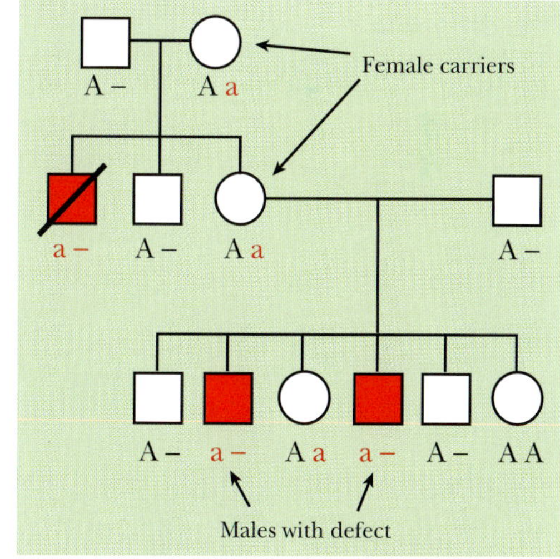

FIGURE 16.13 X-Linked Recessive Disease in a Family

The gene "A" is located on the X chromosome. Males have only one X chromosome and their Y chromosome has no corresponding copy of the gene (symbolized by "–"). So any male who gets one copy of the defective allele ("a") will get the disease.

One Duchenne's muscular dystrophy patient had a deletion large enough to locate to the Xp21 region of the X chromosome using a light microscope. Using the subtractive hybridization method, DNA from this patient was used to clone the DNA that was missing in the deletion. In this approach, the *DMD* gene to be cloned is present in the DNA from a healthy person. The sample of DNA with the *DMD* deletion is used to remove other, unwanted, genes from the healthy DNA, by hybridization. This leaves behind the healthy copy of the *DMD* gene, which is then cloned onto a suitable vector for further analysis. (See Chapter 3 for details of cloning by subtractive hybridization.)

Several types of heritable muscular dystrophies are known and have been reasonably well characterized at the genetic level.

GENETIC SCREENING AND COUNSELING

As explained earlier, close relatives are more likely to produce malformed children because of harmful recessive alleles coming together. It is also true that two people from otherwise unrelated families that both have a history of the same hereditary disease might be wise to avoid having children. Today we can do more than give general advice. Testing of prospective parents for high-risk genetic diseases can now be done for an increasing number of genes. The typical approach is to use a hybridization probe (see Chapter 3) or PCR analysis (see Chapter 4). This will reveal whether a mutant copy of the gene is present in a DNA sample from the person tested. If marital partners both test positive for a recessive defect, they will have to decide whether or not to take the risk of having children. For a recessive defect, where both parents are carriers, one in four children will get the disease.

The basic problem with genetic screening is that our ability to detect genetic defects has far outrun our ability to cure them. Thus, screening may reveal a defect about which nothing can be done, and this may cause severe psychological distress to the individuals concerned. For example, Huntington's disease is an autosomal dominant condition that results in movement disorder, dementia, and ultimately death. The symptoms usually appear only in the 30s, although they may be delayed as late as the 60s in some patients. Genetic screening helps those who are afflicted decide whether or not to have children. It also provides them with the knowledge that they will eventually develop an incurable and fatal disease. Some cancers and coronary heart disease have a genetic component (often multigenic). Screening may allow affected individuals to avoid factors that tend to trigger the disease. On the other hand, especially if the treatment is complicated and/or only partially effective, such knowledge may be a burden.

Genetic screening sometimes allows successful intervention by modifying diet or lifestyle. The classic case is **phenylketonuria**. The absence of the enzyme **phenylalanine hydroxylase** causes a buildup of phenylalanine and a deficiency of the product, tyrosine (Fig. 16.14). Excess phenylalanine is neurotoxic; if untreated, this condition results in severe damage to the central nervous system, and patients require permanent care for life. However, a diet low in phenylalanine avoids the buildup of this amino acid and largely alleviates the problems. Phenylketonuria occurs in 1 in 10,000 live births and is routinely screened for in developed nations using blood samples taken a week or so after birth. The diet may be relaxed somewhat later in life after development of the nervous system is largely complete.

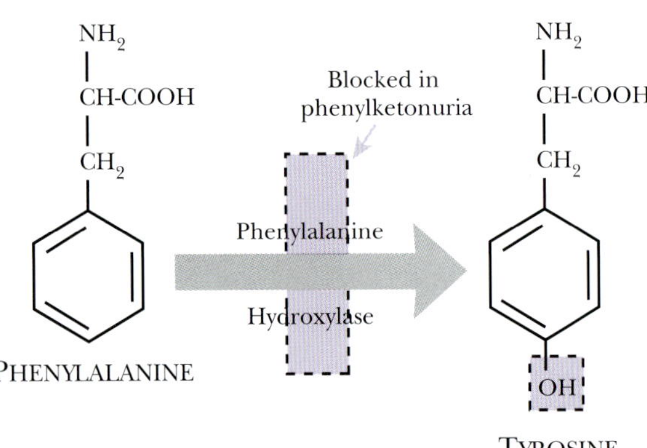

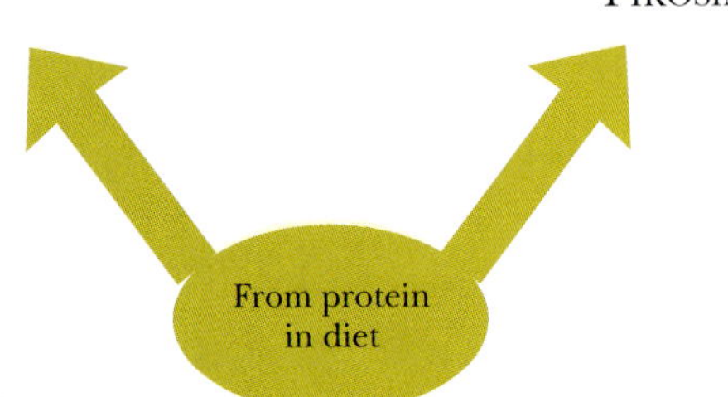

FIGURE 16.14
Phenylketonuria
Phenylalanine hydroxylase catalyzes the conversion of phenylalanine to tyrosine. If the enzyme is absent, phenylalanine accumulates and is toxic to the nervous system.

Given that a reliable test is available, it is generally agreed that neonatal screening is justified if the disease is reasonably frequent (say 1 in 10,000 or more), the disease is severely damaging, and that treatment is available that significantly improves the condition.

It is also possible to examine embryos during early pregnancy by drawing samples of amniotic fluid which contain some cells from the fetus, a procedure called **amniocentesis**, or by examining extraembryonic fetal tissue. These cells may be genetically screened to see, for example, if the fetus is homozygous for a recessive defect. Embryos doomed to grow up with hereditary defects could then be aborted early if the parents wish to do so. It is also now possible to screen embryos obtained by *in vitro* fertilization before they are implanted. This avoids aborting of an unwanted fetus should a serious genetic defect be discovered.

Genetic screening is possible for a wide and growing number of hereditary defects. Unfortunately, most genetic defects cannot yet be cured.

Summary

A variety of genetic defects affect the human population. Some are due to mutations in single genes and are reasonably well understood. Others involve multiple genes and are very complex. Many mutations have minor effects, whereas others are life-threatening. At present, our ability to analyze hereditary defects has far outrun our ability to treat them.

End-of-Chapter Questions

1. Why might a defective gene not have an affect on an organism's genotype?
   - **a.** because the organism is haploid and the other copy compensates for the defective copy
   - **b.** because the mutation is often lethal
   - **c.** because mutations can never be passed to progeny
   - **d.** because the organism is diploid and the other copy compensates for the defective copy
   - **e.** none of the above
2. Which of the following is the first molecular disease to be identified?
   - **a.** sickle cell anemia
   - **b.** phenylketonuria
   - **c.** hemophilia
   - **d.** cystic fibrosis
   - **e.** Duchenne's muscular dystrophy
3. What causes Down syndrome?
   - **a.** an extra Y chromosome
   - **b.** an extra X chromosome
   - **c.** an extra copy of chromosome 21
   - **d.** an extra copy of each chromosome
   - **e.** none of the above
4. What are dominant negative mutations?
   - **a.** a mutation that activates a gene when it would not normally be active
   - **b.** a mutant protein that loses its own function and also interferes with the function of another, non-mutated, protein
   - **c.** a mutation that causes a protein to lose function
   - **d.** a mutation that produces no detrimental effects on the protein
   - **e.** none of the above
5. What are dynamic mutations?
   - **a.** tandem repeats of bases within the coding regions of DNA that increase in number each generation
   - **b.** mutations that affect the coding sequence of the DNA
   - **c.** mutations that cause defects in methylation of DNA
   - **d.** mutations that affect mitochondrial DNA
   - **e.** none of the above

(Continued)

6. When does imprinting occur?
   a. when DNA is methylated at every base
   b. when the enzymes in the DNA methylation pathways are mutated
   c. when the mitochondrial DNA is not methylated
   d. methylation patterns in gametes survive and affect gene expression in the new organism
   e none of the above
7. Why are mitochondrial defects surprisingly prevalent?
   a. Damage to DNA due to reactive oxygen species is much higher.
   b. Mitochondrial DNA is not protected by histones.
   c. Mitochondria contain few repair systems.
   d. Mitochondrial DNA has a much higher mutation rate than nuclear DNA.
   e. all of the above
8. How are mitochondrial disorders inherited?
   a. maternally
   b. paternally
   c. during fetal development
   d. all of the above
   e. none of the above
9. In what way can the presence of a genetic defect on a segment of DNA be screened after positional cloning?
   a. sequencing of the DNA
   b. hybridization of mRNA extracted from the tissues most affected by the disease to suspect DNA
   c. by the presence of open reading frames
   d. identification of CpG islands
   e. all of the above
10. Which ion channel is defective in people with cystic fibrosis?
   a. sodium transporter
   b. potassium transporter
   c. chloride transporter
   d. calcium transporter
   e. zinc transporter
11. What advantage does a △F508 heterozygote have over a patient with cystic fibrosis (i.e., homozygous for △F508)?
   a. △F508 heterozygotes are more susceptible to malaria.
   b. They are more resistant to the effects of enteric diseases, such as typhoid and cholera.
   c. They have no advantage since only one defective copy is needed to cause cystic fibrosis.
   d. The △F508 mutation does not cause disease, so there is no advantage.
   e. none of the above

12. From a genetic standpoint, why does Duchenne's muscular dystrophy primarily occur in males?
  a. because the defective gene is located on the X chromosome, which has no backup allele in men, who are XY, since the Y chromosome is shorter than the X
  b. affects both males and females similarly
  c. because it is on the X chromosome
  d. because the defective gene is located on the Y chromosome, which is passed down to males only
  e. none of the above

13. Which protein must be defective for Duchenne's muscular dystrophy to occur?
  a. insulin
  b. CFTR
  c. dystrophin
  d. synthetase
  e. ion channel proteins

14. Why is mating between close relatives not a good idea, genetically?
  a. Recessive genes within the family have a chance to come together and cause disease.
  b. Dominant alleles within the family have a chance to come together and cause disease.
  c. The mitochondrial DNA is not genetically diverse.
  d. DNA methylation patterns have more tendency to create imprinting.
  e. none of the above

15. What is a specific case in which genetic screening has saved and improved the quality of life for an individual through diet modification?
  a. Duchenne's muscular dystrophy
  b. cystic fibrosis
  c. Prader-Willi syndrome
  d. phenylketonuria
  e. Down syndrome

Further Reading

Antonarakis SE, Epstein CJ (2006). The challenge of Down syndrome. *Trends Mol Med* **12**, 473–479.

Butland SL, Devon RS, Huang Y, Mead CL, Meynert AM, Neal SJ, Lee SS, Wilkinson A, Yang GS, Yuen MM, Hayden MR, Holt RA, Leavitt BR, Ouellette BF (2007). CAG-encoded polyglutamine length polymorphism in the human genome. *BMC Genomics* **8**, 126.

Edwards CA, Ferguson-Smith AC (2007). Mechanisms regulating imprinted genes in clusters. *Curr Opin Cell Biol* **19**, 281–289.

Feinberg AP (2007). Phenotypic plasticity and the epigenetics of human disease. *Nature* **447**, 433–440.

Gadsby DC, Vergani P, Csanády L (2006). The ABC protein turned chloride channel whose failure causes cystic fibrosis. *Nature* **440**, 477–483.

Green NS, Dolan SM, Murray TH (2006). Newborn screening: Complexities in universal genetic testing. *Am J Public Health* **96**, 1955–1959.

Hore TA, Rapkins RW, Graves JA (2007). Construction and evolution of imprinted loci in mammals. *Trends Genet* **23**, 440–448.

Jirtle RL, Skinner MK (2007). Environmental epigenomics and disease susceptibility. *Nat Rev Genet* **8**, 253–262.

Krokan HE, Kavli B, Slupphaug G (2004). Novel aspects of macromolecular repair and relationship to human disease. *J Mol Med* **82**, 280–297.

Li SH, Li XJ (2004). Huntingtin-protein interactions and the pathogenesis of Huntington's disease. *Trends Genet* **20**, 146–154.

Nithianantharajah J, Hannan AJ (2007). Dynamic mutations as digital genetic modulators of brain development, function and dysfunction. *Bioessays* **29**, 525–535.

Ollero M, Brouillard F, Edelman A (2006). Cystic fibrosis enters the proteomics scene: New answers to old questions. *Proteomics* **6**, 4084–4099.

Orr HT, Zoghbi HY (2007). Trinucleotide repeat disorders. *Annu Rev Neurosci* **30**, 575–621.

Paul S (2007). Polyglutamine-mediated neurodegeneration: Use of chaperones as prevention strategy. *Biochemistry (Moscow)* **72**, 359–366.

Ranum LP, Cooper TA (2006). RNA-mediated neuromuscular disorders. *Annu Rev Neurosci* **29**, 259–277.

Ranum LP, Day JW (2004). Myotonic dystrophy: RNA pathogenesis comes into focus. *Am J Hum Genet* **74**, 793–804.

Reik W (2007). Stability and flexibility of epigenetic gene regulation in mammalian development. *Nature* **447**, 425–432.

Sangiuolo F, D'Apice MR, Gambardella S, Di Daniele N, Novelli G (2004). Toward the pharmacogenomics of cystic fibrosis—an update. *Pharmacogenomics* **5**, 861–878.

Taylor RW (2005). Gene therapy for the treatment of mitochondrial DNA disorders. *Expert Opin Biol Ther* **5**, 183–194.

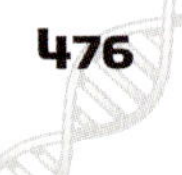

Gene Therapy

Gene Therapy or Genetic Engineering?
General Principles of Gene Therapy
Gene Patching by Oligonucleotide Crossover
Aggressive Gene Therapy
Adenovirus Vectors in Gene Therapy
Cystic Fibrosis Gene Therapy by Adenovirus
Adeno-Associated Virus
Retrovirus Gene Therapy
Retrovirus Gene Therapy for SCID
Nonviral Delivery in Gene Therapy
Liposomes and Lipofection in Gene Therapy
Aggressive Gene Therapy for Cancer
Antisense RNA and Other Oligonucleotides
Aptamers—Blocking Proteins with RNA
Ribozymes in Gene Therapy

GENE THERAPY OR GENETIC ENGINEERING?

Genetic engineering means that we alter an organism permanently so that the changes will be stably inherited. For multicellular organisms this implies deliberate alteration of the DNA in the germline cells. In contrast, **gene therapy** (occasionally called **genetic surgery**) is less permanent. The patient is cured, more or less, by altering the genes in only part of the body. For example, cystic fibrosis patients might be partially cured by introducing the wild-type gene into the lungs. However, these changes are not inherited, and the alleles in the germline cells remain defective.

True human genetic engineering is still in the future. At present, genetic engineering is restricted to nonhumans and has resulted in the creation of transgenic plants and animals as described in Chapters 14 and 15. **Eugenics** refers to deliberate improvement of the human race by selective breeding. Early eugenic proposals were based on choosing superior parents by visual inspection or medical screening and breeding them in much the same way as for prize pigs and pedigreed dogs. Today we have reached the position where direct alterations of the human genome at the DNA level are technically feasible, although still clumsy.

> Genetic engineering may create organisms with changes that are stably inherited. Gene therapy uses genetics to cure a disease but does not alter the germline cells. Consequently the changes are not heritable.

GENERAL PRINCIPLES OF GENE THERAPY

The most straightforward use of gene therapy is to deal with a hereditary defect due to a single gene and that occurs only when both copies of the gene are defective—that is, a recessive condition. Introducing a single good copy of the gene can then cure the defect. This is sometimes known as **replacement gene therapy**. Furthermore, it would obviously simplify treatment if the disease mostly affects just one or a few organs. The main steps involved in replacement gene therapy are as follows:

(a) Identification and characterization of gene
(b) Cloning of gene
(c) Choice of vector
(d) Method of delivery
(e) Expression of gene

The first step is to identify the genetic defect and to clone a good copy of the gene involved (see Chapter 10). The gene must then be delivered to the patient. This involves choosing a vector together with a suitable method of delivery. In addition, the vector/gene construct must be designed to allow proper expression of the gene, once inside the patient. Delivery may be performed in a variety of ways. The vector/gene construct may be injected into the bloodstream or other tissue. It may be aerosolized and sprayed into the nose and airways. In some cases, cells are removed from the patient, engineered while growing in culture, and then returned to the patient. This approach is known as ***ex vivo* gene therapy** because the actual genetic engineering takes place outside the patient. (The direct delivery of the vector/gene construct to the patient is sometimes called "*in vivo* gene therapy" to contrast with this.)

In the laboratory, most manipulations are done with genes carried on bacterial plasmids. Although gene therapy has occasionally been performed directly with plasmid DNA carrying a therapeutic gene, more often specialized delivery systems are used. In most cases a modified virus is used as the vector. Because viruses cause disease, they first need to be genetically

Table 17.1 Delivery Systems Used in Gene Therapy Trials

Vector Used	Trials Number	%
Retrovirus	273	25.4
Adenovirus	269	25
Other viruses	201	18.7
Naked DNA or RNA	187	17.4
Liposomes	93	8.6
Gene gun	5	0.5
Not classified	42	3.9
Total	1076	100

Data as of 2005. Current data on the numbers and types of gene therapy trials are available from the website of the *Journal of Gene Medicine*:http://www.wiley.co.uk/wileychi/genmed/clinical/

disarmed in order to be used in gene therapy. About 70% of human gene therapy trials have used viral vectors (Table 17.1). Two main groups of viruses have been used, retroviruses and adenoviruses. In addition, in a smaller proportion of cases DNA has been delivered inside liposomes or projected into tissues by the gene gun.

In gene replacement therapy, a functional copy of the gene responsible for the hereditary defect is inserted. The most popular approach is for the gene to be carried on an engineered virus vector.

GENE PATCHING BY OLIGONUCLEOTIDE CROSSOVER

During replacement gene therapy, we normally think of replacing the whole defective gene with a complete functional copy. However, some genetic defects consist of just a single base change, or perhaps a cluster of closely linked base alterations. In this case, the defective gene could be "patched" rather than replaced. This may be done by crossing over with a relatively short double-stranded oligonucleotide, which carries the wild-type sequence (Fig. 17.1).

The crossover frequency may be improved by using an RNA-DNA hybrid oligonucleotide. Such hybrids are known to take part in crossover formation more readily than double-stranded DNA. In addition, designing hairpin bends protects the ends of the oligonucleotide and prevents degradation by exonucleases. This technique is still in the experimental phase.

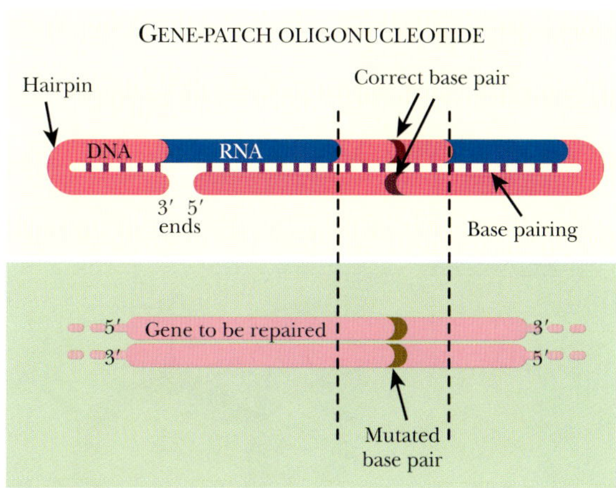

FIGURE 17.1 Patching Defective Gene by Oligonucleotide Crossover
A special gene-patch oligonucleotide can be synthesized to provide a corrected copy of a short specific region of a defective gene. The oligonucleotide is designed to promote a crossover in the defective region of the gene.

Small localized genetic defects could be patched by relatively short oligonucleotides rather than complete functional genes.

AGGRESSIVE GENE THERAPY

The original idea behind gene therapy was curing hereditary defects by replacing defective genes. However, there is no inherent reason why gene therapy must only be "defensive" and suppress defects. We can go on the offensive and provide genes whose products may cure a disease even though the genes we use were not responsible for the problem in the first place. The best examples of such **aggressive gene therapy** are not in curing hereditary defects but in the treatment of cancer. Here the objective is to kill cancer cells. In fact, as shown in Table 17.2, the majority of gene therapy trials are now directed against cancer.

> Aggressive gene therapy uses genes to kill or destroy unwanted cells and is especially useful against cancer.

Table 17.2 Target Conditions for Gene Therapy Trials

Disease Targeted	Trials: Number	Trials: %
Cancer	715	66.4
Single gene defects	95	8.8
Vascular diseases	92	8.6
Infectious diseases	72	6.7
Other diseases	33	3.1
Investigative	69	6.4
Total	1076	100

FIGURE 17.2
Adenovirus Structure
Adenoviruses have an icosahedral capsid with 20 faces and 12 vertices. Each face is composed of hexons (yellow balls). At each of the vertices are a penton base (red ball) and a fiber (black strand). Proteins IIIa, VI, VIII, and IX stabilize the capsid. A strand of double-stranded DNA is packaged inside the capsid, with two terminal proteins to protect the ends.

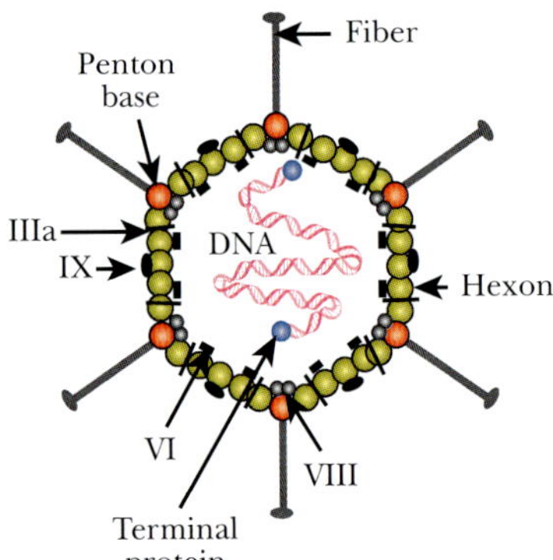

ADENOVIRUS VECTORS IN GENE THERAPY

Adenoviruses are relatively simple, double-stranded DNA viruses that infect humans and other vertebrates. The virus particle consists of a simple icosahedral shell, or capsid, containing a single linear dsDNA molecule of approximately 36,000 base pairs (Fig. 17.2). A terminal protein protects each end of the DNA. The capsid is made of 240 **hexons**, each of which is a trimer of the hexon protein. Hexons are named for their sixfold symmetry and are each surrounded by six neighboring hexons. The hexon protein has loops that project outward from the virus surface. Adenoviruses are subdivided into about 50 serotypes by their response to antibody binding. This variation is largely due to variation in the loops of the hexon protein.

Five faces of the virus particle converge at each vertex. Here is found a **penton**, which consists of a base (a pentamer) plus a fiber (a trimer). The fiber varies greatly in length in different subgroups of adenovirus and its tip binds to the receptor on the host cell surface (Fig. 17.3). The adenoviruses share the same receptor as B-group coxsackieviruses. This protein is therefore known as **CAR** (or **coxsackievirus adenovirus receptor**), but its normal physiological role is unknown.

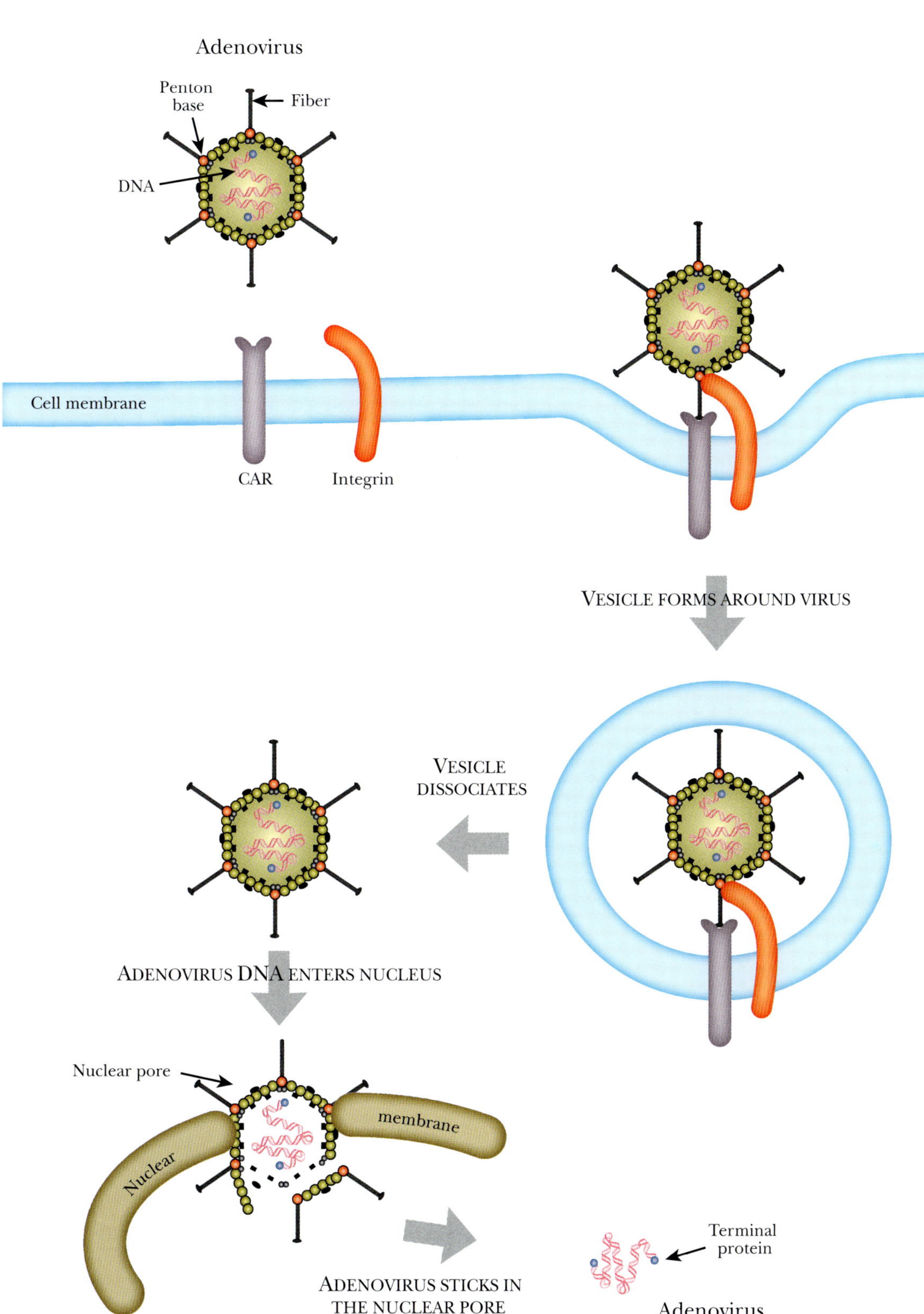

FIGURE 17.3 Adenovirus Entry
Adenoviruses enter human cells by recognizing two receptors, CAR and integrin. The virus is taken into the cell attached to the receptors and surrounded by a membrane vesicle that dissociates in the cytoplasm. The adenovirus injects its DNA into the nucleus through a nuclear pore.

After the fiber tip binds to CAR, the penton base binds to integrins on the host cell surface. (Integrins are transmembrane proteins involved in adhesion.) Next the membrane puckers inward and forms a vesicle that takes the adenovirus inside the cell. The virus is then released into the cytoplasm and travels toward the nucleus. The virion is disassembled outside the nucleus, and only the DNA (with its terminal proteins) enters the nucleus.

Adenoviruses were among the first viruses chosen as vectors for use in human gene therapy. Their advantages are as follows:

(a) Adenoviruses are relatively harmless. They cause mild infections of epithelial cells, especially those lining the respiratory or gastrointestinal tracts.
(b) Adenoviruses are nononcogenic (i.e., they do not cause tumors).
(c) Adenoviruses are relatively easy to culture and can be produced in large quantities.
(d) The life cycle and biology of adenovirus are well understood.
(e) The function of most adenovirus genes is known.
(f) The complete DNA sequence is available, in particular for adenovirus serotype 5 of subgroup C.

Although mild, adenoviruses do cause inflammation and can cause serious illness in patients with damaged immune systems. Therefore, when designing an adenovirus vector for gene therapy, the virus needs to be disarmed by crippling its replication system. This is done by deleting the gene for **E1A protein**, a virus protein made immediately on infection. E1A has two functions (Fig. 17.4). First, it promotes transcription of other early virus genes. Second, it binds to host cell Rb protein, which normally prevents the cell from entering S-phase. This prompts the host cell to express genes for DNA synthesis, which the virus utilizes for its own replication. In the lab, crippled adenovirus is grown in genetically modified host cells that have the viral E1A gene integrated into host cell DNA. The virus particles generated by this approach cannot replicate in normal animal cells.

The DNA of adenovirus is packaged by a headful mechanism. If the DNA is more than 5% shorter or longer than wild type, packaging fails (Fig. 17.5). Insertion of a therapeutic gene into an adenovirus will make the DNA longer. If the inserted gene is much longer than the

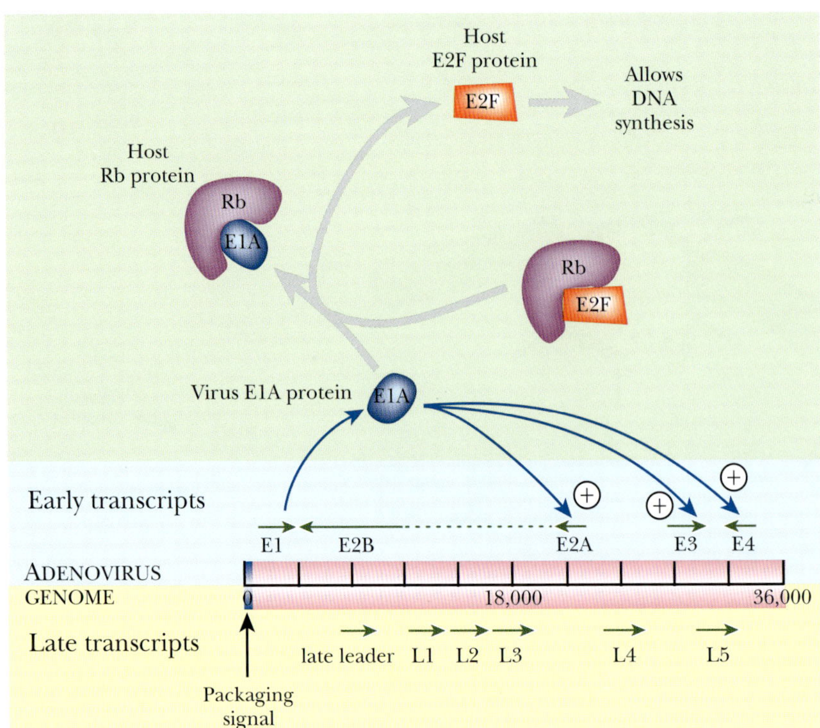

FIGURE 17.4 Role of Adenovirus E1A Protein
Eliminating E1A prevents adenovirus replication. E1A is a transcription factor that activates other adenovirus genes, such as E2A, E3, and E4. E1A also binds host cell Rb protein so releasing host E2F protein to activate DNA synthesis.

deletion used to cripple virus replication, the virus will fail to assemble correctly. Deleting other nonessential virus DNA solves this problem.

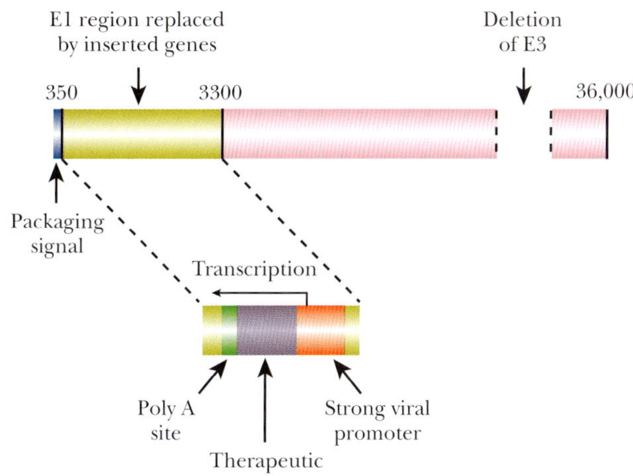

FIGURE 17.5 Engineered Adenovirus
The length of DNA in a virus particle must be very close to the natural adenovirus length of 36,000 base pairs for proper packaging. The therapeutic gene replaces the E1 region. If the gene is much longer than E1A, then a region containing gene E3 is deleted to keep the overall length of the DNA constant.

Although genes carried on engineered adenovirus have been expressed successfully in both animal and human tissues, there are problems. The major difficulty is that adenovirus infections are short-lived. Thus the therapeutic gene is expressed for only a few weeks before the immune system eliminates the virus. Furthermore, the patient develops immunity to the virus so that a second treatment with the same engineered virus will fail. Thus adenovirus vectors cannot be used for long-term gene therapy for hereditary diseases.

Even with the limitations just described, adenovirus vectors may help deliver a deadly gene to cancer cells. Here only a short period of expression should be needed. The CAR receptor is normally only expressed highly by epithelial cells, which limits adenovirus entry to these cell types. However, many cancers also express the CAR receptor at high levels. Consequently, the majority of gene therapy trials using adenovirus are now aimed at cancer cells (see later discussion).

Adenovirus has been widely used as a gene therapy vector. Eventually the immune system eliminates the virus, restricting its use in long-term therapy for inherited conditions.

CYSTIC FIBROSIS GENE THERAPY BY ADENOVIRUS

Because the lungs are relatively accessible to viral infection, cystic fibrosis (see Chapter 16) has been a prime candidate for gene therapy. The healthy version of the cystic fibrosis gene has been cloned and inserted into a crippled adenovirus. Aerosols containing the engineered adenovirus with the cystic fibrosis gene have been sprayed into the noses and lungs, first of rats, and then of humans. In some instances the healthy cystic fibrosis gene was expressed and normal chloride ion movements were also restored. Unfortunately, expression falls off over a 30-day period and repeated doses of virus have little effect, largely because of recognition and destruction of the virus by the immune system.

It is hoped that in the near future, improved vectors will allow cystic fibrosis to be cured by nasal sprays containing genetically engineered viruses. Note, however, that this sort of gene therapy cures only the symptoms in the lungs; it does not correct the genetic defect in the germline cells. The defect will still be passed on to the next generation.

Cystic fibrosis has been targeted for gene therapy because the lungs are readily accessible. Early attempts using adenovirus vectors were partially successful, but only for a brief period.

ADENO-ASSOCIATED VIRUS

Because of the problems with using adenovirus discussed above, other DNA viruses have been considered as vectors. Although none are yet widely used, the **adeno-associated virus (AAV)** shows considerable promise. AAV is a defective or "satellite" virus that depends on adenovirus (or some herpes viruses) to supply some necessary functions. Consequently, it is usually found in cells that are infected with adenovirus. Unlike adenovirus, AAV seems to be entirely harmless.

The benefits of using AAV are as follows:

(a) It does not stimulate inflammation in the host.
(b) It does not provoke antibody formation and can therefore be used for multiple treatments.
(c) It infects a wide range of animals, as long as an appropriate helper virus is also present. It can therefore be cultured in many types of animal cells, including those from mice or monkeys.
(d) It can enter nondividing cells of many different tissues, unlike adenovirus.
(e) The unusually small size of the virus particle allows it to penetrate many tissues of the body effectively.
(f) AAV integrates its DNA into a single site in the genome of animal cells (the *AAVS1* site on chromosome 19 in humans). This allows the therapeutic gene to be permanently integrated.

One drawback is that the AAV genome is small (4681 nucleotides of single-stranded DNA) and the virus can carry only a relatively short segment of DNA. (AAV is unusual in packaging both plus and minus strands into virus particles. Although each virus particle contains only one ssDNA molecule, a virus preparation contains a mixture of particles, half with plus and half with minus strands.) On entering a host cell, the DNA is converted to the double-stranded replicative form, or RF, which is used for both replication and transcription. In the absence of helper virus, AAV integrates into the host chromosome and becomes latent.

FIGURE 17.6 Double AAV System
Two AAV vectors are used to provide erythropoietin to laboratory animals. The regulatory vector (*top*) has two genes, *FRB/RBD* and *ZFHD1/RBD*, which encode two halves of a transcription factor each linked to a rapamycin binding domain (RBD). The therapeutic vector (*bottom*) has the erythropoietin gene (*EPO*) downstream of the transcription factor binding sites. When rapamycin is added, the transcription factor assembles and activates the *EPO* gene.

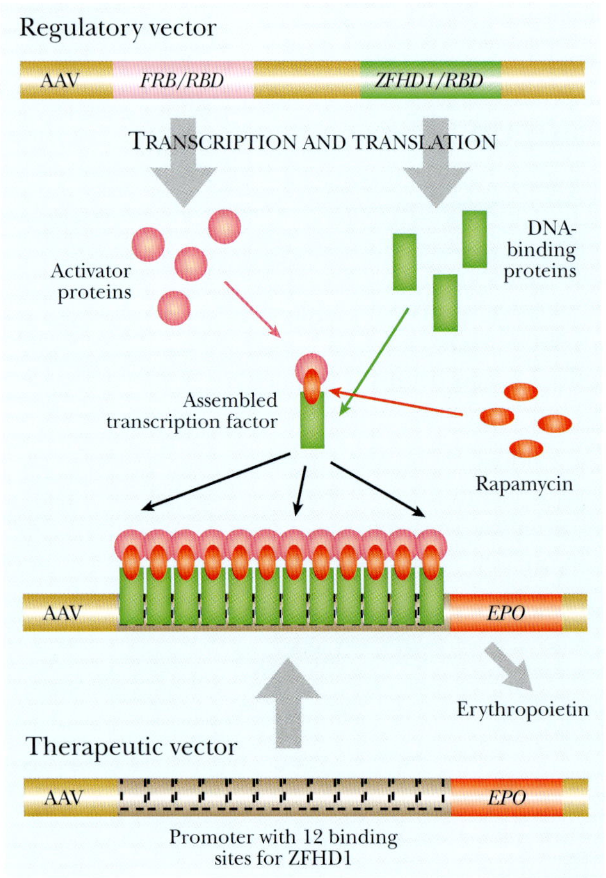

Genes that are permanently integrated need to be carefully regulated. This may be tackled by using two AAV vectors. Mice and monkeys have been experimentally treated with a double AAV system that provides **erythropoietin**, a protein required for development of red blood cells. One AAV vector carries the gene for erythropoietin with a promoter that must be activated by a transcription factor. The second AAV vector carries an artificial regulatory system (Fig. 17.6). This consists of two genes encoding hybrid proteins, each with one domain of the transcription factor. The other domain binds rapamycin (used as an immunosuppressant). In the presence of rapamycin, the two hybrid proteins associate via their rapamycin binding domains to form a functional transcription factor. This activates erythropoietin expression.

After delivery of the two vectors to mice, there was no production of erythropoietin. But when the animals were injected with rapamycin, the transcription factor was assembled and the erythropoietin gene was activated. The levels of erythropoietin increased up to 100-fold and the number of red blood cells rose. Even after several months, injection of rapamycin triggered a sharp rise in erythropoietin levels. Preliminary studies are now being performed in cystic fibrosis patients with AAV vectors carrying the *CFTR* gene.

Adeno-associated virus (AAV) is being developed as a gene therapy vector. A major advantage over adenovirus is that AAV does not provoke antibody formation and can be used for multiple treatments.

RETROVIRUS GENE THERAPY

Retroviruses infect many types of cells in mammals. They need dividing cells for successful infection, and will not infect many tissues where host cell growth and division have come to a standstill. Moreover, the genetic material of retroviruses passes through both DNA and RNA stages. This means that introns must be removed from any therapeutic genes before they are cloned into a retrovirus. Despite these extra technical difficulties, a retrovirus has the distinction of carrying the first gene in successful human gene therapy (see later discussion).

The retrovirus particle has an inner nucleocapsid consisting of an RNA genome inside a protein shell and an outer envelope, derived from the cytoplasmic membrane of the previous host cell. The basic retrovirus genome consists of three genes (*gag, pol,* and *env*) enclosed between two **long terminal repeats (LTRs)**, although more complex retroviruses such as HIV have extra genes involved in regulation. The LTR sequences are needed for integration of the DNA version of the virus genome into the host cell DNA. Between the upstream or 5′ LTR and the *gag* gene is the **packaging signal** (Fig. 17.7), which is essential for packaging the RNA into the virus particle.

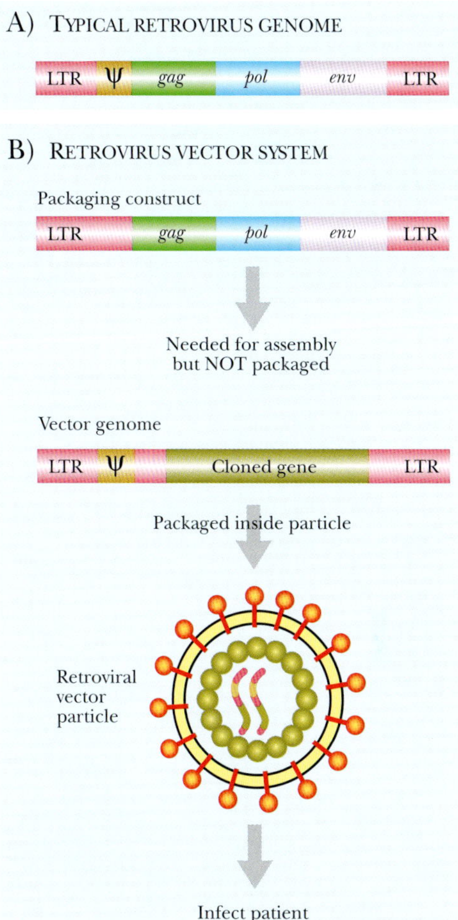

FIGURE 17.7
Retrovirus Genome and Vector System
(A) The retrovirus genome has a packaging signal (ψ) and the genes *gag, pol,* and *env* flanked by two direct repeats known as LTRs. (B) Retrovirus gene therapy uses two virus constructs. The therapeutic vector carries the cloned gene and packaging signal flanked by two LTRs. The packaging vector has the three genes necessary for virus assembly and packaging: *gag, pol,* and *env.* Because the packaging vector does not carry the packaging signal, it is never packaged and does not infect the patient. When both constructs are present, the therapeutic vector plus cloned gene is packaged into the capsid.

Vectors for gene therapy have been derived from the simpler retroviruses, especially **murine leukemia virus (MuLV)**. The vectors have all the retrovirus genes removed, and as a result they are completely defective in replication. They retain only the packaging signal and the two LTRs (Fig. 17.7) and can carry approximately 6 to 8 kb of inserted DNA. A virus promoter in the 5′ LTR drives expression of the cloned gene. Because the vector lacks *gag, pol,* and *env* it cannot make virus particles. Hence these functions must be provided by a **packaging construct**, a defective provirus that is integrated into the DNA of the producer cell (see Fig. 17.7). The packaging construct lacks the packaging signal, so although it is responsible for manufacture of virus particles, it is not packaged itself. The virus particles generated contain only the retrovirus vector carrying the cloned gene.

After infection of the patient, the RNA inside the retroviral vector is reverse transcribed to give a DNA copy. (Although the retroviral vector does not carry a copy of the reverse transcriptase gene, a few molecules of reverse transcriptase enzyme are packaged in retrovirus particles.) Ideally, the cloned gene, enclosed between the two LTR sequences, is then integrated into host cell DNA.

Because the retroviral vectors are completely devoid of genes for retrovirus proteins, they do not cause an immune response or significant inflammation. Furthermore, their ability to integrate into host cell DNA means that the therapeutic gene will become a permanent part of the host cell genome. In principle the retrovirus could integrate into a harmful location, thus disrupting the function of regulation of a host cell gene. In practice, because most DNA in animal cells is noncoding, the chances are low, and only occasional cells would be damaged.

More serious problems are that retroviral vectors can carry only small amounts of DNA (about 8 kb) and cannot infect nondividing cells. However, the lentivirus family of retroviruses (to which HIV belongs) is unusual in being able to infect some nondividing cells. Naturally, using HIV itself is risky, but a future possibility is to transfer this ability into other, safer retroviruses. Alternatively, lentiviruses that infect other mammals, such as FIV (feline immunodeficiency virus), might be used to derive vectors.

Engineered retroviruses are the most frequently used viral vectors in gene therapy. Defective retrovirus vectors are grown in cells with an integrated helper virus to allow formation of virus particles.

RETROVIRUS GENE THERAPY FOR SCID

Severe combined immunodeficiency (SCID) occurs when both the B cells and T cells of the immune system are defective and results in an almost totally defective immune response. Children with SCID have to be shielded from all contact with other people and are kept inside special sterile plastic bubble chambers. Without immune protection any disease, even a cold, could prove fatal. Several genetic defects are known that cause SCID. About 25% are due to mutations in the ***Ada* gene** that encodes the enzyme **adenosine deaminase**. This is needed for the metabolism of purine bases, and its absence prevents development of lymphocytes (white blood cells including both the B cells and T cells).

The first successful instance of human gene therapy used a retroviral vector to provide a functional copy of the *Ada* gene to a child with SCID. The cells affected by SCID are the lymphocytes that circulate in the blood, where they carry out immune surveillance. They are produced by division of bone marrow cells (Fig. 17.8). Gene therapy involves removing bone marrow cells from the patient and maintaining them in cell culture outside the body. Because bone marrow cells constantly replenish the blood supply, they divide often and are suitable for retroviral infection. While in culture, the bone marrow cells are infected with genetically engineered retrovirus carrying the *Ada* gene and are then returned to the body.

FIGURE 17.8
***Ex Vivo* Retroviral Gene Therapy for *Ada* Deficiency**
Gene therapy for SCID requires the removal of bone marrow cells from the patient. The cells are cultured and the mutant *Ada* gene is replaced with a functional copy. The bone marrow culture is treated with neomycin to kill nontransformed cells. The bone marrow cells are then replaced in the patient and repopulate the blood supply with normal blood cells.

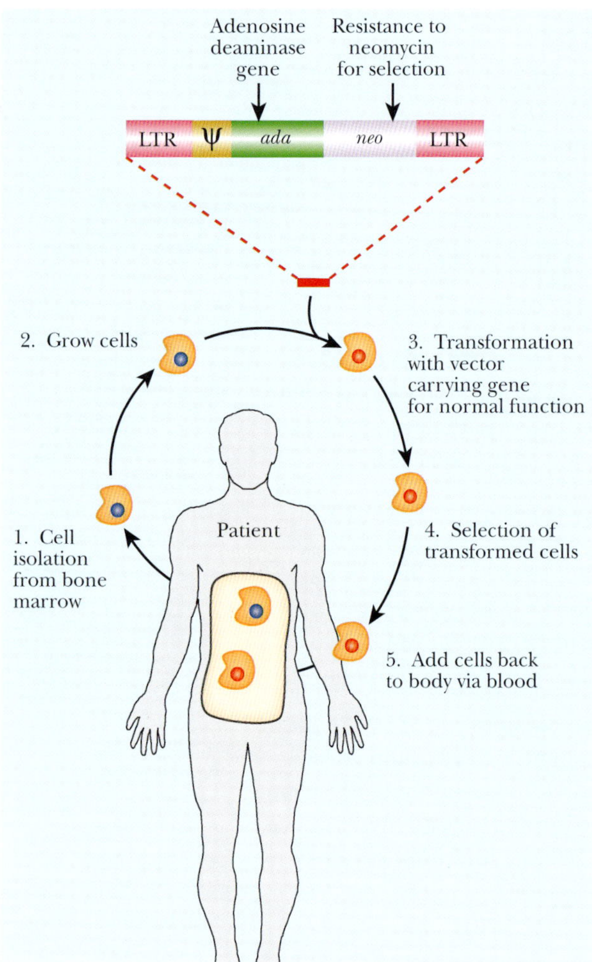

Since 1991, several children have been treated by this approach. However, because T cells live for only 6 to 12 months, the procedure must be repeated at intervals. This problem has been tackled by using blood **stem cells**, which divide to provide the precursors to all types of blood cells and also give rise to more stem cells. These are found in bone marrow but only in very small numbers. However, umbilical cord blood has a much higher proportion of stem cells. So in 1993, blood stem cells were obtained from the umbilical blood of several newborn babies who were known to carry homozygous defects in *Ada*. The *Ada* gene was introduced into the stem cells on a retroviral vector, which has resulted in a long-term supply of healthy T cells.

In all the cases just discussed, the patients have been receiving injections of purified ADA enzyme as well as gene therapy. It is therefore unclear how much of their improvement is due to the gene therapy, even though these patients now have functional T cells in their blood. However, a clear-cut result has been obtained more recently with another variant of SCID. This variant is due to defects in the receptor for several **interleukins**, including IL-7, a protein that promotes development of T cells from stem cells. Because B cells need helper T cells in order to function, both B and T cells are inactive. In this case, the gene for the missing subunit of the IL-7 receptor was inserted into cultured stem cells using a retroviral vector.

Because purified protein is not used in treating this type of SCID, the patients had to depend on gene therapy alone. The patients did develop normal numbers of T cells and were successfully vaccinated against several infectious diseases, indicating a proper immune response. Consequently they were able to leave their bubbles and enjoy normal lives. Although most of the dozen patients treated so far are still doing well, one developed leukemia, presumably due to retrovirus insertion into other genes. Thus the use of retrovirus for gene therapy needs further safeguards before it can be used more routinely.

Some types of severe combined immunodeficiencies have been successfully treated by gene therapy using a retrovirus vector.

NONVIRAL DELIVERY IN GENE THERAPY

However sophisticated a viral delivery system may be, nonviral vectors are inherently safer. Nonetheless, they have been relatively neglected because viruses were more efficient. However, several unfortunate incidents have occurred with viral vectors, especially retroviruses, including the occasional appearance of leukemia-like disease. This has resulted in renewed interest in nonviral delivery systems.

About 75% of gene therapy trials have used viral vectors. A variety of alternative approaches have also been investigated, though few have been effective or widely used so far. These include:

- **(a)** Use of naked nucleic acid (DNA or less often RNA). Many animal cells can be transformed directly with purified DNA. The therapeutic gene may be inserted into a plasmid and the plasmid DNA used directly. Some 10% to 20% of gene therapy trials have used unprotected nucleic acid.
- **(b)** Particle bombardment. DNA is fired through the cell walls and membranes on metal particles. This method was originally developed to get DNA into plants and is therefore discussed in Chapter 14. However, it has also been used to make transgenic animals and is occasionally used for humans.
- **(c)** Receptor-mediated uptake. DNA is attached to a protein that is recognized by a cell surface receptor. When the protein enters the cell, the DNA is taken in with it.
- **(d)** Polymer-complexed DNA. Binding to a positively charged polymer, such as polyethyleneimine, protects the negatively charged DNA. Such complexes are often taken up by cells in culture and may in principle be used for *ex vivo* gene therapy.
- **(e)** Encapsulated cells. Whole cells engineered to express and secrete a needed protein may be encapsulated in a porous polymeric coat and injected locally. Foreign cells excreting nerve growth factor have been injected into the brains of aging rats. The rats showed some improvement in cognitive ability, suggesting that this approach may be of value in treating conditions such as Alzheimer's disease.
- **(f)** Liposomes are spherical vesicles composed of phospholipid. They have been used in around 10% of gene therapy trials (see later discussion).

A variety of approaches, other than viruses, can be used to get foreign DNA into target cells. These include using naked DNA, DNA bound to artificial polymers or proteins, particle bombardment, and liposomes.

LIPOSOMES AND LIPOFECTION IN GENE THERAPY

About 10% of gene therapy trials have used **liposomes**. These are hollow microscopic spheres of phospholipid, and can be filled with DNA or other molecules during assembly. The liposomes will merge with the membranes surrounding most animal cells and the contents of the liposome end up inside the cell (Fig. 17.9), a process known as **lipofection**. Although lipofection works reasonably well, it is rather nonspecific, because liposomes tend to merge with the membranes of any cell.

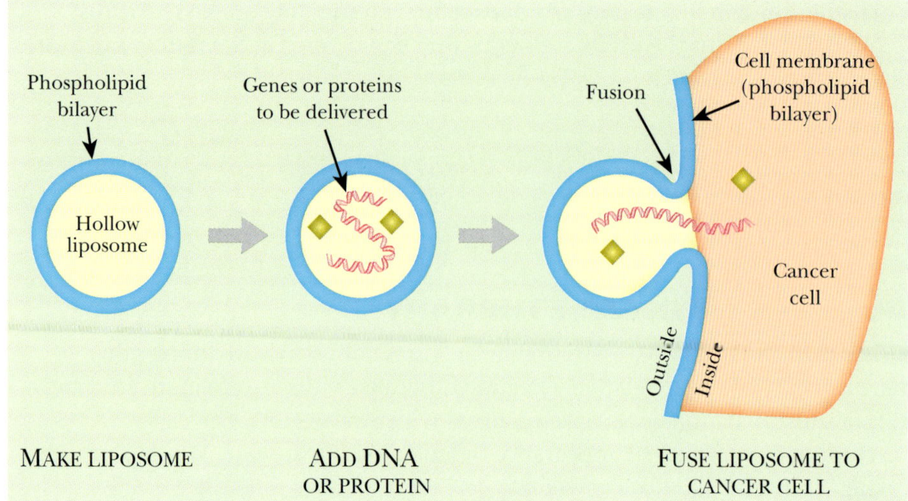

FIGURE 17.9
Delivery of Anticancer Agent by Lipofection
Liposomes are hollow spheres surrounded by a lipid bilayer. They can be filled with DNA or proteins such as TNF. Liposomes interact with cell membranes because both have hydrophobic lipid layers. Once in contact, the liposome merges with the cell membrane, delivering its payload into the cell.

A formidable problem during anticancer gene therapy is how to get foreign DNA specifically into cancer cells (see later discussion). Lipofection is a promising approach because "armed" liposomes can be injected directly into tumor tissue. In fact, liposomes are probably of more use in delivering proteins than DNA, something not feasible when using viruses as genetic engineering vectors. For example, toxic proteins such as **tumor necrosis factor (TNF)** can be packaged inside liposomes and injected into tumor tissue. The liposomes merge with the cancer cell membranes, and the lethal proteins are then released inside the cancer cells.

Liposomes are hollow microscopic spheres of phospholipid. They can be used to carry foreign DNA or proteins to a target tissue.

AGGRESSIVE GENE THERAPY FOR CANCER

Although most cancers are not inherited via the germ line, cancer is nonetheless a genetic disease. In the case of hereditary disease we may attempt to replace the defective component, thus preventing cell death. In contrast, when dealing with a cancer we need to destroy the

Box 17.1 Embryonic Stem Cell Therapy

Rather than replace defective genes, it may be possible in the future to replace whole cells and manipulate them to grow back and regenerate defective tissues or organs. The present scenario involves the use of embryonic stem cells. These are ancestral stem cells, found in early embryos, that differentiate into other less generalized stem cells, which in turn give rise to the final tissues and organs of the body.

The use of embryonic stem cells has given rise to a moral, religious, political, and even occasionally scientific debate that has been blown out of proportion. Claims that stem cells will cure every defect and disease are wildly exaggerated.

At the moment it is possible to maintain many lines of stem cells in culture fairly reliably. Preliminary experiments have shown that stem cells can be implanted into animals and divide and differentiate. What is missing is the ability to precisely control how the stem cells differentiate. In particular, it is difficult to stop them from dividing and changing further once they have replaced damaged tissue. For example, attempts using stem cells to replace insulin-secreting cells of the pancreas resulted in a period of insulin production followed by further uncontrolled growth forming a teratoma.

The great paradox of the stem cell debate is that once we understand cellular differentiation well enough to manipulate stem cells, we will probably no longer need them. If we can control the regulatory circuits for cell differentiation, then we should be able to take adult cells and dedifferentiate them back into pseudo–stem cells. At this point the ethical problems with using embryonic tissues should evaporate.

cancer cells, or at least inhibit their growth and division. Several strategies have been used and may be classified as follows:

(a) Gene replacement
(b) Direct attack
(c) Suicide
(d) Immune provocation

Gene replacement therapy for cancer is analogous to its use in correcting hereditary defects. The cancer is analyzed to identify the mutant gene(s) that are responsible. The wild-type version of the oncogene or tumor suppressor gene is then inserted into the cancer cells. For example, the wild-type version of the p53 gene has been delivered to p53-deficient cancer cells. The delivery method is usually via an adenovirus vector, but sometimes liposomes have been used.

In the direct plan of attack, a gene that helps kill cancer cells is used. For example, the ***TNF* gene** encodes tumor necrosis factor. This is produced by white blood cells known as **tumor-infiltrating lymphocytes (TILs)**. These cells normally infiltrate into tumors where they release TNF, which is fairly effective at eradicating small cancers. To attack a large cancer that is out of control, TNF production must be increased. First the *TNF* gene is cloned. Then white blood cells are removed from the patient and cultured. Multiple copies of the *TNF* gene—or perhaps an improved *TNF* gene with enhanced activity—are introduced into the white cells. Then the white cells are injected back into the patient.

Although TNF is very effective in killing cancer cells, it is also toxic to other cells. Thus high levels of TNF are dangerous to the patient. There are two sides to this problem. One is limiting TNF or other toxic agents to the cancer cells. The other is getting the toxic agent to the relatively inaccessible cells on the inside of a tumor. A variety of modifications are being tested to solve these problems—for example, putting the TNF gene under control of an inducible promoter and using adenovirus to transfer the gene into cancer cells. The chosen promoter is designed to be induced by agents already used in treating cancer cells, such as radiation or cisplatin.

The suicide strategy is actually a hybrid of anticancer drug therapy with gene transfer therapy. A harmless compound, or **prodrug**, is chosen that can be converted to a toxic anticancer drug by a specific enzyme. If the enzyme is present, the cell expressing it will commit suicide when the prodrug is available (Fig. 17.10). Consequently, an enzyme that is not present in normal human cells must be chosen for this approach. The gene encoding the suicide enzyme must be delivered to the target cancer cells, usually by a viral vector or in liposomes. If the enzyme is successfully expressed in the cancer tissue, then the toxic drug will be generated inside the cancer cells. Thus the prodrug can be administered to the patient by normal means but is specifically lethal for the cancer cells.

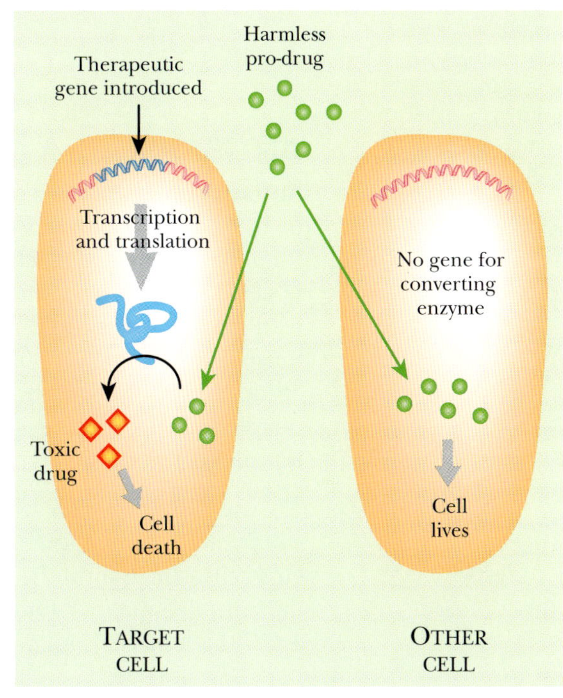

FIGURE 17.10 Suicide Gene Therapy
Suicide gene therapy begins by delivering a therapeutic gene into the cancer cells. The gene encodes an enzyme that will convert a nontoxic prodrug into a toxic compound. Because noncancerous cells do not have the suicide enzyme, they are not affected.

In practice two major suicide enzyme/prodrug combinations have been used. Gene therapy has been used to deliver the enzyme **thymidine kinase**, originally from herpes virus, to cancer cells. The nontoxic prodrug, the nucleoside analog **ganciclovir**, is converted to its monophosphate by thymidine kinase (hence its clinical use in treating herpes virus infections). Because only the cancer cells have thymidine kinase, all the noncancerous cells are unaffected. Normal cellular enzymes then convert the monophosphate to ganciclovir triphosphate (GCV-TP). This acts as a **DNA chain terminator** (Fig. 17.11). DNA polymerase incorporates

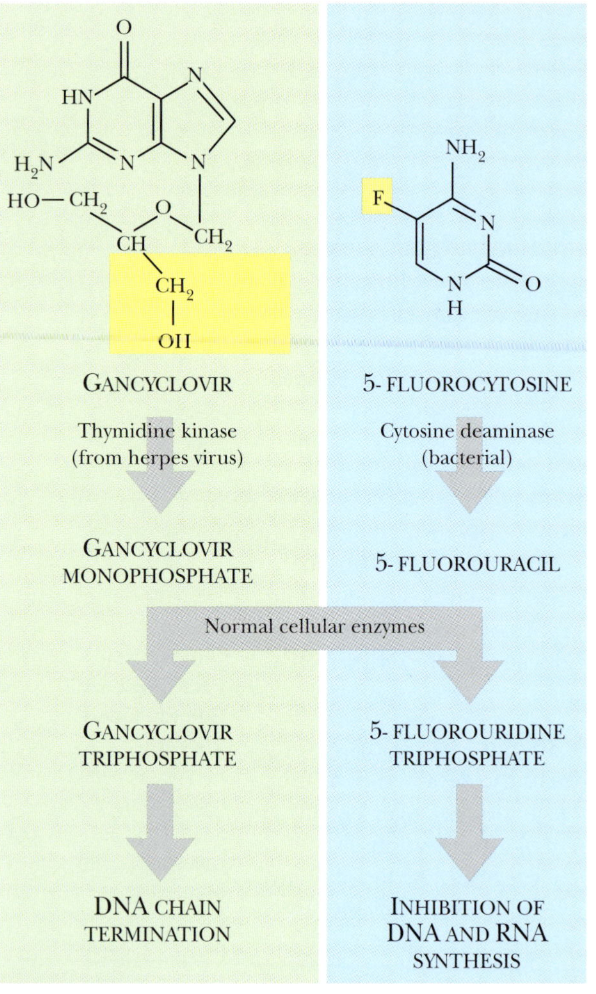

FIGURE 17.11
Ganciclovir and 5-Fluorocytosine
Ganciclovir is converted to a DNA chain terminator in cells containing thymidine kinase (from herpes virus). Similarly, 5-fluorocytosine is converted to a DNA and RNA synthesis inhibitor in cells that have cytosine deaminase (from bacteria). The groups highlighted in yellow prevent addition of further nucleotides. Genes for thymidine kinase or cytosine deaminase must be inserted into the cancer cells via gene therapy.

GCV-TP into growing strands of DNA. However, lack of a 3′-OH group prevents further elongation of the nucleic acid strand. DNA synthesis is thus inhibited and the cell is killed. A similar scheme involves the conversion of 5-fluorocytosine to 5-fluorouracil by **cytosine deaminase**, originally from bacteria. Again, cellular enzymes finish the job by making the phosphorylated nucleoside that actually inhibits DNA and RNA synthesis.

A more indirect approach relies on the body's natural defenses. Our immune systems are effective at killing cancers, provided they identify them while still small. To survive, a cancer has to somehow evade the body's immune surveillance. In this approach, gene therapy inserts a gene that attracts the attention of the immune system to the tumor cells. For example, the HLA (= MHC) proteins are exposed on the surfaces of mammalian cells where they act in cell recognition. Different individuals have different combinations of **HLA genes**, which act as molecular identity tags so that cells of the body are recognized as "self." If HLA genes that are not originally present in that particular individual are inserted into the cancer cells, the tumor appears alien, and the immune system will now mount an assault.

A related approach is to use **cytokines**. These are short proteins that attract immune cells and stimulate their division and development. The genes for several cytokines of the interleukin family (especially IL2, IL4, and IL12) have been used to provoke immune attacks on cancer cells.

Aggressive gene therapy uses engineered genes to attack cancer rather than to replace defects. Sometimes therapeutic genes are used to increase the activity of the body's own anticancer systems. In other cases, the product of the therapeutic gene is used to specifically activate a toxic drug within cancer cells.

ANTISENSE RNA AND OTHER OLIGONUCLEOTIDES

Antisense RNA binds to the corresponding messenger RNA and prevents its translation by the ribosome. Antisense RNA and related antisense nucleic acids are now being tested for possible therapeutic effects. In most cases, the objective is to inhibit the expression of a target gene. Two main alternatives exist when using antisense RNA. First, it is possible to use a full-length **anti-gene** that is transcribed to give a full-length antisense RNA. The anti-gene must be carried on a suitable vector and expressed in the target cells. Second, much shorter artificial RNA oligonucleotides may be used. An antisense RNA of 15 to 20 nucleotides is often capable of binding specifically to part of the complementary mRNA and preventing translation.

The use of a full-length anti-gene has been tested experimentally against cancer cells in culture. Malignant glioma is the most common form of human brain cancer. Here the glial cells, which form the interstitial tissue between nerve cells, grow and divide in an uncontrolled manner. Often this is due to overproduction of **insulin-like growth factor 1 (IGF1)**, which stimulates the growth and division of these cells. A vector carrying an

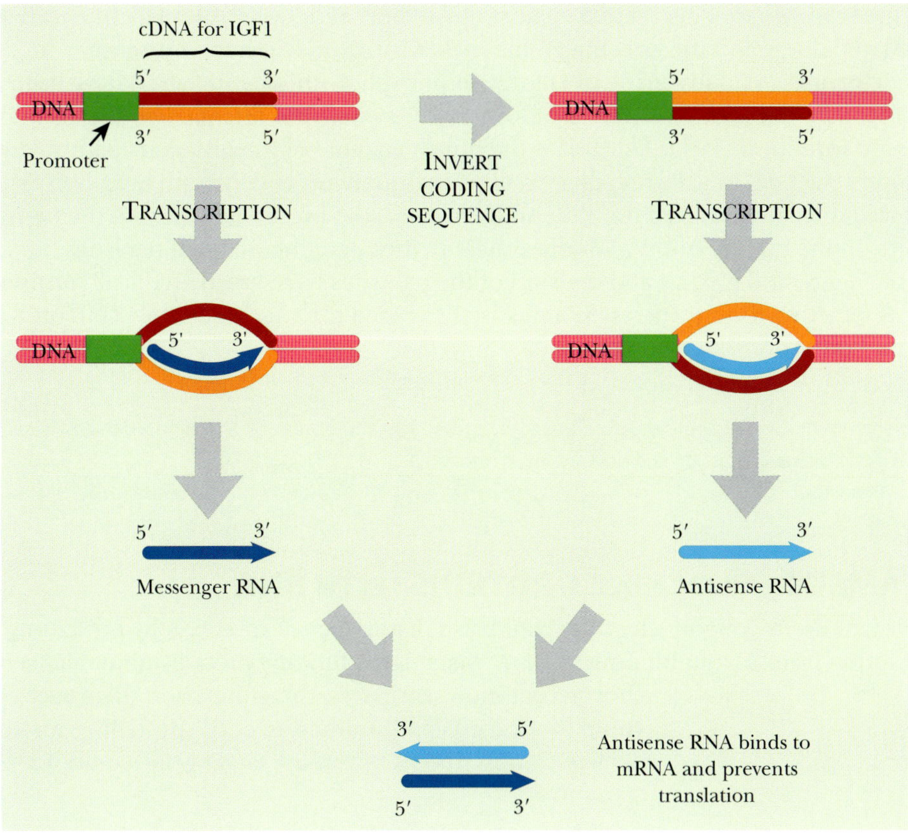

FIGURE 17.12
Full-Length Antisense Therapy for Cancer
The cDNA gene for IGF1 was inverted so that the RNA transcribed from it was complementary to the mRNA from the wild-type IGF1 gene. When this antisense construct was transformed into glial cells, the antisense RNA bound all the wild-type IGF1 mRNA, preventing production of IGF1 protein. The antisense IGF1 gene was controlled by a metallothionein promoter that turns on only in the presence of heavy metals.

FIGURE 17.13
Antisense Oligonucleotide Targets
Antisense oligonucleotides (red) bind splice sites and 3′ UTR in order to prevent pre-mRNA from being processed into mRNA. Antisense oligonucleotides can also bind to the 5′ UTR and the translational start site on mRNA in order to prevent protein translation.

anti-IGF1 gene was constructed by using the cDNA version of the ***IGF1* gene** and simply reversing its orientation (Fig. 17.12). The *anti-IGF1* gene was placed under control of the **metallothionein** promoter, which responds to trace levels of heavy metals. The vector was inserted into glioma cells in culture. Adding a low concentration of zinc sulfate turned on the *anti-IGF1* gene, and the cancer cells stopped dividing and no longer caused tumors if injected into rats.

Artificial RNA oligonucleotides may be aimed at a variety of possible sites on the target RNA. They may be designed to block the 5′-end of a transcript, thus preventing translation from beginning. However, other sites within the mRNA are often just as effective. It is also possible to make an antisense oligonucleotide that blocks the splice site of the primary transcript and so prevents formation of the mature mRNA by the spliceosome (Fig. 17.13).

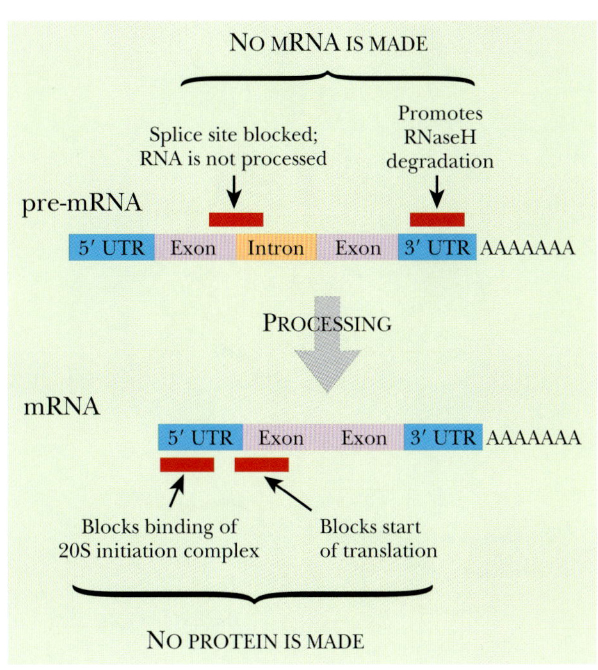

Antisense RNA oligonucleotides are not very stable and are taken up poorly by many cells. Use of liposomes to deliver antisense RNA may overcome the uptake problem. Another way to overcome the first problem is to use modified oligonucleotides. Antisense segments made of DNA rather than RNA are just as effective, are easier to synthesize, and are more stable than RNA. Oligonucleotides with phosphorothioate linkages are more stable than those with phosphodiester linkages. Those made of peptide nucleic acid are extremely stable.

FIGURE 17.14
Selection of Aptamer against Enzyme
To find an oligonucleotide that binds to a specific enzyme, a pool of oligonucleotides with random sequences is synthesized. Short primer binding sequences are added to each end for later PCR amplification. The pool of random oligonucleotides is passed over a column with the target enzyme. Those oligonucleotides that recognize the enzyme stick to the column while everything else passes through. The sequences that bound to the column can be isolated and amplified by PCR.

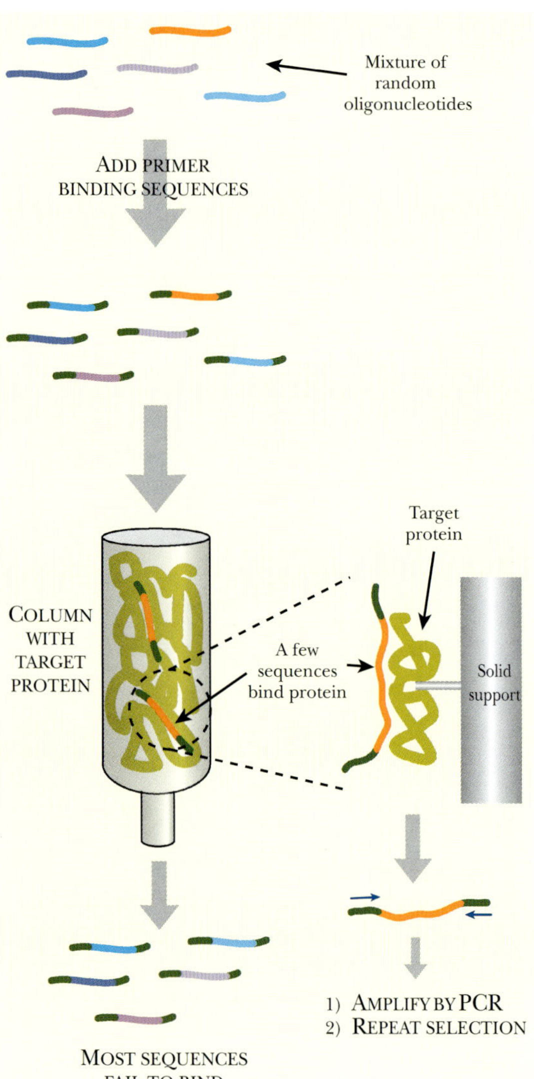

Antisense reagents are being tested against malaria because of the shortage of effective antimalarial drugs and the growing global incidence of this disease. Antisense oligonucleotides (actually made of DNA with phosphorothioate linkages) have been used experimentally to block synthesis of aldolase by ***Plasmodium falciparum***, the agent of the malignant form of malaria. Aldolase is the fourth enzyme of the glycolytic pathway, and the blood stages of *P. falciparum* depend almost entirely on energy from glycolysis. The antisense molecules targeted both the splice sites on the precursor RNA and the translation start site on the mature mRNA. Aldolase activity dropped, the malarial parasites ran short of energy, and growth and division of the parasites were greatly reduced. Antisense reagents targeting topoisomerase II and Clag9 (needed for attachment to CD36 protein of red blood cells) are also being tested.

Antisense RNA oligonucleotides are capable of preventing expression of specific mRNAs. Clinical use requires problems of stability and delivery to be overcome.

APTAMERS—BLOCKING PROTEINS WITH RNA

In theory, antisense oligonucleotides could also disrupt gene expression by preventing transcription factors from binding to DNA. Many DNA binding proteins have relatively short recognition sequences; therefore, short oligonucleotides could block the DNA binding site of the protein. Without an active transcription factor, the target gene would not be transcribed.

In principle, an artificial oligonucleotide could be designed to fit into a binding site of any shape. Even if an enzyme does not normally bind nucleic acids, an oligonucleotide could be designed that would block the active site of the protein. Such oligonucleotides do not correspond to natural nucleic acid sequences and are known as **aptamers**. This procedure has been demonstrated experimentally for the enzyme **thrombin**, a protease involved in the blood clotting cascade. The antithrombin aptamer is short-lived because it is degraded rapidly *in vivo*. It might perhaps be useful as an anticlotting agent, such as during bypass surgery.

Aptamers may be made by a combination of molecular selection and PCR amplification. Random oligonucleotides (of 50 or 60 bases) are artificially synthesized by solid-state synthesis, using a mixture of all four bases at each step. The pool of random sequences is then ligated at each end to primer binding sequences (approximately 20 bases long). The mixture is poured through a column to which the target protein is attached (Fig. 17.14). Some oligonucleotides will bind to the protein and are therefore retained by the column. These are dissociated and then amplified by PCR. The bound oligonucleotides will be a mixture, some of which bind weakly and some strongly. Therefore the binding and PCR steps are repeated two or three times to select oligonucleotides with high binding affinity.

Recently, pegaptanib (Macugen; from Eyetech Pharmaceuticals, Pfizer, New York) has been approved for treatment of blindness due to age-related macular degeneration. Pegaptanib is an aptamer that inhibits the action of an isoform of vascular endothelial growth factor (VEGF) that is responsible for abnormal vascularization of the eyes. Binding of pegaptanib to VEGF prevents VEGF from binding to its receptor.

Short artificial oligonucleotides known as aptamers can be designed to block protein activity.

RIBOZYMES IN GENE THERAPY

RNA molecules with catalytic ability are known as **ribozymes**. Some ribozymes act by cleaving other RNA molecules. Such ribozymes have catalytic domains that carry out the reaction fused to domains that recognize the substrate RNA by base pairing (Fig. 17.15). If the sequence of the substrate recognition domain is changed, the ribozyme will recognize a different target RNA molecule. Thus ribozymes may be engineered to recognize and destroy any target messenger RNA molecule. So far this use of ribozymes is still in the experimental phase. In practice, a vector carrying a gene encoding the ribozyme would presumably be used for delivery to the target cell. Transcription of this gene would result in production of the ribozyme RNA. The ribozyme would then bind and cleave the target mRNA.

All naturally occurring nucleic acid enzymes are made of RNA. However, there is no chemical reason why DNA cannot act catalytically. Indeed, **deoxyribozymes** may be synthesized artificially, using sequences equivalent to those of RNA ribozymes. The DNAzymes are less easily degraded and may be manufactured in multiple copies by a PCR-based technique.

Most DNAzymes presently under investigation are targeted at cancer cells or infectious microorganisms. Antibiotic resistance among bacteria is becoming a major problem. One interesting possibility is to use DNAzymes to cleave the mRNA for enzymes that make bacteria resistant to antibiotics. The bacteria could then be killed with the antibiotics despite carrying a resistance gene.

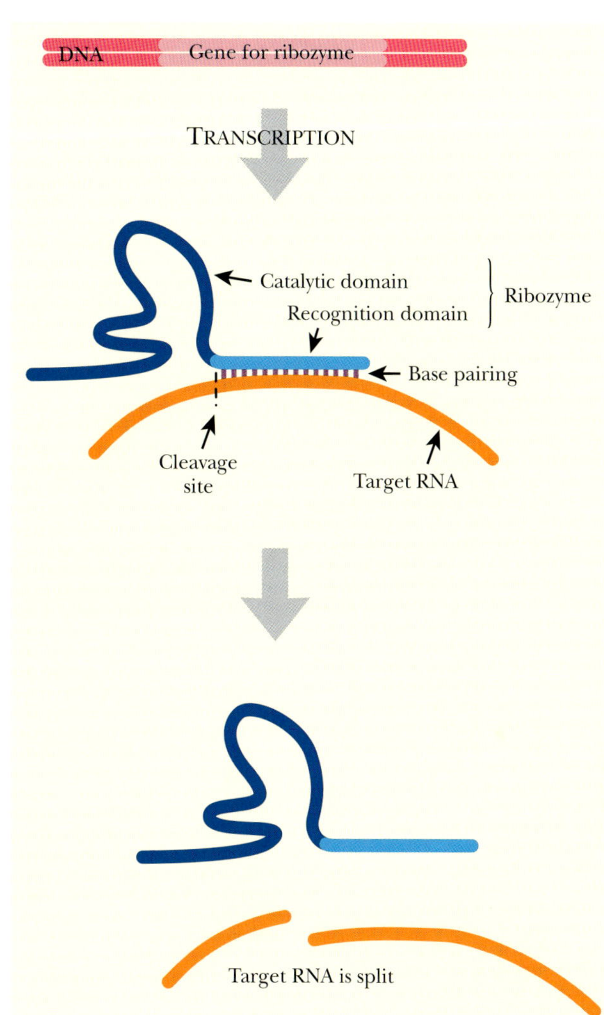

FIGURE 17.15
Targeted Ribozymes
Gene therapy can also be used to transfer a gene for a ribozyme into a defective cell. The ribozyme has two domains; the catalytic domain acts as an enzyme, and the recognition domain binds to a specific target RNA. Once bound, the enzyme domain cleaves the target RNA.

Ribozymes that act by cleaving RNA target molecules have been proposed as therapeutic agents.

Summary

Genetic engineering can be used to combat hereditary defects. This approach is mostly still under development, but a few successes have been achieved. Modified viruses, especially retroviruses, are the favorite vectors for inserting genes into human patients, although other vectors such as liposomes are being tested. Aggressive gene therapy versus cancer uses genes and proteins for attack rather than replacement of genetic defects and has shown considerable promise. Using oligonucleotides to silence genes or to inhibit protein activity is an area under rapid development.

End-of-Chapter Questions

1. What is a difference between genetic engineering and gene therapy?
   - **a.** Genetic engineering produces inheritable changes, but changes through gene therapy are never inherited.
   - **b.** Gene therapy produces inheritable changes, but changes through genetic engineering are never inherited.
   - **c.** Changes from both gene therapy and genetic engineering are passed on to subsequent generations.
   - **d.** Changes from both gene therapy and genetic engineering are never passed on to future generations.
   - **e.** None of the above.
2. Which of the following is the most popular approach to delivering gene therapy?
   - **a.** bacterial vectors
   - **b.** naked DNA
   - **c.** liposomes
   - **d.** viral vectors
   - **e.** gene gun
3. Which one of the following statements about oligonucleotide crossover is not correct?
   - **a.** A defective gene could be patched with double-stranded oligonucleotides.
   - **b.** The frequency of the crossover is improved by the use of RNA-DNA hybrid oligonucleotides.
   - **c.** There is no way to protect the ends of the oligonucleotide from exonucleases after delivery.
   - **d.** Hairpin structures on the ends of the oligonucleotide protect the ends from degradation.
   - **e.** All of the above are correct.
4. What disease are the majority of aggressive gene therapy trials directed against?
   - **a.** infectious diseases
   - **b.** HIV
   - **c.** cancer
   - **d.** vascular diseases
   - **e.** single gene defects
5. Which of the following statements is considered an advantage to using adenoviruses for human gene therapy?
   - **a.** They do not cause cancer.
   - **b.** The complete DNA sequences are available and almost every gene has a known function.
   - **c.** They are easily cultured and can be produced in large quantities.
   - **d.** They only cause minor illnesses, if any illness at all.
   - **e.** All of the above are advantages.

6. Why are adenoviruses not used for long-term gene therapy of inherited diseases?
   a. Adenovirus infections are short-lived.
   b. The patient produces immunity against the virus, and that particular virus cannot ever be used again.
   c. The virus is limited by which cell types express the receptor for the virus.
   d. All of the above are reasons to not use adenoviruses.
   e. None of the above is a reason not to use adenoviruses.
7. What is a benefit of using AAV over adenoviruses for delivery of genes?
   a. AAV does not provoke antibody development and can be used multiple times.
   b. AAV has a wide host and tissue range.
   c. AAV does not stimulate inflammation in the host.
   d. AAV integrates its DNA into the host genome, which allows the therapeutic gene to remain in the host.
   e. All of the above are benefits of using AAV over adenoviruses.
8. What is a drawback to using AAV?
   a. AAV does not provoke antibody development.
   b. The virus is small and can only package small amounts of DNA.
   c. AAV has a wide host and tissue range.
   d. AAV integrates its DNA into the host genome thus causing cancer.
   e. None of the above is a drawback.
9. Which was the first to succeed in carrying a gene during gene therapy?
   a. retrovirus
   b. adenovirus
   c. herpes virus
   d. adeno-associated virus
   e. cytomegalovirus
10. A mutation in which gene accounts for 25% of SCID cases?
   a. CFTR
   b. ZFHD1
   c. Ada
   d. EPO
   e. FRB
11. What disease has resulted after the use of retroviruses in gene therapy?
   a. AIDS
   b. leukemia
   c. HPV
   d. SCID
   e. none of the above
12. What can liposomes do that other methods for delivering genes cannot?
   a. Liposomes can degrade extracellular DNA.
   b. Liposomes can only deliver DNA to specific cells.
   c. Liposomes have no other function than delivering DNA.
   d. Liposomes can deliver proteins to cells, such as tumor necrosis factor.
   e. none of the above

(Continued)

13. When referring to aggressive gene therapy, which of the following is a strategy for destroying cancer cells?

- **a.** direct attack of cancer cells by using a gene that helps kill the cells, such as *TNF*
- **b.** inserting the wild-type copy of the gene that is defective and causing the cancer through gene replacement therapy
- **c.** creation of a hybrid system composed of both gene therapy and cancer drug therapy
- **d.** insertion of a gene that, when expressed, attracts the host's immune system to the cancer
- **e.** all of the above

14. Which of the following cancers has been shown in tissue culture to respond to "anti-gene" therapy?

- **a.** malignant glioma
- **b.** neuroblastoma
- **c.** breast cancer
- **d.** ovarian cancer
- **e.** prostate cancer

15. How might ribozymes be used in gene therapy after delivering the gene to target cells?

- **a.** The ribozyme could cleave ribosomal RNA.
- **b.** The ribozyme could cleave a specific mRNA.
- **c.** The ribozyme could cleave transcription factors for specific mRNAs.
- **d.** The ribozyme could cleave the RNA of retroviruses.
- **e.** None of the above are uses for gene therapy.

Further Reading

Aartsma-Rus A, van Ommen GJ (2007). Antisense-mediated exon skipping: A versatile tool with therapeutic and research applications. *RNA* **13**, 1609–1624.

Athanasopoulos T, Graham IR, Foster H, Dickson G (2004). Recombinant adeno-associated viral (rAAV) vectors as therapeutic tools for Duchenne muscular dystrophy (DMD). *Gene Ther* **11**(Suppl 1), S109–S121.

Bagheri S, Kashani-Sabet M (2004). Ribozymes in the age of molecular therapeutics. *Curr Mol Med* **4**, 489–506.

Conese M, Boyd AC, Di Gioia S, Auriche C, Ascenzioni F (2007). Genomic context vectors and artificial chromosomes for cystic fibrosis gene therapy. *Curr Gene Ther* **7**, 175–187.

Devi GR (2006). siRNA-based approaches in cancer therapy. *Cancer Gene Ther* **13**, 819–829.

Duan D (2006). Challenges and opportunities in dystrophin-deficient cardiomyopathy gene therapy. *Hum Mol Genet* **15**, R253–R261.

Fichou Y, Férec C (2006). The potential of oligonucleotides for therapeutic applications. *Trends Biotechnol* **24**, 563–570.

Foster K, Foster H, Dickson JG (2006). Gene therapy progress and prospects: Duchenne muscular dystrophy. *Gene Ther* **13**, 1677–1685.

Fritz JJ, Gorbatyuk M, Lewin AS, Hauswirth WW (2004). Design and validation of therapeutic hammerhead ribozymes for autosomal dominant diseases. *Methods Mol Biol* **252**, 221–236.

Gleave ME, Monia BP (2005). Antisense therapy for cancer. *Nat Rev Cancer* **5**, 468–479.

Jang JH, Lim KI, Schaffer DV (2007). Library selection and directed evolution approaches to engineering targeted viral vectors. *Biotechnol Bioeng* **98**, 515–524.

Lee JE, Choi JH, Lee JH, Lee MG (2005). Gene SNPs and mutations in clinical genetic testing: Haplotype-based testing and analysis. *Mutat Res* **573**, 195–204.

Li SD, Huang L (2006). Gene therapy progress and prospects: Non-viral gene therapy by systemic delivery. *Gene Ther* **13**, 1313–1319.

Loewen N, Poeschla EM (2005). Lentiviral vectors. *Adv Biochem Eng Biotechnol* **99**, 169–191.

Lu PY, Xie F, Woodle MC (2005). *In vivo* application of RNA interference: From functional genomics to therapeutics. *Adv Genet* **54**, 117–142.

Pelletier R, Caron SO, Puymirat J (2006). RNA based gene therapy for dominantly inherited diseases. *Curr Gene Ther* **6**, 131–146.

Warrington KH Jr, Herzog RW (2006). Treatment of human disease by adeno-associated viral gene transfer. *Hum Genet* **119**, 571–603.

Wilton SD, Fletcher S (2006). Modification of pre-mRNA processing: Application to dystrophin expression. *Curr Opin Mol Ther* **8**, 130–135.

Wu Z, Asokan A, Samulski RJ (2006). Adeno-associated virus serotypes: Vector toolkit for human gene therapy. *Mol Ther* **14**, 316–327.

Molecular Biology of Cancer

CANCER IS GENETIC IN ORIGIN

Both inherited diseases and cancers are genetic in origin. Inherited diseases are genetic defects that are passed on from one individual to another via the germline. Mutations must occur in the germline cells, which give rise to the eggs and sperm, to be passed on to the descendants of a multicellular organism. In contrast, each occurrence of cancer is limited to a single multicellular organism and is not passed on to the next generation. If a mutation does occur in somatic cells—those making up the rest of the body—a variety of possibilities may result (Fig. 18.1). A mutation that occurs early on in embryonic development may be highly detrimental, because

A) Somatic mutation

Early embryo

Somatic cells of body

Mutation

Germline cells (no defect passed on)

Eye cells

Liver cells

Muscle cells

Normal cells

Mutant spleen cells

Egg or sperm

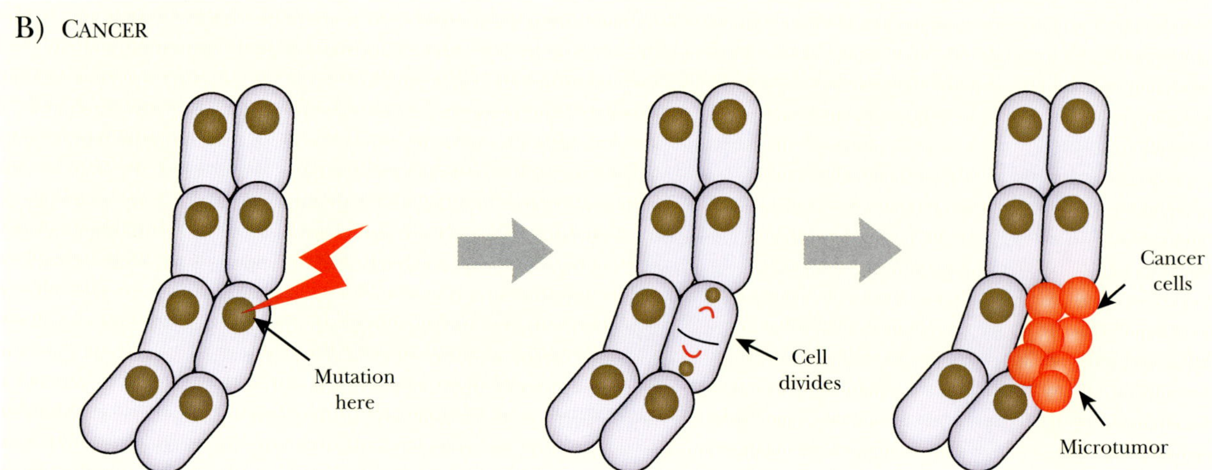

FIGURE 18.1 Cancer Is Caused by Somatic Mutations
(A) Early embryos have germline cells, which become eggs and sperm, and somatic cells that make up the rest of the organism. Depending on the timing, a somatic mutation usually affects only a small portion of the entire organism, such as the spleen. (B) Cancer occurs when a somatic cell mutates because of errors during DNA replication or exposure to a carcinogen. The mutation allows division in a cell that should no longer divide. The mutant cells keep dividing to form a microtumor.

each cell of the embryo gives rise to many cells during development. If the single precursor cell for a major organ or tissue suffers a serious mutation, the results may be serious or fatal. Most **somatic mutations** occurring later in development will affect only one or a few cells and will be of little major significance.

Some somatic mutations occurring after the organism has reached maturity are still dangerous. **Cancers** are the result of somatic mutations that damage the regulatory system controlling cell growth and division. Cancer often starts with a single cell starting to grow and divide again, long after it is supposed to have ceased (Fig. 18.1). Cancers develop in several stages and require multiple mutations. First, a cell's normal control of cell division is lost. Second, the abnormal (mutant) cell divides to form a microtumor. Many microtumors are formed throughout our bodies. Most of these are destroyed by the immune system, whereas others lie quiescent for many months to years.

Small microtumors of about a million cells lie dormant because they have no way of obtaining nutrients. But if further mutations occur, the tumor may begin to acquire blood vessels. The mutant cancer cells emit signaling molecules to the surrounding tissues, attracting vascular endothelial cells to organize into blood vessels, a process called **angiogenesis**. Once the microtumor gets its own blood supply, the tumor can continue to grow into a large mass. If this stays in one place, it is known as a benign tumor and can often be cut out by a surgeon, resulting in complete recovery. Finally, as a result of further mutations, a cancer may gain the ability to invade other tissues and form secondary tumors and is then said to be malignant. Cancers that have spread are much more difficult to cure.

Cancers are somatic mutations that disturb the control of cell division and cell death. Multiple somatic mutations are needed to form microtumors. Microtumors grow into large tumors when the tumor grows blood vessels.

ENVIRONMENTAL FACTORS AND CANCER

It is true that smoking cigarettes, harmful chemicals in our environment, and nuclear radiation all cause cancer. Nonetheless, all of these factors first act by causing mutations. Because cancers are primarily the result of mutations in somatic cells, those chemicals that cause mutations, that is, mutagens, can also cause cancer. Radiation that causes DNA damage similarly leads to mutations and cancers. Not every mutation will actually cause cancer. Most mutations do not even affect transcribed regions of DNA, and even if they do affect a particular gene, it is not likely to be involved in controlling cell division.

In practice, some cancer-causing mutations are due to environmental factors, whereas others occur spontaneously as a result of mistakes made during replication of the cell's DNA. Cancer-causing agents are often called **carcinogens**. Almost all carcinogens are also mutagens. Occasional discrepancies occur due to metabolism within the body, usually in the liver. Certain chemicals that do not react with DNA themselves may be altered by the body, giving rise to derivatives that do react with DNA. In this case, the original compound is by definition a carcinogen but not, strictly speaking, a mutagen.

Approximately 80% of cancers are derived from the epithelial cells that form the outer covering of tissues. Epithelial cells are the surface cells of the skin and are also found lining the intestines and the lungs. Because the outermost layers are constantly worn away, the underlying layers must keep dividing. Cells from tissues where cell division is rare only occasionally become cancerous. (Nerve and muscle cancers account for only 3% to 4% of the total.) In addition, the surface cells are much more likely to suffer exposure to dangerous chemicals and harmful radiation.

Carcinogens are cancer-causing agents. Mutagens are agents that change the genetic code in a cell. Almost all carcinogens are mutagens.

NORMAL CELL DIVISION: THE CELL CYCLE

To understand further how cancer occurs, we must consider the process of normal cell division. The eukaryotic cell cycle has four stages (see Chapter 4 and Fig. 4.9):

1. **G_1 phase**—the cell grows
2. **S (synthesis) phase**—the DNA and chromosomes are duplicated
3. **G_2 phase**—the cell grows and prepares to divide
4. **M (mitosis) phase**—the cell and its nucleus divide.

In addition, cells may exit from the growth and division cycle into the **G_0 phase**. Most nondividing cells are in G_0. Many of these will differentiate and rarely divide again, under normal circumstances (Fig. 18.2).

To move from one stage to another requires the permission of proteins called **cyclins**, one for each major stage. The cyclins act as security checkpoints. They monitor the environment and also check to make sure that the previous stage of the cell cycle has been finished properly before moving on. The cyclins work in conjunction with the **cyclin-dependent kinases (CDKs)**. When the cyclin for a particular step in the cell cycle senses that conditions are appropriate, it binds to the appropriate CDK (Fig. 18.3). This activates the CDK, which then adds phosphate groups to a series of other proteins. These are the enzymes and structural proteins that actually carry out the process of cell division. These proteins are on standby until the added phosphate group activates them. Several anti-oncogenes act by blocking the action of the cyclins (see later discussion).

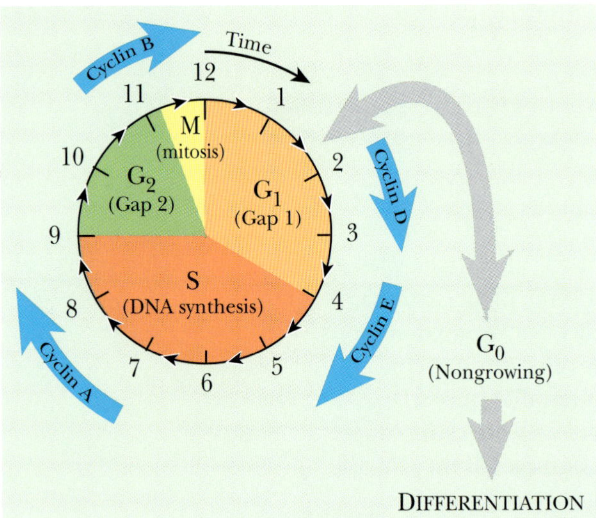

FIGURE 18.2
Eukaryotic Cell Cycle: Division versus Differentiation
The cell cycle normally consists of the four stages G_1, S, G_2, and M. However, if the conditions are right, rather than going from G_1 into the S phase, the cell may differentiate and enter G_0. If the cell does not differentiate, a signal is received from cyclin D and E, and the cell enters S phase and replicates its DNA. After about 5 hours another signal from cyclin A triggers the cell to enter G_2. After cyclin B becomes active, the cell enters mitosis and divides.

Perhaps the most critical checkpoint is the transition between the G_1 and S phases, which is controlled by two transcription factors, **E2F** and p53 together with the **pRB** protein (product of the retinoblastoma gene, an anti-oncogene). E2F promotes the expression of several genes involved in DNA replication. It also increases synthesis of the cyclins E and A that control the cell cycle beyond G_1 (Fig. 18.4). Binding to pRB inactivates E2F until the cell receives a signal from the outside. External **growth factors** cause synthesis of cyclin D. This activates CDK4, which in turn phosphorylates pRB. Phosphorylated pRB releases E2F, which is then free to activate its target genes. The negative side of this regulation system is dominated by p53 protein (see later discussion).

Eukaryotic cells use cyclins to control the progression through the stages of cell division. At each stage, the cell reaches a checkpoint to ensure all the proteins and DNA are in the correct position before moving to the next stage. E2F and pRB control the checkpoint from G_1 to S.

CELL DIVISION RESPONDS TO EXTERNAL SIGNALS

A large number of extracellular growth factors and hormones exist. Many are specific for particular cells or tissues; for example, epidermal growth factor (EGF) stimulates epithelial cells and fibroblast growth factor (FGF) stimulates muscle cells. Growth factors are bound by specific cell surface receptors. Binding activates the inner domain of the receptor, which is often a protein kinase itself or else activates an associated protein kinase (Fig. 18.5). Typically, the protein kinase activates a protein of the Ras family and this in turn stimulates a phosphotransfer cascade. This consists of three **mitogen-activated protein (MAP) kinases**. MAP kinase kinase kinase (MAPKKK) activates MAP kinase kinase (MAPKK), which activates MAP kinase (MAPK). Finally, MAPK phosphorylates transcription factors that activate genes needed early in cell division, including the gene for cyclin D.

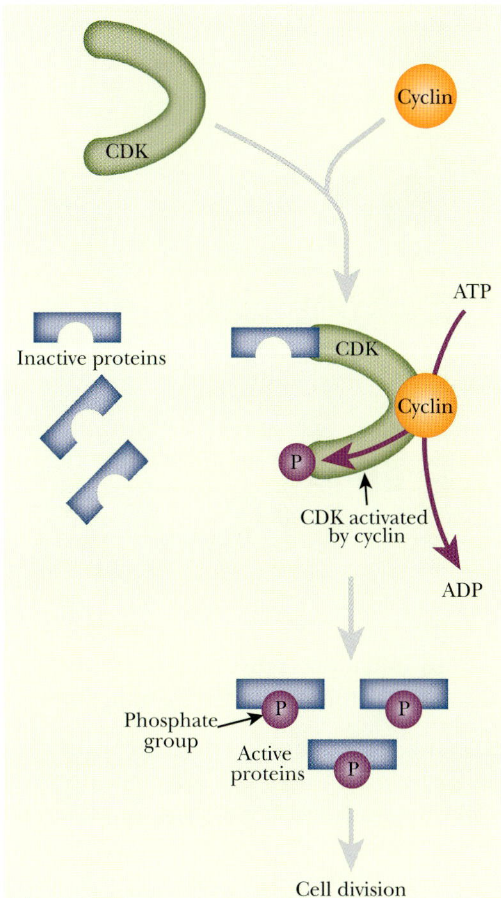

FIGURE 18.3 Cyclins Operate via Cyclin-Dependent Kinases
Before each cell can enter a new phase of the cell cycle, the cyclin must complex with a cyclin-dependent kinase (CDK). Addition of a phosphate from ATP activates the cyclin/CDK complex. The active complex then transfers the phosphate to other proteins that execute cell division.

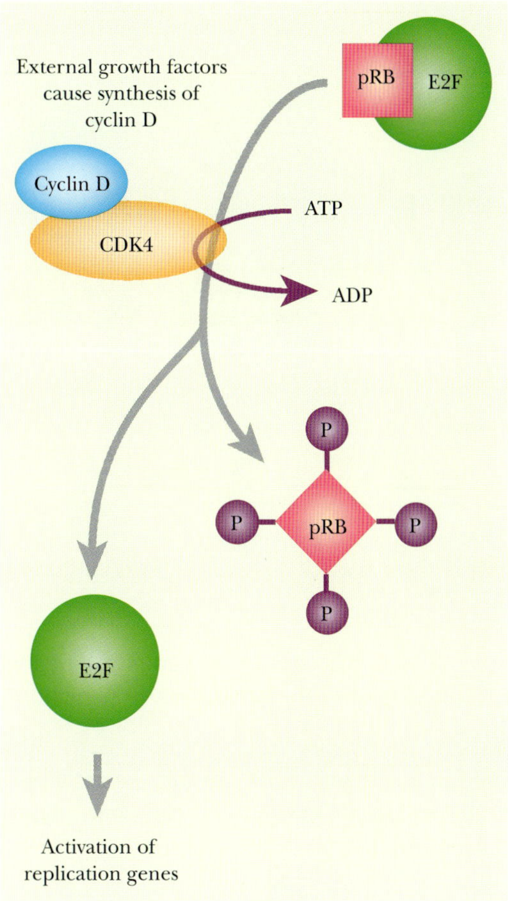

FIGURE 18.4 Control of G_1 to S Transition by pRB and E2F
At the end of G_1, cyclin D must be activated to initiate transition into S phase. This requires binding of cyclin D to its partner, CDK4. Once they are together, ATP is hydrolyzed and a phosphate is transferred to the cyclinD/CDK4 complex. This phosphate is then transferred to pRB, which releases E2F to transcribe genes needed to initiate DNA replication.

Cell division is activated by growth factors, which send a signal from the outside of the cell to the nucleus via a cascade of protein kinases.

GENES THAT AFFECT CANCER

It is important to distinguish two general types of genes that affect cancer. Some genes affect the susceptibility to develop cancer. Increased susceptibility to cancer may be inherited (see later discussion) just like any other genetic defect. In addition, there are two classes of genes directly involved in producing cancers as the result of somatic mutations. These are the **oncogenes** and the **tumor-suppressor genes** or **anti-oncogenes**.

A major difference between oncogenes and anti-oncogenes depends on the fact that animals are diploid and possess two copies of each gene. Oncogenes are cancer-causing genes, and mutant oncogenes promote the development of cancer cells. Mutations in oncogenes are dominant, and so a single oncogenic mutation in just one of a pair of genes is sufficient to give an effect. The second, wild-type copy of the gene cannot make up for the defect.

In contrast, tumor suppressor genes have a negative effect on cancer development. As their name suggests, they normally suppress division of cancer cells. To allow cancers to grow, both

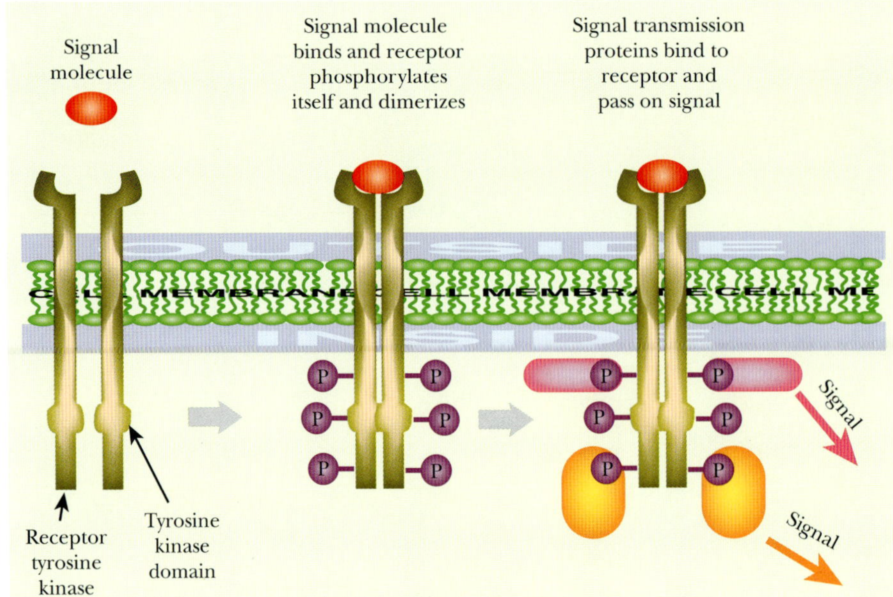

FIGURE 18.5
Activation of Growth Factor Receptor
Growth factor receptors have three domains, the extracellular binding site, the transmembrane region, and the intracellular tyrosine kinase domain. When a signal molecule binds to the extracellular domain, two receptor molecules bind together. This prompts the tyrosine kinase domains to activate each other by cross phosphorylation. The phosphates are then used to activate downstream signaling molecules, which in turn activate the transcription factors for cell growth.

copies of a tumor suppressor gene must be inactivated by mutation. A defective mutation in just one copy of a tumor suppressor gene has no effect; that is, these are recessive mutations.

Because cancers are due to mutations, the techniques used in genomic analysis are widely used in analyzing cancers at the genetic level. These include the use of DNA sequencing, PCR and microarrays, and have already been described in Chapter 4, DNA Synthesis, and Chapter 8, Genomics.

Two types of genes can cause cancer, oncogenes and tumor-suppressor genes. Oncogenes form when a single dominant mutation occurs in a proto-oncogene. Tumor-suppressor genes protect cells from cancer, and when both copies of these genes are mutated, the cell is no longer protected from cancer.

ONCOGENES AND PROTO-ONCOGENES

Oncogenes were first discovered on cancer-causing viruses, but they are found in all normal cells as well. The original, unmutated wild-type allele of an oncogene is sometimes called the **proto-oncogene**. The wild-type proto-oncogene promotes growth and division of the cell. During development of a multicellular organism, cell division must be closely controlled. Once an organ or tissue has reached its correct size, it should stop growing. In other words, its cells should stop dividing. Clearly, mutations in genes that control cell division are potentially very dangerous. These mutant versions are the cancer-causing oncogenes (Fig. 18.6).

The genomes of cancer-causing viruses also contain oncogenes. However, these oncogenes were originally derived from those of the host cells that the viruses infect. Certain types of viruses occasionally pick up cellular DNA and incorporate it into the viral genome. Sometimes they pick up an oncogene and the result is a cancer-causing or **oncogenic virus** (Fig. 18.7). The virus-borne version of an oncogene is sometimes written ***v-onc*** to distinguish it from the cellular version, ***c-onc***. Although quite a few cancer viruses are known, most human cancers are not due to viruses, but are due to new mutations of cellular proto-oncogenes to the oncogene form.

Oncogenes are mutated forms of the proto-oncogene, which is the term for the original unmutated gene.
Oncogenic viruses acquire oncogenes from one host cell and pass the oncogene onto other hosts during infection.

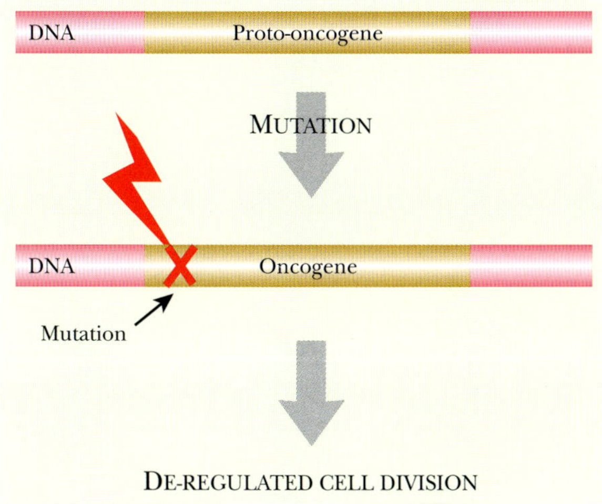

FIGURE 18.6 Oncogenes Are Mutant Alleles of Proto-Oncogenes
Within our DNA are genes called proto-oncogenes, which usually control cell division. When these proto-oncogenes mutate the cell no longer regulates cell division, and may ultimately grow into a tumor.

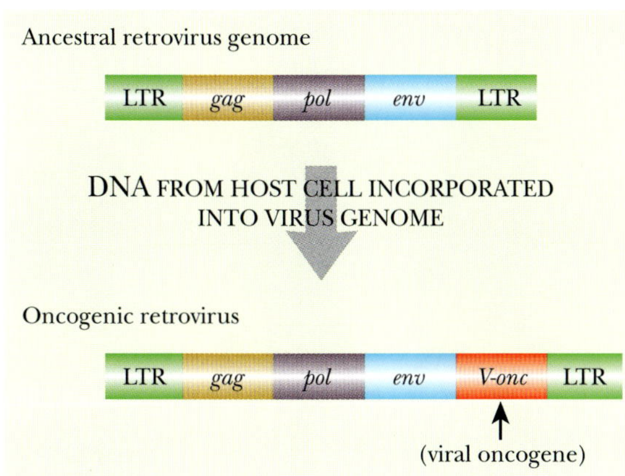

FIGURE 18.7 Acquisition of Viral Oncogene from Host Cell
Comparison of an ancestral viral genome with its modern equivalent shows that new genes have been added. In this example, the virus picked up the *v-onc* gene, which is a copy of a proto-oncogene in the host cell DNA. When the virus enters a new host, overexpression of the virus-borne oncogene deregulates cell division in the host, ultimately causing cancer.

DETECTION OF ONCOGENES BY TRANSFORMATION

If DNA is extracted from cancer cells and is then inserted into healthy cells, it may change them into cancer cells. This is known as **transformation**. (Note that *transformation* has a related but different meaning in bacterial genetics, where it refers to the uptake and incorporation of any foreign DNA.) Thus, if the presence of an oncogene is suspected, this can be tested by adding a sample of the suspect DNA to suitable cells in culture.

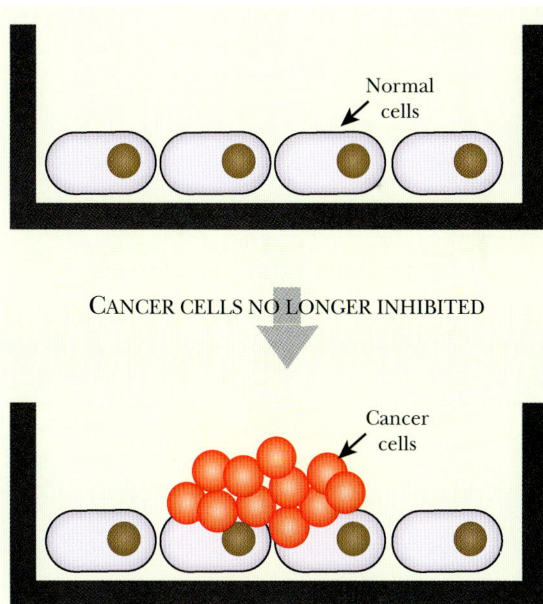

FIGURE 18.8 Contact Inhibition Disappears in Cancer Cells
In culture, normal cells grow until they touch other cells on all sides. They then stop dividing. If mutations are present that affect cell division genes, the cells keep dividing, forming small microtumors in the dish.

Normal cultured animal cells usually grow as a thin monolayer on the surface of a culture dish. They do not normally crawl on top of each other or pile up into heaps (Fig. 18.8). Once the available surface is covered and they are touching neighboring cells on all sides, they stop dividing, a phenomenon known as **contact inhibition**. In contrast, cancer cells are uninhibited and they continue to divide and to aggregate into heaps (see Fig. 18.8). Such miniature heaps of cells can be seen when DNA containing an oncogene is added to normal cells in culture. These heaps may be regarded as tiny tumors. If cells from these heaps are injected into an experimental animal such as a mouse, a real tumor will form (Fig. 18.9).

A related feature of cancer cells is *anchorage independence.* Unlike most normal cells, cancer cells can grow and divide in the absence of binding to proteins of the extracellular matrix. As a consequence, such cancer cells can move around and proliferate in "incorrect" locations in the body. This behavior is especially prominent in malignant (as opposed to benign) tumors.

Normal cells that are cultured *in vitro* form a single layer on a dish and stop growing. Cancer cells continue to grow and are not inhibited by contact.

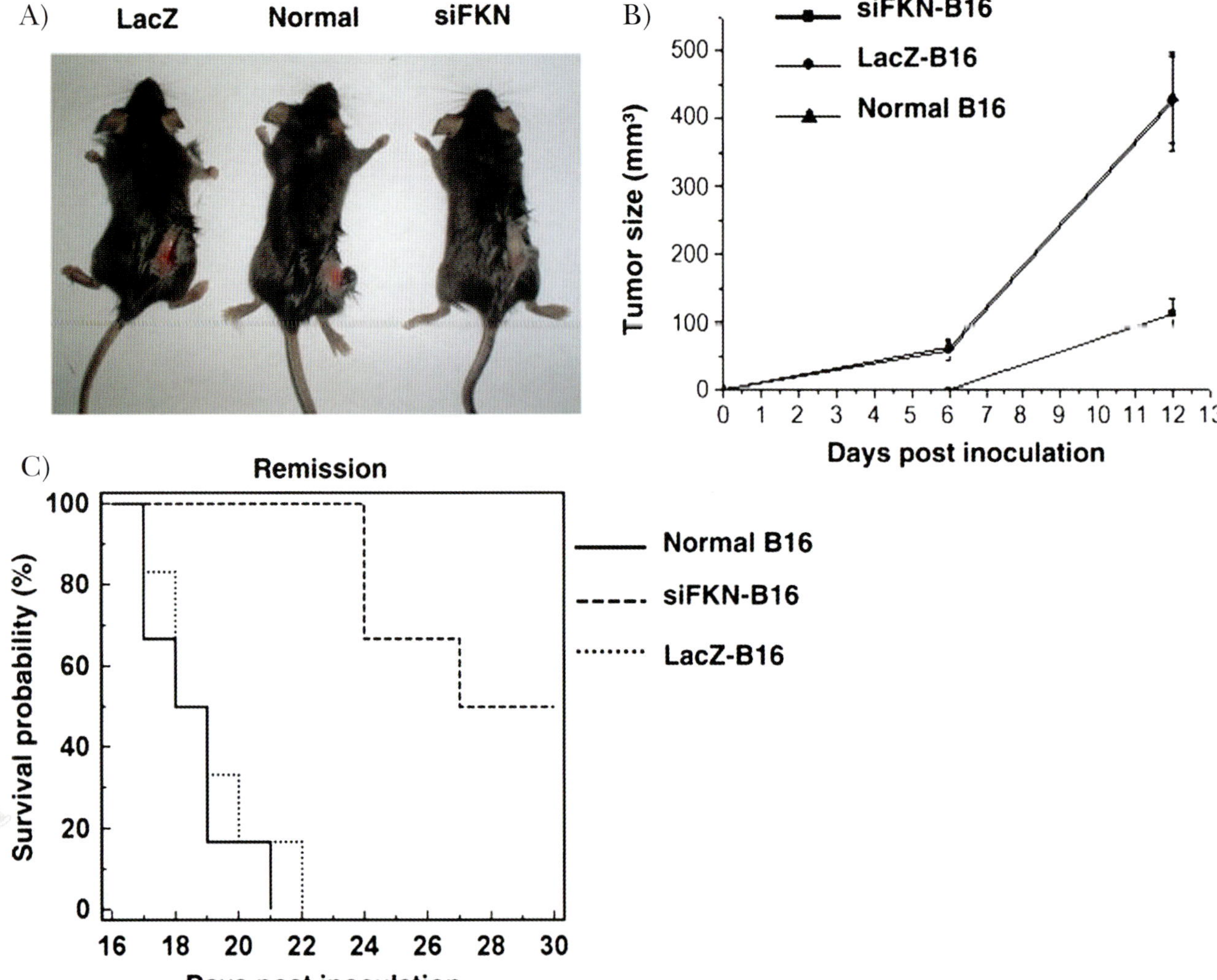

FIGURE 18.9 Human Cancer Cells Cause Tumors in Mice

Mouse melanoma cell line B16-F0 cells were grown *in vitro* with DMEM medium and fetal bovine serum, and then 1×10^6 live cells were injected into the right-side flank of C57BL/6 mice. Tumors with blood vessels developed within 3 weeks (middle mouse). In the control mouse on the left, the B16-F0 cells were transfected with a plasmid vector containing the gene for β-galactosidase before implantation in the mouse. In the mouse on the right, the B16-F0 cells were transfected with a plasmid vector containing a short-interfering RNA (siRNA) that blocks the expression of CX_3CL1/fractalkine, a cell adhesion molecule. The tumor is much smaller and has much less vascularization when the adhesion protein is blocked. From: Ren, Chen, Tian, Wei (2007). *Biochem Biophys Res Commun* **364**, 978–984. Reprinted with permission.

TYPES OF MUTATIONS THAT GENERATE ONCOGENES

The mutations that create oncogenes are dominant and result in increased cell division. This is because oncogenic mutations do not result from loss of activity but from increased activity of the product encoded by the oncogene. Such gain-of-function mutations may be due either to alteration of the protein encoded by the proto-oncogene or to its increased production (Fig. 18.10). Consequently, the wild-type duplicate copy of the proto-oncogene cannot overcome the effect of the oncogene. Like any other gene, a proto-oncogene has a regulatory region plus a structural region that encodes a protein. Some oncogenes are the result of changes in the regulatory region that increase gene expression. Others are due to mutations in the structural portion of the proto-oncogene that produce a protein with increased activity.

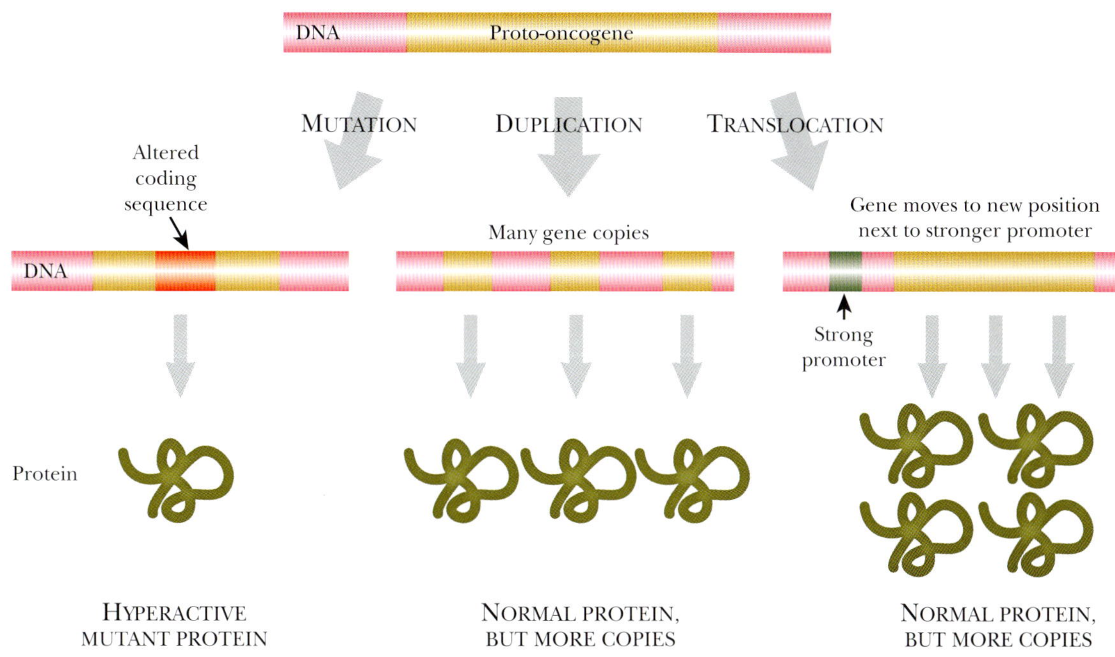

FIGURE 18.10 Oncogenic Mutations Result in Increased Activity
Three different possible mutations can cause a proto-oncogene to become oncogenic. First, the actual protein sequence may be altered by a mutation, resulting in a hyperactive protein. Second, the entire proto-oncogene may be duplicated one or more times. The extra copies cause an abnormally high amount of protein to be produced. Third, the proto-oncogene may be moved next to a strong promoter due to chromosomal rearrangements. The new promoter increases gene expression and hence the level of protein is increased.

The pathway for activating cell growth and division has several stages, and proto-oncogenes encode the proteins taking part in this scheme (Fig. 18.11). Not surprisingly, mutations that result in hyperactivation of any of these components can turn proto-oncogenes into oncogenes. The major components involved in cell division are:

1. Growth factors: These are proteins or other chemical messengers circulating in the blood that carry signals for promoting growth to the cell surface.
2. Cell surface receptors: These proteins are found in the cell membrane where they receive chemical messages from outside the cell. They pass the signal on, often by activating other proteins, such as G-proteins.
3. Signal transduction proteins: These pass on the signal from outside the cell to proteins or genes involved in cell division. Many of these are protein kinases that activate or inactivate other proteins by addition of a phosphate group.
4. Transcription factors: These proteins bind to and switch on genes in the cell nucleus. This results in the synthesis of new proteins, as opposed to the activation of those already present.

Proto-oncogenes become oncogenes after a mutation makes the protein more active, the gene is duplicated, or a mutation in the promoter increases the expression of the protein.

Cellular growth is signaled by a growth factor that binds and activates its cell surface receptor. The intracellular portion of the receptor activates intracellular proteins that move to the nucleus. Inside the nucleus, the proteins activate a transcription factor that turns on genes for growth.

THE *RAS* ONCOGENE—HYPERACTIVE PROTEIN

Ras protein transmits signals concerning cell division in humans, flies, and even yeast cells (Fig. 18.12). Growth signals are received from outside the cell by receptors at the cell

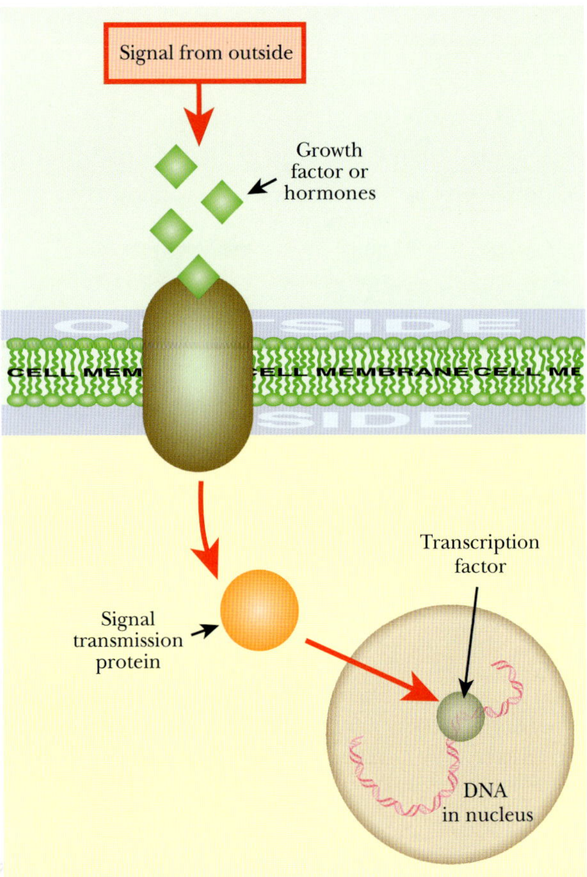

FIGURE 18.11 Components That Promote Cell Growth and Division
The execution of a cell growth signal involves a variety of steps. First, the signal to grow involves the production of growth factors. These bind and activate a cell surface receptor, which in turn activates a variety of intracellular proteins that transmit the signal from the cell membrane to the nucleus. Inside the nucleus, a transcription factor activates genes necessary for cell division and growth.

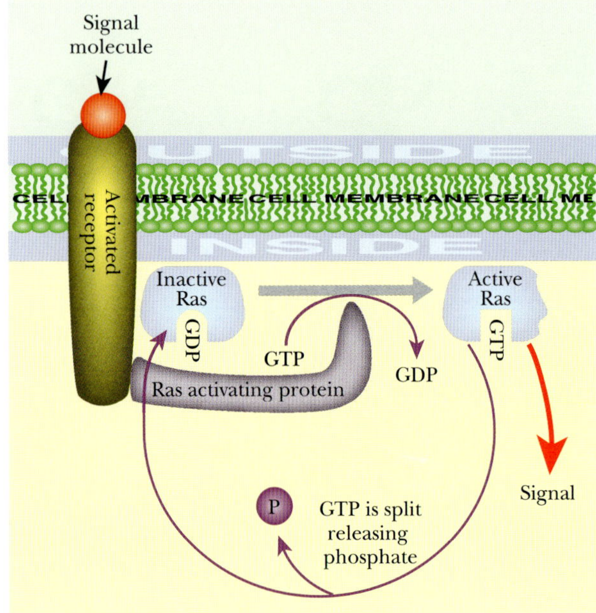

FIGURE 18.12 Function of Ras Protein
When a cell surface receptor receives a signal to grow, it transmits the signal to Ras. Activated Ras binds GTP and sends the growth signal to the nucleus. After the signal has been sent, GTP is hydrolyzed to GDP and Ras becomes inactive once again.

surface. The activated receptor transfers the signal to intracellular Ras protein. After receiving a signal, normal Ras protein binds guanosine triphosphate (GTP) and enters signal-emitting mode. After transmitting a brief pulse of signals, Ras then splits the GTP into guanosine diphosphate (GDP) plus phosphate and relapses into standby mode again. The cancer-causing form of the Ras protein is locked permanently into the signal-emitting mode and never splits the GTP. Therefore it constantly floods the cell with signals urging cell division, even when none is received from outside.

The ***ras* oncogene** is the result of a single base change in the structural region of the gene. This causes an alteration in a single amino acid in the encoded protein (Fig. 18.13). Most *ras* mutations alter the amino acid at position 12; others affect position 13 or 61. Only a few very specific mutations can create a *ras* oncogene from the proto-oncogene. The 3D structure of the Ras protein has been solved by x-ray crystallography. Those few amino acid residues that are changed by oncogenic mutations are all directly involved in the binding and splitting of the GTP. The consequence of hyperactivation of Ras is uncontrolled cell division and the beginnings of a possible cancer. Mutations of *ras* are frequently detected and have been analyzed in detail, in cancers of the lung, colon, pancreas, and thyroid.

Ras becomes an oncogene when a mutation alters amino acids at position 12, 13, or 61. The oncogenic form of ras protein continually sends signals to the nucleus for growth. Normal Ras protein signals growth only when a growth factor binds to its cell surface receptor.

THE *MYC* ONCOGENE—OVERPRODUCTION OF PROTEIN

Some oncogenes are due to mutations that alter the structure of a protein such as Ras. Instead of a hyperactive mutant protein, many other oncogenes suffer changes that vastly increase the amount of the protein formed, although the protein itself is not changed.

A well-known example is the ***myc* oncogene**, which encodes a transcription factor involved in switching on several other genes involved in cell division. A **Myc protein** overdose can occur in two ways. Some *myc*-dependent cancers result from chromosomal changes in which the *myc* gene is duplicated many times. Instead of the normal two copies, 50 to 100 copies may occur because of mistaken duplication of the segment of DNA carrying the *myc* gene (Fig. 18.14). The Myc protein will then be overproduced by 50- to 100-fold, too.

Alternatively, the number of copies of the *myc* gene may remain unchanged, but their regulation may be altered. In Burkitt's lymphoma, a rare chromosomal translocation swaps segments of two unrelated chromosomes. This separates the *myc* structural gene from its own regulatory region and fuses it to the highly active promoter of another gene (Fig. 18.15). The Myc protein is now produced continuously in substantial amounts instead of being strictly regulated as before.

Mutations in *myc* are among the most common in mammalian cancers. Most types of human cancers show overexpression of Myc, although the frequency differs considerably from one type of tumor to another. In particular, mutations in *myc* tend to occur during the later progression of many cancers, including those of lung, breast, and ovary.

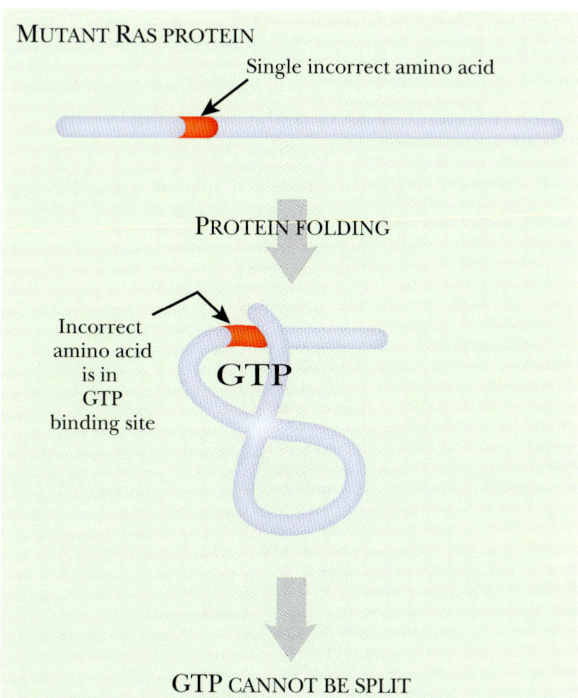

FIGURE 18.13 Mutations Creating the *ras* Oncogene Are Highly Specific
The transition from the normal proto-oncogene to the *ras* oncogene involves a single amino acid substitution. This amino acid change prevents the hydrolysis of GTP by Ras; therefore the protein is always in its signal-emitting mode.

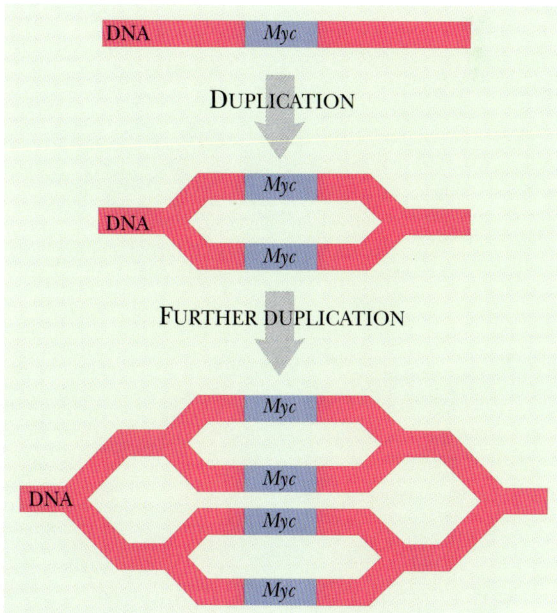

FIGURE 18.14 Overproduction of Myc by Gene Dosage Effect
Normally, the gene for Myc is present in two copies. Because of aberrant chromosomal replication, the *myc* gene may be duplicated. Once this occurs, the number of *myc* oncogenes grows exponentially. Since all these copies are expressed, the amount of Myc protein increases and cancer results.

FIGURE 18.15 Overproduction of Myc by Changing Promoters
Chromosomal translocation of the *myc* oncogene may enhance the amount of protein by fusing the structural portion of the gene to a highly active promoter.

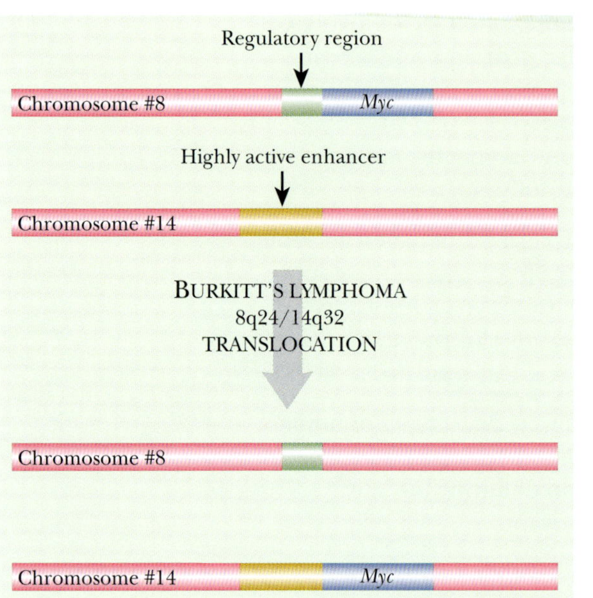

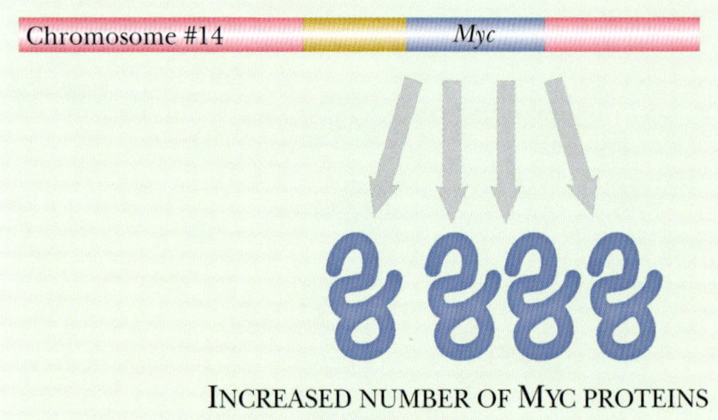

Myc is a common oncogene. In some cancers, the number of *myc* genes increases due to aberrant DNA replication. In other cancers, the *myc* gene is translocated to a strong promoter. Both mutations increase the amount of Myc protein, resulting in too much cell growth.

TUMOR-SUPPRESSOR GENES OR ANTI-ONCOGENES

Tumor-suppressor genes, or anti-oncogenes, normally suppress cell division. Consequently, it is necessary to inactivate both copies in order to initiate cancerous growth. A mutation in just one copy of a tumor-suppressor gene has no effect, that is, these are recessive mutations. When both copies of a gene have been inactivated by null mutations, this is known as the **nullizygous** state. The two most common anti-oncogenes are Rb and p53 (see later discussion). Almost all human tumors inactivate either Rb or p53 and often both.

There are two ways to end up with both copies of a gene inactivated (Fig. 18.16). First, during division of the cells that form the body, two successive somatic mutations may occur. First one copy of the gene is inactivated, and then a second mutation strikes the second copy of the same gene. Although this seems highly unlikely, it will occasionally happen given enough time and a sufficiently high number of cells.

The second route is more frequent. The first mutation occurs in one copy of the gene in a germline cell of an ancestor. If this defective copy is passed on, a newborn individual will start life with one copy of the gene already inactivated in every cell. A somatic mutation that inactivates the second copy may then occur during cell division. In other words, one defective allele is inherited and the other is acquired by somatic mutation. Individuals who inherit a single defective anti-oncogene do not always develop cancer. What they inherit is an increased chance of doing so.

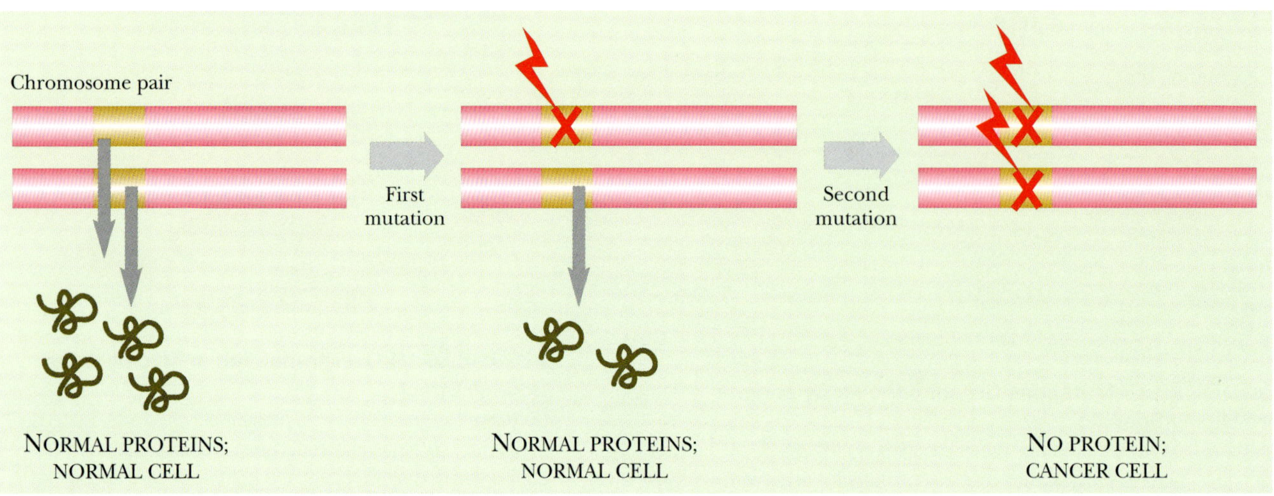

FIGURE 18.16 Mutations in Tumor-Suppressor Genes Are Recessive
In a normal situation, two wild-type anti-oncogenes are present. Two mutation events must occur to completely inactivate the anti-oncogene.

This scheme applies to a dozen or more tumor-suppressor genes. Some of them are involved in a wide range of cancers, such as the **retinoblastoma (Rb) gene**, originally discovered and named for a rare cancer of the retina of the eye, but also involved in many other tumors. Other anti-oncogenes are very tissue specific and inactivation of both copies triggers cancer of a particular organ, such as the Wilms' tumor gene responsible for a kidney cancer.

Anti-oncogenes encode proteins whose normal role is to inhibit growth or prevent cell division. Some signals that promote growth and division come from outside the cell. In other words, hormones circulating in the blood control growth and development. Some proteins encoded by anti-oncogenes are involved in detecting these external hormonal signals and blocking their effect on cell division (Fig. 18.17). An example is the PTEN protein that antagonizes insulin signaling and thus restricts cell growth and protein synthesis.

More recently, evidence has accumulated that in addition to external control, many somatic cell lines are also preprogrammed. They have a preset number of allowed cell divisions. An internal "generation clock" of some sort counts off the number of permitted divisions left, and when it reaches zero, growth and division stop. Some anti-oncogenes are part of this system, and when they are defective the cell fails to stop dividing. Indeed, many of the proteins encoded by anti-oncogenes are DNA binding proteins, often with zinc fingers.

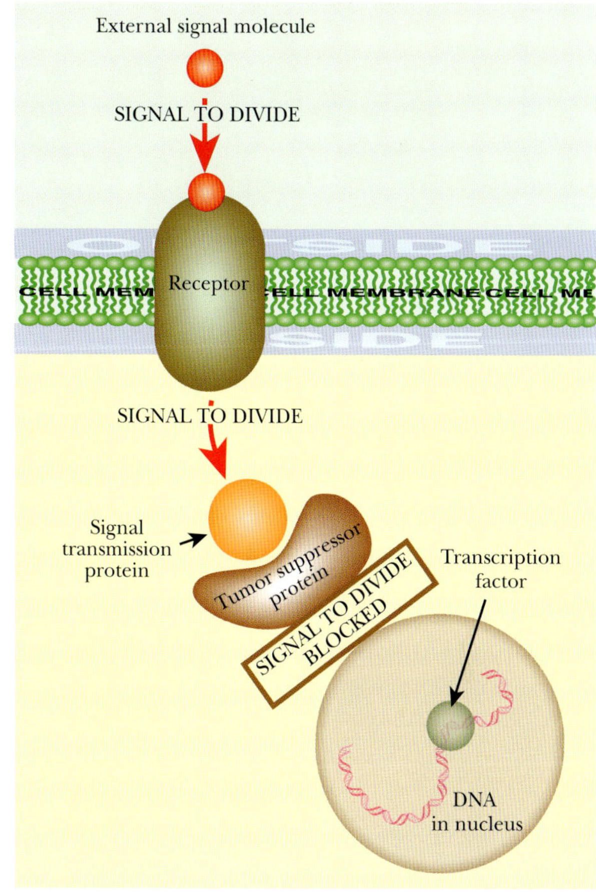

FIGURE 18.17
External Signal for Cell Division
Because growth-promoting signals often circulate in the blood, most cells in the body will be exposed to the growth cue. Only some cells divide, because proteins encoded by anti-oncogenes block transmission of the external stimulus. Anti-oncogene proteins may bind to signal transmission proteins and prevent them from activating transcription factors.

> Tumor-suppressor genes require a mutation in both copies in order to cause cancer. Sometimes both mutations occur in the same cell at different times. Other cancers form because the offspring received one mutated copy of the gene from the parent and therefore need only one mutation to occur.

THE p16, p21, AND p53 ANTI-ONCOGENES

Best known of the human anti-oncogenes is the infamous ***p53* gene**, found on the short arm of chromosome 17. The **p53 protein** (also known as TP53) is a DNA binding protein that, together with proteins p16 and p21, acts as an emergency braking system for the cell cycle. The *p53* gene is involved in a very large number of diverse cancers because its behavior differs from that of a standard anti-oncogene. Mutant alleles of typical anti-oncogenes are recessive and a single mutation does not allow cell division. In contrast, a single defective p53 allele does show such effects, even in the presence of a second, normal copy of the gene. Thus p53 mutations are dominant negatives. Well over half of all human cancers are defective in p53.

The reason for this aberrant behavior is the formation of mixed tetramers. The protein encoded by the *p53* gene assembles into groups of four (Fig. 18.18). When a cell has one good copy of the *p53* gene and one bad copy, it will produce a mixture of functional and defective p53 protein subunits. These will assemble into mixed tetramers, and even if a tetramer contains some good subunits, a single defective subunit will cripple the whole assembly. The likelihood of assembling a wholly good

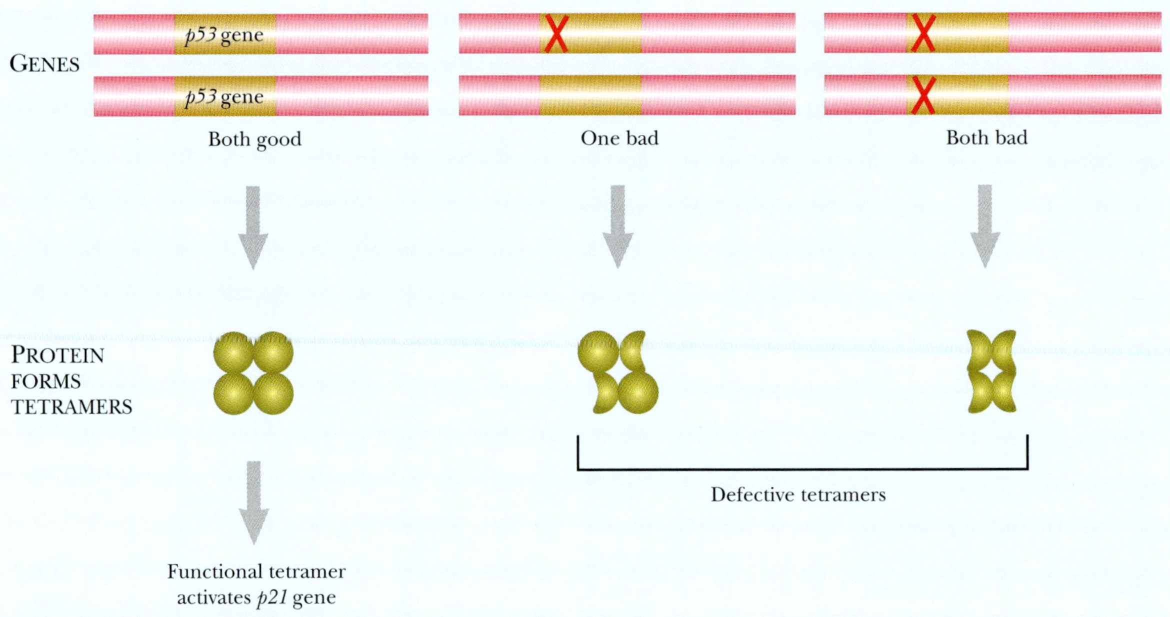

FIGURE 18.18 Mixed Tetramers of p53 Protein
Mutations in the *p53* gene are detrimental whether one or both copies are mutated. p53 protein works as a tetramer, so even if only one of the alleles has a mutation, most of the tetramers are defective. p53 only functions correctly if all four subunits of the tetramer are normal.

tetramer from such a 50:50 mixture of functional and defective subunits is $\frac{1}{2} \times \frac{1}{2} \times \frac{1}{2} \times \frac{1}{2} = \frac{1}{16}$. Thus, only one-sixteenth of the p53 protein assemblies will function correctly even though half of the individual protein subunits are normal.

If a cell's DNA is damaged in any way, the normal p53 protein activates the gene for p21. The **p21 protein** then blocks the action of all of the cyclins and freezes the cell wherever it is in the cell cycle until the damage can be repaired (Fig. 18.19). The p16 protein acts similarly, but just blocks cyclin E. It stops division at the critical point just before the cell enters the S-phase in which the DNA is replicated.

FIGURE 18.19 p53 and p21 Block the Cell Cycle
When a cell senses DNA damage, the p53 protein forms active tetramers that bind to the control region of the *p21* gene. p53 stimulates the transcription and translation of p21 protein. The p21 protein then binds to and inactivates the cyclins, preventing progression of the cell cycle.

The p53 protein is not necessary for normal cell division. Mice with both copies of the *p53* gene knocked out grow normally to start with. However, they all die of cancer after 3 or 4 months. The role of p53 is to shut down cell division in DNA emergencies, such as ultraviolet radiation damage. The p53 pathway also responds to a shortage of nucleotides or a lack of oxygen. In severe cases, p53 may initiate programmed cell death, via Bax protein (see Chapter 20) rather than merely arresting cell division (Fig. 18.20). The p53 protein is inactive when first synthesized and cannot bind to its recognition sequence in the upstream region of those genes it controls. It may be activated to form the tetramer by binding to single-stranded DNA or by phosphorylation.

Both alleles of the *p53* gene are expressed into protein so when one allele has a mutation, the p53 protein tetramer is not active. This is one example where a single mutation in a tumor-suppressor gene can cause cancer.

p53 protein prevents defective cells from progressing through the cell cycle and, in extreme damage, triggers the cell to commit suicide via apoptosis.

FORMATION OF A TUMOR

The actual generation of a real tumor requires several steps. In practice multiple somatic mutations are necessary for the production of most cancers (Fig. 18.21). For example, many colon cancers carry the following defects:

1. Inactivation of the *APC* anti-oncogene (both copies)
2. Activation of the *Ras* oncogene
3. Inactivation of the *DCC* anti-oncogene (both copies)
4. Mutation of a single copy of the *p53* gene

Cancer development needs half a dozen mutational steps before a full-blown tumor results. Even then, the cancer cells will all stay in one place,

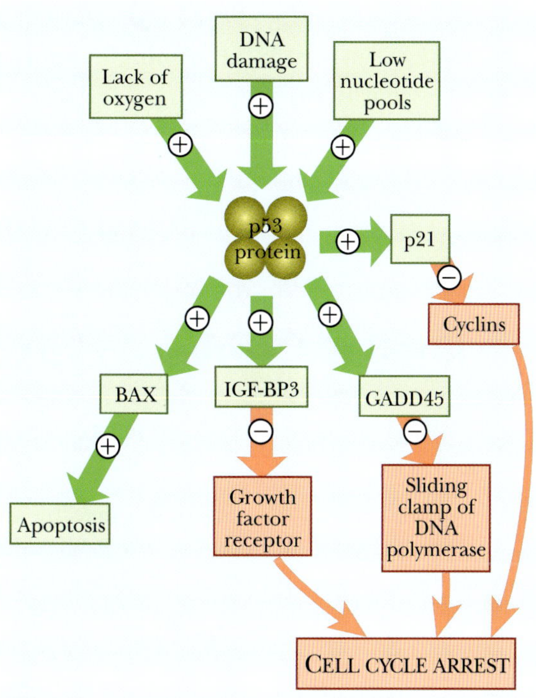

FIGURE 18.20
Central Role of p53
Three possible cues activate p53 protein: lack of oxygen, DNA damage, and low nucleotide pools. Activated p53 affects many different targets. If there is severe damage, p53 activates Bax protein and initiates programmed cell death. Otherwise, p53 activates cell cycle arrest by turning on p21, which in turn blocks the cyclins. Active p53 can also block synthesis by DNA polymerase through the action of GADD45. Active p53 also binds to growth factor receptors to block any further growth signals.

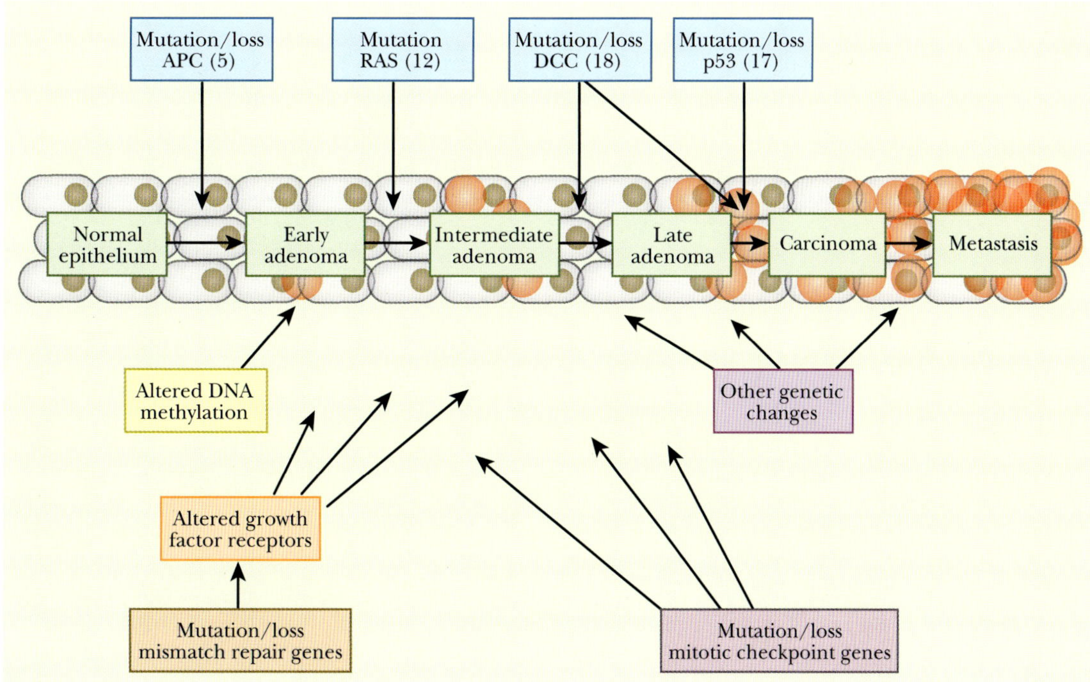

FIGURE 18.21 Steps in Tumor Development and Metastasis
The complexity of cancer is demonstrated by the number of mutations necessary before a metastatic colon cancer emerges. First, the loss of *APC* on chromosome 5 begins the process, followed by mutations in *ras*, inactivation of *DCC*, and finally a defect in *p53*. Chromosome locations of genes are shown in parentheses. Each mutation leads the cells further from normality, causing them to become more and more dysfunctional.

and if the tumor is cut out by surgery, all may still be well. However, cancers do not stay put forever. Eventually they bud off cancer cells that travel around the body, settle down in other tissues, and grow into secondary tumors. This is known as **metastasis**, and once things reach this stage it is virtually impossible to find and remove all of the cancers. Other mutations, which aren't fully understood, are necessary for cancer cells to start traveling. These include mutations that result in:

1. Loss of adhesion to neighboring cells in the home tumor
2. Ability to penetrate the membranes surrounding other tissues
3. Vascularization of the tumor, which not only provides nutrients but also allows mobile cancer cells access to the circulatory system

A cancerous tumor requires multiple genetic mutations to form. For it to become metastatic, even more mutations are required.

INHERITED SUSCEPTIBILITY TO CANCER

It is thought that 5% to 10% of cancers may be largely due to inherited defects. Many of the genes involved in this are poorly understood, but they may be divided into three general categories.

First, as we have already discussed, it is possible to inherit one defective copy of an anti-oncogene. In this case every somatic cell starts life with one faulty copy and only a single somatic mutation is needed to completely inactivate the pair of anti-oncogenes. (Note: Inheriting two defective copies of an anti-oncogene is normally lethal. Artificially engineered mice that are doubly negative for such genes generally die before birth.)

Secondly, mutations in certain special genes affect the rate at which further mutations occur during cell division. Such genes are known as **mutator genes**. These include genes involved in DNA synthesis, such as the genes encoding DNA polymerase. Some mutator genes are more subtle and are involved in DNA repair. These have been analyzed in detail in bacteria, although less is known for humans. Nonetheless, it appears that certain inherited forms of colon cancer are due to defects in genes involved in mismatch repair. This, in turn, increases the rate of mutation in all other genes including the tumor-suppressor genes. Defects in mutator gene are generally recessive, like those in other tumor-suppressor genes.

Box 18.1 Cancer Revisionism and Lead Time Bias

Since the "War on Cancer" was started, most of the major advances have been in the diagnosis and understanding of cancer, rather than its cure. Because of the massive vested interests in the cancer research industry, cancer statistics are often presented in an overly optimistic way. One frequently used measure of misinformation is what is known as *lead time bias*.

Consider a patient who is diagnosed with a serious cancer and dies 5 years later. Now consider what happens if the ability to diagnose cancer improves and the same cancer is detected 5 years earlier. If treatment has not improved, the patient will die 10 years after diagnosis. In the second case, the patient survived for 10 years with cancer as opposed to only 5 years in the first case. However, no real increase has occurred in survival. The patient did not live any longer but was merely diagnosed earlier and had the burden of worrying about cancer for 5 years longer. By using this bogus measure of survival time, it is sometimes made to look as if great advances have been made in curing cancer. In fact (using data from 1950 to 1982) massive improvements in genuine life expectancy have been made in only a few relatively rare forms of cancer.

Inherited breast cancer falls into this category. Inheriting a single defective copy of either the ***BRCA1*** or ***BRCA2*** **genes** (breast cancer A genes) predisposes women to breast cancer. About 0.5% of U.S. women carry mutations in *BRCA1*, which also predisposes to cancer of the ovary. As with other tumor-suppressor genes, the second copy must mutate during division of somatic cells for a cancer to arise. Individuals with two defective *BRCA* alleles die as embryos. Both BRCA1 and BRCA2 proteins are involved in DNA repair. Both bind to RAD51, which takes part in mending double-stranded DNA breaks (Fig. 18.22). In addition BRCA1 has a dual role as a transcriptional regulator of other components needed for DNA repair.

Finally, there are indirect effects on cancer frequency due to genetic differences between races or within populations. For example, some skin cancers result from mutations caused by ultraviolet radiation from the sun. White people, especially those exposed to high levels of sunshine in the tropics or Australia, develop skin cancer much more often than darker-skinned people. The reason is obvious: the more pigment, the less UV radiation penetrating to your DNA.

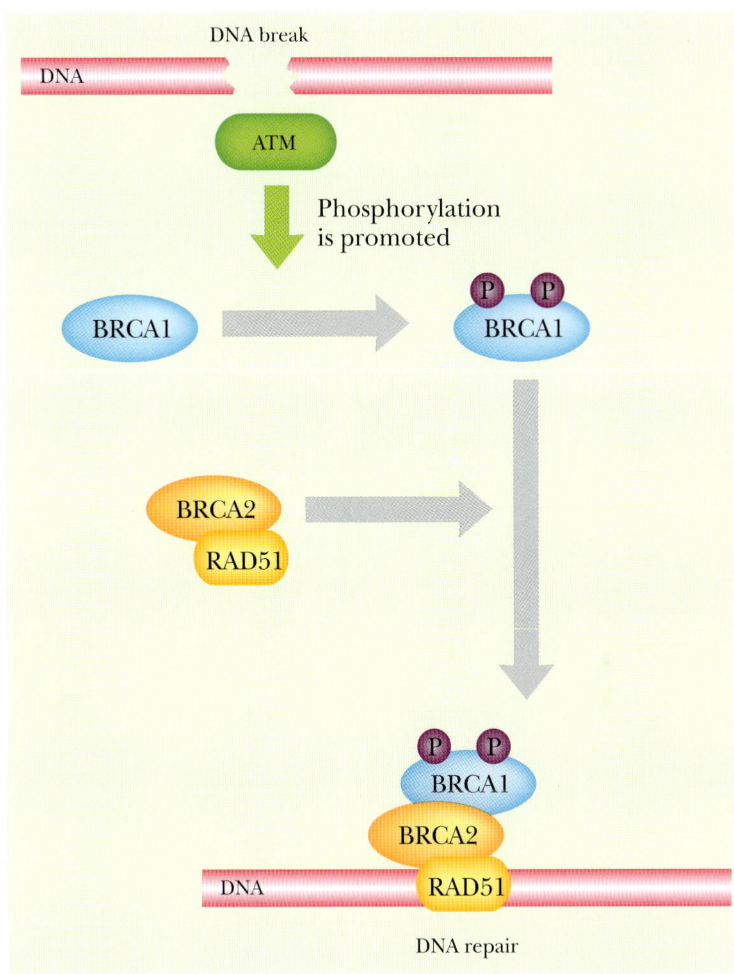

FIGURE 18.22
BRCA1 and BRCA2 Bind to RAD51
The presence of a double-stranded DNA break triggers the phosphorylation of BRCA1 via the ATM protein. BRCA2 plus RAD51 then bind to phospho-BRCA1 and take part in the DNA repair process.

Inherited defects in anti-oncogenes, components of DNA replication, and DNA repair systems can also predispose a person to cancer. Some genetic differences make people more susceptible to certain cancers.

CANCER-CAUSING VIRUSES

Although retroviruses that cause cancer in chickens and mice were important in the discovery of oncogenes, few human cancers are due to retroviruses. The most famous human retrovirus is AIDS (see Chapter 22). Although some AIDS patients die of cancers, the AIDS virus does not cause cancer directly. What it does is cripple the immune system that would otherwise kill off most cancer cells before they get too far out of control. Kaposi's sarcoma, often seen in AIDS patients, is caused by secondary infection with another virus, **human herpesvirus 8 (HHV8)**.

The first genuine cancer-causing retrovirus to be discovered was **Rous sarcoma virus (RSV)**. Long ago, the ancestor of this virus picked up a copy of the chicken *src* **oncogene**. The genome inside a retrovirus particle is RNA. When RSV infects chickens the RNA is converted into a DNA copy by reverse transcriptase (see Chapter 22). The viral DNA is then integrated into the host cell DNA along with the *src* oncogene that it carries. Expression of this copy of the *src* oncogene causes muscle tumors, known as **sarcomas**.

Only about 15% of human cancers are due to viruses, and most of these are due to DNA viruses (**papillomaviruses** and **herpesviruses**). The major human cancer-causing viruses are listed in Table 18.1. **Simian virus 40 (SV40)**, which causes cancer in monkeys, has been studied most. Such DNA tumor viruses act by blocking the cell's tumor-suppressor genes. First the DNA tumor virus integrates its DNA into the host cell's chromosome. Second, it makes a virus protein that binds to the cell's Rb and p53 proteins (Fig. 18.23). This activates cell division and may lead to a tumor.

Table 18.1 Oncogenic Viruses of Humans

Virus	Examples of Cancer
DNA viruses	
Human papillomavirus	Cervical cancer, skin cancer
Epstein-Barr virus	Burkitt's lymphoma, Hodgkin's lymphoma
Human herpesvirus 8	Kaposi's sarcoma
RNA viruses	
HTLV-1	Adult T-cell leukemia
Hepatitis B virus	Liver cancer

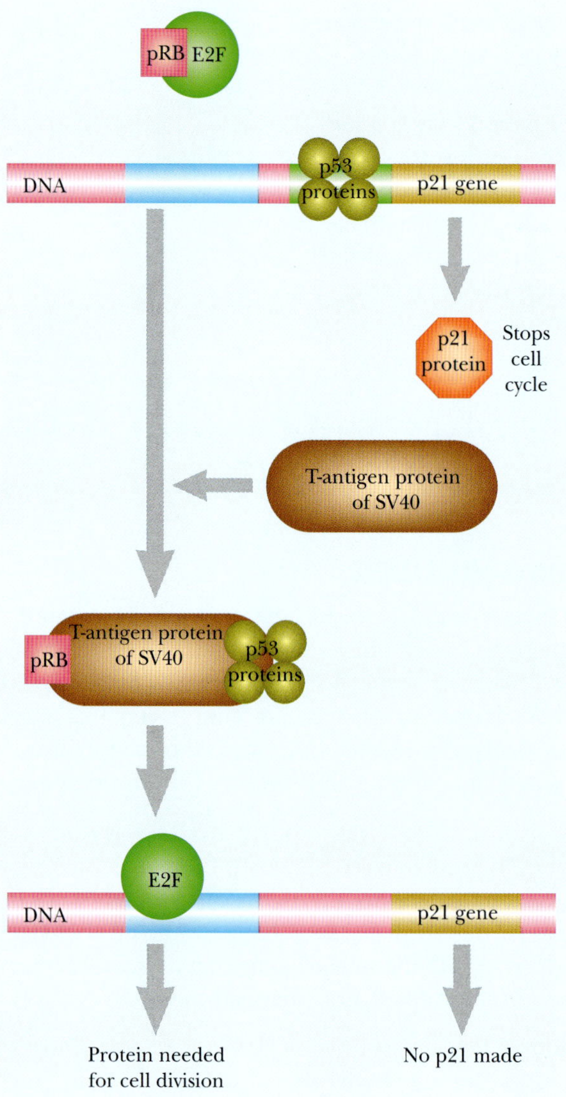

FIGURE 18.23 DNA Tumor Virus Inactivates pRb and p53
Normally, p53 and pRB act together to stop the cell cycle. If the cell is infected with SV40, the virus produces a protein called the T-antigen. This protein binds to pRB and p53, thus freeing E2F to stimulate cell division.

Human papillomaviruses act in a similar manner but usually produce only benign growths, that is, warts. Most papillomaviruses rarely cause dangerous tumors. However, there is a subgroup of human papillomaviruses that are sexually transmitted. These are responsible for virtually all human cervical cancers and, occasionally, cancers of other organs.

Only 15% of human cancers are caused by viruses; most of them are DNA viruses.

ENGINEERED CANCER-KILLING VIRUSES

Lytic viruses infect and kill their target cells (as opposed to integrating or lying latent). If lytic viruses happen to infect cancer cells, then they kill these, too. If such viruses could be made specific for cancer cells, they could provide an effective therapy against cancer.

Several viruses are being considered as the basis for engineered **oncolytic viruses**. At least for now, measles seems most convenient in many ways. Most people in advanced nations are already immunized against measles, providing a level of protection against nonspecific tissue infection and other problems.

Indeed, certain lineages of measles virus have inherent strong antitumor activity. However, they are not cancer specific and infect other cell types also. To make them cancer specific, it is necessary to change their receptor specificity. The natural receptors for measles are the human cell surface proteins CD46 and SLAM (signaling lymphocyte activation molecule; CD150) which are recognized by the H-protein of measles virus (Fig. 18.24).

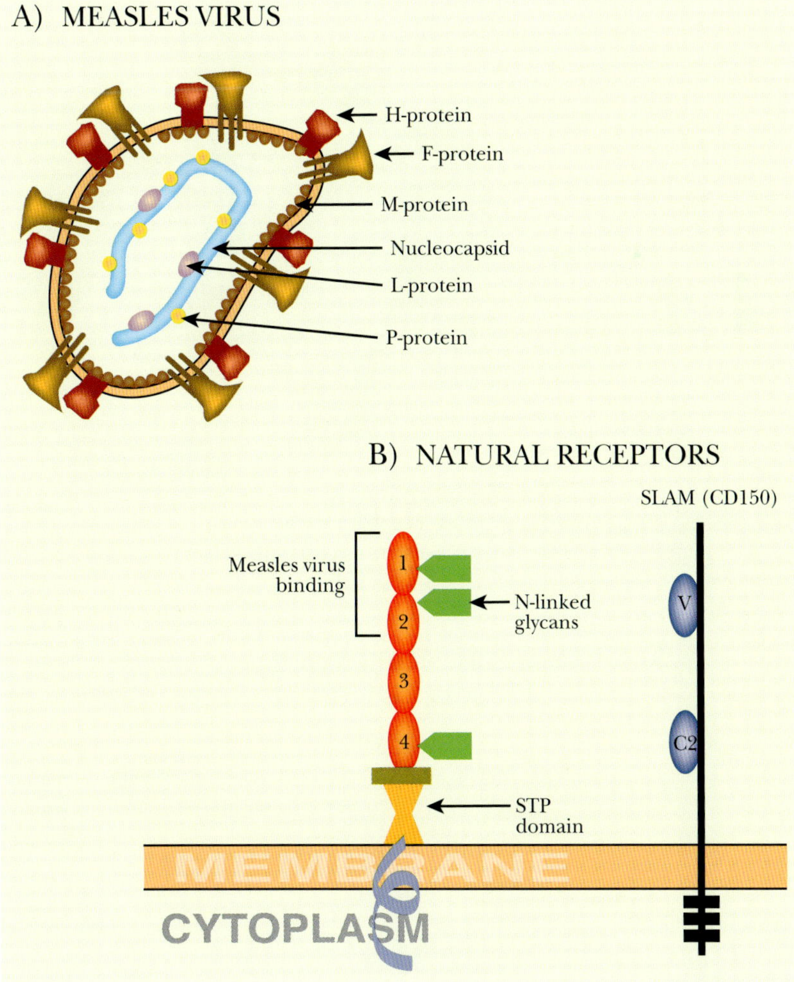

FIGURE 18.24
Natural Receptors for Measles Virus
Measles virus uses two cell surface proteins, CD46 and SLAM, as receptors. (A) Measles virus is an enveloped virus whose DNA genome is inside a nucleocapsid. The glycoproteins, H and F, are on the surface of the virus particle and are required for entry into the host cell. The internal L and P proteins take part in virus replication, whereas M is a structural protein. (B) Domain structures of CD46 and SLAM (signaling lymphocyte activation molecule).

Unlike many other virus families, the paramyxoviruses, which include measles, use two separate virus proteins for recognizing the host cell (protein H) and entering the host cell (protein F). Consequently, it is possible to radically modify the recognition protein, H, without crippling virus entry and infection. Recently, measles viruses with a range of engineered H-proteins have been made. The modified H-proteins have mutations that prevent binding to CD46 or SLAM plus an extra C-terminal domain that recognizes other host cell surface proteins. The new recognition domain was generated as a single-chain antibody Fv fragment or scFv (see Chapter 6 for structure and generation of scFv chains).

Retargeted measles viruses have been designed to bind to CD38 or EGFR (epidermal growth factor receptor), proteins often expressed at high levels by cancer cells (Fig. 18.25). The new viruses were specific for cultured cancer cells with CD38 or EGFR on the cell surface. Better yet, when mice with implanted human tumors displaying CD38

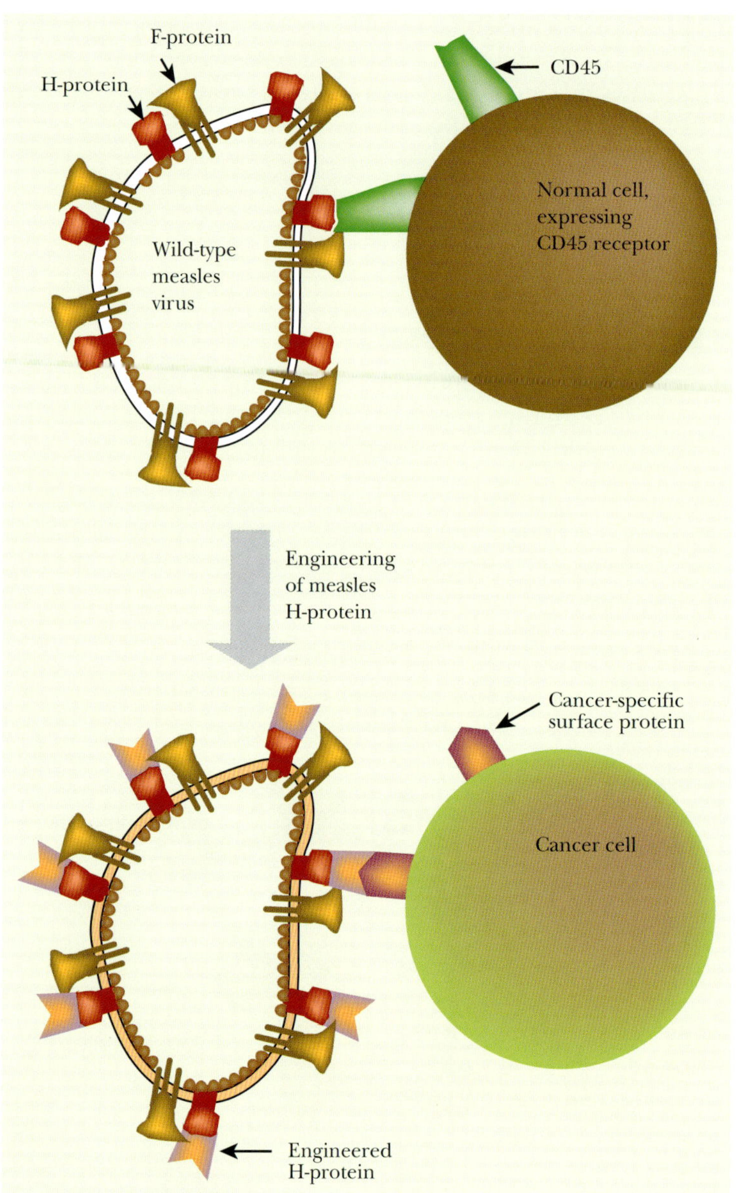

FIGURE 18.25 Cancer Retargeted Measles Virus
The external glycoprotein H may be engineered to recognize a different host cell protein. Instead of binding to CD46 or SLAM, it now binds to cancer-specific surface proteins such as CD38 or EGFR.

or EGFR proteins were inoculated with the retargeted viruses, the tumors were efficiently destroyed. It is hoped that this approach will eventually work on human cancer patients.

Changing the viral protein that binds to the human host cell so it only recognizes cancer cells makes a normal virus oncolytic, that is, the virus will only infect and kill cancer cells.

Summary

Cancer is a genetic disease, that is, a mutation in the DNA must occur for cancer to form. In normal cells, growth begins when a growth factor binds to the external portion of its cell surface receptor. The internal portion of that receptor then activates a cascade of proteins that trigger the nucleus. In the nucleus, transcription factors turn on genes that cause the cell to divide. Cell division is a highly regulated process, and progression from G_1 to S to G_2 to M occurs only when all the proper proteins are in the correct position. When the DNA of a normal cell is damaged by a carcinogen or mutagen, various proteins halt the progression through the cell cycle until the damage is repaired. These checkpoints are controlled by cyclins and transcription factors such as E2F and pRB. In cancerous cells, a mutation in the proteins that control the checkpoints allows the defective cell to progress, even though the cell is damaged. Most cancers usually have multiple proteins that are damaged along the pathway because there are so many internal controls. A large tumor mass begins to form only after blood vessels form to provide nutrients to the internal cells. Without vascular tissue and a blood supply, the cancerous region stays a microtumor. Finally, some tumor cells release from the primary tumor and travel to different places in the body, a process called metastasis.

Cancer is caused by mutations in proto-oncogenes and tumor-suppressor genes. A single dominant mutation changes a proto-oncogene into an oncogene, which in turn causes cancer. In contrast, two mutations, one in each allele, must occur for a tumor-suppressor gene to cause cancer. Sometimes, cancers occur because a virus transmits an oncogene to its host.

Oncogenic mutations occur as a result of three different changes. In the first change, proteins become hyperactive by mutations in the actual amino acid sequence. For example, mutations in the amino acid sequence of Ras cause the protein to signal the nucleus to undergo cell division without any type of signal from a growth factor. Other mutations occur due to aberrant DNA replication, which increases the number of genes by duplication. The extra oncogene copies make much more protein than should be produced. Finally, some oncogenes are formed when a normal gene is translocated to a different, stronger promoter. Many cancers are caused by Myc protein overproduction, which causes the transcription factor to trigger cell division too often. In some cancers, the *myc* gene is duplicated multiple times during DNA replication, and in other cancers, the *myc* gene is translocated behind a strong promoter.

Tumor-suppressor genes encode proteins whose normal role is to inhibit growth or prevent cell division. When both copies of these genes are mutated, the cell grows into a cancer. An exception, *p53* can actually cause cancer when only one copy of the gene is mutated since defective proteins interfere with normal *p53*. *p53* normally prevents cells from progressing through the cell cycle when it detects cellular damage, so when the protein is defective, the cell divides even though it has some sort of damage.

The goal for curing cancer is finding a method of killing the cancer cells without harming any of the normal tissue. One new treatment for cancer may occur by creating oncolytic viruses, that is, viruses that infect and kill cancer cells. The oncolytic virus is created by changing the protein that binds to the human cells so that it binds only to cancer cells. Because the virus only recognizes the cancer, it binds to these cells and destroys the host cell by viral lysis.

End-of-Chapter Questions

1. What causes cancer?
  - **a.** somatic mutations that disrupt normal cell division and death
  - **b.** germline mutations that result in cell death
  - **c.** somatic mutations that cause cells to die prematurely
  - **d.** germline mutations that disrupt normal cell division and death
  - **e.** none of the above

(Continued)

2. How might a carcinogen not be a mutagen as well?
   a. A carcinogen is always a mutagen because it will always react with DNA.
   b. A carcinogen may be converted in the body to a benign substance, which does not interact with DNA.
   c. A chemical that does not react with DNA could be converted in the body to give rise to a derivative that does interact with DNA.
   d. A carcinogen never reacts with DNA, but instead alters protein folding.
   e. all of the above

3. Which protein regulates the cell cycle by monitoring the environment of each stage and making sure the previous cycle is complete?
   a. cyclin synthases
   b. cyclin-dependent kinases
   c. E2F
   d. cyclins
   e. p53

4. What activates cell division?
   a. Ras
   b. E2F
   c. growth factors
   d. cell surface receptors
   e. phosphotransferase

5. Which of the following statements about oncogenes and anti-oncogenes is not true?
   a. Mutations in oncogenes are dominant.
   b. Both copies of tumor-suppressor genes must be inactivated to give rise to cancer.
   c. Mutations in oncogenes are recessive and can often be overcome by the non-affected, wild-type allele.
   d. Tumor-suppressor genes normally protect cells from cancer.
   e. All of the above statements are true.

6. How do oncogenic viruses form?
   a. Upon excision from the genome, the virus "picks up" an oncogene from the host.
   b. A naturally occurring oncogene within a virus becomes mutated.
   c. A virus is exposed to carcinogens, which mutates DNA.
   d. Because viruses are haploid, they do not have a backup copy of the mutated oncogene, and thus become oncogenic.
   e. none of the above

7. How might oncogenes be detected?
   a. transduction of suspect DNA into animal cells using a virus
   b. transformation of suspect DNA into animal cells and observation of cell growth (i.e., lack of contact inhibition)
   c. using phage-display technology
   d. by cloning the suspect gene onto an expression vector and assaying for the presence of the defunct protein in bacterial cells
   e. none of the above

8. Which of the following does not cause a proto-oncogene to become an oncogene?
- **a.** a mutation in the coding region that changes the amino acid sequence to make the protein hyperactive
- **b.** a mutation within the regulatory region of the proto-oncogene that increases expression of that gene
- **c.** a duplication in the proto-oncogene
- **d.** a genetic rearrangement of the proto-oncogene that places the gene behind a strong promoter, thus increasing expression
- **e.** all of the above

9. How does *Ras* become an oncogene?
- **a.** by a mutation in the coding region of the gene that creates a loss of an important function in cell division
- **b.** by a mutation in the genetic code that alters one of three amino acid residues involved in the binding and splitting of GTP
- **c.** by a duplication of the *Ras* proto-oncogene that results in more protein expressed
- **d.** by a genetic rearrangement that places *Ras* proto-oncogene downstream of a strong promoter
- **e.** none of the above

10. A mutation in which oncogene tends to occur in the later progression of lung, breast, and ovarian cancers?
- **a.** *Ras*
- **b.** cyclin-dependent kinase
- **c.** *Rb*
- **d.** *Myc*
- **e.** none of the above

11. What is nullizygous?
- **a.** when both copies of an anti-oncogene have been inactivated by mutations
- **b.** when one copy of an anti-oncogene is inactivated but the other is still active
- **c.** when both copies of an anti-oncogene are not active, but not due to a mutation, rather the anti-oncogenes are not needed in the cell at that time
- **d.** when both copies of a anti-oncogene are active and overproducing protein
- **e.** none of the above

12. Why does a mutation in only one p53 anti-oncogene allele cause a defect in the tetrameric protein?
- **a.** A mutation in only one of the p53 anti-oncogene alleles does not cause a defect in the protein.
- **b.** Because p53 is expressed from both alleles and one mutant copy causes half of the tetramer to be defective.
- **c.** Because only one p53 allele is active at any given time.
- **d.** Because each allele is expressed, making both good and bad p53 proteins to assemble into mixed tetramers.
- **e.** none of the above

(*Continued*)

13. Which of the following is a mutation associated with colon cancer?
- **a.** inactivation of both copies of the *DCC* anti-oncogene
- **b.** activation of the *Ras* oncogene
- **c.** mutation in one copy of the *p53* gene
- **d.** inactivation of both copies of the *APC* anti-oncogene
- **e.** all of the above

14. Which of the following inherited defects predisposes a person to cancer?
- **a.** DNA repair systems
- **b.** anti-oncogenes
- **c.** components of DNA replication
- **d.** genetic differences between races (i.e., skin color and melanoma)
- **e.** all of the above

15. How is a virus made oncolytic?
- **a.** by integrating the genes for tumor-suppressor and necrosis factors into the viral genome under a constitutive promoter
- **b.** by changing the location of a viral DNA integration into the host cell so that it causes a gene knockout in the defective gene that is causing the cancer
- **c.** changing the viral protein that binds to the host cells so that it only recognizes and infects, then kills, cancer cells
- **d.** by cloning genes for apoptosis into the viral genome, which integrates into the host cancer cells and kills the cells
- **e.** none of the above

Further Reading

Boulton SJ (2006). Cellular functions of the BRCA tumour-suppressor proteins. *Biochem Soc Trans* **34**, 633–645.

Esquela-Kerscher A, Slack FJ (2006). Oncomirs—microRNAs with a role in cancer. *Nat Rev Cancer* **6**, 259–269.

Gudmundsdottir K, Ashworth A (2006). The roles of BRCA1 and BRCA2 and associated proteins in the maintenance of genomic stability. *Oncogene* **25**, 5864–5874.

Jass JR (2006). Colorectal cancer: A multipathway disease. *Crit Rev Oncog* **12**, 273–287.

Karakosta A, Golias Ch, Charalabopoulos A, Peschos D, Batistatou A, Charalabopoulos K (2005). Genetic models of human cancer as a multistep process. Paradigm models of colorectal cancer, breast cancer, and chronic myelogenous and acute lymphoblastic leukaemia. *J Exp Clin Cancer Res* **24**, 505–514.

Maddika S, Ande SR, Panigrahi S, Paranjothy T, Weglarczyk K, Zuse A, Eshraghi M, Manda KD, Wiechec E, Los M (2007). Cell survival, cell death and cell cycle pathways are interconnected: Implications for cancer therapy. *Drug Resist Updat* **10**, 13–29.

Nagaraju G, Scully R (2007). Minding the gap: The underground functions of BRCA1 and BRCA2 at stalled replication forks. *DNA Repair (Amsterdam)* **6**, 1018–1031.

Ren T, Chen Q, Tian Z, Wei H (2007). Down-regulation of surface fractalkine by RNA interference in B16 melanoma reduced tumor growth in mice. *Biochem Biophys Res Commun* **364**, 978–984.

Takahashi C, Ewen ME (2006). Genetic interaction between Rb and N-ras: Differentiation control and metastasis. *Cancer Res* **66**, 9345–9348.

Wang W (2007). Emergence of a DNA-damage response network consisting of Fanconi anaemia and BRCA proteins. *Nat Rev Genet* **8**, 735–748.

Zhang B, Pan X, Cobb GP, Anderson TA (2007). microRNAs as oncogenes and tumor suppressors. *Dev Biol* **302**, 1–12.

Zhang H (2007). Molecular signaling and genetic pathways of senescence: Its role in tumorigenesis and aging. *J Cell Physiol* **210**, 567–574.

Noninfectious Diseases

Many of the more complex diseases of higher organisms are due to aberrations in signaling and regulation. These may be the result of hereditary defects, or they may result from the injuries and insults accumulated during growth and development. The complexity of such issues as obesity, diabetes, and behavioral problems often results from the interaction of genetic predisposition with environmental effects. We will begin by considering the basics of communication between the cells of higher, multicellular organisms.

CELLULAR COMMUNICATION

Multicellular organisms depend on coordinating the activities of many different cells, and this requires constant communication. Sometimes signals are sent by cell-to-cell contact. More often, signals are sent by chemical means, whether between neighboring cells or over relatively long distances. **Local mediators** are molecules that carry signals between nearby cells, whereas **hormones** carry signals to remote tissues and organs (Fig. 19.1). In addition to internal signals between cells, organisms may also send signals from one organism to another. Signal molecules that are passed from one organism to another are called **pheromones**. Even single-celled microorganisms such as yeast and bacteria communicate with each other by pheromones. Yeast secrete pheromones to signal readiness for mating. Note that pheromones are used for signaling between organisms, whereas hormones circulate internally inside multicellular organisms.

I LOCAL MEDIATOR

Signaling cell

Signal molecule

Receptor

Target cells

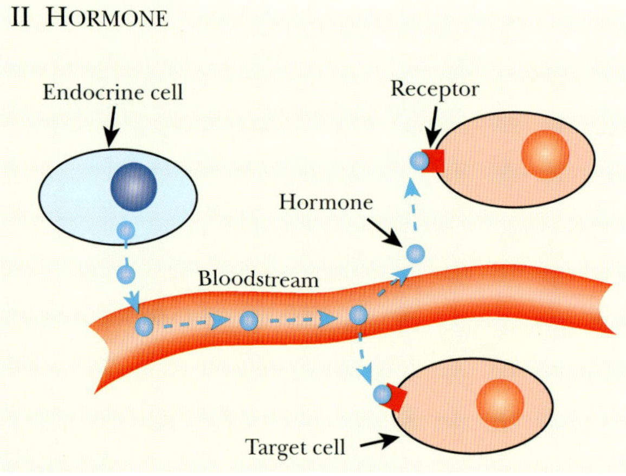

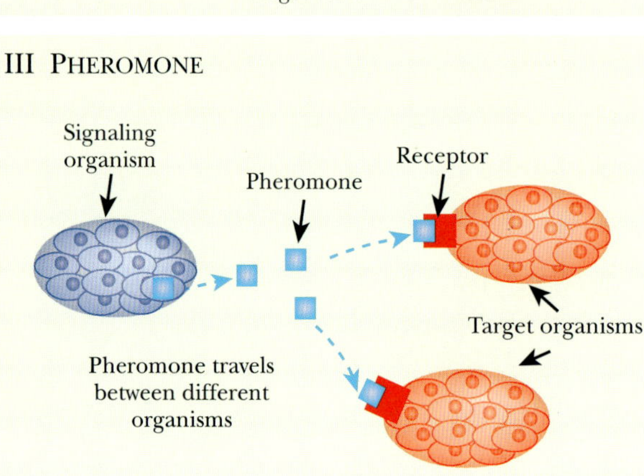

FIGURE 19.1
Local Mediators, Pheromones and Hormones
Three different types of chemical signals are used to stimulate cells. Neighboring cells may communicate via local mediators. Endocrine cells secrete hormones that enter the bloodstream and carry signals to other cells far away from the original cell. Finally, single or multicellular organisms can emit pheromones that travel through the environment rather than within the organism.

Hormones and pheromones carry messages over long distances by animals, plants, and fungi. In contrast to plants and fungi, multicellular animals also possess nervous systems to send messages. In this case, the signal travels in electrical mode along an extremely elongated cell, known as a **neuron** until it reaches a junction, or **synapse** (Fig. 19.2). Here the signal must traverse the gap between two cells by chemical means. Each arriving electrical impulse stimulates the release of a pulse of **neurotransmitter** molecules from the nerve cell. This chemical signal diffuses across the narrow gap to the next cell. Nervous impulses travel much faster than hormones, and in addition, each neuron delivers its signal to only a single target cell. In contrast, hormones act on multiple recipient cells and travel relatively slowly.

Cellular communication is important in multicellular organisms. Local mediators send signals from one cell to its neighbors. Hormones trigger different cells at long distances because they travel in the bloodstream. Pheromones travel from one organism to the next via the air, water, or soil. Vertebrates have neurons that facilitate cellular communication by neurotransmitters.

RECEPTORS AND SIGNAL TRANSMISSION

A wide variety of molecules are used in signaling, but in all cases the recipient cell needs a **receptor**. Usually the receptor is a protein situated in the cytoplasmic membrane with the site for binding the messenger facing the outside. When a signal molecule appears, it binds to the receptor causing a conformational change. The receptor then passes the signal on to other proteins, known generally as **signal transmission proteins**. Both receptors and signal transmission proteins often dimerize or dissociate during the process of signal transmission. There are three main types of signal transmissions.

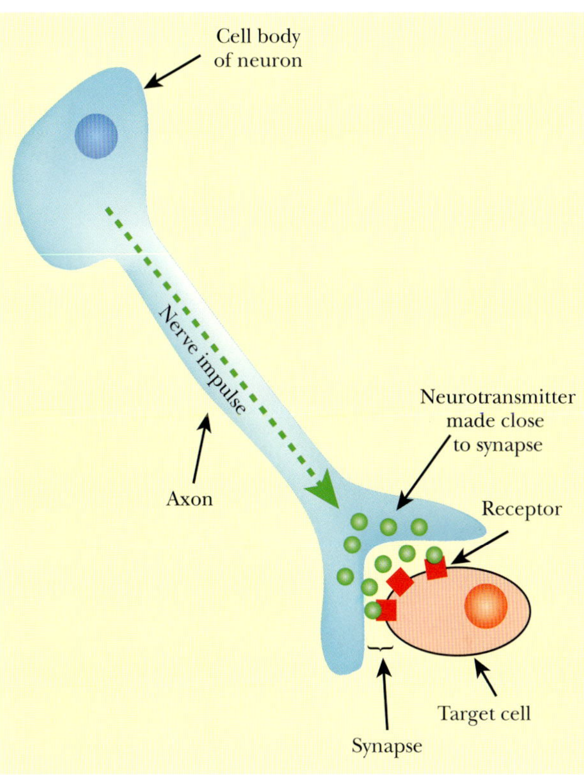

FIGURE. 19.2
Neurons and Neurotransmitters
Neurons are specialized signaling cells used by multicellular animals. The neurons produce a neurotransmitter that activates the target cell by traversing the synapse.

(a) Phosphotransferase systems. Activation of the receptor causes the phosphorylation of signal transmission proteins (Fig. 19.3). **Protein kinases** transfer phosphate groups from ATP to other proteins. The activated receptor may act as a protein kinase itself, or it may bind to and activate a separate protein kinase. Often several proteins take part in a phosphotransfer cascade that allows the signal to be amplified or modulated in a variety of ways.

(b) Second messengers. Activation of the receptor results in synthesis of a small intracellular signal molecule known as the second messenger (Fig. 19.4). This may directly activate or inhibit various enzymes and may also enter the nucleus and affect gene expression. Often a GTP-binding **G protein** links the receptor to the enzyme that makes the second messenger.

(c) Ion channel activation. In this case, the receptor itself acts as an ion channel (Fig. 19.5). On receiving the external message it either opens or closes. The altered movement of ions through the channel then mediates further signaling.

Signal transmission takes the external signal and transmits the information to the nucleus. Signal transmission occurs via phosphotransfer systems, second messengers and ion channels.

STEROIDS AND OTHER LIPOPHILIC HORMONES

Although most hormones and pheromones rely on binding to cell surface receptors, this is not always the case. Certain lipophilic hormones can diffuse across the cell membrane. These include **thyroxine (thyroid hormone)**, retinoic acid, and the **steroid hormones** (Fig. 19.6). Once inside the cell, they bind to internal receptors that are soluble proteins. Indeed, these receptors do double duty as transcription factors and bind directly to DNA. However, because they are poorly soluble in water, lipophilic hormones rely on carrier proteins to transport them from cell to cell via the bloodstream. In contrast, water-soluble hormones diffuse freely through the blood but cannot cross the cell membrane and so do not actually enter the cell. Instead they bind to and activate receptors on the cell surface.

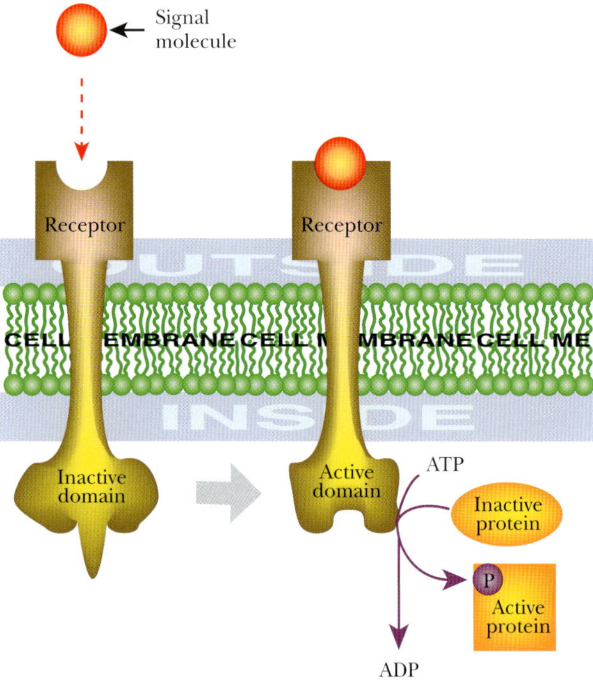

FIGURE 19.3 Phosphotransferase System
A signal molecule induces a conformational change in the receptor that induces self-phosphorylation of the receptor. The receptor then transfers the phosphate group to downstream proteins.

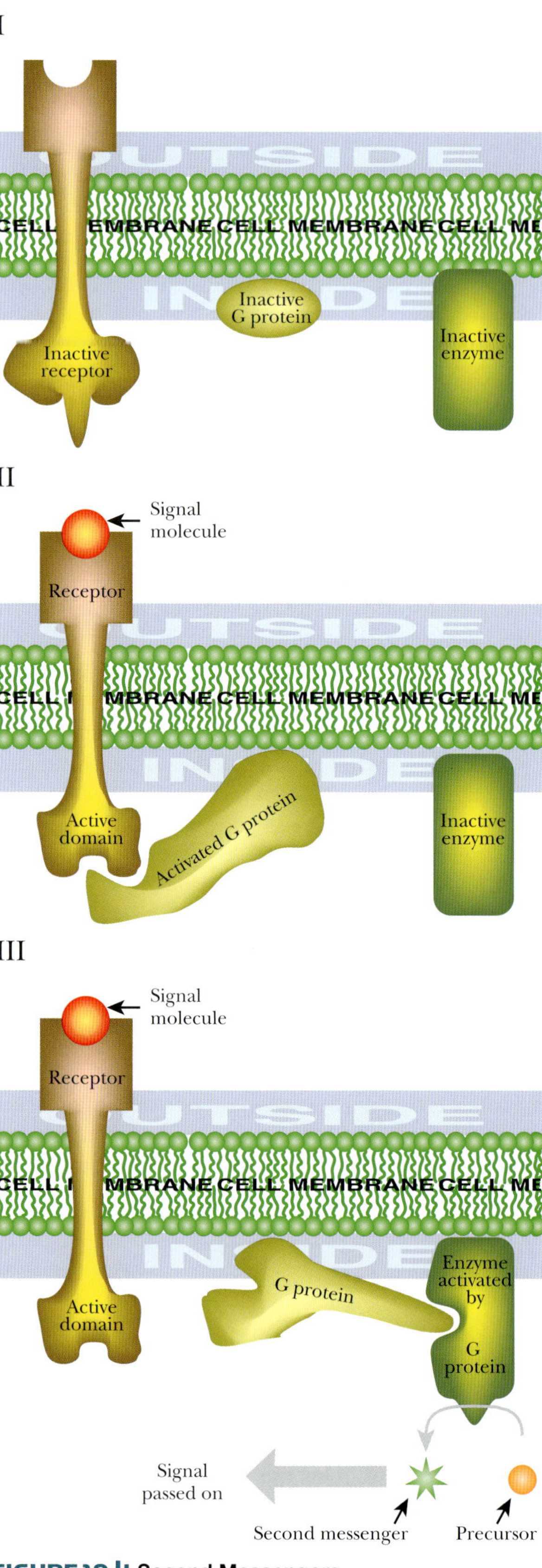

FIGURE 19.4 Second Messengers
Small intracellular signaling molecules called second messengers are produced in response to an activated receptor. The active receptor activates a G protein, which in turn activates a membrane-bound enzyme, which in turn synthesizes the second messenger.

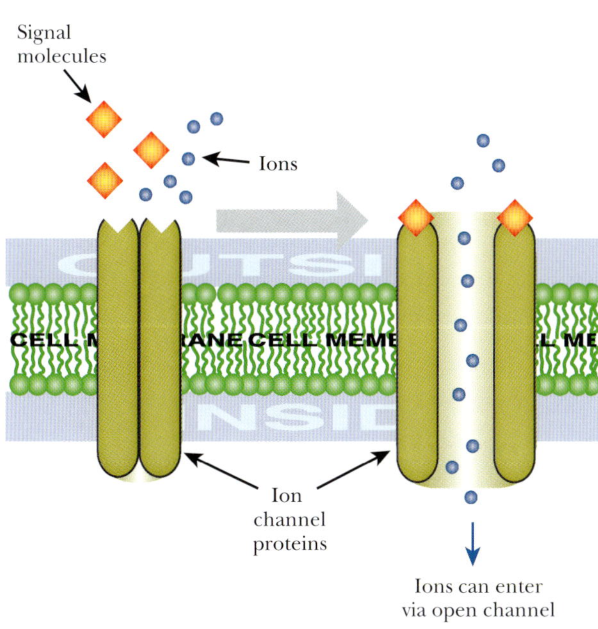

FIGURE 19.5 Ion Channel Activation
Binding of a signal molecule to an ion channel induces a conformational change that opens the channel. Ions may now pass through the open channel freely.

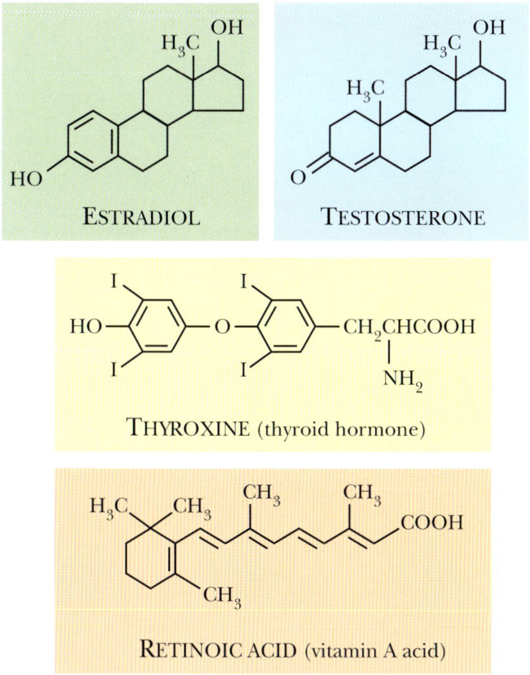

ESTRADIOL

TESTOSTERONE

THYROXINE (thyroid hormone)

RETINOIC ACID (vitamin A acid)

FIGURE 19.6 Lipophilic Hormone Structures
Lipophilic hormones are able to diffuse across cellular membranes because they are small and relatively hydrophobic.

The receptors for the lipophilic hormones belong to the **nuclear-receptor superfamily** of proteins (Fig. 19.7). They all act as transcription factors in addition to hormone receptors and bind to recognition sequences in the DNA of the upstream region of their target genes. They may be subdivided into two classes:

(a) Most **steroid receptors**, such as the estrogen receptor or glucocorticoid receptor, are found in the cytoplasm. After binding the steroid, they enter the nucleus (Fig. 19.8). These receptors form dimers that recognize inverted repeats of 6 bp in the DNA.

(b) The receptors for androgens, vitamin D (a steroid), thyroxine, and retinoic acid are permanently located in the nucleus. Apparently, the hormones migrate via the intracellular membranes to the nuclear membrane and join their receptors inside the nucleus. These receptors bind to the related **RXR protein** to form mixed dimers (see Fig. 19.7). These recognize direct repeats of 6 base pairs in the DNA. The same 6 base-pair sequence may be recognized by different receptors; however, in this case the spacing between the 6 base-pair repeats differs.

Industrial pollution has contaminated the environment with a variety of polycyclic molecules resembling steroids, to a greater or lesser extent. These may enter the body (via ingestion or through the skin) and can diffuse through membranes due to their lipophilic nature. Some of these, known as **xenoestrogens**, bind to steroid receptors and mimic the action of estrogens. This may cause reproductive problems or promote cancer of the uterus. It has even been suggested that the major drop in sperm counts seen over the past half century in the industrial nations, especially the United States, may be due to such estrogen mimics, though this remains speculative.

Lipophilic hormones travel from the extracellular region, through the cell membrane, and into the cytoplasm by diffusion.

Lipophilic hormones bind to receptors found in the cytoplasm or nucleus. The activated hormone/receptor complex binds directly to DNA and activates transcription of genes involved in growth.

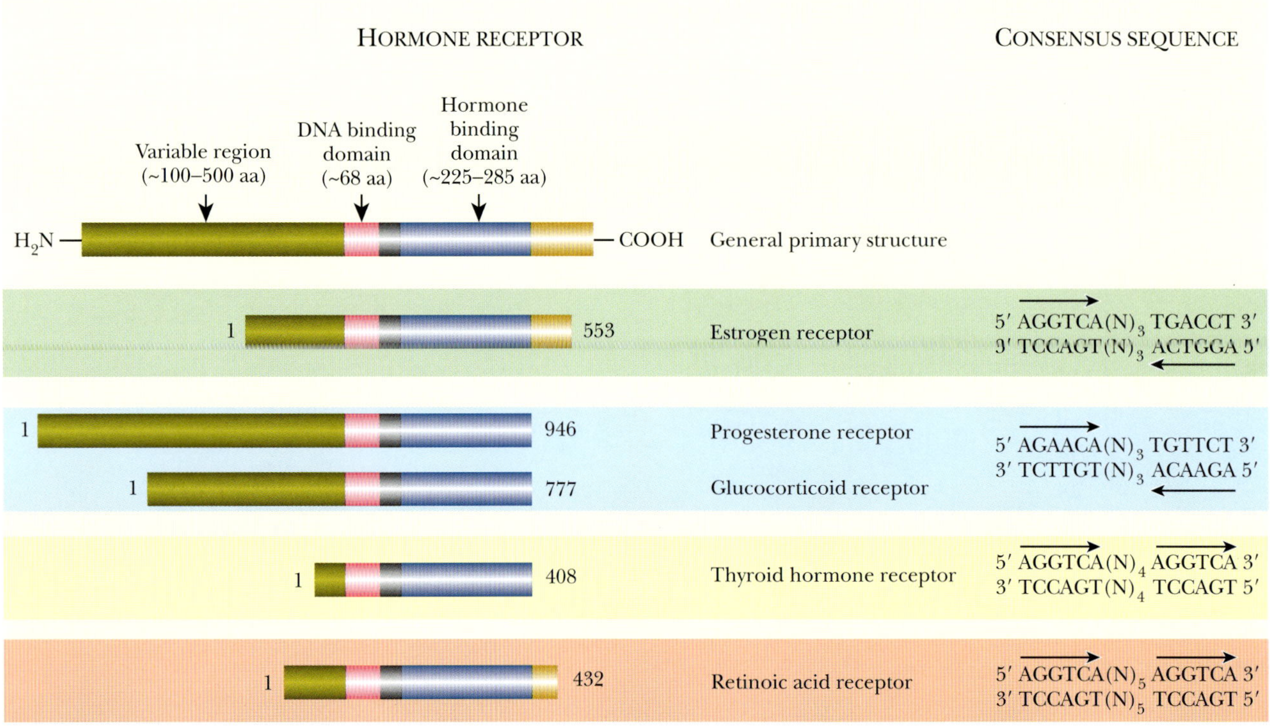

Hormone receptor
Consensus sequence
Variable region (~100–500 aa)
DNA binding domain (~68 aa)
Hormone binding domain (~225–285 aa)
H2N
COOH
General primary structure
1
553
Estrogen receptor
5′ AGGTCA(N)3 TGACCT 3′
3′ TCCAGT(N)3 ACTGGA 5′
1
946
Progesterone receptor
1
777
Glucocorticoid receptor
5′ AGAACA(N)3 TGTTCT 3′
3′ TCTTGT(N)3 ACAAGA 5′
1
408
Thyroid hormone receptor
5′ AGGTCA(N)4 AGGTCA 3′
3′ TCCAGT(N)4 TCCAGT 5′
1
432
Retinoic acid receptor
5′ AGGTCA(N)5 AGGTCA 3′
3′ TCCAGT(N)5 TCCAGT 5′

FIGURE 19.7 Hormone Receptor Family
Hormone receptors have a variable region, a DNA binding domain, and a hormone binding domain. The various receptors recognize different 6 base-pair repeats on the DNA in order to activate transcription of hormone responsive genes.

Steroid hormone
Steroid receptor
Steroid receptor
Inhibitory protein
Activated receptor
Activated receptor
Binds DNA

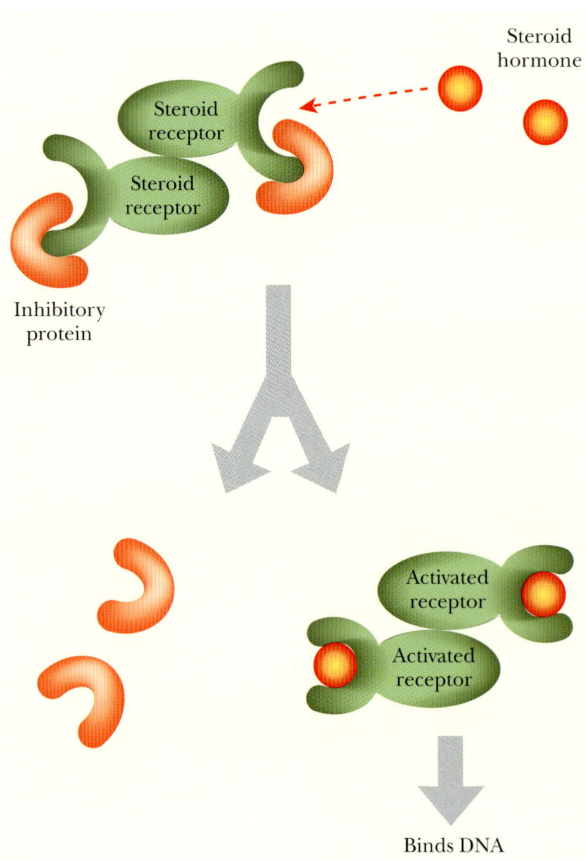

FIGURE 19.8
Activation of Steroid Receptors
Steroid receptors are dimers that have two binding sites for the hormone. In the inactive state, inhibitory proteins block the binding sites. Activation releases the inhibitory proteins, allowing the hormone to bind. The active receptors are able to bind to their recognition sequence on the DNA, thus activating the appropriate genes.

CYCLIC AMP AS SECOND MESSENGER

The regulatory nucleotide **cyclic AMP (cAMP)** is often used as a second messenger by multicellular animals. Comparison of the way in which cyclic AMP is used by bacteria, single-celled eukaryotes, and higher organisms illustrates how the interplay among signaling, multicellularity, and reproduction has developed as we ascend the evolutionary ladder.

Cyclic AMP is synthesized from ATP by the enzyme **adenylate cyclase**, which is located in the cytoplasmic membranes of both bacteria and eukaryotes. Cyclic AMP is widespread in nature, but its precise regulatory function varies widely. In bacteria cAMP often regulates gene expression in response to nutrient availability, although the nutrient may vary from one type of bacteria to another. In animals, cAMP is often used as a second messenger, transmitting a signal from a receptor to the nucleus.

Despite the apparent differences, there are fundamental similarities between the prokaryotic and eukaryotic signaling systems. The level of cAMP depends on the activity of adenylate cyclase, which in turn depends on signals received from other proteins. In both bacteria and animals, this original signal originates from outside the cell via a membrane protein.

In the case of *Escherichia coli*, adenylate cyclase responds to the external presence of glucose (or other highly favored sugars). The proteins of the **phosphotransferase system (PTS)**, which transports glucose and other sugars, transmit the signal to adenylate cyclase. As glucose enters the bacterial cell, it receives a phosphate group from the glucose transporter and is converted to glucose 6-phosphate. When glucose is plentiful, it consumes so many phosphate groups that the proteins of the PTS are mostly in their nonphosphorylated state (Fig. 19.9). When glucose is scarce, the phosphorylated forms accumulate. In particular, phosphorylated enzyme IIA^{glc} is involved in signal transmission. This binds to and activates adenylate cyclase to make cyclic AMP.

In animals, adenylate cyclase often responds to hormonal messages that reveal the nutritional status of the whole animal. This is frequently mediated by the coupling of a hormone receptor to a G protein in the cell membrane. G proteins were so named because their mechanism depends on binding GTP when activated. They go

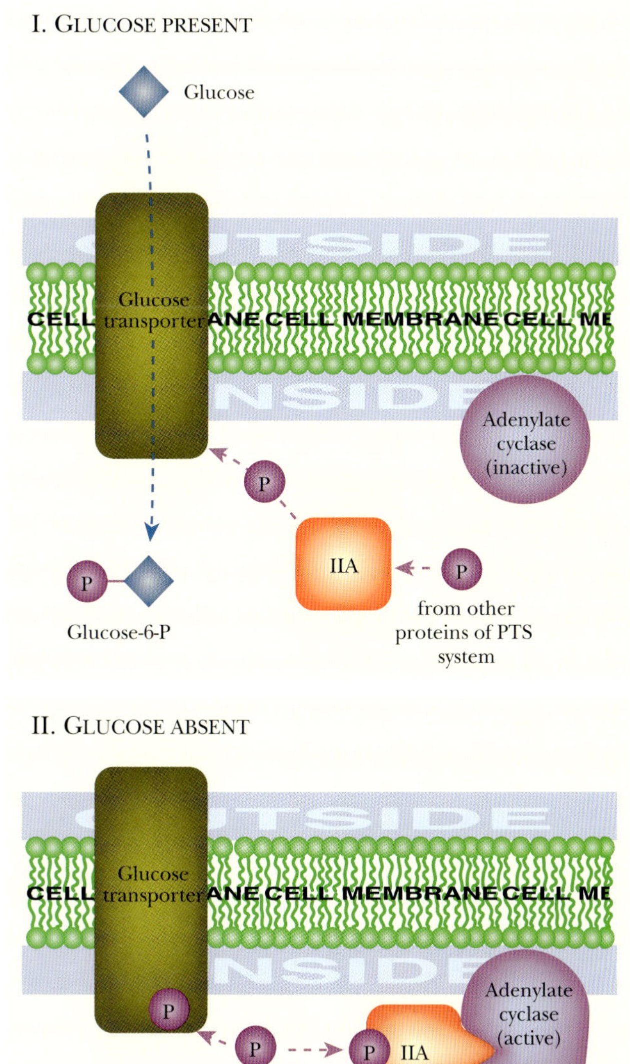

FIGURE 19.9
Control of Adenylate Cyclase by Glucose Transport
E. coli prefer to use glucose as an energy source. The glucose transporter converts glucose to glucose 6-phosphate to initiate its metabolism. The glucose transporter is phosphorylated by enzyme IIA^{glc}, and when glucose is present, this enzyme is mostly not phosphorylated. When the glucose supply is depleted, the cell must signal that other carbon sources should be used. Because enzyme IIA^{glc} has no place to transfer its phosphate molecule, the phosphorylated form accumulates and activates adenylate cyclase. This enzyme converts ATP into cyclic AMP, the second messenger that signals the cell to use another carbon source.

through a cycle in which the bound GTP is split to give GDP and phosphate (Fig. 19.10). Thus, cyclic AMP acts as a second messenger in response to signals delivered to the cell by a hormone.

In both animals and bacteria, cyclic AMP increases transcription of certain genes. In bacteria, cAMP binds to **CRP protein**. This is a DNA binding protein that acts as a global gene activator. In animal cells, cyclic AMP not only affects gene expression but also directly affects enzyme activity (Fig. 19.11). In animals, accumulation of cyclic AMP activates **protein kinase A (PKA)**. In its inactive form this enzyme consists of two regulatory (R) subunits and two catalytic (C) subunits. When cyclic AMP binds to the R-subunits, the C-subunits are released and proceed to phosphorylate a variety of target proteins, depending on the cell type. Some of these are enzymes and are located in the cytoplasm. Phosphorylation may cause activation or deactivation, depending on the particular enzyme. In addition, a few active C-subunits enter the nucleus where they phosphorylate the **CREB protein (cyclic AMP response element binding protein)**. The phosphorylated form of CREB binds to a specific DNA sequence, the **CRE (cyclic AMP response element)** found in front of those genes that are activated by cyclic AMP.

Adenylate cyclase makes cAMP in response to external signals. cAMP is the second messenger that transmits the external signal to the nucleus.

Bacteria preferentially use glucose as an energy source, and it enters via the phosphotransferase system. When glucose is scarce, phosphorylated enzyme IIA^{glc} activates adenylate cyclase to make cAMP. cAMP turns on genes to metabolize other sugars.

Animal cells use cAMP as a second messenger to activate genes in the nucleus. cAMP also is used to activate cytoplasmic proteins via phosphorylation.

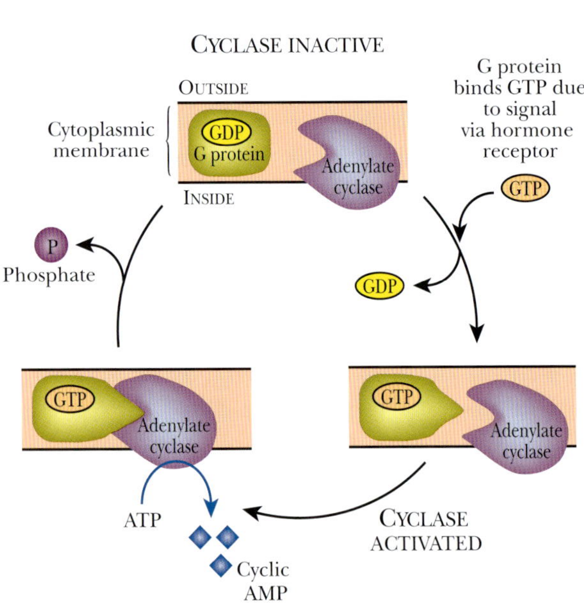

FIGURE 19.10 Control of Adenylate Cyclase by G Proteins

A mammalian hormone receptor activates the G protein to bind GTP. The GTP-bound form can now bind to adenylate cyclase, which is activated and converts ATP into cyclic AMP. The G protein then splits GTP and releases adenylate cyclase so that the signal is no longer produced.

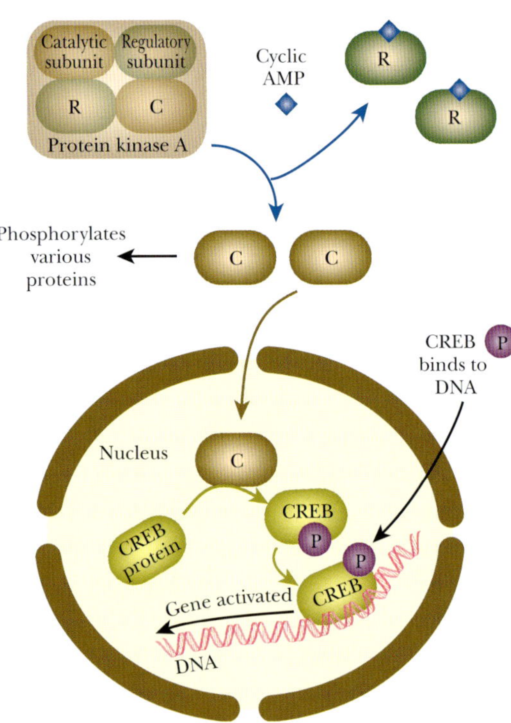

FIGURE 19.11 The Cyclic AMP Cascade in Animal Cells

In animal cells, PKA responds to the presence of cyclic AMP. When no cyclic AMP is present, PKA is an inactive tetramer with two regulatory subunits and two catalytic subunits. Cyclic AMP binds to the two regulatory subunits and the tetramer dissociates, allowing the catalytic subunits to phosphorylate other proteins in the cytoplasm and nucleus. One of the proteins activated by PKA is CREB, a transcription factor that activates genes with a CRE binding site.

NITRIC OXIDE AND CYCLIC GMP

Eukaryotic cells use several other second messengers. These include **cyclic GMP (cGMP)**, Ca^{2+} ions, inositol triphosphate, and other products derived from the membrane lipid phosphatidylinositol. Cyclic GMP is made from GTP by the enzyme **guanylate cyclase**. There are two different types of guanylate cyclases. One is membrane bound and also functions as a receptor for peptide hormones. The other is a soluble protein found in the cytoplasm that is activated by **nitric oxide** (Fig. 19.12). Cyclic GMP controls ion channel opening and closing, and it activates gene expression via protein kinases. Cyclic GMP-dependent protein kinase is capable of entering the nucleus where it regulates gene transcription by phosphorylating certain transcription factors, such as Oct-1.

Nitric oxide is unusual in being a gaseous signal molecule. Whether it should be regarded as a hormone or a neurotransmitter is debatable, because its action has characteristics of both types of signals. Nitric oxide acts via cyclic GMP. Nitric oxide (NO) is short-lived and is involved in control of a variety of local cellular activities, including the contraction of blood vessel walls. In response to nerve cell signals, NO is made from arginine and oxygen by **NO synthase** in endothelial cells and diffuses into nearby muscle cells. Here it binds to a heme cofactor attached to guanylate cyclase and triggers cGMP production (Fig. 19.13). This leads to muscle relaxation and dilation of blood vessels. (Nitric oxide at higher levels is toxic and is made as an antibacterial agent by activated immune cells.)

Another second messenger, cGMP, is made by guanylate cyclase. Hormones activate the membrane-bound form of guanylate cyclase, and a gaseous signaling molecule, nitric oxide, activates guanylate cyclase in the cytoplasm.

FIGURE 19.12 Two Types of Guanylate Cyclase
Both the receptor form and the intracellular form of guanylate cyclase convert GTP into cyclic GMP. Peptide hormones activate the receptor form, and nitric oxide activates the intracellular form.

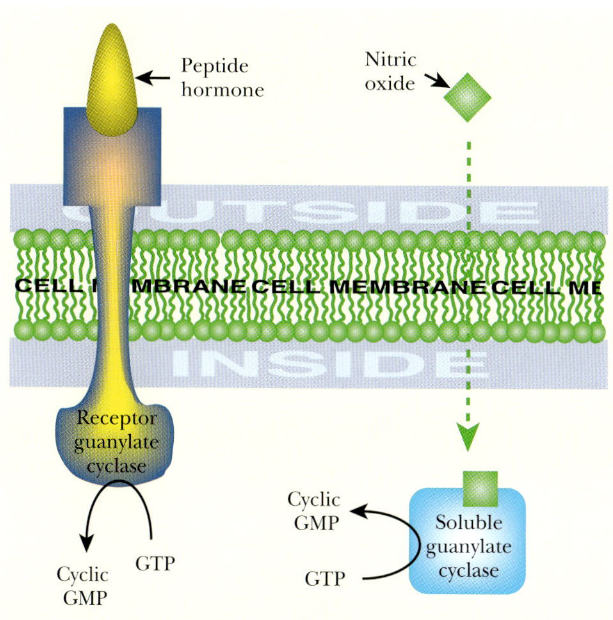

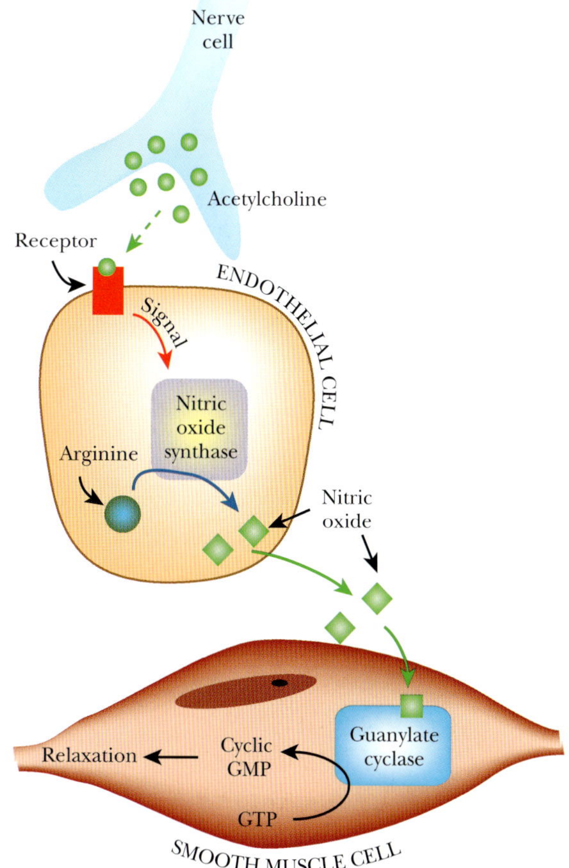

FIGURE 19.13 Nitric Oxide Control of Guanylate Cyclase
Endothelial cells manufacture NO in response to signals from neighboring nerve cells. NO diffuses across the endothelial cell membrane and into nearby muscle cells. Once inside, NO activates guanylate cyclase to manufacture cyclic GMP, which in turn makes the muscle cell relax.

CYCLIC PHOSPHODIESTERASE AND ERECTILE DYSFUNCTION

The cyclic nucleotides are inactivated by the enzyme **cyclic phosphodiesterase** that converts them to the corresponding 5′-nucleoside monophosphates (5′-AMP or 5′-GMP). At least 10 phosphodiesterase isoenzymes are found in humans. They are distributed in different tissues, and some are specific for cAMP, some for cGMP (PDE5, PDE6, and PDE9), and others work on both.

Inhibition of **phosphodiesterase 5 (PDE5)**, which is specific for cyclic GMP, keeps cyclic GMP levels up and so keeps blood vessels dilated for longer. This is the basis of the treatment of male erectile dysfunction by the drug **Viagra** (**sildenafil;** Fig. 19.14), an inhibitor of PDE5. An estimated 20 to 30 million American men suffer from some degree of sexual dysfunction, and perhaps half are estimated to benefit from Viagra treatment.

More recently two new PDE5 inhibitors vardenafil (Levitra) and tadalafil (Cialis) have come to market. These act in a very similar manner to sildenafil but are somewhat more potent. Moreover, tadalafil has a prolonged half life. Consequently, whereas the duration of action of sildenafil and vardenafil is about 4 hours, that of tadalafil is about 36 hours.

Sildenafil inhibits PDE5, a phosphodiesterase that blocks the second messenger, cGMP. Without a signal from cGMP, the blood vessels stay dilated and therefore treat male erectile dysfunction.

CYCLIC GMP

SILDENAFIL

FIGURE 19.14
Cyclic GMP and Its Analog Sildenafil
PDE5 is found in platelets, visceral smooth muscle, and lungs as well as the smooth muscle of blood vessels. Sildenafil inhibits PDE5 in these tissues as well, causing the side effects of Viagra treatment. In addition, sildenafil is not totally specific for PDE5. It also mildly inhibits PDE6, which is located in the retina of the eye. This may be the cause of abnormalities in color vision (blue, green) that have been observed as side effects.

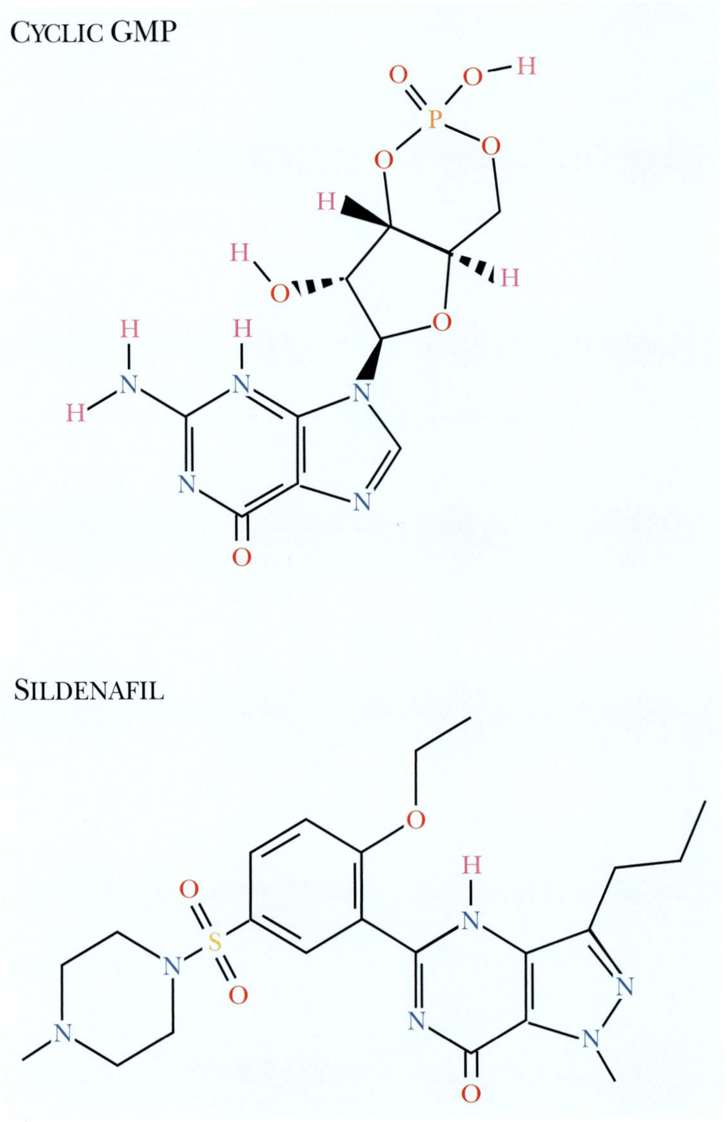

INSULIN AND DIABETES

Diabetes mellitus is actually a group of related diseases in which the level of glucose in the blood and/or urine is abnormally high. There is considerable variation in the detailed symptoms, and multiple genes are involved. Many cases of diabetes are due to the absence of **insulin**, a small protein hormone made by the pancreas, which controls the level of sugar in the blood. Lack of insulin results in high blood sugar, and this causes a variety of complications. In patients with insulin-dependent diabetes mellitus (IDDM), injections of insulin keep blood sugar levels down to near normal. Other defects affect the **insulin receptor**, and so these will not respond to insulin treatment.

Insulin is a protein made of two separate polypeptide chains, the A- and B-chains (Fig. 19.15). Disulfide bonds hold the two chains together. Although the final protein has two polypeptide chains, insulin is actually encoded by a single gene. The original gene product, **preproinsulin**, is a single polypeptide chain, which contains both the A- and B-chains together with the C- (or connecting) peptide and a signal sequence. Preproinsulin itself is not a hormone but must first be processed to give insulin. The signal sequence at the N-terminal end is required for secretion and is then removed by signal peptidase. This leaves **proinsulin**. Removal of the **C-peptide** requires **endopeptidases** that cut within the polypeptide chain. These recognize pairs of basic amino acids at the junctions of the C-peptide with the A- and B-chains. Finally, the terminal Arg and Lys residues are trimmed off by **carboxypeptidase H**.

Insulin is a hormone produced as preproinsulin. The signal sequence is removed by signal peptidase, the C-peptide is removed by endopeptidase, and the final arginine and lysine are trimmed by carboxypeptidase H. After processing, insulin has two chains, A and B, linked by disulfide bonds.

Some diabetics do not produce any insulin and require the insulin as a shot. Other diabetics do not have the insulin receptor, so insulin cannot act on the target cells.

THE INSULIN RECEPTOR

The insulin receptor is a tetramer of two alpha and two beta chains (Fig. 19.16). Just as in insulin, a single gene produces a single protein product that is processed to release the signal sequence and then to give the separate alpha and beta chains of the receptor. The alpha chains are exposed on the cell surface and bind the hormone. Disulfide bonds form between the cysteine-rich regions of the two alpha chains and hold them together. The paired beta chains are embedded in the membrane and possess an internal signal-transmitting domain with protein kinase activity. A variety of genetic defects may affect the insulin receptor. Lowered expression of the receptor gene, defects in processing, defective insulin binding, and defects in protein kinase activity are all known. The various syndromes that result are resistant to insulin treatment.

In normal insulin receptors, binding of insulin triggers a conformational change that activates the protein kinase domain. This initiates two alternative phosphorylation cascades. The protein kinase B pathway is responsible for short-term control of glucose metabolism and involves no new protein synthesis (Fig. 19.17). The other pathway proceeds via the signal transmission proteins Ras and MAPK (see Chapter 18). This leads to gene activation and the synthesis of new proteins, over a longer time scale. The details of insulin action vary between different types of target cells (e.g., liver, fat, or muscle). Overall, insulin lowers blood glucose and promotes its storage as glycogen. If blood sugar is too low, the hormone glucagon acts in opposition

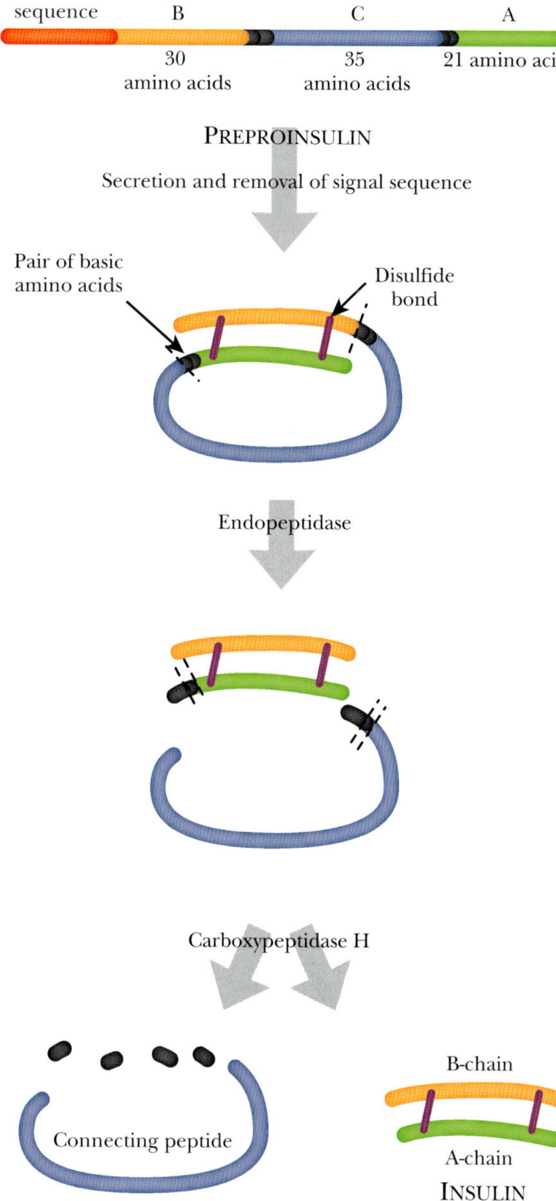

FIGURE 19.15
Processing of Insulin
The gene for insulin produces one transcript that is translated into a single protein called preproinsulin. The signal sequence of preproinsulin is removed after secretion. Next an endopeptidase removes the C-peptide. This leaves the A- and B-chains held together by disulfide bonds. Last, carboxypeptidase H trims terminal Arg and Lys residues, leaving active insulin.

to insulin, promoting the breakdown of glycogen and the release of more glucose into the bloodstream. Glucagon uses cyclic AMP as second messenger.

The insulin receptor has two alpha chains where insulin binds, and two beta chains where the intracellular kinase domain activates the intracellular proteins.

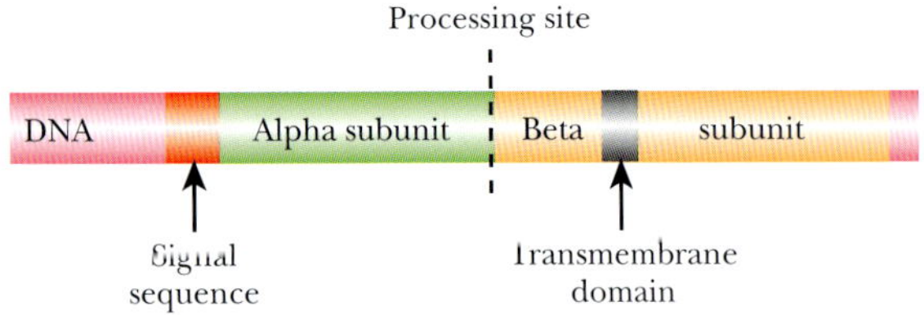

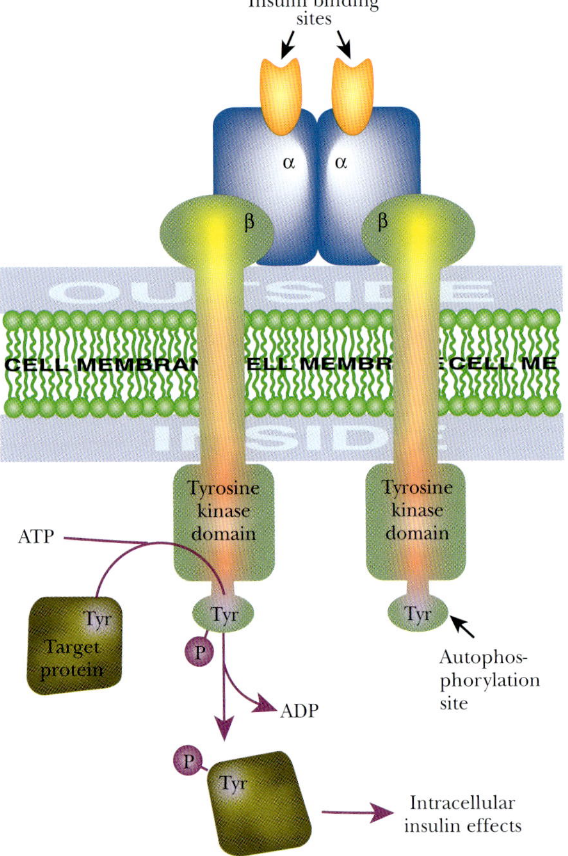

FIGURE 19.16 The Insulin Receptor
(A) The gene for the insulin receptor has three domains that encode the signal sequence used for secretion, the alpha subunit and the beta subunit, respectively. (B) The insulin receptor protein has four subunits. The two alpha subunits bind insulin. The beta subunits are tyrosine kinases that autophosphorylate on a tyrosine residue and transfer the phosphate to tyrosine residues on the target protein. The receptor is held together by disulfide bonds between the two alpha subunits and between the alpha and beta subunits.

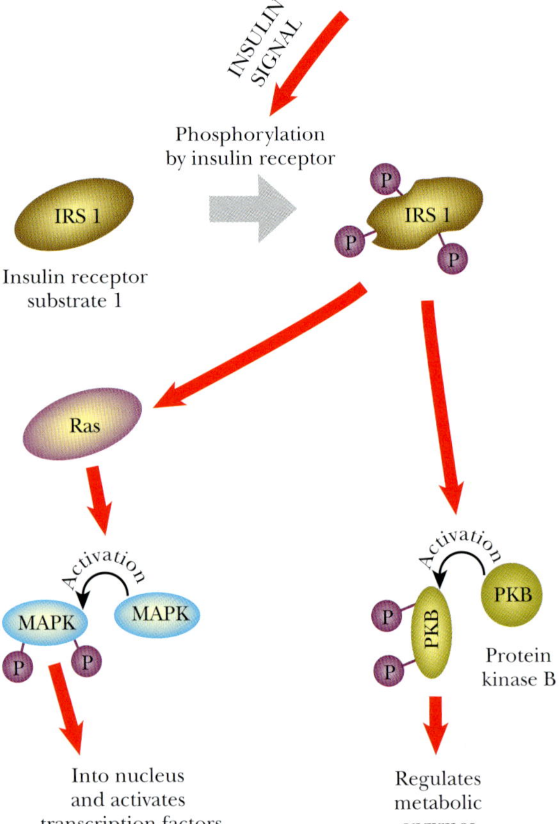

FIGURE 19.17 Insulin-Triggered Regulatory Cascades
IRS-1 is a downstream target for the activated insulin receptor. Once phosphorylated, IRS-1 activates the Ras and MAPK pathway to increase transcription of genes involved in converting glucose to glycogen. IRS-1 also regulates metabolic enzymes by phosphorylating protein kinase B.

CLONING AND GENETIC ENGINEERING OF INSULIN

Insulin was the first genetically engineered hormone to be made commercially available for human use. Before cloned human insulin was available, people with diabetes had to give themselves injections of insulin extracted from the pancreas of animals such as cows or pigs. Although this worked well on the whole, occasional allergic reactions occurred, usually to low-level contaminants in the extracts. Today, genuine human insulin (Humulin, marketed by Eli Lilly Inc.) made by recombinant bacteria is available.

If the insulin gene is cloned and directly expressed in bacteria, preproinsulin would be made. Because bacteria lack the mammalian processing enzymes, the preproinsulin would not be converted into insulin (see Fig. 19.15). In practice, there are two possible solutions to this problem. The first is to purify the preproinsulin and then treat it with enzymes that convert it into insulin. This means the processing enzymes must be manufactured as well. Clearly this is overly complex. The solution chosen was to make two artificial **mini-genes**, one for the insulin A-chain and the other for the insulin B-chain (Fig. 19.18). Two pieces of DNA, encoding the two insulin chains, were synthesized chemically. The two DNA molecules were inserted into plasmids that were put into two separate bacterial hosts. Thus, the two chains of insulin were produced separately by two bacterial cultures. They were then mixed and treated chemically to generate the disulfide bonds linking the chains together.

The approach just described gives insulin that works well. Nonetheless, natural insulin, even natural human insulin, is not perfect, and we can improve on nature. The problem is that natural insulin tends to form hexamers. This clumping covers up the

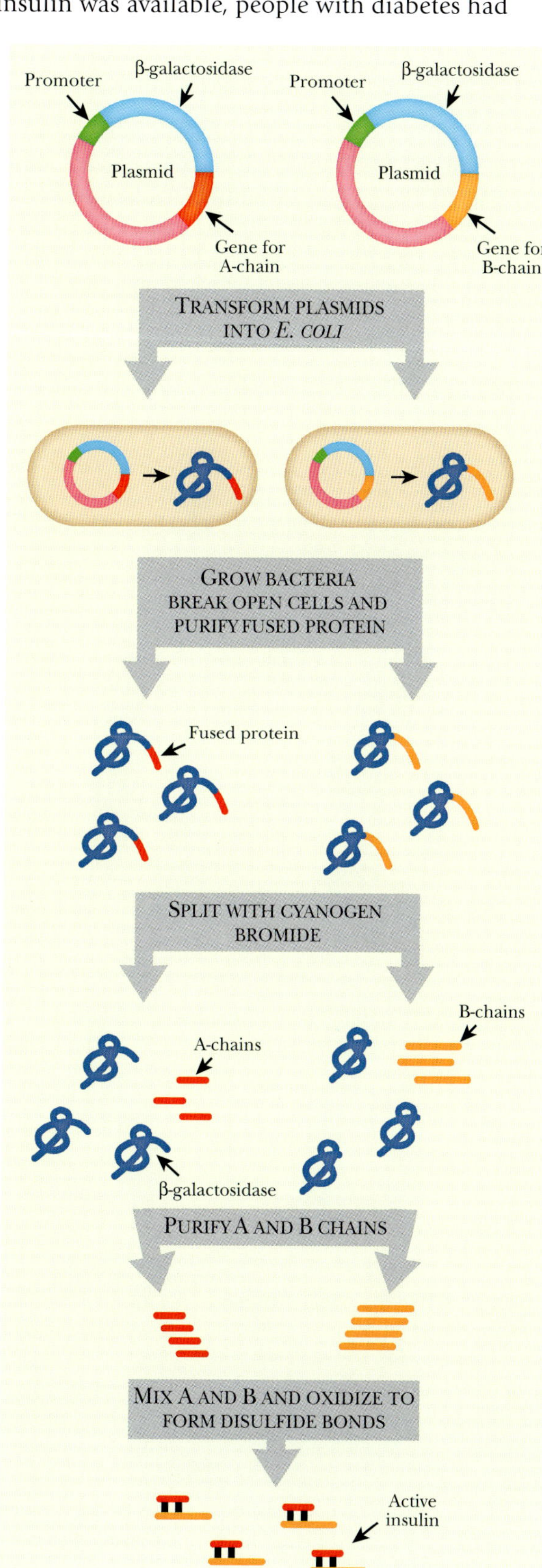

FIGURE 19.18
Cloning of Insulin as Two Mini-Genes
The genes for the A- and B-chains of insulin were cloned on two separate plasmids. Both mini-genes were fused to β-galactosidase because this protein is easy to purify. The plasmids were transformed into bacteria and expressed in separate cultures. The bacteria from each culture were harvested and the fusion proteins purified. The A- and B-chains were cleaved from β-galactosidase by cyanogen bromide, then mixed under oxidizing conditions to form the disulfide bonds, thus making human insulin.

surfaces by which the insulin molecule binds to the insulin receptor, thus preventing most of the insulin from activating its target cells (Fig. 19.19). *In vivo,* insulin is secreted from the pancreas as a monomer and is distributed rapidly by the bloodstream before it gets a chance to clump. However, when insulin is injected, a high concentration of insulin is present in the syringe and clumping occurs. After injection, it takes a while for the hexamers to dissociate, and it may take several hours for the patient's blood glucose to drop to normal levels.

Insulin can be genetically engineered to prevent clumping. The DNA sequence of the insulin gene is altered to change the amino acid sequence of the resulting protein. A proline located at the surface where the insulin molecules touch each other when forming the hexamer is replaced with aspartic acid, whose side chain carries a negative charge. So when two modified insulin molecules approach each other, they are mutually repelled by their negative charges and no longer clump (Fig. 19.20). The altered insulin causes a faster drop in blood sugar than native insulin. In 1999 the Danish pharmaceutical company Novo received approval from the European Union, and the improved insulin may eventually replace the natural product.

Insulin used for treating diabetes is now produced as a recombinant protein in bacteria. Recombinant insulin is expressed as two mini-genes rather than expressing preproinsulin. Changing the proline to aspartic acid prevents recombinant insulin from clumping.

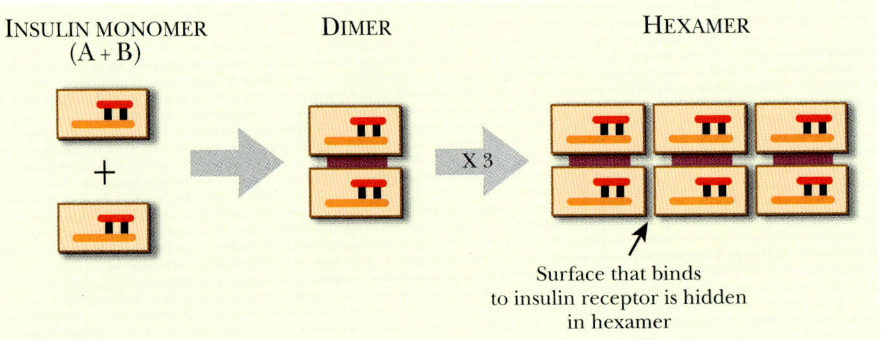

FIGURE 19.19 Insulin Forms Hexamers
High concentrations of insulin cause the monomers to clump into hexamers. The proteins stick to one another by their receptor binding sites.

Box 19.1 Stem Cell Therapy for Diabetes

Today there is considerable interest in using embryonic stem cells to regenerate damaged or diseased tissues. In addition to the religious and political controversy, there are complex technical problems in stem cell therapy—not least being the source of appropriate stem cells. Assuming that all these issues are eventually settled, one obvious application of stem cell therapy is for type 1 diabetes.

At the cellular level, this condition is due to destruction of the insulin-secreting β cells of the pancreas. Implanting new pancreatic β cells should cure type 1 diabetes by providing an internal source of insulin. Considerable work is being done to control the differentiation of stem cells into insulin-secreting cells. As of 2005, mouse cells that secrete insulin in response to glucose had been generated. On implantation into mice, they initially cured diabetes. However, after a few weeks, the cells formed a teratoma—a randomly differentiated tissue mass—and rescue from diabetes failed. Eventually, if such problems are solved and the technology is applied to human cells, it may be possible to actually cure diabetes as opposed to suppressing the symptoms by constant insulin injections.

OBESITY AND LEPTIN

Our word *obesity* comes from the Latin *obedere*, meaning "to devour." Overeating and the accompanying consequences have become epidemic in the industrialized nations. Americans spend more than $12 billion a year on slimming products (diets, exercise machines, liposuction, etc.), yet they are the most overweight people of any industrial nation and are getting fatter. About 35 million Americans are sufficiently overweight to seriously affect their health. The increased health costs amount to some $40 billion per year, more than three times the amount spent on slimming products! Obesity is not just a matter of diet and (lack of) exercise, but depends greatly on genetics.

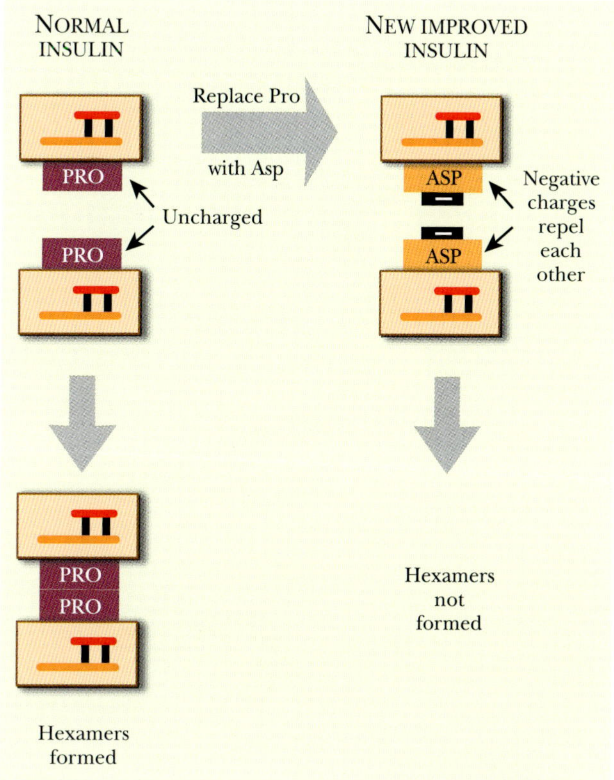

The ***ob* (obese) gene** encodes a protein hormone called **leptin**. Mice defective in both copies of the *ob* gene lack leptin and weigh up to three times as much as normal mice. When obese mice are treated with leptin, they not only eat much less, but also burn off fat much faster and may lose 30% of their body weight. The genes for human leptin and mouse leptin have both been cloned into the bacterium *E. coli*, allowing the hormones to be made in large quantities. Leptin from both species is very similar and acts in the same way on obese mice.

NORMAL INSULIN

NEW IMPROVED INSULIN

Replace Pro with Asp

PRO

PRO

Uncharged

ASP

ASP

Negative charges repel each other

PRO

PRO

Hexamers formed

Hexamers not formed

FIGURE 19.20
Engineered Fast-Acting Insulin
Natural insulin has a sticky patch around a proline residue, which causes two insulin molecules to dimerize and eventually form a hexamer. Using genetic engineering, the proline was replaced with a negatively charged aspartic acid residue. The negative charges repel each other and prevent hexamer formation.

Human trials using cloned human leptin have shown only moderate effects. Some test subjects lost a dozen pounds over 6 months. However, leptin has little effect on most very obese people, indicating that other factors are important. Leptin defects cause gross obesity, which appears almost from birth and is technically known as *morbid obesity*. This is a relatively rare condition in humans. Unlike obesity in mice, few humans have been found who have defective *ob* genes and whose obesity is due to lack of leptin. Another reason why leptin is unlikely to be widely used is that digestive enzymes degrade it. So leptin cannot be taken by mouth but has to be injected daily.

Other obese humans have defects in other components of the leptin regulatory circuit. Many obese humans actually have higher than normal leptin levels and may suffer from defects in the **leptin receptor**. Mice lacking leptin receptor, due to mutation of the ***db* gene**, have two to three times the normal levels of leptin in their blood. The leptin receptor is found on the surface of brain cells, especially in the hypothalamus. It binds leptin that is circulating in the bloodstream and so passes the signal on to the brain (Fig. 19.21).

When leptin binds to its receptor, it turns off the release of **neuropeptide Y** by the brain. Neuropeptide Y increases feeding and so makes animals fatter. In addition, leptin stimulates the release of **noradrenalin (= norepinephrine)** by the nervous system. This binds to **β3-adrenergic receptors** on the surfaces of fat cells and signals them to burn off fat. Partially defective β3-adrenergic receptors have been found in some overweight people.

The leptin receptor has a cytoplasmic tail that acts as a protein kinase and is responsible for transmitting the signal. When leptin binds, it initiates a phosphotransfer cascade that ultimately triggers release of noradrenaline. However, the mRNA for the leptin receptor undergoes alternative splicing to give proteins of different lengths (Fig. 19.22). The longer **isoform** is found in the hypothalamus and activates the multiple signaling pathways described earlier. The shorter isoform lacks the cytoplasmic kinase domain and its role is unclear, although it may bind leptin and carry it across the blood-brain barrier. As noted earlier, *db* mutants lack receptor completely. However, in *fa* mutants the receptor is present

A)
Hypothalamus
Secretion of neuropeptide Y (turned off by leptin)
Leptin receptor
Bloodstream
Fat cell
Leptin secreted into blood by fat cell

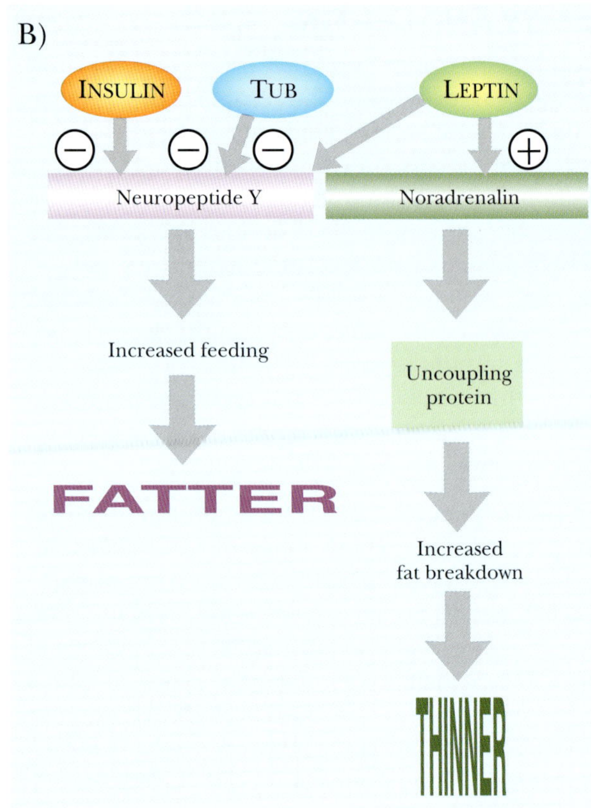

FIGURE 19.21 Leptin and the Leptin Receptor
(A) Fat cells make leptin and secrete it into the bloodstream. The hormone binds its receptor in the hypothalamus, blocking the secretion of neuropeptide Y and increasing the production of noradrenalin. (B) Neuropeptide Y stimulates appetite, whereas noradrenalin increases fat breakdown. Insulin and Tub (for "tubby"; see later discussion) also inhibit neuropeptide Y secretion.

but has a damaged signaling tail. (The *fa* mutation, found originally in rats, is in the *db* gene, encoding the leptin receptor, *not* in the *fat* gene.)

Leptin, a hormone encoded by the *ob* gene, binds to the leptin receptor (encoded by the *db* gene), which activates the intracellular kinase domain to block the release of neuropeptide Y. When leptin or the leptin receptor are defective, neuropeptide Y is made continually and stimulates appetite.

Box 19.2 Obesity Due to Virus Infection

Some are born fat, some achieve fatness by their own efforts, and, bizarre though it may seem, some have fatness thrust upon them by virus infection. This was first noticed among overweight hens in India that were infected with an often fatal adenovirus. After infection, surviving chickens gained up to 75% more fat. A related virus, **adenovirus-36**, infects both humans and chickens. In humans it causes coldlike symptoms and diarrhea. A survey of humans weighing more than 250 lb showed that around 15% had antibodies to adenovirus-36 in their blood, implying that they had been infected with the virus at some time in the past. Obese people usually have higher levels of cholesterol and fats in their blood. However, the adenovirus-36–positive subgroup had lower than normal blood cholesterol and fat, implying that their obesity had a different mechanism from the majority. Very few cases of prior adenovirus-36 infection were detected among people of normal size. The mechanism by which adenovirus-36 promotes obesity is unknown.

A MULTITUDE OF GENES AFFECT OBESITY

Fat accumulation is affected by many factors, both genetic and environmental. Most cases of obesity are not due to defects in leptin or its receptor. Nonetheless, the discovery that at least some cases of obesity are largely genetic has created a major research effort in this area. In addition, there are the potential profits from drugs for controlling appetite and body weight. An assortment of genes and proteins now known to affect body weight and fat accumulation is listed in Table 19.1.

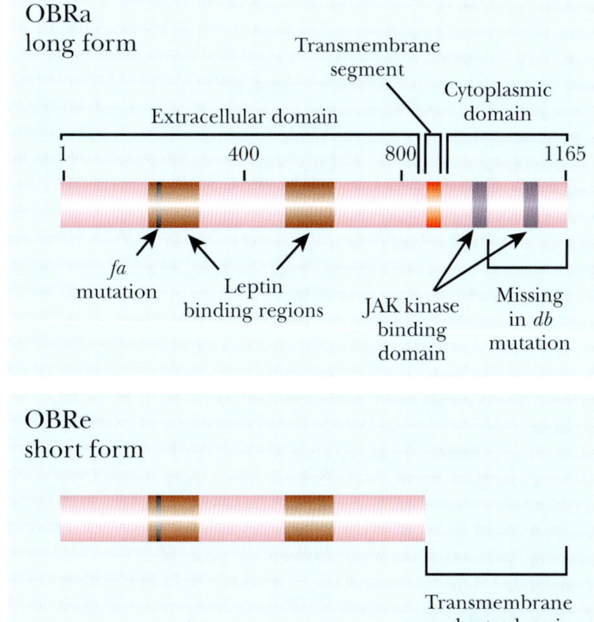

FIGURE 19.22
Alternative Forms of Leptin Receptor
The leptin receptor has two isoforms, which arise by alternate splicing of the *db* gene. The long form has a cytoplasmic domain with protein kinase activity. The short form lacks this domain.

This list gives a confusing choice of genes on which to blame obesity. It also gives a large number of possible targets for drug development. Unfortunately, a change in any one factor tends to affect the level of other components of the regulatory system. Consequently, the development of a "magic bullet" to control obesity is unlikely. Exactly how the overall control network is arranged remains for future research to uncover.

Insulin not only controls sugar levels in the blood but also plays a role in fat metabolism. Furthermore, insulin receptors are present on some brain cells, thus linking insulin to the control of eating and body fat. Insulin decreases feeding by regulating neuropeptide Y. As discussed earlier, insulin is secreted as the inactive precursor proinsulin, which must be processed to insulin in order to act. Carboxypeptidase E is an enzyme that processes proinsulin into active insulin. In mice referred to technically as "fat," there is a genetic defect in the gene for carboxypeptidase E. These mice therefore accumulate proinsulin. They also resemble many overweight people in growing fatter as they age, illustrating how lowered insulin levels contribute to fat accumulation.

Another set of regulators that affect fat metabolism is the **melanocortins** (Fig. 19.23). Pro-opiomelanocortin (POMC) is a precursor protein that is cleaved to give several different melanocortins. Of these, α-melanocyte stimulating hormone (αMSH) affects fat accumulation. There are also several different **melanocortin receptors**. Of these, melanocortin-4 receptor (MC4R) is needed by αMSH to promote fat breakdown. Defects in the *POMC* or *MC4R* genes may result in obesity. How melanocortin control interacts with other regulatory factors is still being debated.

Another defect is found in "tubby" mice, defective in the *tub* gene whose function is still uncertain. Defects in *tub* cause moderate accumulation of fat later in life. The Tub protein is present in the hypothalamus, the same area of the brain targeted by leptin. It also seems to regulate the level of neuropeptide Y, but by a mechanism different from leptin or melanocortin.

Ghrelin is a recently discovered 28-amino-acid peptide that has major effects on feeding behavior. The levels of ghrelin in blood rise as mealtimes approach and fall afterward, suggesting that ghrelin plays a major role in signaling hunger. Ghrelin was originally named as a "growth hormone releasing factor" but has other effects and appears to block leptin action. Injecting rats with ghrelin stimulates them to eat more and gain weight. Obesestatin is a related peptide hormone that opposes the action of ghrelin. Fascinatingly, both ghrelin and obesestatin are derived from processing of the same prohormone, proghrelin.

Many proteins and enzymes regulate the body mass index. The pathways and effects of these defects are still under investigation.

Insulin, melanocortin, melanocortin receptor, and ghrelin are all linked to fat accumulation and metabolism.

Table 19.1 Genes and Proteins Involved in Obesity

Gene	Protein Encoded
ob (for obesity)	leptin
db (for diabetic; in mice)	leptin receptor
fa (in rats)	leptin receptor
NPY	neuropeptide Y
NPY1R-NPY5R	neuropeptide Y receptors; the NPY5R receptor is probably most important in feeding
tub (for tubby)	brain protein, probably transcription factor
POMC	pro-opiomelanocortin, the precursor of αMSH (α-melanocyte stimulating hormone)
MC4R	melanocortin-4 receptor
AGRP	agouti-related protein, antagonist of melanocortin receptors
CART	cocaine- and amphetamine-regulated transcript
fat (in mice)	carboxypeptidase E (CPE)
PCSK1	prohormone convertase 1 (protease)
GHRL	ghrelin
GHRL	obesestatin
GHSR	G-protein coupled receptor for ghrelin
GPR39	G-protein coupled receptor for obestatin
ADRB3	β3-adrenergic receptor
UCP1	uncoupling protein 1 or thermogenin
UCP2	uncoupling protein 2
UCP3	uncoupling protein 3
AQP7	aquaporin 7

FAT DEGRADATION

A final key issue is the degradation of fat. Stored fat consists mostly of triglycerides (i.e., glycerol linked to three fatty acids). In adipose cells, triglycerides are hydrolyzed to release fatty acids and glycerol. The fatty acids are oxidized to CO_2 in the mitochondria, yielding energy. Normally, fat oxidation is coupled to energy conservation by the respiratory chain. Energy is used to generate a proton gradient across the mitochondrial inner membrane, which may be used to synthesize ATP. If the body needs less energy, then fat oxidation is normally decreased. Burning off surplus fat involves wasting the energy released, and this requires deliberate **uncoupling** of the respiratory chain (Fig. 19.24). This results in the energy being wasted as heat.

Uncoupling proteins (UCPs) are located in the inner membranes of mitochondria, especially in brown fatty tissue. Brown fat mitochondria are uncoupled when animals get too cold and need more body heat. Three related uncoupling proteins are known. UCP1 is normally found only in brown fatty tissue and was named **thermogenin** because it helps keep the body warm (Fig. 19.25). When noradrenalin binds to β3-adrenergic receptors on the surfaces of brown fat cells, preexisting UCP1 is rapidly activated. In addition, the same

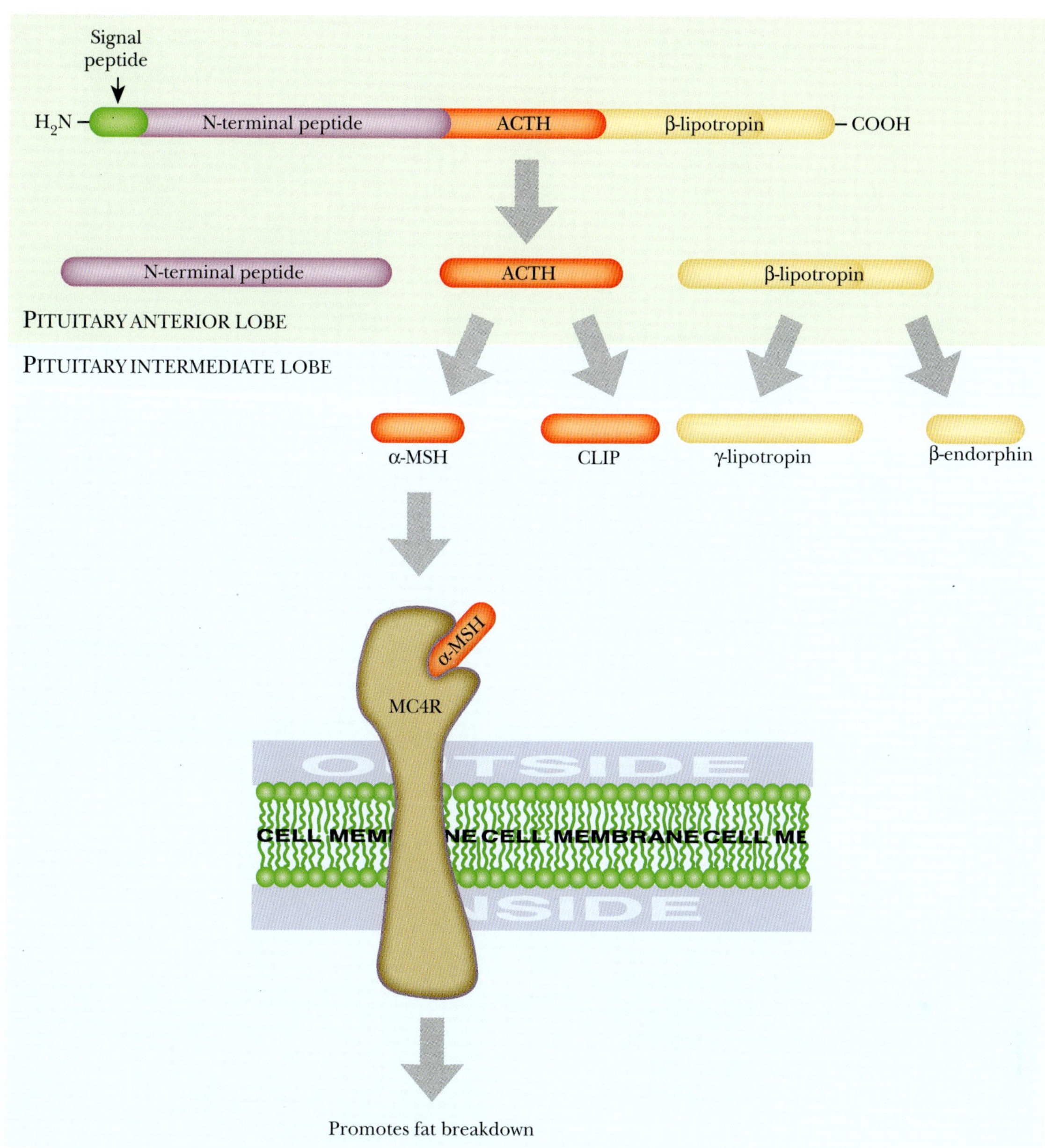

FIGURE 19.23 Melanocortins and Melanocortin Receptors

Pro-opiomelanocortin (POMC) has four domains: the N-terminal signal sequence promotes secretion; the N-terminal peptide, which has an unknown function; the adrenocorticotropin (ACTH) domain; and the β-lipotropin domain, which promotes lipolysis. POMC is processed into four different melanocortins as it passes from the anterior lobe of the pituitary to the intermediate lobe. The α-MSH peptide binds to the melanocortin receptor, MC4R, and promotes fat breakdown. The others are CLIP, γ-lipotropin, and β-endorphin.

stimulus over a longer time period increases expression of the *UCP1* gene, thus increasing the level of UCP1 protein. However, knockout mice lacking UCP1 are cold-sensitive but not obese.

It has been known for a long time that smoking makes people thinner. The active factor is the **nicotine** in tobacco. Nicotine not only acts to raise UCP1 levels in brown fatty tissue, but also induces the synthesis of UCP1 in white fat cells, where it is normally absent. Nicotine also decreases the appetite.

UCP2 is present in most tissues and UCP3 mainly in muscle. The amount of UCP2 goes up if mice eat a high-fat diet, and this presumably helps burn off the fat. Mice unable to make higher levels of UCP2 do put on weight. Certain mutations in the *UCP2* and *UCP3* genes,

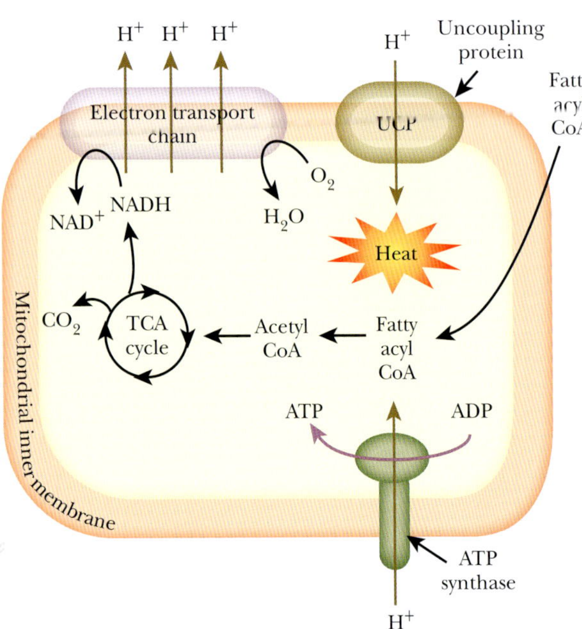

FIGURE 19.24 Fat Metabolism and UCP
Fatty acids are converted into acetyl CoA, which is fed into the TCA cycle for oxidation. The energy released ultimately allows the electron transport chain to pump protons into the intermembrane space. This creates a proton gradient that drives the synthesis of ATP. Alternatively, the protons may move back across the membrane via an uncoupling protein (UCP), which generates heat instead of ATP.

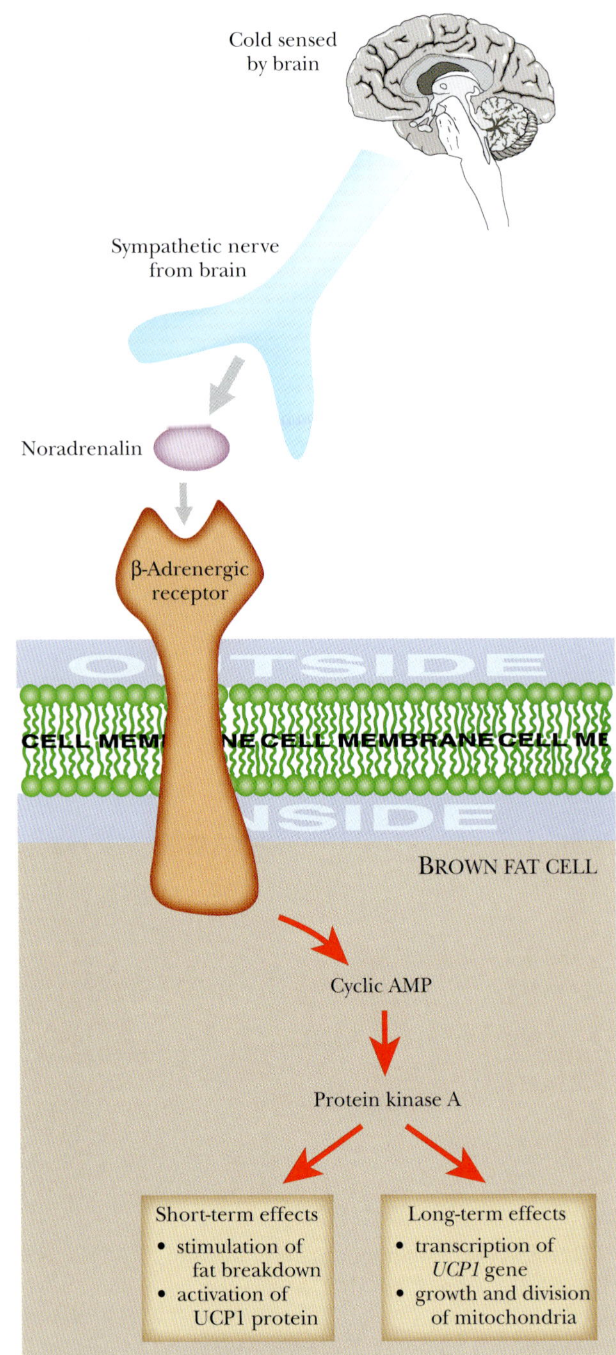

FIGURE 19.25 Cold and UCP1
When the brain senses that the body is cold, noradrenalin is produced. The hormone binds to the β-adrenergic receptor, which signals via cyclic AMP and protein kinase A. This prompts the cell to increase the production and activation of UCP1. The UCP1 uses the proton gradient of the mitochondria to generate heat.

which are both located close together on chromosome 11, have been linked to human obesity. However, the overall regulation and role of UCP2 and UCP3 are still poorly understood.

A curious recent finding is that a deficiency of aquaporin 7 promotes obesity. Although water and glycerol can diffuse across cell membranes, aquaporins greatly speed up this movement. In the absence of aquaporin 7, glycerol cannot exit fat-degrading cells rapidly enough. Consequently, triglyceride levels build up and fat is not broken down efficiently.

Fat metabolism creates a proton gradient in the inner mitochondrial membrane. Uncoupling proteins use the gradient to create body heat. Certain mutations in *UPC2* and *UPC3* are linked to obesity.

Box 19.3 Fierce Mice Cured by Human Brain Gene

Tackling human behavioral abnormality at the genetic level is difficult. A promising approach is to find or create genetic defects in mice that give rise to behavioral changes similar to those of humans. Mouse genes corresponding to human genes suspected of behavioral effects can be analyzed directly by making gene knockouts. This may be followed by testing the cloned human gene for possible reversal of defects seen in the knockout mice.

Mutant mice showing pathological aggression have been called "fierce mice." Deletion of both copies of the *NR2E1* gene (encoding nuclear receptor 2E1) results in mice with abnormal brain development that display extreme aggression. Their aggressive behavior is constant and maladaptive—fierce mouse males often kill their mates. The cloned human version of the *NR2E1* gene has been shown to cure the defects in brain development and restore normal behavior to fierce mice. Whether defects in the *NR2E1* gene are responsible for mental disorders in humans is uncertain. There is a correlation of bipolar disorder with the 6q21-22 region where the *NR2E1* gene is located. However, other candidate genes share this chromosomal location.

MONOAMINE OXIDASE AND VIOLENT CRIME

The interaction of genetics and behavior is complex. Note also that metabolism and behavior overlap. As discussed earlier, the genetic effects on obesity are not merely physiological. Feeding behavior is also affected by mutations in several of the genes controlling body weight.

Previous information in the area of behavioral genetics has been largely statistical in nature, but connections at the molecular level are beginning to emerge. For example, there appears to be a connection between violence and uptake of the neurotransmitter serotonin. Different alleles of the gene for the serotonin transporter are known and the lower activity variants correlate with aggressive behavior.

Another association with violent criminal behavior involves **monoamine oxidase (MAO)** activity in the brain. It has been known for many years that there is a statistical correlation between violence and the levels of MAO assayed in blood platelets. Lower MAO activity is found in violent criminals, both male and female. More recently, the underlying genetic defects have been characterized for a small number of extreme cases.

Signals are transmitted between nerve cells by a variety of chemical neurotransmitters that must be degraded once the signal has been received. Monoamine oxidase A is located in the outer membrane of the mitochondria of neurons and initiates degradation of neurotransmitters of the **catecholamine** family—dopamine, adrenalin (= epinephrine), and noradrenalin (= norepinephrine; Fig. 19.26). The gene for monoamine oxidase A (MAO-A) is located on the X chromosome, and individuals with deletions and point mutations are known. Alterations in the ***MAO-A* gene** result in

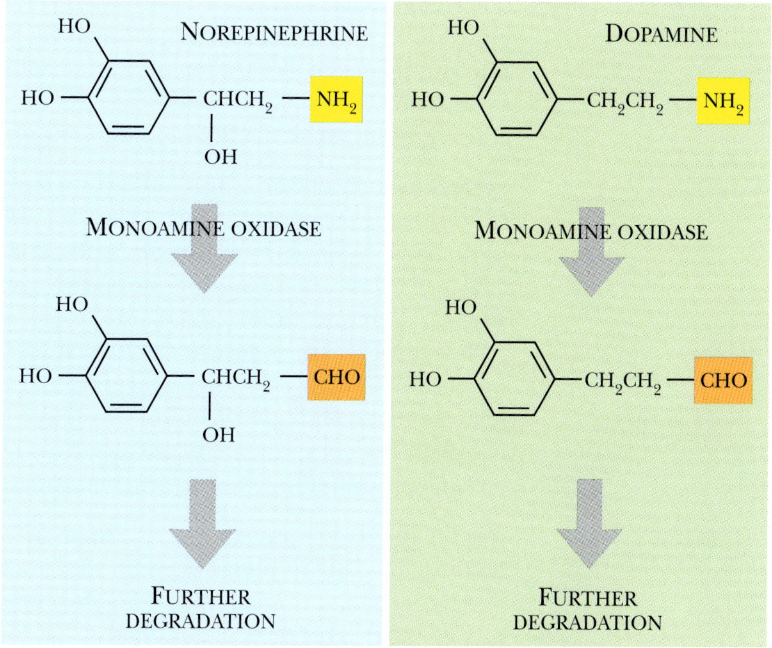

FIGURE 19.26
Monoamine Oxidase A Degrades Neurotransmitters
Monoamine oxidase A degrades norepinephrine by converting the amine group into the aldehyde group.

marked changes in monoamine metabolism and are associated with variable cognitive deficits and behavioral changes in both humans and transgenic mice.

A monoamine oxidase A defect has been linked to violence among males in a family in the Netherlands. More than a dozen males who showed marginal mental retardation and uncontrolled violent outbursts proved to have a point mutation in the eighth exon of the *MAO-A* gene. This mutation changed a codon (CAG) for Gln to a stop codon (TAG) and resulted in a truncated protein. No mutation was found in the *MAO-A* gene of normal brothers of the violent males. Females who gave birth to violent males were shown to have one normal and one mutant allele, as expected for carriers of a sex-linked defect. In affected individuals neurotransmitter levels would rise well above normal levels, especially under stressful situations. It seems likely that minor stress triggers a disproportionate response.

The incidence of severe MAO-A defects is extremely low and is unlikely to account for more than a tiny proportion of criminal behavior. On the other hand, it is conceivable that the wider correlation between MAO levels and violent crime is due to genetic alterations that result in moderate reduction of monoamine oxidase activity. This is a matter for further investigation.

Some people with violent behavior may have a genetic basis for their actions. Defects in the monoamine oxidase A gene have been linked to violent behavior and slight mental retardation.

Summary

Bacteria or viruses do not cause all the diseases found in humans. Many conditions are caused by defects in cellular communication. Normal cellular communication occurs via local mediators that affect only cells in the local vicinity of the signal, hormones that affect all the cells in the body with the proper receptors, and pheromones that affect nearby organisms. Vertebrates also have neurons that transmit signals throughout the body. Once the cell receives one of these signals (i.e., local mediator, hormone, pheromone, or neurotransmitter), the cell must transmit the signal to the nucleus. Some signals are transmitted via a phosphotransfer relay, where one phosphorylated protein passes the phosphate to another, which activates the second protein to pass the phosphate to another, and so on. Other signals are transmitted via second messenger molecules such as cAMP and cGMP. Still other signals are transmitted via ion channels. Some lipophilic hormones do not transmit their signals via a membrane receptor. Rather, these hormones readily diffuse into the nucleus or cytoplasm, where they bind a receptor. The receptor has a second domain where it binds DNA and activates transcription directly.

Second messengers are critical molecules for bacteria as well as humans. In bacteria, cAMP controls gene transcription by turning on operons associated with sugar metabolism. In humans, cAMP activates various genes, but also activates many different cytoplasmic enzymes by phosphorylation. Another second messenger, cGMP, mediates the signal for vascular dilation. Nitric oxide is a small gaseous molecule that activates cellular guanylate cyclase, which in turn makes cGMP. When NO is produced in endothelial cells, it diffuses into nearby muscle cells and triggers cGMP production, and the muscles of the blood vessels relax. In males with erectile dysfunction, the blood flow is restricted and blood vessels do not dilate. The drug sildenafil, or Viagra, increases the amount of cGMP by inhibiting the enzyme phosphodiesterase 5, which normally breaks down the cGMP.

Insulin and its receptor are the two main defects in diabetes. Insulin-dependent diabetes requires the patient to inject insulin so that glucose levels are kept low. Insulin is now made as a recombinant protein in bacteria. It is made by expressing two mini-genes and then mixing the two proteins in conditions that facilitate disulfide bonds. The final recombinant insulin is identical to human insulin, except that at high concentrations, the insulin tends to clump. This does not happen in the body because the insulin is never concentrated. To help reduce the clumping, a proline residue has been changed to arginine, which repels the other insulin proteins.

Obesity is a disease with a very complex etiology. Many different proteins and enzymes are involved with the production, storage, and metabolism of fat in the body. The environment as well as the genetics of a person triggers the regulators, so determining one cause for obesity is difficult. Some of the genes associated with obesity include *ob*, which encodes the hormone leptin, and *db*, which encodes the leptin receptor. Leptin and its receptor turn off the release of neuropeptide Y in the brain, which stimulates appetite and feeding. Other proteins and enzymes that control obesity include insulin, melanocortin, melanocortin receptor, ghrelin, and uncoupling proteins such as UPC2 and UPC3.

End-of-Chapter Questions

1. Which of the following is not involved in cellular communication?
   - **a.** pheromones
   - **b.** local mediators
   - **c.** hormones
   - **d.** mRNA
   - **e.** All of the above are involved in communication between cells.
2. How does signal transmission occur?
   - **a.** phosphotransfer systems
   - **b.** ion channel activation
   - **c.** second messengers
   - **d.** all of the above
   - **e.** none of the above
3. To which family do lipophilic receptors belong?
   - **a.** nuclear-receptor superfamily
   - **b.** Ras family
   - **c.** Xeno family
   - **d.** cAMP family
   - **e.** none of the above
4. Which of the following statements about cAMP is not true?
   - **a.** cAMP is made by adenylate cyclase.
   - **b.** In animals, cAMP binds to the CRP protein to regulate transcription.
   - **c.** cAMP is used in animals as a second messenger.
   - **d.** In bacteria, cAMP is made in response to nutrient availability.
   - **e.** Accumulation of cAMP in animals activates PKA.
5. Which of the following triggers the production of cGMP?
   - **a.** calcium ions
   - **b.** chloride ions
   - **c.** nitric oxide
   - **d.** cAMP
   - **e.** all of the above

(Continued)

6. What health issue do PDE5 inhibitors treat?
   a. blue-green color blindness
   b. erectile dysfunction
   c. anxiety
   d. diabetes mellitus
   e. angina

7. Which of the following is not involved in the manufacturing of insulin?
   a. endopeptidases
   b. carboxypeptidase H
   c. signal peptidase
   d. phosphodiesterase 5
   e. All of the above are involved in the manufacturing of insulin.

8. Which of the following reacts to low blood sugar by breaking down glycogen?
   a. glucagon
   b. insulin
   c. MAPK
   d. Ras
   e. PDE5

9. Which of the following statements about insulin manufacturing is not true?
   a. Diabetics used to have to inject themselves with insulin that had been purified from the pancreas of either cows or pigs.
   b. Bacteria are able to express preproinsulin and process this peptide into insulin.
   c. Natural insulin forms hexamers, which causes the hormone to clump.
   d. A proline converted to an aspartic acid keeps insulin from clumping.
   e. Recombinant insulin is expressed as two mini-genes.

10. In experimental mice, what has been the issue with using stem cells to treat diabetes?
   a. The injected stem cells eventually continue to grow and divide, causing formation of a teratoma.
   b. The injected stem cells never produce insulin in the pancreas.
   c. Ethical issues have prevented researchers from using stem cells in mice.
   d. The injected stem cells fail to produce enough insulin to make any difference.
   e. Diabetic mice are caused by defects in the insulin receptor instead of a lack of insulin, so this procedure has no affect.

11. For what does the *ob* gene code in humans?
   a. leptin
   b. collagen
   c. leptin receptor
   d. insulin receptor
   e. none of the above

12. For what does the *db* gene code in humans?
   a. leptin
   b. insulin receptor
   c. collagen
   d. leptin receptor
   e. none of the above

13. What is ghrelin?
   a. a hormone that controls the level of insulin
   b. a signal molecule that notifies the brain that the body does not need food
   c. a recently discovered peptide that affects feeding behavior
   d. a hormone that controls fat accumulation
   e. none of the above

14. Which of the following has not been linked to obesity?
   a. UCP2
   b. UCP3
   c. aquaporin
   d. MAO
   e. All of the above have been linked to obesity.

15. A mutation in which of the following was linked to violence in males within a family from The Netherlands?
   a. monoamine oxidase A
   b. UCP1
   c. melanocortin receptor
   d. ghrelin
   e. *ob* gene

Further Reading

Altucci L, Leibowitz MD, Ogilvie KM, de Lera AR, Gronemeyer H (2007). RAR and RXR modulation in cancer and metabolic disease. *Nat Rev Drug Discov* **6**, 793–810.

Atkinson RL (2007). Viruses as an etiology of obesity. *Mayo Clin Proc* **82**, 1192–1198.

Bian K, Ke Y, Kamisaki Y, Murad F (2006). Proteomic modification by nitric oxide. *J Pharmacol Sci* **101**, 271–279.

Cary SP, Winger JA, Derbyshire ER, Marletta MA (2006). Nitric oxide signaling: No longer simply on or off. *Trends Biochem Sci* **31**, 231–239.

Corthésy-Theulaz I, den Dunnen JT, Ferré P, Geurts JM, Müller M, van Belzen N, van Ommen B (2005). Nutrigenomics: The impact of biomics technology on nutrition research. *Ann Nutr Metab* **49**, 355–365.

Craig IW (2007). The importance of stress and genetic variation in human aggression. *Bioessays* **29**, 227–236.

Hofbauer KG, Nicholson JR, Boss O (2007). The obesity epidemic: Current and future pharmacological treatments. *Annu Rev Pharmacol Toxicol* **47**, 565–592.

Josselyn SA, Nguyen PV (2005). CREB, synapses and memory disorders: Past progress and future challenges. *Curr Drug Targets CNS Neurol Disord* **4**, 481–497.

Jun HS, Yoon JW (2005). Approaches for the cure of type 1 diabetes by cellular and gene therapy. *Curr Gene Ther* **5**, 249–262.

Nogueira FT, Borecký J, Vercesi AE, Arruda P (2005). Genomic structure and regulation of mitochondrial uncoupling protein genes in mammals and plants. *Biosci Rep* **25**, 209–226.

Palamara KL, Mogul HR, Peterson SJ, Frishman WH (2006). Obesity: New perspectives and pharmacotherapies. *Cardiol Rev* **14**, 238–258.

Rankinen T, Zuberi A, Chagnon YC, Weisnagel SJ, Argyropoulos G, Walts B, Pérusse L, Bouchard C (2006). The human obesity gene map: The 2005 update. *Obesity (Silver Spring)* **14**, 529–644.

Reif A, Rösler M, Freitag CM, Schneider M, Eujen A, Kissling C, Wenzler D, Jacob CP, Retz-Junginger P, Thome J, Lesch KP, Retz W (2007). Nature and nurture predispose to violent behavior: Serotonergic genes and adverse childhood environment. *Neuropsychopharmacology* **32**, 2375–2383.

Seftel AD (2005). Phosphodiesterase type 5 inhibitors: Molecular pharmacology and interactions with other phosphodiesterases. *Curr Pharm Des* **11**, 4047–4058.

Yang W, Kelly T, He J (2007). Genetic epidemiology of obesity. *Epidemiol Rev* **29**, 49–61.

CHAPTER 20

Aging and Apoptosis

Aging is a process we all start at birth and continue until death. Some people seem to age very gracefully, whereas others show signs of aging very early. Even ignoring the effects of infection or accident, quite a large range of life spans occurs among humans. Interestingly, there is a dramatic difference in the average life span of similar sized mammalian species. For example, mice live about 2-3 years, whereas other rodents of similar size average about 5–10 years. So age is a relative term with differences within and between species. Until recently, there was no clear molecular reason for aging other than slow accumulation of wear and tear on our cells, tissues, organs, and ultimately on our bodies as a whole. Until mammalian cells were cultured in the laboratory, understanding the molecular changes that occur over time was simply impossible. Many people try to offset the signs of aging, and spend vast amounts of money on products that claim to repair or mask the visible signs. Not surprisingly, the biotechnology industry hopes to tap into this flow of money by finding biological solutions to aging.

CELLULAR SENESCENCE

When normal adult cells from a mouse or a human are grown in laboratory dishes, the cells divide for a certain number of generations and then stop dividing. Even the addition of growth inducers has no effect. However, the cells do not die as long as they are fed and maintained properly. This is called **cellular** or **replicative senescence**. When cells are isolated from a mammal, they have an internal clock that controls when senescence will occur. Human fibroblasts from a fetus will divide 60 to 80 times in culture, whereas fibroblasts from an older person only divide 10 to 20 times. Replicative senescence depends on the number of cell divisions, not the calendar. The allowed number of divisions is programmed into the cell, rather than being controlled by circulating hormones or surrounding tissues. Additionally, cells from animals with short life spans divide fewer times than cells from animals with long life spans; therefore, replicative senescence correlates with the life span of the organism itself.

There are three main characteristics associated with replicative senescence. First, the senescent cells arrest their cell cycle in G_1 and never enter S phase (see Chapter 4). The cells are metabolically active, that is, they produce proteins, generate energy, and function in their normal capacity, yet the cells do not replicate their DNA or divide. Second, many become **terminally differentiated**. In other words, once cell divisions are over, the cell specializes in a particular function. For example, immature melanocytes divide until the alarm bell rings on the senescence clock; the melanocytes then stop dividing and produce melanin to protect skin tissue from sun damage. Both terminally differentiated and senescent cells no longer divide, but the terminally differentiated cells have also changed their physiology. Cells can senesce without changing their physiological role. Finally, senescent cells become resistant to apoptosis or programmed cell death (see later discussion).

Interestingly, cellular senescence varies among species. Often mouse tissue is used in laboratory settings because it is easy to obtain. Because mice are mammalian, many parallels are made to humans. However, when mouse cells are cultured *in vitro*, a small percentage of the cells will never senesce. Eventually, the nonsenescent cells will outnumber the senescent ones, and a dish of immortal cells is obtained. Such an escape from senescence is never seen in cultured human cells.

If all our cells entered a senescent state, then we would never be able to heal a wound or recover from damage due to bacterial or viral attack. Therefore, some of our cells never enter a senescent state. Obviously during embryogenesis and development very few cells are senescent. Developing bodies are filled with **stem cells**, immature or precursor cells from which differentiated cells originate. As we age, the number of stem cells decreases, but some remain to replenish various tissues, especially the skin, intestinal lining, immune system, and blood cells. Beside stem cells, germline cells do not enter senescence and always maintain the ability to divide when necessary. Tumor cells are another example of nonsenescent cells. Unlike stem cells and the germline, tumor cells have mutated in a way that overrides the entry into senescence.

Cellular senescence is a state where a cell is alive, but no longer divides. Some senescent cells are terminally differentiated to perform a function for the organism. Senescent cells are resistant to apoptosis.

FACTORS THAT ACTIVATE SENESCENCE

Many different factors activate senescence in cells and have been implicated in the aging process. In normal cells, the length of the **telomeres** controls whether or not cells senesce. During the replication of chromosomes, DNA polymerase cannot replicate the DNA at the distal tips of the chromosomes. Consequently, each chromosome shortens each time the cell divides (Fig. 20.1). When the telomeres reach a critically short length, the cell enters senescence.

Another factor that activates senescence and may contribute to aging is DNA damage. Normal metabolic activity produces a variety of **reactive oxygen metabolites (ROM)** that attack various proteins, lipids, and the DNA. This damage accumulates until it reaches a threshold. After that point, the cells may initiate senescence or apoptosis, depending on the severity and extent of the damage (Fig. 20.2). Consequently, highly defective cells do not accumulate in the body.

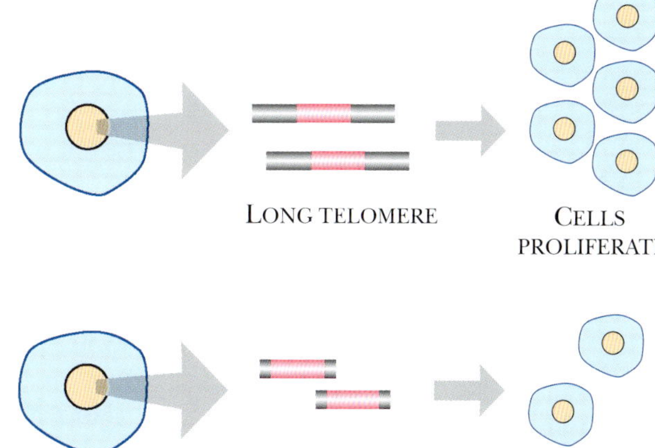

FIGURE 20.1 Telomere Shortening Causes Cellular Senescence A cell with long telomeres is able to continually divide and multiply. A cell with short telomeres may divide only one more time before entering senescence.

Cellular senescence is linked to the length of telomeres and the amount of DNA damage due to reactive oxygen metabolites.

LINKS BETWEEN CANCER AND AGING

The third major factor that triggers cellular senescence is the presence of oncogenic mutations, that is, mutations that promote cancer (see Chapter 18 for details). To defend the organism, tainted cells sacrifice themselves by starting the genetic program for senescence or apoptosis, thus entering **premature senescence**. This brings us to the question of why cells have a deliberate senescence program. The most prevalent theory is simply to avoid cancer.

There are three key pieces of evidence for this theory. First, malignant tumors have cells that are immortal and grow indefinitely because they have bypassed the normal senescence program. Second, oncogenes that induce a normal cell to become malignant often bypass or block senescence, thus extending the life span of the cell. Third, loss of **p53** or **retinoblastoma protein (pRb)** prevents a cell from entering senescence. Many cancers have mutations in one or both of these two proteins, which are both involved in control of the cell cycle (see Chapter 18).

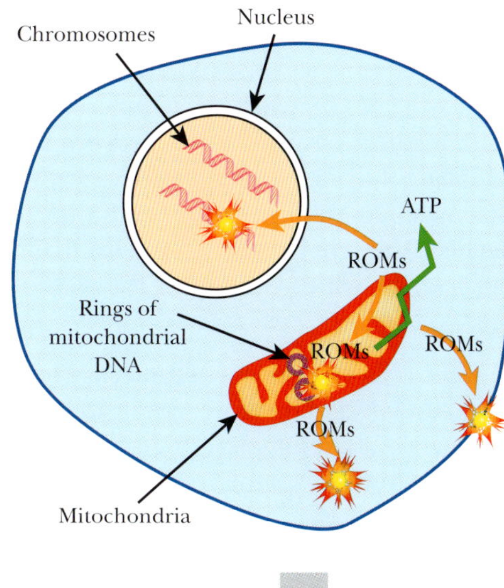

FIGURE 20.2 DNA Damage Activates Senescence Mitochondrial respiration produces ATP, but also generates reactive oxygen metabolites that damage cell components. If the damage is extensive, the cell will senesce or die.

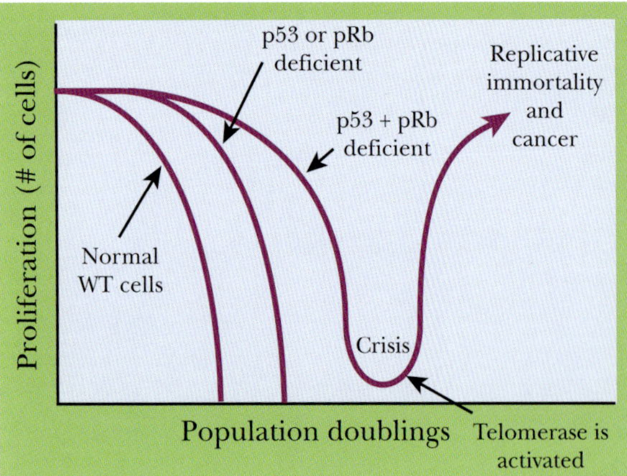

FIGURE 20.3
Defects in p53 and pRb Extend Cellular Life span
Normal cells proliferate for a specific time and then stop dividing. Cells with mutations in p53 or pRb continue to divide and replicate longer that normal. If both p53 and pRb are mutated, the cells divide for even longer. Eventually, the mutants do stop dividing unless some further oncogenic mutation occurs.

If either of p53 or pRb is defective, cells replicate for more cycles than usual (Fig. 20.3). If cells are defective for both p53 and pRb, the effect is additive and the replicative life span is longer than with either mutation alone. Nonetheless, such double mutants do eventually enter senescence, after an extended number of cell divisions. Thus cells have other mechanisms to trigger senescence. The p53 and pRb proteins are also implicated as key regulators of cellular senescence because they are the targets for several viral oncogenes.

p53 activates several cellular programs in response to stress, including cellular senescence, apoptosis, growth arrest, and DNA repair. Its expression is tightly regulated, because both a surplus and a shortage of p53 are detrimental. The p53 protein is a transcription factor that regulates its own gene as well as a variety of other genes. In particular p53 activates transcription of the genes for two other regulators, **PUMA (p53-upregulated modulator of apoptosis)** and **SLUG**. PUMA is a member of the Bcl-2 protein family that activates Bax, thus promoting apoptosis (see later section on Mammalian Apoptosis). Conversely, SLUG protein blocks the transcription of PUMA and opposes apoptosis. Thus, the balance between PUMA and SLUG determines the choice between trying to save a damaged cell via growth arrest plus DNA repair or eliminating the cell via apoptosis. The role of p53 in the death or survival of a damaged cell is very complex and is still being investigated.

When a cell becomes damaged, it enters senescence early. Severely damaged cells commit suicide via apoptosis. Damaged cells that bypass the senescence program and block apoptosis are called cancer.

p53 and pRb are proteins that regulate senescence. Mutations in these genes make the cells resistant to senescence and often cause cancer.

TELOMERES SHORTEN DURING AGING

That all cells normally go senescent implies that there is an internal clock that drives this phenomenon. The clock alarm may be triggered early by mutant oncogenes, but usually the alarm goes off when the telomeres get dangerously short. Thus, normal human chromosomes have telomeres between 5 and 15 kb long, whereas senescent cells have telomeres 4 to 7 kb long. Telomeres are highly repetitive sequences at the ends of eukaryotic chromosomes that get shorter each time the chromosomes replicate (see Chapter 4 for details). They are bound by proteins that protect the ends of the DNA from double-stranded break repair, which would fuse the ends of different chromosomes together—a disastrous mistake.

In some cells, the enzyme **telomerase** repairs the shortened ends of the chromosome (Fig. 20.4) by adding more repeats to the ends. It is particularly active in reproductive cells. Telomerase is a nontypical reverse transcriptase with two components, protein and RNA. The protein catalyzes the synthesis of new DNA, and the RNA acts as a template.

In most human cells, telomerase activity is very low, and so these cells are highly susceptible to senescence. In contrast, mice have significant telomerase activity in most cells and have much longer telomeres (approximately 40 to 60 kb). These differences are evident when cells are cultured. Mouse cell lines divide many more times before entering senescence and sometimes become immortal spontaneously. Mice have been engineered so as to delete the RNA component of telomerase. These mice appear normal for a few generations, although the telomeres shorten with each generation, because they can no longer be extended. At a critical point in telomere length, the mice begin to show chromosomal fusions, higher cellular senescence, and increased apoptosis. Thus, the mouse cells come to resemble human cells (which normally have shorter telomeres).

The human inherited disease dyskeratosis congenita (DKC) is linked to lower telomerase activity. People with DKC have blocked tear ducts, learning difficulties, pulmonary disease, graying and loss of hair, and osteoporosis, to name just a few of the problems. Interestingly, many of these are also symptoms of aging. The gene responsible for DKC encodes a protein that processes rRNA precursors in the nucleolus. In DKC patients the level of the RNA component of telomerase was much lower than normal and the telomeres were unusually short. Some of the symptoms of DKC are probably due to improper telomere length.

Telomere length is a critical issue in cancer also. Because cancer cells divide beyond the normal stopping point, their telomeres become extremely short. This contributes to the genetic instability associated with cancer. If a tumor progresses, its cells will eventually turn on telomerase activity, which restabilizes the genome. Of course, many rearrangements may have already occurred by then.

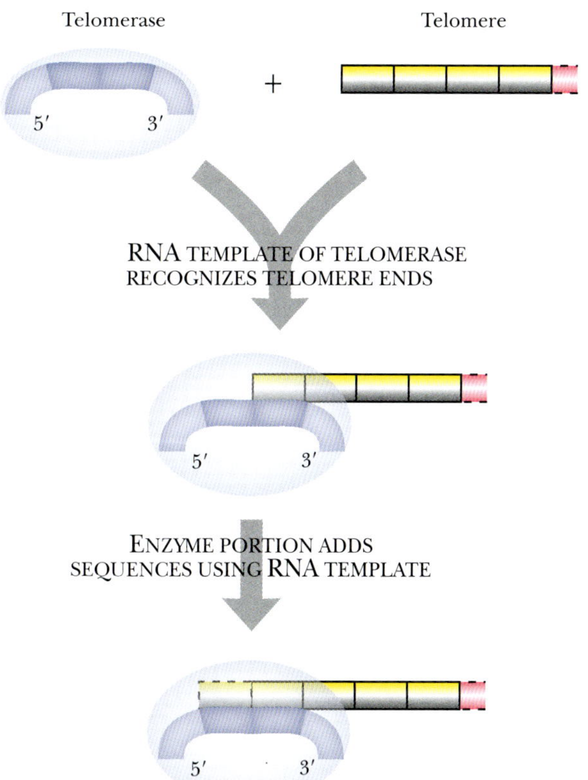

FIGURE 20.4 Telomerase Increases the Length of Telomeres
Telomerase consists of a protein plus an RNA component. The RNA recognizes the telomere repeats, and the protein elongates the telomeres.

Telomeres shorten because DNA polymerase cannot synthesize the end of the chromosome. The enzyme telomerase can lengthen telomeres in some cells.

When telomeres become too short, cells enter into senescence. If senescence is blocked, as in cancer, the chromosomes become unstable, triggering rearrangements and fusions.

MITOCHONDRIA AND AGING

As described earlier, cellular senescence and aging can be attributed to oxidative stress. Indeed, many people consider oxidative damage the main culprit in aging. Because the majority of oxidizing radicals are formed by aerobic respiration, the mitochondria play a central role here. In brief, mitochondrial electron transport, which manufactures most of the ATP, uses about 85% of the oxygen. Partial reduction of the O_2 molecule creates highly reactive species including superoxide ions ($\bullet O_2^-$), peroxides (H_2O_2), and hydroxyl radicals ($OH\bullet$). These reactive oxygen metabolites damage protein, lipid, and DNA nonspecifically (Fig. 20.5).

Mitochondrial DNA (mtDNA) is a main target for ROM because it is not protected by histones and is close to the electron transport chain. If the mtDNA is oxidized, mutations may accumulate in the genes for energy production that are carried on the mtDNA. This may give rise to a defective electron transport chain. Hence the mitochondria produce less energy and cellular processes slow down. Moreover, defective electron transport proteins may produce even more ROM, which damages the mtDNA further, so starting a vicious cycle. Remember that each mitochondrion has multiple copies of its genome, each cell has many mitochondria, and each tissue has many cells. Damage from many events must accumulate for its effects to be noticeable. Cells that are highly metabolically active tend to accumulate oxidative damage faster.

Antioxidant enzymes can alleviate the effects of oxidant stress. There are several of these with varied roles. The two most familiar are **superoxide dismutase (SOD)** and **catalase**.

ROMs attack mtDNA, nuclear DNA, lipids and cellular proteins

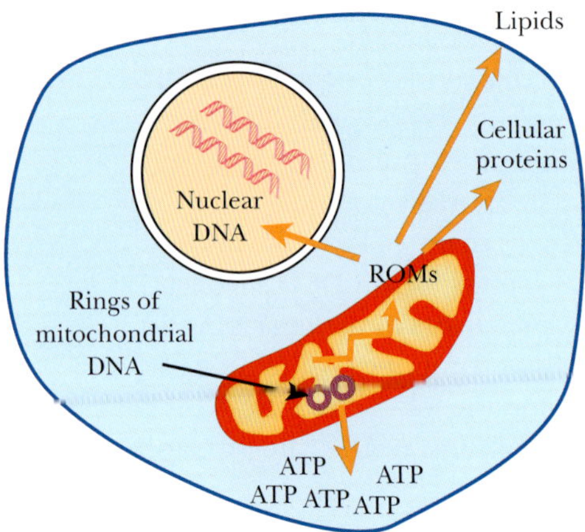

Mutated mtDNA make defective electron transport proteins

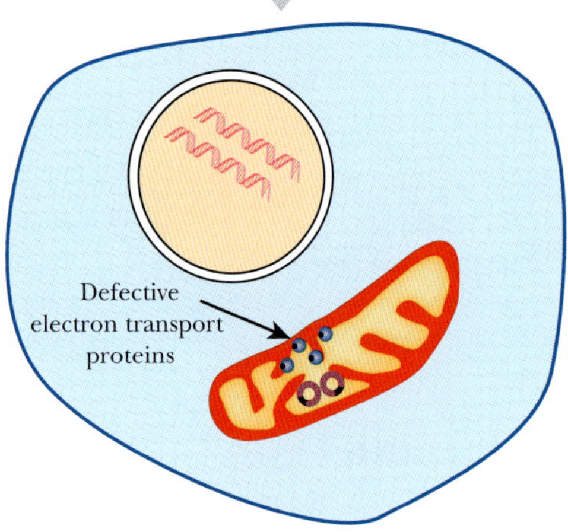

Mutated electron transport proteins make more ROMs and less ATP

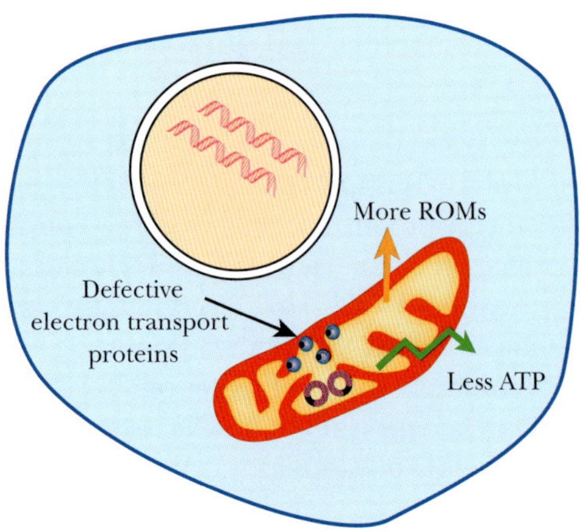

FIGURE 20.5 Oxidative Stress Damages Surrounding Molecules
Reactive oxygen metabolites (ROM) attack mitochondrial DNA, cellular proteins, lipids, and nuclear DNA. Damaged mitochondrial DNA may produce mutant respiratory chain proteins, which in turn make more ROM and/or less ATP.

There are two forms of SOD with different metal ions at the active site. Cu/Zn SOD is found primarily in the cytoplasm, whereas Mn SOD is found in the mitochondria. SOD converts superoxide ions to oxygen plus hydrogen peroxide. Catalase converts hydrogen peroxide to oxygen plus water.

Superoxide dismutase: $2 \bullet O_2^- + 2H^+ \rightarrow O_2 + H_2O_2$

Catalase: $2H_2O_2 \rightarrow O_2 + 2H_2O$

If the microscopic transparent worm *Caenorhabditis elegans* is treated with antioxidants, its mean life span increases significantly—by as much as 54%. These worms did not have a smaller body size or reduced fertility, suggesting there was little nonspecific reduction in metabolism. Other studies with the fruit fly, *Drosophila melanogaster*, support this view. Adding antioxidants to the diet of *Drosophila* has varied effects on the average life span. For example, vitamin E (alpha-tocopherol, a free radical scavenger) increased the life span by 13%, but vitamin C had no effect. Three studies have shown that the antioxidant thioproline increases life span, but the effect was highly variable (6% to 30% increase). Genetically engineered flies with one extra gene for SOD or catalase gene showed no change in life span. However, flies with three copies of SOD and three copies of catalase gained a one-third increase in life span due, presumably, to less oxidative damage.

Because metabolism and oxidation are linked to aging, if the metabolic rate decreases, less damage should accumulate, and the organism should live longer. Decreasing the body temperature lowers the metabolic rate, as seen in cold-blooded or hibernating animals. Such animals are able to survive on their body fat for extended periods of time by lowering the heart rate and slowing metabolism. Decreased physical activity also lowers the metabolic rate, as does limiting the diet. To test this theory, mice were put on a calorie-restricted diet. This decreased blood glucose and insulin levels. The mice also had improved insulin sensitivity and lower body temperature. In support of the theory that calorie restriction limits oxidative damage, the rodents had fewer reactive oxygen metabolites in the mitochondria and accumulated less oxidative damage over their life span. The greatest effect was seen in the brain and skeletal muscle, both tissues with high metabolic rates.

Although these results are promising, the jury is still out on whether antioxidants can lengthen the life span of all organisms. Many factors can contribute to the oxidation of proteins or DNA and subsequent aging of the organism. Further research must be done to determine the exact role each factor plays.

Reactive oxygen species are formed from the partial reduction of oxygen. These attack the mitochondrial DNA and the electron transport chain enzymes, because these are in closest proximity to oxygen reduction.

Less cellular damage may occur if ROMs are lessened by caloric restriction or by adding antioxidant enzymes.

LIFE SPAN AND METABOLISM IN WORMS

The role of metabolism and nutrition in aging has been revealed by studying the nematode worm *C. elegans*, which normally lives for 2 to 3 weeks. Several mutations are known that increase the life span and/or the worm's entry into a type of hibernation called **dauer** (Fig. 20.6). Normally these worms go through four larval stages before they become adults. When there is not enough food or water, a worm goes into the dauer stage. It produces a thicker cuticle to prevent dehydration and closes off its mouth and anus so it cannot eat. It does not feed or move until stimulated with more food or water. Some mutations in the genes *daf-2* and *age-1* cause the worms to enter the dauer stage. Many of these mutations are **temperature sensitive**, that is, the mutation is fully expressed only at higher temperature. Other *daf-2* or *age-1* mutations cause an increased life span, of up to 2 months, without inducing the dauer stage.

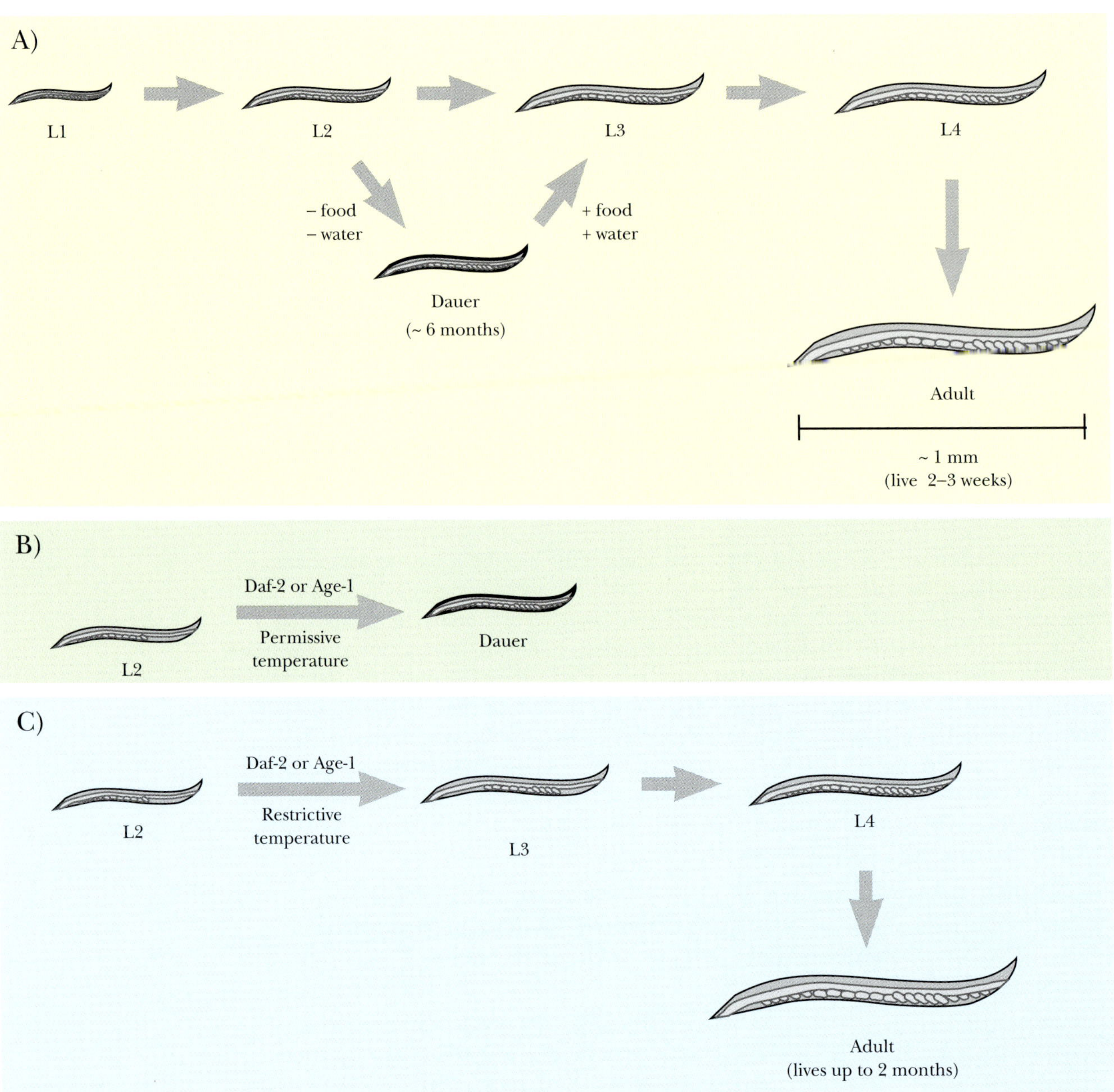

FIGURE 20.6 Effects of *daf-2* and *age-1* on *C. elegans* Life Span and Hibernation
(A) Normal worms go through four larval stages before becoming adults, which live about 2 to 3 weeks under normal conditions. If worms encounter starvation while in L2 they can hibernate as a dauer for up to 6 months. (B) Some *daf-2* or *age-1* mutations cause the worms to enter the dauer state no matter what their nutritional status. (C) Other *daf-2* or *age-1* mutations do not induce the dauer state but result in greatly increased life span.

The Daf-2 and Age-1 proteins both regulate metabolic rate (Fig. 20.7). According to sequence data, *Age-1* encodes a worm homolog to the catalytic subunit of phosphatidylinositol-3-kinase (PI3K), and *daf-2* encodes a homolog to the insulin receptor. The mammal homologs of these two proteins are involved in insulin signaling (see Chapter 19). When this pathway is impaired in humans, we develop type II diabetes (adult-onset). The worm mutations in *daf-2* and *age-1* block the signal transmitted by the insulin-like receptor. By analogy with the human system, the worm mutations probably affect glucose levels, lower the metabolic rate, and hence extend life span. In support of this, *daf-2* and *age-1* worms show lower metabolic rates than wild-type and more resistance to oxidative damage.

Another set of mutations in *C. elegans* also increases the average life span of the worm. These genes are called *clk*, for abnormal function of biological clocks, and *gro*, for growth rate

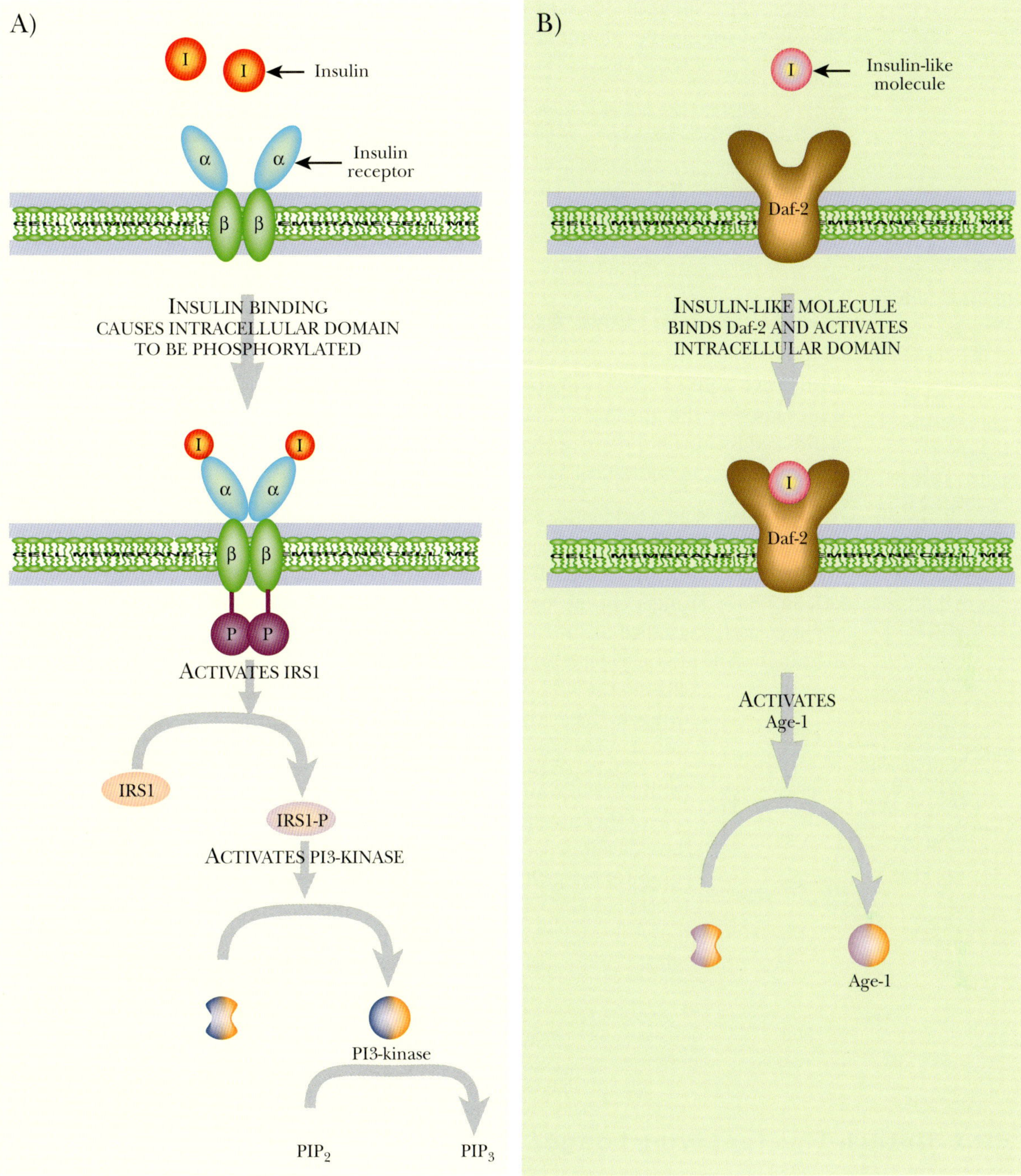

FIGURE 20.7 Insulin Signaling in Humans and *C. elegans*

(A) When insulin binds to the human insulin receptor, the intracellular domain is activated by phosphorylation. The phosphate is transferred via IRS-1 to PI3-kinase (see Chapter 19), which converts the lipid molecule phosphatidylinositol 3,4-bisphosphate (PIP_2) to phosphatidylinositol 3,4,5-trisphosphate (PIP_3) by adding a phosphate. PIP_3 trigger the intracellular response to insulin. (B) In worms Daf-2 protein binds an insulin-like molecule and activates Age-1 protein. This protein is similar to the PI3-kinase of mammals.

abnormal. These mutations lengthen the time it takes for the worm to go through embryonic development and postembryonic development, ultimately increasing the life span. The whole developmental process is delayed. Unlike *daf-2* and *age-1*, mutations in *clk* or *gro* have absolutely no effect on dauer formation. The closest homolog of CLK-1 is a yeast protein that is involved in growth regulation and allows the cells to switch from a fermentable carbon source like glucose to a nonfermentable one. The yeast protein controls the synthesis of ubiquinone, an electron carrier involved in mitochondrial respiration. So, once again, the extension of

C. elegans life span boils down to an altered rate of metabolism. The mitochondria of *clk* mutants release smaller amounts of reactive oxygen metabolites, and aging is slowed.

Daf-2 and *age-1* are genes that encode insulin metabolism-like genes in *C. elegans*. Mutations in these increase life span of the worm, or send the worm into hibernation. These may extend life span because the mutations slow the worm's metabolic rate.

Mitochondria of *clk* mutant *C. elegans* release smaller amounts of reactive oxygen metabolites, and the worms stay at each stage of development for longer than average.

SIRTUINS, HISTONE ACETYLATION, AND LIFE SPAN IN YEAST

Despite being a single-celled eukaryote, yeast undergoes aging of two different types, **replicative aging** and **chronological aging**. Replicative aging applies to dividing cells and is measured by the number of daughter cells one mother can produce. Chronological aging is induced by nutrient deprivation. The length of time the cell can live without dividing determines its chronological age.

Aging in yeast is affected by the Sir2 protein, a member of the sirtuin family of histone deacetylases. (Remember that eukaryotic gene expression involves remodeling of the nucleosomes, which in turn is affected by the level of histone acetylation.) Yeast cells that lack Sir2 have a longer chronological life span. They have a decreased rate of DNA mutation and increased resistance to stressful conditions such as heat shock and oxidative stress. The mammalian homolog to Sir2, SirT1, controls glucose and fat metabolism. Mammalian cells lacking the *sirt1* gene survive longer in culture, and transgenic mice lacking this gene have lower body fat and live longer. Sir2 is thus implicated in the caloric restriction pathway discussed earlier.

FIGURE 20.8 Sirtuin Inhibitors

Several compounds based on the indole ring structure have been identified as possible sirtuin inhibitors by high-throughput screening. The indole nucleus consists of the benzene ring fused to the five-membered ring containing nitrogen. The R group may be H, CH_3, or Cl in different members of the series.

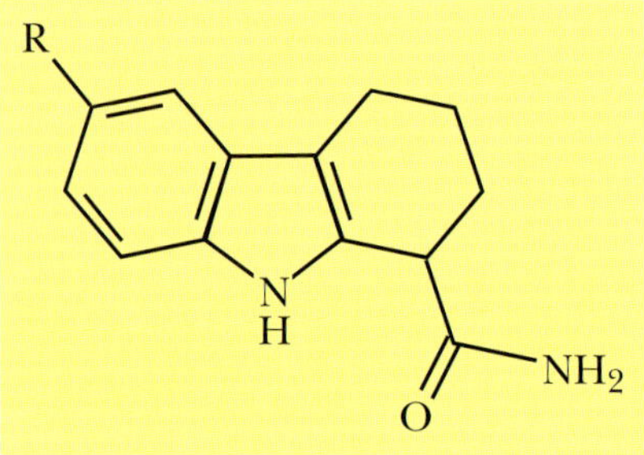

R N H NH_2 O

Inhibitors of sirtuins are being developed as possible anti-aging drugs. Indeed, one of the biotech companies involved is named Elixir Pharmaceuticals. They have developed a series of modified indoles as specific and potent sirtuin inhibitors (Fig. 20.8). Sirtuins act by transferring the acetyl group from the histone to NAD, giving *O*-acetyl-ADP-ribose plus nicotinamide. The indole-based sirtuin inhibitors bind after the release of nicotinamide and prevent the release of the histone and the *O*-acetyl-ADP-ribose.

Sirtuins are deacetylases that remove acetyl groups from histones. When these proteins are defective, yeast cells, cultured mammalian cells, and mice have longer life spans.

Box 20.1 Biotech Tips for Living Longer

Although we cannot increase our own life spans by gene knockouts, there are a variety of less drastic measures that can increase the odds of living longer.

Eat less overall to reduce the production of reactive oxygen intermediates, the buildup of fat in blood vessels, and other problems. Animals on restricted diets have shown significant increases in longevity, and humans who are overweight have shortened life expectancies.

Eat more antioxidants to combat reactive oxygen intermediates. Brightly colored fruits and vegetables usually contain high levels of these, as do coffee, dark chocolate, and red wine. Try purple potatoes and other unusual-colored versions of standard vegetables.

Drink moderate amounts of alcohol—those who consume two to three drinks per day (one to two for females) live longer than both nondrinkers and heavy drinkers. One theory is that alcohol simply acts as a solvent to help dissolve away fat deposits in blood vessels. Other theories suggest complex metabolic and regulatory effects (so far unproven).

Get a pet—even nontransgenic cats and dogs lower the blood pressure of their owners. (Yes, this has been directly measured—it is genuine science, not urban myth!)

APOPTOSIS IS PROGRAMMED CELL DEATH

Have you ever wondered what happens to a tadpole's tail when it turns into a frog, or what happens to all the extra white blood cells after you are over an infection? What happens to senescent cells that are defective and beyond repair?

All of these processes involve **apoptosis,** or **programmed cell death**. The tadpole's tail and fins do not simply fall off. Rather, each cell in the tail undergoes programmed cell death, which allows all the proteins, lipids, and nucleotides to be reused. When you get sick, your immune system makes new cells to fight off the infection. If these extra cells just burst and died when the battle was over, the body would have quite a mess to clean up! Instead, the cells calmly activate genes that initiate a controlled and deliberate death program. Apoptosis occurs in an orderly manner with genes regulating each step of the process. Few to none of the cytoplasmic components are released, and neighboring cells engulf the neatly packaged debris. All the components are recycled.

During apoptosis, the dying cell undergoes a process that is morphologically distinct and recognizable by sight. This was first described in 1972, but not until the past 10 years has the process been analyzed genetically. First, the cell membrane starts to form blebs, or regions that balloon out (Fig. 20.9). The nucleus shrinks, condenses, and divides into smaller

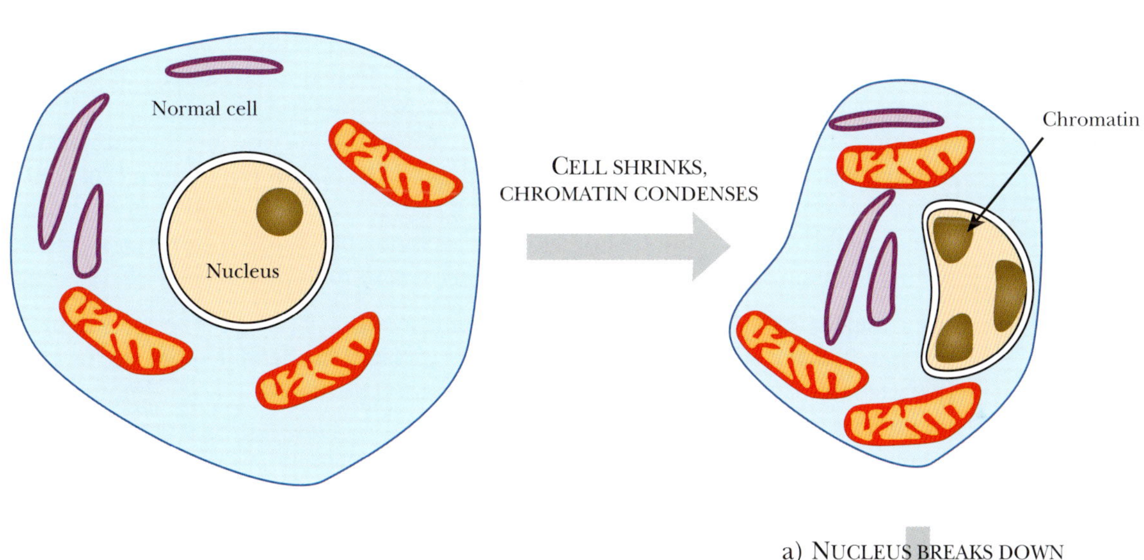

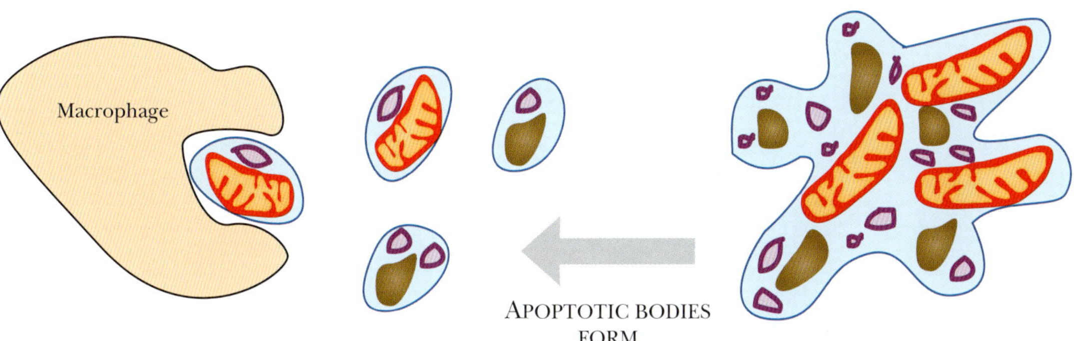

FIGURE 20.9 Apoptosis
The cell shrinks and chromatin condenses as apoptosis begins. Then the nucleus fragments and other cellular components are degraded by enzymes called caspases. Finally, the condensed apoptotic bodies are eaten by macrophages.

fragments. Finally the entire cell shrinks and divides into large condensed fragments called **apoptotic bodies**. Other cells engulf the debris by phagocytosis, thus cleaning up the remnants. The proteins, lipids, and nucleic acids are all digested and recycled.

Apoptosis is very distinct from other types of cell death. When cells are damaged from external injury, oxygen starvation, or energy depletion, they undergo **necrosis**. Necrotic cells swell because their osmotic balance is perturbed. Their proteins are denatured and degraded. Finally, the cell ruptures and dies. Necrosis is a messy way to die and often elicits an immune response. Apoptosis has many advantages over these other types of cell death. The stepwise program allows the cell to halt the process if necessary. In addition, the final result does not alter the physiology of the entire organism. In necrosis, bursting cells can cause collateral damage to surrounding tissue, making the situation worse. The immune system must clean up the mess and the tissues must be repaired and replaced. In contrast, apoptosis does not elicit an immune response unless the cells are dying in response to a viral or bacterial attack.

Apoptosis is an organized and efficient process of cell death.

APOPTOSIS INVOLVES A PROTEOLYTIC CASCADE

The first well-documented account of apoptosis was seen in the nematode *Caenorhabditis elegans*. This small worm is normally found in soil and eats soil bacteria. The worm is used as a model organism for developmental genetics because it is easy to maintain, has four developmental stages, and is transparent. Using a special microscope, the **Nomarski** or **differential interference microscope**, every cell of *C. elegans* can be seen (Fig. 20.10). Using this technique, scientists have been able to trace each cell division from the single-celled egg through the entire adult. During development, *C. elegans* generates 1090 cells, but the final adult worm has only 959. The other 131 cells die via apoptosis.

Many mutant worms were identified in which the number of cells in the adult was more or less than 959, the wild-type number. Four genes involved in aberrant apoptosis have been characterized. Three of these are *ced-3*, *ced-4*, and *ced-9* ("ced" = cell death abnormal), and the fourth is *egl-1* (for egg laying defective). When the *ced-3*, *ced-4*, or *egl-1* genes are defective, there are more than 959 cells in the adult worm. Thus these genes initiate or execute the apoptotic program. When *ced-9* is defective, there are fewer than 959 cells, indicating more apoptosis than normal. Thus, CED-9 protein inhibits apoptosis.

The CED and EGL proteins work in a cascade that initiates cell death. That is, the action of one protein activates the next protein, which in turn activates further proteins. Genetic and biochemical experiments have given the following model for apoptosis in *C. elegans* (Fig. 20.11). The three CED proteins form an inactive complex in the membrane of the mitochondrion. A signal from surrounding cells activates the synthesis of EGL-1 protein. EGL-1 binds to CED-9 and removes it from the complex. This activates CED-4, which is a protease that specifically cleaves a small inhibitory domain from the end of CED-3. Activated CED-3 forms a heterotetramer of two small and two large domains. This in turn digests various cellular proteins by cutting after aspartic acid residues. This type of enzyme is known as a **caspase** (cellular aspartate-specific protease—see Box 20.2). Once CED-3 is active, it cleaves inhibitory domains off other proteases, nucleases, and other caspases, thus executing the apoptotic program.

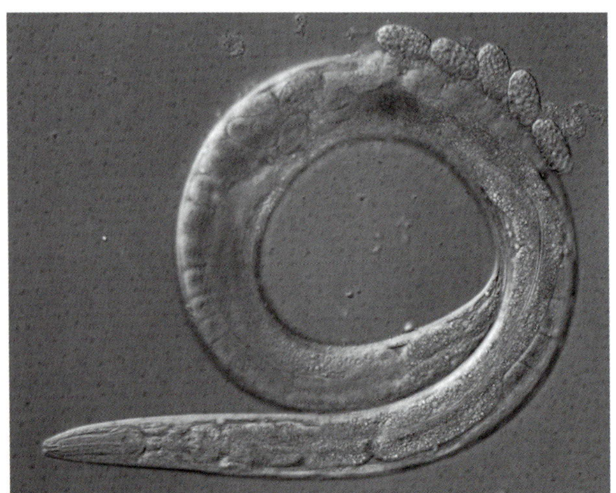

FIGURE 20.10
Nomarski Image of Adult *Caenorhabditis elegans*
Nomarski optics transforms differences in density into differences in height, giving an image that looks like a three-dimensional relief. The technique is useful in visualizing tissues that are several cells deep. Courtesy of Jill Bettinger, Virginia Commonwealth University, Richmond, VA.

During development, *C. elegans* generates more cells than are found in the adult worm. The extras die via apoptosis.

The CED and EGL proteins of *C. elegans* work in a cascade that initiates cell death. Ced-3 and Ced-4 are special proteases called caspases, which are key regulators to apoptosis.

MAMMALIAN APOPTOSIS

Comparison of the genome sequence of *C. elegans* with those of mice and humans has allowed the discovery of many mammalian genes that are homologs of worm apoptosis genes. Because apoptosis in humans occurs throughout development and adulthood, there are significantly more proteins and regulators associated with apoptosis than in nematode worms. There are more than 15 caspases with varied tissue specificity and functions. In addition, there are more than 20 mammalian proteins that are similar to CED-9. The precise roles of proteins involved in regulating mammalian apoptosis are still being elucidated.

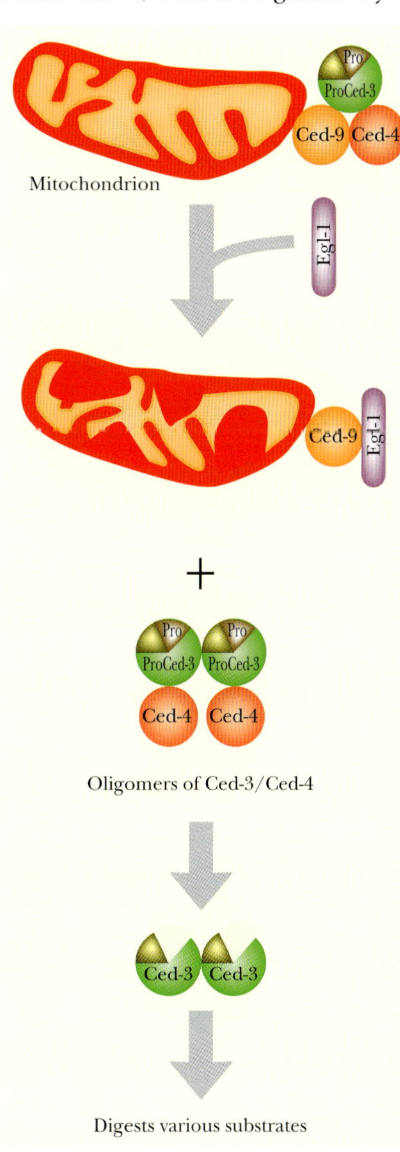

FIGURE 20.11 Programmed Cell Death in *C. elegans*
In the membrane of *C. elegans* mitochondria, a complex of CED-9, proCED-3, and dimers of CED-4 remains inactive. When apoptosis is triggered, EGL-1 protein binds to CED-9, releasing CED-4 and proCED-3. Tetramers of CED-4 are formed and cleave the inhibitory domain from proCED-3. Activated CED-3 is the central enzyme that activates the remaining caspases and executes the apoptosis program.

Mammalian cells have two major pathways to trigger apoptosis (Fig. 20.12). These eventually converge on the caspases. One pathway is designated the **death receptor pathway** because the death signal originates at a cell surface receptor protein, or **death receptor**. An external signal molecule binds to the extracellular domain on the receptor. The receptor transmits the signal to its intracellular domain and then recruits a variety of internal proteins to the membrane. This ultimately activates the caspases. The other pathway is the **mitochondrial death pathway**. This is usually triggered by intracellular catastrophe, such as irreparable DNA damage. This activates proteins in the mitochondria, which release different effector molecules to activate the caspases.

The death receptor pathway can function independently of the mitochondrial pathway, although both pathways are often activated simultaneously. The signal molecules that bind to the death receptor are quite varied, and the biological significance of most is still uncertain. The best understood is CD95, or Fas ligand (Fig. 20.13). This is a highly glycosylated cell surface protein found in immune cells. When CD95 binds to its receptor on the target cell (the CD95 receptor—CD95R), it causes the receptor to trimerize. The intracellular domain of each receptor is activated by proximity to the others. Activated CD95R recruits a complex of proteins known as **DISC** (**death-inducing signaling complex**) to the cell membrane. One component of DISC is caspase-8, which cleaves its own pro-domain, releasing the active form (see Box 20.2). The main target for caspase-8 is caspase-3, the direct homolog of CED-3 in *C. elegans*. Once caspase-3 is activated in mammalian cells, there is no turning back. The cell will die via apoptosis.

The mitochondrial pathway is usually activated by internal stimuli such as irreparable DNA damage. The signal converges on a family of mitochondrial proteins named after its founding member, **Bcl-2**, first identified in a B-cell lymphoma. Bcl-2 is homologous to the

C. elegans CED-9 protein. Unlike *C. elegans*, which has only a single protein of this class, mammals have a family of many proteins, some of which induce apoptosis, whereas others protect cells from apoptosis. The members of the Bcl-2 family are active as dimers, and the relative number of each protein in the dimer state controls the death pathway (Fig. 20.14). Thus, if there are more dimers of Bcl-2, which is anti-apoptotic, then the cell will not die. If there are more dimers of Bax, a pro-apoptotic family member, then the cell dies.

If a pro-apoptotic signal is received, the pro-apoptotic members of the Bcl/Bax family allow **cytochrome *c*** (cyt *c*) to escape from the mitochondria through pores in its outer membrane (Fig. 20.15). Once released, cyt *c* induces formation of the **apoptosome**, a signaling complex containing cyt *c*, caspase-9, and Apaf-1. Apaf-1 is the mammalian homolog to CED-4 of

Box 20.2 Caspases

Caspases are proteases that cleave their target proteins after certain specific aspartate residues. Specificity results from recognition of the four amino acid residues following the aspartate. There are many different caspases. They are highly conserved throughout evolution and are found in mammals, flies, nematodes, and even hydra. Scientists regard them as central executioners because they carry out the death sentence of apoptosis. Indeed, inhibiting caspases prevents apoptosis.

Caspases are thought to work as heterotetramers, that is, an assembly of four domains (Fig. A). Most mammalian caspases have two p20 domains and two p10 domains. The p20 domain has the enzyme active site; therefore, each active caspase can cleave two different proteins at the same time. The inactive form of a caspase has three different domains: the prodomain, which blocks the active site, a p20 domain, and a p10 domain.

There are three ways to activate caspases (Fig. B). The first method requires another, previously activated caspase to cleave between the prodomain and p20, and between p20 and p10. When two molecules have been digested, the p20 and p10 domains associate into their final structure. The second method of caspase activation occurs via self-association. For example, when three molecules of caspase-8 trimerize, one caspase-8 cleaves its neighbor. When two molecules of the triad have been cut, they dissociate from the third to form the heterotetramer. This mechanism is used to activate the caspase cascade in the death receptor pathway of apoptosis (see later discussion). The third method involves specific regulatory proteins that are not caspases. When the apoptosome forms in the mitochondrial pathway of apoptosis (see later discussion), two regulatory proteins, Apaf-1 and cytochrome *c*, bind to and initiate cleavage of caspase-9.

There are more than a dozen different caspases in humans, each with its own exclusive set of targets. In some cases, digesting the target protein inactivates it. In some cases cleavage releases part of the protein that has a new function. In other cases, the caspase cleaves an inhibitory domain from an inactive protein, releasing the active form.

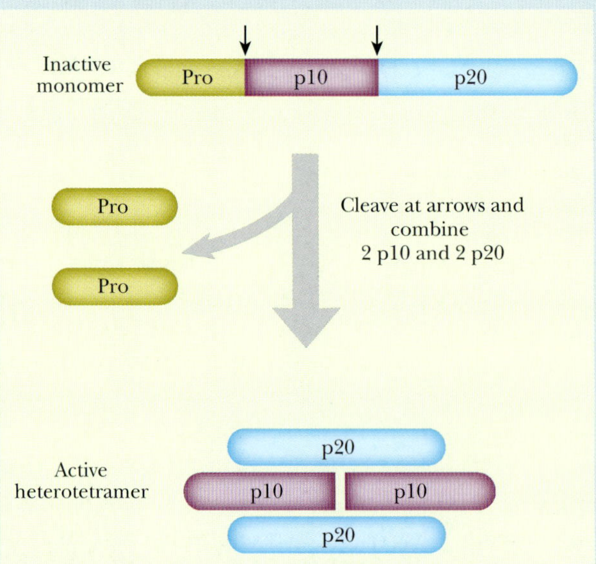

FIGURE A Active Caspases are Heterotetramers
Caspases are usually found as inactive monomers with three domains: the prodomain, which inhibits activity, the p10 domain, and the p20 domain with the active site. When the prodomain is removed, two p10 domains and two p20 domains come together as a heterotetramer.

Box 20.2 Caspases—cont'd

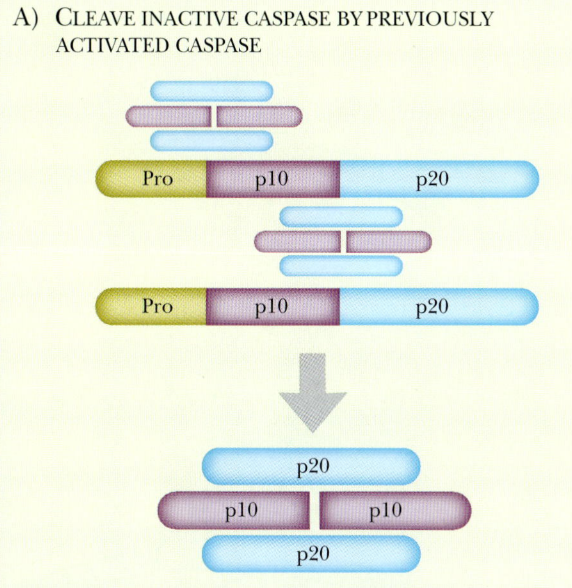

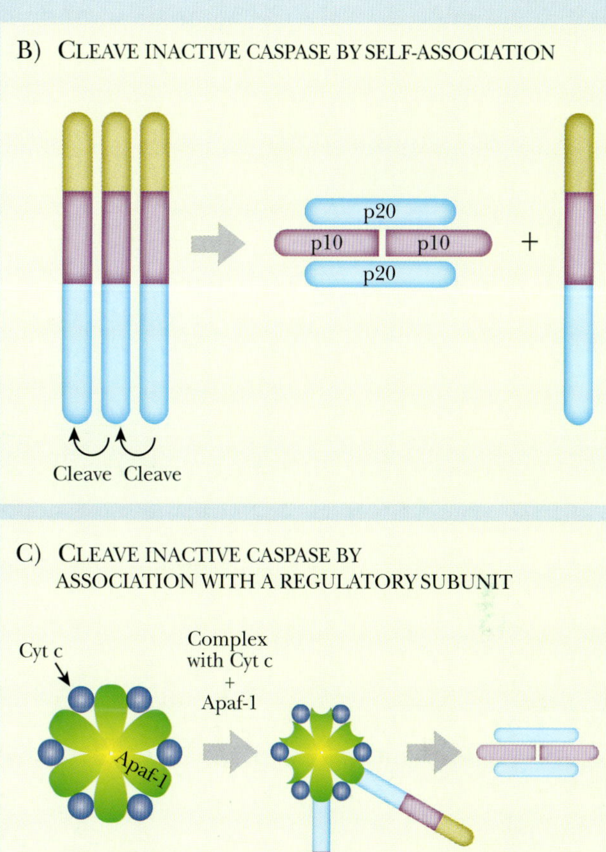

FIGURE B Three Methods to Activate a Caspase
(A) Inactive caspase monomers can be cleaved by previously activated caspases. (B) Association of three inactive caspase monomers can trigger one enzyme to cleave its neighbors, resulting in two active p10 and p20 domains. (C) Other enzymes than a caspase can also cleave the prodomain from inactive caspase monomers.

C. elegans and caspase-9 corresponds to CED-3. However, whereas CED-4 is blocked by CED-9, Apaf-1 has an auto-inhibitory domain that is part of the Apaf-1 protein itself. Furthermore, CED-4 forms a tetramer when activated but Apaf-1 forms a heptamer (Fig. 20.15). The Apaf-1 heptamer in the apoptosome activates caspase-9, which in turn activates caspase-3 as in the death receptor pathway.

Two different pathways activate apoptosis, the death receptor pathway and the mitochondrial death pathway.

In the death receptor pathway, CD95 or Fas ligand binds to its receptor, CD95R, which aggregates with two other activated CD95Rs. The intracellular domains activate DISC, which in turn, activates caspase-3.

The ratio of pro-apoptotic (Bax) and anti-apoptotic (Bcl-2) dimers controls whether cells die via the mitochondrial pathway. The pro-apoptotic signal triggers the release of cytochrome *c* from the mitochondria. Cytochrome *c* binds to Apaf-1 and caspase-9 to form the apoptosome.

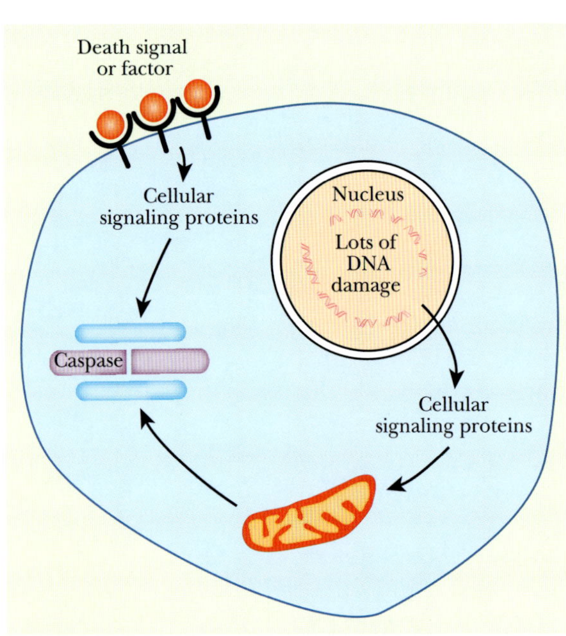

FIGURE 20.12 Activation of Apoptosis
Apoptosis in mammals has two different ways of being activated. An external death signal or factor can bind to a cell surface receptor and initiate an intracellular signal to active caspases. An internal signal such as massive DNA damage can also initiate a signal cascade in the mitochondria, which also activates the caspases.

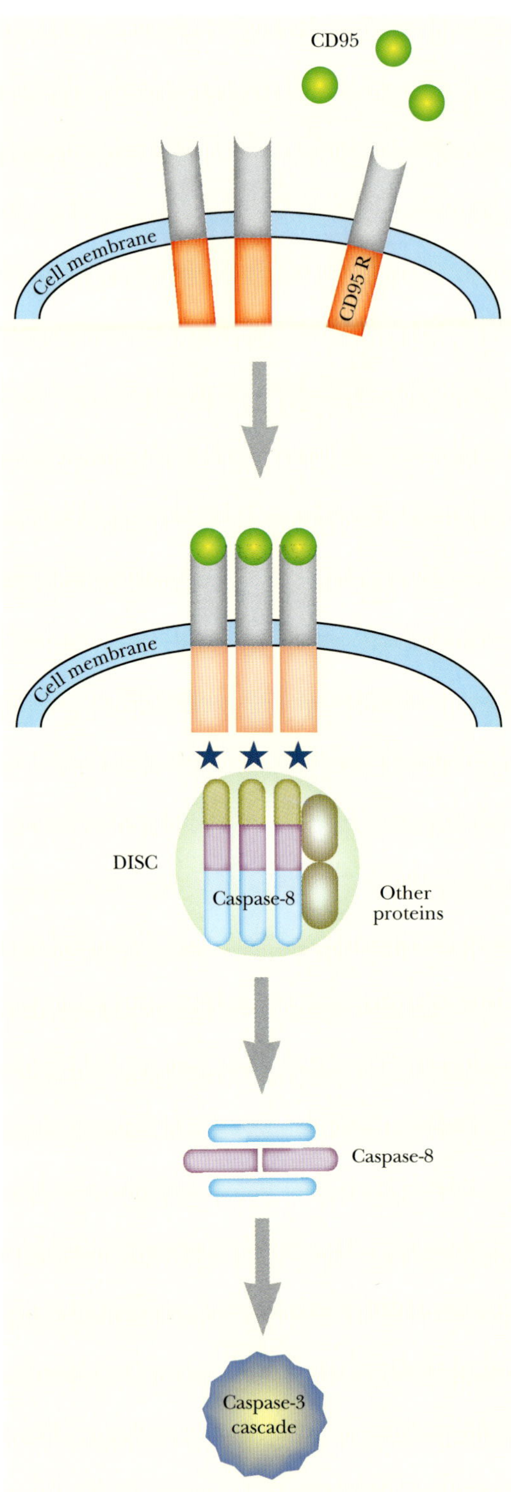

FIGURE 20.13 Death Receptor Pathway
An external signaling molecule called CD95 can initiate cell death. When three molecules of CD95 bind to the CD95 receptor, the three receptors come together to form one complex, which activates the intracellular domains. The DISC complex then binds to the receptor. The key component of DISC is caspase-8. When three molecules of caspase-8 come together, the pro-domains are cleaved off and the remaining domains form a hetero-tetramer that activates caspase-3, which in turn activates the other caspases.

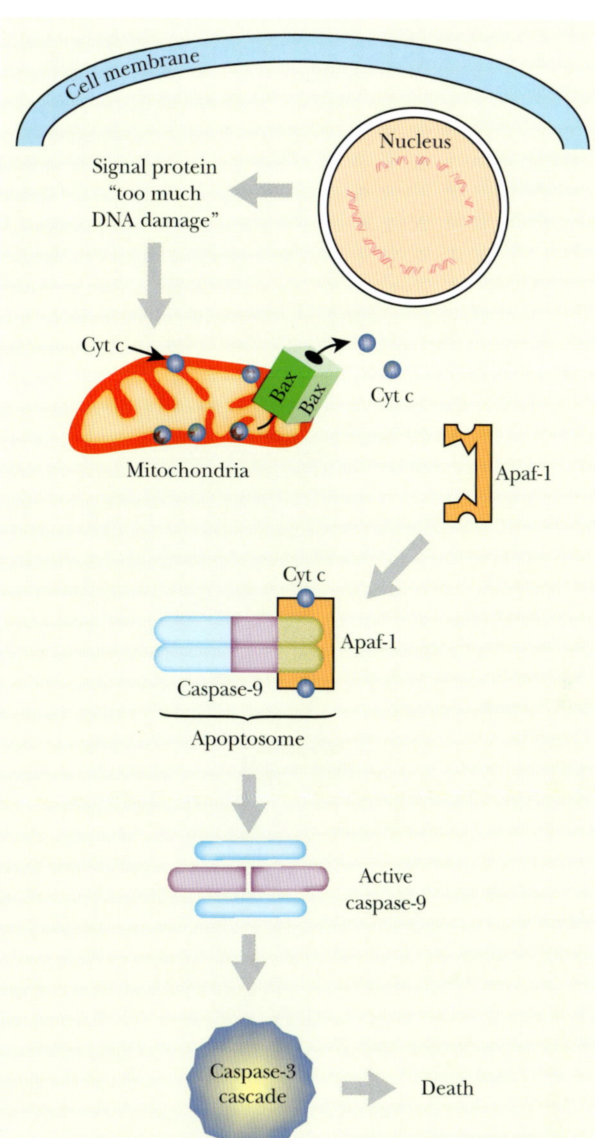

FIGURE 20.15 Mitochondrial Pathway of Apoptosis
Severe DNA damage signals for an increase in Bax/Bax dimers. These form a channel through which cytochrome *c* escapes from the mitochondria. Cytochrome *c* binds to Apaf-1 and induces a conformational change. Activated Apaf-1 assembles into a heptamer which binds to pro-caspase-9 and cleaves off the pro-domain. Activated caspase-9 activates caspase-3 and the rest of the caspases carry out the death sentence.

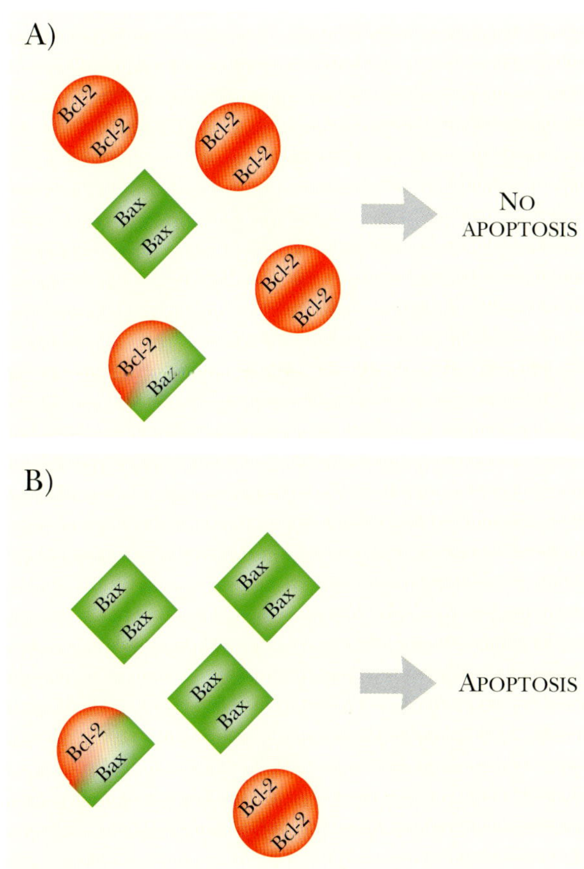

FIGURE 20.14 Bax/Bcl-2 Control of Apoptosis
The ratio of pro-apoptotic (Bax) and anti-apoptotic (Bcl-2) dimers controls whether cells die via apoptosis.

EXECUTION PHASE OF APOPTOSIS

Activating CED-3 in *C. elegans*, and caspase-3 in humans commits the cell to apoptosis, which is executed by a caspase cascade. One of the more famous targets is **CAD**, for **caspase-activated DNase**, a nuclease that cuts nuclear DNA between the nucleosomes (Fig. 20.16). The fragments of DNA are about 180 base pairs in length, which is the length of DNA wound around a nucleosome. To determine if cells are in the process of apoptosis, scientists isolate the genomic DNA and look at the fragment sizes. If they are in multiples of approximately 200 base pairs, this implies the cells were going through apoptosis.

There are many other substrates of caspases. Proteins that maintain nuclear structure are digested so that the nucleus shrinks and breaks into small pieces. The cytoskeleton is cleaved and the cell architecture disintegrates and compacts. Other organelles are digested so they

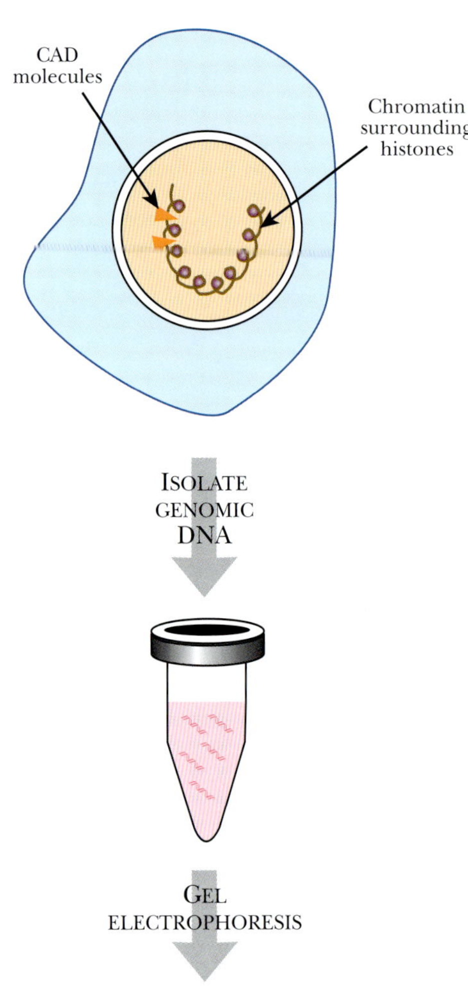

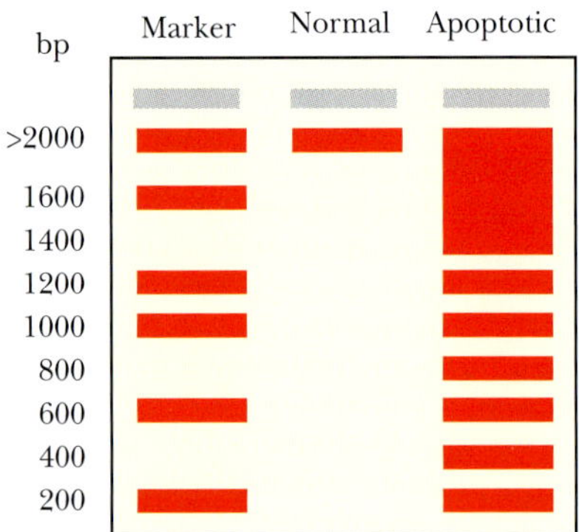

FIGURE 20.16 CAD Cleaves Nuclear DNA During Apoptosis
One of the targets for caspases is CAD, which digests the nuclear chromatin. Because about 200 base pairs of DNA is wound around each nucleosome, and CAD digests only between these particles, DNA from an apoptotic cell exists in multiples of 200 base pairs. When genomic DNA is isolated from a cell in apoptosis and electrophoresed through an agarose gel, the DNA forms a ladder.

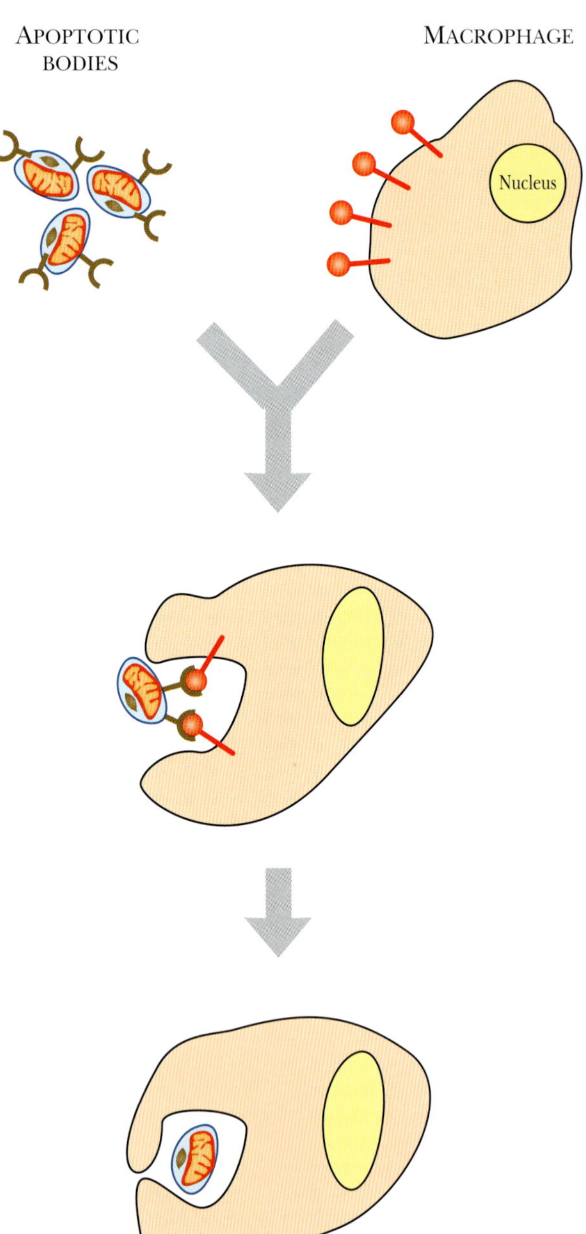

FIGURE 20.17 Removal of Apoptotic Bodies
In mammals most apoptotic bodies are removed by macrophages. These recognize cell surface receptors on the apoptotic body and ingest the entire cellular fragment. Any remaining proteins are broken down and recycled by the macrophage.

lose their structure, which leads to the final result of small compact granules of cellular material called apoptotic bodies.

> During the execution phase, caspases digest cellular components such as chromatin, nuclear architecture, cytoskeleton, and organelles.

CORPSE CLEARANCE IN APOPTOSIS

The apoptotic bodies are removed by phagocytosis. In *C. elegans*, neighboring cells take up the apoptotic bodies. In mammals, the situation is more complex. In a few areas of the body, neighboring cells engulf the apoptotic bodies. In most cases though, apoptotic bodies are engulfed by macrophages (Fig. 20.17). Macrophages are cells of the immune system whose primary role is to digest anything foreign, such as invading bacteria. Normally macrophages recruit other immune cells to the site of infection. However, when dealing with apoptotic bodies, they do not recruit other immune cells. Thus macrophages can distinguish an apoptotic body from a foreign invader.

How does the macrophage know that apoptotic bodies do not require an immune response? Normally, when macrophage ingest bacteria, they secrete soluble proteins to recruit other immune cells. Apparently apoptotic bodies have molecules on the surface that trigger the macrophage to digest them but without secreting the immune signaling factors. The numbers and mechanisms of "eat me" receptors are quite complex since different mammalian tissues may have different mechanisms to activate the macrophage. One such molecule appears to be phosphatidylserine, a phospholipid normally found only in the inner leaflet of the cell membrane (Fig. 20.18). During apoptosis phosphatidylserine may be translocated to the

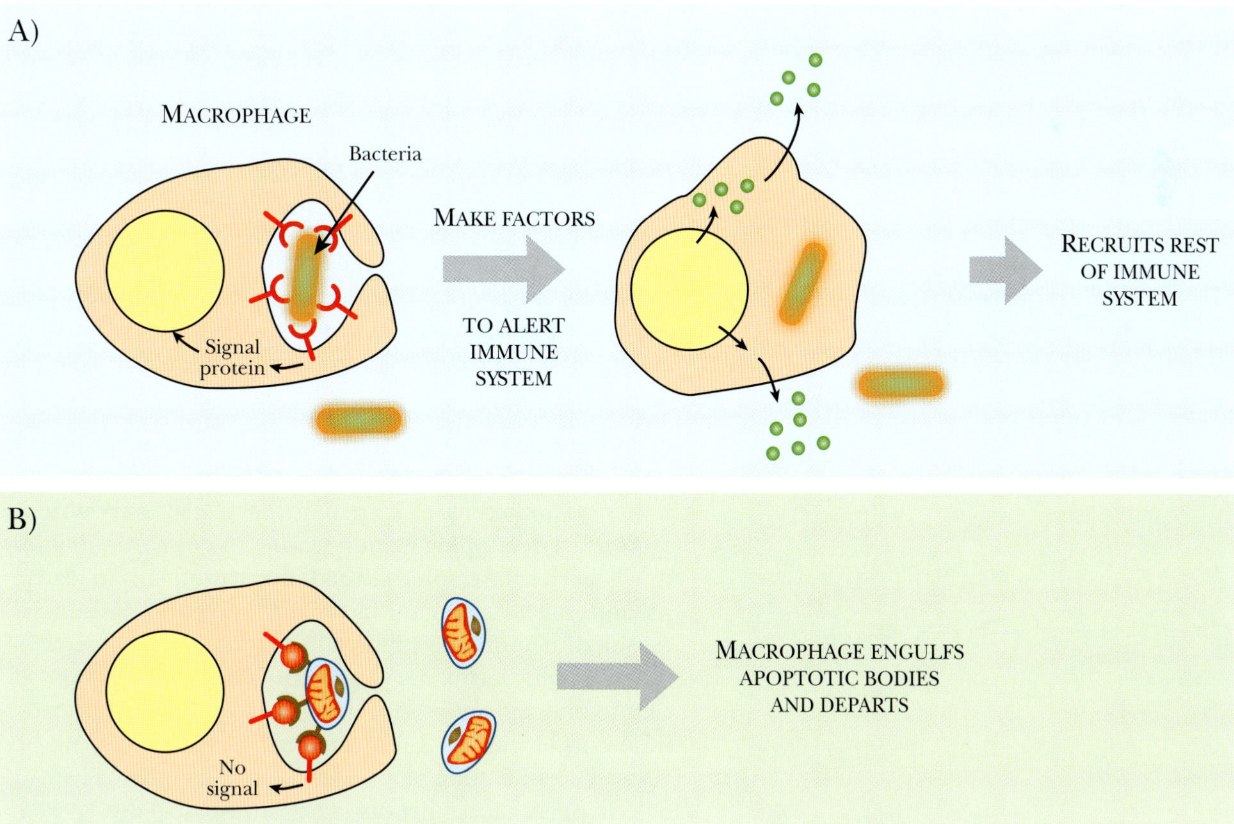

FIGURE 20.18 Apoptotic Bodies Do Not Trigger an Immune Response by Macrophages
(A) Normally, macrophages ingest invading pathogens such as bacteria. This triggers the macrophage to release factors that attract other immune cells. (B) In contrast, the cell surface signal on apoptotic bodies does not trigger the macrophage to release these factors, so other immune cells are not recruited to the site.

outer leaflet. Macrophages have receptors for phosphatidylserine, allowing them to recognize the apoptotic body as "self."

Macrophages digest apoptotic bodies and recycle the cellular proteins without triggering other immune cells.

CONTROL OF APOPTOTIC PATHWAYS IN DEVELOPMENT

Although controlling the onset of apoptosis is very complex, the ramifications of losing control are dire. Too much apoptosis or inappropriate activation of apoptosis can destroy fully functioning cells. Death of too much tissue can kill a developing organism. Not enough apoptosis, especially during development, can create surplus tissue and disrupt the normal operation of tissues and organisms. Many disease states may arise from inappropriate or defective apoptosis.

In *C. elegans* sex development depends on apoptosis. *C. elegans* comes in two "sexes," males, which produce only sperm, and hermaphrodites, which produce both sperm and eggs. No true females are produced. The decision to become male or hermaphrodite hinges on apoptosis (Fig. 20.19). Two neurons control the muscles around the vulva so that eggs can be laid. If the neurons are present, the worm is a hermaphrodite. If the neurons are absent, the worm cannot lay eggs, essentially making it a male. The presence or absence of these hermaphrodite-specific neurons (HSN) depends on apoptosis. Defects in *ced-3* or *elg-1* affect whether the worm is male or hermaphrodite. Without *elg-1*, too much apoptosis occurs; therefore, all worms are missing the HSN and no egg-laying worms are produced. Conversely, without *ced-3*, apoptosis cannot occur and the HSN survive in all worms.

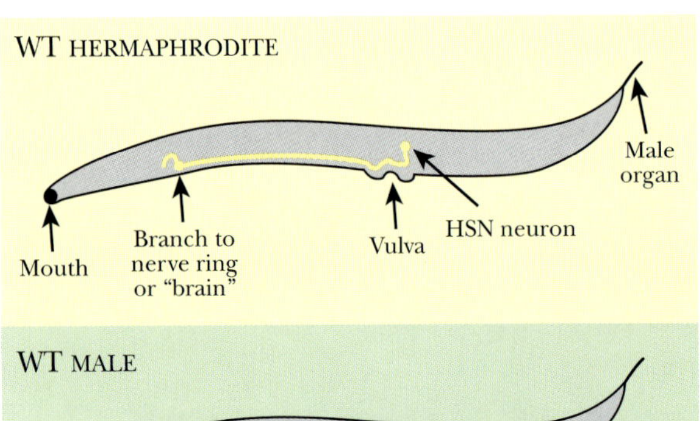

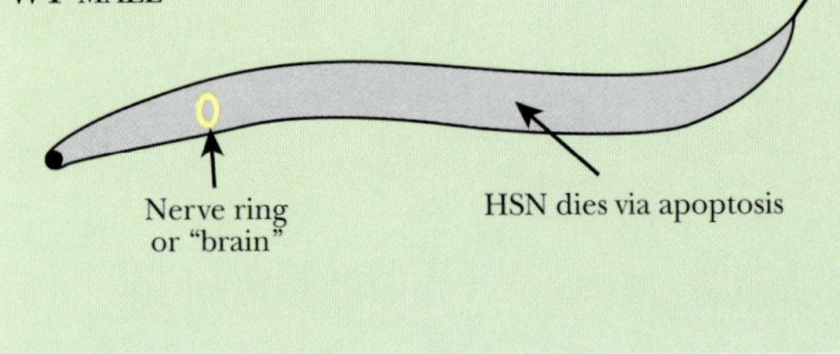

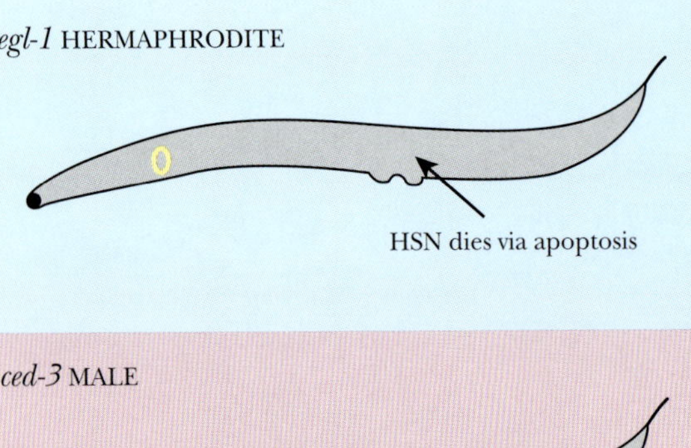

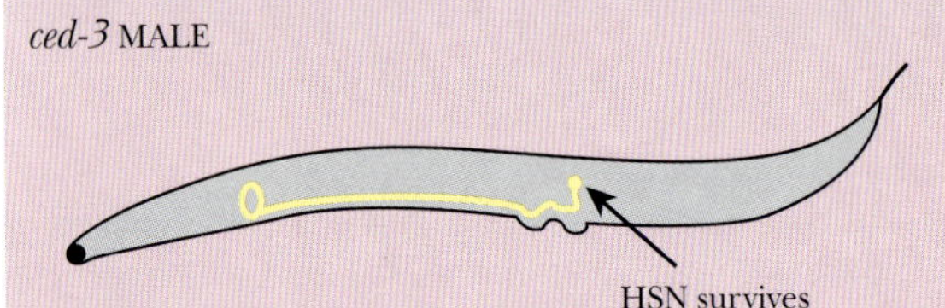

FIGURE 20.19 Apoptosis Controls *C. elegans* Sex Determination
C. elegans has two sexes, males that produce sperm and hermaphrodites that produce both sperm and eggs. One nerve, the HSN, determines whether the worm will lay eggs or not. The presence or absence of HSN depends on apoptosis in development. EGL-1 mutations induce too much apoptosis; therefore, the HSN nerve is killed during development and *egl-1* hermaphrodites cannot lay eggs. In contrast, *ced-3* mutations prevent apoptosis; therefore, the HSN nerve is never missing. Even *ced-3* males are able to lay eggs.

Apoptosis is tightly regulated during an immune response. During infection, the body responds to the attack by increasing the number of white cells of the immune system. When the infection is past, the body eliminates the surplus immune cells via apoptosis. Immune cells use the death receptor pathway to trigger apoptosis (see earlier discussion). Too much apoptosis would deplete our immune system of essential cells and disable the immune response. During HIV infection, the number of T-cells plummets to dangerously low levels, leaving the patient open to many secondary infections (see Chapter 22). One theory is that HIV kills T cells by inducing apoptosis.

The nervous system is another tissue that is highly sensitive to apoptosis. During development a large number of neurons undergo apoptosis. One theory suggests that neurons die if they do not receive a "keep on living" signal or **trophic factor**. If a developing neuron reaches its correct destination, it will receive the trophic factor. If the neuron fails to reach its target it gets no trophic factor and enters apoptosis by default. If neurons are cultured in a laboratory dish, removal of one trophic factor, **nerve growth factor**, induces the cells to undergo apoptosis (Fig. 20.20). Addition of caspase inhibitors blocks cell death, proving the cells were dying via apoptosis. During mouse development, embryos with defective genes for either caspase-3 or caspase-9 die. Lack of apoptosis in the developing neural system is the main cause for death in both cases. In the adult brain, apoptosis of neurons causes irreparable damage because neurons do not regenerate. Extensive apoptosis may play a role in many diseases, such as Alzheimer's (see later discussion), Parkinson's disease, Huntington's disease, and amyotrophic lateral sclerosis (ALS). The exact role of apoptosis in these diseases is still being investigated.

- Apoptosis controls the presence or absence of the HSN, which controls whether *C. elegans* becomes a male or a hermaphrodite.
- When neurons do not receive trophic factor signals, the cells die via apoptosis.

ALZHEIMER'S DISEASE

Alzheimer's disease is identified by two pathological hallmarks in the brain: **senile neuritic plaques** and **neurofibrillary tangles**. The senile neuritic plaques form when degenerating neurons aggregate into large clumps. The aggregates contain protein fragments known as **amyloid β (Aβ)** that are derived from amyloid precursor protein (APP). Normally, the precursor protein is digested by an enzyme in the endoplasmic reticulum to form a 40-amino-acid form called $A\beta_{40}$. In Alzheimer's, the precursor is digested in the wrong spot, and an aberrant 42-amino-acid form ($A\beta_{42}$) is made instead (Fig. 20.21). This form tends

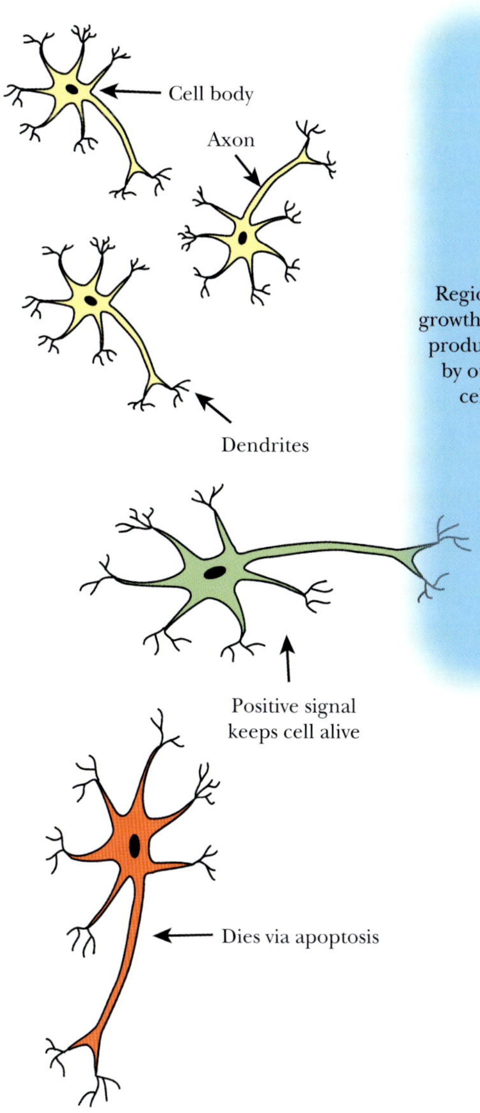

Cell body
Axon
Dendrites
Region of growth factor production by other cells
Positive signal keeps cell alive
Dies via apoptosis

FIGURE 20.20 (left) Growth Factors Keep Neurons Alive
Nerve cells undergo apoptosis if their dendrites do not receive trophic factors such as nerve growth factor.

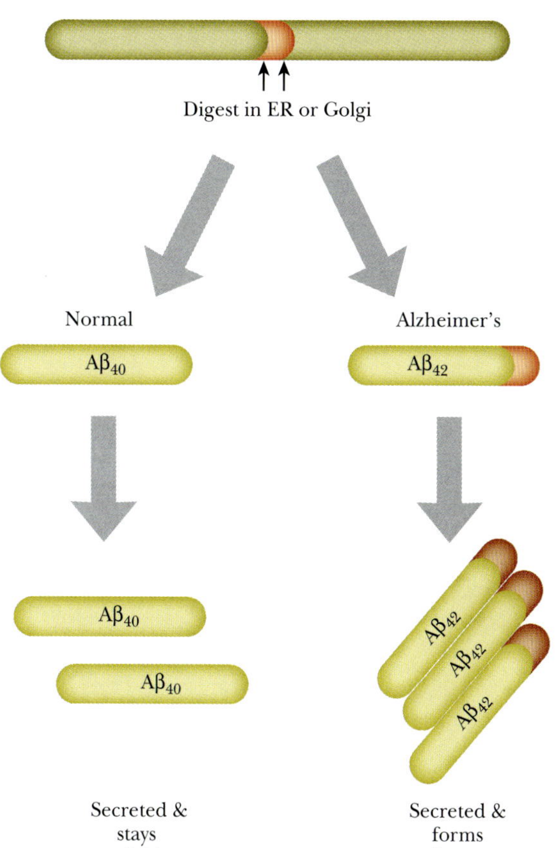

FIGURE 20.21 (right) Amyloid Precursor Protein Processing
The APP protein is normally digested into a 40-amino-acid protein as it travels through the endoplasmic reticulum (ER). In Alzheimer's patients, APP is processed into a 42-amino-acid form that aggregates into large clumps.

to aggregate into clumps. In addition to the clumps of neurons, neurofibrillary tangles form inside otherwise normal neurons. This occurs when **tau**, a protein associated with microtubules, starts forming helical aggregates. These tangles eventually kill the neuron, which relies on the microtubules (and tau) to ship neurotransmitters from the cell body to the dendrites.

Beside tau and amyloid β protein, several other proteins are implicated in the pathology of Alzheimer's disease. The **presenilins** are transmembrane proteins found in the endoplasmic reticulum and Golgi apparatus. Mutations in presenilins may increase the secretion of the $A\beta_{42}$ form of amyloid β. Whether or not the presenilins are the actual proteases that digest the amyloid precursor protein remains to be proven. In addition, mutant presenilins make the nerve cell more sensitive to apoptosis, although the mechanism is not understood. Mutations in the presenilin genes, *presenilin-1* and *presenilin-2*, are the most common defects in patients with familial Alzheimer's disease. These defects also cause the disease to start earlier than usual. Another protein implicated in Alzheimer's is apolipoprotein E. A specific allele of the gene for this (ε4) promotes the formation of plaques. If someone inherits this allele, he or she is more likely to develop Alzheimer's after the age of 60.

The formation of senile neuritic plaques causes the nerve cells to die, and the patient loses the function of that region of the brain. When a plaque forms, the immune system is alerted by soluble signals from nearby microglial cells. These cells are activated by the aggregation of amyloid β protein. Once the immune system is activated, the immune cells will kill more neurons by secreting proteins that activate the death receptor pathway of apoptosis. (The immune system plays a role in the progress of many neurodegenerative diseases.)

Because the role of amyloid β in plaque formation is a central aspect of Alzheimer's disease, recent work to establish a treatment has focused on this molecule. Scientists have been able to produce an antibody that breaks up aggregates of amyloid β in mice. The antibody can be given to Alzheimer's patients as a vaccine, and early clinical trials have shown that it has few to no side effects. Although the exact mechanism is unknown, clinical trials are under way to determine how effective the vaccine is in human Alzheimer's patients. Potentially in 5 to 10 years a real treatment will exist for this devastating disease.

Alzheimer's disease is identified by two pathological hallmarks in the brain: senile neuritic plaques and neurofibrillary tangles.

Amyloid β, tau, presenilins, and apolipoprotein E are proteins implicated in causing Alzheimer's disease.

PROGRAMMED CELL DEATH IN BACTERIA

Although apoptosis has not been seen in single-celled organisms, a genetic system that kills *Escherichia coli* when under extreme stress does exist. Morphologically, the death does not resemble apoptosis, but like apoptosis, the death system is genetically encoded. In *E. coli*, an **addiction module** of two genes controls the death-inducing system (Fig. 20.22). One gene encodes a toxin, MazF, which is quite stable. The second gene encodes the antitoxin, MazE, which prevents the toxin from killing the bacteria. The antitoxin is unstable and degrades very fast after translation. If its transcription or translation is stopped or slowed in any way, the level of antitoxin plummets and the toxin kills the bacteria.

The MazF toxin is a specific endoribonuclease that degrades messenger RNA. It recognizes the sequence ACA and cleaves to the 5′-side. Such enzymes have been named mRNA interferases and have now been found in a variety of bacteria. The net result is the destruction of mRNA followed by a halt in protein synthesis. Cell death rapidly follows.

When bacteria are depleted of nutrients, transcription and translation slow down, and this may trigger the MazEF suicide system. Perhaps some *E. coli* commit suicide for the good of the rest, because the proteins, lipids, and nucleic acids of the dead cell could provide food for nearby cells. Despite being unicellular, bacteria have genetic programs for the good of the population. Another theory is that the MazEF suicide system is designed to limit the multiplication of bacterial viruses. Indeed, mutants of *E. coli* deleted for the whole *mazEF* operon give higher yields of bacteriophage when they burst. In wild-type cells, the MazEF system kills the cell before virus replication is complete and thus reduces the number of viruses produced.

Addiction modules also exist in bacteriophages such as P1 and lambda that are maintained in a lysogenic state. The toxin/antitoxin pair of proteins prevents the bacteriophage genome from being destroyed or lost during *E. coli* growth. The bacteriophage genome encodes both the toxin and the antitoxin. Just as in the MazEF system, the toxin protein is very stable, whereas the antitoxin degrades quickly and must be produced continually. If the P1 or lambda genome is lost, the stable toxin protein will kill the bacteria. Interestingly, the toxin produced by P1 does not kill *E. coli* directly but acts by activating the bacterium's own MazEF system (Fig. 20.23). The P1 toxin inhibits translation of the MazE antitoxin, which activates the MazF toxin, which in turn kills the cell.

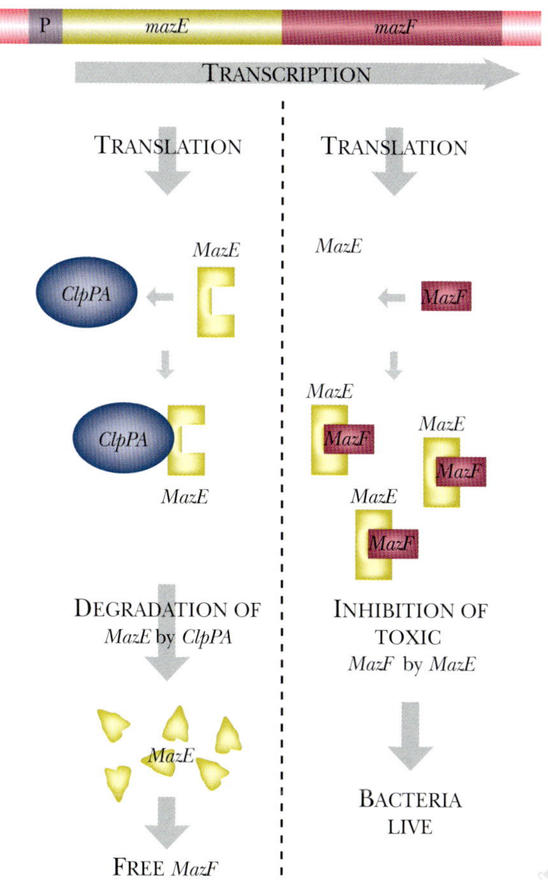

FIGURE 20.22
MazEF System of *E. coli*
Two genes, *mazE* and *mazF*, control whether or not *E. coli* self-destruct. MazF protein is a toxin that kills the bacteria. It is very stable and is produced continually. The MazE antitoxin protects *E. coli* by inhibiting the MazF toxin. However, MazE is degraded very fast by the ClpA protease. Under normal conditions, *E. coli* continually makes the antitoxin, but if the bacteria encounter stress, they may halt protein synthesis. The antitoxin is then no longer made and the toxin is free to initiate the suicide program.

E. coli have a programmed cell death system that is controlled by two proteins, MazE and MazF. MazE is the antitoxin and requires constant synthesis; MazF is a toxin that kills the cell when MazE is missing.

Another two-gene system, Phd/Doc, of P1 bacteriophage, controls whether its host cell lives or dies. If the cell loses the P1 genome, then the Doc toxin induces bacterial cell death via MazE and MazF.

USING APOPTOSIS TO TREAT CANCER

Scientists have been trying to induce death in cancer cells for many years, usually with radiation or chemotherapy. The drawback to most current therapies is that they are nonspecific. They kill noncancerous cells as well as cancer cells. However, every mammalian cell contains the genetic information to undergo apoptosis—even cancer cells, although they have overridden the normal controls. Recently, scientists have focused on deliberately inducing apoptosis in cancer cells.

One strategy is to induce the mitochondrial apoptotic pathway. A high level of Bcl-2 protects cells from apoptosis. So if the amount of Bcl-2 were reduced in cancer cells, perhaps they would become more sensitive to apoptosis. (Mice with the *Bcl-2* gene deleted die from too much apoptosis, particularly in the lymphoid tissue, where immune **B cells** mature.) The synthesis of Bcl-2 protein can be reduced by using a single-stranded **antisense DNA** of 18 nucleotides. Its sequence is complementary to the translation start site of the *Bcl-2* gene. When the **antisense DNA oligonucleotide** binds to the Bcl-2 mRNA, translation is blocked, and less protein is made (Fig. 20.24). In a recent clinical trial of patients with B-cell

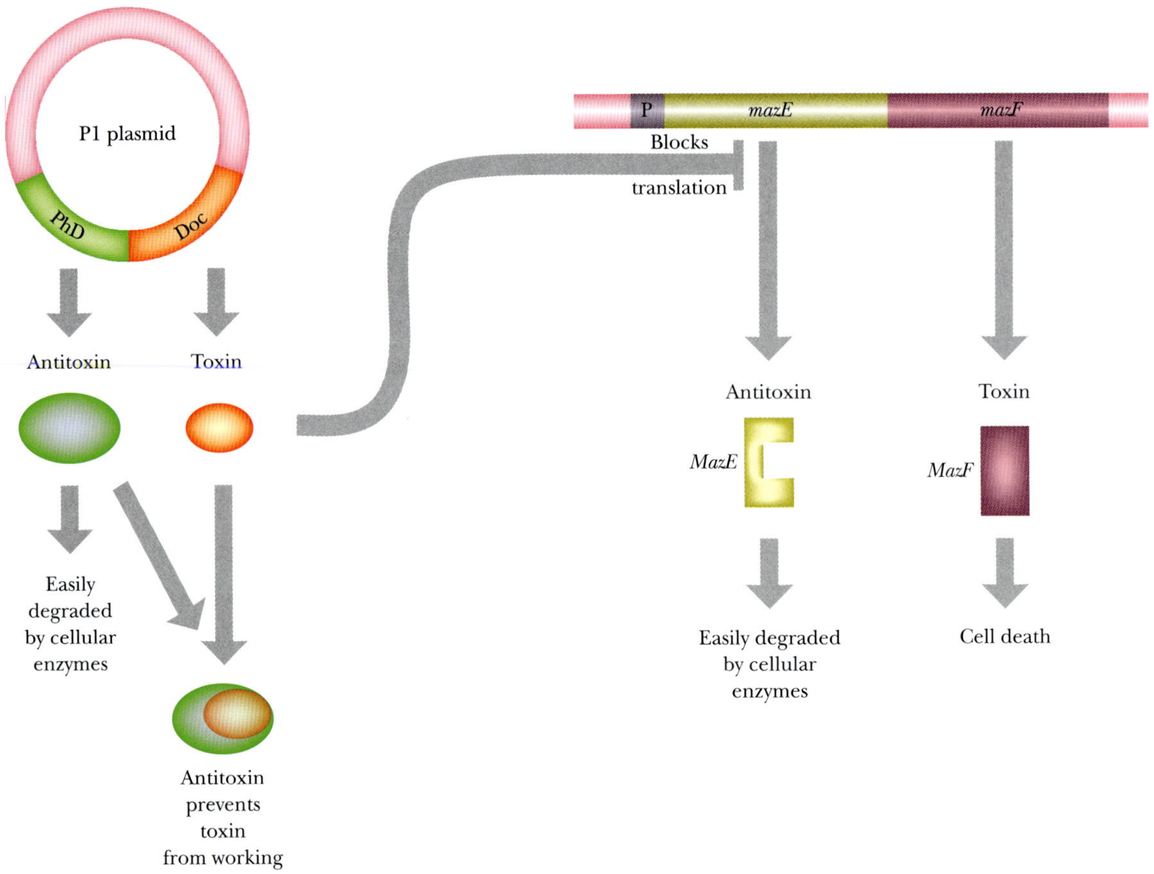

P1 plasmid
PhD
Doc
Antitoxin
Toxin
Easily degraded by cellular enzymes
Antitoxin prevents toxin from working
P
mazE
mazF
Blocks translation
Antitoxin
MazE
Toxin
MazF
Easily degraded by cellular enzymes
Cell death

FIGURE 20.23 Phd/Doc System of P1
PhD/Doc is an addiction module carried by P1 that encodes an antitoxin/toxin pair. The antitoxin is unstable and must be synthesized continually, or else the toxin will become active. If for some reason the PhD protein is not produced, Doc inhibits translation of MazE. Without MazE, the MazF toxin activates the suicide program.

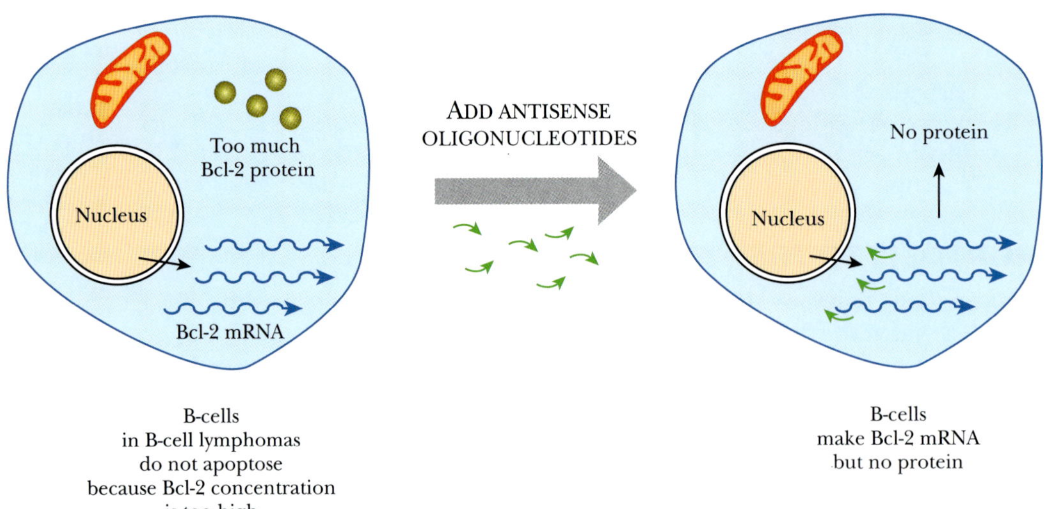

FIGURE 20.24 Antisense Oligonucleotides Block Apoptosis in Cancer Cells
The anti-apoptotic protein, Bcl-2, protects B cell lymphomas from apoptosis. Short antisense DNA with a sequence complementary to the *Bcl-2* gene may be added to cancer cells. The antisense DNA binds to the translation start site and blocks Bcl-2 protein synthesis. The cancer cells are no longer protected from apoptosis.

lymphomas, six of 14 patients had a much lower level of Bcl-2 and showed improved anti-cancer responses.

Antisense DNA oligonucleotides to Bcl-2 inhibit its translation and therefore decrease the amount of Bcl-2 in cancer cells. Without the anti-apoptotic Bcl-2, the cancer cells are more likely to die from apoptosis.

Summary

All organisms age, some more slowly than others, but every living organism slowly accumulates mutations and other genetic changes over time. Cells growing *in vitro* also age. Human cells grow and continue to divide until they reach cellular senescence, a state where they are alive, but do not divide and multiply. Certain types of cells are terminally differentiated, meaning they do not divide, but change their physiology to fulfill an important function for the organism. The type of cell and the organism from which the cells originate determine how long the cells continue to divide *in vitro*. In some areas of the body, stem cells maintain a supply of undifferentiated cells that are able to divide.

Senescence is a genetic program that is triggered by critically short telomeres and damage due to reactive oxygen metabolites. Damaged cells activate the program for premature senescence. If the damage is extreme, then the cell initiates apoptosis, or programmed cell death. In cancers, the genes that activate senescence or apoptosis are defective, allowing damaged cells to continue to grow and form a tumor. Two of these genes are *p53* and *pRb*. Components of the mitochondrial death pathway are being targeted by antisense oligonucleotides to make cancer cells more sensitive to apoptosis.

Studies in *C. elegans* have found genes that affect the average life span of the worm, called *daf-2* and *age-1*. The mutant genes increase the life span of the worm for up to 2 months. These proteins regulate the metabolic rate of the worm and are homologous to the mammalian insulin signaling pathway that controls the sugar levels in our bloodstream. Other experiments in *C. elegans* have identified *clk* mutations that control the rate at which *C. elegans* develops. These mutants also have fewer reactive oxygen metabolites. In yeast cells, aging is controlled by sirtuins, which are deacetylases that regulate histones.

Apoptosis occurs when cellular telomeres become too short or the cell is irrevocably damaged. Apoptosis is the organized method of suicide that begins with the membrane forming blebs, the nucleus shrinking, condensing, and fragmenting, and finally the formations of condensed fragments called apoptotic bodies. Phagocytes digest these fragments, and the components are recycled.

Apoptosis is an important process to development of all organisms because apoptosis controls tissue remodeling and physiology. In *C. elegans*, apoptosis reduces the number of cells from 1090 to 959 during development. Apoptosis of the HSN during development controls whether the worm becomes a male or a hermaphrodite. Apoptosis keeps human immune systems in balance, deleting the excess immune cells when the bacteria or virus is no longer a threat. Neurons that stop receiving trophic factors commit suicide via apoptosis. Control of apoptosis is critical, and too much apoptosis may be a component of many diseases, such as Alzheimer's and AIDS.

Apoptosis occurs via a cascade of different enzymes from two different pathways. In the death receptor pathway, a small factor called Fas binds to its receptor in the membrane of a cell. Fas causes the receptors to trimerize, activating the intracellular enzymatic domains. The intracellular portion activates proteins associated with the DISC, death-inducing signaling complex, which has caspase-8 (cellular aspartate-specific protease). Caspase-8 removes its own pro-domain and activates caspase-3. Caspase-3 is homologous to CED-3 of *C. elegans*, which has other genes similar to those found in humans. In the alternate pathway, the mitochondrial death pathway, two proteins, Bcl-2 and Bax, control whether or not apoptosis is activated. The ratio of pro-apoptotic (Bax), and anti-apoptotic (Bcl-2) dimers determines whether cytochrome *c* is released from the mitochondrial membrane. Cytochrome *c* activates a complex of proteins called the apoptosome. During the execution phase of apoptosis, the caspases activate other cellular proteases and digest the cellular proteins to create the apoptotic bodies.

Many organisms besides humans have apoptosis, including *E. coli*. The apoptosis in *E. coli* is not morphologically similar, but does rely on two genes that work cooperatively to control whether or not the cell dies. Healthy *E. coli* produce MazF, a very stable toxin, and MazE, the unstable antitoxin. As long as the cell stays healthy, the antitoxin prevents the toxin from killing the cell. As soon as the health of the cell is compromised, transcription and translation slow down, making less antitoxin. The toxin is free to kill the compromised cell. The Phd/Doc system of bacteriophage P1 is also a toxin/antitoxin pair. Here, bacteriophage P1 makes the antitoxin to prevent the toxin from killing its host cell. If the host cell loses the P1 genome, then the Doc toxin induces bacterial cell death via MazE and MazF.

End-of-Chapter Questions

1. What is cellular senescence?
   - **a.** when cells are alive but are no longer dividing
   - **b.** when cells finally die
   - **c.** when the cells are alive, but are no longer expressing genes
   - **d.** when cells divide uncontrollably
   - **e.** none of the above

2. What are the ends of the chromosomes called that do not get fully replicated by DNA polymerase?
   - **a.** ROMs
   - **b.** stem-loop structures
   - **c.** poly A tails
   - **d.** telomeres
   - **e.** none of the above
3. What is the most prevalent theory to explain why cells might prematurely senesce?
   - **a.** because there are no signals coming from the environment to induce or repress gene expression
   - **b.** to avoid cancer
   - **c.** because the cell has acquired holes in the membrane and is about to die
   - **d.** because the cell has already become cancerous
   - **e.** none of the above
4. Which enzyme lengthens the telomeres in some cells?
   - **a.** DNA polymerase
   - **b.** polyadenylate polymerase
   - **c.** telomerase
   - **d.** RNA polymerase
   - **e.** none of the above because telomeres cannot be lengthened

5. Which two enzymes are responsible for decreasing the effects of oxidants on cellular components?
   a. catalase and telomerase
   b. superoxide dismutase and telomerase
   c. PUMA and telomerase
   d. catalase and superoxide dismutase
   e. catalase and PUMA
6. Besides *age-1*, a mutation in which gene of *C. elegans* has been shown to increase lifespan and induce hibernation?
   a. Daf-2
   b. catalase
   c. superoxide dismutase
   d. telomerase
   e. peroxidase
7. A defect in which of the following causes yeast cells, cultured mammalian cells, and mice to live longer?
   a. insulin receptors
   b. histones
   c. sirtuins
   d. nicotinamide binding proteins
   e. none of the above
8. Which of the following statements about apoptosis is not correct?
   a. Apoptosis is an organized and efficient process of cell death.
   b. During apoptosis, the cell shrinks and divides into apoptotic bodies.
   c. All of the components of an apoptotic cell are recycled.
   d. An apoptotic cells undergo necrosis.
   e. All of the above statements are true.
9. Which enzymes are considered the key regulators of apoptosis?
   a. caspases
   b. telomerases
   c. catalases
   d. superoxide dismutases
   e. none of the above
10. Which of the following apoptotic pathways is usually triggered by intracellular catastrophe, such as irreparable DNA damage?
   a. death receptor pathway
   b. apoptotic caspase activation pathway
   c. death-inducing pathway
   d. mitochondrial death pathway
   e. none of the above
11. Which of the following is digested by caspase?
   a. chromatin
   b. cellular organelles
   c. cytoskeleton
   d. nuclear architecture
   e. all of the above

(*Continued*)

12. What enables macrophages to recognize apoptotic bodies as "self"?
- **a.** phosphatidylserine
- **b.** outer surface proteins of apoptotic bodies
- **c.** cAMP
- **d.** Fas ligands
- **e.** none of the above

13. Which of the following is implicated during Alzheimer's?
- **a.** $A\beta_{42}$
- **b.** tau
- **c.** apolipoprotein E
- **d.** presenilins
- **e.** all of the above

14. Which two proteins control programmed cell death in *E. coli*?
- **a.** CED-3 and CED-2
- **b.** MazE and MazF
- **c.** caspase and catalase
- **d.** Bax and Bcl-2
- **e.** none of the above

15. Which one of the following antisense DNA oligonucleotides has been used in clinical trials to induce apoptosis in cancer cells?
- **a.** CED-3
- **b.** Bax
- **c.** Bcl-2
- **d.** caspase
- **e.** none of the above

Further Reading

Anderton BH (1999). Alzheimer's disease: Clues from flies and worms. *Curr Biol* **9**, R106–109.

Campisi J (1996). Replicative senescence: An old lives' tale? *Cell* **84**, 497–500.

Finch CE, Ruvkun G (2001). The genetics of aging. *Annu Rev Genomics Hum Genet* **2**, 435–462.

Hazan R, Sat B, Reches M, Engelberg-Kulka H (2001). Postsegregational killing mediated by the P1 phage "addiction module" phd-doc requires the *Escherichia coli* programmed cell death system mazEF. *J Bacteriol* **183**, 2046–2050.

Hekimi S, Lakowski B, Barnes TM, Ewbank JJ (1998). Molecular genetics of life span in *C. elegans*: How much does it teach us? *Trends Genet* **14**, 14–20.

Hodgkin J (1999). Sex, cell death, and the genome of *C. elegans*. *Cell* **98**, 277–280.

Itahana K, Dimri G, Campisi J (2001). Regulation of cellular senescence by p53. *Eur J Biochem* **268**, 2784–2791.

Kenyon C (1996). Ponce d'elegans: Genetic quest for the fountain of youth. *Cell* **84**, 501–504.

Krammer PH (2000). CD95's deadly mission in the immune system. *Nature* **407**, 789–795.

Le Bourg E (2001). Oxidative stress, aging and longevity in *Drosophila melanogaster*. *FEBS Lett* **498**, 183–186.

Leevers SJ (2001). Growth control: Invertebrate insulin surprises. *Curr Biol* **11**, R209–212.

Marciniak RA, Johnson FB, Guarente L (2000). Diskeratosis congenital, telomeres and human ageing. *Trends Genet* **16**, 193–195.

Meier P, Finch A, Evan G (2000). Apoptosis in development. *Nature* **407**, 796–801.

Melov S, Ravenscroft J, Malik S, Gill MS, Walker DW, Clayton PE, Wallace DC, Malfroy B, Doctrow SR, Lithgow GJ (2000). Extension of life-span with superoxide dismutase/catalase mimetics. *Science* **289**, 1567–1569.

Metzstein MM, Stanfield GM, Horvitz HR (1998). Genetics of programmed cell death in *C. elegans*: Past, present and future. *Trends Genet* **14**, 410–416.

Miller RA (1996). The aging immune system: Primer and prospectus. *Science* **273**, 70–74.

Nicholson DW (2000). From bench to clinic with apoptosis-based therapeutic agents. *Nature* **407**, 810–816.

Orrell RW, Habgood JJ, Gardiner I, King AW, Bowe FA, Hallewell RA, Marklund SL, Greenwood J, Lane RJ, deBelleroche J (1997). Clinical and functional investigation of 10 missense mutations and a novel frameshift insertion mutation of the gene for copper-zinc superoxide dismutase in UK families with amyotrophic lateral sclerosis. *Neurology* **48**, 746–751.

Parrish J, Metters H, Chen L, Xue D (2000). Demonstration of the *in vivo* interaction of key cell death regulators by structure-based design of second-site suppressors. *Proc Natl Acad Sci USA* **97**, 11916–11921.

Pawelec G, Hirokawa K, Fülöp T (2001). Altered T cell signaling in ageing. *Mech Ageing Dev* **122**, 1613–1637.

Raloff J (2001). Coming to terms with death: Accurate descriptions of a cell's demise may offer clues to diseases and treatments. *Science News* **159**, 378–380.

Rich T, Allen RL, Wyllie AH (2000). Defying death after DNA damage. *Nature* **407**, 777–783.

Sat B, Hazan R, Fisher T, Khaner H, Glaser G, Engleberg-Kulka H (2001). Programmed cell death in *Escherichia coli*: Some antibiotics can trigger mazEF lethality. *J Bacteriol* **183**, 2041–2045.

Sat B, Reches M, Engelberg-Kulka H (2003). The *Escherichia coli* mazEF suicide module mediates thymineless death. *J Bacteriol* **185**, 1803–1807.

Savill J, Fadok V (2000). Corpse clearance defines the meaning of cell death. *Nature* **407**, 784–788.

Serrano M, Blasco MA (2001). Putting the stress on senescence. *Curr Opin Cell Biol* **13**, 748–753.

Shepherd PR, Withers DJ, Siddle K (1998). Phosphoinositide 3-kinase: The key switch mechanism in insulin signaling. *Biochem J* **333**, 471–490.

Sherr CJ, DePinho RA (2000). Cellular senescence: Mitotic clock or culture shock. *Cell* **102**, 407–410.

Smith JR, Pereira-Smith OM (1996). Replicative senescence: Implications for *in vivo* aging and tumor suppression. *Science* **273**, 63–67.

Sohal RS, Weindruch R (1996). Oxidative stress, caloric restriction, and aging. *Science* **273**, 59–63.

Takahashi Y, Kuro OM, Ishikawa F (2000). Aging mechanisms. *Proc Natl Acad Sci USA* **97**, 12407–12408.

Wright WE, Shay JW (2000). Telomere dynamics in cancer progression and prevention: Fundamental differences in human and mouse telomere biology. *Nat Med* **6**, 849–851.

Yuan J, Yankner BA (2000). Apoptosis in the nervous system. *Nature* **407**, 802–809.

Bacterial Infections

INTRODUCTION

Infections of humankind as well as of animals and plants are caused by a diverse assortment of microorganisms, including viruses, bacteria, and various single-celled eukaryotes. The mechanisms of infection range from the simple approach of certain filamentous fungi that merely grow on unprotected organic matter to the highly sophisticated schemes for invasion and survival of specialized pathogens such as bubonic plague or malaria. Here we are concerned with applying modern molecular biology to understand and combat infection.

The molecular mechanisms used by infectious microorganisms are best understood for pathogenic bacteria, especially those closely related to the molecular biologist's model organism, *Escherichia coli*. While most strains of *E. coli* are harmless, a few virulent strains exist that illustrate many of the principles of infection at the molecular level. Although viruses have smaller genomes than bacteria, they are obligate intracellular parasites and depend on many host-cell gene products for replication. Consequently, viruses interact in a complex manner with the host-cell genome, which often makes analysis more complicated than for bacteria (see Chapter 22). Most difficult of all to understand are infections such as malaria or sleeping sickness that are due to single-celled eukaryotes.

Modern genetic analysis can be applied to any organism, whether invading microorganism or victim of infection, that has DNA or RNA that can be extracted, sequenced, and manipulated. Molecular approaches have greatly aided both the understanding and the diagnosis of infectious disease. Today they are starting to provide new methods to protect against disease.

Pathogenic bacteria are responsible for many human infections. Genetic approaches have yielded faster diagnostic methods and are providing the foundation for creating novel antibacterial agents.

MOLECULAR APPROACHES TO DIAGNOSIS

A major contribution of molecular biology has been the development of improved methods for diagnosis. In practice *diagnosis* usually means identifying the agent of disease, whether a bacterium, virus, or protozoan. Some pathogenic bacteria grow slowly or not at all when cultured outside their host organisms. Viruses are obligate parasites and can only be grown in the laboratory by infecting appropriate cultured host cells. Furthermore, different microorganisms require different culture media and culture conditions. All these factors make traditional methods of identification laborious. In contrast, molecular approaches analyze macromolecules such as DNA, RNA, or protein rather than attempting to grow the disease agents. Some new methods involve the use of antibody technology and are dealt with in Chapter 6. Here we will consider nucleic acid–based approaches.

Molecular methods usually start with extraction of DNA from either cultured pathogens or an infected patient. The DNA is often amplified by PCR (see Chapter 4) and then analyzed by sequencing or hybridization. Thus the same reagents and procedures may be used for many different microorganisms. Moreover, molecular methods are often quicker, more accurate, and more sensitive than classical microbiological techniques. Most of the molecular techniques used in clinical diagnosis have already been described elsewhere in this book. Here we will consider their applications.

Every species of organism has a different **small-subunit ribosomal RNA** sequence (**ssu rRNA**; 16S rRNA in bacteria, 18S rRNA for eukaryotes). Hence bacteria and eukaryotic parasites may be identified by analysis of their ssu rRNA sequences. In practice, the gene encoding the ssu rRNA is sequenced, rather than the RNA itself.

(a) **Ribotyping** may be done by detailed restriction analysis of the rRNA genes. DNA from a bacterial strain is digested with several different restriction enzymes and the fragments are separated by gel electrophoresis. The fragments are transferred to a membrane and subjected to Southern blotting to identify bands. A probe that corresponds to part of the 16S rRNA sequence is used. Many bacteria have more than one 16S rRNA gene and so several bands will be seen for each digestion. This approach needs significant amounts of DNA.

(b) Amplification of DNA by PCR followed by DNA sequencing allows identification using only tiny amounts of DNA. Primers that recognize the conserved region of 16S rRNA are used to generate a segment of the 16S gene by PCR. The PCR fragment is then sequenced and compared with a database of known DNA sequences. Note that the same procedure and reagents are applicable to all bacterial infections. Furthermore, this method works with bacteria that cannot be cultured. Variants of PCR, such as RAPD (randomly amplified polymorphic DNA; for details see Chapter 4), may also be used to identify the pathogen. For bacterial RAPD-PCR, a random mixture of six-base primers is often used. This allows different strains of the same bacterial species to be distinguished and has been used to track the source and spread of contaminating bacteria in food or water.

(c) **Checkerboard hybridization** is a technique that allows multiple bacteria to be detected and identified simultaneously in a single sample. A series of probes corresponding to different bacteria are applied in horizontal lines across a hybridization membrane (Fig. 21.1). PCR is used to amplify a portion of the 16S gene from the target bacteria or clinical samples, which may contain a mixture of bacteria. The PCR fragments are then labeled with a fluorescent dye and applied vertically to the membrane. After denaturation and annealing to allow hybridization, the membrane is washed to remove unbound DNA. Those samples that hybridize to the probes appear as bright fluorescent spots.

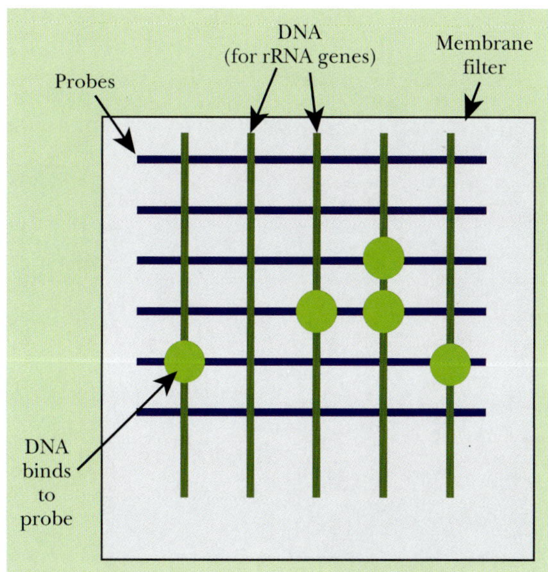

FIGURE 21.1
Checkerboard Hybridization
Probes corresponding to 16S rRNA for each candidate bacterium are attached to a membrane filter in long horizontal stripes (one candidate per stripe). To quickly identify a group of unknown pathogens, mixed DNA is extracted from a sample and amplified by PCR using primers for 16S rRNA. The PCR fragments are tagged with a fluorescent dye and applied in vertical stripes. Each sample is thus exposed to each probe. Wherever a 16S PCR fragment matches a 16S probe, the two bind, forming a strong fluorescent signal where the two stripes intersect.

Diagnosing pathogenic bacteria by molecular approaches, especially using ribosomal RNA sequences, is faster than traditional culture techniques.

VIRULENCE GENES ARE OFTEN FOUND ON MOBILE SEGMENTS OF DNA

Many infectious diseases are caused by bacterial invasion of the human body. Some pathogenic bacteria penetrate the interior of host cells, whereas others remain outside and inhabit extracellular spaces. The molecular mechanisms involved in infectious disease vary greatly in detail. Nonetheless, invading microorganisms face similar problems and so share many of the same general abilities. Properties that allow microorganisms to cause infections are called **virulence factors** and may be subdivided into three main groups: those required for invasion of the host, those for survival inside the host, and those for aggression against the host.

The blocks of DNA encoding virulence factors are often mobile. In some cases (e.g., anthrax, bubonic plague) they are borne on **virulence plasmids**. In other cases (e.g., cholera toxin, diphtheria toxin), they may be carried on lysogenic bacteriophage and inserted into the chromosomes of certain bacterial strains. In yet other cases, they are grouped together in regions of the chromosome flanked by inverted repeats and known as **pathogenicity**

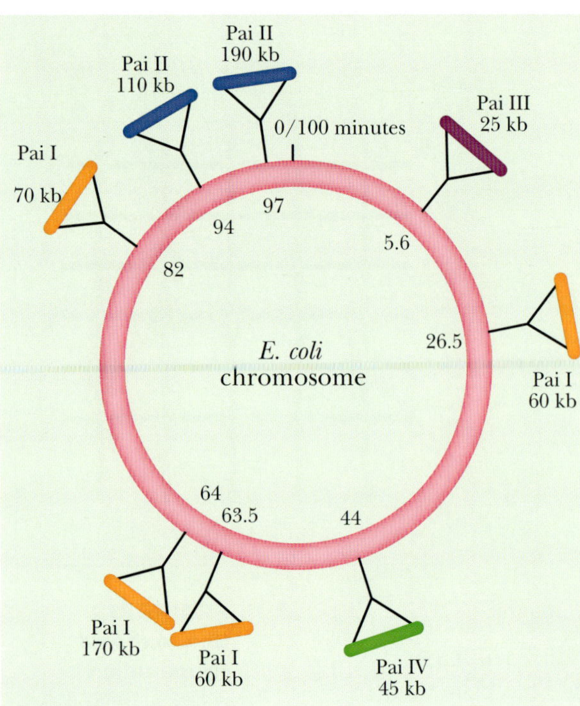

FIGURE 21.2
Pathogenicity Islands of *Escherichia coli*
Different strains of *E. coli* vary greatly in their abilities to cause disease. Pathogenic *E. coli* have unique regions of DNA that are not found in nonpathogenic strains, called pathogenicity islands (PAI). The regions are designated I–IV, where I encodes alpha-hemolysin; II encodes alpha-hemolysin and fimbriae; III encodes fimbriae; and IV encodes the yersiniabactin iron-chelating system.

islands (Fig. 21.2). The inverted repeats imply that the pathogenicity islands may move as a unit by transposition. The same or closely related virulence factors may be chromosomally integrated in one bacterial strain and plasmid-borne in others.

The mobility of virulence factors has several implications. First, closely related bacterial strains vary enormously in their capacity to cause disease. Harmless bacterial strains that are closely related to cholera, plague, diphtheria, anthrax, and so on are widely distributed, but normally draw little attention. Second, virulence factors may be transferred to hitherto harmless strains, thus creating novel pathogens in one or a few steps. If the harmless strain is a close relative, we may merely get a new variant of an old disease (e.g., the new biotypes of cholera that have appeared over the past century are due to transfer of the CTXphi phage into harmless strains of *Vibrio*). However, if the recipient for a block of virulence factors is unrelated to the previous "owner," a genuinely novel bacterial pathogen may be created. It seems likely that bubonic plague, *Yersinia pestis*, arose in this way several thousand years ago. Apparently an ancestral *Yersinia* that caused a mild intestinal infection was transformed by acquiring two virulence plasmids into a devastating disease injected directly into the bloodstream by fleas.

The genes for bacterial virulence factors are often carried by mobile DNA. That is, they are present on plasmids, integrated viruses, transposable elements, or pathogenicity islands rather than being a permanent feature of bacterial chromosomes.

ATTACHMENT AND ENTRY OF PATHOGENIC BACTERIA

The first step in many infections is the binding of bacteria to the surface of cells of the host animal. This is mediated by proteins known as adhesins that usually bind to sugar residues of glycoproteins or glycolipids on the animal cell surface. There are two major types of adhesins, fimbrial adhesins and nonfimbrial adhesins (Fig. 21.3). **Pili** (singular, *pilus*) or **fimbriae** (singular, *fimbria*) are thin filaments that protrude from the surface of bacteria (Fig. 21.4). The shaft is composed of helically arranged subunits of protein (**pilin**). Several specialized proteins, including **adhesins**, are carried at the very tip. Nonfimbrial adhesins are found on the surface of bacterial cells. In many cases, pili make first contact with the host cell and the nonfimbrial adhesins are responsible for a later and closer stage of binding.

A second common step of infection is entering an animal cell. Not all bacteria that adhere to animal cell surfaces possess the ability to invade. Some pathogens remain permanently outside the host cells. A classic example is cholera. Here the bacteria remain in the lumen of the intestine attached to the outside of intestinal cells. Only the toxin enters the intestinal cells and is responsible for the symptoms of the disease (see later discussion). Other pathogens possess various strategies for entering host cells. In some cases animal cells that are normally phagocytic, such as many ameboid cells of the immune system, ingest the bacteria but fail to destroy them. In other cases, bacteria provoke animal cells to swallow them by means of proteins known as **invasins**. For example, the invasin of *Yersinia* binds to the integrin proteins on mammalian cell surfaces and promotes internalization of the bacteria.

The spread of antibiotic resistance has provoked scientists to consider alternative approaches to treating infections. Several proposals have been made that take advantage of adhesins

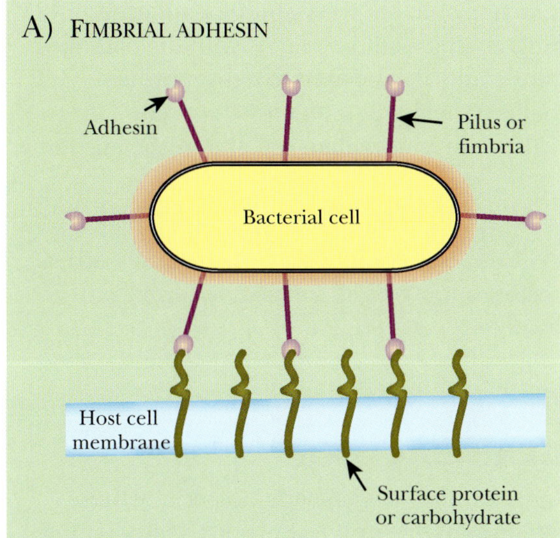

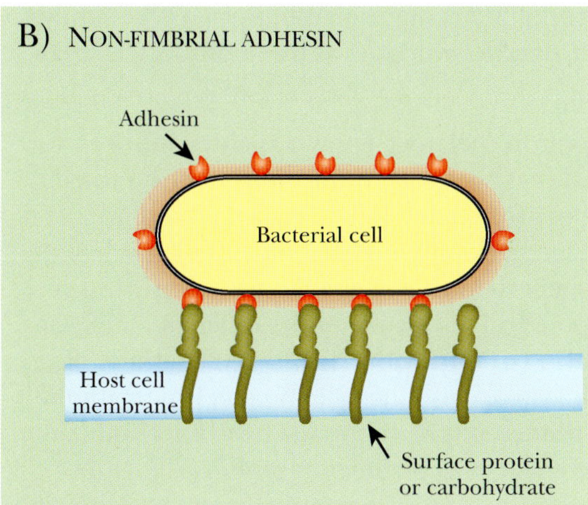

FIGURE 21.3 Bacterial Adhesins
(A) The surface of some bacterial cells is covered with pili (fimbriae), composed of helically arranged pilin protein. At the tip of the pili are adhesins, which recognize the surface glycoproteins of the host cell. (B) Nonfimbrial adhesins are found on the surface of the bacterial cell.

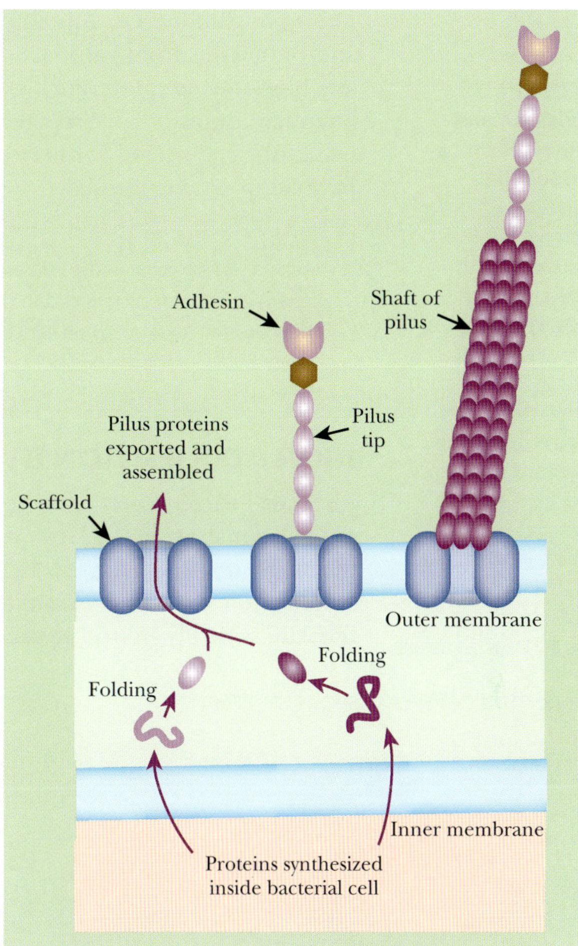

FIGURE 21.4 Assembly of Bacterial Pilus
The pilus has two segments, the tip and the shaft, which are assembled on the outside of the bacterium. The protein subunits of the pilus are synthesized in the cytoplasm and exported across both membranes. The proteins are folded in the periplasmic space. The pilus is assembled from the tip to the base by starting with the adhesin protein and other tip proteins and then adding further layers of pilin protein beneath.

and invasins to turn the tables on bacterial invaders. Because these are still mostly in the experimental phase, we will outline a few examples briefly.

Binding studies combined with x-ray crystallography can reveal details of the molecular targets for adhesins. Thus, the FimH adhesin of pathogenic *Escherichia coli* binds mannose residues on the surface of mammalian glycoproteins. Several alkyl- and aryl-mannose derivatives bind with extremely high affinity to the adhesin and can block attachment to the natural receptor. Mannose derivatives might form the basis for designing antiadhesin drugs that prevent bacterial binding.

A further suggested step is to genetically engineer harmless gut bacteria, such as nonpathogenic strains of *E. coli*, to express the target oligosaccharide for adhesins on their cell surfaces. Pathogenic bacteria would then bind to these decoys instead of to mammalian cells. This would avoid the need for administration of expensive sugar derivatives, because the decoy strains of *E. coli* would multiply naturally in the intestine. Furthermore, one engineered decoy strain could carry multiple adhesin targets.

A third possibility is to equip nonpathogenic strains of *E. coli* with genes for adhesins and/or invasins from pathogens. These harmless strains would compete with the pathogens and block the receptors. Such engineered strains could also be used to deliver protein pharmaceuticals or large segments of DNA for gene therapy into mammalian cells. Once inside, the engineered *E. coli* would be digested by the mammalian cell and its therapeutic payload would be released.

Bacteria use proteins known as adhesins to bind to the surface of host cells. Other proteins, known as invasins, promote entry into the host cell. Both adhesins and invasins are potential targets for novel antibiotics but have yet to be successfully exploited.

IRON ACQUISITION BY PATHOGENIC BACTERIA

Almost all bacteria need iron because it is a cofactor for many enzymes, especially those of the respiratory chain. However, the concentration of free iron in the body, including the bloodstream, is kept low by a variety of specialized proteins that bind iron very tightly. Surplus iron is bound by **transferrin** and **lactoferrin**, which act as iron transporters, or by **ferritin**, which is an iron storage protein.

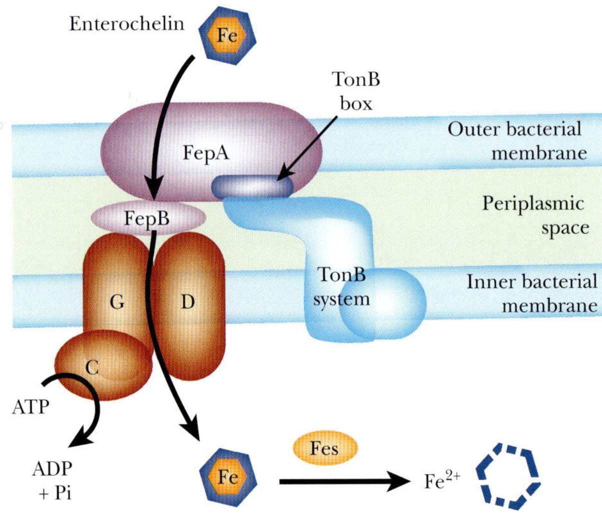

FIGURE 21.5
Acquisition and Uptake of Iron by Enterochelin
FepA protein is the outer membrane receptor for enterochelin. Energy for crossing the outer membrane requires the TonB system, which uses the proton motive force. The FepB protein gets enterochelin from FepA and passes it to the inner membrane permease, consisting of FepG and FepD. The FepC protein uses ATP to supply energy to FepGD for transport across the inner membrane.

Bacteria use iron chelators, known as **siderophores**, to bind iron and, if necessary, extract it from host proteins. Siderophores are excreted by the bacteria, bind iron, and are then taken back into the bacterial cell by specialized transport systems. Most bacteria have a variety of such iron transport systems for use under different conditions. **Enterochelin (or enterobactin)** is a cyclic trimer of 2,3-dihydroxybenzoylserine (DHBS) and is perhaps the best known siderophore. It is made by *E. coli* and many enteric bacteria. The Fep transport system (Fig. 21.5) conveys the enterochelin/Fe complex across both outer and inner membranes. Enterochelin binds iron so tightly that once inside the bacterial cell it must be destroyed in order to release the iron. Fes protein is an esterase that hydrolyzes Fe^{3+}-enterochelin into DHBS monomers, releasing the iron. Despite this, enterochelin does not bind iron well enough to extract it from transferrin, the major iron binding protein of blood.

Pathogenic bacteria often possess more potent siderophores (plus their receptor systems) that are capable of extracting iron from transferrin. Two examples are mycobactin, made by *Mycobacterium tuberculosis*, and yersiniabactin, named after *Yersinia pestis*, the plague bacterium, in which it was first discovered. Yersiniabactin is widespread among pathogenic bacteria of the enteric family, and in *Yersinia* itself the genes for synthesis and uptake are clustered on a mobile pathogenicity island. Other virulent enteric bacteria carry plasmid-borne genes for **hemolysin**, a toxin that lyses red blood cells, thus allowing access to the hemoglobin.

Bacteria need a source of iron. They use iron chelators known as siderophores to compete for iron with other organisms, including the host in the case of bacterial infections.

Box 21.1 Deliberate Design of Novel Antibiotics

Absence of the high-potency siderophores largely abolishes virulence in both plague and tuberculosis. Because siderophores are not made by mammals, their unique biosynthetic pathways provide an attractive target for development of novel antibiotics. Several high-potency siderophores, including mycobactin and yersiniabactin, have a salicyl group capping the siderophore structure (Fig. A). The salicylate is activated by using ATP to generate salicyl-AMP. Deliberate chemical synthesis of substrate analogs in which salicylate is linked to adenosine via a sulfamoyl group (instead of phosphate) results in highly active and specific inhibitors of siderophore synthesis. These prevent growth of both *Mycobacterium* and *Yersinia* under iron-limiting conditions. It is hoped that this approach will lead to novel inhibitors for use against antibiotic-resistant pathogens.

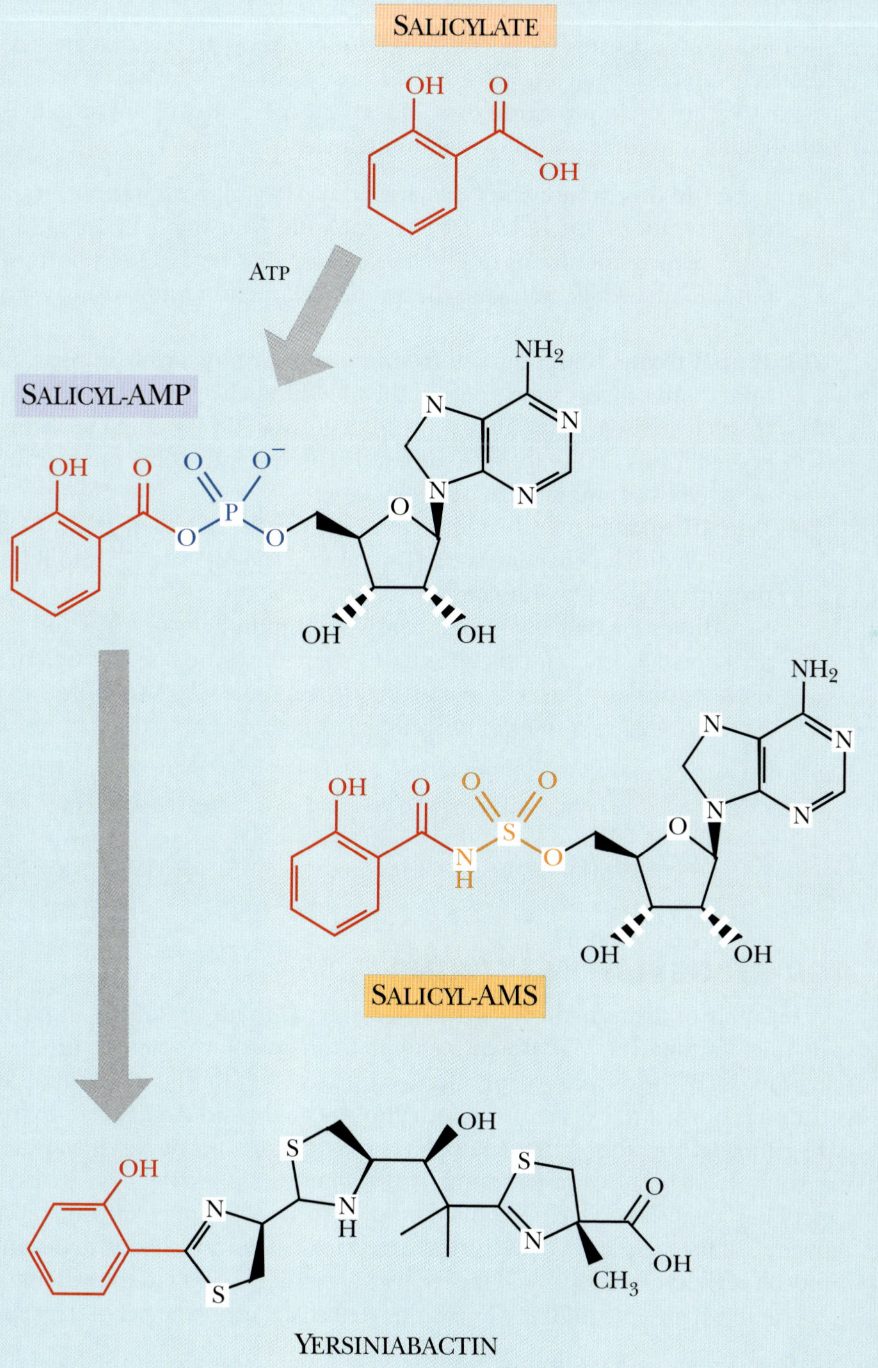

FIGURE A
Yersiniabactin and Salicyl-AMP
The structure of yersiniabactin shows the salicyl group in red. The precursor salicyl-AMP is made by activating salicylate with ATP. The sulfamoyl analog, salicyl-AMS, inhibits the incorporation of the salicyl group into yersiniabactin.

BACTERIAL TOXINS

In addition to invasion and survival, many pathogenic bacteria also mount aggressive attacks against eukaryotic cells by making **toxins**. In its broadest sense, the term *toxin* includes any molecule that damages eukaryotic cells. Some toxic effects may be regarded as "accidental," whereas others are deliberate. For example, bacterial **endotoxin** is actually the lipid A component of lipopolysaccharide (LPS) that forms part of the outer membrane of gram-negative bacteria. If such bacteria are killed by the immune system, their cell walls will disintegrate and release LPS. The CD14 receptor on immune cells binds the LPS, and this triggers the release of cytokines. Simultaneous destruction of many bacteria will release large amounts of LPS and may result in septic shock.

Nonetheless, most pathogenic bacteria make one or more toxins that are designed to deliberately kill or damage the cells of their host. Because these are secreted from still-living bacteria, they are known as **exotoxins**. Most bacterial exotoxins are proteins, and they may be subdivided into three groups based on their site of action:

(a) **Type I toxins** do not enter the target cell. They trigger a harmful response by binding to a receptor on the cell surface. For example, heat-stable toxin a (STa) is made by some pathogenic strains of *E. coli* that cause diarrhea. STa is a hormone analog that binds to guanylate cyclase in the animal cell membrane, causing overproduction of cyclic GMP.

(b) **Type II toxins** act on the cell membrane of the target cell. Some degrade membrane lipids, and others create pores in the membrane. For example, hemolysin A is made by some pathogenic strains of *E. coli* that cause kidney damage. Hemolysin A was named for its ability to lyse red blood cells, but it disrupts the membranes of many other types of animal cells also.

(c) **Type III toxins** enter the target cell. They consist of a toxic factor (A-protein) together with a delivery system (B-protein). The A ("active") and B ("binding") proteins may be separate molecules or may be different domains of a single protein. Sometimes the delivery system consists of multiple B-subunits and, conversely, cases are known where multiple A-proteins share the same delivery system (Table 21.1). These toxins are the most interesting from a molecular viewpoint, and we will discuss a few selected examples in more detail.

Many different bacterial toxins are known. Endotoxin is the lipid A fragment of lipopolysaccharide. Protein toxins are classified according to whether they act from outside the target cell, damage the membrane, or enter the cell to act. Most toxins that enter the target cell consist of toxic subunits plus binding proteins.

ADP-RIBOSYLATING TOXINS

A large family of toxins hydrolyzes the cofactor NAD to nicotinamide and **ADP-ribose**. The toxin then transfers the ADP-ribose fragment to an acceptor molecule, usually a protein that binds GTP. The target protein is locked into its GTP-binding conformation and cannot perform its normal role (Fig. 21.6). Both **cholera toxin** and **diphtheria toxin** work by **ADP-ribosylation**, although the targets are different. Cholera toxin inactivates the G protein that controls adenylate cyclase in animal cells, whereas diphtheria toxin attacks elongation factor EF-2, a translation factor required for protein synthesis in eukaryotic cells. It was originally thought that the toxic effect of such NAD-using toxins was merely due to destruction of NAD. Although high levels of toxin will indeed hydrolyze all the NAD in the cell, the true lethal effect occurs at much lower toxin levels and is due to the ADP-ribosylation of target proteins.

The genes for cholera toxin and diphtheria toxin are not part of the bacteria genome. Rather, the genes are carried on lysogenic viruses whose DNA is integrated into the bacterial chromosome. Only bacteria that carry the bacteriophage produce toxin, whereas those that lack it are nonpathogenic.

Table 21.1 Examples of A-B–Type Bacterial Toxins

Toxin and Organism	Structure and Mode of Action
Cholera toxin *Vibrio cholerae*	AB_5 ADP-ribosylation of G-protein that controls adenylate cyclase
Pertussis toxin *Bordetella pertussis*	$A'B'_5$ (five B-subunits are nonidentical) ADP-ribosylation of G-protein that controls adenylate cyclase
Shiga toxin *Shigella*	AB_5 Removal of adenine in 28S rRNA; inhibition of protein synthesis
Diphtheria toxin *Corynebacterium diphtheriae*	A-B (A = N-terminal domain) ADP-ribosylation of translation factor EF2; inhibition of protein synthesis
Exotoxin A *Pseudomonas aeruginosa*	A-B (A = C-terminal domain) ADP-ribosylation of translation factor EF2; inhibition of protein synthesis
Botulinum toxin *Clostridium botulinum*	A-B Zn-dependent protease that cleaves proteins in nerve synapses so blocking transmission in peripheral nerves
Edema toxin *Bacillus anthracis*	$A\text{-}B_7$ (edema factor plus protective antigen) Adenylate cyclase activity
Lethal toxin *Bacillus anthracis*	$A\text{-}B_7$ (lethal factor plus protective antigen) Protease that cleaves mitogen-activated protein kinase kinases (MAPKKs)

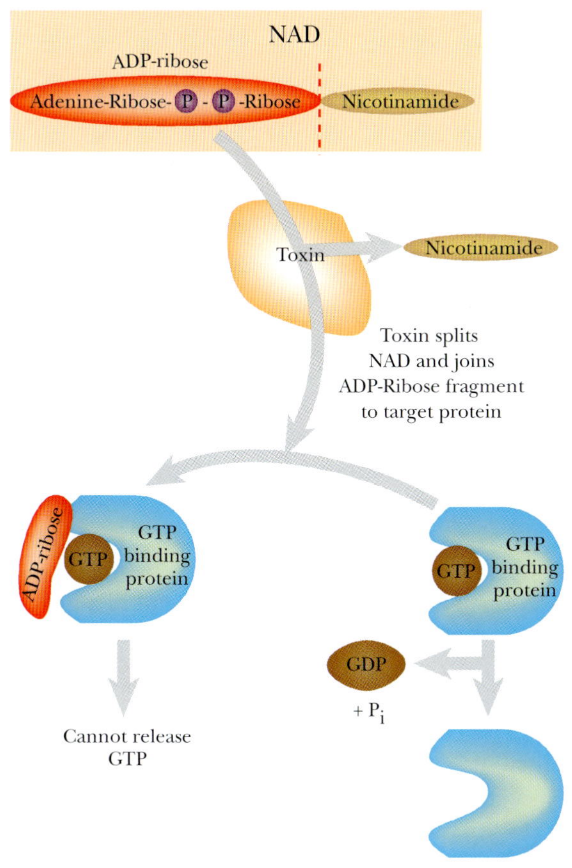

FIGURE 21.6 ADP-Ribosylating Toxins
Nicotinamide adenine dinucleotide (NAD) consists of ADP-ribose linked to nicotinamide. These are split by some bacterial toxins and the ADP-ribose is attached to a GTP-binding protein, thus preventing it from splitting GTP.

Certain other bacteriophages are known to possess enzymes that use NAD to ADP-ribosylate proteins of their bacterial hosts. Usually several bacterial proteins are modified, so which is the intended target is often uncertain. The blockade of key cellular enzymes may cripple host metabolism. A more sophisticated effect of ADP-ribosylation is the modification of host DNA or RNA polymerases to change their specificity. Examples include bacteriophage T4, which ADP-ribosylates the RNA polymerase of its host, *E. coli*. The RNA polymerase thereupon loses much of its ability to transcribe *E. coli* genes but still works well with T4 genes. Like the targets of cholera and diphtheria toxins, RNA polymerase also binds GTP, although admittedly as just one of several alternative nucleotides.

> Many toxins act by adding an ADP-ribose group, taken from NAD, to a target protein. This target is generally a GTP-using enzyme, and this activity is blocked by the presence of ADP-ribose.

CHOLERA TOXIN

The cholera bacterium, *Vibrio cholerae*, does not invade the tissues of the host. It attaches to the exterior of cells lining the small intestine and stays there. The damage is due to the secretion of cholera toxin. This attacks intestinal epithelial cells, causing them to lose sodium ions and then water into the intestinal tract. The clinical symptoms of cholera are loss of body fluids by massive diarrhea and subsequent death by dehydration.

The virulence proteins of *Vibrio cholerae* include cholera toxin as well as both pilus-borne and cell-surface adhesins for binding to intestinal cells. The genes for cholera toxin are carried by the **CTXphi** filamentous bacteriophage that lysogenizes *Vibrio cholerae.* Synthesis of both adhesins and toxin is co-regulated by the ToxR protein found in the bacterial inner membrane. This senses whether the bacterium is inside an animal intestine (temperature, pH, and bile salts are all involved) and activates the virulence genes accordingly (Fig. 21.7). The internal domain of ToxR protein binds directly to the promoters of these genes.

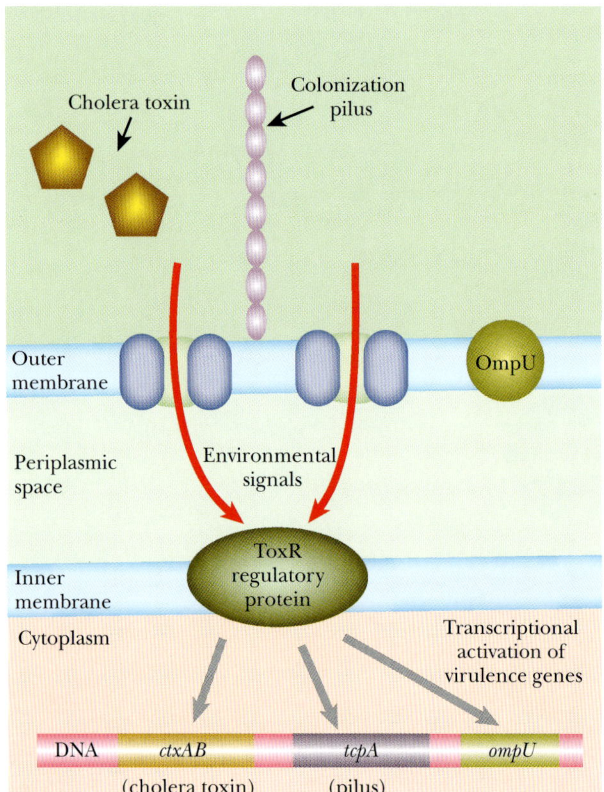

FIGURE 21.7 Regulation of *V. cholerae* Virulence Genes
ToxR of *V. cholerae* sits in the cytoplasmic membrane, where it senses that the cell is in a human host and directly activates the genes for cholera toxin and for attachment.

Cholera toxin consists of two proteins, A and B, encoded by the *ctxAB* genes (Fig. 21.8). The original A protein is split by a protease to give A1 (23.5 kDa) and A2 (5.5 kDa), which remain linked by a single disulfide bond. Five B proteins (11.6 kDa) form a ringlike structure. The A1-S-S-A2 protein protrudes through the ring of five B-subunits. Cholera toxin binds to a glycolipid, known as **ganglioside GM1**, in eukaryotic cell membranes. Each B subunit binds to the galactose end of a ganglioside molecule. After binding of the A1A2/B_5 complex to the cytoplasmic membrane, A1 is released from A2 by reduction of the disulfide bond. A1 then enters the cell. Free B protein protects eukaryotic cells against toxin by competing for GM1 binding sites.

After entering the intestinal cell, cholera toxin splits NAD into nicotinamide and ADP-ribose, which is then used to ADP-ribosylate target molecules. Cholera toxin A1 protein can actually ADP-ribosylate a variety of acceptors, including free arginine and its derivatives as well as many proteins. A1 protein ADP-ribosylates itself, increasing its activity by 50%. The true *in vivo* lethal target is a **G protein** involved in regulation of **adenylate cyclase** (Fig. 21.9).

Normally, when the G protein is activated by a signal from outside the cell, it binds GTP and then binds to and activates adenylate cyclase. GTP hydrolysis allows the release of the G protein from the adenylate cyclase and results in deactivation. ADP-ribosylation of an arginine residue by cholera toxin prevents the hydrolysis of GTP and locks the G protein in its GTP binding state. This causes hyperactivation of adenylate cyclase and overproduction of cyclic AMP, which results in the loss of sodium ions and water. GTP analogs that cannot be hydrolyzed (such as GMP-P-NH-P or GMP-P-CH_2-P) show effects similar to cholera toxin.

Cholera toxin and the heat-labile **enterotoxins** of other enteric bacteria are actually variants of the same toxin, although cholera toxin is more potent. The heat-labile enterotoxins found in some pathogenic strains of *E. coli* are encoded on the Ent-plasmid, which may be transferred between strains. These toxins have similar amino acid sequences to cholera toxin and cause loss of intestinal fluids by the same mechanism.

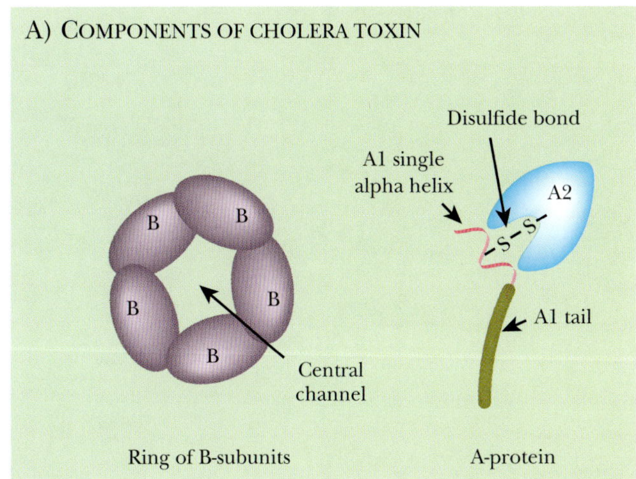

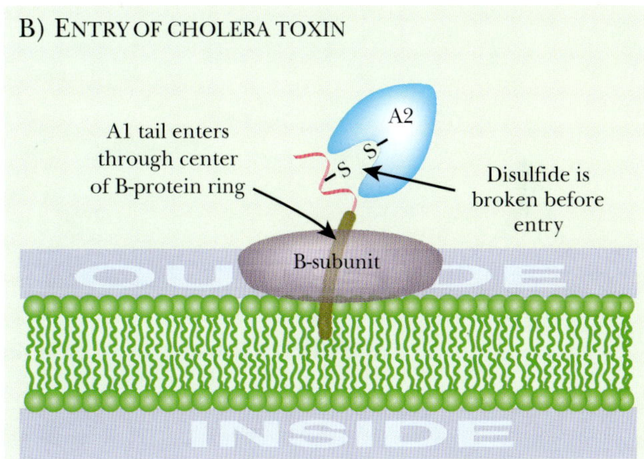

FIGURE 21.8
Structure and Entry of Cholera Toxin
(A) Cholera toxin consists of an A protein plus five copies of B protein. The A protein is split into two halves (A1 and A2), held together by a disulfide bond. The B protein forms a ring with a central channel for the A1-S-S-A2 protein. (B) Cholera toxin binds to the host cell when the five B-subunits recognize ganglioside GM1. The disulfide bond in A1-S-S-A2 breaks, allowing A1 to enter the cell.

The genes for cholera toxin are carried on a virus inserted into the bacterial chromosome. The toxin adds ADP-ribose to a host G protein that regulates adenylate cyclase. Overproduction of cyclic AMP leads to dehydration.

ANTHRAX TOXIN

Anthrax is caused by the gram-positive bacterium *Bacillus anthracis*, the first bacterium proven to be the cause of a disease. In 1877, Robert Koch grew this organism in pure culture, demonstrated its ability to form spores, and produced anthrax experimentally by injecting it into animals. Virulence factors of anthrax include the exotoxins and the capsule, both plasmid-borne. There are two plasmids: pXO1 carries the regulatory and structural genes for the exotoxins and pXO2 carries the genes for the capsule. The capsule is made from poly-D-glutamic acid and protects against attack by cells of the immune system. Chromosomal sequencing has shown that, apart from its virulence plasmids, *Bacillus anthracis* is remarkably similar to other "species" of *Bacillus*, such as *Bacillus cereus* (a common soil bacterium) and *Bacillus thuringiensis* (well known for making the insecticidal toxins used in transgenic plant engineering—see Chapter 14).

The genes for known virulence factors of *B. anthracis* comprise only a small percentage of the DNA of pXO1 and pXO2. For example, plasmid pXO1 (184 kb) has 143 predicted genes and carries a pathogenicity island of 45 kb bordered by inverted repeats. This region contains the three genes *pag, lef,* and *cya*, which encode the binding protein and the two exotoxins,

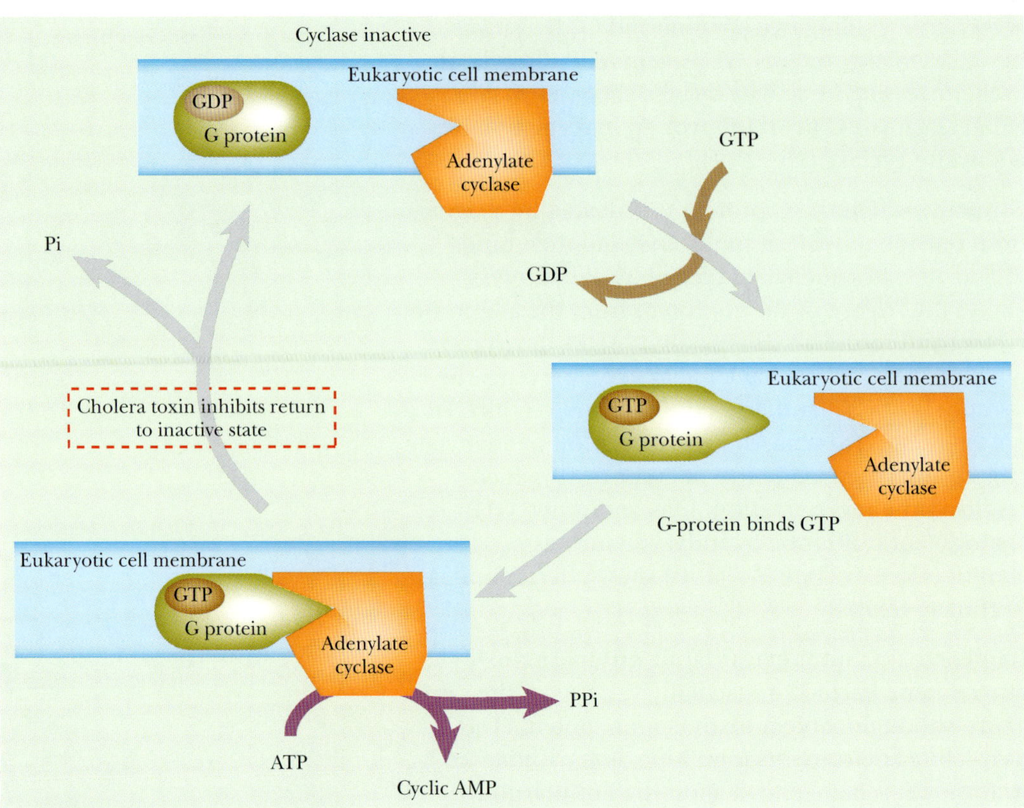

FIGURE 21.9 Mechanism of Action of Cholera Toxin
In their inactive state G proteins bind GDP. When an external signal activates the G protein, the GDP is exchanged for GTP. The G protein then activates adenylate cyclase. Normally, the GTP is hydrolyzed and the G protein returns to its inactive state. Cholera toxin cleaves NAD and attaches the ADP-ribose group to an arginine in the G protein. This prevents the G protein from splitting GTP. Consequently adenylate cyclase does not get turned off and continues to produce cyclic AMP.

respectively. Transcription of the toxin genes is under the control of two *trans*-acting regulatory genes, *atxA* and *atxR*. Toxin synthesis responds to levels of bicarbonate/carbon dioxide and temperature. These factors indicate whether the anthrax bacteria have reached the interior of a warm-blooded animal.

Anthrax makes two separate toxin proteins, **edema factor (EF)** and **lethal factor (LF)**, which both share the same binding protein, known as **protective antigen (PA)**. Edema toxin consists of edema factor plus PA and lethal toxin is lethal factor plus PA. Thus the protective antigen corresponds to the B protein of other toxins because it recognizes and binds to the host cell. PA forms a ring of seven subunits with a central cavity through which the toxic factors enter the target cell. When first synthesized, PA protein is an inactive monomer (83 kDa). Activation involves cleavage by a host protease to give the 63-kDa form (PA63), which has the ability to oligomerize and insert itself into the host cell membranes (Fig. 21.10).

Edema factor acts as an adenylate cyclase. It is not fatal alone but increases the toxic effects of lethal factor. As its name implies, lethal factor is fatal alone, although higher doses are needed in the absence of edema factor. Lethal factor is a protease that cleaves several mitogen-activated protein kinase kinases (MAPKKs) such as MEK1, MEK2, and MKK3, which are involved in the control of animal cell division. Cleavage occurs within the N-terminal proline-rich region preceding the kinase domain. This disrupts the domain involved in protein-protein interactions needed for signaling. The lethal effects of anthrax toxin are due to the lysis of macrophages, which are especially susceptible to this toxin. This in turn causes the excessive release of interleukins, resulting in shock leading to respiratory failure and/or cardiac failure.

In anthrax toxin, two toxic proteins share the same receptor binding subunit. Edema factor is an adenylate cyclase, and lethal factor cleaves host regulatory proteins. Genes for the anthrax toxin subunits are plasmid borne.

ANTITOXIN THERAPY

Even if infections occur and toxins are secreted by the invading bacteria, it may be possible to protect the patient against the toxins. Traditional antitoxin treatment has relied on antibodies against bacterial toxins. However, new gene-oriented approaches are emerging.

One approach relies on **dominant-negative mutations** in the binding subunit of the toxin. Defective mutations typically result in proteins that are inactive. However, occasional mutations give rise to proteins that not only are inactive themselves, but interfere with the functional version of the protein. The presence of such a mutation in the same cell as the wild-type version of the gene results in absence of activity—hence the term *dominant-negative.*

The mechanism usually involves the binding of a defective protein subunit to functional subunits resulting in a complex that is inactive overall. Consequently, most dominant-negative mutations affect proteins with multiple subunits. The multisubunit B proteins of A and B toxins such as cholera toxin and anthrax toxin are good examples. Dominant-negative mutations have been deliberately isolated in the B protein (i.e., the protective antigen, PA) of anthrax toxin. Mixing mutant subunits with active ones resulted in the assembly of inactive heptamers that bind the lethal or edema factor proteins (i.e., the A subunits) but cannot transport them into target cells (Fig. 21.11). Treatment with the dominant-negative PA protein protected both cultured human cells and whole mice or rats from death by lethal levels of anthrax toxin.

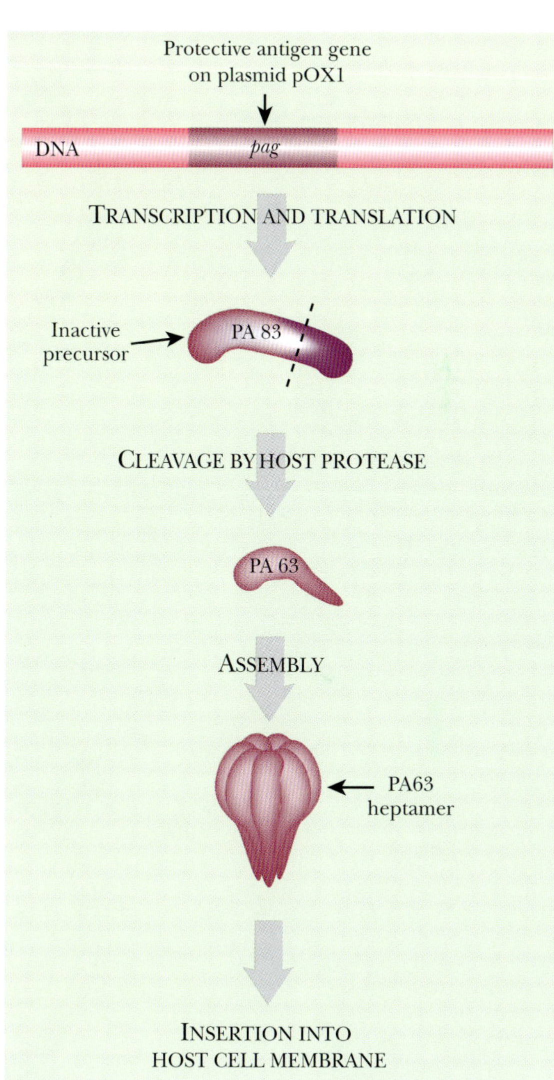

FIGURE 21.10 Activation of Protective Antigen
The *pag* gene of the pOX1 plasmid encodes the protective antigen (PA) of *B. anthracis.* PA is synthesized as an inactive precursor that is cleaved and assembled into a ring structure.

Another approach is the use of phage display (see Chapter 9) to isolate nonnatural peptides that bind to bacterial toxins. Such peptides typically bind rather weakly to single proteins. However, if several copies of the peptide are attached to a flexible backbone, this gives what is known as a **polyvalent inhibitor**. Binding to multiple target proteins will occur, which results in massively increased overall binding affinity. As discussed earlier, for this to work, the target must be a multisubunit protein such as the heptameric PA protein of anthrax toxin. Polyvalent peptide inhibitors with a polyacrylamide backbone have proven successful in protecting animals against anthrax toxin.

Dominant-negative mutations in toxin binding subunits can interfere with the binding of normal toxins. Polyvalent peptide inhibitors are also effective for toxins with multiple receptor binding subunits.

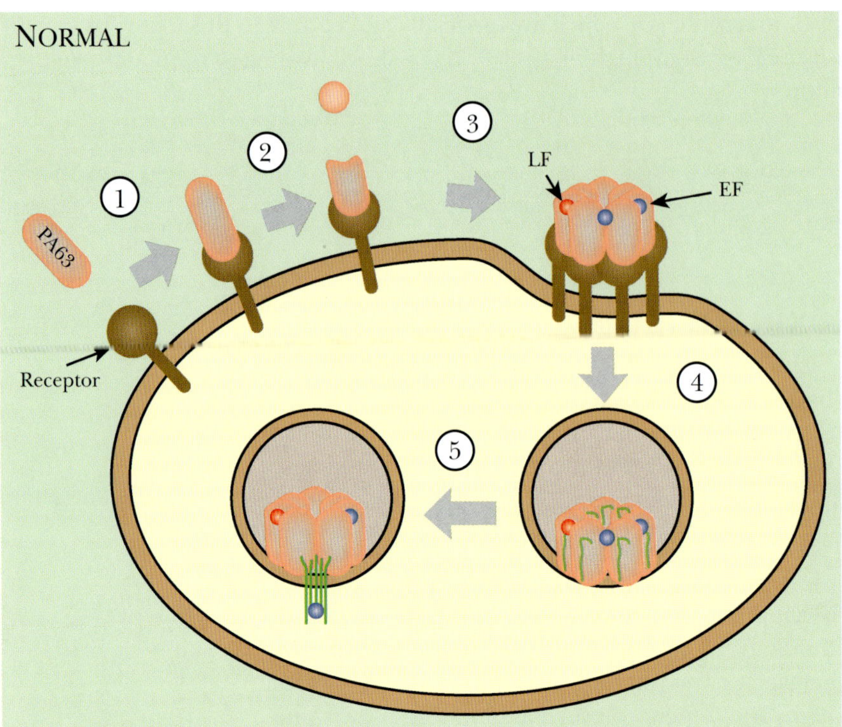

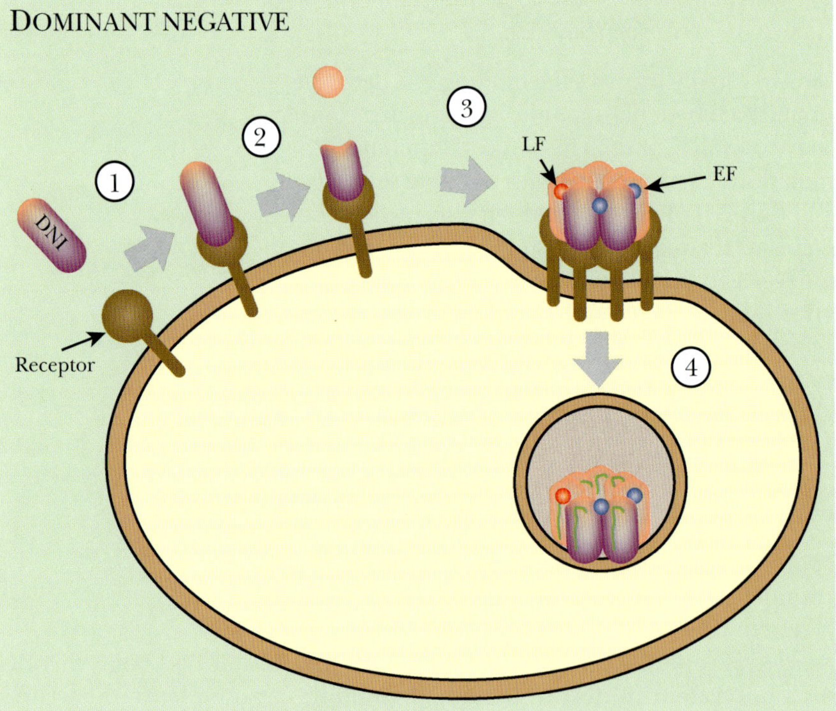

FIGURE 21.11 Dominant-Negative Toxin Mutations

The PA63 protein (protective antigen) binds the lethal factor (LF) and edema factor (EF) and transports them into the target cell cytoplasm via an endocytotic vesicle. The dominant-negative inhibitory (DNI) mutant of the PA63 protein (purple) assembles together with normal PA63 monomers (pink) to give an inactive complex that cannot release the LF and EF toxins from the vesicle into the cytoplasm.

Summary

Most bacterial infections can be treated successfully by antibiotics. Indeed, this is one of the major triumphs of modern science. Modern challenges in this area are largely due to mobile DNA. Plasmids, bacterial viruses, and transposons move antibiotic resistance genes and virulence genes between different bacterial species and can thus generate novel pathogens. Detailed analysis of bacterial toxins may yield some possible novel approaches.

End-of-Chapter Questions

1. What advantage do molecular methods have over traditional methods when identifying infections in patients?
   - **a.** The molecular methods are more sensitive.
   - **b.** Many different organisms can be identified with the same reagents and procedures.
   - **c.** Molecular methods are more accurate than traditional methods.
   - **d.** Molecular methods are often quicker than traditional methods.
   - **e.** All of the above are advantages of molecular methods over traditional methods of identifying a pathogen.
2. Which of the following molecular techniques can be used to identify pathogens?
   - **a.** DNA sequencing
   - **b.** ribotyping
   - **c.** RAPD
   - **d.** checkerboard hybridization
   - **e.** all of the above
3. Where might virulence factors be located?
   - **a.** pathogenicity islands
   - **b.** virulence plasmids
   - **c.** integrated viruses
   - **d.** transposable elements
   - **e.** all of the above
4. What mediates the binding of pathogenic bacteria to the host's cells?
   - **a.** integrins
   - **b.** adhesins
   - **c.** MHC class II
   - **d.** invasins
   - **e.** none of the above
5. Which of the following is used by bacteria to incite the host cell to "swallow" the bacterial cell?
   - **a.** integrins
   - **b.** adhesins
   - **c.** invasins
   - **d.** MHC class II
   - **e.** none of the above

(Continued)

6. Which of the following is used by bacteria to bind iron and/or extract it from proteins?
   a. transferrins
   b. siderophores
   c. lactoferrins
   d. ferritin
   e. all of the above

7. Which one of the following is capable of extracting iron from transferrin?
   a. mycobactin
   b. lactoferrin
   c. ferritin
   d. hemolysin
   e. none of the above

8. Which of the following sentences about toxins is not true?
   a. Some Type II toxins degrade cell membranes.
   b. Type III toxins kill the immune response system of the eukaryotic cell.
   c. Type I toxins do not enter eukaryotic cells.
   d. Some Type II toxins create pores in the cell membranes of eukaryotic cells.
   e. None of the above is true.

9. Which of the following is not an A-B type toxin?
   a. shiga toxin
   b. botulinum toxin
   c. hemolysin A toxin
   d. cholera toxin
   e. diphtheria toxin

10. Which of the following processes account for the lethal mechanism of diphtheria and cholera toxins?
   a. ADP-ribosylation
   b. adenylate cyclase cleavage
   c. iron acquisition from the host
   d. uncoupling of the electron transport chain
   e. blocking of transmission in peripheral nerves

11. Which one of the following statements about cholera toxin is not correct?
   a. The cholera toxin genes are carried on the filamentous bacteriophage CTXphi.
   b. ToxR regulates the synthesis of both adhesins and toxin.
   c. Cholera toxin is produced after the *Vibrio* cells are taken up into the host cell.
   d. Cholera toxin attacks intestinal cells and causes them to release large amounts of sodium ions and water.
   e. All of the above statements are correct.

12. To enter the host cell, to what does cholera toxin bind?
   a. ganglioside GM_1
   b. adenylate cyclase
   c. G proteins
   d. ADP-ribose
   e. none of the above

13. Which protein do both edema factor and lethal factor bind from *B. anthracis*?
- **a.** ganglioside GM1
- **b.** adenylate cyclase
- **c.** protective antigen
- **d.** ADP-ribosylated proteins
- **e.** none of the above

14. Where are the genes for anthrax toxin subunits located?
- **a.** virulence plasmids
- **b.** pathogenicity islands
- **c.** transposable elements
- **d.** integrated bacteriophage
- **e.** none of the above

15. Which of the following would be most affected by dominant-negative mutations?
- **a.** DNA
- **b.** single subunit proteins
- **c.** mRNA
- **d.** proteins with multiple subunits
- **e.** none of the above

Further Reading

Arnold DL, Jackson RW, Waterfield NR, Mansfield JW (2007). Evolution of microbial virulence: The benefits of stress. *Trends Genet* **23**, 293–300.

Baldari CT, Tonello F, Paccani SR, Montecucco C (2006). Anthrax toxins: A paradigm of bacterial immune suppression. *Trends Immunol* **27**, 434–440.

Clarridge JE 3rd (2004). Impact of 16S rRNA gene sequence analysis for identification of bacteria on clinical microbiology and infectious diseases. *Clin Microbiol Rev* **17**, 840–862.

Diep BA, Gill SR, Chang RF, Phan TH, Chen JH, Davidson MG, Lin F, Lin J, Carleton HA, Mongodin EF, Sensabaugh GF, Perdreau-Remington F (2006). Complete genome sequence of USA300, an epidemic clone of community-acquired methicillin-resistant *Staphylococcus aureus*. *Lancet* **367**, 731–739.

Gal-Mor O, Finlay BB (2006). Pathogenicity islands: A molecular toolbox for bacterial virulence. *Cell Microbiol* **8**, 1707–1719.

Hung DT, Shakhnovich EA, Pierson E, Mekalanos JJ (2005). Small-molecule inhibitor of *Vibrio cholerae* virulence and intestinal colonization. *Science* **310**, 670–674.

Jolley KA, Brehony C, Maiden MC (2007). Molecular typing of meningococci: Recommendations for target choice and nomenclature. *FEMS Microbiol Rev* **31**, 89–96.

McLeod SM, Kimsey HH, Davis BM, Waldor MK (2005). CTXphi and *Vibrio cholerae*: Exploring a newly recognized type of phage-host cell relationship. *Mol Microbiol* **57**, 347–356.

Miethke M, Marahiel MA (2007). Siderophore-based iron acquisition and pathogen control. *Microbiol Mol Biol Rev* **71**, 413–451.

Mulvey MR, Boyd DA, Olson AB, Doublet B, Cloeckaert A (2006). The genetics of *Salmonella* genomic island 1. *Microbes Infect* **8**, 1915–1922.

Palomino JC (2006). Newer diagnostics for tuberculosis and multi-drug resistant tuberculosis. *Curr Opin Pulm Med* **12**, 172–178.

Shakhnovich EA, Hung DT, Pierson E, Lee K, Mekalanos JJ (2007). Virstatin inhibits dimerization of the transcriptional activator ToxT. *Proc Natl Acad Sci USA* **104**, 2372–2377.

Waldor MK, Friedman DI (2005). Phage regulatory circuits and virulence gene expression. *Curr Opin Microbiol* **8**, 459–465.

Yates SP, Jørgensen R, Andersen GR, Merrill AR (2006). Stealth and mimicry by deadly bacterial toxins. *Trends Biochem Sci* **31**, 123–133.

Viral and Prion Infections

Viral Infections and Antiviral Agents
Interferons Coordinate the Antiviral Response
Influenza Is a Negative-Strand RNA Virus
The AIDS Retrovirus
Chemokine Receptors Act as Co-receptors for HIV
Treatment of the AIDS Retrovirus
Infectious Prion Disease
Detection of Pathogenic Prions
Approaches to Treating Prion Disease
Using Yeast Prions as Models

VIRAL INFECTIONS AND ANTIVIRAL AGENTS

Many human diseases are due to viruses. These are less well understood than bacterial diseases, to a large extent because viruses cannot be grown alone in culture but depend on a host cell. Until recently, protection against virus diseases relied on public health measures and vaccination. Only recently have a significant number of specific antiviral agents become available.

Pathogenic bacteria contain many unique components not found in eukaryotic cells, which can be targeted by antibiotics. In contrast, because viruses rely on the host cell for almost all of their metabolic reactions, they usually have few unique components apart from the structural proteins of the virus particle. Consequently, most chemical agents that prevent virus metabolism are also toxic to the host cells.

Like pathogenic bacteria, viruses must also attach to and invade host cells. Recognition proteins on the surface of the virus capsid bind to specific receptors on the surface of the host cell. After entry, viral replication occurs at the expense of the host cell, which supplies not only raw material and energy, but also the ribosomes needed for synthesis of viral proteins and often many of the enzymes required for synthesis of viral nucleic acids as well. Finally, new virus particles are assembled and exit the cell. These stages, and corresponding antiviral agents are listed in Table 22.1. Most modern antiviral agents have been developed to combat HIV and are discussed in the later section on AIDS.

In principle, soluble fragments of the host receptor protein should compete for binding virus particles and prevent them attaching to host cells. Such approaches (e.g., use of CD4 fragments for HIV) are presently being investigated. **WIN compounds** (named for Sterling Winthrop pharmaceuticals) prevent the attachment and entry of many **picornaviruses**. This virus family includes **enteroviruses** (e.g., poliovirus) and **rhinoviruses** (one of the major virus groups causing the common cold). Many of these viruses bind to the protein **ICAM-1 (intercellular adhesion molecule 1)** on the surface of animal cells. This fits into a dimple or "canyon" in the coat of the virus particle (Fig. 22.1). At the bottom of the canyon is a hydrophobic pocket that normally contains a lipid molecule. WIN compounds displace this lipid and block changes in the conformation of the virus capsid protein, VP1, which are needed for binding to the receptor. A few antiviral agents are known that prevent entry of the host cell by the virus, such as amantadine in the case of influenza (see later discussion).

> Relatively few antiviral agents are available compared to the number of antibiotics for treating bacterial infections. Moreover, most antivirals have harmful side effects.

Table 22.1 Virus Life Cycle and Antagonists

Stage in Life Cycle	Possible Antiviral Agents;
Binding to receptor	Solubilized receptor fragments, WIN compounds (picornavirus)
Entry into host cell	Amantadine (influenza)
Gene expression	Interferons α and β
Reverse transcription (retroviruses only)	Reverse transcriptase inhibitors including nucleoside analogs
Replication	Nucleoside analogs
Assembly of virions	Protease inhibitors
Release from cell	Neuraminidase inhibitors (influenza)

INTERFERONS COORDINATE THE ANTIVIRAL RESPONSE

Interferons are a class of proteins induced in animal cells in response to virus infection. Clinical treatment with interferons is used to treat viral infections in a few cases (e.g., against hepatitis B and hepatitis C infections). **Interferons α** and **β** (**INF α** and **INF β**) block the spread of viruses by interfering with virus replication. (Although **interferon γ** shares the same name, it is quite distinct and is not induced directly by virus infection. It has a regulatory role in response to intracellular pathogens.) Interferons α and β are secreted in response to double-stranded RNA, which is symptomatic of the replication of most RNA viruses. They bind to the interferon receptors of both the infected cell itself and its neighbors. Locally, this triggers a phosphorelay signal pathway that activates several genes involved in opposing virus infection (Fig. 22.2).

Antiviral proteins induced by interferon include oligoadenylate synthetase, which converts ATP into 2′-5′-linked poly(A). This removes the ATP required as an energy source for viral replication. In addition, 2′-5′-poly(A) activates an endonuclease that cleaves viral RNA. P1 kinase is also activated and phosphorylates initiation factor eIF2, halting protein synthesis. The **Mx**

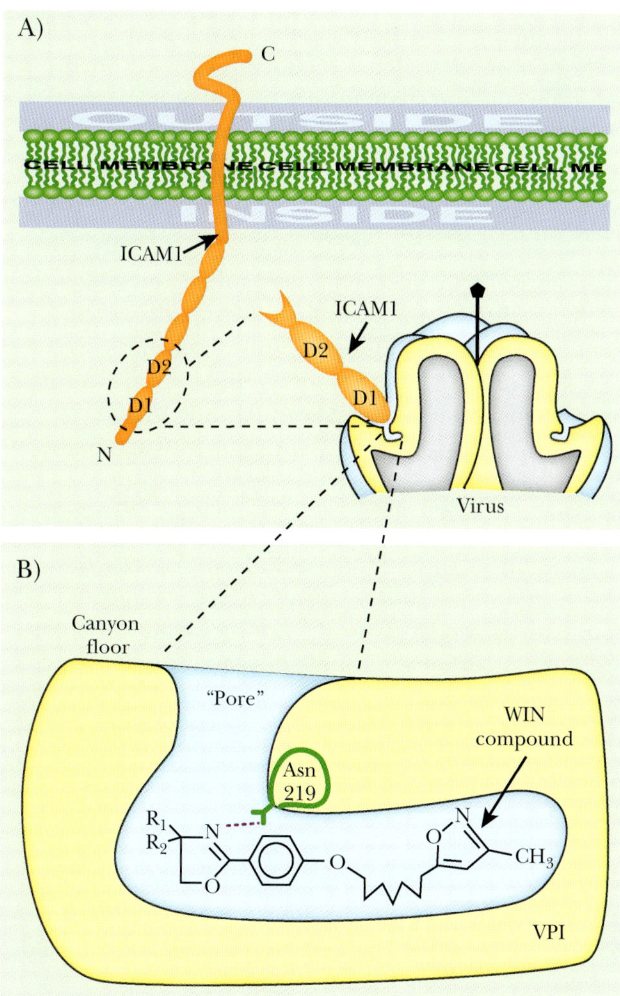

FIGURE 22.1 WIN Compounds and Picornavirus Capsid
(A) Picornavirus gain entry into a cell by binding to the D1 and D2 regions of ICAM-1. The D1 region binds to a canyon on the surface of the virus. (B) WIN compounds bind to the lipid binding site below the ICAM-1 recognition site. WIN compounds bind tightly to Asn 219, which displaces the lipid and prevents virus protein VP1 from binding ICAM-1.

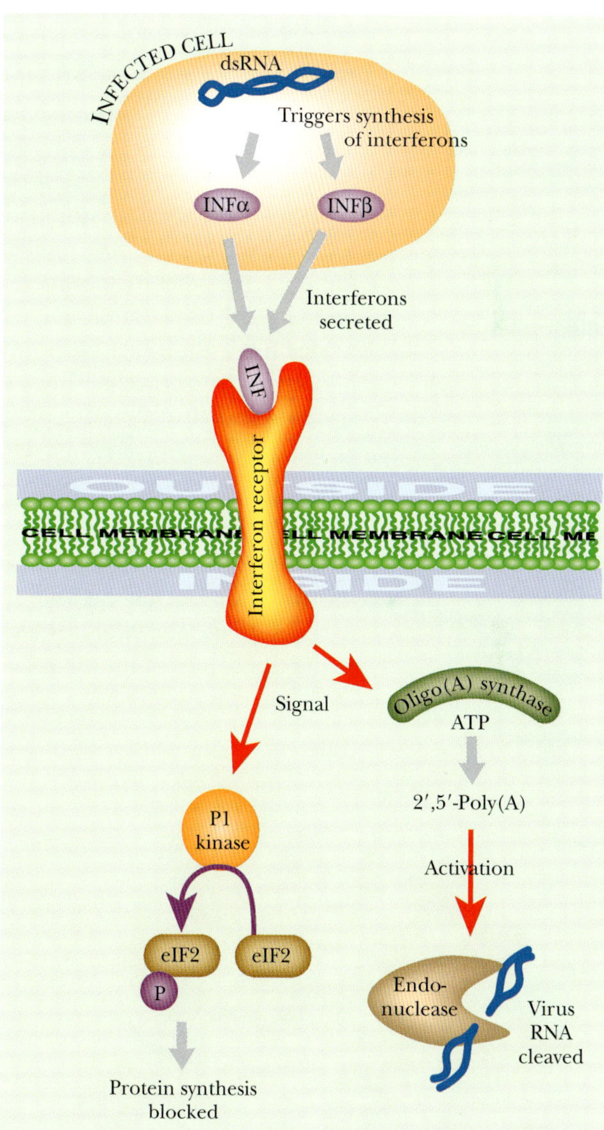

FIGURE 22.2 Interferons and Antiviral Proteins
The presence of dsRNA inside an infected cell triggers production of INFα and INFβ. These are secreted to neighboring cells, bind to the interferon receptor, and activate various antiviral proteins. P1 kinase blocks protein synthesis by phosphorylating eIF2 (an elongation factor). Oligo(A) synthetase converts ATP to 2′,5′-poly(A), which activates an endonuclease to digest dsRNA and depletes the ATP supply. Without ATP and protein synthesis, the virus cannot survive in the host cell.

proteins are GTPases that probably interfere with the RNA polymerases of negative-strand RNA viruses (e.g., influenza, parainfluenza). Interferons also help activate immune system cells, such as NK cells, which selectively destroy virus infected cells.

Interferon alpha was one of the first mammalian proteins to be manufactured via genetic engineering. However, its clinical effects have been disappointing except in a few cases. Modern attempts at antiviral therapy have tended to move away from interferons and toward stimulating the RNA interference system (see Box 22.1).

Interferons are animal proteins that promote the antiviral response by inducing synthesis of a range of enzymes with specific antiviral activities.

INFLUENZA IS A NEGATIVE-STRAND RNA VIRUS

Influenza virus, an **orthomyxovirus**, is an example of a negative-strand single-stranded RNA virus. In other words, the virus genome is present in the virus particle as noncoding (= antisense = negative-strand) RNA. The flu virus particle contains a segmented genome consisting of eight separate pieces of single-stranded RNA ranging from 890 to 2341 nucleotides long. These are each packed into an inner **nucleocapsid** and are surrounded by an outer envelope (Fig. 22.3). Although the outer membrane is derived from host-cell material, it contains virus-encoded proteins such as neuraminidase, hemagglutinin, and ion channels. These viral proteins are made on the ribosomes of the infected host cell and are involved in virus recognition and entry into successive host cells. The hemagglutinin and neuraminidase of influenza differ slightly but significantly between strains of flu. The variants are designated by H and N numbers. Thus the Spanish flu of 1918 was H_1N_1 and the avian flu presently spreading worldwide is H_5N_1.

FIGURE 22.3
Structure of the Influenza Virus
The influenza virus has an outer envelope containing neuraminidase, hemagglutinin, and ion channels. Several individual negative-strand ssRNA molecules are packaged within the outer membrane. Each strand is coated with nucleocapsid proteins. An RNA replicase molecule is also included with each ssRNA strand to ensure expression.

When a flu virus comes in contact with an appropriate host cell it is engulfed and ends up inside a vesicle. Both the vesicle and the outer coat of the virus particle are dissolved, releasing the nucleocapsids, which enter the nucleus. The nucleocapsids disassemble inside the nucleus releasing the RNA molecules (Fig. 22.4). Replication of the influenza RNA occurs in the nucleus. The viral mRNA exits the nucleus just like normal cellular mRNA and travels to the ribosomes in the cytoplasm. Here the proteins for the new virus particles are made.

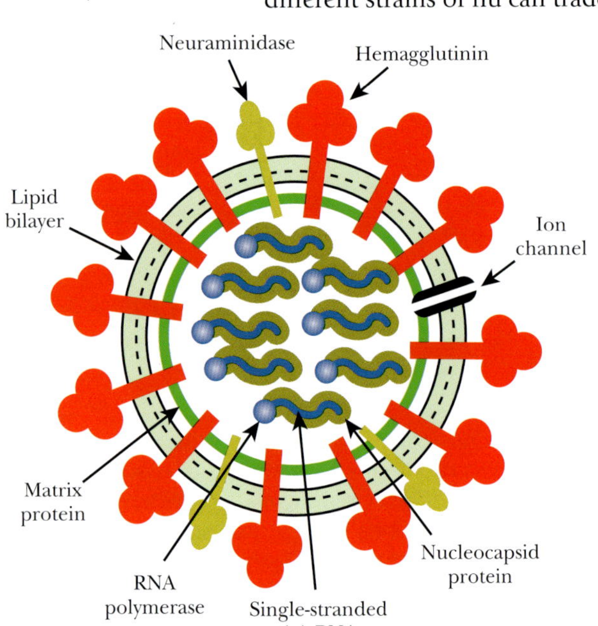

Because influenza virus has its genes scattered over eight separate molecules of RNA, different strains of flu can trade segments of RNA and form new genetic combinations (Fig. 22.5). In addition, mutations occur at a higher rate during RNA replication than in DNA. These two mechanisms result in a lot of genetic diversity. Consequently, different strains of flu emerge every couple of years. The changing surface antigens of the virus allow it to avoid immune recognition. These different flu strains vary greatly in their apparent virulence. However, this depends as much on the immune history of the human population as on genetic changes in the virus.

Influenza viruses fall into two major groups, influenza A and B. Mutation of both A and B causes annual epidemics due to slow antigenic drift. Influenza B is largely restricted to humans and has less genetic variation. Influenza A has a wider host range ("people, pigs, and poultry"). As a result, influenza A gives rise to severe but less common epidemics due to reassortment of viruses from different hosts during mixed infections.

The Spanish flu of 1918–1919 was the worst influenza A pandemic so far and is estimated to have killed around

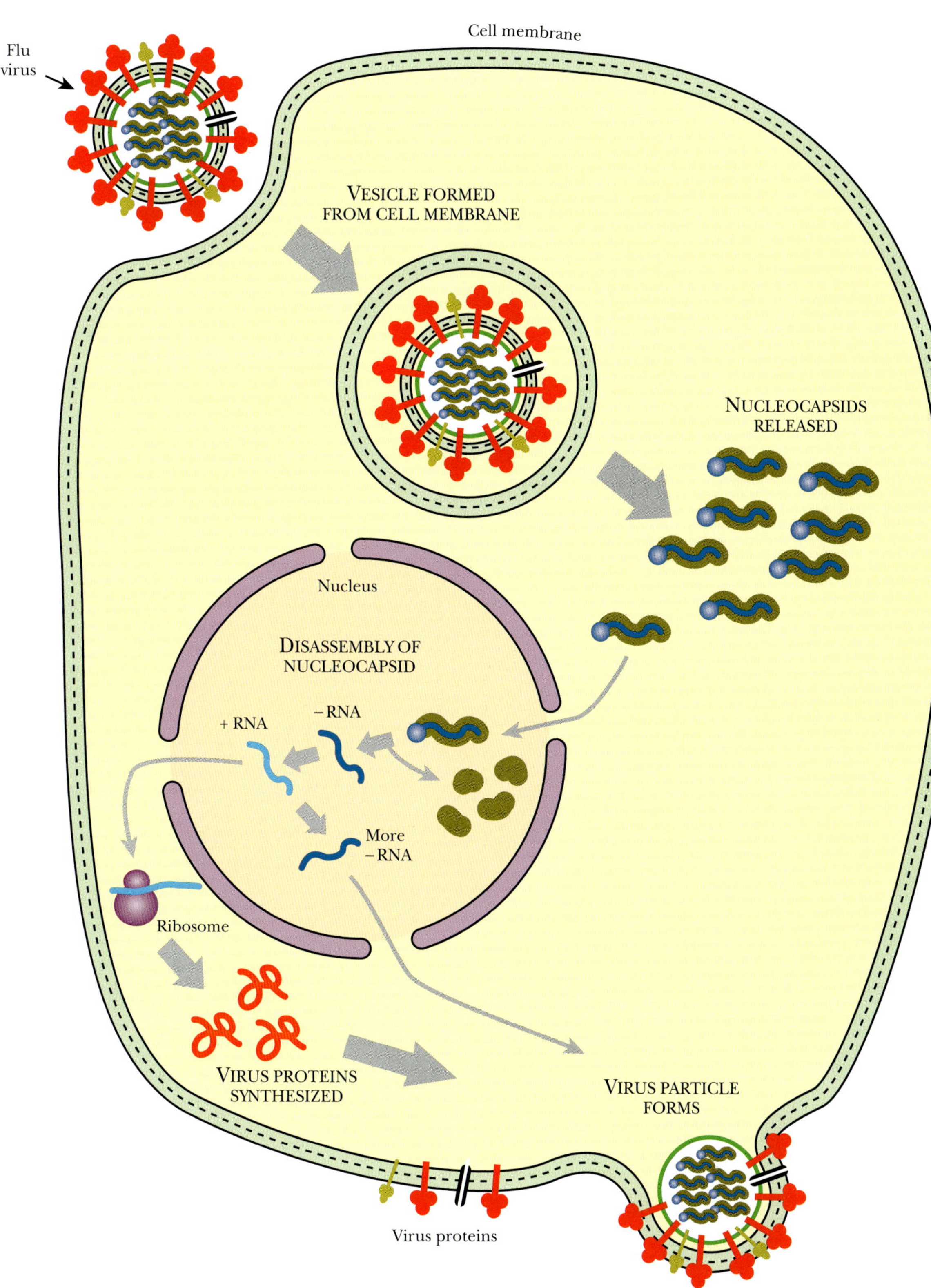

FIGURE 22.4 Life Cycle of the Influenza Virus
After entry into the host cell, the nucleocapsids enter the nucleus before disassembly. There the viral replicase makes positive RNA strands and more negative strands. The (+) RNA strands are exported to the ribosomes where they act as mRNA and are translated. The resulting viral proteins are assembled into more virus particles, together with the (–) RNA strands.

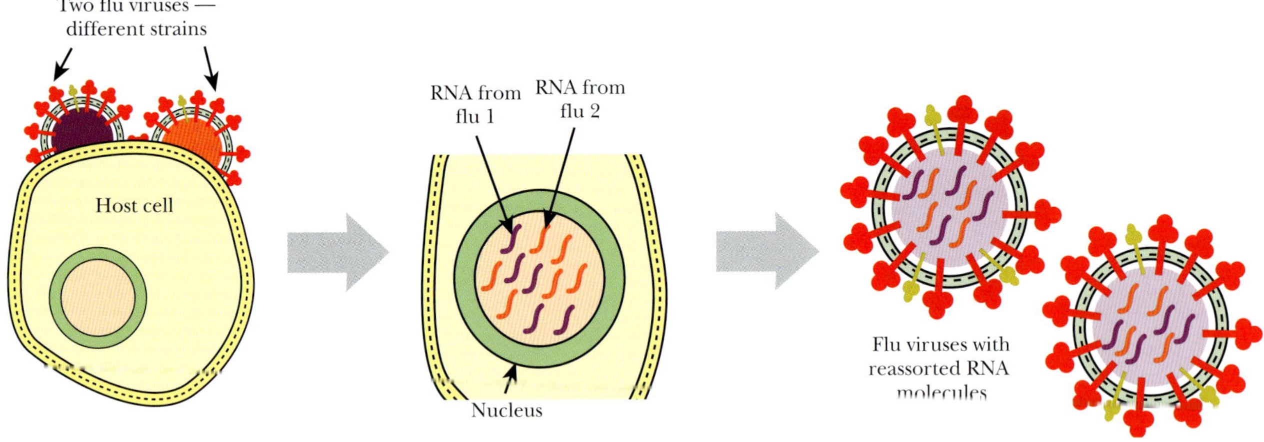

FIGURE 22.5 Influenza Viral Genomes Can Switch RNA Segments
If two different influenza strains infect the same host cell, the genomes of both will enter the nucleus. When new virus particles are formed, some nucleocapsids from strain 1 may be packaged with strain 2, and vice versa. Thus complete ssRNA molecules from different influenza strains may be reshuffled to generate new assortments. Such reshuffling more often happens in pigs and birds than in human hosts.

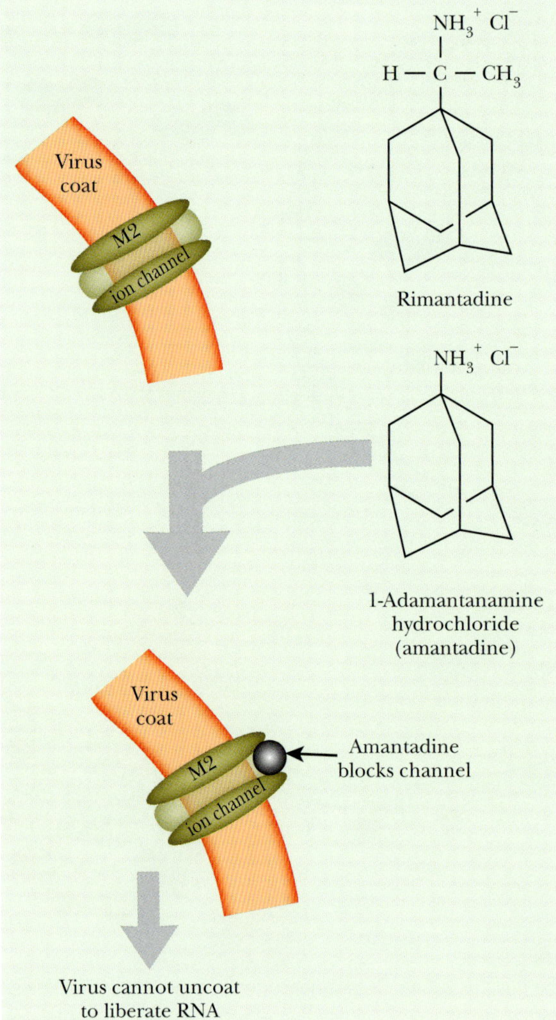

FIGURE 22.6 Amantadine Blocks M2 Ion Channel
The amantadine molecule blocks ions from passing though the M2 channel in the virus coat, thus preventing uncoating and RNA molecule release.

50 million people (more than World War I). Will there be another major flu pandemic soon? The World Health Organization is worried about the unusually virulent H_5N_1 avian flu emerging in Asia. So far a few humans have caught this from birds. The major threat is that this will mutate further and become readily transmissible from person to person.

Amantadine is a tricyclic amine that binds to the M2 protein, one of the transmembrane ion channels found in the outer envelope of influenza A virus. M2 is not expressed by influenza B and consequently amantadine works only against type A strains of influenza. Amantadine blocks the M2 ion channel, and this stops entry of protons, which prevents uncoating of the virus particle (Fig. 22.6). Thus entry of the virus is prevented. Amantadine must be given very early in infection. Amantadine was actually the first specific and effective antiviral agent to be discovered, although its mode of action has only recently been elucidated.

Influenza (both A and B) may also be treated with neuraminidase inhibitors, such as oseltamivir (= Tamiflu) or zanamivir. These are analogs of *N*-acetylneuraminic acid. Neuraminidase normally cleaves this from the virus receptor, allowing progeny virus particles to be released. If neuraminidase is inhibited, progeny virus is trapped in infected cells.

Influenza is an extremely common viral infection of humans and some other animals. Its genome consists of eight pieces of RNA of negative complementarity. As a result it shows a high rate of both mutation and recombination. Very few drugs are available to treat influenza.

Box 22.1 Virus Therapy Using RNA Interference

The number of effective antiviral agents is low, and most have side effects. Antisense RNA and ribozyme therapy have been proposed as alternative approaches, but neither has proven effective so far. However, using RNA interference (RNAi) to treat virus infections looks promising.

RNA interference is discussed in detail in Chapter 5 and is in fact used by host cells to protect themselves against invasion by RNA viruses. RNAi targets double-stranded RNA (dsRNA) derived from RNA virus replication and destroys both the dsRNA and corresponding single-stranded RNA (in practice this will usually be viral mRNA). As described before, RNAi is triggered by short dsRNA molecules of just over 20 nucleotides known as short-interfering RNA (siRNA).

Not surprisingly, many viruses have evolved mechanisms to avoid destruction by RNAi. However, in mammals, administration of artificially synthesized siRNA around 17 to 21 nucleotides long will provoke a strong RNAi response even against viruses with protection mechanisms. The sequence of the siRNA is designed to represent conserved regions of the RNA virus genome.

RNAi therapy is especially useful for viruses infecting the respiratory tract. This is because the siRNA can be administered easily by inhalation. Today the RNAi approach is being developed for influenza, parainfluenza, and respiratory syncytial viruses. The siRNA sequences can be screened for effectiveness in cell culture before being used on whole organisms. Tests with mice have shown good protection against several respiratory viruses. Phase I clinical trials in humans have begun using siRNA against respiratory syncytial virus, and so far the results are promising.

THE AIDS RETROVIRUS

AIDS (acquired immunodeficiency syndrome) is caused by human immunodeficiency virus (HIV), which damages the immune system. Most AIDS patients die of **opportunistic infections**. These are infections seen only in patients with defective immune systems and are caused by assorted viruses, bacteria, protozoans, and fungi that are normally relatively harmless but may take the opportunity to attack when host defenses are down. In addition, without immune surveillance, cancers caused by other viruses or somatic mutations often grow out of control.

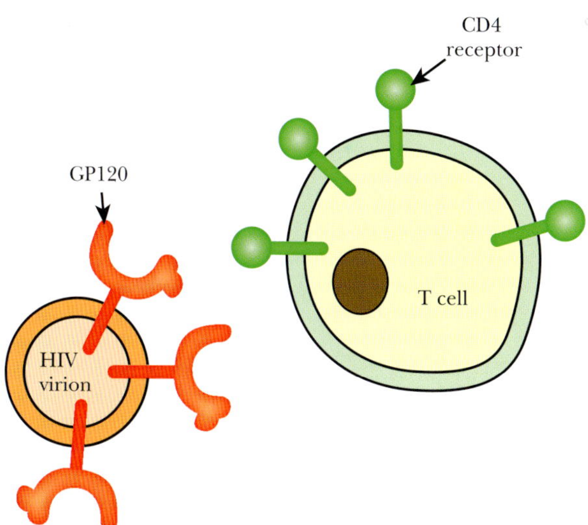

FIGURE 22.7 HIV Uses CD4 Protein as Receptor
HIV particles are coated with gp120, which recognizes the T cells of the immune system. The viral glycoprotein gp120 binds to protein CD4, on the surface of the T cell. The viral particle is then taken into the T cell, where it takes over the cellular machinery to produce more virus.

HIV infects circulating white blood cells belonging to the immune system, the **T cells**. The **CD4 protein** is found on the surface of many T cells, where it acts as an important receptor during the immune response (see Chapter 6). HIV also uses the CD4 protein as a receptor (Fig. 22.7). The **gp120** protein in the outer envelope of HIV is a glycoprotein of 120-kDa molecular weight. It recognizes and binds to CD4, which is needed for entry of the virus.

The CD4 protein is also found on the surface of some other immune system cells, the monocytes and macrophages. HIV does not seriously harm these two cell types, but the cells become reservoirs to spread the virus to more T cells. It is the damage to the T cells that is most critical to immune function. Once HIV has entered the T cell, the DNA form of the retrovirus genome integrates into the host chromosome and begins to express virus genes. Viral proteins are manufactured on host ribosomes. In particular, gp120 is made in large amounts and is inserted into the host cell membrane. Consequently, infected T cells carry the gp120 protein from the HIV particle in their surface membranes. This binds to the CD4 protein on other T cells. The result is that several T cells clump together and fuse (Fig. 22.8). The giant, multiple cell soon dies. About 70% of the body's T cells carry the CD4 receptor. As they gradually die off, the immune response fades away over a 5- to 10-year period.

> AIDS is caused by a retrovirus that uses the CD4 protein on the surface of T cells as a receptor. Damage to T cells cripples the immune response, leaving the body open to other infections.

CHEMOKINE RECEPTORS ACT AS CO-RECEPTORS FOR HIV

The entry of HIV into T cells requires binding of virus to both the CD4 protein and one of several chemokine receptors, which act as **co-receptors**. The **chemokine receptors** are membrane proteins with seven *trans*-membrane segments. They bind **chemokines**, a group of approximately 50 small messenger peptides that activate the white blood cells of the immune system and attract them to the site of infections. The most important chemokine receptors for HIV entry are **CCR5** and to a lesser extent CXCR4.

Mutations in CCR5 are largely responsible for the small proportion of the population who are naturally resistant to infection by the AIDS virus. The *CCR5Δ32* allele has a deletion of 32 base pairs and results in nonfunctional CCR5 protein. Individuals homozygous for *CCR5Δ32* are vastly less susceptible to infection by HIV (though not totally resistant). In addition, if the individual is infected, the disease progresses much more slowly. About 2% of Europeans are homozygous for *CCR5Δ32* and 14% are heterozygous. Heterozygotes appear to be mildly protected and show slower progression, in accord with the lower levels of CCR5 protein on the surfaces of their T cells. The origin of the *CCR5Δ32* allele has been traced back to around 700 years ago in northwest Europe, at about the time of the Black Death. Conceivably, the defects in CCR5 were selected by providing resistance against the bubonic plague. Variations in susceptibility to AIDS also result from alterations in the DNA sequence of the promoter for the *CCR5* gene. Presumably these cause variations in the level of CCR5 protein expressed.

Preexisting receptors that direct the uptake of important molecules into animal cells are often the targets for viruses. It is quite possible for the same host cell protein to be used as a receptor by unrelated infectious agents, including both viruses and bacteria. Thus, the myxoma poxvirus, which causes immune deficiency in rabbits, also uses the CCR5 and CXCR4 chemokine receptors. Which receptors are used by smallpox or other poxviruses is still unknown. Other pathogens, including the malaria parasite, also target chemokine receptors, although not CCR5 and CXCR4. Scientists are presently trying to identify the functions of the various receptors on immune cells in the hope of understanding how viruses exploit them for their own use.

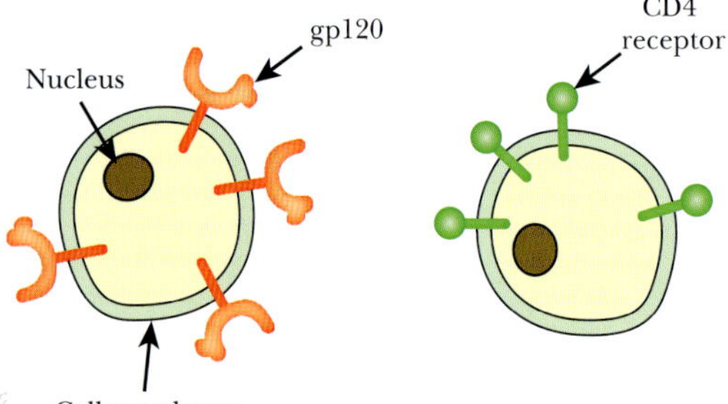
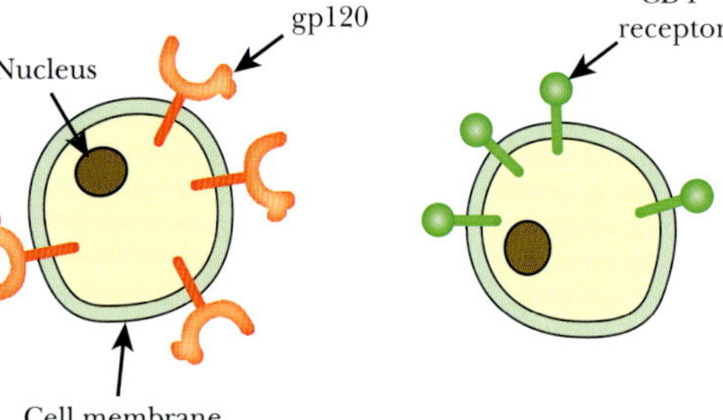

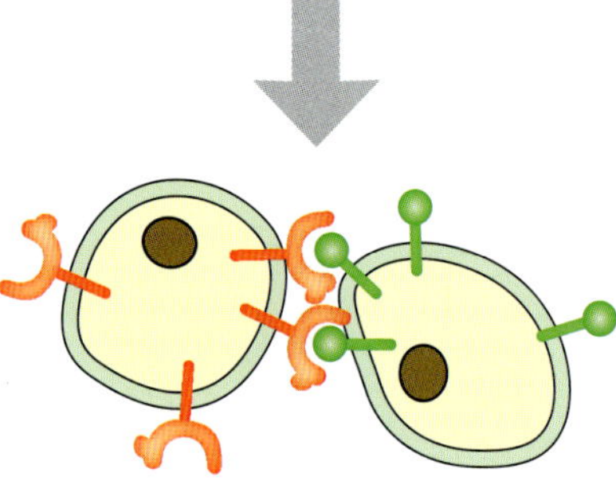
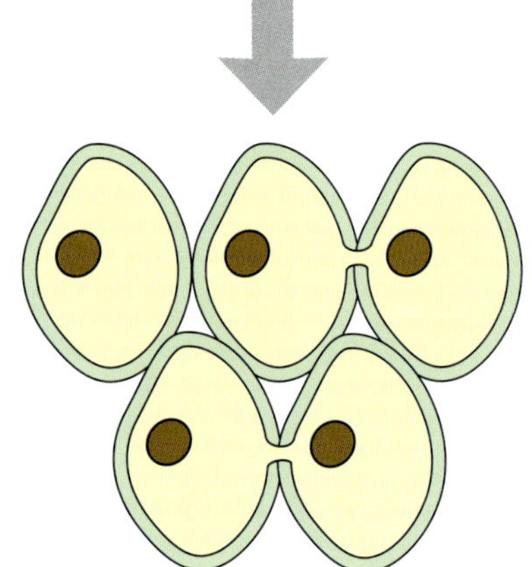

FIGURE 22.8 Fusion of Infected T Cells
Once HIV has entered the T cell, gp120 is made in large amounts and is inserted into the host cell membrane. T cells with gp120 in their membranes bind to other T cells via the CD4 receptor, which causes the cells to fuse. The process continues until large clumps of T cells form. These soon die, crippling the immune system.

Entry of HIV into target cells requires co-receptors. Natural resistance to AIDS results from defects in co-receptors, especially the CCR5 chemokine receptor.

TREATMENT OF THE AIDS RETROVIRUS

AIDS has recently overtaken malaria as the leading cause of death from infectious disease in the world. No complete cure or effective vaccine yet exists for AIDS, although several treatments are now available that significantly extend patients' lives. The fundamental problem with all anti-AIDS drugs is that HIV is an RNA virus and so has a relatively high mutation rate. HIV mutates at a rate of approximately one base per genome per cycle of replication. Even within a single patient, HIV exists as a swarm of closely related variants known as a **quasi-species**. Consequently, strains of HIV resistant to individual drugs appear at a relatively high frequency. Attempts to control AIDS (Fig. 22.9), whether by using vaccines, protein processing inhibitors, or antisense RNA, all face the same problem: HIV will mutate to produce resistant variants. In practice this problem may be partially overcome by simultaneous treatment with several drugs that hit different targets.

Azidothymidine (**AZT**, or zidovudine) was one of the first drugs used against AIDS. It is an analog of thymidine that lacks the 3′-hydroxyl group. A variety of other **nucleoside analogs** that lack the 3′-hydroxyl group are also in use. AZT and other 3′-deoxy base analogs are converted to the 5′-triphosphate by the cell and then incorporated into the growing DNA chain during reverse transcription (Fig. 22.10). Because AZT lacks a 3′-hydroxyl group, the DNA chain cannot be extended. AZT is thus a DNA **chain terminator**. Although AZT is incorporated more readily by the viral reverse transcriptase than by most host-cell DNA polymerases, it is not completely specific. Thus one major drawback is that AZT partially inhibits host DNA synthesis in uninfected cells of the body. In particular it is toxic to bone marrow cells (B cells), which are another part of the immune system. Mutations in the HIV reverse transcriptase may cause resistance to base analogs. For example, Met41Leu (i.e., replacement of methionine at position 41 with leucine) increases resistance to AZT by fourfold and a second mutation of Thr215Tyr gives an overall 70-fold resistance.

Certain drugs that do not bind at the active site can also inhibit reverse transcriptase. These are referred to as **nonnucleoside reverse transcriptase inhibitors** (**NNRTI**; Fig. 22.11). They bind to the enzyme at a separate site, relatively close to the active site. This distorts the structure of reverse transcriptase and inhibits its activity. Unfortunately, mutations that alter the NNRTI binding site occur quite frequently, and these give rise to resistant reverse transcriptase enzyme. These drugs are therefore generally used in combination with nucleoside analogs.

Most individual HIV proteins are joined together as polyproteins when first made and must therefore be cut apart by HIV protease. For example, the *env* gene is transcribed and translated to give gp160, which is cleaved to gp41 and gp120. The *gag* gene encodes a polyprotein that includes the proteins of the virus core. Consequently, inhibition of polyprotein cleavage will prevent the assembly of the virus particle. The HIV protease recognizes and binds a stretch of seven amino acids around the cleavage site. This step may be blocked with **protease inhibitors** that are analogs of several amino acid residues around the cut site (Fig. 22.12). For example, saquinavir is an analog of Asn-Tyr-Pro.

At present the favored approach in AIDS therapy is to use three drugs with different mechanisms in combination. Note that different drugs should *not* be used one after the other, because this allows resistance to develop to each drug in turn. If three are used simultaneously, emerging virus mutants that are resistant to one drug will be killed by the others. A typical cocktail consists of a reverse transcriptase inhibitor, hydroxyurea, and a protease inhibitor. Since 1995, when protease inhibitors became available, deaths from AIDS

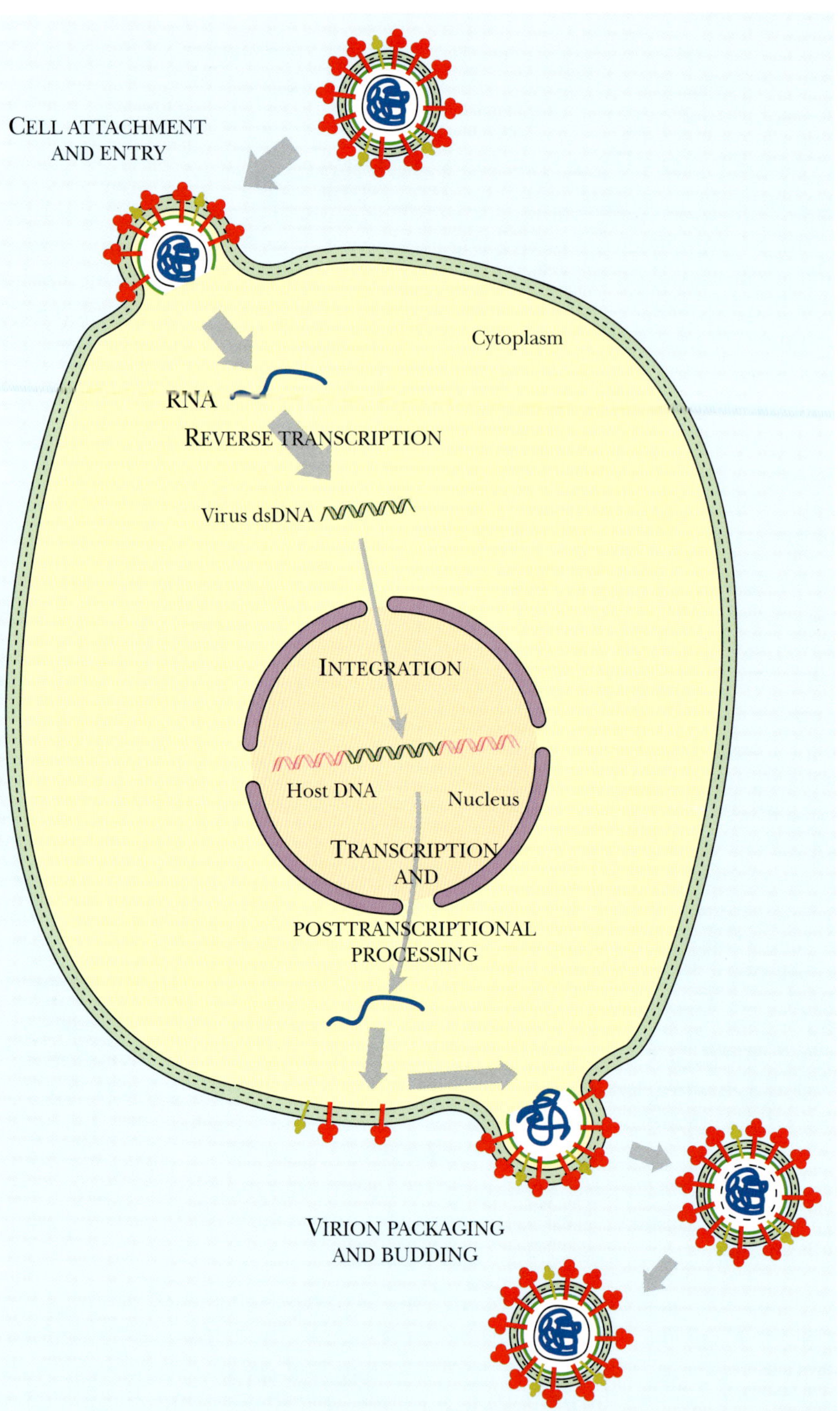

FIGURE 22.9 Possible Steps for HIV Inhibition
HIV infections could be stopped at the following steps: (1) at the cell surface, competing molecules could prevent virus attachment; (2) enzyme inhibitors may block the action of reverse transcriptase; (3) integration of the viral genome could be prevented; (4) transcription and translation could be blocked; (5) finally, blocking virion packaging and budding would protect other cells from becoming infected.

have dropped 60% to 80% in those nations whose citizens can afford expensive long-term treatment with costly pharmaceuticals. In 2000 about a third of HIV-positive Americans were receiving treatment and about 80% received a cocktail including protease inhibitors. In the United States, treatment with such a cocktail may cost from $800 to $1500 per month, although the cost keeps dropping.

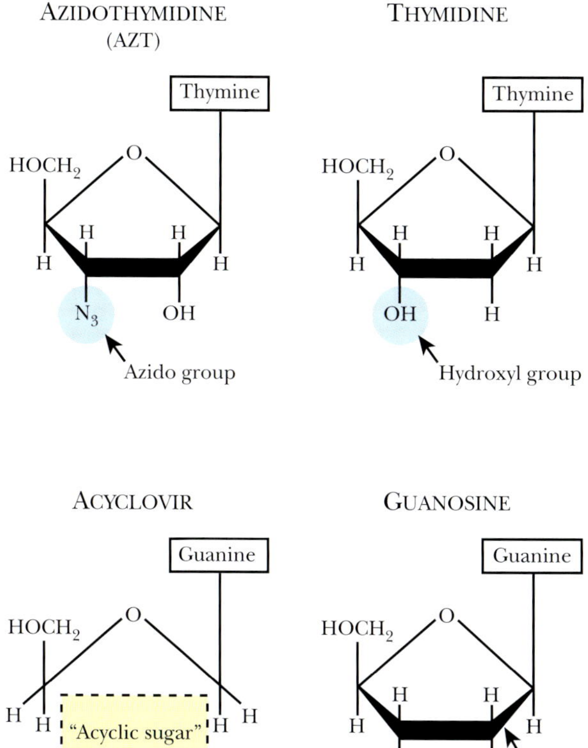

FIGURE 22.10 Nucleoside Analogs Act as Chain Terminators

Two examples of chain terminators are azidothymidine (AZT) and acyclovir, which replace thymine and guanine, respectively. AZT has an azido group on the 3′ position of the deoxyribose ring rather than a hydroxyl. The entire deoxyribose ring is altered in acyclovir. In both cases, the analogs are incorporated into DNA during the reverse transcriptase reaction. Once the analog has been inserted, reverse transcriptase cannot elongate the DNA chain any further because the analogs lack the 3′-OH group to which the next nucleotide would be added.

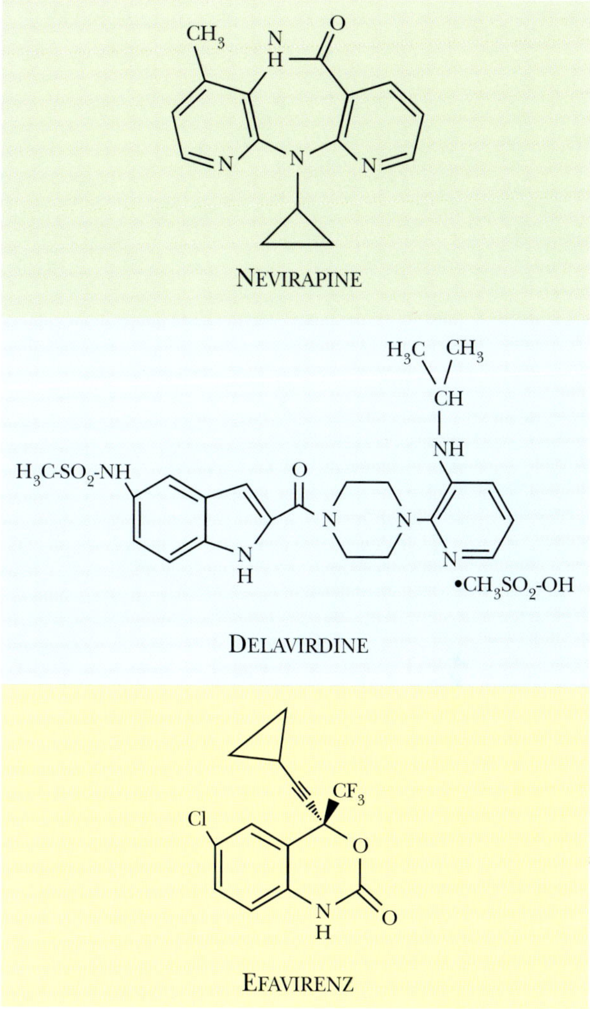

NEVIRAPINE

DELAVIRDINE

EFAVIRENZ

FIGURE 22.11 Nonnucleoside Reverse Transcriptase Inhibitors

Chemical structures of three NNRTIs that are presently in use: nevirapine, delavirdine, and efavirenz. These are specific for HIV-1 and have no effect on HIV-2.

The inclusion of hydroxyurea is, in some ways, the cleverest part of the cocktail scheme. Hydroxyurea inhibits enzymes of the human host cell that are needed for the AIDS virus to replicate. Because human genes encode these proteins, the virus cannot mutate to produce hydroxyurea-resistant enzymes. Both the advantage and the problem with hydroxyurea is that it inhibits human cell DNA replication. Therefore, it cannot be given in dosages so high they eliminate *all* DNA replication.

Most recently developed antiviral agents were designed to treat AIDS. They include nucleoside analogs (chain terminators), nonnucleoside reverse transcriptase inhibitors, and protease inhibitors.

INFECTIOUS PRION DISEASE

Prions are proteins with unique properties that are capable of causing inherited, spontaneous, or infectious disease. The **prion protein (PrP)** exists in two conformations, the normal harmless or "cellular" form **(PrP^c)** and the pathogenic **(PrP^{Sc})** form, named after

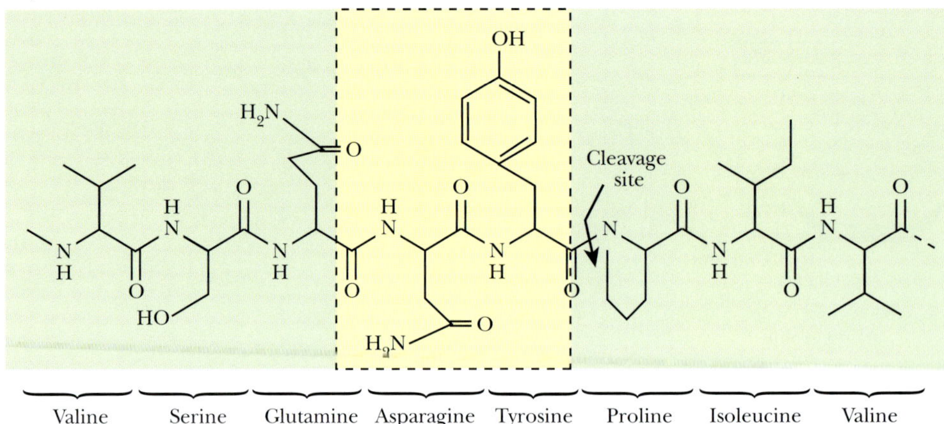

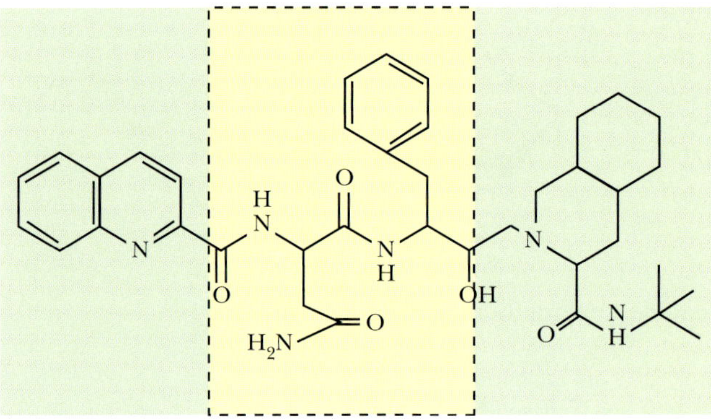

FIGURE 22.12 Protease Inhibitors

(A) HIV-1 protease recognizes Asn-Tyr-Pro, cleaving the protein between the tyrosine and proline. (B) Saquinavir has a structure that mimics these three amino acids. HIV-1 proteinase binds to saquinavir, but cannot cleave or release it because the cleavage site is missing.

scrapie, a disease of sheep (Fig. 22.13). Rogue prion proteins bind to their normal relatives and induce them to refold into the disease-causing conformation. Thus, a small number of misfolded prions will eventually subvert the population of normal proteins. Over time, this leads to neural degeneration and eventually death.

Mutations within the ***Prnp* gene** that encodes the prion protein may result in prions with a greatly increased likelihood of misfolding. This causes hereditary prion disease. Several clinically different variants are known, depending on the precise location of the mutation within the prion protein and the nature of the amino acid alteration. The most common is Creutzfeldt-Jakob disease (CJD). Even normal prions occasionally misfold. The result is spontaneous prion disease, which occurs at a rate of about one per million of the human population.

If misfolded prions are transmitted to another susceptible host, the result is infectious prion disease, also known as **transmissible spongiform encephalopathy (TSE)**. Such an infection can be passed from one cell to another and one animal to another by entry of the PrP^{Sc} form of the prion. The two individuals may be of the same or different species. Infection of a new victim by prions is relatively difficult. It requires uptake of rogue prion proteins from infected nervous tissue, especially brain, but the details of infection remain obscure. The best known infectious prion diseases are:

1. Scrapie, a disease of sheep and goats
2. **Kuru**, a disease of cannibals
3. **Mad cow disease**, officially known as **bovine spongiform encephalopathy (BSE)**
4. **Chronic wasting disease (CWD)** of deer and elk. This can apparently be transmitted via saliva, unlike the other TSEs.

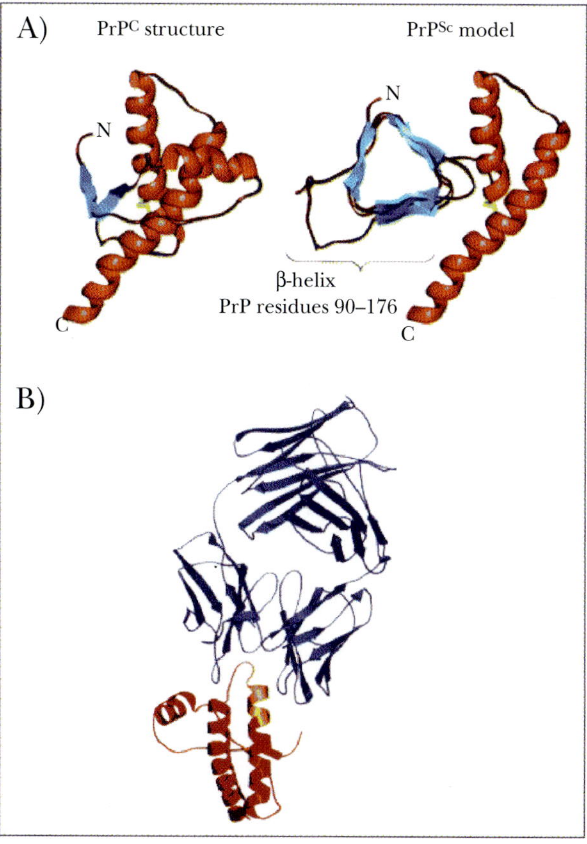

FIGURE 22.13
Normal and Pathogenic Forms of the Prion Protein
The PrP^c structure is on the right and the PrP^{Sc} structure is on the left. Note the greatly increased proportion of beta-sheet in the PrP^{Sc} structure. From: Eghiaian (2005). Structuring the puzzle of prion propagation. *Curr Opin Struct Biol* **15**, 724–730. Reprinted with permission.

Scrapie is a disease of sheep and related animals that has been recorded going back several hundred years in Europe. The name comes from the behavior of infected sheep that constantly scrape themselves against fences, trees, or walls and often seriously injure themselves. Only certain breeds of sheep are susceptible, because of the slight differences in prion sequence between breeds. Dead and decomposing sheep may contaminate the grass of their fields with prion proteins. These are unusually stable and long-lived and may be eaten by healthy sheep.

Kuru was transmitted by ritual cannibalism and used to be endemic among the Fore tribe of New Guinea. The women had the honor of preparing the brains of dead relatives and participating in their ritual consumption. As a result, 90% of the victims were women together with younger children who accompanied them. It may take 10 to 20 years for symptoms to develop, but once they do, the progression from headaches to difficulty walking to death from neural degeneration took from 1 to 2 years. No one born since 1959, when cannibalism stopped, has developed kuru.

Brain degeneration, or spongiform encephalopathy, due to misfolded prions is possible in any mammalian species. In addition to scrapie, BSE, and CWD, a variety of less well-characterized prion diseases are known in other animals. In a way, these are really all the same disease, because there is a single prion gene encoding a single prion protein that is found in the brain of all mammals. Symptoms vary slightly from species to species, but after a long incubation period, the result is degeneration and death of cells of the central nervous system. As the popular name *mad cow disease* indicates, progressive degeneration of the brain and nervous system causes the infected animals to behave bizarrely during later stages of the disease.

Mad cow disease was spread by overly intensive farming practices. Animal remains, including the brains, were ground up and incorporated into animal feed. Because sheep remains were included in feed for cows, the epidemic of mad cow disease, which began in England in 1986, was originally blamed on sheep with scrapie. However, people in England and other European countries have eaten sheep with scrapie since the 1700s without any noticeable ill effects. Nor have any other domestic animals, including cows, ever caught scrapie, despite sharing the same fields. Moreover, sheep prions are not infectious for cows. It is now thought that a random flip-flop event converted a normal prion into the rogue form inside a cow's brain somewhere in England in the late 1970s or early 1980s. The rogue cow prions were recycled in animal feed and spread, eventually causing an epidemic. After mad cow disease broke out in England, the recycling of animal remains in feed was prohibited and infected herds were destroyed.

Mad cow disease can be transmitted to humans, but the rate of infection is extremely low. The first human cases were confirmed in 1996 and were named **variant CJD** in an attempt to obscure their origin. However, when the rogue cow prion infects humans, the misfolded prions are characteristic of mad cow disease, not genuine CJD. In humans with CJD or kuru, the precise conformation of the misfolded prions is different. The human victims of mad cow disease are scattered randomly throughout the population, suggesting that relatively few humans are actually susceptible to infection. As of 2006, about 200 people, mostly in England, have come down with BSE. Calculations based on the history and age distribution of BSE in humans since the outbreak started suggest an average incubation period of about 15 years and that the number of cases will eventually reach around 300. These estimates are much lower than many earlier and rather emotional predictions and reflect the extremely low infectivity of prions when crossing from one species to another.

Prion diseases (scrapie, Creutzfeldt-Jakob disease, mad cow disease) are due to misfolding of the prion protein that is expressed at high levels in nervous tissue, especially brain. Inherited, infectious, and spontaneous variants of prion disease are found.

DETECTION OF PATHOGENIC PRIONS

The emergence of mad cow disease (BSE) has created the need to screen cows and their products for the presence of the pathogenic form of the prion protein (PrP^{Sc}). This is presently done by immunological detection. Unfortunately, even though the 3D folding of the normal (PrP^{c}) and pathogenic (PrP^{Sc}) forms is different, it has not yet proven possible to obtain antibodies specific to each form. Because the pathogenic form of the prion is protease resistant, samples are first treated with protease to destroy the normal (PrP^{c}) form and then subjected to immunological testing by Western blotting (see Chapter 6). The overall procedure is tedious and of only moderate sensitivity. In particular, it would be valuable to have a test that reveals prion disease well before symptoms develop to allow time for possible treatments.

The **protein misfolding cyclic amplification (PMCA)** procedure amplifies the levels of misfolded prion in a manner analogous to the use of PCR for amplifying DNA (Fig. 22.14). This allows greatly increased sensitivity of detection of PrP^{Sc} in clinical samples. Small samples suspected of containing PrP^{Sc} are mixed with normal brain homogenate containing a surplus of the normal PrP^{c}. The PrP^{c} is converted to PrP^{Sc} and incorporated into the growing PrP^{Sc} aggregates. The sample is then sonicated to break up the aggregates. This procedure is repeated for several cycles. Increases of around 60-fold over five cycles are typical.

Detection of prions is technically difficult. Cyclic amplification of prions has greatly increased the sensitivity of detection.

APPROACHES TO TREATING PRION DISEASE

At present there is no effective treatment for any of the prion diseases, although a variety of agents are being tested. Relatively few drugs cross the blood-brain barrier effectively. Nonetheless, random screening of those known to do so revealed that both quinacrine and chlorpromazine eliminate prions from infected animal brain cells in culture. (Quinacrine is a rarely used antimalarial drug, and chlorpromazine is widely used to treat schizophrenia.) Unfortunately, they do not cure the disease in whole animals. Several high-tech approaches to dealing with prion diseases are in progress, although none has yet reached clinical use.

Removal of prions from infective material is an alternative approach. Filters have recently been developed that remove prions. These were developed by screening combinatorial libraries (see Chapter 11) for ligands that bound prion protein. The ligands are attached to resins and placed in columns for filtration of blood or other liquids that might contain active prions. When scrapie-infected hamster blood was injected into hamsters, 15 of 99

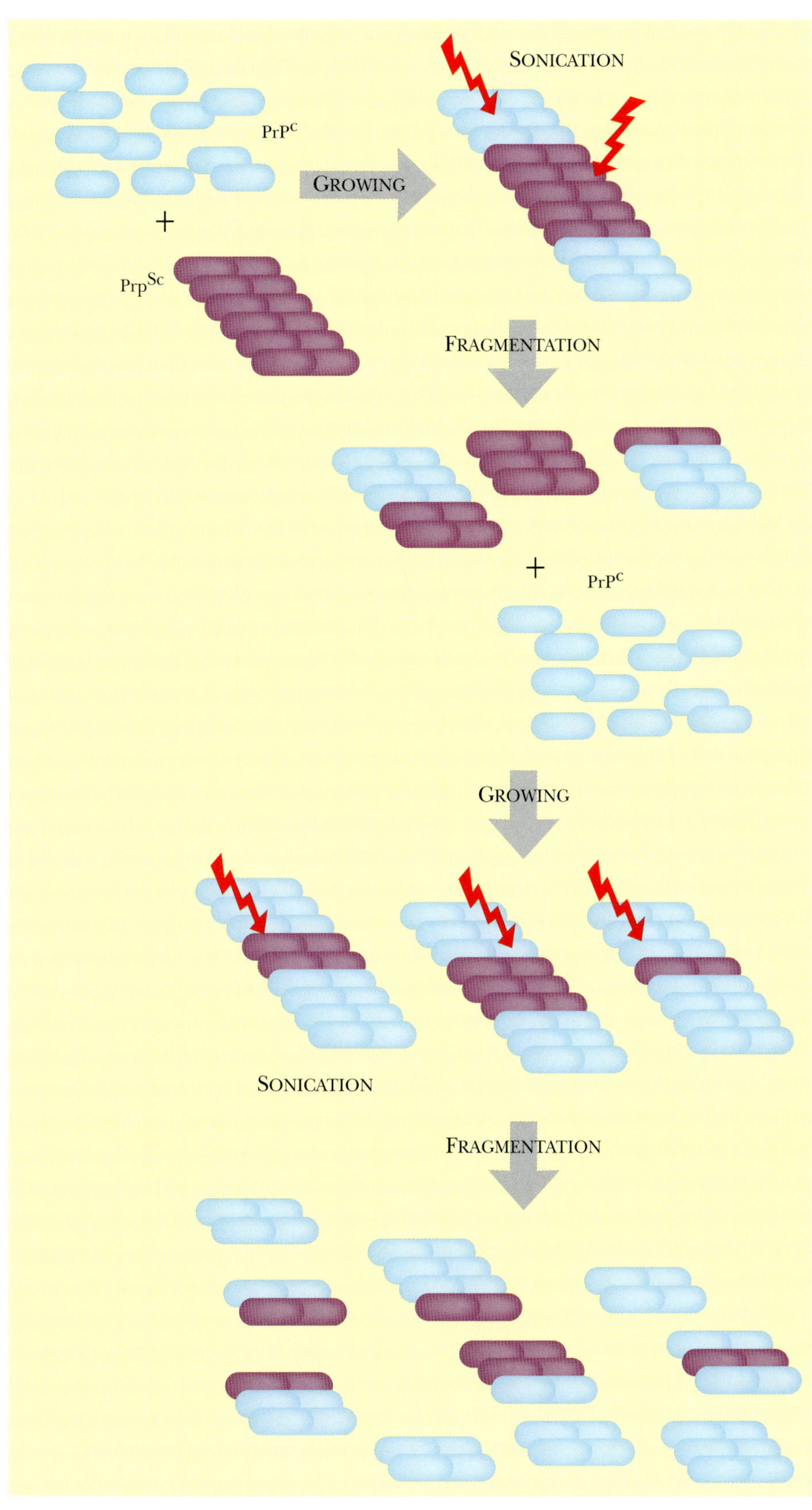

FIGURE 22.14 Protein Misfolding Cyclic Amplification (PMCA)
Amplification involves multiple cycles of incubating PrP^{Sc} in the presence of excess PrP^{c} followed by sonication. During the incubation periods, the size of PrP^{Sc} aggregates (purple) increases because of incorporation of normal prion protein (blue). During sonication the aggregates are disrupted, producing more pathogenic conversion units.

were infected, but if the blood was first filtered through a column of affinity resin L13, none of 96 hamsters were infected.

RNA interference (see Chapter 5) is widely used to suppress gene expression in laboratory studies. It is possible to generate siRNA that will suppress expression of the *Prnp* gene in mice. In order to generate the siRNA in prion-infected cells, a retrovirus vector was used that expresses short hairpin RNAs. These are processed in the target cells by Dicer to give the siRNA. This in turn triggers RNA interference directed against *Prnp* mRNA, which is degraded. Retroviral vectors were chosen because they can infect nongrowing cells, such as those of the nervous system. At least in mice, intracranial injection of the vectors was able to reduce prion levels and prolong survival.

Knocking out the prion gene in livestock is another approach to eliminating prion disease. Transgenic mice lacking both copies of the *Prnp* gene were engineered several years ago. They grow and develop normally. They are unable to make prion protein and are resistant to infection by pathogenic prions. This confirmed that new (and normal) prion proteins are first made by the host cell, before changing conformation during infectious prion disease. Although the prion gene is not needed for survival and its role is still unclear, it does appear to be involved in long-term memory and spatial learning.

Recently, cattle lacking both copies of the *Prnp* gene have been engineered and after 2 years are normal in growth and development. Brain cells from such animals are resistant to prion infection. *Prnp* knockout livestock could be used to provide prion-free products, if transgenic animals are approved as a source of human food.

There is presently no treatment for prion disease, although several lead compounds have been found with partial activity in cell culture.

USING YEAST PRIONS AS MODELS

A new approach to screening compounds for use in prion therapy is based on the use of **yeast prions**. Some weird genetic behavior in yeast turned out to be due to proteins that behave like the prions of mammals. The yeast prions are nonlethal and are cytoplasmic rather than membrane proteins. Nonetheless, they do show nucleic acid-free inheritance and their misfolded forms catalyze conversion of the normal protein into the misfolded version. Furthermore, the misfolded versions of yeast prions form insoluble **amyloid aggregates**, like those of mammals. However, although they do display similar structural domains, the mammalian and yeast prion proteins show no sequence homology.

The two best known yeast prions are [URE3] and [PSI+], which are the misfolded forms of the Ure2p and Sup35p proteins. Ure2p is involved in nitrogen regulation and Sup35p is a translation termination factor. In yeast, prions are detected by loss of function. For example, the prion form, [PSI+], of the Sup35p protein is insoluble and inactive. This results in increased read-through of stop codons. This forms the basis for a clever and quick genetic screening system for possible antiprion drugs.

Yeast mutants defective in adenine biosynthesis turn red because of accumulation of metabolic by-products. A yeast mutant with a nonsense mutation (i.e., a premature stop codon) in the ADE1 gene thus forms red colonies. If the Sup35p protein is in its prion form, read-through of stop codons occurs and enough full-length protein is made to allow adenine synthesis, that is, the mutation is suppressed. Consequently, prion-positive strains form white colonies. If the prions are lost, the yeast goes back to forming red colonies. This allows rapid color-based screening of chemical compounds simply by adding them to the medium and looking for those causing a white-to-red color shift of the yeast colonies. As of 2003 several candidates had been identified (Fig. 22.15). All were tricyclic, like quinacrine and

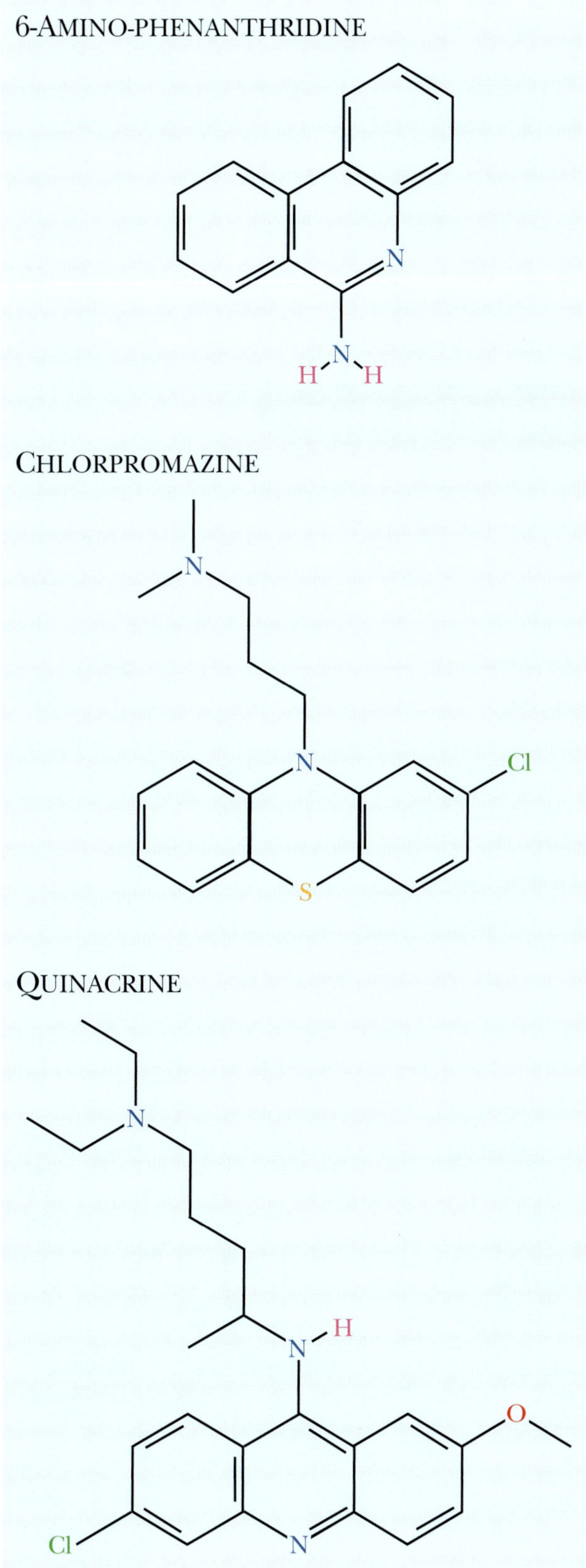

6-AMINO-PHENANTHRIDINE

CHLORPROMAZINE

QUINACRINE

FIGURE 22.15 Antiprion Agents
Structures of 6-aminophenanthridine, chlorpromazine, and quinacrine.

chlorpromazine. The most active were six-amino derivatives of phenanthridine. It is hoped that this approach will lead to an effective treatment for prion disease.

Prions have been discovered in yeast. This has allowed systematic screening for antiprion agents.

Summary

Viruses rely on a host cell for growth and replication. As a consequence it is often difficult to stop viral replication without damaging the host. The recent increased development of antiviral agents has largely been driven by the AIDS epidemic and, more recently, by the threat of pandemic flu. Prions are infectious proteins that cause neurodegenerative diseases. So far no effective therapy has been found to combat prion infection.

End-of-Chapter Questions

1. Which of the following is a possible antiviral agent?
  - **a.** interferon ß
  - **b.** amantadine
  - **c.** nucleoside analogs
  - **d.** protease inhibitors
  - **e.** all of the above

2. What is ICAM-1?
  - **a.** an adhesin molecule on the surface of animal cells that some viruses bind
  - **b.** an antiviral agent that prevents binding of the virus to the animal cell
  - **c.** a ligand on the surface of viral particles
  - **d.** an interferon that inhibits viral particle attachment
  - **e.** none of the above

3. What prompts interferon ß to be secreted?
  - **a.** ssDNA
  - **b.** dsRNA
  - **c.** ssRNA
  - **d.** dsDNA
  - **e.** mtDNA

4. Which of the following is not a component of influenza virus?
  - **a.** nucleocapsid
  - **b.** neuraminidase
  - **c.** negative ssRNA
  - **d.** outer envelope
  - **e.** caspase

5. Which influenza group causes the most severe outbreaks?
  - **a.** influenza A
  - **b.** avian influenza
  - **c.** influenza B
  - **d.** influenza AB
  - **e.** all of the above

6. What is the mode of action for the drug Tamiflu?
   a. hemagglutinin protease
   b. ion channel blocker
   c. neuraminidase inhibitor
   d. reverse transcriptase blocker
   e. none of the above
7. Which HIV protein binds to CD4?
   a. ganglioside GM_1
   b. neuraminidase
   c. hemagglutinin
   d. gp120
   e. chemokines
8. A mutation in which gene is responsible for natural resistance to HIV infection?
   a. CD4
   b. CCR6
   c. CCR5
   d. CXCR4
   e. gp120
9. What is the main problem with treatment of HIV/AIDS?
   a. availability of antiviral drugs
   b. the high mutation rate of the virus
   c. socioeconomic issues within populations
   d. adequate testing facilities for the disease
   e. adequate education about HIV/AIDS prevention
10. What is the mode of action for AZT?
   a. a DNA chain terminator
   b. reverse transcriptase inhibitor
   c. gp120 analog
   d. CCR5 analog
   e. none of the above
11. What is the favored method for HIV/AIDS therapy?
   a. treatment with a two-drug cocktail
   b. treatment with a three-drug cocktail
   c. treatment with a reverse transcriptase inhibitor only
   d. treatment with hydroxyurea only
   e. treatment with a protease inhibitor only
12. How do prions cause disease?
   a. Prion proteins bind to cells and induce apoptosis.
   b. Prion proteins bind to DNA polymerase and prevent replication.
   c. Prion proteins induce normal cellular proteins to refold into the prion form.
   d. Prion proteins induce an immune response against the "self."
   e. none of the above

(*Continued*)

13. Which of the following diseases is not caused by a prion?
- **a.** kuru
- **b.** BSE
- **c.** scrapie
- **d.** HIV
- **e.** CJD

14. Which method is used to identify prion infections?
- **a.** PMCA
- **b.** PCR
- **c.** RT-PCR
- **d.** brain biopsy
- **e.** none of the above

15. How might prion diseases be treated?
- **a.** RNAi
- **b.** *Prnp* knockouts
- **c.** removal of prions using a filter
- **d.** quinacrine or chlorpromazine treatment
- **e.** all of the above

Further Reading

Balzarini J (2007). Carbohydrate-binding agents: A potential future cornerstone for the chemotherapy of enveloped viruses? *Antivir Chem Chemother* **18**, 1–11.

Crusat M, de Jong MD (2007). Neuraminidase inhibitors and their role in avian and pandemic influenza. *Antivir Ther* **12**, 593–602.

D'Cruz OJ, Uckun FM (2006). Dawn of non-nucleoside inhibitor-based anti-HIV microbicides. *J Antimicrob Chemother* **57**, 411–423.

De Clercq E (2005). Recent highlights in the development of new antiviral drugs. *Curr Opin Microbiol* **8**, 552–560.

De Clercq E (2006). Antiviral agents active against influenza A viruses. *Nat Rev Drug Discov* **5**, 1015–1025.

Haasnoot J, Berkhout B (2006). RNA interference: Its use as antiviral therapy. *Handb Exp Pharmacol* **173**, 117–150.

Haller O, Staeheli P, Kochs G (2007). Interferon-induced Mx proteins in antiviral host defense. *Biochimie* **89**, 812–818.

Hovanessian AG, Justesen J (2007). The human 2′-5′ oligoadenylate synthetase family: Unique interferon-inducible enzymes catalyzing 2′-5′ instead of 3′-5′ phosphodiester bond formation. *Biochimie* **89**, 779–788.

Karpala AJ, Doran TJ, Bean AG (2005). Immune responses to dsRNA: Implications for gene silencing technologies. *Immunol Cell Biol* **83**, 211–216.

Kocisko DA, Caughey B (2006). Searching for anti-prion compounds: Cell-based high-throughput *in vitro* assays and animal testing strategies. *Methods Enzymol* **412**, 223–234.

Lori F, Foli A, Kelly LM, Lisziewicz J (2007). Virostatics: A new class of anti-HIV drugs. *Curr Med Chem* **14**, 233–241.

Pestka S (2007). The interferons: 50 years after their discovery, there is much more to learn. *J Biol Chem* **282**, 7–51.

Pinto LH, Lamb RA (2007). Controlling influenza virus replication by inhibiting its proton channel. *Mol Biosyst* **3**, 18–23.

Reece PA (2007). Neuraminidase inhibitor resistance in influenza viruses. *J Med Virol* **79**, 1577–1586.

Reiche EM, Bonametti AM, Voltarelli JC, Morimoto HK, Watanabe MA (2007). Genetic polymorphisms in the chemokine and chemokine receptors: Impact on clinical course and therapy of the human immunodeficiency virus type 1 infection (HIV-1). *Curr Med Chem* **14**, 1325–1334.

Rottinghaus ST, Whitley RJ (2007). Current non-AIDS antiviral chemotherapy. *Expert Rev Anti Infect Ther* **5**, 217–230.

Rusconi S, Scozzafava A, Mastrolorenzo A, Supuran CT (2007). An update in the development of HIV entry inhibitors. *Curr Top Med Chem* **7**, 1273–1289.

Sakudo A, Nakamura I, Ikuta K, Onodera T (2007). Recent developments in prion disease research: Diagnostic tools and *in vitro* cell culture models. *J Vet Med Sci* **69**, 329–337.

Scherer L, Rossi JJ, Weinberg MS (2007). Progress and prospects: RNA-based therapies for treatment of HIV infection. *Gene Ther* **14**, 1057–1064.

Soto C, Estrada L, Castilla J (2006). Amyloids, prions and the inherent infectious nature of misfolded protein aggregates. *Trends Biochem Sci* **31**, 150–155.

Souvignet C, Lejeune O, Trepo C (2007). Interferon-based treatment of chronic hepatitis C. *Biochimie* **89**, 894–898.

Vana K, Zuber C, Nikles D, Weiss S (2007). Novel aspects of prions, their receptor molecules, and innovative approaches for TSE therapy. *Cell Mol Neurobiol* **27**, 107–128.

Wang X, Jia W, Zhao A, Wang X (2006). Anti-influenza agents from plants and traditional Chinese medicine. *Phytother Res* **20**, 335–341.

Wickner RB, Edskes HK, Shewmaker F, Nakayashiki T (2007). Prions of fungi: Inherited structures and biological roles. *Nat Rev Microbiol* **5**, 611–618.

Biowarfare and Bioterrorism

INTRODUCTION

Biological warfare involves not only choosing a disease agent and perhaps modifying it genetically, but also manufacturing and storage, weaponization, and delivery. This chapter is intended to address the molecular biology of biowarfare and makes no attempt to cover the more physical and military aspects.

In addition, the subject is plagued by a lack of verifiable information. Accusations of using biological warfare or developing biowarfare agents are often made as propaganda ploys in the absence of any real evidence. Suspicious outbreaks of disease have frequently led to accusations of biowarfare, especially during the Cold War period. Accusing any nation that irritates the U.S. government of harboring "weapons of mass destruction" has become virtually routine. The official list of naughty nations who are supposedly guilty of developing biowarfare agents changes with the constantly shifting political situation. In consequence, reliable information on alleged germ warfare programs is hard to ascertain.

Biological warfare among humans is politically controversial, and hard facts are often in short supply.

BACTERIA MAKE LETHAL PROTEINS TO KILL OTHER ORGANISMS

Before discussing human biological combat, let's remember that biological warfare has been practiced throughout evolutionary history by organisms at all levels on the evolutionary scale. Many of the following examples have already been discussed under different headings; here, we will summarize them briefly before talking about deliberate use of biological agents in human conflict.

When related bacteria compete for the same habitat or same resources, they often kill each other, using toxic proteins known as **bacteriocins**. Generally speaking, bacteria are most likely to kill their close relatives because the more closely related two strains of bacteria are, the more likely they will compete for the same resources. For example, many strains of *Escherichia coli* deploy a wide variety of bacteriocins, referred to as **colicins**, intended to kill other strains of the same species. The genes for colicins are normally carried on plasmids. The colicin plasmids of *E. coli* have been used as the basis of many plasmids commonly used in molecular biology and genetic engineering (see Chapter 1). *Yersinia pestis*, the plague bacterium, also makes bacteriocins, called pesticins in this case, designed to kill competing strains of its own species.

When proteins produced by bacteria act against higher organisms we refer to them as **toxins**. The difference in terminology between *bacteriocin* and *toxin* is thus really a matter of perspective. Bacteria deploy bacteriocins against their fellow bacteria with the deliberate intention of killing them. In contrast, pathogenic bacteria do not usually "intend" to kill the people they infect. The longer the host organism stays alive, the longer it provides a home for the infecting bacteria. Bacteria also infect insects and make toxins that kill insects but are harmless to vertebrates. The bacterium *Bacillus thuringiensis* has become famous for its production of a toxin that kills insect pests. The use of Bt toxin in creating genetically modified crop plants has been covered in Chapter 14.

Bacteriocins are toxic proteins made by bacteria to kill other related bacteria that are competing for the same natural resources. Bacteria also make proteins that kill insects or higher animals.

GERM WARFARE AMONG THE LOWER EUKARYOTES

Paramecium is a well-known ciliated protozoan. Many strains of *Paramecium* carry symbiotic bacteria (*Caedibacter*), known as **kappa particles**, that grow and divide inside the larger, eukaryotic, *Paramecium* cell (Fig. 23.1). Those strains of *Paramecium* with kappa particles are known as **killers**. Killing occurs when kappa particles are released by a killer and eaten by a sensitive *Paramecium* (i.e., one lacking kappa particles). Digestion of the kappa particle results in release of a protein toxin and death of the sensitive *Paramecium*. Thus germ warfare is practiced by single-celled eukaryotes.

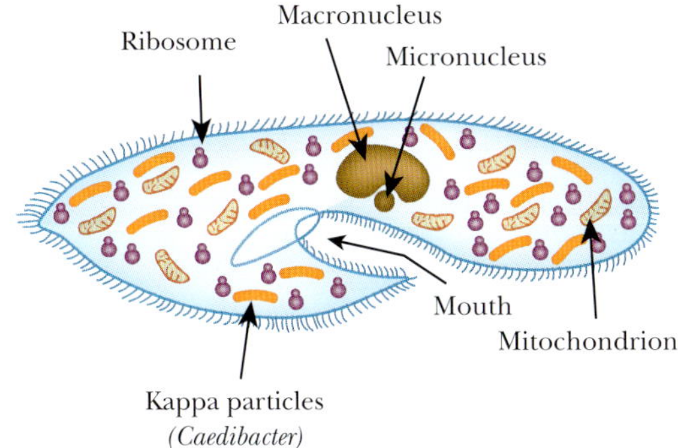

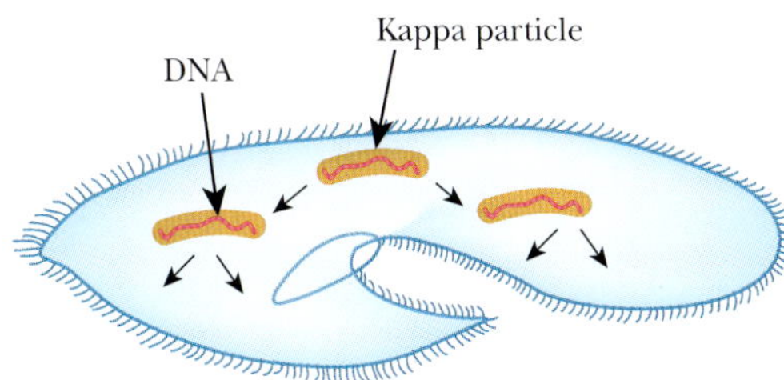

FIGURE 23.1 Killer *Paramecium* Uses a Bacterial Toxin
(A) The kappa particles are found in the cytoplasm of the *Paramecium*.
(B) Kappa particles are symbiotic *Caedibacter* that are found in many strains of *Paramecium*, yet they have their own DNA and divide like typical bacteria.

The toxin gene is borne not on the chromosome of the symbiotic bacteria, but on a bacterial plasmid derived from a defective bacterial virus. So a toxin encoded by a virus infecting the kappa particle bacteria has been diverted to the purpose of killing other strains of *Paramecium*. This appears to be a general principle. Many of the toxins used by pathogenic bacteria that infect humans are actually encoded by DNA of nonchromosomal origin—viruses, plasmids, or transposons. These elements are often integrated into the chromosome of pathogenic strains of bacteria.

Some insects rely on viruses to wage biological warfare. Certain parasitic wasps inject their eggs into plant-eating maggots (i.e., the larvae of plant-eating insects). After the eggs hatch, the newborn wasps eat the living maggots from inside. The maggots are eventually killed and a new generation of wasps is released to continue their life cycle.

The secret to the wasp's success is injecting a virus along with their eggs (Fig. 23.2). The virus, a member of the adenovirus family, targets the maggot's "fat body" (vaguely equivalent to the liver of higher animals). The virus cripples both the maggot's developmental control system and its primitive immune system. The maggot loses its appetite for plants and is prevented from molting and turning into a pupa, the next stage on the way to an adult insect capable of laying more eggs. There are many types of plant-eating maggots, including major agricultural pests such as the tobacco budworm and cotton bollworm, and many types of wasps that attack them, and consequently many types of viruses, each designed to soften up a particular insect larva.

Lower eukaryotes often use viruses or bacterial toxins to attack competitors or cripple prey.

HISTORY OF HUMAN BIOLOGICAL WARFARE

Burning crops was probably the earliest form of warfare aimed at undermining an enemy's survival by biological means. Early in history, the water supply was also a prime biological target for feuding nomads, especially in areas where water was scarce. Presumably tossing dead or rotting animals into waterholes poisoned the drinking water and proved to be reasonably effective in driving the enemy away.

Throughout history there have been occasional sporadic attempts to deliberately spread infection for military purposes. However, these have mostly been ineffective or irrelevant. During the **Black Death** epidemic of the mid-1300s, the Tartars catapulted plague-ridden corpses over the walls into cities held by their European enemies.

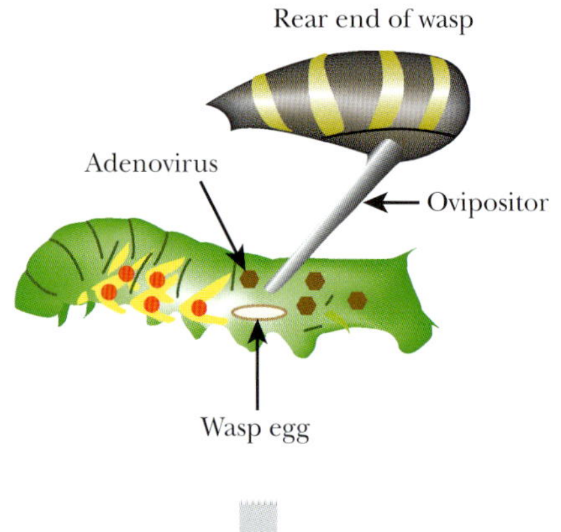

FIGURE 23.2 Wasps Use Viruses against Maggots
Certain types of wasps lay their eggs inside tobacco hornworm larvae. The wasp lands on the back of the larva and injects the eggs plus adenovirus into the maggot through the ovipositor. The adenovirus prevents the larva from eating and therefore developing into a pupa. When the eggs hatch, the young use the insides of the larva as a food source, to grow and develop into adult wasps.

Although this is sometimes credited with spreading the plague, in reality, rats and their fleas spread **bubonic plague**, not contact with corpses. Catapulting bodies into a city may deserve points for enthusiasm, but it doesn't earn an A in microbiology. In medieval Europe, dead or sick animals were hurled over the walls into castles or walled cities to break sieges by spreading disease. Nonetheless, given the state of hygiene in most medieval towns or castles, there was often little need to provide an outside source of infection. With plague, typhoid, **smallpox**, dysentery, and diphtheria already around, all that was usually necessary was to let nature take its course. Similarly, attempts of white settlers to spread smallpox among the American Indians were not only rather ineffective but also largely irrelevant because smallpox had already spread by itself.

The reason why germ warfare has been of little account until recently is that plenty of dangerous infections were already in circulation. If an army was crowded and unhygienic, some natural disease would undoubtedly attempt a biological assault without waiting for artificial prompting. Until recently, armies, like civilian populations, were so dirty and disease-ridden that practicing germ warfare was rather like trying to kill a shark by drowning it. Only in our modern disinfected age has spreading disease deliberately become a meaningful threat.

Although biological warfare is historically old, it has rarely been effective. This is probably due to the massive effect of natural infectious disease.

EXPECTANCY AND EXPENSE

During the Vietnam War, the Viet-Cong guerillas dug camouflaged pits as booby traps. Within these they often positioned sharpened bamboo stakes or splinters smeared with human waste. Although it was possible to contract a nasty infection from these, the main purpose was psychological. Worrying about possible booby traps hampered the movements of American troops out of all proportion to actual casualties. Thus the threat of chemical or biological warfare may have great psychological effect. An example is the recent anthrax bioterrorism in the United States. There have been several times as many fatalities from the naturally spreading **West Nile virus** as from the deliberate delivery of anthrax spores. Yet response to the anthrax scare has involved colossal disruption of the postal service and massive expense.

Taking protective measures against a possible biological attack is costly and inconvenient. Vaccinating soldiers against all possible diseases that might be used is impractical. In addition, vaccines sometimes have side effects, especially if they have been developed under emergency conditions without thorough testing. Consider the anthrax vaccine used by the U.S. army that was approved in 1971. It has been thoroughly tested and is regarded as relatively safe. Vaccination requires six inoculations plus annual boosters. It produces swelling and irritation at the site of injection in 5% to 8% and severe local reactions in about 1% of those inoculated. Major systemic reactions are "rare." Although it works against "natural" exposure, it is uncertain whether it would protect against a concentrated aerosol of anthrax spores. Dressing infantry in protective clothing and respirators hampers their mobility, making them easier targets for conventional weaponry. In hot climates extra clothing may also promote heat stress.

Even without deliberate germ warfare, troops from hygienic temperate nations are at a major disadvantage when operating in tropical Third World situations. Drugs given to ward off malaria and other endemic tropical infections are costly, rarely 100% effective, and may damage health if taken over a long period. Constant exposure to insecticides to kill mosquitoes, lice, and so forth may damage the nervous system. An additional factor is that the inhabitants of rich Western nations expect to live into their seventies or eighties nowadays. Consequently, those sent into backward areas of the world demand ever-increasing levels of protective equipment and medication. Military actions in Third World nations are thus becoming ever more expensive. In contrast, troops belonging to a poverty-stricken local regime will go unencumbered by the extra protective gear that they cannot, in any case, afford. Moreover, human casualties are of much less importance to the regimes of overcrowded nations where life expectancy is much lower. Over the past half century, armed interference in the Third World by nations such as the Soviet Union and the United States has steadily become less enthusiastic and less effective, at least in part because of these trends.

Much of the effect of biological warfare is due to its psychological threat.

IMPORTANT FACTORS IN BIOLOGICAL WARFARE

The basic strategy of germ warfare is to choose a human disease and use it to eliminate or frighten away human competitors. Perhaps the disease agent could even be improved by genetic engineering, as discussed later. However, several major factors need to be considered. The relative importance of these factors will depend on whether biowarfare agents are intended for use on a military scale or by terrorists.

Incubation Time

One major problem with biological warfare is that death from infectious diseases is slow. Even the most virulent pathogens, such as Ebola virus or the pneumonic form of Black Death, take a few days to kill. Thus infected personnel would still be capable of fighting for a significant period. In contrast, conventional weapons kill or incapacitate rapidly.

Dispersal

Another drawback to germ warfare is the problem of delivery. Most protocols for spreading infections deliberately involve some sort of airborne delivery. This tends to place delivery at the whim of the weather. Not only do you need a breeze, but also the wind needs to blow in the right direction! During the 1950s the British government conducted field tests with harmless bacteria. When the wind blew them over "healthy" farmland, many of the airborne bacteria survived the trip and reached the ground alive. In contrast, when the wind blew the bacteria over industrial areas, especially oil refineries or similar installations, the airborne bacteria were almost all killed. In the event of aerial delivery, even if the wind is in the right direction, most of the population of an industrial nation will be found in cities where they will be at least partly protected from airborne bacteria by air pollution!

In addition, many infectious agents are sensitive to desiccation and become inactive if exposed to air for significant periods of time. Moreover, natural UV radiation from the sun also inactivates many bacteria and viruses. Thus most biowarfare agents must be protected from this "open air factor" before use and then dispersed as rapidly as possible.

Dispersal by terrorists involves rather different considerations. Delivery of small batches of biowarfare agents can be performed individually in a variety of ways—for example, by mailing letters containing anthrax spores. The minimalist dispersal scheme is to infect a volunteer and have him or her use public transport among crowded target populations.

Persistence of the Agent

Once biological weapons have been successfully used, the victor presumably wants to move in and take over the enemy territory. This may involve exposing the invading troops to the infectious agent, depending how long it persists in the environment. **Anthrax** is a favorite biological weapon because it acts rapidly and is highly lethal. Unfortunately, ***Bacillus anthracis,*** which causes this disease, spreads by forming spores that are tough, difficult to destroy, and last for an extremely long time. When suitable conditions return, the spores germinate and resume growth as normal bacterial cells. In contrast, many viruses last only a few days, if that, outside their animal or human hosts. However, infections due to these agents may persist among the local population, and this too may prove a threat to an invading army.

Storage of the Agent

An infectious agent that persists may be inconvenient from the perspective of a military occupation. However, an infectious agent that decays very rapidly has the opposite problem—it is difficult to store.

Preparation of the Agent

Some pathogenic microorganisms are relatively easy to grow in culture, whereas others are extremely difficult and/or expensive to manufacture in sizeable quantities. One major drawback with viruses is that they can grow only inside host cells. Culturing animal cells is far more difficult than growing bacteria and the yields are much lower. Large-scale manufacture of viruses is expensive and difficult. Many bacteria (e.g., plague, tularemia, typhoid) are relatively easy to culture. In the years just after World War II, the British kept large-scale cultures of *Yersinia pestis*, the agent of plague, growing continuously in case of need. Doubtless other nations did much the same.

Another factor is the issue of formulation. That is, the disease agent must be prepared in a manner that facilitates storage and dispersal. Both bacterial cells and spores tend to clump together spontaneously. Consequently, they must be **weaponized** to disperse them effectively by aerosols and other delivery systems. This issue is highly technical, and the details are beyond the scope of this book.

High-Containment Laboratories

In addition to inherent problems with culturing bacteria or viruses, there is the need for protecting personnel who work with infectious agents. High-containment laboratories are needed for research and development in such areas.

Biological containment is rated on a scale with four levels. Laboratories with a **biosafety level** 4 rating are needed for working with agents such as Ebola virus. Operations are done inside safety cabinets with glove ports. The whole laboratory is sealed off and kept at a little under normal atmospheric pressure. That way, if there is a leak, air will leak in (not out!). To enter a high-containment facility involves exchanging outside clothes for a separate set of lab clothes—in extreme cases not just lab coats but lab underwear, too! Spacesuit-style protective clothing may be used at the highest biosafety levels (Fig. 23.3). Entry is via an air lock, and exits are provided with disinfectant showers and ultraviolet lights. Lab clothes are left behind. Some high-containment labs are designed so that the only exit is via total submersion in a pool of disinfectant. Ultraviolet lights are used to sterilize both the laboratories themselves and the air locks, especially when working with viruses.

Using high-containment facilities not only is expensive but is very time consuming. High containment is also needed for any bioweapons manufacturing facility. Here everything is on a large industrial scale and the problems are correspondingly worse. On the other hand, terrorist groups may care only about secrecy and be willing to take risks that are unacceptable to an industrialized nation. Similarly, Third World dictatorships may not care about casualties,

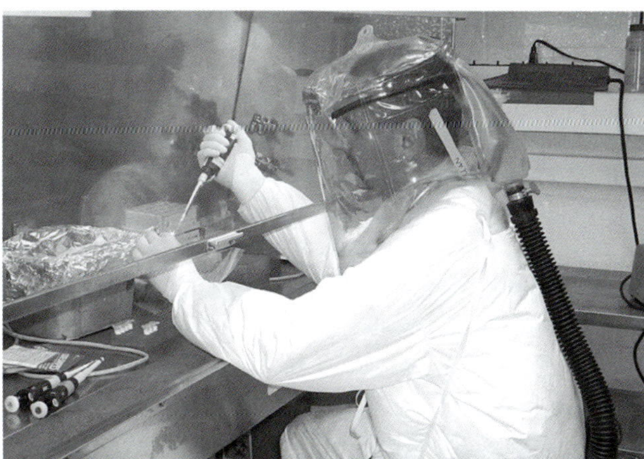

FIGURE 23.3 Biohazard Clothing, Then and Now
(A) Even during the bubonic plague, doctors wore protective clothing to prevent exposure to the deadly pathogens, as illustrated in Bartholin, Thomas Hafniae, 1654–1661: *Historiarum anatomicarum.* Courtesy U.S. National Library of Medicine. (B) Today's suits are more streamlined, but serve the same purpose. Laboratory worker wearing BSL-4 protective gear. Courtesy of USAMRIID, DoD, and the NIAID Biodefense Image Library.

but if a suspected biowarfare agent escapes into the surrounding population it will draw unwelcome attention.

A wide range of physical factors affect biological warfare in addition to the nature and properties of the infectious agent to be used.

WHICH DISEASE AGENTS ARE SUITABLE FOR BIOLOGICAL WARFARE?

Which disease? Bacteria, virus, or eukaryote? Most bacterial diseases can, in principle, be cured by antibiotics, whereas viral diseases cannot. On the other hand, bacteria are easier to grow than viruses. Many bacteria can be grown in relatively simple and cheap culture media. However, the production of virus particles requires culturing host cells for the virus to infect and consequently is more difficult because animal cells have complex growth requirements. Pathogenic eukaryotes such as *Plasmodium* (malaria) or *Entamoeba* (amoebic dysentery) have rarely even been considered as possible biowarfare agents because of the difficulty of culturing them on a large scale. Pathogenic fungi are a possible exception: some can be grown relatively easily. Although viruses might be grown in quantity by a government, large-scale virus culture is probably impractical for small groups of bioterrorists.

Outside the tropics, most incurable infectious diseases are due to viruses. The fundamental issue is that viruses are not themselves living cells but rely on the host cells they infect to assemble new virus particles. Consequently, chemical agents that prevent virus replication usually kill the host cells, too. A small and growing range of specific antiviral agents are available (see Chapter 22); nonetheless, no cure yet exists for most viral diseases. Vaccination may protect against catching many viral diseases such as mumps, measles, and smallpox. However, protection against a viral biowarfare agent would require an effective vaccine against the particular strain of virus being used and vaccinating a sufficiently large proportion of the target population to prevent spread of the disease.

Box 23.1 Requirements for Biowarfare Agents

According to the U.S. Army, a biological warfare agent should fulfill the following requirements:

1. It should consistently produce death, disability, or damage.
2. It should be capable of being produced economically and in militarily adequate quantities from available materials.
3. It should be stable under production and storage conditions, in munitions, and in transportation.
4. It should be capable of being disseminated efficiently by existing techniques, equipment, or munitions.
5. It should be stable after dissemination from a military munition.

Among the bacterial diseases, anthrax, bubonic plague, brucellosis, tularemia, glanders, and melioidosis have all been suggested as possible biological weapons. The properties of these agents are summarized in Table 23.1. A variety of viruses have been suggested, including both tropical diseases such as dengue and yellow fever and emerging diseases such as Lassa fever and Ebola virus, but the only consistent choice among viruses seems to be smallpox.

A variety of viruses, bacteria, and toxins have been proposed as effective biowarfare agents. These are classified into three major categories by the Centers for Disease Control and Prevention (CDC) according to their level of risk.

Table 23.1 CDC-Listed Agents Relevant to Biowarfare

CATEGORY A AGENTS INCLUDE ORGANISMS THAT POSE A RISK BECAUSE:	
• they can be easily disseminated or transmitted person-to-person • they cause high mortality • they might cause public panic and social disruption • they require special action to protect public health	
Bacteria	
Anthrax	*Bacillus anthracis*
Plague	*Yersinia pestis*
Tularemia	*Francisella tularensis*
Viruses	
Smallpox	*Variola major*
Filoviruses	Ebola hemorrhagic fever Marburg hemorrhagic fever
Arenaviruses	Lassa fever Junin virus (Argentine hemorrhagic fever)
Toxins	
Botulinum toxin from *Clostridium botulinum*	
CATEGORY B AGENTS INCLUDE THOSE THAT:	
• are moderately easy to disseminate • cause moderate morbidity and low mortality • require improved diagnostic capacity and enhanced surveillance	
Bacteria	
Brucellosis	*Brucella* (several species)
Glanders	*Burkholderia mallei*

Table 23.1 CDC-Listed Agents Relevant to Biowarfare—cont'd

Melioidosis	*Burkholderia pseudomallei*
Q fever	*Coxiella burnetti*
Several food- or waterborne enteric diseases including:	*Salmonella, Shigella dysenteriae, Vibrio cholerae*
Viruses	
Alphaviruses	Venezuelan encephalomyelitis Eastern and western equine encephalomyelitis
Toxins	
Ricin toxin from *Ricinus communis* (castor bean)	
Epsilon toxin from *Clostridium perfringens*	
Enterotoxin B from *Staphylococcus*	
CATEGORY C AGENTS:	
Emerging pathogens that could possibly be engineered for mass dissemination in the future, such as Nipah virus, hantaviruses, flaviviruses (yellow fever, dengue fever), multidrug-resistant tuberculosis	

ANTHRAX

Anthrax is a virulent disease of cattle that infects humans quite easily. It is caused by the bacterium *Bacillus anthracis*, which is relatively easy to culture and forms spores, which can survive harsh conditions that would kill most bacteria. The spores may lie dormant in the soil for years and then germinate on contact with a suitable animal victim. Three main forms of the disease occur. Cutaneous anthrax, that is, infection of the skin, is rarely dangerous. **Inhalational anthrax**, in which the spores enter via the lungs, gives a high death rate. Gastrointestinal anthrax is relatively rare and occurs from ingestion of bacteria or spores, mostly by eating contaminated meat. Two major toxins are responsible for the symptoms of anthrax and, as described in Chapter 21, genes carried on the plasmid pOX1 encode the toxins. In some ways, anthrax is the ideal biological weapon—lethal, highly infectious, and cheap to produce, with spores that store well.

The problem with anthrax is that the spores are so tough and long-lived that getting rid of them after hostilities are over is almost impossible. During World War II the British tested anthrax (using sheep as the targets) on the tiny island of Gruinard, which lies off the coast of Scotland. Although it was fire-bombed and disinfected, the island remained uninhabitable because of anthrax spores still surviving in the soil until 1987, when it was treated with seawater and formaldehyde.

The germ warfare facility of the now-defunct Soviet Union is supposed to have concentrated on anthrax and smallpox. For the Soviet Union, a nation that had vast expanses of thinly populated land, anthrax was a good choice for defense against possible intrusion by massively overpopulated Asian neighbors such as China or India.

In many ways, anthrax is one of the best biological weapons. It is lethal, highly infectious, and easy to produce, and it has long-lived spores.

THE 2001 ANTHRAX ATTACK IN THE UNITED STATES

Shortly after the terrorist attack on the World Trade Center in September 2001, anthrax spores were distributed via the U.S. Postal Service. The anthrax attack was notable in two respects. First, it killed only a small handful of victims, supporting the contention that biological warfare is not usually very effective in practice. Second, it generated a vastly disproportionate reaction, illustrating the importance of the psychological aspects of bioterrorism. Undoubtedly governmental overreaction and public panic did far more damage than the anthrax attack itself.

The anthrax attack came from inside the United States. Furthermore, although the case has not been solved at the time of writing, it is thought likely that the perpetrator was an insider in America's own biological warfare research establishment. The attacker used the Ames strain of *Bacillus anthracis*, which is widely used in laboratories in the United States.

A major problem with tracing the origin of anthrax outbreaks is that all of the strains of *Bacillus anthracis* are closely related and difficult to tell apart. No differences in either 16S rRNA or 23S rRNA sequence occur between different strains. In practice analysis is done using variable number tandem repeats or VNTRs (see Chapter 24 for details). For example, the *vrrA* gene of *Bacillus anthracis* contains from two to six copies of the sequence CAATATCAACAA within the coding region for a protein of unknown function (Fig. 23.4). These repeats were probably originally generated by slippage of DNA polymerase during replication. The repeats do not alter the reading frame, but result in corresponding repeats of the four-amino-acid sequence Gln-Tyr-Gln-Gln within the encoded protein. Several other VNTRs are also now used, including some on the pOX1 virulence plasmid. The greatest diversity of *Bacillus anthracis* strains, as assessed by multiple VNTR analysis, comes from Southern Africa, which is therefore regarded as the probable homeland of anthrax.

Different strains of anthrax are identical in ribosomal RNA sequence. They may be identified by using VNTRs.

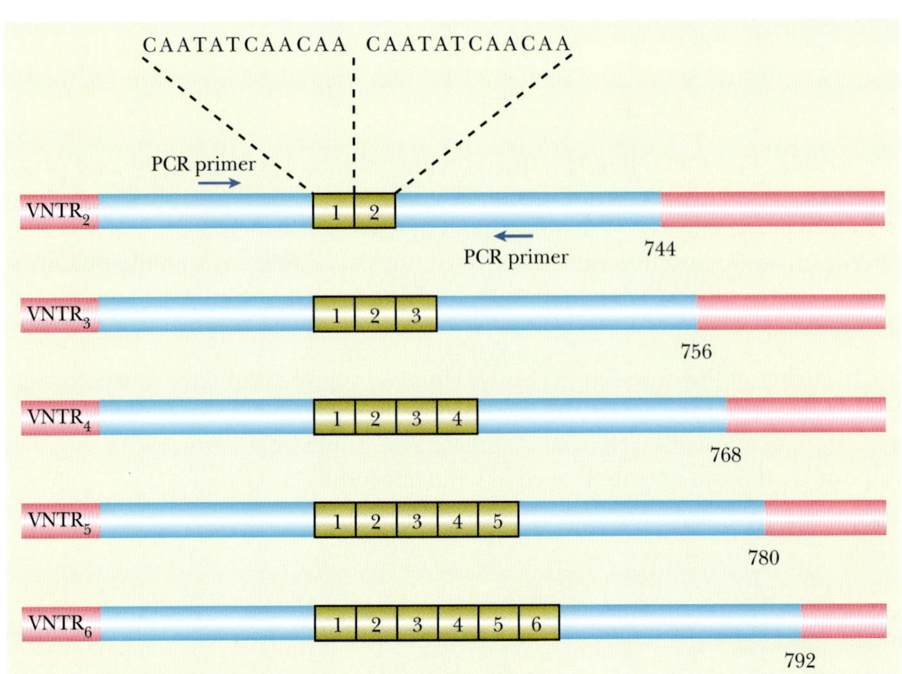

FIGURE 23.4 The *vrrA* VNTR of Anthrax The *vrrA* gene of anthrax (blue) has a stretch of repeats in the coding region. Different strains of anthrax have different numbers of repeats (green) due to polymerase slippage and can therefore be traced by comparing the number of repeats. PCR is used to amplify the region containing the repeats. The length of the PCR product reveals the number of repeats.

OTHER BACTERIAL AGENTS

Yersinia pestis, the causative agent of plague, was responsible for the notorious Black Death epidemics of the Middle Ages. It is highly infectious, the death rate is high, and it kills fast. Bubonic plague is spread by fleas; the more dangerous pneumonic form is spread from person to person directly via the air. Around World War II, plague was in vogue. The Japanese apparently tested bubonic plague on prisoners and also tried small-scale airborne dispersal of infected fleas carrying plague on the Chinese. This had little effect—partly because plague was endemic in China and already in natural circulation. The British biological warfare center at Porton Down maintained large-scale plague cultures for several years following World War II. In the 1960s the United States apparently experimented with spreading plague among rodents in Viet Nam, Laos, and Cambodia, again with little effect. Needless to say, this is officially denied. Since then, plague seems to have gone out of fashion.

Other bacteria qualify as potential bacterial agents. Brucellosis, caused by *Brucella*, is a disease of cattle, camels, goats, and related animals. Brucellosis was developed as a biological weapon by the United States from 1954 to 1969, though the choice seems curious. In humans it behaves erratically, both in the time for symptoms to emerge and the course of the disease. Though human victims often fall severely ill for several weeks, it is rarely fatal, even if untreated. **Tularemia**, caused by ***Francisella tularensis***, is a disease of rodents or birds that has a death rate of 5% to 10% in humans, if untreated. It is highly infectious and generally regarded as an incapacitating rather than a lethal agent. It is still considered a possible threat. Melioidosis, caused by the bacterium *Burkholderia pseudomallei*, is related to glanders (*Burkholderia mallei*). Glanders is a disease of horses, and melioidosis is a rare disease of rodents from the Far East that is spread by rat fleas. Despite the *pseudo-* in its name, melioidosis is more virulent than glanders and is fatal some 95% of the time in humans.

Plague is extremely lethal, and its pneumonic form can be spread from person to person. Tularemia is rarely fatal but is extremely infectious and acts as an incapacitating agent. Both are possible biowarfare agents.

SMALLPOX VIRUS

Smallpox virus is a member of the **poxvirus** family. These are large viruses that contain double-stranded DNA (Fig. 23.5). Poxviruses are the most complex animal viruses and are so large they may just be seen with a light microscope. They measure approximately 0.4 by 0.2 microns, compared to 1.0 by 0.5 microns for bacteria such as *E. coli*. Unlike other animal DNA viruses, which all replicate inside the cell nucleus, poxviruses replicate their dsDNA in the cytoplasm of the host cell. They build subcellular factories known as inclusion bodies, inside which virus particles are manufactured. Poxviruses have 185,000 nucleotides encoding 150 to 200 genes, about the same number as the T4 family of complex bacterial viruses.

Variola virus infects only humans, which allowed its eradication by the World Health Organization, a task completed by 1980. Smallpox is highly infectious and exists as two variants, *Variola major* with a fatality rate of 30% to 40% and *Variola minor* with a fatality rate of around 1%. The two differ by about 2% of their base sequence. *Vaccinia* virus is a related but very mild poxvirus which is used as a live vaccine. Immunity is effective against several closely related poxviruses including smallpox and monkeypox.

For an industrial nation, the preparation of large amounts of a virus whose particles are fairly stable and long-lived, such

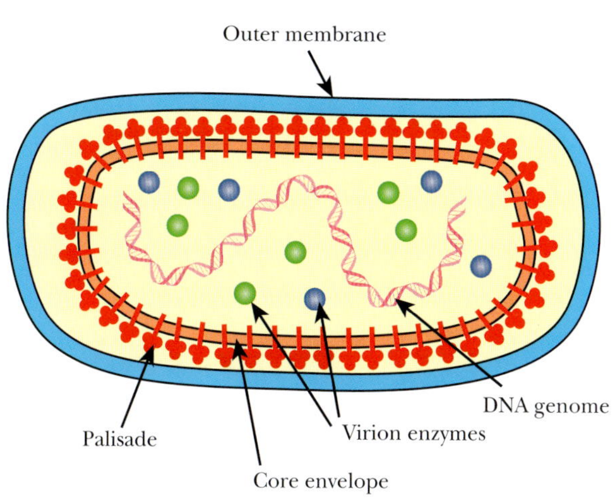

FIGURE 23.5
Smallpox Virus
Poxviruses, including smallpox, are closely related in structure and DNA sequence. They have genomes of dsDNA surrounded by two envelope layers. A protein layer, known as the palisade, is embedded within the core envelope. Premade viral enzymes are also packaged with the genome to allow replication immediately on infection. Poxviruses infect animals, and the outermost viral membrane is derived from the membrane of the previous host cell.

as smallpox, is feasible. For a Third World nation, it might be possible. A suggested terrorist scenario is that suicidal volunteers would be deliberately infected and then travel to densely populated target areas. They would deliberately mingle with as many people as possible, by attending mass events and traveling the subways in large cities.

Smallpox is the most likely virus to be used as a biowarfare agent. It is highly infectious and the death rate may reach 30% to 40%.

OTHER VIRAL AGENTS

A variety of other viruses have been considered as possible biowarfare agents, but none seems to have achieved a consensus. As noted above, viruses need host cells and so most viruses are difficult to culture in large amounts. In addition, many virus particles are unstable during storage.

Filoviruses, including Ebola and Marburg viruses, are a family of negative single-stranded RNA viruses that form long, thin filaments. The virus spreads through the blood system and causes vomiting of blood in doomed patients. Ebola virus has caused outbreaks in Sudan and Zaire with 80% to 90% fatality. Nonetheless, many closely related virus strains exist that are nowhere near so virulent. For example, in 1989, an Ebola outbreak occurred among monkeys in a research facility in Reston, Virginia. The monkeys, long-tailed macaques, had been freshly imported from the Philippines. Although the incident received a great deal of publicity, the Reston strain of Ebola virus was not even lethal to humans. Furthermore, most filoviruses are relatively difficult to catch by casual exposure, and transmission usually requires substantial exposure to infected body fluids. In practice this makes them relatively poor choices for biowarfare agents.

Dengue fever and yellow fever are both caused by members of the Flavivirus family. Yellow fever is frequently lethal, whereas dengue is rarely fatal, but is highly painful and incapacitates its victims for several days. However, both are spread by insect bites, which would greatly impede their possible use as biological weapons.

Lassa fever is an emerging virus disease, which appeared in the Lassa River region of Nigeria in the late 1960s. Lassa is a member of the Arenavirus family and contains segmented single-stranded RNA. Rodents spread the virus. Lassa virus has extremely high mortality and was considered briefly as a possible biowarfare agent.

Several other viruses resemble smallpox in having a very high death rate. However, effectively spreading the virus presents major technical problems in most of these cases.

PURIFIED TOXINS AS BIOWARFARE AGENTS

Another approach to biological warfare is to use purified toxins rather than a living infectious agent. A variety of toxins are known that may be purified in substantial quantities. Bacteria, primitive eukaryotes such as algae or fungi, higher plants, and animals all make toxins (Table 23.2).

Purified highly potent toxins are possible biowarfare agents. Some toxins come from bacterial sources, others are made by plants or animals, and some toxins are wholly artificial.

BOTOX—BOTULINUM TOXIN

Botulinum toxin is the most toxic substance known. It is made by the anaerobic bacterium, ***Clostridium botulinum***, and is the cause of botulism, a severe form of food poisoning. It has been proposed as a biowarfare agent but has actually found its most frequent application in

Table 23.2 Toxins Relevant to Biowarfare

Toxin	LD50 (μg/kg)	Producer Organism
Botulinum toxin A	0.001	Bacterium (*Clostridium*)
Enterotoxin B	0.02	Bacterium (*Staphylococcus*)
Ciguatoxin P-CTX-1	0.2	Marine dinoflagellate
Batrachotoxin	2	Poison arrow frog
Ricin (RIP)	3	Castor bean (*Ricinus communis*)
Tetrodotoxin	8	Pufferfish
VX	15	Synthetic nerve agent
Anthrax lethal toxin	50	Bacterium (*B. anthracis*)
Aconitine	100	Plant (monkshood = wolfsbane)
Mycotoxin T-2	1200	Fungus (*Fusarium*)

cosmetics, under the name **Botox**. (Perhaps there's hope for the human race after all!) It is also used to treat a few clinical conditions where a muscle relaxant is needed.

Botulinum toxin is a **neurotoxin** that blocks transmission of signals from nerves to muscles, thus causing muscular paralysis. The incredible potency of botulinum toxin is due to its enzyme activity—it is a zinc proteinase that cleaves proteins of the neuromuscular junction that are required for release of the neurotransmitter acetylcholine. Death is generally due to paralysis of the lungs and respiratory failure.

Clostridium botulinum almost never causes infections but will grow in canned food that has not been thoroughly sterilized. Botulinum toxin then accumulates in the contaminated food and, if ingested, may cause botulism. The bacteria are actually widespread, and traces are often present in food. The problem arises when contaminated food is not sterilized and is stored under anaerobic conditions that allow *Clostridium* to multiply. If the food is not heated thoroughly (which destroys the toxin), poisoning may occur.

Botulinum toxin lacks a leader sequence, and instead of being secreted it is released by bacterial lysis. The toxin is synthesized as a single inactive precursor protein. This is then cleaved to give a heavy chain (the binding protein; MW approximately 100,000) plus a light chain (the actual toxin; MW approx 50,000). These are held together by a disulfide bond.

Botulinum toxin is actually produced in a complex with a hemagglutinin, which causes blood cells to clump or agglutinate, and other associated proteins. The genes encoding these components are clustered together in two operons, transcribed in opposite directions (Fig. 23.6). They are controlled by an activator gene, *botR*, which lies between the two structural gene operons. The toxin gene cluster is found on the chromosome, on plasmids, or on bacteriophages integrated into the chromosome, depending on the strain of *Clostridium botulinum*.

The toxin binds very strongly to peripheral nerve synapses, in particular those of neuromuscular junctions, and blocks the release of the neurotransmitter acetylcholine. The toxin binds specifically to receptors in the synapse

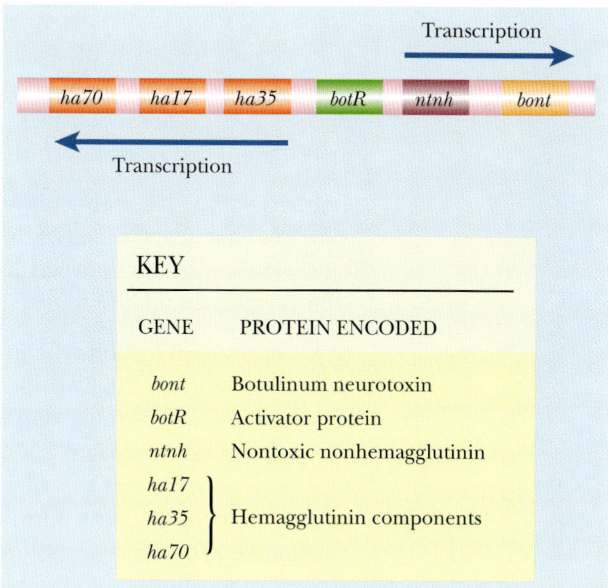

FIGURE 23.6
Botulinum Toxin Gene Cluster
BotR protein regulates the expression of two operons and so controls the expression of botulinum toxin.

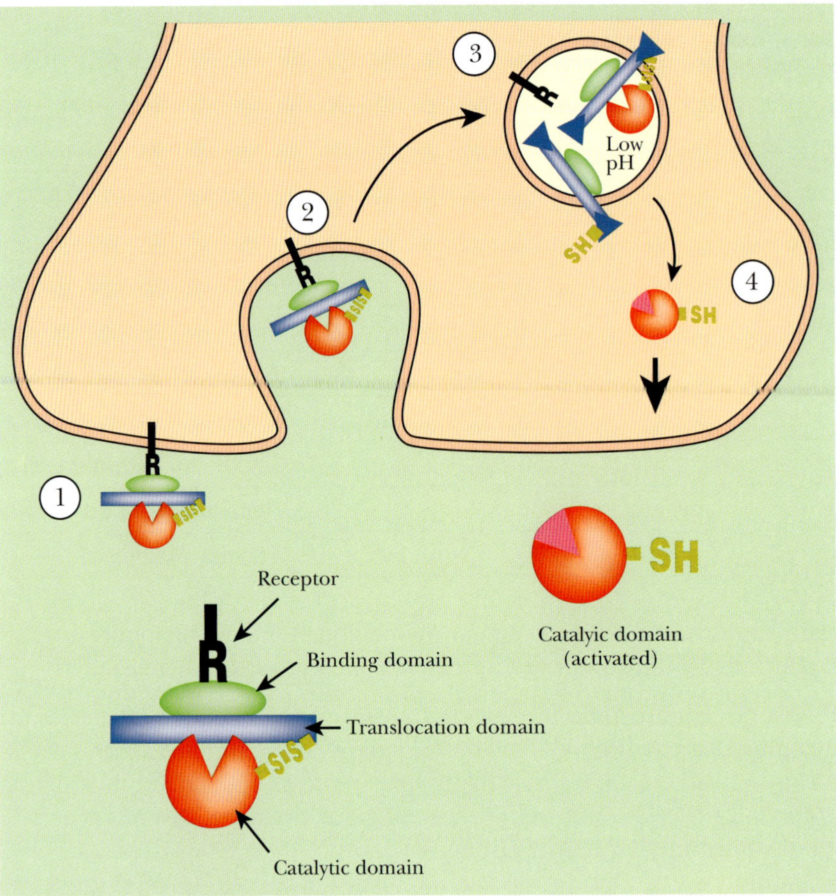

FIGURE 23.7 Uptake of Botulinum Toxin

Current view of botulinum toxin entry at nerve terminals. (1) Toxin binds to the nerve cell membrane at an as-yet-unidentified receptor. (2) Internalization inside endocytotic vesicles, whose interior is acidified by an ATP-driven proton pump. (3) At low pH, the toxin changes conformation and inserts into the vesicle membrane. The light chain (red) is released into the cytosol. (4) In the cytosol the light chain degrades one of the three SNARE proteins.

membrane and is taken up as a whole into vesicles (Fig. 23.7). At this point, the two chains separate and, as is typical for A and B toxins, only the toxic component (light chain) enters the eukaryotic cytoplasm and the binding protein (heavy chain) is left behind in the vesicle.

Botulinum toxin is a Zn-dependent protease that cleaves the SNARE proteins involved in acetylcholine release by the synapse. In actuality, several different variants of the toxin exist, made by different strains of *Clostridium botulinum*. Each toxin variant acts at a different site on one of the three SNARE proteins (SNAP25, synaptobrevin, and syntaxin). **Botulinum toxin type A** is the toxin variant that is used both clinically and for cosmetic treatment. It cleaves SNAP25.

Terrorists of the Japanese cult Aum Shinrikyo have attempted to use botulinum toxin. Aerosols were dispersed at various sites in Tokyo and at U.S. military installations in Japan on several occasions between 1990 and 1995. The attacks failed, mainly because they used strains of *Clostridium botulinum* that failed to produce toxin.

Botulinum toxin inhibits the muscle contraction responsible for wrinkles and frown lines. Treatment with extremely dilute preparations of botulinum toxin A (Botox) is now used to remove wrinkles from the skin of the old and ugly—provided they can afford the expense. Botox injections cost from $300 to $500 and the results last for about 5 months. They are the most rapidly growing form of cosmetic medicine and more than 1.6 million people received injections in 2001. Botox parties cater to this trend and reduce the price by selling in bulk. The patients (or customers?) mingle, chat, and consume refreshments. During the festivities each of them slips away individually for a few minutes for the injections.

Botulinum toxin is the most potent toxin known. It acts by blocking nerve transmission. It is responsible for occasional cases of food poisoning and is now widely used in cosmetics to eliminate wrinkles.

RIBOSOME-INACTIVATING PROTEINS

Many higher plants make **ribosome-inactivating proteins (RIPs)**. These split the N-glycosidic bond between adenine and ribose, releasing adenine from a specific sequence in the large-subunit ribosomal RNA (28S rRNA in eukaryotes, 23S rRNA in prokaryotes). In rat 28S rRNA, the target nucleotide is adenine 4324 (Fig. 23.8). Clipping adenine from the rRNA totally inactivates the ribosome. RIPs are enzymes and can therefore catalyze many successive reactions. Consequently, a single RIP molecule is sufficient to inactivate all the ribosomes and kill a whole cell. The Shiga toxins of *Shigella dysenteriae* (causative agent of bacterial dysentery) and the related toxins made by some pathogenic strains of *E. coli* share the same mechanism.

The target adenosine nucleotide lies within a highly conserved loop close to the 3′ end of rRNA. Removal of the single adenine makes the large subunit of the ribosome incapable of binding elongation factor 2 (EF2). Protein synthesis is therefore prevented (see Chapter 2 for structure and activity of the ribosome). The RIPs attack purified rRNA from both bacteria and higher organisms. However, intact ribosomes from different groups of organisms differ greatly in their sensitivity to RIPs. Mammalian ribosomes are by far the most sensitive. In particular, the activity of many RIPs against the bacterial ribosome is low or negligible, and this has allowed the genes for some RIPs to be cloned and expressed in *E. coli*.

FIGURE 23.8
Site of Action of Plant Ribosome-Inactivating Proteins
The large subunit of the ribosome is a complex of protein and rRNA. RIPs recognize this ribosomal subunit and cleave a specific adenine from the 28S rRNA component. Without the adenine, the ribosome becomes nonfunctional.

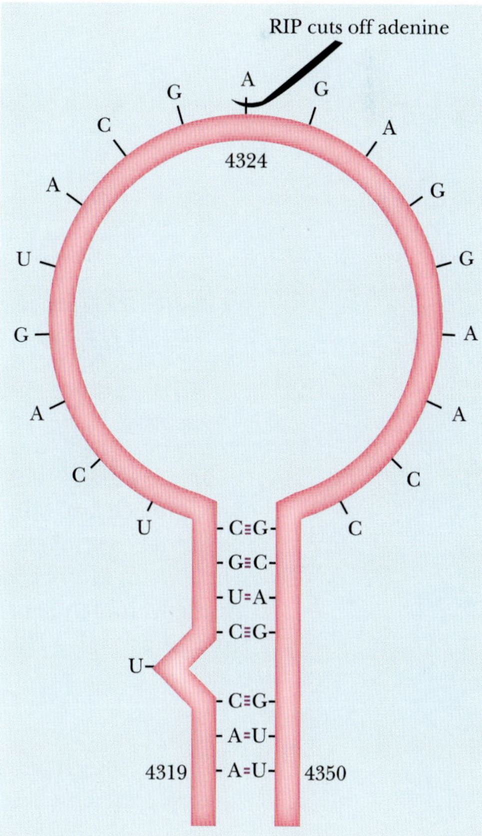

The RIPs are divided into two classes. Type I RIPs consist of a single protein chain and are relatively nontoxic because of their inability to enter mammalian cells effectively. In contrast, type II RIPs consist of two chains, A and B, and are some of the most toxic agents known. Like many bacterial toxins (see Chapter 21) the A chain expresses the toxic enzyme activity and the B chain mediates entry into the target cell. The most famous type II RIP, **ricin**, is lethal at around 3 micrograms/kilogram body weight, meaning that half a milligram should kill a large human. Though less well known, Abrin, from the rosary pea, is 10-fold more toxic. Ricin achieved international notoriety in 1978 when the Bulgarian defector Georgi Markov was assassinated in a London street by ricin. The communist assassin wielded a modified umbrella that injected a hollow 0.6-mm-diameter metal sphere, filled with ricin, into Markov's leg.

The RIPs are synthesized as precursor proteins (pre-RIPs) that are processed only after exiting from the cytoplasm of the plant cell responsible for making them; therefore, the toxin does not kill the plant. Ricin is encoded by a single gene and synthesized as a single precursor polypeptide. This is processed in two stages. After export the signal sequence is cleaved from the N terminus. Next an intervening peptide is cut out to leave two shorter polypeptide chains held together by a disulfide bond (Fig. 23.9). The A chain is very similar to the type I RIPs, and the B chain is a **lectin** that binds to galactose residues in cell surface glycoproteins. (Lectins are plant carbohydrate binding proteins.)

Ricin is extracted from the seeds of the castor bean plant, *Ricinus communis* (Fig. 23.10). This is widely grown, both for ornamentation and on a large scale for castor oil production. Because of its

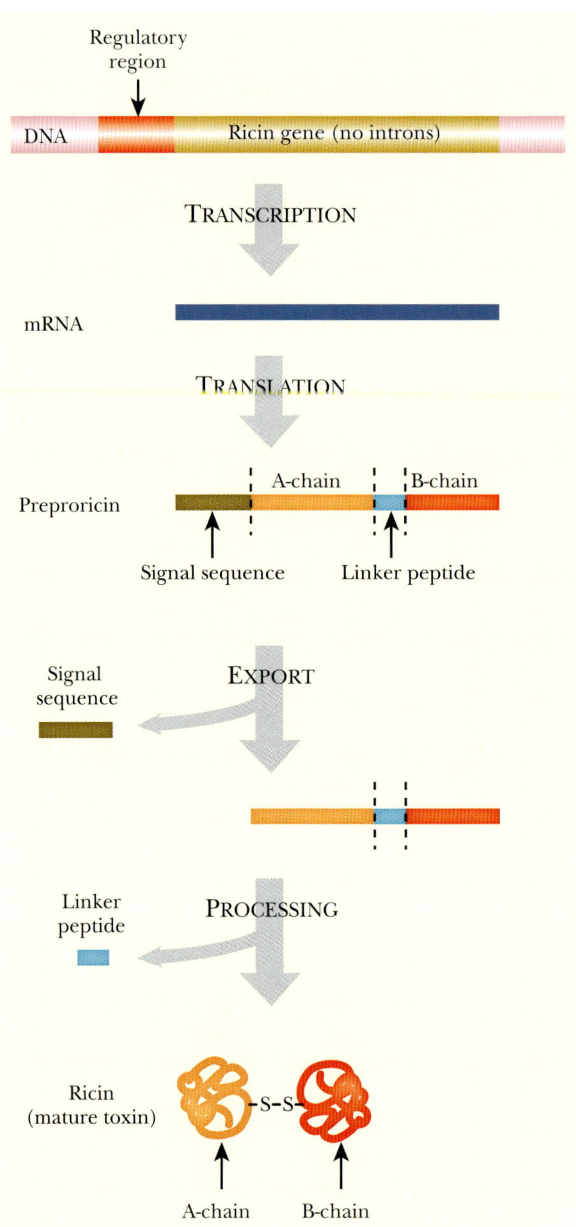

FIGURE 23.9 Processing of Type II RIP
Ricin is a type II RIP that is encoded by a single gene. After transcription and translation, the preproricin is signaled for export from the cell by its N-terminal signal sequence. Removal of the signal sequence gives proricin. After export, the proricin is processed by removing the linker peptide from between the A and B chains. The A and B peptides remain linked via a disulfide to form the mature toxin.

high toxicity, stability, the lack of any antidote, and the ready availability of the raw material, several extremist groups have attempted to purify and use ricin. For example, in 1991 in Minnesota, four members of the Patriots Council, an extremist group with an antigovernment, antitax ideology, purified ricin in a home laboratory. They were arrested for plotting to kill a U.S. marshal with ricin.

> Ribosome-inactivating proteins are made by certain plants. They inactivate the rRNA of target cells and are especially active against mammals. The best known example is ricin from the castor bean.

AGRICULTURAL BIOWARFARE

An alternative form of biological warfare is the destruction of the food supply. Instead of using pathogenic microorganisms that target humans directly, infectious agents that target food animals or crop plants might be used. Because domestic animals also rely on crops to eat and humans can survive on a vegetarian diet, in practice the obvious targets are staple crops such as cereals or potatoes. A wide variety of diseases exist that destroy such crop plants and might be used for biological warfare. Primitive fungi, such as rusts, smuts, and molds, cause many plant diseases. Their spores are often highly infectious, dispersed by wind or rain, and in many cases there is no effective treatment.

FIGURE 23.10 Seeds of the Castor Bean, *Ricinus communis* Courtesy of Dan Nickrent, Department of Plant Biology, Southern Illinois University, Carbondale, IL.

Soybean rust and wheat stem rust are examples of pathogenic fungi that could destroy major crops. In addition to destruction of the crop, certain fungi may produce toxins. For example, when the **ergot** fungus grows on rye or other cereals, it produces a mixture of toxins that cause a syndrome referred to as ergotism, which causes sickness and death in people and livestock (Fig. 23.11). An entire crop might have to be screened even if only a small part was infected, causing major disruption and economic losses. Spores from pathogenic fungi could be sprayed into the air from a single small plane. Another approach would be the deliberate infection of seeds. Many seeds planted in the United States today are now grown in other countries, where they might be accessed for contamination before importing. Modern agriculture is especially vulnerable to biowarfare because large acreages are planted at high density with genetically identical cultivars of major crops. This reduces the genetic variability that makes natural populations relatively resistant to many infections. It also makes it relatively easy for any infection to spread rapidly. Another advantage of agricultural biowarfare is that the agents used are harmless to humans and so would pose no danger to those using them.

FIGURE 23.11 Ergot on Quackgrass Fungi such as *Claviceps purpurea* infect various grains such as wheat or rye as well as grasses such as quackgrass, shown here. The mature fungus forms purple to black bodies, called ergot bodies or sclerotia, where the grain would normally be positioned. Courtesy of David Barker, Department of Horticulture and Crop Science, Ohio State University, Columbus, OH.

The spores of highly infectious fungi could be used as biowarfare agents to target crops rather than humans.

GENETIC ENGINEERING OF INFECTIOUS AGENTS

It is often suggested that genetic engineering could be used to create more dangerous versions of infectious agents. Although there is some truth in this assertion, there is also a great deal of exaggeration. Suppose we take a harmless laboratory bacterium such as *E. coli* and engineer it. The bacteria could be engineered to invade the intestinal wall after being swallowed and to invade other cells. Other genes could be added for ripping vital supplies of iron away from blood cells. And finally, we could add genes for potent toxins that kill human cells. What we would actually achieve is to convert *Escherichia coli* into its near relative, *Yersinia pestis*, which causes the bubonic plague—a disease that already exists and is endemic today in India, China, and California!

The reason we are not all dying of virulent epidemics is not any lack of dangerous pathogens, but the existence of modern hygiene and vaccination programs. Given the right conditions, diseases that already exist are quite capable of decimating human populations. From this perspective, the improvement of diseases by genetic engineering is a relatively minor threat.

As regards actual engineering of human infectious diseases, the Soviet germ warfare facility is known to have modified smallpox virus and generated a variety of artificial mutants and hybrids. The details are largely unavailable. However, certain recent experiments with **mousepox (*Ectromelia* virus)** have given disturbing results. Mousepox is related to smallpox, but it only infects mice. Its virulence varies greatly depending on the strain of mouse. Genetically resistant mice rely on cell-mediated immunity, rather than antibodies. Natural killer cells and cytotoxic T cells both destroy cells infected with mousepox virus, thus clearing the virus from the body.

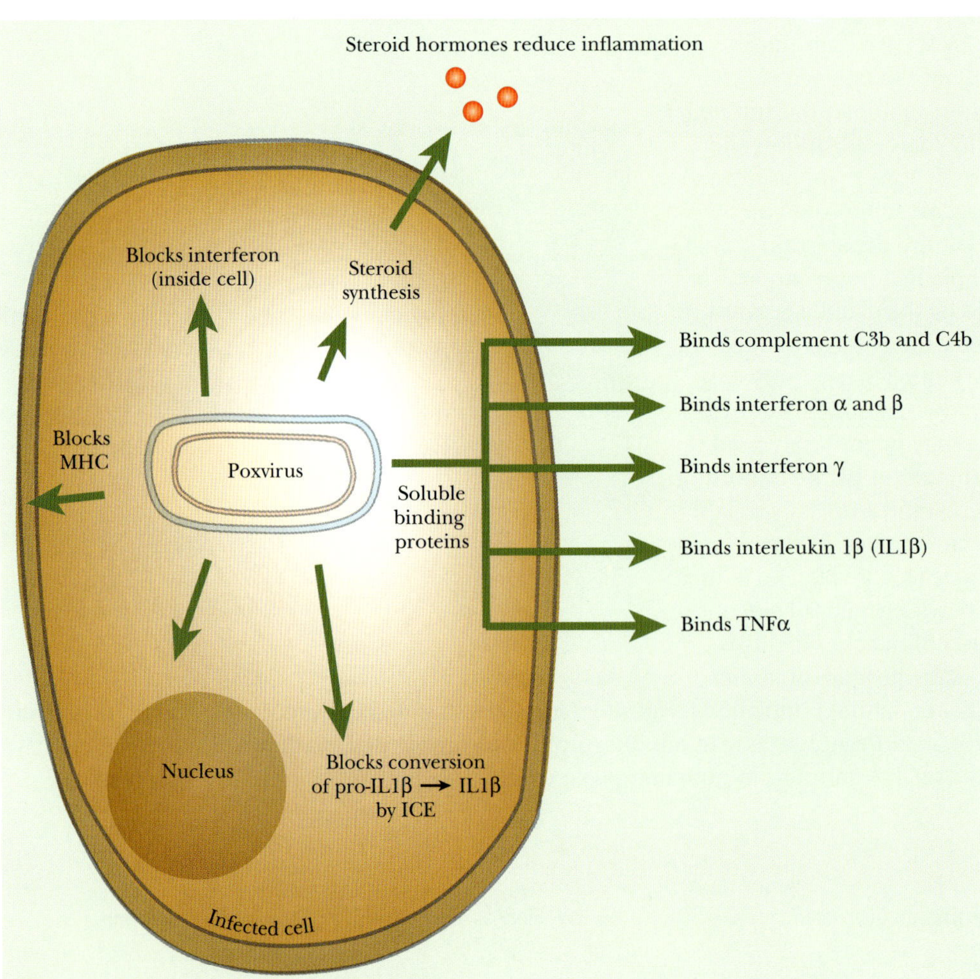

FIGURE 23.12
Poxvirus Immune Evasion
Poxvirus deploys many different proteins to prevent the infected cell from being attacked by the host's immune system.

Researchers modified mousepox virus by inserting the human gene for the cytokine **interleukin 4 (IL-4)**. IL-4 is known to stimulate the division of B cells, which synthesize antibodies. The rationale for engineering the virus was that IL-4 would stimulate the production of antibodies and lead to an improved and more balanced immune response. What actually happened was the opposite of what was expected: the creation of a virus with vastly greater virulence. Not only did it kill all of the genetically resistant mice, but it also killed 50% of mice that had been vaccinated against mousepox. The expression of excess IL-4 suppressed the NK cells and cytotoxic T cells. Furthermore, it failed to increase the antibody response. The reasons are not fully understood at present, but do serve to remind us that the immune system is under extremely complex control.

Similar results have been seen with strains of *Vaccinia* virus, used for vaccination. Whether insertion of IL-4 or other immune regulators into smallpox itself would lead to increased virulence by undermining the immune response is unknown. Poxviruses already carry genes designed to protect the virus by interfering with the action of NK cells and cytotoxic T cells (Fig. 23.12). These are the **cytokine response modifier genes (*crm* genes)**, and they vary in effectiveness between different poxviruses. Thus it is possible that one reason why smallpox is so virulent is that it already subverts the body's cell-mediated immune response. In this case, addition of IL-4 would be expected to give little or no further increase in virulence.

Genetic engineering of certain poxviruses has increased their virulence. Whether this approach would work with smallpox is uncertain.

CREATION OF CAMOUFLAGED VIRUSES

Another worrying achievement of genetic engineering is the ability to hide a potentially dangerous virus inside a harmless bacterium. This is not really novel, because bacteriophages naturally insert their genomes into bacterial chromosomes or plasmids and later reemerge to infect other hosts. Using standard molecular biology, it is possible to clone the entire genome of a small virus that normally infects animals or plants and then insert it into a bacterial plasmid, essentially hiding a pathogenic virus inside harmless bacteria. To accommodate the genomes of larger viruses, bacterial artificial chromosomes may be used instead of plasmids. In the case of RNA viruses, a cDNA copy of the virus genome must first be generated by reverse transcriptase before cloning it into a bacterial vector. There are three main technical issues to face when cloning complete virus genomes:

(a) The fidelity of reverse transcriptase and of the polymerase used during PCR. The enzymes that were originally available introduced too many errors. Nowadays, high-fidelity reverse transcriptase and PCR polymerases are available. Hence the length of RNA or DNA that can be generated error-free has greatly increased.

(b) Suitable vectors to carry large inserts of DNA. Vectors able to carry extremely large inserts, such as bacterial or yeast artificial chromosomes (BACs and YACs), have been developed to clone and sequence large segments of eukaryotic genomes.

(c) Certain base sequences found in virus genomes are not stably maintained or replicated on plasmids in bacterial hosts. These are referred to as **poison sequences**. For example, the cDNA version of yellow fever virus could not be cloned in one piece. Instead it was cloned as two segments that were replicated separately in a bacterial host. To generate a complete, functional cDNA, the two fragments had to be ligated *in vitro*. This problem can sometimes be solved by using a suitable low-copy vector.

Many cell types, both bacterial and eukaryotic, can take up DNA or RNA under certain circumstances by transformation (see Chapter 3). Consequently the naked nucleic acid genomes of many viruses (both DNA and RNA) are infectious even in the absence of their protein capsids. Once a virus genome is cloned, DNA molecules containing the virus genome can be generated by replication of the plasmid inside the bacterial host cell. Although such

DNA contains extra plasmid sequences, it still may be infectious if transformed into the appropriate host cell.

Amazingly, this is sometimes even true for RNA viruses—that is, the cDNA version of an RNA virus can successfully infect host cells and give rise to a new crop of RNA-containing virus particles. For this to occur, the cDNA must enter the nucleus of the host cell and be transcribed to give an RNA copy of the virus. The viral RNA then proceeds through its normal replication cycle. This has been demonstrated for RNA viruses such as poliovirus, influenza, and coronavirus (one of the causative agents of common colds).

An improved strategy for generating RNA viruses is to clone the cDNA version of their genomes onto a bacterial plasmid downstream of a strong promoter (Fig. 23.13). In this case, the natural RNA version of the virus genome will be generated by transcription. This may be done inside the bacterial host cell by using a bacterial promoter. Alternatively, a eukaryotic promoter may be used to improve transcription of cDNA into viral RNA once the cDNA has entered the eukaryotic host cell. The technology thus exists to create bacteria carrying "hidden" plasmid-borne animal viruses. If the complete cDNA from an RNA virus is placed under control of a strong bacterial promoter, the bacterial cell could generate large amounts of infectious viral RNA internally by transcription. When the bacterial cell dies and disintegrates, the viral RNA would be liberated. If a dangerous human RNA virus was loaded into a harmless intestinal bacterium under control of a promoter designed to respond to conditions inside the intestine, this could pose a serious threat.

Certain RNA viruses can be converted to DNA forms and inserted into bacterial cells. Infectious viral particles or RNA may then be generated by several possible approaches.

BIOSENSORS AND DETECTION OF BIOWARFARE AGENTS

Biosensors are devices for detection and measurement of reactions that rely on a biological mechanism, often an enzyme reaction adapted to generate an electrical signal. Biosensors have been traditionally used in clinical diagnosis and in food and environmental analysis. By far the biggest use has been the clinical monitoring of glucose levels using the enzyme glucose oxidase. In particular, knowing the concentration of glucose is critical to proper care of diabetics. Glucose oxidase is unusually stable, a major reason for its widespread use. The enzyme catalyzes the following reaction:

$$\text{glucose} + O_2 \rightarrow \delta\text{-gluconolactone} + H_2O_2$$

The glucose biosensor consists of a thin layer of glucose oxidase attached to the bottom of an oxygen electrode (Fig. 23.14). The electrode detects oxygen released by the enzyme reaction. The current generated provides a measure of the glucose concentration. A potential of about 0.6 volts is applied between the central positive platinum electrode and the surrounding negative silver/silver chloride electrode. The electrolyte solution is saturated potassium chloride. The negative electrode (cathode) is covered by a thin Teflon membrane, which allows oxygen to diffuse through but keeps out other molecules that might react. The electrode reactions are:

Platinum cathode (electrons consumed)	$O_2 + 4H^+ + 4\,e^- \rightarrow 2H_2O$
Ag/AgCl anode (electrons released)	$4Ag + 4Cl^- \rightarrow 4AgCl + 4e^-$

There is growing interest today in using biosensors to detect toxins, viruses, and perhaps other possible biowarfare agents. In particular a handheld device giving a rapid response would be highly useful. Several proposals exist that would use specific antibodies or antibody fragments as detectors for biowarfare agents (see Chapter 6 for antibody engineering).

Each B cell carries antibodies specific for one antigen and one proposal is to use whole B cells in the biosensor. When an antigen binds to the antibody on the surface of a B cell,

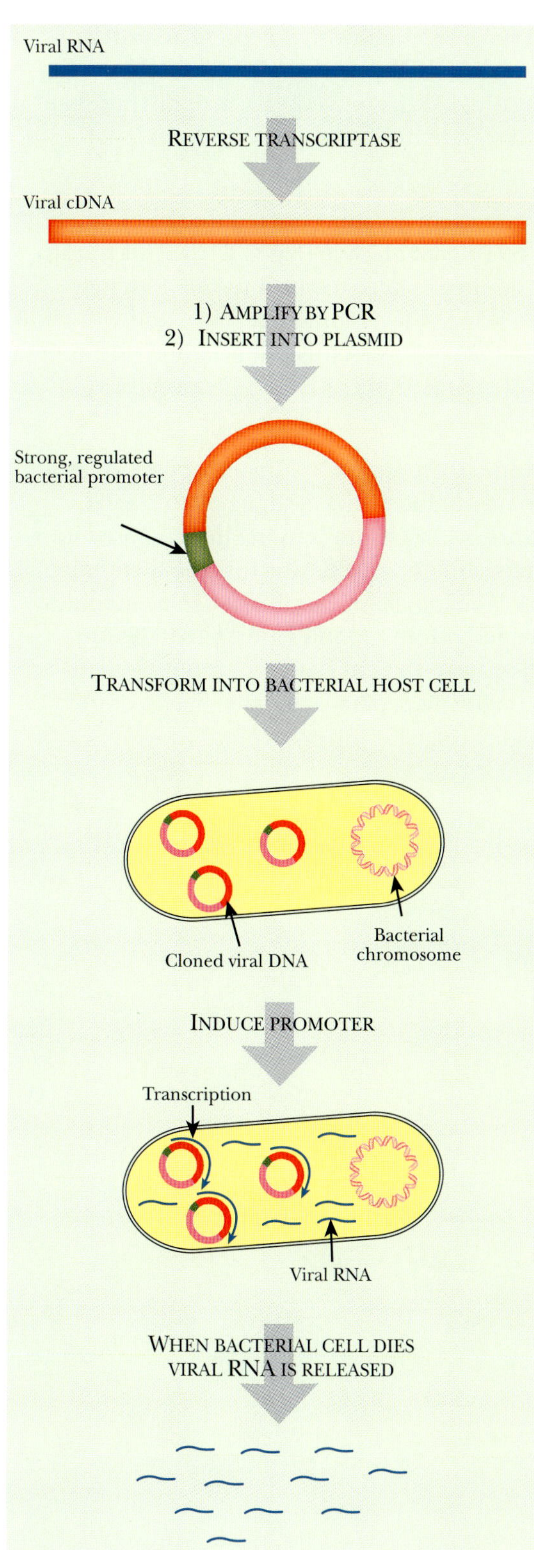

FIGURE 23.13 Expression of Cloned RNA Virus
Cloning an RNA virus requires making a double-stranded DNA copy using reverse transcriptase. The cDNA is inserted into an appropriate bacterial plasmid and transformed into the bacterial cells. To control the expression of the viral DNA, a strong promoter is placed upstream of the viral cDNA. If the promoter is inducible, when the bacteria are given the appropriate stimulus, the viral cDNA will be expressed, resulting in production of viral particles that could infect many people.

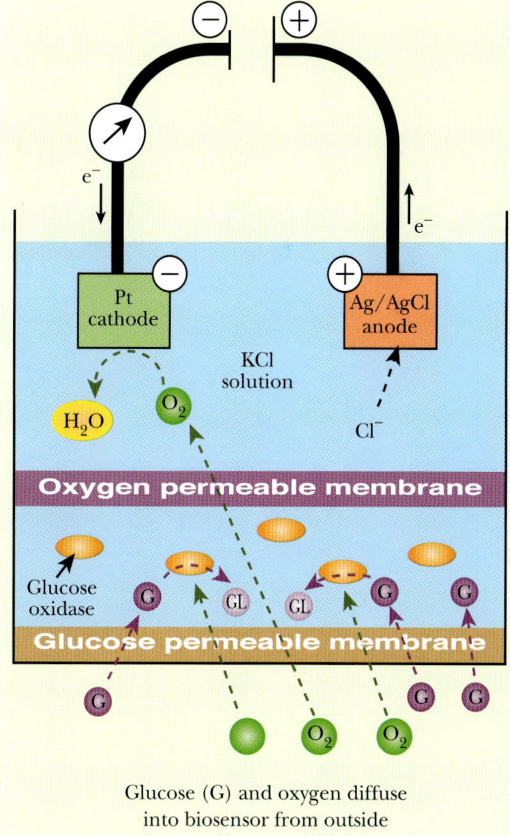

FIGURE 23.14 Glucose Oxidase Biosensor
This simple biosensor measures the concentration of glucose in a sample (*bottom of figure*). The glucose and oxygen molecules diffuse through the membrane. Glucose (G) enters only the bottom chamber, but oxygen can enter either chamber. Glucose oxidase converts glucose plus oxygen to δ-gluconolactone (GL) and hydrogen peroxide. This lowers the oxygen concentration reaching the cathode and hence causes a change in the voltage differential between the cathode and anode.

it triggers a signal cascade. Engineered B cells have been made that express **aequorin**, a light-emitting protein from the luminescent jellyfish *Aequorea victoria*. Aequorin emits blue light when triggered by calcium ions (Fig. 23.15). Living jellyfish actually produce flashes of blue light, which are transduced to green by the famous green fluorescent protein (GFP).

In the biosensor, when a B cell detected a disease agent (or any targeted microbe), the cell would release calcium ions due to activation of a signal cascade (Fig. 23.16). This in turn triggers light emission by aequorin. The light emitted is detected by a sensitive charge-coupled device (CCD) detector. This approach is in its developmental stages and will allow detection of 5 to 10 particles of a pathogenic agent such as a virus or bacterium. Multiple patches of around 10,000 B cells specific to different pathogens would be assembled in array fashion onto the same chip.

Another scheme, under development by the Ambri Corporation of Australia, uses antibody fragments mounted on an artificial biological membrane, which is attached to a solid support covered by a gold electrode layer. Channels for sodium ions are incorporated into the membrane. When the ion channels are open, sodium ions flow across the membrane and a current is generated in the gold electrode. The ion channels consist of two modules, each spanning half the membrane. When top and bottom modules are united, the ion channel is open. When the top module is pulled away, the ion channel cannot operate. Binding of biowarfare agents by the antibody fragments separates the two halves of the channels, which in turn affects the electrical signal (Fig. 23.17).

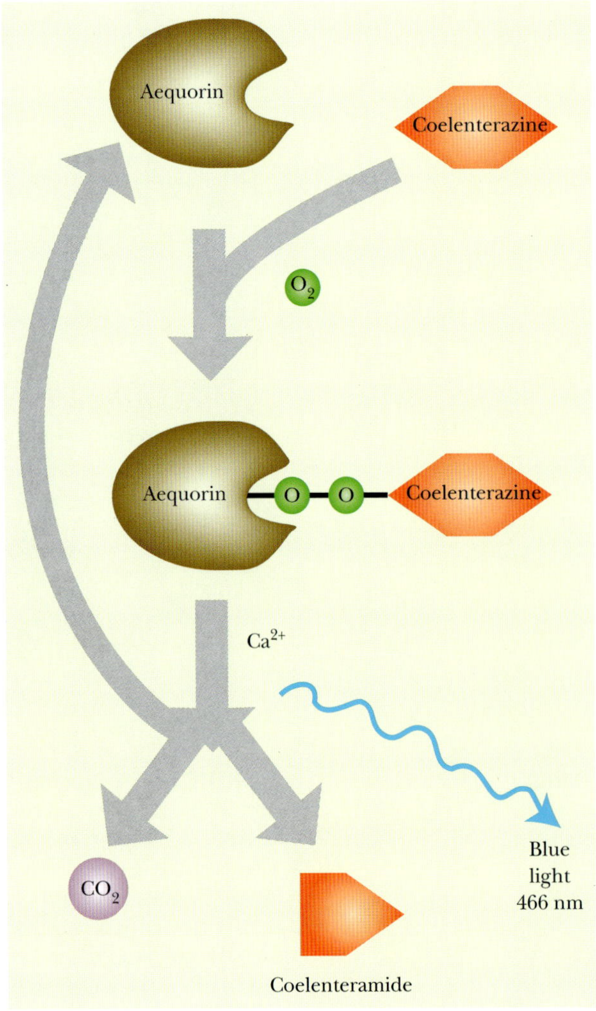

FIGURE 23.15
Light Emission by Aequorin
Aequorin, from *Aequorea victoria*, emits blue light when provided with its substrate, coelenterazine, plus oxygen and calcium. The enzyme binds to aequorin via the oxygen, and when calcium is present, the complex emits blue light, degrades the substrate to coelenteramide, and releases carbon dioxide.

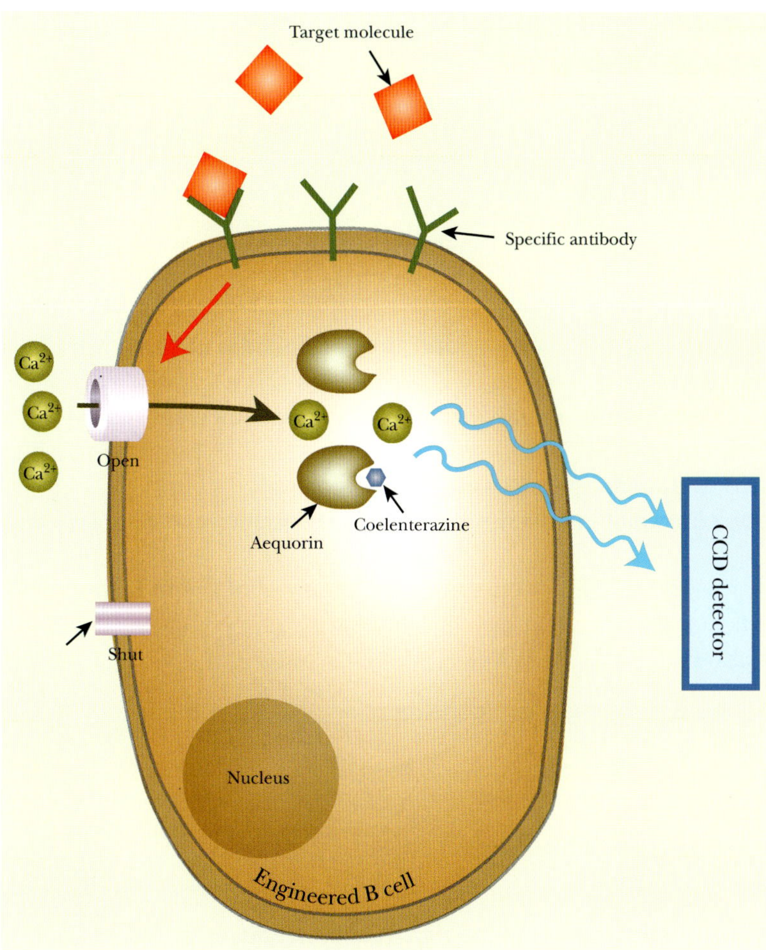

FIGURE 23.16 B-Cell Optical Biosensor
Expressing aequorin in a B cell would provide a detection system for B-cell activation. When a trigger molecule, such as a biowarfare agent, binds to receptors on the B cell, the calcium channels are opened and calcium floods the cell. The high calcium levels would activate aequorin to emit blue light. A CCD detector would measure the emissions and warn the user of a biological agent.

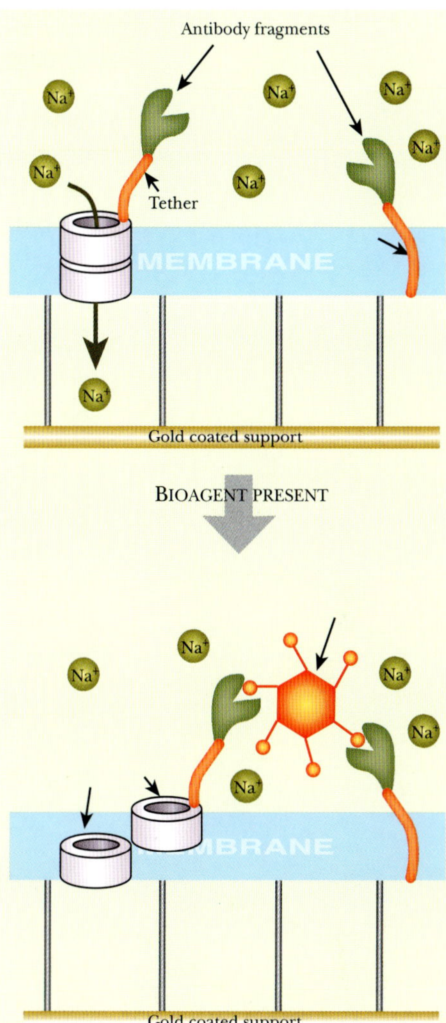

FIGURE 23.17 Antibody Ion-Channel Biosensor
Antibody fragments that bind specific biological weapons can be engineered and tethered to a fixed location on an artificial membrane. Another molecule of the same antibody fragment is tethered to a sodium channel. The artificial membrane is carried on a gold-coated solid support that acts as an electrode. This detects sodium ions that pass through the ion channel. When a biological agent is present, the antibody fragments bind it, pulling the top half of the sodium channel out of alignment with the bottom half. Sodium ions no longer pass to the gold electrode, decreasing the signal.

Summary

Natural competition has generated a vast range of both aggressive and defensive strategies among living organisms. Humans have often attempted to use biological agents in warfare, although with little overall success so far. Several highly virulent infectious agents including anthrax, plague, and smallpox as well as certain biological toxins are regarded as likely biowarfare agents. Whether or not genetic engineering can create "improved" bioweapons is as yet uncertain.

End-of-Chapter Questions

1. What can bacterial toxins kill?
   a. insect cells
   b. human cells
   c. other bacterial cells
   d. protozoa
   e. all of the above
2. Even though biological warfare has been around for hundreds of years, why was it still ineffective?
   a. because there were already massive amounts of infectious diseases circulating
   b. because there was more hygiene hundreds of years ago
   c. because people had already been exposed to the agent and developed immunity
   d. biological warfare hundreds of years ago was largely effective at killing mass quantities of people
   e. none of the above
3. Which of the following is an important consideration of germ warfare?
   a. dispersal
   b. persistence of the agent
   c. incubation time
   d. storage and preparation of the agent
   e. all of the above
4. According to the U.S. Army, which of the following is a requirement for biological weapons?
   a. It should be able to be produced economically.
   b. It should consistently produce death, disability, or damage.
   c. It should be stable from production through delivery.
   d. It should be easy to disseminate quickly and effectively.
   e. All of the above are requirements for biological agents.
5. Which one of the following has rarely been considered as a possible biological warfare agent?
   a. viruses
   b. bacteria
   c. pathogenic eukaryotes
   d. pathogenic fungi
   e. none of the above
6. According to the text, which of the following is one of the best biological weapons?
   a. anthrax
   b. malaria
   c. amoebic dysentery
   d. smallpox
   e. none of the above

7. Since *B. anthracis* strains are closely related, how is it possible to differentiate between the strains?
- **a.** 16S rRNA sequencing
- **b.** VNTRs
- **c.** 23s rRNA sequencing
- **d.** gene expression profiles
- **e.** none of the above

8. How does bubonic plague spread?
- **a.** ticks
- **b.** person-to-person
- **c.** fleas
- **d.** rodents
- **e.** birds

9. Which of the following is used as a live vaccine for smallpox?
- **a.** *Variola* major
- **b.** monkeypox
- **c.** *Variola* minor
- **d.** *Vaccinia* virus
- **e.** none of the above

10. To what virus family do dengue fever and yellow fever belong?
- **a.** Flaviviruses
- **b.** Poxviruses
- **c.** Filoviruses
- **d.** Variola viruses
- **e.** Arenaviruses

11. Which of the following is the most toxic substance of all?
- **a.** ricin
- **b.** mycotoxin T-2
- **c.** enterotoxin B
- **d.** botulinum toxin
- **e.** aconitine

12. What is the mode of action for ricin?
- **a.** inactivation of transcription
- **b.** inactivation of rRNA
- **c.** activation of the apoptosis pathway
- **d.** inactivation of the immune system
- **e.** creation of pores in cell walls

13. Which of the following could be used as a biological agent against agriculture crops?
- **a.** viruses
- **b.** bacteria
- **c.** pathogenic fungi spores
- **d.** prions
- **e.** none of the above

(Continued)

14. Which of the following gained virulence upon introduction of the IL-4 gene?
    a. monkeypox
    b. smallpox
    c. chickenpox
    d. mousepox
    e. camelpox

15. How are pathogens detected by using biosensors?
    a. by antibodies that are connected to components to give electrical signals or trigger light emission
    b. by isolating the pathogen directly from the sample
    c. biosensors detect antibodies against specific pathogens, similar to a Western blot
    d. by using PCR to amplify variable regions of the pathogen's genome
    e. none of the above

Further Reading

Domaradskij IV, Orent LW (2006). Achievements of the Soviet biological weapons programme and implications for the future. *Rev Sci Tech* **25**, 153–161.

Fang Y, Rowland RR, Roof M, Lunney JK, Christopher-Hennings J, Nelson EA (2006). A full-length cDNA infectious clone of North American type 1 porcine reproductive and respiratory syndrome virus: Expression of green fluorescent protein in the Nsp2 region. *J Virol* **80**, 11447–11455.

Fokin SI (2004). Bacterial endocytobionts of ciliophora and their interactions with the host cell. *Int Rev Cytol* **236**, 181–249.

Halverson KM, Panchal RG, Nguyen TL, Gussio R, Little SF, Misakian M, Bavari S, Kasianowicz JJ (2005). Anthrax biosensor, protective antigen ion channel asymmetric blockade. *J Biol Chem* **280**, 34056–34062.

Jeblick J, Kusch J (2005). Sequence, transcription activity, and evolutionary origin of the R-body coding plasmid pKAP298 from the intracellular parasitic bacterium *Caedibacter taeniospiralis*. *J Mol Evol* **60**, 164–173.

Li Y, Sherer K, Cui X, Eichacker PQ (2007). New insights into the pathogenesis and treatment of anthrax toxin-induced shock. *Expert Opin Biol Ther* **7**, 843–854.

Neumann G, Kawaoka Y (2004). Reverse genetics systems for the generation of segmented negative-sense RNA viruses entirely from cloned cDNA. *Curr Top Microbiol Immunol* **283**, 43–60.

Osborne SL, Latham CF, Wen PJ, Cavaignac S, Fanning J, Foran PG, Meunier FA (2007). The Janus faces of botulinum neurotoxin: Sensational medicine and deadly biological weapon. *J Neurosci Res* **85**, 1149–1158.

Paterson RR (2006). Fungi and fungal toxins as weapons. *Mycol Res* **110**, 1003–1010.

Pinton P, Rimessi A, Romagnoli A, Prandini A, Rizzuto R (2007). Biosensors for the detection of calcium and pH. *Methods Cell Biol* **80**, 297–325.

Prentice MB, Rahalison L (2007). Plague. *Lancet* **369**, 1196–1207.

Puissant B, Combadière B (2006). Keeping the memory of smallpox virus. *Cell Mol Life Sci* **63**, 2249–2259.

Sliva K, Schnierle B (2007). From actually toxic to highly specific—novel drugs against poxviruses. *Virol J* **4**, 8.

Stirpe F (2004). Ribosome-inactivating proteins. *Toxicon* **44**, 371–383.

Trull MC, du Laney TV, Dibner MD (2007). Turning biodefense dollars into products. *Nat Biotechnol* **25**, 179–184.

van Belkum A (2007). Tracing isolates of bacterial species by multilocus variable number of tandem repeat analysis (MLVA). *FEMS Immunol Med Microbio* **49**, 22–27.

Forensic Molecular Biology

The Genetic Basis of Identity

Blood, Sweat, and Tears

Forensic DNA Testing

DNA Fingerprinting

Using Repeated Sequences in Fingerprinting

Using the Polymerase Chain Reaction (PCR)

Probability and DNA Testing

The Use of DNA Evidence

Tracing Genealogies by Mitochondrial DNA and the Y Chromosome

Identifying the Remains of the Russian Imperial Family

THE GENETIC BASIS OF IDENTITY

DNA technology has many practical uses. Because every individual has a unique DNA sequence, DNA samples can be used for identification. The legal system is now using DNA evidence to determine guilt or innocence. The application of DNA technology began in Britain in the mid-1980s and appeared in America shortly afterward. Today many societies have reached the point of compiling DNA databases of known criminals—especially serious offenders. However, the most frequent use of DNA evidence is actually in cases of unknown or disputed paternity.

Identity can also be established in other ways. Just a casual glance reveals major differences among people. Geneticists refer to this outward appearance as the *phenotype*. Most physical differences between people are due to complex interactions of several genes during development. Some are obvious at a glance; others require close observation. Fingerprints are the classic example of a phenotype used in law enforcement. They are due to variations in the pattern of dermal ridges, small skin elevations on our fingers. Fingerprint patterns depend on more than one gene (i.e., they are **multigenic**). This creates the huge genetic diversity underlying this phenotype. Although you might expect the fingerprints of identical twins to be the same, they are not identical. Minor variations in fingerprint patterns occur as a result of environmental factors affecting development. Fingerprints were being used for identification by the late 1800s.

Retinal scans provide a more high-tech form of unique identification. These take advantage of the unique pattern of blood vessels on the retina at the back of the eyes. Scanning typically takes about a minute, because several scans are needed. The subject must place the eye close to the scanner, keep the head still, and focus on a rotating green light. Infrared light is used for scanning because blood vessels on the retina absorb this better than the surrounding tissues. A computer algorithm is then used to convert the scan into digital data. There is about 10-fold more information in a retinal scan than in a fingerprint.

Previously, retinal scanning was mostly used in high-security situations—for example, by the CIA, the FBI, NSA, and some prison systems. However, it is now being used for animal identification (Fig. 24.1). The Optibrand corporation now makes a handheld scanner for livestock, especially cattle. The setup incorporates a Global Positioning System time, date, and location stamp. The Optibrand system can identify and track individual animals through the food processing chain. This is of special relevance in tracing any animals suspected of being exposed to mad cow disease.

Moving to the molecular level, another set of identifying features are the proteins and polysaccharides made by all cells. Good examples of individual differences that involve proteins are the various blood types found in human populations. But for ultimate identification at the molecular level, we must examine the genes themselves to determine the genotype. This is what is meant by DNA typing or **DNA fingerprinting**, a technique that is described later.

> Fingerprints and retinal scanning use highly detailed biological data for identification of individual humans or animals.

BLOOD, SWEAT, AND TEARS

All kinds of body tissues and fluids may be used to establish identity. Although DNA technology is relatively new, it is a logical outgrowth of the work on **blood typing** that has been used in the courtroom for more than 50 years. Although blood analysis is most common, other body fluids such as sweat, tears, urine, saliva, and semen also have cells with surface proteins that can be analyzed.

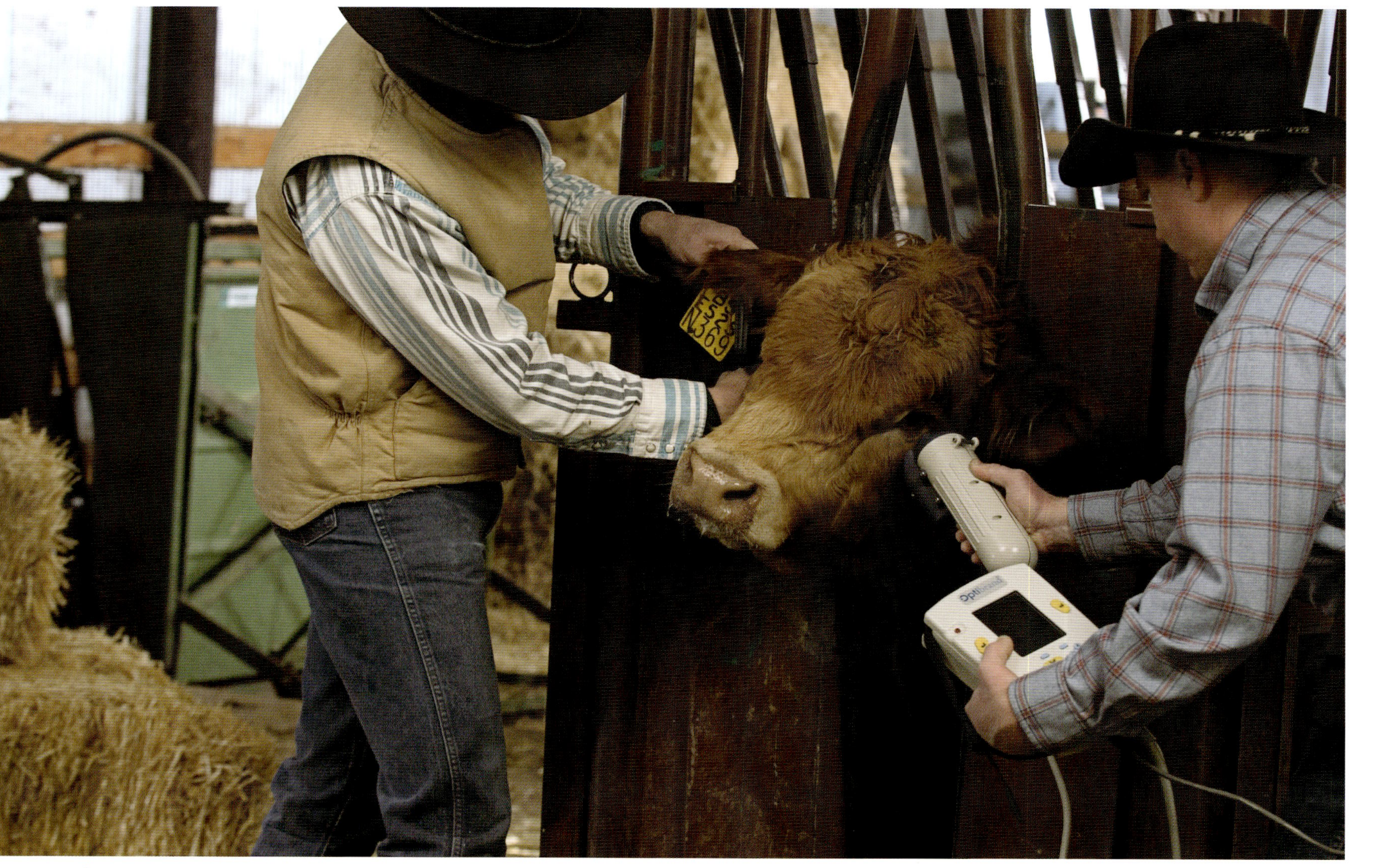

FIGURE 24.1 Optibrand Retinal Scanning of Cattle
Retinal scanning is now being used for cattle identification. Courtesy of Optibrand Ltd. LLC, Fort Collins, Colorado.

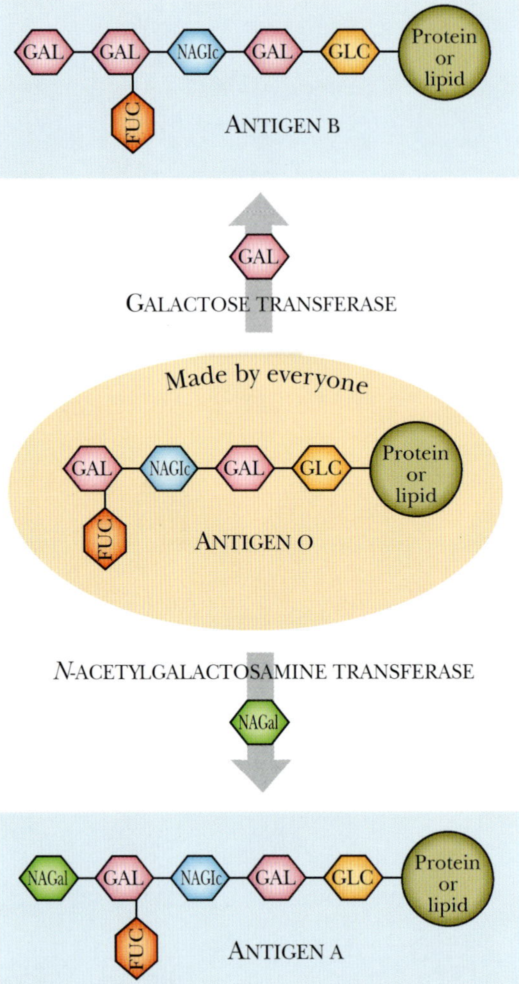

FIGURE 24.2
ABO Blood System Molecules
Antigenic glycolipids determine each individual's blood type. The enzyme *N*-acetylgalactosamine transferase makes A antigen by adding *N*-acetylgalactosamine to the end of the O antigen. Galactose transferase makes B antigen by adding galactose to the end of the O antigen. Abbreviations: NAGal, *N*-acetylgalactosamine; GAL, galactose; NAGlc, *N*-acetylglucosamine; FUC, fucose; GLC, glucose.

The membranes of red blood cells contain several proteins and lipids with attached carbohydrate portions that are exposed on the outside of the cell. These are highly antigenic, and in blood typing, they are referred to as **blood antigens**. Binding of an antibody to the corresponding antigen is highly specific. Consequently, two related blood antigens with only relatively small shape differences can be told apart because different antibodies will bind them.

Several groups of blood antigens are routinely used in identification. The best known is the **ABO blood group** system. Three different glycolipids, A, B, and O, are involved. These consist of different carbohydrate structures attached to the same lipid. The A antigen is made by adding *N*-acetylgalactosamine to the end of the O antigen, and the B antigen by adding galactose (Fig. 24.2). Antibodies are made against the A and B antigens, but the shorter O "antigen" is poorly antigenic and invokes little antibody production. The closely related enzymes that make the A and B antigens are encoded by different alleles of the same gene. Absence of this enzyme gives the O allele.

Thus there are three alleles—A, B, and O—present in the population. Because we all have two copies of each gene, we all have two alleles for the ABO system. These may be identical or different in any given person. The alleles for the A and B antigens are both dominant; therefore, if at least one allele for either A or B is present, that antigen will be expressed (Table 24.1). A person with one A and one B allele will express both antigens on the surfaces of his or her blood cells (AB blood group).

People do not make antibodies against those antigens present on their own red blood cells. Each individual makes antibodies against foreign antigens, that is, those present only in other individuals. Consequently, people with type A blood will express antibodies against

Table 24.1 ABO Blood System

Alleles Present		Blood Type Expressed	Antigens Made	Antibodies
A	A	A	A	anti-B
A	O	A	A	anti-B
B	B	B	B	anti-A
B	O	B	B	anti-A
A	B	AB	A and B	neither
O	O	O	neither	anti-A and anti-B

the B antigen, whereas those with type B blood will express antibodies to the A antigen. People with type O blood will express antibodies to both A and B antigens and those with type AB have neither antibody (see Table 24.1). If a person is given blood of the wrong type, the person's own antibodies cause the foreign blood cells to clump together or agglutinate. People with type AB blood are considered universal acceptors, because they will not react against any type of blood. Conversely, type O people are universal donors but can only accept type O themselves, because they have antibodies to both the A and B antigens.

Because humans are diploid, a child may belong to a different blood type than either of its parents. For example, an AO mother will make A antigen and a BO father will make B antigen. Despite this, they can have a type O child, because the child may inherit the O gene on one chromosome from the heterozygous mother and another O gene from the heterozygous father (Fig. 24.3).

FIGURE 24.3
Inheritance of ABO Blood Type
If parents with two different blood types have children, the offspring may differ in blood type from either parent. In this example, parent 1 has the alleles for type O and A antigens. Parent 2 has the alleles for type B and O antigens. Mendelian analysis shows that the child could have type AB, type B, type A, or type O—that is, all the possibilities.

Parent 1 AO

gametes	A	O
B	AB	BO
O	AO	OO

Parent 2 BO

Usually we know who the mother of a child is. Mostly it is the father who may be difficult to identify. If you are accused of being a father in a paternity suit, ABO typing will only exclude you in around 15% to 20% of cases, assuming you are innocent. Consider, for example, a mother who has blood type A with a daughter of type AB. We know the father must have contributed the B allele to the daughter. Therefore the father must be BO, BB, or AB but cannot be AA, AO, or OO. Therefore, anyone accused of paternity who has type A or type O blood is innocent. Individuals with type B or AB blood might be guilty. Because large numbers of people share each blood group, ABO blood typing alone cannot provide proof of guilt, although it can prove innocence.

There are several other blood and tissue antigen systems similar in principle to the ABO system that are also used in forensic medicine. Using the HLA system of white blood cells, the chance of exclusion is over 90%. When the HLA and ABO systems are combined, the chances of exclusion are about 97%. Including the analysis of blood serum proteins with the others makes exclusion almost certain, if the accused is indeed innocent.

The following formula is used to determine the combined probability (P) of exclusion from multiple tests with individual probabilities P1, P2, P3, and so on:

$$P = 1 - (1-P1)\ (1-P2)\ (1-P3)\ (1-P4) \text{ etc.}$$

Blood typing relies on identifying antigens present on the surface of blood cells. The widely used ABO system consists of three different alleles distributed among the human population.

FORENSIC DNA TESTING

In many criminal cases, blood typing is the primary evidence. Juries have sometimes convicted suspects on ABO typing combined with other blood antigens giving overall probabilities as low as 25% to 50% that the suspect and the blood evidence matched. DNA evidence can do much better and provide probabilities that are almost 100%.

No two individuals have the same DNA. During gamete development and fertilization, sets of individual chromosomes are distributed to offspring in so many possible combinations that it is incredibly unlikely that any two individuals will have the same DNA. Identical twins are the exception that proves the rule, because they occur when the egg divides after fertilization has already happened.

DNA tests alone, without supportive evidence, have sometimes been sufficient for conviction, where identity was the key issue. DNA evidence is now almost always sufficient for exoneration of misidentified individuals who were wrongly convicted of committing a crime. DNA evidence can be obtained from any body tissue or secretion that has cell nuclei that contain DNA. There are two major types of testing used to determine whether DNA found at the scene of a crime matches that of the suspect or the victim. One is popularly known as *DNA fingerprinting*, and the other is *polymerase chain reaction (PCR) amplification* followed by hybridization or sequencing. Tissue samples taken from suspects may be compared with evidence obtained from a crime scene.

DNA evidence can provide virtually unambiguous identification of individuals if carried out properly.

DNA FINGERPRINTING

DNA fingerprinting relies on the unique pattern made by a series of DNA fragments after separating them according to length by gel electrophoresis. DNA samples from different suspects, the victim, and samples from the crime scene are first purified. Restriction enzymes cut the DNA samples into fragments of different lengths. Consequently, the variation in the size of the fragments and hence of their positions on an agarose gel is due to differences in where cutting occurs. Thus differences in patterns between individuals are due to differences in the base sequence of their DNA. The nucleotide differences that cause the fragment lengths to vary are called *restriction fragment length polymorphisms* (RFLPs; see Chapter 3). There is believed to be approximately one difference in every 1000 nucleotides between nonrelated individuals.

The steps involved in DNA fingerprinting are as follows (Fig. 24.4):

1. The DNA is cut with a restriction enzyme.
2. The DNA fragments are separated according their length or molecular weight by gel electrophoresis.
3. The fragments are visualized by Southern blotting. After transfer of the separated fragments from the gel to nylon paper a radioactively labeled DNA probe is added. The probe will bind to those DNA fragments whose DNA sequences are complementary to the probe.
4. An autoradiograph is made by covering the blot with radiation-sensitive film. This will show the location of those DNA fragments that reacted with the radioactive probe.

There are many different restriction enzymes, most with unique cutting properties. In practice several different enzymes are used with the same DNA samples, giving different sets of fragments for different people. These can be compared with DNA samples taken from a victim or found at a crime scene. Because there is so much genetic diversity, RFLP patterns from different people can vary a lot. Even if mutations have changed a small percentage of the target sequence around the cut site, there will usually still be enough similarity for binding of the probe to occur. The entire process requires several weeks to finish.

The final product of a DNA fingerprint is an autoradiograph that contains at least five essential lanes (Fig. 24.5). The markers are standardized DNA fragments of known size, which have been radioactively labeled. These help determine the size of the various fragments. The "control" is DNA from a source known to react positively and reliably to the DNA probes and shows whether the test has worked as expected. The experimental lanes have samples from the victim, the defendant, and the crime scene. In this example, blood from the defendant's clothing was compared with his/her own blood and the victim's blood. The DNA from the clothing actually matches that of the victim.

Two variants of DNA fingerprinting have been used—**single-locus probing (SLP)** and **multiple-locus probing (MLP)**. In SLP, a probe is used that is specific for a single site, that is, a single locus, in the genomic DNA. Because humans are diploid, an SLP probe will therefore

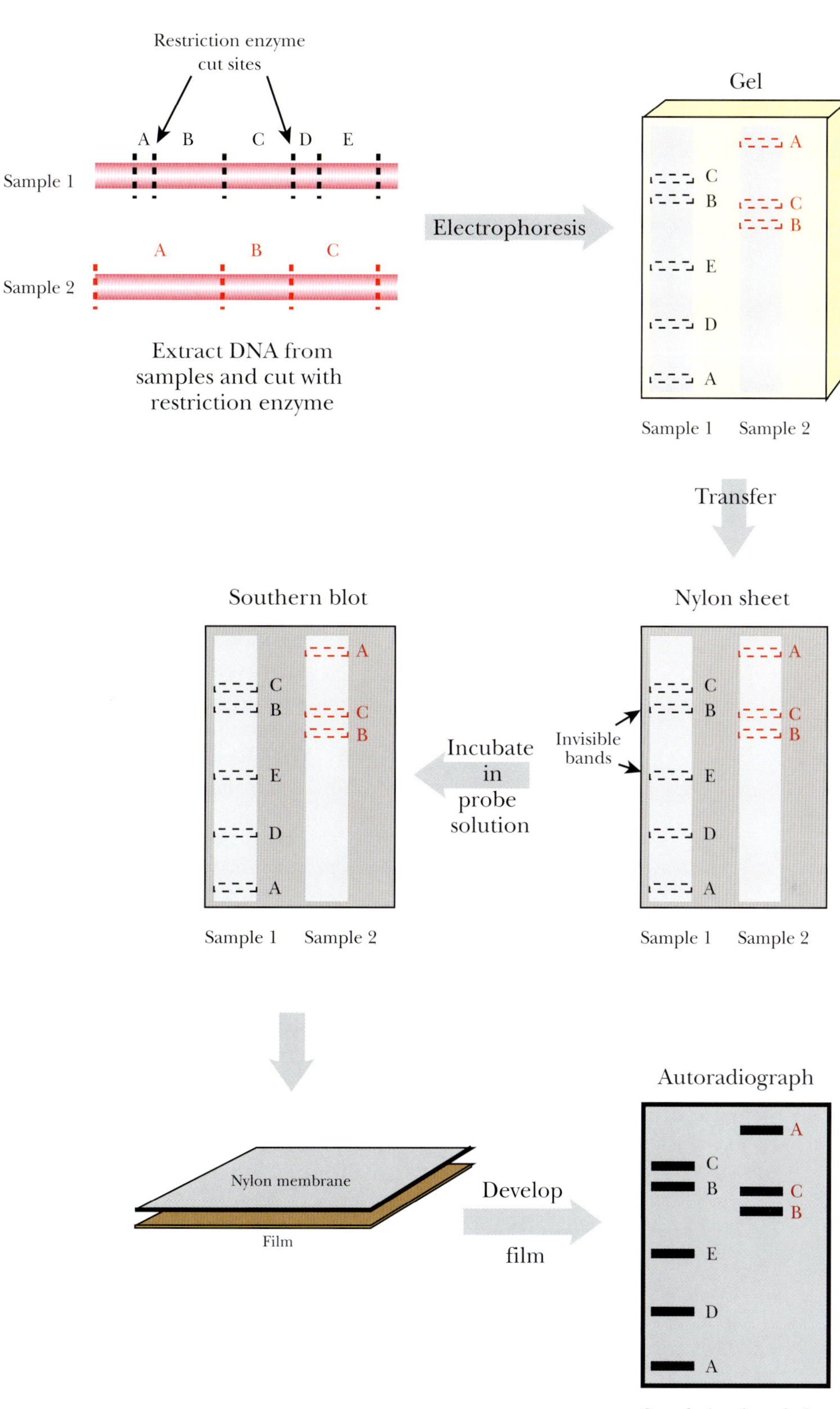

FIGURE 24.4 Outline of DNA Fingerprinting Procedure
A pattern of different-sized DNA fragments is generated during DNA fingerprinting. The sequence of the DNA determines where the restriction enzyme cuts in the first step; thus different people will have a different pattern of fragments for the same restriction enzyme. These differences may be used to identify people.

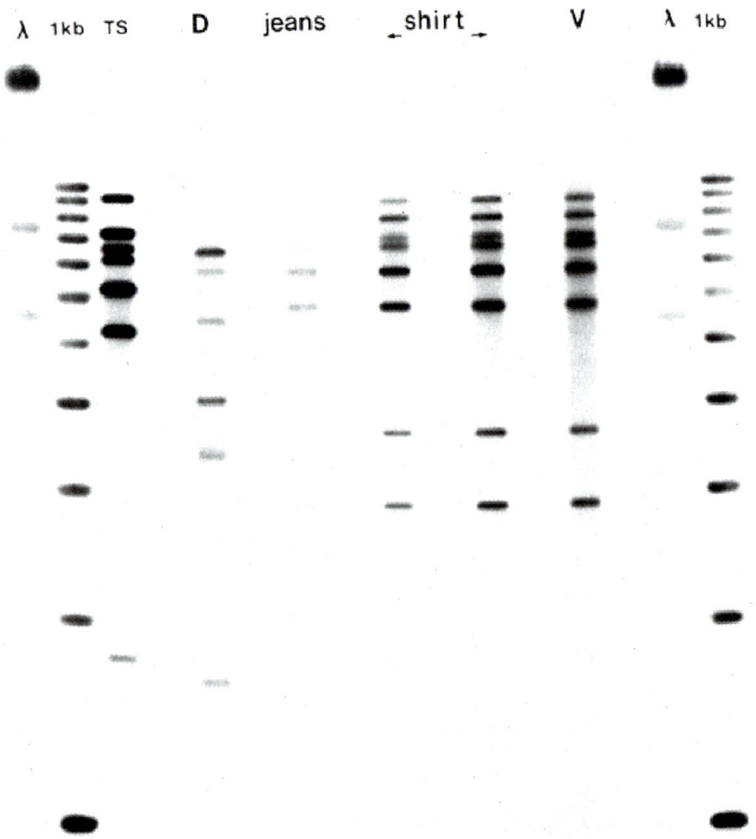

normally give rise to two bands from each person for each particular locus. This assumes that the chosen locus shows substantial allelic variation. Occasional persons will be homozygous and hence show only a single band. For full identification using SLPs, it is necessary to run several reactions, each using a different SLP probe. SLP analyses use smaller amounts of material than MLP and are easier to interpret and compare. Statistical analysis and population frequencies are possible using SLP data.

Historically, MLP was used before SLP. In MLP, a probe is used that binds to multiple sites in the genome. Consequently, an MLP probe generates multiple bands from each individual. Because it is not known which particular band comes from which particular locus, interpretation is difficult. Furthermore, statistical analysis is impractical and data cannot be stored reliably in computer databases. In practice, fingerprints generated by MLP probes must be directly compared with others run on the same gel. Consequently, MLP methods have largely been displaced by SLP analysis.

FIGURE 24.5
DNA Fingerprint
Actual DNA fingerprint showing that the pattern of DNA fragments of the victim (V) were found on the defendant's clothing (jeans/shirt). The first two and last two lanes are the standard size markers (labeled λ and 1 kb). The lane marked TS is a positive control showing that the fingerprint technique was successful. The lane marked D is the defendant's DNA pattern. Reprinted with permission from Quick Publishing, LC.

DNA fingerprinting relies on the unique pattern found in different individuals when a series of DNA fragments is separated according to length.

USING REPEATED SEQUENCES IN FINGERPRINTING

A variation of DNA fingerprinting is to look at regions of the DNA that contain **variable number tandem repeats (VNTRs)**. As discussed in Chapter 8, this means that sequences of DNA are repeated multiple times and that different people have different numbers of repeats. Repeat sequences vary greatly in length; however, for forensic purposes, relatively short repeated sequences are now generally used and are known as **short tandem repeats (STRs)**—see later discussion. VNTRs usually occur in noncoding regions of DNA. Hence, using VNTRs protects privacy in the sense that an individual's coding DNA is not revealed during forensic investigations. VNTRs may be visualized by using restriction enzymes to cut out the DNA segment containing the VNTR, followed, as before, by separation of DNA bands by gel electrophoresis and visualization by Southern blotting. Alternatively, the DNA fragments for VNTR analysis may be generated by PCR (see later discussion). Figure 24.6 shows corresponding DNA fragments from three individuals who differ in the number of repeats in the same VNTR. Consequently the length of the fragment differs from person to person.

There is often an enormous variation between people in the number of repeats at any particular VNTR site. So there is a very low likelihood of two people matching exactly, or, if you prefer, a high probability they will differ. Some VNTRs have 100 to 200 different variants, making them very useful for forensic analysis. Although VNTRs are not genuine genes, their variants are regarded as alleles and so VNTRs are considered to be "multiallelic" systems.

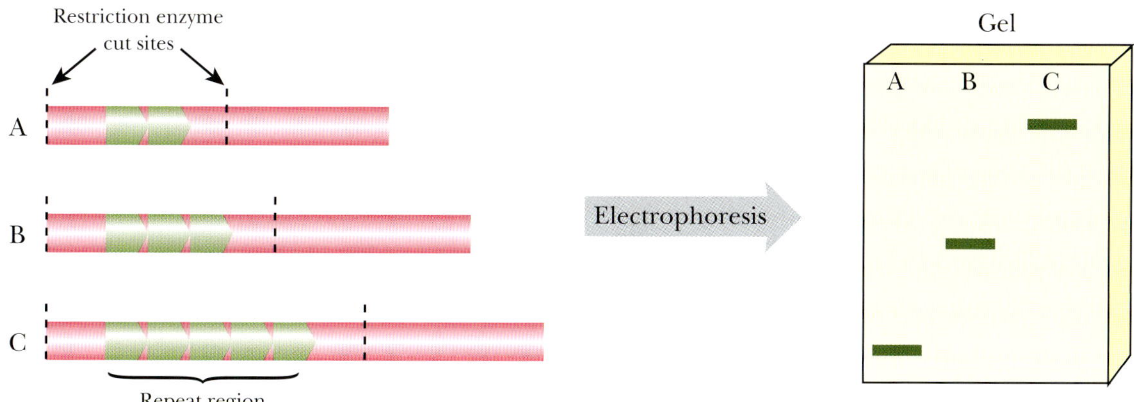

FIGURE 24.6 VNTR Fingerprinting
Genomic DNA has regions with repeated sequences. In each individual, the number of repeats varies, and therefore the lengths of these regions can be compared to distinguish identities. The repeated region is isolated using restriction enzymes from three individuals marked A, B, and C. The fragments are run on agarose gels to compare the lengths.

One practical problem here is that multiple tandem repeats give so many closely packed bands that standard agarose gel electrophoresis cannot discriminate the different fragments. Different types of gels such as polyacrylamide may resolve closely spaced bands. Alternatively, the fragments can be separated on a gradient gel.

The original DNA fingerprints, invented by Alec Jeffreys in England in 1985, used highly variable VNTRs with long repeat sequences. DNA was isolated and cut with restriction enzymes because PCR had not been invented. The cut fragments were probed using MLP to generate the early DNA fingerprints.

The STR (short tandem repeat) is a subcategory of VNTR in which the repeat is from two to six nucleotides long. Most STRs are not as variable in the number of repeats as VNTRs with longer unit sequences. Many STR sequences have only 10 to 20 alleles and hence cannot provide unique identification alone. However, many STR loci are available, and if several are analyzed simultaneously, this will provide enough data that the pattern would be unique for each individual. Today, STR analysis is done using PCR (see later discussion) to generate the DNA fragments.

Present-day DNA fingerprinting uses repeated sequences. In particular, STR (short tandem repeat) sequences are used because they are convenient for analysis.

USING THE POLYMERASE CHAIN REACTION (PCR)

The polymerase chain reaction (PCR) is a procedure for amplifying tiny amounts of DNA and is used when there is too little DNA, or the DNA is too degraded for DNA fingerprinting via the RFLP approach. The details of PCR have already been discussed in Chapter 4. PCR machines can amplify a segment of DNA (100 to 3000 bp long) in a few hours, starting from only a picogram (10^{-12} g), although microgram (10^{-6} g) quantities or larger are better. In fact, PCR can be used successfully to analyze DNA from a single cell.

Whereas classical DNA fingerprinting requires relatively long strands of DNA, PCR can be used on short segments of DNA. PCR is most useful for regions of the DNA with high individual variability. Small regions with high person-to-person variability are the best to amplify. If the sequences of two samples match in several highly variable regions, they are probably from the same person. In current practice, forensic DNA analysis is almost all done by PCR-based methodology.

Once the DNA from the forensic sample has been amplified by PCR, it is compared with DNA from the suspect, or suspects. Spots of both DNA samples are bound to a membrane and tested for binding to a DNA probe that is either radioactive or tagged with a fluorescent dye. The probe either binds or doesn't bind, so any spot is either positive or negative. This kind of test is known as a dot blot (Fig. 24.7). Thus, the major difference between PCR and DNA fingerprinting is that DNA fingerprinting looks for differences in fragment sizes, whereas PCR tests for the presence or absence of specific stretches of DNA with identical (or almost identical) sequences. If necessary, segments of DNA that have been amplified by PCR can be fully sequenced. In this case, we do not need to rely just on hybridization as an indication of related sequences. In the future, such sequence identification will most likely be done by DNA array technology (see Chapter 8), which allows simultaneous analysis of multiple short sequences.

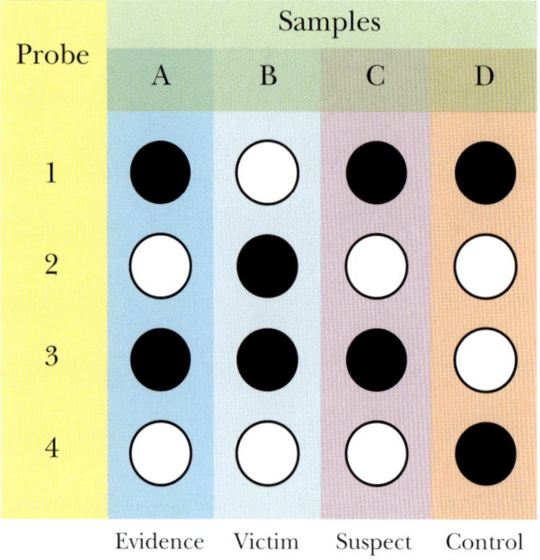

FIGURE 24.7 PCR plus Dot Blots for Identification
This example compares whether or not evidence from a crime scene matches the victim or suspect. DNA from the victim, the suspect, and the evidence was isolated. PCR was used to amplify a short segment of the DNA using primers flanking a highly variable region. The PCR products from the evidence, victim, suspect, and a control were each applied as four separate spots to a nylon membrane. The membrane was treated with four different probes, such that each probe was in contact with each of the different PCR products. Probes bound specifically to those spots that had sequences matching the probe. Notice that the pattern of the suspect and the evidence are identical. Therefore, the suspect's DNA was present at the scene of the crime.

It is possible to analyze several STR loci simultaneously by running several PCR reactions in the same tube using different primers (**multiplex PCR**; Fig. 24.8). This requires that the multiple sets of primers do not interfere with each other, which is often difficult to achieve when running six or more amplifications together. Nonetheless, commercial kits are now available that can run up to 13 STR analyses in one reaction tube. Such a multiplex analysis gives a gel track with up to $2N$ bands (where N is the number of loci analyzed). For this example, 13 STR loci would produce 26 bands. Fewer bands will be seen in individuals who are homozygous at any of the chosen loci. Despite the apparent complexity, the STRs that are used derive from known sequences at known chromosomal locations, and hence the individual bands can be identified and entered into computer databases. Multiplex STR analysis is the basis of the national database set up in the UK in 1995. Similar databases are now used in other European nations and, since 1998, in the United States.

The polymerase chain reaction (PCR) is now used to generate the DNA fragments for DNA fingerprinting. Multiple standardized PCR reactions can be run in the same tube.

Box 24.1 Heroic Cockatoo Provides DNA Evidence

A fascinating example of DNA evidence was presented in September 2002 to a grand jury in Dallas, Texas. A pet bird pecked and clawed intruders while trying to protect its owner. Blood found at the crime scene apparently came from a wound the bird pecked on the head of one of the suspects. The bird, a white-crested cockatoo, stands 18 inches tall and has a beak powerful enough to snap thin tree branches. Sadly, the intruders killed both the bird and its owner. Both the murdered owner and the bird have been autopsied and the bird's beak and claws are being checked for blood in a search for additional DNA evidence. When confronted with the condemning DNA evidence, one of the suspects confessed—but blamed his partner for the actual killing.

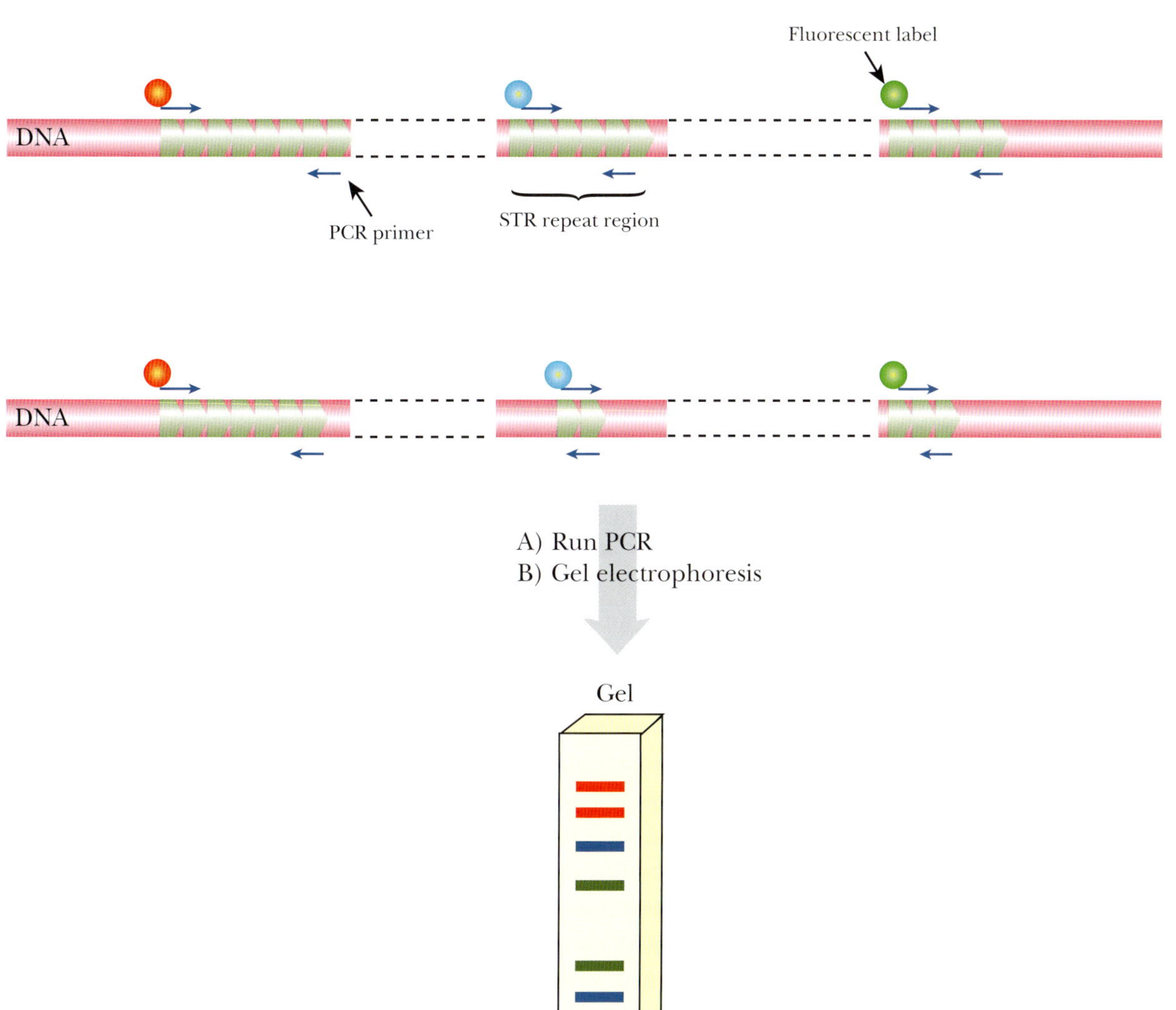

FIGURE 24.8 Multiplex STR Fingerprinting
Three different STR loci are amplified using PCR primers. Each set of primers is labeled with a different fluorescent label to distinguish each locus from the other. The PCR products are run on an agarose gel to determine the length of the fragment, and hence the number of repeats. This individual is heterozygous for each locus because there are two different-sized bands for each PCR primer set.

PROBABILITY AND DNA TESTING

If two DNA samples are different, then they must have come from different people. Hence, DNA testing can readily exclude an individual from being suspected. But what if two DNA samples match for whatever tests we have run? Positive identification requires the use of probability. Inclusion depends on the assumption that it is highly improbable that the DNA of the suspect and the DNA from the evidence match merely by chance. Using DNA profiling it is now possible to achieve probabilities of less than 1 in the total world population of a chance match. We should be cautious if close relatives are suspects in criminal proceedings, because the probability of a match is obviously much greater than for the general population.

The following general steps are important when determining the probability of a match:

1. From the same population of which the suspect is a member, select a random sample of individuals.

2. Determine the genotype of these randomly selected individuals and estimate the frequency of the alleles at the loci used in DNA typing.
3. Calculate the probability of finding the genotype of the suspect by assuming that this individual's alleles at each single locus represent a random selection from the population in general. (We also assume that the alleles tested are not linked but are independent of each other.)
4. Multiply together the frequencies that are determined from the various loci. The figure obtained represents the overall probability that the suspect's DNA would match the evidence by chance.

The details of population genetics used to establish probabilities for genetic screening, whether DNA or blood groups, are beyond the scope of this book. However, the probabilities from DNA testing are now sufficiently good in practice to make identification virtually certain, provided that the tests are carried out properly on reasonably good samples of DNA.

Interestingly, convictions have been obtained using DNA evidence where the probability of a chance match was one in 100, but with the addition of supporting evidence. However, in cases where the evidence is primarily based on DNA testing, juries more and more often expect astronomical odds such as one in a million or billion. This was the case in the notorious O. J. Simpson trial, but the DNA evidence was ignored. However good the scientific evidence, it cannot overcome the corruption of justice by wealth and politics.

Modern DNA analyses using multiple STR sequences can provide almost total certainty of unique identification.

THE USE OF DNA EVIDENCE

The Frye rule (*Frye v. United States*, 1923) concerns the admissibility of scientific evidence. The principle states that new scientific tests must be generally accepted in appropriate scientific circles before evidence from them is admissible in courts. In addition, a "helpfulness" standard is applied in some states, which involves the use of expert witnesses to assist the court in interpreting facts from scientific evidence. Recent court cases have almost all allowed DNA testing to be admitted into evidence, although there have been occasional exceptions. By 1996, DNA evidence had been admitted in more than 2500 criminal cases in the United States. There are relatively few U.S. government labs doing DNA testing. Accredited private labs perform most forensic work, and these services are available to both the prosecution and defense.

The main impact of DNA technology has been the far greater certainty with which individuals can be associated with or excluded from a particular crime than was possible with traditional blood tests. Experience has shown that if DNA testing is given as evidence, there is a higher probability of conviction than without DNA testing. DNA evidence is commonly used in cases of rape. However, in most cases of rape, the accused admits knowing the alleged victim and identity is not an issue. DNA testing can also be used by law enforcement to narrow the number of possible suspects, given that simple DNA testing protocols can determine sex and racial characteristics.

The British judicial system has led the way in DNA testing. In one early case, DNA testing was used to exclude the person first suspected of the sexual assault and murder of a young girl. But to find the real perpetrator, the police screened more than 5000 men in the village by blood testing (ABO and phosphoglucomutase), only to find no match with anyone. Ironically, the murderer was discovered because it was revealed that he had paid another man to give blood for testing. DNA testing subsequently confirmed, with high probability, that his DNA matched that of the semen sample taken from the victim. A conviction was finally obtained. Overall, there is considerable public support in Britain for maintaining DNA profiles on the entire population.

In the United States, a national DNA data bank is presently maintained by the FBI and is often screened by computer searches to find suspects. Although some worry about invasion of privacy, those who feel that liberty includes the freedom to walk down the street without being assaulted regard these developments favorably. Can DNA information be misused? The answer, of course, is that any information can be abused. People with different racial or genetic characteristics have been persecuted in the past. Then again, detailed DNA sequence information is hardly necessary for identifying people by race. In practice, the vastly increased accuracy of DNA testing compared to, say, blood group analysis means that unique individual identification is usually possible and hence racial bias is largely excluded in DNA testing.

DNA testing has become widespread in industrial nations, and many countries are setting up national DNA data banks.

TRACING GENEALOGIES BY MITOCHONDRIAL DNA AND THE Y CHROMOSOME

Mitochondrial DNA sequences have been very useful in tracing the recent evolution of the human species at the molecular level. Analysis of mitochondrial DNA (mtDNA) can also be used in forensics. The main advantage is that mitochondrial DNA is present in multiple copies per cell and so is relatively easier to obtain in sufficient amounts for analysis. The sequence of mtDNA varies by 1% to 2% between unrelated individuals.

The major disadvantage is that mitochondrial DNA does not vary between closely related individuals. Mitochondria are inherited maternally, and mitochondrial DNA sequences are therefore shared among groups of people derived from the same maternal lineage. If two samples of DNA show different mitochondrial sequences, this indicates that they come from different people. However, the opposite is not true. Identical mitochondrial sequences are found in people related on the mother's side.

Mitochondrial DNA has been used to derive family ancestries. Indeed, it is now possible to submit personal samples of DNA for analysis to companies such as Oxford Ancestors. Their MatriLine service allows persons of European descent to trace their maternal ancestry back to one of seven ancestral females (Fig. 24.9). Almost everyone in Europe, or whose maternal roots are in Europe, is descended from one of only seven women whose descendants make up well over 95% of modern Europeans. For genealogical purposes, each of these seven women may be regarded as the founder of a "maternal clan." For

Box 24.2 DNA Tackles Fishy Business

DNA can be used to identify animals as well as people. What happens when a sleazy operator mixes low-quality bonito in with premium tuna? The taste test may tell us something is wrong, but this is not sufficient for legal action. DNA analysis can reveal the species of fish present, even if mixed in with others. Sequence differences and RFLP profiles of PCR fragments generated from the cytochrome *b* gene, carried on mitochondrial DNA, can be used to distinguish multiple closely related fishes. The food regulatory agencies of the European Union are now using this procedure.

those whose maternal roots lie outside Europe, a similar analysis is available, but is not yet so detailed.

In contrast to mitochondrial DNA, the Y chromosome follows a paternal pattern of inheritance. The Y chromosome contains many STR sequences in noncoding regions. However, most have few different alleles, and so only a few are suitable for forensic analysis. One advantage of using Y-linked STR loci is that any sequence specific to the Y chromosome must have come from a male. This is often useful in cases of sexual assault.

> Genealogies are increasingly being traced by DNA sequencing. Mitochondrial DNA is maternally inherited and is often used to specifically trace maternal ancestry. Y chromosome sequences can be used to trace male ancestry.

IDENTIFYING THE REMAINS OF THE RUSSIAN IMPERIAL FAMILY

An interesting example of forensics concerns the identification of the remains of the Russian royal family. Analysis of both short tandem repeats in chromosomal DNA and sequencing of mitochondrial DNA were involved. The last Tsar of Russia, Nicholas Romanov II, was executed in July 1918 together with his family and a handful of servants. After execution by a firing squad of Bolshevik soldiers, the bodies were buried in a hidden mass grave. The burial site was rediscovered in 1989, and in 1991 nine skeletons were excavated. Thorough forensic analysis of the bones, clothing, and personal possessions from the grave provided evidence that some of the skeletons belonged to the tsar and his family. American and British teams, at the invitation of the Russian government, then carried out DNA testing.

Nuclear and mitochondrial DNA tests were performed on the nine bone samples. Five of the bodies were clearly related, as demonstrated by STR analysis at five different genetic loci (Fig. 24.10). These were Tsar Nicholas, his wife, the Tsarina Alexandra, and three of their four daughters. The fourth daughter and their son, Prince Alexei, the heir to the throne, were missing—their bodies had apparently been destroyed completely by burning before the mass burial. The other four remains were those of servants who were unrelated to the royal family.

The identity of the remains of the tsarina was confirmed by sequencing mitochondrial DNA. Tsarina Alexandra was the granddaughter of Queen Victoria of England. Alexandra's sister, Princess Victoria of Hesse, was the grandmother of Prince Philip, Duke of Edinburgh, husband of the present queen of England. A sample of blood provided by Prince Philip showed an mtDNA sequence that was identical to that of the remains presumed to belong to Tsarina Alexandra.

The mtDNA of the tsar himself proved more intriguing. Two distant maternal relatives of the tsar, Countess Xenia Sfiri and the Duke of Fife, contributed samples for

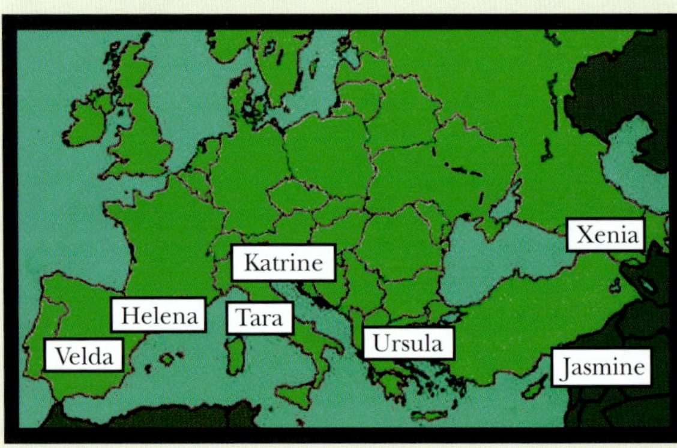

FIGURE 24.9 Seven Daughters of Eve
Oxford Ancestors refers to these so-called "Seven Daughters of Eve" as Ursula (Latin for "she-bear"), Xenia (Greek for "hospitable"), Helena (Greek for "light"), Velda (Scandinavian for "ruler"), Tara (Gaelic for "rock"), Katrine (Greek for "pure"), and Jasmine (Persian for "flower"). People of European descent can trace their lineage back to one of these seven women by comparing the mitochondrial DNA sequences. Used with permission from Oxford Ancestors (http://www.oxfordancestors.com).

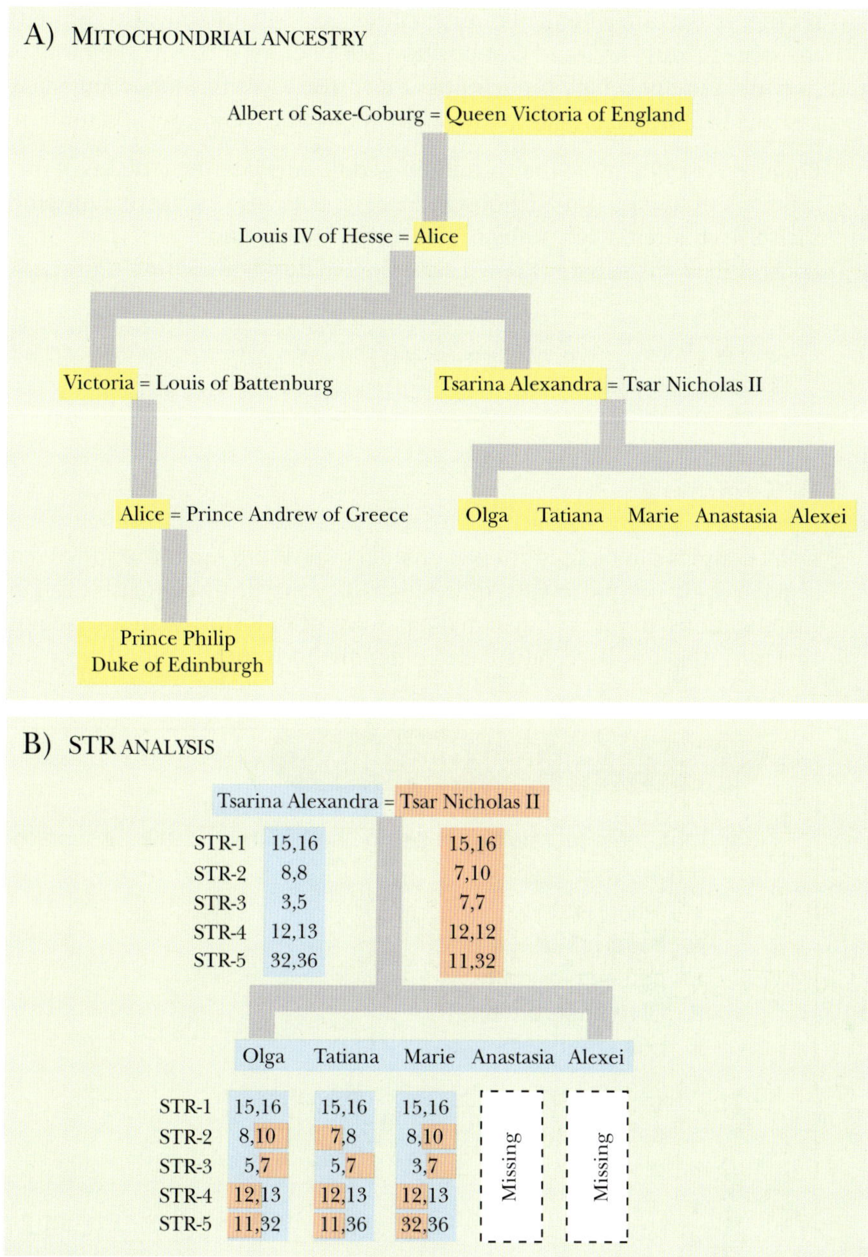

FIGURE 24.10 Russian Royal Family

(A) Family tree showing the ancestry of Tsarina Alexandra, Tsar Nicholas II, and their children. The yellow highlights show people with identical mitochondrial DNA. (B) STR analyses of the skeletal remains of Tsar Nicholas II and his family. The remains were examined with PCR primers for five different STRs (labeled 1–5). The three children had combinations of STR fragments found in either parent. The number of the STR is color coded, with red from Tsar Nicholas II and blue from Tsarina Alexandra. Two children's remains were missing from the grave.

comparison. Their mtDNA sequences were identical to each other. The tsar's mtDNA was identical to the relatives except at position 16169. Here both relatives had T but the tsar had a mixture, with 70% T and 30% C at position 16169. This suggested either that the sample was contaminated or that he had a rare condition known as **heteroplasmy**.

In a few individuals, there are two populations of mitochondria with slight differences in mtDNA sequence—that is, heteroplasmy. This condition is sometimes inherited via the maternal line. However, often the minority population of mitochondria is not present in all descendants. The matter was settled by exhumation of the body of Georgij Romanov, younger brother to the tsar, who died of tuberculosis in 1899. Georgij also showed heteroplasmy with the same mixture of T and C at position 16169 of his mtDNA. The rarity of heteroplasmy provides extremely high probabilities for correct identification when a match is found. In this case the likelihood ratio for authenticity was estimated at over 100 million!

On July 17, 1998, more than a million people attended the reburial of the last Imperial monarch of Russia, Tsar Nicholas II, together with his wife Tsarina Alexandra, and three of their five children, Olga, Tatiana, and Maria. The ceremony took place in the Peter and Paul Fortress in St. Petersburg (known as Leningrad during the communist period).

> Analysis of both nuclear and mitochondrial DNA sequences successfully identified the remains of the Russian royal family.

Summary

DNA samples can be used for personal identification. Apart from identical twins, DNA sequences are unique to each individual. They may be used to establish identity in criminal investigations, in cases of disputed paternity, or in historical or archaeological research. DNA fingerprinting has largely displaced older methods such as blood typing in these areas.

End-of-Chapter Questions

1. What new technology is being used to track cattle through the food processing chain?
   - **a.** retinal scans
   - **b.** fingerprints
   - **c.** cattle color patterns
   - **d.** DNA fingerprinting
   - **e.** none of the above
2. Which type of body fluid can be analyzed?
   - **a.** semen
   - **b.** blood
   - **c.** urine
   - **d.** tears
   - **e.** all of the above
3. In the ABO blood group, how many alleles are present in the population?
   - **a.** 2
   - **b.** 4
   - **c.** 6
   - **d.** 3
   - **e.** 1
4. Which one of the following statements about ABO blood grouping is not correct?
   - **a.** People with type A blood will express antibodies against type B blood.
   - **b.** A person with AB blood type is the universal donor.
   - **c.** A child may not have the same blood type as either one of his/her parents because humans are diploid.
   - **d.** A person with O blood type is the universal donor.
   - **e.** A type O person is only able to accept type O blood.

5. Which of the following techniques is not involved in the identification of DNA at crime scenes against possible suspects?
   a. PCR
   b. Western blot
   c. sequencing
   d. hybridization
   e. DNA fingerprinting

6. Which one of the following is not a step in DNA fingerprinting?
   a. DNA fragments are separated according to their molecular weights during gel electrophoresis.
   b. Autoradiography is used to identify the location of the radioactive-labeled probe after hybridization.
   c. A cDNA copy of the mRNA is made.
   d. Southern blotting is used to visualize the DNA fragments.
   e. The DNA is cut with restriction enzymes.

7. Why have MLP methods been displaced by SLP methods?
   a. Interpretation is difficult since it is unknown which band corresponds with which locus.
   b. In order to make comparisons, fingerprints from MLP must be run on the same gel.
   c. It is difficult to store the data from MLP methods in a database.
   d. Statistical analysis of MLP is difficult to obtain.
   e. All of the above are reasons SLP has replaced MLP.

8. What is a problem associated with using VNTRs to identify persons of interest in criminal cases?
   a. Multiple tandem repeats create densely packed bands during agarose gel electrophoresis, so other separation methods must be used.
   b. VNTRs are usually the same from person-to-person, so very little information is obtained from this technique.
   c. VNTRs occur within coding regions; thus privacy is not well protected.
   d. There have been no problems associated with using VNTRs to identify people.
   e. none of the above

9. What is the major difference between DNA fingerprinting and PCR?
   a. Both techniques provide adequate data for forensic science.
   b. DNA fingerprinting looks for differences in sizes, whereas PCR tests for the presence or absence of specific regions.
   c. DNA fingerprinting identifies regions that are absent or present, whereas PCR identifies different sized fragments.
   d. There are no major differences between the two techniques.
   e. none of the above

10. Which of the following terms describes running several PCR reactions simultaneously in one tube?
    a. DNA fingerprinting
    b. multiplex PCR
    c. VNTR
    d. RFLP
    e. STR PCR

(Continued)

11. When close relatives are suspects, what happens to the probability of a match during DNA testing?
    a. The probability increases.
    b. The probability decreases.
    c. The probability remains the same.
    d. The probability becomes 100%.
    e. The probability becomes 0.

12. What has been the main impact of DNA technology on the criminal justice system?
    a. uncertainty between the crime scene sample and the suspect's DNA
    b. providing loopholes in laws regarding DNA testing and admissibility in courts
    c. not much impact
    d. greater certainty when matching crime scene DNA with suspect DNA
    e. none of the above

13. What technique has been used to prevent low-quality bonito from being included with premium tuna?
    a. RFLP
    b. VNTR
    c. RT-PCR
    d. Western blot
    e. none of the above

14. How have mitochondrial DNA sequences been used?
    a. to identify differences between closely related people
    b. to roughly determine how many mitochondria are in one cell
    c. to derive family ancestries
    d. mitochondrial DNA sequences have not been used in forensics
    e. none of the above

15. What method was used to identify the remains of the Russian Imperial Family?
    a. PCR
    b. mitochondrial DNA sequencing
    c. VNTR
    d. RFLP
    e. Western blot

Further Reading

Butler JM (2006). Genetics and genomics of core short tandem repeat loci used in human identity testing. *J Forensic Sci* **51**, 253–265.

Chaix R, Austerlitz F, Khegay T, Jacquesson S, Hammer MF, Heyer E, Quintana-Murci L (2004). The genetic or mythical ancestry of descent groups: Lessons from the Y chromosome. *Am J Hum Genet* **75**, 1113–1116.

Gusmão L, Butler JM, Carracedo A, Gill P, Kayser M, Mayr WR, Morling N, Prinz M, Roewer L, Tyler-Smith C, Schneider PM (2006). DNA Commission of the International Society of Forensic Genetics (ISFG): An update of the recommendations on the use of Y-STRs in forensic analysis. *Forensic Sci Int* **157**, 187–197.

Horsman KM, Bienvenue JM, Blasier KR, Landers JP (2007). Forensic DNA analysis on microfluidic devices: A review. *J Forensic Sci* **52**, 784–799.

Jobling MA, Gill P (2004). Encoded evidence: DNA in forensic analysis. *Nat Rev Genet* **5**, 739–751.

Katoh T, Munkhbat B, Tounai K, Mano S, Ando H, Oyungerel G, Chae GT, Han H, Jia GJ, Tokunaga K, Munkhtuvshin N, Tamiya G, Inoko H (2005). Genetic features of Mongolian ethnic groups revealed by Y-chromosomal analysis. *Gene* **346**, 63–70.

Redd AJ, Chamberlain VF, Kearney VF, Stover D, Karafet T, Calderon K, Walsh B, Hammer MF (2006). Genetic structure among 38 populations from the United States based on 11 U.S. core Y chromosome STRs. *J Forensic Sci* **51**, 580–585.

Sahoo S, Singh A, Himabindu G, Banerjee J, Sitalaximi T, Gaikwad S, Trivedi R, Endicott P, Kivisild T, Metspalu M, Villems R, Kashyap VK (2006). A prehistory of Indian Y chromosomes: Evaluating demic diffusion scenarios. *Proc Natl Acad Sci USA* **103**, 843–848.

Schneider PM (2007). Scientific standards for studies in forensic genetics. *Forensic Sci Int* **165**, 238–243.

Sobrino B, Brión M, Carracedo A (2005). SNPs in forensic genetics: A review on SNP typing methodologies. *Forensic Sci Int* **154**, 181–194.

Varsha A (2006). DNA fingerprinting in the criminal justice system: An overview. *DNA Cell Biol* **25**, 181–188.

Woolfe M, Primrose S (2004). Food forensics: Using DNA technology to combat misdescription and fraud. *Trends Biotechnol* **22**, 222–226.

Bioethics in Biotechnology

APPROACH TO BIOETHICS

The purpose of this book is to explain the science of molecular biology and its applications in biotechnology. This chapter does not attempt to teach bioethics as such, but attempts to survey briefly the moral issues arising from advances in biotechnology. In each section, we have indicated problems and possibilities. We have asked a lot of questions but given few definite answers, because we feel that moral decisions are for the reader to make, not the authors. On the other hand, we have not attempted to artificially hide our own biases, and these are fairly obvious in several cases. In practice, many moral questions have been decided, either by general public acceptance or by the imposition of laws. Nonetheless, laws may be repealed and what is viewed as morally acceptable constantly changes.

Much of what is regarded as "official" bioethics derives from the clinical arena, including the practice of medicine as well as clinical trials and experiments. Although these principles still apply to the testing and use of clinical protocols that involve genetically engineered materials, they do not cover many of the newer issues in genetics and biotechnology. Instead of dealing with traditional, clinically oriented ethics, we have chosen to consider issues that have arisen from recent scientific advances. The list of topics was generated, in part, from an informal survey given to students. Before dealing with the individual scientific topics, some general issues are discussed.

Most standard bioethics derives from medicine. Here we consider novel issues arising from modern genetic technology.

POWER, PROFIT, POVERTY, AND ACCESS

Morality is a private and costly luxury.

Henry B. Adams, 1838–1918

Many of the questions raised by modern biology and genetic engineering are well worn in other arenas. Who should control technology? What should be banned or permitted and who should decide? Who should profit? Should access to novel and expensive technology be provided to those who cannot afford it? If so, who should pay, the government or private individuals? A related issue is that of access to technology for those who live in poor Third World nations. The answers to these questions differ according to personal beliefs and cultural outlook. In any case, they are neither novel nor restricted to biology, let alone biotechnology, and we will make no attempt to answer them here.

Over the past half century, the gap between the rich and poor has widened. This is true both between the industrial nations and the Third World and, more recently, within the industrial nations themselves. This is partly due to advancing technology. Mechanization has made unskilled human labor less necessary. For example, less than 5% of the population is needed to grow food for everyone in the advanced nations (compared to some 90% in medieval times). Automation has also reduced the number of people needed to produce industrial goods. Consequently, a significant proportion of the population has increasing difficulty finding worthwhile employment. Whether biotechnology will merely add to the prevailing trend toward automation or whether it will alleviate some of the problems of poverty is still unclear.

Many clinical procedures involving novel technology are expensive and are beyond the reach of the poor. This is the old question of the distribution of wealth, in a technological guise, and has no special link to genetic engineering. The rich have always had greater access to expensive health care, whether drugs, surgery, or simply high-quality nursing. For example, treatment with botulinum toxin (Botox) is now used to remove wrinkles from the skin of the old and ugly (Fig. 25.1)—provided they can afford it. Botox injections cost from $300 to $500

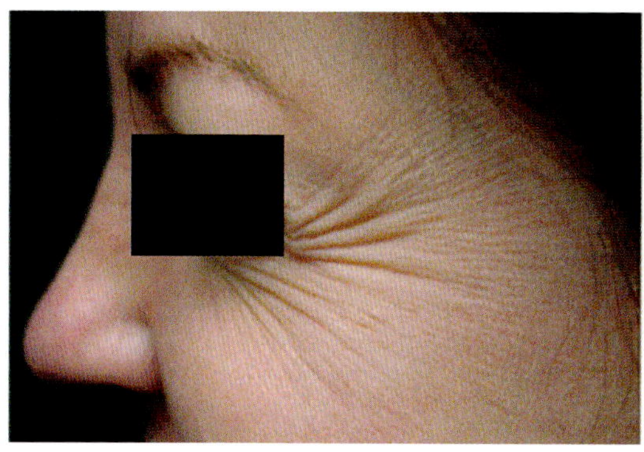
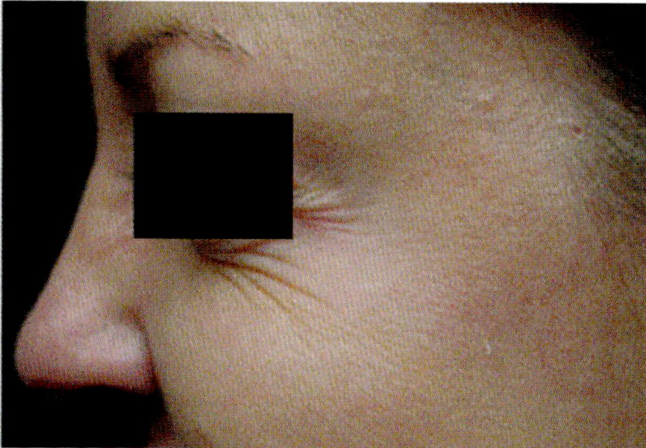

FIGURE 25.1 Botox Treatment: Before and After
Treatment of crow's feet with 12 U of Botox in each periorbital area. Three injection points were used. The patient is shown before and 2 weeks after treatment. From: Flynn TC (2006). Update on botulinum toxin. *Semin Cutan Med Surg 25*, 115. Reprinted with permission.

(more than a month's wages in many Third World nations) and the results last for about 5 months. More than 1.6 million people received injections in 2001. Botulinum toxin type A is a protein toxin made by the bacterium *Clostridium botulinum*, which causes food poisoning. The toxin blocks the release of the neurotransmitter acetylcholine by nerve cells that control muscle contraction; hence its use in very low amounts to inhibit the muscle contraction responsible for wrinkles and frown lines (see Chapter 23 for more details).

Those who discuss the social aspects of gene therapy and other new technology often mention the question of access by Third World nations. When millions die every year from malaria, tuberculosis, and AIDS, there is little point in discussing the benefits of expensive high technology. Most inhabitants of Third World nations cannot afford basic antimalarial drugs, let alone AIDS cocktails. Many do not even have pure drinking water. Some scientific advances may indeed benefit the poor nations. Mass immunization against infections with cheaper, more effective vaccines is an example. Transgenic crops able to grow in poor soils and give higher yields without fertilizers may also help. Huge strides in traditional plant breeding increased crop yields drastically between the 1940s and 1980s (the so-called Green Revolution). But merely saving lives from starvation causes population expansion. This, in turn, causes overcrowding, thus promoting the spread of infections. Unless the world can control its population, we will merely enter a futile spiral of trading one problem for another.

Advancing technology sometimes benefits the poor. In other cases it increases the gap between rich and poor.

IGNORANCE, NOVELTY, AND CULTURAL VIEWPOINT

Another general point is the effect of ignorance and novelty on supposed morality. People are frightened of the unfamiliar, and these fears are often expressed in moralistic terms. Whenever some new advance occurs in science and technology it is almost certain to be greeted with claims that it is either immoral or hazardous to society or both. For example, when the steam locomotive appeared, the flames in the firebox were compared to the fires of Hell by overly enthusiastic preachers. This aspect of morality has nothing to do with science in particular and is also seen in other areas. For example, novel fashions in clothing, music, or entertainment are frequently condemned as evidence of moral decline.

Advances in science are subject to fear stemming from novelty and compounded by the effects of fear stemming from ignorance. Genetic engineering and human cloning are very complex issues, and relatively few members of the general public understand the science behind them. Because these topics are understood by so few of the general public, misinformation, both deliberate and accidental, becomes a real issue in the debate. The entertainment industry often portrays genetically engineered or cloned organisms as evil and scary.

The other side of this phenomenon is that once a new fashion or technique has become familiar, complaints against its supposed immorality fade away. When rock music emerged, it was widely condemned as immoral by preachers. Today, many churches feed their congregations rock music to keep in touch with youth. Many rock bands use their lyrics to help spread the word of God. Saving lives by giving smallpox vaccinations was once regarded as "playing God," and hence blasphemous. Yet today every preacher and his or her kids are vaccinated against a dozen diseases. Today human cloning is often criticized as "playing God." Will tomorrow see the College of Cardinals advocating the cloning of Mother Teresa?

When the first animal, Dolly the sheep, was cloned, it hit the headlines and gave rise to a heated moralistic debate. Today there is a big furor about human cloning and stem cells. Meanwhile, animal cloning has largely been accepted and the cloning of yet another species scarcely qualifies as front-page news anymore. When "test tube babies" (*in vitro* fertilization) first appeared, they were the subjects of intense ethical debate by self-appointed leaders in human morality. Louise Brown, the world's first test tube baby, was born on July 25, 1978, in England. Since then there have been more than a million others, and today the procedure is covered by most health insurance plans. In December of 2006, Louise Brown gave birth to a naturally conceived son, Cameron John, illustrating that test tube babies reproduce normally as adults. The "morality" of this topic is rarely even discussed any more. Indeed, it is now possible to buy test tube baby necklaces (Fig. 25.2). The fact that most "fundamental ethical issues" fade away at much the same rate as fashions in women's clothing brings their deep significance into question.

FIGURE 25.2 Test Tube Baby Jewelry Sealed air-tight in its tube, the Onch Test Tube Baby rests peacefully in its special formulated nutrient! Courtesy of Onch Movement, onch@onchmovement.com.

Another factor that makes most ethical decisions highly subjective is cultural viewpoint. It is rather noticeable that Americans tend to oppose genetic engineering on religious grounds. Europeans are much more likely to question the possible dangers to human health or the environment. Thus the furor about the dangers of genetically modified crop plants and food arose first in Western Europe. In contrast, the ethics of using human stem cells has been muddled into the American religious controversy surrounding abortion. Consider that abortion was legalized in most Western European nations in the 1950s, but in America not until the 1970s. Was it moral to have an abortion in England in 1960 but immoral to have one in the United States the same year? (Note that rich American women would cross the Atlantic to get abortions in Europe during this period, thus buying themselves out of their own society's legal situation.) Does this suggest that much morality is merely local custom? If so, whose brand of ethics should be used? Alternatively, one might argue that abortion is (or is not) immoral and that local legal differences do not reflect ethics.

Ethical viewpoints vary both over time and among cultures. Ignorance and fashion tend to affect novel discoveries, but their effect fades over time.

POSSIBLE DANGERS TO INDIVIDUALS, SOCIETY, OR NATURE

What of real dangers as opposed to a vague fear of the unknown? History shows that most technologies can be used for a variety of purposes, both helpful and destructive. Dynamite is used in mining and quarrying and in ammunition. Improved nutrition gives healthier children and stronger soldiers to kill others. Box-cutters can be used to open a carton of Red Cross supplies or hijack an aircraft. And so on. Obviously biotechnology can be abused, just like any other technology. Should we stop making advances because they can be abused?

What about the accidental or incidental hazards of biotechnology? All improvements in human health and prosperity have side effects that we cannot predict. Increased life expectancy results in more lonely old people, which in turn burdens the health care system. A greater proportion of retired people perturbs the distribution of wealth. Decreased infant mortality exacerbates overpopulation and affects the environment both by consuming scarce resources and causing pollution. Overcrowding also promotes the emergence and spread of novel infectious diseases. Other technologies also cause unwanted side effects. Modern transportation has speeded getting the sick to hospital, yet large numbers of people are killed in car accidents. Should we get rid of our cars, trains, and planes because some people die using them?

A widespread fear is that genetic engineering will result in the creation of monsters, mutant humans, or virulent new diseases. Will some genetic construct escape from a laboratory and crossbreed with some wild organism, forming a fearsome hybrid monster (Fig. 25.3)? How many experiments are necessary to examine whether or not these things will actually happen? How can you predict all possible outcomes or side effects? Should these technologies be banned until we can predict all possibilities? In reality, very few people are extremists who would ban all technological advances. Most people tend to examine each case on an individual basis.

Indeed, in 1975, during the early days of recombinant DNA research, the molecular biology community itself met at Asilomar, California, and called for a moratorium on those experiments that were seen as potentially hazardous. This respite allowed the NIH to generate guidelines to oversee recombinant DNA research.

More mundane, but also more realistic, is the worry that improved characteristics engineered into useful crop plants may be genetically transferred to weeds. The improved weeds would gain similar advantages such as resistance to drought, insects, or herbicides. The potential

FIGURE 25.3
Ancient Greek Chimera
The Chimera was a hybrid monster of ancient Greek legend that was thought to live in southwest Anatolia (present-day Turkey). It combined a lion, goat, and fire-breathing dragon. This modern representation was created by Rebecca Kemp, http://www.wildlife-fantasy.com/.

for escape of engineered organisms or clusters of engineered genes and their possible effects on the natural world is a major source of contention. Of course, ecosystems can be damaged simply by the entry of new species from elsewhere (without any genetic engineering). The classic case is the introduction of rabbits to Australia.

Accurately predicting possible dangers from genetic engineering, or any other novel technology, is extremely difficult.

HEALTH CARE AND RELATED ISSUES

A variety of specific ethical issues may be loosely grouped under the heading of health care. These issues largely affect the present generation as opposed to altering their descendents genetically. Few of these issues are truly new from a moral perspective. Once the novelty has worn off, the attached moral issues will fade as the technology becomes accepted. Besides novelty, safety and expense are issues that must be addressed when focusing on health care.

Bioterrorism and Germ Warfare

Objectively, the likelihood of surviving a biological attack is much better than surviving a nuclear strike (see Chapter 23 for details of germ warfare). Despite this, those who are unfamiliar with microbiology tend to find biological warfare very frightening and often regard it as more immoral than chemical or nuclear warfare. This disproportionate response to germ warfare can be seen in the hysterical response of the United States to the anthrax attacks of 2001–2002 that followed the terrorist destruction of the World Trade Center. The actual number of casualties was low, yet the associated fear was widespread and became a hot media topic. Perhaps one major reason for the fear is lack of visibility. Guns and bombs are highly visible. Infectious microbial agents are invisible to the naked eye. The fear of invisible dangers can become quite obsessive.

Whether or not research on germ warfare should be done is hotly debated. Knowing how to protect against an infectious disease inevitably provides information that would help in using the disease against an enemy. This conundrum is true in other areas. For example, the technology to build a nuclear power station is closely related to that needed to develop nuclear weapons. The same body of knowledge can often be applied to both positive and negative objectives.

Another issue is that of the Third World versus the industrial nations. Germ warfare has been described as the "poor man's nuclear weapon." Nations too poor to develop costly high-tech weapons could throw together crude biological weapons relatively easily and cheaply. Germ warfare thus represents a possible means by which Third World nations could protect themselves against the rich nations. This aspect is compounded by the fact that soldiers from rich countries have higher life expectancies and a better quality of living than do the poor inhabitants of the Third World. Thus a poor dictatorship might be tempted to release a biological agent within its own borders and accept casualties to its own people, knowing that this would frighten off a rich invader. There is some historical precedent for this. In World War I typhus epidemics were common on the Eastern front. The Serbians lost 150,000 men to typhus in the first 6 months of the war, including more than half of their 60,000 Austrian prisoners of war. Paradoxically, this actually aided the Serbs, because the Austrians were so frightened by the typhus epidemic that they kept their armies out of Serbia for fear of infection. Third World nations are also much more accustomed to death and illness due to extreme poverty. Perhaps this is one reason why the rich nations are so eager to ban germ warfare while keeping more expensive weapons of mass destruction in circulation.

Questions:

Does the method of killing large numbers of people affect the morality of doing so?

Is research intended to develop germ warfare agents more or less immoral than research into nuclear or chemical weapons?

Should preventative research be regarded with suspicion because the same, or closely related, technology can be used both for attack and defense?

Should biological weapons be banned by international law while more expensive weapons of mass destruction are allowed? Is it wrong for poor countries to possess biological weapons but OK for "responsible" advanced nations to do so?

Biological warfare generates emotions out of proportion to its effects. Perhaps this is due to the invisible nature of infectious agents.

Gene Therapy

The technology of gene therapy has been discussed in Chapter 17. Here we are excluding heritable changes to the human germline (i.e., transgenic humans) and considering only somatic gene therapy. The issues involved are mostly the same as for any other novel technology—safety and cost. In practice, gene therapy is still mostly in the experimental stages, and most questions have to do with the safety of the procedures and avoiding possible harm to the patients. Several individual incidents have occurred over the past few years in which patients receiving experimental gene therapy died or suffered serious injury. On the other hand, most of these patients have incurable conditions and hence little expectation of living long, healthy lives.

Questions:

If you know you are going to die within a year or two, why not risk a novel treatment that might kill you or might give you a dozen years of healthy life?

Do excessive safety regulations passed in response to one or two unfortunate cases slow down overall the development of treatments that might otherwise save many lives?

To what extent should the taxpayer bear the burden of expensive novel technology for those who cannot afford it?

Should public support be restricted to funding research or include clinical applications once these are successful?

Gene therapy is a high-cost, high-risk technology that is still largely experimental.

Organ Replacement, Artificial Parts, and the Bionic Man

The current system of organ transplants relies on people donating their organs after death. Too few people volunteer, resulting in a shortage. To alleviate this, it has been proposed to develop human clones as a source of replacement organs (as opposed to the creation of new individual persons). Artificial tissues (i.e., of nonbiological origin) are also being developed, and eventually wholly artificial organs may become available. One can also imagine replacement organs that are mixtures of artificial and biological parts. Yet another alternative is nanotechnology, the use of engineering on a microscopic scale (see Chapter 7).

This may eventually provide tiny devices that can replace biological components, although using different mechanisms, such as miniature filtration units to replace defective kidneys, or photosensors for defective vision (Fig. 25.4). However, all these options are somewhat futuristic, and at present, stem cells seem the most likely basis for organ regeneration (see later discussion).

Creating and transplanting new body parts is very complex, and thus expensive. This raises the old issue of access to medical care by the poor. A related issue is how to distribute a limited supply of spare organs among those who need replacements. Currently, organs are distributed on a first-come-first-served basis combined with urgency. Even here, the classic question still exists: "If only one liver is available, should it be given to the alcoholic (who will abuse it), the aging textbook writer (whose best work is over), or a young child (future unknown)?" Until there is an unlimited supply of organs, this debate will continue and make for good storylines for books and movies.

From the 1980s onward, several European countries have introduced presumed consent laws for organ donation. Under this system, it is presumed that an individual is willing to contribute organs upon death, unless he or she has registered as a nondonor or there is other evidence to the contrary (e.g., from relatives). In most cases this has greatly increased the supply of organs.

Questions:

Is human cloning for spare parts morally different from human cloning to produce new individuals?

Is developing wholly artificial organs a good way to avoid moral problems associated with using natural organs? Or does the prospect of the "Bionic Man" raise other moral questions?

When a technological resource is scarce, how should the few lucky recipients be chosen from those who need the therapy?

Should the United States switch to presumed consent for organ donation?

Organ replacement is a relatively old issue. New technology may allow creation of novel and improved artificial replacements.

Antibiotics and Antiviral Agents

When bacteria are exposed to antibiotics, they may gain resistance. This may be due to mutation or to gaining mobile genetic elements such as plasmids, bacteriophage, and transposons, which already carry antibiotic resistance genes. Overuse and improper use of antibiotics have led to the spread of antibiotic resistance. It is consequently getting difficult to find effective antibiotics to treat certain infections that used to be susceptible. It is estimated that about two people die per hour (or 17,000 per year) in American hospitals

Box 25.1 Nanophobia and Gray Goo

The emergence of nanotechnology has inevitably given rise to the parallel appearance of nanophobia. Early writings on nanotechnology suggested the use of microscopic self-replicating robots (nanobots) to build and repair both industrial products and damaged human tissues. This in turn gave rise to the suggestion that the nanobots might replicate out of control and convert all available resources into copies of themselves—thus turning the world into "gray goo."

In fact, organisms capable of replicating in less than half an hour and continuing until they have used up all the available resources already exist. They are the bacteria.

Perhaps nanophobia will divert the minds of the gloom-and-doom brigade away from biology, and genetic engineers will be able to breathe more freely and get on with cloning and stem cell research.

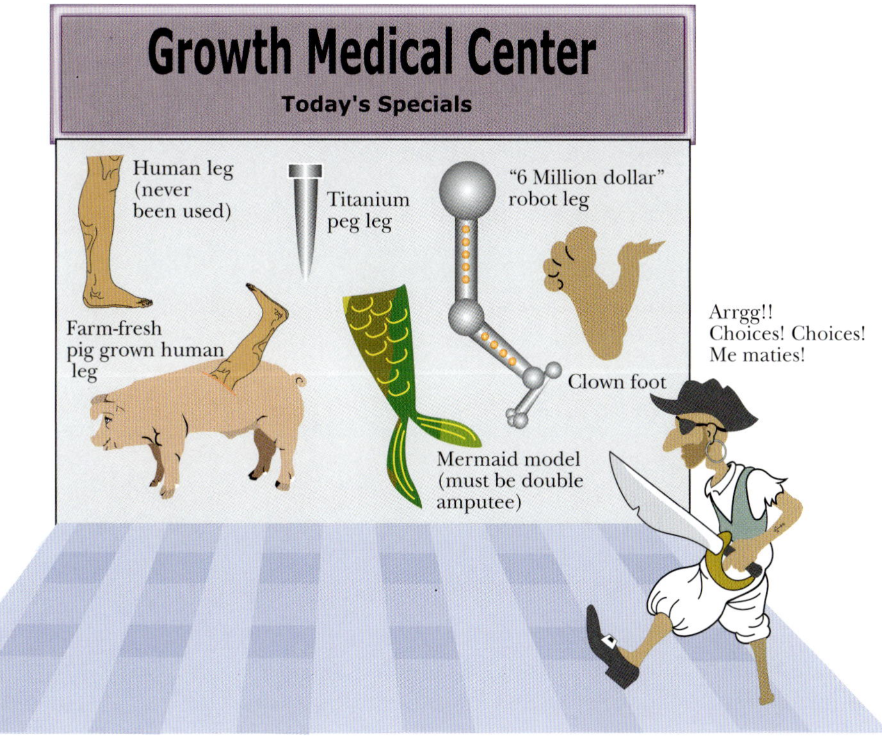

FIGURE 25.4
Alternative Replacement Parts
A futuristic vision of possible replacement parts.

as a result of infection with drug-resistant bacteria. Many practices lead to the spread of antibiotic-resistant bacteria—greed, ignorance, and poverty are all involved.

Certain antibiotics are widely used in agriculture to promote growth of domestic animals and improve meat yields. The United States leads the world in antibiotic consumption, using 23 to 25 million pounds in 2001. Only 10% of this was used to treat humans, and the other 90% was used on animals. Not surprisingly, bacteria resistant to such antibiotics are now widespread. Unlike the United States, this practice has been greatly restricted in Europe, where only 40% of the antibiotics are now given to animals. There has been a major drop in antibiotic-resistant bacteria in Europe, especially in Denmark, which has instituted a total ban on antibiotics in animal feed. Furthermore, the prices of pork and poultry have not risen because of lowered animal yields.

Overprescription of antibiotics for minor ailments, even when these may not even be caused by bacteria, is another factor in the emergence and spread of antibiotic resistance. Another problem is that antibiotic treatment is often discontinued too early. If the infecting bacteria are not totally destroyed by completing the course of antibiotic treatment, the survivors may gain resistance and spread. Poorly educated patients tend to stop taking medication as soon as the symptoms disappear. Poverty also has a major effect. In poor countries, the dosage and length of antibiotic treatment are decreased in order to save money.

Similar considerations apply to antiviral agents, such as those used to treat AIDS. In this case the virus mutates so fast (see Chapter 22) that treatment with just one antiviral agent will almost certainly result in resistant HIV mutants appearing in most patients. In practice AIDS patients are given cocktails of at least three antiviral agents so that the other antiviral agents will kill mutants resistant to one. However, patients who cannot afford expensive cocktails may be treated with a single agent. This allows development of resistant virus to one antiviral agent at a time. If such HIV strains are transmitted from one person to another, they may gradually pick up multiple resistance. Discontinuation of AIDS medication when patients begin to feel healthy again is also a serious problem, especially among the poor and uneducated. This is exacerbated by the high cost of AIDS therapy and the inconvenience of taking multiple different pills at different times of the day.

Questions:

Should use of antibiotics in agriculture be forbidden? Or should certain antibiotics be restricted to human use only?

Should use of antibiotics on humans be restricted to serious illness?

Should the rich countries forbid the export of certain antibiotics to poor countries in order to keep them safe from underprescription?

Should those unable to afford AIDS cocktails be forbidden to take a single antiviral agent because of the risk of producing a resistant virus?

Overuse of antibiotics has led to increased antibiotic resistance. Several European nations have begun to restrict antibiotic use in animal feed.

INTERFERENCE WITH THE NATURAL WORLD

To some people, particularly in Europe, the creation of transgenic animals and plants is seen as hazardous meddling with nature. Of course, humans have been tinkering with nature by improving livestock and crop plants for thousands of years. However, it is now possible to move blocks of genetic information across major taxonomic boundaries (e.g., from bacteria to animals or plants), rather than merely selecting for reassortment of genetic variation within a population or hybridizing closely related species.

Transgenic animals and plants, together with their uses, have been described in Chapters 14 and 15. Most transgenic animals, especially those designed for production of large amounts of clinically or industrially useful proteins, are unlikely to compete well in the wild. Furthermore, transgenic animals can be contained fairly efficiently. On the other hand, plants with improved resistance to drought, disease, or insect pests may well have an advantage in the wild and might spread naturally. In addition, plants are much more likely to hybridize with related natural species, possibly transferring the transgenes into wild populations. Movement of transgenes into wild relatives has already been observed with transgenic corn in Mexico. Thus we already have cases of genetically engineered organisms that have escaped into the wild. How this will alter the balance of nature is unknown.

Humans have been interfering with nature for thousands of years. However, genetic engineering allows us to make bigger changes much faster.

Transgenic Crop Plants

There has been considerable controversy over the use of transgenic plants in agriculture. Although the term *genetically modified organism* (GMO) is often used, we should remember that all domesticated plants and animals have been genetically modified by more traditional methods and consequently differ greatly from their wild ancestors. There are three main issues to consider for transgenic crops. First is whether the food product is safe for human consumption. Second is the question of containment. Third is the question of hazard to the environment.

Containment of transgenic plants is unrealistic on an agricultural scale. In practice, seeds from different batches of corn are impossible to keep wholly separate, and mixing of GMO with natural corn has occurred (e.g., the Starlink case in 2000—see Chapter 14). DNA of transgenic origin has been detected in wild plants. For example, wild maize (corn) in Mexico examined

Table 25.1 Percentage of GM Crops by Region (2004 Data)

Country	%
USA	59
Canada	6
Argentina	20
Brazil	6
China	5
Others	4

in 2001 contained transgenic DNA, even though planting transgenic corn plants was stopped in 1998. Worrisome possibilities include genes for herbicide resistance moving from crop plants to weeds. This would make weed control more difficult. Similarly, insecticide toxins expressed in pollen grains might harm bees, impairing the pollination of crops that depend on the bees. Events like these could decrease agricultural productivity.

Perspectives on GMO food vary greatly (Table 25.1) but seem rather predictable, based on known vested interests. Those who grow, export, and profit from GMO crops claim that they are safe and that the controversy is largely an emotional overreaction. Originally, the corporations and farmers were overall pro-GMO. However, the terminator controversy caused a rift between these two groups (see Box 25.2). Environmental and consumer groups tend to oppose GMO, as with any new technology.

Questions:

If you believe that interference with Nature is wrong, dangerous, or even blasphemous ("playing God"), where do you draw the line?

Should development of transgenic crops capable of flourishing under poor conditions, without need for expensive fertilizers, insecticides, weed-killers, and so forth, be developed because this would especially benefit poor countries?

Because advanced nations such as the United States have a food surplus using traditional methods, why not just play safe and avoid transgenic crops?

> Transgenic crops are widely used in some countries but viewed with suspicion in others. Safety to the consumer and to the environment where the crop is grown are common concerns.

Loss of Biodiversity

Genetic engineering followed by cloning to distribute many identical animals or plants is sometimes seen as a threat to the diversity of nature. However, humans have been replacing diverse natural habitats with artificial monoculture for millennia. Most natural habitats in the advanced nations have already been replaced with some form of artificial environment based on mass production or repetition. The real threat to biodiversity is surely the need to convert ever more of our planet into production zones to feed the ever-increasing human population. The cloning and transgenic alteration of domestic animals makes little difference to the overall situation.

Conversely, the renewed interest in genetics has led to a growing awareness that there are many wild plants and animals with interesting or useful genetic properties that could be used

Box 25.2 Terminator Genes in Seeds

One divisive aspect of the GMO controversy was the development of "terminator" technology. Crop plants were engineered so that their seeds would be sterile. The pretense was that this would prevent escape of GMO plants into the wild. The underlying motive was mere greed. Farmers would be forced to buy a new supply of seeds each year instead of planting seeds saved from the previous year's harvest. This would both increase the profits of the seed corporation and make farmers dependent on their seed suppliers. The attempted use of terminator technology to blackmail farmers caused a great deal of ill feeling.

The terminator scheme involves three transgenes:

(a) A gene for a toxin that is lethal only in developing seeds. The toxin gene is otherwise inactive due to a DNA spacer flanked by *loxP* sites inserted between the promoter and the coding sequence.

(b) A gene for Cre recombinase (see Chapter 14), which recognizes the *loxP* sites and recombines them so deleting the spacer sequence. This allows expression of the toxin gene.

(c) A gene encoding a variant of the TetR repressor (see Chapter 15) that prevents expression of the Cre recombinase gene.

Before sale, the seeds are soaked in a solution of tetracycline that binds to and inactivates the repressor. This allows the Cre recombinase to become active and remove the spacer sequence. The toxin gene is now expressed. Because the toxin does not harm the growing plant, except for the developing seeds, the crop grows normally except that the seeds are sterile.

for a variety of as-yet-unknown purposes. This has led in turn to a realization that we should avoid destroying natural ecosystems because they may harbor tomorrow's drugs against cancer, malaria, or obesity.

Our planet's biodiversity is under threat from human agriculture. Livestock cloning may damage the ecosystem, but renewed interest in natural products tends to have the opposite effect.

Animal Testing

The morality of animal testing applies to traditional products and medications as much as to new genetically engineered ones. It should be noted that vastly more animals are used for product testing and quality control than for actual research. On the one hand, each novel product or procedure means more animal testing. On the other hand, advances in molecular biology mean that many tests can now be performed on cells in culture, avoiding the use of living animals. For example, potential carcinogens can be screened initially by the Ames test, which uses bacteria to detect mutagenic agents. Furthermore, advances in genomics and proteomics now allow vast amounts of data to be garnered from testing the responses of multiple genes in cultured cells rather than from measuring the response of a single enzyme in whole organisms.

A generation ago, social activists demanded that medicines, cosmetics, shampoos, foodstuffs, and every other product that might come into contact with a human being should be rigorously tested for safety using animals. This led to massive government legislation mandating such testing. Today's animal rights activists are demanding less animal testing. There are even suggestions that such products as soap and shampoo, with familiar ingredients and properties, do not really need to be constantly tested by rubbing them into the eyes of animals to demonstrate that they will cause discomfort. This is a good example of how moral fashions vary, and even reverse, over time.

Animal testing is an old ethical issue. Opinions have varied considerably, even within recent history.

Transgenic Animals and Animal Cloning

Humans have meddled with nature since time immemorial. Historically humans have altered animals and plants by deliberate selective breeding and hybridization. In addition, human activity has led to unconscious genetic modification of many organisms. For example, we have undoubtedly selected alterations in the mice that infest our fields and grain storage facilities and the insects that rely on human crops. The novelty of genetic engineering is not in what we are doing, but in how we do it. Today we generate transgenic organisms by direct manipulation of their genetic material.

Even if you devote a whole field to growing a crop plant that is natural, you are eliminating the natural inhabitants of that patch of land. Moreover, you will select for life forms—both weeds and insect pests—that adapt to croplands. The European corn borer is a huge threat to the corn crop, but if we did not grow so much corn, these insects would be rare. Whether we want to or not, whether we are aware of it or not, we are imposing genetic selection on many other organisms, whatever we do.

Genetic manipulations could create future organisms that are truly bizarre by today's standards. By manipulating the homeobox genes, which control body plans and segmentation, maybe a "chickapede"—a chicken with multiple legs and body segments—could be created. Perhaps more grotesque would be to develop feed animals lacking most of the brain. This would avoid the suffering of domestic animals that are kept for slaughter. The controversy surrounding such future creations is yet to arise.

Questions:

Is interference with nature worse when dealing with animals versus plants? What about "pets" versus "livestock"?

Should people be allowed to clone their pets?

Is prescientific selective breeding OK?

Is applying mendelian genetics acceptable?

Is genetic engineering OK as long as no foreign DNA is introduced from another species?

Would it be OK to develop a chicken with, say, 10 legs for food?

Humans have genetically manipulated animals for thousands of years. Today, transgenics allows animals to be altered much faster than selective breeding. Ethical opinions on this issue vary greatly.

Box 25.3 Cloned Pets from Genetic Savings & Clone

Julie (last name withheld by request) of Texas became the first paying client to receive a pet clone, when the "twin" of her deceased cat Nicky—dubbed "Little Nicky"—was presented to her at a December 10 holiday party thrown by Genetic Savings & Clone (GSC) at a San Francisco restaurant.

As the first clone delivered to a paying client, Little Nicky made a huge splash in the media when GSC announced the delivery in December 2004. "He looks identical, his personality is extremely similar, they are very close," said Julie, an airline employee from Dallas who placed the order, during an interview on *Good Morning America* on December 23, 2004. Little Nicky, who was born October 17 in Austin, Texas, is a clone of Nicky, a Maine Coon who died in November 2003 at age 17.

In December 2006 Genetic Savings & Clone shut down due to lack of demand.

Transgenic Animals for Art and Amusement

A rather frivolous use of biotechnology is the creation of transgenic animals for artistic reasons—"**transgenic art**." The insertion of the marker gene *gfp*, which encodes green fluorescent protein (GFP), is now routine in genetic engineering. Other fluorescent proteins with other colors are also increasingly used. Consequently, green fluorescent animals are now quite common. Such animals appear normal in daylight, but if illuminated in the dark with near-UV or blue light, the whole animal fluoresces green. It is necessary to use animals with white fur or naked skins to visualize the color clearly. Thus albino strains of mice and rabbits show this effect well, whereas in animals with dark fur the green fluorescence is obscured.

Alba the green fluorescent rabbit—"GFP Bunny"—has been claimed as genuine transgenic art. Alba was born in France in the spring of 2000. Alba is a transgenic albino rabbit expressing high levels of GFP. She is surrounded by controversy. The artist, Eduardo Kac, claims that Alba was engineered at his request, whereas the scientists claim that she was made for research purposes and have not released the rabbit to the artist! But what if red and green fluorescent mice are marketed as children's pets or even as Christmas décor? What about creating red, white (i.e., nontransgenic), and blue fluorescent mice for patriotic purposes? Will NASA send a trio of red, white, and blue rhesus monkeys into space? Genetic art based on bacteria has also been displayed recently (Fig. 25.5).

Humans have bred animals for art and entertainment for a long time. Many breeds of dogs fall in this category, and a variety of other animals are also bred for show rather than for food or work. Transgenic technology has merely speeded up this process and allowed more drastic alterations than the traditional crossbreeding methods.

Question:

Should transgenics be restricted to "serious" areas such as health care and agriculture?

Transgenics can be used for relatively trivial purposes, such as art or entertainment. This is presumably less justifiable than improving livestock.

A)

B)

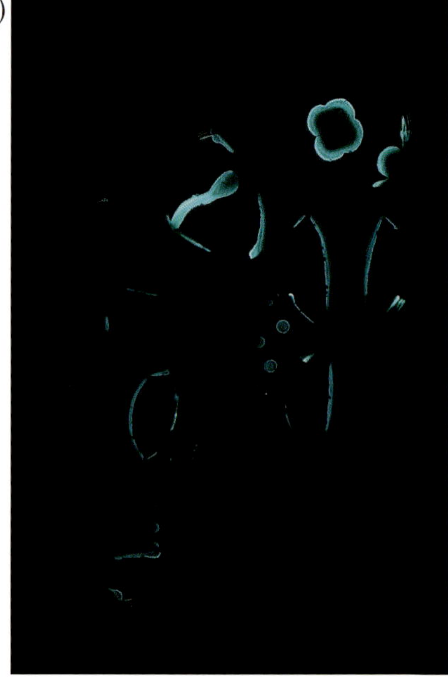

FIGURE 25.5 Genetic Art: Plant Embryo Stages 1 and 5
Hunter O'Reilly creates controlled line drawings using bioluminescent bacteria. The bacteria grow and become collaborators in the art. The bacteria in the drawing are photographed at intervals. First appearing with bright light, they fade away over a 2-week period as available nutrients are used. Courtesy of Hunter O'Reilly (http://www.artbyhunter.com/artgallery/livingbacterialdrawings/index.html).

ALTERING THE HUMAN GERMLINE

Genetic Screening in Pregnancy and Abortion

Genetic screening of newborn babies has been practiced for some time. Such information is used to allow early treatment of newborn infants, the classic case being phenylketonuria (PKU). People with PKU lack the enzyme that converts phenylalanine to tyrosine, and excessive amounts of phenylalanine causes permanent brain damage (see Chapter 16). Newborn screening allows infants with PKU to be given a diet low in phenylalanine, greatly reducing the damage. More recently it has become possible to screen developing fetuses for a variety of genetic defects long before birth. Analytical techniques are constantly advancing and an ever-increasing list of inherited defects can be monitored, at ever earlier stages of development.

However, prenatal genetic screening could also be used to decide whether to abort a fetus destined to suffer from an inherited defect. As understanding of the human genome increases, it will become possible to deduce such things as the probable future height, eye color, IQ, and beauty of the developing fetus. Most parents would like to have smart, healthy, and attractive children, and the temptation to have abortions based on these characteristics will soon become a reality.

The abortion issue is of course a peculiarly American obsession. Most European nations legalized abortion in the 1950s, and few Europeans take seriously the moralistic pronouncements on this issue that come from the other side of the Atlantic. The central question of the abortion issue is "When does human life start?" From a biological perspective, life does not start at one particular point but is a continuum. Sperm cells are alive, and so are the eggs they fertilize. Fusion of egg and sperm to create a zygote forms a new living individual with a unique genetic constitution. Rather than the beginning of life, the issue is perhaps really about consciousness. When do we actually become a conscious being? This is impossible to answer because no one yet understands consciousness, let alone has the ability to measure it.

Because society has arbitrarily decided that abortion is legal until the end of the first trimester, who should decide if an abortion is to be performed? From a genetic viewpoint, both father and mother have an equal share in the new individual—except for the mitochondrial DNA that is maternal in origin. From the viewpoint of biological resources, the mother has more invested and has traditionally been allowed to make the decision. Thus, the father often has fewer rights over the children. Although this outlook was not deliberately based on evolutionary considerations, it does in fact coincide with Darwinian logic.

Questions:

Are European views on abortion (and related issues) more advanced or more degenerate than American attitudes?

Should prenatal genetic screening be allowed for inherited diseases?

Should defective fetuses be terminated by abortion? Who should make the decision whether a defective fetus lives or dies?

Should we enforce paternity tests to make sure that the true genetic parents of a child are notified of any decisions about the child's welfare?

Genetic screening of both fetus and newborn is widespread and expanding. The difficult decisions lie in how to apply the knowledge.

Stem Cell Research

Another issue that has become entangled with the abortion controversy is stem cell research. Stem cells are the precursors to the differentiated cells that make up the body. Different types of stem cells correspond to different types of tissues. Embryonic stem cells are found in the developing embryo and retain the ability to develop into any body tissue. Embryonic stem cells can be maintained in culture and may be used to create transgenic animals by insertion of DNA.

It is hoped that engineered stem cells will eventually be used to regenerate damaged tissues or organs. One controversy concerns the source of the embryonic stem cells. In particular, should they be taken from discarded fetuses? One side claims that stem cell research will encourage abortions just to provide material. The other side claims that stopping research will deny patients medical improvements such as organ replacements. A related issue is the use of stem cells from leftover embryos in fertility clinics. Because few stem cells are needed for research, and vast numbers of aborted fetuses already exist, increased numbers of abortions seems unlikely. In addition, no one has yet grown an actual human organ from a stem cell. Thus this controversy is based on possibilities, not realities. One suspects that if technology advances far enough to grow organs in culture, it will also allow the use of stem cells from the patient's own body and embryonic tissue will no longer be required.

Stem cell research merges into other areas of biotechnology. If scientists are not allowed to use existing aborted tissue, can they create their own embryos *in vitro*? How far should such embryos be allowed to develop? Should brain tissue be used, since that is where people believe our consciousness lies?

Questions:

Should any research be allowed using embryonic stem cells?

Should researchers be allowed to harvest embryonic stem cells from aborted fetuses or only be allowed to subculture stem cell lines already in existence?

What if researchers in less repressive countries than the United States continue with stem cell research and are successful in inventing new therapies? Will Americans be allowed to benefit from the therapy?

> Using stem cells has raised concerns about the sanctity of life and generated a heated controversy. Attitudes and regulations vary greatly in different countries.

Human Cloning

Human cloning is another contentious issue. We have discussed the technology behind cloning animals (Chapter 15) and have already noted some ethical issues in animal cloning (discussed earlier). Not surprisingly, human cloning is generally seen as more unethical than animal cloning. Two uses have been proposed for human cloning. One is the generation of new individuals, identical to the donor of the genetic information. The other is the use of cloning to produce new parts for organ and tissue replacement.

The technical problems in cloning have been discussed in Chapter 15. Assuming that these are solved and cloning becomes safe and reliable, is there any reason to prohibit human cloning? It has been suggested that parents could replace a child who died prematurely or that dictators could keep a series of backup clones for emergencies. However, cloned humans would have to be carried by a surrogate mother. After proceeding through normal development, they would only reach adulthood after many years. Thus the idea of "replacing" individuals with cloned replicas faces the problem of a massive time lag. Furthermore, environmental and developmental influences would mean that although genetically identical, the clone would not be a true "behavioral replica." Remember that although identical twins are genetically identical—"natural clones"—they still show considerable divergence in personality, behavior, and ability.

Cloning incomplete humans for spare parts seems less upsetting to most people today, though technologically it might actually be more difficult because modifications would be needed to avoid developing complete individuals. In particular it would be necessary to engineer a humanoid with no conscious mind, presumably by greatly reducing development of the brain. Will we end up in a world where everyone has a brainless backup clone in the basement to provide spare parts? Or will a central facility grow generic clones with elongated bodies, missing heads, and 20 kidneys? At a more mundane level, the presumed use of stem cells for organ regeneration has probably made cloning for replacement parts an obsolete approach.

If such creatures are engineered, will a further step be to use cloned humanoids lacking higher consciousness as food? Such cannibalism would avoid harming animals and would provide food with exactly the right composition. Although this may sound bizarre, remember that it has been suggested that the Aztecs practiced cannibalism for nutritional (rather than cultural) reasons and on a scale large enough to have contributed significantly to the diet, at least of the upper classes. This of course is controversial, because accurate numbers for both the total populations and numbers of victims are lacking. In a fascinating new twist, it has been shown that diets based almost solely on corn (maize) cause serotonin deficiency in the brains of rats. Moreover, low serotonin levels have been correlated with bizarre and violent behavior in both experimental animals and humans. This has led to the suggestion that it was their high-maize diet that promoted cannibalism among the Aztecs. Thus, it is conceivable that morality can be affected by diet!

Questions:

Assuming technical problems are solved, is human cloning to create new individuals acceptable?

Is cannibalism inherently unethical, or is it the source of the body (sacrifice, murder, nonbrained cloning) that determines its morality?

If it is convincingly proven that a particular diet promotes unacceptable behavior, is it unethical to follow such a diet?

> Human cloning is also controversial. Reproductive cloning to create new individuals and therapeutic cloning to provide replacement organs tend to be viewed differently.

Eugenics and Selective Breeding

> Men are Generally More Careful of the Breed of their Horses and Dogs than of their Children.
>
> **—William Penn, 1693**

The ultimate in playing God is not merely to clone humans but to improve them genetically also. This is Frankenstein's monster in its 21st-century guise. In works of fiction, engineered humans always display some unforeseen and fatal flaw (an exaggerated fear of fire in the case of Frankenstein's monster). The implication is that we do not know enough to tinker safely and should therefore refrain from meddling with nature. Another common theme, at least in grade B movies, is the emergence of a group of superior humans, often with psychic powers, higher IQs, and stronger bodies. Sometimes this is due to accidental exposure to radiation, sometimes the result of mysterious "upward evolution," and sometimes caused by deliberate selective breeding or genetic engineering. The hero typically eradicates these cliques of super-people, presumably demonstrating that it is morally preferable to be weaker and less intelligent.

Occasional half-baked attempts have been made at eugenics—the deliberate genetic improvement of the human species. For example, between the two world wars, the American eugenics movement promoted the sterilization of the mentally subnormal and violent criminals. Such schemes were based on selective breeding (as used on domestic animals over the centuries) rather than genetic engineering. In practice, institutionalization of criminals and mental patients greatly reduces the number of offspring, so the overall results of sterilization programs were probably marginal.

Those opposed to modern genetic engineering often mention these old eugenics programs. Although rarely included in such discussions, providing welfare to those unable to feed their own children also has an effect on the future genetic constitution of the species. Welfare may increase the number of dependent people who do not contribute to society as a whole. For that matter, almost any major social or political change affects the human gene pool. Any major epidemic favors those individuals with greater resistance to disease and hence carries out genetic selection on the human population under attack. Mutations providing resistance to a variety of virulent infections (e.g., smallpox, malaria, dysentery/cholera) are now being discovered in the human genome (see Chapter 16).

Mutation and selection have also resulted from cultural changes. For example, the domestication of cattle has resulted in the genetic modification of those humans who chose this way of life. Lactase is the enzyme that breaks down the sugar lactose present in milk. Ancestral humans expressed the gene for lactase only during infancy. Modern cultures that drink large amounts of milk in adulthood show lactase expression in adults. This is due to a T-to-C base change in a regulatory sequence about 14 kb upstream of the structural gene encoding lactase. Thus dairy farming has led to the selection of lactase expression in adults. Those able to efficiently use the new food supply had a selective advantage. Such genetic changes were the unwitting side effects of cultural choices. Whether we are aware of the changes or not, humans are constantly modifying themselves genetically.

Returning to Man's Best Friend, Fig. 25.6 demonstrates the remarkable changes that have been brought about by selective breeding without the help of genetic engineering.

Questions:

One major issue in eugenics is control—who should decide on deliberate genetic selection, the government or individuals?

Is it wrong to offer violent criminals shorter sentences in exchange for accepting sterilization?

Is it wrong ("undue influence") to suggest to the mentally subnormal that they should avoid having children?

If it is morally wrong to sterilize the "unfit," is it OK to promote the increase of the unfit by welfare, while deliberately ignoring the genetic implications?

Should we attempt to improve the human species by deliberate genetic engineering? If so, who is to choose what improvements to incorporate?

Transgenic Humans and Designer Children

Future technology may allow humans to be engineered by choosing variants of existing human genes (e.g., resistance to disease, improved intelligence, blue eyes). This can already

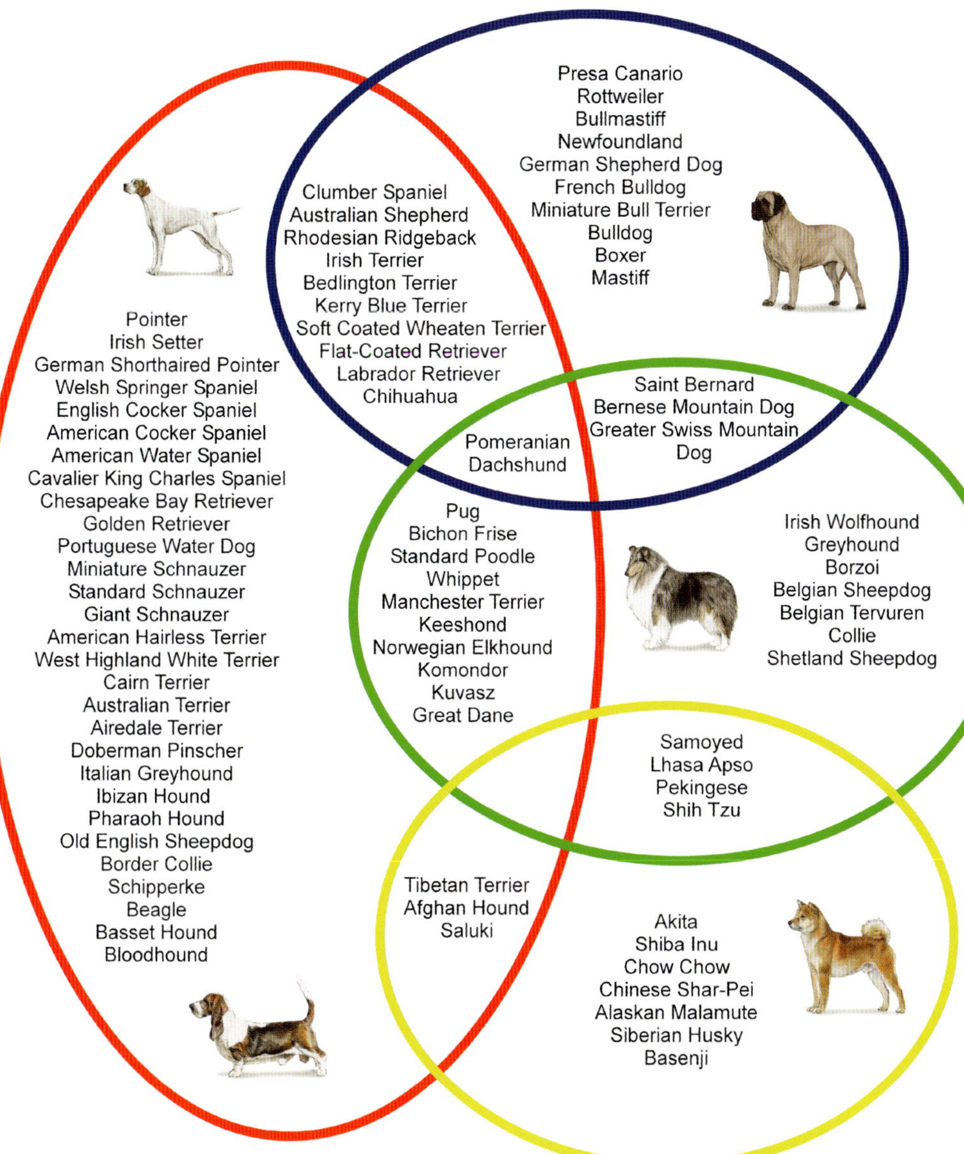

FIGURE 25.6
Population Structure of Dog Breeds
Five unrelated dogs from each of 85 breeds were genotyped using 96 (CA)n repeat-based microsatellites (VNTRs) that spanned the dog genome at an average density of 30 megabases. Four main genetically related clusters of breeds exist, as indicated by the colored circle, together with a variety of intermediate breeds. From: Parker and Ostrander (2005). Canine genomics and genetics: Running with the pack. *PLoS Genet* 1, e58. Public domain and courtesy of Elaine Ostrander.

Box 25.4 Eugenics or Dysgenics?

The Napoleonic Wars provided a fascinating but accidental example of dysgenics (negative eugenics). Napoleon, who was himself short, deliberately recruited tall men into the French Imperial Army. Although Napoleon won many battles, his casualties were enormous, and even when victorious, he often lost more men than his enemies because he used columns many soldiers deep. Bullets and shells often passed through men in the front rows, killing those behind, too. The result of constantly selecting tall men and subjecting them to massive casualties was that the average height of the French nation decreased significantly in the generation that followed!

Since industrialization began, welfare programs of one kind or another have enabled the poor to raise children at the expense of social programs. On the other hand, overcrowding due to urbanization has greatly aided the spread of infectious disease, especially among the crowded and badly housed poor. Have we been selecting for or against the less able? No one really knows. Nonetheless, our actions are changing the human gene pool, whether we care to admit it or not. Today we possess better techniques not only for screening genetic defects but also for artificial manipulation of individual genes and whole organisms. Will we use this knowledge sensibly or will this issue remain taboo while we continue to make changes to society that alter our genetic heritage in an ignorant and semirandom manner?

be done in a primitive way by genetic screening followed by abortion of an undesired fetus, as noted earlier. However, transgenic technology may eventually allow deliberate engineering of offspring for desired characters.

Such technology would also allow the insertion of foreign DNA into the human genome, creating transgenic humans. Most parents hope their children will do well, and one can easily imagine parents wanting "new and improved" children, if available. We have discussed engineering animals to make their own essential amino acids, thus avoiding the need to provide these in the diet (see Chapter 15). What about combating malnutrition and vitamin deficiency in humans by inserting DNA encoding metabolic pathways for essential amino acids or vitamin C? One can also imagine fringe elements wanting children with novel or cosmetic genetic enhancements derived from nonhuman life forms. Will we eventually see green fluorescent children to add to the string of GFP animals mentioned earlier?

Genetic engineering allows not only replacement of defective genes but also the incorporation of genes from other organisms. One can only imagine the future controversy if this is ever applied to humans.

KNOWLEDGE, IDENTITY, AND IDEOLOGY

Privacy and Personal Genetic Information

It may become possible to predict future health problems by analyzing an individual's DNA. At present this is true for a few inherited defects, mostly with major and easily observed effects. Such information might be of interest not only to the individual (see later discussion) but also to the health care system, insurance companies, employers, the military, and so on. This brings up the question of invasion of privacy.

Does the health insurance company have a right to know about your potential future health problems? Life insurance providers currently screen for age, weight, and cholesterol levels and charge higher premiums for those at risk. How will insurance companies react if they can screen for any health issue using genetic screening? Obviously, the insurance companies will have to provide most people with insurance to stay in business, but how much will rates vary depending on genomic data?

Other privacy issues include such possibilities as a national health care database. (Remember, databases for convicted criminals, especially sex offenders, are now current in many countries.) This might be used to provide prospective marital partners with information about possible inherited defects that might arise from a particular combination of parents. Other suggestions include a DNA-based identity card that would provide not only unique identity but also health care data useful in an emergency (such as allergies and blood group data) (Fig. 25.7). How will these affect the government's role in our lives?

Box 25.5 Gene Doping—A Future Dilemma?

The advent of improved transgenic mice (Mighty Mouse, Marathon Mouse—see Chapter 15) has led to wishful thinking on the part of some people. Could the same improvements be applied to human athletes? Researchers involved in these areas have received quite a few inquiries from athletes looking for muscle enhancement.

The term *gene doping* refers to the provision of extra copies of genes that offer a competitive advantage (as opposed to doping with illegal steroids, for example). Greater muscle mass, increased endurance, or higher red cell counts are all possible. The extra proteins would be very similar or identical to the body's own. Hence, detection would be extremely difficult.

At the moment gene doping is still a future worry. But given the pace of scientific development and the willingness of athletes to try performance-enhancing drugs, even if illegal, the issue may emerge into reality soon.

Questions:

How much personal genetic information should be available to health care systems, insurance companies, governments, employers, marriage partners, and so forth?

Will our "genetic identity" become a commodity much like our financial identity? Will people steal genetic information cards to obtain cheaper life insurance, marry a superior partner, or get a good job?

> The invasion of privacy is an old issue. However, the increased availability of DNA sequence data raises the question of how to protect genetic privacy.

Foreknowledge or Deliberate Ignorance

Is ignorance bliss, or is forewarned forearmed? Genetic screening can reveal the presence of genetic defects in adults, children, or unborn fetuses. Unfortunately, most of these defects cannot be cured. Suppose that a family has a history of some genetic defect—should the children be examined and informed of the results? For example, suppose that a child has a 50:50 chance of inheriting some severely crippling defect whose symptoms appear only later in life. If the child were free of the defect, they would most likely be greatly relieved to know. But suppose they carry the genetic defect—would they want to live with the knowledge that later in life they would become an invalid? Or would they rather not know? (Note that you cannot carry out the analysis and then withhold the information from the unlucky ones—patients who were not informed would realize that they had the defect.) Examples of this situation are relatively rare, and Huntington's disease (see Chapter 16) is perhaps the best known.

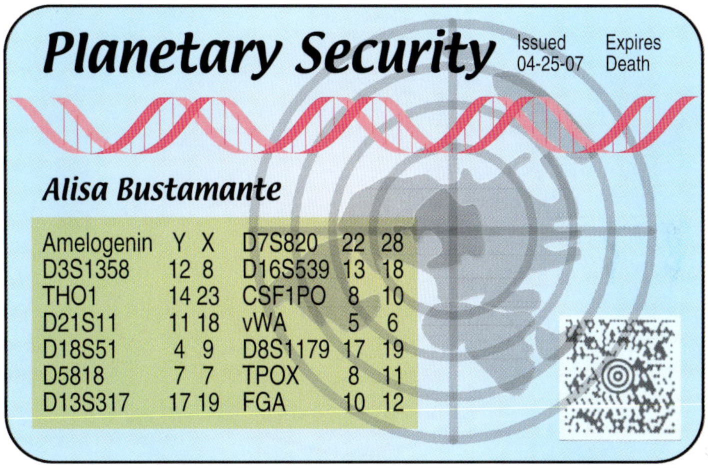

FIGURE 25.7
Genetic Identity Card
DNA fingerprinting that relies on a dozen STR sequences is the state-of-the-art method in forensic analysis. The amelogenin marker identifies the sex chromosomes. Such a fingerprint is sufficiently specific to distinguish more than 100 million million (10^{14}) individuals.

One can argue that it is best to know for several reasons:

You can take measures to avoid passing the defect on by refraining from having children.
Future advances may produce a cure that you can take advantage of, providing you are informed.
You can plan your life to minimize the distress.
If you didn't know one way or the other, you would worry anyway.

Question:

Should genetic information be used to plan marriages, careers, and make health-oriented decisions (e.g., moving to a better climate to avoid potential asthma) before any symptoms appear?

> Genetic screening may reveal severe defects that cannot be treated. Should such information be revealed, or does this cause more harm than good?

Forensics and Crime

Obviously any technology that is used to combat crime can be abused. As discussed in Chapter 24, Forensic Molecular Biology, the use of DNA for identification in both criminal investigations and civil cases (mostly paternity suits) is now widely accepted (Fig. 25.8). Early

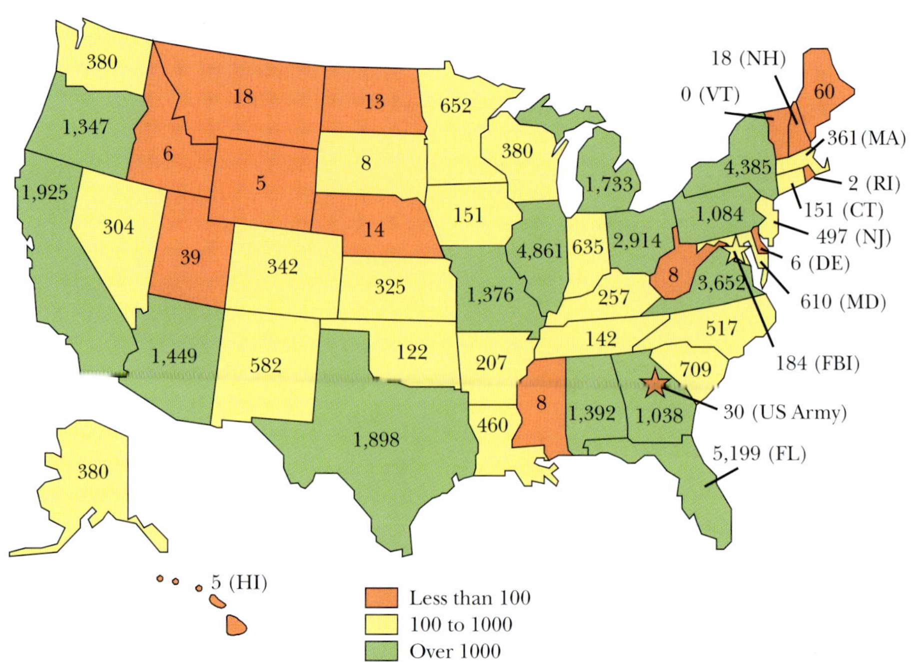

FIGURE 25.8
Acceptance of DNA Evidence in Today's Culture
This map shows the number of cases aided by the FBI's CODIS system for DNA evidence. Data are through December 2006 and include 43,156 investigations aided in 49 states and 2 federal laboratories.

technical problems have been ironed out, and the probabilities are now so overwhelming that, when properly done, DNA analysis can give a reliable and essentially unique identification. In fact, one major result of using DNA evidence has been the release of significant numbers of suspects who were wrongly convicted based on less reliable means of identification.

The remaining ethical issues concern such matters as setting up national or international databases of criminal DNA records. Who will be included? Who will have access? At the moment, DNA sequences used in identification are located in regions of noncoding DNA. However, it may eventually be possible to deduce a person's physical appearance and mental characteristics from DNA. If this is feasible, will DNA left at a crime scene be further analyzed to provide this information about possible suspects?

DNA evidence is now widely accepted in criminal cases. Compiling databases of criminal DNA and associated privacy issues remain somewhat controversial.

Conflict of Science with Traditional Religion

Religious and political establishments have been in conflict with Western science throughout history. In particular, modern advances in biology frequently conflict with traditional morality. Although genetic engineering has drawn considerable attention in recent years, few of the moral issues it raises are either truly new or unique. Germ warfare is one example. Using genetically modified viruses for germ warfare seems horribly immoral to some. Yet, is it any more (or less) immoral for a biologist to design a lethal virus than for a physicist to design a rocket with a nuclear warhead? And what of the caveman who designed a better axe to kill his competitors for food and shelter?

Although topics such as bioterrorism, evolution versus creationism, abortion, and human cloning are most familiar to modern readers, it is important to view the struggle between biology and the religious establishment historically. Long before Charles Darwin wrote *The Origin of Species*, religion and biology were locked in a power struggle. This was the clash between traditional religious belief and secular medicine. Both traditional Christianity and other religions have regarded life and health as given by God. Consequently, the established religion of many societies has claimed control over birth, reproduction, healing, and death.

This has often resulted in opposition to secular intervention in health care. For example, when vaccination first emerged, its use was widely condemned by the Christian church, especially the papacy, on the grounds that it interfered with God's will. Epidemics of infectious diseases such as plague or smallpox were seen as God's judgment on sinful man. If God intended you to live, then you would survive the epidemic, and if God wanted you to die, then being vaccinated to avoid this was tantamount to blasphemy.

From this perspective, issues such as abortion or human cloning may simply be regarded as modern episodes in a long-running power struggle between religious and secular control of human life. For that matter, modern biology conflicts just as much with almost all new religious cults and New Age ideology. Molecular biology affects these areas mostly in adding more detail to the picture of evolution. The basic ideological disagreements remain much as before and are beyond the scope of this book.

Questions:

Was there supernatural involvement in the origin of life or the universe?

If God created the world, what are the implications for genetic engineering and other human meddling with the natural world?

Is evolution compatible with religious belief?

If there is no God, where does morality come from? For example, is evolutionary ethics, based on competition and the struggle for survival, the only basis for judging behavior?

Should research in areas that clash with religious beliefs be suppressed?

Has life evolved on other planets in the universe, and if so, what are the implications?

If we are alone in the universe, do we have some moral obligation to colonize it and populate it with earthly life forms?

Should we use genetic engineering to create life forms that can inhabit planets where unaltered earthly life could not survive?

Science and religion have frequently clashed. Genetic engineering does not really affect the fundamental disagreements.

Clash with Egalitarian Ideologies

Just as modern biology clashes with religious ideas, it also conflicts with most secular ideologies. Whole families of secular ideologies are based on the notion of equality. But humans are not equal; they differ greatly in their capabilities, and many of these are partly or largely inherited. Both Soviet communism and American liberalism try to explain the problem of differences in human ability, especially between different demographic groups, by blaming environmental influences. Similar claims are made for the differences in male and female behavior. Just as the Christian establishment tried to censor embarrassing areas of science, so too leftist ideologies have attempted to suppress investigation and free discussion in areas such as the inheritance of IQ or the effect of hormones in determining "gender roles." For a considerable while, the Soviet Union allowed access to "capitalist genetics" only for trusted party members, and this area was forbidden to the general population. Today genes involved in behavior and brain function are beginning to be identified, and molecular details of their mechanisms and roles are starting to accumulate. These findings increasingly undermine the whole concept of human equality.

One fascinating example is the case of homosexuality. Some 20 years ago the suggestion that homosexuality was due to a genetic defect was highly "politically incorrect" and would have drawn accusations of "homophobia" from American liberals. Today, some gay organizations themselves maintain that homosexuality has a genetic basis. So far there is no convincing evidence one way or another, although it is hard to believe that genetics is not involved.

Questions:

The whole "social engineering" scenario (affirmative action, quotas, etc.) that has dominated American society for the past half century is based on the claim that differences in ability and/or behavior between different groups are due to environmental influences and "discrimination." What if they are mostly genetic?

Should research in areas that clash with political correctness be suppressed or censored? Should "sensitive information" be restricted to those trusted for their political loyalty?

If behavioral defects or tendencies have a biological basis, should attempts be made to cure them?

What if it becomes possible to detect genetic predisposition to violent crime while the affected individuals are still children? Should potential murderers and rapists be sent to jail before they have committed any crimes as a preventative measure?

Will analysis of fetal DNA allow predictions as to the likely future IQ of the child? If so, will parents choose to abort low-IQ fetuses? Will high-IQ donors provide (or sell) sperm for artificial fertilization?

Science and nonreligious ideologies are also in conflict. Modern genetics is steadily undermining many modern ideologies.

Ownership of Genetic Information

The ownership of knowledge is an age-old question. It is generally accepted in the Western industrial nations that if you invent something, you are entitled to patent it and profit from it. But how far do such rights extend? If you invent a camera, it is clear that you own the rights to the camera. But what if you use the camera to take a photo of the moon? Do you then own the moon, or just the photo? This sounds silly as stated. But what if you invent a DNA sequencing apparatus and proceed to sequence a gene (perhaps from some other human)—do you own the sequence information? Should you be able to patent preexisting information that new technology has revealed, or should patents be restricted to novel methods, machines, or techniques?

Genetically modified bacteria and other transgenic organisms have been patented in considerable numbers. They are presumably regarded as new inventions. The question of ownership of DNA sequences is still disputed, and different countries have different attitudes. In practice it is selling drugs and diagnostic kits that brings in the money. So far, at least, no DNA sequence has allowed the direct prediction of a drug or treatment that will cure a disease. Consequently, the actual financial value of knowing a DNA sequence is debatable.

Should scientists working at universities with public funding be allowed to patent their discoveries and profit personally? In practice, most universities have schemes whereby profits are split, say, 50:50 between investigator and institution. Another viewpoint is that all knowledge from research funded by the taxpayer should be public property, or even that all information should be shared freely by all humankind. Although the last attitude may sound excessively idealistic, keeping information secure is a lot more difficult in practice than guarding physical possessions. The copying of videotapes and music CDs or the downloading of MP3s over the Internet are cases in point.

Questions:

If you are paid by a company or by the taxpayers to develop a new technology, do you or the company/public own that technology? What if the "development" was an idea—who owns it then?

Would you sign a release form that entitles a company to be sole owner of any ideas, processes, or inventions you develop?

Who owns genetic information? The researcher? Those who funded the research? The person whose genome the data came from? The public?

LONG-TERM BIOLOGICAL PROBLEMS

Many of the bioethics issues mentioned earlier are fashionable because of their technological novelty and seem likely to fade from public awareness relatively soon. What will mostly be left are underlying issues, such as access to health care and privacy, that apply both to new advances and to previous technology. However, there are several biological issues that are less romantic but may well be of more real importance. We will briefly mention these as a counterweight to topics such as human cloning.

Two centuries of advancing medical technology have increased life expectancy from the mid-thirties to the mid-seventies in the industrial nations. Infant mortality has dropped from nearly 50% to less than 1%. The result is a population explosion that is far more hazardous to the planetary environment that any high-tech tinkering with nature. Although antipollution measures and recycling may help slightly, the growth of the human population inevitably consumes more resources and encroaches on the natural world.

Increased life expectancy also means that the average age of the human population is increasing. The ever greater proportion of old and retired people is putting a major strain on the health care systems of the advanced nations. Predictions of the coming collapse of American Medicare or the British National Health System are heard with increasing frequency. These trends are exacerbated by the high cost of much novel medical technology. In the United States some 20% of expenditure is now in the general area of health care, and a vastly disproportionate amount is spent keeping old people alive for their last few months. Another factor is obesity. More and more the inhabitants of the advanced nations are getting fatter. This causes major health problems, many of which, like diabetes, need expensive long-term treatment.

Population growth means increased crowding. Modern transport has led to increased mobility. The combination of these two factors has resulted in the rapid spread of infections around the world. From major pandemics such as AIDS and tuberculosis down to lesser epidemics such as cholera and West Nile virus, there are ominous signs that infectious disease is making a comeback. At the same time we have the spread of genetic resistance: to antibiotics among bacteria, to antivirals among viruses, and to insecticides among the insects that carry many infections or ravage crops. On the one hand, fending off novel or resistant infections is becoming ever more expensive in the rich nations. On the other hand, the spread of lethal infections is counteracting the population explosion to some extent in the poorer nations. This is especially evident in Africa, where actual population declines are predicted, largely as a result of AIDS.

Listing problems tends to create a gloomy atmosphere. So let us end by saying that most problems today are the problems of success. Western science is responsible for today's overpopulation precisely because it solved yesterday's problems of famine and disease. We believe that technology will solve many of the new generation of problems. The foregoing list of issues should be viewed more as a to-do list than a forecast of gloom and doom.

Many of today's problems are the results of yesterday's success. More surviving people and higher living standards inevitably create a need for more resources.

Summary

New advances in both knowledge and technical capability bring new ethical and regulatory problems. Closer inspection suggests that many new ethical issues are merely old issues in a new guise. Other issues do not involve ethics so much as familiarity. Nonetheless, some aspects of genetic engineering, such as transgenics and human cloning, do pose questions that are at least partly new.

End-of-Chapter Questions

1. What percentage of the population is needed to grow food for everyone in advanced nations?
   - **a.** 5%
   - **b.** 10%
   - **c.** 90%
   - **d.** 2%
   - **e.** 50%

2. Which of the following statements about ethical issues is not correct?
   - **a.** New advances in science and technology are usually associated with being immoral or detrimental to society.
   - **b.** Ignorance of scientific milestones, such as Dolly, fuel fear in a society that generally does not understand the issue.
   - **c.** Smallpox vaccinations were once considered as "playing God."
   - **d.** Fundamental ethical issues tend to fade away over time.
   - **e.** All of the above statements are true.

3. Which of the following is an incidental side effect of technology?
   - **a.** Increased life-expectancy burdens the health care system.
   - **b.** Decreased infant mortality contributes to overcrowding and more use of finite resources.
   - **c.** Overcrowding because of increased life expectancy or decreased infant mortality promotes the emergence of infectious diseases.
   - **d.** Cars can enable a sick person to reach the hospital much faster than horses, but many people die in car accidents each year.
   - **e.** All of the above are side effects of technology.

4. What is one theory that describes why biological warfare creates more emotion than what is needed?
   - **a.** because of a fear of the invisible (i.e., microorganisms)
   - **b.** does not cause fear
   - **c.** because of multi-drug resistant bacteria
   - **d.** because of a lack in sufficient control mechanisms to contain an outbreak
   - **e.** none of the above

5. How has the supply of donated organs increased in European nations?
    a. Upon death, people are assumed to be donors unless pre-registered as non-donors.
    b. People are generally more friendly in European nations and are more willing to give.
    c. Payment for the organs is allowed in European nations; therefore, people are willing to donate more.
    d. The supply of donated organs has not increased in European nations.
    e. none of the above
6. How has poverty contributed to the emergence of some antibiotic resistance bacteria?
    a. The dosage and length of antibiotic treatment are decreased to save money.
    b. Poverty has not contributed to the emergence because there are social programs in place that assure all individuals have proper medications.
    c. People in impoverished countries have the same access to health care as rich individuals.
    d. Antibiotics are more widely available in impoverished countries.
    e. none of the above
7. If humans have been interfering with nature for thousands of years, why has genetic engineering been hotly debated?
    a. Genetic engineering is not hotly debated and has been widely accepted by most nations.
    b. Any interference, even by natural means, is considered "playing God."
    c. Genetic engineering allows us to make bigger changes much faster.
    d. Genetic engineering ensures that biodiversity is not compromised.
    e. none of the above
8. Which of the following is an example of how humans have genetically selected organisms without performing genetic engineering?
    a. Growing entire fields of crops eliminates the natural vegetation on the land.
    b. Selectively breeding animals to produce desired looks and features.
    c. Growing large amounts of a certain crop invites insect pests which would normally have not inhabited that land before the crop was planted.
    d. Selectively breeding crops for desired characteristics.
    e. All of the above are examples.
9. How might prenatal genetic screening be used now and in the future?
    a. to test for inherited defects in the fetus
    b. for early detection of diseases such as PKU
    c. to test for particular facial features
    d. to test for IQ
    e. all of the above
10. What would be the most practical use to cloning humans?
    a. to use as food
    b. to produce new organs or tissues for transplants
    c. to have a behavioral replica of one's self
    d. all of the above
    e. none of the above

(Continued)

11. What is a major issue with eugenics?
    - a. who makes the decision about deliberate eugenics
    - b. who pays for the process of genetically engineering humans
    - c. who is responsible for the outcome of all eugenics research
    - d. who chooses which genes to alter and for what outcome
    - e. none of the above
12. Who might be interested in a person's individual genetic information?
    - a. military
    - b. health care system
    - c. insurance companies
    - d. employers
    - e. all of the above
13. What is a major ethical concern of DNA obtained during a criminal or civil case?
    - a. privacy issues of DNA records
    - b. if the procedures used to exonerate or implicate a suspect are accurate
    - c. if the DNA obtained in the investigation matches the suspect's DNA
    - d. entering DNA evidence into a trial is common practice and therefore, there are no ethical concerns
    - e. none of the above
14. What undermines the concept of human equality?
    - a. the involvement of genes in behavior and brain function and the investigation of their roles
    - b. the role that religion plays in society and moral issues
    - c. the role that the government plays in society and moral issues
    - d. the ideas about race and gender roles in society
    - e. none of the above
15. What is a debate surrounding genetic information ownership?
    - a. Should the knowledge gained from taxpayer money be free information?
    - b. Should researchers that are funded by taxpayers be able to patent and profit from their research?
    - c. Should someone be able to patent preexisting information?
    - d. If a researcher patented a particular apparatus, does he or she own the information obtained from the equipment?
    - e. All of the above are debated.

Further Reading

Adcock M (2007). Intellectual property, genetically modified crops and bioethics. *Biotechnol J* **2**, 1088–1092.

Baoutina A, Alexander IE, Rasko JE, Emslie KR (2007). Potential use of gene transfer in athletic performance enhancement. *Mol Ther* **15**, 1751–1766.

Birnbacher D (2005). Human cloning and human dignity. *Reprod Biomed Online* **10**(Suppl 1), 50–55.

Caulfield T, Cook-Deegan RM, Kieff FS, Walsh JP (2006). Evidence and anecdotes: An analysis of human gene patenting controversies. *Nat Biotechnol* **24**, 1091–1094.

Devolder K (2006). What's in a name? Embryos, entities, and ANTities in the stem cell debate. *J Med Ethics.* **32**, 43–48.

Fiester A (2005). Ethical issues in animal cloning. *Perspect Biol Med* **48**, 328–343.

Fischbach GD, Fischbach RL (2004). Stem cells: Science, policy, and ethics. *J Clin Invest* **114**, 1364–1370.

French AJ, Wood SH, Trounson AO (2006). Human therapeutic cloning (NTSC): Applying research from mammalian reproductive cloning. *Stem Cell Rev* **2**, 265–276.

Melo EO, Canavessi AM, Franco MM, Rumpf R (2007). Animal transgenesis: State of the art and applications. *J Appl Genet* **48**, 47–61.

Murray TH (2005). Ethical (and political) issues in research with human stem cells. *Novartis Found Symp* **265**, 188–211.

Roche PA, Annas GJ (2006). DNA testing, banking, and genetic privacy. *N Engl J Med* **355**, 545–546.

Strong C (2005). Reproductive cloning combined with genetic modification. *J Med Ethics* **31**, 654–658.

Taymor KS, Scott CT, Greely HT (2006). The paths around stem cell intellectual property. *Nat Biotechnol* **24**, 411–413.

Thomas SM (2004). Society and ethics—the genetics of disease. *Curr Opin Genet Dev* **14**, 287–291.

Unal M, Ozer Unal D (2004). Gene doping in sports. *Sports Med* **34**, 357–362.

Wolpert L (2007). The public's belief about biology. *Biochem Soc Trans* **35**, 37–40.

Yoshimura Y (2006). Bioethical aspects of regenerative and reproductive medicine. *Hum Cell* **19**, 83–86.

Biotechnology Multiple-Choice Answer Key

CHAPTER 1

1:A, 2:C, 3:D, 4:E, 5:C, 6:E, 7:C, 8:B, 9:E, 10:E, 11:A, 12:E, 13:E, 14:A, 15:E

CHAPTER 2

1:E, 2:A, 3:B, 4:D, 5:E, 6:C, 7:C, 8:E, 9:D, 10:C, 11:E, 12:D, 13:A, 14:B, 15:C

CHAPTER 3

1:C, 2:E, 3:A, 4:E, 5:B, 6:A, 7:B, 8:E, 9:C, 10:A, 11:E, 12:A, 13:D, 14:E, 15:D

CHAPTER 4

1:E, 2:A, 3:B, 4:D, 5:C, 6:C, 7:A, 8:C, 9:A, 10:A, 11:E, 12:D, 13:E, 14:D, 15:E

CHAPTER 5

1:E, 2:C, 3:E, 4:A, 5:E, 6:D, 7:B, 8:E, 9:A, 10:B, 11:C, 12:D, 13:E, 14:A, 15:E

CHAPTER 6

1:A, 2:D, 3:C, 4:B, 5:E, 6:A, 7:A, 8:E, 9:E, 10:A, 11:E, 12:B, 13:C, 14:D, 15:E

CHAPTER 7

1:A, 2:E, 3:C, 4:B, 5:A, 6:E, 7:E, 8:D, 9:C, 10:D, 11:C, 12:A, 13:E, 14:B, 15:C

CHAPTER 8

1:E, 2:A, 3:B, 4:D, 5:E, 6:E, 7:B, 8:C, 9:B, 10:A, 11:C, 12:D, 13:E, 14:E, 15:C

CHAPTER 9

1:A, 2:E, 3:B, 4:C, 5:A, 6:D, 7:A, 8:A, 9:B, 10:D, 11:E, 12:C, 13:E, 14:A, 15:B

CHAPTER 10

1:E, 2:E, 3:C, 4:A, 5:B, 6:D, 7:E, 8:B, 9:A, 10:B, 11:C, 12:C, 13:B, 14:A, 15:E

CHAPTER 11

1:A, 2:D, 3:A, 4:B, 5:C, 6:D, 7:A, 8:B, 9:E, 10:C, 11:A, 12:B, 13:D, 14:E, 15:D

CHAPTER 12

1:B, 2:E, 3:A, 4:C, 5:E, 6:B, 7:D, 8:A, 9:E, 10:E, 11:E, 12:B, 13:C, 14:A, 15:A

CHAPTER 13

1:B, 2:A, 3:D, 4:A, 5:C, 6:A, 7:D, 8:C, 9:D, 10:A, 11:E, 12:E, 13:B, 14:D, 15:B

CHAPTER 14

1:E, 2:A, 3:C, 4:E, 5:A, 6:D, 7:B, 8:C, 9:E, 10:A, 11:E, 12:D, 13:E, 14:D, 15:B

CHAPTER 15

1:B, 2:A, 3:A, 4:C, 5:E, 6:A, 7:E, 8:E, 9:C, 10:A, 11:D, 12:A, 13:E, 14:C, 15:A

CHAPTER 16

1:D, 2:A, 3:C, 4:B, 5:A, 6:D, 7:E, 8:A, 9:E, 10:C, 11:B, 12:A, 13:C, 14:A, 15:D

CHAPTER 17

1:A, 2:D, 3:C, 4:C, 5:E, 6:D, 7:E, 8:B, 9:A, 10:C, 11:B, 12:D, 13:E, 14:A, 15:B

CHAPTER 18

1:A, 2:C, 3:D, 4:C, 5:C, 6:A, 7:B, 8:E, 9:B, 10:D, 11:A, 12:D, 13:E, 14:E, 15:C

CHAPTER 19

1:D, 2:D, 3:A, 4:B, 5:C, 6:B, 7:D, 8:A, 9:B, 10:A, 11:A, 12:D, 13:C, 14:D, 15:A

CHAPTER 20

1:A, 2:D, 3:B, 4:C, 5:D, 6:A, 7:C, 8:D, 9:A, 10:D, 11:E, 12:A, 13:E, 14:B, 15:C

CHAPTER 21

1:E, 2:E, 3:E, 4:B, 5:C, 6:B, 7:A, 8:B, 9:C, 10:A, 11:C, 12:A, 13:C, 14:A, 15:D

CHAPTER 22

1:E, 2:A, 3:B, 4:E, 5:A, 6:C, 7:D, 8:C, 9:B, 10:A, 11:B, 12:C, 13:D, 14:A, 15:E

CHAPTER 23

1:E, 2:A, 3:E, 4:E, 5:C, 6:A, 7:B, 8:C, 9:D, 10:A, 11:D, 12:B, 13:C, 14:D, 15:A

CHAPTER 24

1:A, 2:E, 3:D, 4:B, 5:B, 6:C, 7:E, 8:A, 9:B, 10:B, 11:A, 12:D, 13:A, 14:C, 15:B

CHAPTER 25

1:A, 2:E, 3:E, 4:A, 5:A, 6:A, 7:C, 8:E, 9:E, 10:B, 11:A, 12:E, 13:A, 14:A, 15:E

β3-adrenergic receptor Receptor found on fat cells that binds noradrenaline (norepinephrine)

β-galactosidase Enzyme that splits lactose and related disaccharides into simple sugars

β-lactamase Enzyme that degrades antibiotics of the β-lactam family

β-lactams Large family of antibiotics including both penicillins and cephalosporins

2-micron circle (2μ circle) A small multicopy plasmid found in the yeast *Saccharomyces cerevisiae*, whose derivatives are widely used as vectors

3′ untranslated region (3′ UTR) Sequence at the 3′-end of mRNA, downstream of the final stop codon that is not translated into protein

30-nanometer fiber Chain of nucleosomes that is arranged helically, approximately 30 nm in diameter

30S initiation complex Initiation complex for translation that contains only the small subunit of the bacterial ribosome

4S pathway Pathway of four steps for removing sulfur from dibenzothiophene and related molecules

5′ untranslated region (5′ UTR) Region of an mRNA between the 5′-end and the translation start site

59-base element (59-be) (see also *attC*) A repeated genetic sequence that flanks an integron gene cassette and contains a seven-base-pair core that is highly conserved

5-bromo-2-deoxyuridine (BrdU) A base analog for thymine that is easily incorporated into DNA during replication

6HIS Six tandem histidine residues that are fused to proteins so allowing purification by binding to nickel ions that are attached to a solid support. Also known as polyhistidine tag

70S initiation complex Initiation complex for translation that contains both subunits of the bacterial ribosome

a factor Yeast mating factor that acts as a pheromone, attracting yeast cells of the opposite mating type

ABO blood group Set of blood antigens on surface of red cells consisting of three different but related glycolipids

acceptor stem Base-paired stem of tRNA to which the amino acid is attached

acetylserine sulfhydrylase Enzyme that converts *O*-acetylserine plus hydrogen sulfide to cysteine

achondroplasia Inherited defect due to mutation of the *FGFR3* gene that encodes the fibroblast growth factor (FGF) receptor #3. Also known as short-limbed dwarfism

acquired immunity Type of immunity that responds to specific antigens over a course of days, has the ability to distinguish self from non-self, remembers past invading pathogens, and generates new antibodies

acquired immunodeficiency syndrome (AIDS) A disease caused by the HIV retrovirus, which slowly undermines the immune system by destroying helper T cells

activator proteins Protein that switches a gene on

***Ada* gene** Gene that encodes the enzyme adenosine deaminase

addiction module Set of genes that control cell death in bacteria such as *E. coli*

adeno-associated virus (AAV) A defective or satellite virus that depends on adenovirus (or certain herpesviruses) to supply necessary functions

adenosine deaminase Enzyme involved in purine metabolism whose absence prevents development of B cells and T cells

adenovirus-36 Strain of adenovirus that causes obesity

adenoviruses Family of small, spherical, double-stranded DNA viruses that infect animals

adenylate cyclase Enzyme that synthesizes cyclic AMP

adherent cell lines Cultured cells that grow attached to the culture dish

adhesin Bacterial protein that binds to glycoproteins or glycolipids on animal cell surfaces

adjuvant (or **carrier)** An agent used to increase the antigenicity of a vaccine

ADP-ribose Molecular fragment consisting of adenosine diphosphate plus ribose that is normally obtained by cleavage of NAD

ADP-ribosylation Addition of ADP-ribose group to a protein, thereby altering its activity or completely inactivating it

adverse drug reaction (ADR) Term used to describe serious side effects or lack of drug efficacy in certain patients

Aedes aegypti A mosquito species that transmits yellow fever

aequorin A protein from the luminescent jellyfish *Aequorea victoria* that emits blue light when triggered by calcium ions

affinity HPLC Chromatography technique where the stationary phase has a binding site for a specific molecule (e.g., an antibody to a particular protein)

A-form An alternative form of the double helix, with 11 base pairs per turn, often found for double-stranded RNA, but rarely for DNA

agarose A polysaccharide from seaweed that is used to form gels for separating nucleic acids by electrophoresis

aggressive gene therapy Therapy of cancers or infections by introducing genes or gene products that act to kill cancer cells or infectious agents

agouti Brown coat color in mice due to a dominant allele

AIDS (acquired immunodeficiency syndrome) A disease caused by the HIV retrovirus, which slowly undermines the immune system by destroying helper T cells

aligned (in reference to DNA sequence comparison) matching the most similar DNA sequences between two or more different genes

alkaline phosphatase An enzyme that cleaves phosphate groups from a wide range of molecules

allele One particular version of a gene, or more broadly, a particular version of any locus on a molecule of DNA

alpha (α) factor Yeast mating factor that acts as a pheromone, attracting yeast cells of the opposite mating type

alpha complementation Assembly of functional β-galactosidase from N-terminal alpha fragment plus rest of protein

alternative splicing library Library of gene or protein sequences generated by randomized inclusion or exclusion of exons from an original protein

Alu element An example of a SINE, a particular short DNA sequence found in many copies on the chromosomes of humans and other primates

amantadine Antiviral agent that blocks ion channels found in the outer envelope of influenza A virus

aminoacyl tRNA synthetases Enzyme that attaches an amino acid to tRNA

aminopeptidase A protease that removes amino acids from a polypeptide chain starting at the amino-terminal end

amniocentesis Procedure for drawing samples of amniotic fluid that contains some fetal cells

amyloid β A peptide fragment of a protein found in neuritic senile plaques

amyloid aggregate Insoluble clump of misfolded proteins found in several disease conditions including Alzheimer's and prion disease

amylopectin Branched component of starch

amylose Linear component of starch

analyte Mixture of molecules (such as proteins) that is being studied by chromatography

Angelman's syndrome Inherited defect resulting in loss of function of genes on the maternally derived copy of chromosome 15 that are subject to imprinting

angiogenesis The development of new blood vessels

Anopheles gambiae A mosquito species that transmits malaria

anthrax Virulent bacterial disease of cattle that readily infects humans and is caused by *Bacillus anthracis*

antibodies Proteins of the immune system that recognize and bind to foreign molecules (antigens)

anticodon Group of three complementary bases on tRNA that recognize and bind to a codon on the mRNA

antigen capture immunoassay Method of protein detection where an antibody is linked to a solid support and captures its cognate protein from a mixture

anti-gene Gene that gives a full-length antisense RNA when transcribed

antigens Molecules that cause an immune response and which are recognized and bound by antibodies

anti-oncogene Gene that acts to prevent unwanted cell division (same as tumor-suppressor gene)

antiparallel Parallel, but running in opposite directions

antisense (or **noncoding**) The strand of DNA that has complementary sequence to mRNA

antisense DNA oligonucleotide Short antisense DNA

antisense DNA Single-stranded DNA with a sequence complementary to a specific target molecule of DNA or RNA, usually mRNA; binding of mRNA by antisense DNA may block its translation

antisense RNA RNA complementary in sequence to messenger RNA and which therefore base-pairs with it

anti-Shine-Dalgarno sequence Sequence on 16S rRNA that is complementary to the Shine-Dalgarno sequence of mRNA

antiterminator proteins Protein that allows transcription to continue through a transcription terminator

AP-1 (activator protein-1) Eukaryotic transcription factor that activates a variety of different genes

apoptosis Genetic program that eliminates damaged cells or cells that are no longer needed without activating the immune system

apoptosome A complex of signaling proteins that activates mammalian caspase-3 during apoptosis

apoptotic bodies Dense granular fragments of an apoptotic cell

aptamer Oligonucleotide that binds to another molecule, often a protein, but is not encoded by a naturally occurring DNA sequence

artificial chromosome A self-replicating element used to clone large fragments of DNA that has three main components: an origin of replication, a centromere, and telomeres

ascospores Type of spore made inside an ascus by fungi of the ascomycete group, including yeasts and molds

ascus Specialized spore-forming structure of ascomycete fungus

aspartic protease A class of protease that has two aspartic acid residues in its active site

assembler A general-purpose device for molecular manufacturing capable of guiding chemical reactions by positioning molecules

atomic force microscope (AFM) An instrument able to image surfaces to molecular accuracy by mechanically probing their surface contours

ATP synthetase Enzyme that synthesizes of ATP using energy from the respiratory chain

attenuated vaccine A live pathogen that no longer causes a disease state, but still stimulates the immune system to create antibodies

attenuation riboswitches Type of riboswitch that causes premature termination by inducing an alternative stem and loop structure in the leader region of the mRNA when triggered by a signal

autogenous regulation Self-regulation, i.e., when a DNA binding protein regulates the expression of its own gene

automatic DNA sequencer Machine that separates a dye-labeled dideoxy sequencing reaction into increasing larger fragments, analyzes the final nucleotide of each fragment by recording the color, and prints the results in a graphical format to determine the sequence

autoradiography Allowing radioactive materials to take pictures of themselves by laying them flat on photographic film

avidin A protein from egg white that binds biotin very tightly

azidothymidine (AZT) Nucleoside analog that acts as a chain terminator. It is used against AIDS and is also known as zidovudine

B cell The type of immune system cell that produces antibodies

Bacillus anthracis Bacterium that causes anthrax and that readily infects humans

back-crossed Procedure in which pollen from the progeny of two parents is cross-pollinated onto one of the original parents

bacmid Hybrid cloning vector made from a baculovirus and a plasmid

bacteriocin Toxic protein made by bacteria to kill closely related bacteria

bacteriophage Virus that infects bacteria

bacteriophage vector A bacteriophage genome where nonessential bacteriophage genes are replaced with a multiple cloning site where other DNA fragments can be cloned. The vector can be maintained within a bacterial cell since it contains all the necessary genes for replication, growth, and lysis

baculoviruses Family of DNA viruses that infect insects and related invertebrates and are widely used as vectors

bait The fusion between the DNA binding domain of a transcriptional activator protein and another protein as used in two-hybrid screening

barcode Unique 20-nucleotide sequence added to each clone in a library that marks or identifies each different clone

base Alkaline chemical substance, in molecular biology especially refers to the cyclic nitrogen compounds found in DNA and RNA

base peak Peak for the most intense ion detected in a sample

base substitution Mutation in which one base is replaced by another

basic peptide A short length of basic amino acids that facilitate translocation across biological membranes. These are derived from known proteins or made synthetically

Bcl-2 Mitochondrial protein that helps control entry into apoptosis

beta-galactosidase or **β-galactosidase (LacZ)** Enzyme that splits lactose and related molecules to release galactose

B-form The normal form of the DNA double helix, as originally described by Watson and Crick

bifurcation Division of two different taxonomic groups from each other during evolution

bioaugmentation Adding degrading microorganisms (normal or engineered) plus nutrients to remove a contaminant from the environment

bioinformatics The computerized analysis of large amounts of biological sequence data

biopanning Method of screening a phage display library for a desired protein by binding to a bait molecule attached to a solid support

bioreactors Large chambers or vats used to grow organisms such as yeast or bacteria so that the biotechnological product can be harvested on a large scale

bioremediation Using any microorganism, fungus, plant, or enzyme to clean up or restore the environment to its original condition

biosafety level Rating system (BL1–BL4) for high-containment laboratories

biosensor Detection or monitoring device that relies on a biological mechanism

biosensory system A genetic circuit that is controlled by a specific environmental cue

biostimulation Using different chemical compounds to stimulate naturally occurring microorganisms to remove a contaminant from its environment

Black Death Bubonic plague, or more specifically, the great plague epidemic of the mid-1300s

blastocyst A very early stage of the embryo

blood antigens Glycoproteins and glycolipids on the surface of blood cells that are used for antigenic analysis

blood typing Use of differences in blood antigens to determine identity

blunt ends Ends of a double-stranded DNA molecule that are fully base-paired and have no unpaired single-stranded overhang

Bohr radius The natural preferred distance between positive and negative charges in a material

Botox Botulinum toxin, especially as used for cosmetic purposes

botulinum toxin Protein toxin made by *Clostridium botulinum* that blocks nerve transmission

botulinum toxin type A Specific type of botulinum toxin used clinically and for cosmetic purposes

bovine spongiform encephalopathy (BSE) Official name for mad cow disease, a brain degeneration disease of cattle caused by prions

BRCA1 Breast cancer A1 gene, involved in DNA repair. Defects in this gene predispose women to breast cancer

BRCA2 Breast cancer A2 gene, involved in DNA repair. Defects in this gene predispose women to breast cancer

Bt toxin Toxin found in the soil bacterium *Bacillus thuringiensis*, which kills certain caterpillars that destroy crops

bubonic plague Virulent bacterial disease spread by fleas and primarily found in rodents

bud The new asexual daughter cell of a yeast that forms as bulge on the surface of the mother cell

bZIP proteins Family of transcription factors that each have a leucine zipper domain

CAD (caspase-activated DNase) A nuclear DNase that cleaves DNA between histones; activated by caspases during apoptosis

callus An unorganized, proliferating mass of undifferentiated plant cells

callus culture Dedifferentiating a mass of plant tissue *in vitro*, and then growing the undifferentiated tissue in a petri dish

calmodulin A small calcium-binding protein of animal cells

cancer Tissue or cluster of cells resulting from uncontrolled cell growth and division due to somatic mutations affecting cell division

candidate cloning Approach to cloning that involves making an informed guess as to what kind of protein is likely to be involved in a hereditary disorder

CAP (catabolite activator protein) A transcription enhancer protein that activates the transcription of sugar metabolizing operons when bound by cAMP (also CRP-cyclic AMP receptor protein)

cap Structure at the 5′-end of eukaryotic mRNA consisting of a methylated guanosine attached in reverse orientation

capsid Shell or protective layer that surrounds the DNA or RNA of a virus particle

CAR (coxsackievirus adenovirus receptor) Receptor on animal cells shared by adenoviruses and B-group coxsackieviruses

carboxypeptidase Enzyme that removes C-terminal amino acid residue from a protein

carboxypeptidase H A particular carboxypeptidase that is needed during the processing of insulin

carcinogen Cancer-causing agent

carrier (or **adjuvant**) An agent used to increase the antigenicity of a vaccine

caspase Enzyme that carries out the protein degradation characteristic of apoptosis; caspases cleave other proteins after aspartic acid residues

caspase-activated DNase (CAD) A nuclear DNase that cleaves DNA between histones; activated by caspases during apoptosis

catalase Enzyme that converts hydrogen peroxide to oxygen plus water

catecholamines Family of neurotransmitters including dopamine, adrenalin (= epinephrine) and noradrenalin (= norepinephrine)

catenated Structure in which two or more circles of DNA are interlocked

CCR5 A protein that acts as a co-receptor for entry of HIV into host cells. Its natural role is as a chemokine receptor

CD4 protein Protein found on the surface of many T cells that is used a receptor by the AIDS virus

cell cycle Series of stages that a cell goes through from one cell division to the next

cell-mediated immunity Type of immunity that relies on T cells and cell-cell interactions (rather than on antibodies made by B cells)

cellular senescence Arrested cell division of cultured mammalian cells; same as replicative senescence

cellulose Structural polymer of β-1,4-linked glucose found in plant cell walls

central dogma Basic plan of genetic information flow in living cells which relates genes (DNA), message (RNA), and proteins

centromere Region of eukaryotic chromosome where the microtubules attach during mitosis and meiosis

centromere (Cen) sequence Sequence at centromere of eukaryotic chromosome that is needed for correct partition of chromosomes during cell division

cephalosporins Sub-family of antibiotics belonging to the β-lactam family of antibiotics

CFTR protein Cystic fibrosis transporter—protein encoded by the cystic fibrosis gene and found in the cell membrane where it acts as a channel for chloride ions

CG islands Region of DNA in eukaryotes that contains many clustered CG sequences that are used as targets for cytosine methylation

chain termination sequencing Method of sequencing DNA by using dideoxynucleotides to terminate synthesis of DNA chains. Same as dideoxy sequencing

chain terminator Nucleoside analog that is incorporated into a growing DNA chain but prevents further extension of the DNA chain

chaperonin Protein that oversees the correct folding of other proteins

checkerboard hybridization Technique for mass screening of DNA samples using probes to 16S rRNA genes corresponding to many different bacteria

chemokine receptor Cell surface protein required for uptake of chemokines into animal cells

chemokines Group of approximately 50 small messenger peptides that activate the white cells of the immune system

chimeric animal Animal in which different cells vary genetically

ChIP-chip Technique to identify the sequence of transcription factor binding sites by whole-genome arrays after the binding sites are crosslinked to the transcription factors and isolated by immunoprecipitation

chloroplast transit peptide A small peptide added to the N-terminus of a protein that directs the protein from the ribosome into the chloroplast

cholera toxin Protein toxin made by *Vibrio cholerae* that hyperactivates the adenylate cyclase of animal cells

cholesterol Sterol derivative found in humans and other animals that has a major effect on atherosclerosis

chromatin Complex of DNA plus protein which constitutes eukaryotic chromosomes

chromatin immunoprecipitation (ChIP) Immunoprecipitation of transcription factors crosslinked to their binding sites on DNA

chromatography Assorted techniques that separate a mixture of molecules by size or chemical properties

chromatography column A long tube that contains the stationary phase for chromatography

chromosome walking Method for cloning neighboring regions of a chromosome by successive cycles of hybridization using overlapping probes

chronic wasting disease (CWD) Prion disease of deer and elk in the northern United States

chronological aging Length of time the yeast cell can live without dividing

circum-sporozoite protein Protein on surface of sporozoite stage of malarial parasite

cistron Segment of DNA (or RNA) that encodes a single polypeptide chain

clade A group of organisms that share a common evolutionary ancestor

cladistics A system of biological classification based on using quantitative analysis of physical traits to construct evolutionary relationships

clamp-loading complex Group of proteins that loads the sliding clamp of DNA polymerase onto the DNA

class I major histocompatibility complex (class I MHC) MHC proteins consisting of two chains of unequal length and found on the surface of all cells. Their role is to display fragments of proteins originating inside the cell

class II major histocompatibility complexes (class II MHC) MHC proteins consisting of two chains of equal length and found only on the surface of certain immune cells. Their role is to display fragments of proteins from digested microorganisms

ClfA (clumping factor A) A protein found on the surface of *Staphylococcus aureus* that allows the bacteria to adhere to fibrinogen of the host cell

cloned animals Genetically identical animals derived from the same original cell line

cloning vectors Any molecule of DNA that can replicate itself inside a cell and is used for carrying cloned genes or segments of DNA. Usually a small multicopy plasmid or a modified virus

Clostridium botulinum Bacterium that makes botulinum toxin

cluster of differentiation (CD) antigen Cell surface proteins found on leukocytes and used to classify the different types of leukocytes

coding strand (see nontemplate or sense) The strand of DNA equivalent in sequence to the messenger RNA (same as plus strand)

codon bias Some organisms use only a subset of redundant codons for a particular amino acid, whereas other organisms will use a different subset. This bias can hinder expression of foreign proteins in genetic engineering

codon usage The favoring of different alternative codons for the same amino acid in different organisms

codon Group of three RNA or DNA bases that encodes a single amino acid

co-immunoprecipitation Method of identifying protein interactions by using antibodies to one of the proteins

"cold" Slang for nonradioactive

colicin Toxic protein or bacteriocin made by *Escherichia coli* to kill closely related bacteria

combinatorial library Large series of related molecules that have been systematically generated by combining chemical groups and/or molecular motifs

combinatorial screening Screening of a large series of related molecules for those with useful properties

comparative genomics Comparing genome sequences of different organisms, especially in attempts to determine potential functions for genes or proteins by comparison with characterized examples

competent Cell that is capable of taking up DNA from the surrounding medium

complementarity determining region (CDRs) Short segment that forms a loop on the surface of the variable region of an antibody, thus forming part of the antigen-binding site

complementary DNA (cDNA) DNA copy of a gene that lacks introns and therefore consists solely of the coding sequence. Made by reverse transcription of mRNA

complex transposon A transposon that moves by replicative transposition

c-onc Cellular version of an oncogene

conditional mutation Mutation whose phenotypic effects depend on environmental conditions such as temperature or pH

conjugation Process in which genes are transferred by cell-to-cell contact in bacteria

conservative substitution Replacement of an amino acid with another that has similar chemical and physical properties

conservative transposition Same as cut-and-paste transposition

constant region Region of an antibody protein chain that remains constant in sequence

constitutive heterochromatin Regions of more or less permanent heterochromatin found on both copies of homologous chromosomes, especially around centromeres

constitutive promoter A promoter that functions in all tissues at all times

constitutive Term used to describe genes that are expressed during all conditions

construct Term used to describe a cloning vector derived from various different genes and DNA segments assembled into one

contact inhibition Inhibition of cell division that occurs due to contact with neighboring cells

contig A stretch of known DNA sequence, built up from smaller cloned fragments, that is contiguous and lacks gaps

contig map A genome map based on contigs

controlled pore glass (CPG) Glass with pores of uniform sizes that is used as a solid support for chemical reactions such as artificial DNA synthesis

Coomassie blue A blue dye used to stain proteins

copy number The number of copies of a gene or plasmid found within a single host cell

core enzyme The part of DNA or RNA polymerase that synthesizes new DNA or RNA (i.e., lacking the recognition and/or attachment subunits)

co-receptor Protein required for virus entry into host cell in addition to the main virus receptor

co-repressor In prokaryotes—a small signal molecule needed for some repressor proteins to bind to DNA; in eukaryotes—an accessory protein, often a histone deacetylase, involved in gene repression

cos **sequences (lambda cohesive ends)** Complementary 12 base-pair-long overhangs found at each end of the linear form of the lambda genome

cosmid Small multicopy plasmid that carries lambda *cos* sites and can carry around 45 kb of cloned DNA

co-suppression An alternate term used to describe RNA interference-like phenomena

coxsackievirus adenovirus receptor (CAR) Receptor on animal cells shared by adenoviruses and B-group coxsackieviruses

C-peptide Connecting peptide that originally links the A- and B-chains of insulin but which is absent from the final hormone

CRE (cyclic AMP response element) A specific DNA sequence found in front of genes that are activated by cyclic AMP in higher organisms

Cre A recombinase from bacteriophage P1, which directs recombination at specific sites (*loxP* sites)

Cre recombinase Enzyme encoded by bacterial virus P1 that catalyzes recombination between inverted repeats (*loxP* sites)

Cre/*loxP* A system allowing a specific gene to be added or removed during development by exploiting Cre recombinase to excise DNA sequences flanked by *loxP* recombination sites

CREB protein (cyclic AMP response element binding protein) Protein that regulates genes in response to cyclic AMP in animal cells by binding to CRE sequences in promoters

Creutzfeldt-Jakob disease (CJD) Inherited or spontaneous brain degeneration disease of humans caused by prions

crossover Structure formed when the strands of two DNA molecules are broken and rejoined with each other

cross-pollination Taking the pollen from one plant and placing it on the stigma of another in order to direct the exchange of genetic information

CRP protein (cyclic AMP receptor protein) Bacterial protein that binds cyclic AMP and then binds to DNA (see CAP, catabolite activator protein)

Cry protein Crystalline protein found in *Bacillus* bacterial spores that breaks into delta endotoxin (Bt toxin)

CTXphi Filamentous bacteriophage that carries genes for cholera toxin and lysogenizes *Vibrio cholerae*

cut-and-paste transposition Type of transposition in which a transposon is completely excised from its original location and moves as a whole unit to another site

cycle sequencing Technique that combines PCR and chain termination sequencing to determine the sequence of a template DNA

cyclic AMP (cAMP) (cyclic adenosine monophosphate) A signal molecule used in global regulation (in bacteria) and as a second messenger (in higher organisms). A cyclic mononucleotide of adenosine that is formed from ATP

cyclic AMP response element (CRE) A specific DNA sequence found in front of genes that are activated by cyclic AMP in higher organisms

cyclic AMP response element binding protein (CREB protein) Protein that regulates genes in response to cyclic AMP in animal cells by binding to CRE sequences in promoters

cyclic GMP (cyclic guanosine monophosphate) A signal molecule used as a second messenger by eukaryotic cells

cyclic phosphodiesterase Enzyme that degrades cyclic nucleotides, including cyclic AMP and cyclic GMP

cyclin-dependent kinase (CDK) Specialized protein kinase that is activated by a cyclin and participates in control of cell division

cyclins Family of proteins that controls the cell cycle

cysteine protease A class of protease with cysteine in its active site

cystic fibrosis An inherited disease in which the major symptom is the accumulation of fibrous tissue in the lungs, ultimately due to defects in trans-membrane chloride channel

cytochrome *c* Mitochondrial protein that functions in electron transport; released from mitochondria during apoptosis

cytogenetic map A visual chromosome map based on light and dark bands due to staining and seen under the light microscope

cytokine response modifier genes (*crm* genes) Genes of poxviruses that interfere with the action of NK cells and cytotoxic T cells

cytokines Short peptides that stimulate cell growth and division, especially of immune cells

cytosine deaminase Enzyme, usually from bacteria, that converts cytosine into uracil

data mining The use of computer analysis to find useful information by filtering or sifting through large amounts of data

dauer Stage of *C. elegans* life cycle initiated by low food and water; characterized by lack of movement and eating

***db* gene** Gene that encodes the receptor for the hormone leptin

***de novo* methylase** An enzyme that adds methyl groups to wholly nonmethylated sites

death receptor A cell surface receptor that transmits an external signal to die to the intracellular proteins responsible for killing the cell

death receptor pathway Program of apoptosis that involves activating membrane-bound receptors with extrinsic factors

death-inducing signaling complex (DISC) A complex of proteins activated by binding to a trimerized death receptor; the proteins initiate apoptosis

defensin A Anti-bacterial peptide made by mosquito

degenerate primer Primer with several alternative bases at certain positions

degradome The complete set of proteases that are expressed at one time, under a defined set of conditions

dehydrogenase Type of enzyme that removes hydrogen atoms from its substrate

deletion Mutation in which one or more bases is lost from the DNA sequence

delta endotoxin Actual toxin released from the Cry protein into the gut of caterpillars that digest *Bacillus* bacterial spores

demethylases An enzyme that removes methyl groups

deoxyribose The sugar with five carbon atoms that is found in DNA

deoxyribozyme Artificial DNA molecule that acts as an enzyme

***DHFR* gene** Gene that encodes the enzyme dihydrofolate reductase

diabetes mellitus Group of related diseases causing inability to control level of blood sugar due to defect in insulin production

diabody Artificial antibody construct made of two single-chain Fv (scFv) fragments assembled together

dibenzothiophene (DBT) Thiophene fused to two benzene rings, widely regarded as a model compound for organic sulfur in coal and oil

Dicer An enzyme that cuts double-stranded RNA into small segments of 21 to 23 nucleotides in length (siRNA)

dicots (see dicotyledonous) Plants with broad leaves with netlike veins, whose seedlings have two cotyledons, or seed leaves

dicotyledonous plants (see dicots) Plants with broad leaves with netlike veins, whose seedlings have two cotyledons, or seed leaves

dideoxynucleotide Nucleotide whose sugar is dideoxyribose instead of ribose or deoxyribose

differential fluorescence induction (DFI) Method to identify genes active in infectious agents that might be used for potential vaccines. Fluorescently labeled library clones are used to determine which genes are active after the infectious agent enters the cell

differential interference microscopy (see Nomarski optics) Type of polarization microscopy that transforms differences in density into differences in height so that the image looks like a three-dimensional relief

dihydrofolate reductase Enzyme that takes part in one carbon metabolism and is needed for the synthesis of thymine and adenine

dimethoxytrityl (DMT) group Group used for blocking the 5′-hydroxyl of nucleotides during artificial DNA synthesis

dioxygenase Enzyme that inserts two oxygen atoms into its substrate so yielding a diol

diphtheria toxin Protein toxin made by *Corynebacterium diphtheriae* that inactivates elongation factor EF-2 of animal cells

diploid Having two copies of each chromosome and hence of each gene

direct immunoassay (or reverse-phase array) Method of protein detection, where the protein is bound to a solid support and an antibody with a detection system is added

directed evolution Technique for enhancing or altering the original activity of an enzyme by randomly mutating the gene and then screening for the new activity

directed mutagenesis Deliberate alteration of the DNA sequence of a gene by any of a variety of artificial techniques

DISC (death-inducing signaling complex) A complex of proteins activated by binding to a trimerized death receptor; the proteins initiate apoptosis

***DMD* gene** Gene located on the X-chromosome that encodes dystrophin and is defective in Duchenne's muscular dystrophy

DNA (deoxyribonucleic acid) Nucleic acid polymer of which the genes are made

DNA adenine methylase (Dam) A bacterial enzyme that methylates adenine in the sequence GATC

DNA cassette Deliberately designed segment of DNA that is flanked by convenient restriction or recombinase sites

DNA chain terminator A nucleotide analog that is incorporated into a growing DNA chain and stops further elongation

DNA chip technology Method of hybridization where a chip is used to simultaneously detect and identify many short DNA fragments. Also known as DNA array technology or oligonucleotide array technology

DNA cytosine methylase (Dcm) A bacterial enzyme that methylates cytosine in the sequences CCAGG and CCTGG

DNA fingerprinting Individual identification by means of differences in DNA sequence generally visualized as the pattern of DNA fragments after separation by gel electrophoresis

DNA gyrase An enzyme that introduces negative supercoils into DNA, a member of the type II topoisomerase family

DNA helicase Enzyme that unwinds double helical DNA

DNA invertase An enzyme that recognizes specific sequences at the two ends of an invertible segment and inverts the DNA between them

DNA ligase Enzyme that joins DNA fragments covalently, end to end

DNA microarray Chip used to simultaneously detect and identify many short DNA fragments by DNA-DNA hybridization. Also known as DNA array or oligonucleotide array detector, same as DNA array or DNA chip

DNA polymerase An enzyme that elongates strands of DNA, especially when chromosomes are being replicated

DNA polymerase III (Pol III) Enzyme that makes most of the DNA when bacterial chromosomes are replicated

DNA SELEX Method where catalytically active DNA sequences are altered by mutations to use new substrates

DNA shuffling Method of artificial evolution in which genes are cut into segments, which are mutagenized, mixed, shuffled, and rejoined

DNA synthesizer Machine that adds one nucleotide after another in a 3′ to 5′ direction to assemble a length of DNA

DNA vaccine Vaccine that consists of DNA that encodes a specific antigenic protein of the disease agent. After entering the host, the DNA is expressed, and the immune system reacts to the foreign protein

dominant-negative mutation Mutation giving a gene product that not only is functionally defective but also inactivates the wild-type gene product. Usually occurs with multisubunit proteins

dot blot Hybridization technique where various DNA samples are attached to a filter in small spot. The blot can then be probed with a gene of interest to find matching sequences

double helix Structure formed by twisting two strands of DNA spirally around each other

Down syndrome Defective development, including mental retardation, resulting from an extra copy of chromosome 21

driver Normal DNA used in suppressive subtraction hybridization that is used to "subtract" or remove the common sequences

***dszABC* operon** Group of genes encoding pathway for removing sulfur from dibenzothiophene

Duchenne's muscular dystrophy One particular form of muscular dystrophy, degenerative muscle disease

duplication Mutation in which a segment of DNA is duplicated

dynamic mutation Mutation consisting of multiple tandem repeats that increase in number over successive generations

dystrophin Protein encoded by the *DMD* gene whose malfunction causes muscular dystrophy. Dystrophin helps attach muscle fibrils to the membranes of muscle cells

E1A protein An adenovirus early protein that promotes transcription of other early virus genes and binds to host cell Rb protein

E2F Regulatory protein that controls synthesis of cyclins E and A

early genes Genes expressed early during virus infection and that mainly encode enzymes involved in virus DNA (or RNA) replication

ecdysone Steroid hormone from insects that is involved in molting

edema factor (EF) Protein toxin made by anthrax bacteria that acts as an adenylate cyclase

edible vaccine Vaccine that is expressed in the edible part of a plant. Eating an edible vaccine would confer resistance to the disease state

elastin Protein encoded by the *ELN* gene and found in the elastic tissues of skin, lung, and blood vessels

elastin-like polypeptide (ELP) A protein created in the laboratory that has elastin-like properties

electrochemical detector A device that detects changes in oxidation and reduction of proteins (or other molecules) in the mobile phase during chromatography

electroporation Technique that uses high-voltage discharge to make cells competent to take up DNA

electrospray ionization (ESI) Type of mass spectrometry in which gas-phase ions are generated from ions in solution

ELISA See enzyme-linked immunosorbent assay

elongation (in reference to protein translation) The process of adding amino acids onto a growing polypeptide chain

elongation factors Proteins that are required for the elongation of a growing polypeptide chain

elution Removing bound molecules from a chromatography column by passing through buffer

embryonic stem cell Stem cell derived from the blastocyst stage of the embryo

endomesoderm Mesoderm that derives from the endoderm of a two-layered blastodisc

endopeptidase Protease that cuts within a polypeptide chain

endotoxin Lipopolysaccharide from the outer membrane of gram-negative bacteria

enhancer Regulatory sequence outside, and often far away from, the promoter region that binds transcription factors

enterochelin (or **enterobactin**) A siderophore used by *E. coli* and many enteric bacteria

enterotoxin Protein toxin secreted by enteric bacteria

enteroviruses Group of small RNA viruses that infect animal intestines and includes poliovirus

Enviropig™ Transgenic pig with much less phosphorus in its waste

enzyme-linked immunosorbent assay (ELISA) Sensitive method to assay the amount of a specific protein in a sample using enzyme-linked antibodies

epigenetic change 1. Refers to inherited changes that are not due to alterations in the DNA sequence. 2. Changes in gene regulation that persist after tissue culture, but are not inherited by any progeny of the plant

epitopes Localized regions of an antigen to which an antibody binds

EPSPS An enzyme essential for making amino acids of the aromatic family in plants and bacteria

ergot Fungus that grows on cereals, especially rye, and produces a mixture of toxins that causes a syndrome called ergotism

error-prone PCR Type of PCR where mutations are introduced at random during the amplification steps

erythromycin An antibiotic of the macrolide family that inhibits protein synthesis

erythropoietin A protein required for proper development of red blood cells

Escherichia coli A species of bacterium commonly used in genetics and molecular biology

ethidium bromide A stain that specifically binds to DNA or RNA and appears orange if viewed under ultraviolet light

eugenics Deliberate improvement of the human race (or other species) by selective breeding

event A term used to distinguish different insertions of the same transgene. The transgene is identical, but the integration sites will vary

***ex vivo* gene therapy** Gene therapy by removal, engineering, and return of cells derived from the same original patient

exon Segment of a gene that codes for protein and that is still present in the messenger RNA after processing is complete

exopeptidase A protease that cuts starting at either the carboxy- or amino-terminal end of the polypeptide chain

exotoxin Protein toxin secreted by bacteria

explant Tissue taken from an original site in a plant and transferred to artificial medium for growth and maintenance

expressed (as pertaining to genes) Converting a DNA region into RNA and/or protein to be used by the living cell

expressed sequence tag (EST) A special type of STS derived from a region of DNA that is expressed by transcription into RNA

expression library A library where each cloned piece of DNA is expressed into a protein by the vector. See also expression vector

expression vector Vector designed to enhance gene expression, usually by providing a strong promoter that drives expression of the cloned gene

Fab fragments Antigen-binding fragments of an antibody

facultative heterochromatin Heterochromatin that has the ability to return to normal euchromatin

family (in relationship to gene sequences) Group of closely related genes that arose by successive duplication and perform similar roles

fast neutron mutagenesis Mutagenesis technique where seeds are exposed to fast neutrons that induce small deletions

fast neutrons Free neutrons with kinetic energy around 1 MeV that are used to create DNA deletions in plant seeds

Fc fragment The stem region of an antibody, i.e., the fragment that does not bind the antigen

ferritin An iron storage protein of animals

Fierce Mouse Transgenic mouse with abnormal aggression

filoviruses Family of negative-strand ssRNA viruses that includes Ebola virus and Marburg virus

fimbria (plural **fimbriae**) Thin helical protein filaments found on the surface of bacteria; same as pilus

FLAG tag A short peptide tag (AspTyrLysAspAspAspAspLys) that is bound by a specific anti-FLAG antibody and that may be attached to proteins

flight tube Tube or channel in which ions generated during mass spectroscopy are separated according to size or charge

flippase (same as Flp recombinase and Flp protein) Enzyme encoded by the 2-micron plasmid of yeast that catalyzes recombination between inverted repeats (FRT sites)

floral dip (also called *in planta transformation*) Method in which a transgene is carried by *Agrobacterium* and transformed into *Arabidopsis* by dipping a developing flower bud into a solution of *Agrobacterium* plus a surfactant

flow cytometry Analyzing (without sorting) different cells as they flow past a fluorescent detector by observing the attached fluorescent antibodies

Flp protein Enzyme encoded by the 2-micron plasmid of yeast that catalyzes recombination between inverted repeats (FRT sites)

Flp recombinase (same as flippase and Flp protein) Enzyme encoded by the 2-micron plasmid of yeast that catalyzes recombination between inverted repeats (FRT sites)

fluorescence activated cell sorting (FACS) Technique that sorts cells (or chromosomes) based on fluorescent labeling

fluorescence detector A detector that records fluorescence emitted by the passing mobile phase after it has been excited by light

fluorescence *in situ* hybridization (FISH) Using a fluorescent probe to visualize a molecule of DNA or RNA in its natural location

***Fok*I** A particular type II restriction endonuclease with separate recognition and nuclease domains

foster mother (as used in genetics) Female animal that carries engineered embryos

founder animal Animal that is the original host for a transgene and maintains it stably

fragile X syndrome Inherited defect resulting in a fragile site within the long arm of the X chromosome that can be seen by microscopic observation

Francisella tularensis Bacterium that causes tularemia

FRT site Flp recombination target, the recognition site for Flp recombinase

functional cloning Approach to cloning that starts with a known protein that is suspected of involvement in a hereditary disorder

functional genomics The study of the whole genome and its expression

G_0 phase Resting phase in which eukaryotic cells no longer grow or divide

G_1 phase First stage in the eukaryotic cell cycle, in which cell growth occurs

G_2 phase Third stage in the eukaryotic cell cycle, in which the cell prepares for division

G-418 Aminoglycoside antibiotic that kills animal cells by blocking protein synthesis; also known as genetecin

Gal4 protein Transcriptional activator from yeast that has a DNA binding domain and a transcription activating domain

ganciclovir A nucleoside analog that is converted by cells to a DNA chain terminator. Used against virus infections and in cancer therapy

ganglioside GM_1 Glycolipid found in eukaryotic cell membranes that is used as a receptor by cholera toxin

GC ratio The amount of G plus C relative to all four bases in a sample of DNA. The GC ratio is usually expressed as a percentage

gel electrophoresis Electrophoresis of charged molecules through a gel meshwork in order to sort them by size

gene cassette Simple genetic element that has one or two open reading frames flanked by repeats called 59-be or *attC* and inserts into integrons

gene creature Genetic entity that consists primarily of genetic information, sometimes with a protective covering, but without its own machinery to generate energy or replicate macromolecules

gene fusion Structure in which parts of two genes are joined together, in particular when the regulatory region of one gene is joined to the coding region of a reporter gene

GENE impedance (GENEi) An alternate term used to describe RNA interference. Proposed to encompass all the different phenomena that induce the degradation of a specific RNA transcript homologous to a short interfering RNA

gene library Collection of cloned segments of DNA that is big enough to contain at least one copy of every gene from a particular organism. Same as DNA library

gene superfamily Group of related genes that arose by several stages of successive duplication. Members of a superfamily have often diverged so far that their ancestry may be difficult to recognize

gene testing Analyzing DNA sequences to identify potential disease states in a patient

gene therapy Medical treatment that involves use of recombinant DNA technology; originally referred to curing hereditary defects by introducing functional genes

general secretory system (Sec system) Standard system for exporting proteins across membranes that is found in most organisms

general transcription factors Proteins that work to enhance or repress gene expression for all genes

genetecin (G-418) Aminoglycoside antibiotic that kills animal cells by blocking protein synthesis

genetic changes Changes in plants after tissue culture that are inherited by the progeny, including alterations in chromosomes

genetic circuits Combinations of genes, promoters, enhancers, and repressors that control the output or expression of the final gene product

genetic maps Maps of genetic markers and/or genes ordered by using linkage information but without exact base-pair distances

genetic markers Physical landmarks used to construct genomic maps that are genetic in origin, such as genes, single nucleotide polymorphisms

genetic surgery Less common name for gene replacement therapy

genome browser A computer program that allows the scientist to look graphically at the entire genome of an organism to identify regions or genes of interest

genome map A graphical representation of an organism's genome

germline Reproductive cells producing eggs or sperm that take part in forming the next generation (in eukaryotic organisms)

ghrelin Peptide hormone that signals hunger

glucocorticoid hormones Group of steroid hormones in mammals that are involved in water and ion balance

glutathione-*S*-transferase (GST) Enzyme that binds to the tripeptide glutathione; often used in making fusion proteins

glycogen Polymer of glucose linked by α-1,4 bonds and used as a storage polysaccharide by animals and bacteria

glyphosate A weed killer that inhibits synthesis of amino acids of the aromatic family in plants

gp120 Glycoprotein of 120 kDa found in outer envelope of HIV that binds to CD4 protein on host cell, resulting in entry of the virus

G-proteins Class of GTP-binding eukaryotic proteins involved in signal transmission

gratuitous inducer A molecule (usually artificial) that induces a gene but is not metabolized like the natural substrate; the best known example is the induction of the *lac* operon by IPTG

green fluorescent protein (GFP) A jellyfish protein that emits green fluorescence and is widely used in genetic analysis

growth factor Protein or other chemical messenger circulating in the blood that carries signals for promoting growth to the cell surface

guanylate cyclase Enzyme that synthesizes cyclic GMP

hairpin A double-stranded base-paired structure formed by folding a single strand of DNA or RNA back on itself

hairpin ribozyme Small catalytic RNA molecule with four helices around two internal loops

hammerhead ribozyme Small catalytic RNA containing three helices around one core loop

haploid spores A fungal sexual offspring that carries a single copy of all the genes (generally used of organisms that have two or more sets of each gene)

haploinsufficiency Situation where a defect in one of the two copies of a diploid gene causes significant phenotypic defects

heat-stable oral vaccines Vaccines that are made by expressing a disease antigen in plant leaf tissue. Once the plant tissue is grown, it can be freeze-dried and encapsulated so that refrigeration is unnecessary

heavy chains The longer of the two pairs of chains forming an antibody molecule

helper component proteinase (Hc-Pro) A polyviral protein that inhibits accumulation of plant siRNAs, but has no effect on the spread of RNA interference to other parts of the plant

helper virus A virus that provides essential functions for defective viruses, satellite viruses, and satellite RNA

hemicellulose Mixture of polymers found in plant cell walls that consist of several sugars, especially mannose, galactose, xylose, arabinose, and glucose

hemimethylated Methylated on only one strand

hemolysin Protein toxin that lyses red blood cells

hepatitis delta virus (HDV) A single-stranded RNA satellite virus that has ribozyme activity

Herceptin Trade name of trastuzumab, a monoclonal antibody to the HER2 receptor on metastatic breast cancer cells that is used as a therapeutic treatment to block growth and spread of the cancer cells

heritable A trait that is passed from the parent to the offspring

herpesviruses Family of DNA-containing viruses that cause a variety of diseases and sometimes cause tumors. They contain dsDNA and an outer envelope surrounding the nucleocapsid

heterochromatin A highly condensed form of chromatin that cannot be transcribed because it cannot be accessed by RNA polymerase

heterologous Derived from a different species

heteroplasmy Presence of different mitochondrial (or chloroplast) genomes in a single individual

hexon Protein subunit of virus capsid with sixfold symmetry that is therefore surrounded by six neighboring subunits

high-pressure liquid chromatography (HPLC) A type of chromatography where a liquid phase containing a mixture of molecules is passed over a solid phase under high pressure

histone acetyl transferase (HAT) Enzyme that adds acetyl groups to histones

histone deacetylases (HDACs) Enzyme that removes acetyl groups from histones

histones Special positively charged protein that binds to DNA and helps to maintain the structure of chromosomes in eukaryotes

HIV (human immunodeficiency virus) The member of the retrovirus family that causes AIDS

HLA genes Family of genes for proteins found on the cell surface and acting in cell recognition

hole The absence of an electron from an atom. Used in combination with electrons to create current in a semiconductor

homologous co-suppression A type of RNAi where multiple copies of a transgene decrease the expression of related host genes

homologous recombination Recombination or genetic exchange between two lengths of DNA that are identical, or nearly so, in sequence

homoplasmy Condition where the population of mitochondria within an individual are all identical

hormone Molecule that carries signals inside multicellular organisms

"hot" Slang for radioactive

housekeeping genes Genes that are switched on all the time because they are needed for essential life functions

human chorionic gonadotropin (hGC) Hormone produced by the placenta during pregnancy

human factor IX One of the proteins involved in blood clotting

human herpesvirus 8 (HHV8) Virus of the herpesvirus family that causes Kaposi's sarcoma, often seen in AIDS patients

human immunodeficiency virus (HIV) The member of the retrovirus family that causes AIDS

human leukocyte antigens (HLAs) Another name for the proteins of the major histocompatibility complex (MHC) of humans, due to their presence on the surface of white blood cells

humanized (of an antibody) Replacing all of a protein, except the antigen binding region, with human encoded sequences

humoral immunity Type of immunity that relies on B cells to produce antibodies

Huntington's disease Inherited defect affecting nerve cells that results in loss of control of the limbs, impaired cognition and dementia

hybrid dysgenesis A genetic mechanism causing abnormally low frequency of viable offspring in a mating between two parents

hybridization Pairing of single strands of DNA or RNA from two different (but related) sources to give a hybrid double helix

hybridoma A cell made by researchers in which an antibody-producing cell (B cell) is fused with a myeloma cell to form a self-proliferating cell that produces one specific monoclonal antibody

hydrolase Type of enzyme that degrades its substrate by hydrolysis

ICAM1 (intercellular adhesion molecule 1) Protein found on the surface of animal cells that is used as a receptor by many viruses of the picornavirus family

ice nucleation factor Protein that acts as a seed for ice crystals to form

***IGF1* gene** Gene that encodes insulin-like growth factor 1

immune memory Memory of antigens previously encountered by the immune system due to specialized memory B cells

immunity protein Protein that provides immunity. In particular bacteriocin immunity proteins bind to the corresponding bacteriocins and render them harmless

immunocytochemistry Technique of visualizing specific cellular proteins by using antibodies. When the antibody binds the protein, the label reveals its position

immunoglobulin G (IgG) The class of antibody with a γ (gamma) heavy chain

immunoglobulins Another name for antibody proteins

immunohistochemistry Technique of visualizing specific proteins in a tissue section using labeled antibodies

imprinting When the expression of a particular allele depends on whether it originally came from the father or the mother (imprinting is a rare exception to the normal rules of genetic dominance)

***in planta Agrobacterium* transformation** (see floral dip transformation) Method in which a transgene is carried by *Agrobacterium* and transformed into *Arabidopsis* by dipping a developing flower bud into a solution of *Agrobacterium* plus a surfactant

***in vivo* induced antigen technology (IVIAT)** Method used to identify genes active after an infectious agent enters the host cell. The technique identifies antibodies from the patient that react with intracellular proteins of the disease agent

***inaZ* gene** Gene encoding ice nucleation protein of *Pseudomonas syringae*

inclusion bodies Dense crystals of misfolded, nonfunctional proteins found in host cells that are expressing a foreign protein

indigo Bright blue pigment based on the indole ring system

indole Ring system containing nitrogen and found in tryptophan and indigo

inducer (signal molecule) Molecule that exerts a regulatory effect by binding to a regulatory protein

inducible promoter A promoter that functions only under special circumstances

influenza virus Member of the orthomyxovirus family with eight separate ssRNA molecules

inhalational anthrax Form of anthrax in which the spores of *Bacillus anthracis* enter via the lungs and the death rate is high

initiation factors Proteins that are required for the initiation of a new polypeptide chain

initiator box Sequence at the start of transcription of a eukaryotic gene

insertion Mutation in which one or more extra bases are inserted into the DNA sequence

insertional inactivation Inactivation of a gene by inserting a foreign segment of DNA into the middle of the coding sequence

insulator binding protein (IBP) Protein that binds to insulator sequence and is necessary for the insulator to function

insulator sequence A DNA sequence that shields promoters from the action of enhancers and also prevents the spread of heterochromatin

insulators DNA sequences that shield promoters from the action of enhancers and also prevent the spread of heterochromatin

insulin receptor Protein on cell surface that acts as a receptor for insulin

insulin Small protein hormone made by the pancreas that controls the level of sugar in the blood

insulin-like growth factor 1 (IGF1) Peptide that stimulates the growth and division of certain animal cells

integrase (*intI*) Enzyme that inserts a segment of dsDNA into another DNA molecule at a specific recognition sequence. In particular, lambda integrase inserts lambda DNA into the chromosome of *E. coli*

integron analysis Identifying the genes embedded within integrons in order to identify useful new genes from the environment

integron Genetic element consisting of an integration site (for gene cassettes) plus a gene encoding an integrase

intein Self-splicing intervening sequence that is found in a protein

interferon γ (INF γ) Protein induced in animal cells in response to intracellular pathogens

interferons (INF) Family of proteins induced in animal cells in response to virus infection or intracellular pathogens

interferons α and β (INF α and INF β) Proteins induced in animal cells in response to virus infection and that induce antiviral responses

interleukin 4 (IL-4) A cytokine that (among other effects) stimulates the division of B cells, which synthesize antibodies

interleukins Subclass of cytokines involved in development of immune system cells

internal ribosomal entry site (IRES) Sequence allowing the translation of multiple coding sequences on the same message in a eukaryotic cell. IRES sequences are found on some animal viruses

intron Segment of a gene that does not code for protein but is transcribed and forms part of the primary transcript

invasin Bacterial protein that provokes an animal cell to swallow the bacteria

inverse PCR Method for using PCR to amplify unknown sequences by circularizing the template molecule

inversion Mutation in which a segment of DNA has its orientation reversed, but remains at the same location

invertase (strictly, DNA invertase) An enzyme that recognizes specific sequences at the two ends of an invertible segment and inverts the DNA between them

ion-exchange HPLC Chromatography technique that separates noncharged from charged molecules. The stationary phase has a net charge that attracts charged molecules, but neutral molecules are not retained

IPTG (isopropyl-thiogalactoside) A gratuitous inducer of the *lac* operon

IRES Internal ribosomal entry site

isoelectric focusing Technique for separating proteins according to their charge by means of electrophoresis through a pH gradient

isoform One of several alternative forms of a protein that are encoded by the same gene. They differ because of alternative splicing of mRNA or alternative processing of precursor protein

JNK (Jun amino-terminal kinase) A eukaryotic protein that transfers a phosphate group from itself to AP-1 in order for AP-1 to bind and activate gene transcription

jumping gene Popular name for a transposable element

junk DNA Term used to describe defective selfish DNA that is of no use to the host cell it inhabits and which can no longer move or express its genes

kappa particles Symbiotic bacteria (*Caedibacter*) that grow inside killer *Paramecium* and make toxin

killer *Paramecium* *Paramecium* carrying kappa particles and therefore capable of killing sensitive paramecia

knockout mice Mice containing genes that have been inactivated by genetic engineering, usually by insertion of a DNA cassette to disrupt the coding sequence

kuru Brain degeneration disease of cannibals caused by prion

LacI repressor Repressor protein that controls *lac* operon

lactate dehydrogenase (LDH) Enzyme that interconverts pyruvate and lactate

lactoferrin An iron transport protein of animals

lactose acetylase Protein product of the LacA gene that has unknown function in lactose metabolism

lactose permease (LacY) The transport protein for lactose

***lacZ* gene** Gene encoding β-galactosidase; widely used as a reporter gene

lagging strand The new strand of DNA that is synthesized in short pieces during replication and then joined later

large-offspring syndrome Imprinting defect that causes abnormal body size

late genes Genes expressed later in virus infection and that mainly encode enzymes involved in virus particle assembly

latency State in which a virus replicates its genome in step with the host cell without making virus particles or destroying the host cell. Same as lysogeny, but generally used to describe animal viruses

lawn A uniform growth of bacteria that coats the entire surface of the growth medium. Single colonies of bacteria are not visible

LCR (same as locus control region) Regulatory sequence in eukaryotes found in front of a cluster of genes that it controls

leading strand The new strand of DNA that is synthesized continuously during replication

lectin Plant protein that specifically binds carbohydrates

leptin A protein hormone that controls the appetite and the burning of fat by the body

leptin receptor Receptor for leptin, encoded by the *db* gene

lethal factor (LF) Protein toxin made by anthrax bacteria that is a protease and cleaves several host cell mitogen-activated protein kinase kinases (MAPKKs)

ligated In biotechnology, joining up DNA fragments end to end using an enzyme such as DNA ligase

light chains The shorter of the two pairs of chains forming an antibody molecule

lignin Insoluble polymer of crosslinked aromatic residues found in plant cell walls

linkers Short, blunt-ended, double-stranded DNA segments that contain recognition sequences for restriction enzymes used to make cohesive ends on DNA

lipofection Use of liposomes to transfer DNA or proteins into a target cell

liposomes Vesicles with an aqueous core surrounded by a phospholipid shell that may be used to deliver oligonucleotides, drugs, or other molecules across the cell membrane

live cell microarrays A method used to analyze siRNA library clones where the library DNA is spotted onto a glass slide and eukaryotic cells are grown over the slide. The siRNA library clones are taken up and the cells are assessed for physical differences

local mediator Molecule that carries signals between nearby cells

locus control region Regulatory sequence in eukaryotes found in front of a cluster of genes that it controls

long interspersed element (LINE) Long sequence found in multiple copies that makes up much of the moderately repetitive DNA of mammals

long terminal repeats (LTRs) Direct repeats of several hundred base pairs found at the ends of retroviruses and some other retroelements required for insertion into host DNA

***loxP* site** Specific sequence that is recognized by Cre recombinase

***luc* gene** Gene encoding luciferase from eukaryotes

luciferase Enzyme that emits light when provided with a substrate known as luciferin

luciferin Chemical substrate used by luciferase to emit light

Lumi-Phos An artificial substrate that is split by alkaline phosphatase, releasing an unstable molecule that emits light

***lux* gene** Gene encoding luciferase from bacteria

lysogeny Type of virus infection in which the virus becomes largely quiescent, makes no new virus particles, and duplicates its genome in step with the host cell. Same as latency but used of bacterial viruses

lysozyme An enzyme found in many bodily fluids that degrades the peptidoglycan of bacterial cell walls

lytic phase Type of growth in which a virus generates many virus particles and destroys the cell

M (mitosis) phase Fourth phase of the eukaryotic cell cycle, in which the cell divides; also known as mitosis

macrolides Class of antibiotics derived from the polyketide pathway

mad cow disease Brain degeneration disease of cattle caused by prions, also known as bovine spongiform encephalopathy (BSE)

magnetosomes Prokaryotic organelle that contains mineralized magnetic crystals of Fe_3O_4 or Fe_3S_4 surrounded by a protein layer

maintenance methylase Enzyme that adds a second methyl group to the other DNA strand of half-methylated sites

MalE protein Carrier protein for maltose found in the periplasmic space of *E. coli*

maltose-binding protein (MBP) Protein of *E. coli* that binds maltose during transport; often used in making fusion proteins

***MAO-A* gene** Gene that encodes monoamine oxidase A

MAR proteins Proteins that form the connection between the chromosomes and the nuclear matrix

Marathon Mouse Transgenic mouse that runs farther

MAT* locus** Chromosomal locus in yeast that controls the mating type and exists as two alternative forms, *MATa** or *MAT*α

matrix attachment regions (MAR) Site on eukaryotic DNA that binds to proteins of the nuclear matrix or of the chromosomal scaffold—same as SAR sites

matrix-assisted laser desorption-ionization (MALDI) Type of mass spectrometry in which gas-phase ions are generated from a solid sample by a pulsed laser

mediator complex A protein complex that transmits the signal from transcription factors to the RNA polymerase in eukaryotic cells

melanocortin receptor Receptor for one of several hormones of the melanocortin family

melanocortins Family of peptide hormones all derived from the same precursor protein, pro-opiomelanocortin (POMC)

melt When used of DNA, refers to its separation into two strands as a result of heating

melting temperature, Tm Temperature at which 50% of a protein is denatured

memory cells Specialized B cells that wait for possible infections instead of manufacturing antibodies

metabolic fingerprinting Characterizing all the metabolites that are present under a certain set of conditions or at a certain time

metabolome The total complement of small molecules and metabolic intermediates of a cell or organism

metagenomic library A library of DNA sequences isolated from an environmental sample. Contains gene sequences from many different organisms and genetic elements

metagenomics The study of all the genomes in a particular environment

metalloprotease A protease that uses a metal cofactor to facilitate protein digestion

metallothionein A metal-binding protein whose synthesis is induced only when certain heavy metals are present

metallothionein promoter Promoter of gene encoding metallothionein; used in genetic engineering because it is very strong and is induced by traces of zinc or other metals

metastasis Process in which cancer cells from a primary tumor move around the body and form secondary cancers

methionine sulfoximine Toxic analog of methionine

methotrexate Antibiotic that inhibits the enzyme dihydrofolate reductase in animal cells

methyl *tert*-butyl ether (MTBE) A fuel additive that oxygenates gasoline so the engine runs with less knocking and the gasoline burns more completely

methylcytosine binding proteins Type of protein in eukaryotes that recognizes methylated CG islands

microfluidics Manipulation of fluids on a micrometer scale

microRNAs (miRNAs) Small regulatory RNA molecules of eukaryotic cells

microsatellite polymorphisms Genetic markers that consist of very short repeated sequences of 2 to 5 nucleotides in length

Mighty Mouse Transgenic mouse with colossal muscle development

mini-gene Miniature gene encoding part of a protein, usually made by artificial DNA synthesis

minisatellite Another term for a VNTR (variable number tandem repeat)

Minos A transposon originally found in insects

minus (–) strand The noncoding strand of RNA or DNA

mismatch repair system DNA repair system that recognizes mispaired bases and cuts out part of the DNA strand containing the wrong base

missense mutation Mutation in which a single codon is altered so that one amino acid in a protein is replaced with a different amino acid

mitochondrial death pathway Program of apoptosis that involves activating mitochondrial proteins to kill the cell; often activated by internal factors such as DNA damage

mitogen-activated protein (MAP) kinases Family of signal transmission proteins that form part of a phosphotransfer cascade in animal cells

mobile DNA Segment of DNA that moves from site to site within or between other molecules of DNA

mobile phase The liquid or solvent containing the mixture of molecules that moves over the stationary phase in column chromatography

modification enzyme Enzyme that binds to the DNA at the same recognition site as the corresponding restriction enzyme but methylates the DNA

modular design Taking various useful domains of different proteins and combining them into a new engineered protein

molar absorptivity The absorbance of a 1 molar solution of pure solute at a given wavelength. The higher it is, the more light is absorbed

molecular chaperone Protein that oversees the correct folding of other proteins

molecular phylogenetics Study of evolutionary relationships using DNA or protein sequences

molecular weight standards Mixture of varying sized DNA or proteins of known size used to compare to unknown proteins

monoamine oxidase (MAO) Enzyme involved in degradation of neurotransmitters of the catecholamine family

monocistronic mRNA mRNA carrying the information of a single cistron that is a coding sequence for only a single protein

monoclonal antibody A pure antibody with a unique sequence that recognizes only a single antigen and which is made by a cell line derived from a single B cell

morpholino-antisense oligonucleotides Synthetic oligonucleotides with morpholino rings instead of ribose and phosphorodiamidate linkages between nucleotides

mousepox (*Ectromelia* virus) Relative of smallpox virus that infects mice

mRNA or messenger RNA The class of RNA molecule that carries genetic information from the genes to the rest of the cell

multigenic Due to interaction of several genes

multiple cloning site (MCS) (see **polylinker**) A stretch of artificially synthesized DNA that contains cut sites for seven or eight widely used restriction enzymes

multiple locus probing (MLP) Variant of DNA fingerprinting where multiple probes are used

multiple nuclear polyhedrosis virus (MNPV) A particular baculovirus widely used as a cloning vector

multiplex PCR Running several PCR reactions with different primers in the same tube

murine leukemia virus (MuLV) A simple retrovirus frequently used to construct vectors for gene therapy

muscular dystrophy Several diseases that result in the wasting away of muscle tissue and cause premature death

mutation An alteration in the DNA (or RNA) that composes the genetic information

mutation breeding Using mutagens to induce genetic changes in plants in order to make or increase a desirable trait

mutation hot spot Region of DNA where alterations in DNA due to mutation are common

mutator gene Gene that affects the rate at which mutations occur

Mx proteins Antiviral proteins of animal cells that interfere with RNA polymerase of negative strand RNA viruses

myc A small DNA segment that encodes for a peptide epitope that can be recognized by antibodies. The tag is used to mark uncharacterized proteins for analysis

***myc* oncogene** Oncogenic version of gene encoding Myc protein

Myc protein Transcription factor involved in switching on several genes involved in cell division

myeloma cells Cancer cells derived from B cells, which therefore express immunoglobulin genes

NAD Nicotinamide adenine dinucleotide; a cofactor that carries reducing equivalents during dehydrogenations; NAD usually acts in degradative pathways

NADP Nicotinamide adenine dinucleotide phosphate; a cofactor that carries reducing equivalents during dehydrogenations; NADP usually acts in biosynthetic pathways

nano- A prefix meaning one billionth (1/1,000,000,000)

nanocarpets Structure formed by stacking a large number of nanotubes together, with their cylindrical axes aligned vertically. The carpet has antibacterial qualities and the ability to change color

nanoparticles Particles of submicron scale—in practice from 100 nm down to 5 nm in size—that can be constructed in a variety of shapes

nanorods Nanoparticles that have a long cylindrical shape. Only the diameter must be in the nanoscale range

nanoscale ion channel Small nanoscale channel created using biological molecules that puncture small holes in the membrane and allow passage of ions under controlled conditions

nanoshells Hollow nanosized particles that can hold different materials

nanotechnology Control of the structure of matter based on molecule-by-molecule control of products and by-products; the products and processes of molecular manufacturing, including molecular machinery

nanotubes Cylinders made of pure carbon with diameters of 1 to 50 nanometers that have novel properties that make them potentially useful in a wide variety of applications

nanowires Wires of dimensions in the order of a nanometer (10^{-9} meters) range, which can be metallic, semiconducting, and insulating. These can be designed of repeating organic units such as DNA

naphthalene oxygenase Enzyme that carries out the first step in naphthalene breakdown by inserting oxygen into the aromatic ring

necrosis Death of a cell characterized by cellular swelling and rupture; elicits an immune response

negative (–) strand The noncoding strand of RNA or DNA

negative regulation Regulatory mode in which a repressor keeps a gene switched off until the repressor is removed

neomycin phosphotransferase Enzyme that confers resistance to antibiotics such as kanamycin and neomycin

nerve growth factor A soluble trophic factor that is required to keep neurons alive

neurofibrillary tangles Intracellular clumps of protein found in neurons of Alzheimer's patients

neuron Nerve cell

neuropeptide Y (NPY) Peptide in brain that increases feeding and so makes animals fatter

neurotoxin Toxin that attacks nerve cells

neurotransmitter Molecule that carries signals across synapses between cells within the nervous system

***N*-formylmethionine (fMet)** Modified methionine used as the first amino acid during protein synthesis in bacteria

nick A break in the backbone of a DNA or RNA molecule (but where no bases are missing)

nicotine Alkaloid in tobacco that raises levels of uncoupling protein 1 in brown fatty tissue hence promoting fat metabolism

nitric oxide (NO) Gaseous molecule used in signaling by animal cells

nitric oxide synthase (NO synthase) Enzyme that synthesizes nitric oxide

nitrocellulose A pulpy or cotton-like substance formed into sheets and used to attach proteins in Western blotting

NO synthase (nitric oxide synthase) Enzyme that synthesizes nitric oxide

Nomarski optics Type of polarization microscopy that transforms differences in density into differences in height so that the image looks like a three-dimensional relief

noncoding (or **antisense**) The strand of DNA that has complementary sequence to mRNA

nonnucleoside reverse transcriptase inhibitor (NNRTI) Antiviral agent that inhibits the reverse transcriptase of viruses such as HIV but is not a nucleoside analog

nonsense mutation Mutation due to changing the codon for an amino acid to a stop codon

nontarget organism Any organism exposed to a specific insecticide, herbicide, or transgenic plant which that product was not intended to harm

nontemplate The strand of DNA equivalent in sequence to the messenger RNA (same as plus strand)

noradrenalin Neurotransmitter of the catecholamine family (also known as norepinephrine)

norepinephrine Neurotransmitter of the catecholamine family (also known as noradrenalin)

Northern blots Hybridization technique in which a DNA probe binds to an RNA target molecule

nosocomial infections Those infections acquired in a hospital setting

npt The gene that encodes neomycin phosphotransferase

nuclear envelope Two concentric membranes that surround the nucleus of eukaryotic cells

nuclear microinjection Technique for insertion of foreign DNA into a host cell nucleus

nuclear pore Pore in nuclear membrane that allows proteins, RNA, and other molecules into and out of the nucleus

nuclear transplantation Technique in which nuclei from one cell are transplanted into another cell from which the nucleus has previously been removed

nuclear-receptor superfamily Large family of proteins that act as transcription factors; includes receptors to many hormones that are poorly water soluble

nucleocapsid Inner structure of certain viruses that consists of RNA or DNA surrounded by protein

nucleoside analog Molecule that mimics a nucleoside well enough to be incorporated into a growing chain of DNA by synthetic enzymes

nucleosome Subunit of a eukaryotic chromosome consisting of DNA coiled around histone proteins

nucleotide Monomer or subunit of a nucleic acid, consisting of a pentose sugar plus a base plus a phosphate group

nullizygous When both copies of a gene are fully inactivated

***ob* (obese) gene** Gene that encodes the protein hormone leptin

Okazaki fragments The short pieces of DNA that make up the lagging strand

oligonucleotide A molecule of 25 nucleotides or fewer used to prime *in vitro* DNA replication, sequencing, or PCR reactions

oncogene Mutant gene that promotes cancer

oncogenic virus Cancer-causing virus

oncolytic virus Cancer-destroying virus

oocyte Female egg cell

open reading frame (ORF) Sequence of bases (either in DNA or RNA) that can be translated (at least in theory) to give a protein

operator Site on DNA to which a repressor protein binds

operon A cluster of prokaryotic genes that are transcribed together to give a single mRNA (i.e., polycistronic mRNA)

opportunistic infection Infections seen in patients with defective immune systems that are caused by normally harmless microorganisms

oral vaccine A vaccine that is taken by mouth, as either a liquid or a pill

origin Site on a chromosome or any other DNA molecule where replication begins

origin of chromosome (*oriC*) Origin of replication of a chromosome

origin of replication (*ori*) Site on a chromosome or any other DNA molecule where replication begins

orthomyxoviruses Family of negative-strand ssRNA viruses that includes influenza

overlap PCR PCR technique that uses overlapping primers to match small regions of two different gene segments and is used in joining segments of DNA from different sources

oxygenase Enzyme that inserts one or more oxygen atoms into its substrate

P elements Transposons found in *Drosophila* and other insects

p21 protein A protein that blocks cell division by binding to and inhibiting the cyclins

***p53* gene** A notorious anti-oncogene often mutated in cancer cells

p53 protein (also known as TP53) DNA-binding protein, encoded by the *p53* gene, that acts to stop cell division; causes some cancer cells to enter senescence

packaging construct Defective provirus that is integrated into the DNA of a producer cell to manufacture virus particles but which is not packaged itself because of lack of the packaging signal

packaging signal Site on retrovirus genome that is essential for packaging the RNA into the virus particle

papillomaviruses Family of small DNA-containing viruses that sometimes cause tumors

pathogenicity island Region of bacterial chromosome flanked by inverted repeats that carries many genes involved in virulence and pathogenicity

pathway engineering The assembly of a new or improved biochemical pathways, using genes from one or more organisms

penicillins Subfamily of antibiotics belonging to the β-lactam family of antibiotics

Penicillium notatum The original mold that makes penicillin

penton Protein subunit of virus capsid with fivefold symmetry that is found at vertexes surrounded by five neighboring subunits

pentose A five-carbon sugar, such as ribose or deoxyribose

peptidase Same as proteinase; an enzyme that degrades polypeptides by hydrolysis

peptide nucleic acid (PNA) Artificial analog of nucleic acid with a polypeptide backbone

peptide vaccines Vaccines that contain a single epitope or peptide attached to a carrier protein. The immune system forms antibodies to the peptide and gains immunity to the disease agent

peptidyl transferase Enzyme activity on the ribosome that makes peptide bonds; actually 23S rRNA (bacterial) or 28S rRNA (eukaryotic)

PEST sequence Region of 10–60 amino acids that is rich in P (proline), E (glutamate), S (serine), and T (threonine) and is recognized by proteases

phage display Fusion of a protein or peptide to the coat protein of a bacteriophage whose genome also carries the cloned gene encoding the protein. The protein is displayed on the outside of the virus particle and the corresponding gene is carried on the inside

phage Short for bacteriophage, a virus that infects bacteria

pharmacogenetics Study of inherited differences in drug metabolism and response

pharmacogenomics Study of genes involved in drug metabolism and efficacy

phase variation Reversible inversion of a segment of DNA leading to differences in gene expression

phenol Organic chemical with the formula C_6H_5OH consisting of a hydroxyl group attached to a phenyl ring. It is used to remove protein from nucleic acids by dissolving the proteins

phenotype catalogue A listing of proteins that cause the same phenotype when their function is disrupted

phenotypic signature A set of physical features that are categorized together to classify a particular cell function, such as adherence, motility, or cell division

phenylalanine hydroxylase The enzyme that converts the amino acid phenylalanine into tyrosine

phenyl-boronate Resin used to bind beta-lactamases

phenylketonuria Inherited defect causing lack of the enzyme phenylalanine hydroxylase and hence a buildup of phenylalanine

pheromone Molecule that carries signals between organisms

***phoA* gene** Gene encoding alkaline phosphatase; widely used as a reporter gene

phosphate group Group of four oxygen atoms surrounding a central phosphorus atom found in the backbone of DNA and RNA

phosphodiester bond The linkage between nucleotides in a nucleic acid that consists of a central phosphate group esterified to sugar hydroxyl groups on either side

phosphodiesterase 5 (PDE5) One particular cyclic phosphodiesterase of animal cells that degrades cyclic GMP

phospholipase Enzyme that degrades phospholipids

phosphorelay Term to describe transferring one phosphate from one location to another to successively activate different proteins

phosphorodiamidate An uncharged version of the phosphodiester group linking nucleotides in which one of the oxygen atoms is replaced with an amidate group

phosphorothioate A phosphate group in which one of the four oxygen atoms around the central phosphorus is replaced by sulfur

phosphorothioate oligonucleotide A synthetic oligonucleotide with a sulfur replacing one of the four oxygen atoms in the phosphodiester linkage between nucleotides

phosphotransferase system System found in many bacteria that transports sugars and regulates metabolism. It operates by transferring phosphate groups

photodetector Sensitive instrument that detects an optical signal and converts it into an electrical signal

photolithography Method used to synthesize oligonucleotides directly on a DNA chip where light is passed through a mask to selectively activate some regions and keep other regions blocked

photolyase Enzyme that catalyzes the repair of DNA thymine dimers in response to blue light

phylogenetic tree A diagram showing the evolutionary relationship of different organisms

physical maps Genetic maps that give physical DNA base-pair lengths between features

picornaviruses Virus family that includes enteroviruses (such as poliovirus) and rhinoviruses (which cause the common cold)

piezoelectric ceramics Materials that change shape in response to an applied voltage

piggyBac A transposon originally found in insects

pilin Protein that makes up the main part of a pilus

pilus (plural **pili**) Thin helical protein filaments found on the surface of bacteria; same as fimbria

plaques (when referring to viruses) A clear zone caused by virus destruction in a layer of cultured cells or a lawn of bacteria

plasmid incompatibility The inability of two plasmids of the same family to coexist in the same host cell

plasmid Self-replicating genetic elements that are sometimes found in both prokaryotic and eukaryotic cells. They are not chromosomes or part of the host cell's permanent genome.

Most plasmids are circular molecules of double-stranded DNA, although rare linear plasmids and RNA plasmids are known

Plasmodium falciparum Protozoan parasite that causes the malignant form of malaria

plus (+) strand The coding strand of RNA or DNA

poison sequence Base sequence, often found in virus genomes, that is not stably maintained or replicated on plasmids in bacterial hosts

poly(A) tail A stretch of multiple adenosine residues found at the 3′ end of mRNA

polyacrylamide gel electrophoresis (PAGE) Technique for separating proteins by electrophoresis on a gel made from polyacrylamide

polyadenylation complex Protein complex that adds the poly(A) tail to eukaryotic mRNA

polycistronic mRNA mRNA carrying multiple coding sequences (cistrons) that may be translated to give several different protein molecules; only found in prokaryotic (bacterial) cells

polyclonal antibody Natural antibody that actually consists of a mixture of different antibody proteins that all bind to the same antigen

polyglutamine tract Run of multiple glutamine residues in a protein

polyhedrin Protein that comprises polyhedron structure of baculoviruses

polyhedron In reference to viruses, the packages of virus particles embedded in a protein matrix that are formed by baculoviruses

polyhistidine tag (His6 tag) Six tandem histidine residues that are fused to proteins, thus allowing purification by binding to nickel ions that are attached to a solid support

polyhydroxyalkanoate (PHA) Type of bioplastic polymer made by certain bacteria from hydroxyacid subunits

polyhydroxybutyrate (PHB) Bioplastic polymer made by certain bacteria from hydroxybutyrate subunits

polyketide Class of natural linear polymers whose backbone consists of two carbon repeats with keto groups on every other carbon (when first synthesized)

polylinker (see **multiple cloning site [MCS]**) A stretch of artificially synthesized DNA that contains cut sites for seven or eight widely used restriction enzymes

polymerase chain reaction (PCR) Amplification of a DNA sequence by repeated cycles of strand separation and replication

polyploid Having more than one set of chromosomes per cell

polysome Group of ribosomes bound to and translating the same mRNA

polytene chromosomes Giant chromosomes found in cells that replicate their DNA without dividing into separate cells. Found in *Drosophila* salivary gland cells

polyvalent inhibitor Inhibitor consisting of several linked inhibitor molecules that consequently binds multiple copies of its target, giving a very high overall binding affinity

position effect variation The effect of chromosomal position on the expression of a particular gene. For example, genes embedded within heterochromatin are not expressed, but if positioned in euchromatin, the same gene would be expressed

positional cloning Approach that attempts to locate a gene by its position on a chromosome and is used when the nature of the gene product is unknown

positive (+) strand The coding strand of RNA or DNA

positive regulation Control by an activator that promotes gene expression when it binds

posttranscriptional gene silencing (PTGS) Plant version of RNA interference

poxviruses Family of large animal viruses with dsDNA and approximately 150 to 200 genes

Prader-Willi syndrome Inherited defect resulting in loss of function of genes on the paternally derived copy of chromosome 15 that are subject to imprinting

pRB (retinoblastoma protein) Protein associated with a malignant cancer of the retina; involved with cellular senescence

premature senescence Entering senescence before the normal number of cell divisions

pre-microRNAs Longer precursor molecules that are converted into microRNAs

preproinsulin Insulin as first synthesized, with both a signal sequence and the connecting peptide

presenilins Transmembrane proteins found in the Golgi apparatus and endoplasmic reticulum; associated with Alzheimer's disease

prey The fusion between the activator domain of a transcriptional activator protein and another protein as used in two-hybrid screening

PriA Protein of the primosome that helps primase bind

primary antibody First antibody to the protein of interest used to identify a protein in Western blotting

primary transcript The original RNA molecule obtained by transcription from a DNA template, before any processing or modification has occurred

primase Enzyme that starts a new strand of DNA by making an RNA primer

principle of independent assortment Alleles of a gene sort independently during formation of a gamete

principle of segregation Alleles of a gene segregate or separate into different gametes during gamete formation, but then reunite during fertilization

prion protein (PrP) Brain protein that may exist in two forms, one of which is pathogenic and may cause transmissible prion disease

Prnp Gene that encodes the prion protein

probe molecule Molecule that is tagged in some way (usually radioactive or fluorescent) and is used to bind to and detect another molecule

prodrug A harmless compound that is converted to an active drug by a specific enzyme

programmed cell death Genetic program that eliminates damaged cells or cells that are no longer needed without activating the immune system

proinsulin Precursor to insulin that contains both the A and B chains plus the connecting peptide

promoter Region of DNA in front of a gene that binds RNA polymerase and so promotes gene expression

pronuclei The parental male and female nuclei in a fertilized egg just before nuclear fusion

prophage Bacteriophage genome that is integrated into the DNA of the bacterial host cell

protease inhibitor Inhibitor of protease enzymes, in particular antiviral agent that inhibits the protease of viruses such as HIV

protease Same as proteinase; an enzyme that degrades polypeptides by hydrolysis

protective antigen (PA) Protein that acts as the delivery system for both anthrax toxins

protein A Antibody binding protein from *Staphylococcus* that is often used in making fusion proteins

protein engineering Altering the sequence of a protein by genetic engineering of the DNA that encodes it

protein fusion Hybrid protein made by joining the coding sequences of two proteins together in frame

protein fusion vector Vector designed for fusing cloned proteins to a carrier protein to help expression and/or export

protein interactome The total of all the protein-protein interactions in a particular cell or organism

protein kinase Enzyme that transfers phosphate groups to other proteins, thus controlling their activity

protein kinase A (PKA) One particular protein kinase of animal cells that is activated by cyclic AMP

protein misfolding cyclic amplification (PMCA) Protocol that amplifies misfolded prions in a manner analogous to PCR

proteinase Same as protease; an enzyme that degrades polypeptides by hydrolysis

protein Polymer made from amino acids; may consist of several polypeptide chains

proteome The total set of proteins encoded by a genome or the total protein complement of an organism

proteomics Study or analysis of an organism's complete protein complement

proto-oncogene Original, unmutated wild-type allele of an oncogene

protoplasts Plant cells that have been dissociated and whose cell walls are removed

provirus Virus genome that is integrated into the host cell DNA

PrP^{c} Normal, "cellular" form of the prion protein

PrP^{Sc} Pathogenic, "scrapie" form of the prion protein

pseudogenes Defective copies of a genuine gene

pulsed field gel electrophoresis (PFGE) Type of gel electrophoresis used for analysis of very large DNA molecules and which uses an electric field of "pulses" delivered from a hexagonal array of electrodes

PUMA (p53-upregulated modulator of apoptosis) A Bcl-2 protein family member that activates Bax and promotes apoptosis

purine Type of nitrogenous base with a double ring found in DNA and RNA

pyrimidine Type of nitrogenous base with a single ring found in DNA and RNA

quantum confinement The phenomenon seen in nanoscale structures where an electron-hole pair is kept within a structure that is near its natural Bohr radius. Energy states are discrete in these structures

quantum yield The ratio of photons absorbed to photons emitted during fluorescence

quasi-species Group of related RNA-based genomes that differ slightly in sequence, but which arose from the same parental RNA molecule

quelling Fungal version of RNA interference

radiation hybrid mapping Mapping technique that uses cells (usually from a rodent) that contain fragments of chromosomes (generated by irradiation) from another species

radical replacement Replacement of an amino acid with another that has different chemical and physical properties

radiochemical detector A device that detects radioactively labeled molecules, for example, in the mobile phase during chromatography

random shuffling library Library of gene or protein sequences generated by randomized shuffling and linking of short segments

randomly amplified polymorphic DNA (RAPD) Method for testing genetic relatedness using PCR to amplify arbitrarily chosen sequences

***ras* oncogene** Oncogenic version of gene encoding Ras protein

Ras protein GTP binding protein involved in transmitting signals concerning cell division in animals

raster-scan An image displayed line by line

rBST Acronym for recombinant bovine somatotropin

reactive oxygen metabolites (ROM) Highly reactive oxygen-derived molecules or ions with extra electrons (especially superoxide ions, peroxides, and hydroxyl radicals)

reanneal Renaturation of single-stranded DNA into double-stranded DNA

receptor Molecule that binds another molecule, such as a hormone or a nutrient. In particular receptors are often proteins that participate in signaling and are situated in the cell membrane facing outwards

recombinant bovine somatotropin (rBST) Bovine growth hormone produced in another organism

recombinant human somatotropin (rHST) Human growth hormone produced in another organism

recombinant plasmids Plasmids that contain segments of DNA not originally found in the plasmid, most likely from another organism

recombinant protein Protein expressed from recombinant DNA gene

recombinant tissue plasminogen activator (rTPA) Tissue plasminogen activator produced in another organism

recombinase Enzyme that catalyzes recombination between inverted repeats

refractive index detector A device that records changes in the velocity of light as it passes through liquid such as the eluate of a column

release factor Protein that recognizes a stop codon and brings about the release of a finished polypeptide chain from the ribosome

replacement gene therapy Curing inherited defects by introducing functional copies of the defective gene

replication fork Region where the enzymes replicating a DNA molecule are bound to untwisted, single-stranded DNA

replicative aging The number of daughter cells a yeast cell can produce in its lifetime

replicative form (RF) Double-stranded form of the genome of a single-stranded DNA (or RNA) virus. The RF first replicates itself and is then used to generate the ssDNA (or ssRNA) to pack into the virus particles

replicative senescence (cellular senescence) Arrested cell division of cultured mammalian cells

replicative transposition Type of transposition in which two copies of the transposon are generated, one in the original site and another at a new location

replicator A system able to build copies of itself when provided with raw materials and energy

replicon Molecule of DNA or RNA that contains an origin of replication and can self-replicate

replisome Assemblage of proteins (including primase, DNA polymerase, helicase, SSB protein) that replicates DNA

repressor Regulatory protein that prevents a gene from being transcribed

resolution The sharpness of a peak after separation by chromatography

restriction endonuclease Type of endonuclease that cuts double-stranded DNA at a specific sequence of bases, the recognition site

restriction enzyme-generated siRNA (REGS) A method to generate an RNAi library that uses restriction enzymes to create random double-stranded 21- to 23-nucleotide-long segments of genes

restriction enzyme Type of endonuclease that cuts double-stranded DNA at a specific sequence of bases, the recognition site

restriction fragment length polymorphism (RFLP) A difference in restriction sites between two related DNA molecules that results in production of restriction fragments of different lengths

retinal scan Scan of the unique pattern of blood vessels on the retina

retinoblastoma (*Rb*) gene Anti-oncogene that is responsible for a rare cancer of the retina of the eye

retinoblastoma protein (pRB) Protein associated with a malignant cancer of the retina; involved with cellular senescence

retroviruses A family of animal viruses with single-stranded RNA inside two protein shells surrounded by an outer envelope. Once inside the host cell they use reverse transcriptase to convert the RNA version of the genome to a DNA copy

reverse-phase HPLC Chromatography technique that passes a solution of proteins over a non-polar stationary phase under high pressure. Hydrophobic molecules attach to the stationary phase and hydrophilic molecules elute from the column

reverse transcriptase An enzyme that uses single-stranded RNA as a template for making double-stranded DNA

reverse transcriptase PCR (RT-PCR) Variant of PCR that allows genes to be amplified and cloned as intron-free DNA copies by starting with mRNA and using reverse transcriptase

reverse vaccinology Approach to making new vaccines that uses genomics to identify new epitopes or proteins from infectious agents that are highly antigenic without causing disease

reverse-phase assay (or **direct immunoassay**) Method of protein quantification and detection, where the protein is bound to a solid support, and an antibody with a detection system is added

RFLP (restriction fragment length polymorphism) A difference in restriction sites between two related DNA molecules that results in production of restriction fragments of different lengths

rhinoviruses Group of small RNA viruses that cause a major percentage of cases of the common cold

Rho (ρ) protein Protein factor needed for successful termination at certain transcriptional terminators

Rho-dependent terminator Transcriptional terminator that depends on Rho protein

Rho-independent terminator Transcriptional terminator that does not need Rho protein

rHST Acronym for recombinant human somatotropin

ribonuclease (or **RNase**) Enzyme that cuts or degrades RNA

ribonucleoprotein A protein that has RNA associated with it

ribose The five-carbon sugar found in RNA

ribosomal RNA or rRNA Class of RNA molecule that makes up part of the structure of a ribosome

ribosome binding site (RBS) Sequence close to the front of mRNA that is recognized by the ribosome; only found in prokaryotic cells

ribosome The cell's machinery for making proteins

ribosome-inactivating protein (RIP) Toxic protein that inactivates ribosomes by releasing adenine from a specific site in large-subunit rRNA

riboswitches Domains of messenger RNA that directly sense a signal and control translation by alternating between two structures

ribotyping Identification of bacteria or other living organisms based on the sequence of their small-subunit ribosomal RNA

ribozyme RNA molecule that acts as an enzyme

ricin Highly toxic ribosome-inactivating protein from the castor bean

right-handed helix In a right-handed helix, as the observer looks down the helix axis (in either direction), each strand turns clockwise as it moves away from the observer

RNA (ribonucleic acid) Nucleic acid that differs from DNA in having ribose in place of deoxyribose

RNA interference (RNAi) Response that is triggered by the presence of double-stranded RNA and results in the degradation of mRNA or other RNA transcripts homologous to the inducing dsRNA

RNA polymerase Enzyme that synthesizes RNA

RNA polymerase I Eukaryotic RNA polymerase that transcribes the genes for the large ribosomal RNAs

RNA polymerase II Eukaryotic RNA polymerase that transcribes the genes encoding proteins

RNA polymerase III Eukaryotic RNA polymerase that transcribes the genes for 5S ribosomal RNA and transfer RNA

RNA SELEX Method where catalytically active RNA sequences are altered by mutation to identify new ligands or substrates for their activity

RNA-dependent RNA polymerase (RdRP) RNA polymerase that uses RNA as a template and is involved in the amplification of the RNAi response

RNAi library A library that expresses double-stranded RNA to activate RNA interference. Each library clone inactivates a gene from the organism of interest

RNA-induced silencing complex (RISC) Protein complex induced by siRNA that degrades single-stranded RNA corresponding in sequence to the siRNA

RNA-SIP Method used to enrich the RNA in an environmental sample. A stable isotope is mixed with an environmental sample and becomes incorporated into any life form that is using RNA for protein expression

rolling circle replication Mechanism of replicating double-stranded circular DNA that starts by nicking and unrolling one strand and using the other, still circular, strand as a template for DNA synthesis. Used by some plasmids and viruses

ROM (reactive oxygen metabolites) Highly reactive oxygen derived molecules or ions with extra electrons (especially superoxide ions, peroxides, and hydroxyl radicals)

***rosy* gene** Gene of *Drosophila* that affects eye color

Rous sarcoma virus (RSV) Cancer-causing retrovirus of chickens

rTPA Acronym for recombinant tissue plasminogen activator

RU486 Progesterone antagonist; ingredient of abortion pill

RXR protein Nuclear protein that forms mixed dimers with receptors for androgens, vitamin D, thyroxine, and retinoic acid

S phase Second stage ("synthesis" phase) in the cell cycle, in which chromosomes are duplicated

sarcoma Cancer originating from muscle cells

satellite viruses A defective virus that needs an unrelated helper virus to infect the same host cell in order to provide essential functions

scanning tunneling microscope (STM) An instrument able to image conducting surfaces to atomic accuracy; has been used to pin molecules to a surface

scFv fragment (single-chain antibody) Engineered antibody with VH and VL domains linked together by a short peptide chain

SCID (severe combined immunodeficiency) Immune defect due to lack of both T cells and B cells. About 25% of inherited SCID cases are due to adenosine deaminase deficiency

scintillation counting Detection and counting of individual microscopic pulses of light

scrape-loading A method of getting oligonucleotides into cultured cells by gently scraping cells growing on the surface of a dish. The mechanical scraping introduces small breaks in the cell membrane that allow oligonucleotides into the cytoplasm

scrapie Brain degeneration disease of sheep and goats caused by prion

second messenger Intracellular signal molecule that is made when a cell surface receptor receives a message

secondary antibody Antibody to the primary antibody that has a detection system attached. Used in Western blotting to identify a protein of interest

selective pressure (in data analysis) Continued pressure or refinement of a particular data set based on set criteria

semiconductors Materials that are between a conductor and insulator in electrical conductivity. The electrical current travels via electron-hole pairs

semiconservative replication Mode of DNA replication in which each daughter molecule gets one of the two original strands and one new complementary strand

senile neuritic plaques Aggregates of degenerating neurons in brains of Alzheimer's patients

sense strand The strand of DNA equivalent in sequence to the messenger RNA (same as plus strand)

sensor kinase A protein that phosphorylates itself when it senses a specific signal (often an environmental stimulus, but sometimes an internal signal)

sequence tagged site (STS) A short sequence (usually 100–500 bp) that is unique within the genome and can be easily detected, usually by PCR

serine protease A protease that has a serine residue in its active site

serine transacetylase Enzyme that converts acetyl-CoA plus serine to *O*-acetylserine

severe combined immunodeficiency (SCID) Immune defect due to lack of both T cells and B cells. About 25% of inherited SCID cases are due to adenosine deaminase deficiency

Shine-Dalgarno sequence Same as RBS; sequence close to the front of mRNA that is recognized by the ribosome; only found in prokaryotic cells

short-hairpin RNA (shRNA) Genetically engineered RNA with complementary sequences that fold to become double-stranded. Used to activate RNA interference

short-interfering RNA (siRNA) Double-stranded RNA with 21–22 nucleotides involved in triggering RNA interference in eukaryotes

short interspersed element (SINE) Short repeated sequence that makes up a major fraction of the moderately or highly repetitive DNA of mammals

short tandem repeats (STR) Sub-class of VNTR with short repeated sequences

shotgun sequencing Approach in which the genome is broken into many random short fragments for sequencing. The complete genome sequence is assembled by computerized searching for overlaps between individual sequences

shuttle vector A vector that can survive in and be moved between more than one type of host cell

siderophore Iron chelator used by microorganisms to extract and bind iron from their environments

sigma subunit Subunit of bacterial RNA polymerase that recognizes and binds to the promoter sequence

signal molecule (see inducer) Molecule that exerts a regulatory effect by binding to a regulatory protein

signal transmission protein Protein involved in transmitting signals, often from cell surface receptor to gene regulators

sildenafil Generic name for Viagra

silence In genetic terminology, refers to switching off genes in a relatively nonspecific manner

silver stain A sensitive dye used to stain proteins

simian virus 40 (SV40) Cancer-causing virus of monkeys

similarity search Comparing newly sequenced DNA with other known DNA sequences to determine its identity

single-locus probing (SLP) Variant of DNA fingerprinting where a single probe is used

single nucleotide polymorphism (SNP) A difference in DNA sequence of a single base change between two individuals

single-chain antibody (scFv fragment) Engineered antibody with VH and VL domains linked together by a short peptide chain

single-chain Fv (scFv) Fv fragment of an antibody engineered so that the VH and VL domains are linked together by a short peptide chain

single-stranded binding protein A protein that keeps separated strands of DNA apart

size exclusion chromatography Chromatography technique that separates on the basis of size

sliding clamp Subunit of DNA polymerase that encircles the DNA, thereby holding the core enzyme onto the DNA

SLUG Protein that blocks apoptosis

small subunit ribosomal RNA (SSU rRNA) The 16S rRNA in prokaryotes or 18S rRNA in eukaryotes; target RNA that is isolated from the environment to identify and catalogue all the different organisms

smallpox Virulent virus disease that infects only humans

Smart Mouse Transgenic mouse that has improved learning and memory

sodium bisulfite Chemical that deaminates cytosine to uracil

sodium dodecyl sulfate (SDS) A detergent widely used to denature and solubilize proteins before separation by electrophoresis

somatic cell Cell making up the body, as opposed to the germline

somatic mutation Mutation that occurs in somatic cells and is not passed on to the next generation via the germline

somatotropin A polypeptide hormone that controls cell growth and reproduction in humans and other animals

Southern blots A method to detect single-stranded DNA that has been transferred to nylon paper by using a probe that binds DNA

specific immunity Type of immunity that responds to specific antigens over a course of days, has the ability to distinguish self from nonself, remembers past invading pathogens, and generates new antibodies

specific transcription factors Regulatory proteins that exert their effect on a single gene or operon or on a very small number of related genes

spindle Microtubule-based structure where chromosomes attach and separate during mitosis or meiosis

splicing factors Molecules that remove intervening sequences and rejoin the ends of a molecule; usually refers to removal of introns from RNA

***src* oncogene** Oncogene carried by Rous sarcoma virus and originally derived from chicken cells

stable isotope probing (SIP) Method for enriching the DNA in an environmental sample. The sample is incubated with a stable isotope that is incorporated into any actively dividing life form, and then the labeled DNA is isolated

stanol Sterol derivative whose double bond has been reduced

starch Polymer of glucose linked by α-1,4 bonds and used as a storage polysaccharide in plants

Starlink Controversial transgenic corn that inadvertently entered the human food supply before being fully evaluated by the government

stationary phase The material packed into a column whose physical or chemical properties separate a mixture of molecules such as proteins into separate fractions

stem cell Precursor cell that gives rise to specialized cells of various types as well as to more stem cells

steroid 1. Type of polycyclic lipophilic molecule that includes cholesterol, sex hormones, and corticosterols. 2. Sterol derivative having keto rather than hydroxyl groups

steroid hormones Hormones with a steroid structure, such as the sex hormones

steroid receptor Dual-function protein that acts as receptor for steroid hormone and as transcription factor

sterols Class of lipophilic biological compounds with four fused nonaromatic rings

sticky ends Ends of a double-stranded DNA molecule that have unpaired single-stranded overhangs, generated by a staggered cut

STR (short tandem repeats) Subclass of VNTR with short repeated sequences

strand slippage During replication, DNA polymerase may slip along the template strand at a repeat and reattach at a different site. The daughter strand will have different number of repeats from the parental DNA strand

streptavidin A small biotin binding protein from the bacterium *Streptococcus*

streptolysin O A toxin from *Streptococci pyogenes* that attaches to cholesterol in cellular membranes and aggregates into a circular structure, forming a pore for molecules to enter the cell

structural gene Sequence of DNA (or RNA) that codes for a protein or for an untranslated RNA molecule

subunit vaccines Vaccines that contain a single polypeptide from the disease agent. The immune system creates antibodies to the single polypeptide and inhibits the infectious agent by attacking via the single target protein

subviral agents Infectious agents that are more primitive than viruses and encode fewer of their own functions

supercoiling Higher level coiling of DNA that is already a double helix

superoxide dismutase (SOD) Enzyme that converts superoxide to oxygen plus hydrogen peroxide

suppressive subtraction hybridization (SSH) Culture enrichment technique for metagenomics that identifies differences in DNA content between two different environments

surface-enhanced laser desorption-ionization (SELDI) A form of mass spectroscopy where the protein of interest is in a liquid phase that is placed on a solid surface and then ionized with a laser

suspension cells Cultured cells that grow in a liquid nutrient media, not attached to any surface

suspension culture Dedifferentiating a mass of plant tissue, then growing the cells in a liquid medium

symbiotic theory Theory that the organelles of eukaryotic cells are derived from symbiotic prokaryotes

synapse Junction between cells, across which signals are carried by chemical molecules known as neurotransmitters

syncytium Giant cell with many nuclei

T cells Cells of the immune system that remove virally infected cells and secrete soluble factors that activate other cells in the immune system, particularly B cells. Responsible for cell-mediated immunity; make T-cell receptors rather than antibodies

T lymphocytes (also T cells) Type of immune system cell responsible for cell-mediated immunity and which makes T-cell receptors instead of antibodies. These mature in the thymus

TA cloning Procedure that uses *Taq* polymerase to generate single 3′-A overhangs on the ends of DNA segments that are used to clone DNA into a vector with matching 3′-T overhangs

***tac* promoter** Hybrid promoter containing the ribosome binding site from the *trp* promoter and the operator sequence from the *lac* promoter

tandem mass spectroscopy Mass spectroscopy technique with multiple steps of selection and analysis that use successively smaller fragments for starting material

***Taq* DNA polymerase** Heat-resistant DNA polymerase from *Thermus aquaticus* that is used for PCR

target DNA DNA that is the target for binding by a probe during hybridization or the target for amplification by PCR

target sequence 1. Sequence on host DNA molecule into which a transposon inserts itself. 2. Sequence within the original DNA template that is amplified in a PCR reaction

targeting vector Vector designed to promote the integration of a transgene into a specific location

targeting-induced local lesions in genomes (TILLING) Mutagenesis technique where plant seeds are soaked in a chemical mutagen to induce point mutations and the genomic DNA is analyzed by PCR and hybridization

TATA binding factor or **TATA box factor** Transcription factor that recognizes the TATA box

TATA box Binding site for a transcription factor that guides RNA polymerase II to the promoter in eukaryotes

tau Protein normally associated with microtubules that clumps and forms helical aggregates in Alzheimer's patients

taxonomy Scientific classification of organisms based on physical or genetic relatedness

telomerase Enzyme made of RNA plus protein that re-elongates telomeres by adding DNA to the end of a eukaryotic chromosome

telomeres Special repeated sequences that cap the ends of linear eukaryotic chromosomes

temperature-sensitive mutation Mutation that shows a different phenotype at different temperatures

template Strand of DNA used as a guide for synthesizing a new strand by complementary base-pairing

terminal differentiation Expression of final phenotype for a cell type or tissue

tester DNA from the experimental sample that has sequences unique to that environment. Used in suppressive subtraction hybridization

***tet* operon** Cluster of bacterial genes that confers resistance to the antibiotic tetracycline

***tetO* operator** Site in front of *tet* operon where repressor binds

TetR repressor Repressor protein that controls *tet* operon

tetracyclines Family of antibiotics with four fused rings derived from the polyketide pathway

therapeutic cloning Cloning to obtain tissue for transplantation as opposed to generating a new individual

thermocycler Machine used to rapidly shift samples between several temperatures in a preset order (for PCR)

thermogenin Alternative name for uncoupling protein 1 (UCP1)

thermostable Able to withstand high temperatures without loss of function

theta-replication Mode of replication in which two replication forks go in opposite directions around a circular molecule of DNA

thiophene Five-membered aromatic ring containing sulfur and four carbon atoms

threonine protease A protease that has an active-site threonine

thrombin A protease involved in the blood-clotting cascade

thymidine kinase Enzyme that converts thymidine and related nucleosides to their monophosphate derivatives

thyroid hormone (thyroxine) Hormone made by thyroid gland

thyroxine Hormone made by thyroid gland

time-of-flight (TOF) Type of mass spectrometry detector that measures the time for an ion to fly from the ion source to the detector

Tm Temperature at which 50% of a protein is denatured

***TNF* gene** Gene encoding tumor necrosis factor

TOL plasmid (pTOL) Plasmid-carrying pathway for degradation of toluene

topoisomerase I Enzyme that alters the level of supercoiling or catenation of DNA (i.e., changes the topological conformation)

totipotent The ability of one particular cell to dedifferentiate and redifferentiate into all the different types of cells found in the organism

toxin Poisonous molecule, often a protein, of biological origin; in particular refers to proteins made by pathogenic bacteria

transcription bubble Region where DNA double helix is temporarily opened up so, allowing transcription to occur

transcription factor Protein that regulates gene expression by binding to DNA in the control region of the gene

transcription Process by which information from DNA is converted into its RNA equivalent

transcription start site Starting point where a gene is converted into its RNA copy

transcriptional gene silencing An alternate term used to describe RNA interference-like phenomena

transcriptome The total sum of the RNA transcripts found in a cell, under any particular set of conditions

transfer RNA (tRNA) RNA molecules that carry amino acids to the ribosome

transferrin An iron transport protein of animals

transformation (as used in bacterial genetics) Process in which genes are transferred into a cell as free molecules of DNA

transformation (in cancer biology) Conversion of a normal cell into a cancer cell

transgene A foreign gene that is inserted into an organism using genetic engineering

transgenic An organism with a foreign piece of DNA stably integrated into its genome

transgenic art Art form that involves transgenic animals and plants

transgenic plant A plant containing a gene (transgene) from a different plant or other organism

transition Mutation in which a pyrimidine is replaced by another pyrimidine or a purine is replaced by another purine

translation Making a protein using the information provided by messenger RNA

translational expression vector Vector designed to enhance gene expression at the level of translation

translatome The total set of proteins that have actually been translated and are present in a cell under any particular set of conditions

translocation 1. Transport of a newly made protein across a membrane by means of a translocase. 2. Sideways movement of the ribosome on mRNA during translation. 3. Removal of a segment of DNA from its original location and its reinsertion in a different place

transmissible spongiform encephalopathy (TSE) Infectious form of prion disease

transposable element A mobile segment of DNA that is always inserted in another host molecule of DNA. It has no origin of replication of its own and relies on the host DNA molecule for replication. Includes both DNA-based transposons and retrotransposons

transposase Enzyme responsible for moving a transposon

transposition The process by which a transposon moves from one host DNA molecule to another

transposon A mobile segment of DNA that is always inserted in another host molecule of DNA. It has no origin of replication of its own and relies on the host DNA molecule for replication. Includes both DNA-based transposons and retrotransposons

transversion Mutation in which a pyrimidine is replaced by a purine or vice versa

trehalase Enzyme that degrades trehalose into two glucoses

trehalose A nonreducing storage sugar of plants that protects against dehydration

trehalose-phosphate synthase Enzyme that combines UDP-glucose and glucose 6-phosphate into trehalose 6-phosphate

trehalose 6-phosphate phosphatase Enzyme that removes a phosphate from trehalose 6-phosphate

triplets (see codons) Group of three RNA or DNA bases that encodes a single amino acid

$tRNA_i$ (initiator tRNA) RNA molecule that carries the first amino acid of a protein to the ribosome

$tRNA_i^{fMet}$ Designation for a tRNA charged with *N*-formyl-methionine, which is used as the first amino acid in a prokaryotic protein

trophic factors Soluble molecules that tell neurons, or other cells, to continue to live

tularemia Bacterial disease of rodents or birds that has a death rate of 5% to 10% in humans

tumor necrosis factor (TNF) Protein that kills cancer cells and is produced by tumor-infiltrating lymphocytes

tumor-infiltrating lymphocyte (TIL) White blood cell that secretes TNF

tumor-suppressor gene Gene that acts to prevent unwanted cell division (same as anti-oncogene)

two-component regulatory system A regulatory system consisting of two proteins, a sensor kinase and a DNA binding regulator

two-dimensional polyacrylamide gel electrophoresis (2D-PAGE) A technique used to separate proteins first by size and then by isoelectric focusing (i.e., by electrical charge)

two-hybrid system Method of screening for protein-protein interactions that uses fusions of the proteins being investigated to the two separate domains of a transcriptional activator protein

type I restriction enzyme Type of restriction enzyme that cuts the DNA a thousand or more base pairs away from the recognition site

type I secretory system Specialized export system that spans both inner and outer membranes of gram-negative bacteria such as *E. coli*

type I toxin Toxin that triggers a harmful (internal) response by binding to a cell surface receptor

type II restriction enzyme Type of restriction enzyme that cuts a fixed number of bases away from its recognition site

type II secretory system Specialized export system that spans the outer membrane only

type II toxin Toxin that damages the cell membrane

type III toxin Toxin consisting of a toxic factor (A protein) that enters the target cell together with a delivery system (B protein)

tyrosyl-tRNA synthetase Enzyme that charges tRNA with tyrosine

ultraviolet detector A device that records changes in ultraviolet absorption due to solutes in the mobile phase

uncoupling (of mitochondria) Dissociation of the operation of the respiratory chain from ATP synthesis, which results in the energy being wasted as heat

uncoupling protein (UCP) Protein that uncouples the respiratory chain of mitochondria and releases heat

universal genetic code Version of the genetic code used by almost all organisms

vaccination Artificial induction of the immune response by injecting foreign proteins or other antigens

variable number tandem repeat (VNTR) Cluster of tandemly repeated sequences in DNA, whose number of repeats differs from one individual to another

variable region Region of an antibody whose sequence is varied by gene segment shuffling in order to provide many alternative antigen binding sites

variant CJD (variant Creutzfeldt-Jacob disease) Name used for human cases of mad cow disease

variant Creutzfeldt-Jacob disease (variant CJD) Name used for human cases of mad cow disease

Varkud satellite ribozyme A small ribozyme that initiates self-cleavage and replication of the Varkud satellite virus RNA found in the mitochondria of *Neurospora*

vector vaccine A vaccine where a nonpathogenic virus or bacterium (the vector) is engineered to express a protein or peptide from the disease agent on its surface. The immune system creates antibodies to the expressed protein, which confers immunity to the disease agent

Viagra (sildenafil) Drug used to treat male erectile dysfunction; acts by inhibiting phosphodiesterase 5 and keeping cyclic GMP levels up

virion Virus particle

virosphere All the viruses found within our biosphere

virulence factors Inherited properties that allow pathogenic microorganisms to successfully infect their hosts

virulence plasmid Plasmid that carries genes involved in virulence and pathogenicity

virus-induced silencing An alternate term used to describe RNA interference-like phenomena

VNTR (variable number tandem repeats) Cluster of tandemly repeated sequences in the DNA, whose number of repeats differs from one individual to another

v-onc Virus-borne version of an oncogene

VP16 activator Gene activator protein from herpes simplex virus

weaponized Refers to agents that have been physically prepared for use as biological weapons, such as by making a stable powder

West Nile virus Virus of the flavivirus family originally from the Middle East/Africa that is now spreading in North America

Western blotting Detection technique in which a probe, usually an antibody, binds to a protein target molecule

WIN compound Antiviral agent that prevents the attachment of many picornaviruses

wobble Less rigid base-pairing but only for codon/anticodon pairing during translation

xenobiotic Chemical compound that possesses significant biological activity, but is foreign to the environment

xenoestrogen Foreign polycyclic molecule that binds to steroid receptor and mimics the action of estrogens

X-inactivation The condensation and complete shutting down of gene expression of one of the two X-chromosomes in cells of female mammals

X-Phos 5-bromo-4-chloro-3-indolyl phosphate, an artificial substrate that is split by alkaline phosphatase, releasing a blue dye

xylose Five-carbon sugar that is a major component of various hemicellulose polysaccharides

yeast artificial chromosome (YAC) Single-copy vector based on yeast chromosome that can carry very long inserts of DNA. Widely used in the Human Genome Project

yeast prion Yeast protein that behaves in some ways like the prion protein of mammals

Yersinia pestis Bacterium that causes both the bubonic and pneumonic forms of plague

Z-form An alternative form of double helix with left-handed turns and 12 base pairs per turn. Both DNA and dsRNA may be found in the Z-form

zoo blot Comparative Southern blotting using DNA target molecules from several different animals to test whether the probe DNA is from a coding region

INDEX

E

F

G

U

V

W